Engineering Design Graphics with Autodesk® Inventor® 2015

Engineering Design Graphics with Autodesk® Inventor® 2015

James D. Bethune

Boston Columbus Indianapolis New York San Francisco Hoboken
Amsterdam Cape Town Dubai London Madrid Milan Munich Paris Montreal Toronto
Delhi Mexico City São Paulo Sydney Hong Kong Seoul Singapore Taipei Tokyo

Executive Editor: Lisa McClain
Cover Designer: Sandra Schroeder
Full-Service Project Management: Mohinder Singh/Aptara®, Inc.
Composition: Aptara®, Inc.
Printer/Binder: Edwards Brothers Malloy
Cover Printer: Edwards Brothers Malloy
Text Font: Bookman

Credits and acknowledgments borrowed from other sources and reproduced, with permission, in this textbook appear on the appropriate page within the text. Unless otherwise stated, all artwork has been provided by the author.

Disclaimer:
The publication is designed to provide tutorial information about Inventor® and/or other Autodesk computer programs. Every effort has been made to make this publication complete and as accurate as possible. The reader is expressly cautioned to use any and all precautions necessary, and to take appropriate steps to avoid hazards, when engaging in the activities described herein.

Neither the author nor the publisher makes any representations or warranties of any kind, with respect to the materials set forth in this publication, express or implied, including without limitation any warranties of fitness for a particular purpose or merchantability. Nor shall the author or the publisher be liable for any special, consequential, or exemplary damages resulting, in whole or in part, directly or indirectly, from the reader's use of, or reliance upon, this material or subsequent revisions of this material.

Library of Congress Cataloging-in-Publication Data
Bethune, James D.
Engineering design graphics with Autodesk Inventor 2015 / James D. Bethune.
pages cm
Includes bibliographical references and index.
ISBN 978-0-13-396374-8 (alk. paper) — ISBN 0-13-396374-8 (alk. paper)
1. Engineering graphics. 2. Engineering models—Data processing. 3. Autodesk Inventor (Electronic resource) I. Title.
T386.A974B47 2015
620'.0042028553—dc23

2014035177

10 9 8 7 6 5 4 3 2 1

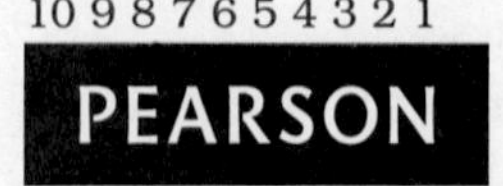

ISBN 10: 0-13-396374-8
ISBN 13: 978-0-13-396374-8

Preface

This book introduces Autodesk® Inventor® 2015 and shows how to use Autodesk Inventor to create and document drawings and designs. The book puts heavy emphasis on engineering drawings and on drawing components used in engineering drawings such as springs, bearings, cams, and gears. It shows how to create drawings using many different formats such as .ipt, .iam, ipn, and .idw for both English and metric units. It explains how to create drawings using the tools located under the **Design** tab and how to extract parts from the **Content Center.**

All topics are presented using a step-by-step format so that the reader can work directly from the text to the screen. There are many easy-to-understand labeled illustrations. The book contains many sample problems that demonstrate the subjects being discussed. Each chapter contains a variety of projects that serve to reinforce the material just presented and allow the reader to practice the techniques described.

Chapters 1 and 2 present 2D sketching tools and the **Extrude** tool. These chapters serve as an introduction to the program. There are 38 Chapter Projects to help students apply the material presented.

Chapter 3 demonstrates the tools needed to create 3D models, including **Shell, Hole, Rib, Split, Loft, Sweep,** and **Coil.** Work points, work axes, and work planes are explained and demonstrated.

Chapter 4 shows how to create orthographic views from 3D models. The creation of isometric views, section views, and auxiliary views is covered. In addition, a comparison between first- and third-angle projection is presented using both ANSI and ISO conventions.

Chapter 5 shows how to create assembly drawings using both the bottom-up and the top-down process. The chapter includes presentation drawings and exploded isometric drawings with title blocks, parts lists, revision blocks, and tolerance blocks. There is an extensive step-by-step example that shows how to create an animated assembly, that is, a drawing that moves on the screen.

Chapter 6 covers threads and fasteners. Drawing conventions and callouts are defined for both inch and metric threads. The chapter shows how to calculate thread lengths and how to choose the appropriate fastener from Inventor's **Content Center.** The **Content Center** also includes an extensive listing of nuts, setscrews, washers, and rivets.

Chapter 7 shows how to apply dimensions to drawings. Both ANSI and ISO standards are demonstrated, but the emphasis is on ANSI standards. Different styles of dimensioning, including ordinate and baseline, and using Inventor's **Hole Table** are presented. Applying dimensions to a drawing is considered an important skill, so many examples and sample problems are included.

Chapter 8 is an extensive discussion of tolerancing, including geometric tolerances. The chapter first shows how to use Inventor to apply tolerances to a drawing. The chapter then shows how to calculate tolerances in various design situations. Positional tolerances for both linear and geometric applications are included. The chapter introduces the **Limits/Fits Calculator** located on the **Power Transmission** panel under the **Design** tab.

Chapter 9 shows how to draw springs using the **Standard.ipt** format and the **Coil** tool. It also shows how to draw springs using the tools on the

Spring panel under the **Design** tab. Compression, extension, torsion, and Belleville springs are included.

Chapter 10 shows how to draw shafts using the **Shaft** tool under the **Design** tab. Chamfers, retaining rings, retaining ring grooves, keys and keyways, splines, pins, O-rings, and O-ring grooves are covered. The chapter contains many exercise problems.

Chapter 11 shows how to match bearings to specific shafts using the **Content Center.** Plain, ball, and thrust bearings are presented. An explanation of tolerances between a shaft and bearing bore and between the bearing's outside diameter and the assembly housing is given. Both ANSI and ISO standards are presented.

Chapter 12 emphasizes how to draw gears and how to mount them into assembly drawings. Spur, bevel, and worm gears are introduced. The chapter shows how to create gear hubs with setscrews, and keyways with keys, and how to draw assembly drawings that include gears. There are two new extensive assembly exercise problems.

Chapter 13 shows how to draw basic sheet metal parts including features such as tabs, reliefs, flanges, cuts, holes, and hole patterns.

Chapter 14 shows how to create and draw weldments. Only fillet and groove welds are covered.

Chapter 15 shows how to design and draw cams. Displacement diagrams and different types of followers are discussed.

Chapter 16 is available online and includes two large project-type problems. They can be used as team projects to help students learn to work together to share and compile files, or they can be used as end-of-the-semester individual projects. This chapter can be found on the web as a supplement to the Instructor's Manual at http://pearsonhighered.com/irc. Instructors may distribute to students.

Download Instructor Resources from the Instructor Resource Center

To access supplementary materials online, instructors need to request an instructor access code. Go to www.pearsonhighered.com/irc to register for an instructor access code. Within 48 hours of registering, you will receive a confirming e-mail including an instructor access code. Once you have received your code, locate your text in the online catalog and click on the Instructor Resources button on the left side of the catalog product page. Select a supplement, and a login page will appear. Once you have logged in, you can access instructor material for all Prentice Hall textbooks. If you have any difficulties accessing the site or downloading a supplement, please contact Customer Service at http://247pearsoned.custhelp.com.

Acknowledgments

I would like to thank the following reviewers for their invaluable input: Rebecca Rosenbauer, Lafayette College; Antigone Sharris, Triton College; Nancy E. Study, Virginia State University; and Marsha Walton, Finger Lakes Community College.

Thanks to my family: David, Maria, Randy, Hannah, Wil, Madison, Jack, Luke, Sam, and Ben.

A special thanks to Cheryl.

James D. Bethune

Style Conventions in *Engineering Design Graphics with Autodesk® Inventor® 2015*

Text Element	Example
Key terms—Bold and italic on first mention in the body of the text. Brief glossary definition in margin following first mention.	Create a ***work axis*** by clicking on the edge of the block.
Inventor tools—Bold and follow Inventor capitalization convention.	Click on the **Line** tool.
Toolbar names, menu items, and dialog box names—Bold and follow capitalization convention in Inventor tab, panel, or pull-down menu (generally first letter capitalized).	The **Design** tab The **Modify** panel The **2D Chamfer** dialog box The **File** pull-down menu
Dialog box controls/buttons/input items—Bold and follow capitalization convention of the name of the item or the name shown in the Inventor tooltip.	Choose the **Metric** tab in the **New File** dialog box. Click on the **Flush** button on the **Place Constraint** dialog box. On the **Assembly** tab, set the **Offset** to **0.000 mm.** **ONG 300/320 sat 76**

Brief Contents

Contents

The following online chapter available at www.pearsonhighered.com/irc.

Engineering Design Graphics with Autodesk® Inventor® 2015

1 chapterone
Get Started

CHAPTER OBJECTIVES

- Learn how to create part drawings
- Understand the different drawing formats
- Learn how to sketch using Inventor
- Draw holes in an object
- Draw angular shapes

Introduction

This chapter presents a step-by-step introduction to Inventor 2015. When the program is first accessed the **Get Started** panel will appear. See Figure 1-1.

Inventor uses a system of tabs and panels to present commands. In Figure 1-1 the tab is **Get Started,** and the panels are **Launch, My Home,** and **New Features.** Within each panel is a series of commands. On the **Launch** panel are the commands **New, Open, Projects,** and **Open Samples.**

To Start a New Drawing

There are two ways to start a **New** drawing; click the **New** tool located under the **Get Started** tab, or click the **Part** tool as shown in Figure 1-1.

Clicking the **Part** tool will let you jump directly to a drawing screen. The drawing will automatically be set to the default values. In this example, the default values are decimal inches; the default values can be edited.

Figure 1-1

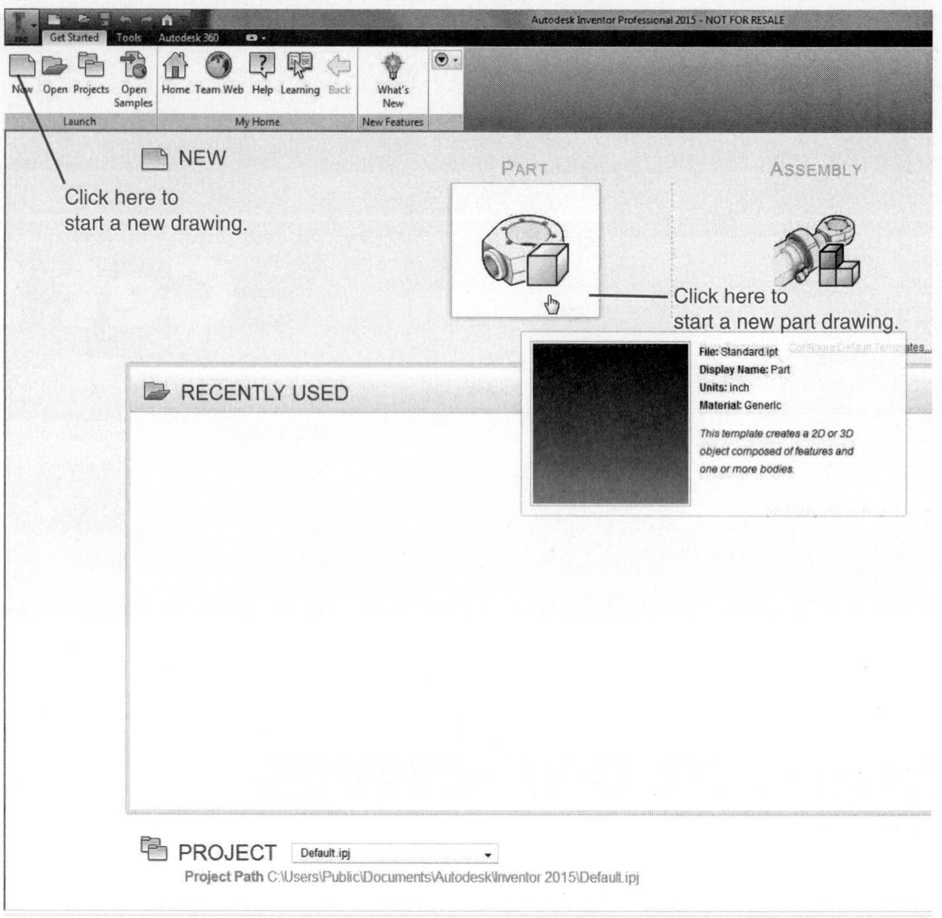

Clicking the **New** tool will allow you to select drawing units, and the type of drawing you want.

1. Click the **New** tool located on the **Launch** panel under the **Get Started** tab.

   The **Create New File** dialog box will appear. See Figure 1-2.

2. Click the type of drawing you want to create.

There are many options for creating drawings with four different types of files. These files are categorized using four different extensions. The extensions are defined as follows:

.ipt: Part files for either 3D model drawings or sheet metal drawings. These files are for individual parts.

.iam: Assembly drawings and weldments. Assembly drawings are formed by combining .ipt files.

.ipn: Presentation files. These files are used to create exploded assembly drawings.

.idw: Drawing layout files. These files are used to create orthographic views from already existing part, assembly, and presentation drawings

The first type of drawing created in this book will be a *Standard (mm).ipt* drawing. The *mm* indicates that the drawing units are millimeters. The first screen to appear will have a colored background. For this book the

Figure 1-2

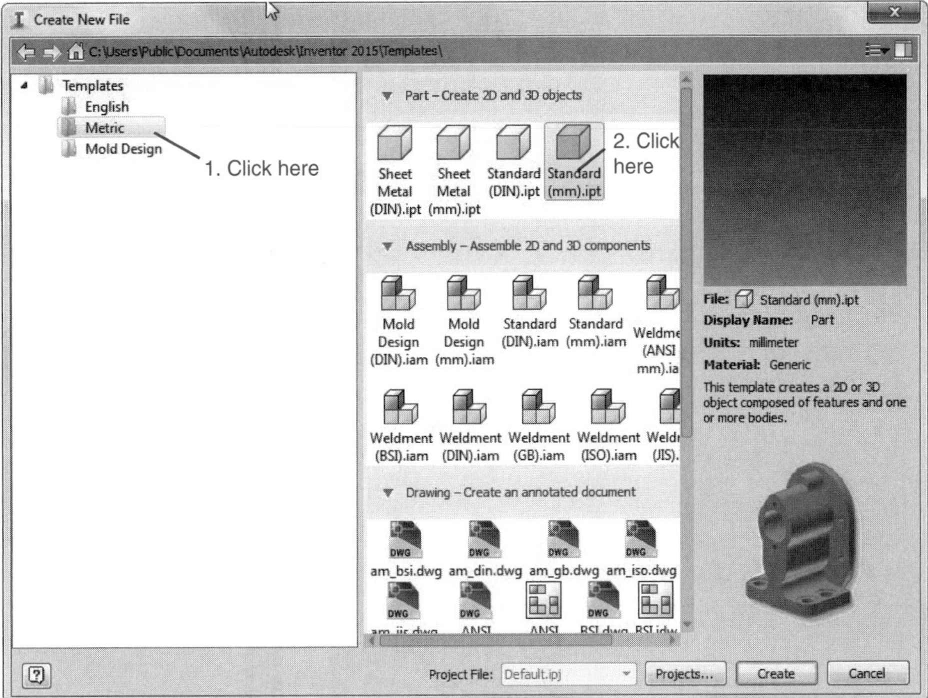

background color was changed to a *Presentation* format. This was done to make the illustrations easier to read. Your screen may have a different background.

To Change the Screen's Background Color

1. Click the **Tools** tab.
2. Click the **Application Options** tool on the **Options** panel.

   The **Application Option** dialog box will appear. See Figure 1-3.
3. Click the **Colors** tab on the **Application Options** dialog box.
4. Select a color scheme.

   In this example the **Presentation** color scheme was selected and a one-color background rather than a gradient.
5. Click **OK.**

Creating a First Sketch

This section shows how to set up, create, and save a first drawing. The intent is to walk through a simple drawing in order to start to understand how Inventor functions.

1. Click the **Get Started** tab.
2. Click the **New** tool on the **Launch** panel.

   The **New File** dialog box will appear. See Figure 1-1.

Figure 1-3

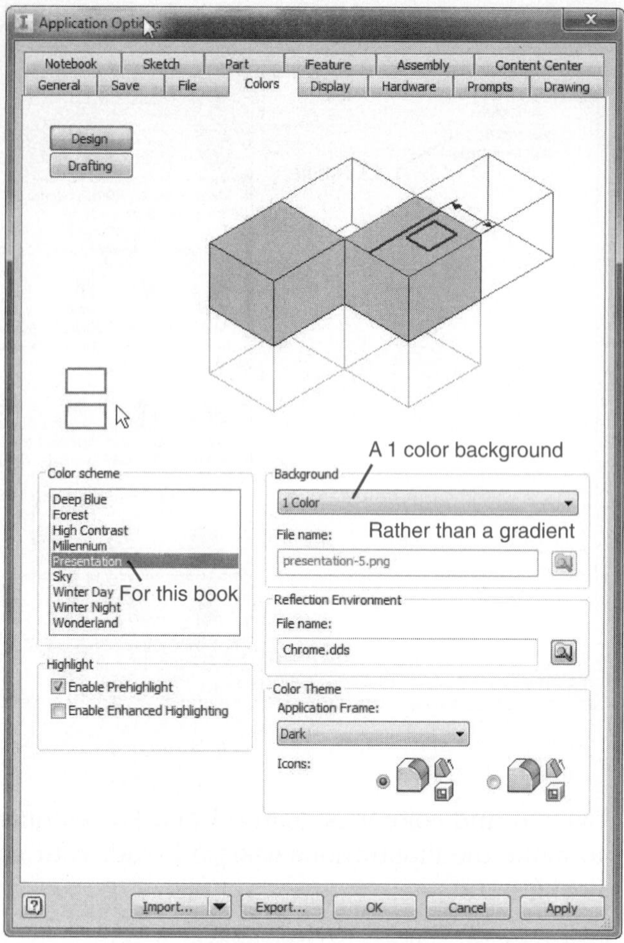

Figure 1-4

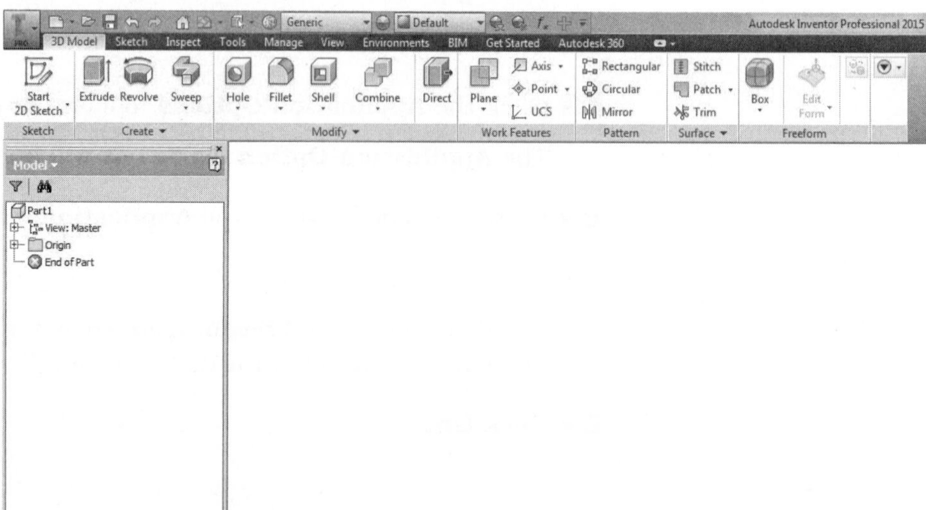

3 Click the **Metric** tab.

The **Metric** tab will define all drawing dimension values in millimeters.

4 Click the **Standard (mm).ipt** tool; click **OK**.

5 Click the **Sketch** tab.

The panels at the top of the screen will change to the **Sketch** panels. See Figure 1-4.

To Sketch a 30 × 40 Rectangle

1 Click the **Line** tool located in the **Draw** panel under the **Sketch** tab.

See Figure 1-5. You must define which plane you want to work on.

Figure 1-5

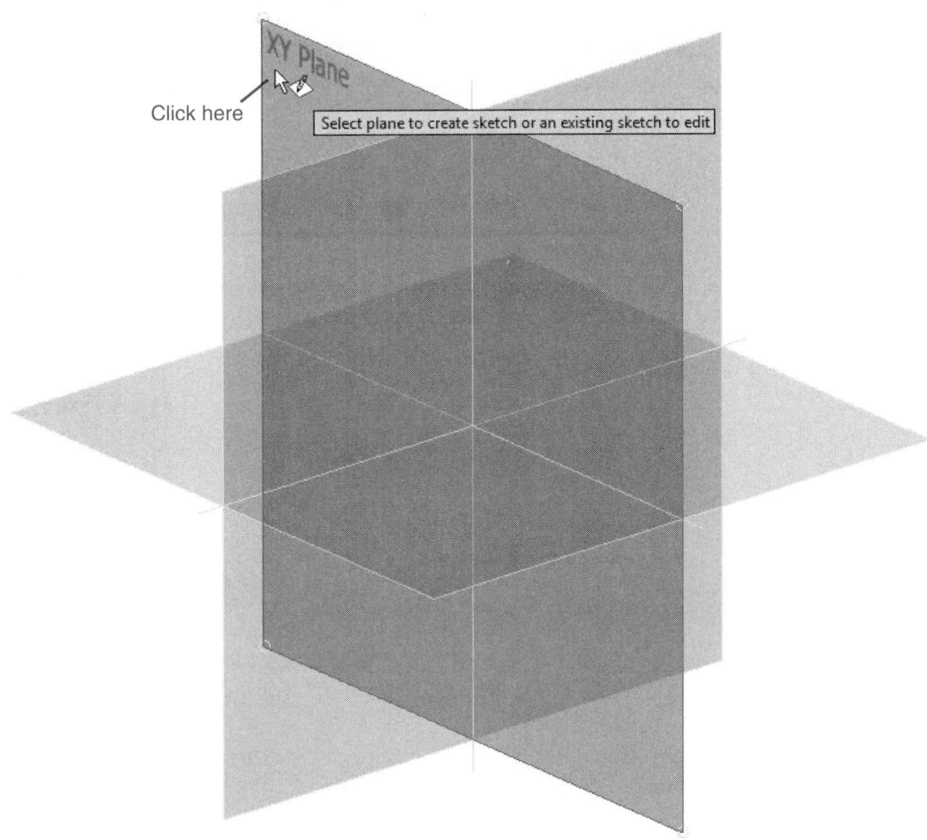

XY Plane
Click here
Select plane to create sketch or an existing sketch to edit

2 Select the XY plane.

The screen will rotate so as to create a direct view of the XY plane. The view is like a blank sheet of drawing paper. We are looking along the Z-axis at the XY plane.

Click a starting point for the line, release the mouse button, and drag the cursor across the drawing screen.

3 Draw a horizontal line, that is, one at 0°.

The length need not be a specific value. See Figure 1-6. The line's length could be defined by entering a numerical value that will appear in the box above the line. The box will disappear once the line's endpoint is defined.

4 Click the mouse to define the end of the line. Drag the cursor downward to create a vertical line, that is, one at 90° to the first line.

5 Draw a second horizontal line parallel to the first line. Continue the line until a broken line appears from the starting point of the first line.

This broken line indicates that the endpoint of the second horizontal line is now aligned with the starting point of the first line.

6 Draw a second vertical line to complete the rectangle.

7 Right-click the mouse and select the **OK** option.

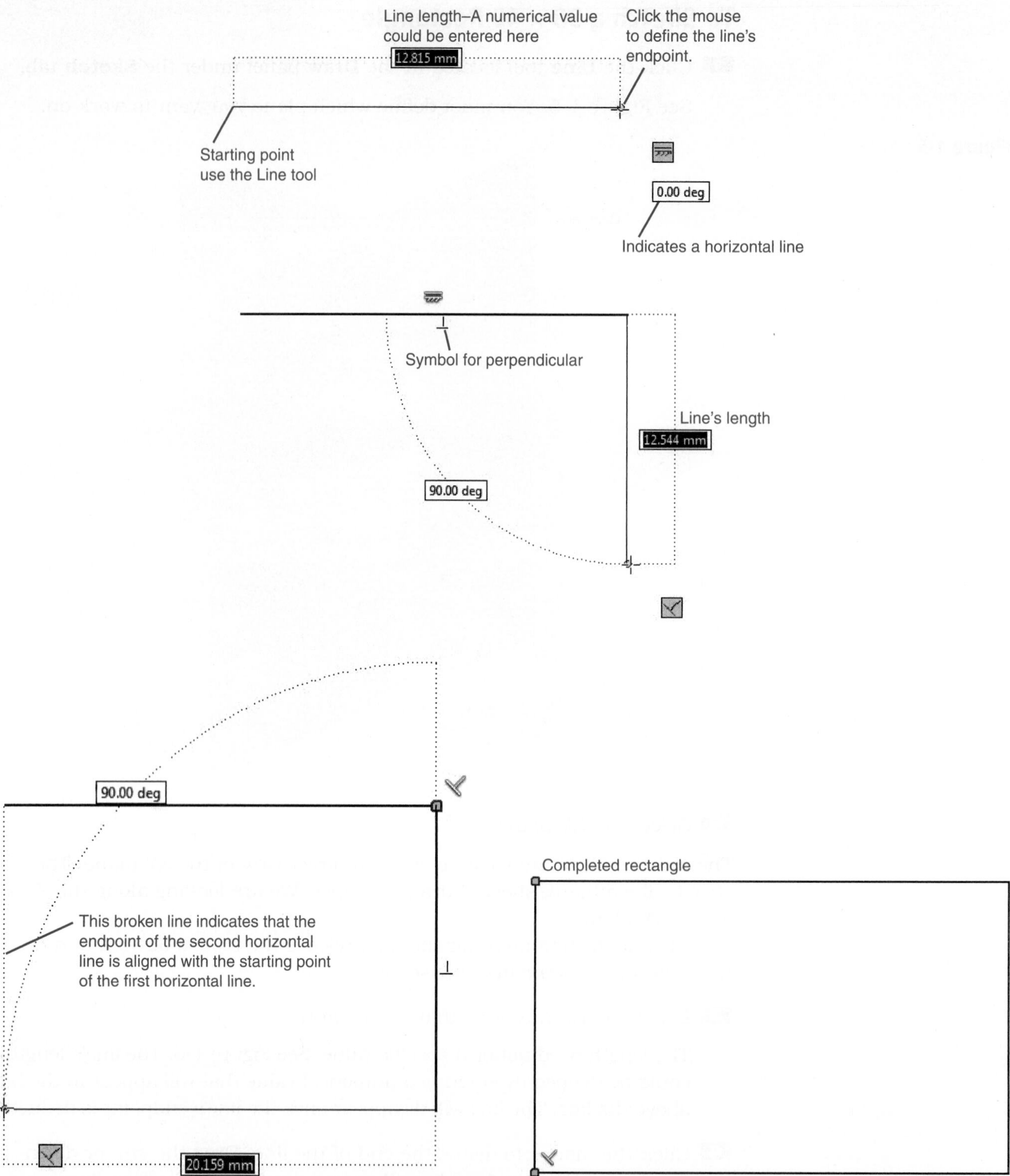
Line length–A numerical value could be entered here
12.815 mm
Click the mouse to define the line's endpoint.
Starting point use the Line tool
0.00 deg
Indicates a horizontal line
Symbol for perpendicular
Line's length
12.544 mm
90.00 deg
90.00 deg
This broken line indicates that the endpoint of the second horizontal line is aligned with the starting point of the first horizontal line.
20.159 mm
Completed rectangle

Figure 1-6

To Size the Rectangle

The rectangle shown in Figure 1-6 was drawn without regard to dimension values; that is, it was sketched. The dimension values must be entered. See Figure 1-7.

NOTE: Do not click the **Finish 2D Sketch**; continue working on the sketch.

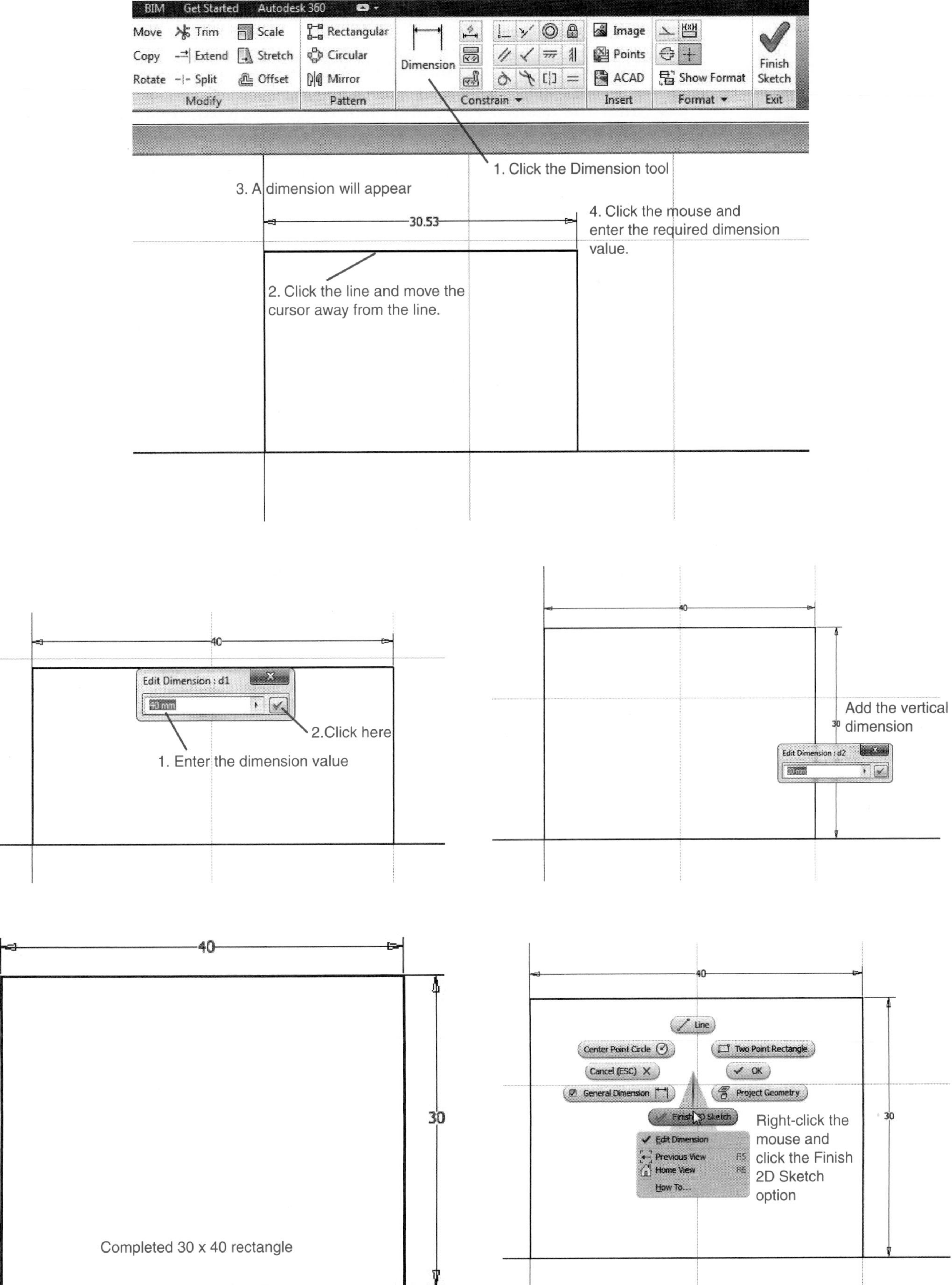
BIM
Get Started
Autodesk 360
Move
Trim
Scale
Rectangular
Copy
Extend
Stretch
Circular
Rotate
Split
Offset
Mirror
Dimension
Image
Points
ACAD
Show Format
Finish Sketch
Modify
Pattern
Constrain
Insert
Format
Exit
1. Click the Dimension tool
3. A dimension will appear
30.53
4. Click the mouse and enter the required dimension value.
2. Click the line and move the cursor away from the line.
40
Edit Dimension : d1
40 mm
2.Click here
1. Enter the dimension value
40
Add the vertical dimension
30
Edit Dimension : d2
40
30
Completed 30 x 40 rectangle
40
Line
Center Point Circle
Two Point Rectangle
Cancel (ESC)
OK
General Dimension
Project Geometry
Edit Dimension
Previous View
F5
Home View
F6
How To...
Right-click the mouse and click the Finish 2D Sketch option
30

Figure 1-7

1. Click the **Dimension** tool, click the top horizontal line, and move the cursor away from the line.

   The **Edit Dimension** dialog box will appear.

2. Enter the desired value and click the **green (OK) check mark** or press the **<Enter>** key.

   The length of the rectangle will change to the entered dimension value.

3. Click the vertical line and move the cursor away from the line.

4. Enter the desired value and click the **green (OK) check mark** or press the **<Enter>** key.

5. Right-click the mouse and select the **Finish 2D Sketch** option.

The screen will change to a 3D view of the rectangle. If needed, click the front surface on the ViewCube to return the 2D view to the XY plane. Use the mouse wheel to size the rectangle to fit the screen.

To Edit an Existing Dimension

The size of an object can be edited by changing the dimensions. For example, say we wish to change the size of the 30 × 40 rectangle to 35 × 50. We can edit the existing drawing. The drawing view was returned to 2D for this example. The dimensions could be edited with the screen in the 3D orientation.

1. Double-click the **30** dimension.

   The dimension value box will appear. See Figure 1-8.

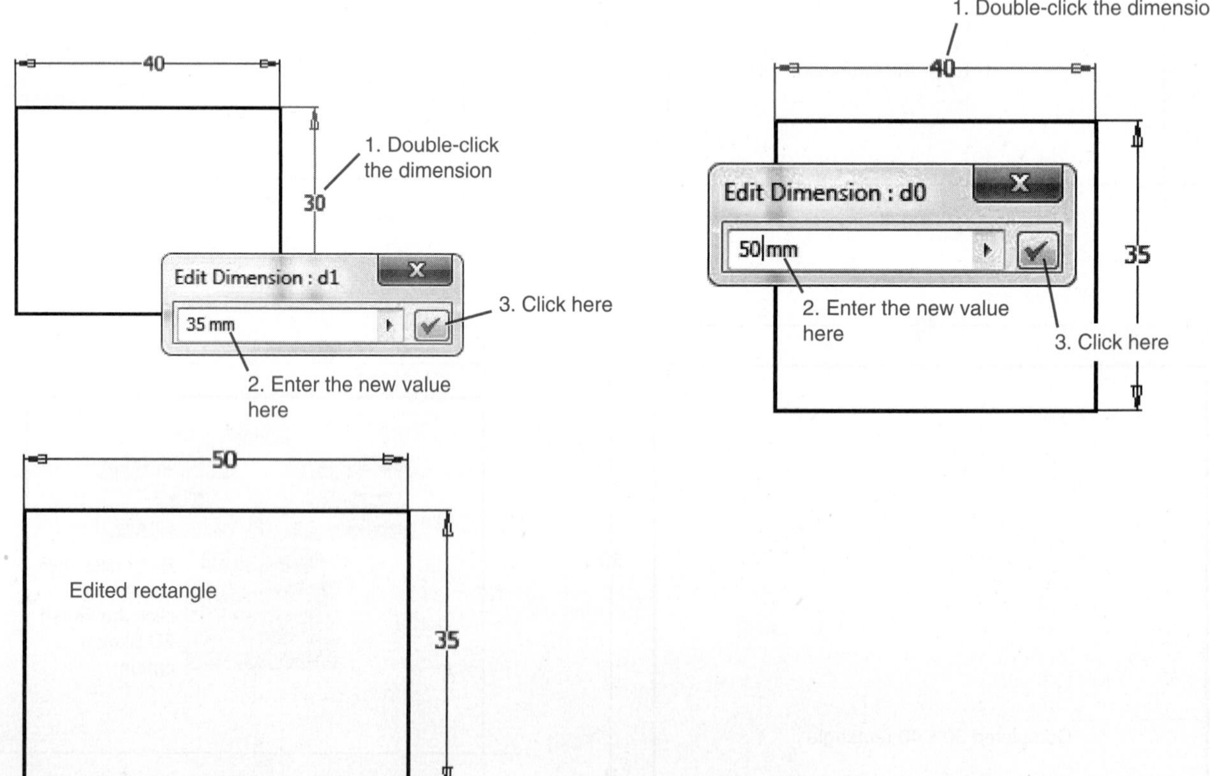

Figure 1-8

2 Change the dimension value to **35**.

3 Click the **green check mark**.

The dimension value will change, as will the shape of the rectangle.

4 Repeat the procedure to change the 40 value to **50**.

The Undo Tool

The **Undo** tool is used to undo commands. The commands will be undone in the reverse order of the sequence in which they were entered.

To Undo a Series of Commands

1 Click the **Undo** tool located at the top of the screen.

See Figure 1-9.

Figure 1-9

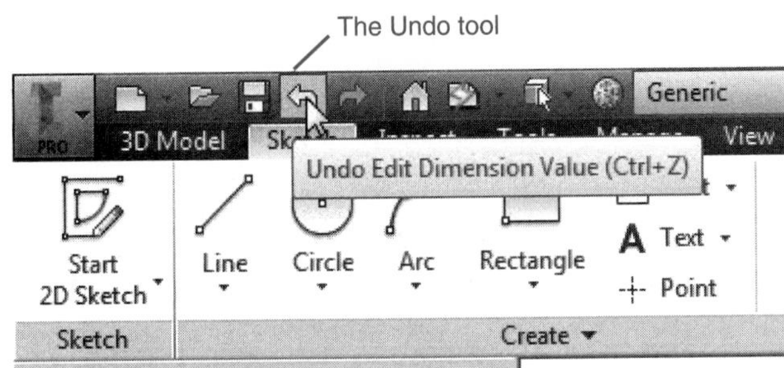

2 Use the **Undo** command to return the 35 × 50 rectangle to the original 30 × 40 rectangle.

Creating a Solid Model

The 30 × 40 rectangle will now be used to create a solid object 15.0 mm thick.

1 Click the **3D Model** tab located near the top left corner of the screen, and, if needed, click the houselike icon next to the ViewCube.

See Figure 1-10. The screen will change view orientation to **3D Isometric** view.

Figure 1-10

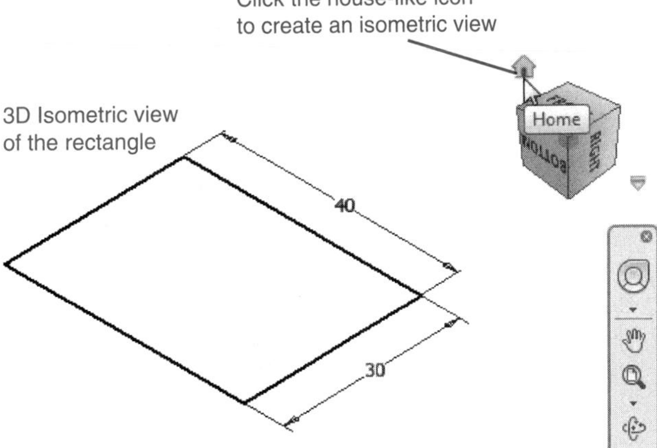

2 Click the **Extrude** tool located in the **Create** panel under the **Model** tab.

The **Extrude** dialog box will appear. See Figure 1-11. In this example the default thickness value is 10 mm.

Figure 1-11

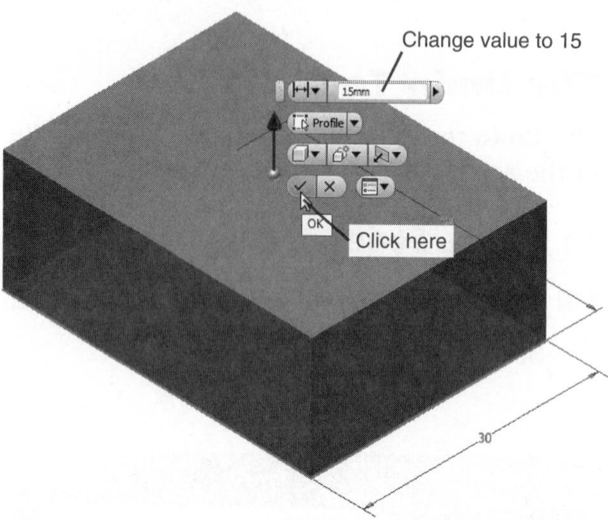

3 Change the thickness value in the **Extrude** dialog box to **15;** click **OK**.

The 30 × 40 rectangle has been used to create a 15 × 30 × 40 box (rectangular prism). This is a solid model.

To Add a Hole

Inventor allows you work on only one plane at a time. To create a hole, create a new drawing plane on the top surface of the box, add a circle to the plane, then use the **Cut Extrude** option to create a hole. See Figure 1-12.

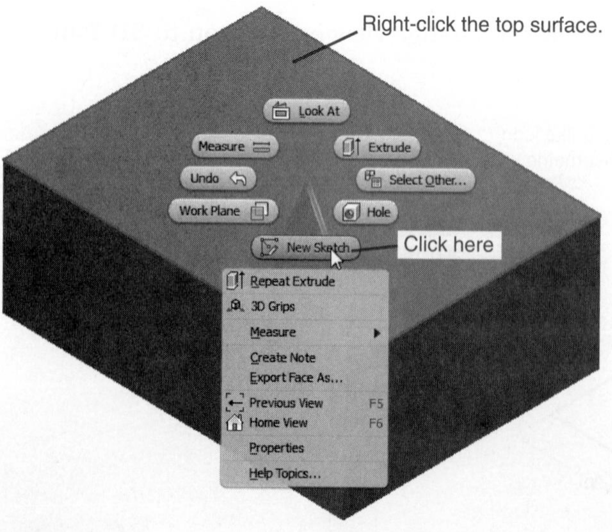

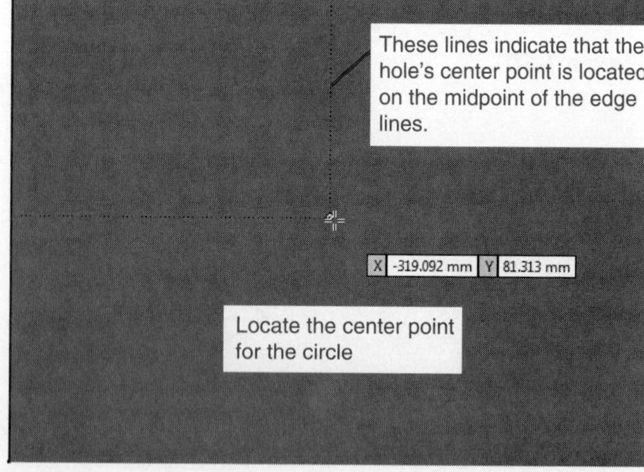

Figure 1-12

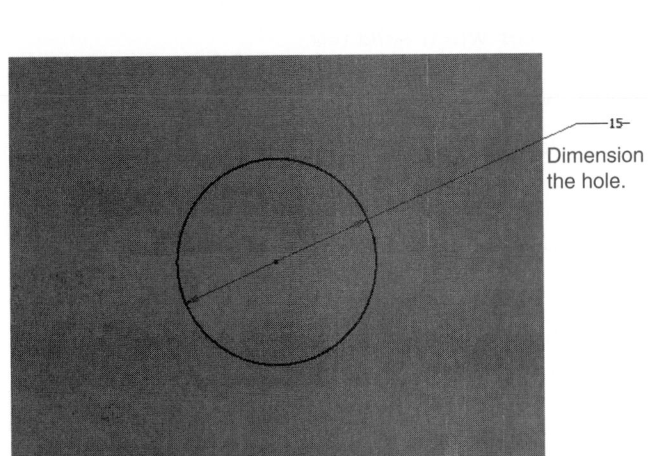

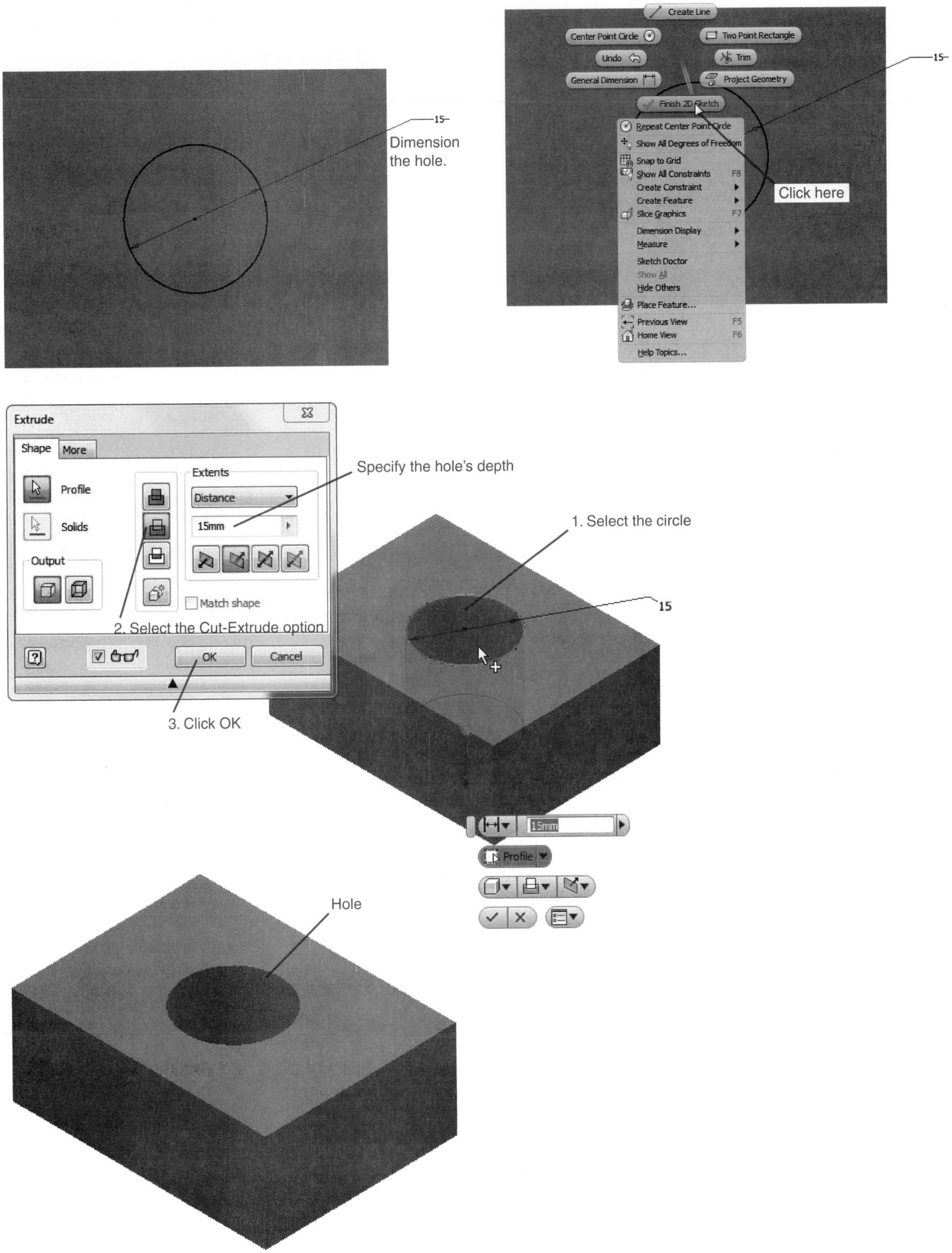

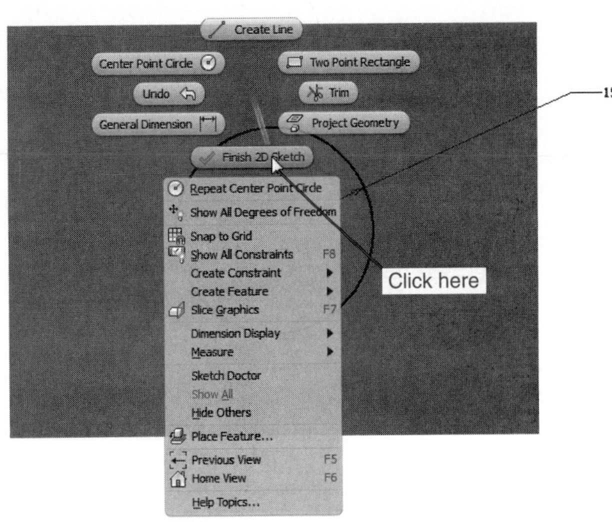

Extrude
Shape More
Profile
Solids
Output
Extents
Distance
15mm
Match shape
OK Cancel
Specify the hole's depth
1. Select the circle
15
2. Select the Cut-Extrude option
3. Click OK
15mm
Profile
Hole

Figure 1-12
(Continued)

1. Move the cursor to the top surface of the box and right-click the mouse.

   The top surface will change color when selected.

2. Click the **New Sketch** option.

   The box will change orientations so that you are looking directly onto the top surface.

3. Click the **Circle** tool on the **Draw** panel under the **Sketch** tab.

4. Locate the circle on the center point of the top surface by moving the cursor around the edge lines until the midpoints are identified. When the cursor intersects an edge line's midpoint, a green circle will appear. Move the cursor away from the midpoint and locate the other edge line midpoint. Move the cursor to the center of the surface. Broken lines will appear from the midpoints of the edge lines of the surface. Locate the center point of the circle on the center point of the top surface of the box.

5. Use the **Dimension** tool from the **Constraint** panel under the **Sketch** tab and define the diameter of the circle as **15.0 mm**.

6. Right-click the top surface and select the **Finish 2D Sketch** option.

7. Click the **Extrude** tool on the **Create** panel under the **Model** tab.

   The **Extrude** dialog box will appear.

8. Click the circle on the top surface.

9. Define the depth of the hole as **15 mm** or greater.

   Be sure to have the extrusion pointing downward into the box.

10. Click the **Cut** option on the **Extrude** dialog box.

11. Click **OK.**

To Save a Drawing

1. Click the **Save As** command located under the **I** icon at the top left of the screen.

   See Figure 1-13. The **Save As** dialog box will appear. See Figure 1-14.

2. Save the drawing as **30x40x15Box;** click **Save.**

There are other ways to create holes. These will be discussed later in the book.

Sample Problem SP1-1

This section shows how to draw the problem presented in Figure P1-2 in the Chapter Project. The dimensions are in inches.

1. Start a **New** drawing. Click the **English** tab, and select the **Standard (in).ipt** format.

   See Figure 1-15.

Figure 1-13

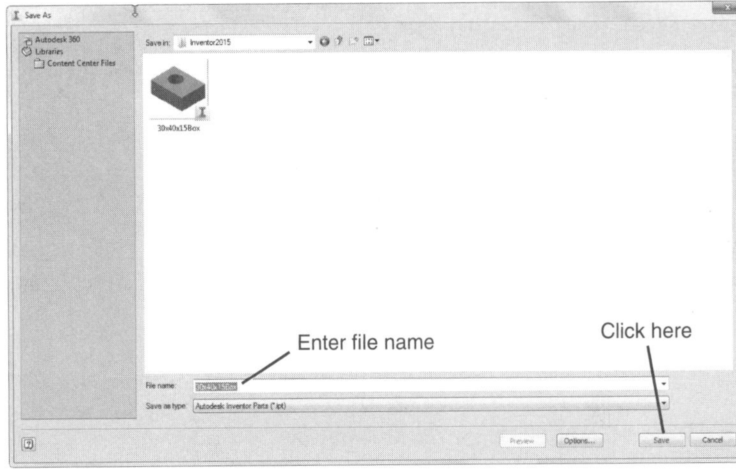

Figure 1-14

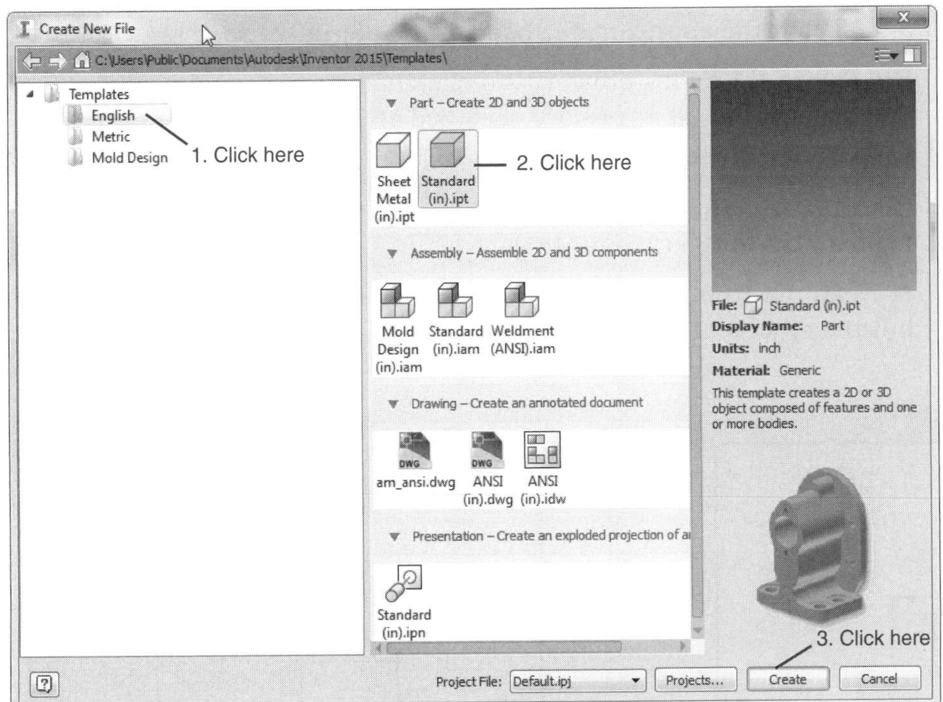

Figure 1-15

2 Click the **Origin** heading in the **Browser Box** and select the **XY Plane** option. Move the cursor into the drawing area, right-click the mouse, and select the **New Sketch** option.

See Figure 1-16.

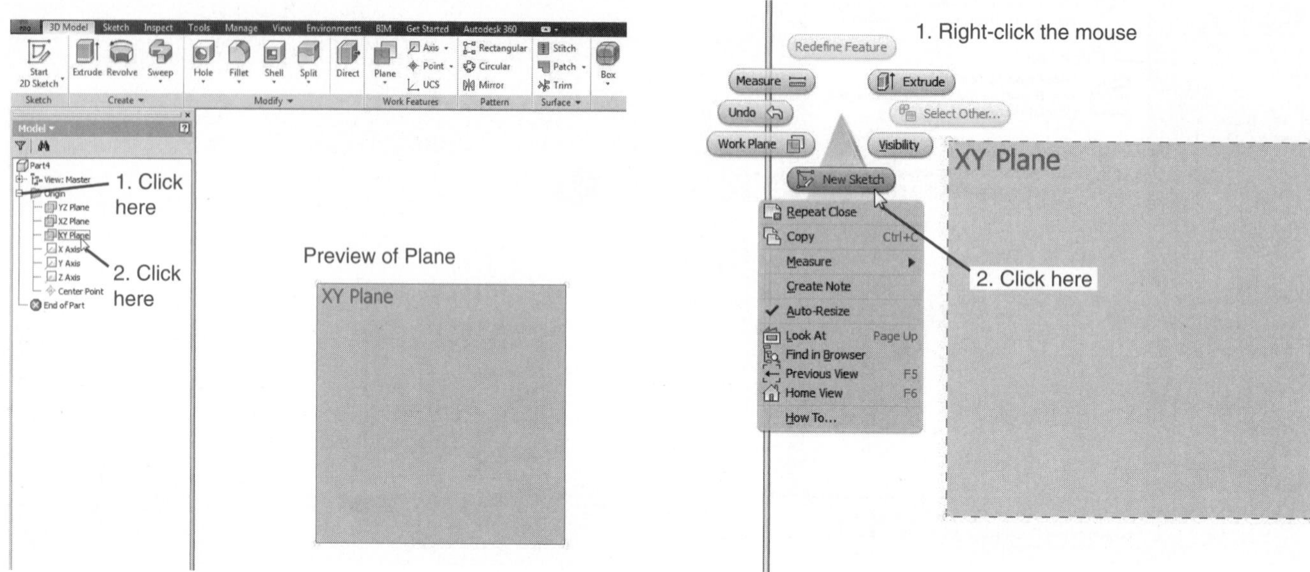

Figure 1-16

3 Click the **Line** tool located on the **Draw** panel under the **Sketch** tab and sketch the approximate shape of the object.

See Figure 1-17. This drawing is a sketch and uses only approximate dimensions, but all angles are sketched at 90°. The lower lines of the sketch were aligned.

4 Click the **Dimension** tool and add the dimensions as presented in Figure P1-2 in the Chapter Project.

If, when the 1.000 dimension is applied, one side of the shape moves, but the other side does not, we could align the two bottom edges by

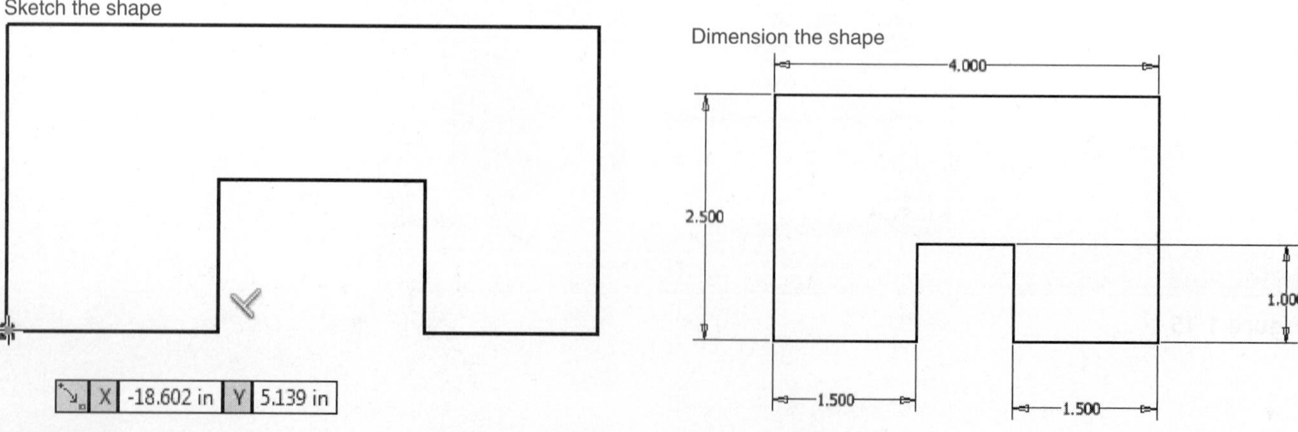

Figure 1-17

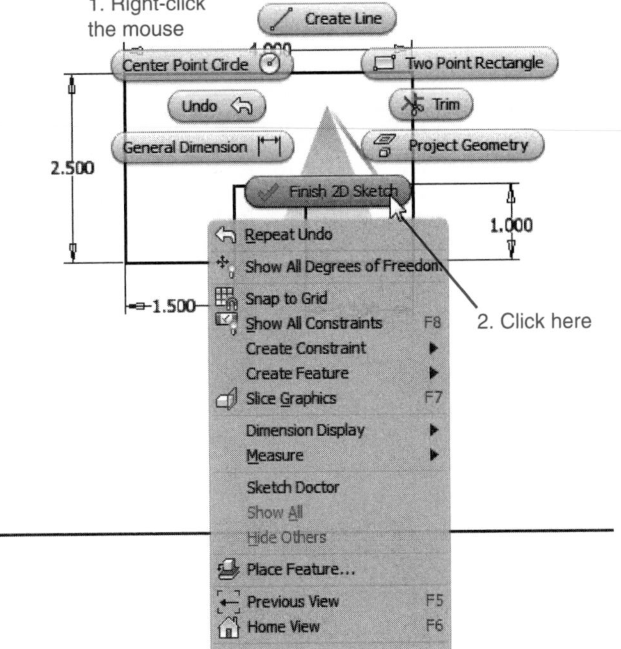

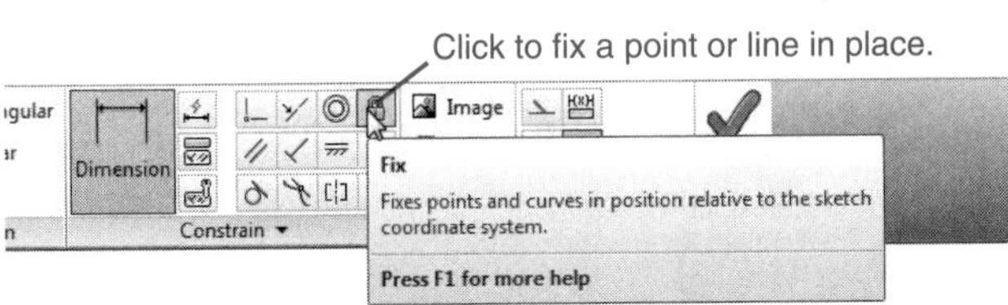

Figure 1-17
(Continued)

adding a second 2.500 dimension to the right edge line, but we can also use a **Constraint** option.

5 Click the **Undo** tool and return the drawing to its original shape without the 1.000 dimension.

6 Click the **Fix** tool located on the **Constrain** panel under the **Sketch** tab.

The **Fix** constraint looks like a padlock.

7 Click one of the lines.

This will fix the line; that is, it will not move when changes are made. The **Fix** constraint will cause the 1.000 dimension to be absorbed by moving the horizontal portion of the slot.

8 Add the **1.000** dimension for the slot.

9 Right-click the mouse and select the **Finish 2D Sketch** option.

Extrude the Shape

See Figure 1-18.

1 Click the **3D Model** tab and click the **Home** (the icon looks like a house) tool next to the ViewCube.

2 Click the **Extrude** tool on the **Create** panel.

3 Enter a thickness value of **.500**.

4 Click **OK**.

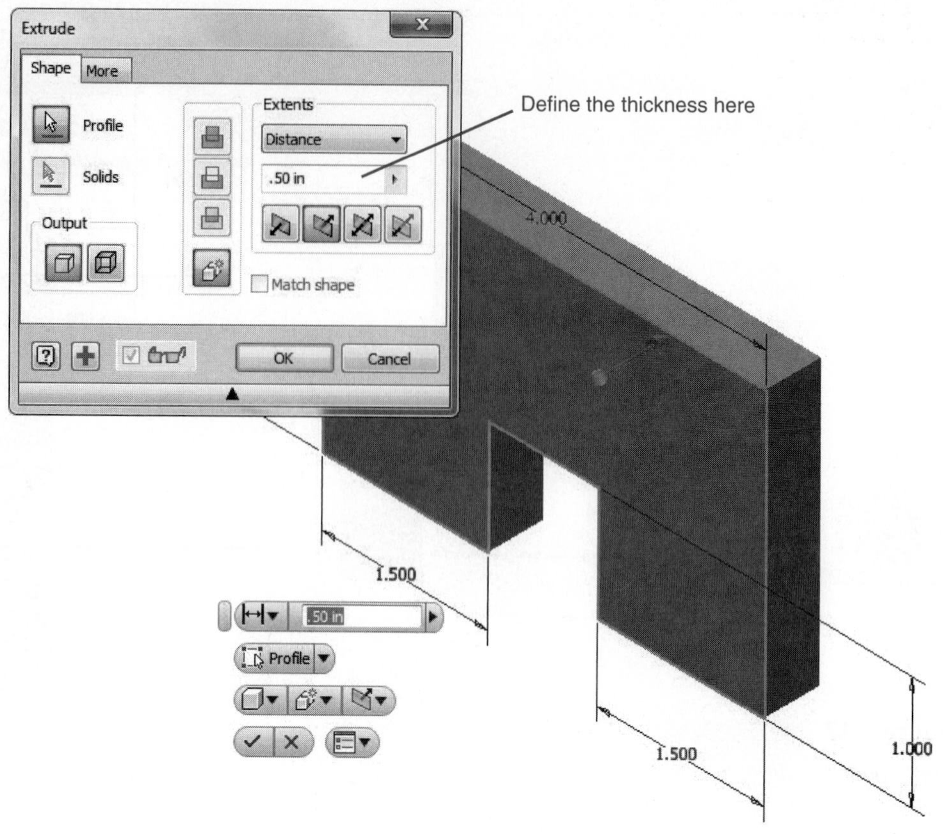
Extrude
Shape
More
Profile
Solids
Output
Extents
Distance
.50 in
Match shape
OK
Cancel
Define the thickness here
4.000
1.500
.50 in
Profile
1.500
1.000

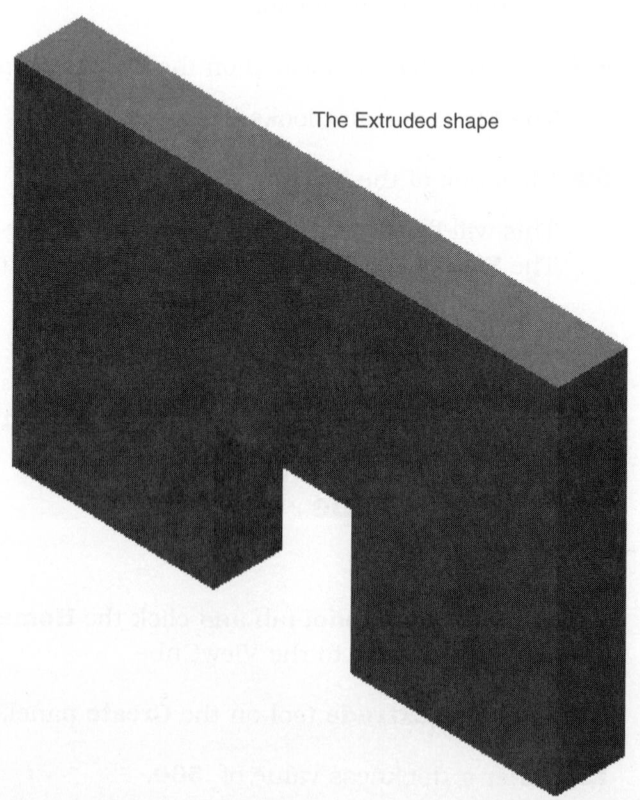
The Extruded shape

Figure 1-18

Add the Holes

See Figure 1-19.

1. Right-click the front surface of the shape and select the **New Sketch** option.

2. Use the **Circle** tool on the **Draw** panel and draw two circles in their approximate position.

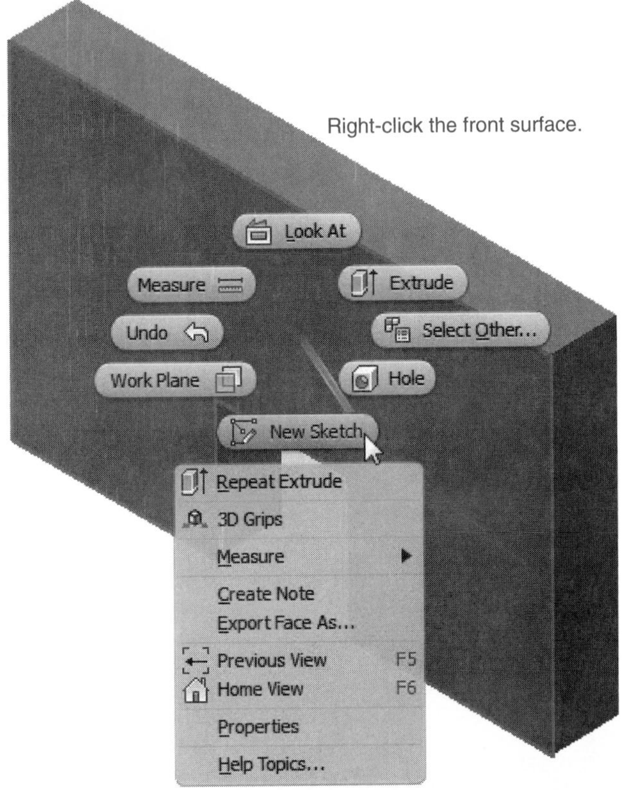

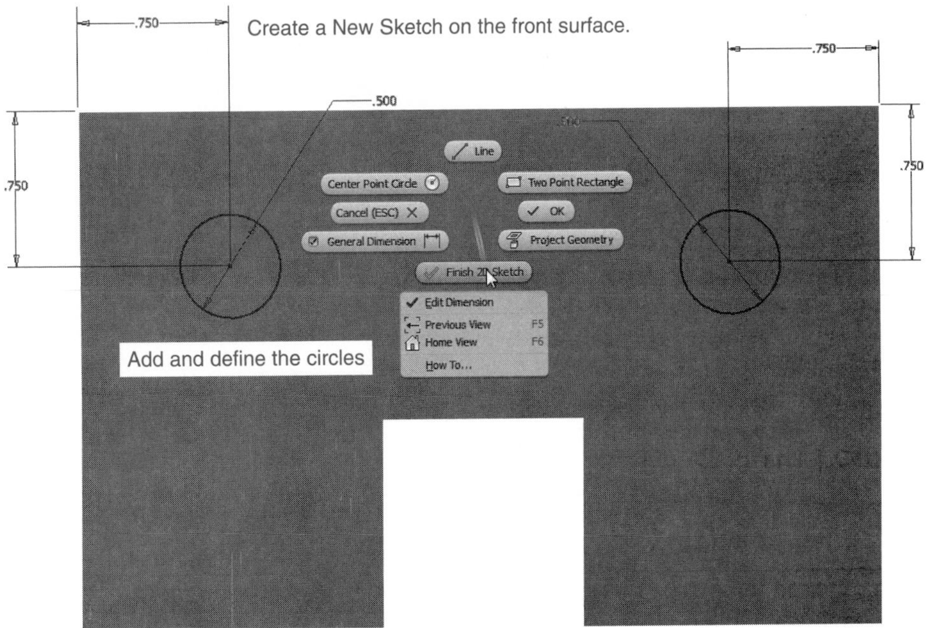

Figure 1-19

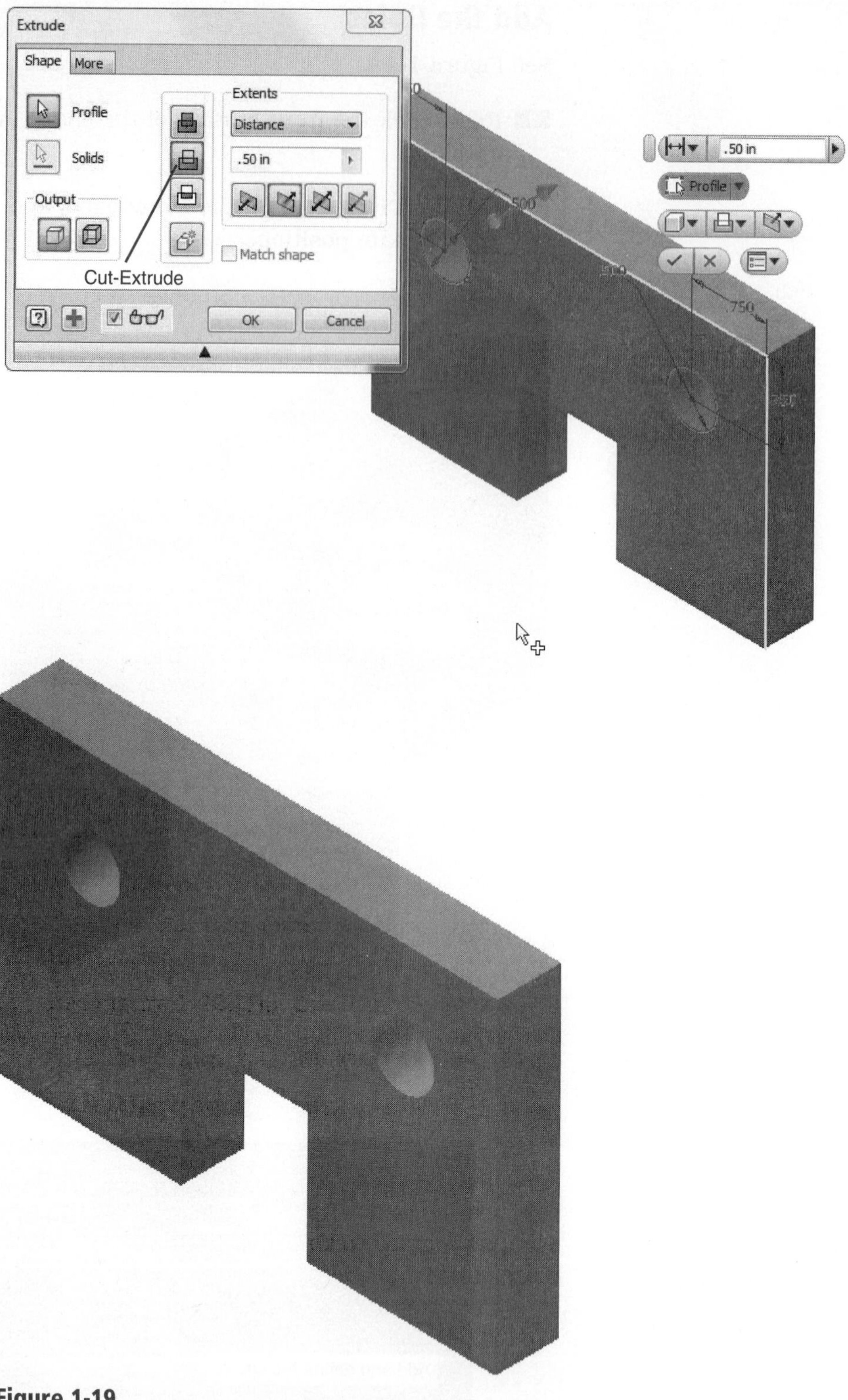

Figure 1-19
(Continued)

3 Use the **Dimension** tool and accurately locate and size the circles using the given dimensions.

4 Click the **3D Model** tab.

5 Click the **Extrude** tool.

6 Click the **Cut** option on the **Extrude** dialog box.

7 Click each circle.

The circles will change color when selected, and a preview will appear. Ensure that the direction of the cut is through the object.

8 Click **OK**.

Sample Problem SP1-2 Angular Shapes

Figure 1-20 shows a shape that has angular edges. The **Dimension** tool can be used to define the required angles.

Figure 1-20

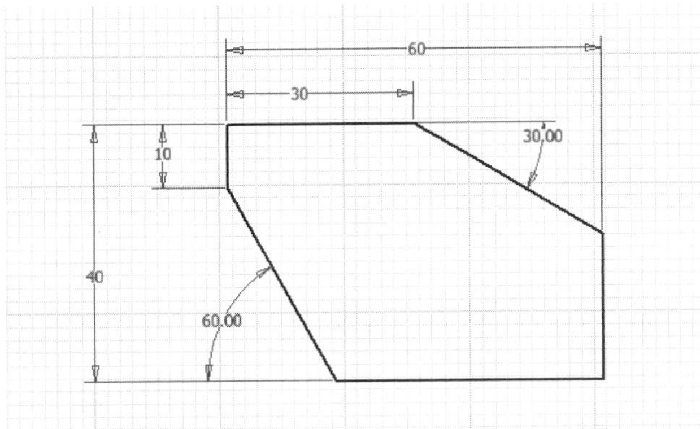

1 Start a **New** drawing. Use the **Standard (mm).ipt** format.

2 Draw an approximate sketch of the object. See Figure 1-21.

3 Use the **Dimension** tool and add the **10, 30,** and **60** dimensions as shown.

4 Use the **Dimension** tool and click the top horizontal line and the right angular line.

An angular dimension will appear.

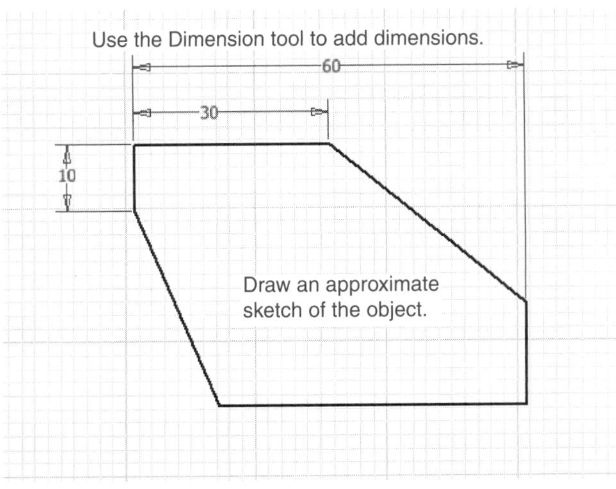

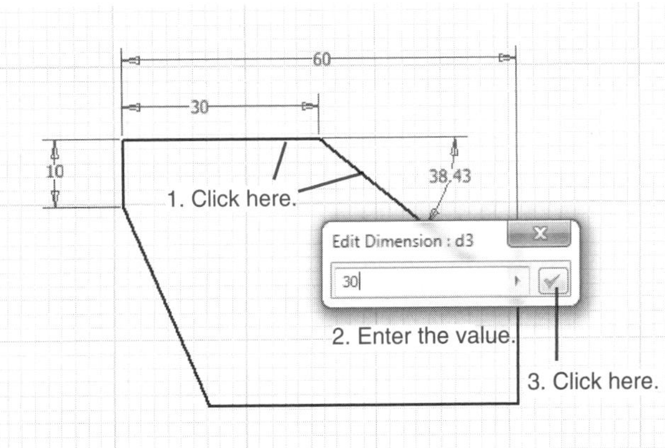

Figure 1-21

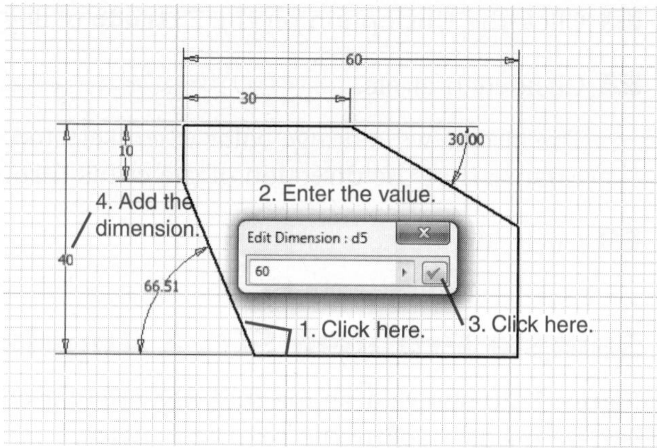

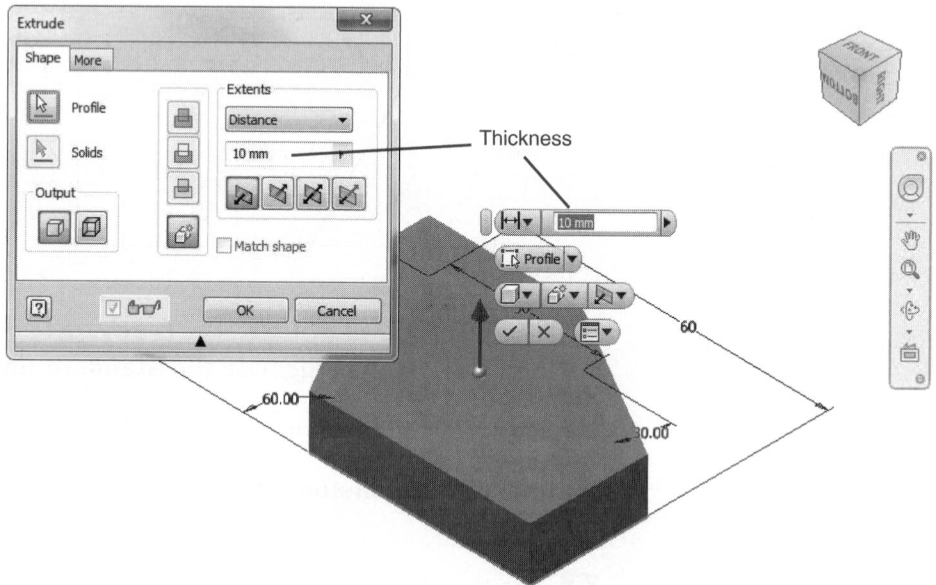

Figure 1-21

(Continued)

5 Enter a value of **30;** click the **green check mark**.

6 Add the overall **40** and the angular **60** dimensions as shown.

7 Use the **Extrude** tool and add a thickness of **10 mm**.

To Rotate the Object

See Figure 1-22.

1 Click the **Free Orbit** tool located along the right edge of the drawing screen.

2 Hold the left mouse button down and rotate the object.

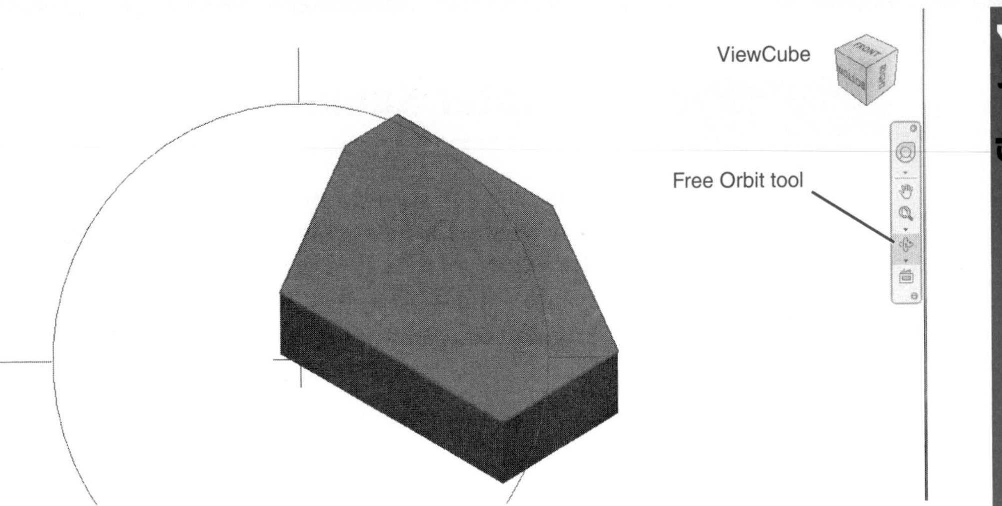

Figure 1-22

To Return the Object to Its Initial Isometric Position

See Figure 1-23.

1. Move the cursor into the area of the ViewCube.

   A houselike icon will appear. This is the **Home** tool.

2. Click the **Home** tool.

   The object will return to its original isometric shape.

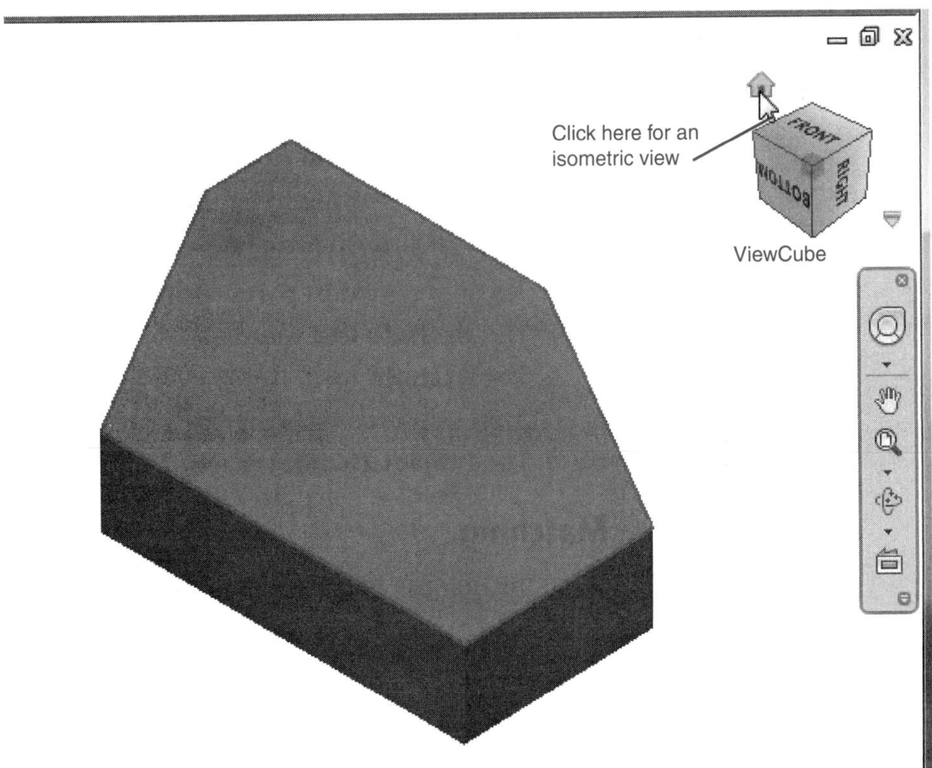

Figure 1-23

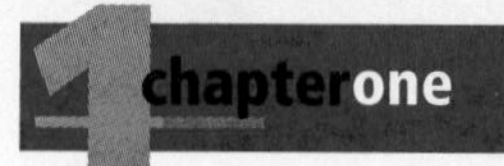

Chapter Summary

This chapter showed how to get started using Inventor 2015. It showed how to start a new drawing, how to create 2D sketches, and how to change the sketches into dimensioned drawings. The drawings were then used to create solid 3D models. Sample problems showed how to add holes and angular edges to 3D models.

Chapter Test Questions

Multiple Choice

Circle the correct answer.

1. Which of the following formats is used to create Inventor sketches?
   a. Weldment.iam b. Standard (in).ipt
   c. Standard.ipn d. ISO.dwg
2. The **Dimension** tool is used to perform which of the following?
   a. Add dimensions to a drawing
   b. Create isometric views
   c. Change the orientation of a drawing
3. The **Extrude** tool is located on the
   a. **Model** panel b. **Sketch** panel
   c. **Draw** panel d. **Create** panel
4. The **Free Orbit** tool is used to
   a. Remove the drawing from the drawing screen
   b. Allow the drawing to be rotated into different orientations
   c. Allow the drawing to be used on different drawings
   d. Free the drawing from its existing file name
5. Holes are created in parts using
   a. The **Extrude Cut** tool
   b. The **Extrude** tool
   c. A presentation drawing
   d. The **Project Geometry** tool

Matching

Write the number of the correct answer on the line.

Column A

a. **Save As** ______
b. **Home** ______
c. Standard (mm).ipt ______
d. **Extrude** ______
e. **Extrude Cut** ______

Column B

1. The format used to start a new metric drawing
2. Used to add thickness to a sketch
3. Used to define the file name of a drawing
4. Used to change a drawing's orientation to isometric
5. Used to create a hole from a circle

True or False

Circle the correct answer.

1. **True or False:** Sketches are created using the ANSI (mm).idw format.
2. **True or False:** Drawings with inch dimensions are created using the Standard (in).ipt format.
3. **True or False:** The ViewCube is used to change the orientation of a drawing.
4. **True or False:** Holes can be created using **Hole** and **Extrude Cut** tools.
5. **True or False:** The **Fix** constraint is used to fix a point in space.

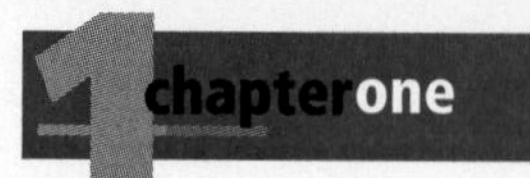

Chapter Project

Project 1-1

Sketch the shapes shown in Figures P1-1 through P1-8, then create solid models using the specified thickness values.

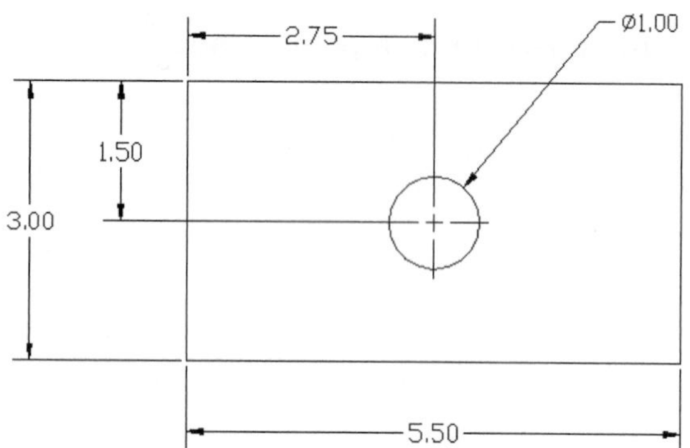

Figure P1-1
Inches

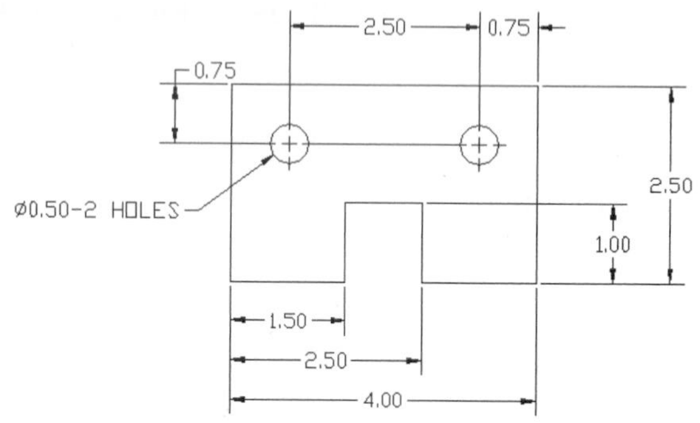

Figure P1-2
Inches

> **NOTE**
> This exercise is presented in Sample Problem SP1-1.

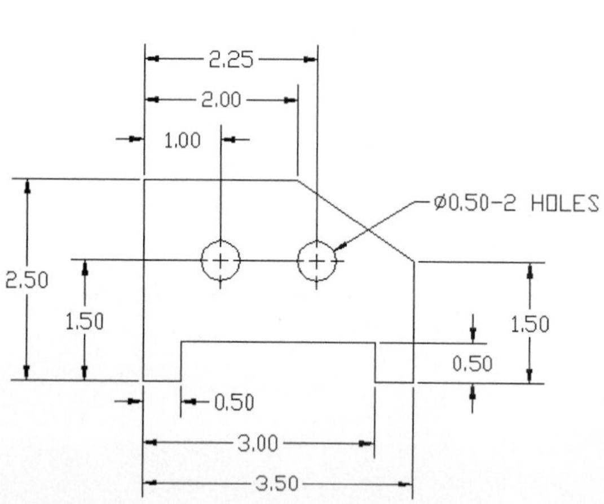

Figure P1-3
Inches

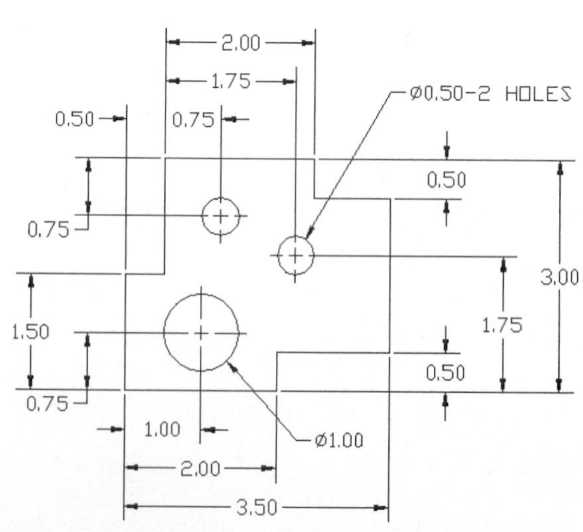

Figure P1-4
Inches

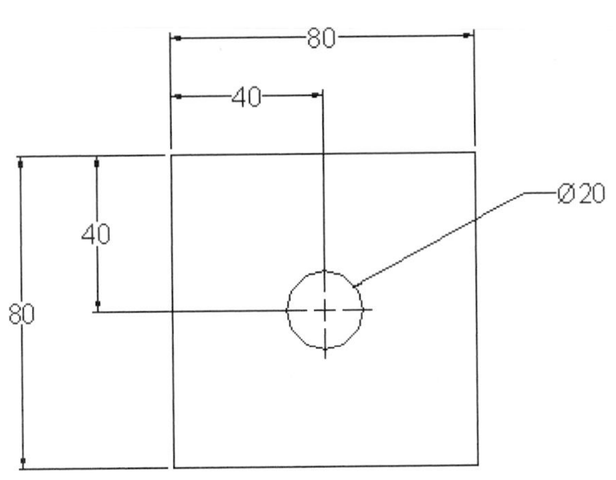

THICKNESS = 10

Figure P1-5
Millimeters

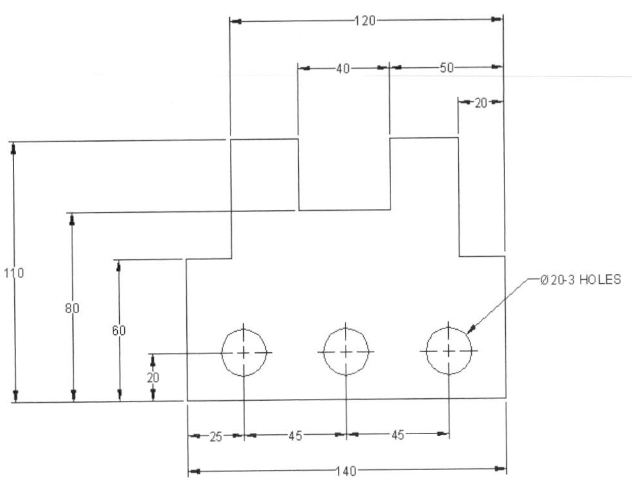

THICKNESS = 15

Figure P1-6
Millimeters

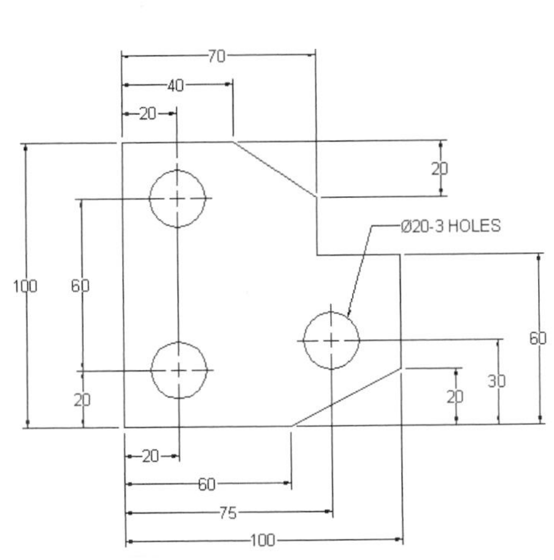

THICKNESS = 5

Figure P1-7
Millimeters

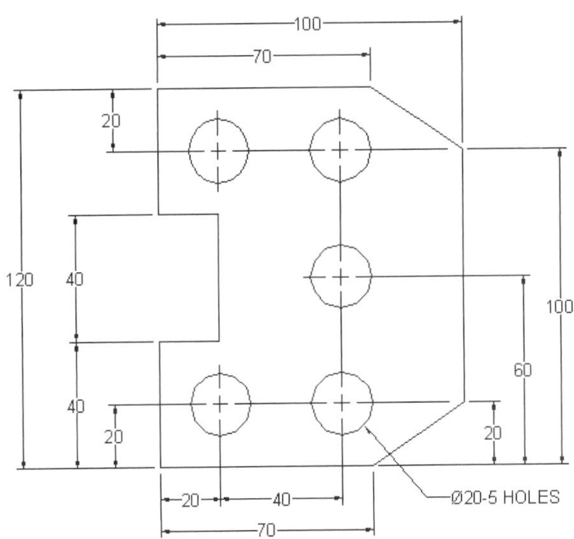

THICKNESS = 12

Figure P1-8
Millimeters

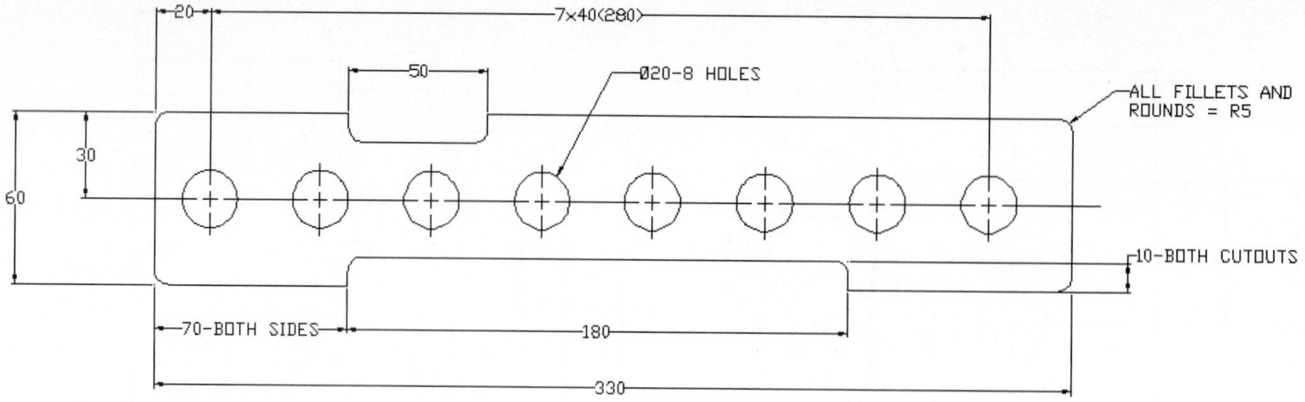

THICKNESS = 2

Figure P1-9
Millimeters

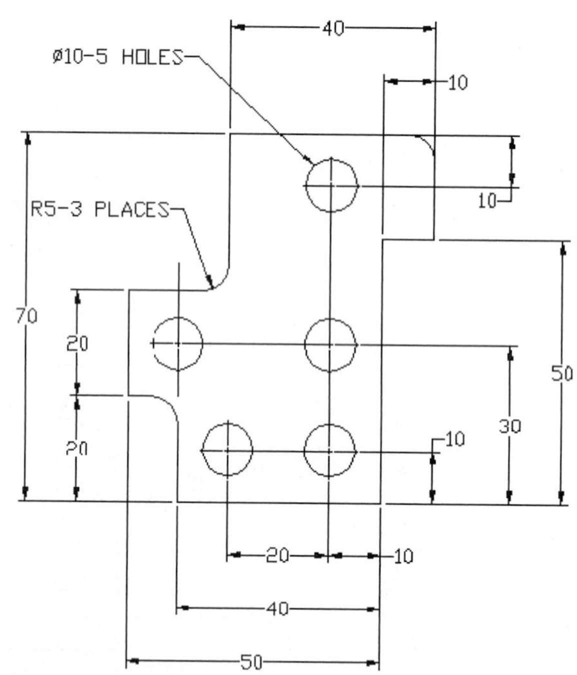

THICKNESS = 8.25

Figure P1-10
Millimeters

2 chapter two

Two-Dimensional Sketching

CHAPTER OBJECTIVES

- Learn about the 2D sketching tools
- Learn how to combine 2D sketching tools to form more complex shapes
- Learn how to edit sketches

Introduction

This chapter introduces most of the tools found on the **Sketch** panel, that is, those tools found under the **Sketch** tab. Inventor 2D models are usually based on an initial 2D sketch that is extended and manipulated to create a final 3D solid model.

Each tool is presented with a short sample application to introduce the tool and show how it can be used.

The Sketch Panel

To Access the Sketch Panel

1. Click the **New** tool.

   The **Create New File** dialog box will appear. See Figure 2-1.

2. Click the **Metric** tab.

3. Scroll down and click the **Standard (mm).ipt** format.

Figure 2-1

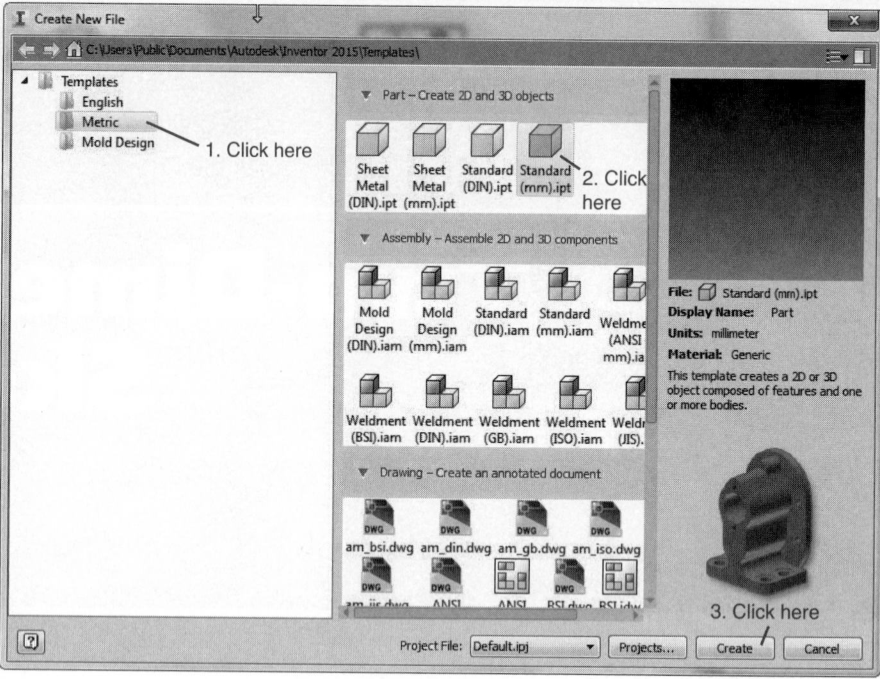

4 Click **Create**.

5 Click the **Create 2D Sketch** tool on the **Sketch** panel.

6 Select the plane for the sketch.

See Figure 2-2.

Figure 2-2

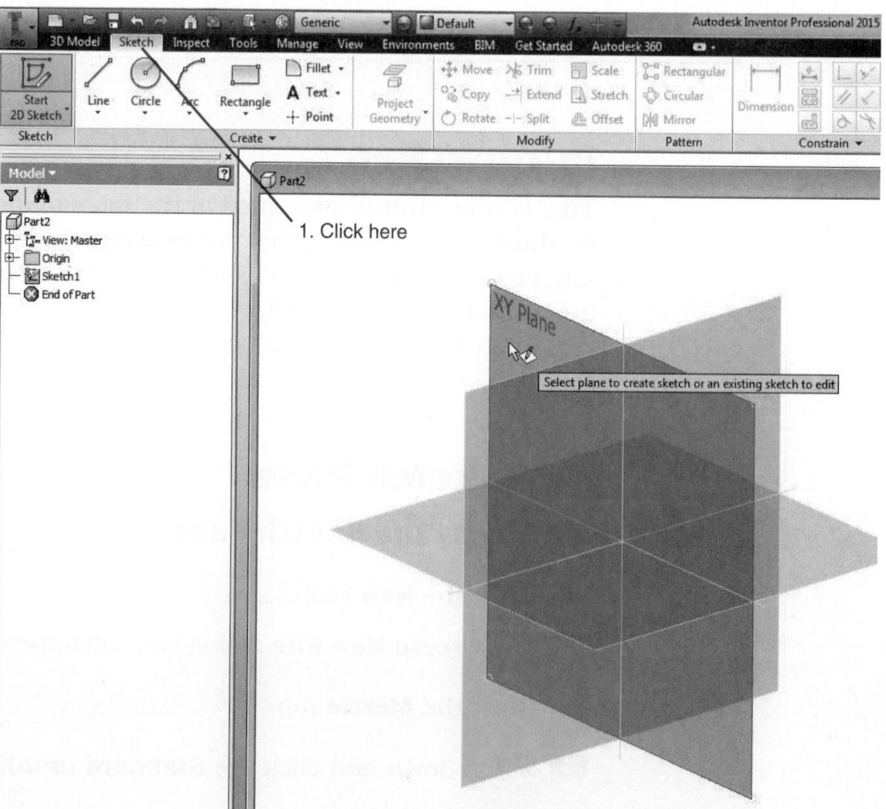

Figure 2-2
(Continued)

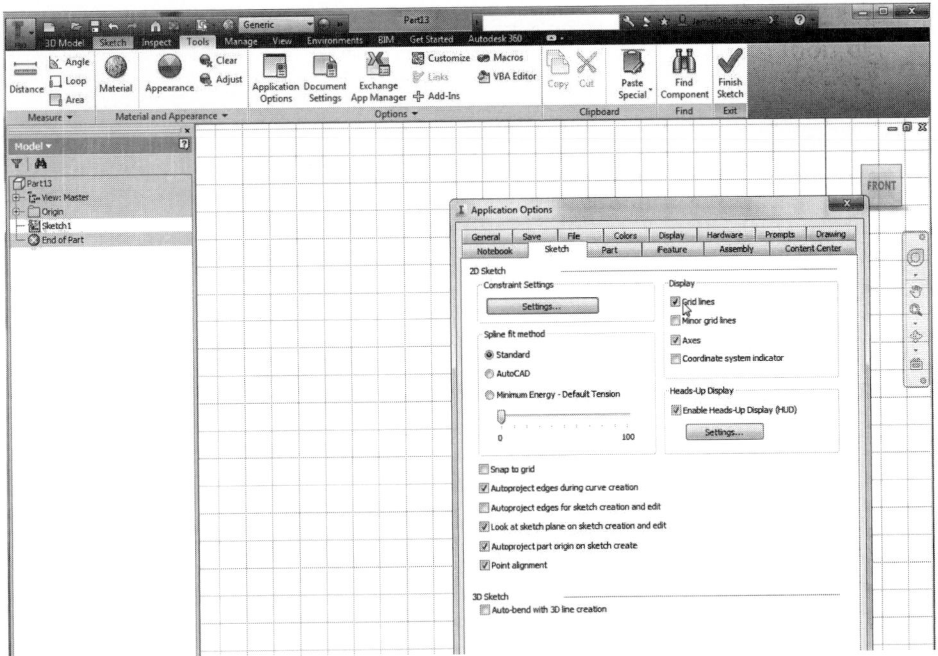

In this example, the XY plane was selected. The one-color grid background is optional. It is included in this chapter for visual referencing and to make it easier to see the figures on the screen.

To Add a Grid Background

1. Click the **Tools** tab and select the **Application Options** tool on the **Options** panel.

2. Click the **Sketch** tab and turn on the **Grid lines** option under the **Display** heading.

   A check mark in the box next to the **Grid lines** heading indicates that the **Grid lines** option is on.

3. Click **Apply** and **Close.**

To Remove the Gradient Background

See Figure 1-3.

1. Click the **Tools** tab and select the **Application Options** tool.

2. The **Application Options** dialog box will appear.

3. Click the **Colors** tab, click the arrowhead to the right of the **Gradient** option located under the **Background** heading, and select the **1 Color** option.

Line

The **Line** tool is used to draw individual straight lines. See Figure 2-3.

1. Click the **Line** tool located on the **Create** panel under the **Sketch** tab.

2. Select a point on the drawing screen, click and release the left mouse button, and drag the cursor across the screen.

Figure 2-3

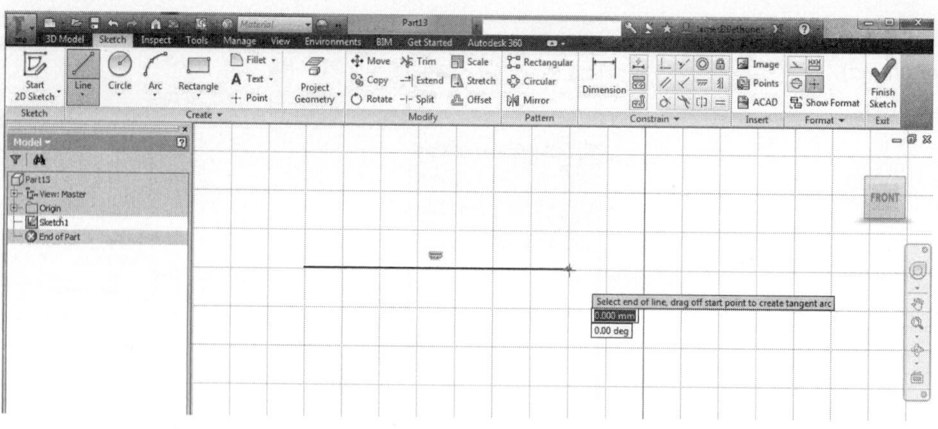

Select end of line, drag off start point to create tangent arc
0.000 mm
0.00 deg

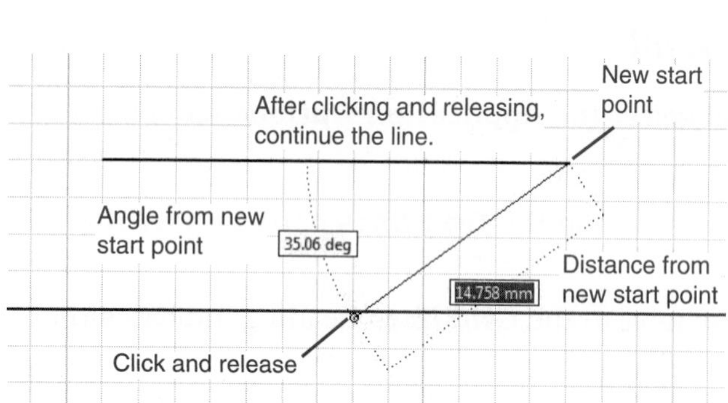

After clicking and releasing, continue the line.
New start point
Angle from new start point
35.06 deg
14.758 mm
Distance from new start point
Click and release

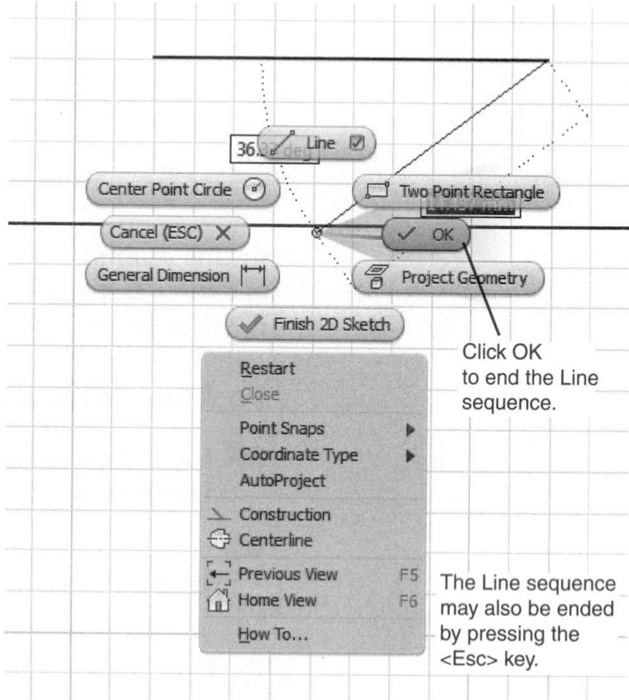

Line
Center Point Circle
Two Point Rectangle
Cancel (ESC)
OK
General Dimension
Project Geometry
Finish 2D Sketch
Click OK to end the Line sequence.
Restart
Close
Point Snaps
Coordinate Type
AutoProject
Construction
Centerline
Previous View F5
Home View F6
How To...
The Line sequence may also be ended by pressing the <Esc> key.

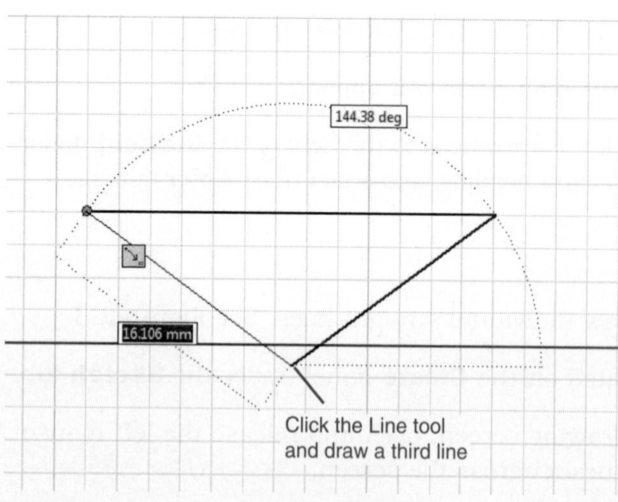

144.38 deg
16.106 mm
Click the Line tool and draw a third line

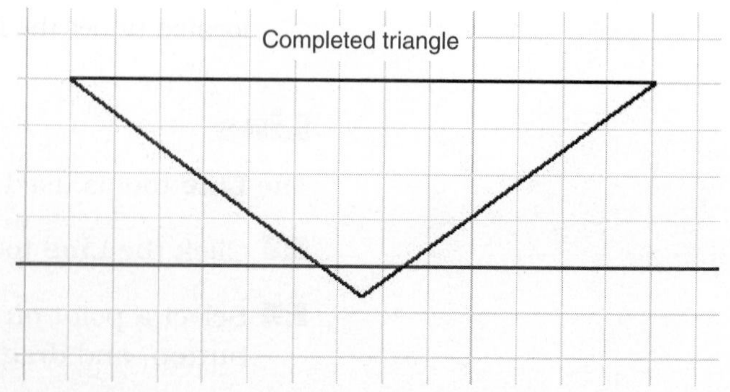

Completed triangle

3 Select an endpoint for the line and again click and release the left mouse button.

As the line is being sketched a box with a dark background will appear above the line, defining the length of the line in real time. Another box, with a light background, will also appear giving the angle of the line relative to the horizontal.

> **NOTE**
> 0° is defined as a horizontal line to the right of the starting point.

4 Continue the line by moving the cursor.

A distance and an angle value will again appear. These values refer to the new starting point. The new starting point will be the same as the end-point of the horizontal line, but the values that are displayed will be based on the new starting point.

5 Select an endpoint and click the left mouse button.

6 Right-click the mouse and click the **OK** option.

This will end the **Line** sequence.

> **NOTE**
> The <Esc> key can also be used to end a command sequence.

7 Start a new line by clicking the **Line** tool and moving the cursor to the endpoint of the second line.

A colored, filled circle will appear on the endpoint when it is selected.

8 Click the point and drag the cursor to the original starting point of the horizontal line.

9 Click the point, right-click the mouse, and select the **OK** option.

To Define the Lengths of the Lines

Figure 2-3 shows an enclosed figure. It was sketched; that is, the lengths and angles of the lines were estimated. The lengths of the lines will now be defined.

1 Click the **Dimension** tool located on the **Constrain** panel under the **Sketch** tab.

2 Click the horizontal line, move the cursor away from the line, and click a location point.

The **Edit Dimension** value box will appear. See Figure 2-4.

3 Enter a value for the dimension and click the **green check mark**.

In this example **25** was entered.

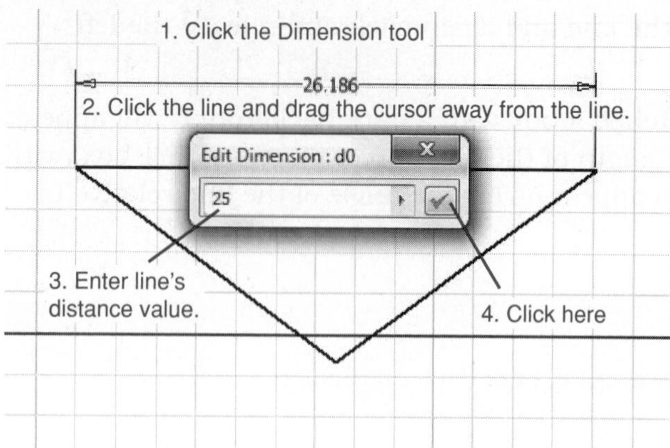

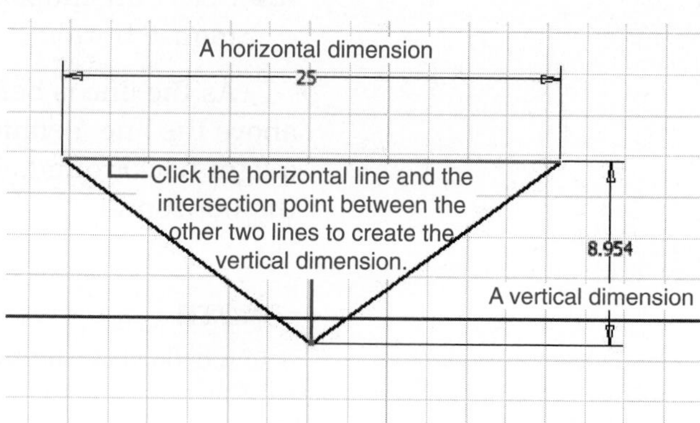

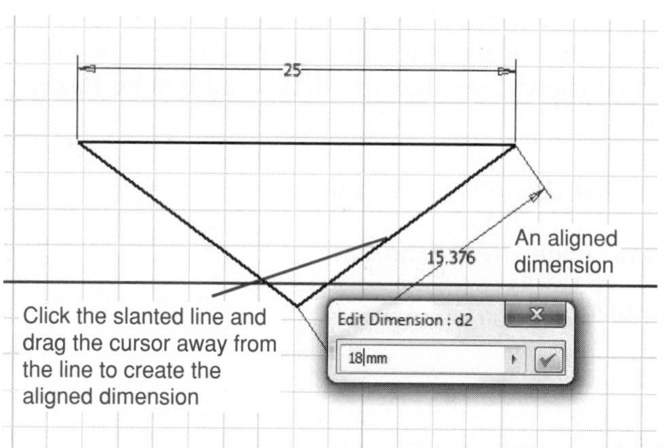

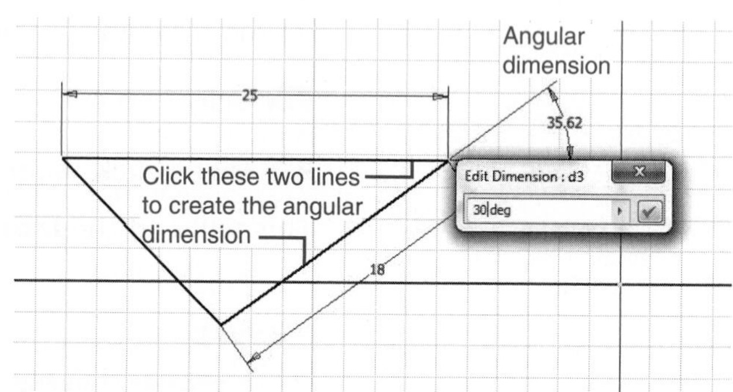

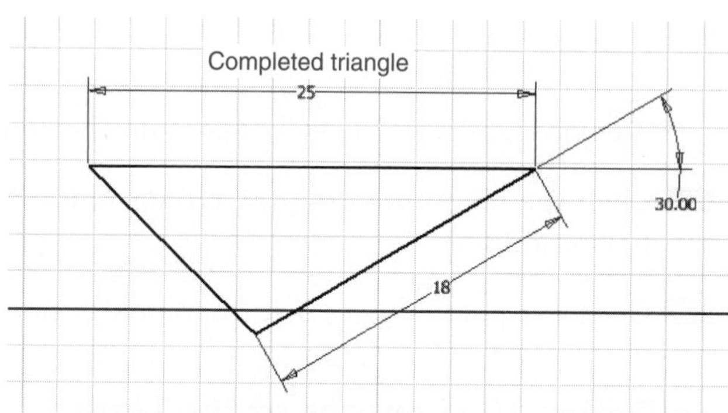

Figure 2-4

4 Click the **Dimension** tool, then click the horizontal line and the intersection point between the two slanted lines, and move the cursor to the right.

A horizontal or a vertical line will appear.

5 Undo the horizontal dimension. Click the **Dimension** tool, click the right slanted line, and move the cursor away from the line.

An aligned dimension will appear with the **Edit Dimension** value box.

6 Enter a value into the **Edit Dimension** value box.

In this example a value of **18** was entered.

7. Click the **green check mark**.

8. Click the **Dimension** tool, click the horizontal line and the right slanted line, and move the cursor up and to the right.

   An angular dimension will appear.

9. Enter an angular value; click the **green check mark**.

Chapter 7 will explain when to use the different types of dimensions to define a drawing.

The object is now completely defined; that is, no more dimensions are needed. If you try to dimension the left slanted line, an error message will appear. See Figure 2-5. If you wished to dimension the length of the left slanted line, you would have to remove one of the other dimensions.

Figure 2-5

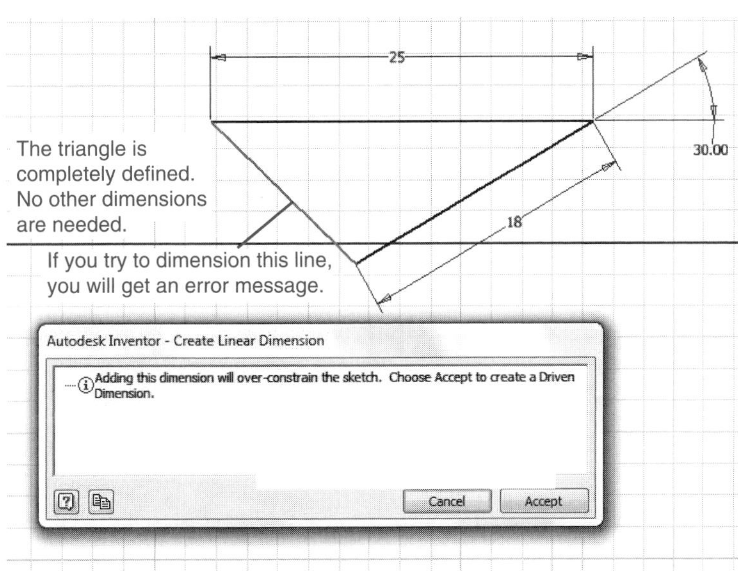

Circle

There are two ways to draw circles in Inventor: by using the **Circle Center Point** tool or the **Circle Tangent** tool.

To Draw a Circle with Circle Center Point

1. Start a new drawing, click the **English** tab, click the **Standard (in).ipt** format, and click **Create**. Select the **Create 2D Sketch** option, and select the XY plane for the sketch.

2. Click the **Circle** tool located on the **Create** panel under the **Sketch** tab.

3. Select a center point and click the left mouse button.

4. Release the mouse button and drag the cursor away from the center point.

5. Select a diameter point and click the left mouse button.

   See Figure 2-6.

Figure 2-6

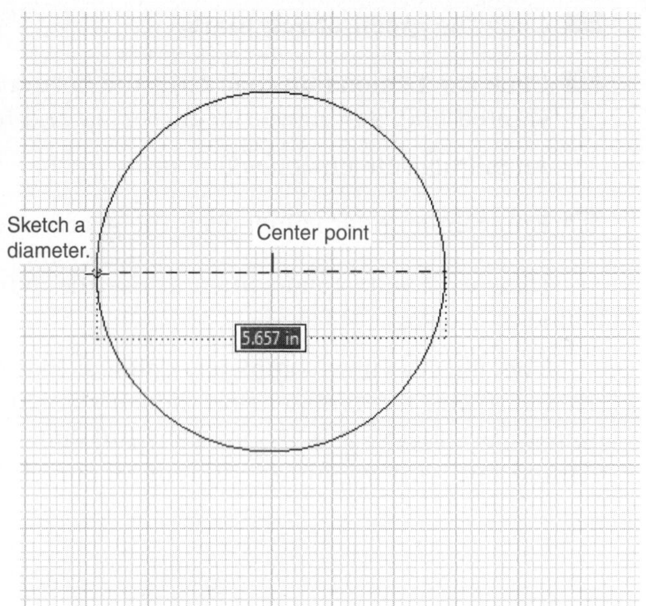

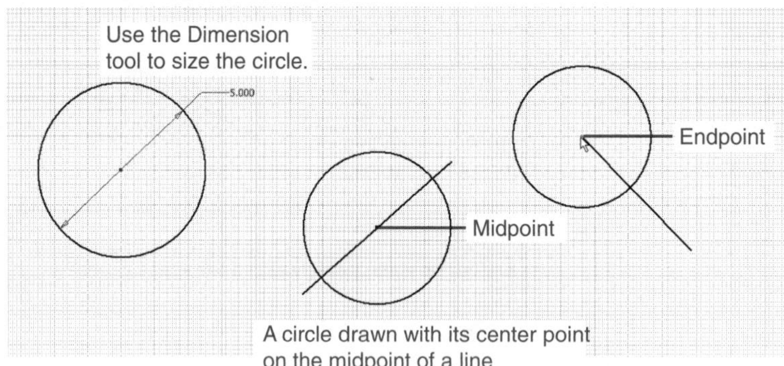

6 Use the **Dimension** tool and size the circle.

In this example a value of **5.00** was selected. The center point for a circle may also be located in association with other features.

To Draw a Circle with Its Center Point on the Midpoint of a Line

1 Click the **Circle** tool and move the cursor to the existing line.

2 Move the cursor up and down the line to locate the line's midpoint.

When the line's midpoint is located and touched, a colored, filled circle will appear.

3 Click the line's midpoint, release the cursor, and draw the cursor away from the point, creating a circle.

4 Click a point to define the circle's diameter.

5 Use the **Dimension** tool to size the circle.

Figure 2-6 also shows a circle created with its center point on the endpoint of a line.

To Draw a Circle Using the Circle Tangent Tool

1 Start a new drawing, click the **Metric** tab, click the **Standard (mm).ipt** format, and click **Create**. Select the **Create 2D Sketch** option, and select the XY plane for the sketch.

2 Sketch an equilateral triangle.

See Figure 2-7. Any length side is acceptable.

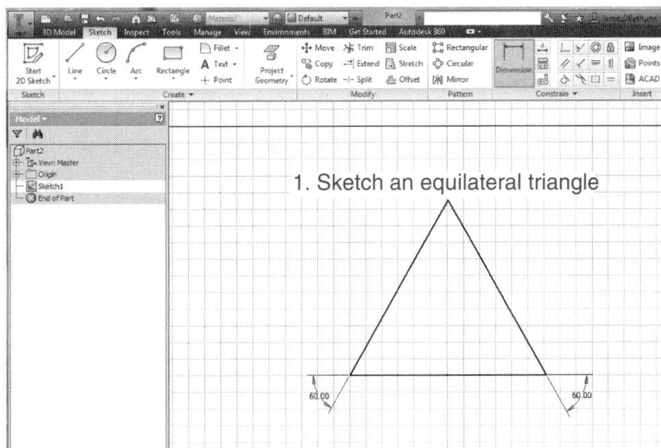

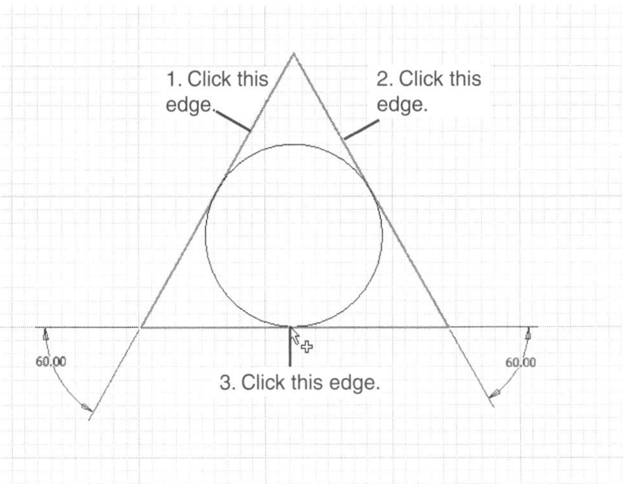

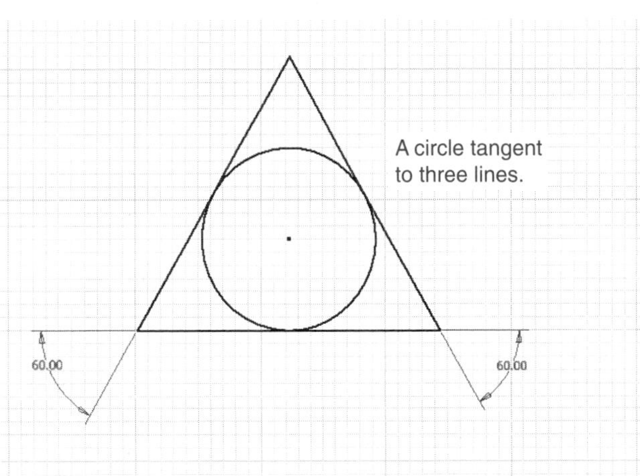

Figure 2-7

3 Click the **Circle Tangent** tool.

4 Click the left edge line, click the right edge line, and click the bottom edge line.

A circle will appear tangent to the three lines.

Arc

There are three types of arcs: three-point, tangent, and center point.

To Draw a Three-Point Arc

1. Start a new drawing, click the **Metric** tab, click the **Standard (mm).ipt** format, and click **Create**. Select the **Create 2D Sketch** option, and select the XY plane for the sketch.

   See Figure 2-8.

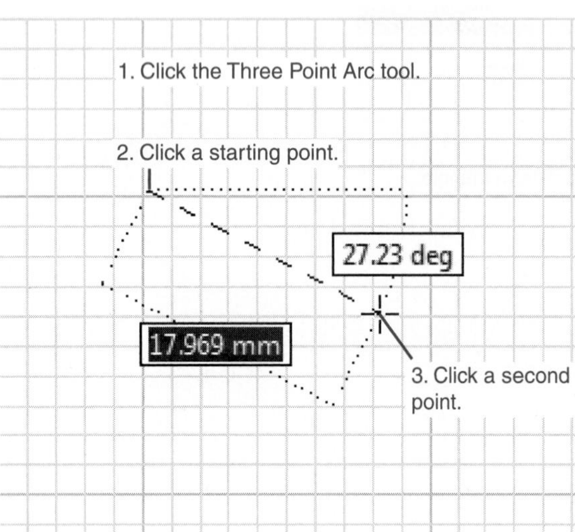

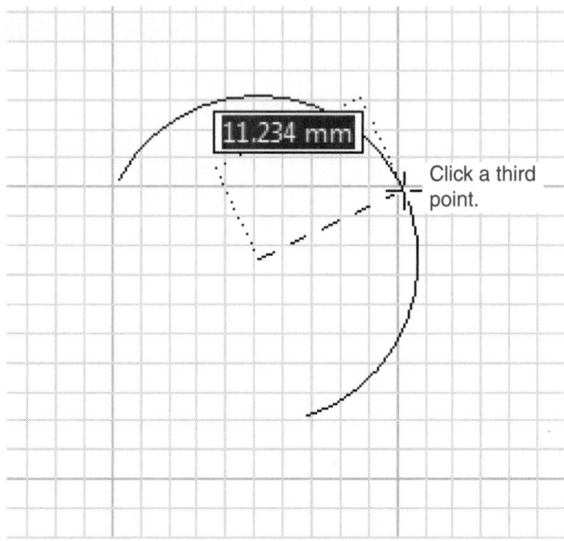

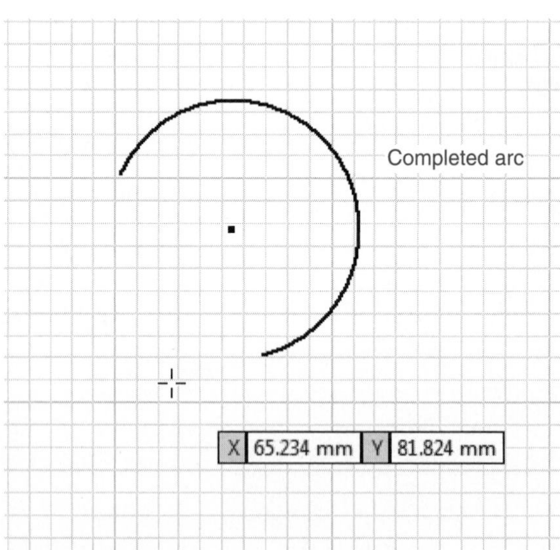

Figure 2-8

2. Click the **Three Point Arc** tool located on the **Create** panel under the **Sketch** tab.
3. Select a starting point for the arc.
4. Select a second point.

5. Select the third point.

6. Right-click the mouse and select the **OK** option.

> **NOTE**
> Arcs are sometimes difficult to control. The **Dimension** and **Trim** tools can be used to edit an arc after it has been sketched.

To Draw a Tangent Arc

A tangent arc requires another entity to be present. In this example a line is first drawn on the screen.

1. Start a new drawing, click the **Metric** tab, click the **Standard (mm).ipt** format, and click **Create**. Select the **Create 2D Sketch** option, and select the XY plane for the sketch.

2. Draw a line.

   See Figure 2-9.

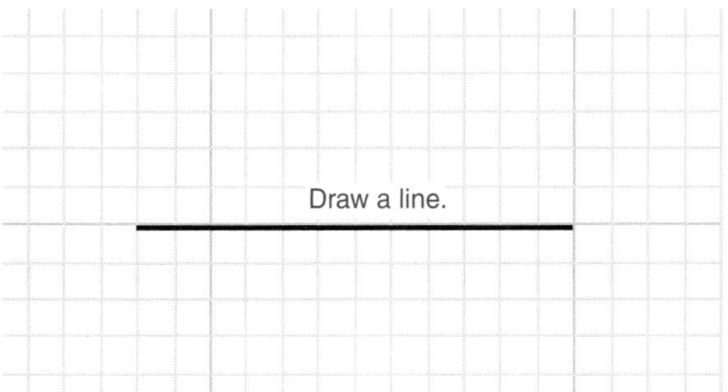

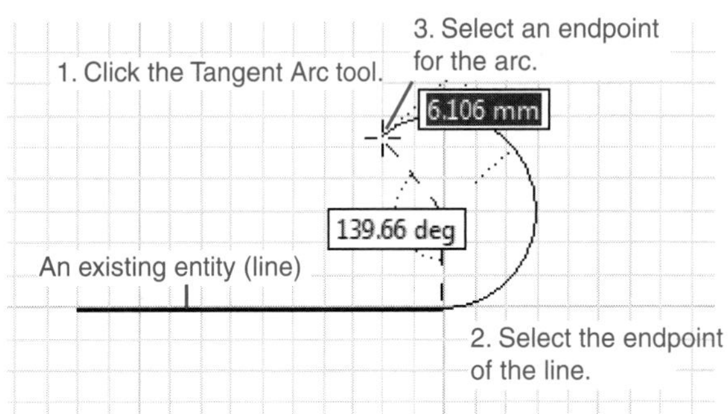

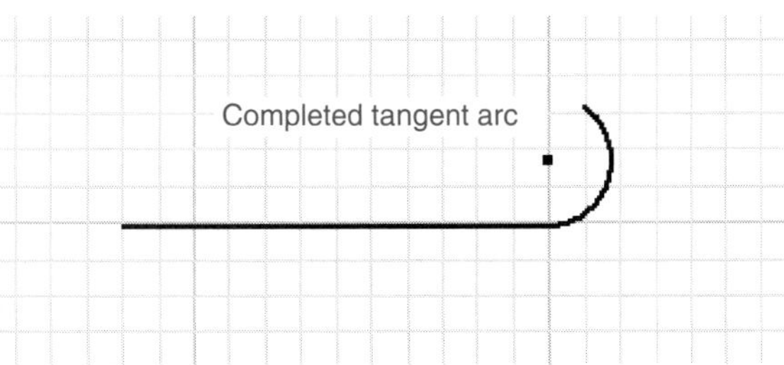

Figure 2-9

3. Click the **Tangent Arc** tool.

4. Click the endpoint of the line.

   The line's endpoint will automatically be selected as long as the line is clicked near the endpoint.

5. Click a second point for the arc.

6. Right-click the mouse and select the **OK** option.

To Draw a Center Point Arc

1. Start a new drawing, click the **Metric** tab, click the **Standard (mm).ipt** format, and click **Create**. Select the **Create 2D Sketch** option, and select the XY plane for the sketch.

2. Click the **Center Point Arc** tool.

   See Figure 2-10.

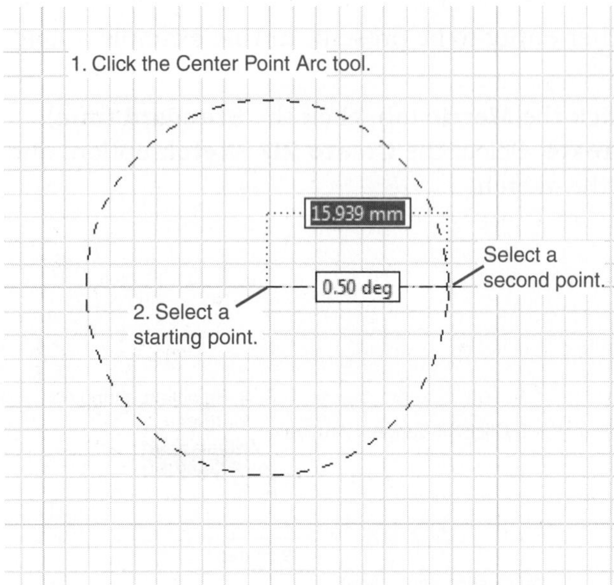

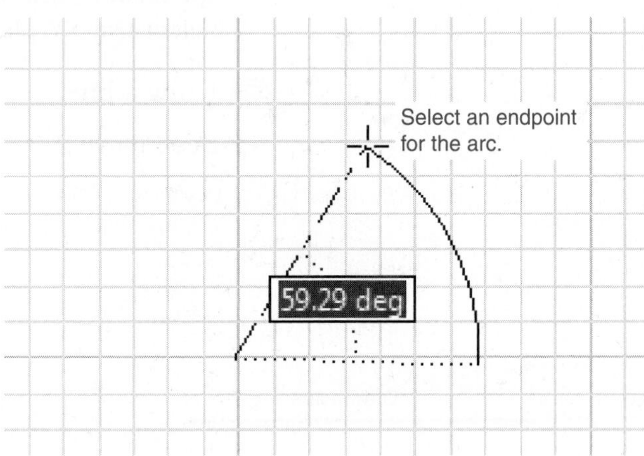

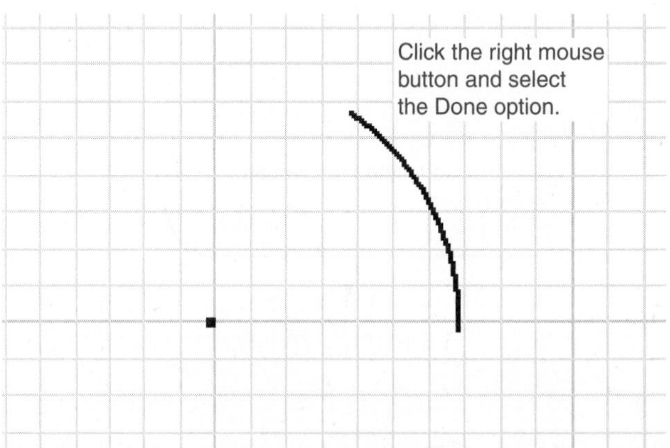

Figure 2-10

3. Select a starting point for the arc, drag the cursor away from the point, and select a second arc point.

4. Drag the cursor and select a third arc point.

5. Right-click the mouse and select the **Done** option.

Spline

spline: A curved line that is defined by a series of vertices.

closed spline: An enclosed curved line on which the start and endpoints are the same point.

open spline: A curved line whose ends do not meet.

A ***spline*** is a curved line that is defined by a series of vertices. A spline that forms an enclosed area is called a ***closed spline***. A spline that does not form an enclosed area is called an ***open spline***.

To Draw an Open Spline

1. Start a new drawing, click the **Metric** tab, click the **Standard (mm).ipt** format, and click **Create**. Select the **Create 2D Sketch** option, and select the XY plane for the sketch.
2. Click the **Spline** tool located on the **Create** panel under the **Sketch** tab. See Figure 2-11.

Figure 2-11

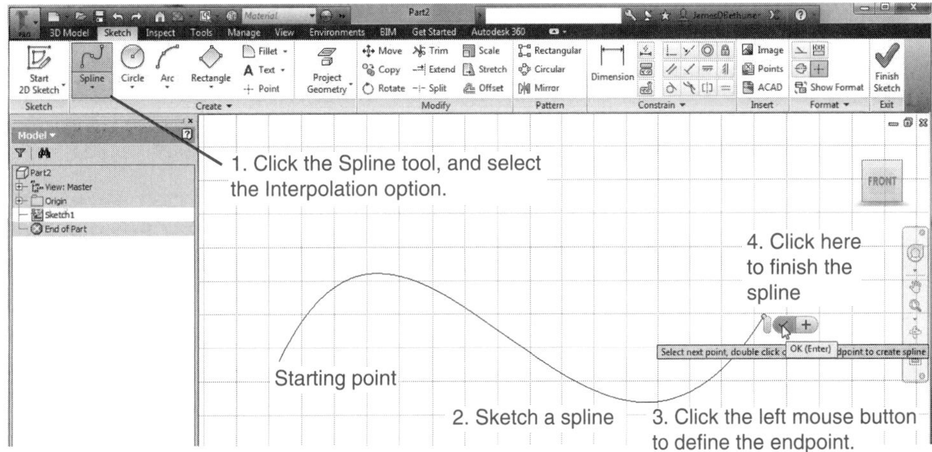
1. Click the Spline tool, and select the Interpolation option.
4. Click here to finish the spline
Starting point
2. Sketch a spline
3. Click the left mouse button to define the endpoint.

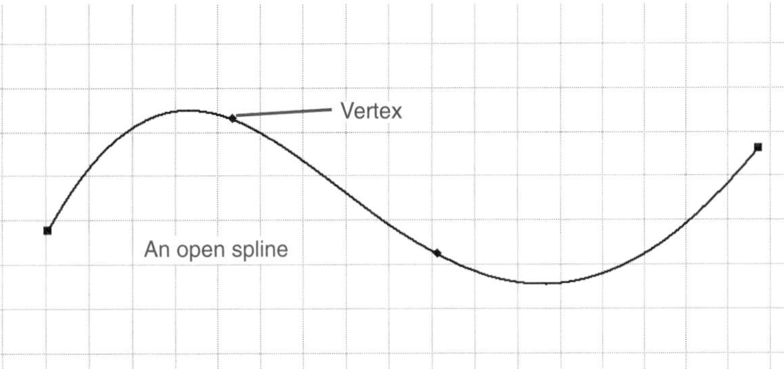
Vertex
An open spline

3. Select the **Spline Interpolation** option.
4. Select a starting point for the spline, select additional points, and select an endpoint.
5. Right-click the mouse and select the **OK** option.

To Edit an Existing Spline

To edit a spline, click and hold one of the vertices and drag it to a new location. See Figure 2-12. Dimensions may also be used to define and edit a spline. Figure 2-13 shows a dimensioned spline. The dimensions locating the second point, 15 and 6, were changed to 10 and 8. Note the difference in shape. Figure 2-14 shows an example of a closed spline.

Figure 2-12

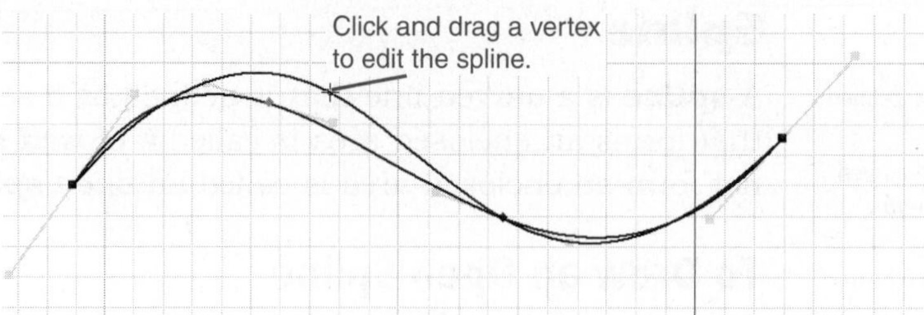

Figure 2-13

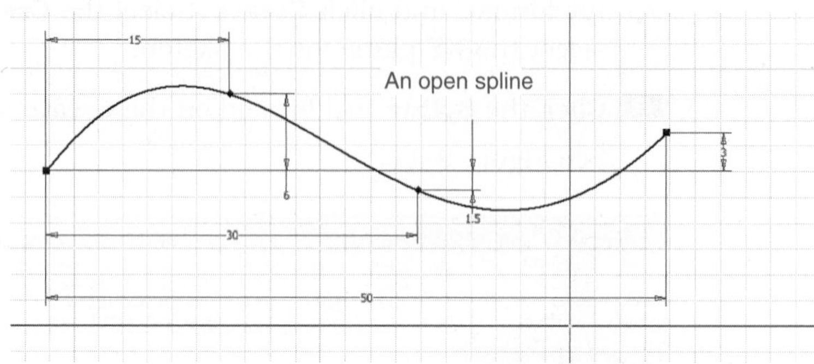

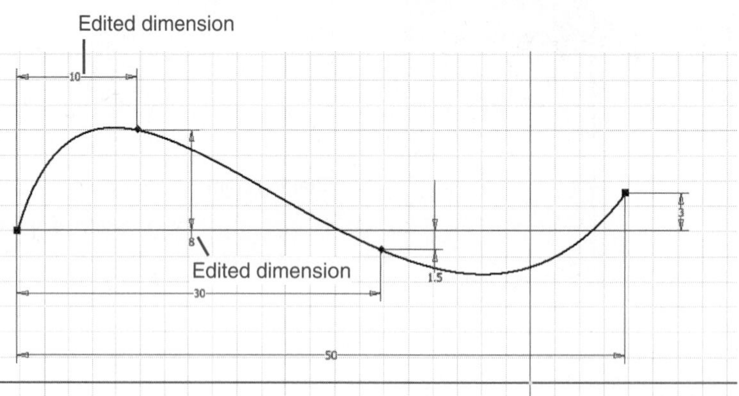

Figure 2-14

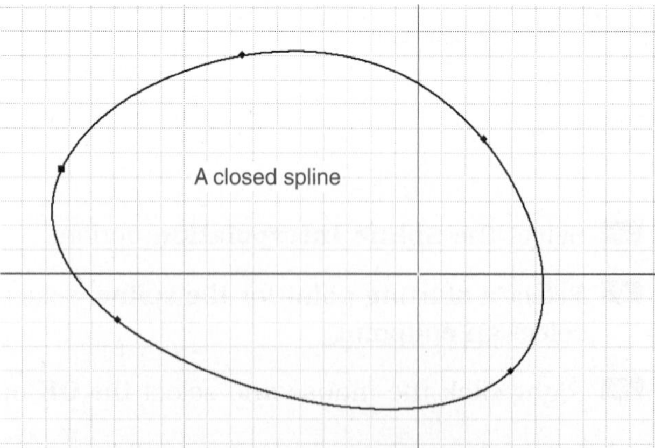

Ellipse

An elliptical shape is defined in one of two ways: by a major and a minor axis, or by a diameter and an angle. Inventor defines an ellipse using a major and a minor axis.

To Draw an Ellipse

1 Start a new drawing, click the **English** tab, click the **Standard (in).ipt** format, and click **Create**. Select the **Create 2D Sketch** option, and select the XY plane for the sketch.

2 Click the **Ellipse** tool located on the **Create** panel under the **Sketch** tab.

See Figure 2-15.

Figure 2-15

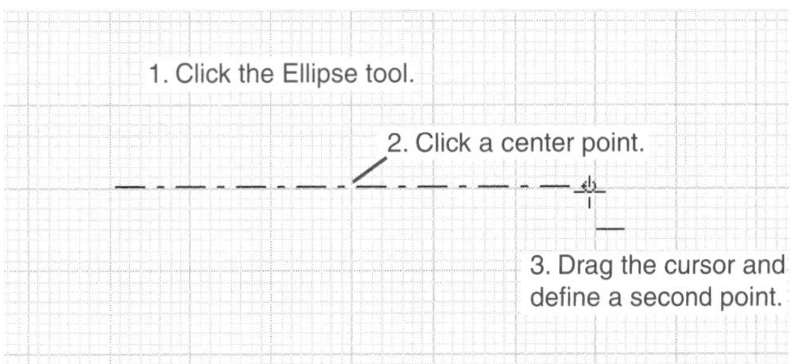
1. Click the Ellipse tool.
2. Click a center point.
3. Drag the cursor and define a second point.

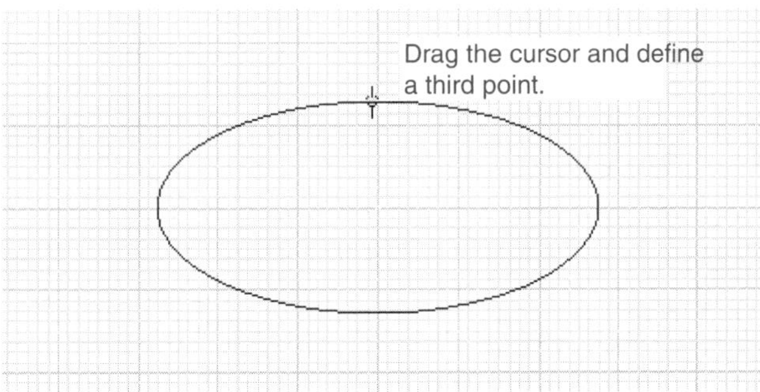
Drag the cursor and define a third point.

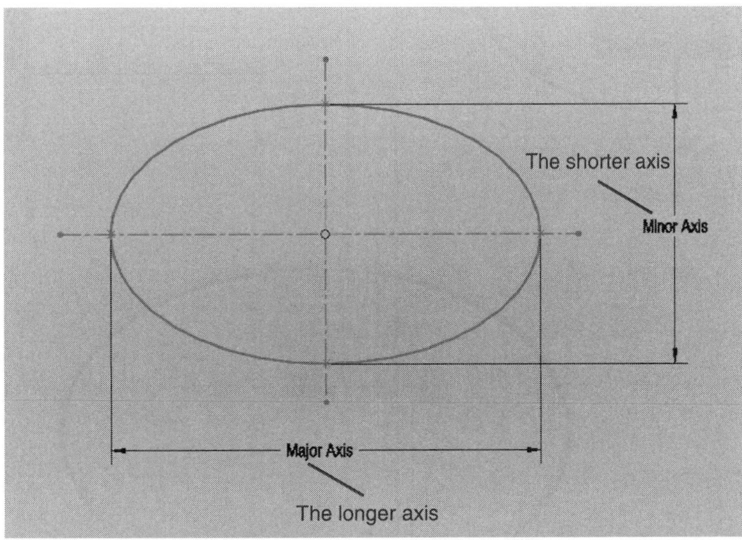
The shorter axis
Minor Axis
Major Axis
The longer axis

3 Select a starting point, click and release the left mouse button, drag the cursor away from the point, and select an endpoint.

The starting point will be the center point of the ellipse. As you drag the cursor, lines will develop in both the left and right directions.

4 Drag the cursor away from the line and click another point.

This point will define either the major or the minor axis, depending on which axis is greater.

An elliptical shape can be drawn to specific dimensions. See Figure 2-16.

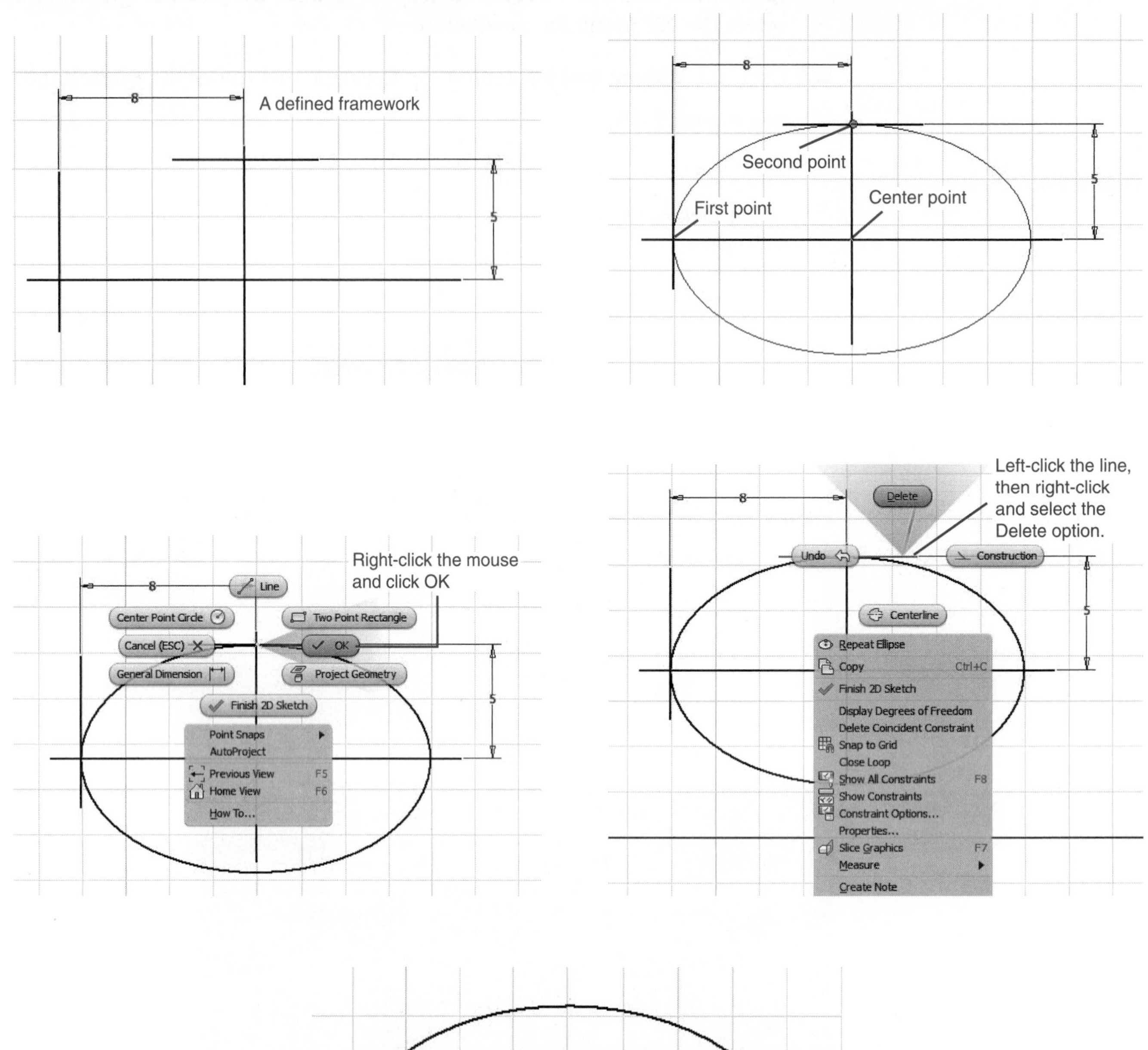

Figure 2-16

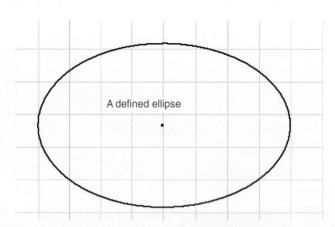

Define and dimension a framework. Use the **Ellipse** tool and define an ellipse based on the intersections of the framework. Use the **Delete** tool to remove the unwanted framework.

Point

The **Point** tool is used to locate points on a sketch. Points are helpful in defining points for a spline, among other uses. Figure 2-17 shows dimensioned points that are then used to define a spline shape.

Figure 2-17

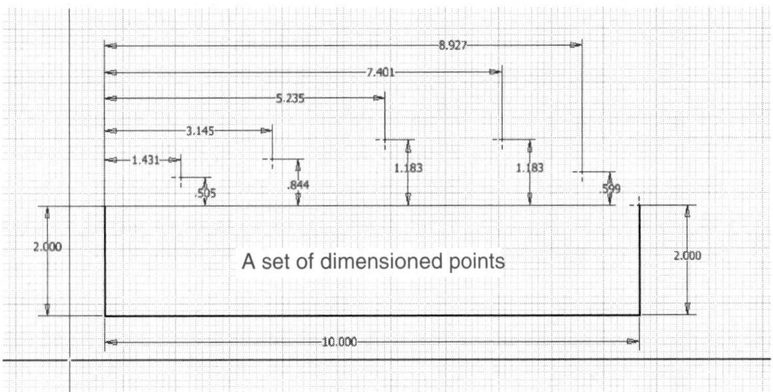

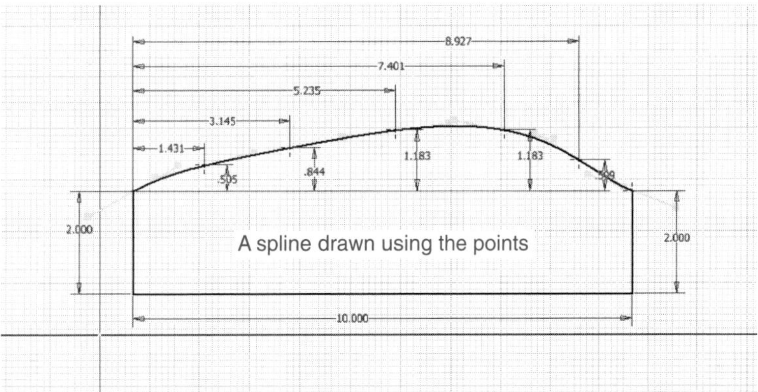

To Create a Point

1. Click the **Point** tool located on the **Create** panel under the **Sketch** tab.
2. Move the cursor to a location and click the left mouse button.

A point will be created at that location. Points can be dimensioned from other drawn features.

Rectangle

There are two ways to draw a rectangle: using either two points or three points.

To Draw a Two-Point Rectangle

1. Start a new drawing, click the **English** tab, click the **Standard (in). ipt** format, and click **Create.** Select the **Create 2D Sketch** option, and select the XY plane for the sketch.
2. Click the **Two Point Rectangle** tool, select a starting point, drag the cursor to a second point, and click the left mouse button.

3. Right-click the mouse and select the **OK** option.

The two points used to define the rectangle are the diagonally opposed corner points of the rectangle. See Figure 2-18.

Figure 2-18

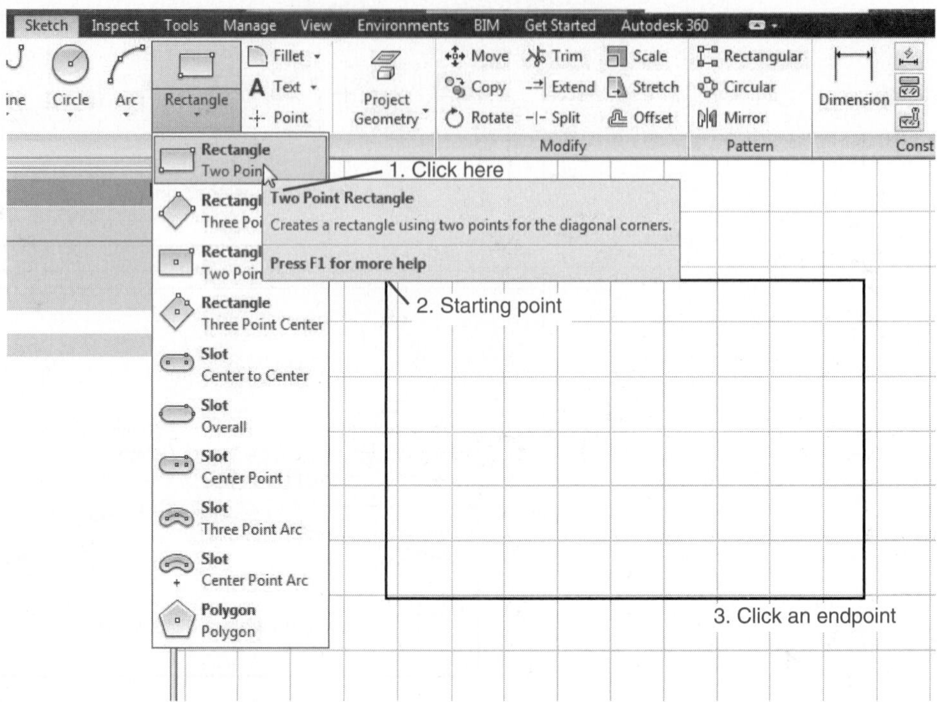

To Draw a Three-Point Rectangle

1. Start a new drawing, click the **English** tab, click the **Standard (in).ipt** format, and click **Create**. Select the **Create 2D Sketch** option, and select the XY plane for the sketch.

2. Click the **Three Point Rectangle** tool, select a starting point, and move the cursor to a second point.

The first two points define the length of one edge of the rectangle. See Figure 2-19.

3. Move the cursor to define a third point.

The distance between the second and third points defines the other edge of the rectangle.

Fillet

fillet: A rounded edge or corner on an entity.

A ***fillet*** is a rounded corner and can be created in either the 2D sketch mode or the 3D model mode. This section presents the 2D commands.

Figure 2-19

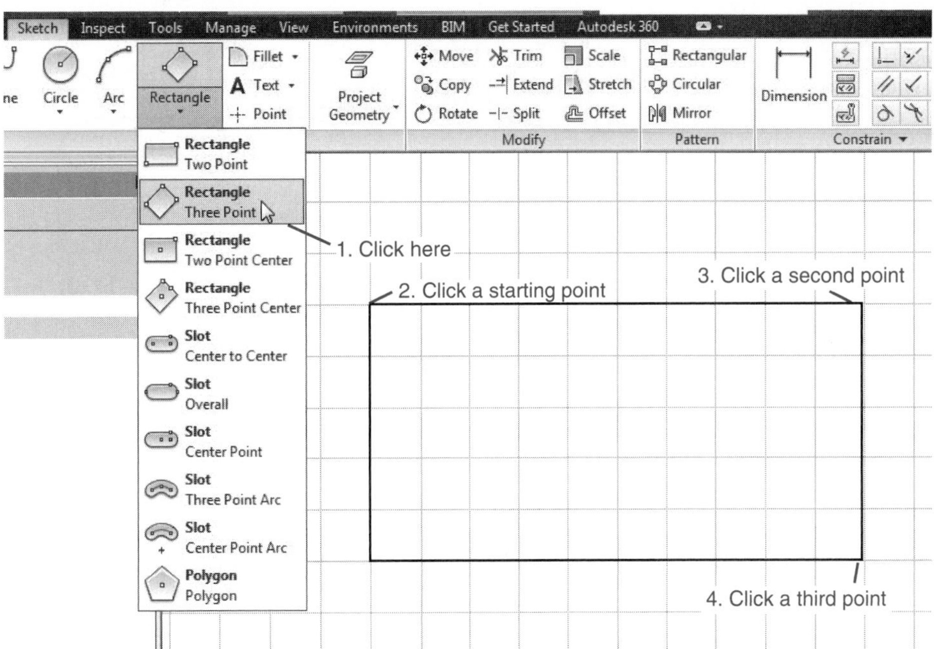

To Draw a Fillet

1. Start a new drawing, click the **Metric** tab, click the **Standard (in).ipt** format, and click **Create**. Select the **Create 2D Sketch** option, and select the XY plane for the sketch.

2. Draw a rectangle.

3. Click the **Fillet** tool located on the **Create** panel under the **Sketch** tab.

   The **2D Fillet** dialog box will appear. See Figure 2-20.

Figure 2-20

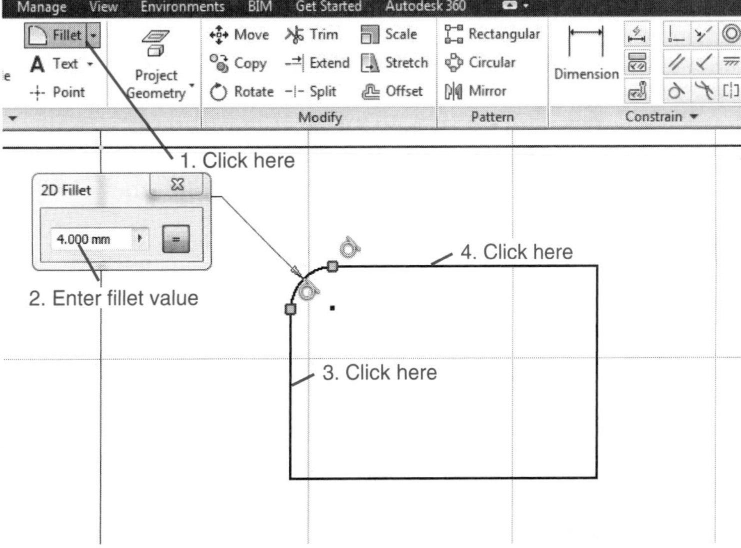

Figure 2-20
(Continued)

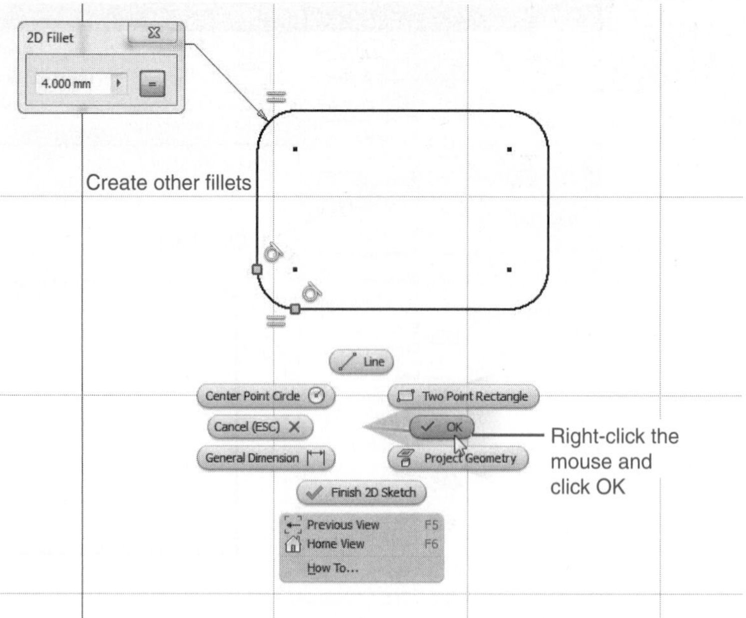

4 Define the radius of the fillet.

In this example a radius value of 4 mm was entered.

5 Click two intersecting lines.

In this example the left vertical line and the top horizontal line were selected.

6 Fillet the other corners of the rectangle.

7 Right-click the mouse and click the **OK** option.

chamfer: An angled edge or corner on an entity.

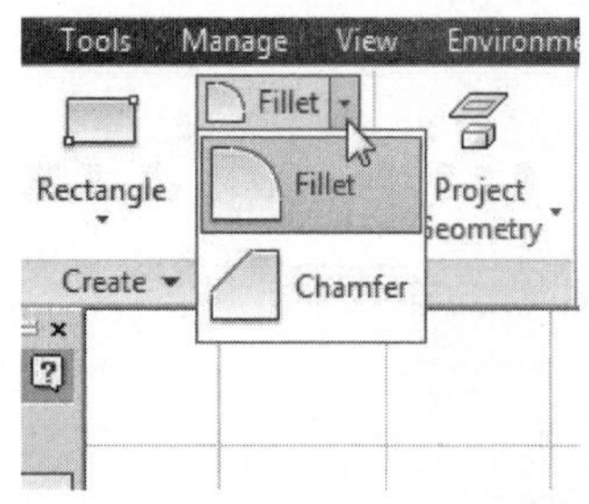

Chamfer

Chamfers are straight-line cuts across corners. They are similar to fillets, but fillets are round, and chamfers are straight.

Figure 2-21 shows the **2D Chamfer** dialog box. There are three different ways to define a chamfer: using two equal distances, two unequal distances, and a distance and an angle.

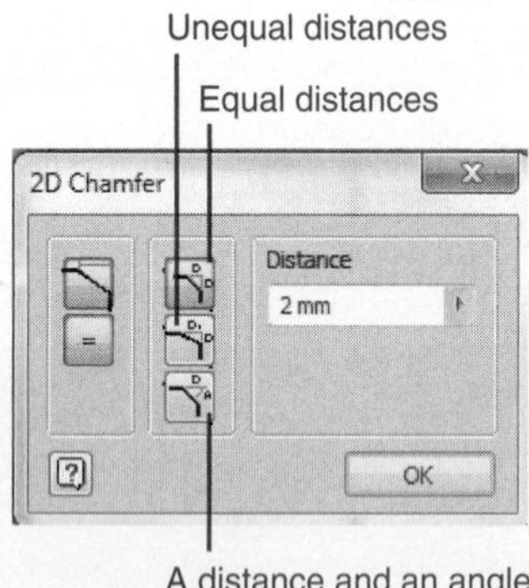

Figure 2-21

To Draw Chamfers

1 Start a new drawing, click the **Metric** tab, click the **Standard (in).ipt** format, and click **Create**. Select the **Create 2D Sketch** option, and select the XY plane for the sketch.

2 Draw a rectangle.

See Figure 2-22.

3 Click the **Chamfer** tool located on the **Create** panel under the **Sketch** tab.

The **Chamfer** tool is a flyout from the **Fillet** tool.

4 Click the **Equal distances** option and enter a value of **5**.

5 Click the left vertical line, then click the top horizontal line.

Figure 2-22

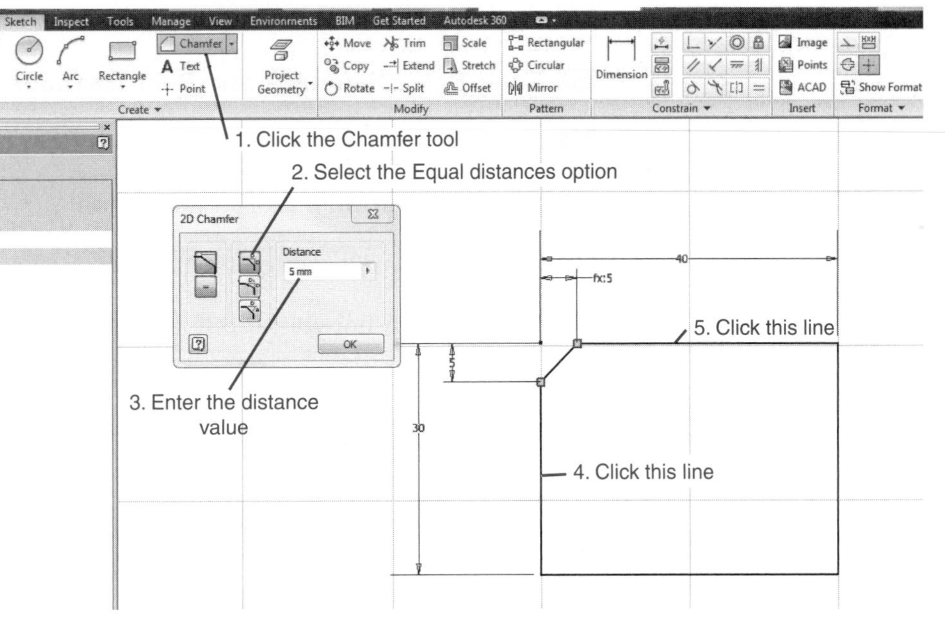
1. Click the Chamfer tool
2. Select the Equal distances option
3. Enter the distance value
4. Click this line
5. Click this line
2D Chamfer
Distance
5 mm
OK

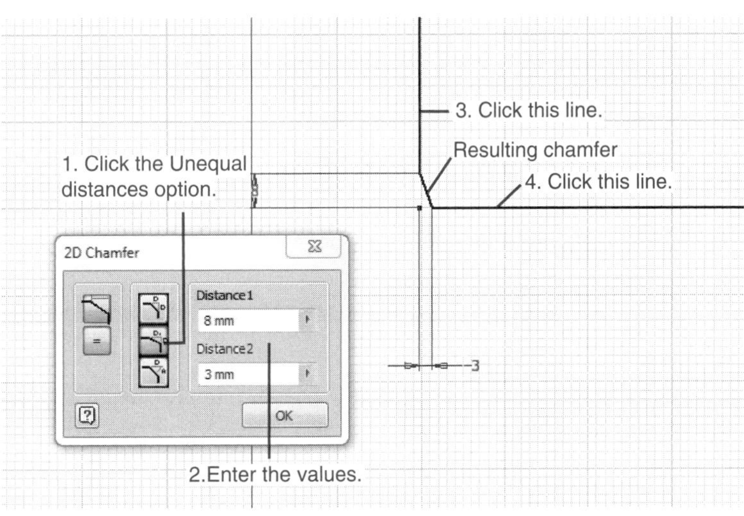
1. Click the Unequal distances option.
2.Enter the values.
3. Click this line.
Resulting chamfer
4. Click this line.
2D Chamfer
Distance1
8 mm
Distance2
3 mm
OK

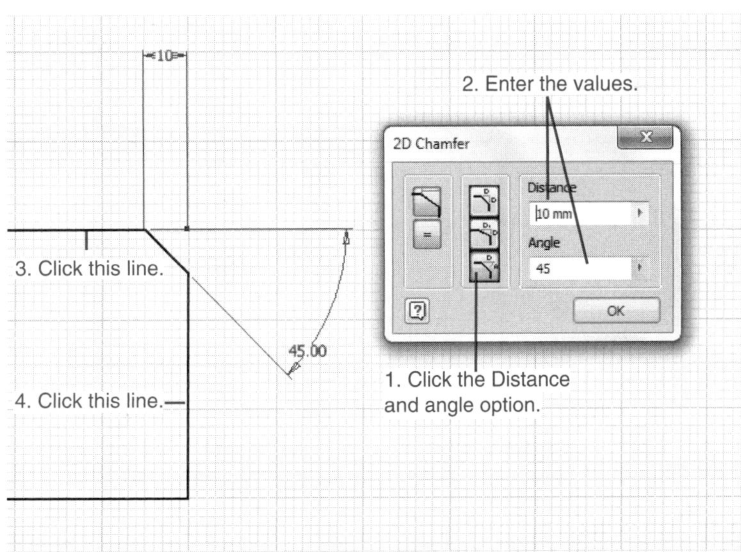
1. Click the Distance and angle option.
2. Enter the values.
3. Click this line.
4. Click this line.
2D Chamfer
Distance
10 mm
Angle
45
OK

6 Click the **Unequal distances** option.

7 Enter **8** for **Distance1** and **3** for **Distance2**.

8 Click the left vertical line and the bottom horizontal line.

9 Click the **Distance and angle** option.

10 Enter a distance value of **10** and an angle value of **45**.

11 Click the top horizontal line and the right vertical line.

Chamfer callouts for drawings are written as follows:

Equal distances: 10 × 10 CHAMFER

Unequal distances: 5 × 10 CHAMFER

Distance and an angle: 10 × 45° CHAMFER

Polygon

polygon: An enclosed figure that has three or more straight-line sides.

A ***polygon*** is an enclosed figure that has three or more straight-line sides. If all the sides are equal, it is called an *equilateral* polygon. Inventor's **Polygon** tool can be used to draw only equilateral polygons. Irregular polygons are created using individual lines.

To Draw a Hexagon

The hexagon shape is commonly used for screw heads and nuts. See Figure 2-23.

Figure 2-23

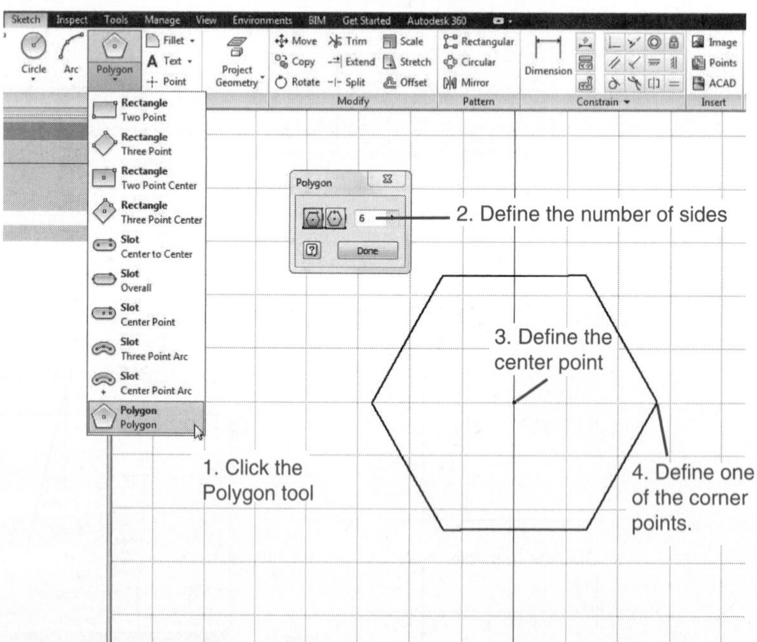
2. Define the number of sides
3. Define the center point
1. Click the Polygon tool
4. Define one of the corner points.

1 Start a new drawing, click the **English** tab, click the **Standard (in).ipt** format, and click **Create**. Select the **Create 2D Sketch** option, and select the XY plane for the sketch.

2 Click the **Polygon** tool located on the **Create** panel under the **Sketch** tab.

The **Polygon** dialog box will appear.

3 Define the required number of sides.

4 Select a center point.

5 Move the cursor away from the center point and define a corner point.

6 Right-click the mouse and click **OK**.

7 Use the **Dimension** tool to define the hexagon's size.

To Define a Hexagon's Size

There are three different dimensions that can be used: the distance across the flats, the distance across the corners, and the edge distance. Each of these distances is defined in Figure 2-24.

Figure 2-24

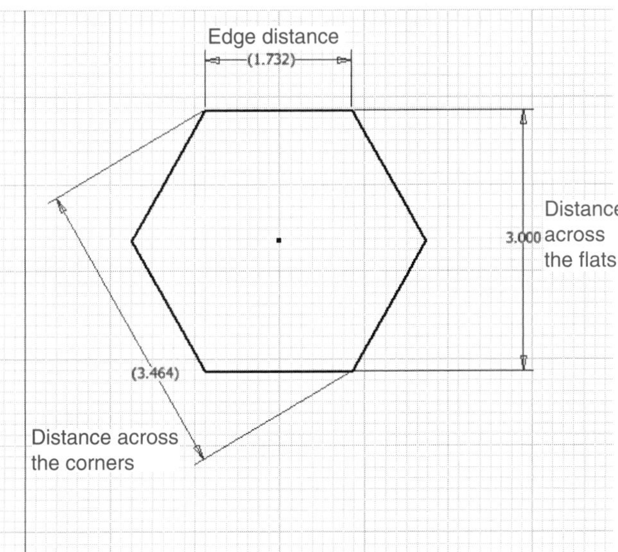

Text

The **Text** tool is used to add text to a sketch. See Figure 2-25.

1 Start a new drawing, click the **English** tab, click the **Standard (in).ipt** format, and click **Create**. Select the **Create 2D Sketch** option, and select the XY plane for the sketch.

2 Click the **Text** tool located on the **Create** panel under the **Sketch** tab.

3 Click the drawing screen.

The **Format Text** dialog box will appear.

4 Type text in the open box as shown.

5 Click **OK**.

Note that the text was created using Tahoma font at a height of 0.120 inch.

Figure 2-25

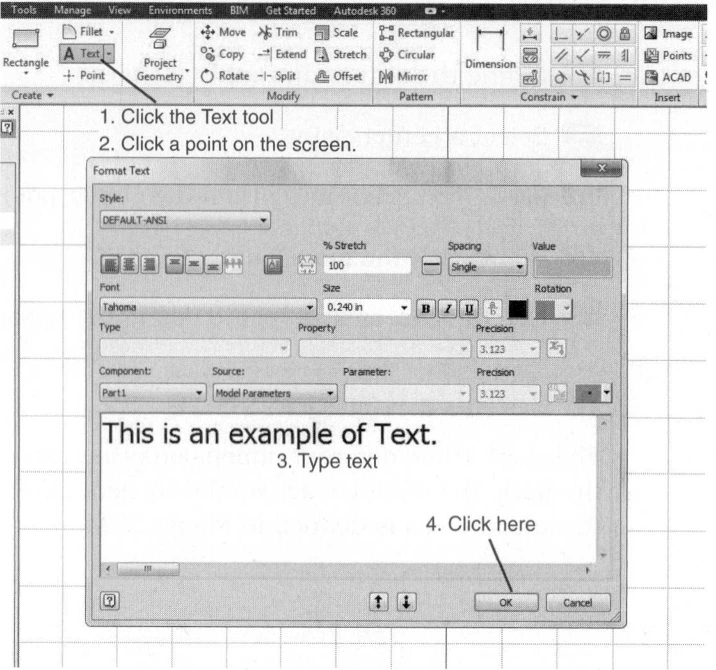

Text on a sketch

This is an example of text.

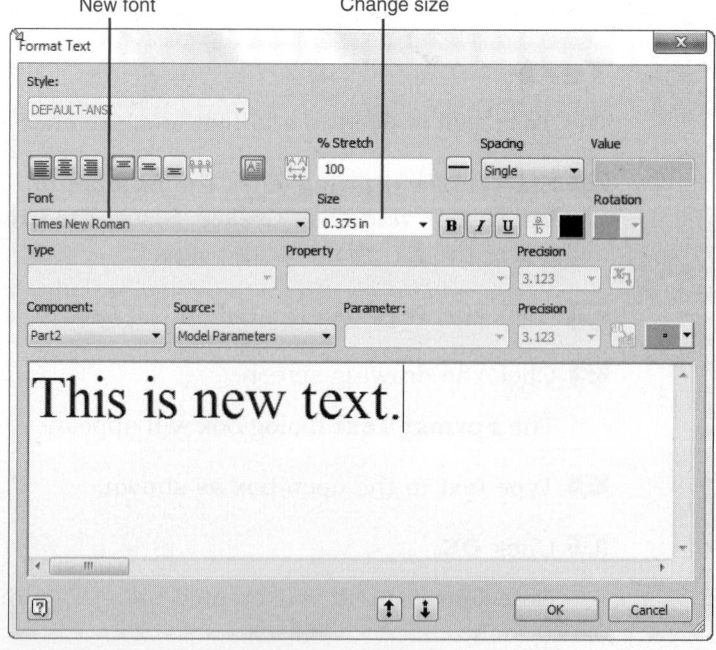

Figure 2-25
(*Continued*)

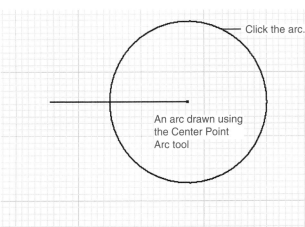

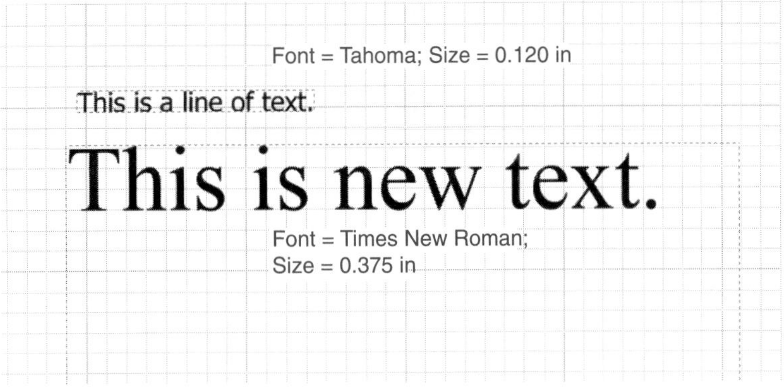

Font = Tahoma; Size = 0.120 in

This is a line of text.

This is new text.

Font = Times New Roman;
Size = 0.375 in

To Change Font and Text Height

1. Start a new line of text by clicking the **Text** tool.
2. Change the **Font** to **Times New Roman** and the **Size** to **0.375 in**.
3. Type in a new line of text.
4. Click **OK.**

   Note the difference between the two lines of text.

Geometry Text

The **Geometry Text** tool is used to write text on a curved line.

1. Draw an arc using the **Center Point Arc** tool.

   See Figure 2-26.

Click the arc.

An arc drawn using the Center Point Arc tool

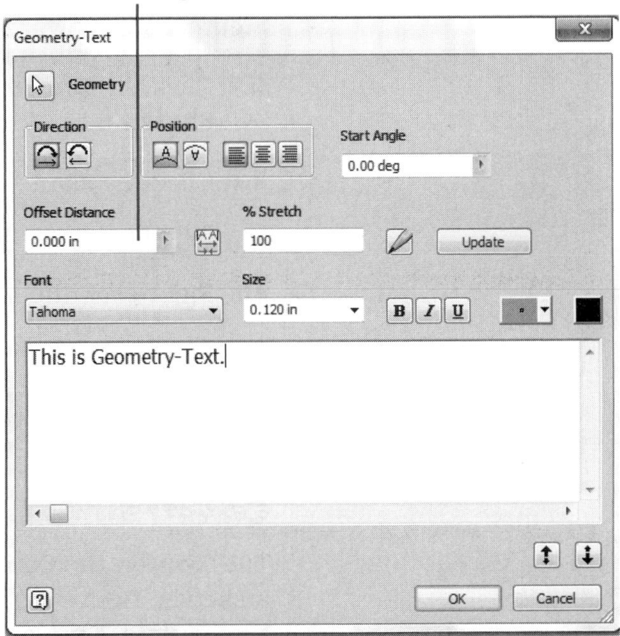

Figure 2-26

Figure 2-26
(Continued)

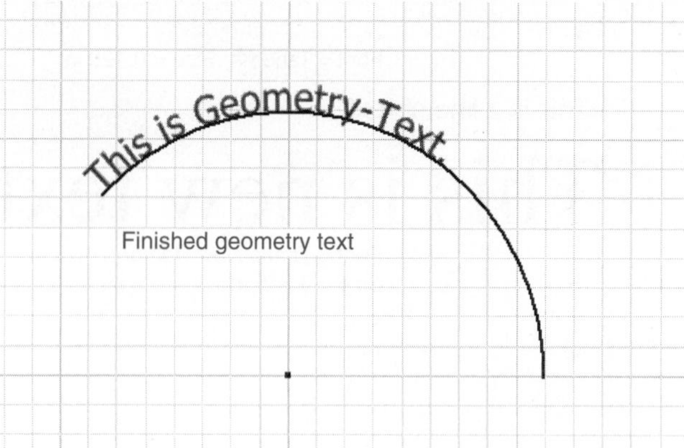

2 Click the **Geometry Text** tool.

The **Geometry Text** tool is a flyout from the **Tex**t tool.

3 Click the arc.

The **Geometry-Text** dialog box will appear.

4 Type in some text.

5 Click **OK.**

Dimension

The **Dimension** tool is used to add size to a sketch. Dimensions will be covered in detail in Chapter 7. This section will show how to create an aligned dimension. Aligned dimensions will be needed for some constructions before Chapter 7.

To Create an Aligned Dimension

Figure 2-27 shows a shape that includes a slanted edge.

1 Click the **Dimension** tool located on the **Constrain** panel under the **Sketch** tab.

2 Click the slanted line and move the cursor away from the line.

A vertical dimension will appear.

3 Right-click the mouse and select the **Aligned** option.

4 The vertical dimension will change to an aligned dimension.

5 Click the mouse and change the dimension value.

Constraints

The tools on the **Constrain** panel are used to orient and limit entities on a sketch. Twelve different constraints are available: **Coincident, Colinear, Concentric, Fix, Parallel, Perpendicular, Horizontal, Vertical, Tangent, Smooth, Symmetric,** and **Equal.** See Figure 2-28.

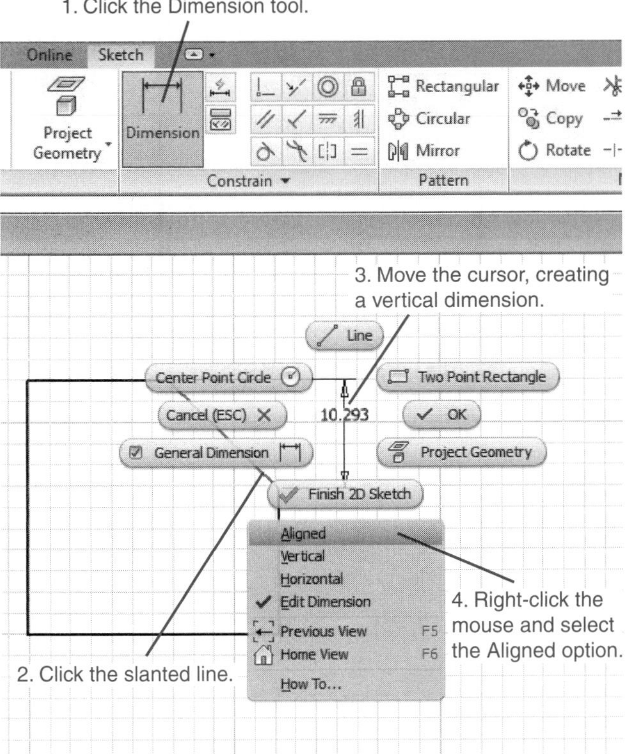

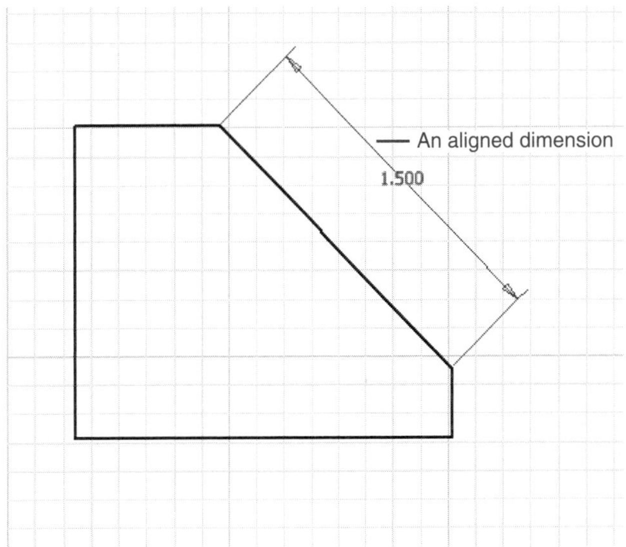

Figure 2-27

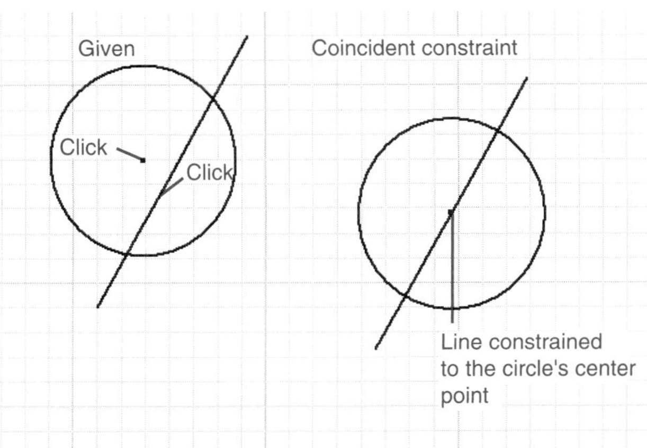

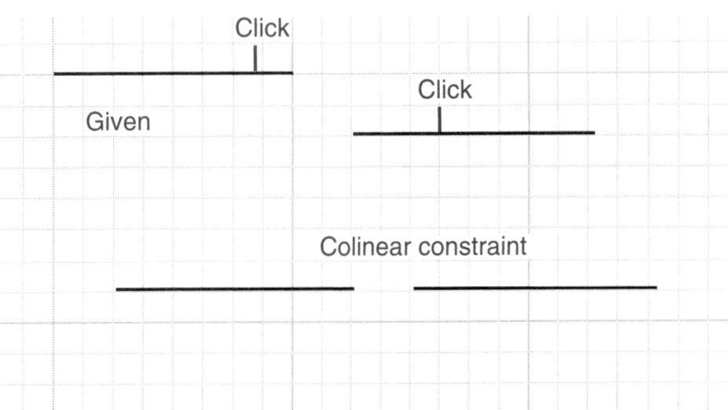

Figure 2-28

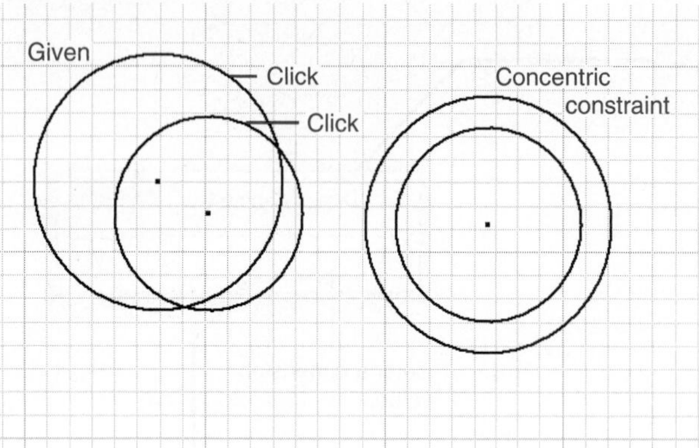

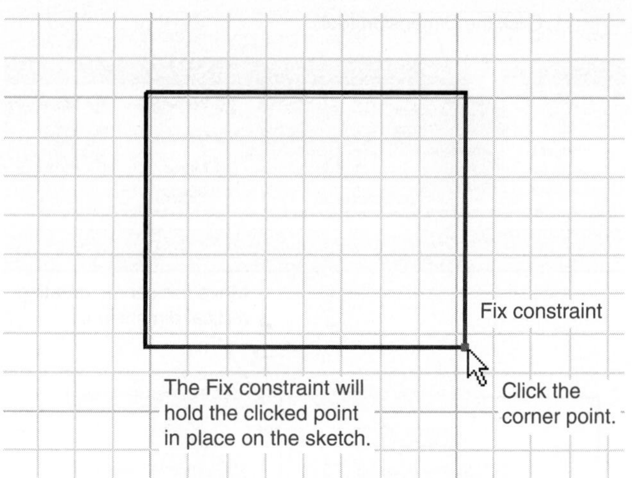

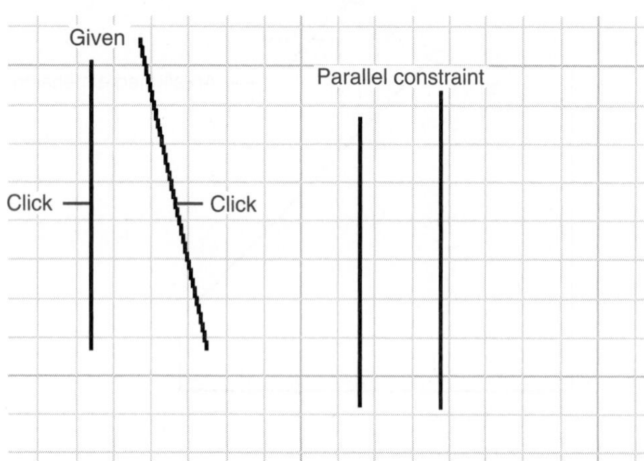

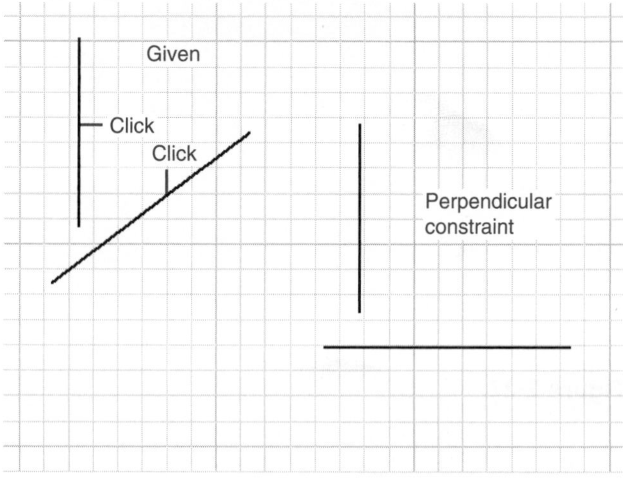

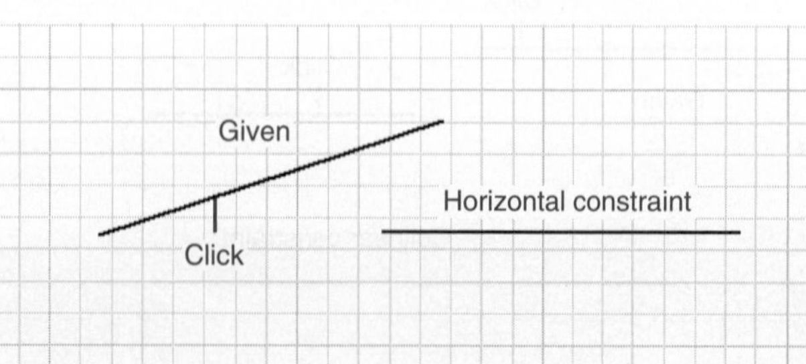

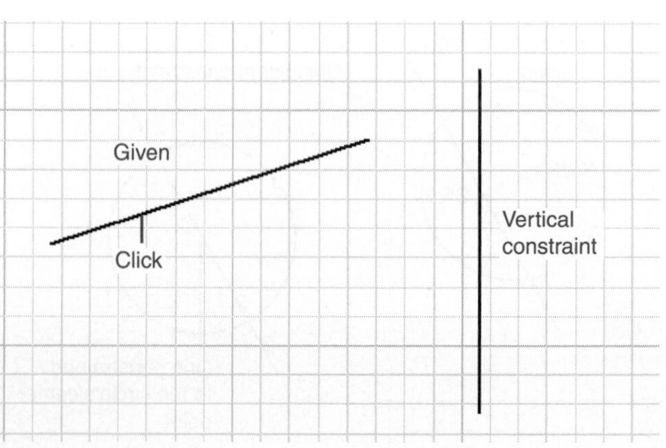

Figure 2-28
(Continued)

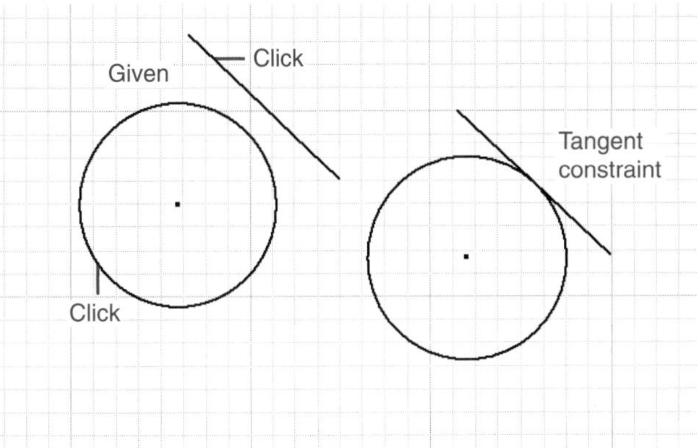

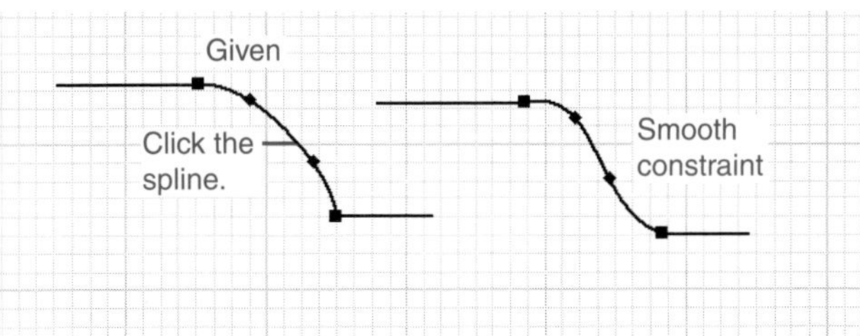

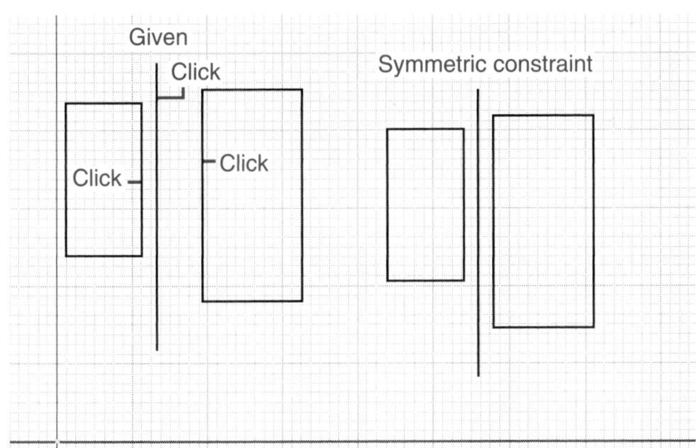

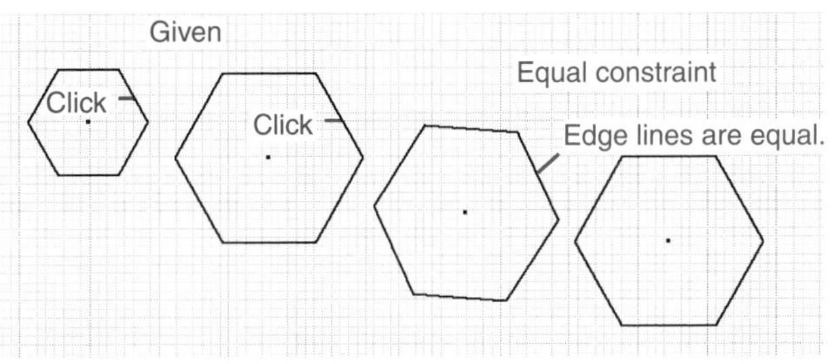

Figure 2-28
(Continued)

Pattern - Rectangular

The **Rectangular** tool is used to create rectangular patterns. See Figure 2-29.

To Create a Rectangular Pattern

1. Start a new drawing, click the **Metric** tab, click the **Standard (in).ipt** format, and click **Create.** Select the **Create 2D Sketch** option, and select the XY plane for the sketch.

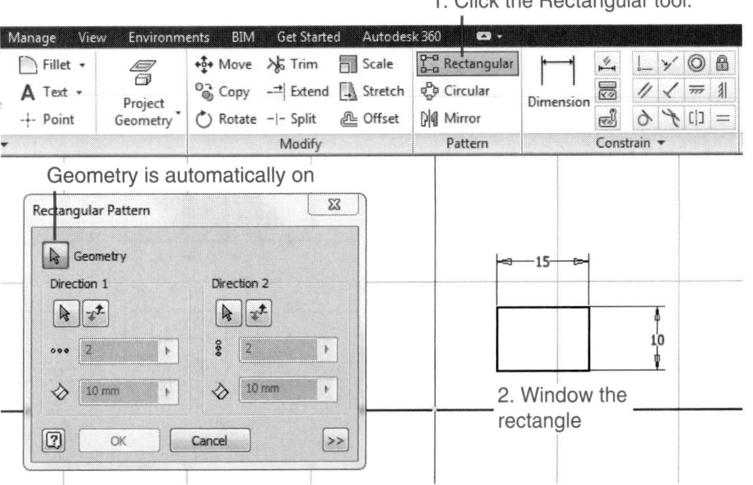

Figure 2-29

Figure 2-29

(Continued)

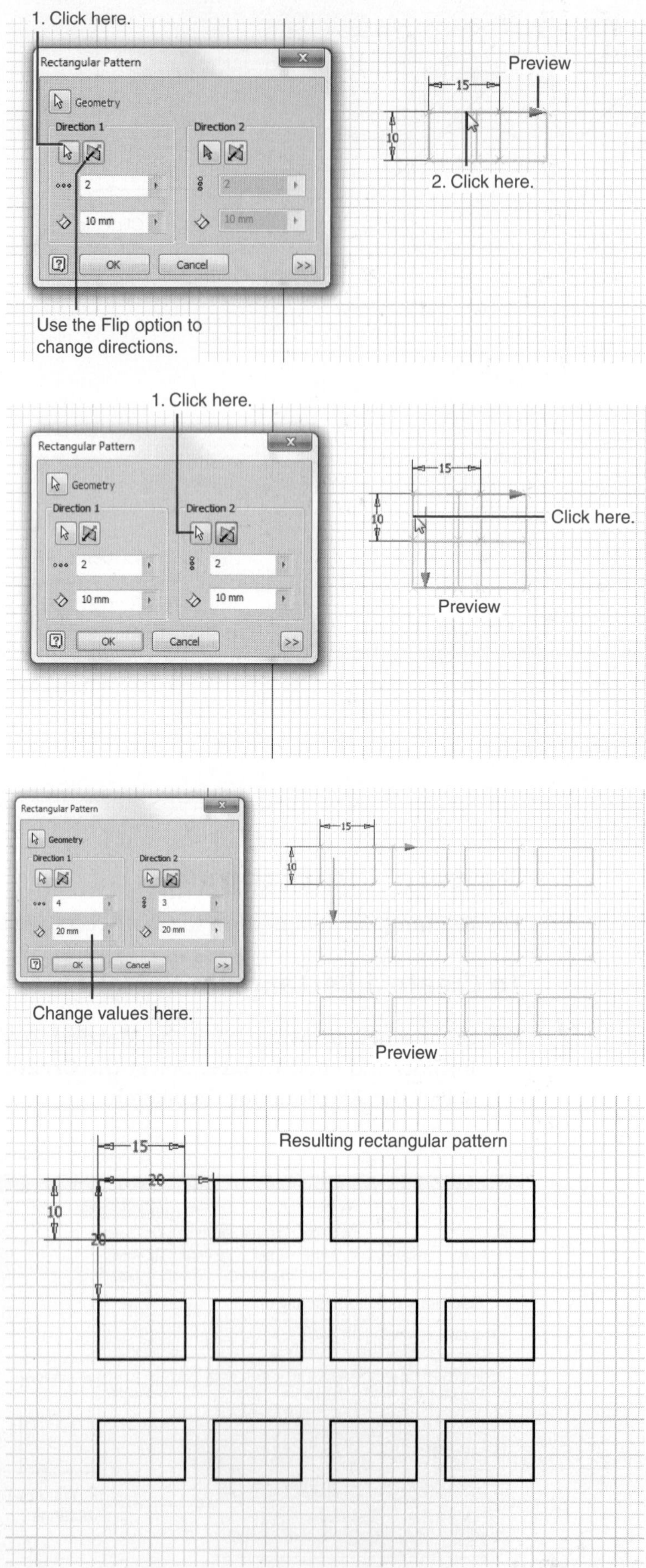

2 Draw a **10** × **15** rectangle.

3 Click the **Rectangular** tool located on the **Pattern** panel under the **Sketch** tab.

The **Rectangular Pattern** dialog box will appear. The **Geometry** box will automatically turn on; that is, the program will ask which entity you wish to pattern.

4 Window the **10** × **15** rectangle. (Click all lines in the rectangle.)

5 Click on the arrow under the **Direction 1** heading.

6 Click the top horizontal line of the rectangle.

NOTE

Use the **Flip** tool located next to the arrow under the **Direction 1** heading if the preview indicates that the pattern is going in the wrong direction.

7 Click the arrow under the **Direction 2** heading and click the left vertical line of the 10 × 15 rectangle.

A preview of the pattern will appear.

8 Enter new values as shown in Figure 2-29.

In this example, the rectangles are spaced 20 mm apart. There are three rows of four rectangles.

9 Click **OK**.

Pattern - Circular

The **Circular** tool is used to create circular patterns, sometimes called *bolt circles*.

1 Start a new drawing, click the **Metric** tab, click the **Standard (in).ipt** format, and click **Create.** Select the **Create 2D Sketch** option, and select the XY plane for the sketch.

2 Draw a **Ø20.0** circle.

3 Draw a point (use the **Point** tool).

See Figure 2-30.

4 Dimension the circle and point as shown.

5 Click the **Circular** tool located on the **Pattern** panel under the **Sketch** tab.

The **Circular Pattern** dialog box will appear. The **Geometry** box will automatically turn on; that is, the program will ask which entity you wish to pattern.

Figure 2-30

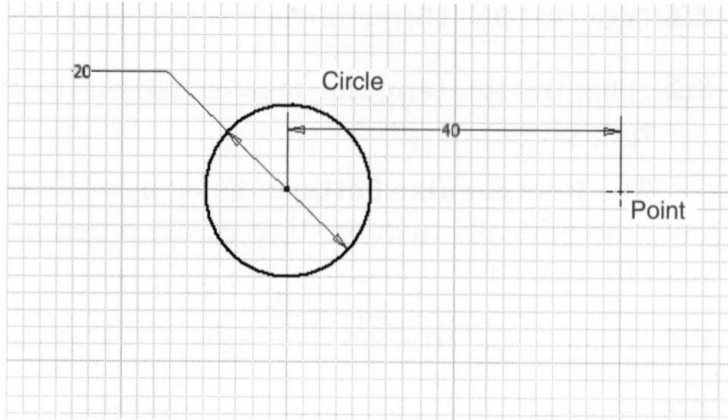
20
Circle
40
Point

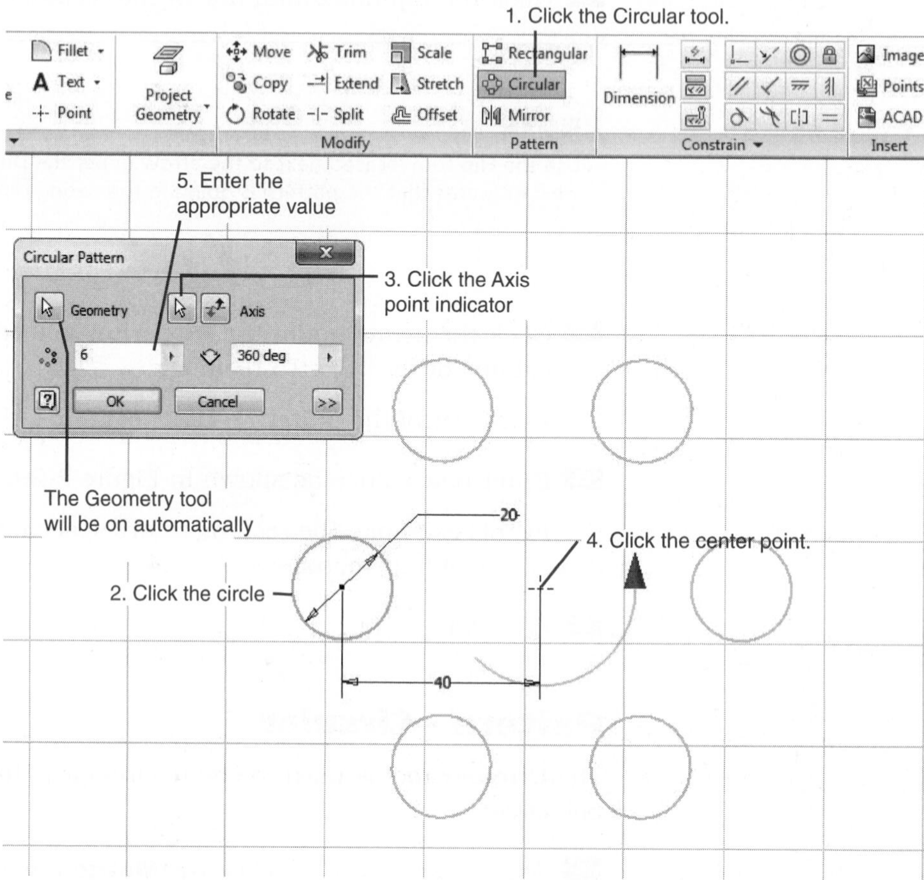
1. Click the Circular tool.
Fillet
Text
Point
Project Geometry
Move
Copy
Rotate
Trim
Extend
Split
Scale
Stretch
Offset
Modify
Rectangular
Circular
Mirror
Pattern
Dimension
Constrain
Image
Points
ACAD
Insert
5. Enter the appropriate value
Circular Pattern
Geometry
Axis
6
360 deg
OK
Cancel
3. Click the Axis point indicator
The Geometry tool will be on automatically
20
4. Click the center point.
2. Click the circle
40

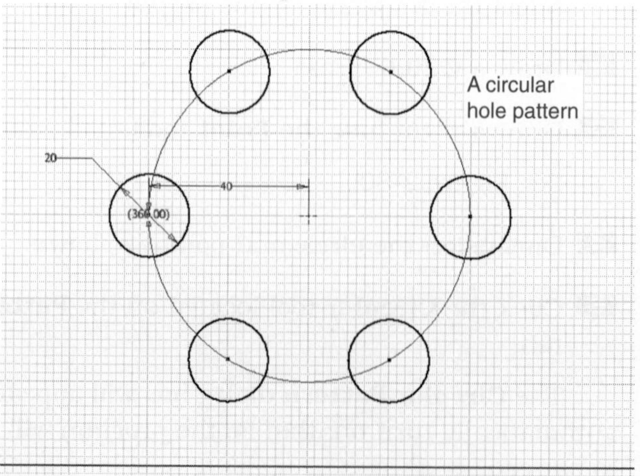
A circular hole pattern
20
40
(360.00)

6 Click the circle.

7 Click the arrow to the left of the **Axis** heading.

8 Select the point.

A preview will appear.

9 Click **OK**.

Pattern - Mirror

The **Mirror** tool is used to create mirror images of an entity. Remember, a mirror image is not the same as a copy. Your hands are approximate mirror images of each other. Figure 2-31 shows an image to be mirrored. The

Figure 2-31

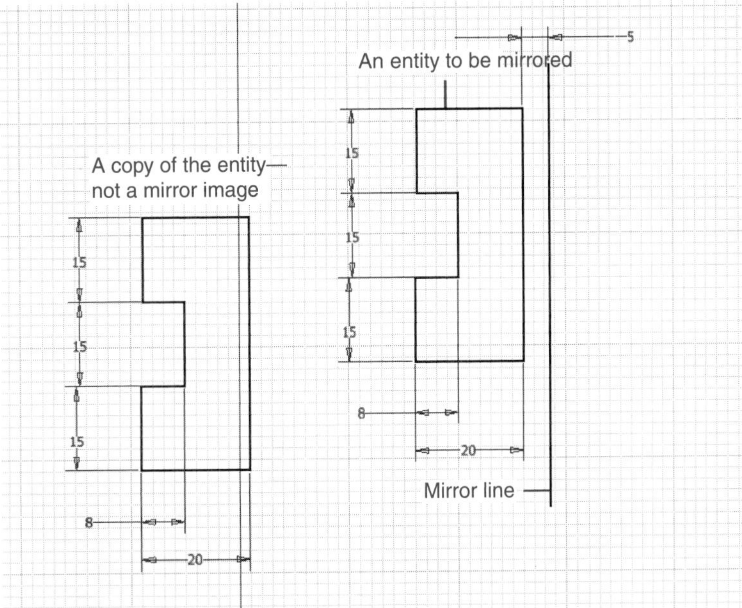

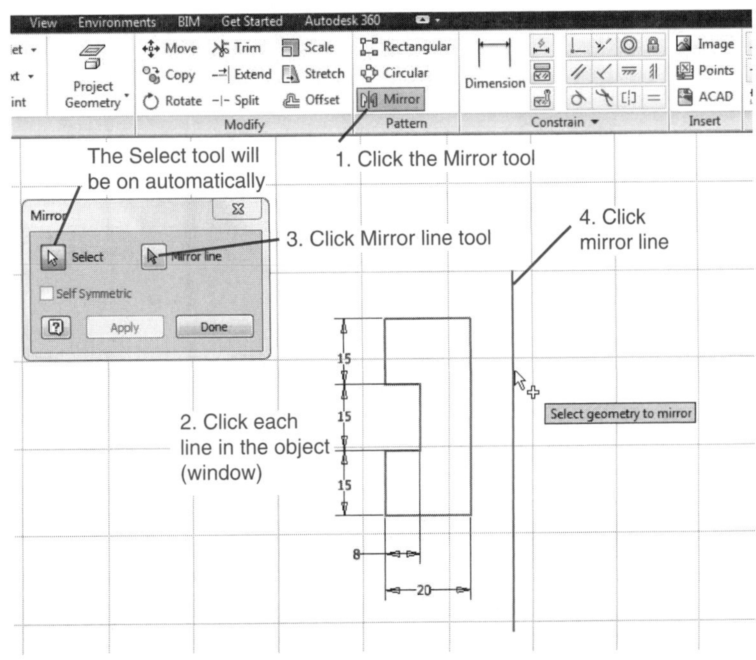

Figure 2-31
(Continued)

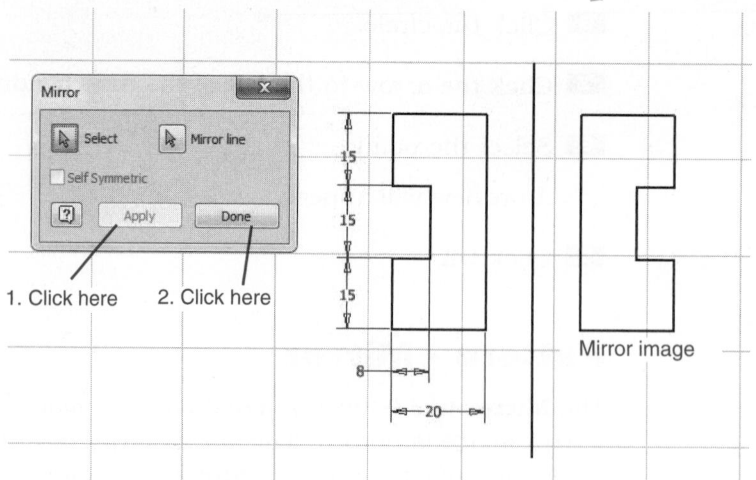

figure also shows a copy of the image. Note the difference between the copy and the mirror image.

1. Start a new drawing, click the **Metric** tab, click the **Standard (in).ipt** format, and click **Create**. Select the **Create 2D Sketch** option, and select the XY plane for the sketch.
2. Draw the shape shown in Figure 2-31.
3. Click the **Mirror** tool located on the **Pattern** panel under the **Sketch** tab.

   The **Mirror** dialog box will appear.
4. Click the **Select** box (the box may be on automatically).
5. Click each line in the shape.
6. Click the **Mirror line** box.
7. Click the long vertical line next to the shape.

   A preview of the mirrored shape will appear.
8. Click the **Apply** box.
9. Click the **Done** box.

Any line within the object can be used as a mirror line. For example, the right vertical edge line could have been used as a mirror line.

Move

The **Move** tool is used to move entities to different locations on the drawing screen. See Figure 2-32.

Note that the **Move** tool creates a different result from that produced by the **Copy** tool. The **Move** tool will move an object but not create a second one or leave the original object in its original location.

1. Start a new drawing, click the **Metric** tab, click the **Standard (in).ipt** format, and click **Create**. Select the **Create 2D Sketch** option, and select the XY plane for the sketch.

Figure 2-32

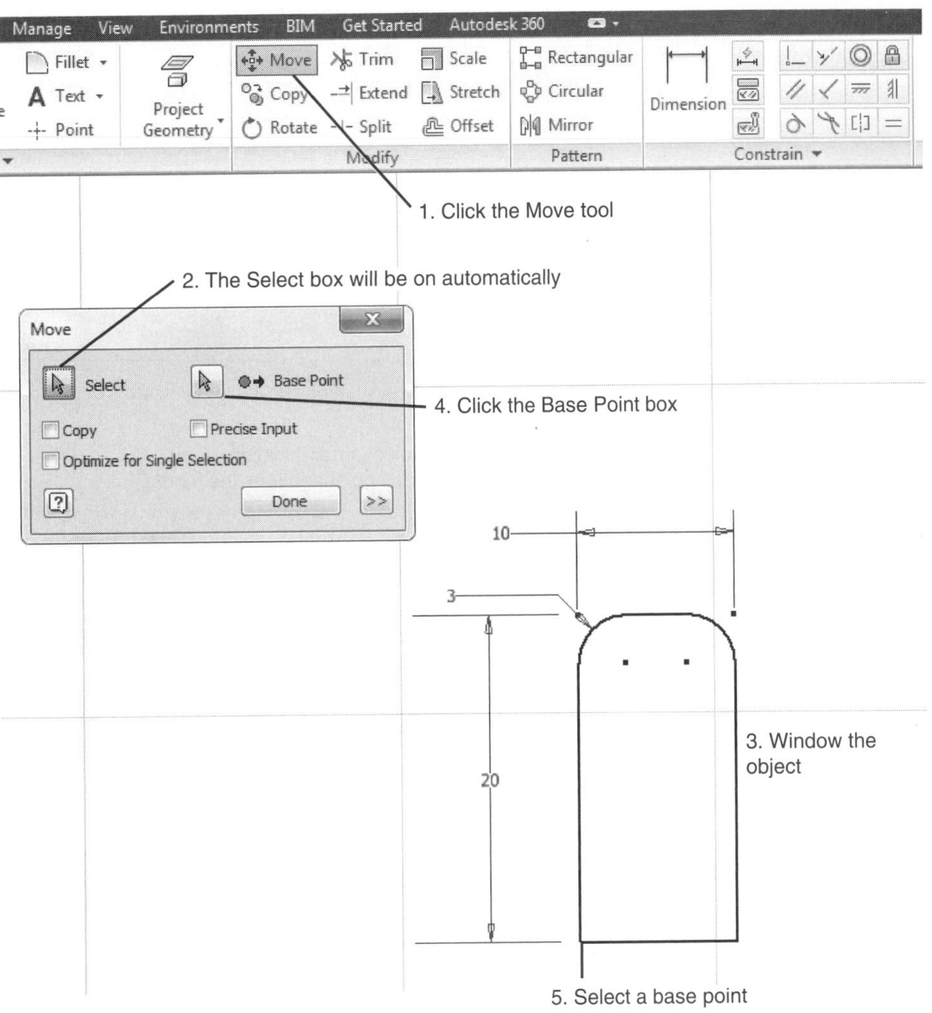

Manage View Environments BIM Get Started Autodesk 360
Fillet
Text
Point
Project Geometry
Move
Copy
Rotate
Trim
Extend
Split
Scale
Stretch
Offset
Modify
Rectangular
Circular
Mirror
Pattern
Dimension
Constrain
1. Click the Move tool
2. The Select box will be on automatically
Move
Select
Base Point
Copy
Precise Input
Optimize for Single Selection
Done
4. Click the Base Point box
10
3
20
3. Window the object
5. Select a base point

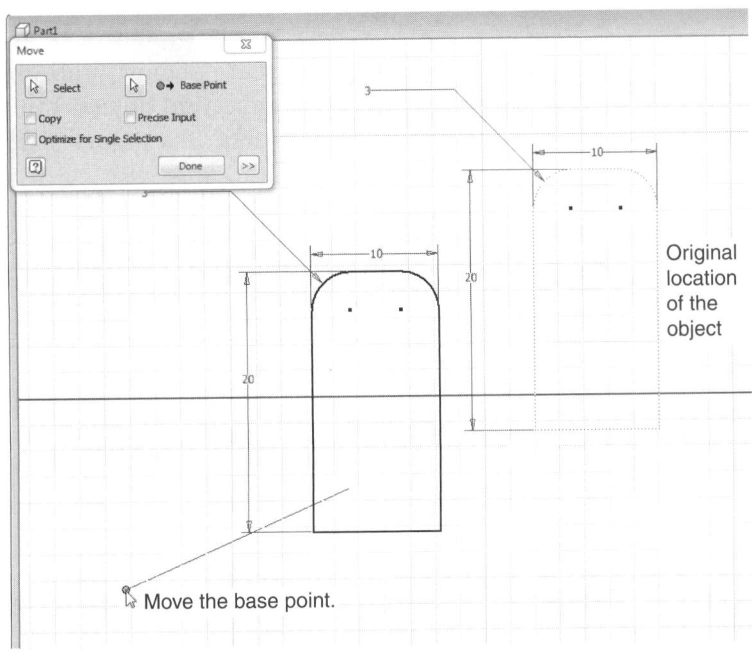

Part1
Move
Select
Base Point
Copy
Precise Input
Optimize for Single Selection
Done
3
10
20
Original location of the object
Move the base point.

Figure 2-32
(*Continued*)

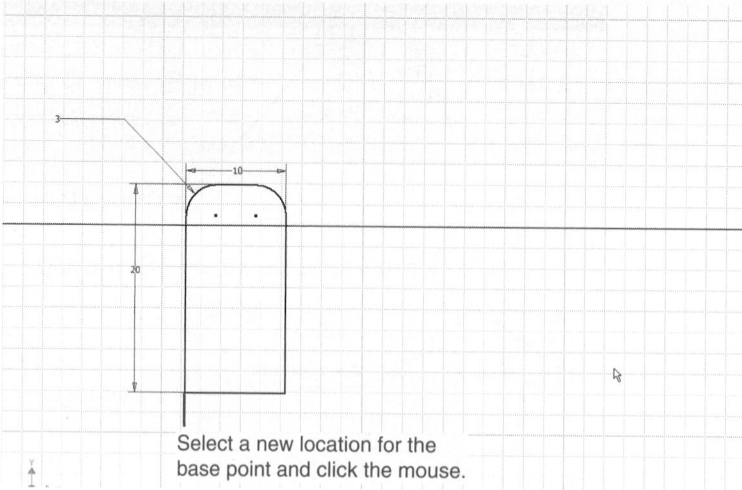

3
10
20
Select a new location for the base point and click the mouse.

2 Draw the object shown.

3 Click the **Move** tool located on the **Modify** panel under the **Sketch** tab.

The **Move** dialog box will appear.

4 Click the **Select** box and window the object.

5 Click the **Base Point** box and select a base point for the move.

6 Click the mouse when a new location has been selected.

7 Click the **Done** box.

In this example the lower left corner was selected. Any point, even one not on the object, may be selected.

Copy

The **Copy** tool is used to create a copy of an object. The copy will not be in the same location as the original object, but the original object will be retained in its original location. See Figure 2-33.

1 Start a new drawing, click the **Metric** tab, click the **Standard (in).ipt** format, and click **Create.** Select the **Create 2D Sketch** option, and select the XY plane for the sketch.

2 Draw the object shown.

3 Click the **Copy** tool located on the **Modify** panel under the **Sketch** tab.

The **Copy** dialog box will appear.

4 Click the **Select** box (the box may be on automatically) and window the object.

5 Click the **Base Point** box and select a base point for the move.

6 Click the mouse when a new location has been selected.

7 Click the **Done** box.

Figure 2-33

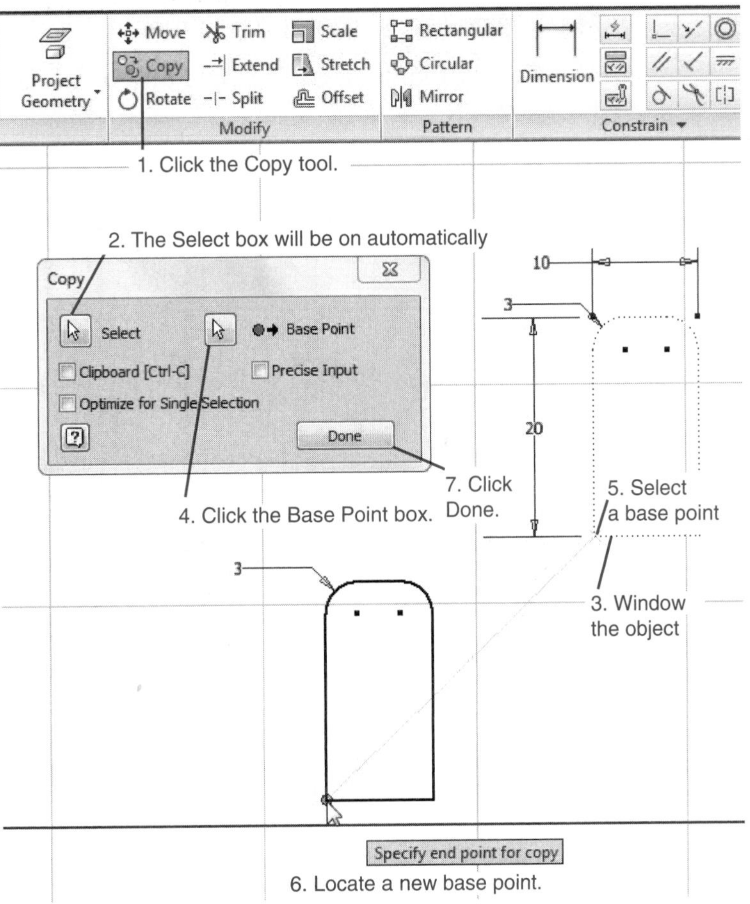
Move
Trim
Scale
Rectangular
Copy
Extend
Stretch
Circular
Dimension
Project Geometry
Rotate
Split
Offset
Mirror
Modify
Pattern
Constrain
1. Click the Copy tool.
2. The Select box will be on automatically
Copy
Select
Base Point
Clipboard [Ctrl-C]
Precise Input
Optimize for Single Selection
Done
7. Click Done.
4. Click the Base Point box.
5. Select a base point
3. Window the object
10
3
20
Specify end point for copy
6. Locate a new base point.

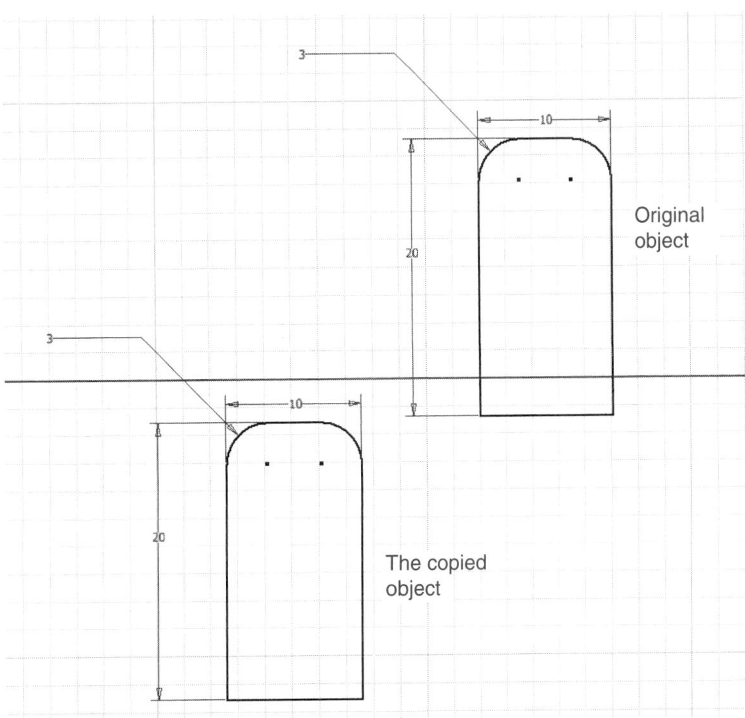
3
10
20
Original object
The copied object

Rotate

The **Rotate** tool is used to rotate an object about a defined point. See Figure 2-34.

Figure 2-34

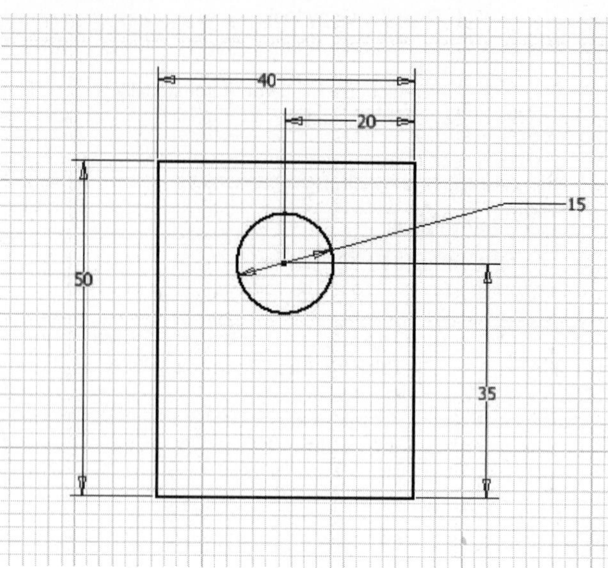

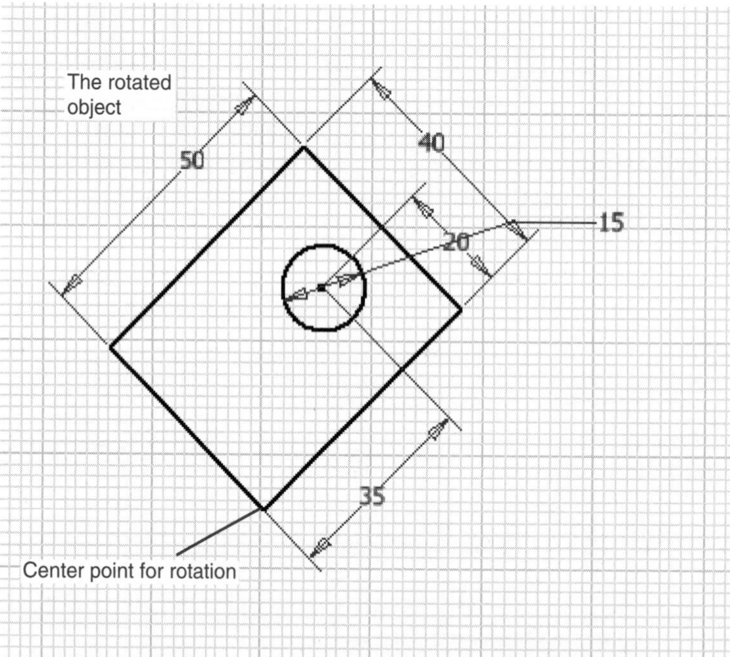

1. Start a new drawing, click the **Metric** tab, click the **Standard (in).ipt** format, and click **Create**. Select the **Create 2D Sketch** option, and select the XY plane for the sketch.

2. Draw the object shown.

3. Click the **Rotate** tool located on the **Modify** panel under the **Sketch** tab.

   The **Rotate** dialog box will appear.

4. Click the **Select** box (the box may be on automatically) and click all the lines in the object.

5 Click the **Select** box and select a center point for the rotation.

6 Enter a value for the angle.

In this example a value of **–45°** was entered. The rotation angle could have been sketched.

> **NOTE**
>
> In Inventor, counterclockwise is the positive direction.

7 Click the **Apply** box.

8 Click the **Done** box.

The original object will disappear. To retain the original object, click the **Copy** box before clicking the **Apply** box.

Trim

The **Trim** tool is used to cut away unwanted parts of an object. Figure 2-35 shows a group of objects that includes a circle, a rectangle, and a line. Trim the line between the left side of the rectangle and the circle.

1 Click the **Trim** tool located on the **Modify** panel under the **Sketch** tab.

2 Locate the cursor on the entity to be trimmed.

The entity will change to a broken line.

3 Click the left mouse button.

The entity will disappear.

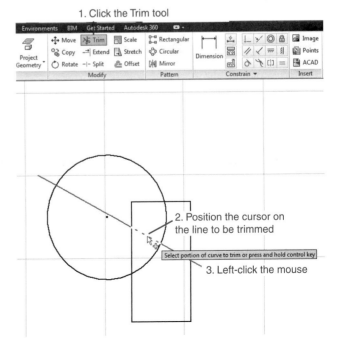

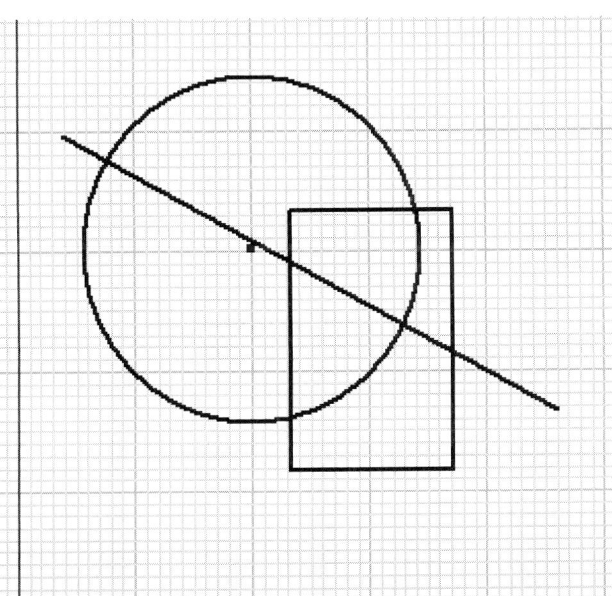

Figure 2-35

Figure 2-35
(*Continued*)

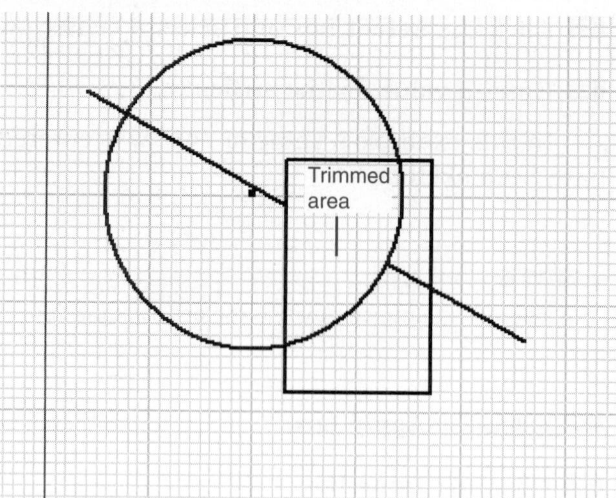

Extend

The **Extend** tool is used to extend entities to known boundaries. Figure 2-36 shows a line and a circle drawn with its center point at the endpoint of the line. Extend the line to the right-hand edge of the circle.

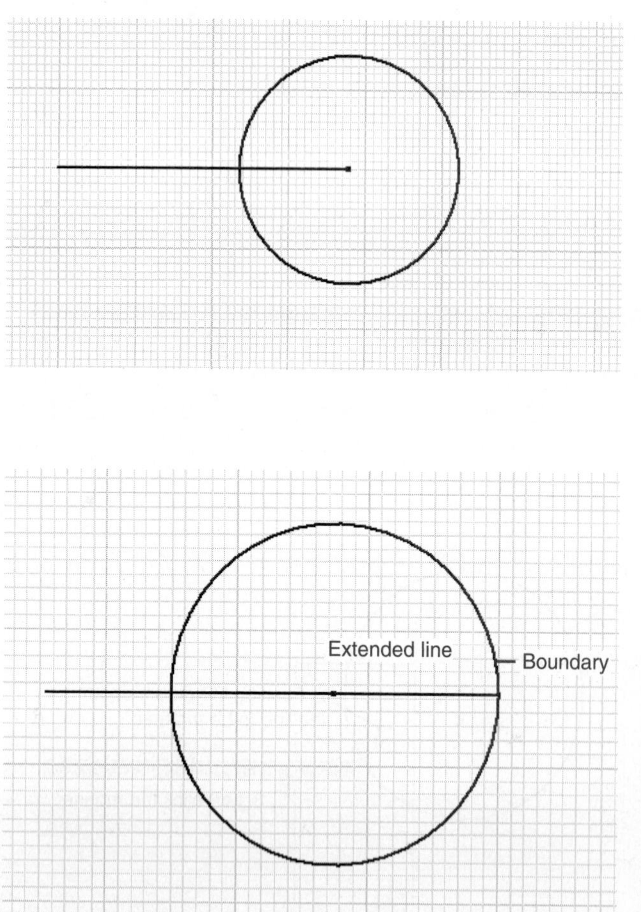

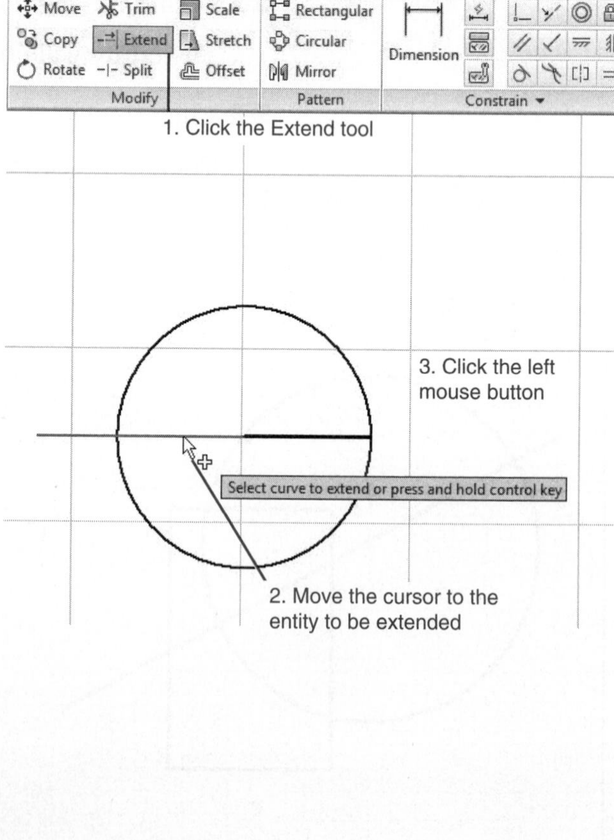

Figure 2-36

1 Click the **Extend** tool.

2 Move the cursor to the line.

The line will automatically extend to the next available entity. In this example that is the edge of the circle.

3 Click the left mouse button.

Offset

The **Offset** tool is used to draw entities parallel to an existing entity. Figure 2-37 shows a line and a circle. Create a new line and a circle offset from the existing entities.

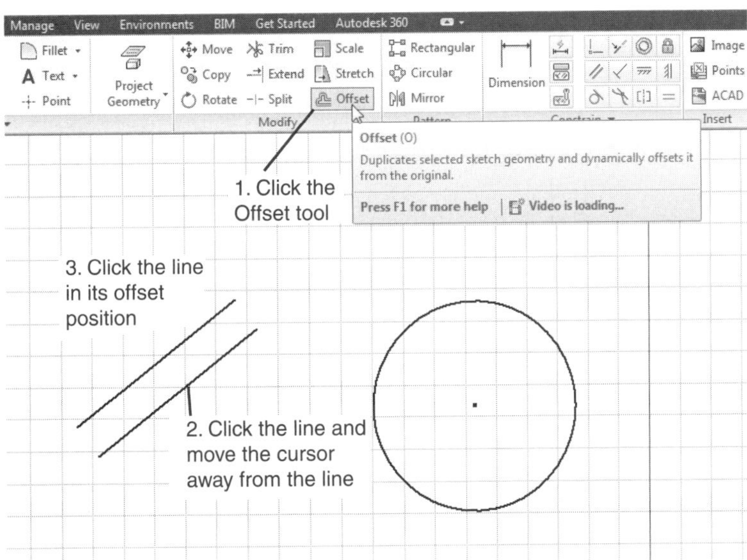

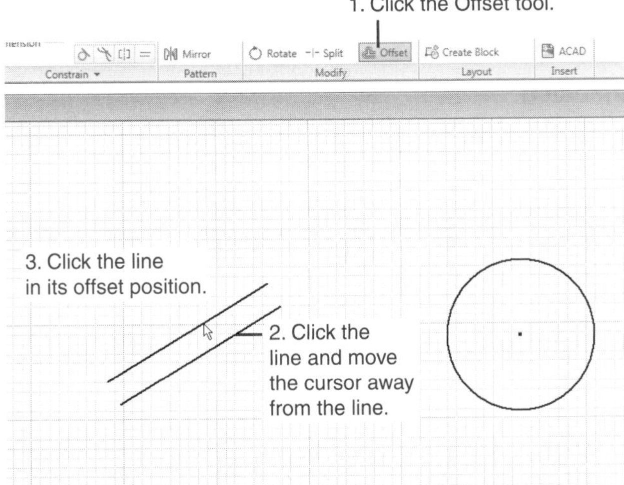

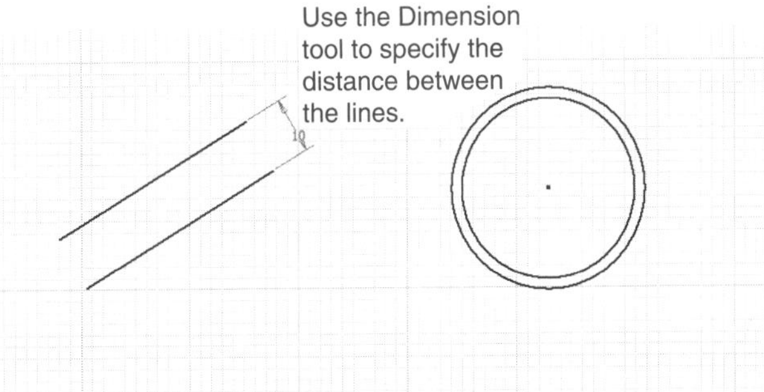

Figure 2-37

1 Click the **Offset** tool.

2 Click the line and move the cursor away from the line.

A new line will appear. It will be parallel and equal in length to the original line.

3 Position the offset line and click the left mouse button.

Use the **Dimension** tool to specify the offset distance, that is, the distance between the lines. Use the **Offset** tool to create a concentric circle relative to the original circle.

Editing a Sketch

Figure 2-38 shows a finished sketch. Any of the features may be changed by editing the dimensions.

Figure 2-38

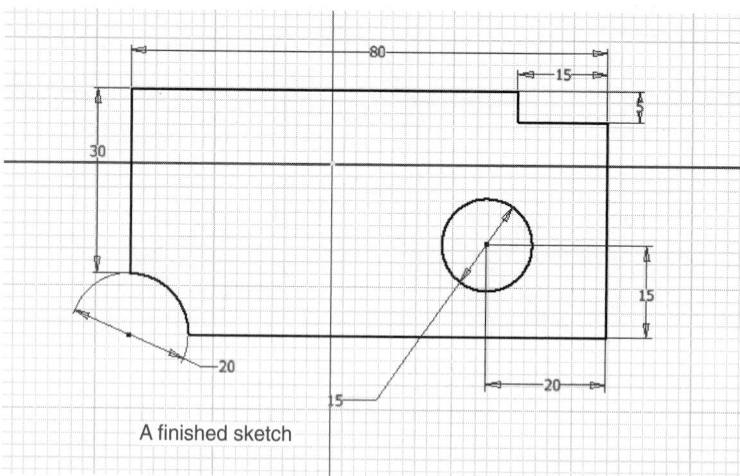

A finished sketch

To Edit the Cutout

Suppose the 5 × 15 cutout located at the top right corner of the object is to be changed to a 5 × 20 cutout.

1 Double-click the **15** dimension.

The **Edit Dimension** dialog box will appear. See Figure 2-39.

2 Change the text value to **20** and click the **green check mark**.

To Change the Location and Size of the Circle

Change the hole to a Ø22, a horizontal location of 25, and a vertical location of 18. See Figure 2-40.

1 Double-click the diameter dimension, **15,** and change it to **22**.

2 Double-click the **20** and **15** dimensions and change them to **25** and **18,** respectively.

Figure 2-39

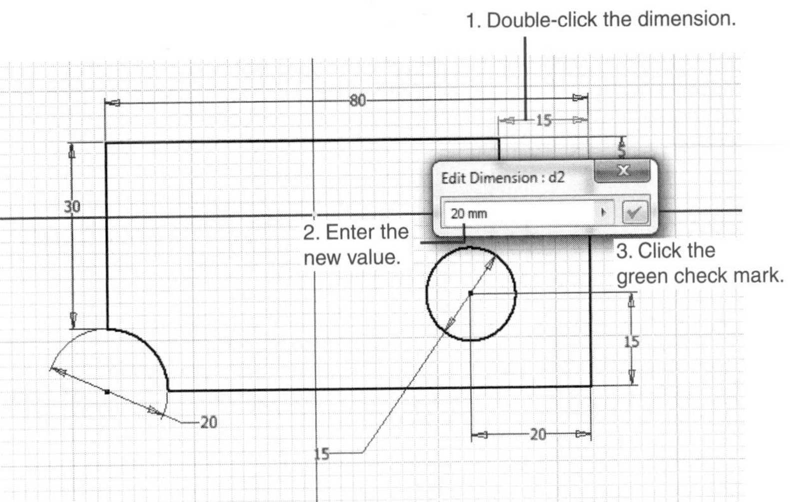
1. Double-click the dimension.
80
15
5
Edit Dimension : d2
20 mm
30
2. Enter the new value.
3. Click the green check mark.
15
20
15
20

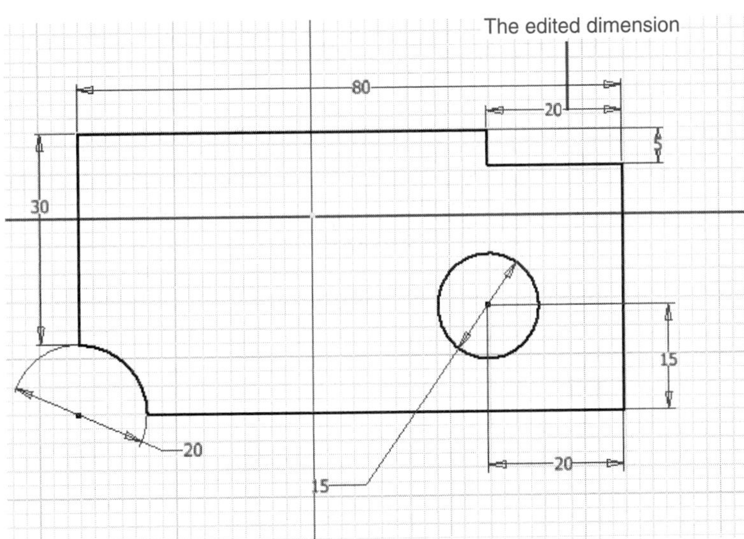
The edited dimension
80
20
5
30
15
20
15
20

Figure 2-40

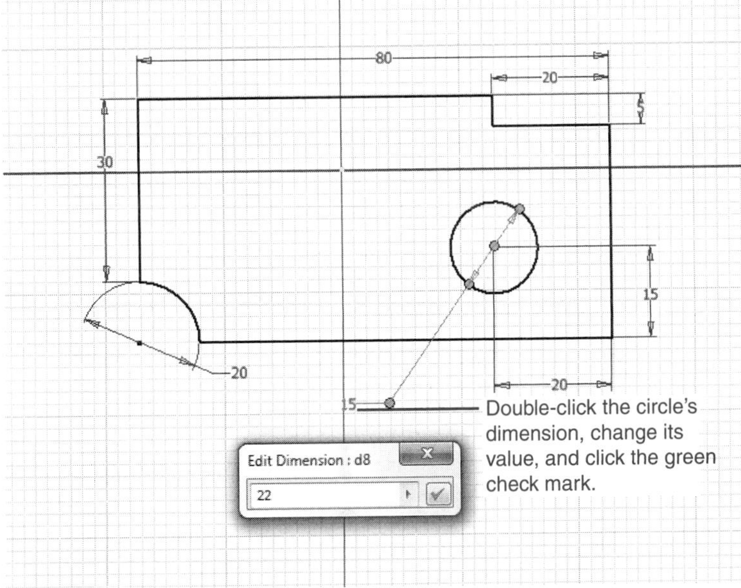
80
20
5
30
15
20
20
15
Double-click the circle's dimension, change its value, and click the green check mark.
Edit Dimension : d8
22

Figure 2-40
(Continued)

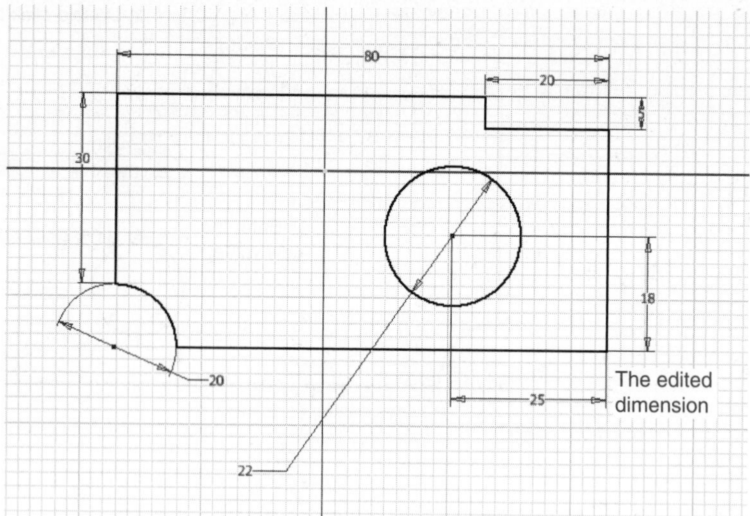

Sample Problem SP2-1

The sample problem presented here is based on Figure P2-8 in the Chapter Project. The dimensions are in millimeters. See Figure 2-41.

1. Start a new drawing using the **Standard (mm).ipt** format.
2. Use the **Rectangle** tool to draw a **100** × **85** rectangle.
3. Use the **Rectangle** tool and add a rectangle aligned with the top horizontal edge.
4. Dimension the location and the depth of the rectangle.

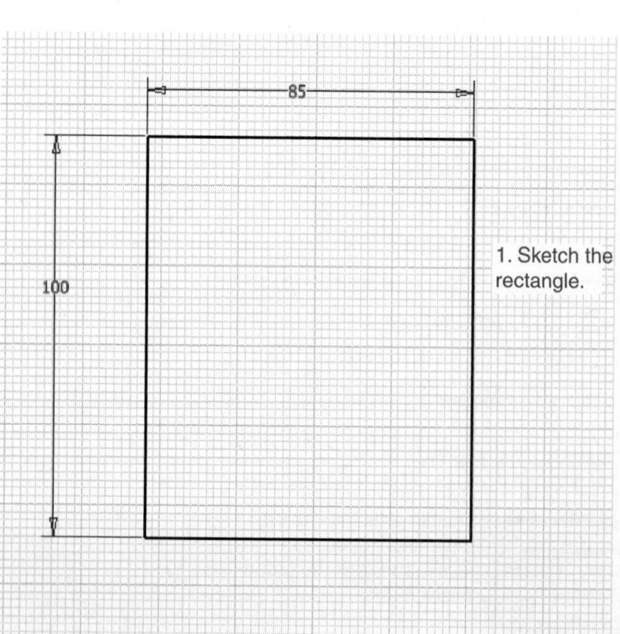

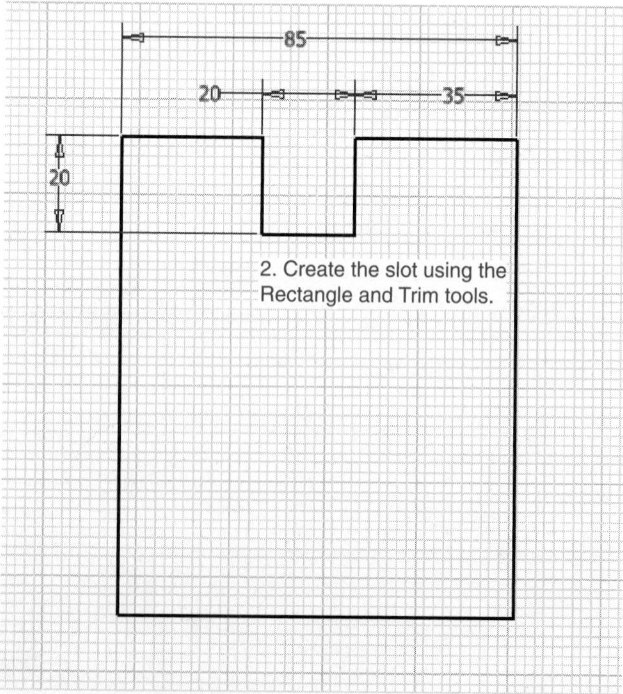

Figure 2-41

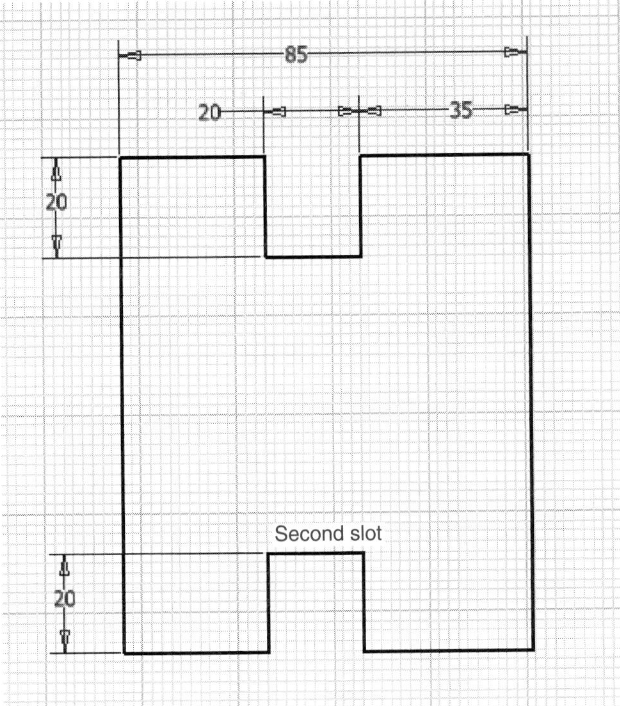
85
20
35
20
Second slot
20

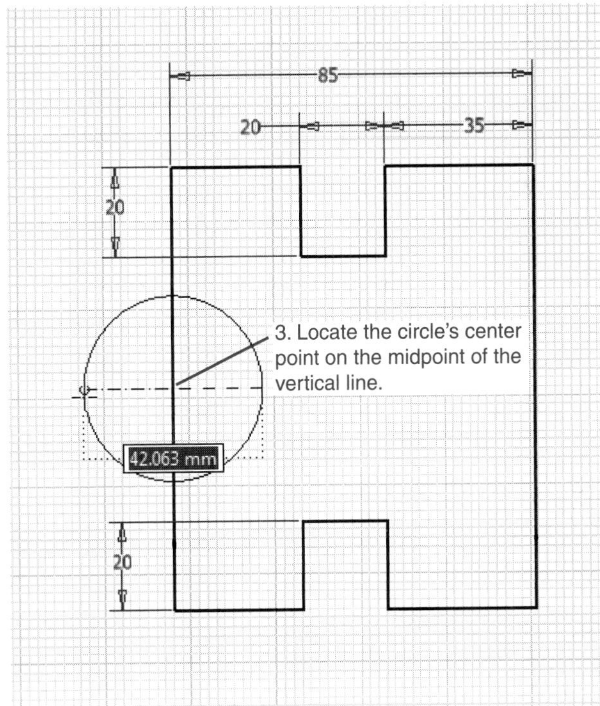
85
20
35
20
3. Locate the circle's center point on the midpoint of the vertical line.
42.063 mm
20

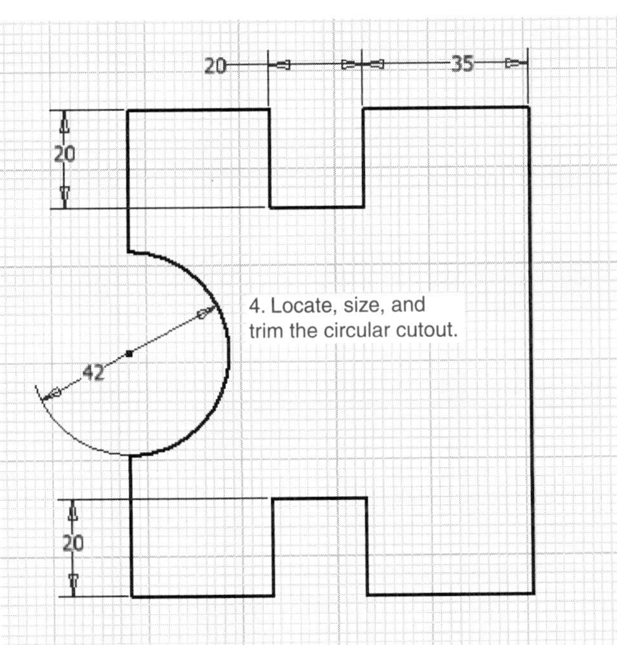
20
35
20
4. Locate, size, and trim the circular cutout.
42
20

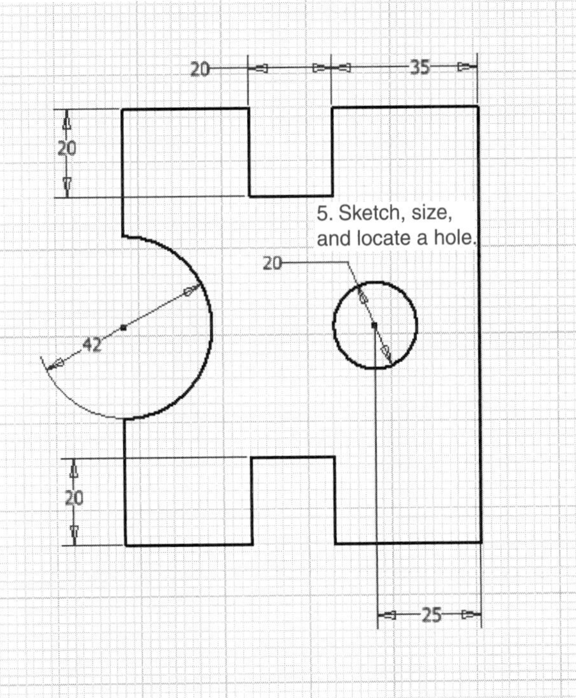
20
35
20
5. Sketch, size, and locate a hole.
20
42
20
25

Figure 2-41

(Continued)

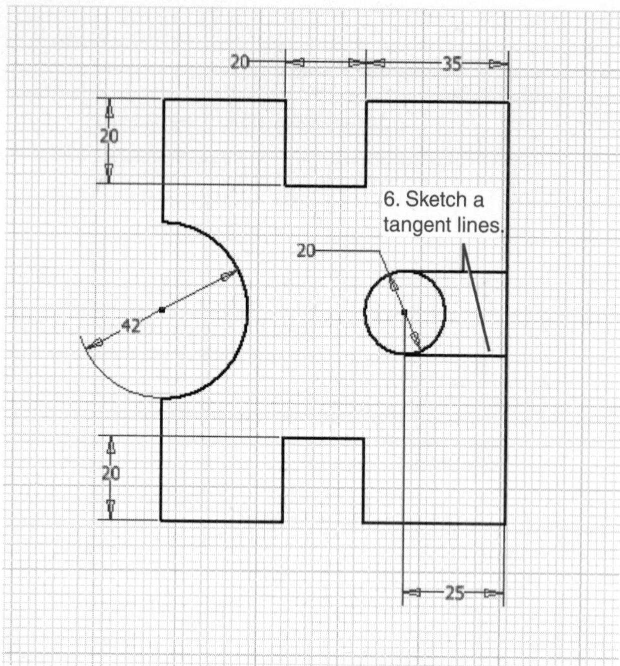

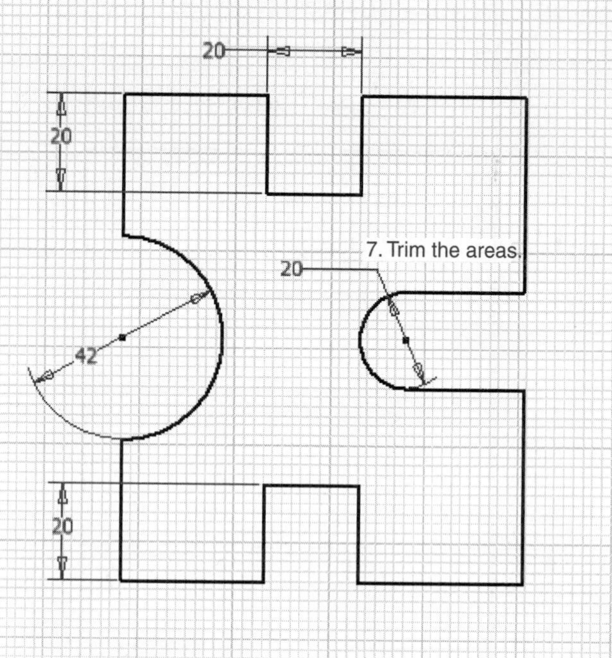

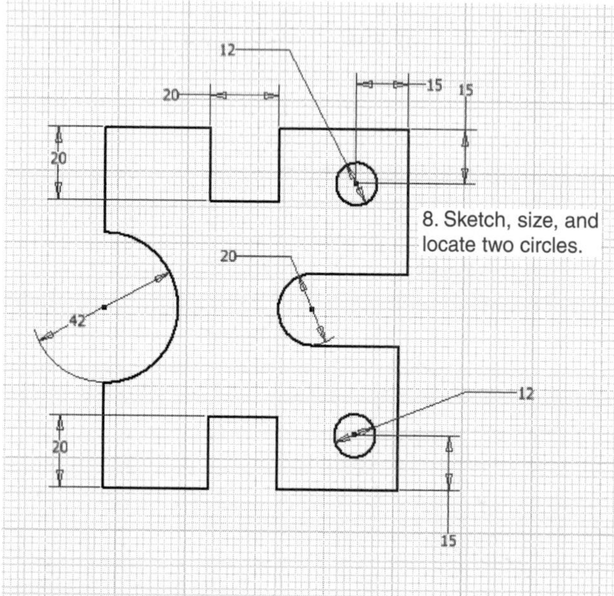

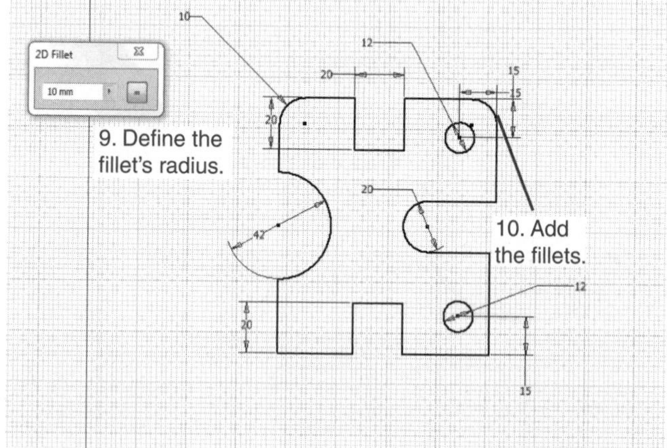

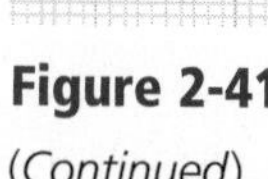

Figure 2-41

(Continued)

Figure 2-41
(Continued)

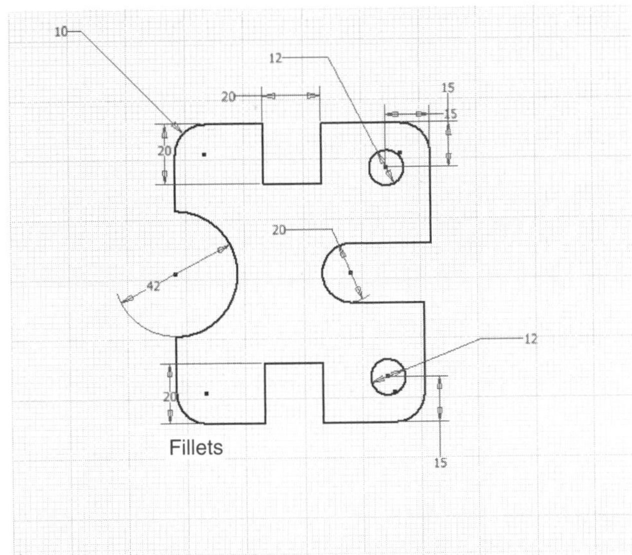

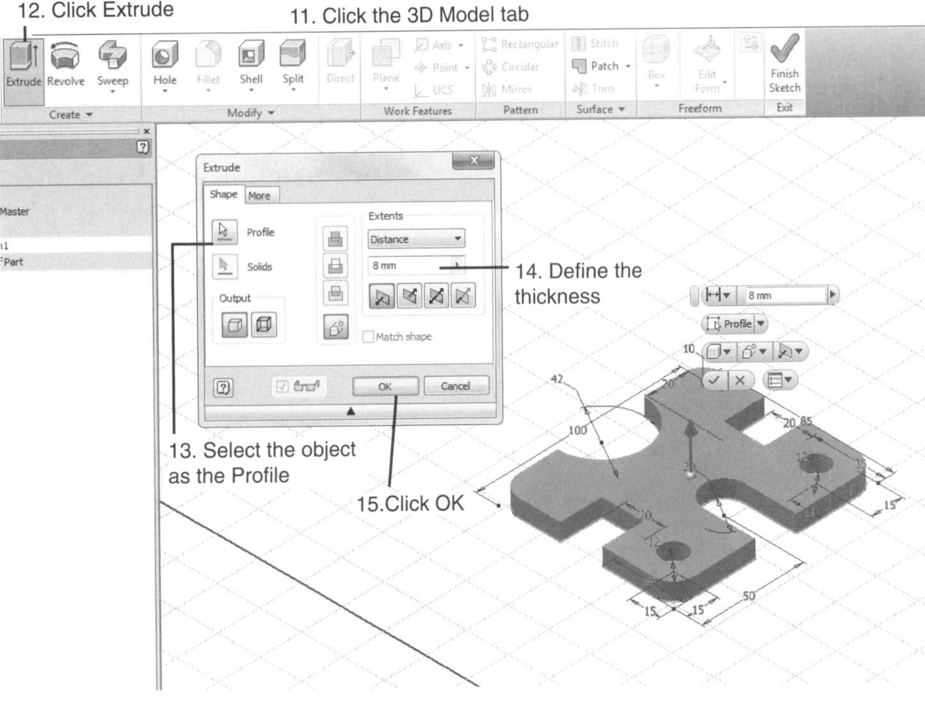

The finished object

5 Use the **Trim** command to remove the horizontal line segment to create a slot.

Remember, there are two lines to trim: the 100 × 85 rectangle and the smaller one added in step 3.

6 Repeat the procedure to create a slot aligned with the bottom horizontal edge.

> **NOTE**
>
> As you create the lower slot, broken lines will appear to help align the edges of the slot with the upper slot. This eliminates the need for locational dimensions for the lower slot.

7 Dimension the height of the slot and trim away the central horizontal line segment.

8 Access the **Circle** tool and scroll the cursor along the left vertical line.

The midpoint of the line will appear as a filled colored circle.

9 Make the midpoint of the line the center of a circle and sketch the circle. Use the **Dimension** tool to create the **Ø42** (R=21) circle. Trim away the excess vertical line.

10 Sketch the circular cutout by first sketching and locating a circle.

11 Sketch lines tangent to the circle to the right edge of the object.

12 Trim the excess lines.

13 Sketch, locate, and dimension the two circles.

14 Use the **Fillet** tool to add the **R=10** rounded corners.

15A Click the **3D Model** tab and select the **Extrude** tool. Select the sketch as the **Profile,** enter an **Extents Distance** value of **8,** and click **OK**.

15B Right-click the mouse and click the **Finish 2D Sketch** option. Select the **Extrude** tool. Select the sketch as the **Profile**, enter an **Extents Distance** value of **8**, and click **OK**.

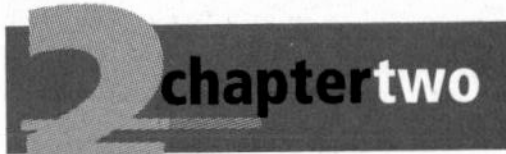

Chapter Summary

This chapter introduced most of the tools found on the panels under the **Sketch** tab, and each of the tools was demonstrated using a short sample exercise. Several of the tools were combined to form more complex drawing shapes. The chapter also showed how to edit sketches and presented a sample problem that required the use of many of the 2D tools.

Chapter Test Questions

Multiple Choice

Circle the correct answer.

1. Once a line has been sketched, which tool is used to add a dimensional value?
   a. **Sketch** c. **General Dimension**
   b. **Line** d. **Auto Dimension**
2. What is a curved line with multiple shape changes called?
   a. Polynomial
   b. Spline
   c. Ogee curve
3. An ellipse is defined by its major axis and which other axis?
   a. Lesser
   b. Second
   c. Minor
4. A chamfer is defined by which of the following?
   a. Two distances
   b. An angle and two distances
   c. Two angles and a distance
5. Which of the following shapes is not a polygon?
   a. Triangle c. Circle
   b. Hexagon d. Square
6. Which tool is used to remove an unwanted portion of a line?
   a. **Erase** c. **Offset**
   b. **Trim** d. **Move**
7. Which tool is used to increase the length of an existing line?
   a. **Scale** c. **Copy**
   b. **Offset** d. **Extend**

8. The symbol Ø is used to define which of the following?

a. Radius
c. Runout
b. Diameter
d. Extension

Matching

Write the number of the correct answer on the line.

Column A

a. Defines a location and size for an entity _____
b. Aligns two entities _____
c. Defines two entities as 90° apart _____
d. Defines a line in the X direction _____
e. Defines an entity equidistant from another entity _____

Column B

1. **Perpendicular**
2. **Parallel**
3. **Colinear**
4. **Fix**
5. **Horizontal**

True or False

Circle the correct answer.

1. **True or False:** The **General Dimension** tool is used to add dimensional values to sketches.
2. **True or False:** A closed spline is a curved line that starts and ends at a common point.
3. **True or False:** Splines cannot be edited once drawn.
4. **True or False:** The **Tangent Arc** and **Center Point Arc** tools are fly-outs from the **Three Point Arc** tool.
5. **True or False:** A fillet is a straight line across a corner.
6. **True or False:** A chamfer is a straight line across a corner.
7. **True or False:** A polygon can have any number of sides.
8. **True or False:** A mirror image is the same as a copied image.
9. **True or False:** A line offset from an original line has a different length than the original line.
10. **True or False:** Many different font styles are available for Inventor text.

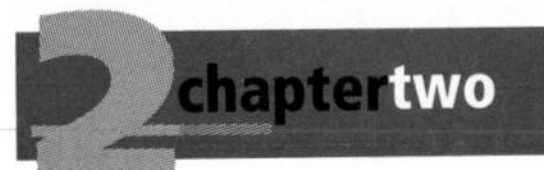

Chapter Project

Project 2-1

Redraw the following objects using the given dimensions. Create solid models of the objects using the specified thicknesses.

GUIDE PLATE

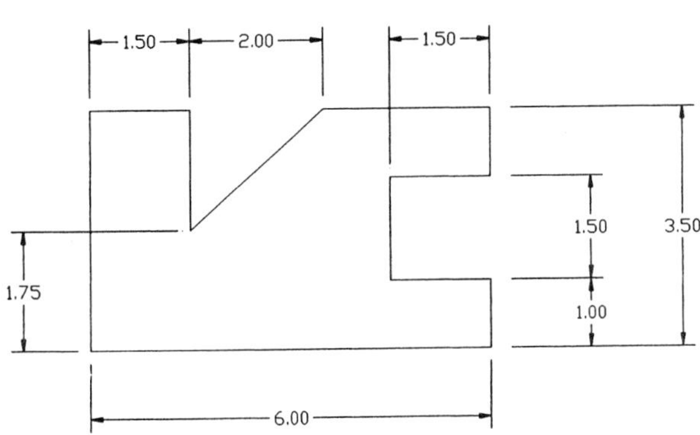

THICKNESS = 1.00

Figure P2-1
INCHES

TOP GASKET

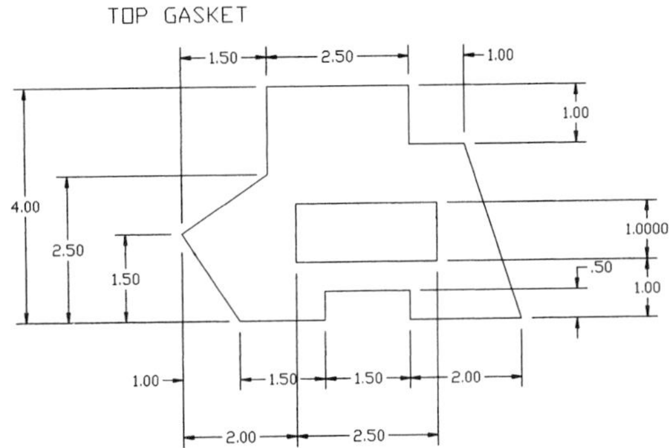

THICKNESS = .625

Figure P2-2
INCHES

BASE PLATE

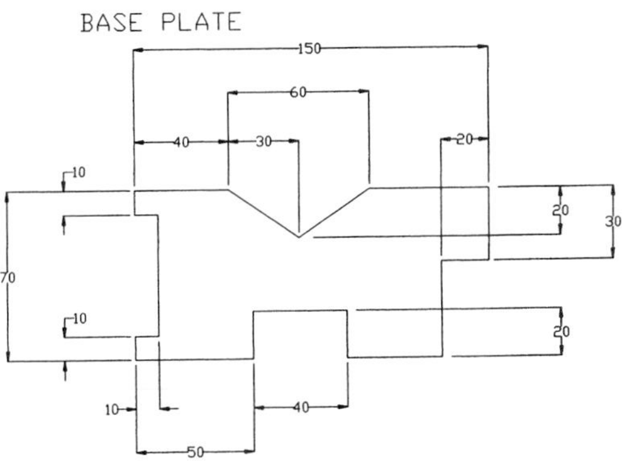

THICKNESS = 12

Figure P2-3
MILLIMETERS

GASKET

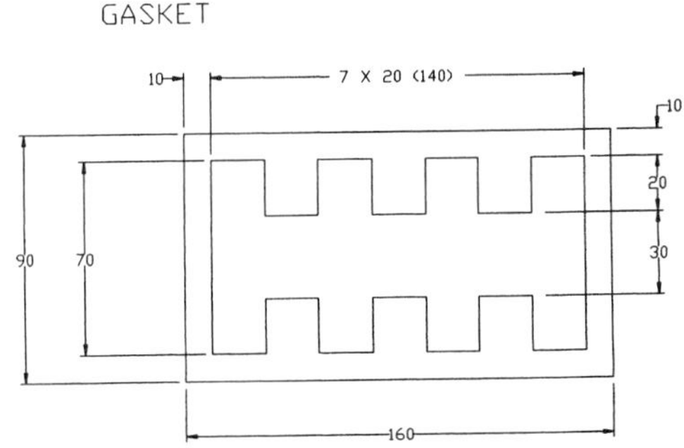

THICKNESS = 2

Figure P2-4
MILLIMETERS

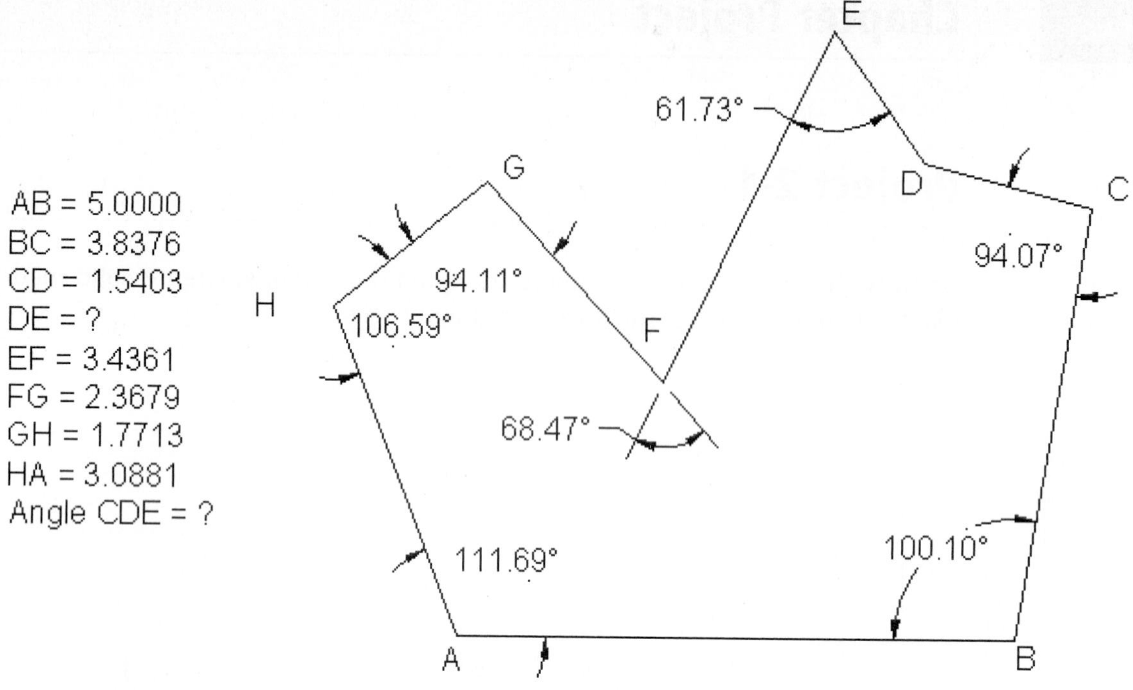

THICKNESS = 1.25

Figure P2-5
INCHES

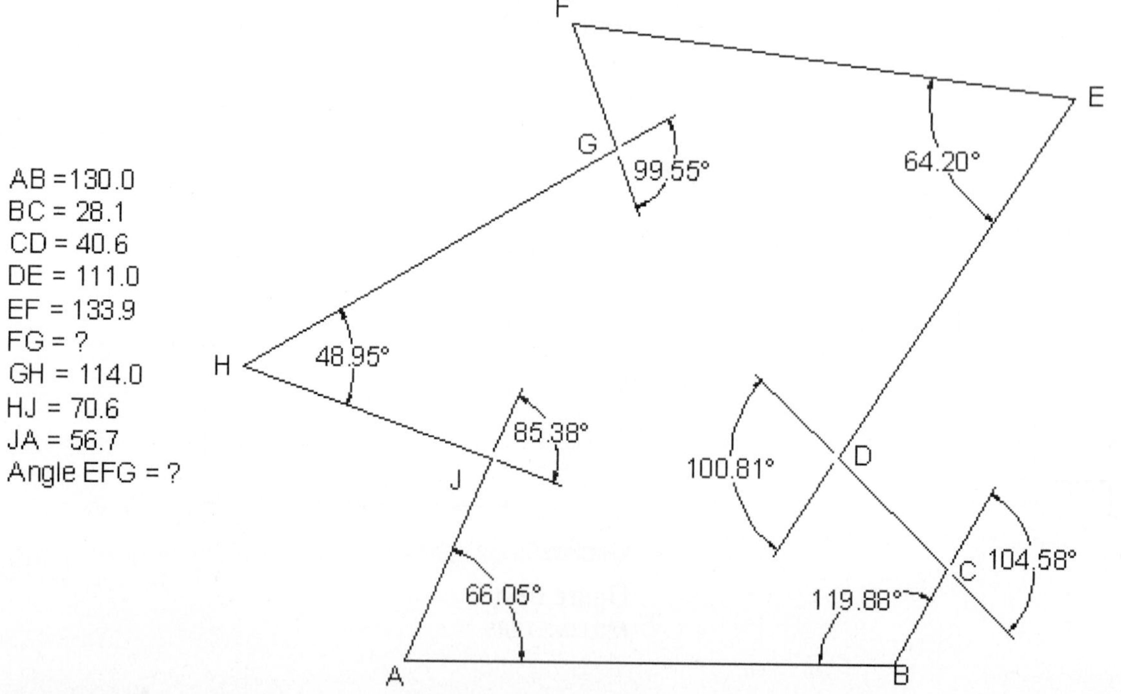

THICKNESS = 12

Figure P2-6
MILLIMETERS

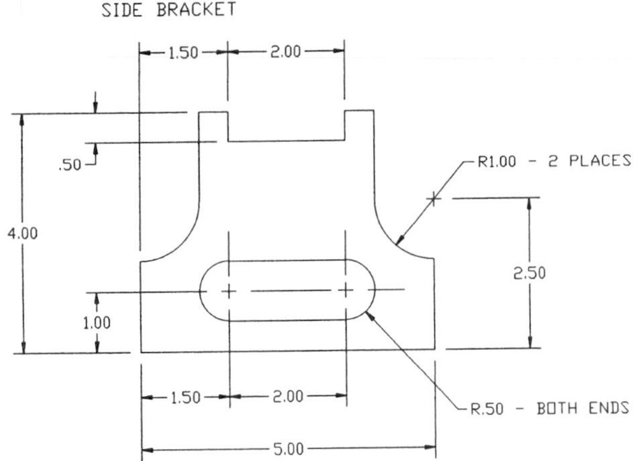

THICKNESS = 1.25

Figure P2-7
INCHES

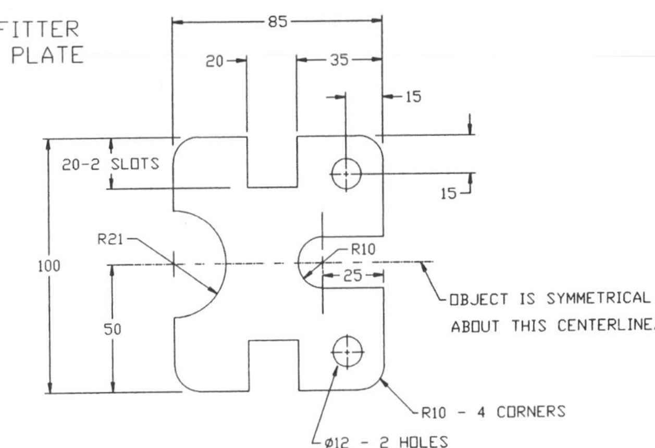

THICKNESS = 8

Figure P2-8
MILLIMETERS

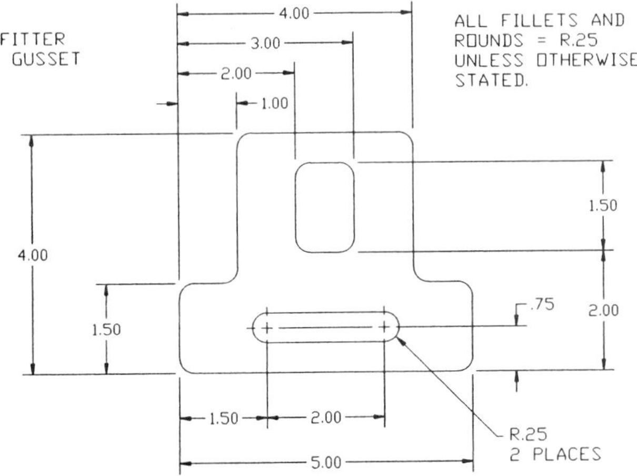

THICKNESS = .375

Figure P2-9
INCHES

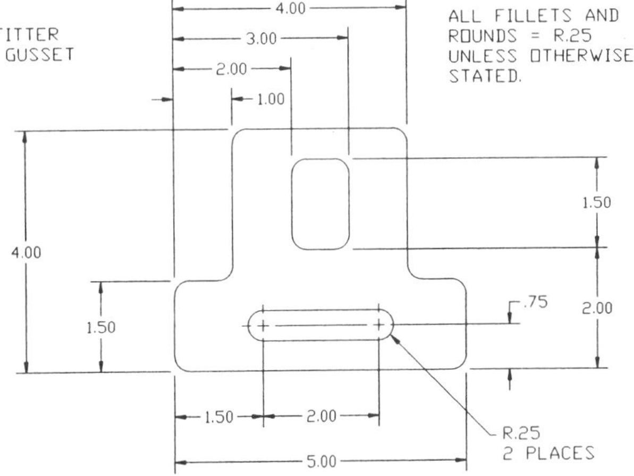

THICKNESS = .375

Figure P2-10
MILLIMETERS

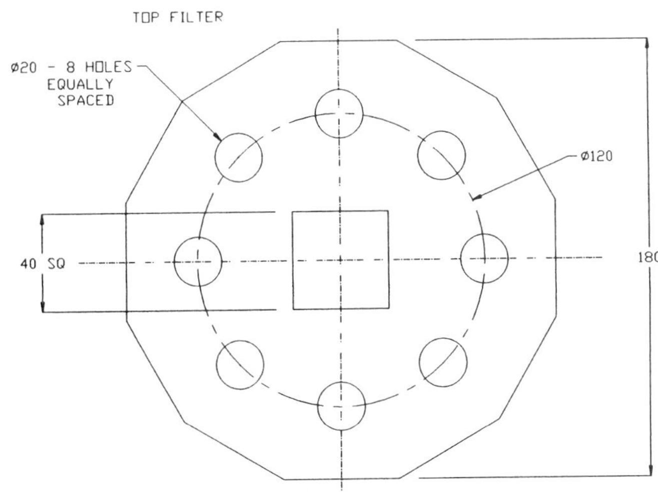

THICKNESS = 6

Figure P2-11
MILLIMETERS

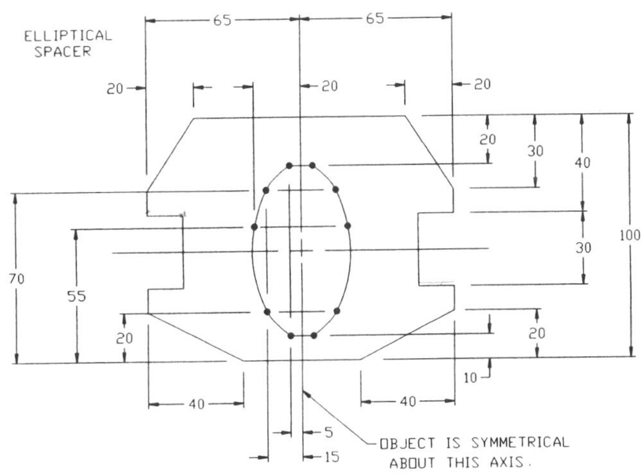

THICKNESS = 10

Figure P2-12
MILLIMETERS

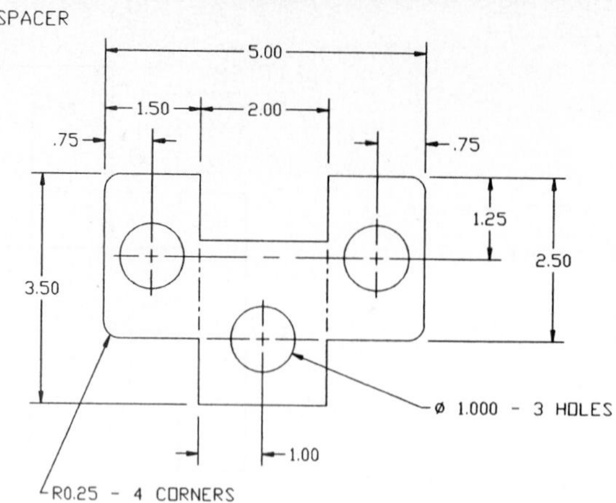

THICKNESS = .75

Figure P2-13
INCHES

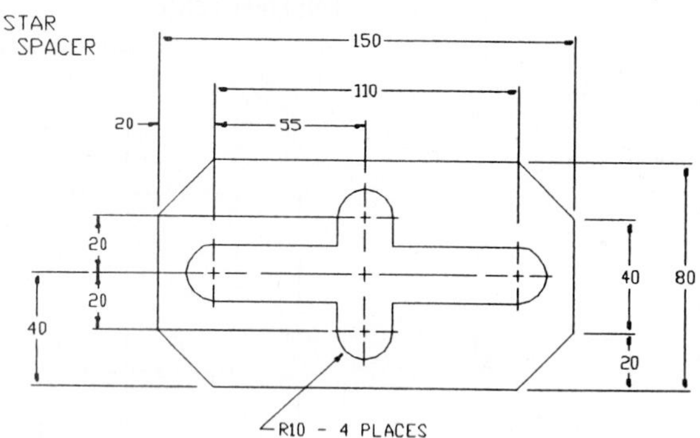

THICKNESS = 7.5

Figure P2-14
MILLIMETERS

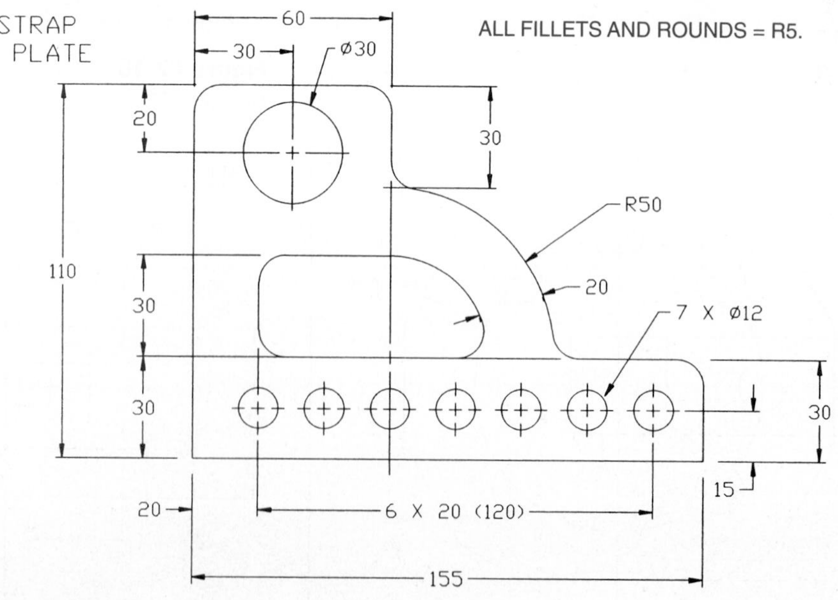

THICKNESS = 5

Figure P2-15
MILLIMETERS

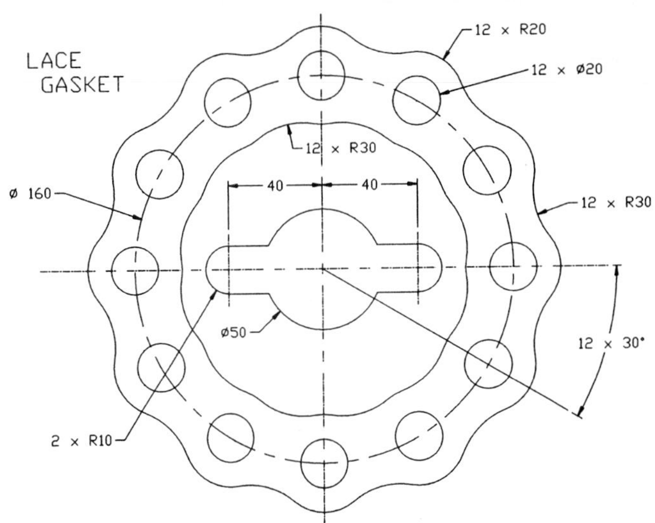

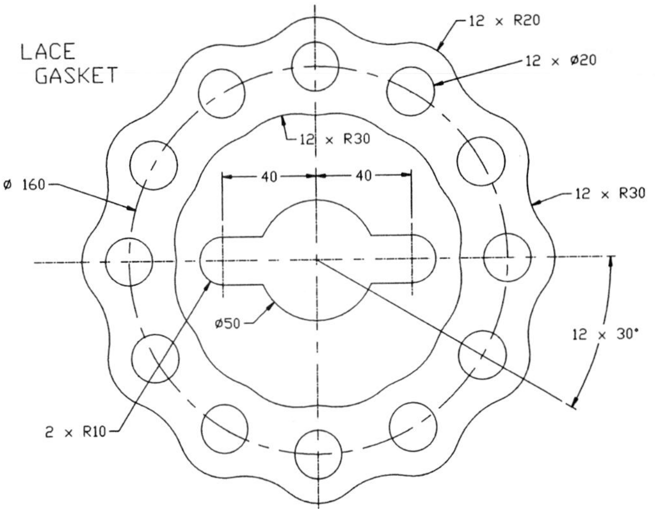

THICKNESS = 6; central area is 4 thick

Figure P2-16
MILLIMETERS

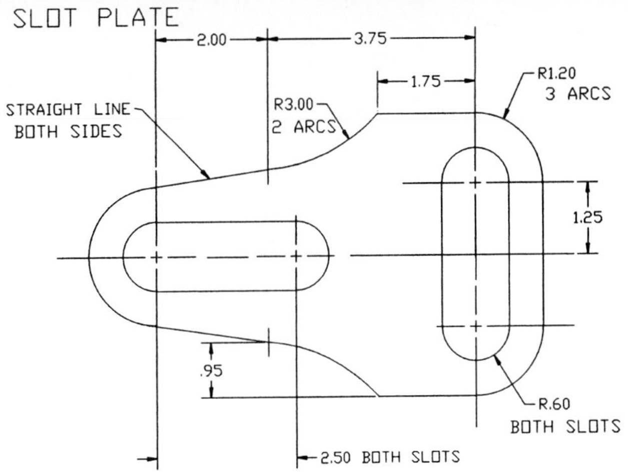

NOTE: Object is symmetrical about its horizontal centerline.

THICKNESS = .875

Figure P2-17
INCHES

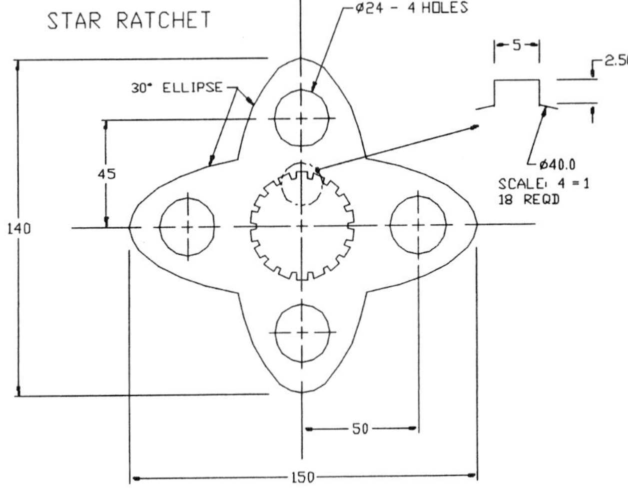

THICKNESS = 6

Figure P2-18
MILLIMETERS

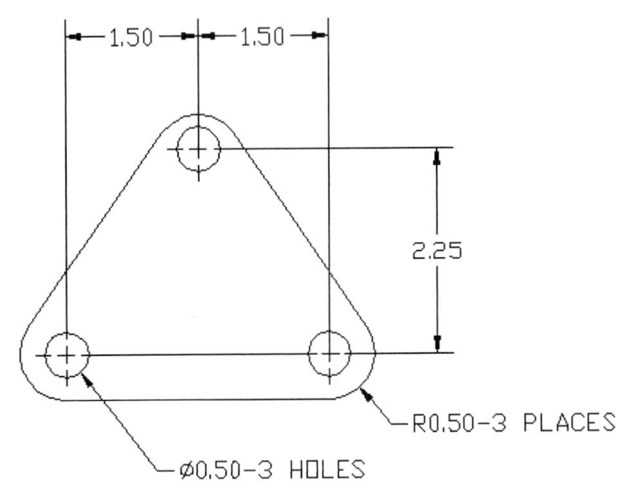

THICKNESS = .250

Figure P2-19
INCHES

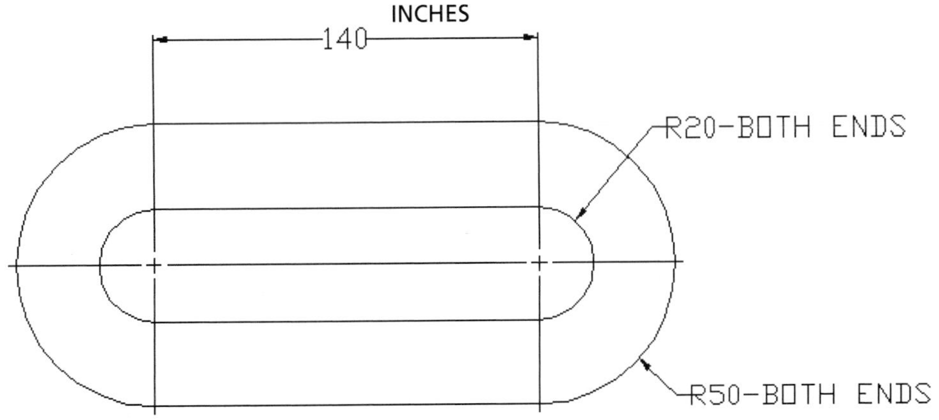

THICKNESS = 6.5

Figure P2-20
MILLIMETERS

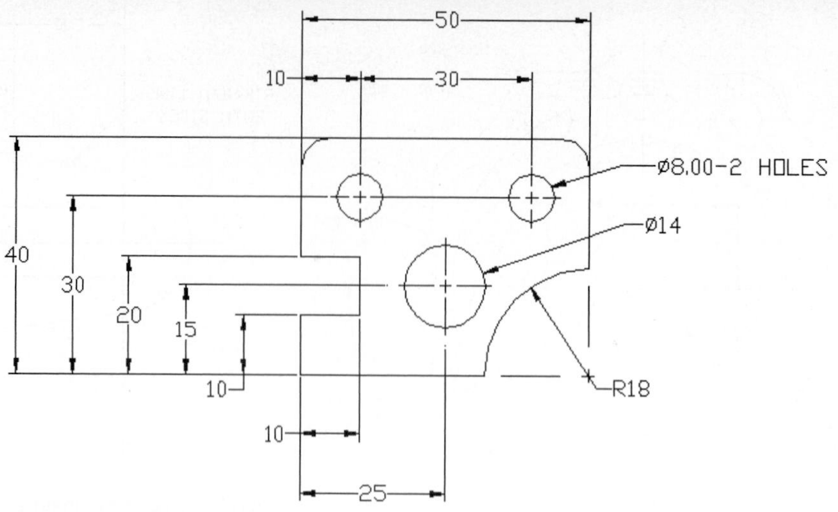

THICKNESS = 16

Figure P2-21
MILLIMETERS

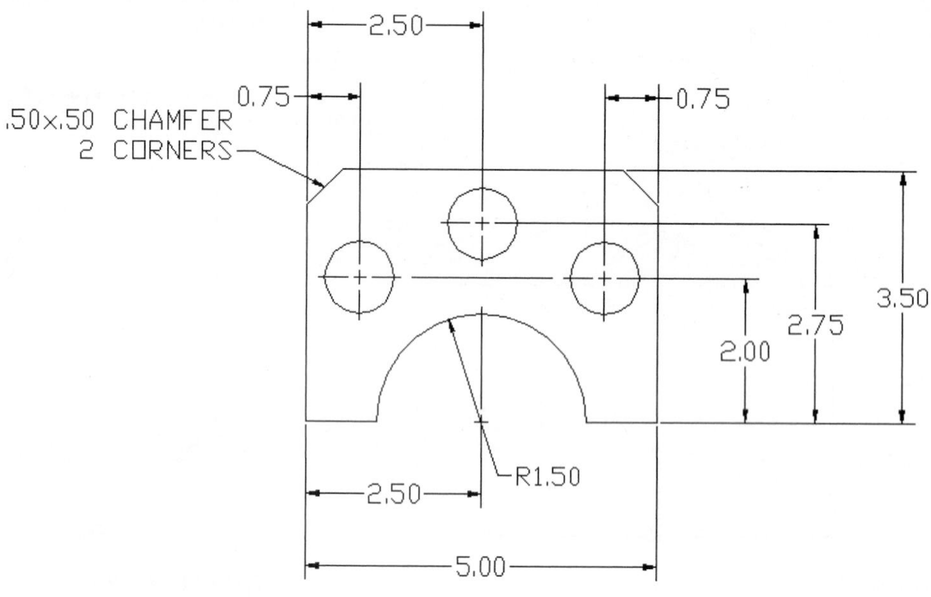

THICKNESS = .50

Figure P2-22
INCHES

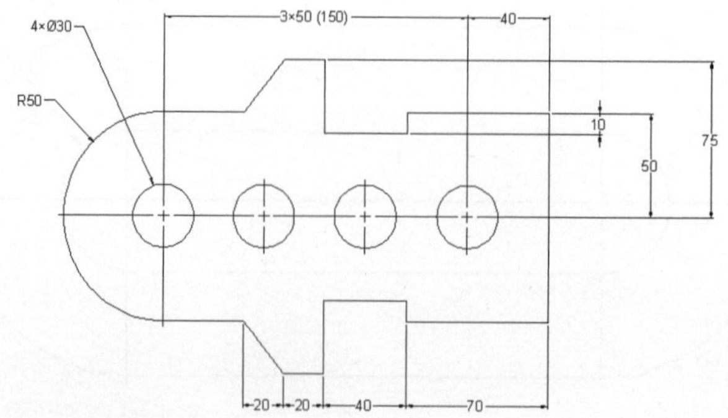

THICKNESS = 12

Figure P2-23
MILLIMETERS

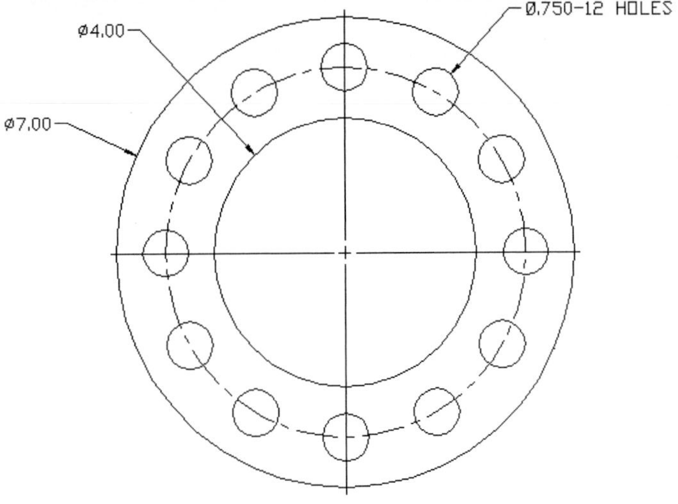

THICKNESS = .1875

Figure P2-24
MILLIMETERS

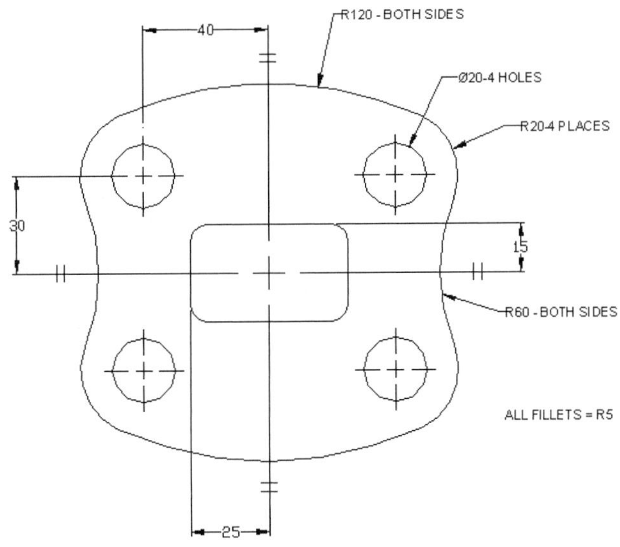

THICKNESS = 7.25

Figure P2-25
MILLIMETERS

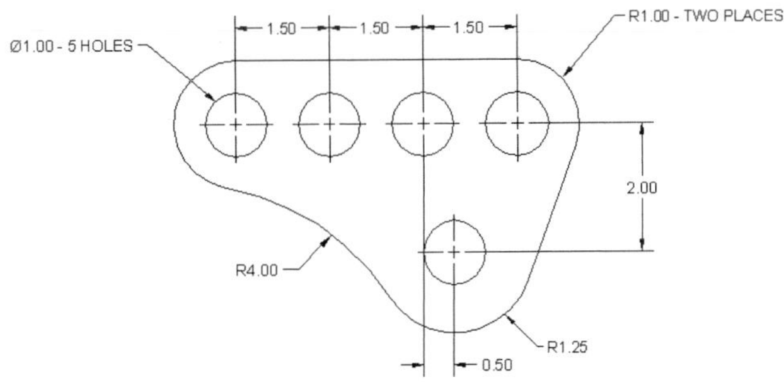

THICKNESS = .25

Figure P2-26
INCHES

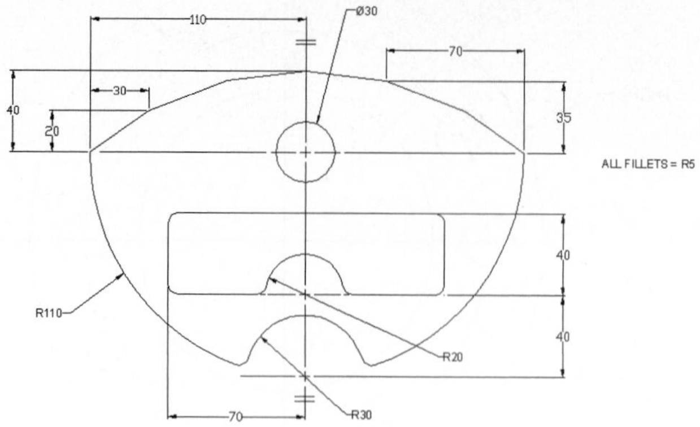

THICKNESS = 5

Figure P2-27
MILLIMETERS

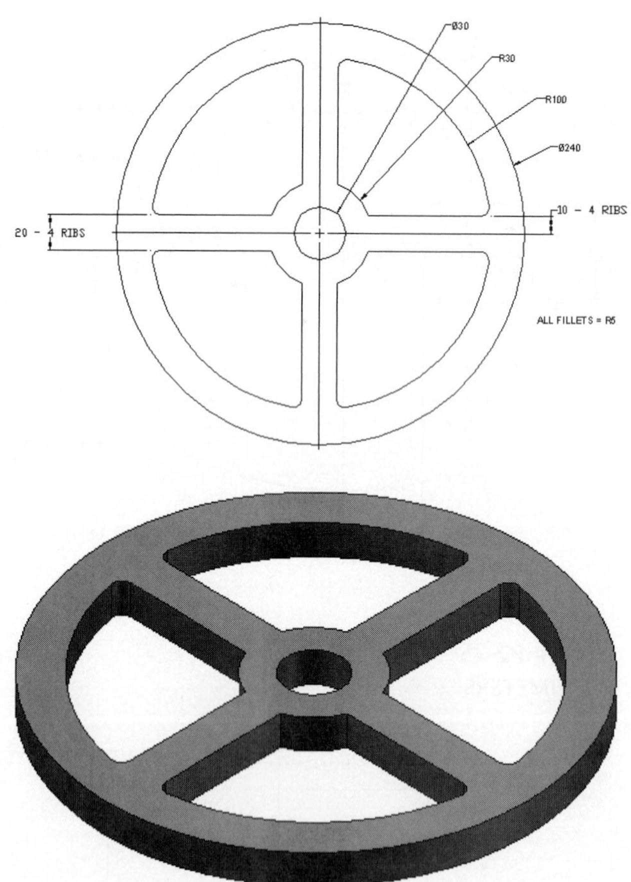

THICKNESS = 16

Figure P2-28
MILLIMETERS

3 chapter three
3D Models

CHAPTER OBJECTIVES

- Learn how to create 3D models
- Learn how to use the tools on the **3D Model** panel
- Learn how to edit 3D models
- Learn how to create and use work planes

Introduction

This chapter introduces and demonstrates how to create 3D models. The tools associated with the **3D Model** panel are presented along with examples of how they are applied to convert 2D sketches into 3D solid models.

Extrude

The **Extrude** tool is used to convert 2D sketches into 3D solid models. See Figure 3-1.

1. Start a new drawing using the **Standard (in).ipt** format.

2. Draw a **3.000** × **6.000** rectangle.

3. Right-click the mouse and select the **Finish 2D Sketch** option. Click the **Extrude** tool located on the **Create** panel under the **3D Model** tab.

The **Extrude** dialog box will appear. As there is only one sketch on the screen, it will automatically be selected as the profile.

Figure 3-1

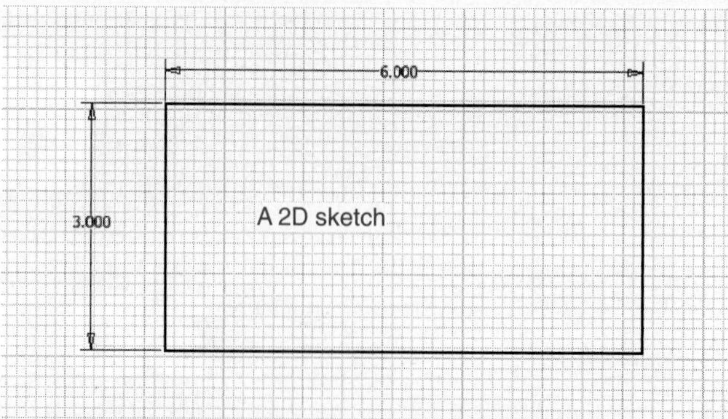
6.000
3.000
A 2D sketch

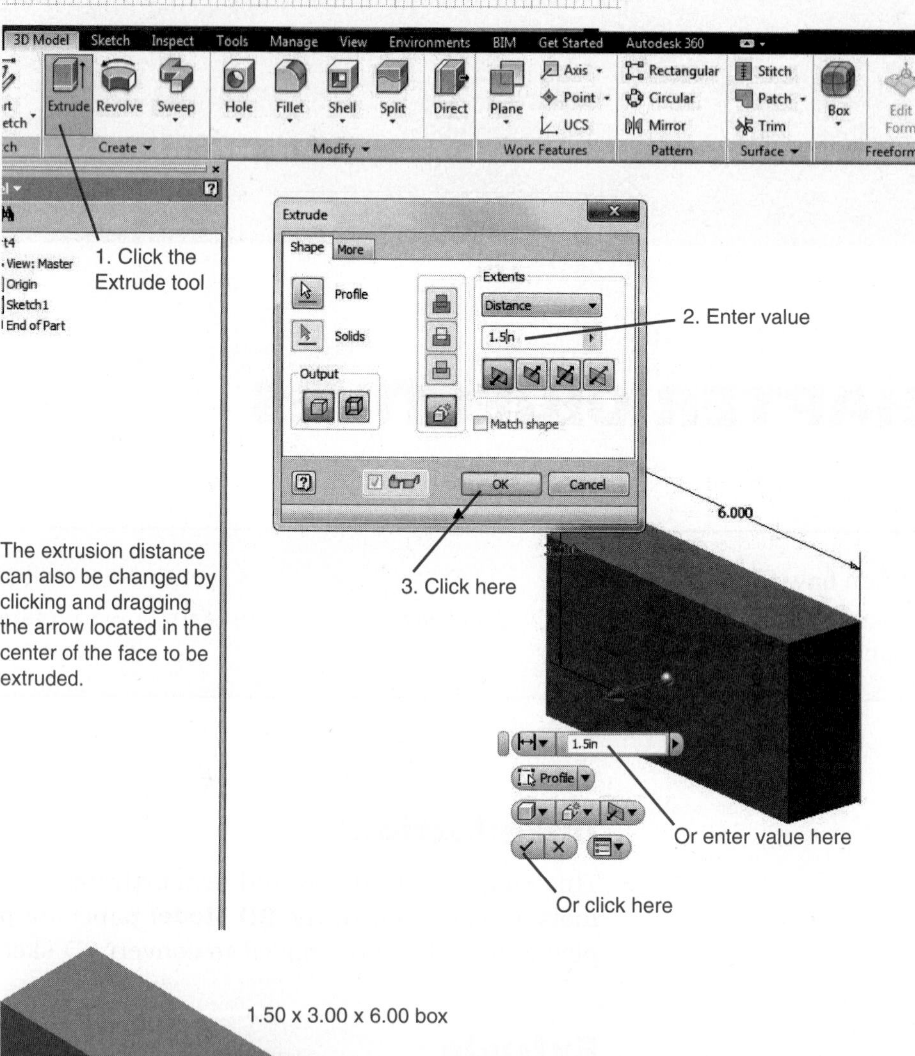
1. Click the Extrude tool
2. Enter value
3. Click here
The extrusion distance can also be changed by clicking and dragging the arrow located in the center of the face to be extruded.
Or enter value here
Or click here
Extrude
Shape
More
Profile
Solids
Output
Extents
Distance
1.5in
Match shape
OK
Cancel
6.000

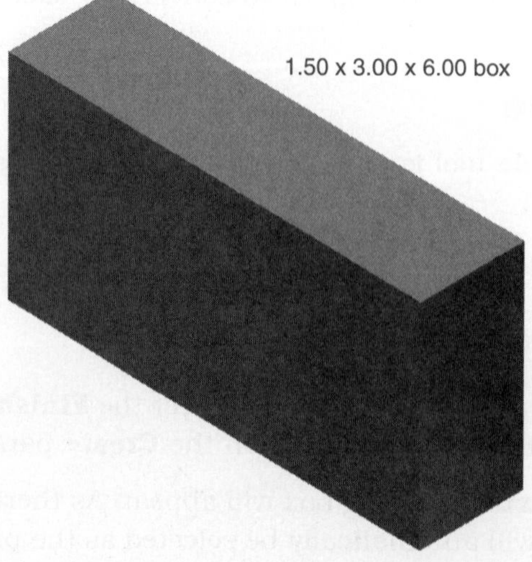
1.50 x 3.00 x 6.00 box

4 Set the thickness value for **1.50**.

5 Click **OK** or the **green check mark**.

> **NOTE**
>
> The extrusion distance can also be changed by clicking and dragging the arrow located in the center of the face to be extruded.

Taper

The sides of the object drawn in Figure 3-1 are at 90° to each other. The **Extrude** tool also allows for tapered sides, that is, sides that are not 90° to each other. See Figure 3-2.

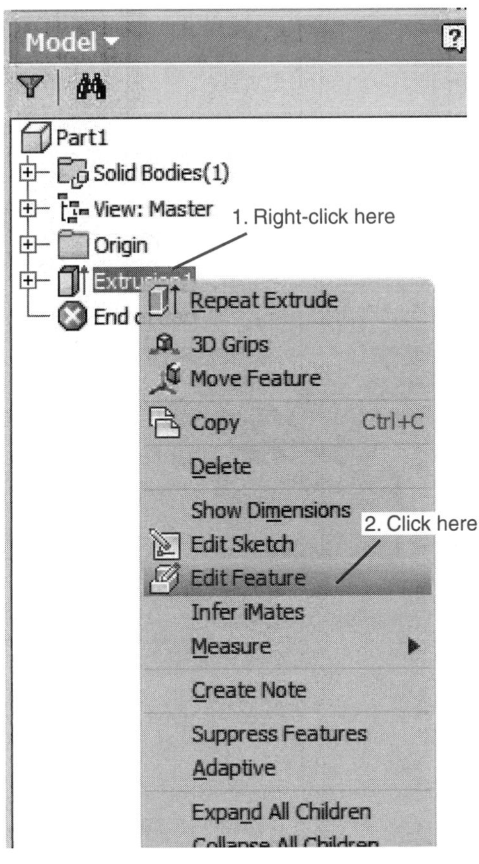

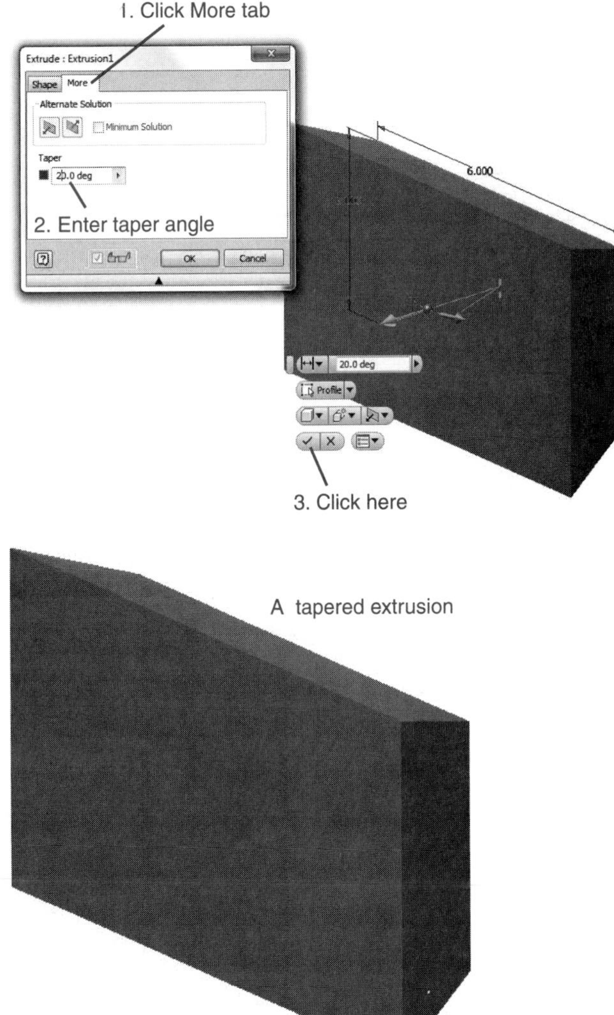

Figure 3-2

1 Right-click the word **Extrusion1** in the browser box.

A dialog box will appear.

2 Click the **Edit Feature** option.

Figure 3-3

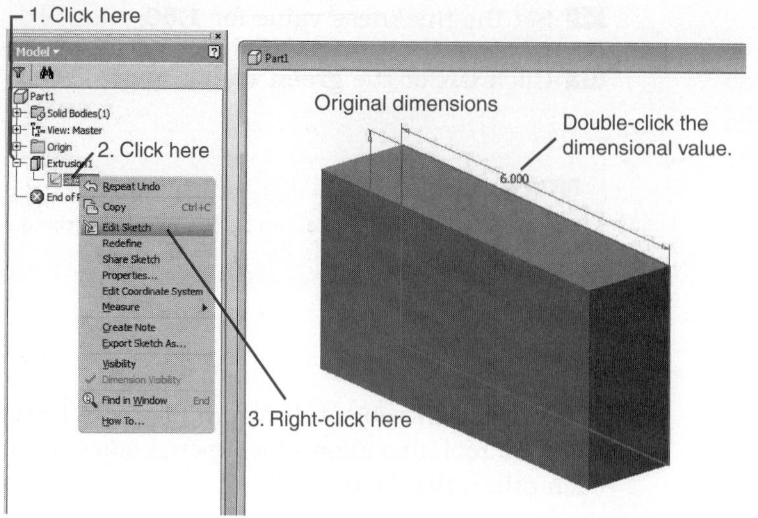
1. Click here
Model
Part1
Solid Bodies(1)
View: Master
Origin
Extrusion1
End of Part
2. Click here
Repeat Undo
Copy Ctrl+C
Edit Sketch
Redefine
Share Sketch
Properties...
Edit Coordinate System
Measure
Create Note
Export Sketch As...
Visibility
Dimension Visibility
Find in Window End
How To...
3. Right-click here
Part1
Original dimensions
Double-click the dimensional value.
6.000

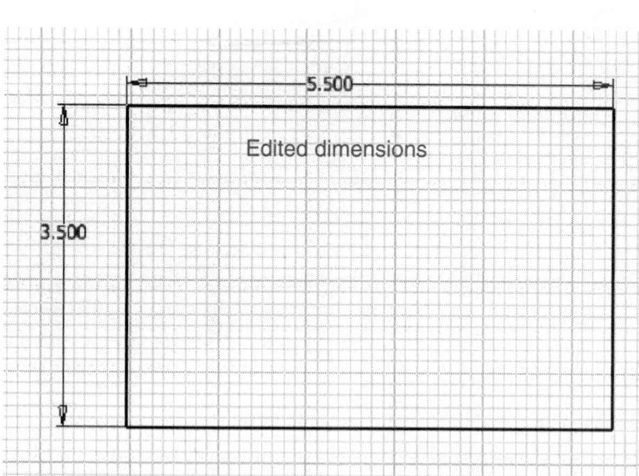
5.500
Edited dimensions
3.500

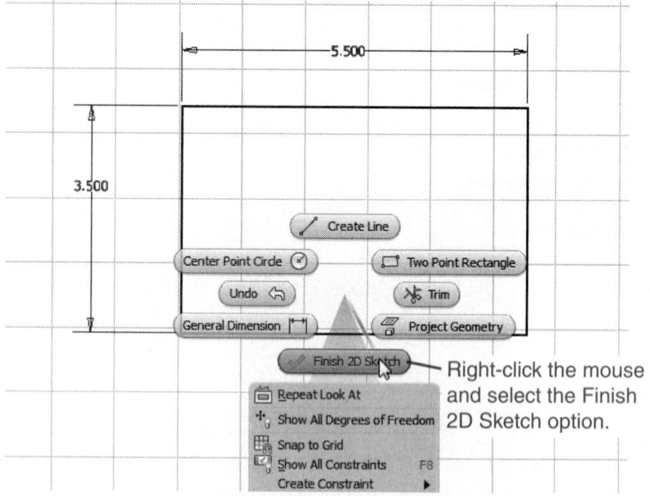
5.500
3.500
Create Line
Center Point Circle
Two Point Rectangle
Undo
Trim
General Dimension
Project Geometry
Finish 2D Sketch
Repeat Look At
Show All Degrees of Freedom
Snap to Grid
Show All Constraints F8
Create Constraint
Right-click the mouse and select the Finish 2D Sketch option.

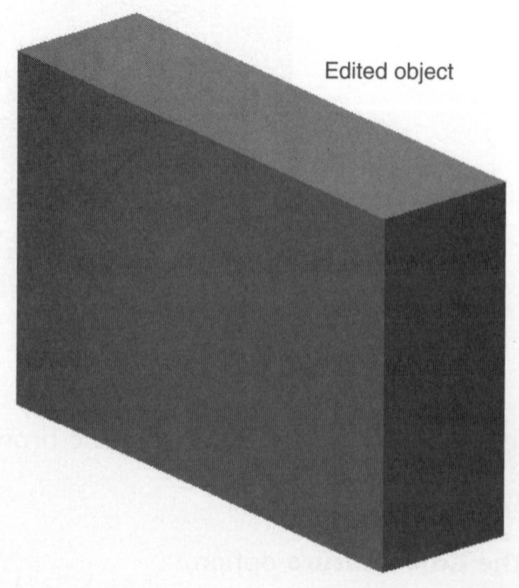
Edited object

3 The **Extrude** dialog box will reappear.

4 Click the **More** tab.

5 Enter a **Taper angle** of **20.0 deg**.

6 Click **OK**.

Editing an Object's Sketch

Click the **Undo** tool to return the tapered object to its original square shape. The object was based on a 3.00 × 6.00 rectangle. Say we wish to change that original size to 3.50 × 5.50. See Figure 3-3.

1 Click the **+ sign** to the left of the word **Extrusion1** in the browser box.

The word **Sketch** will appear under the **Extrusion1** heading.

2 Select the **Edit Sketch** option and right-click the mouse.

The screen will change to show only the original 3.00 × 6.00 sketch.

3 Double-click each dimension and enter the new values.

4 Right-click the mouse and select the **Finish 2D Sketch** option.

The object will reappear using the new dimension values.

Editing an Object's Features

The thickness of the object can also be edited. See Figure 3-4.

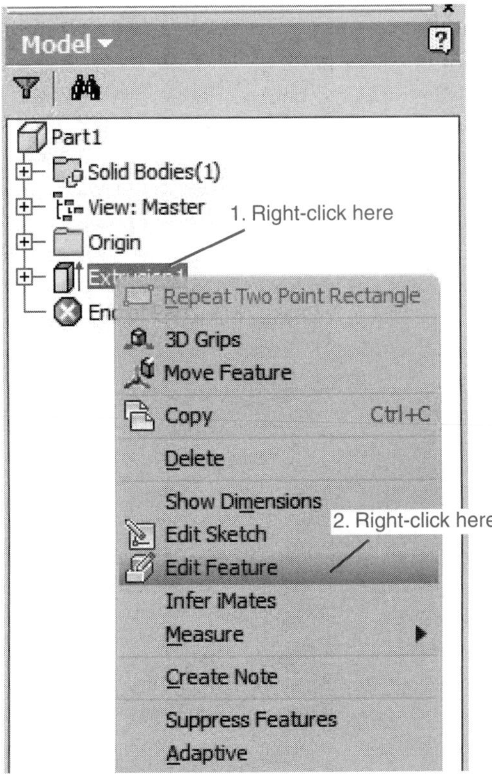

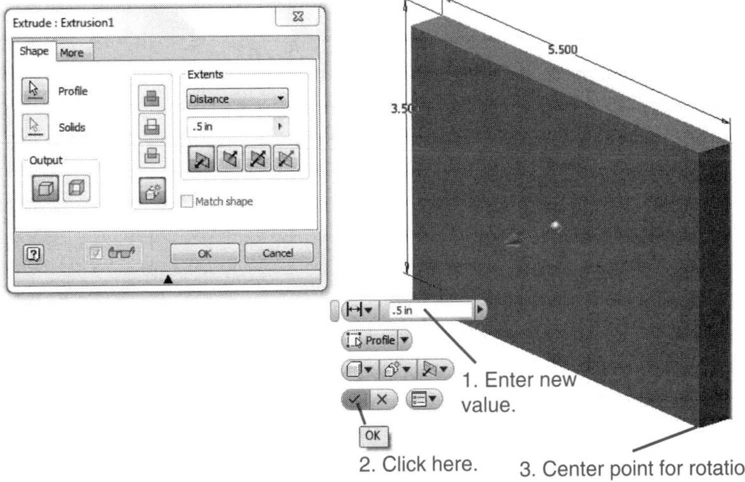

Figure 3-4

1 Right-click the word **Extension1** in the browser box and click the **Edit Feature** option.

The **Extrude** dialog box will reappear.

2 Change the thickness to **.50** and click **OK**.

feature: A shape created with the tools from the **3D Model** panel.

sketch: A shape created with the tools from the **Sketch** panel.

> **NOTE**
>
> Shapes created with tools from the **3D Model** panel are ***features***. Shapes created with the tools from the **Sketch** panel are ***sketches***. The sketches are listed under the features in the browser box. Both sketches and features can be edited in an existing object.

ViewCube

This section will use the .5 × 3.50 × 5.50 box created previously. See Figure 3-5.

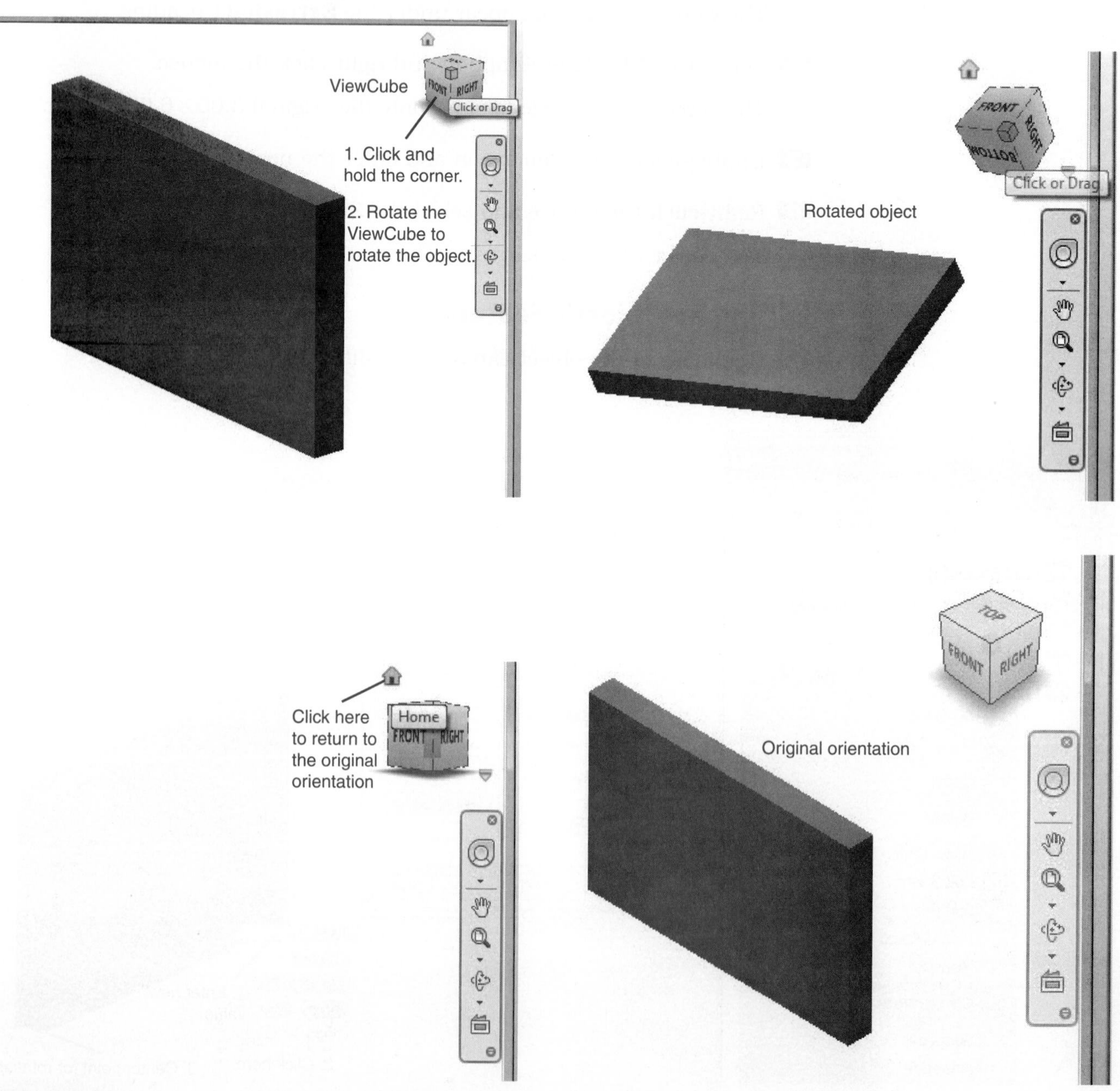

Figure 3-5

The **ViewCube** is located in the upper right of the drawing screen and is used to change the orientation of an object.

1. Move the cursor to the top corner of the **ViewCube** as shown.

   The corner will be highlighted by three small surface planes.

2. Click and hold the corner point.

3. Drag the object to a new orientation using the **ViewCube**.

   In this example, an orientation was chosen that exposed the bottom surface of the object.

4. Click the houselike icon located above the **ViewCube** to return the object to its original orientation.

To Add a Flange

1. Right-click the bottom surface of the object and select the **New Sketch** option.

Use the **ViewCube** to reorient the object so you can see the bottom surface if necessary.

Inventor can draw on only one plane at a time. The **New Sketch** option has now created a new sketching plane (note the grid pattern) aligned with the bottom surface of the object. See Figure 3-6.

2. Use the **Rectangle** tool and draw a rectangle aligned with the bottom surface of the object.

3. Use the **Dimension** tool to define the length of the rectangle as **5.00**.

   The width of the rectangle will be 5.50, as it was aligned with the bottom surface of the object.

4. Right-click the mouse and select the **Finish 2D Sketch** option. Click the **Extrude** tool.

5. Click the sketched rectangle as the profile. Also, click the bottom surface of the existing rectangle.

6. Set the thickness value for **.50** and flip the extrusion to extend up into the existing object.

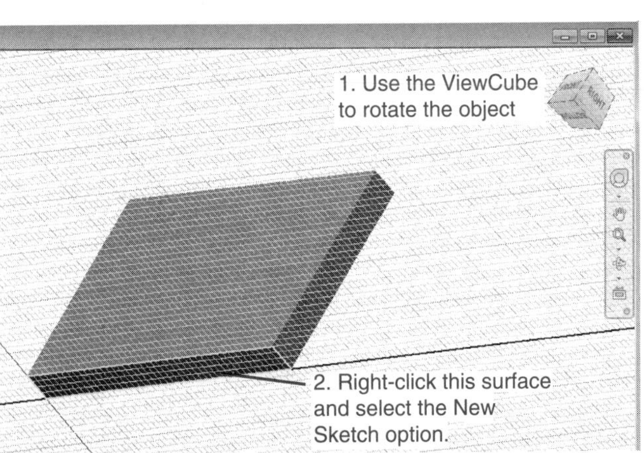

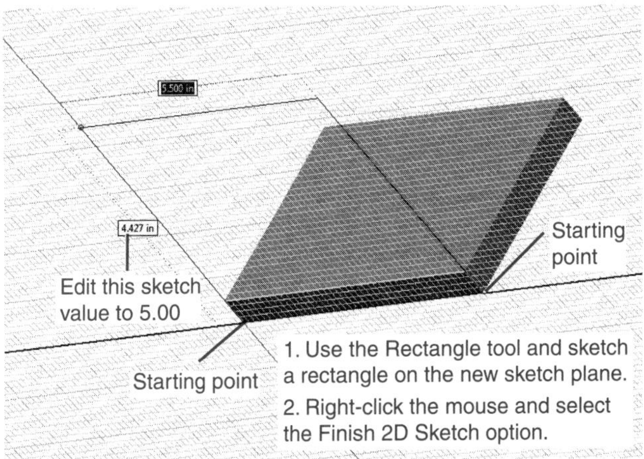

Figure 3-6

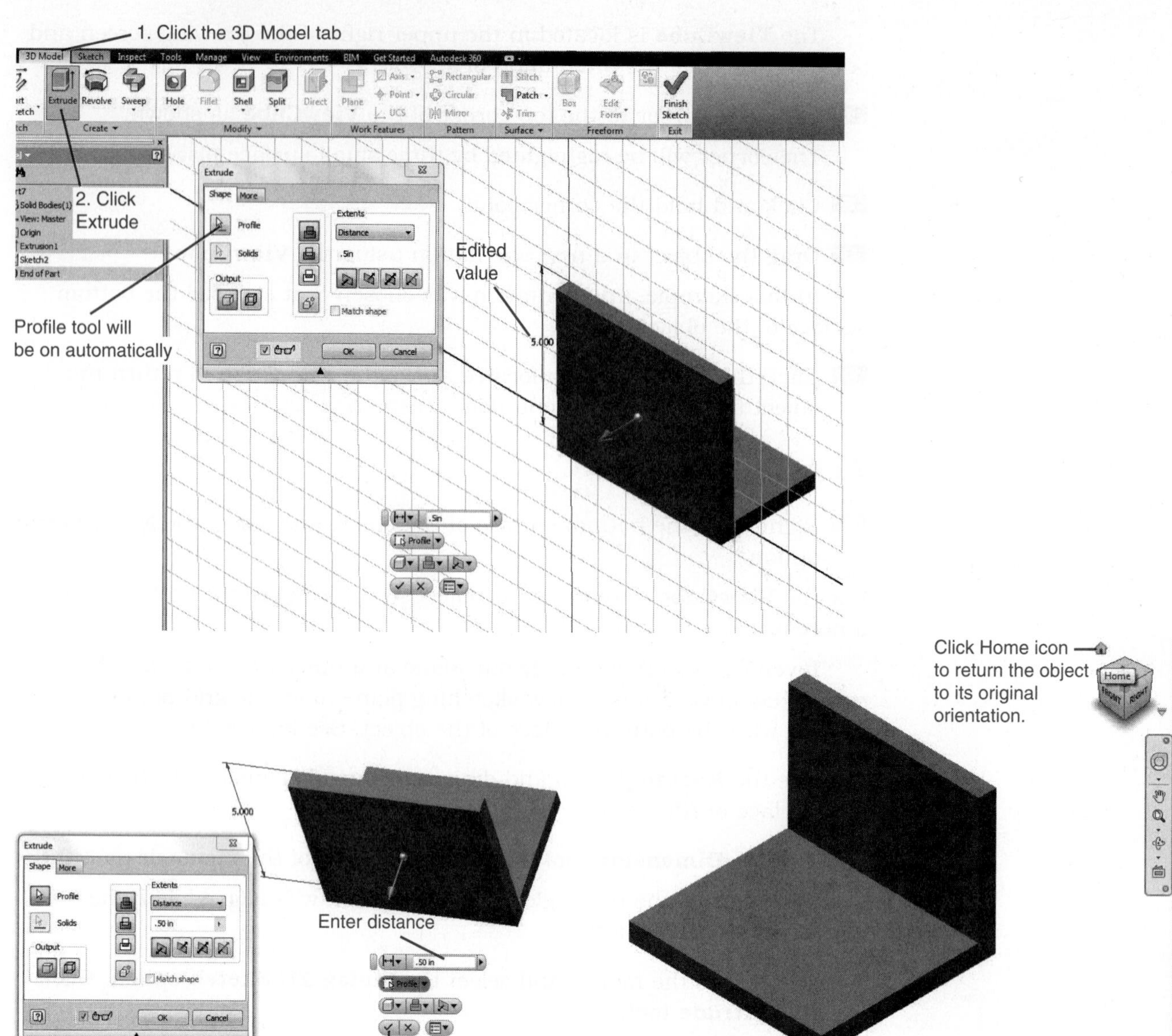

Figure 3-6
(*Continued*)

7 Click **OK**.

8 Move the cursor into the **ViewCube** area and click the **Home** icon (it looks like a small house).

This will return the object to the isometric orientation.

Add another flange to the top surface of the object. See Figure 3-7.

1 Right-click the top surface of the object and select the **New Sketch** option.

2 Align the new sketch with the top surface and extend it a distance of **3.50**.

3 Right-click the mouse and select the **Finish 2D Sketch** option.

4 Extrude the new sketch **.50** into the existing sketch.

5 Click **OK**.

Figure 3-7

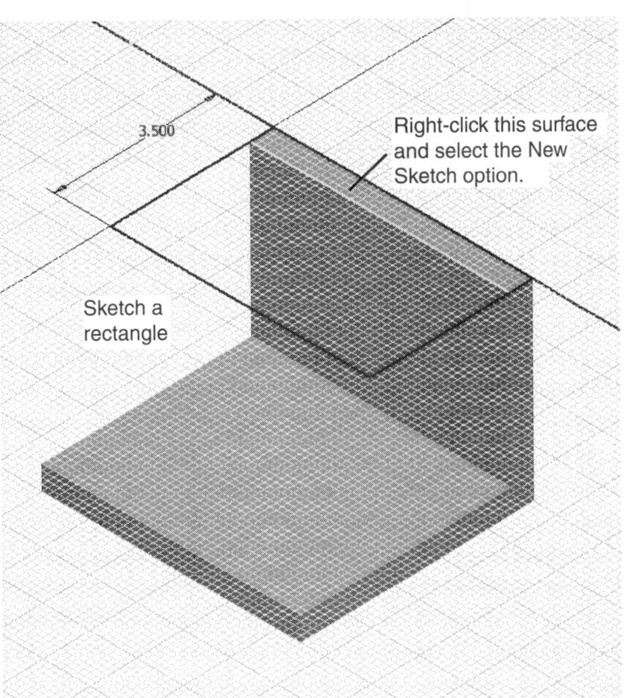
3.500
Right-click this surface and select the New Sketch option.
Sketch a rectangle

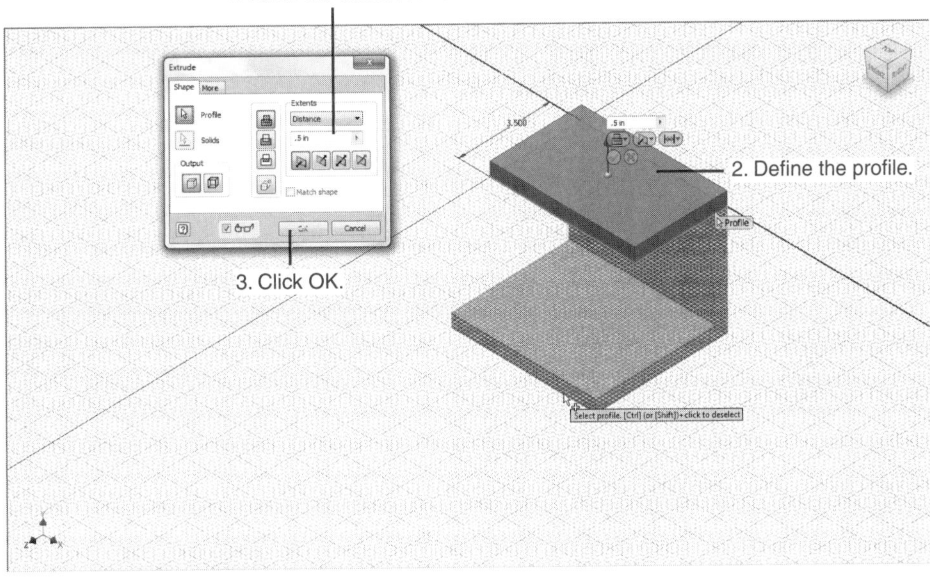
1. Enter the thickness value.
2. Define the profile.
3. Click OK.
Extrude
Shape
More
Profile
Solids
Output
Extents
Distance
.5 in
Match shape
Cancel
3.500
Profile

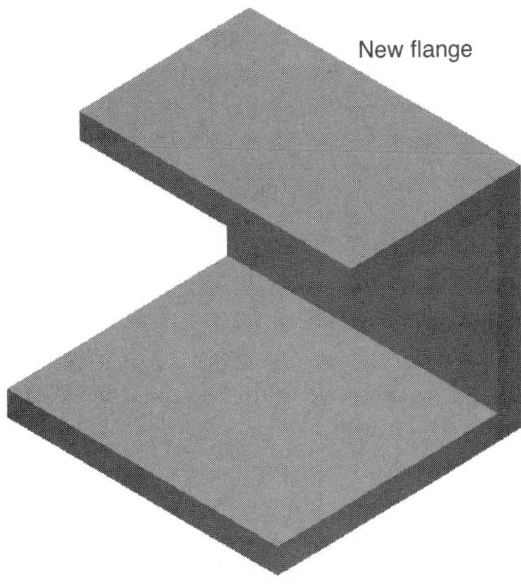
New flange

Revolve

The **Revolve** tool is used to revolve a profile about an axis of revolution. See Figure 3-8.

Figure 3-8

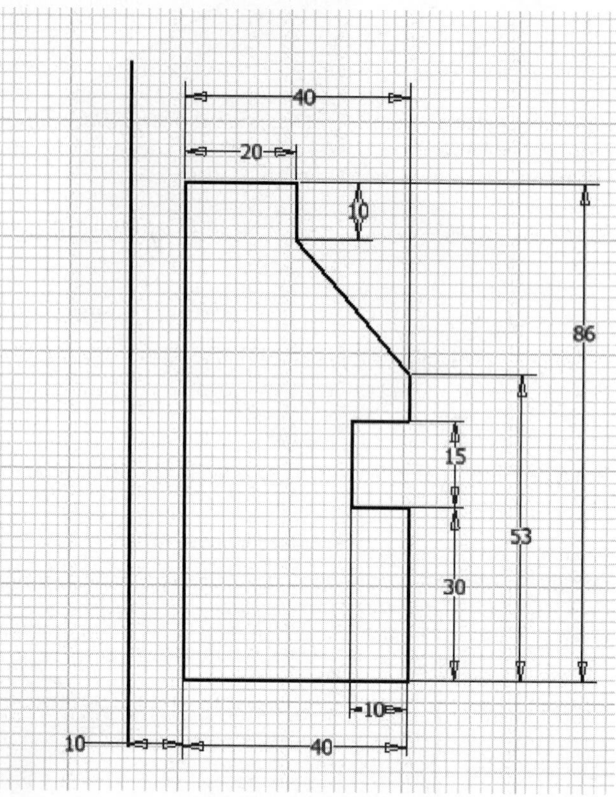

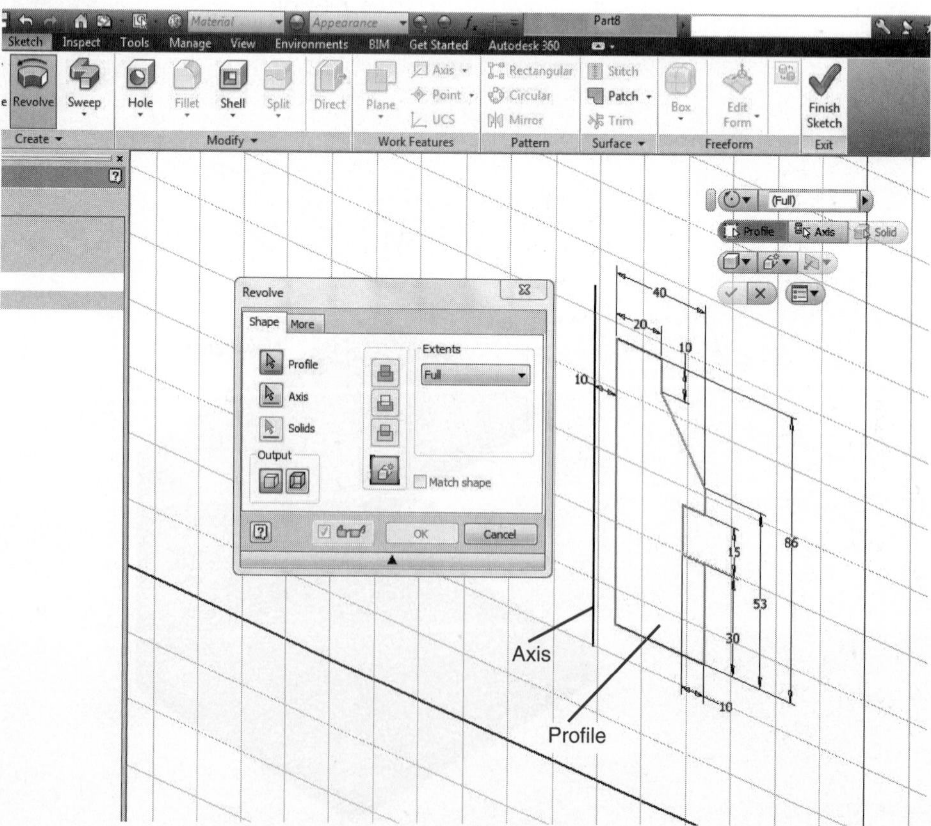

Figure 3-8
(*Continued*)

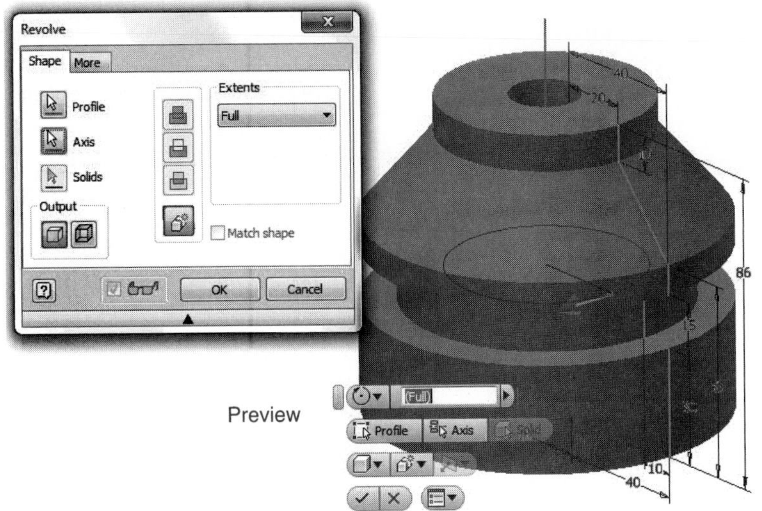

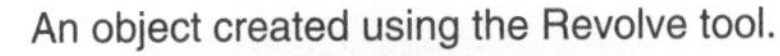

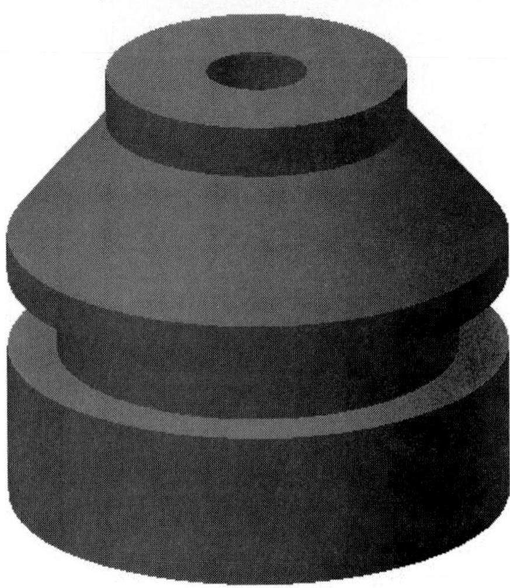

1 Start a new drawing using the **Standard (mm).ipt** format, click the **Create 2D Sketch** tool, and select the XY plane.

2 Sketch the enclosed figure and vertical line shown.

3 Click the **3D Model** tab and select the **Revolve** tool.

The **Revolve** dialog box will appear.

4 Select the enclosed shape as the **Profile**.

5 Select the vertical line as the **Axis**.

6 Click **OK**.

7 Use the **ViewCube** to orient the object.

Figure 3-9 shows a sphere created as a revolved object.

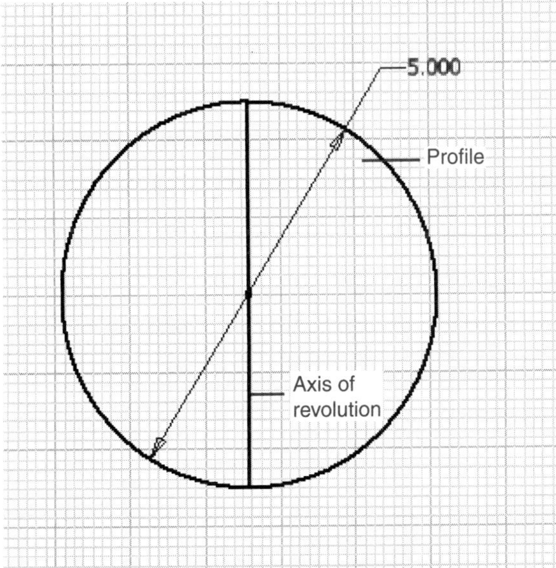

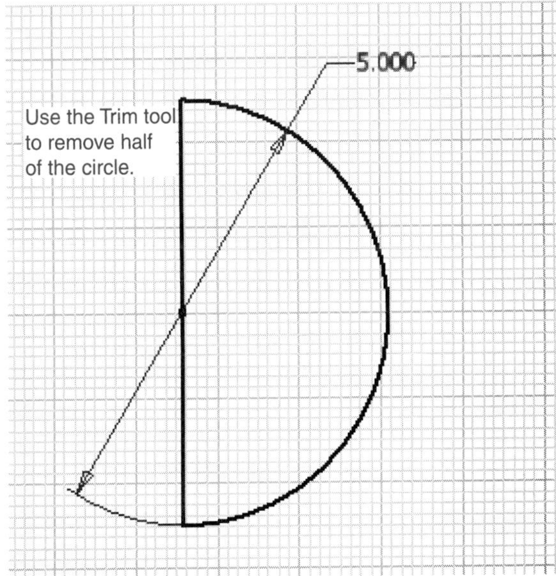

Figure 3-9

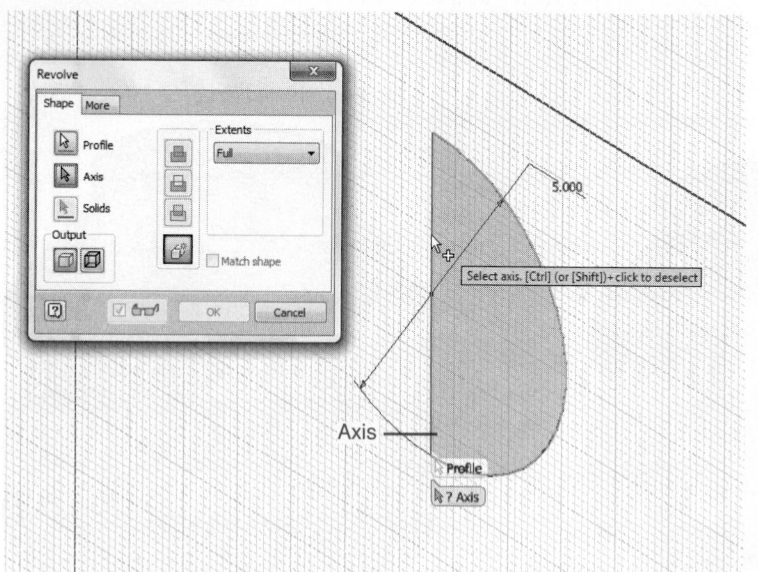

Figure 3-9
(Continued)

Holes

Holes can be created using an extruded cut circle, as was done in the first two chapters, or by using the **Hole** tool located on the **Modify** panel under the **3D Model** tab. The **Hole** tool allows for hole shapes other than straight-through holes, including blind holes, counterbored and countersunk holes, and threaded holes.

To Create a Through Hole

Figure 3-10 shows a 40 × 40 × 60 mm box. Locate a Ø20.0 simple hole through its center.

1. Click the **Hole** tool and select the **Simple Hole** option.

simple hole: A straight hole.

Note that the default value is the **Simple Hole** option. A ***simple hole*** is a straight hole. Note also that the **Face** option is automatically on. The **Face** option defines the surface on which the hole will be located.

2. Define the top surface of the object as the **Face**.
3. Specify the diameter of the hole.
4. Use the **Reference 1** and **Reference 2** options on the **Hole** dialog box to define the location of the hole's center point.
5. Define reference lines 1 and 2 as shown.

Dimensions will automatically appear from the reference lines to the hole's center point.

6. Edit the hole's locating dimensions as needed.
7. Specify the depth of the hole.

In this example the hole is to go completely through the object, so select the **Through All** option.

8. Click **OK**.

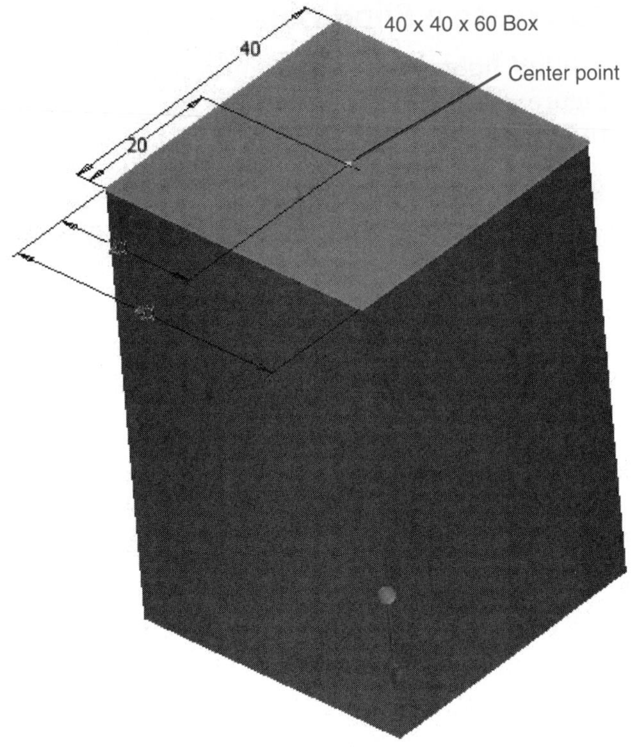

40 x 40 x 60 Box
Center point
40
20

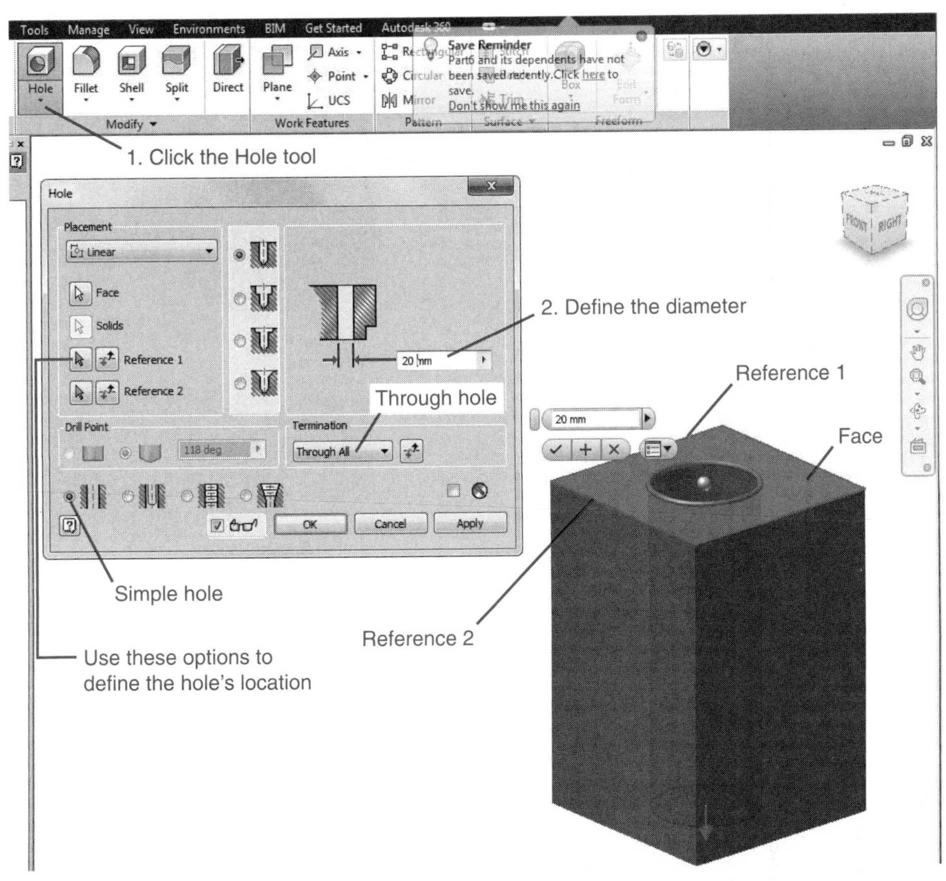

1. Click the Hole tool
2. Define the diameter
Through hole
Reference 1
Face
Reference 2
Simple hole
Use these options to define the hole's location

Ø20 Hole

Figure 3-10

To Create a Blind Hole

blind hole: A hole that does not go completely through an object.

A ***blind hole*** is a hole that does not go completely through an object. Figure 3-11 shows a 40 × 40 × 50 box. Locate a Ø20 hole in the center of the top surface with a depth of 25.

1. Start a new drawing using the **Standard (mm).ipt** format, click the **Create 2D Sketch** tool, select the XZ plane, and sketch the 40 × 40 × 50 box as shown.

2. Select the **Simple Hole** option.

3. Ensure that the **Drill Point** is set on the **118 deg** option.

   Twist drills have a conical endpoint that makes it easier for them to drill holes. This conical shape is included in the drawing. It is not considered part of the hole depth.

4. Define the hole's diameter as **20** and the hole's depth as **25**.

5. Click the **Face** tool and select the top surface of the box as the face.

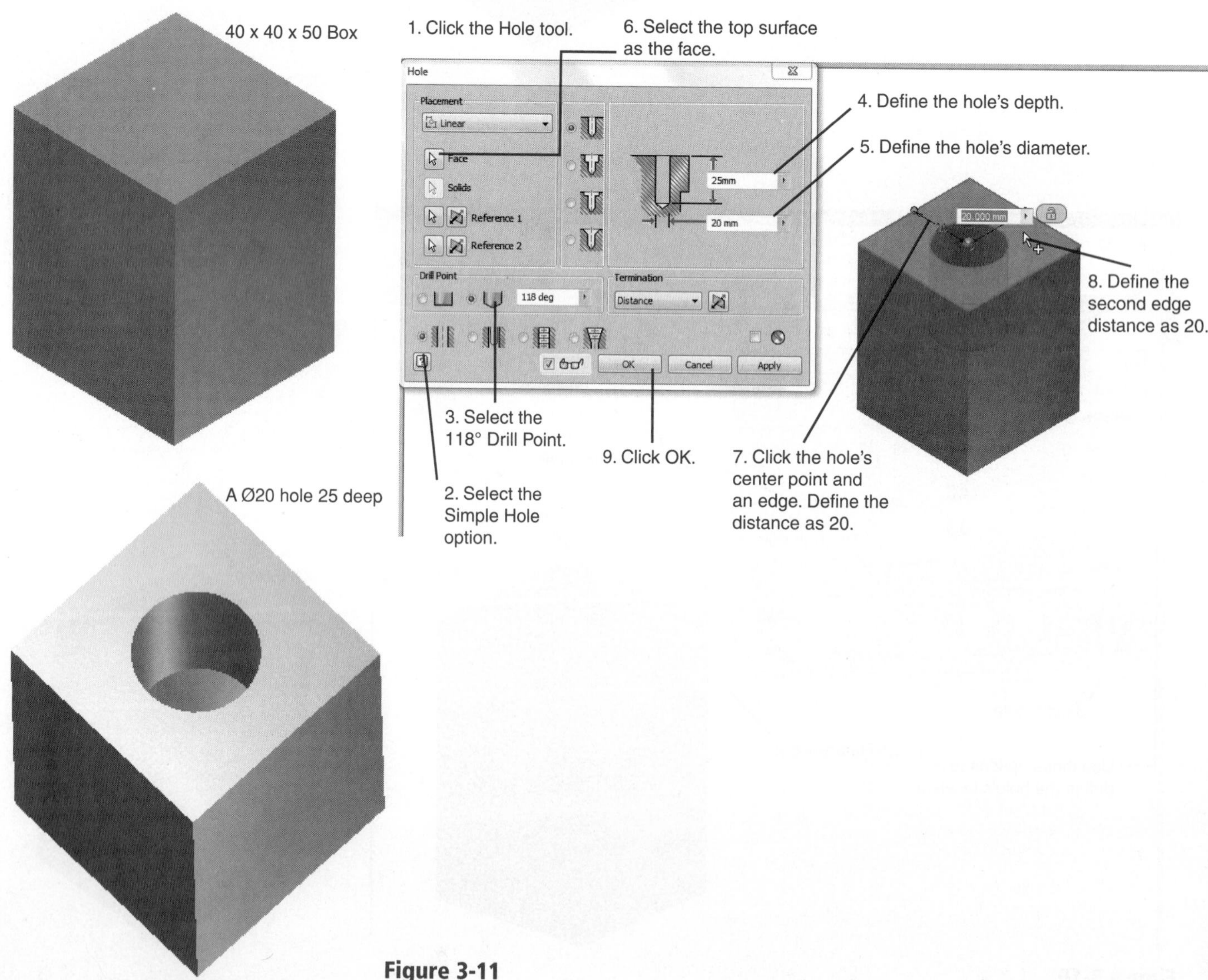

Figure 3-11

6 Click the hole's center point and one of the edges of the top surface.

7 Use the **Reference 1** and **Reference 2** options to locate the hole's center point **20** from each edge.

8 Click **OK**.

Fillet

A fillet is a rounded corner. 2D fillets were covered in Chapter 2; see Figure 2-20. In this section you will create 3D fillets. See Figure 3-12.

1 Use the **Undo** tool and return the box created for the **Hole** tools to a plain 40 × 40 × 50 box.

2 Click the **Fillet** tool located on the **Modify** panel under the **3D Model** tab.

Figure 3-12

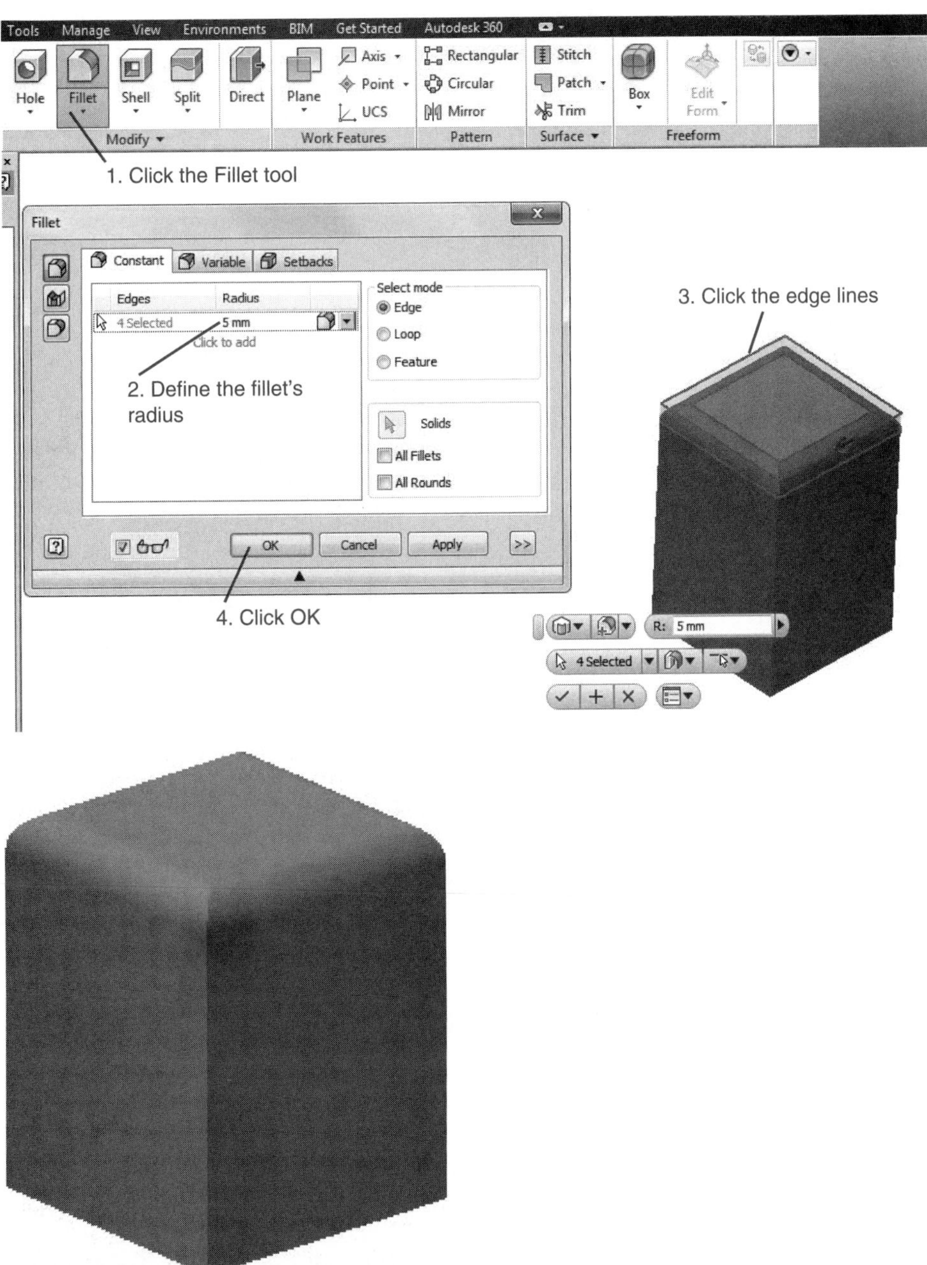

Chapter 3

The **Fillet** dialog box will appear.

3 Enter a fillet **Radius** value of **5**.

4 Move the cursor to the edges to be filleted.

A preview will appear.

5 Click **OK**.

Figure 3-13 shows a fillet of an internal edge.

Figure 3-13

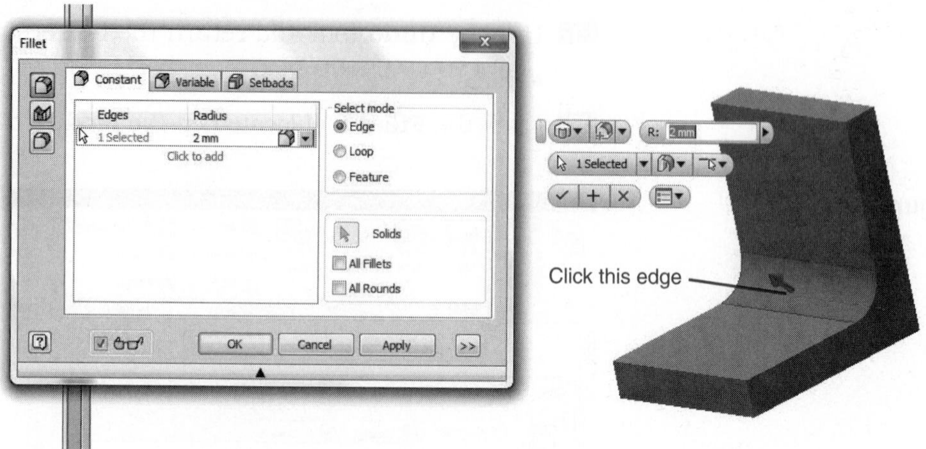

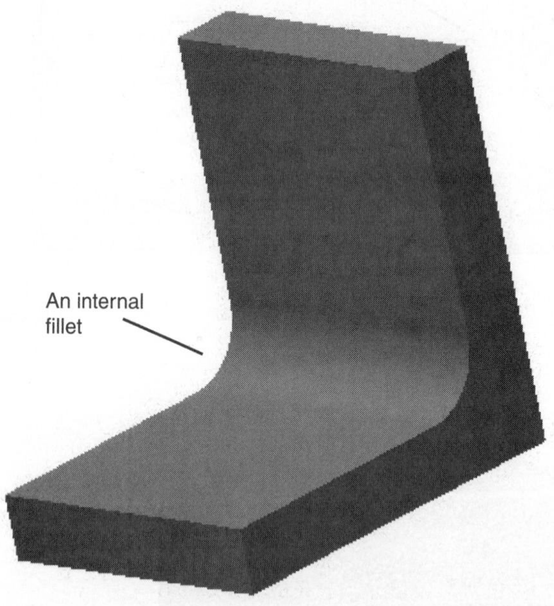

Full Round Fillet

Figure 3-14 shows an L-bracket, 40 × 40, 10 thick, and 40 long. Add a full-round fillet to the top section.

1 Start a new drawing using the **Standard (mm).ipt** format, click the **Create 2D Sketch** tool, and select the XZ plane. Draw the L-bracket.

2 Click the **Fillet** tool located on the **Modify** panel under the **Model** tab.

Figure 3-14

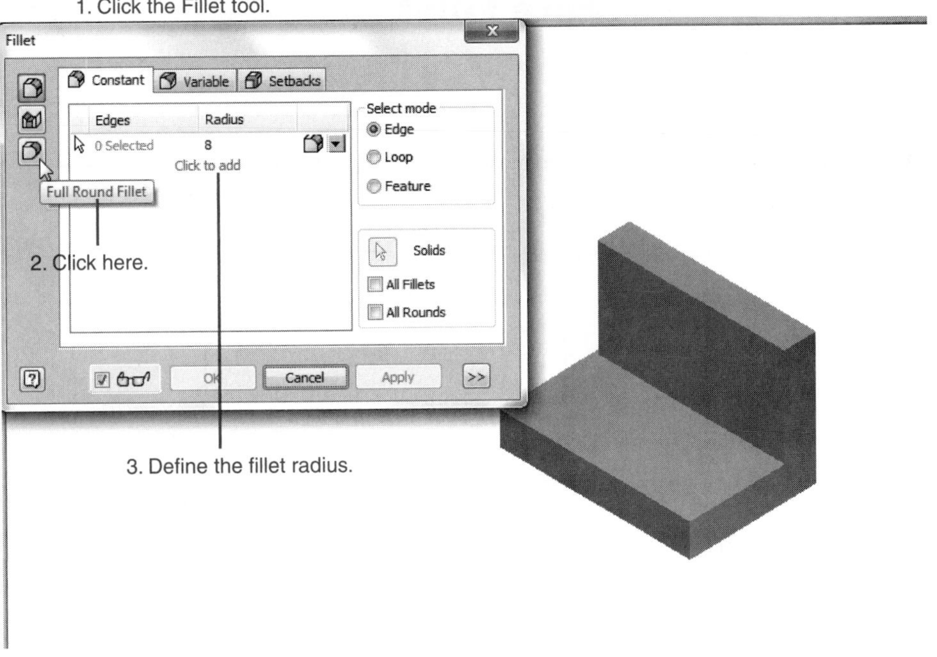

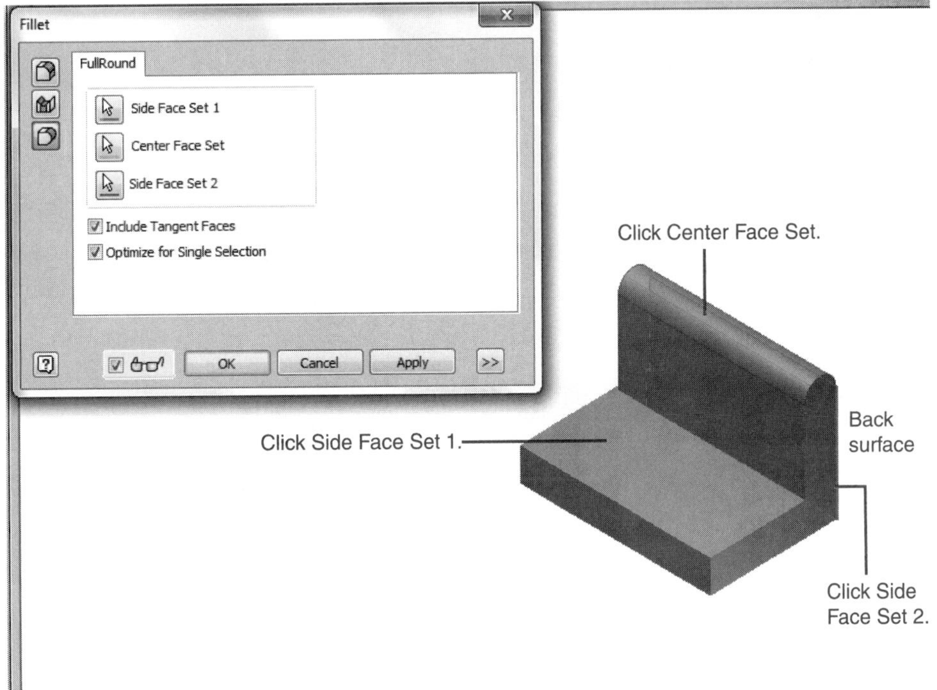

The **Fillet** dialog box will appear.

3 Click the **Full Round** option.

The dialog box will change.

4 Define the faces as shown.

Use the **ViewCube** to rotate the L-bracket so that you can access the back surface and define it as **Side Face Set 2**.

5 Click **OK**.

Face Fillet

See Figure 3-15.

Figure 3-15

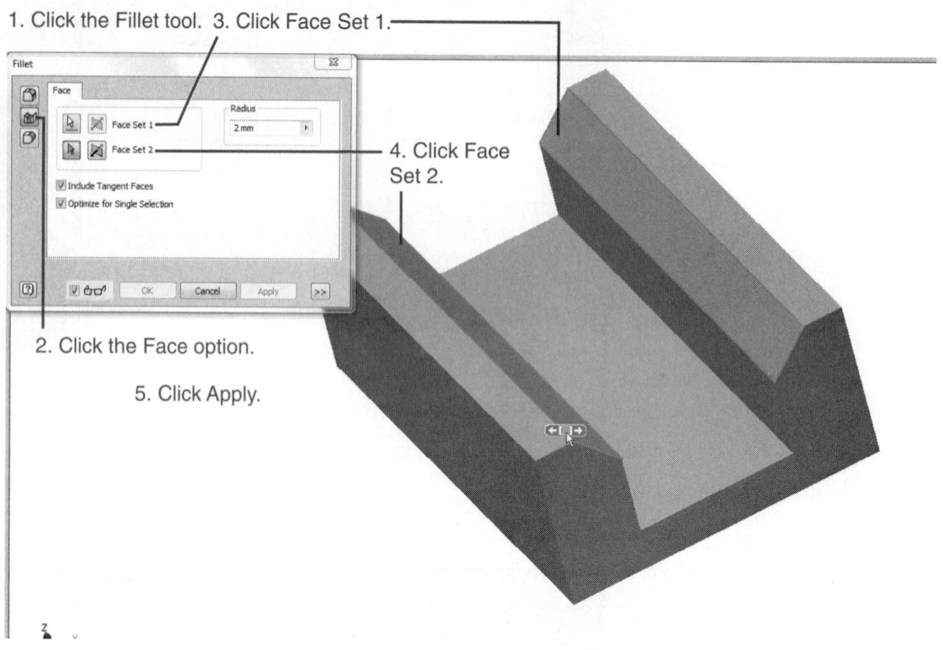

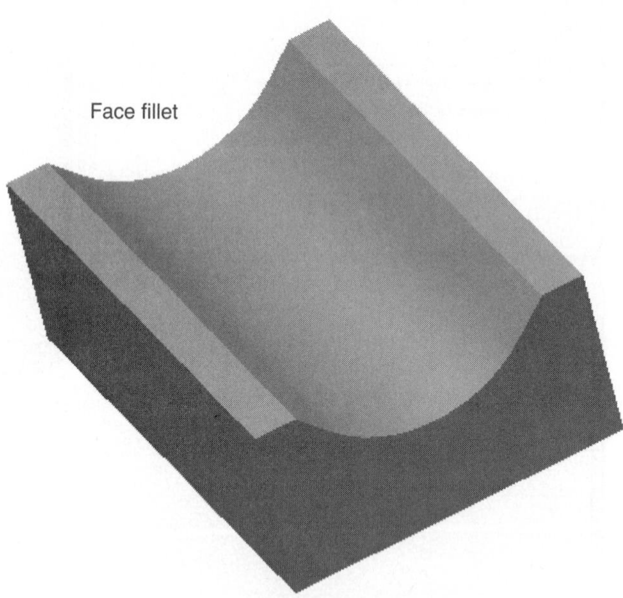

1 Click the **Fillet** tool.

The **Fillet** dialog box will appear.

2 Click the **Face Fillet** option.

The **Fillet** dialog box will change.

3 Define **Face Set 1** and **Face Set 2** as the right and left slanted surfaces as shown.

4 Define the **Radius** value as **2**.

5 Click **Apply**.

Creating a Variable Fillet

See Figure 3-16.

Figure 3-16

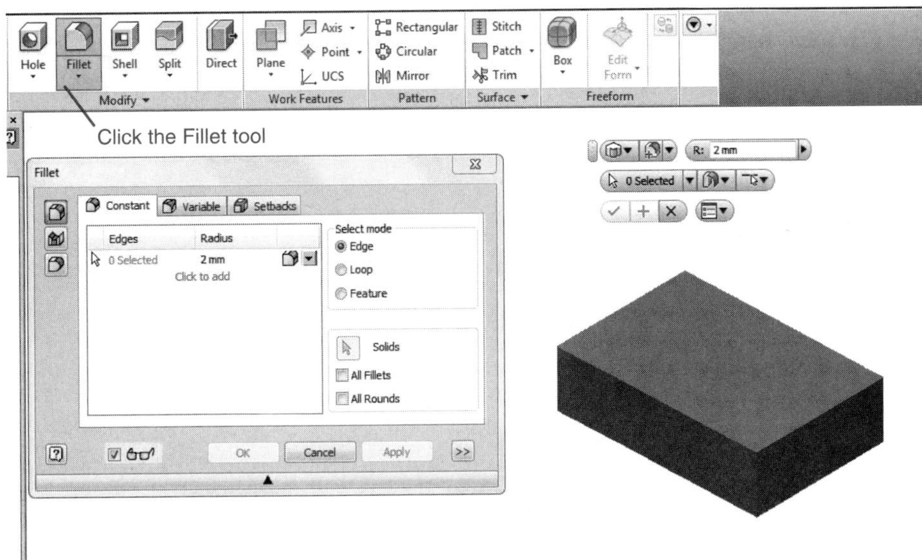
Click the Fillet tool

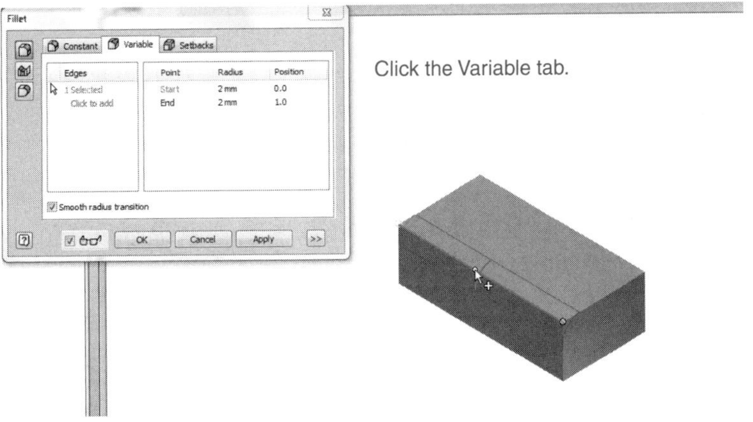
Click the Variable tab.

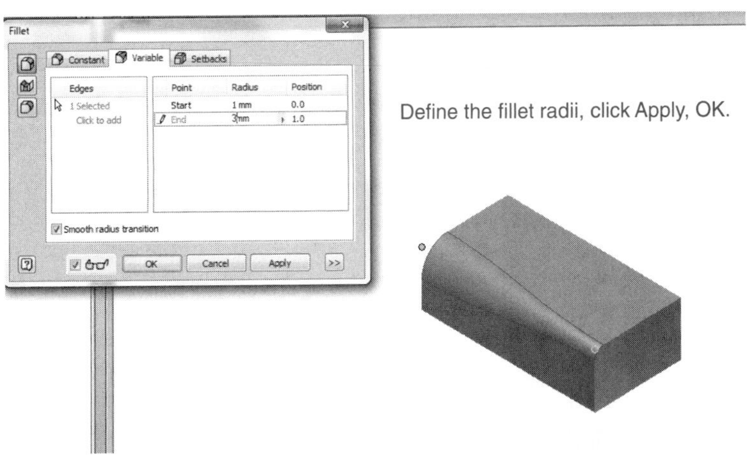
Define the fillet radii, click Apply, OK.

1. Start a new drawing using the **Standard (mm).ipt** format, click the **Create 2D Sketch** tool, and select the XZ plane.

2. Draw a **6** × **10** × **20** box.

3. Select the **Fillet** tool from the **Modify** panel bar under the **3D Model** tab.

4. Click the **Variable** tab.

5. Select the edge for the fillet.

6. Define the **Start** radius.

   In this example a value of **1 mm** was selected.

7. Click the word **End** and define a value.

   In this example a value of **3 mm** was selected.

8. Click **OK.**

Chamfer

The **Chamfer** tool is used to create beveled edges. See Figure 3-17. Chamfers are defined by specifying linear setback distances or by specifying a setback distance and an angle. Most chamfers have equal setback distances or an angle of 45°.

Figure 3-17

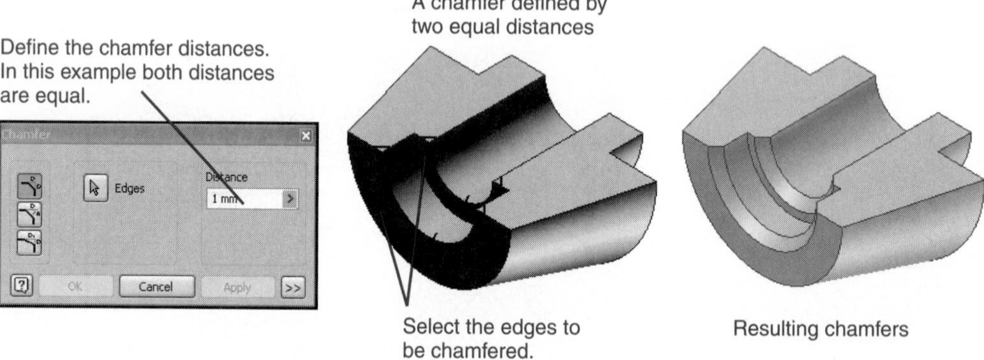
Define the chamfer distances. In this example both distances are equal.
Chamfer
Edges
Distance
1 mm
OK
Cancel
Apply
A chamfer defined by two equal distances
Select the edges to be chamfered.
Resulting chamfers

1. Click the **Chamfer** tool located on the **Modify** panel under the **3D Model** tab.

   The **Chamfer** dialog box will appear. The first option box on the left side of the **Chamfer** dialog box is used to create chamfers with equal distances.

2. Set the distance for **1 mm**.

3. Select the edges to be chamfered.

4. Click **OK**.

The chamfers drawn in Figure 3-17 were defined using two equal distances, creating a 45° chamfer. Chamfers may also be defined using a distance and an angle. Figure 3-18 shows a 2 × 60° chamfer. Chamfers may also be defined using two unequal distances. Figure 3-19 shows a 1 × 4 chamfer.

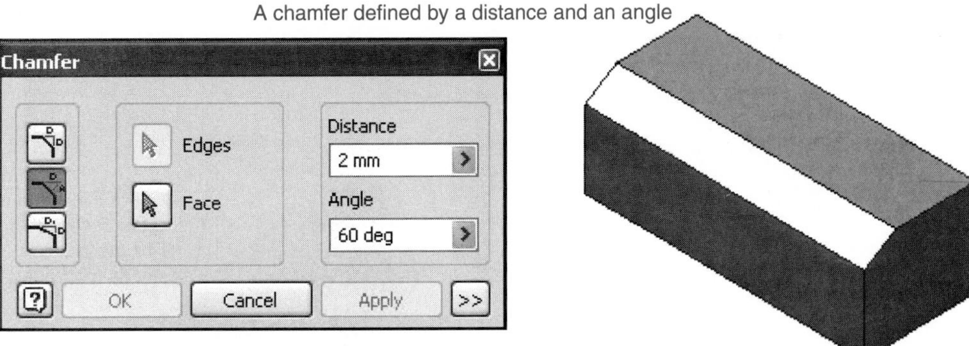

Figure 3-18

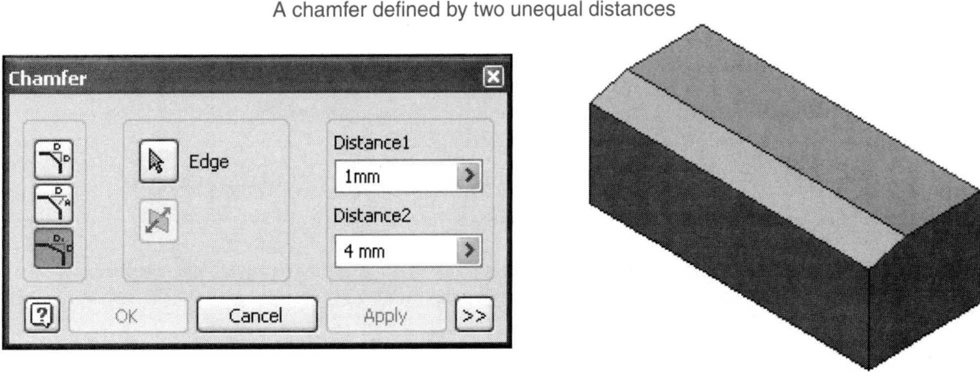

Figure 3-19

The format for the most common chamfer note is 0.25 × 45° CHAMFER. The note specifies a distance and 45° angle, resulting in two equal lengths.

Face Draft

The **Face Draft** tool is used to create angled surfaces. See Figure 3-20. A 5 × 10 × 20 box was used to demonstrate the **Face** tool.

1. Click the **Draft** tool located on the **Modify** panel under the **3D Model** tab.

   The **Face Draft** dialog box will appear.

2. Click the right front surface of the object to define this surface as the **Pull Direction**.

3. Click the **Faces** option and then click the top surface of the box.

Click the surface near or on the front edge of the surface.

4. Enter a **Draft Angle** value of **15 deg**.

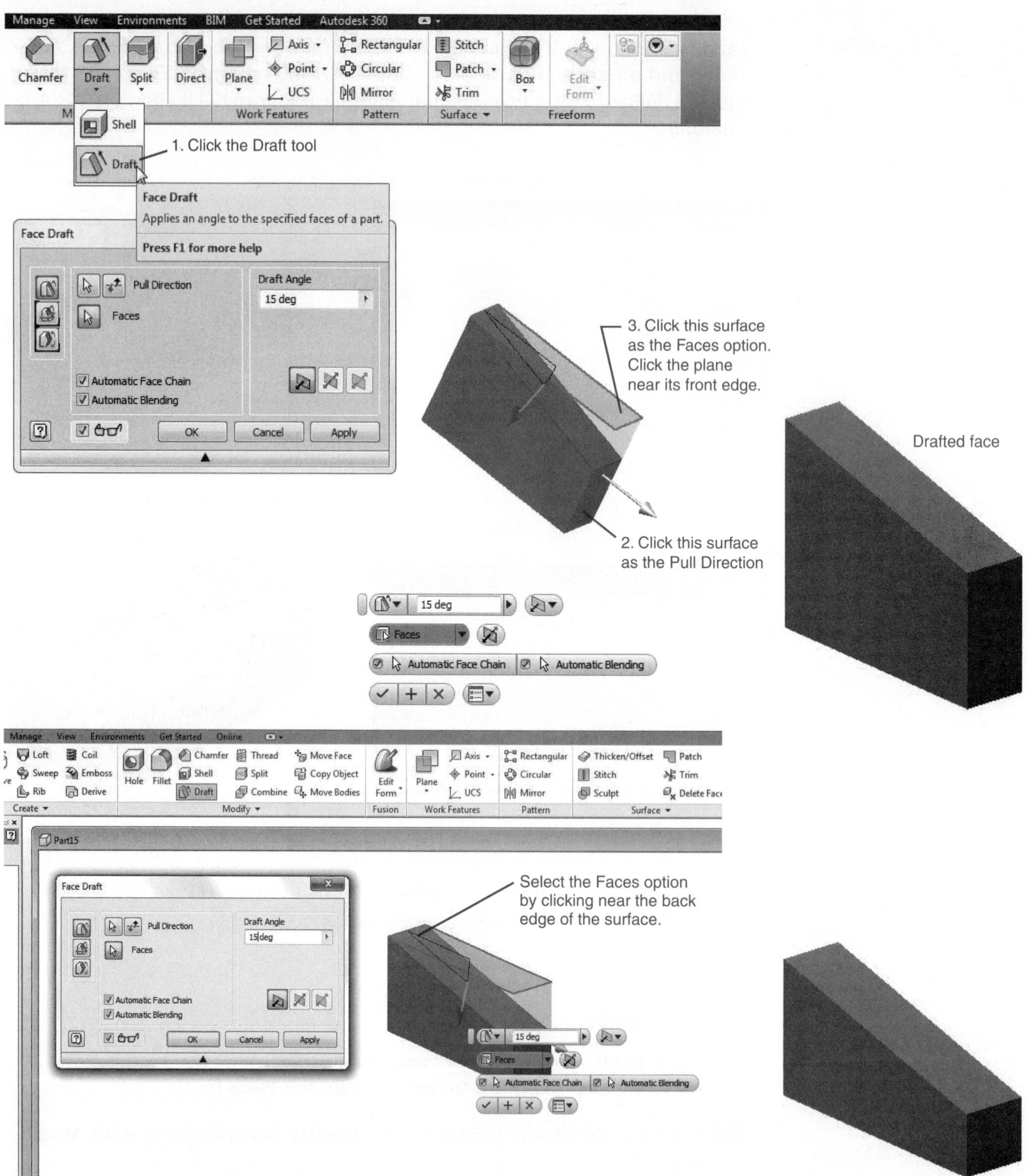

Figure 3-20

5. Click **OK**.
6. Use the **Undo** tool to return the object to its rectangular shape.
7. Click the **Face Draft** tool and again click the right front surface of the object.
8. Click the **Faces** option and then click the top surface of the box.

Click the surface near or on the back edge of the surface.

9 Click **OK**.

Note the differences in the resulting face drafts.
Several surfaces can be drafted at the same time. See Figure 3-21.

Figure 3-21

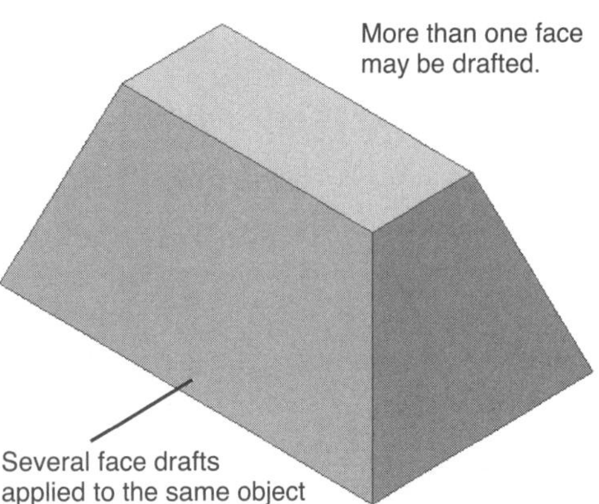
More than one face may be drafted.
Several face drafts applied to the same object

Shell

The **Shell** tool is used to create thin-walled objects from existing models. Figure 3-22 shows a $12 \times 30 \times 20$ model.

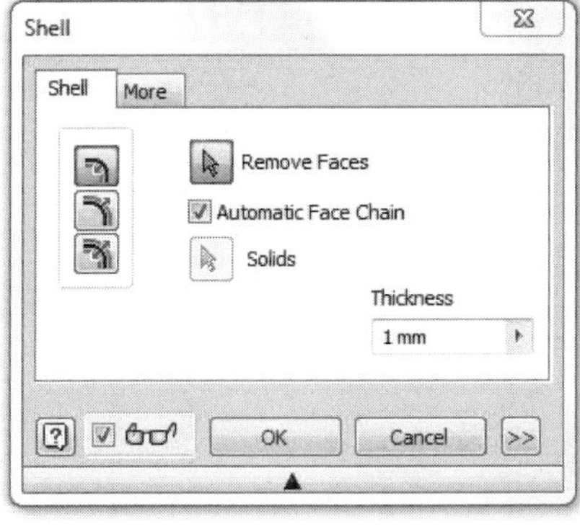
Shell
Shell
More
Remove Faces
Automatic Face Chain
Solids
Thickness
1 mm
OK
Cancel
>>

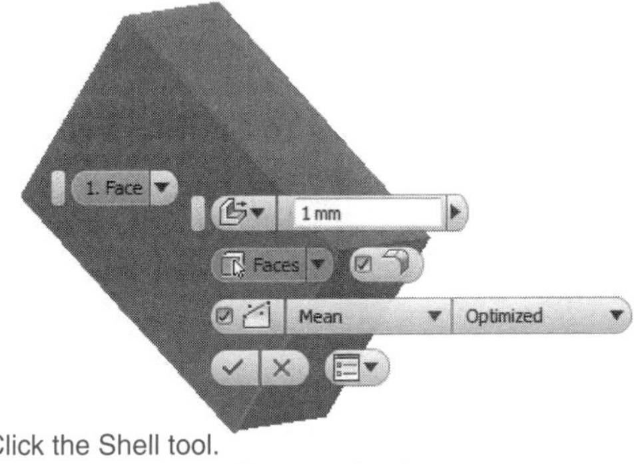
1. Face
1 mm
Faces
Mean
Optimized
1. Click the Shell tool.
2. Ensure the Remove Faces option is on.
3. Click the front face of the object.

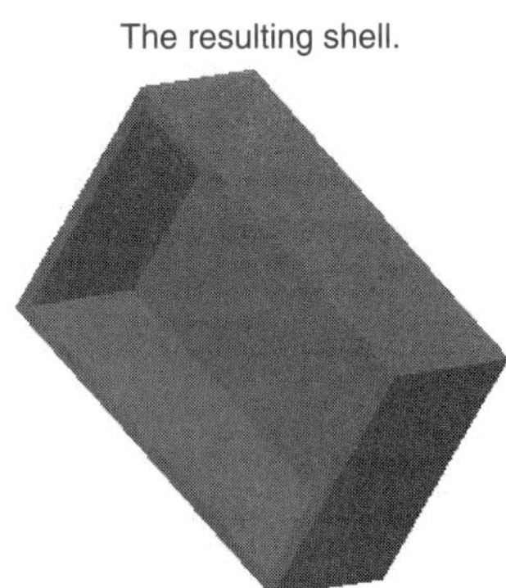
The resulting shell.

Figure 3-22

1 Click the **Shell** tool located on the **Modify** panel under the **3D Model** tab.

The **Shell** dialog box will appear. There are three different ways to define a shell, which are accessed by the three boxes on the left side of the **Shell** dialog box. The options are as follows:

> **TIP**
>
> **Inside:** The external wall of the existing model will become the external wall of the shell.
>
> **Outside:** The external wall of the existing model will become the internal wall of the shell.
>
> **Both Sides:** The existing outside wall will become the center of the shell; half the thickness will be added to the outside and half to the inside.

2. Click the **Inside** option.

3. Click on the front surface of the model, then click **OK**.

Shells may be created from any shape model. Figure 3-23 shows a cone that has been used to create a hollow thin-walled cone.

Figure 3-23

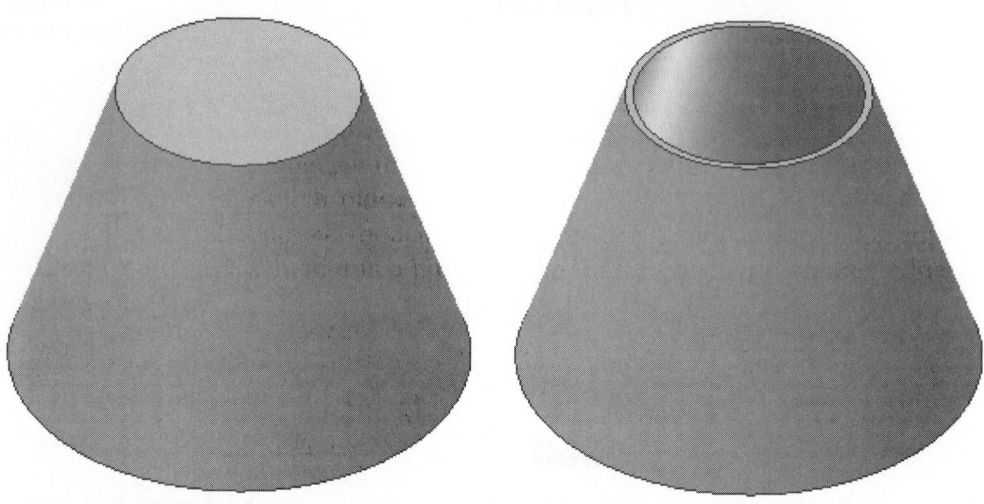

EXERCISE 3-1 Removing More Than One Surface

1. Click the **Shell** tool.

2. Click the surfaces to be removed.

3. Select the **Inside** option.

4. Click **OK**.

See Figure 3-24.

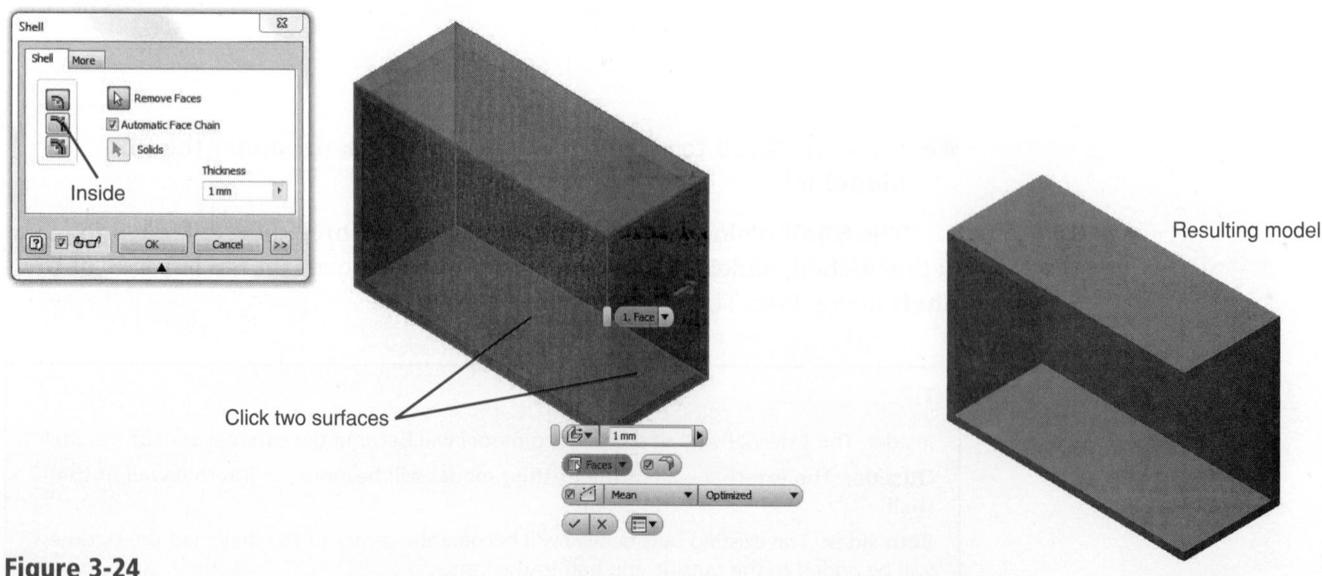

Figure 3-24

Split

The **Split** tool is used to trim away a portion of a model. See Figure 3-25. A sketch line is used to define the location and angle of the split.

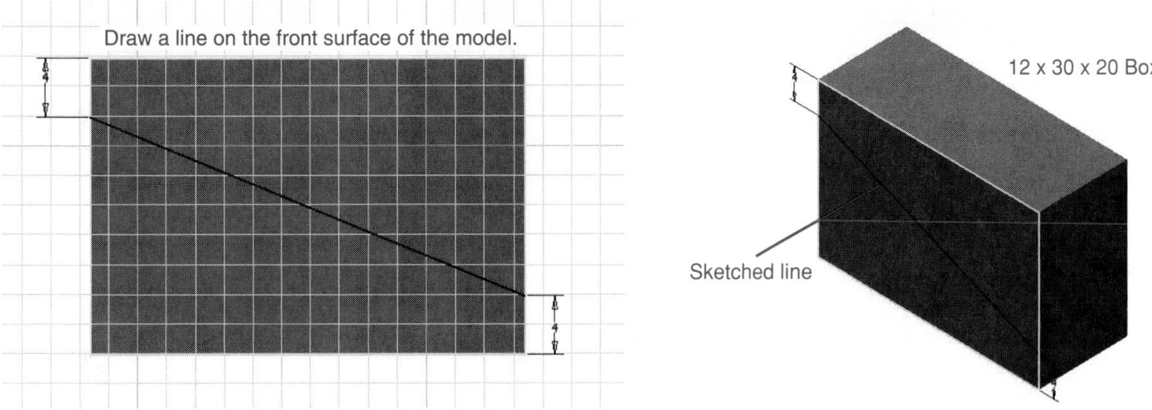

Figure 3-25

EXERCISE 3-2 **Defining the Split Line**

1. Click the left front surface of the 12 × 30 × 20 model.

   The surface will change color.

2. Right-click the mouse and select the **New Sketch** option.

   A grid will appear on the screen oriented to the selected face.

3. Click the **Line** tool and sketch a line across the front left surface.
4. Right-click the mouse and select the **OK** option.
5. Use the **Dimension** tool to locate the line as shown.

EXERCISE 3-3 **Splitting the Model**

1. Right-click the mouse and select the **Finish 2D Sketch** option.
2. Click the **Split** tool located on the **Modify** panel under the **3D Model** tab.

   The **Split** dialog box will appear. See Figure 3-26.

3. Click the **Split Tool** option.
4. Click the sketch line.
5. Select the **Trim Solid** option.

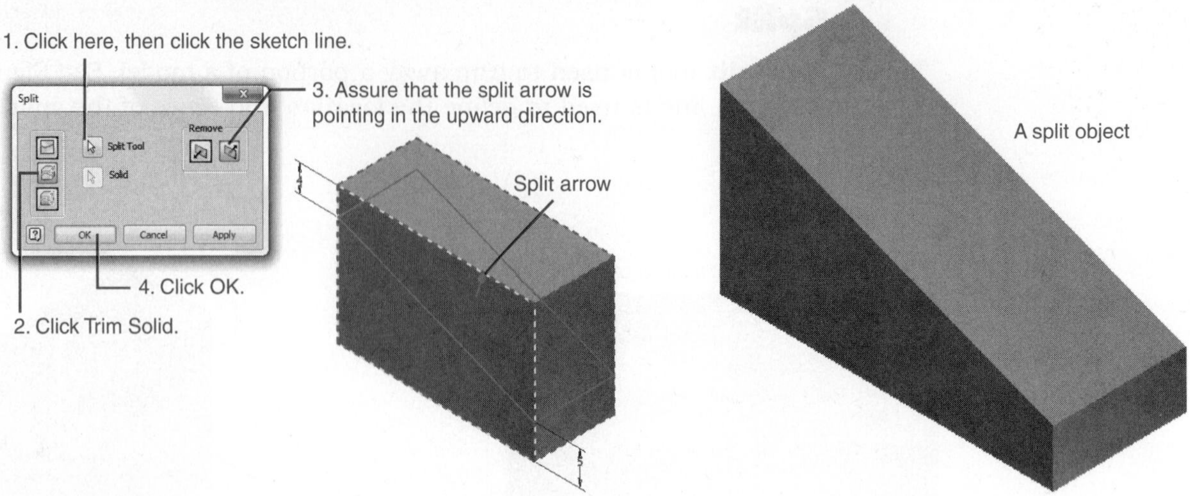

1. Click here, then click the sketch line.

3. Assure that the split arrow is pointing in the upward direction.

Split arrow

A split object

4. Click OK.

2. Click Trim Solid.

Figure 3-26

6 Use the **Remove** option to define which side of the model is to be removed.

The split arrow should be pointing upward.

7 Click **OK**.

Figure 3-27 shows a split that was created using a sketched circle. The arrow that appears on the top surface indicates the direction of removal.

Figure 3-27

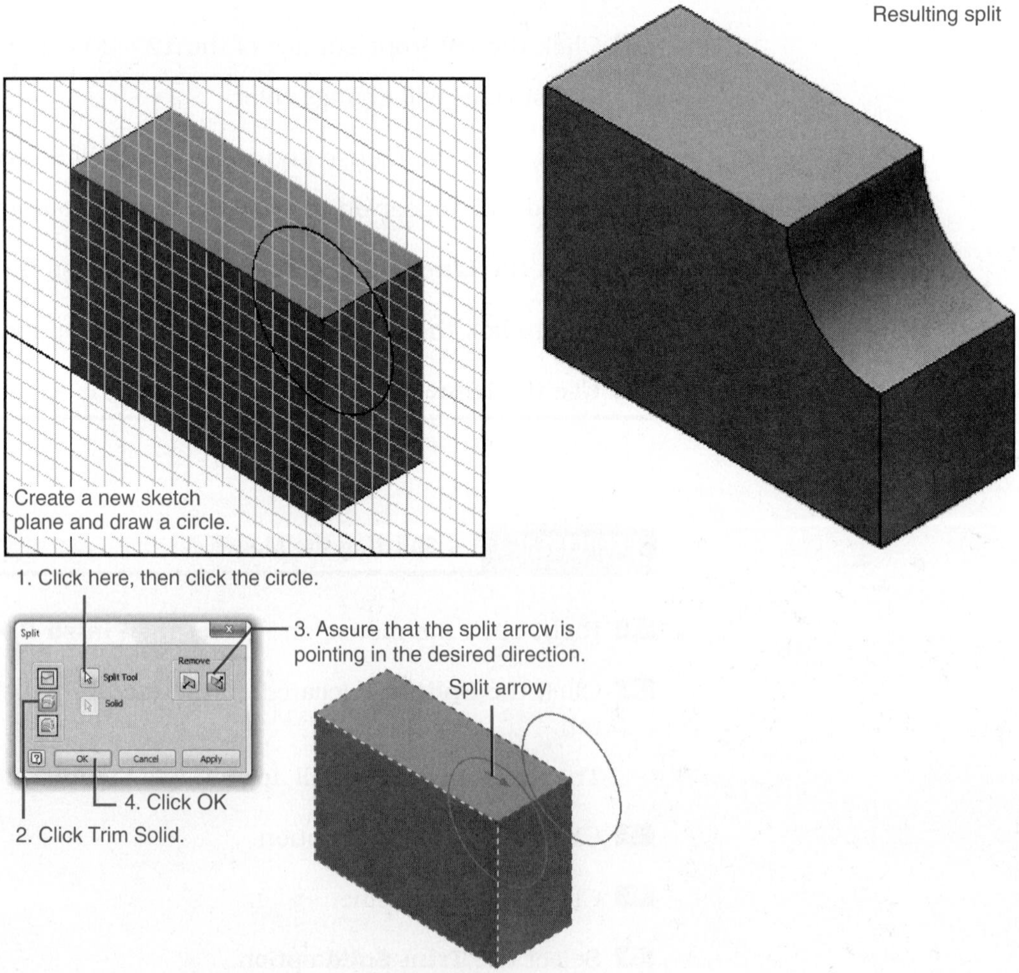

Resulting split

Create a new sketch plane and draw a circle.

1. Click here, then click the circle.

3. Assure that the split arrow is pointing in the desired direction.

Split arrow

4. Click OK

2. Click Trim Solid.

Mirror

The **Mirror** tool is used to create mirror images of an existing model. See Figure 3-28.

Figure 3-28

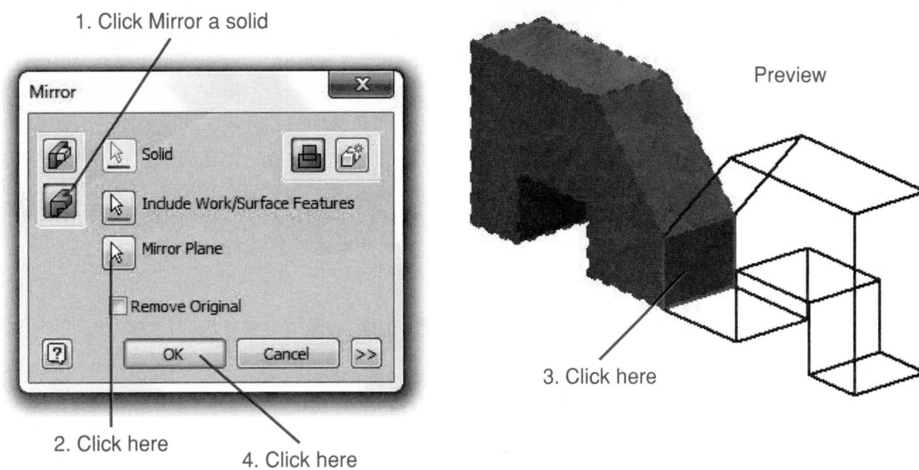

1. Click the **Mirror** tool located on the **Pattern** panel under the **3D Model** tab.

   The **Mirror** dialog box will appear.

2. Click the **Features** option. (It should be on automatically.)
3. Click the model.
4. Click the **Mirror Plane** box.
5. Select one of the model's surfaces as a mirror plane.
6. Click **OK**.

Rectangular Pattern

The **Rectangular Pattern** tool is used to create a rectangular array of an existing model feature. Figure 3-29 shows a 30 × 40 × 5 plate with a Ø5 hole located 5 mm from each edge.

Figure 3-29

1. Assure that the Features option is on, and click the hole.
Rectangular Pattern
Features
Solid
Direction 1
Direction 2
Spacing
OK
Cancel
The feature

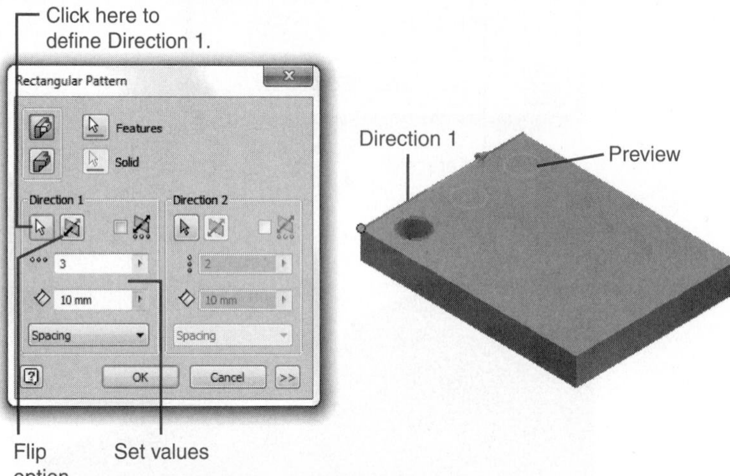
Click here to define Direction 1.
Rectangular Pattern
Features
Solid
Direction 1
Direction 2
3
10 mm
Spacing
OK
Cancel
Direction 1
Preview
Flip option
Set values

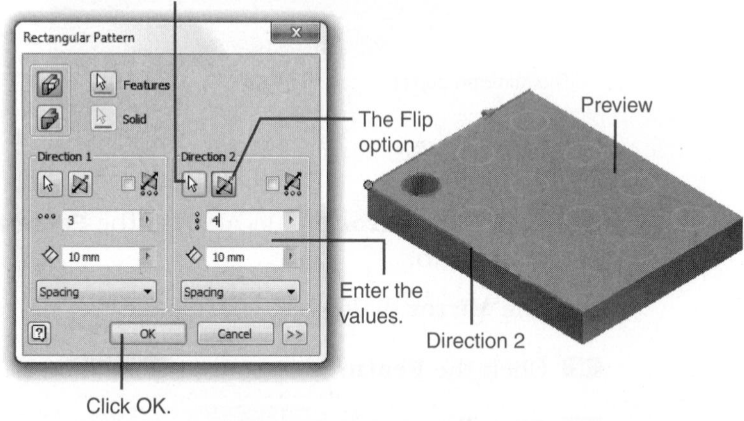
Click here to define Direction 2.
Rectangular Pattern
Features
Solid
Direction 1
Direction 2
3
4
10 mm
10 mm
Spacing
Spacing
OK
Cancel
The Flip option
Preview
Enter the values.
Direction 2
Click OK.

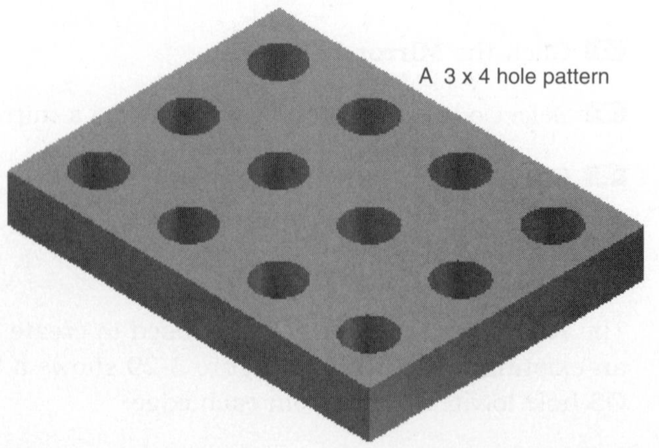
A 3 x 4 hole pattern

1 Click the **Rectangular Pattern** tool located on the **Pattern** panel under the **3D Model** tab.

The **Rectangular Pattern** dialog box will appear. The **Features** box will automatically be active.

2 Click the hole.

The hole is the feature.

3 Click the arrow in the **Direction 1** box, then click the back left edge of the model to define direction 1.

Use the **Flip** option in the **Direction 1** box to reverse the direction if necessary.

4 Set the **Count** value for **3** and the **Spacing** value for **10**.

5 Click the arrow in the **Direction 2** box, then click the front left edge of the model to define direction 2.

Use the **Flip** option in the **Direction 2** box to reverse the direction if necessary.

6 Set the **Count** value for **4** and the **Spacing** for **10**.

7 Click **OK**.

Circular Pattern

The **Circular Pattern** tool is used to create a polar array of an existing model feature. Figure 3-30 shows a Ø40 cylinder 5 mm high with two

Figure 3-30

1. Ensure that the Features option is on, then click the hole.

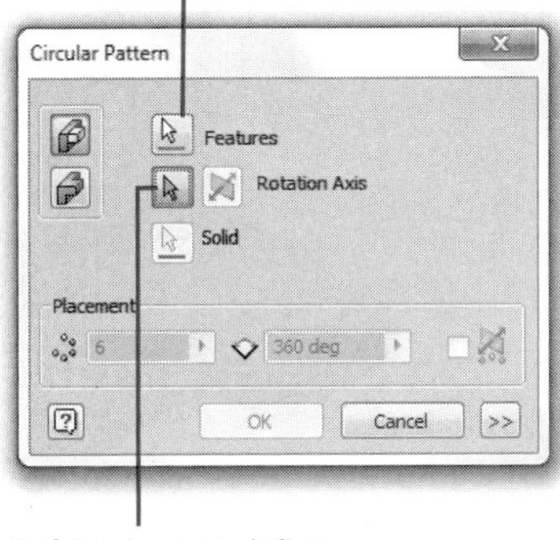

2. Click here to define the axis of revolution.

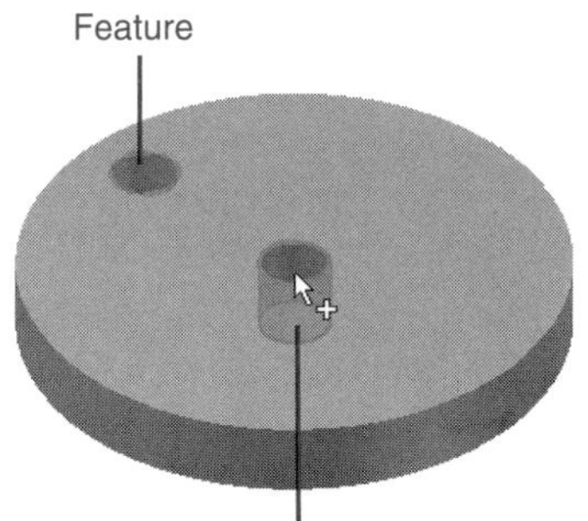

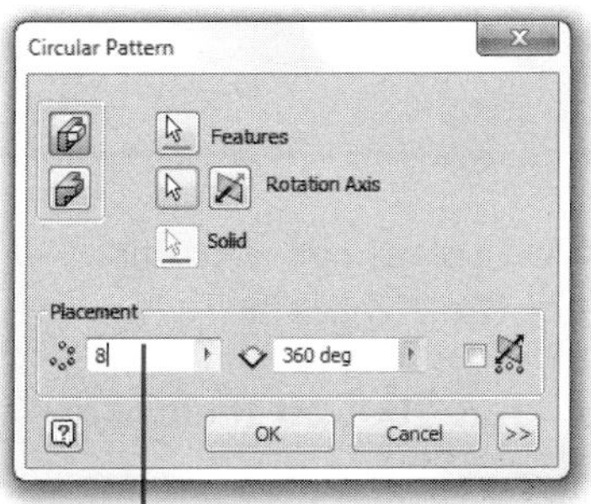

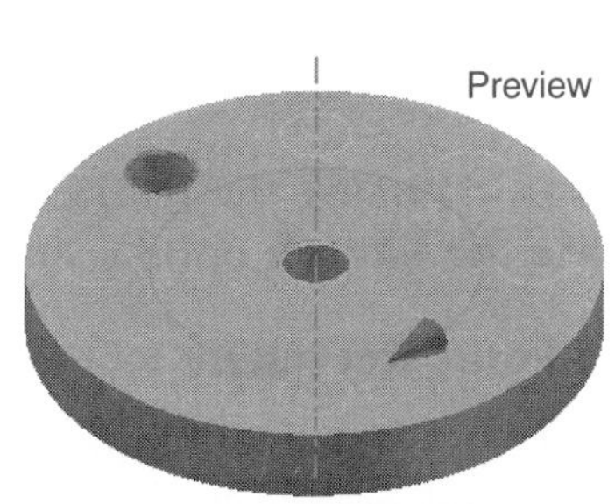

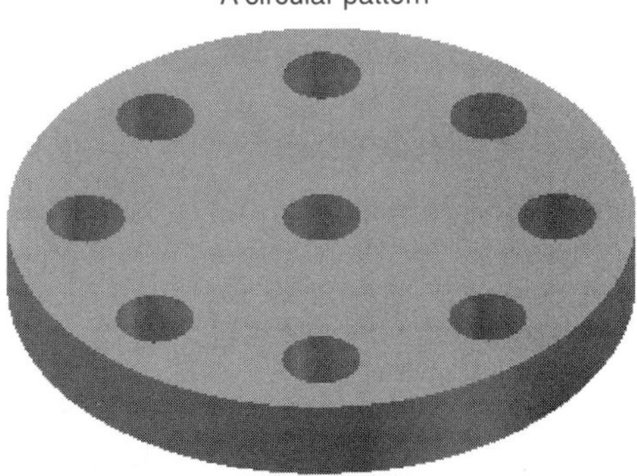

Ø5 holes. One hole is located in the center of the model; the second is located 15 mm from the center.

1. Click the **Circular Pattern** tool located on the **Modify** panel bar under the **3D Model** tab.

The **Circular Pattern** dialog box will appear. The **Features** box will automatically be active.

2. Click the hole to be used to create the circular pattern.

3. Click the **Rotation Axis** button, and click the center hole.

4. Set the **Count** value for **8** and the **Angle** value for **360°**.

5. Click **OK**.

Sketch Planes

sketch plane: A 2D plane drawn on any surface or work plane on a model used for sketching.

Sketches are created on ***sketch planes***. Any surface on a model may become a sketch plane. As models become more complex they require the use of additional sketch planes.

Figure 3-31 shows a model that was created using several different sketch planes. The model is a composite of basic geometric shapes added to one another.

Figure 3-31 A model created using several different sketch planes

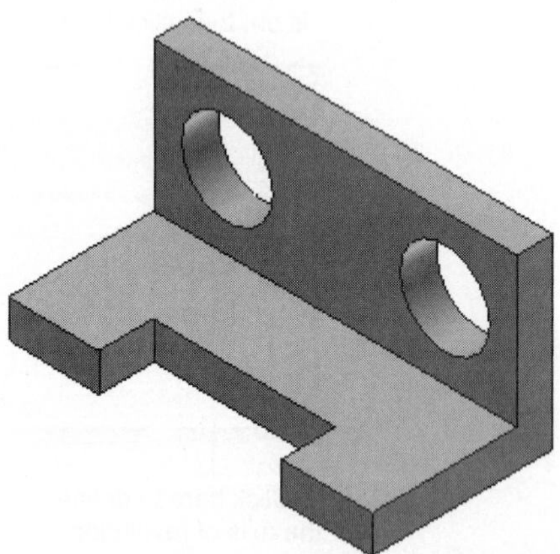

EXERCISE 3-4 Creating the Base

1. Start a new drawing using the metric **Standard (mm).ipt** settings, click the **Create 2D Sketch** option, and select the XZ plane.

2. Click the **Two Point Rectangle** tool and sketch a **10 × 20** rectangle.

See Figure 3-32. The object will automatically be drawn on a sketch plane aligned with the program's XY plane.

Figure 3-32

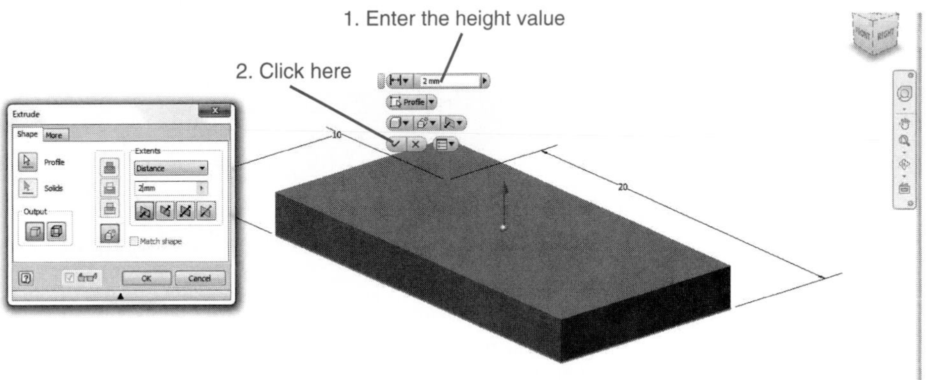

3 Click the **3D Model** tab and click the **Extrude** tool on the **Create** panel. The **Extrude** dialog box will appear.

4 Set the extrusion height for **2 mm** and click **OK**.

EXERCISE 3-5 Creating the Vertical Portion

The rectangular vertical back portion of the model will be created by first defining a new sketch plane on the top surface of the base, then sketching and extruding a rectangle that will be joined to the existing base. See Figure 3-33.

Figure 3-33

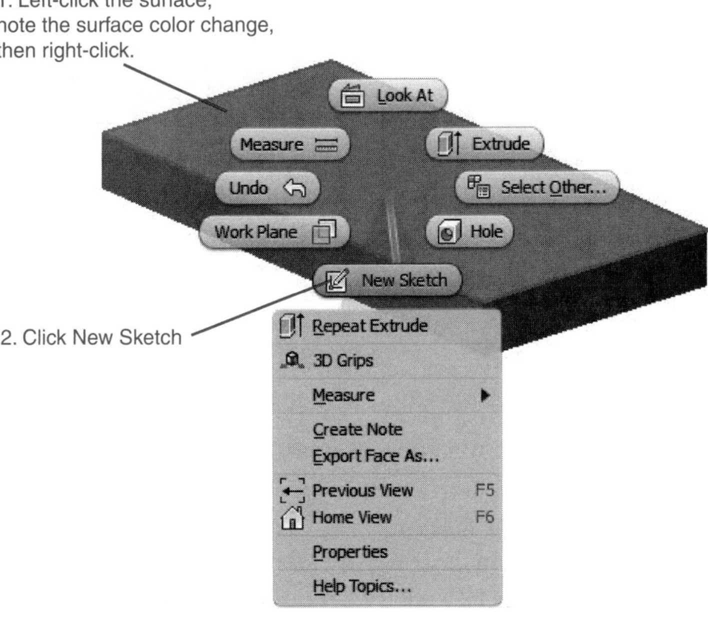

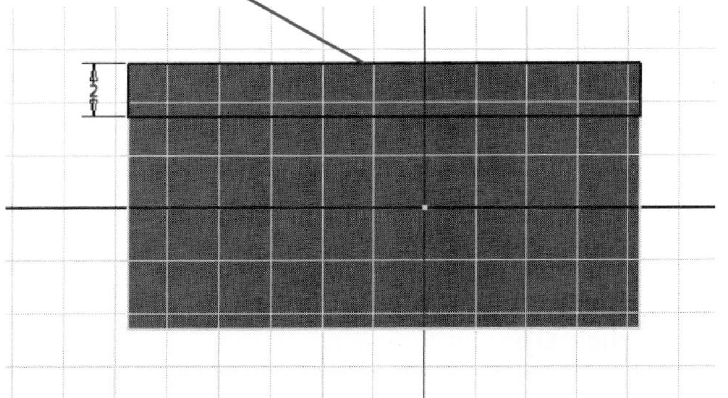

Figure 3-33
(Continued)

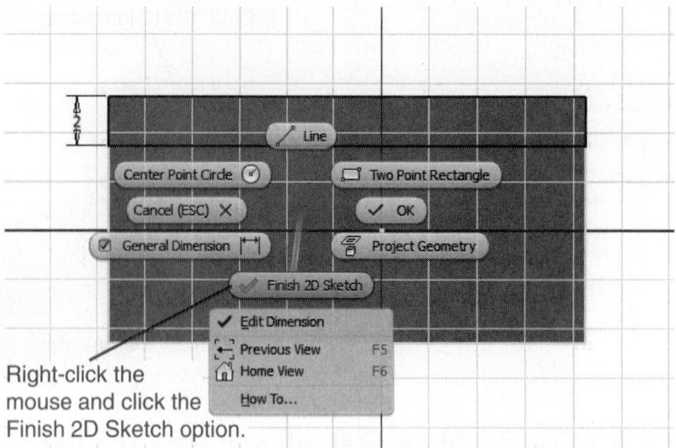

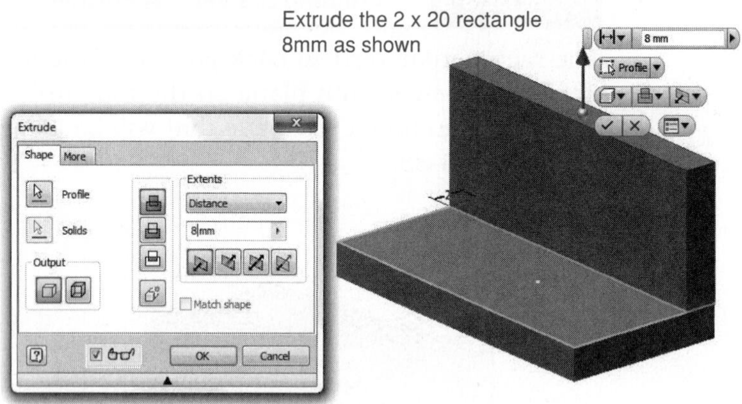

1 Click the top surface of the base.

The surface will change color, indicating that it has been selected.

2 Right-click the mouse and select the **New Sketch** option.

A new grid pattern will appear aligned with the top surface of the base. This is a new sketch plane.

3 Click the **Two Point Rectangle** tool and sketch a **2** × **20** rectangle on the top surface so that it is aligned with the back edge of the base.

Note that the cursor changes from yellow to green when it is aligned with the plane's corner point.

4 Right-click the mouse and select the **Finish 2D Sketch** option. Click the **Extrude** tool.

5 Select the 2 × 20 rectangle and set the extrusion height for **8,** then click **OK**.

Note that the surfaces are unioned together to form one object. See Figure 3-34.

Figure 3-34

EXERCISE 3-6 Adding Holes to the Vertical Surface

1. Click the front edge of the vertical surface.

   The surface will change color, indicating that it has been selected.

2. Right-click the mouse and select the **New Sketch** option.

   A grid will appear on the surface. This is a new sketch plane. The holes are located 4 mm from the top edge and from each of the side edges. See Figure 3-35.

Figure 3-35

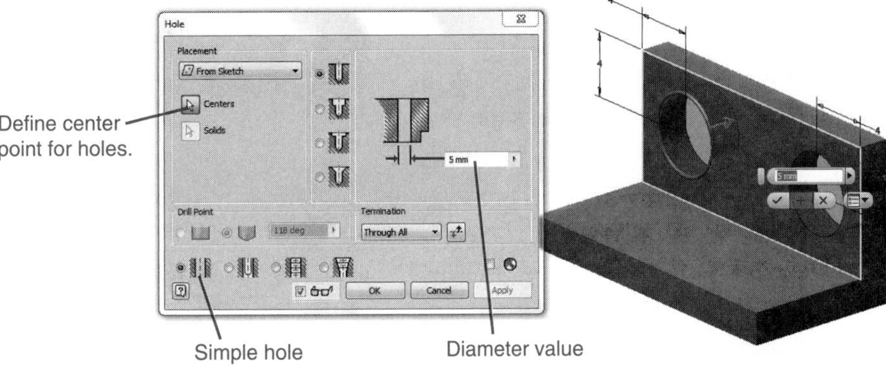

3 Use the **Point, Center Point** and the **Dimension** tools to locate the center points for the two holes.

4 Right-click the mouse and select the **Finish 2D Sketch** option. Click the **Hole** tool.

The **Hole** dialog box will appear.

5 Set the **Termination** for **Through All** and the diameter value for **5,** then click the **Centers** option on the **Hole** dialog box and the center points for the holes.

6 Click **OK**.

EXERCISE 3-7 Creating the Cutout

1 Create a new sketch plane on the top surface of the base.

The cutout is 3 deep with edges 5 from each end of the model.

2 Use the **Two Point Rectangle** and **Dimension** tools to define the cut-out's size.

3 Right-click the mouse and click the **Finish 2D Sketch** option. Click the **Extrude** tool.

The **Extrude** dialog box will appear. See Figure 3-36.

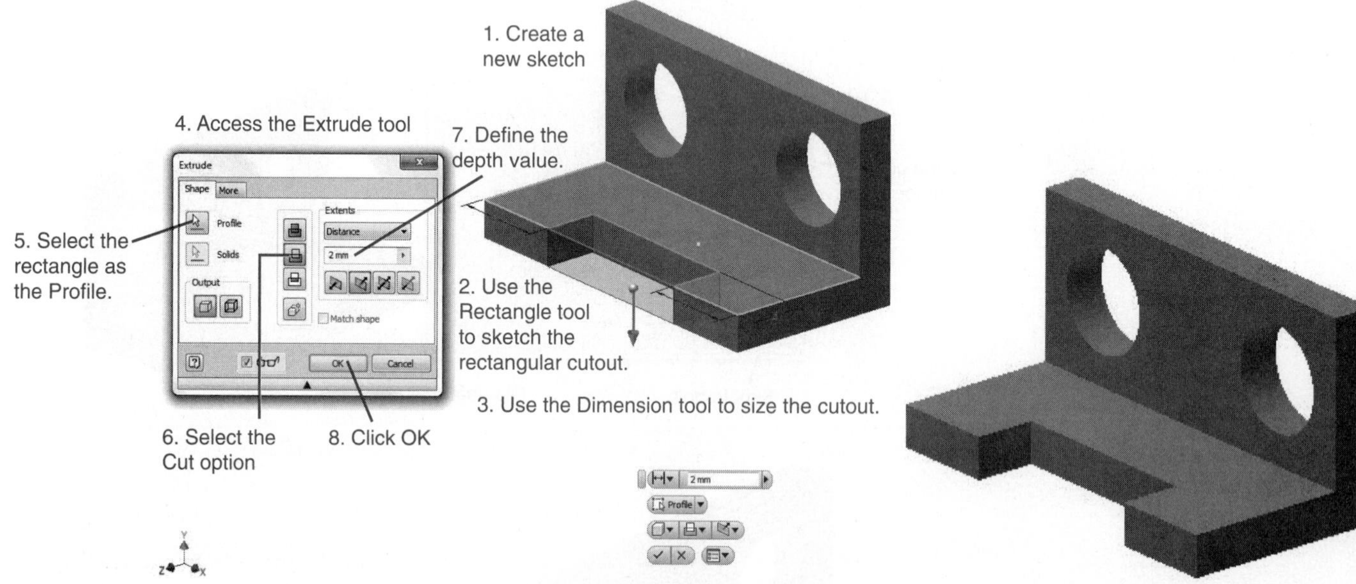

Figure 3-36

4 Select the cutout rectangle as the **Profile,** set the extrusion distance for **2** and the direction arrow for a direction into the model, and select the **Cut** option.

5 Click the **OK** box.

Editing a 3D Model

3D models may be edited; that is, dimensions and features may be changed at any time. For example, suppose the 3D model drawn in the last section and shown in Figures 3-31 through 3-36 requires some changes. The 20 mm length is to be changed to 25, the holes are to be changed from Ø5 to Ø3, and fillets are to be added on the front corners.

There are two types of edits: edit sketch and edit features. The **Edit Sketch** tool applies to shapes created using the **Sketch** panel tools, for example, **Line, Rectangle,** and **Circle**. The **Edit Feature** tool applies to shapes created using the **3D Model** panel bar tools, for example, **Extrude, Hole**, and **Split**.

EXERCISE 3-8 **Changing the Model's Length**

1 Move the cursor into the browser box and click the **+ sign** to the left of **Extrusion1**.

See Figure 3-37. The plus sign will change to a minus sign (−), and **Sketch1** will appear.

Figure 3-37

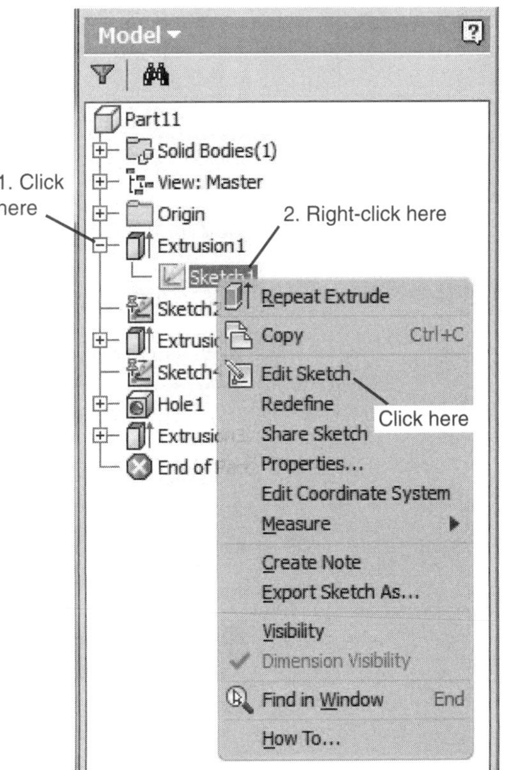

2 Right-click **Sketch1,** then select the **Edit Sketch** option.

See Figure 3-38.

Figure 3-38

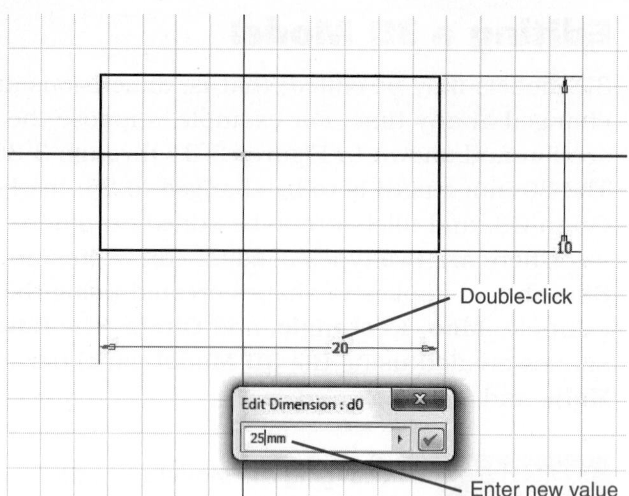

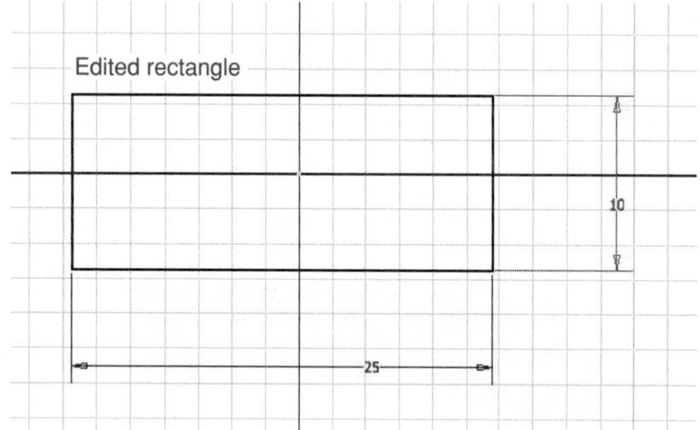

3 Double-click the **20 mm** dimension and enter a value of **25**.

4 Click the check mark.

5 Right-click the mouse and select the **Finish 2D Sketch** option.

TIP

Holes created using the **Hole** tool are *features*. The rectangle face used to create the object is a *sketch*.

EXERCISE 3-9 **Changing the Hole's Diameters**

See Figure 3-39.

1 Right-click **Hole1** in the browser box and select the **Edit Feature** option.

The **Hole: Hole1** dialog box will appear.

2 Change the hole's diameter to **3 mm**.

3 Click **OK**.

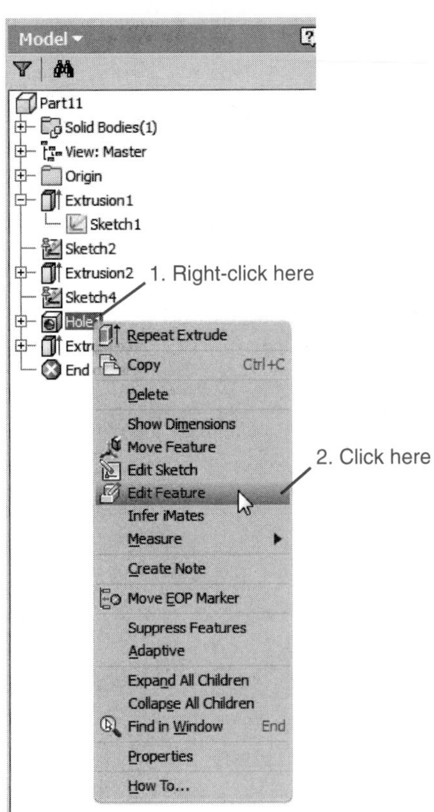

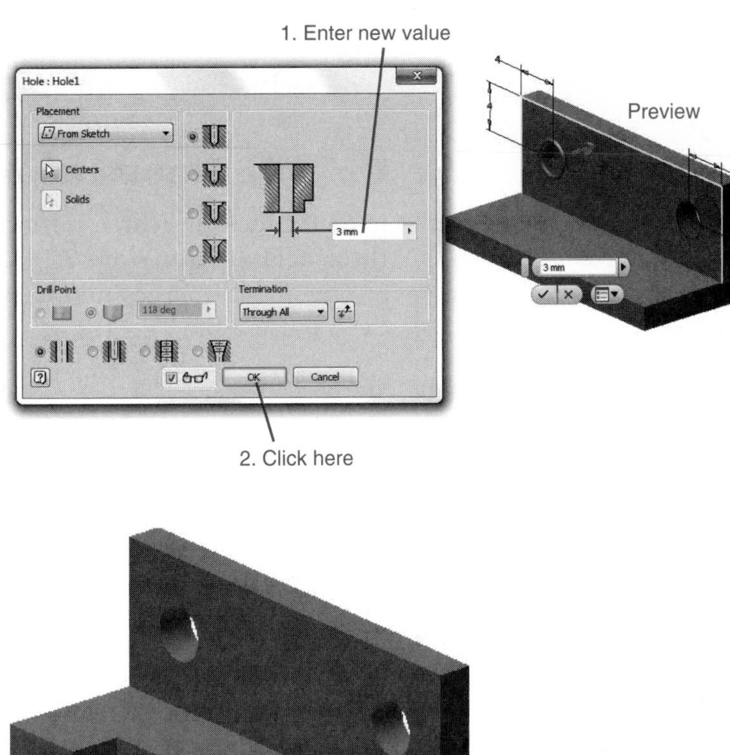

Figure 3-39

EXERCISE 3-10 Adding a Fillet

Fillets and other features may be added to an existing 3D model using the tools on the **3D Model** panels. See Figure 3-40.

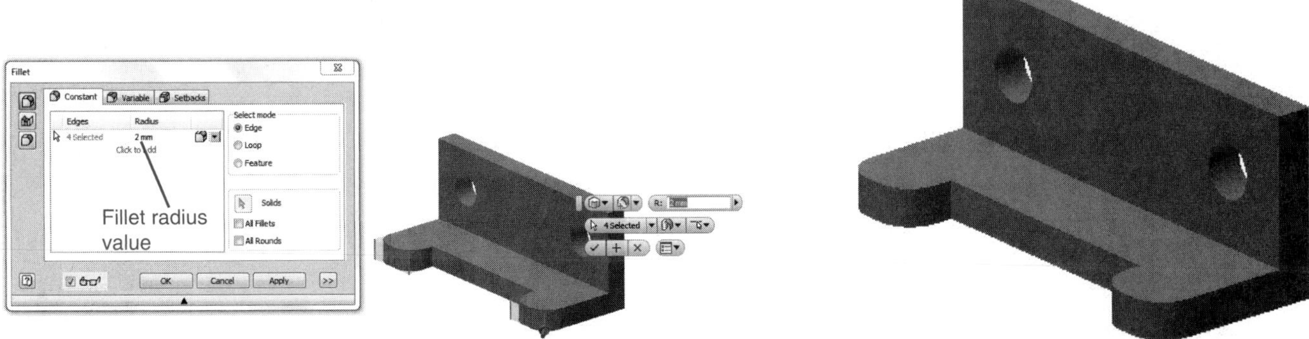

Figure 3-40

1. Click the **Fillet** tool on the **Modify** panel.
2. Set the radius value to **2 mm**.
3. Click the **Edge** box.
4. Click the four edges shown in Figure 3-40. Click **OK**.

> **TIP**
> The default planes are listed in the browser box under the **Origin** heading.

Default Planes and Axes

default planes: In Inventor, the YZ, XZ, and XY planes.

default axes: In Inventor, the X, Y, and Z axes.

Inventor includes three ***default planes*** and three ***default axes***. The three default planes are YZ, XZ, and XY, and the three axes are X, Y, and Z. The default planes and axes tools are accessed through the browser. See Figure 3-41.

Figure 3-41

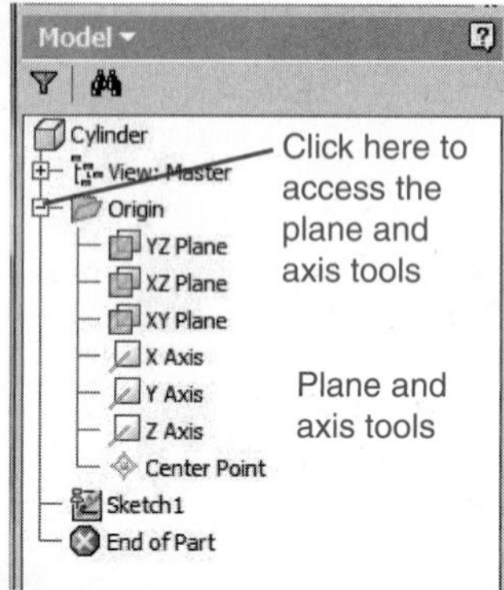

1 Click the + to the left of the **Origin** heading.

The default plane and axis headings will cascade down.

EXERCISE 3-11 Displaying the Default Planes and Axes

Figure 3-42 shows a Ø30 × 16 cylinder that was drawn with its center point on the 0,0,0 origin. The base of the cylinder is on the XY plane. Inventor sketches are automatically created on the default XY axis.

1 Click the + next to **Origin** in the browser box.

2 Move the cursor onto the **XZ Plane** tool.

A plane outline will appear on the screen. It will be red.

3 Click the **XZ Plane** tool.

The plane's color will change to blue.

4 Move the cursor to the **XY Plane** tool.

A red XY plane will appear.

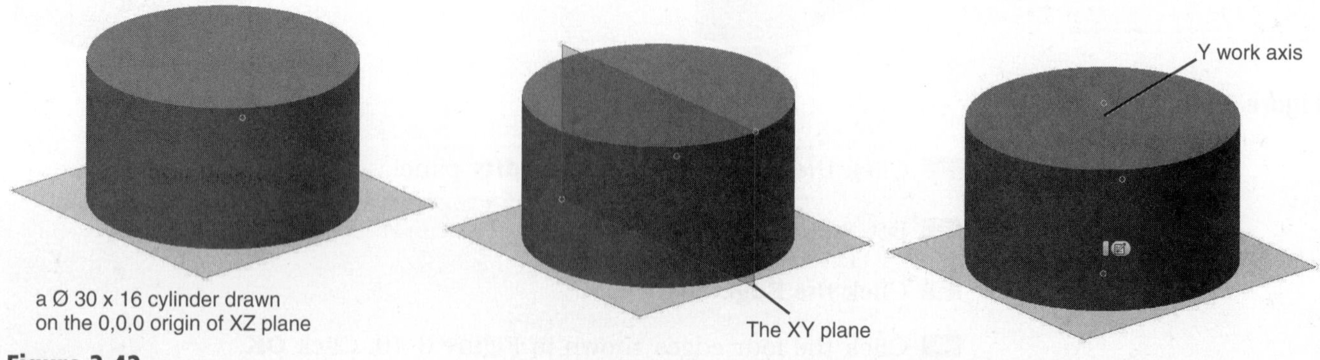

Figure 3-42

5. Move the cursor to the **Y Axis** tool.

   The Y axis will appear.

6. Move the cursor through all the tools and note the planes and axes that appear.

Work Planes

work plane: A plane used for sketching that is created independent of the model.

Work planes are planes used for sketching, but unlike sketch planes, work planes are not created using the surfaces of models. Work planes are created independently of the model. Work planes may be created outside or within the body of a model. Work planes are used when no sketch plane is available.

Work planes may be defined using the parameters shown in Figure 3-43.

Figure 3-43

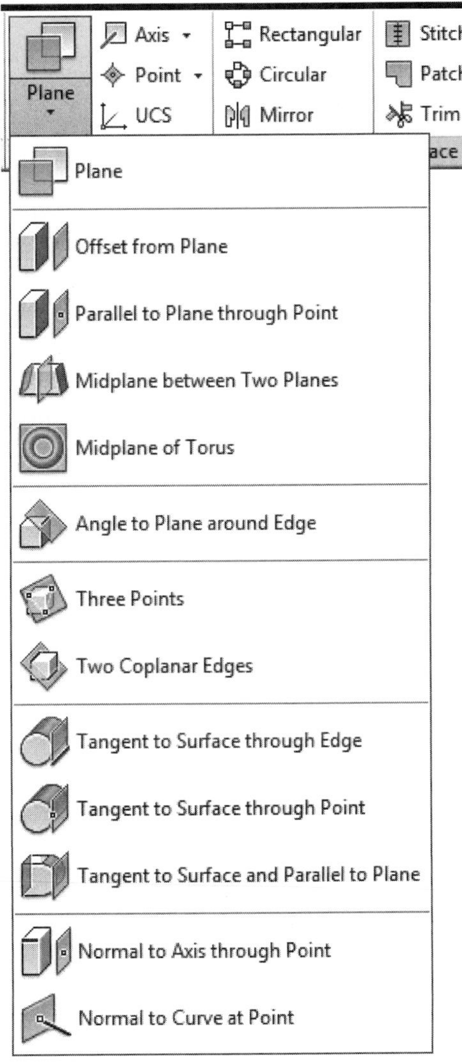

Work Plane Help

If you are not sure how to create a work plane, Inventor includes help features.

1. Type in the keywords "Work Planes". See Figure 3-44.

Figure 3-44

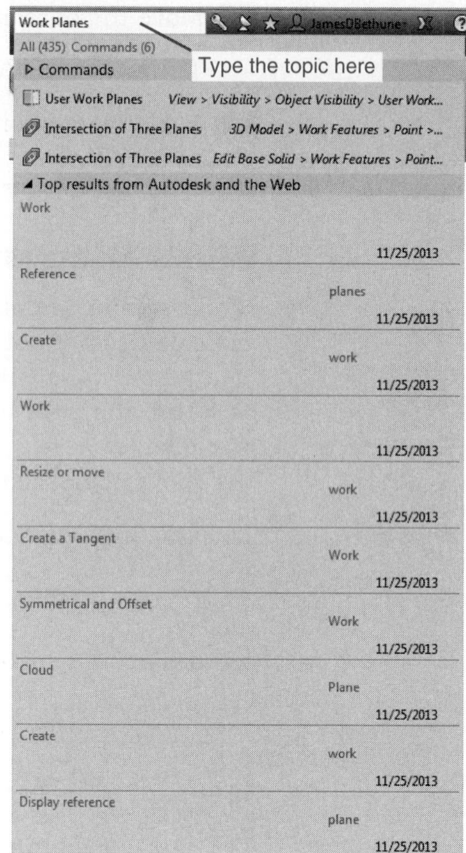

2 A listing of related topics will appear. Click the topic you need.

3 Click the question mark located in the upper right corner of the screen. The AutoDesk Inventor help screen will appear. See Figure 3-45.

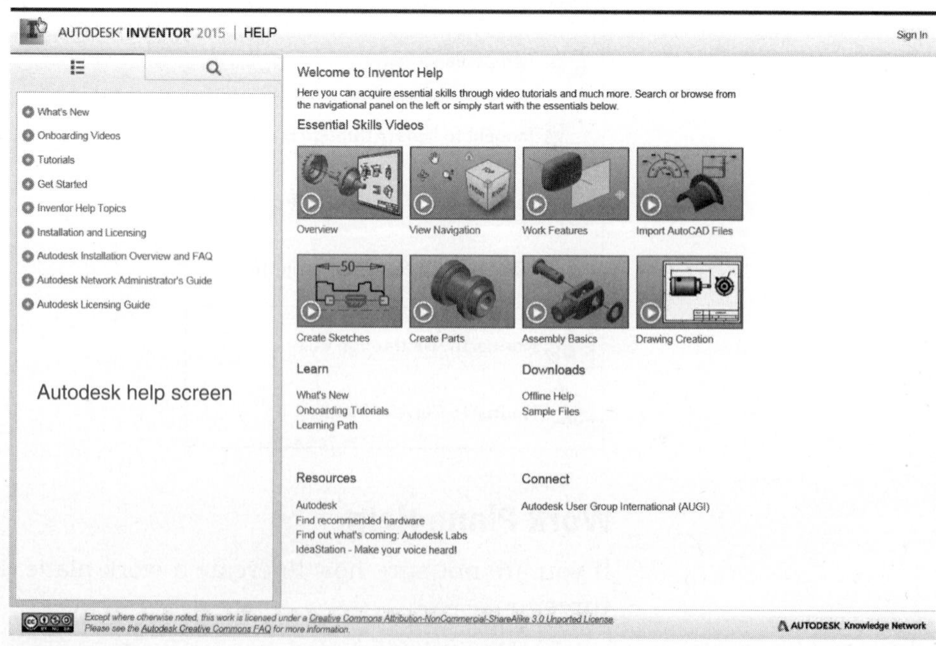

Figure 3-45

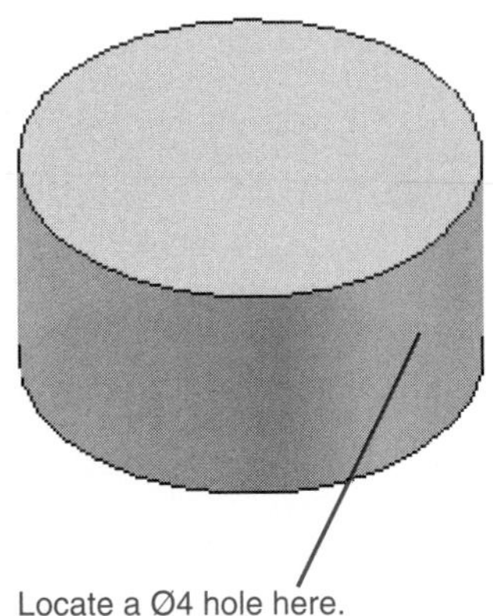

Locate a Ø4 hole here.

Figure 3-46

Sample Problem SP3-1

Figure 3-46 shows a Ø20 × 10 cylinder that was sketched aligned with the system's origin. The sketch was created on the default XZ plane.

Create a Ø4 hole through the cylinder so that its centerline is parallel to the XY plane and 5 above the plane.

The sides of the cylinder cannot be used as a sketch plane, so a work plane is needed. Either the YZ or XZ plane could be used. In this example the YZ plane was used.

EXERCISE 3-12 Creating a Tangent Work Plane

1. Click the **Plane** tool on the **Work Features** panel.
2. Click the **YZ Plane** tool in the browser area.

   A YZ plane will appear on the screen. See Figure 3-47.
3. Move the cursor and click the lower outside edge of the cylinder.

   A work plane will be created tangent to the cylinder.
4. Right-click one of the corners of the work plane (yellow circles will appear) and select the **New Sketch** option.

   A grid will appear.

Figure 3-47

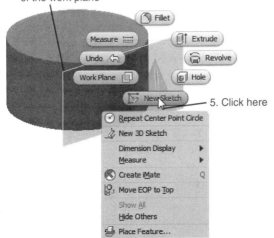

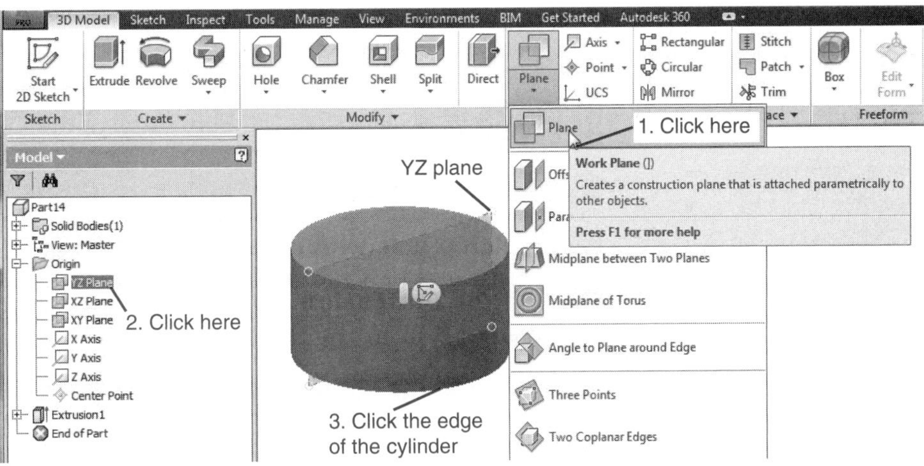

4. Right-click the corner of the work plane

Fillet
Measure
Extrude
Undo
Revolve
Work Plane
Hole
New Sketch
5. Click here
Repeat Center Point Circle
New 3D Sketch
Dimension Display
Measure
Create iMate
Move EOP to Top
Show All
Hide Others
Place Feature...

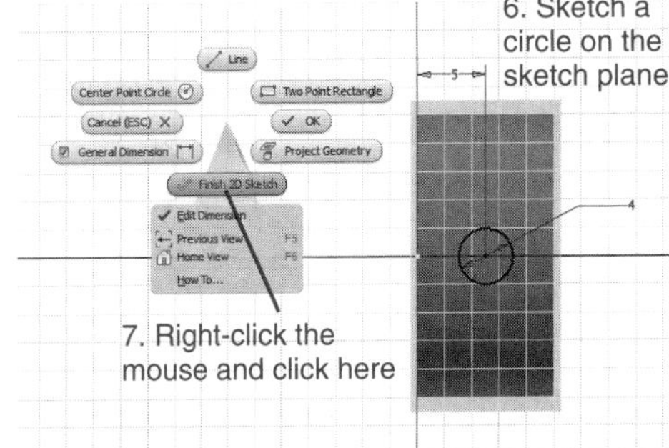

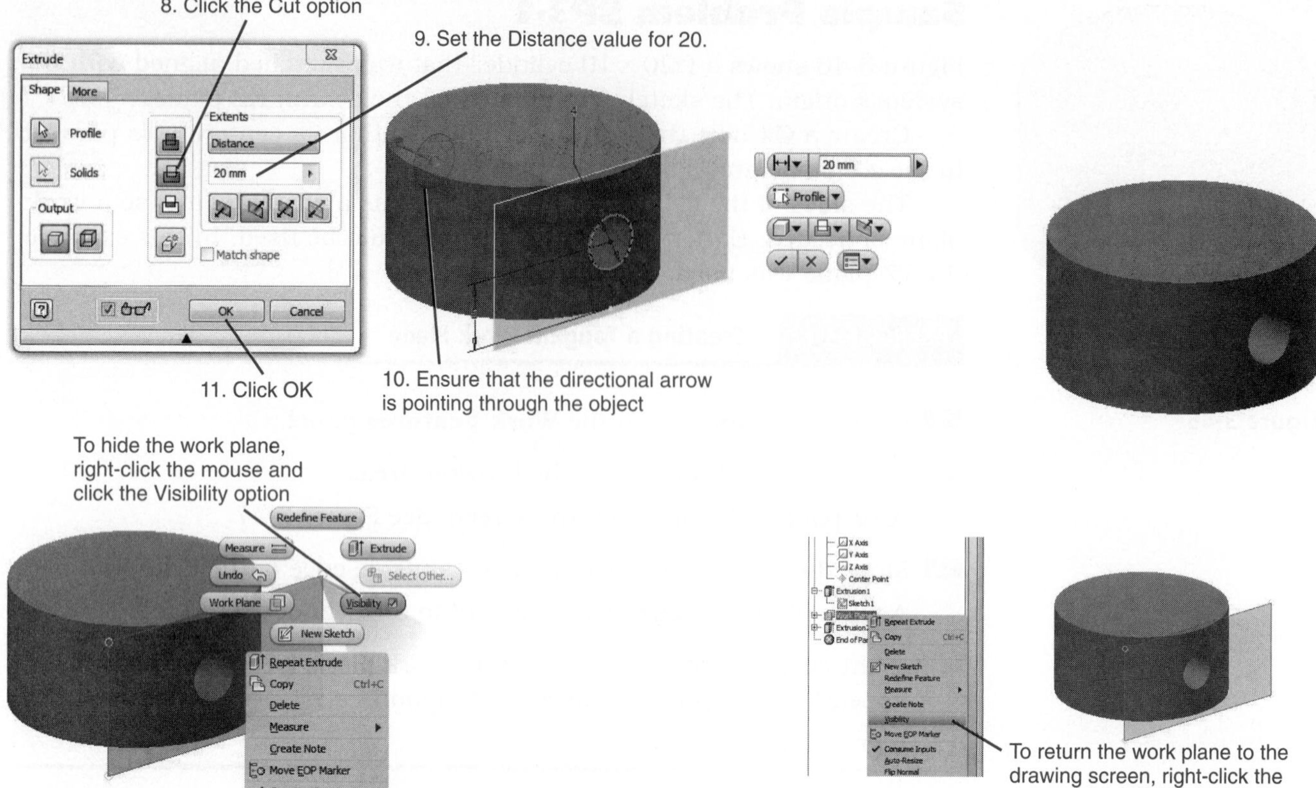

Figure 3-47
(Continued)

EXERCISE 3-13 Creating the Hole through the Cylinder

1. Click the **Sketch** tab and click the **Circle** tool.
2. Sketch a hole with its center point located on the darker vertical line.
3. Use the **Dimension** tool to create a **Ø4** circle with its center point located **5** from the top surface of the cylinder.
4. Click the **3D Model** tab and select the **Extrude** tool.

   The **Extrude** dialog box will appear.
5. Set the extrusion distance for **20** in a direction that passes through the cylinder, and select the **Cut** option.
6. Click **OK**.

The **Point, Center Point,** and **Hole** tools can also be used to create the hole.

Hiding Work Planes

Work planes may be hidden by right-clicking one of the corners of the work plane and selecting the **Visibility** option. See Figure 3-47.

NOTE

Do not delete the work plane, as this will also delete all commands associated with the work plane.

Restoring a Work Plane

To restore a hidden work plane, right-click the work plane's reference in the browser box and select the **Visibility** option.

Angled Work Planes

Work planes may be created at an angle to a model. For example, suppose a Ø10 hole must be drilled through the 30 × 50 × 10 box shown in Figure 3-48 at a 45° angle. Only extrusions perpendicular to a plane can be created, so a plane 45° to the top surface of the box is needed.

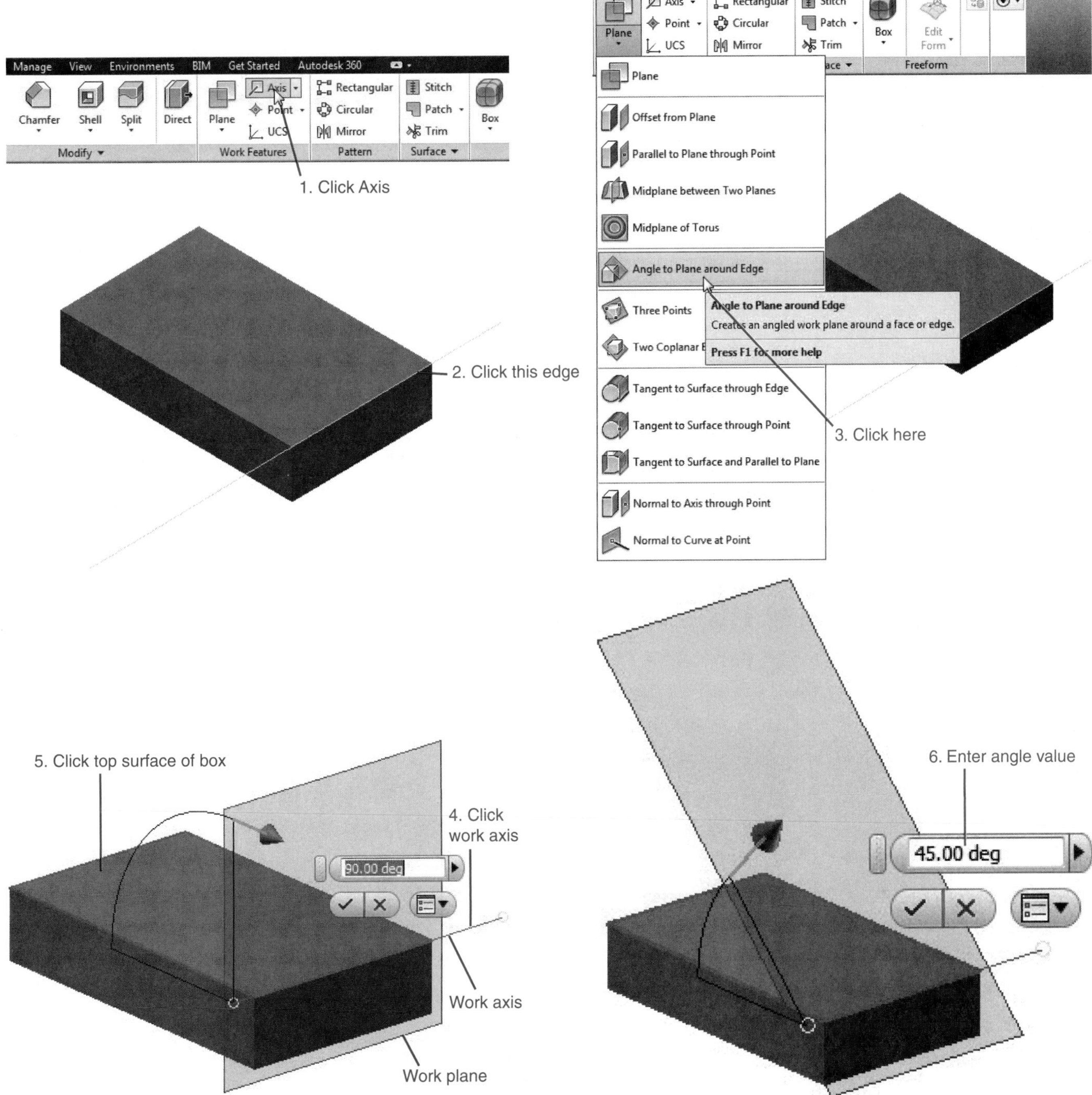

Figure 3-48

Figure 3-48
(Continued)

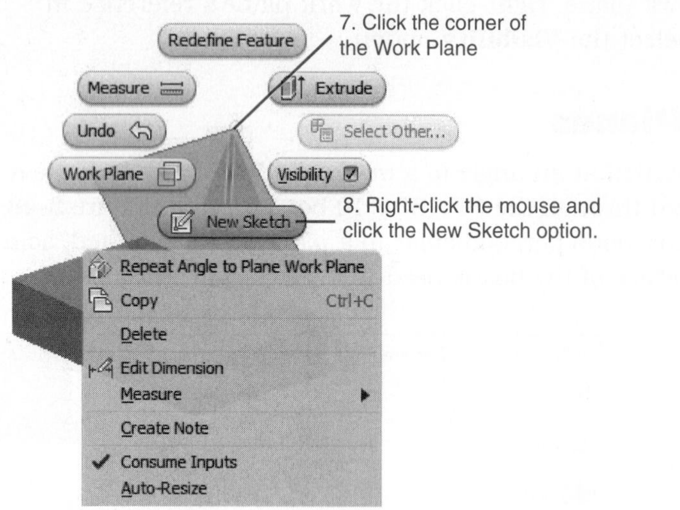

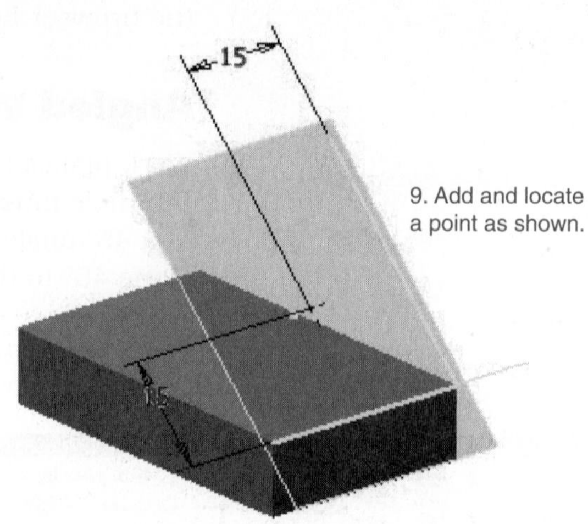

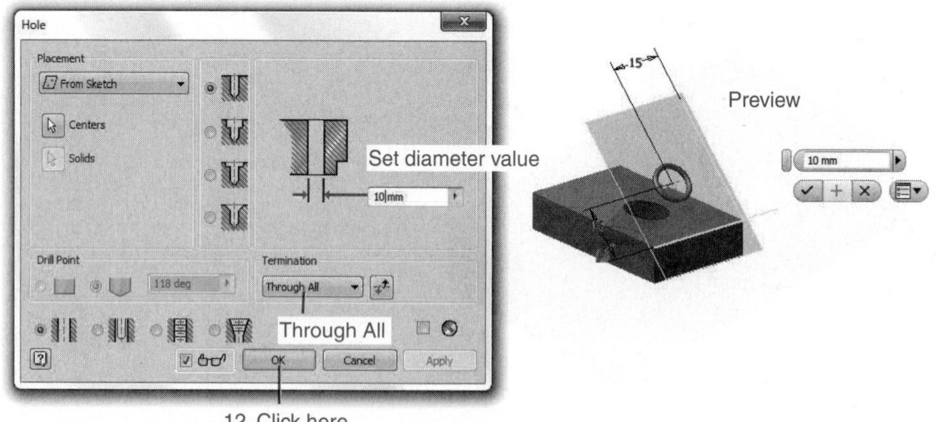

10. Click the Note tool.

11. Create a Ø10 simple hole, through all, using the point created on the angled work plane as the center point for the hole.

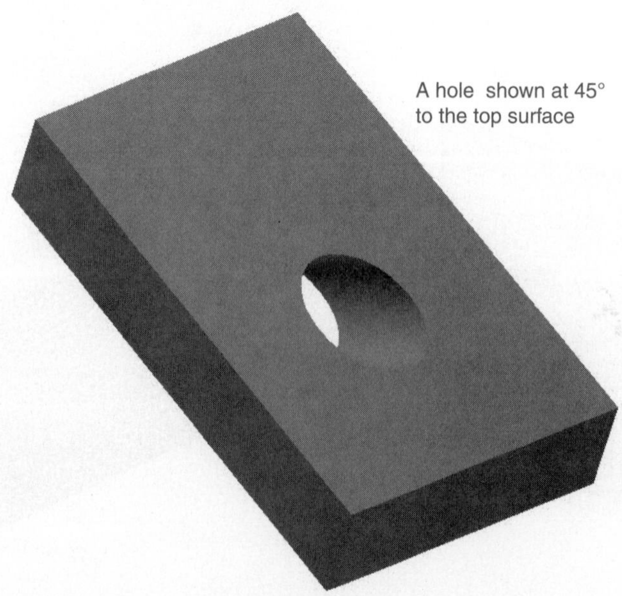

A hole shown at 45° to the top surface

EXERCISE 3-14 **Creating an Angled Work Plane**

work axis: A defined line on a model.

1 Create the 30 × 50 × 10 box, select the **Work Axis** tool located on the **Work Features** panel under the **3D Model** tab, and create a ***work axis*** by clicking on the edge of the block as shown.

See Figure 3-48.

The work axis will appear on the edge of the box.

2 Select the **Angle to Plane around Edge** tool, and click the work axis and the top surface of the box.

The **Angle to Plane around Edge** tool is a flyout from the **Plane** tool located on the **Work Features** panel under the **3D Model** tab.

3 Enter an angle value, and click the **OK** check mark.

In this example, a value of **45**° was entered.

4 Right-click one of the work plane's corner points and click the **New Sketch** option.

5 Use the **Point** and **Dimension** tools to create and locate a point that can be used as the hole's center point.

6 Right-click the mouse and select the **Finish 2D Sketch** option.

7 Select the **Hole** tool and specify the hole's diameter and a length of **Through All**.

In this example a diameter of **10 mm** was entered.

8 Click **OK**.

9 Use the **Visibility** tool to hide the work plane and axis.

Offset Work Planes

Figure 3-49 shows an object in which a small cylinder passes through a larger cylinder. An offset work plane was used to create the object.

Figure 3-49

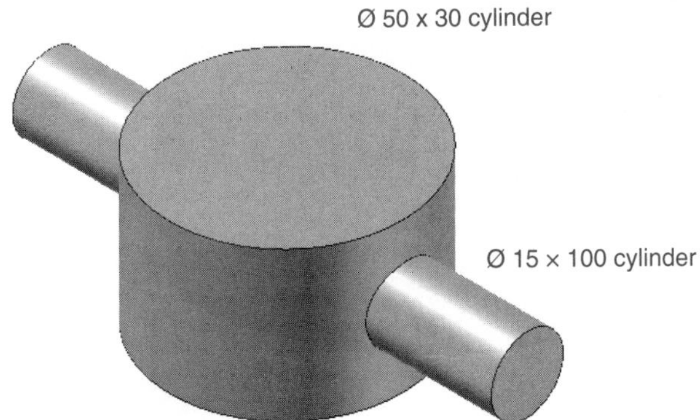

EXERCISE 3-15 Creating an Offset Work Plane

See Figure 3-50.

1. Draw a **Ø50** × **30** cylinder centered on the origin of the XZ plane.
2. Click the arrow under the **Plane** tool on the **Work Features** panel under the **3D Model** tab, and select the **Offset from Plane** option.
3. Click the **YZ Plane** in the browser box.

The work plane will appear at the center of the cylinder.

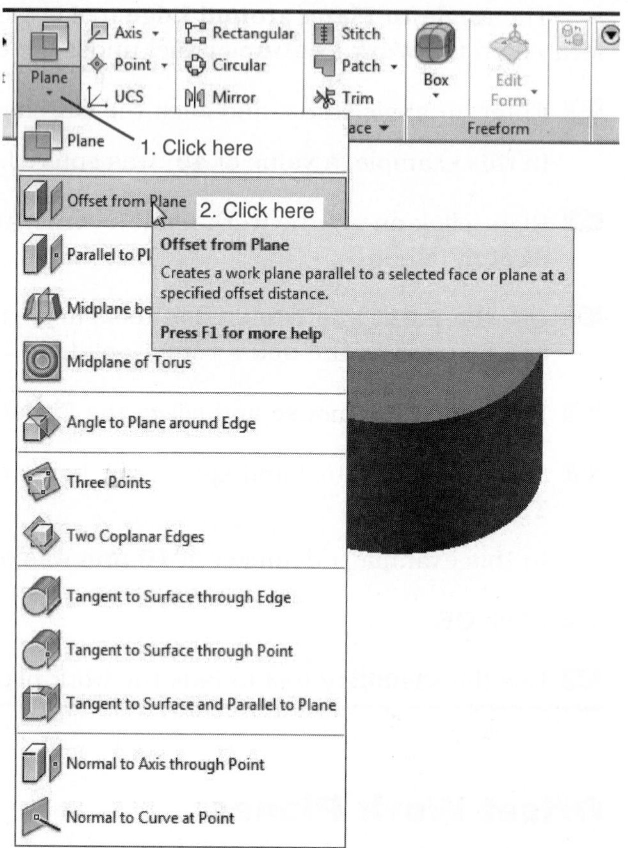

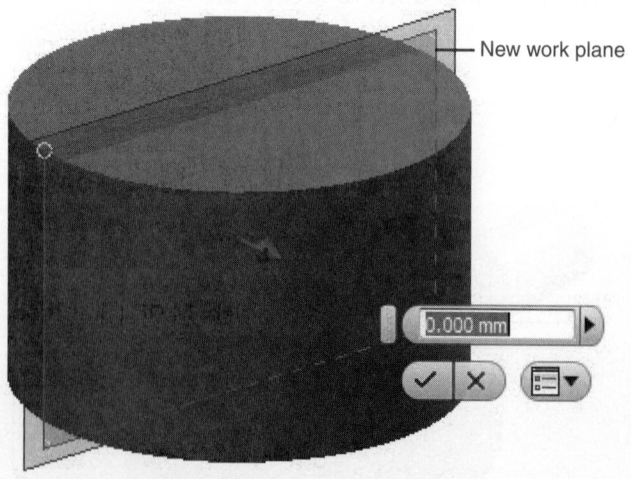

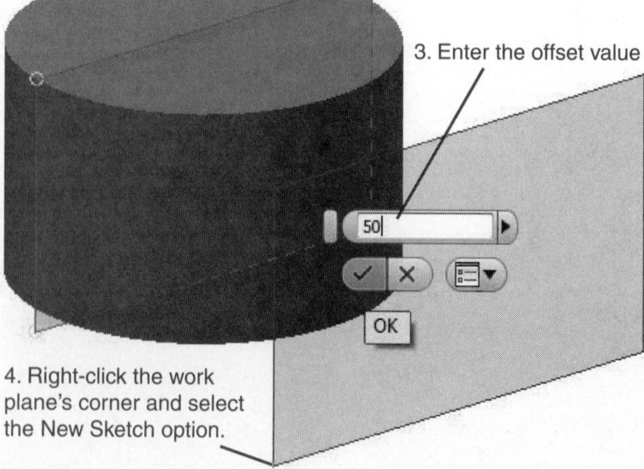

Figure 3-50

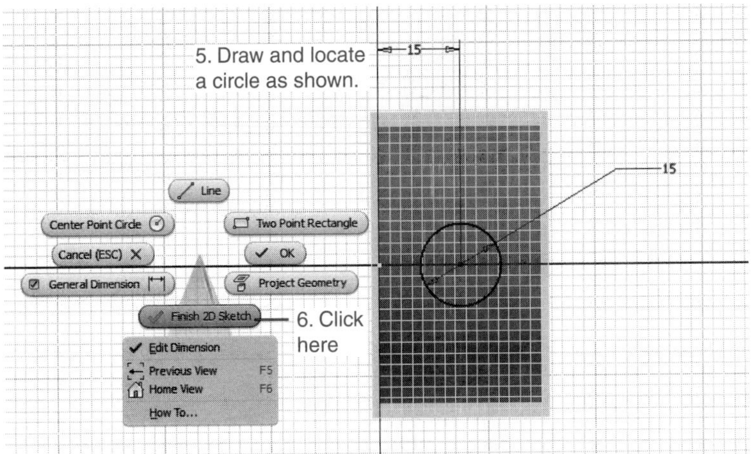

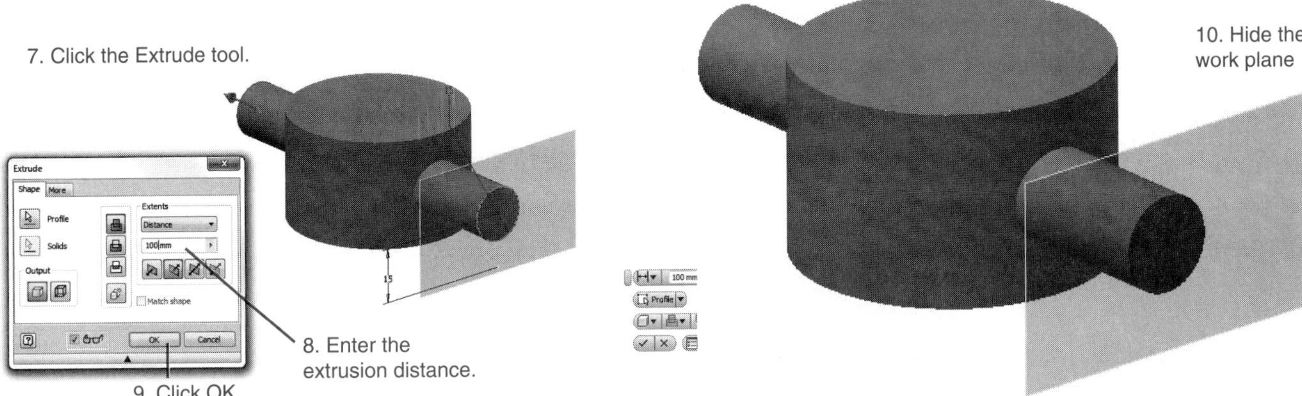

Figure 3-50
(Continued)

4 Specify the offset distance, and click the check mark.

In this example a value of **50** was used.

5 Right-click one of the work plane's corner points and select the **New Sketch** option.

A grid will appear.

6 Draw and locate a **Ø15** circle as shown.

7 Right-click the mouse and select the **Finish 2D Sketch** option.

8 Select the **Extrude** tool, then select the circle as the profile and extrude it **100 mm** through the large cylinder.

9 Click **OK**.

10 Hide the work plane.

Work Points

work point: A defined point on a model used to help locate work planes and work axes.

Work points are defined points on a model. They are used to help locate work planes and work axes. There are nine options associated with the **Work Point** tool. See Figure 3-51.

Figure 3-51

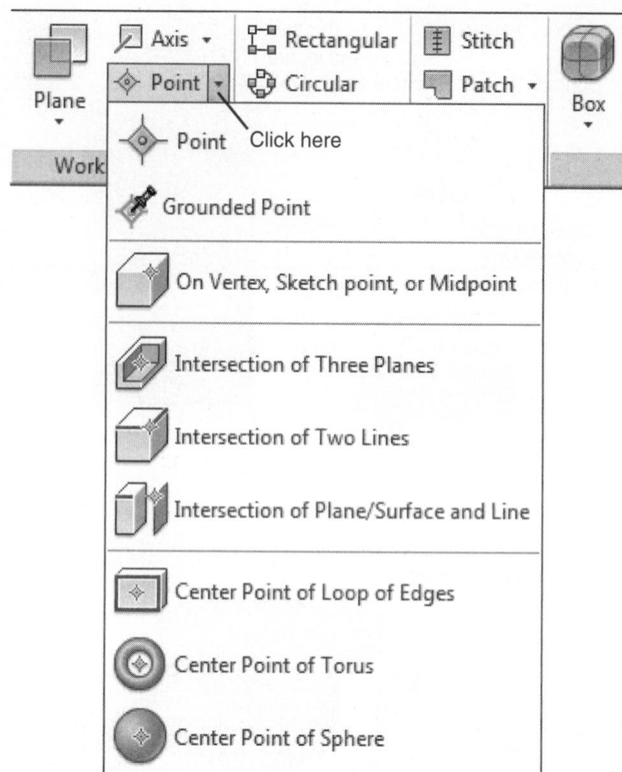

EXERCISE 3-16 **Defining a Work Point**

1. Click on the **Work Point** tool located on the **Work Features** panel under the **3D Model** tab.

2. Select the location for the work points and click the mouse.

In the example, shown in Figure 3-52, the midpoint of the left edge was selected along with the lower front corner. The cursor will snap to the midpoint of the edge.

Figure 3-52

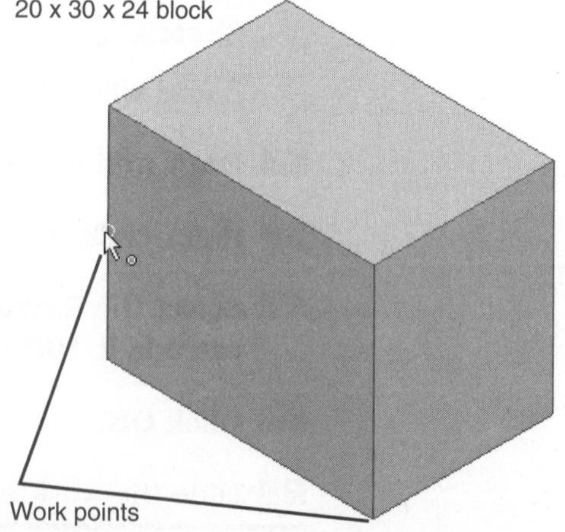

The work points created will be listed in the browser box.

EXERCISE 3-17 **Creating an Oblique Work Plane Using Work Points**

An oblique work plane may be created using work points. Figure 3-52 shows a $20 \times 30 \times 24$ rectangular box.

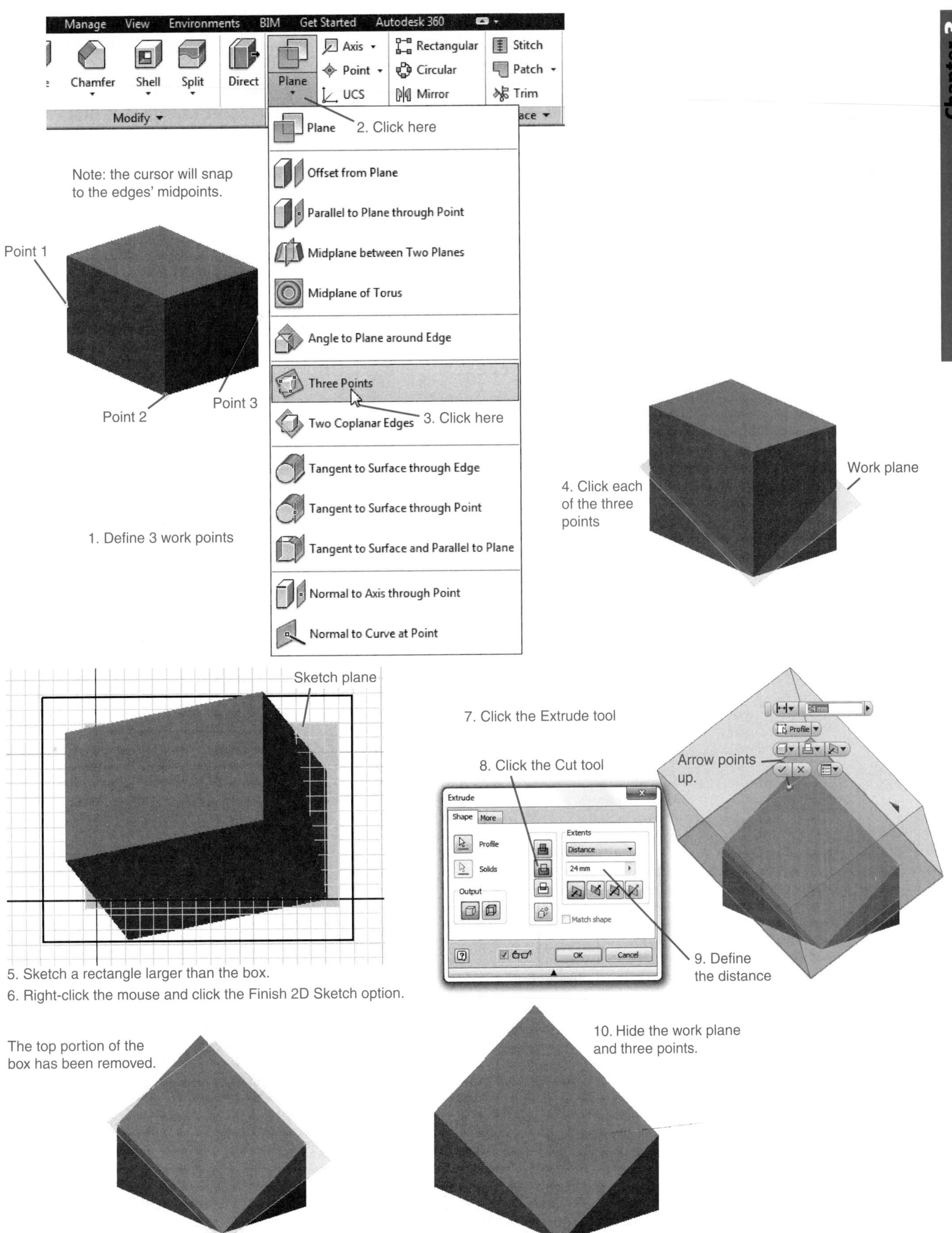
Manage
View
Environments
BIM
Get Started
Autodesk 360
Chamfer
Shell
Split
Direct
Plane
Axis
Point
UCS
Rectangular
Circular
Mirror
Stitch
Patch
Trim
Modify
Plane
2. Click here
Offset from Plane
Parallel to Plane through Point
Midplane between Two Planes
Midplane of Torus
Angle to Plane around Edge
Three Points
Two Coplanar Edges
3. Click here
Tangent to Surface through Edge
Tangent to Surface through Point
Tangent to Surface and Parallel to Plane
Normal to Axis through Point
Normal to Curve at Point
Note: the cursor will snap to the edges' midpoints.
Point 1
Point 2
Point 3
1. Define 3 work points
4. Click each of the three points
Work plane
Sketch plane
5. Sketch a rectangle larger than the box.
6. Right-click the mouse and click the Finish 2D Sketch option.
7. Click the Extrude tool
8. Click the Cut tool
Extrude
Shape
More
Profile
Solids
Output
Extents
Distance
24 mm
Match shape
OK
Cancel
Arrow points up.
9. Define the distance
The top portion of the box has been removed.
10. Hide the work plane and three points.

Figure 3-53

1. Create three work points on the prism: two on the midpoints of the vertical edges, and one at the lower corner as shown in Figure 3-53.

2. Click on the **Work Plane** tool located on the **Work Features** panel under the **3D Model** tab, and click the **Three Points** tool. Click each of the three work points.

3. Right-click one of the work plane's corner points, and click the **New Sketch** option.

4. Click the **Two Point Rectangle** tool and sketch a very large rectangle on the new sketch plane. Right-click the mouse and click the **Finish 2D Sketch** option.

   The rectangle may be any size that exceeds the size of the block.

5. Click the **Extrude** tool, click the **Cut** tool, and remove the top portion of the box.

6. Hide the work plane and the three defining points.

Work Axes

A work axis is a defined line. Work axes are used to help define work planes and to help define the geometric relationship between assembled models. There are eight options associated with the **Work Axis** tool. See Figure 3-54.

Figure 3-54

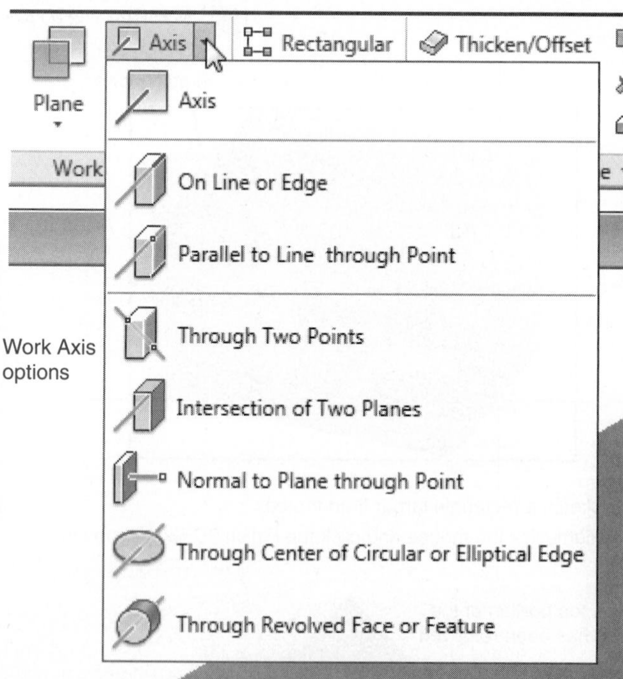

EXERCISE 3-18 **Creating a Work Axis**

1. Click on the **Work Axis** tool located on the **Work Features** panel under the **3D Model** tab.

2. Click the **On Line or Edge** tool.

3 Click the edge line that is to be defined as a work axis.

Figure 3-55 shows a model with two defined work axes.

Figure 3-55

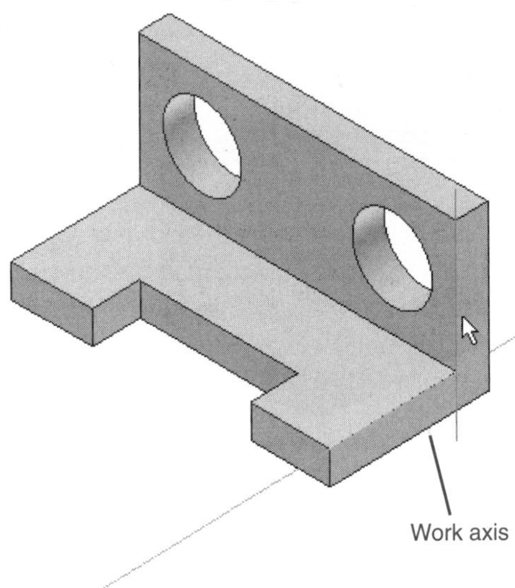

EXERCISE 3-19 Drawing a Work Axis at the Center of a Cylinder

Figure 3-56 shows a cylinder.

1 Click the **Work Axis** tool.

2 Click the lower edge of the cylinder.

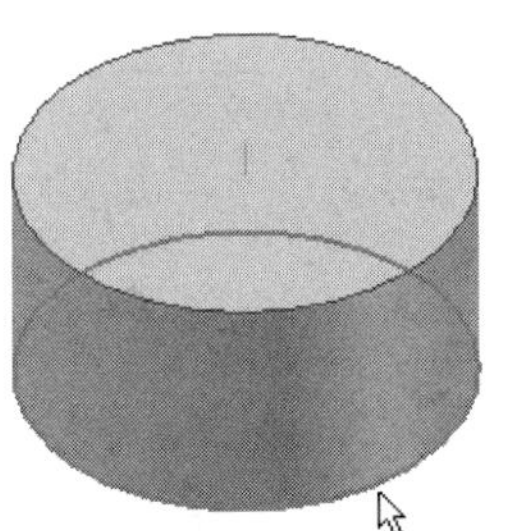

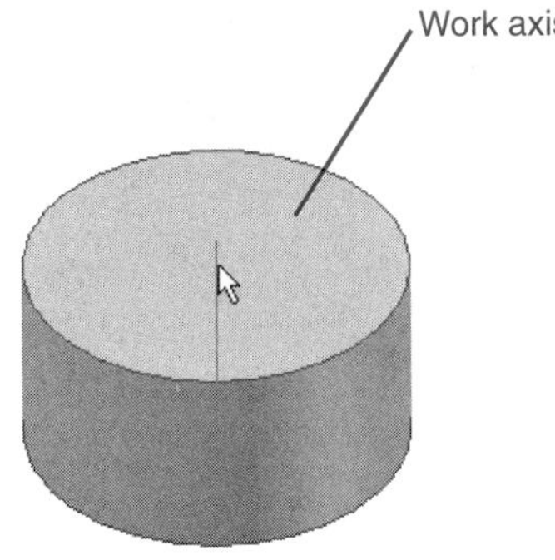

Figure 3-56

A work axis will appear through the center of the cylinder, and the words **Work Axis 1** will appear in the browser. Because there was only one object on the screen, and because the object was a cylinder, the **Through Revolved Face or Feature** will turn on automatically.

> **TIP**
> The relationships among work points, work axis, and work planes will be discussed further in Chapter 5, Assembly Drawings.

Ribs (WEBs)

rib: An element added to a model to give it strength.

A ***rib*** is used to add strength to a part. Ribs or webs are typically used with cast or molded parts. Figure 3-57 shows an L-bracket. The bracket's flanges are 20 × 20, the length is 50, and the thickness is 5. Ribs 5 mm thick are to be added to each end of the bracket.

1 Click the right end surface of the bracket, click the right mouse button, and select the **New Sketch** option.

If needed, add a work plane to the end plane, and create the new sketch.

2 Use the **Line** tool and draw a line across the corner edge points as shown in Figure 3-57.

3 Right-click the mouse and select the **Finish 2D Sketch** option.

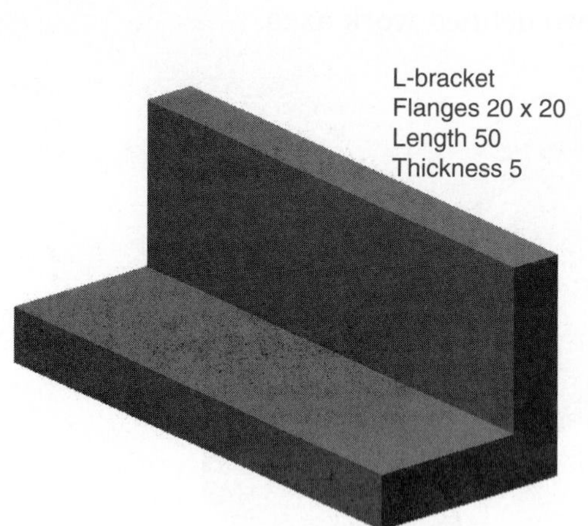

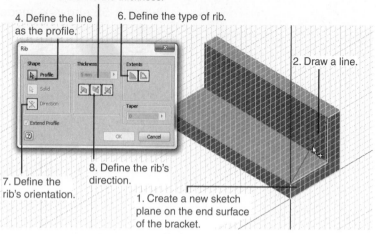

3. Access the Model tools and click the Rib tool.
5. Define the thickness.
4. Define the line as the profile.
6. Define the type of rib.
2. Draw a line.
8. Define the rib's direction.
7. Define the rib's orientation.
1. Create a new sketch plane on the end surface of the bracket.

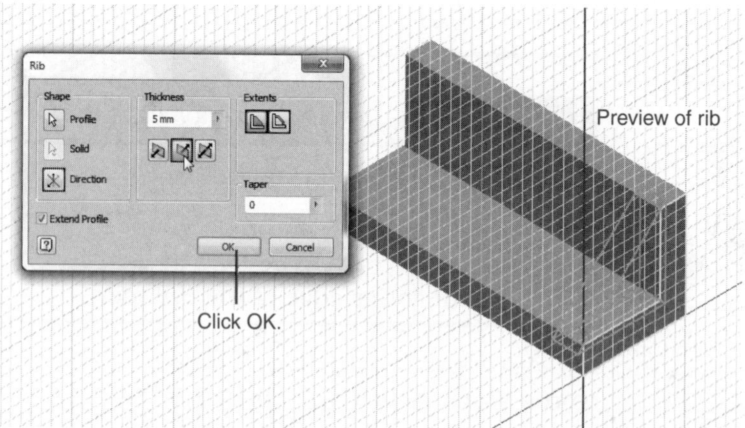

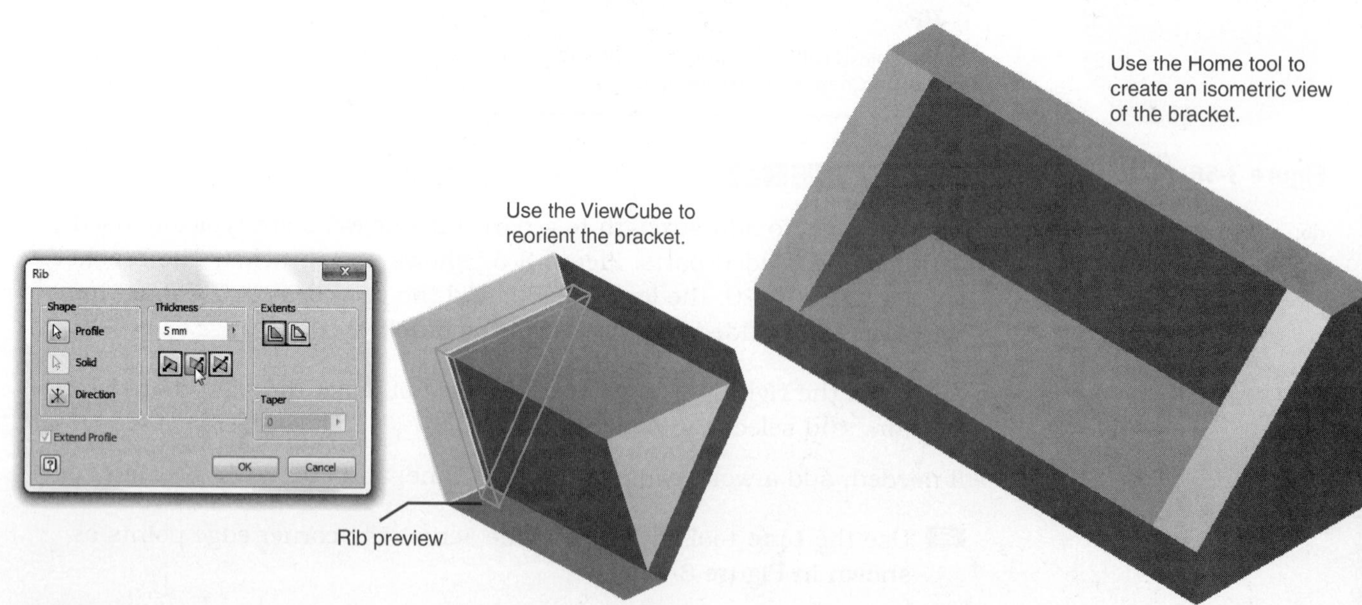

Figure 3-57

4 Click the **Rib** tool located on the **Create** panel under the **3D Model** tab.

The **Rib** dialog box will appear. See Figure 3-57.

5 Define the line as the profile by clicking the line.

6 Enter a **Thickness** value of **5**.

7 Click the left box under the **Extents** heading.

8 Click the middle box under the **Thickness** heading to locate the rib.

The right side of the rib preview should be aligned with the right end surface of the L-bracket.

9 Click **OK**.

The rib will appear on the bracket.

10 Use the **ViewCube** to rotate the bracket so that the other end of the bracket is visible.

11 Right-click the L-shaped surface and create a new sketch plane.

12 Draw a line between the corners as was done for the first rib.

13 Specify the thickness and use the **Direction** tool to specify the rib's orientation.

14 Move the cursor into the rib area and move the rib around until the desired orientation is achieved.

15 Use the direction arrows under the **Thickness** heading to locate the rib.

16 Click **OK**.

17 Use the **Home** tool to create an isometric view of the bracket.

Loft

The **Loft** tool is used to create a solid between two or more sketches. Figure 3-58 shows a loft surface created between a circle and a square. Both the circle and the square are first drawn on the same XY plane. This allows the **Dimension** tool to be used to ensure the alignment between the

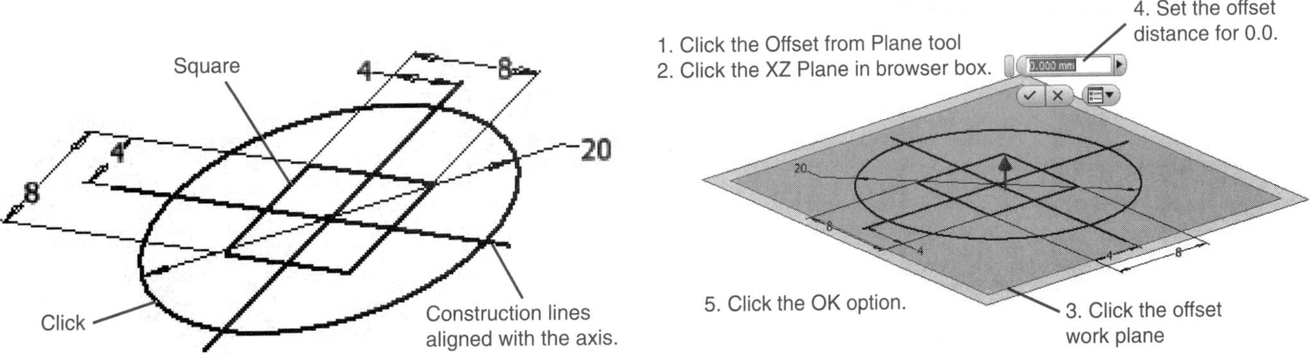

Figure 3-58

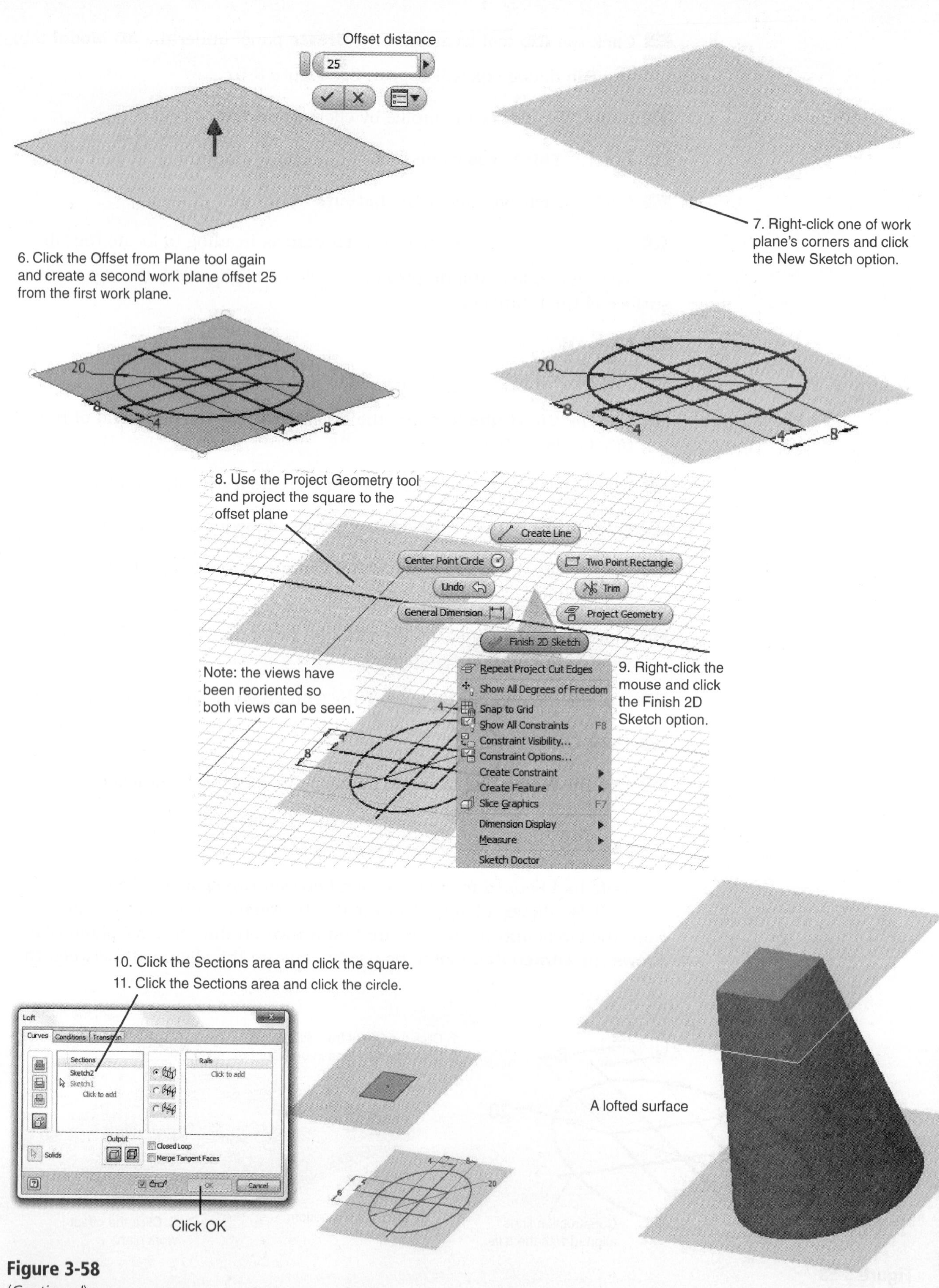

Offset distance
25
6. Click the Offset from Plane tool again and create a second work plane offset 25 from the first work plane.
7. Right-click one of work plane's corners and click the New Sketch option.
20
8
4
4
8
8. Use the Project Geometry tool and project the square to the offset plane
Create Line
Center Point Circle
Two Point Rectangle
Undo
Trim
General Dimension
Project Geometry
Finish 2D Sketch
Repeat Project Cut Edges
Show All Degrees of Freedom
Snap to Grid
Show All Constraints F8
Constraint Visibility...
Constraint Options...
Create Constraint
Create Feature
Slice Graphics F7
Dimension Display
Measure
Sketch Doctor
Note: the views have been reoriented so both views can be seen.
9. Right-click the mouse and click the Finish 2D Sketch option.
10. Click the Sections area and click the square.
11. Click the Sections area and click the circle.
Loft
Curves
Conditions
Transition
Sections
Sketch2
Sketch1
Click to add
Rails
Click to add
Output
Solids
Closed Loop
Merge Tangent Faces
OK
Cancel
Click OK
A lofted surface

Figure 3-58
(Continued)

two sketches. The rectangle is then projected onto another work plane, and the **Loft** tool is used to create a surface between the two planes.

EXERCISE 3-20 Sketching the Circle and the Square

1 Start a new drawing using the **Standard (mm).ipt** format and sketch a **Ø20** circle and an **8 × 8** square aligned to a common center point. Right-click the mouse, and click the **Finish 2D Sketch** option.

See Figure 3-58.

2 Create a work plane aligned with the XY plane, that is, **0** offset, by first clicking the **Offset from Plane** tool on the **Work Features** panel under the **Plane** tool, then clicking the **XY Plane** tool in the browser box.

3 Click and drag one of the new plane's corners upward, but set the offset value to **0.0**.

4 Right-click the mouse and click the **OK** option.

EXERCISE 3-21 Creating an Offset Work Plane

1 Use the **Offset from Plane** tool again, then click the **XY Plane** tool in the browser area.

A new plane will appear aligned with the existing XZ work plane.

2 Click one of the corner points of the new plane and move the cursor upward.

An **Offset** dialog box will appear.

3 Set the offset distance for **25,** right-click the mouse, and select the **OK** option.

Check the browser area to verify that two work planes have been created.

EXERCISE 3-22 Projecting the Square

1 Click one of the corner points of the offset work plane, right-click the mouse, and select the **New Sketch** option.

In this example, the **ViewCube** was used to reorient the model so that both planes can be seen.

2 Click the **Project Geometry** tool located on the **Create** panel under the **Sketch** tab.

3 Select the **8 × 8** square.

Select the square line by line. The square will be projected into the offset work plane.

4 Right-click the mouse and select the **OK** option.

EXERCISE 3-23 Creating a Loft

1. Right-click the mouse and select the **Finish 2D Sketch** option, then select the **Loft** option located on the **Create** panel under the **3D Model** tab.

   The **Loft** dialog box will appear.

2. Click in the **Sections** area of the **Loft** dialog box, and click the 8 × 8 square.

3. Click in the **Sections** area of the **Loft** dialog box again and click the circle.

4. Click **Sketch1** in the **Loft** dialog box.

5. Click **OK**.

   Hide the work planes if desired.

Sweep

The **Sweep** tool is used to project a sketch along a defined path. In this example a shape is created in the XY plane and then projected along a path drawn in the YZ plane. See Figure 3-59.

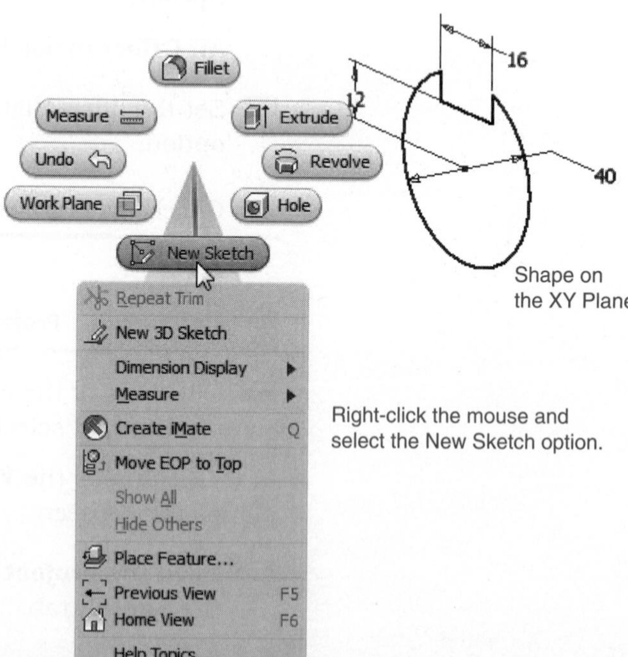

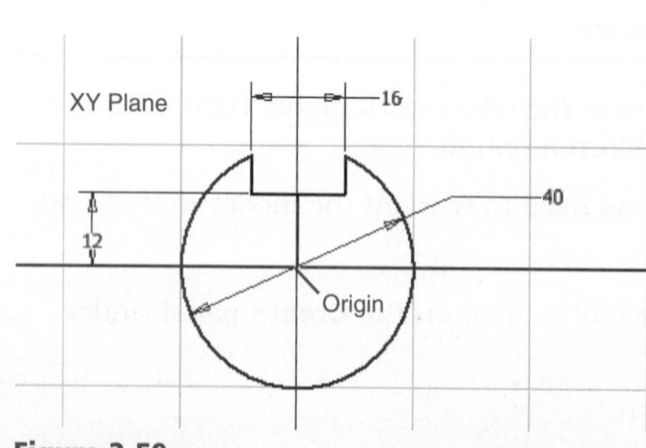

Figure 3-59

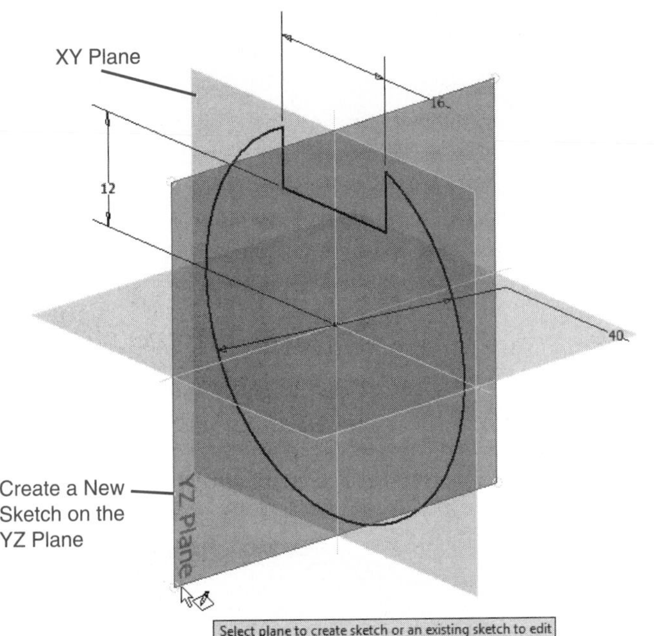
XY Plane
16
12
40
Create a New Sketch on the YZ Plane
YZ Plane
Select plane to create sketch or an existing sketch to edit

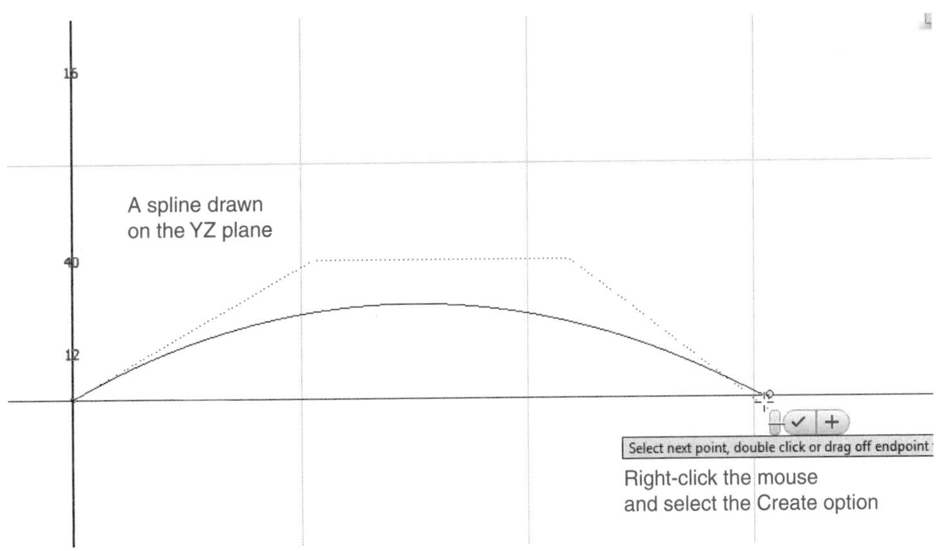
16
A spline drawn on the YZ plane
40
12
Select next point, double click or drag off endpoint
Right-click the mouse and select the Create option

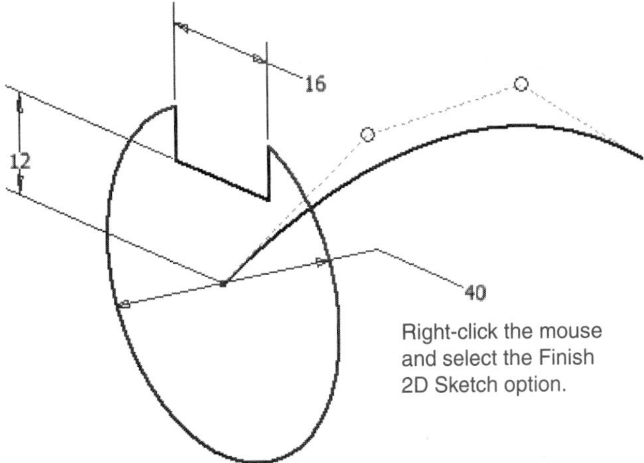
16
12
40
Right-click the mouse and select the Finish 2D Sketch option.

Figure 3-59
(Continued)

Figure 3-59

(Continued)

Click the Sweep tool

Sweep

- Profile
- Path
- Solids
- Type: Path
- Orientation: Path / Parallel
- Taper: 0
- Twist: 0
- Output
- Optimize for Single Selection
- OK
- Cancel

16

12

40

Profile

Path

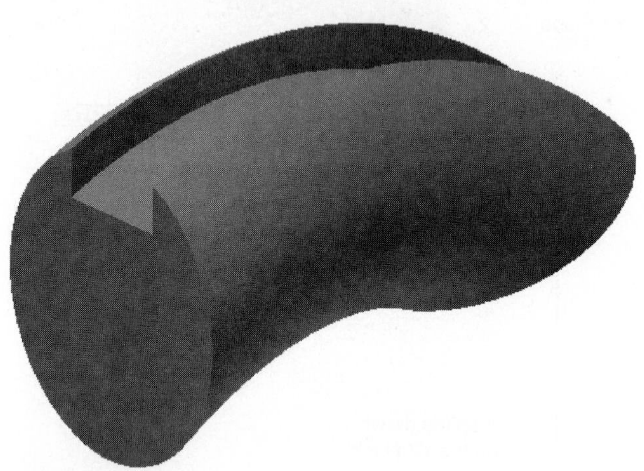

EXERCISE 3-24 Creating the Sketch

1. Start a new drawing using the **Standard (mm).ipt** format.

   Click the **Create 2D Sketch** tool. The drawing will go to the **2D Sketch** tools.

2. Right-click the mouse and select the **New Sketch** option, then select the XY plane.

3. Draw the circular shape shown centered on the origin, and right-click the mouse and select the **Finish 2D Sketch** option.

EXERCISE 3-25 Creating the Path

1. Right-click the mouse and select the **New Sketch** option.

2. Click one of the YZ plane's corner points and create a new sketch plane.

3. Click the **Spline** tool and sketch a spline starting at the hole's center point. Click the right mouse button and select the **Create** option, then right-click the mouse again and select the **OK** option.

   In this example a random spline was used.

EXERCISE 3-26 **Creating the Sweep**

1. Right-click the mouse and select the **Finish 2D Sketch** option, then click the **Sweep** tool.

   The **Sweep** dialog box will appear. The circular sketch will automatically be selected as the **Profile**.

2. Click the spline to define it as the path.

3. Click **OK**.

Coil

A coil is similar to a sweep, but the path is a helix. A sketch is drawn and projected along a helical path.

EXERCISE 3-27 **Creating the Sketch**

1. Start a new drawing using the **Standard (mm).ipt** format.

2. Sketch the shape shown in Figure 3-60 on the XY plane.

Figure 3-60

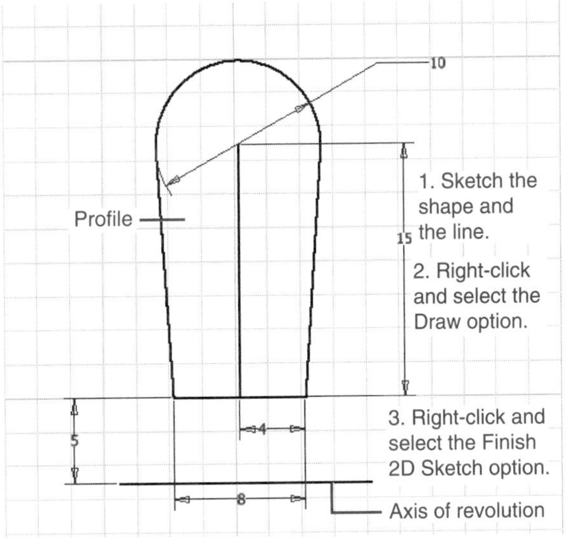

Profile
10
15
5
4
8
1. Sketch the shape and the line.
2. Right-click and select the Draw option.
3. Right-click and select the Finish 2D Sketch option.
Axis of revolution

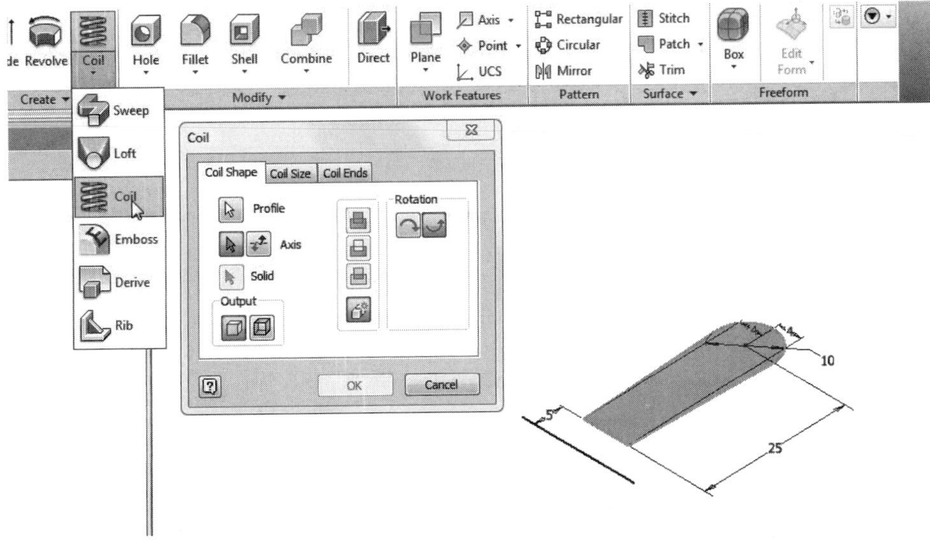

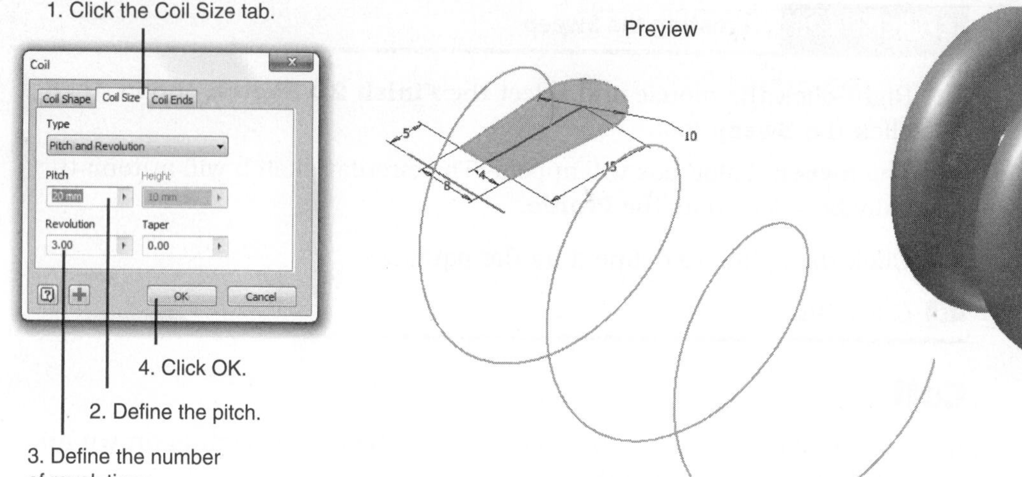

Figure 3-60
(Continued)

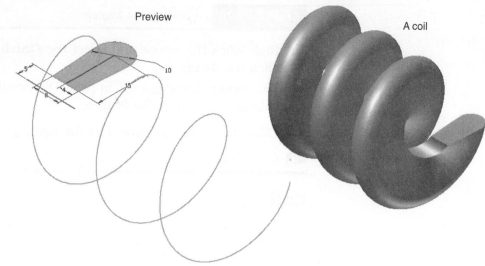

3. Sketch a line below the shape as shown.

   This line will serve as the axis of rotation.

4. Right-click the mouse, and select the **Finish 2D Sketch** option.

   In this example an **Isometric** view orientation was used.

EXERCISE 3-28 Creating the Coil

1. Click the **Coil** tool located on the **Create** panel under the **3D Model** tab.

   The **Coil** dialog box will appear. The sketched profile will be selected automatically.

2. Select the sketch line as the axis.

   A preview will appear.

3. Click the **Coil Size** tab.

   The dialog box will change.

> **NOTE**
> How to draw springs using **Coil** is covered in Chapter 9.

4. Set the **Type** for **Pitch and Revolution,** the **Pitch** for **20,** and the **Revolution** for **3**.

5. Click **OK**.

   See Figure 3-60.

Model Material

A material designation may be assigned to a model. The material designation becomes part of the model's file and will be included on any assembly's parts list that includes the model.

EXERCISE 3-29 **Defining a Model's Material**

1 Right-click on the model's name in the browser box and select the **iProperties** option.

In this example, the name BLOCK was used. See Figure 3-61. The **Block iProperties** dialog box will appear.

2 Select the **Physical** tab and then the scroll arrow on the right side of the **Material** box.

See Figure 3-62.

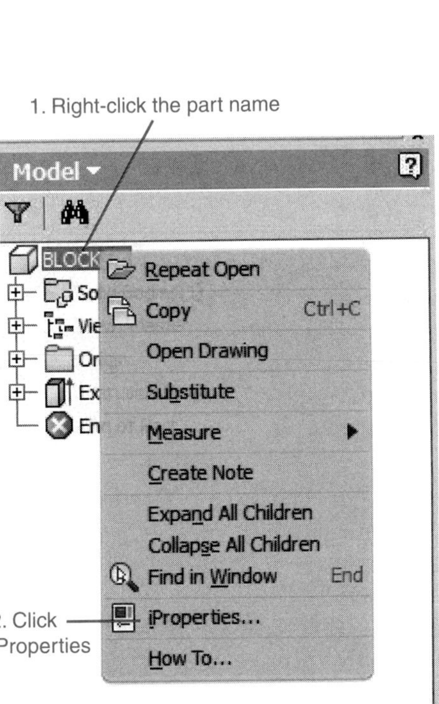

Figure 3-61

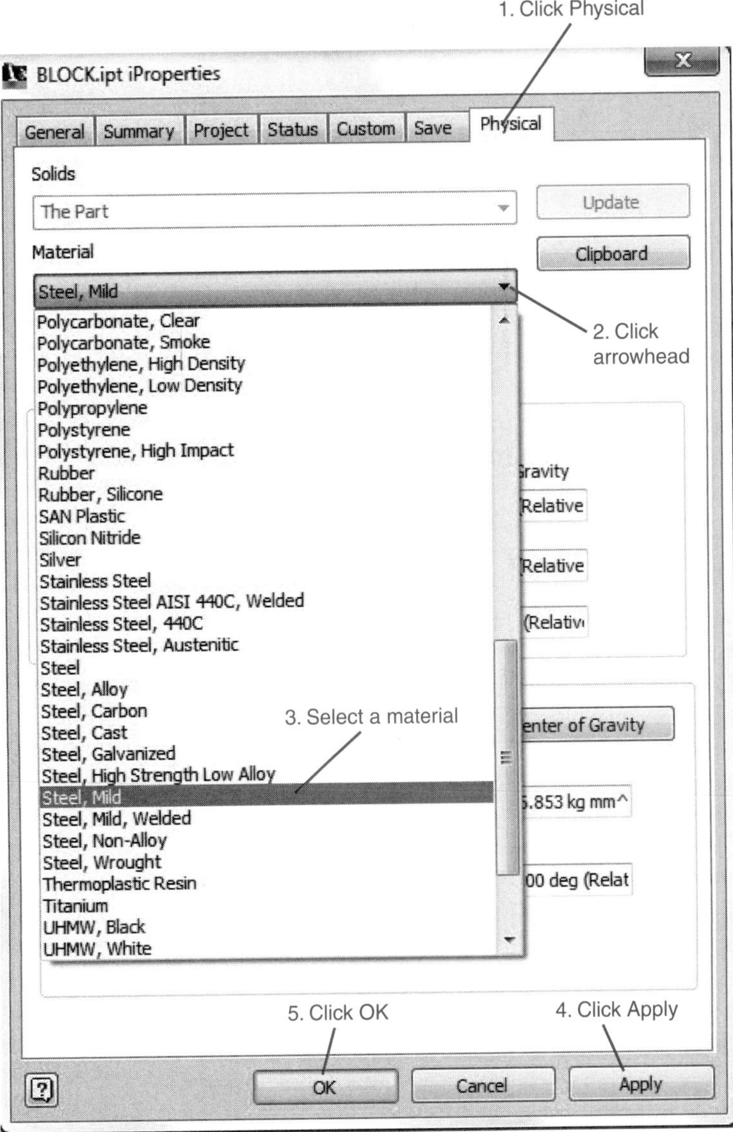

Figure 3-62

3 Select a material.

Figure 3-63 shows the BLOCK using three different materials: mild steel; brass, soft yellow; and glass.

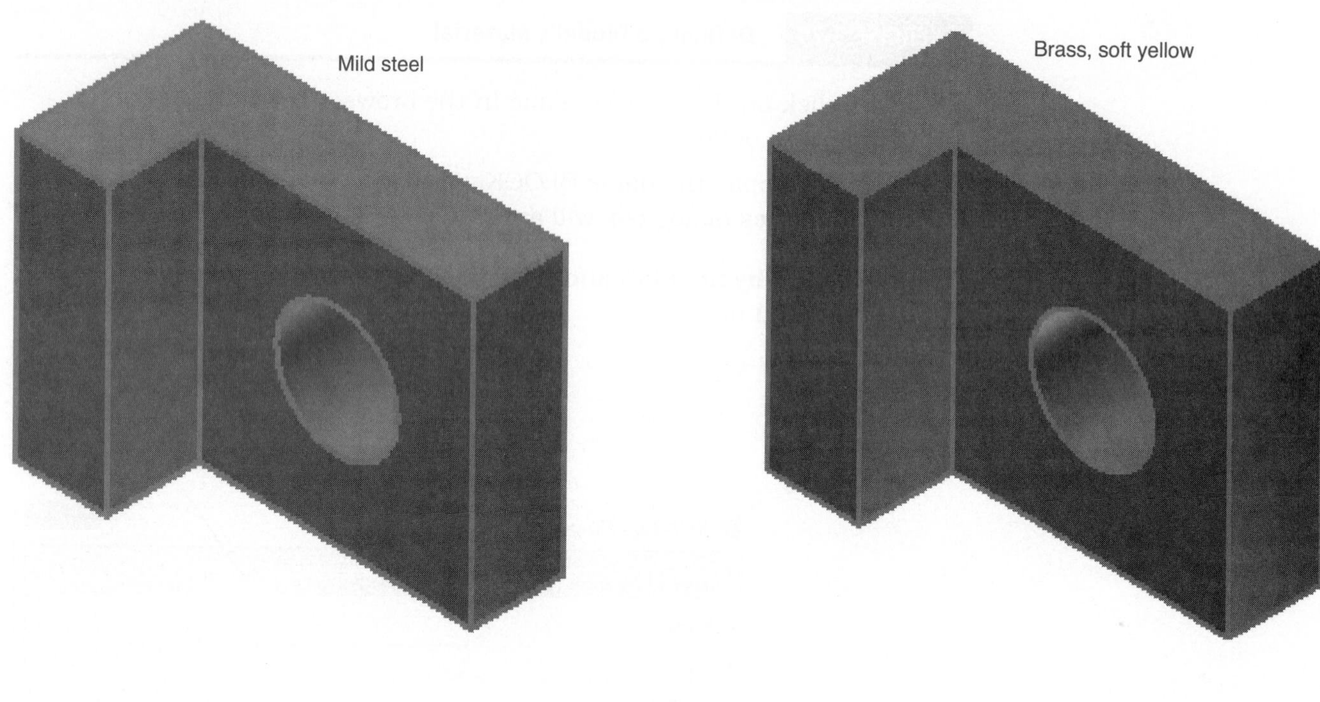

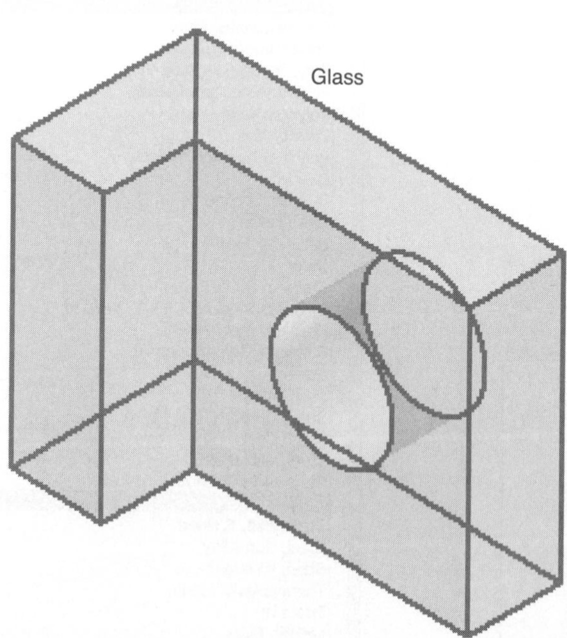

Figure 3-63

4 Click **Apply**.

5 Click **Close**.

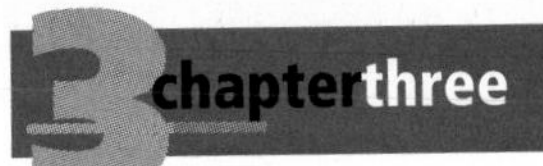

Chapter Summary

The first part of the chapter demonstrated how to convert 2D sketches into 3D models and then modify features using some of the tools in the **3D Model** panel bar. Exercises included extruding, revolving, lofting, and mirroring models as well as trimming away portions, and creating shells. Fillets, chamfers, and holes in both rectangular and circular arrangements were also added to models.

The second part of the chapter introduced sketch and work planes and work axes and demonstrated how to use them to refine 3D models.

Chapter Test Questions

Multiple Choice

Circle the correct answer.

1. Which of the following is not used to define a chamfer?
   a. Angle and a distance
   b. Distance and distance
   c. Two angles and a distance
2. Which tool is used to draw a spring?
   a. **Coil**
   b. **Loft**
   c. **Sweep**
3. The **Edit Sketch** tool can be applied to shapes created with which of the following tools?
   a. **Extrude**
   b. **Rectangle**
   c. **Revolve**
   d. **Hole**
4. The **Edit Feature** tool can be applied to shapes created with which of the following tools?
   a. **Circle**
   b. **Line**
   c. **Point, Center Point**
   d. **Extrude**
5. Which of the following parameters cannot be used to draw a work plane?
   a. Angle to a plane
   b. Point and a tangent
   c. 3-Points
   d. Tangent to a face through

6. Which of the following is a material not listed under the **Physical** tab of the **Properties** dialog box?
   a. Mild Steel
   b. Aluminum Bronze
   c. Glass
7. Sketched shapes can be projected between work planes using which tool?
   a. **Sweep**
   b. **Boundary Patch**
   c. **Move Face**
   d. **Project Geometry**
8. Which of the following will happen if a work plane is deleted?
   a. The work plane will disappear from the screen and all entities will be deleted.
   b. The work plane will disappear from the screen and all entities will remain in place.
9. Why are ribs used on molded parts?
   a. To increases the part's flexibility
   b. To balance the part
   c. To increase the part's strength
10. The **Face Draft** tool is used to
    a. Create airflow
    b. Create an angled surface
    c. Create a current behind a moving object

Matching

Write the number of the correct answer on the line.

Column A

a. **Face Draft** _____
b. **Fillet** _____
c. **Coil** _____
d. **Shell** _____
e. **Work Plane** _____
f. **Rectangular Pattern** _____
g. **Extrude, Cut** _____

Column B

1. The tool used to draw springs
2. The tool used to draw a square pattern of holes
3. The tool used to add a slanted surface to an object
4. The tool used to hollow out an object
5. The tool used to add rounded edges to an object
6. The tool used to remove material from an object
7. The tool used to define planes not located on any surface of an object

True or False

Circle the correct answer.

1. **True or False:** A fillet must always be of constant radius.
2. **True or False:** A chamfer can be defined using a distance and an angle.
3. **True or False:** The **Face Draft** tool is used to create slanted surfaces.
4. **True or False:** Every Inventor drawing includes three default planes and three default axes.
5. **True or False:** The **Shell** tool can be applied to any solid shape.
6. **True or False:** The **Fillet** tool can be applied only to external edges.
7. **True or False:** A sketch plane can be created only on an existing surface.
8. **True or False:** Work planes can be drawn at an angle to an existing object.
9. **True or False:** A work plane can be created using a work point and a face parallel.
10. **True or False:** An object cannot be assigned a material specification of Phenolic.

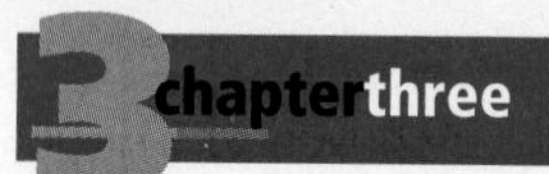

Chapter Project

Project 3-1

Redraw the following objects as solid models based on the given dimensions. Make all models from mild steel.

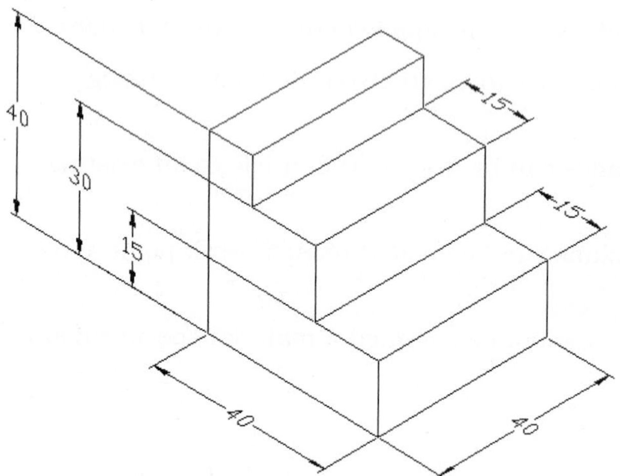

Figure P3-1
MILLIMETERS

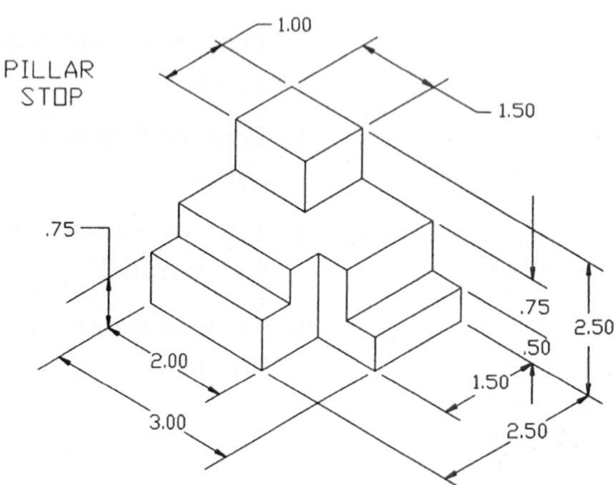

Figure P3-2
INCHES

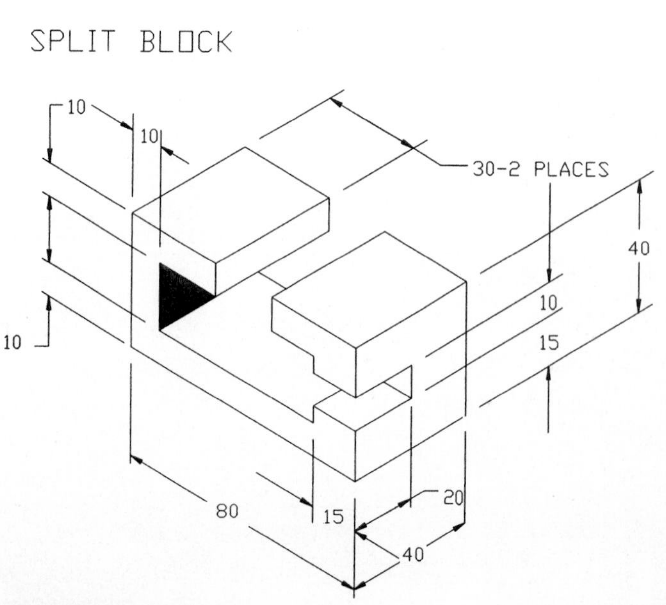

Figure P3-3
MILLIMETERS

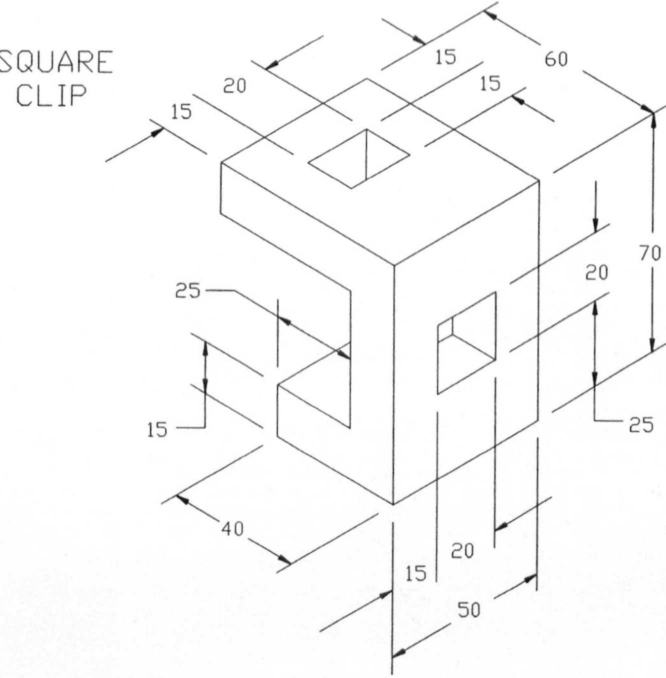

Figure P3-4
MILLIMETERS

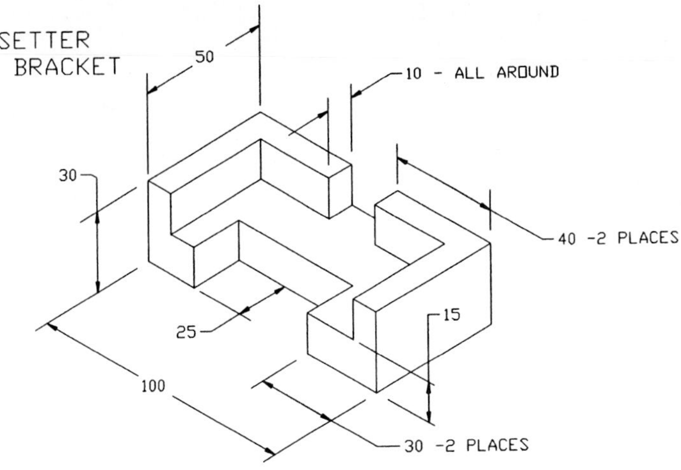

Figure P3-5
MILLIMETERS

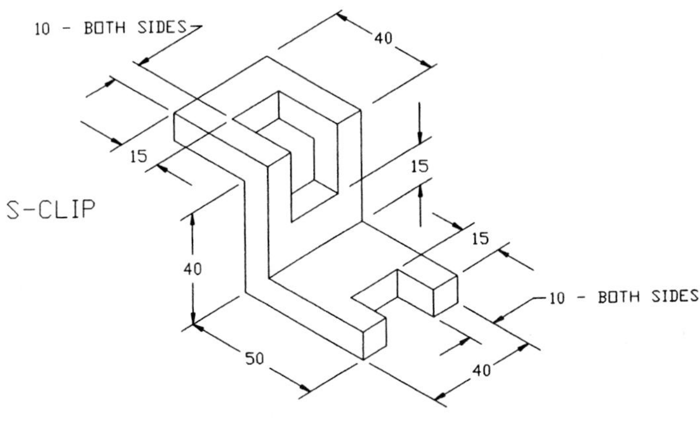

Figure P3-6
MILLIMETERS

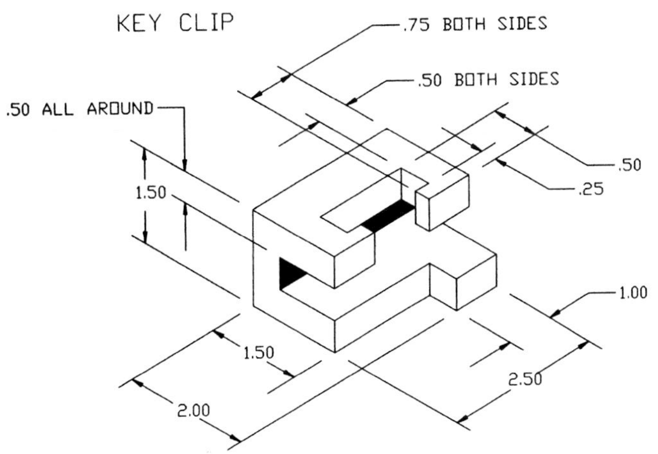

Figure P3-7
INCHES

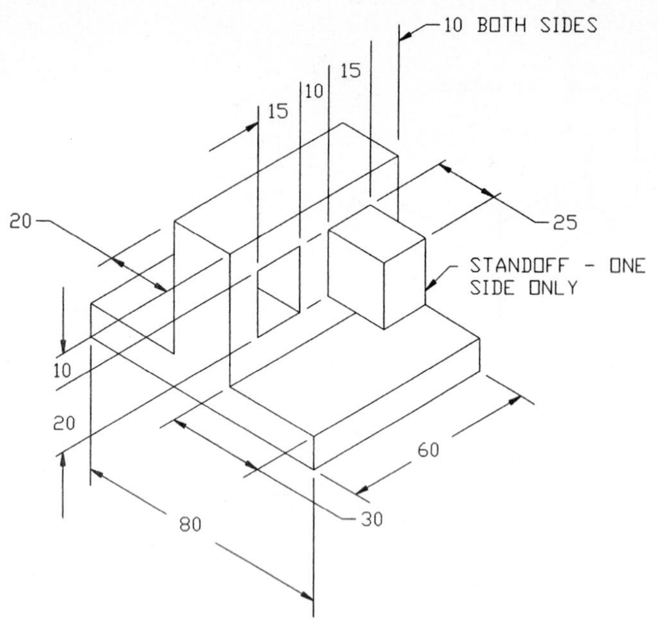

Figure P3-8
MILLIMETERS

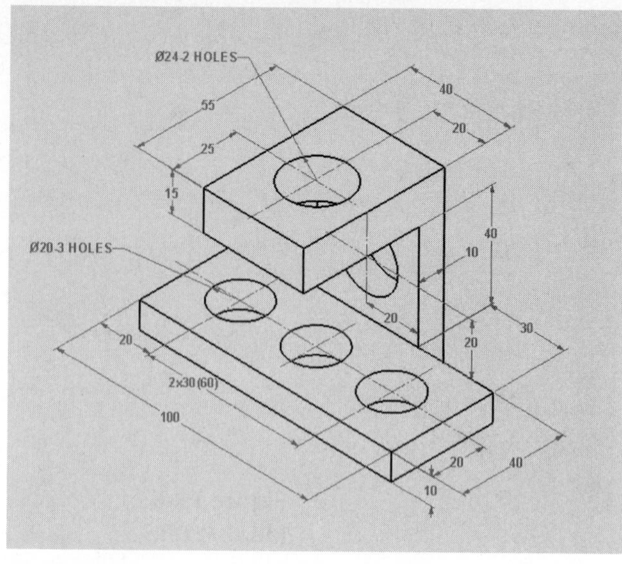

Figure P3-9
MILLIMETERS

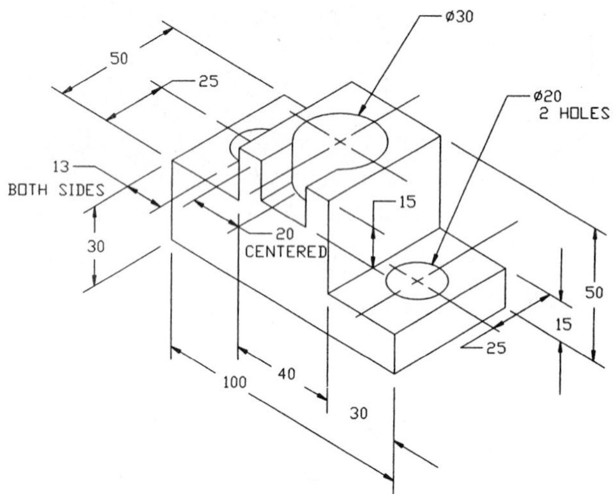

Figure P3-10
MILLIMETERS

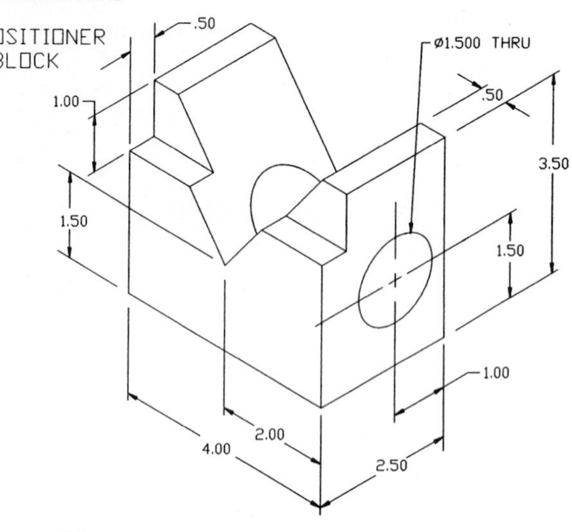

Figure P3-11
INCHES

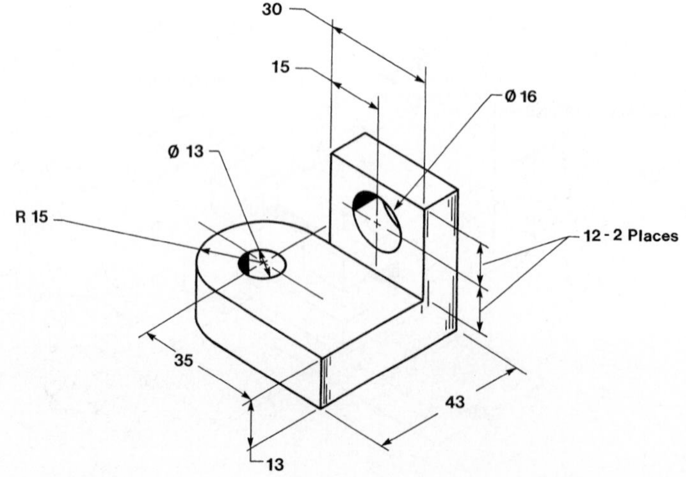

Figure P3-12
MILLIMETERS

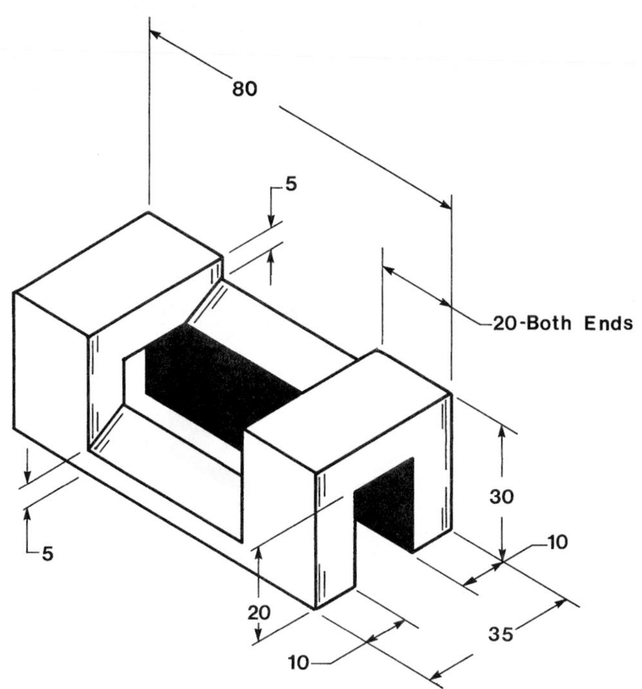

Figure P3-13
MILLIMETERS

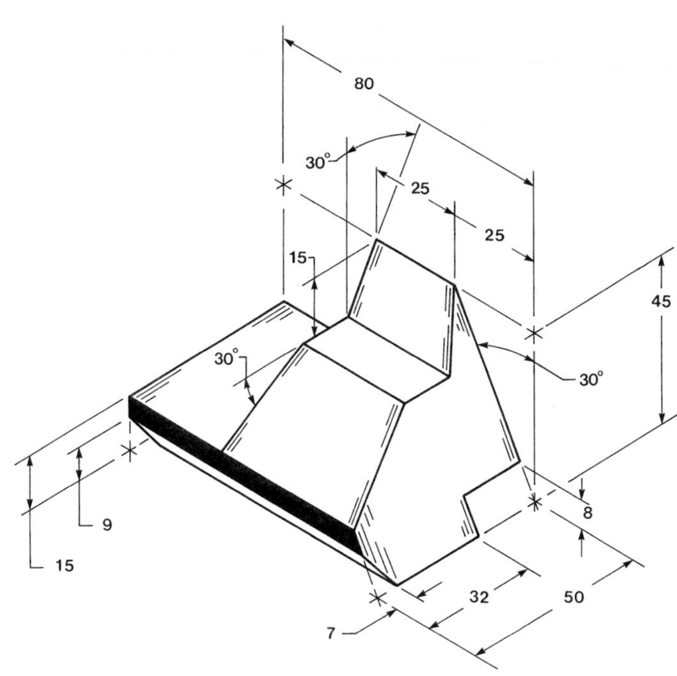

Figure P3-14
MILLIMETERS

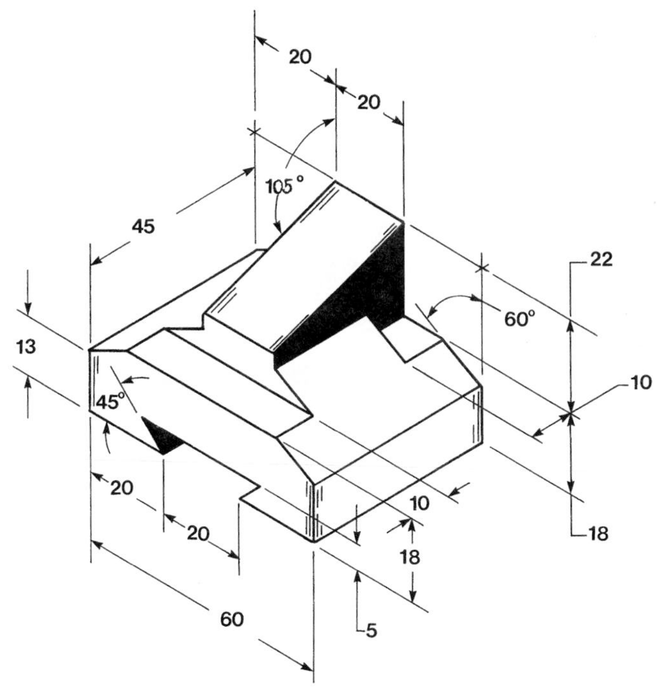

Figure P3-15
MILLIMETERS

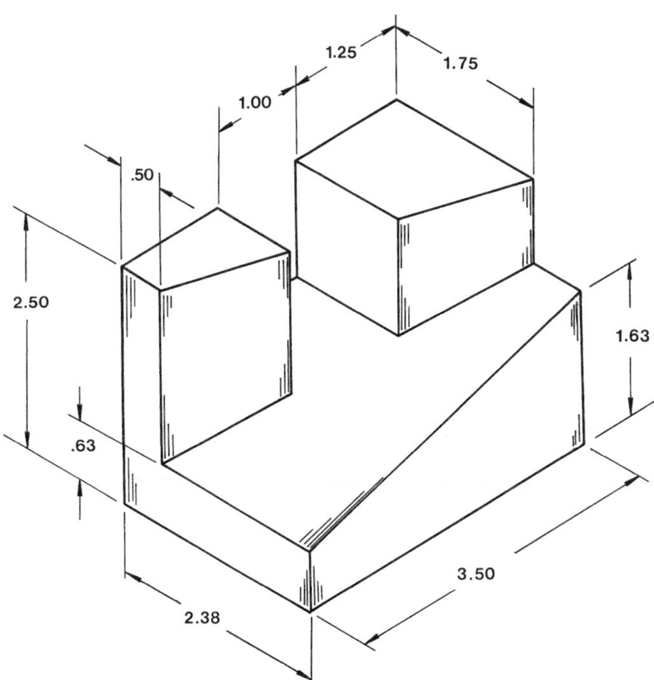

Figure P3-16
INCHES

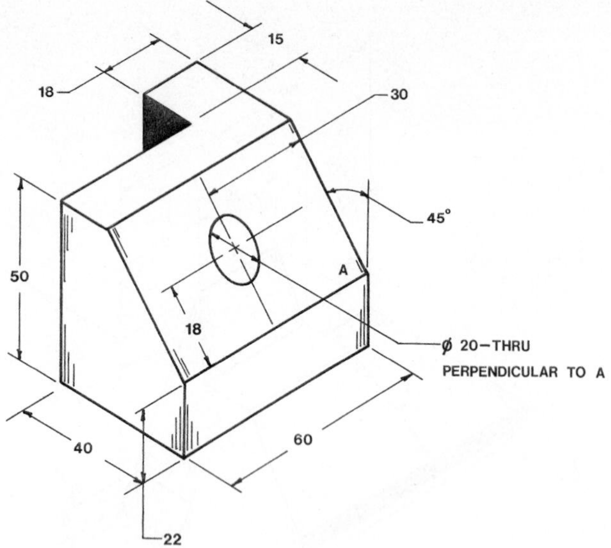

Figure P3-17
MILLIMETERS

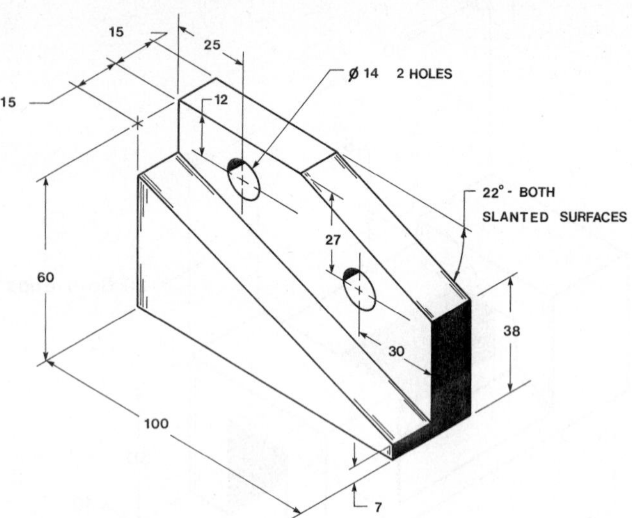

Figure P3-18
MILLIMETERS

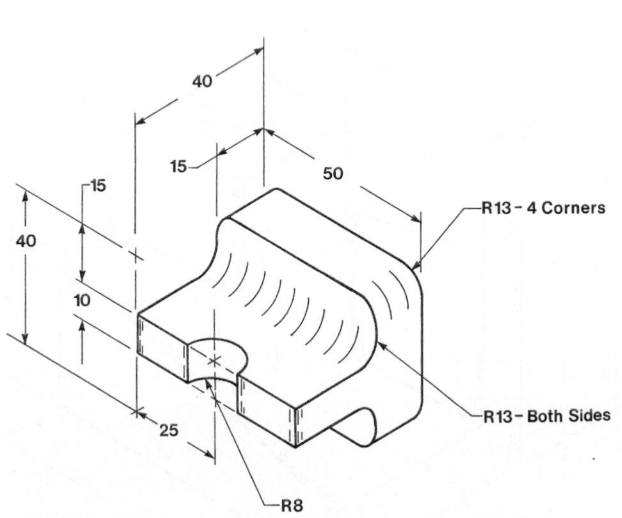

Figure P3-19
MILLIMETERS

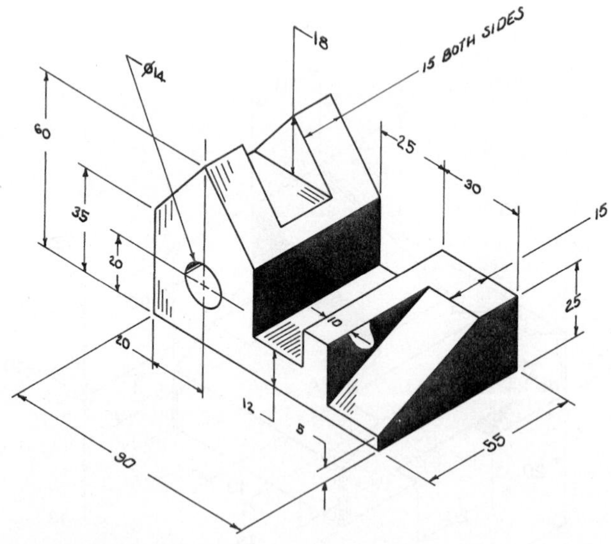

Figure P3-20
MILLIMETERS

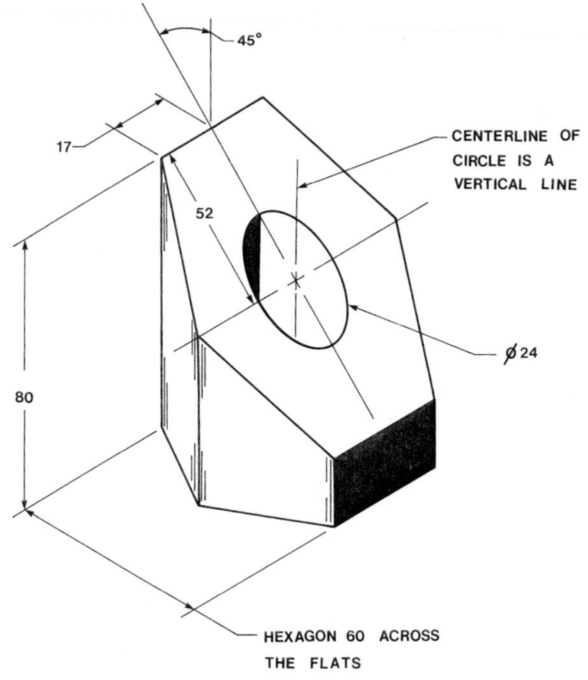

Figure P3-21
MILLIMETERS

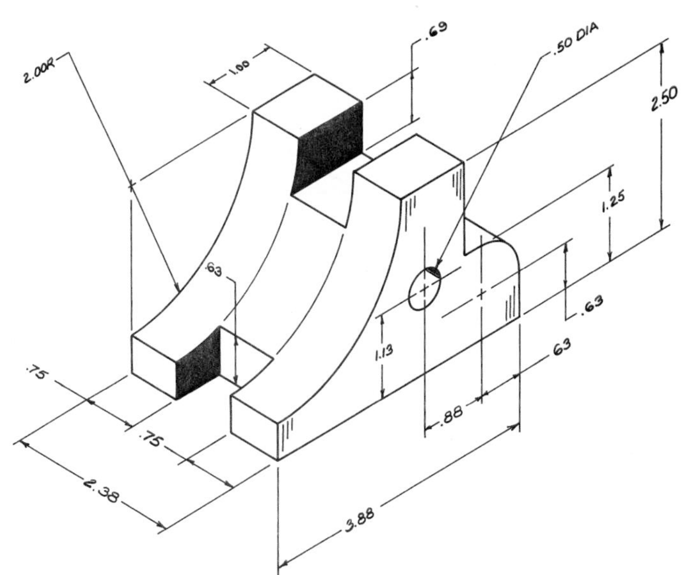

Figure P3-22
INCHES

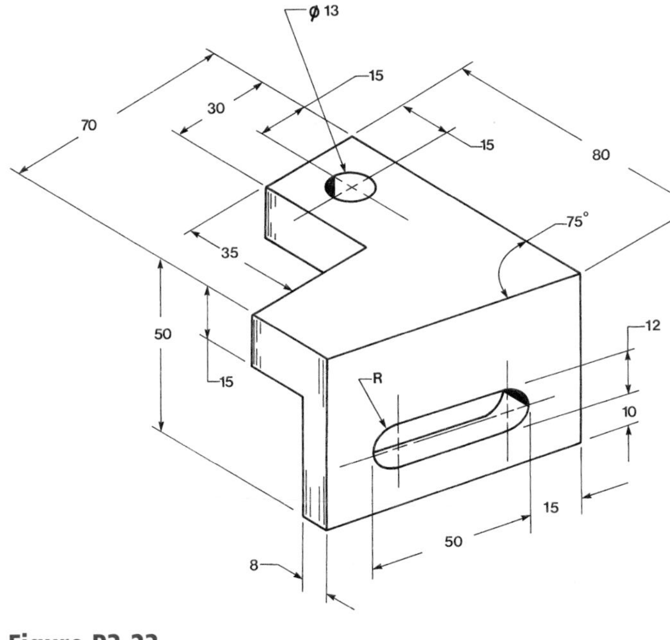

Figure P3-23
MILLIMETERS

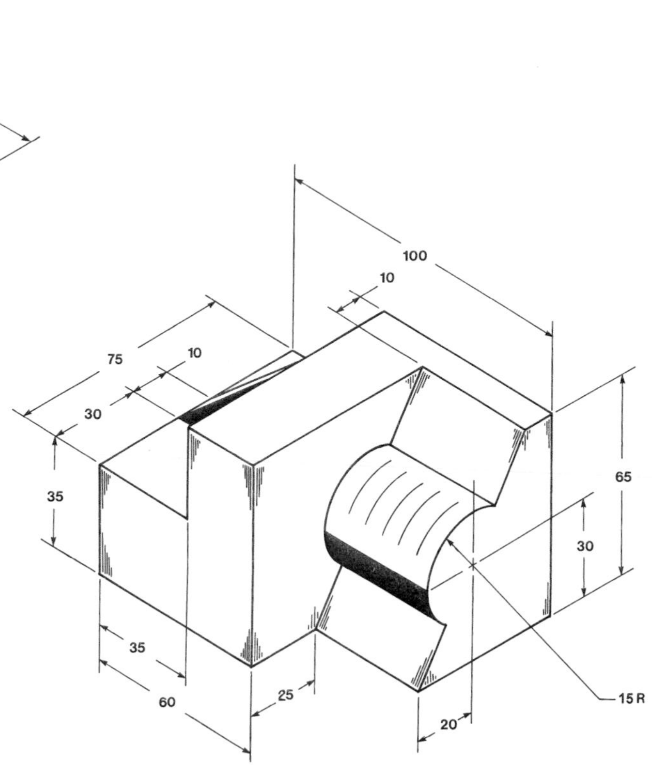

Figure P3-24
MILLIMETERS

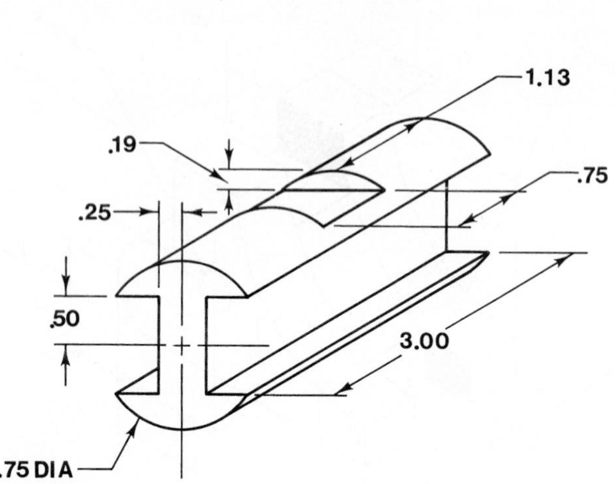

Figure P3-25
INCHES (Scale: 4=1)

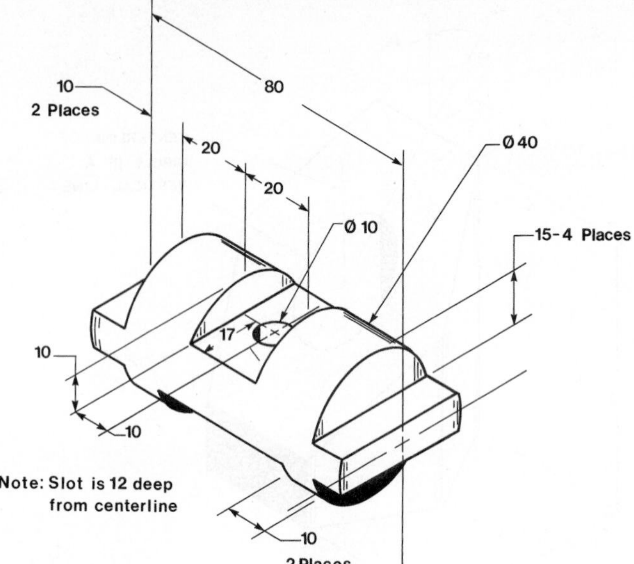

Figure P3-26
MILLIMETERS

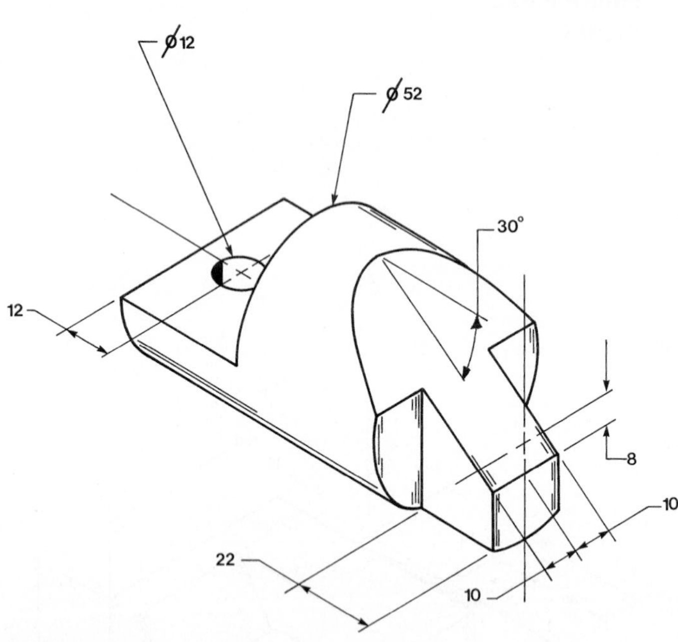

Figure P3-27
MILLIMETERS

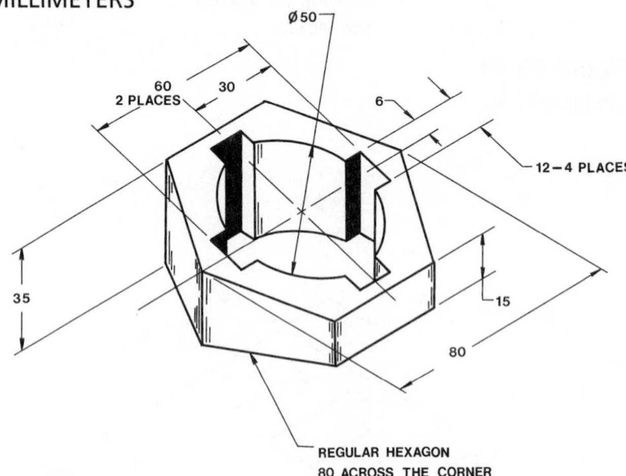

Figure P3-28
MILLIMETERS

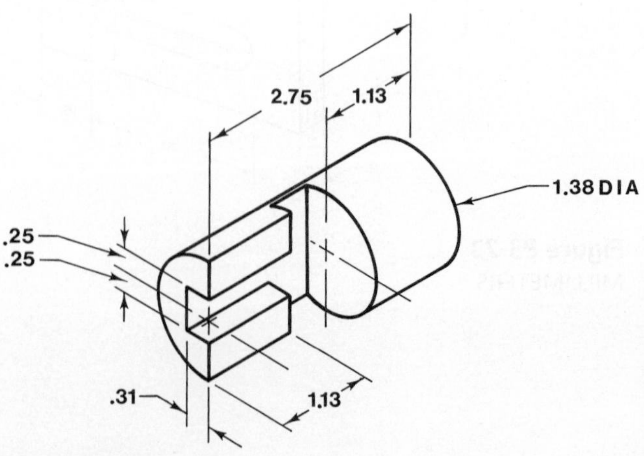

Figure P3-29
INCHES (Scale: 4=1)

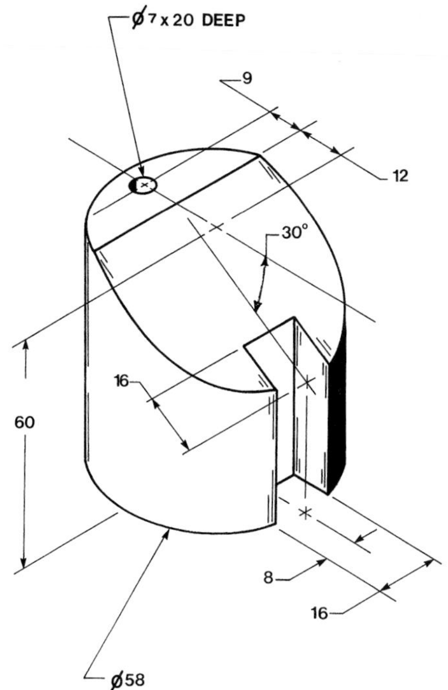

Figure P3-30
MILLIMETERS (Scale: 2=1)

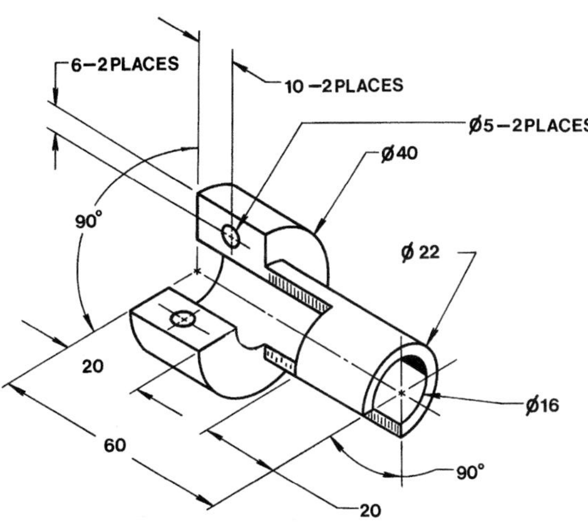

Figure P3-32
MILLIMETERS

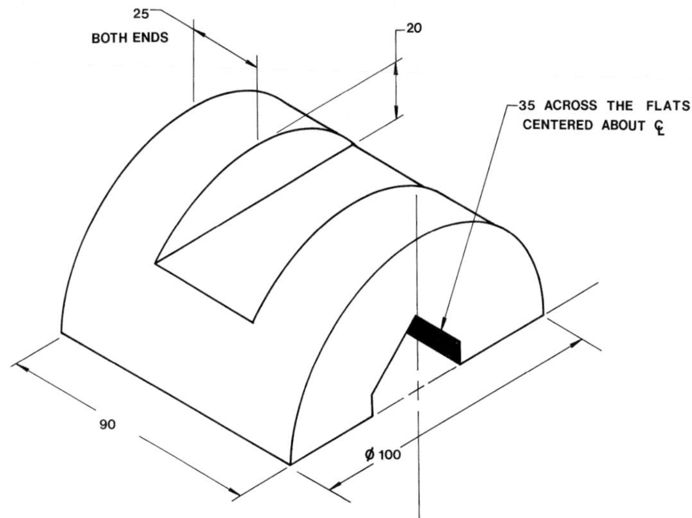

Figure P3-31
MILLIMETERS

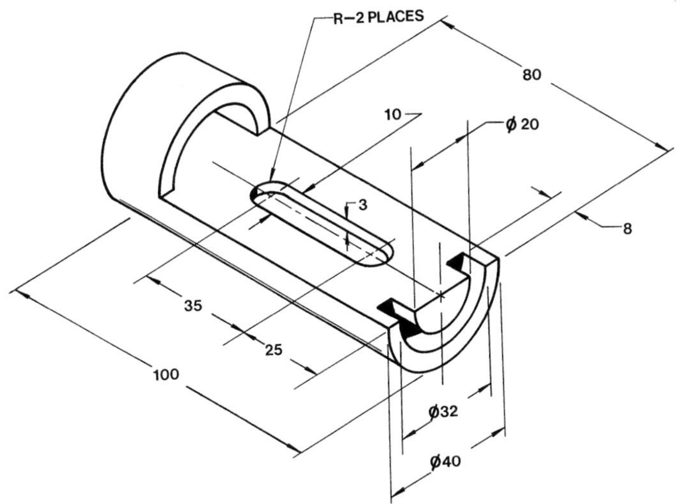

Figure P3-33
MILLIMETERS

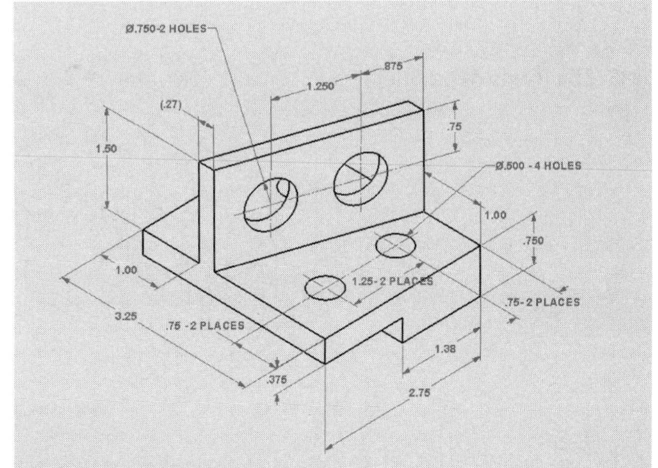

Figure P3-34
MILLIMETERS

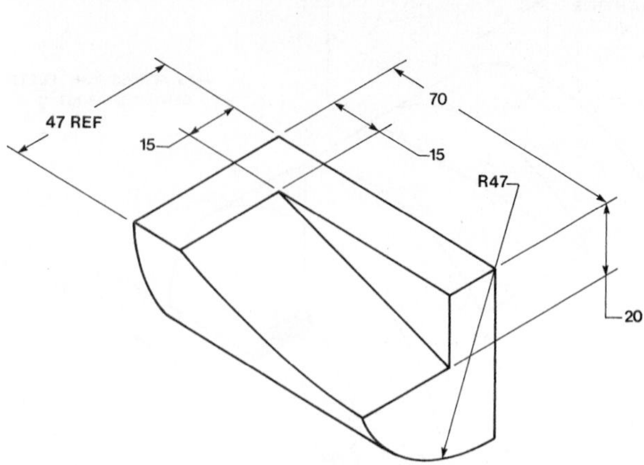

Figure P3-35
MILLIMETERS

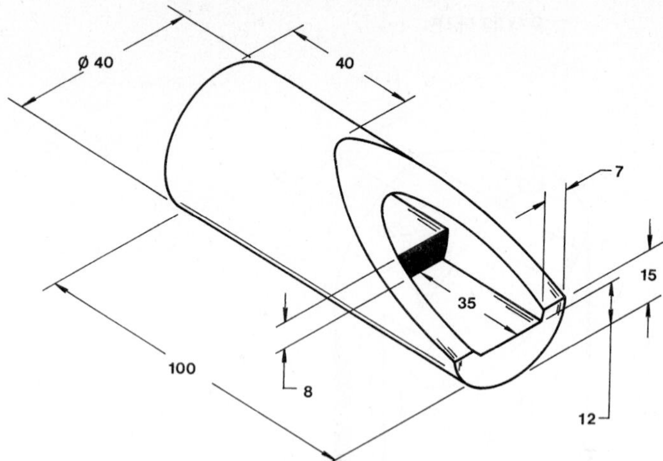

Figure P3-36
MILLIMETERS

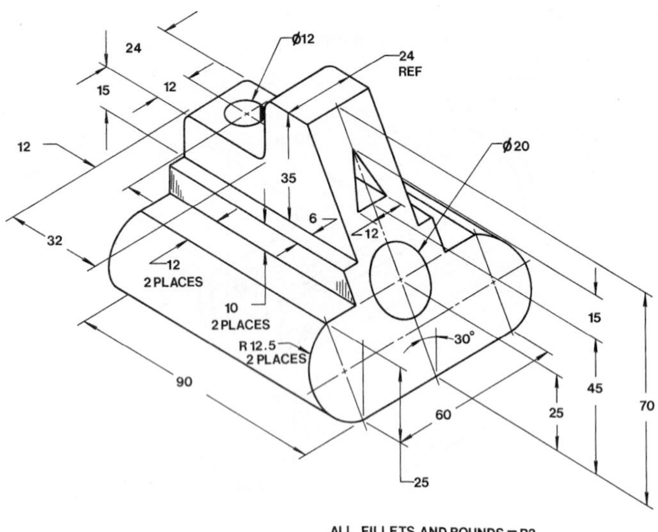

Figure P3-37
MILLIMETERS

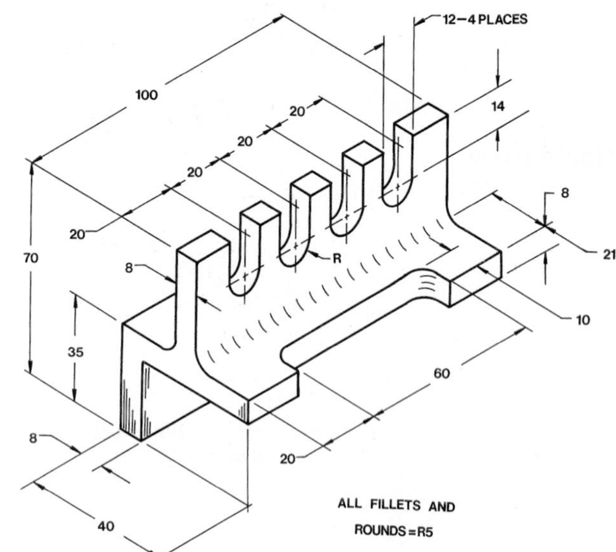

Figure P3-38
MILLIMETERS

Figure P3-39
MILLIMETERS (Consider a Shell)

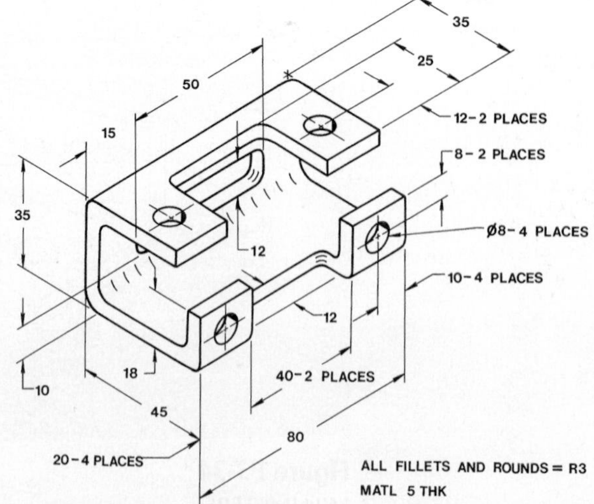

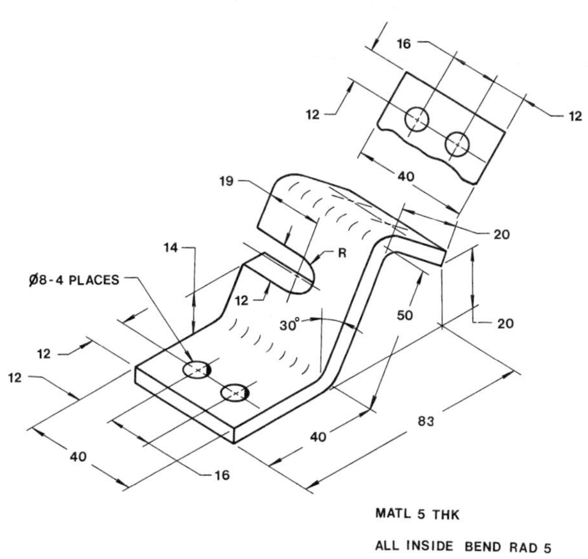

Figure P3-40
MILLIMETERS

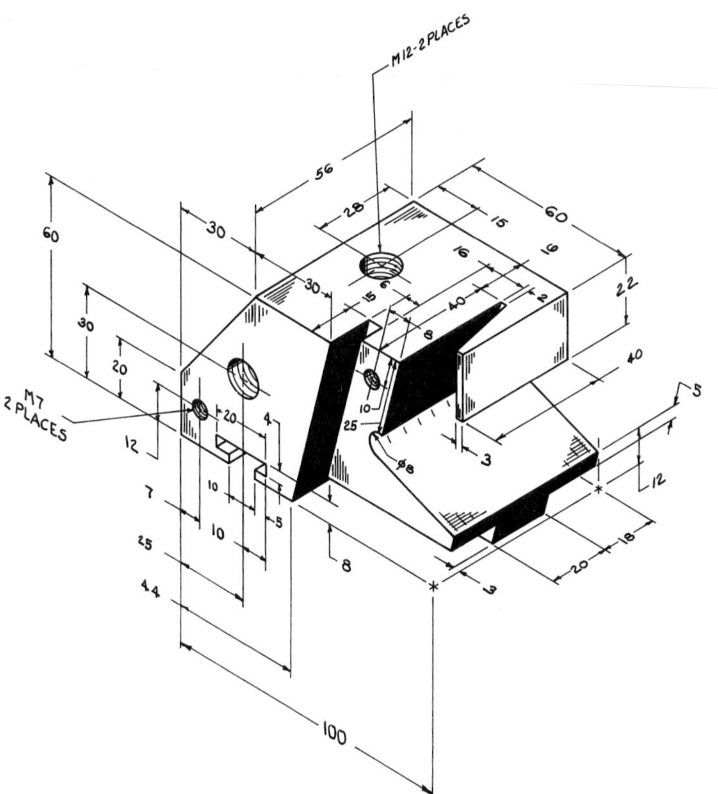

Figure P3-41
MILLIMETERS

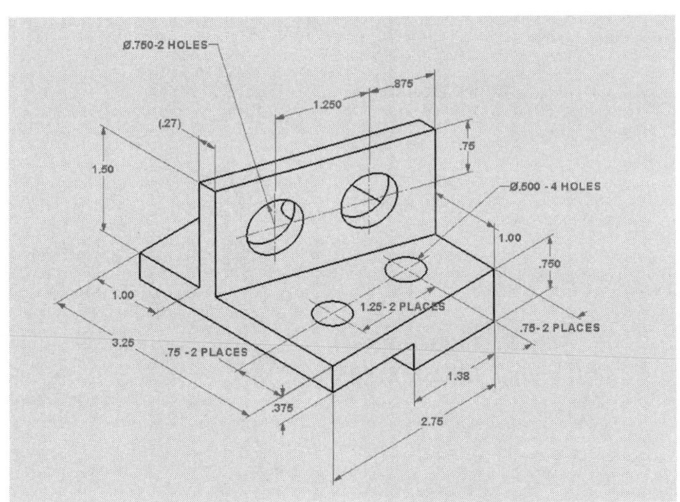

Figure P3-42
INCHES

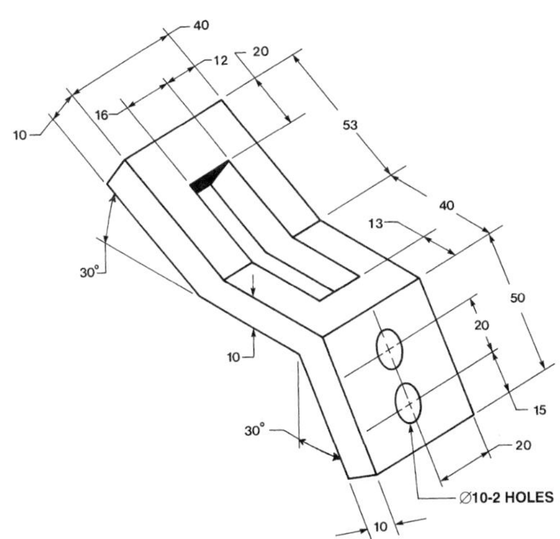

Figure P3-43
MILLIMETERS

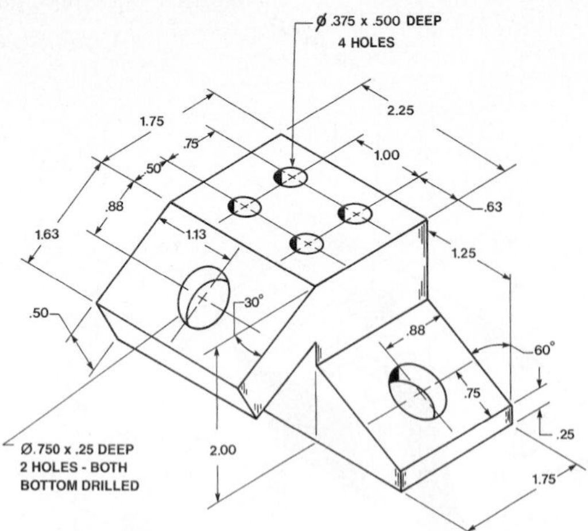

Figure P3-44
INCHES

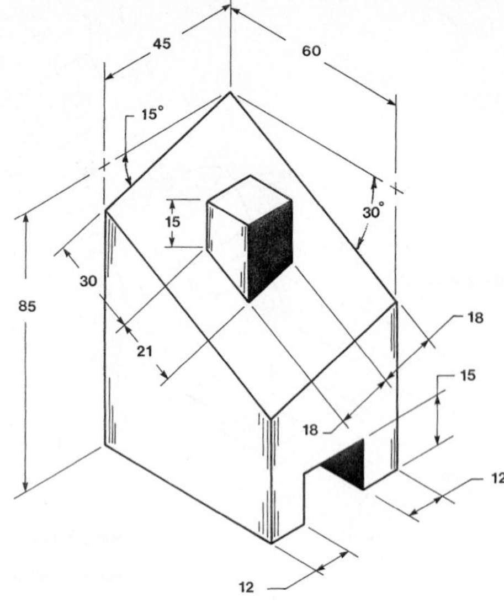

Figure P3-45
MILLIMETERS

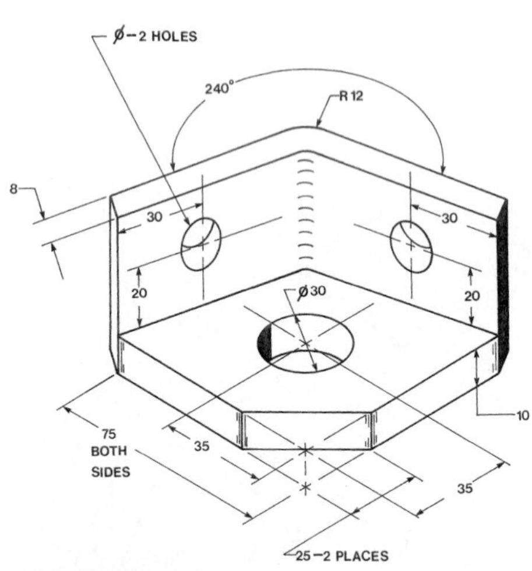

Figure P3-46
MILLIMETERS

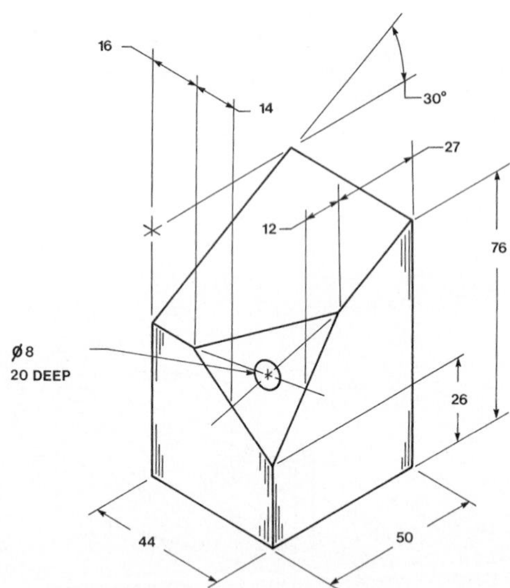

Figure P3-47
MILLIMETERS

Figure P3-48

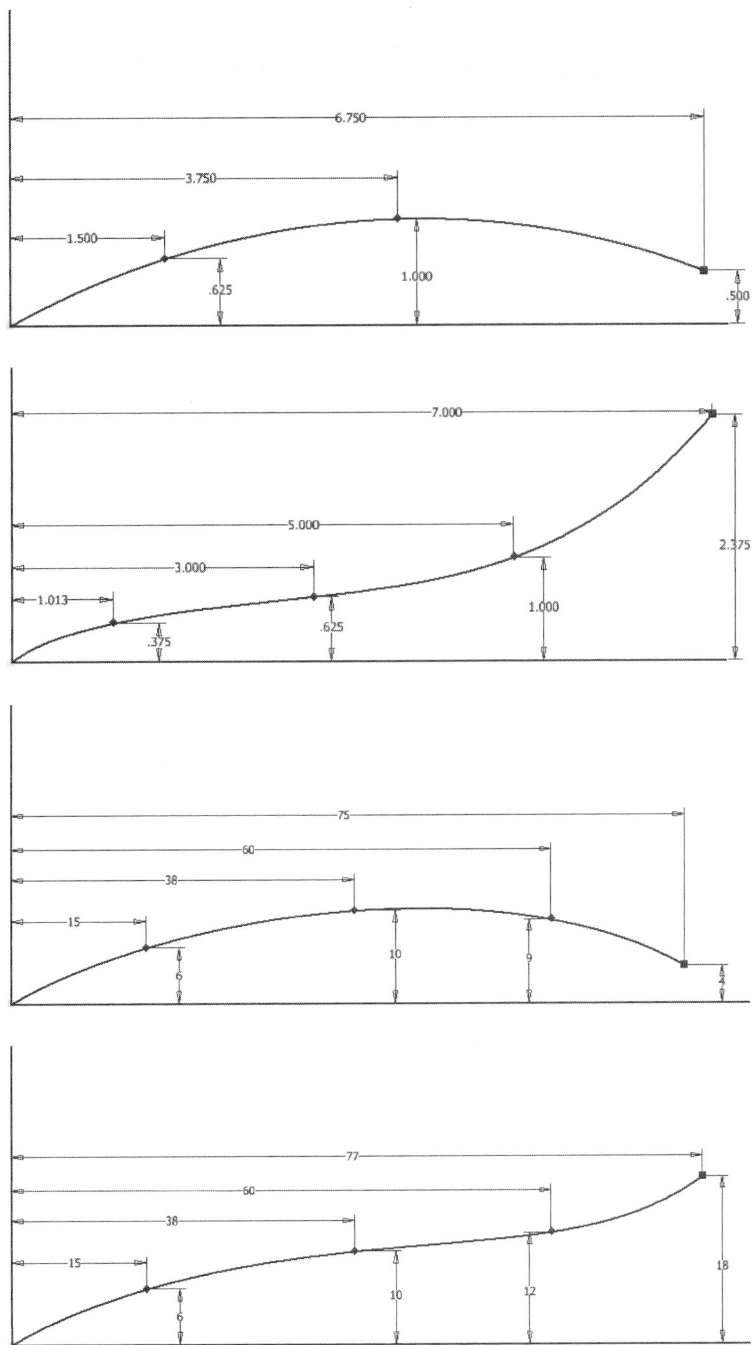

Sweep a circle along a spline.

Circles

1. Ø1.00 inch
2. Ø0.75 inch
3. Ø1.25 inches
4. Ø12.5 millimeters
5. Ø8 millimeters
6. Ø16 millimeters

Splines

A. Spline 1
B. Spline 2
C. Spline 3
D. Spline 4

Create a lofted surface between two of the following surface shapes at one of the specified offset distances.

Circles

1. Ø3.00 inches
2. Ø1.50 inches
3. Ø3.75 inches
4. Ø4.125 inches
5. Ø14.7 millimeters
6. Ø11.0 millimeters
7. Ø21.0 millimeters
8. Ø32 millimeters

Squares

1. 1.50 × 1.50 inches
2. 3.00 × 3.00 inches
3. 5.25 × 5.25 inches
4. 2.875 × 2.875 inches
5. 10 × 10 millimeters
6. 16 × 16 millimeters
7. 23.50 × 23.50 millimeters
8. 33.7 × 33.7 millimeters

Hexagons—Distance across the Flats

1. 2.00 inches
2. 2.875 inches
3. 3.25 inches
4. 1.625 inches
5. 20 millimeters
6. 12 millimeters
7. 23.4 millimeters
8. 28.56 millimeters

Offset Distances

1. 2.00 inches
2. 3.375 inches
3. 4.25 inches
4. 5.53 inches
5. 22 millimeters
6. 18 millimeters
7. 28.75 millimeters
8. 36.75 millimeters

4 chapter four
Orthographic Views

CHAPTER OBJECTIVES

- Learn how to draw orthographic views
- Learn ANSI standards and conventions
- Learn about third-angle projection
- Learn how to draw section and auxiliary views

Introduction

orthographic views: Two-dimensional views used to define a three-dimensional model. (Usually more than one view is needed to define a 3D model.)

Orthographic views may be created directly from 3D Inventor models. ***Orthographic views*** are two-dimensional views used to define a three-dimensional model. Unless the model is of uniform thickness, more than one orthographic view is necessary to define the model's shape. Standard practice calls for three orthographic views: a front, a top, and a right-side view, although more or fewer views may be used as needed.

Modern machines can work directly from the information generated when a solid 3D model is created, so the need for orthographic views—blueprints—is not as critical as it once was; however, there are still many drawings in existence that are used for production and reference. The ability to create and read orthographic views remains an important engineering skill.

This chapter presents orthographic views using third-angle projection in accordance with ANSI standards. ISO first-angle projections are also presented.

Fundamentals of Orthographic Views

Figure 4-1 shows an object with its front, top, and right-side orthographic views projected from the object. The views are two-dimensional, so they show no depth. Note that in the projected right plane there are three rectangles. There is no way to determine which of the three is closest and which is farthest away if only the right-side view is considered. All views must be studied to analyze the shape of the object.

Figure 4-2 shows three orthographic views of a book. After the views are projected they are positioned as shown. The positioning of views relative to one another is critical. The views must be aligned and positioned as shown.

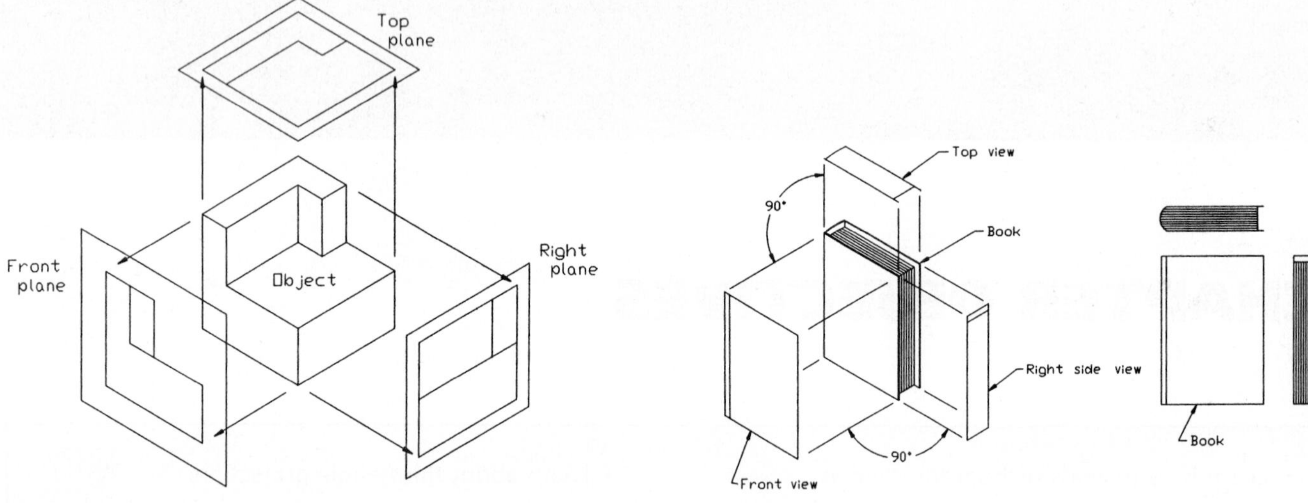

Figure 4-1

Figure 4-2

Normal Surfaces

normal surfaces: Surfaces that are 90° to each other.

Normal surfaces are surfaces that are at 90° to each other. Figures 4-3, 4-4, and 4-5 show objects that include only normal surfaces and their orthographic views.

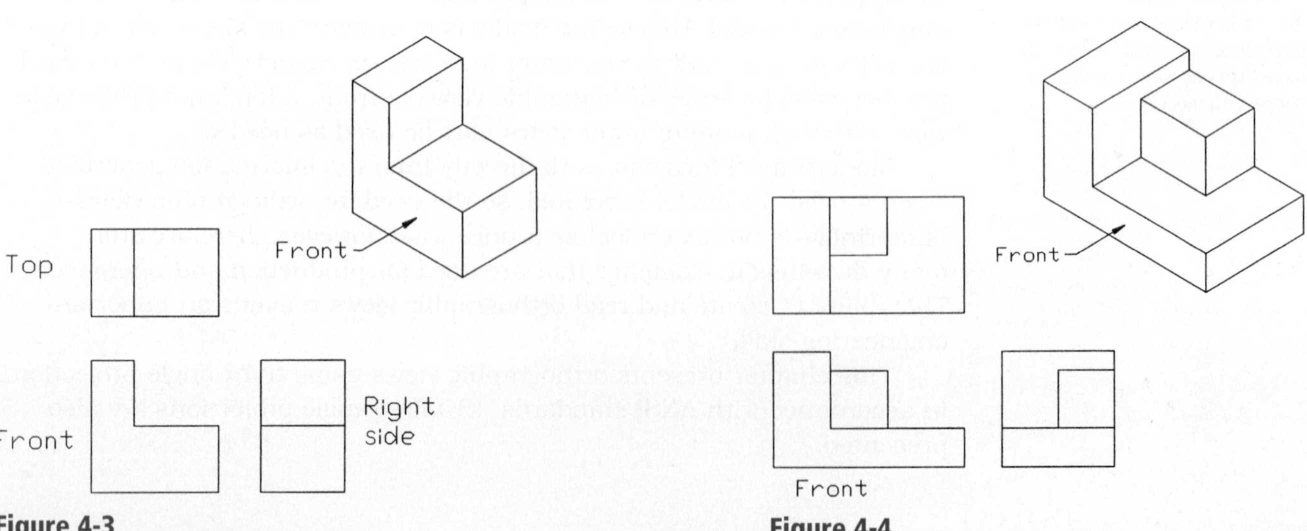

Figure 4-3

Figure 4-4

Figure 4-5

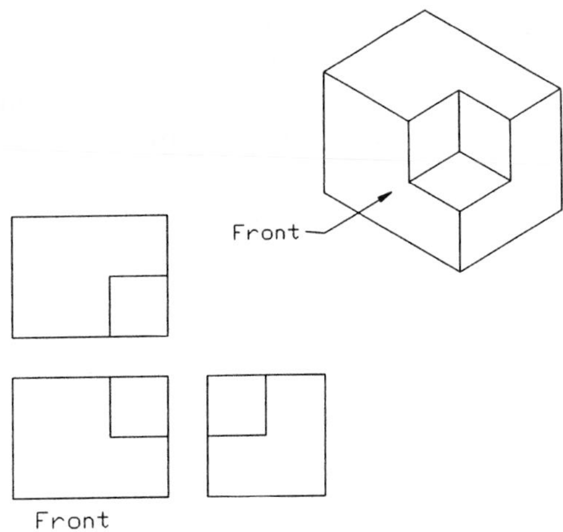

Hidden Lines

Hidden lines are used to show surfaces that are not directly visible. All surfaces must be shown in all views. If an edge or surface is blocked from view by another feature, it is drawn using a hidden line. Figures 4-6 and 4-7 show objects that require hidden lines in their orthographic views.

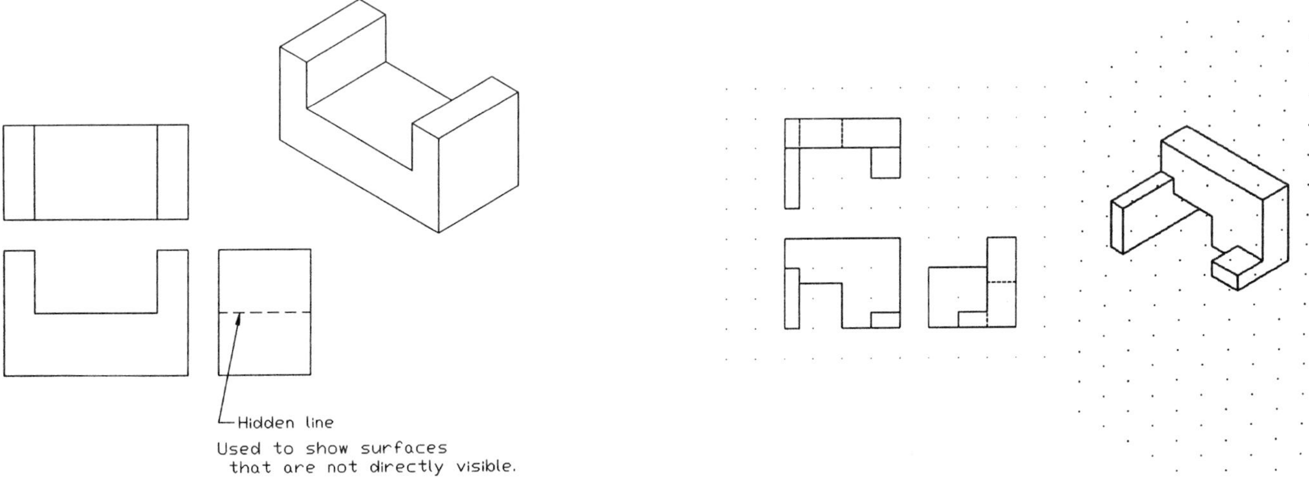

Figure 4-6

Figure 4-7

Figure 4-8 shows an object that contains an edge line, A-B. In the top view, line A-B is partially hidden and partially visible. The hidden portion

Figure 4-8

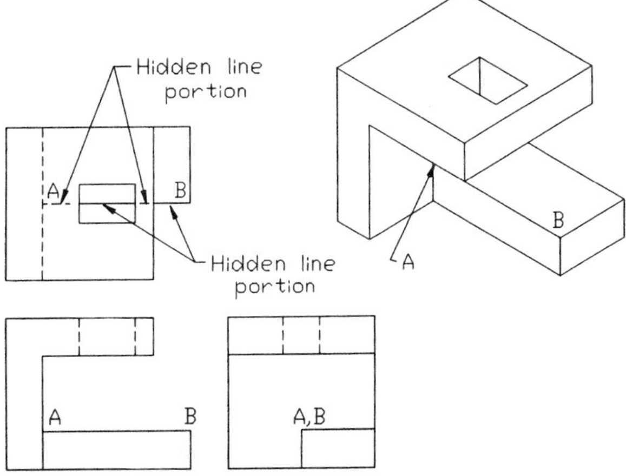

of the line is drawn using a hidden-line pattern, and the visible portion of the line is drawn using a solid line.

Figures 4-9 and 4-10 show objects that require hidden lines in their orthographic views.

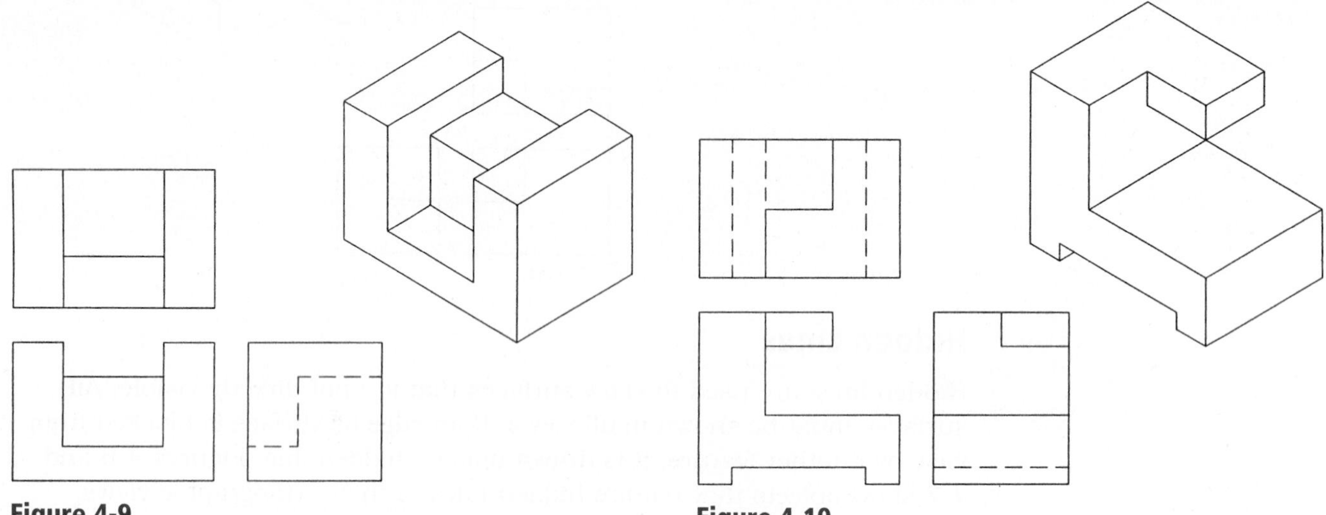

Figure 4-9

Figure 4-10

Precedence of Lines

It is not unusual for one type of line to be drawn over another type of line. Figure 4-11 shows two examples of overlap by different types of lines. Lines are shown on the views in a prescribed order of precedence. A solid

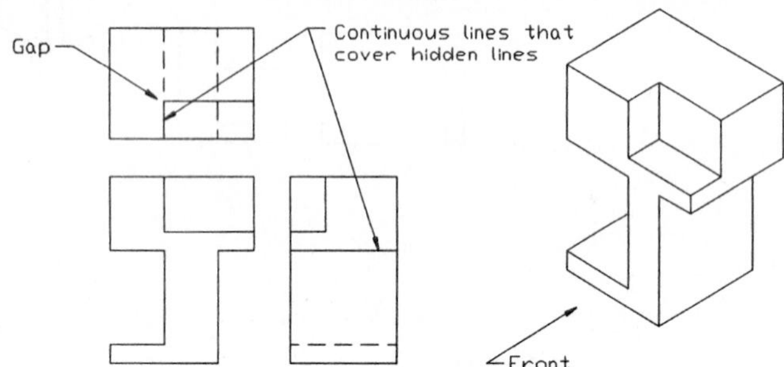

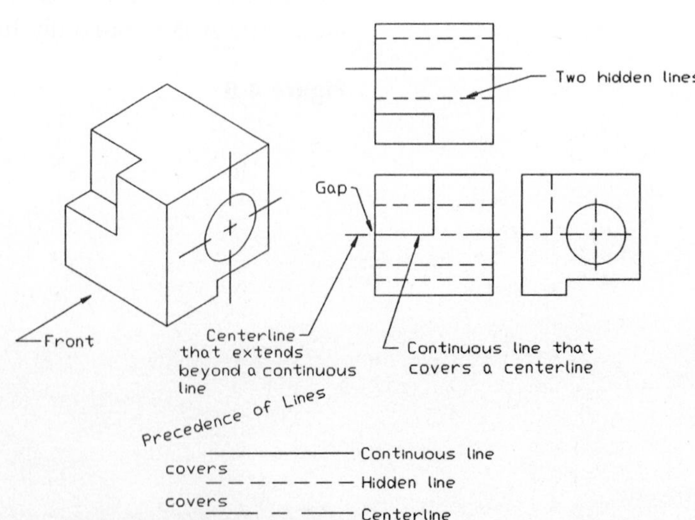

Figure 4-11

line (object or continuous) takes precedence over a hidden line, and a hidden line takes precedence over a centerline.

Slanted Surfaces

slanted surfaces: Surfaces that are at an angle to each other.

Slanted surfaces are surfaces drawn at an angle to each other. Figure 4-12 shows an object that contains two slanted surfaces. Surface ABCD appears as a rectangle in both the top and front views. Neither rectangle represents the true shape of the surface. Each is smaller than the actual surface. Also, none of the views show enough of the object to enable the viewer to accurately define the shape of the object. The views must be used together for a correct understanding of the object's shape.

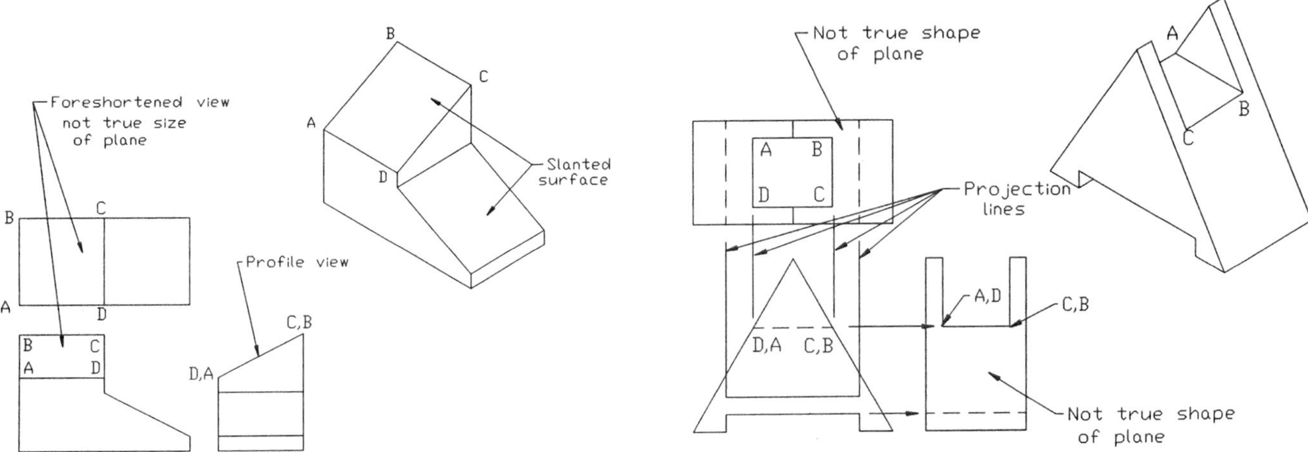

Figure 4-12

Figure 4-13

Figures 4-13 and 4-14 show objects that include slanted surfaces. Projection lines have been included to emphasize the importance of correct view location. Information is projected between the front and top views using vertical lines and between the front and side views using horizontal lines.

Figure 4-14

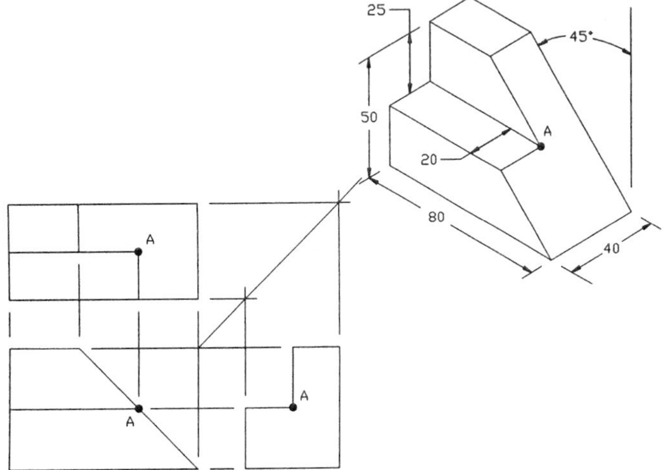

Compound Lines

compound line: A line that is neither perpendicular nor parallel to the X, Y, or Z axis.

A ***compound line*** is formed when two slanted surfaces intersect. Figure 4-15 shows an object that includes a compound line.

Figure 4-15

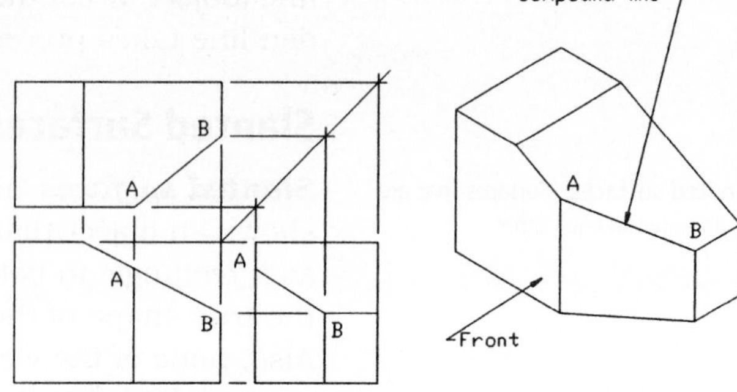

Oblique Surfaces

oblique surface: A surface that is slanted in two different directions.

An ***oblique surface*** is a surface that is slanted in two different directions. Figures 4-16 and 4-17 show objects that include oblique surfaces.

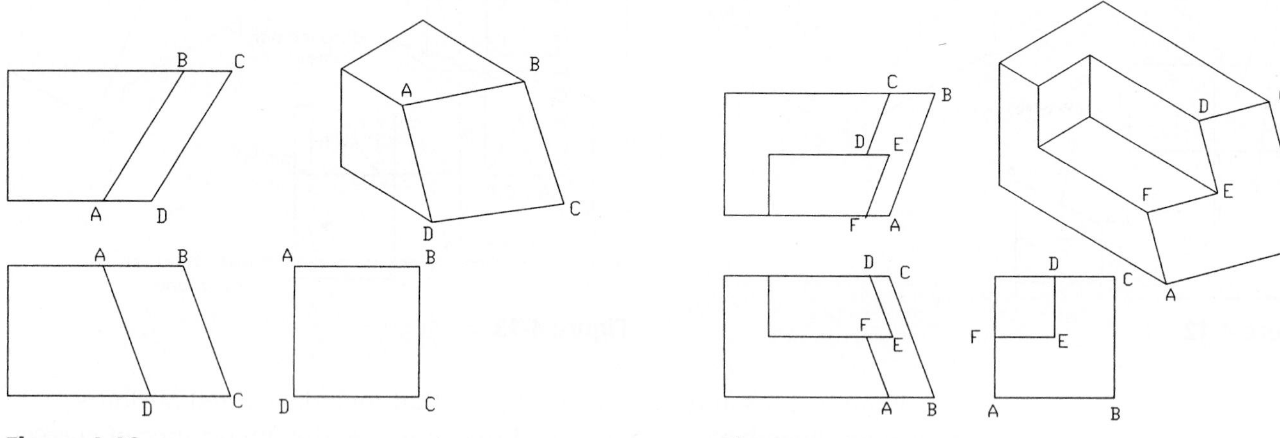

Figure 4-16

Figure 4-17

Rounded Surfaces

Figure 4-18 shows an object with two rounded surfaces. Note that as with slanted surfaces, an individual view is insufficient to define the shape of a surface. More than one view is needed to accurately define the surface's shape.

Figure 4-18

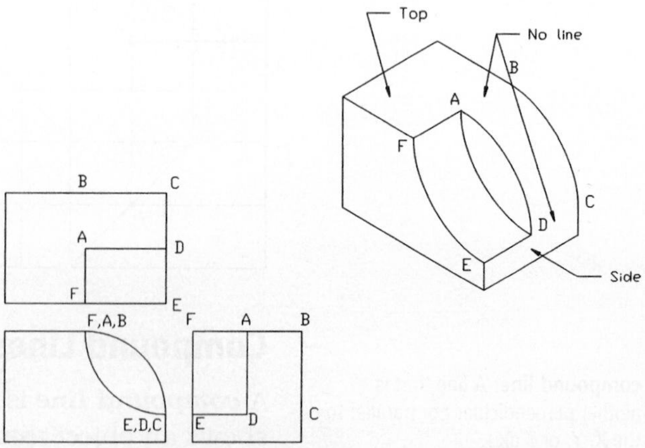

Convention calls for a smooth transition between rounded and flat surfaces; that is, no lines are drawn to indicate the tangency. Inventor includes a line to indicate tangencies between surfaces in the isometric drawings created using the multiview options but does not include them in the orthographic views. Tangency lines are also not included when models are rendered.

Figure 4-19 shows the drawing conventions for including lines for rounded surfaces. If a surface includes no vertical portions or no tangency, no line is included.

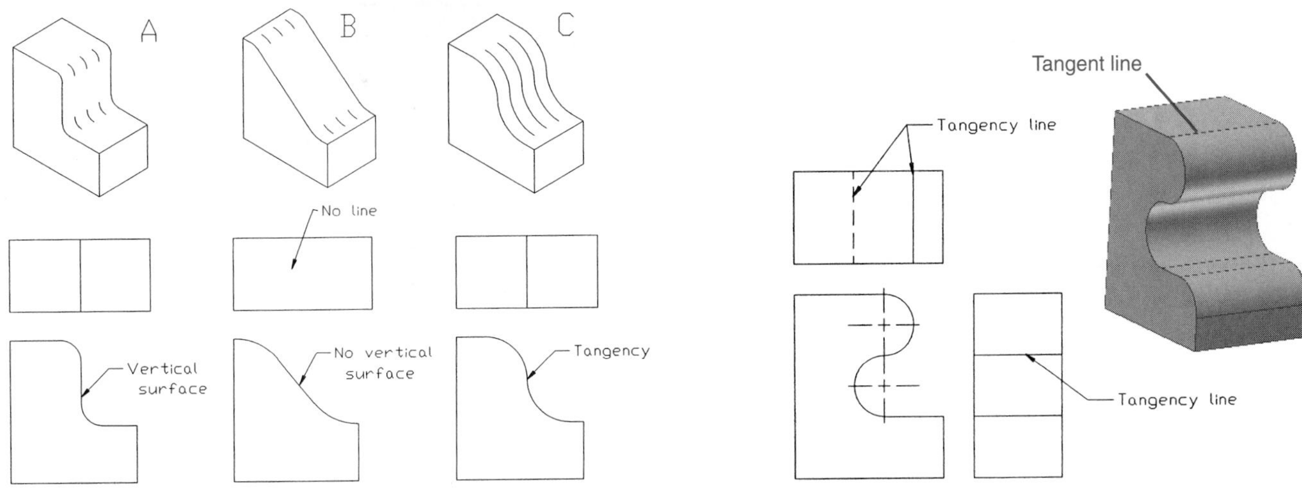

Figure 4-19

Figure 4-20

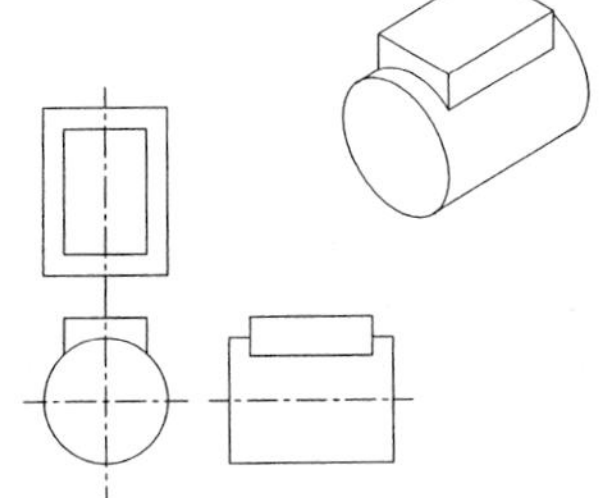

Figure 4-20 shows an object that includes two tangencies. Each is represented by a line. Note in Figure 4-20 that Inventor will add tangent lines to the 3D model. These lines will not appear in the orthographic views.

Figure 4-21 shows two objects with similar configurations; however, the boxlike portion of the lower object blends into the rounded portion exactly on its widest point, so no line is required.

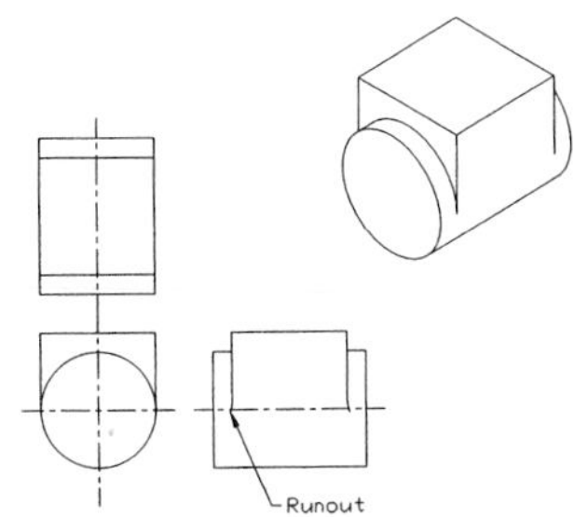

Figure 4-21

Orthographic Views with Inventor

Inventor will create orthographic views directly from models. Figure 4-22 shows a completed three-dimensional model. See Figure P4-7 for the model's dimensions. It was created using an existing file, BLOCK, 3HOLE. It will be used throughout this chapter to demonstrate orthographic presentation views.

Figure 4-22

EXERCISE 4-1 Creating an Orthographic View

1 Start a new drawing, click the **Metric** tab, and select the **ANSI (mm).idw** option.

See Figure 4-23. ANSI stands for American National Standards Institute.

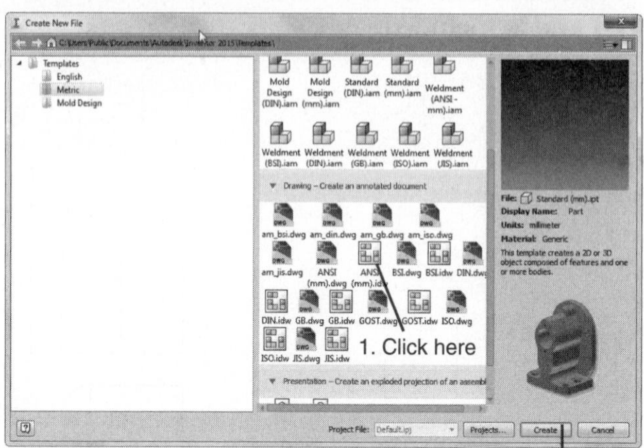

Figure 4-23

2 Click **Create**.

The drawing management screen will appear. See Figure 4-24.

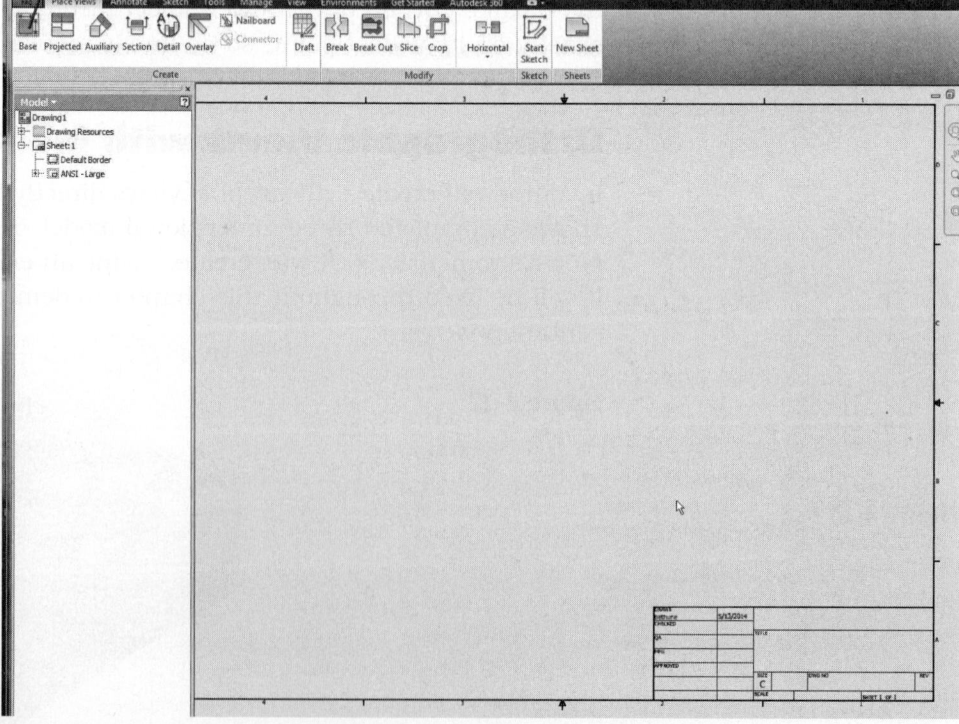

Figure 4-24

3 Click the **Base** tool located on the **Create** panel under the **Place Views** tab.

The **Drawing View** dialog box will appear. See Figure 4-25.

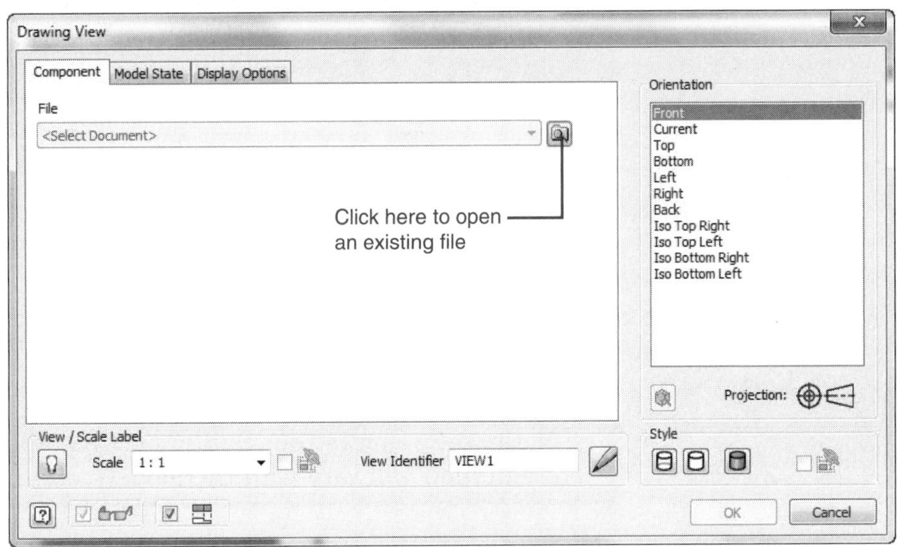

Figure 4-25

4 Click the **Open an existing file** button.

The **Open** dialog box will appear. See Figure 4-26.

Figure 4-26

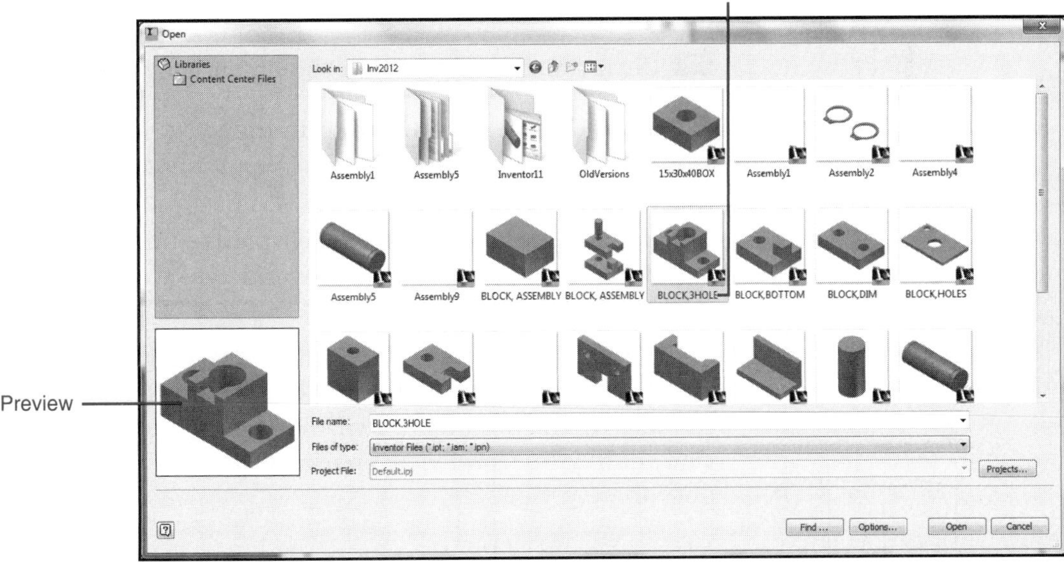

5 Select the desired model. In this example the model's file name is **Block,3HOLE**.

6 Click the **Open** box.

The **Drawing View** dialog box will appear. See Figure 4-27.

Figure 4-27

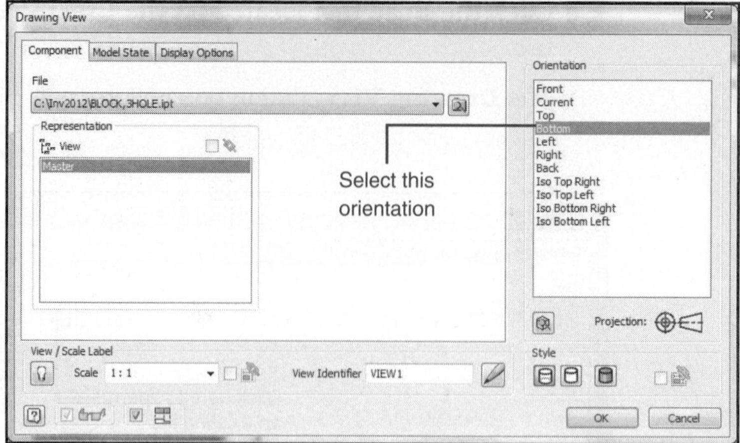

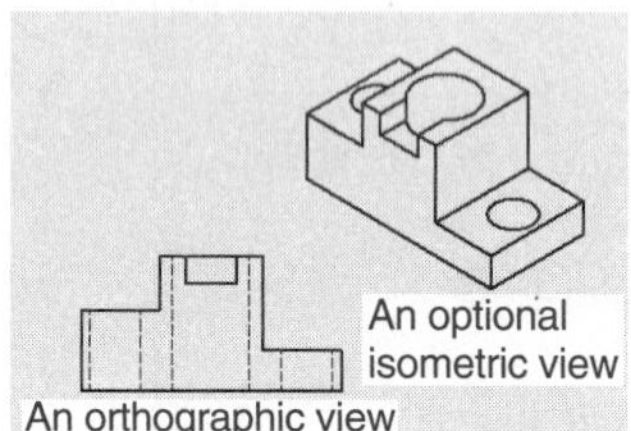

Figure 4-28

7 Select the **Bottom** option, locate the view on the drawing screen, and click the location.

Figure 4-28 shows the resulting orthographic view. The selection of orientation will vary with the model's original orientation.

The screen will include a border and a title block. The lettering in the title block may appear illegible. This is normal. The text will be legible when printed. The section on title blocks will explain how to work with title blocks.

EXERCISE 4-2 Creating Other Orthographic Views

4. Right-click the mouse and click the Create option.

3. Move the cursor upward, select a location, and click the mouse.

1. Click the Projected View tool

2. Click the view

1 Click the **Projected View** tool on the **Create** panel under the **Place Views** tab.

2 Click the view already on the drawing screen.

3 Move the cursor upward from the view.

A second view will appear.

4 Select a location, click the left mouse button to place the view, then click the right mouse button and select the **Create** option.

Top view

Front view

Figure 4-29

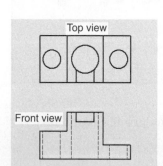

Figure 4-29 shows the resulting two orthographic views. The initial view was created using the **Bottom** option. This is a relative term based on the way the model was drawn. The initial view can be defined as the front view, and the second view created from that front view is also, by definition, the top view.

EXERCISE 4-3 Adding Centerlines

Convention calls for all holes to be defined using centerlines. The views in Figure 4-29 do not include centerlines.

1 Click the **Annotate** tab.

See Figure 4-30.

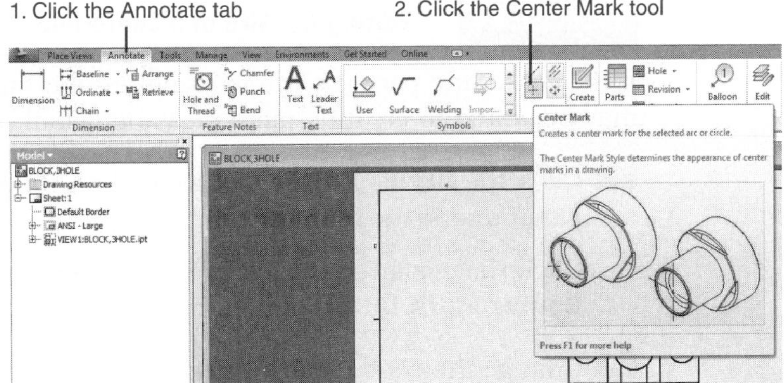

Figure 4-30

2 Click the **Center Mark** tool located on the **Symbols** panel under the **Annotate** tab.

3 Move the cursor into the drawing screen and click the edges of the holes in the top view.

See Figure 4-31.

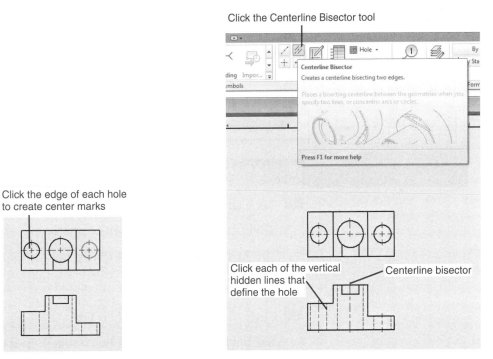

4 Click the **Centerline Bisector** tool located on the **Symbols** panel under the **Annotate** tab.

5 Click each side of the holes' projections in the front view.

Vertical centerlines will appear. See Figure 4-32.

Click the edge of each hole
to create center marks

Figure 4-31

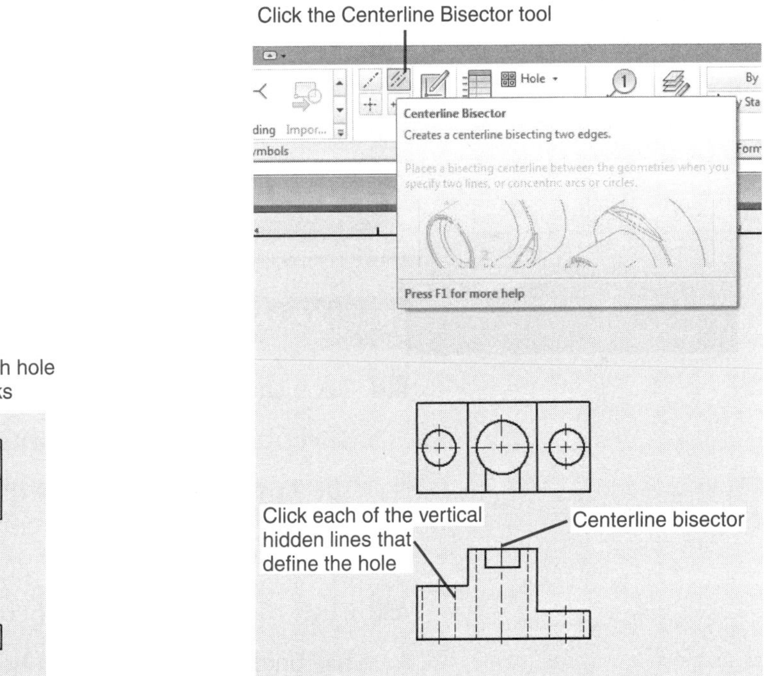

Press F1 for more help
Click each of the vertical
hidden lines that

Figure 4-32

EXERCISE 4-4 Editing the Size of a Centerline

If the centerline patterns are too small or too big for the given feature, they may be edited to create a more pleasing visual picture.

1 Click the **Styles Editor** tool located on the **Styles and Standards** panel under the **Manage** tab.

Click the + sign to the left of the **Center Mark** heading, and select the **Center Mark (ANSI)** option. See Figure 4-33.

2 Change the center mark values as needed.

Figure 4-33

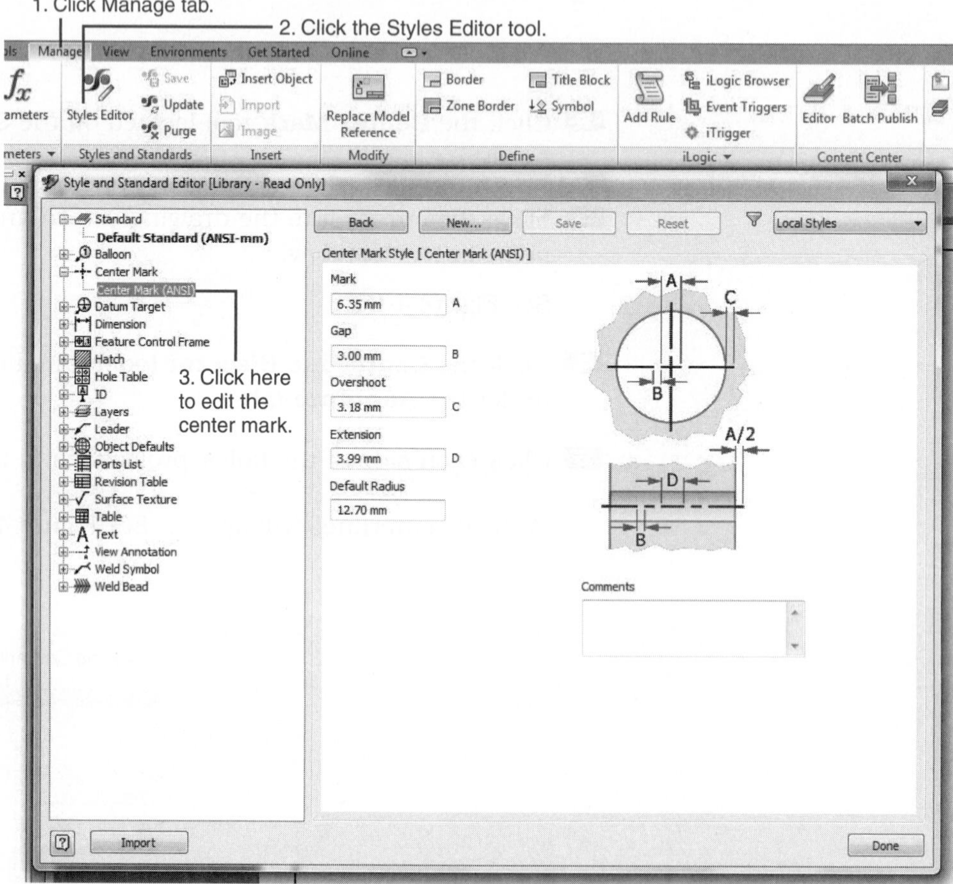

EXERCISE 4-5 Changing the Background Color of the Drawing Screen

1 Click the **Tools** tab at the top of the screen.

Select the **Active Standards** option.

The **Application Options** dialog box will appear. See Figure 4-34.

2 Click the **Colors** tab.

3 Click the desired color, then **OK**.

The background color will be changed. In this example the **Presentation Color scheme** and **1 Color Background** were selected. This format is used throughout the book for visual clarity.

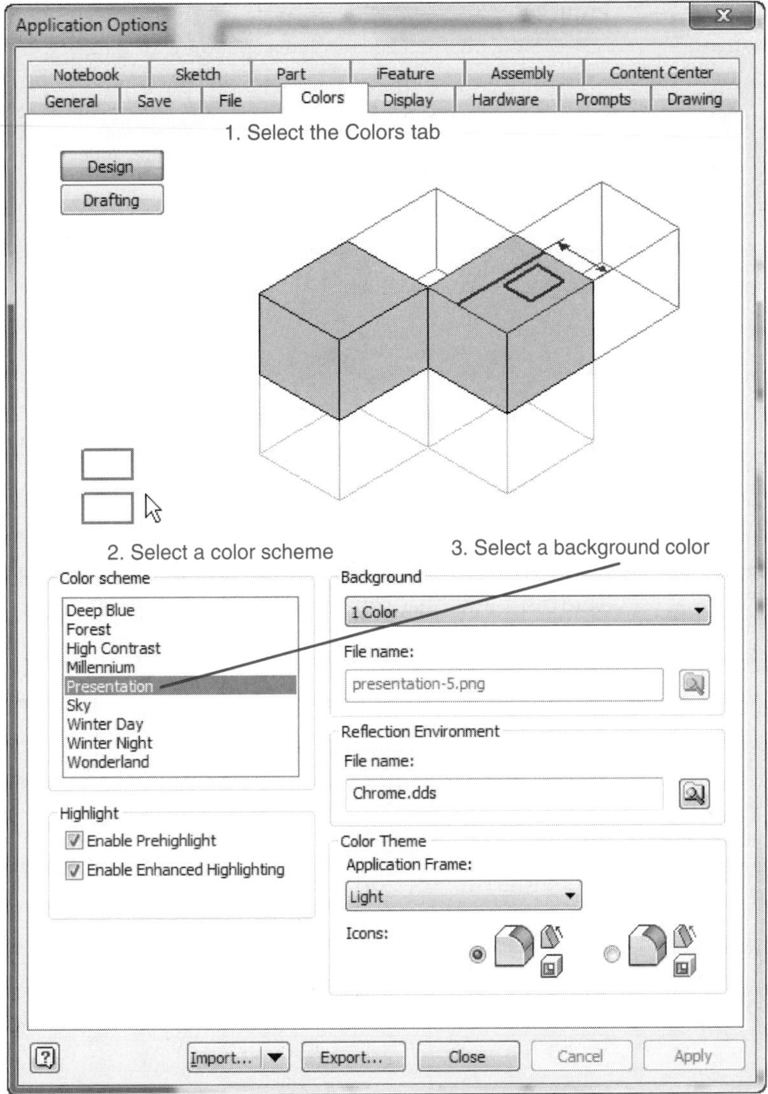

Figure 4-34

Isometric Views

An isometric view may be created from any view on the screen. The resulting orientation will vary according to the view selected. In this example the front view is selected.

1. Access the **Create** panel under the **Place Views** tab and click the **Projected View** tool.

2. Click the **Front** orthographic view.

3. Move the cursor to the right of the front view and select a location for the isometric view by clicking the mouse.

4. Move the cursor slightly and click the right mouse button.

5. Select the **Create** option.

Figure 4-35 shows the resulting isometric view. Isometric views help visualize the orthographic views.

Figure 4-35

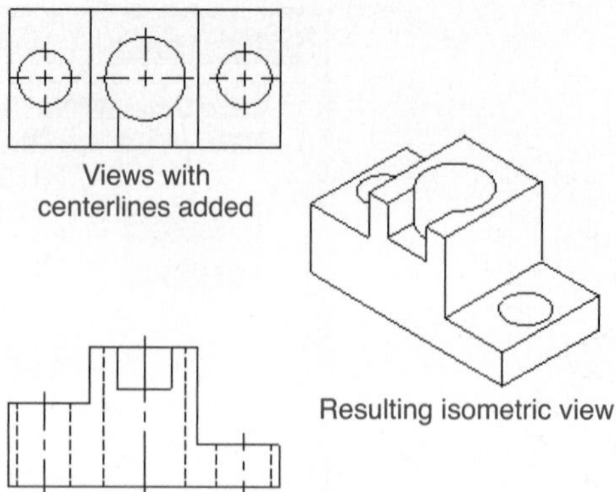

Section Views

section view: A view used to expose an internal surface of a model.

Some objects have internal surfaces that are not directly visible in normal orthographic views. ***Section views*** are used to expose these surfaces. Section views do not include hidden lines.

Any material cut when a section view is defined is hatched using section lines. There are many different styles of hatching, but the general style is evenly spaced 45° lines. This style is defined as ANSI 31 and will be applied automatically by Inventor.

cutting plane: A plane used to define the location of a section view.

Figure 4-36 shows a three-dimensional view of an object. The object is cut by a cutting plane. ***Cutting planes*** are used to define the location of the section view. Material to one side of the cutting plane is removed, exposing the section view.

Figure 4-36

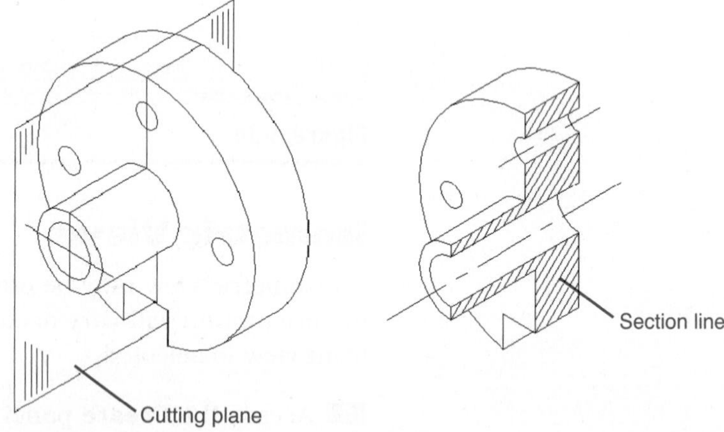

Figure 4-37 shows the same object presented using two dimensions. The cutting plane is represented by a cutting plane line. The cutting plane line is defined as A-A, and the section view is defined as view A-A.

All surfaces directly visible must be shown in a section view. In Figure 4-38 the back portion of the object is not affected by the section view and is directly visible from the cutting plane. The section view must include these surfaces. Note how the rectangular section blocks out part of the large hole. No hidden lines are used to show the hidden portion of the large hole.

Figure 4-37

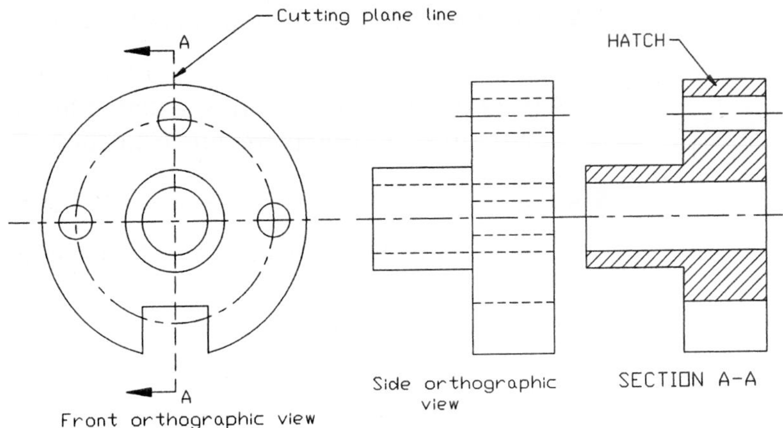

Figure 4-38

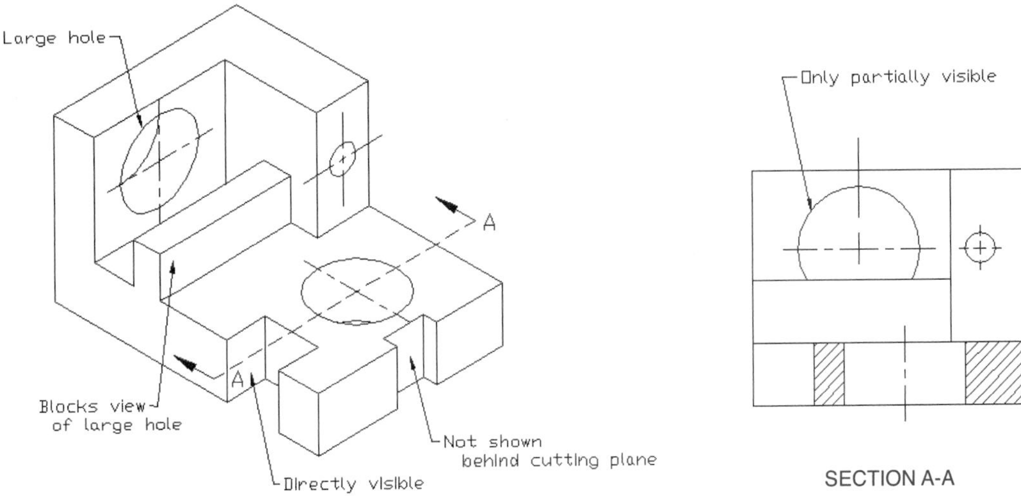

EXERCISE 4-6 Drawing a Section View Using Inventor

Figure 4-39 shows the front and top views of the object defined in Figure P3-10. A section view will be created by first defining the cutting plane line in the top view, then projecting the section view below the front view.

1 Click the **Section View** tool on the **Create** panel under the **Place Views** tab, then click the top view.

The cursor will change to a +-like shape.

Figure 4-39

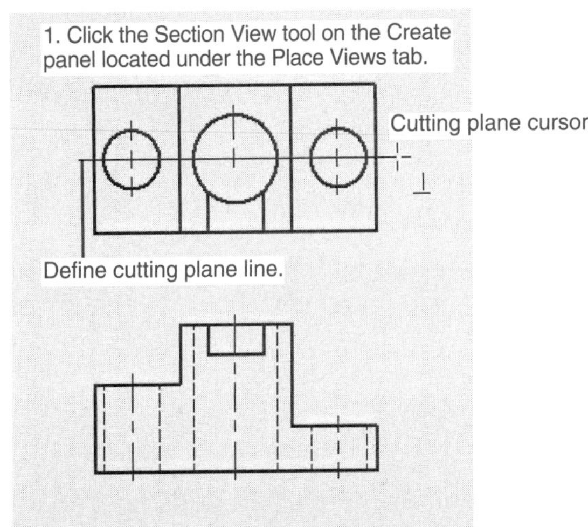

2 Define the cutting plane by defining two points on the top view.

See Figure 4-39. Note that if you touch the cursor to the endpoint of one of the hole's centerlines, a dotted line will follow the cursor, assuring that the cutting plane line is aligned with the holes' centerlines.

3 Right-click the mouse and select the **Continue** option.

The **Section View** dialog box will appear. See Figure 4-40.

Figure 4-40

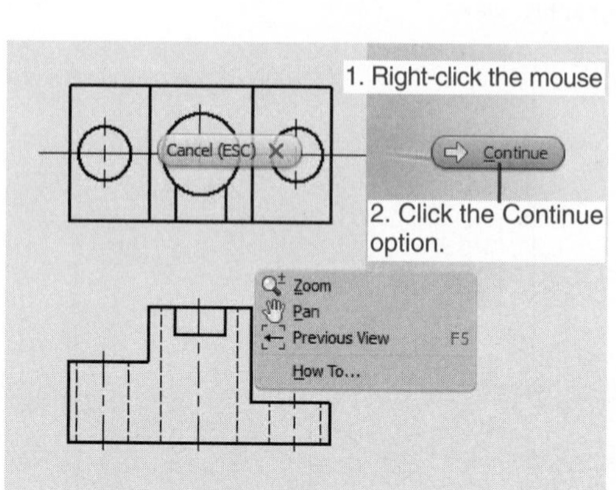

1. Right-click the mouse
Cancel (ESC)
Continue
2. Click the Continue option.
Zoom
Pan
Previous View
F5
How To...

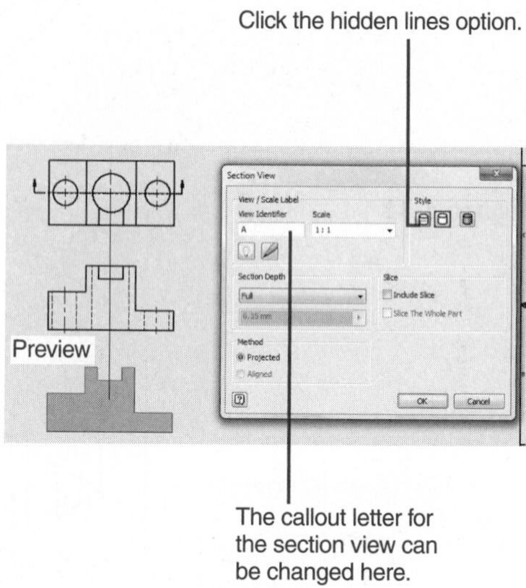

Click the hidden lines option.
Preview
The callout letter for the section view can be changed here.

4 Set the **Label** letter for **A,** the **Scale** for **1:1,** and **Style** for **Hidden Lines.**

5 Move the cursor so as to position the section view below the front view.

6 Click the section view location.

7 Add the appropriate centerlines using the **Centerline Bisector** tool.

Figure 4-41 shows the resulting section view. Notice that the section view is defined as A-A, and the scale is specified. The arrows of the cutting plane line are directed away from the section view. The section view is located behind the arrows.

Figure 4-41

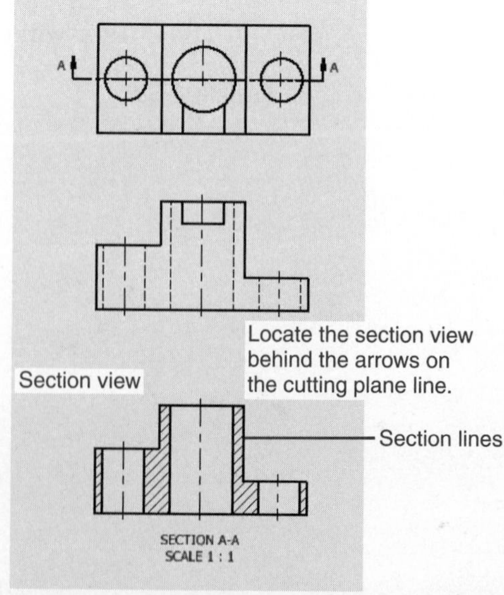

A
A
Locate the section view behind the arrows on the cutting plane line.
Section view
Section lines
SECTION A-A
SCALE 1 : 1

Offset Section Views

Cutting plane lines need not pass directly across an object but may be offset to include several features. Figure 4-42 shows an object that has been cut using an offset cutting plane line.

Figure 4-42

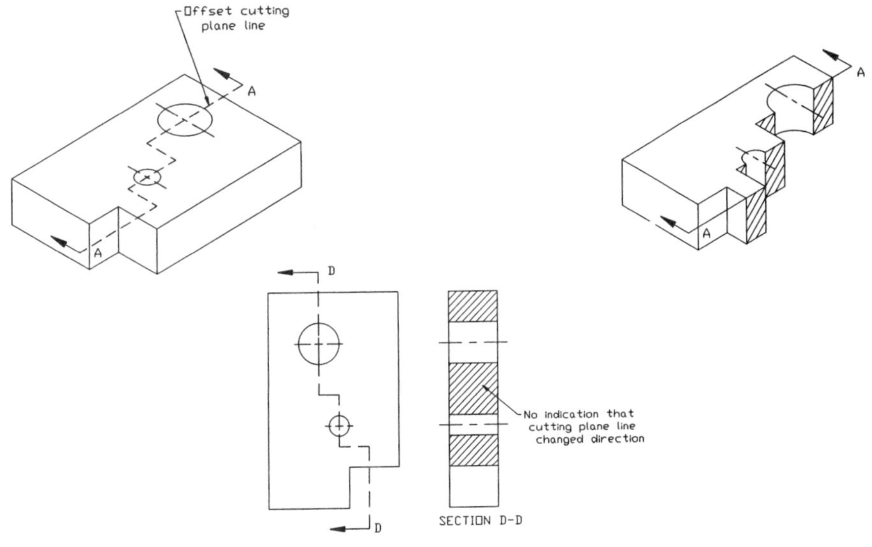
Offset cutting plane line
A
A
A
A
D
D
No indication that cutting plane line changed direction
SECTION D-D

EXERCISE 4-7 Creating an Offset Cutting Plane

Figure 4-43 shows the front and top views of an object. The views were created using the **Create View, Projected View,** and **Centerline** tools.

1. Click the **Section View** tool, and click the top view.
2. Draw a cutting plane across the top view through the centers of each of the three holes.

   When drawing an offset cutting plane line, show the line in either horizontal or vertical line segments.
3. Locate the section view below the front view and add the appropriate centerlines.

Figure 4-43

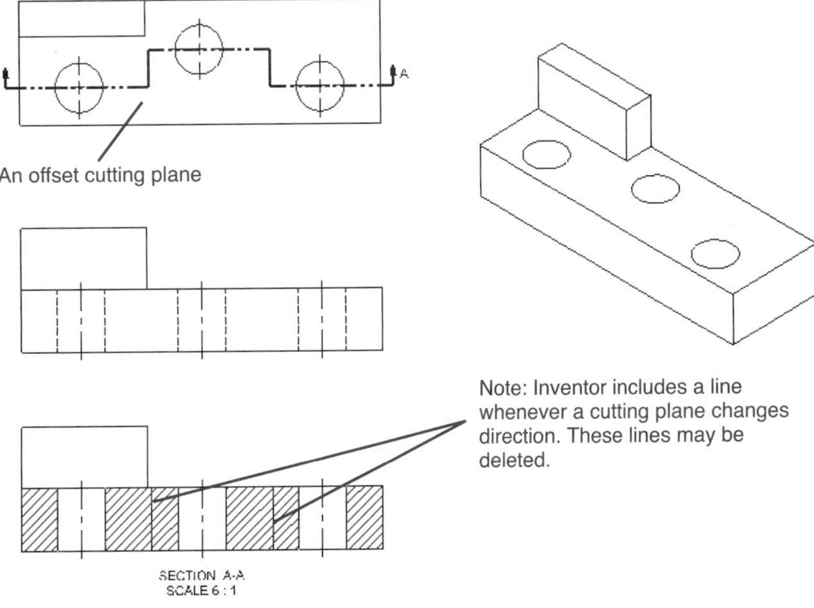
A
A
An offset cutting plane
Note: Inventor includes a line whenever a cutting plane changes direction. These lines may be deleted.
SECTION A-A
SCALE 6 : 1

Aligned Section Views

Figure 4-44 shows an example of an aligned section view. Aligned section views are most often used on circular objects and use an angled cutting plane line to include more features in the section view, like an offset cutting plane line.

Figure 4-44

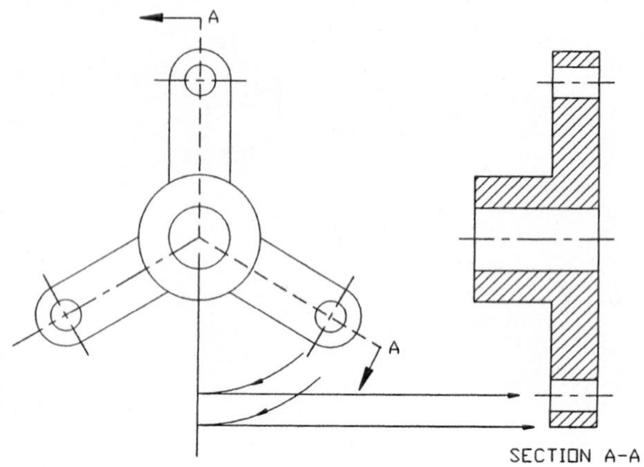

An aligned section view is drawn as if the cutting plane line ran straight across the object. The cutting plane line is rotated into a straight position, and the section view is projected.

Figure 4-45 shows an aligned section view created using Inventor.

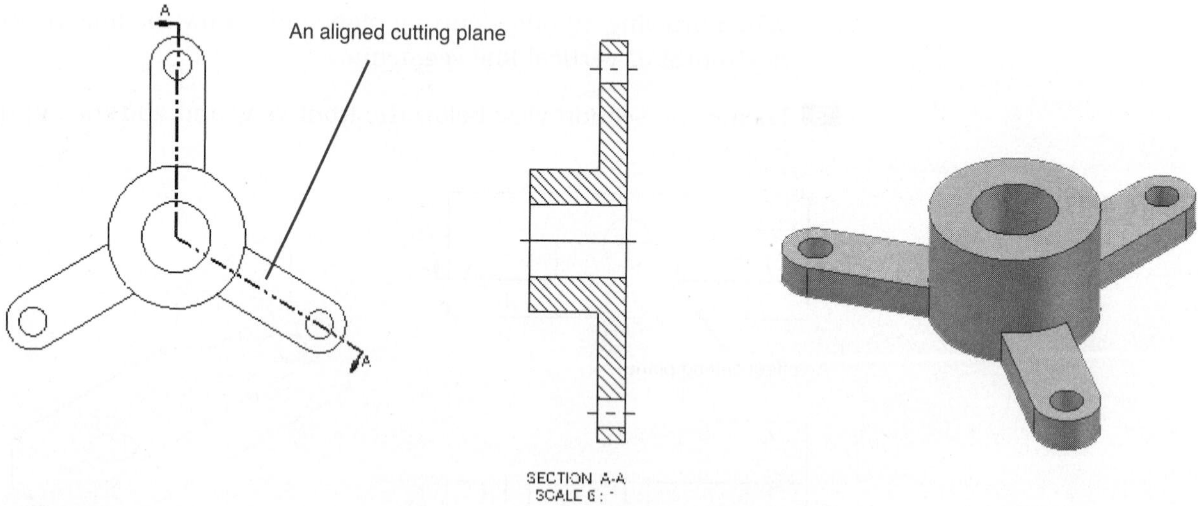

Figure 4-45

Detail Views

detail view: An enlarged view of a portion of a model.

Detail views are used to enlarge portions of an existing drawing. The enlargements are usually made of areas that could be confusing because of many crossing or hidden lines.

EXERCISE 4-8 Creating a Detail View

1 Click the **Detail View** tool on the **Create** panel under the **Place Views** tab, then click the view to be enlarged.

In this example, the top view was selected.

The **Detail View** dialog box will appear. See Figure 4-46.

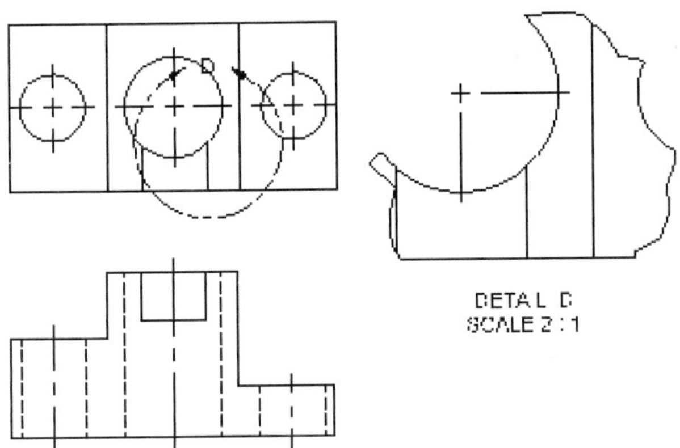

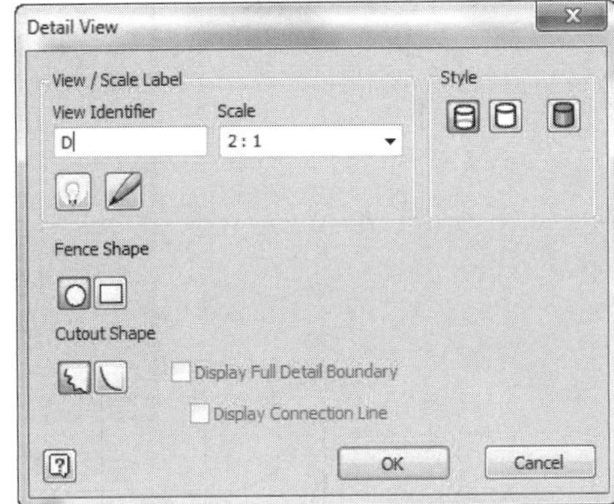

Figure 4-46

2 Set the **Label** letter to **D** and the **Scale** to **2:1,** then pick a point on the view.

3 Move the cursor, creating a circle.

The circle will be used to define the area of the detail view.

4 When the circle is of an appropriate diameter, click the left mouse button and move the cursor away from the view.

5 Locate the detail view and click the location.

Break Views

It is often convenient to break long continuous shapes so that they take up less drawing space. Figure 4-47 shows a long L-bracket that has a continuous shape; that is, its shape is constant throughout its length. Figure 4-48 shows an orthographic view of the same L-bracket.

Figure 4-47

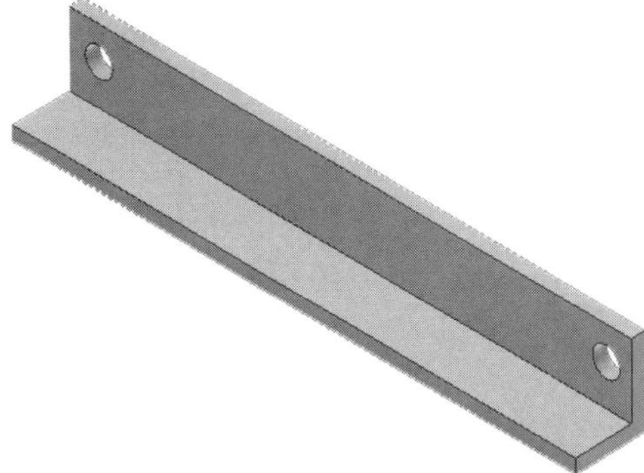

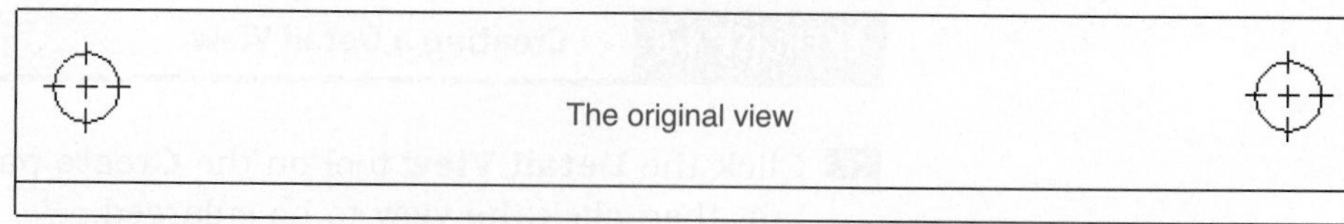

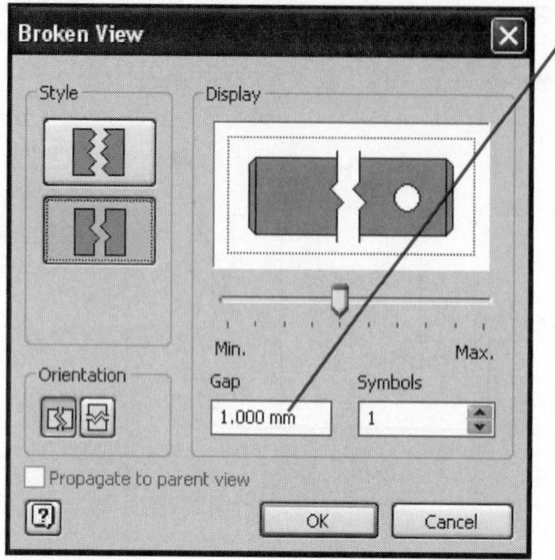

Figure 4-48

EXERCISE 4-9 Creating a Broken View

1 Click the **Break** tool located on the **Create** panel under the **Place Views** tab, then click the orthographic view.

The **Broken View** dialog box will appear.

2 Select the orientation of the break and the gap distance between the two portions of the L-bracket.

In this example the gap distance is 1.00. Do not click the **OK** box. Define the break with the **Broken View** dialog box on the drawing screen.

3 Click a point near the left end of the L-bracket, then move the cursor to the right and click a second point near the right end of the L-bracket.

Figure 4-48 shows the resulting broken view.

Multiple Section Views

It is acceptable to take more than one section view of the same object to present a more complete picture of the object. Figures 4-49 and 4-50 show objects that use more than one section view.

Auxiliary Views

auxiliary view: An orthographic view drawn perpendicular to a slanted or oblique surface.

Auxiliary views are orthographic views used to present true-shaped views of slanted surfaces. Figure 4-51 shows an object with a slanted surface that includes a hole drilled perpendicular to the slanted surface. Note how the right-side view shows the hole as an ellipse and that the surface A-B-C-D is foreshortened; that is, it is not shown at its true size. Surface

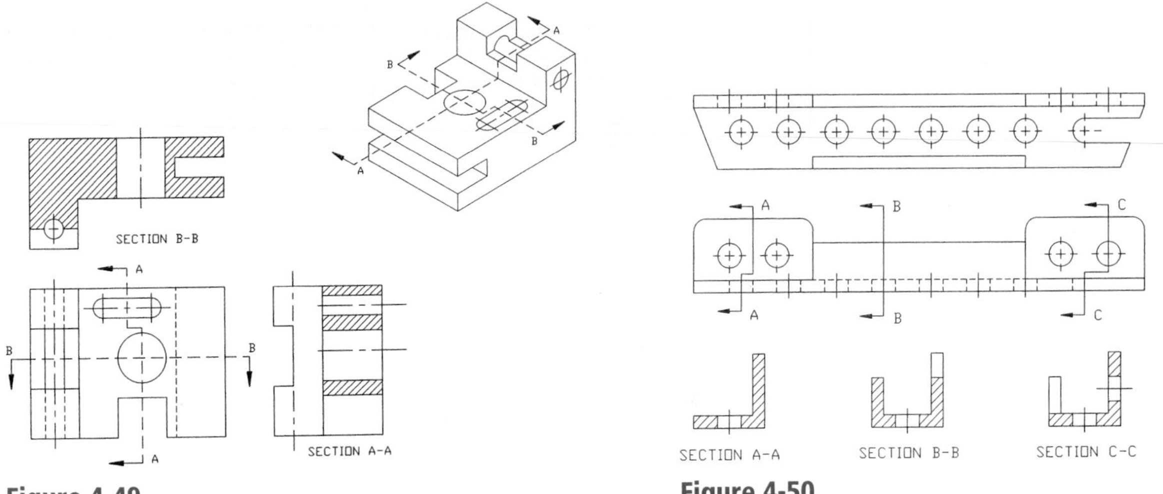

Figure 4-49

Figure 4-50

A-B-C-D does appear at its true shape and size in the auxiliary view. The auxiliary view was projected at 90° from the slanted surface so as to generate a true-shaped view.

Figure 4-52 shows an object that includes a slanted surface and hole.

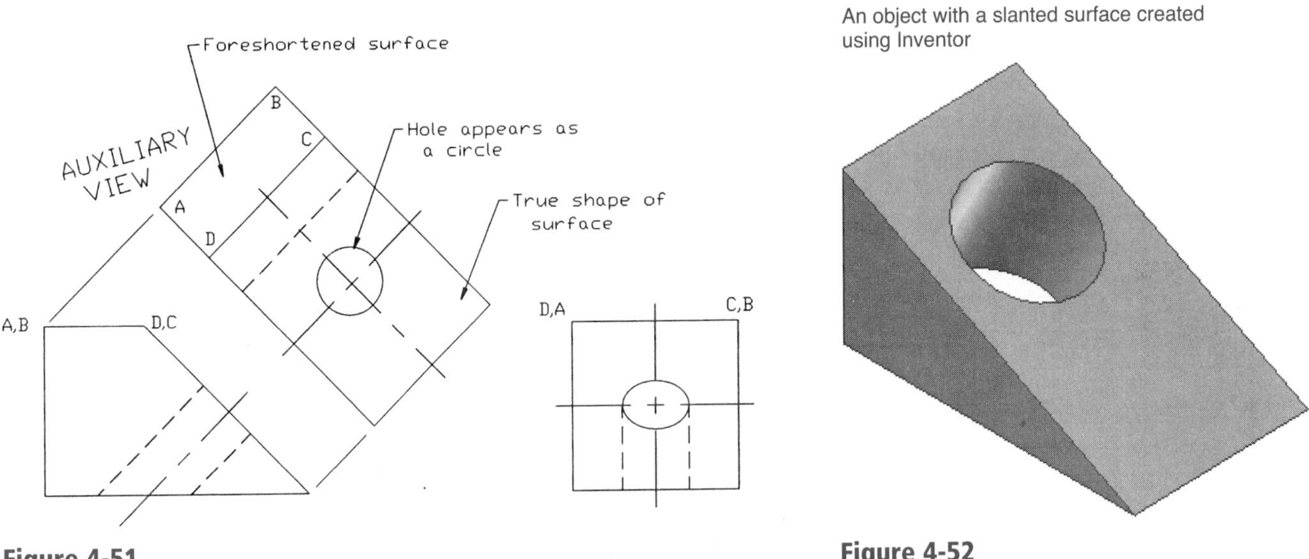

Figure 4-51

An object with a slanted surface created using Inventor

Figure 4-52

EXERCISE 4-10 Drawing an Auxiliary View

1 Create a drawing using the **ANSI (mm).ipt** format. Click the **Base View** and **Projected View** tools on the **Create** panel under the **Place Views** tab and create a front and a right-side view as shown in Figure 4-53.

Click the **Auxiliary View** tool, then the front view.

2 The **Auxiliary View** dialog box will appear.

3 Enter the appropriate settings, then click the slanted edge line in the front view.

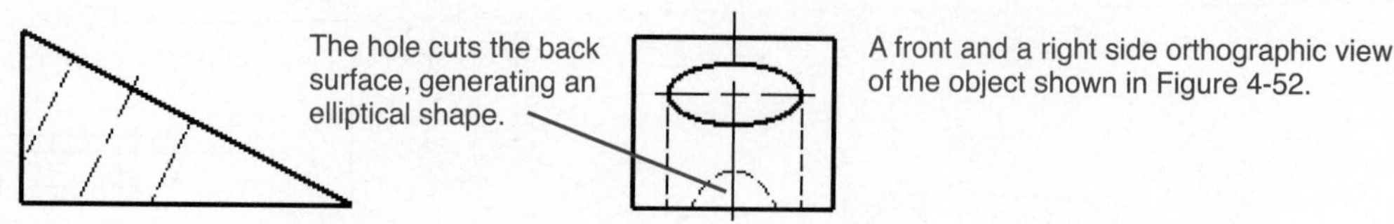

The hole cuts the back surface, generating an elliptical shape.
A front and a right side orthographic view of the object shown in Figure 4-52.

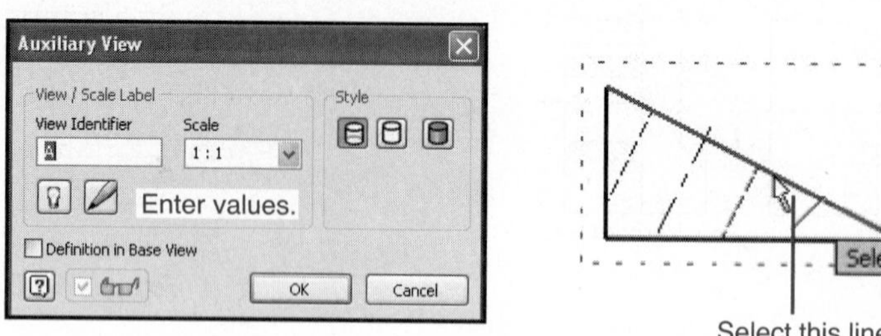

Select a line
Select this line.

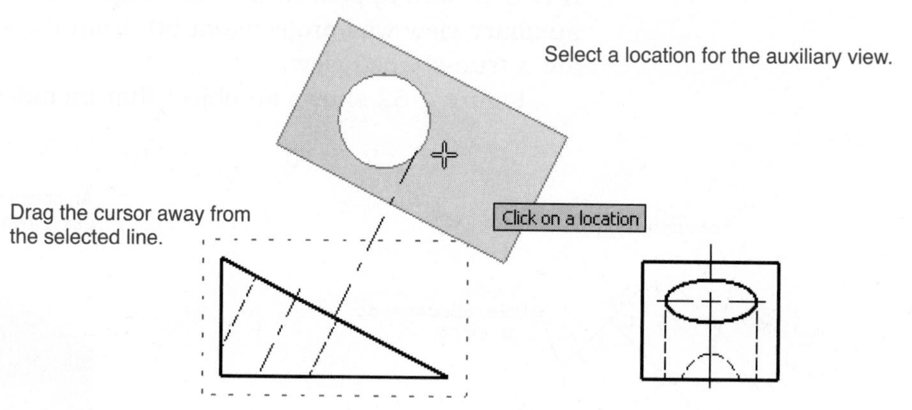

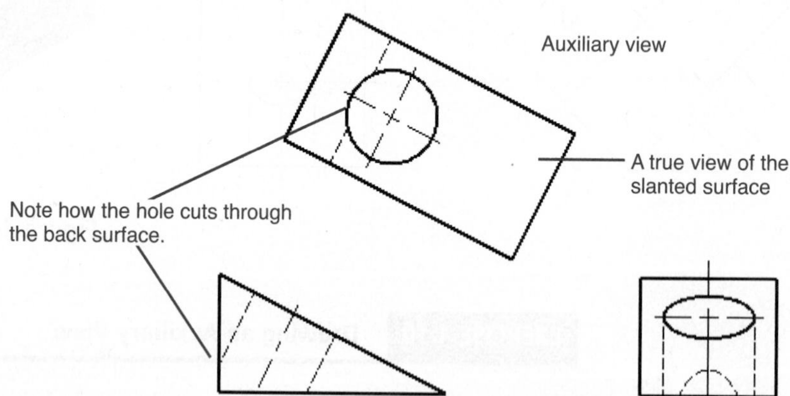

Figure 4-53

In this example, a scale of 1:1 was used.

4 Move the cursor away from the front view and select a location for the auxiliary view.

Click on a location

5 Click the left mouse button and create the auxiliary view.

ASME Y14.3-2003.

Drawing Standards

There are two sets of standards used to define the projection and placement of orthographic views: the American National Standards Institute (ANSI) and the International Organization for Standardization (ISO). The ANSI calls for orthographic views to be created using third-angle projection and is the accepted method for use in the United States. See the American Society of Mechanical Engineers publication ASME Y14-3-2003. Some countries other than the United States use first-angle projection. See ISO publication 128-30.

This chapter has presented orthographic views using third-angle projections as defined by ANSI. However, there is so much international commerce happening today that you should be able to work in both conventions as you should be able to work in both inches and millimeters.

Figure 4-54 shows a three-dimensional model and three orthographic views created using third-angle projection and three orthographic views created using first-angle projection. Note the differences and similarities. The front view in both projections is the same. The top views are the same but are in different locations. The third-angle projection presents a right side view, while the first-angle projection presents a left side view.

Figure 4-54

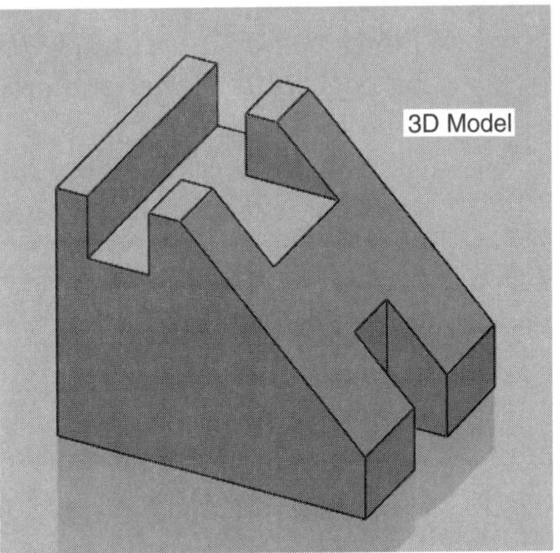

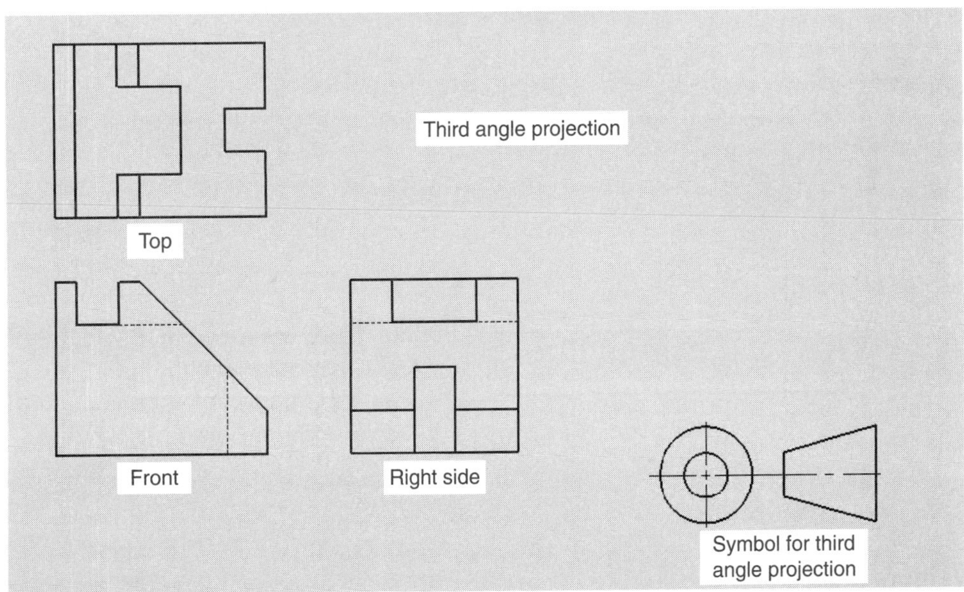

Figure 4-54
(Continued)

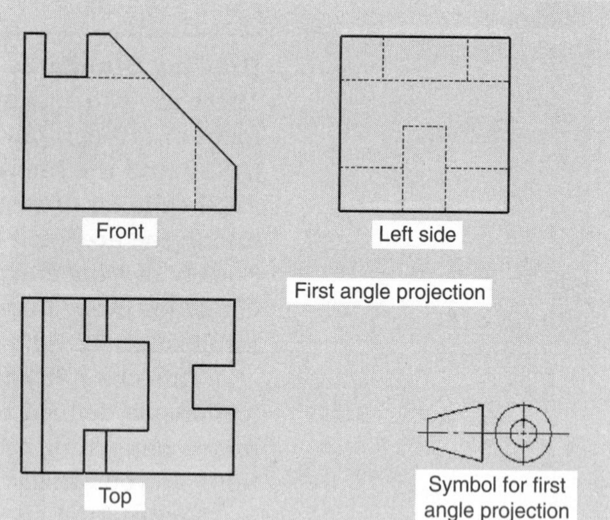

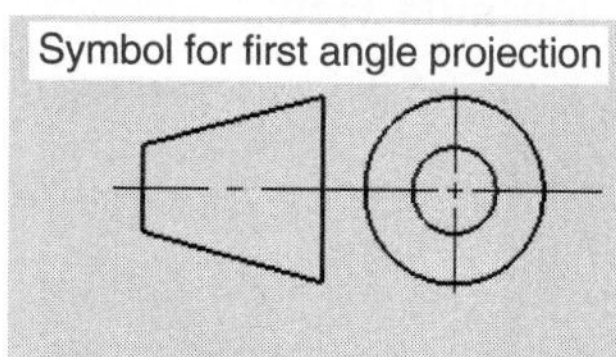

Symbol for first angle projection

Figure 4-55

Figure 4-55 shows the drawing symbols for first- and third-angle projections. These symbols can be added to a drawing to help the reader understand which type of projection is being used. These symbols were included in the projections presented in Figure 4-54.

Third- and First-Angle Projections

Figure 4-56 shows an object with a front orthographic view and two side orthographic views: one created using third-angle projection, and the other created using first-angle projection. For third-angle projections the orthographic view is projected on a plane located between the viewer's position and the object. For first-angle projections the orthographic view is

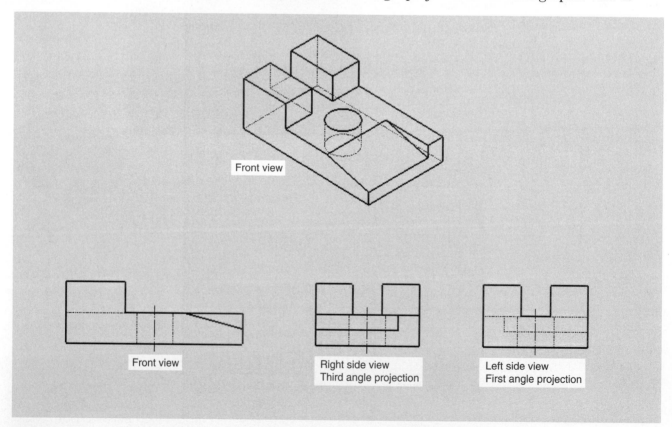

Front view

Front view

Right side view
Third angle projection

Left side view
First angle projection

Figure 4-56

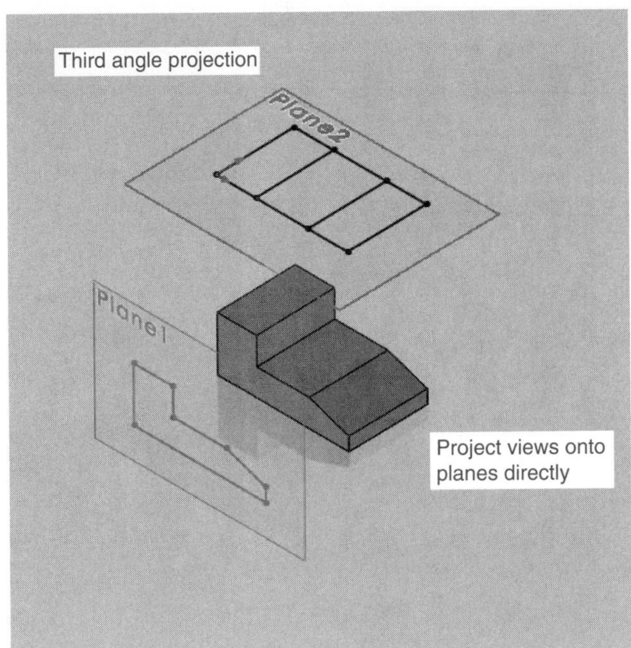

Figure 4-57

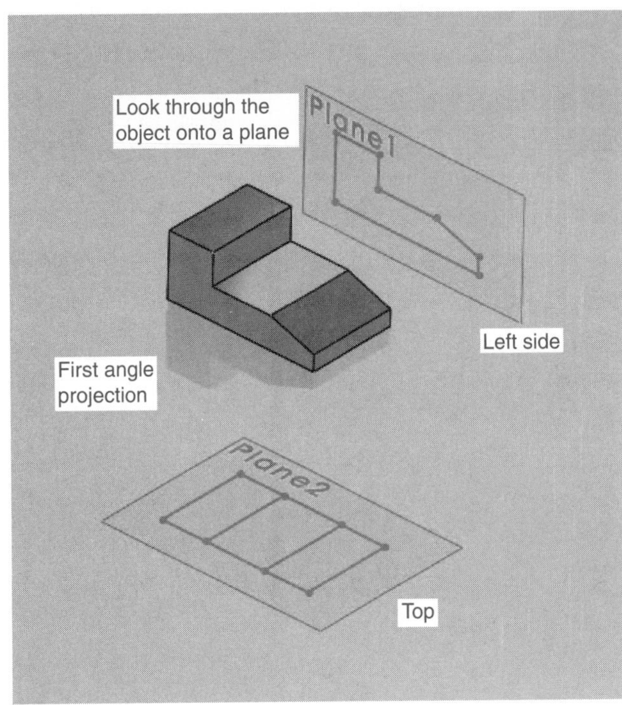

Figure 4-58

projected on a plane located beyond the object. The front and top views for third- and first-angle projections appear the same, but they are located in different positions relative to the front view.

The side orthographic views are different for third- and first-angle projections. Third-angle projection uses a right side view located to the right of the object. First-angle projections use a left side view located to the right of the object. Figures 4-57 and 4-58 show the two different side view projections for the same object. For third-angle projection the viewer is located on the right side of the object and creates the side orthographic view on a plane located between the view position and the object. The viewer looks directly at the object. For first-angle projection the viewer is located on the left side of the object and creates the side orthographic view on a plane located beyond the object. The viewer looks through the object.

To help understand the difference between side view orientations for third- and first-angle projections, locate your right hand with the heel facing down and the thumb facing up. Rotate your hand so that the palm is facing up—this is the third-angle projection orientation. Return to the thumb up position. Rotate your hand so that the palm is down—this is the first-angle view orientation.

To create first-angle projections using Inventor

1. Start a **New** drawing using the **ISO.idw** template.

   This template will automatically create first-angle projection drawings.

2. Click the **Base** tool.

3. Select the appropriate file.

4. Select the orientation.

5. Use the **Projected View** tool to select and position the views (Figure 4-59).

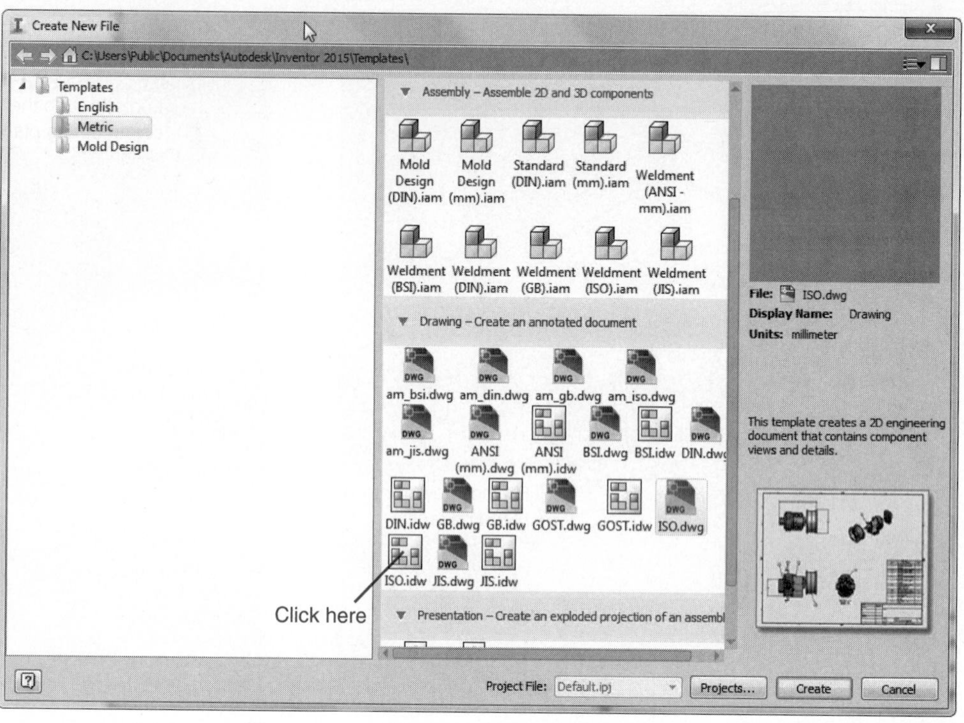
Click here

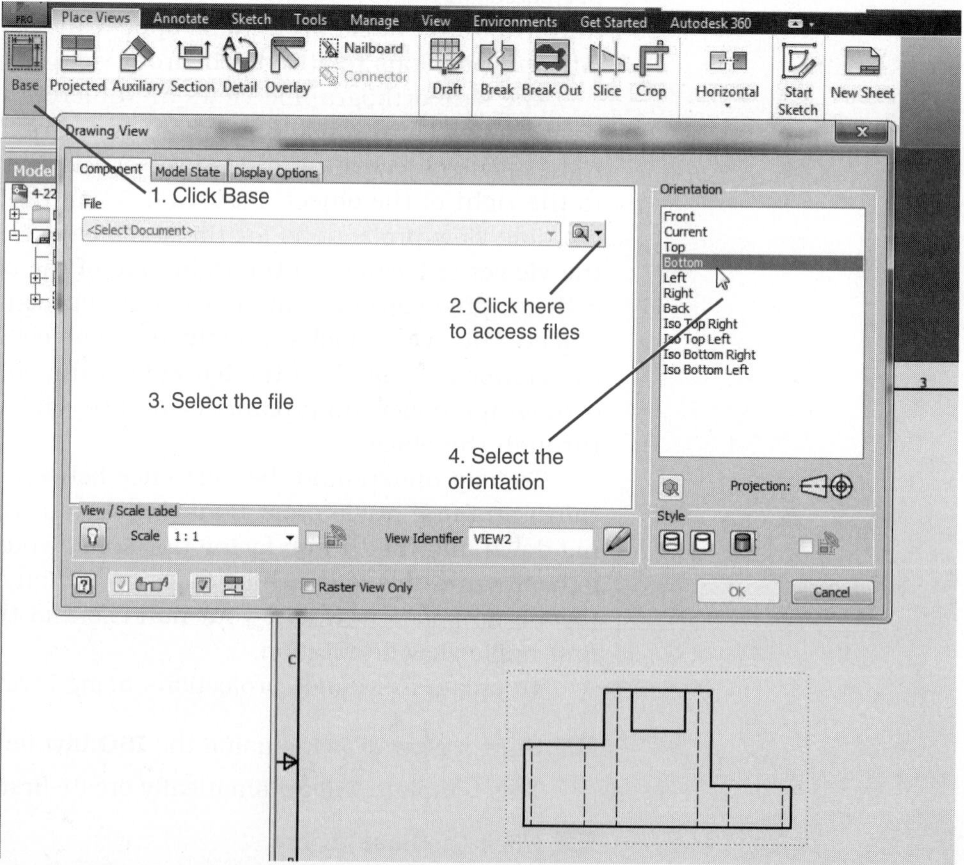
1. Click Base
2. Click here to access files
3. Select the file
4. Select the orientation

Figure 4-59

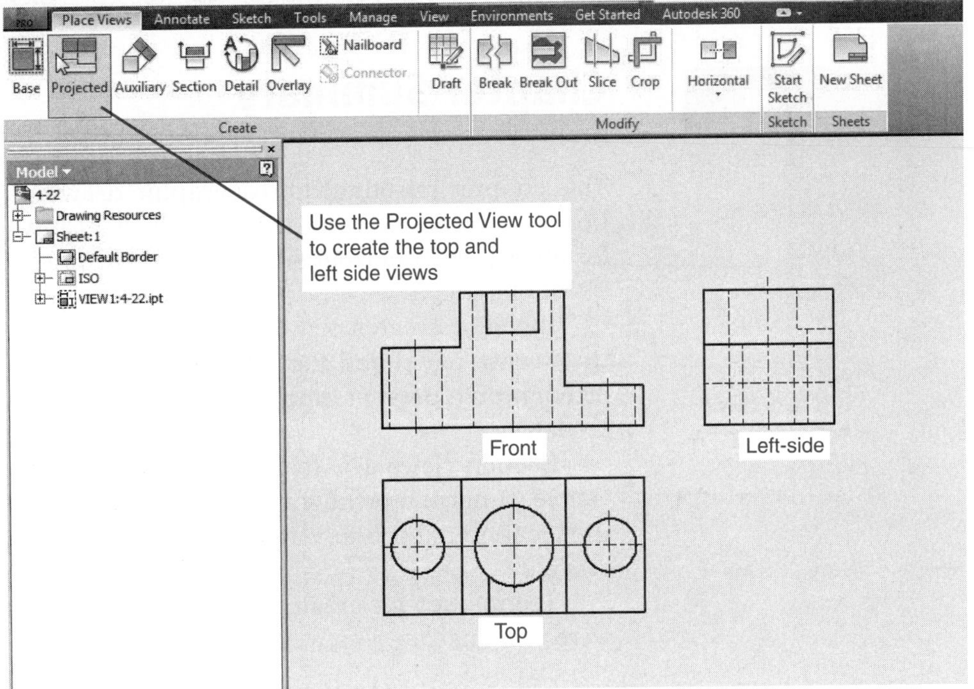

Figure 4-59

(*Continued*)

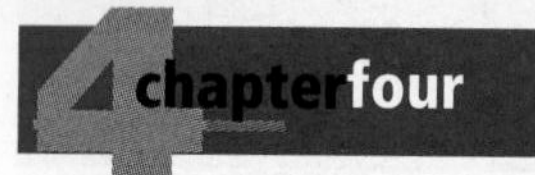

Chapter Summary

This chapter introduced orthographic drawings using third-angle projection in accordance with ANSI standards. Conventions were demonstrated for objects with normal surfaces, hidden lines, slanted surfaces, compound lines, oblique surfaces, and rounded surfaces.

Inventor creates orthographic views directly from models. The tools on the **Create** panel and the **Annotate** panel were introduced for managing orthographic presentation views. Isometric views can also be created from models.

Section views are used to expose internal surfaces that are not directly visible in normal orthographic views. Cutting planes were used to define the location of section views. Offset and aligned section views also were created.

Techniques for creating detail views, broken views, and auxiliary views were demonstrated as well.

Chapter Test Questions

Multiple Choice

Circle the correct answer.

1. Which of the following is not one of the three views generally taken of an object?

a. Front

b. Top

c. Left

d. Right

2. In the precedence of lines, a hidden line covers a(n) _____ line.

a. Continuous

b. Center

c. Compound

d. Oblique

3. Which of the following is used to define a section view?

a. A cutting plane

b. A section line

c. A centerline

4. Section lines are used to define which of the following on a section view?

a. The outside edges of the section cut

b. The location of the section view

c. The areas where the section view passes through solid material

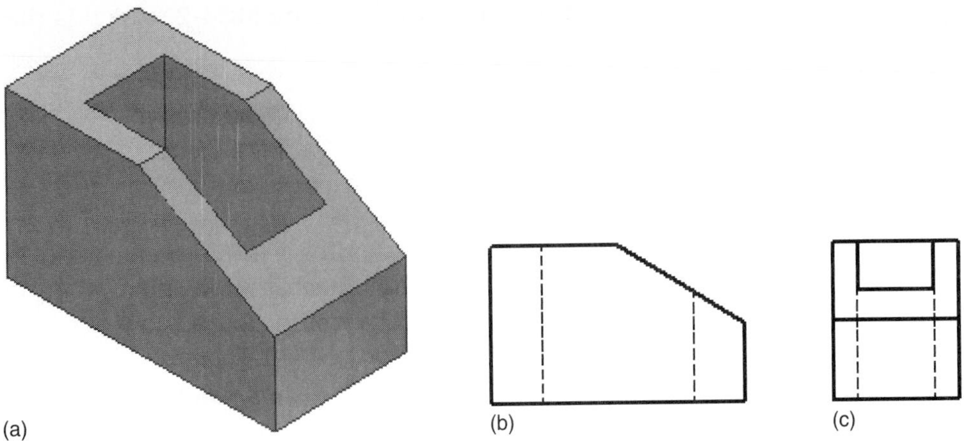

Figure MC4-1

5. Given the model shown in Figure MC4-1, which is the correct top view?

   a. b. c.

6. Given the model shown in Figure MC4-1, which is the correct front view?

   a. b. c.

7. Given the model shown in Figure MC4-1, which is the correct right-side view?

   a. b. c.

8. Given the model shown in Figure MC4-2, which is the correct right-side view?

   a. b. c.

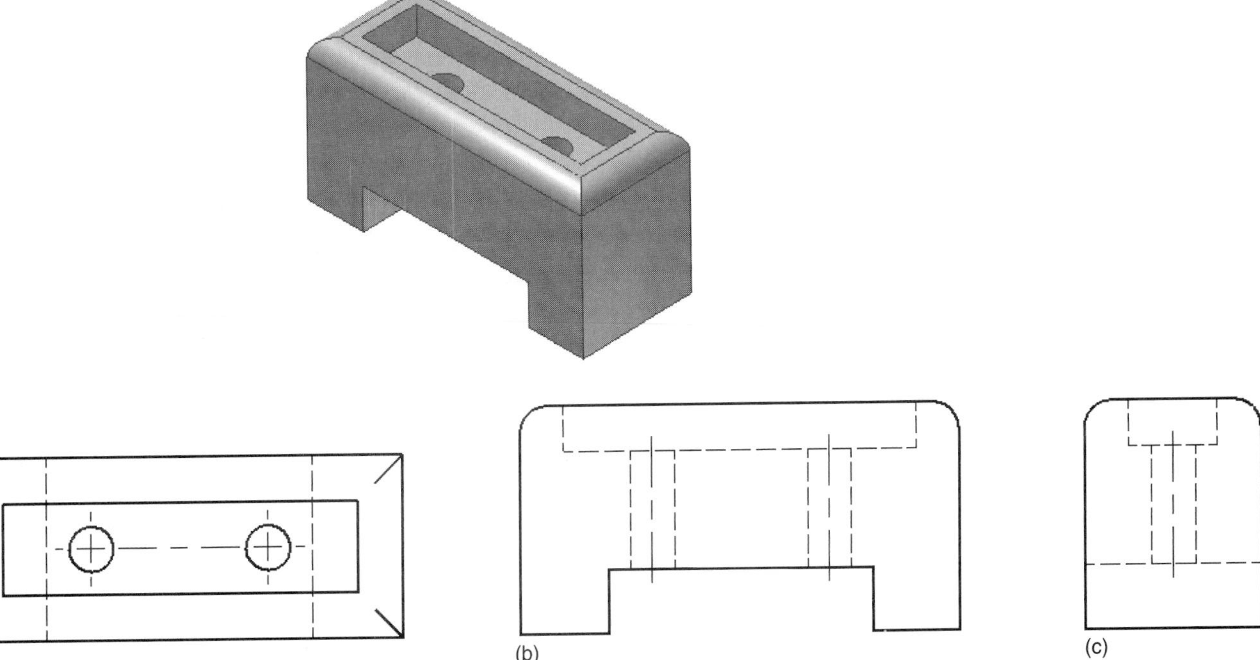

Figure MC4-2

9. Given the model shown in Figure MC4-2, which is the correct top view?

   a. b. c.

10. Given the model shown in Figure MC4-2, which is the correct front view?

   a. b. c.

Matching

Given the drawing shown in MC4-3, identify the types of lines used to create the drawing.

Column A	Column B
a. ____________	1. Centerlines
b. ____________	2. Cutting plane line
c. ____________	3. Continuous line
d. ____________	4. Section line
e. ____________	5. Hidden line

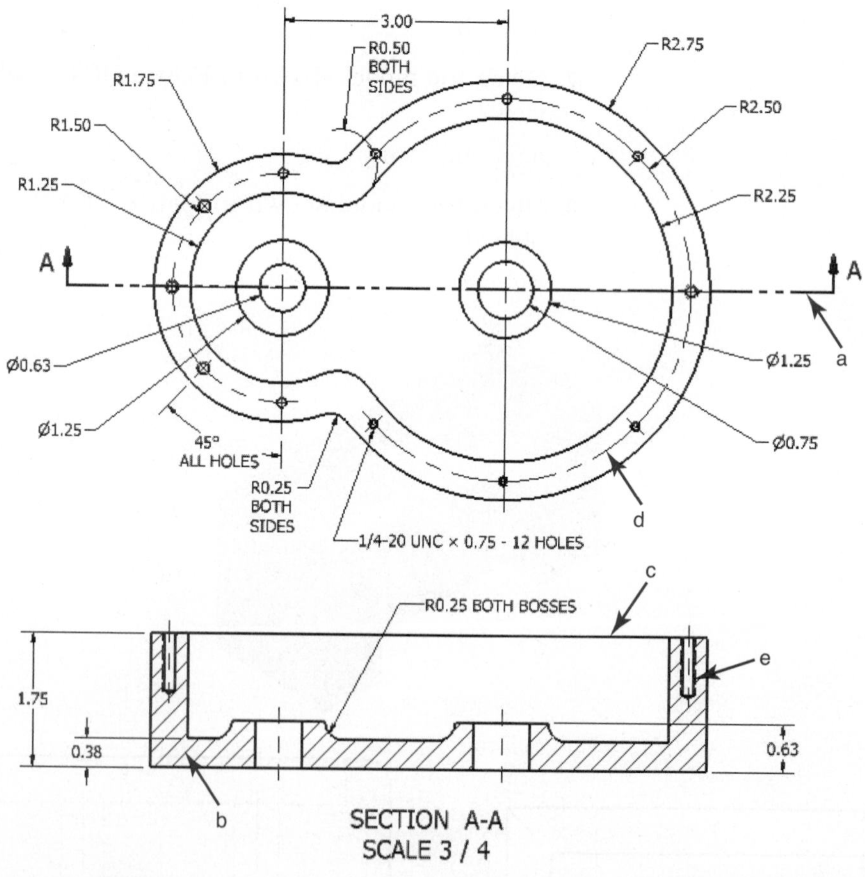

Figure MC4-3

True or False

Circle the correct answer.

1. **True or False:** Orthographic views are two-dimensional views used to define three-dimensional models.
2. **True or False:** Normal surfaces are surfaces located 90° to each other.
3. **True or False:** Hidden lines are not used in orthographic views.
4. **True or False:** A compound line is formed when two slanted surfaces intersect.
5. **True or False:** An oblique surface is a surface that is slanted in two different directions.
6. **True or False:** Center points cannot be edited; they can be used only as they appear on the drawing screen.
7. **True or False:** A section view can be taken only across an object's centerline.
8. **True or False:** Aligned section views are most often used on circular objects.
9. **True or False:** A detail view is used to enlarge portions of an existing drawing.
10. **True or False:** Break views are used to shorten long continuous shapes so they can fit within the drawing screen.

Chapter Projects

Project 4-1

Draw a front, a top, and a right-side orthographic view of each of the objects in Figures P4-1 through P4-24. Make all objects from mild steel.

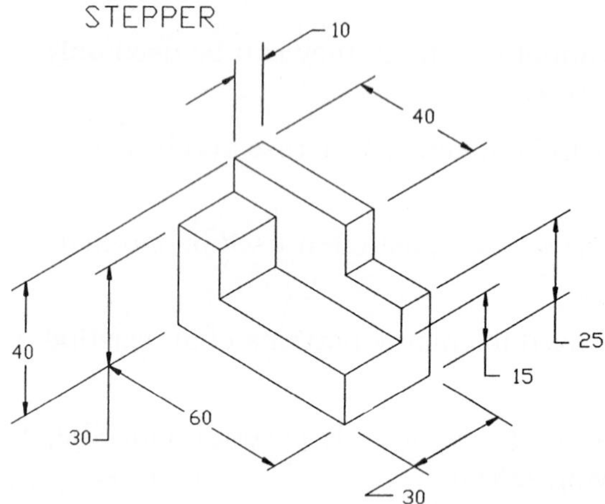

Figure P4-1
MILLIMETERS

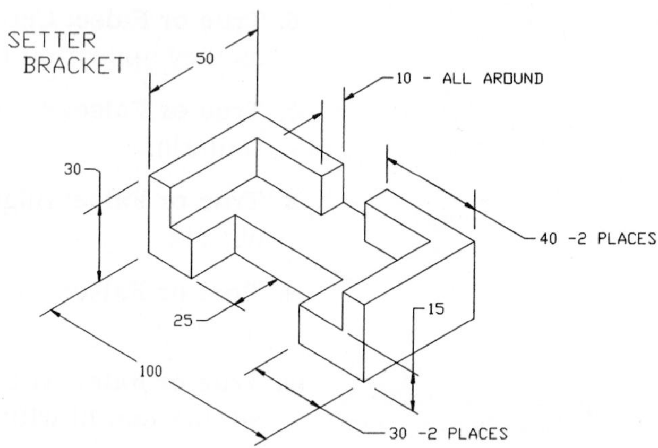

Figure P4-2
MILLIMETERS

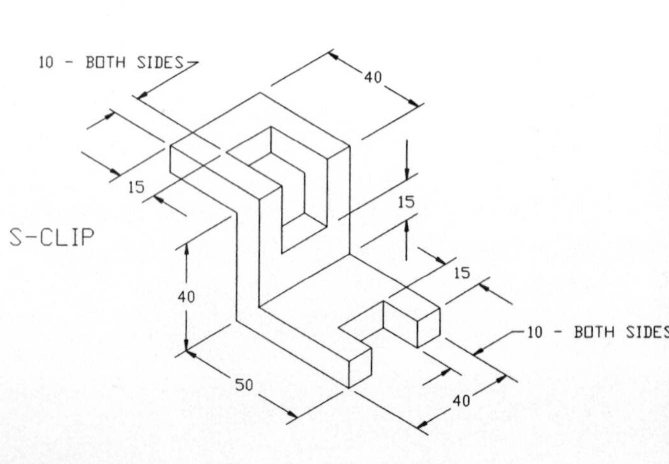

Figure P4-3
MILLIMETERS

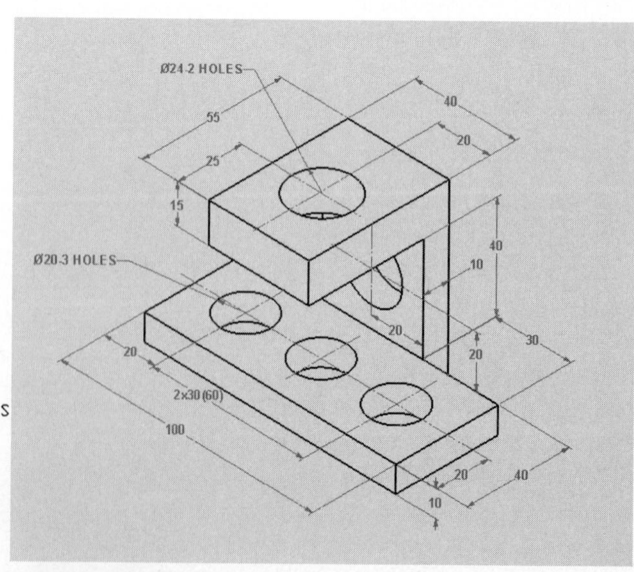

Figure P4-4
MILLIMETERS

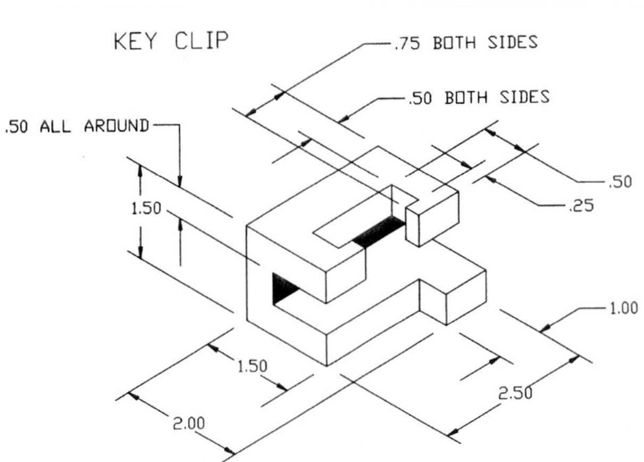

Figure P4-5
INCHES

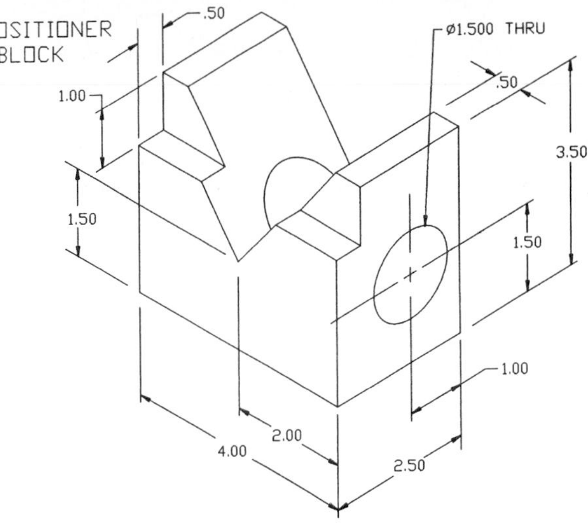

Figure P4-6
INCHES

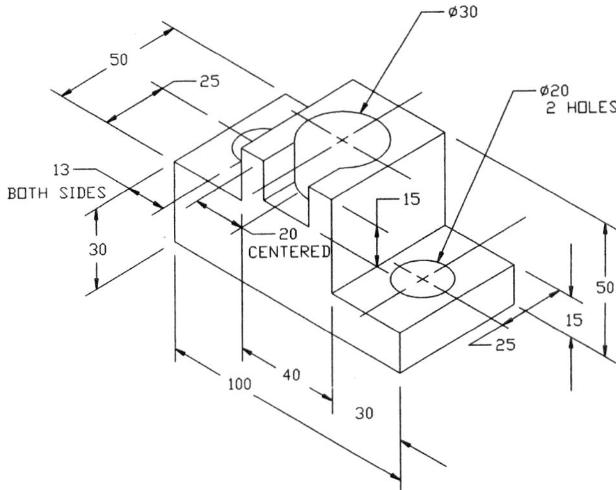

Figure P4-7
MILLIMETERS

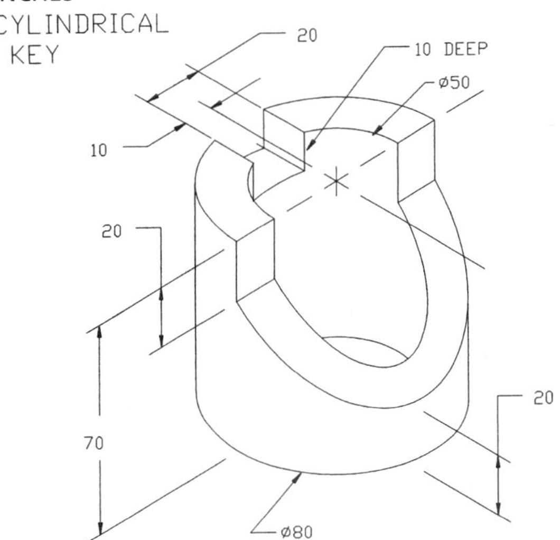

Figure P4-8
MILLIMETERS

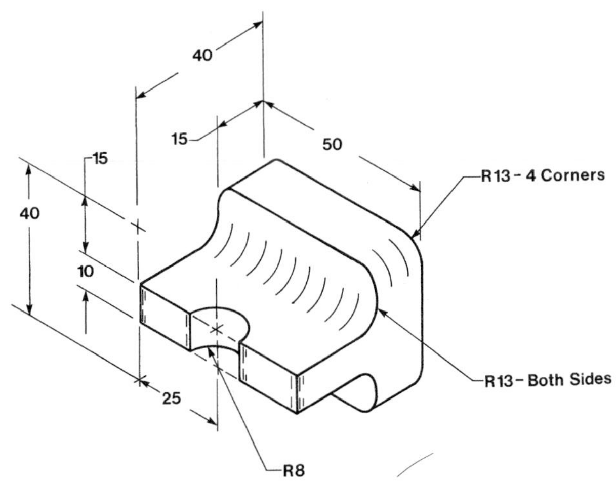

Figure P4-9
MILLIMETERS

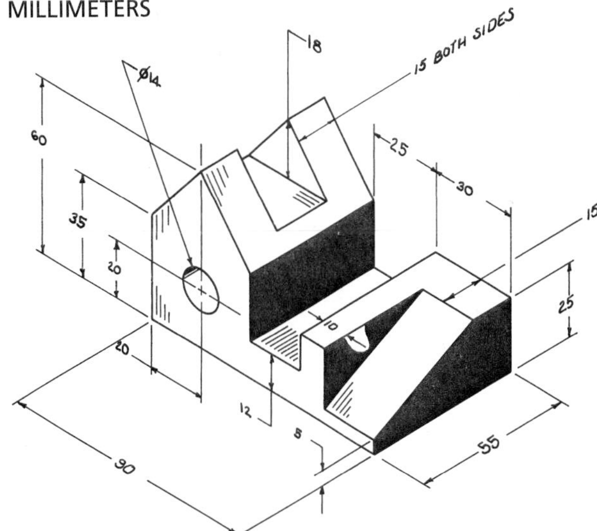

Figure P4-10
MILLIMETERS

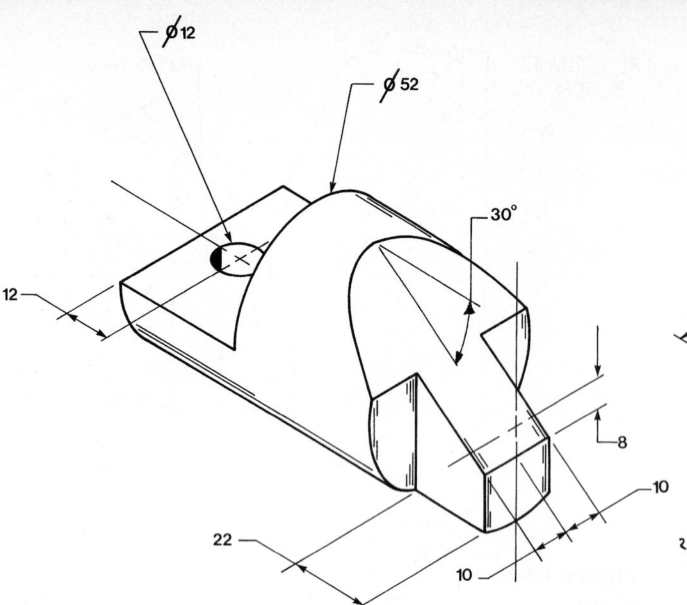

Figure P4-11
MILLIMETERS

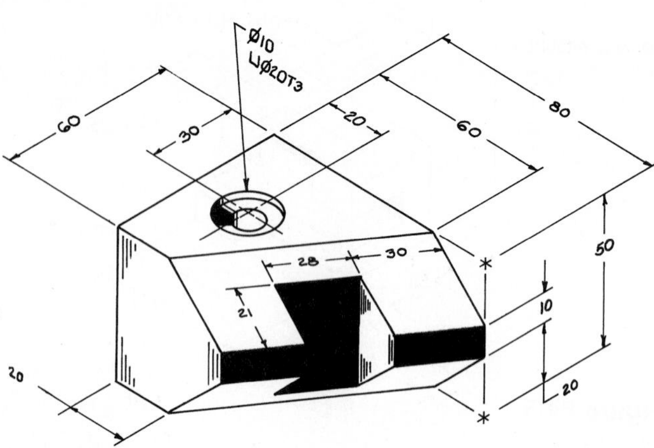

Figure P4-12
MILLIMETERS

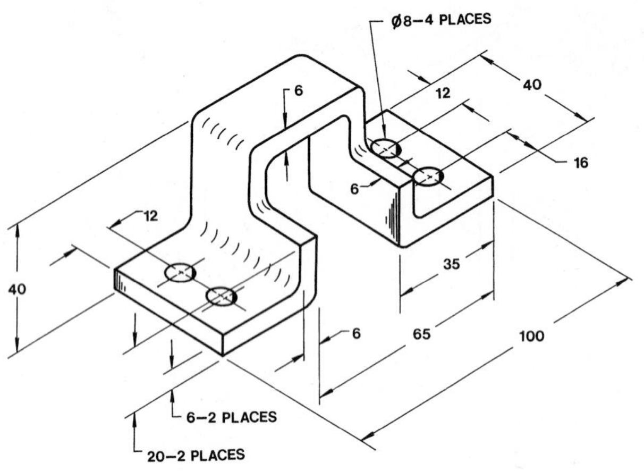

NOTE: ALL FILLETS AND ROUNDS=R3

Figure P4-13
MILLIMETERS

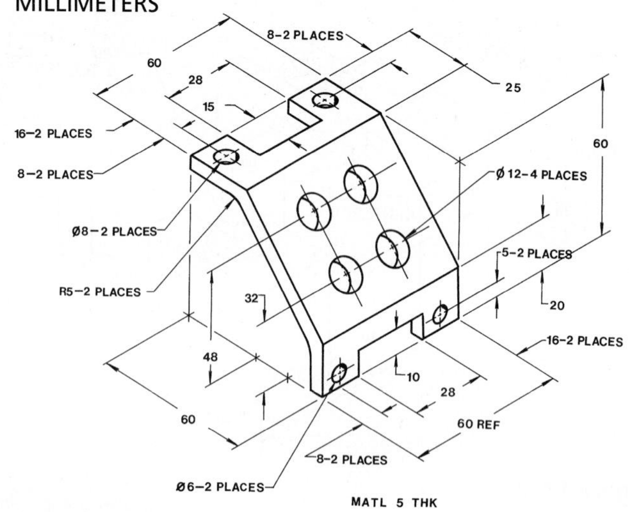

Figure P4-14
MILLIMETERS

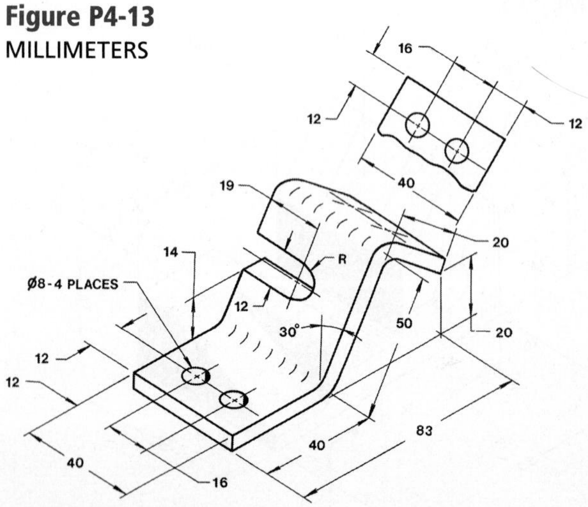

16
12
12
19
40
20
R
14
Ø8-4 PLACES
12
30°
50
20
12
12
40
40
83
16

MATL 5 THK
ALL INSIDE BEND RAD 5

Figure P4-15
MILLIMETERS

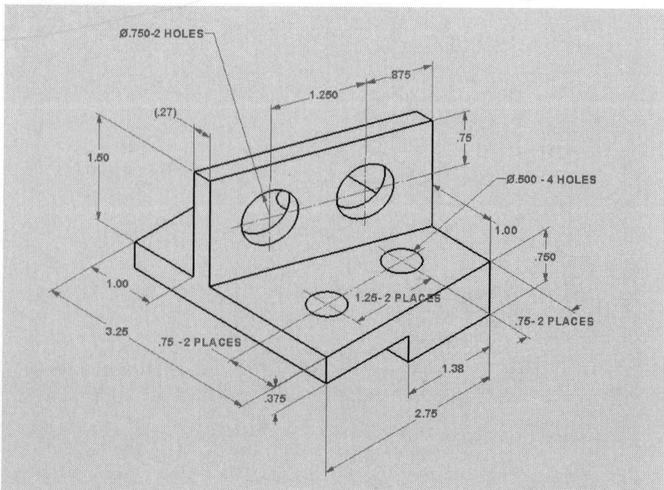

Figure P4-16
INCHES

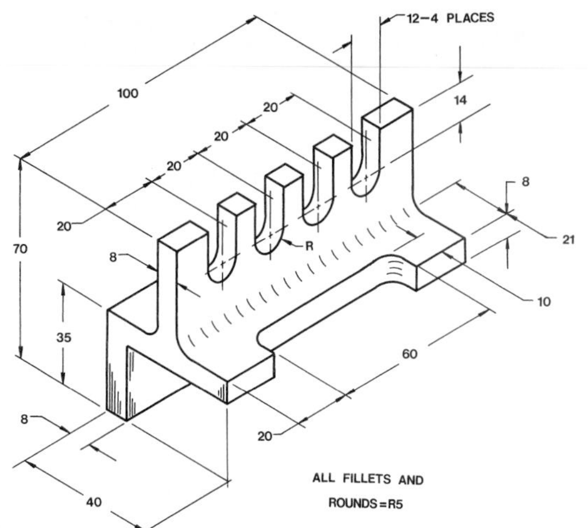

Figure P4-17
MILLIMETERS

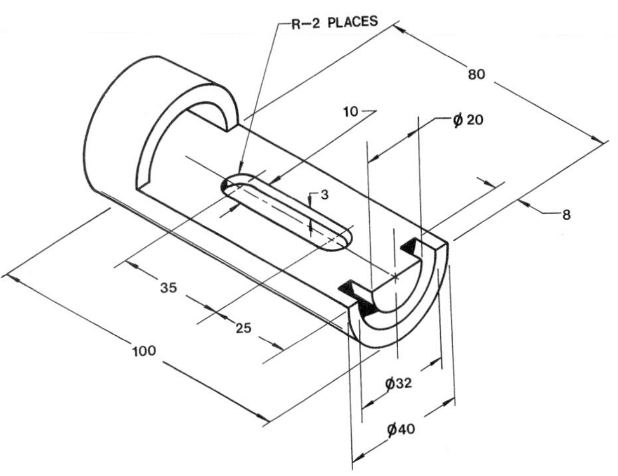

Figure P4-18
MILLIMETERS

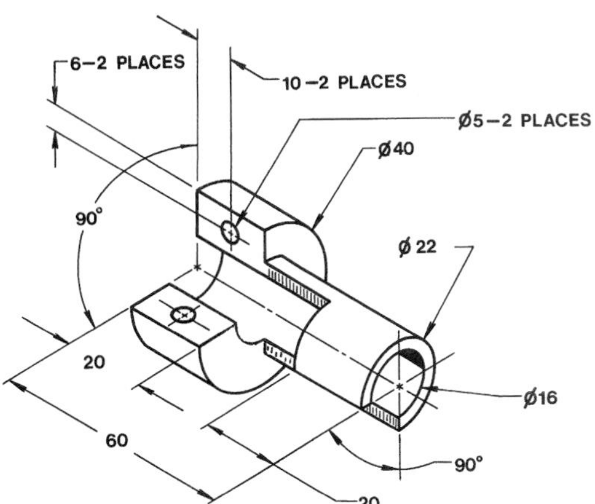

Figure P4-19
MILLIMETERS

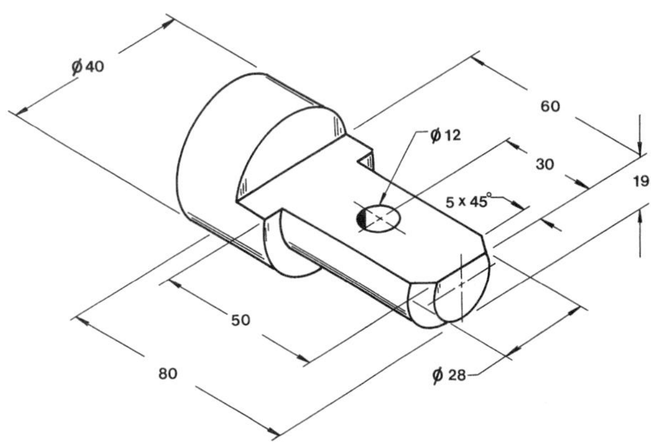

Figure P4-20
MILLIMETERS

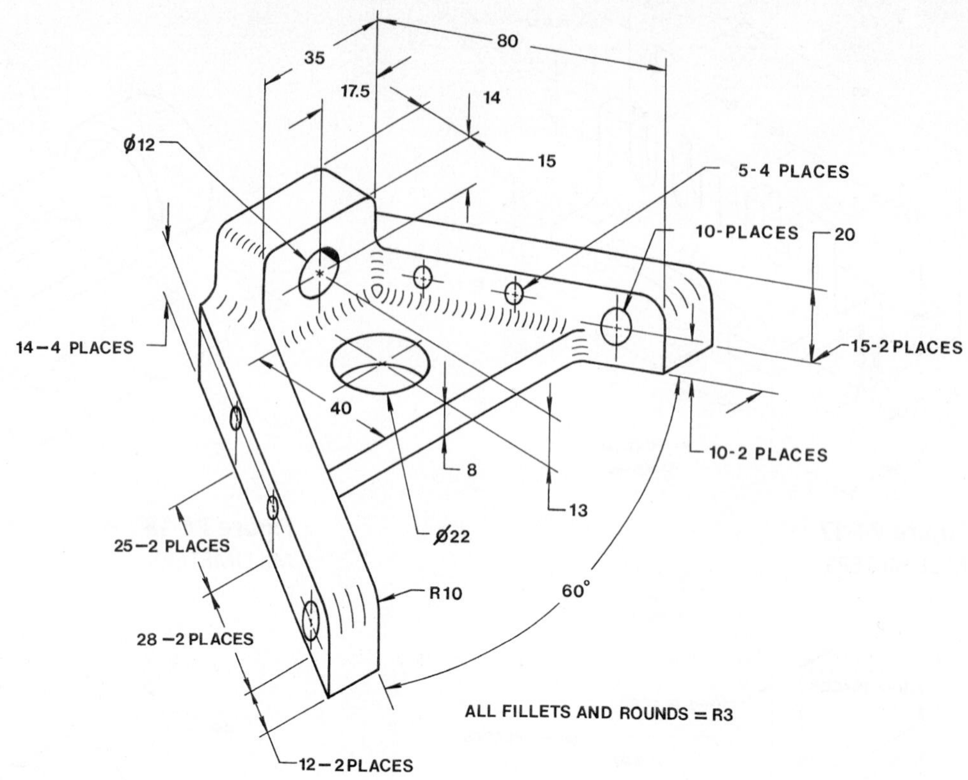

Figure P4-21
MILLIMETERS

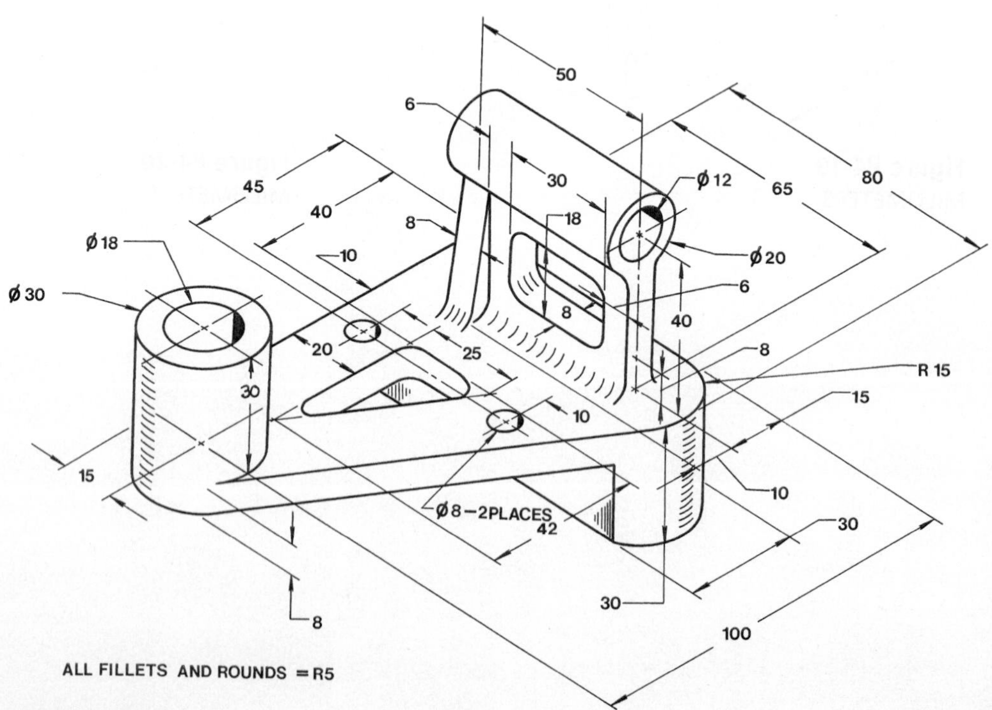

Figure P4-22
MILLIMETERS

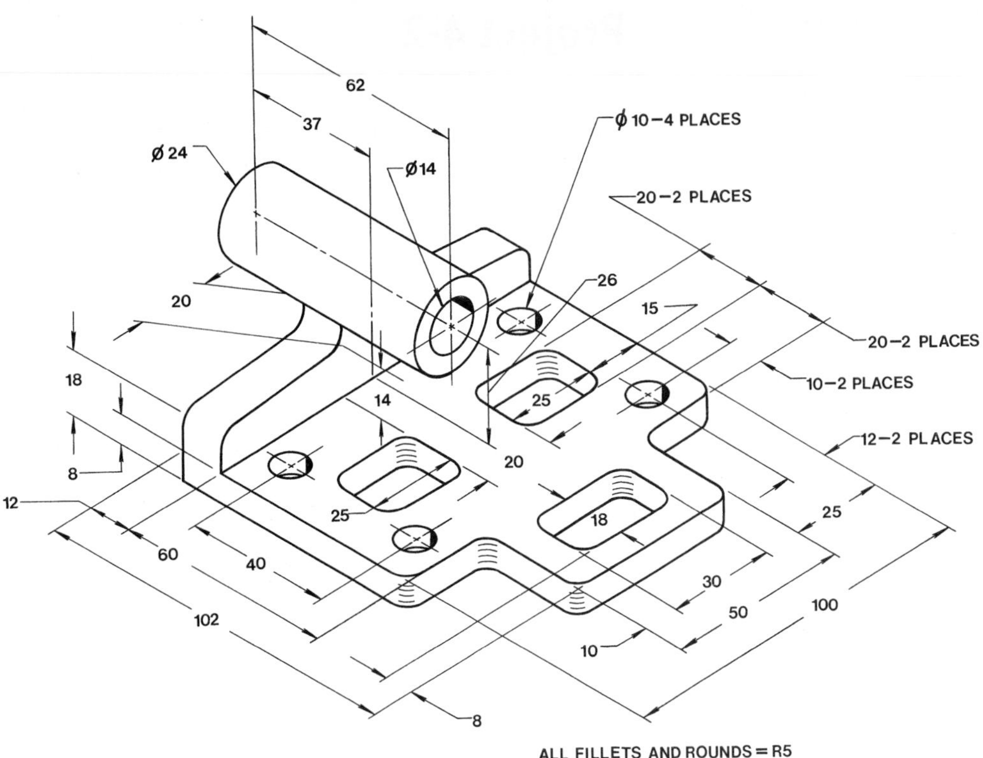

Figure P4-23
MILLIMETERS

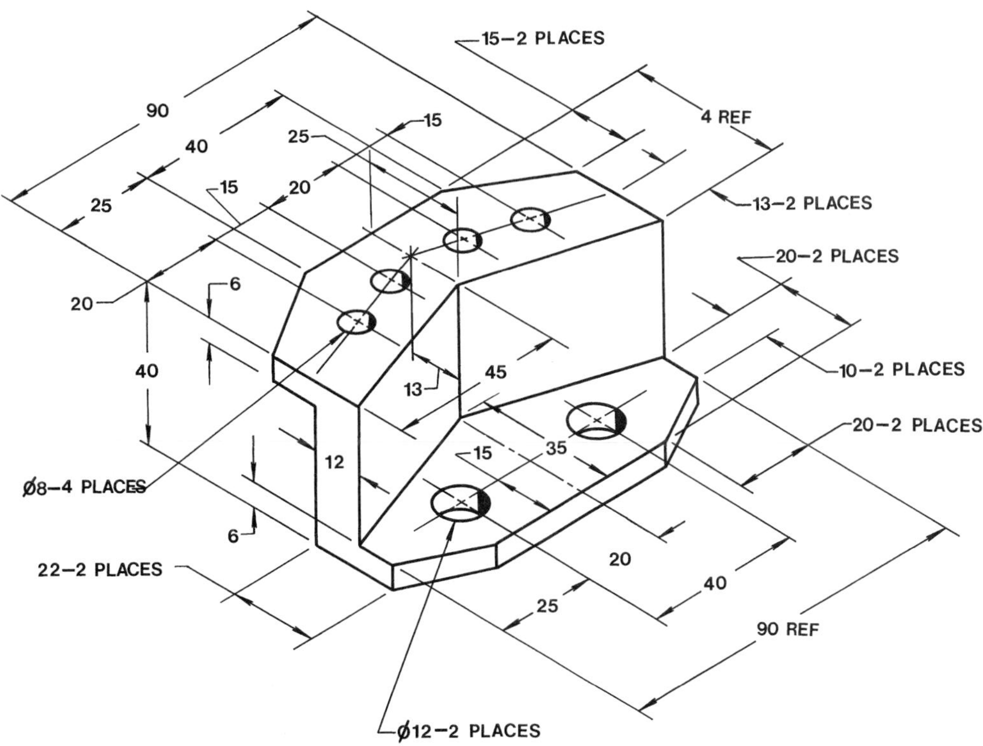

Figure P4-24
MILLIMETERS

Project 4-2

Draw at least two orthographic views and one auxiliary view of each of the objects shown in Figures P4-25 through P4-36.

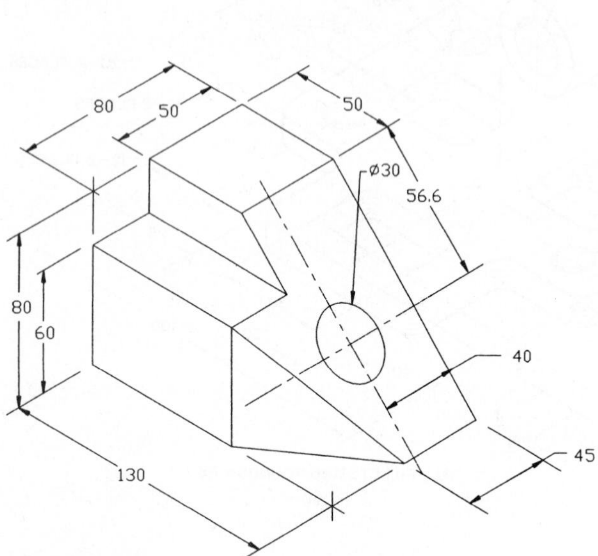

Figure P4-25
MILLIMETERS

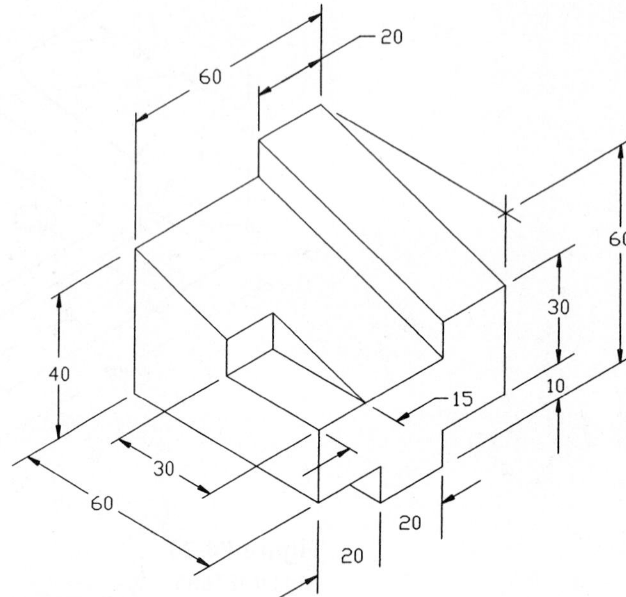

Figure P4-26
MILLIMETERS

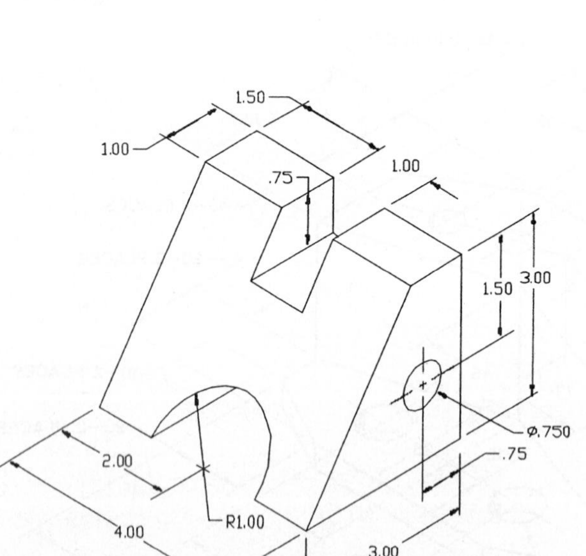

Figure P4-27
INCHES

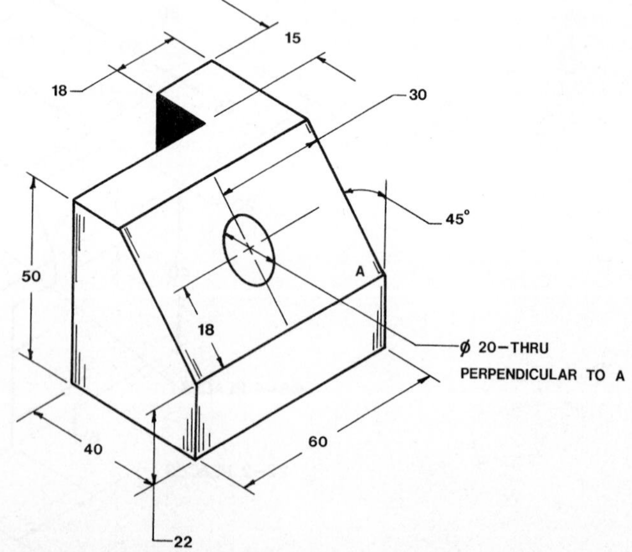

Figure P4-28
MILLIMETERS

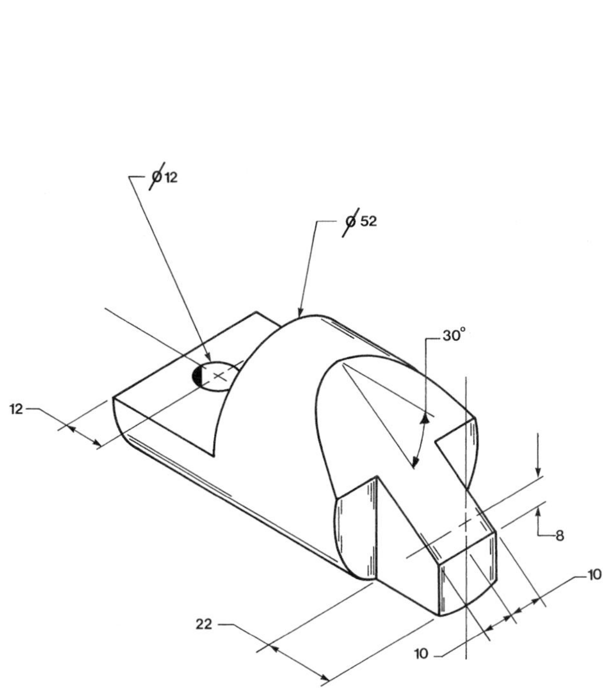

Figure P4-29
MILLIMETERS

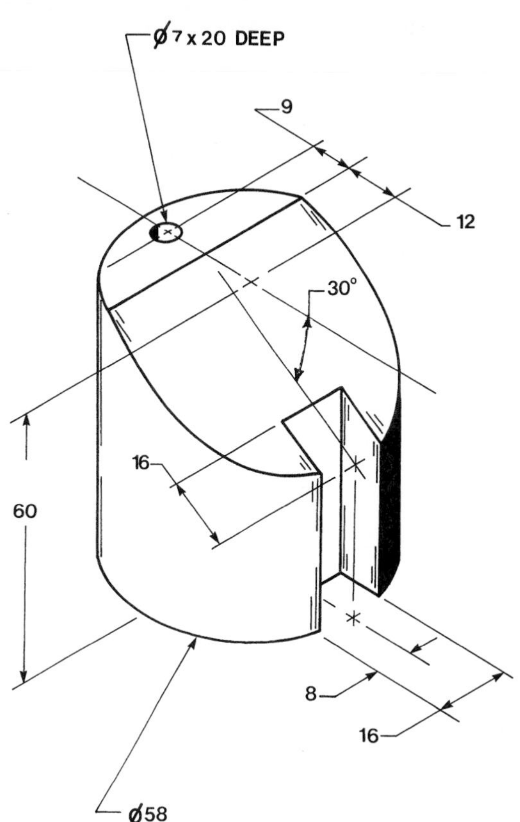

Figure P4-30
MILLIMETERS

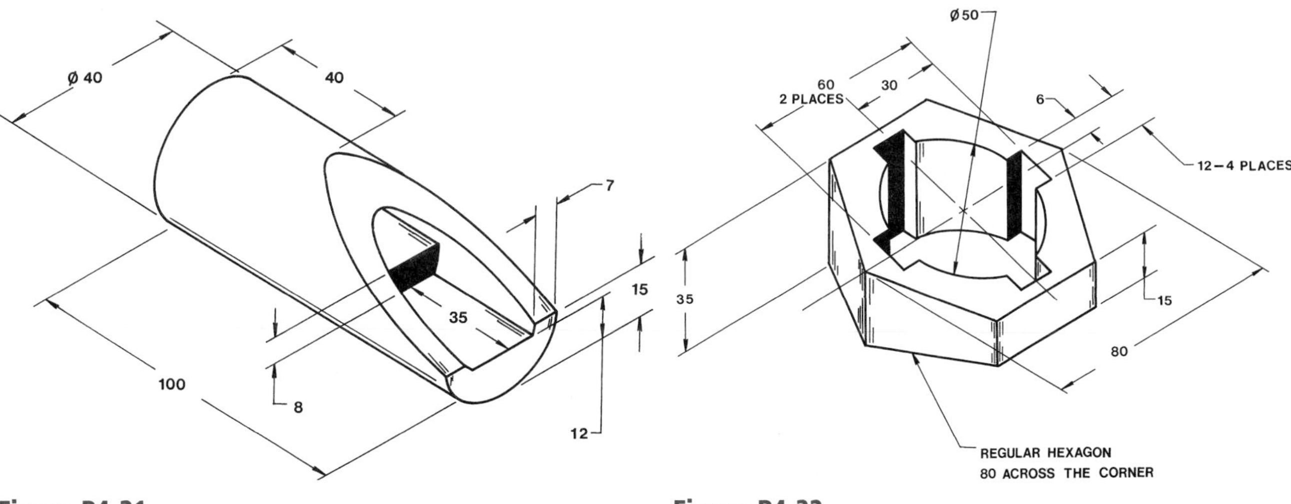

Figure P4-31
MILLIMETERS

Figure P4-32
MILLIMETERS

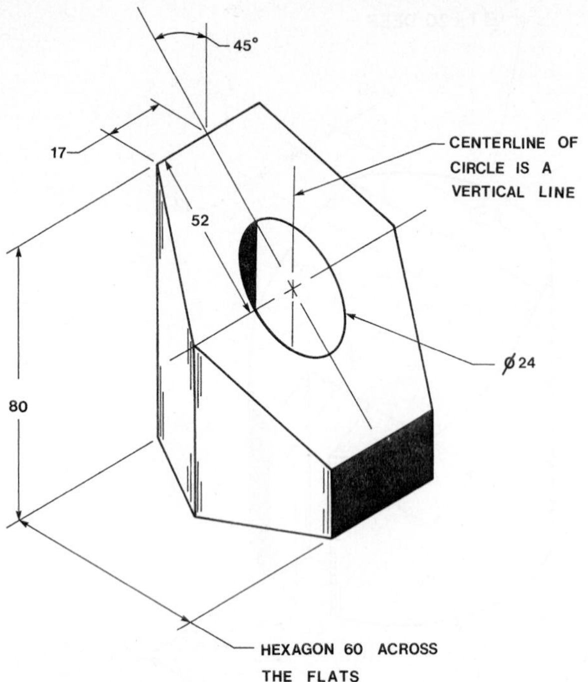

Figure P4-33
MILLIMETERS

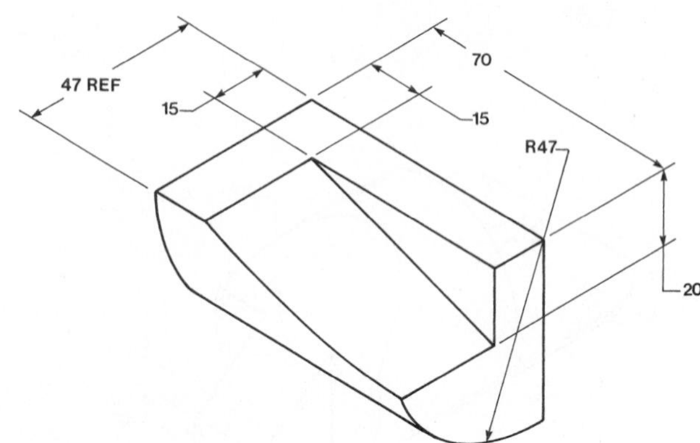

Figure P4-34
MILLIMETERS

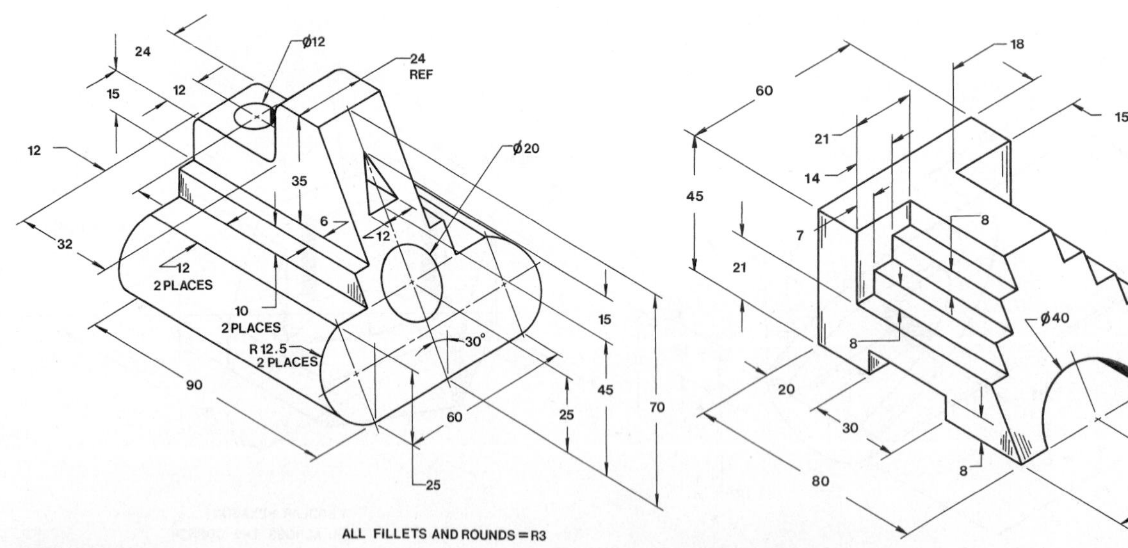

Figure P4-35
INCHES

Figure P4-36
MILLIMETERS

Project 4-3

Define the true shape of the oblique surfaces in each of the objects shown in Figures P4-37 through P4-40.

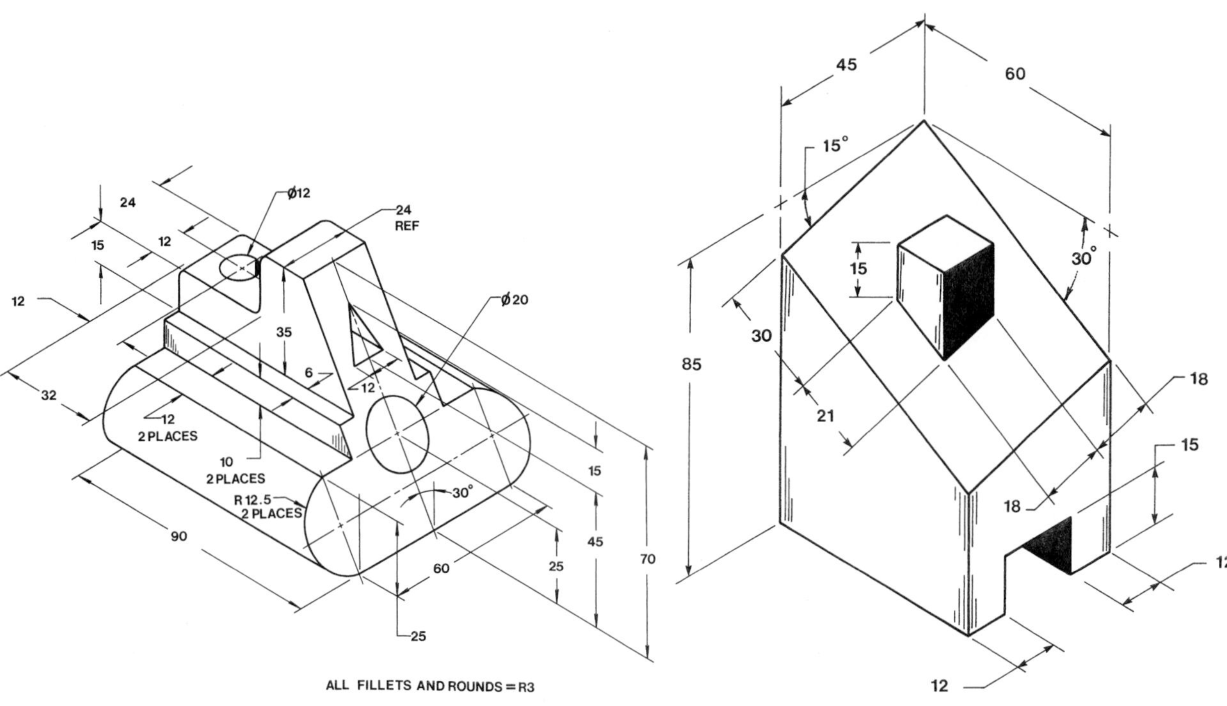

Figure P4-37
INCHES

Figure P4-38
MILLIMETERS

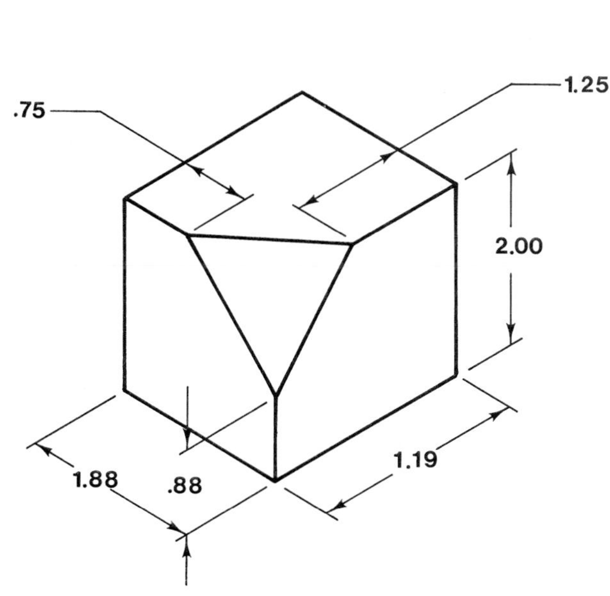

Figure P4-39
INCHES

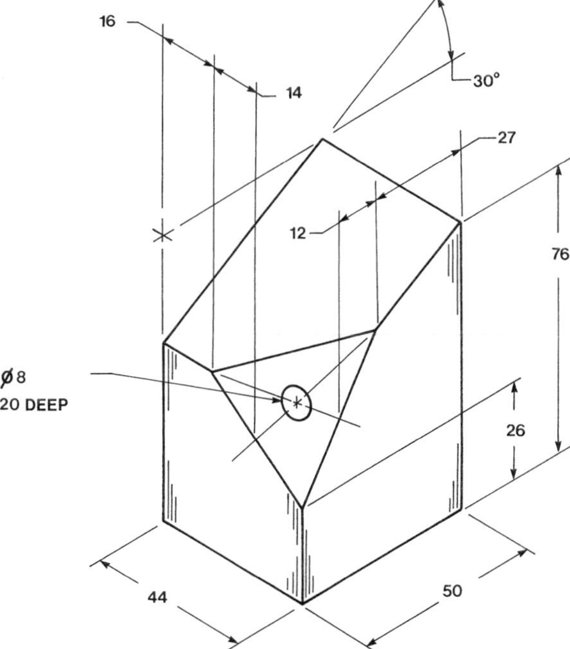

Figure P4-40
MILLIMETERS

Project 4-4

Draw each of the objects shown in Figures P4-41 through P4-44 as a model, then draw a front view and an appropriate section view of each.

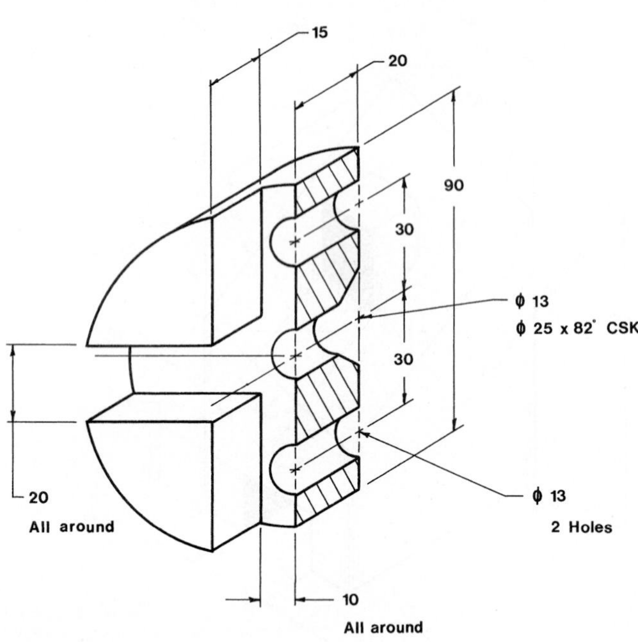

Figure P4-41
MILLIMETERS

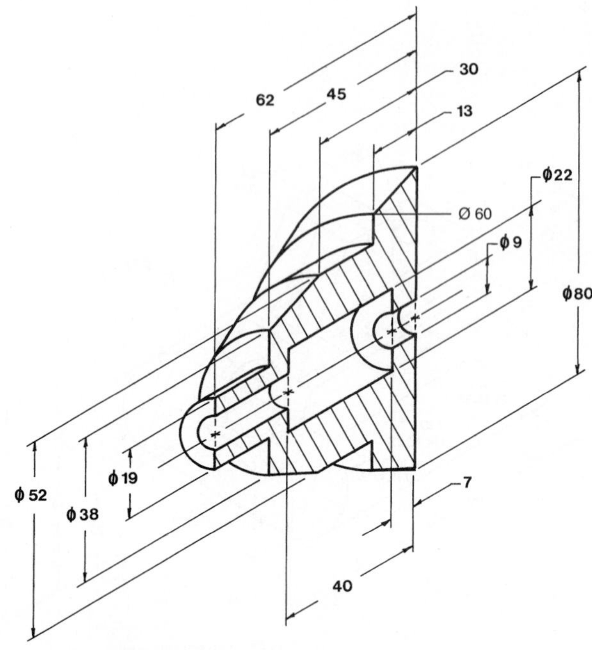

Figure P4-42
MILLIMETERS

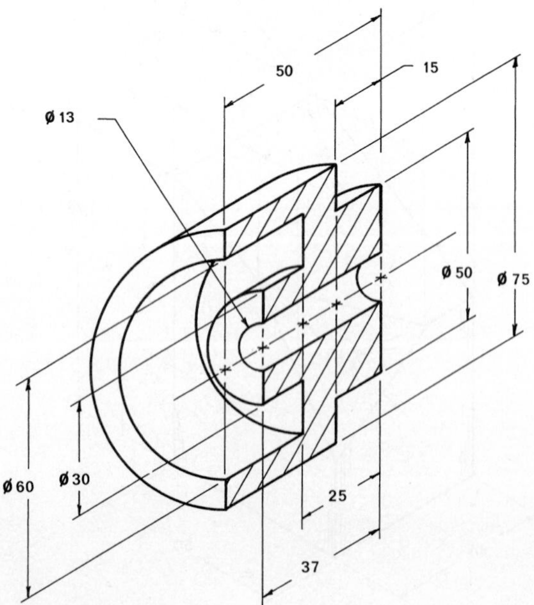

Figure P4-43
MILLIMETERS

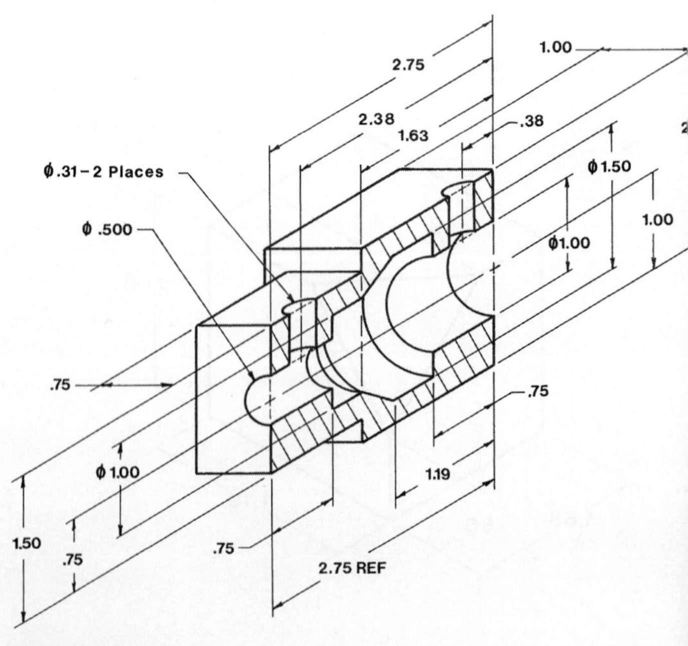

Figure P4-44
INCHES

Project 4-5

Draw at least one orthographic view and the indicated section view for each object shown in Figures P4-45 through P4-50.

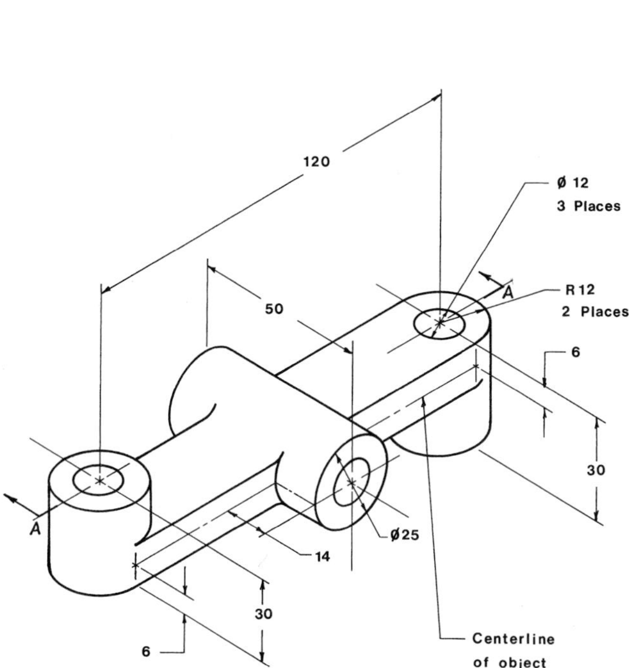

Figure P4-45
MILLIMETERS

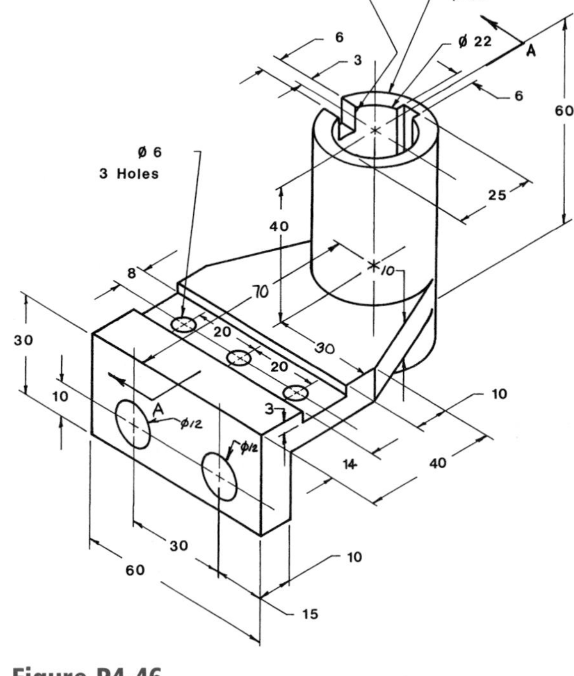

Figure P4-46
MILLIMETERS

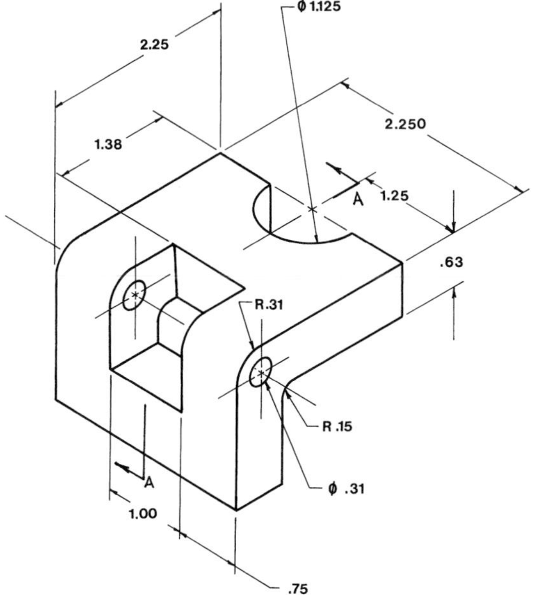

Figure P4-47
INCHES

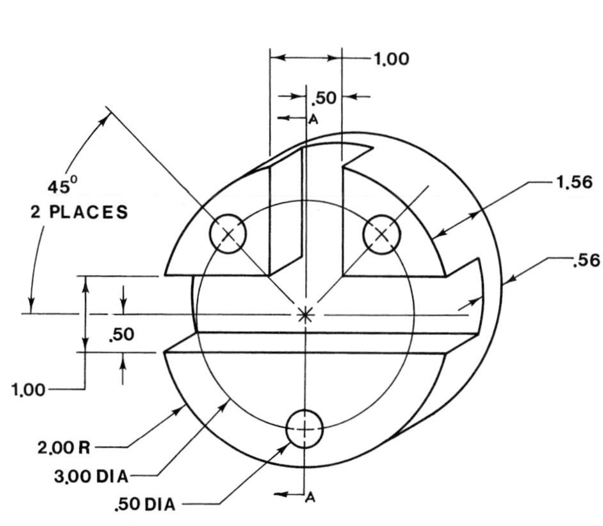

Figure P4-48
INCHES

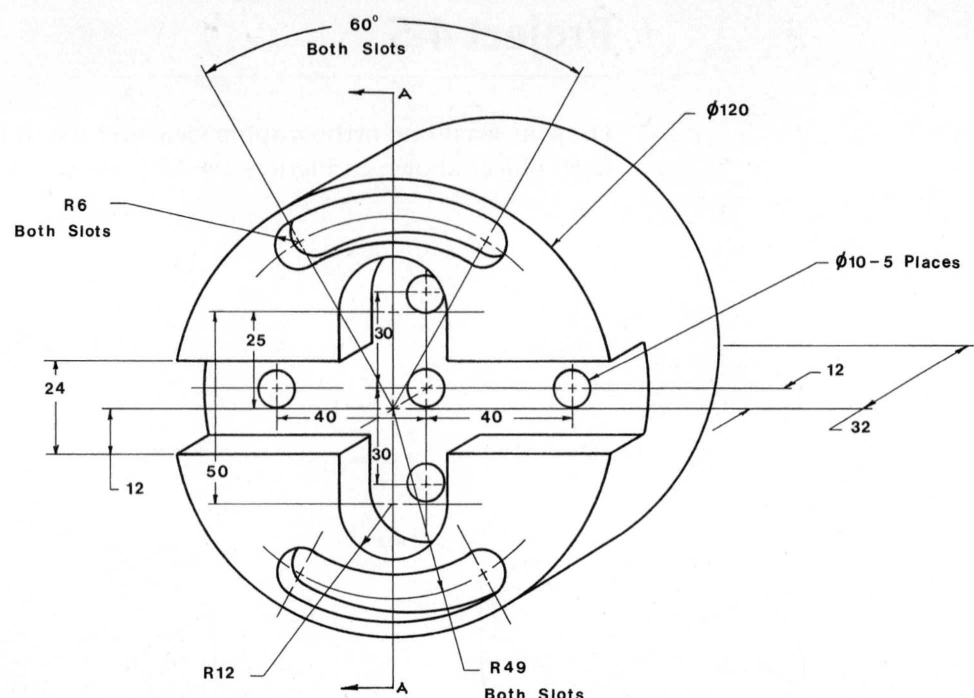

Figure P4-49
MILLIMETERS

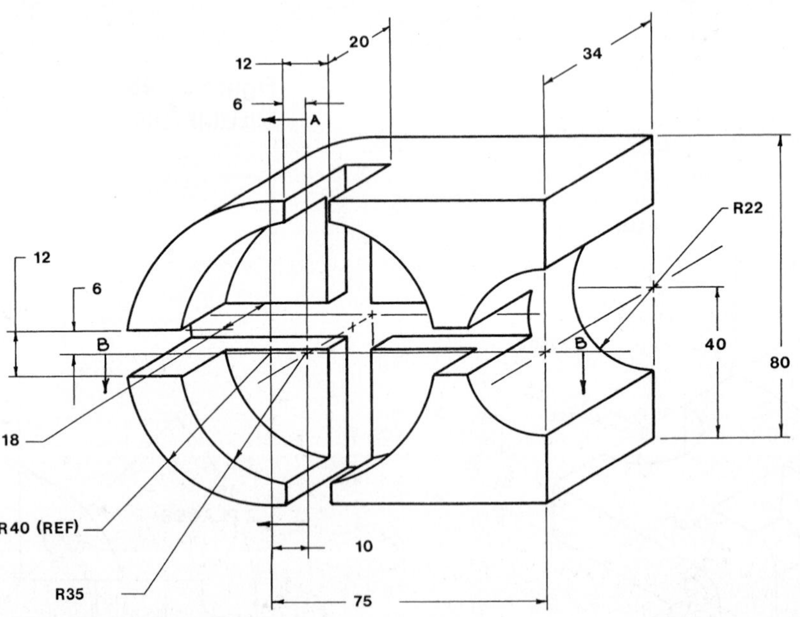

Figure P4-50
MILLIMETERS

Project 4-6

Given the orthographic views in Figures P4-51 and P4-52, draw a model of each, then draw the given orthographic views and the appropriate section views.

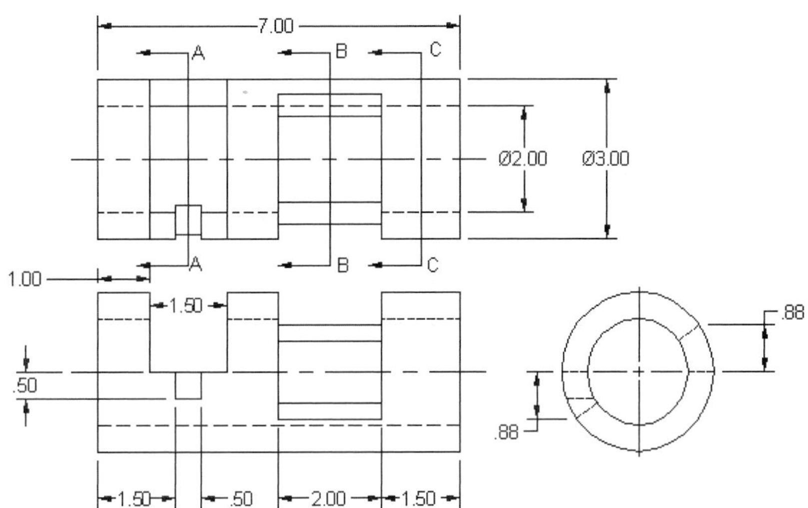

Figure P4-51
INCHES

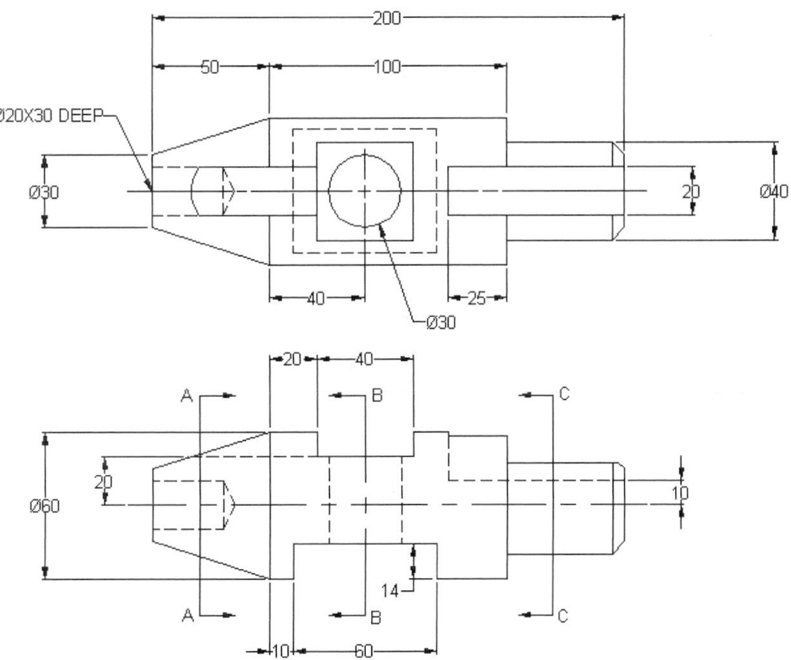

Figure P4-52
MILLIMETERS

Project 4-7

Draw a 3D model and a set of multiviews for each object shown in Figures P4-53 through P4-60.

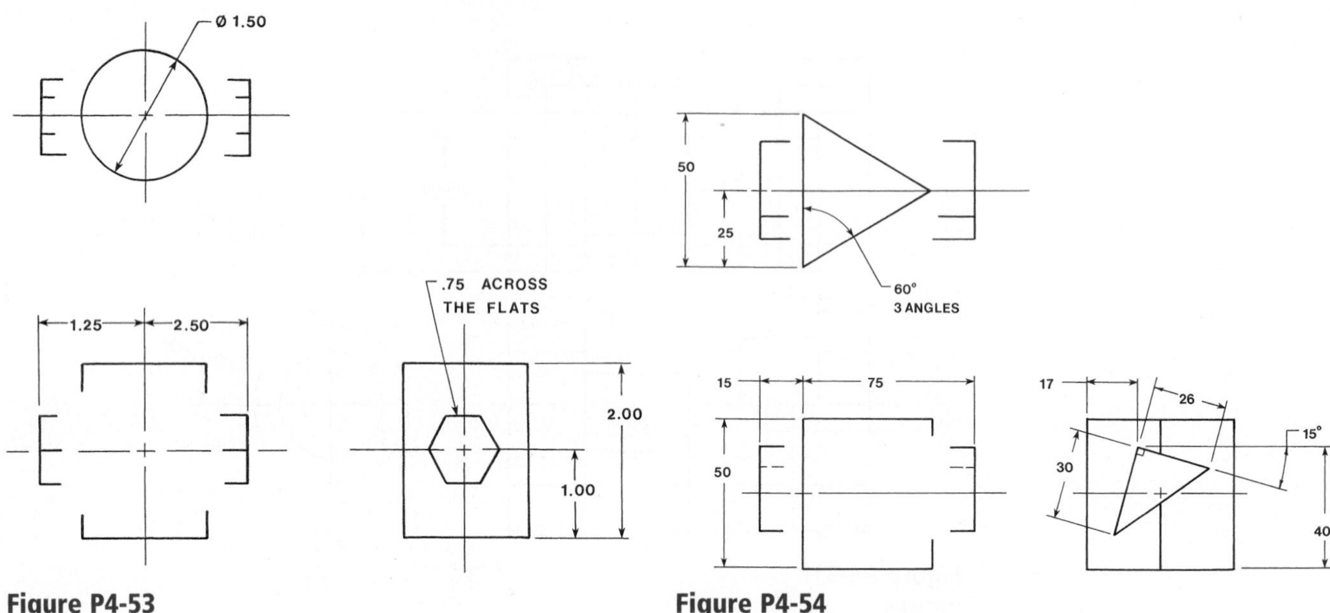

Figure P4-53
INCHES

Figure P4-54
MILLIMETERS

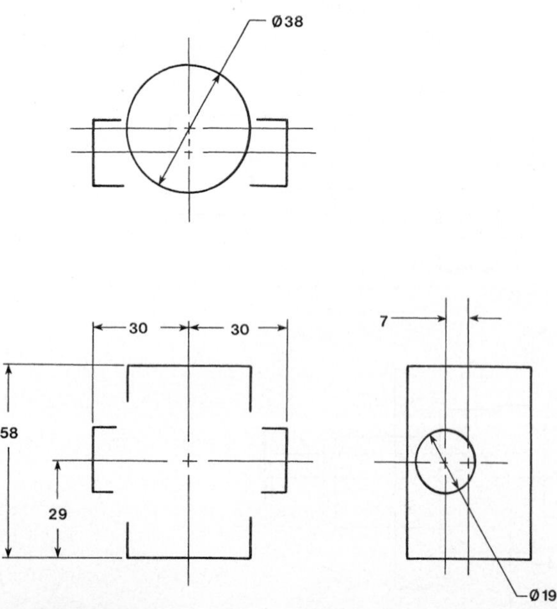

Figure P4-55
MILLIMETERS

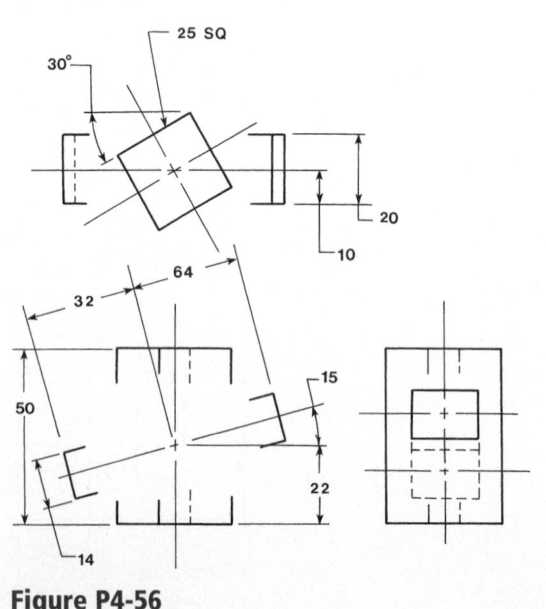

Figure P4-56
MILLIMETERS

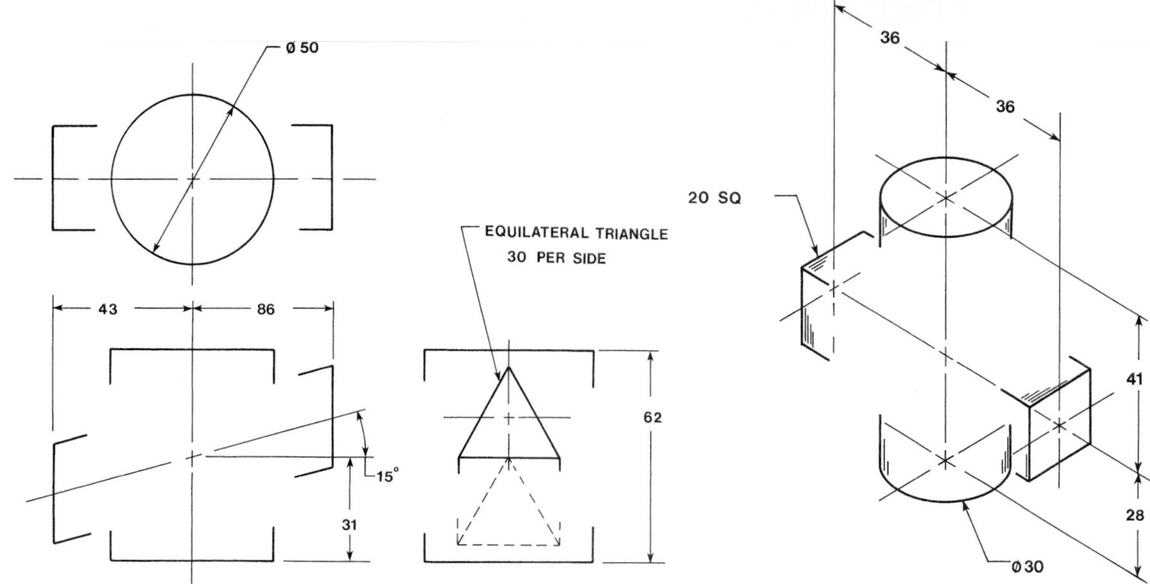

Figure P4-57
MILLIMETERS

Figure P4-58
MILLIMETERS

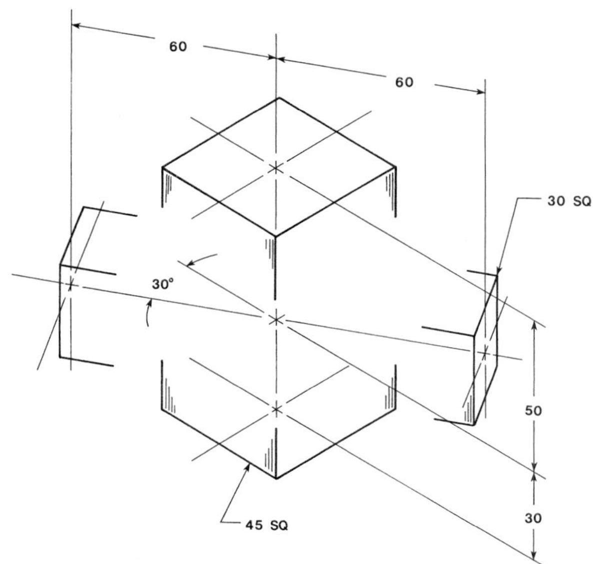

Figure P4-59
MILLIMETERS

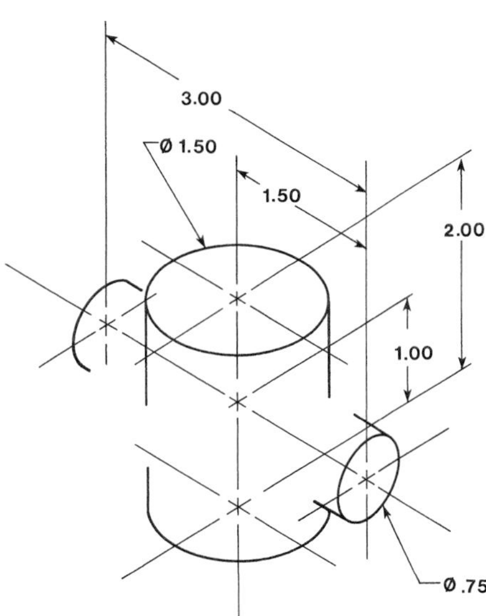

Figure P4-60
INCHES

Project 4-8

Figures P4-61 through P4-66 are orthographic views. Draw 3D models from the given views. The hole pattern defined in Figure P4-61 also applies to Figure P4-62.

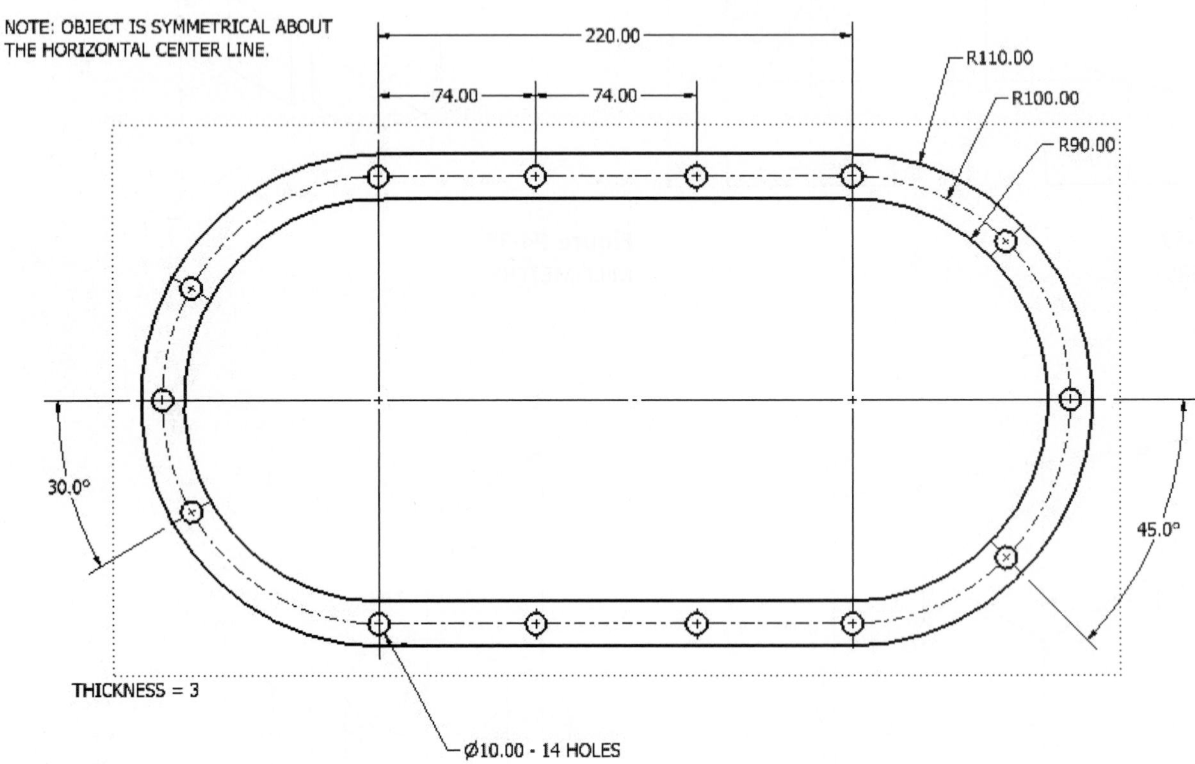

Figure P4-61
MILLIMETERS

Figure P4-62
MILLIMETERS

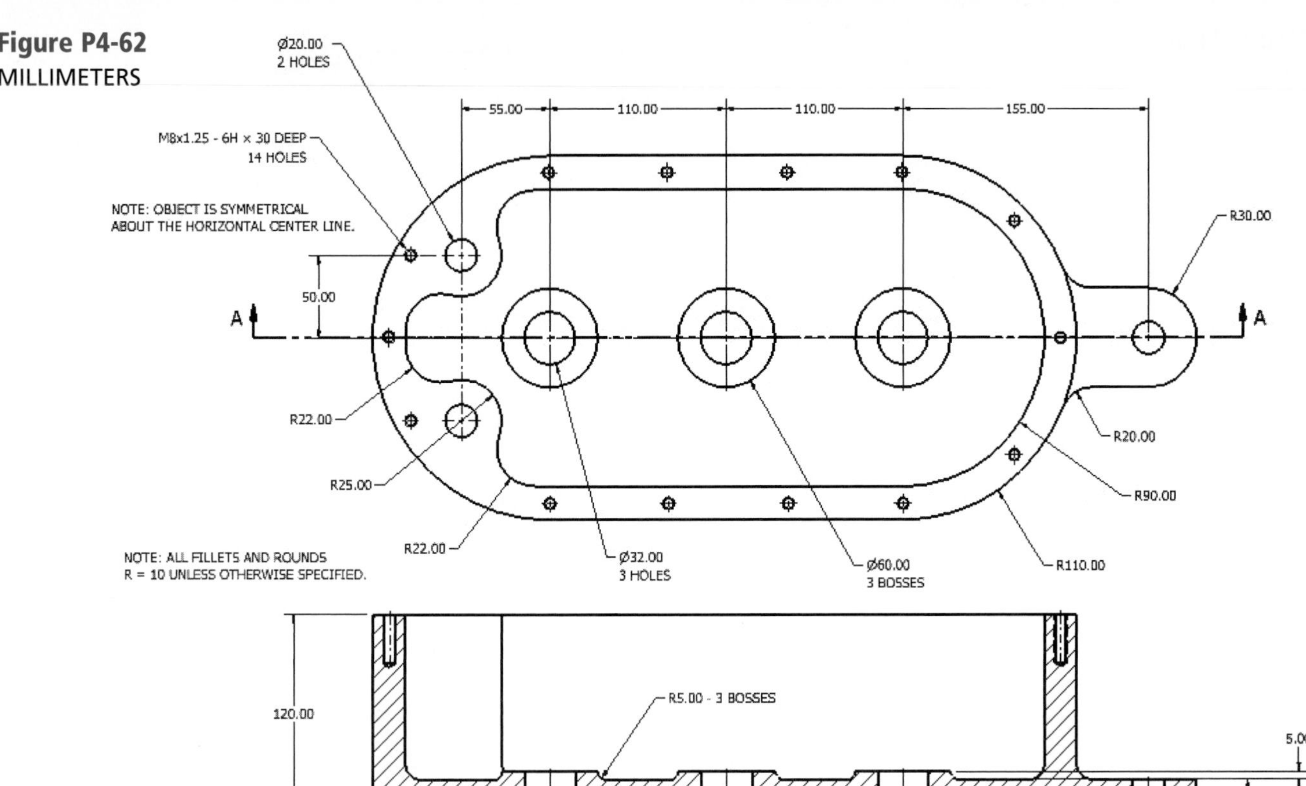
Ø20.00
2 HOLES
M8x1.25 - 6H × 30 DEEP
14 HOLES
NOTE: OBJECT IS SYMMETRICAL ABOUT THE HORIZONTAL CENTER LINE.
55.00
110.00
110.00
155.00
R30.00
50.00
A
A
R22.00
R25.00
R22.00
R20.00
R90.00
R110.00
NOTE: ALL FILLETS AND ROUNDS R = 10 UNLESS OTHERWISE SPECIFIED.
Ø32.00
3 HOLES
Ø60.00
3 BOSSES
120.00
R5.00 - 3 BOSSES
5.00
20.00
SECTION A-A
SCALE 1 / 2

Figure P4-63
INCHES

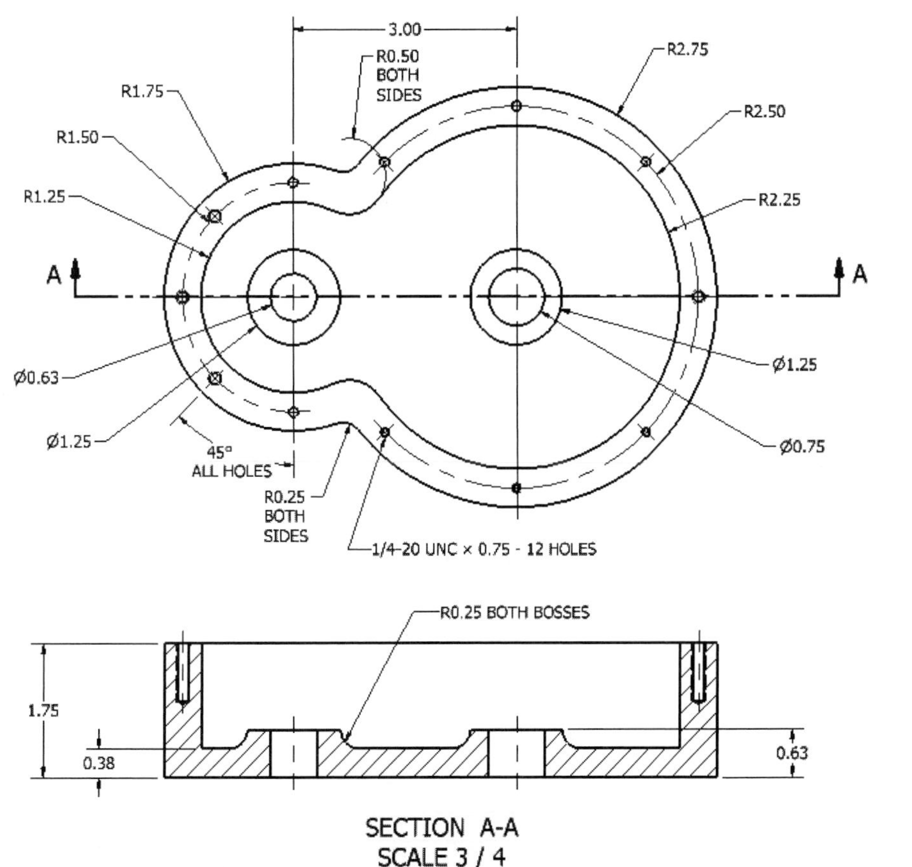
3.00
R0.50
BOTH
SIDES
R2.75
R1.75
R2.50
R1.50
R1.25
R2.25
A
A
Ø0.63
Ø1.25
Ø1.25
Ø0.75
45°
ALL HOLES
R0.25
BOTH
SIDES
1/4-20 UNC × 0.75 - 12 HOLES
R0.25 BOTH BOSSES
1.75
0.38
0.63
SECTION A-A
SCALE 3 / 4

Figure P4-64
INCHES

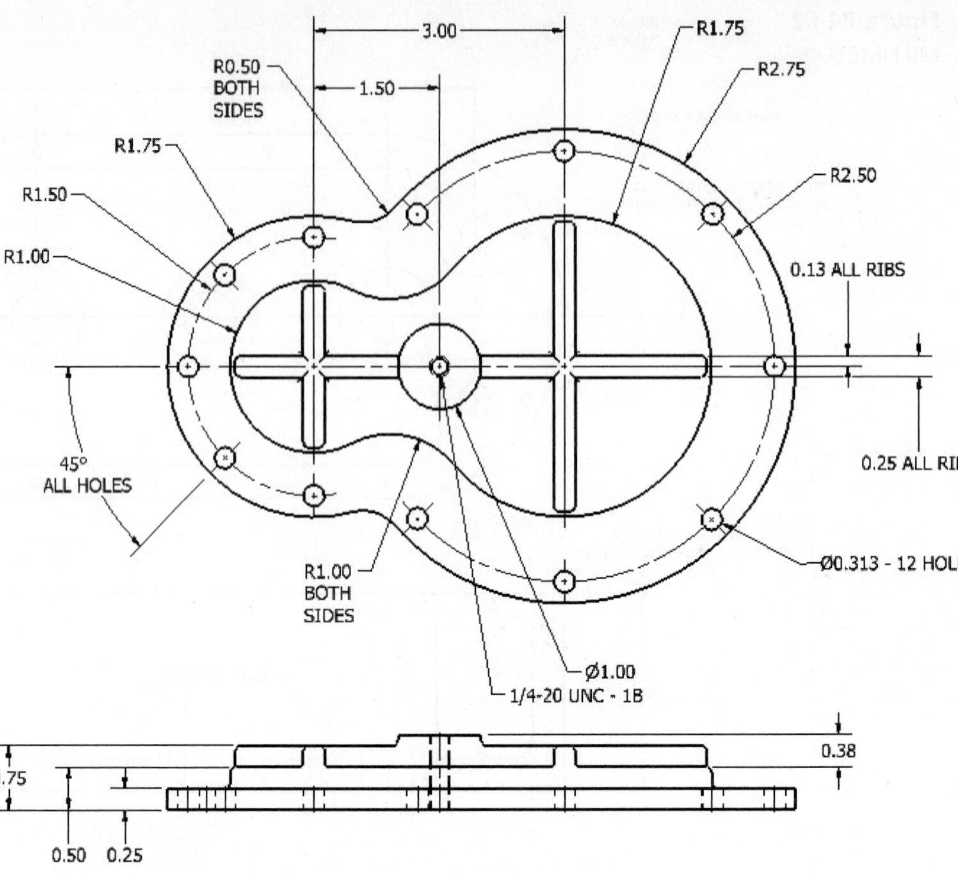

Figure P4-65
MILLIMETERS

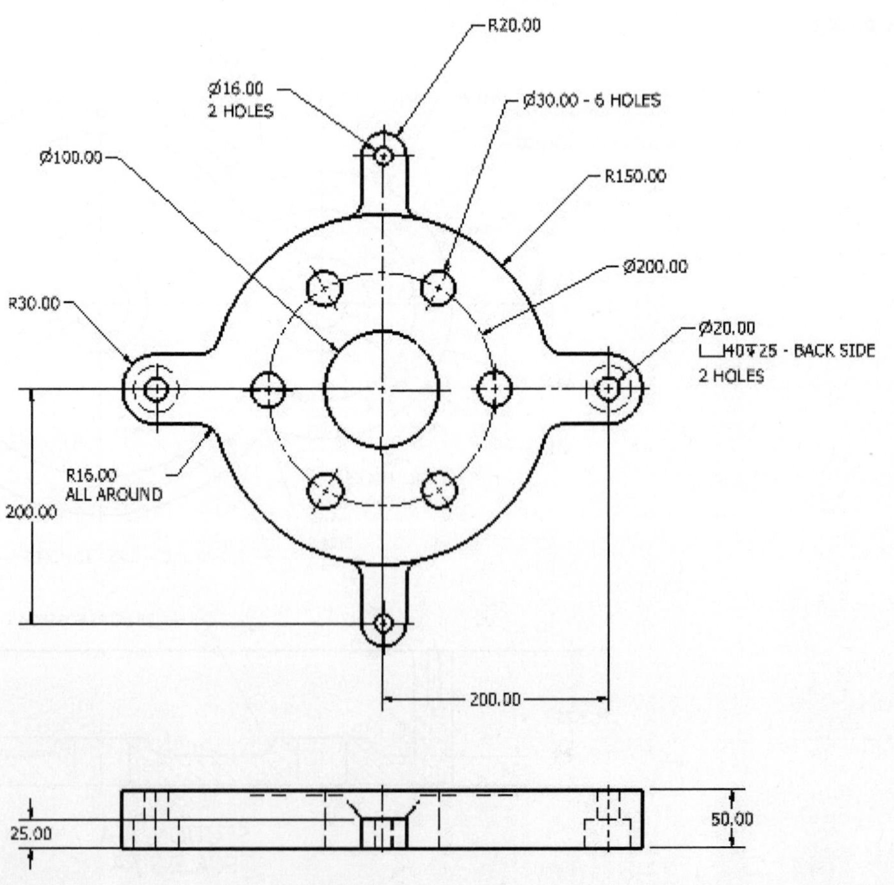

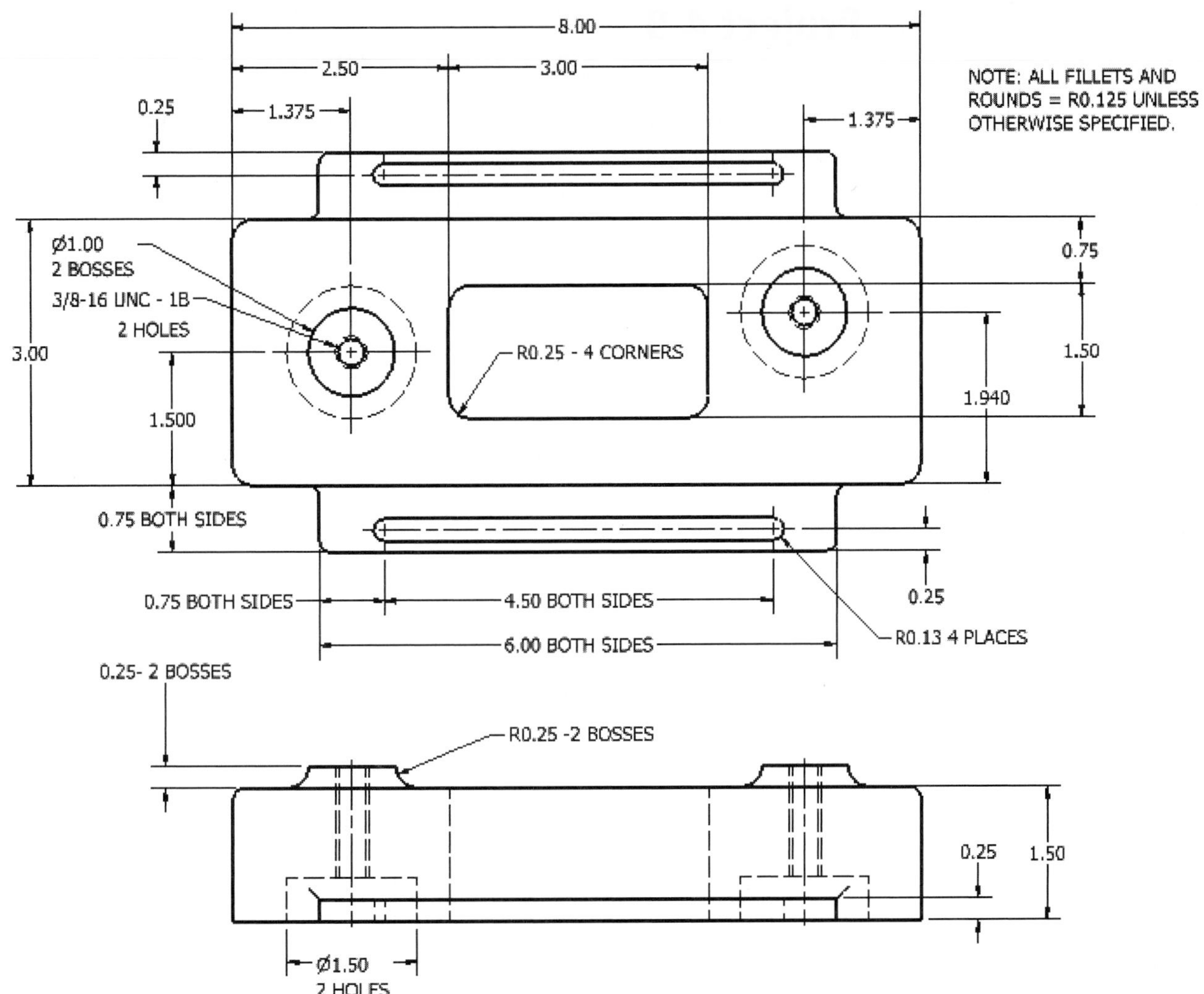

Figure P4-66
INCHES

Project 4-9

Figures P4-67 through P4-71 are presented using first-angle projection and ISO conventions.

A. Create a solid model from the given orthographic views.

B. Draw front, top, and right-side orthographic views of the objects using third-angle projection and ANSI conventions.

Figure P4-67
MILLIMETERS

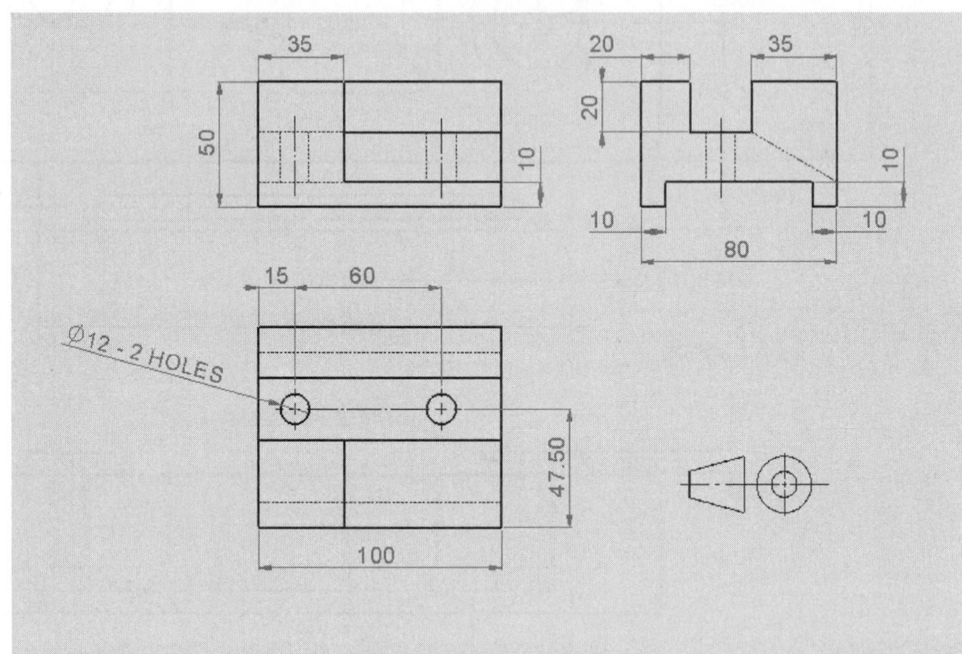

Figure P4-68
MILLIMETERS

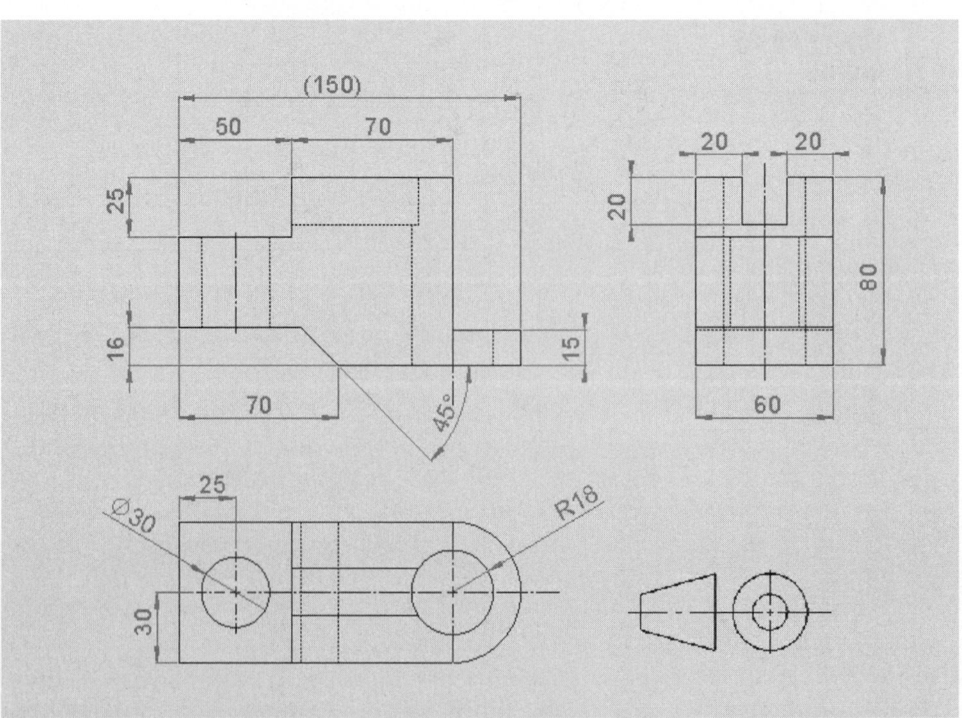

Figure P4-69
MILLIMETERS

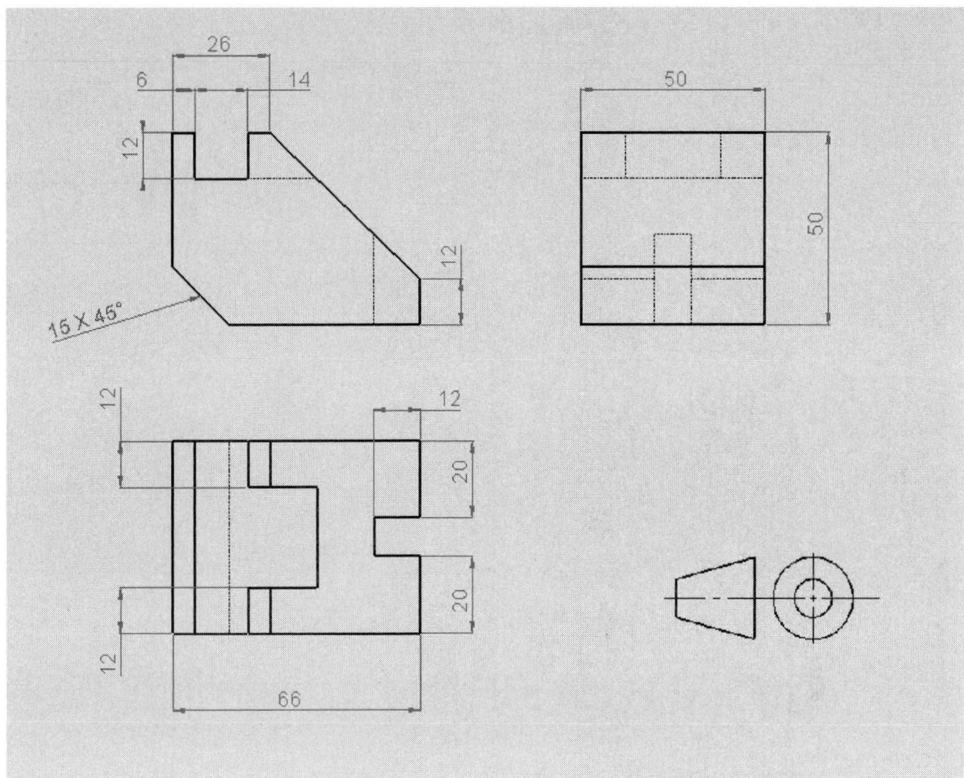

Figure P4-70
MILLIMETERS

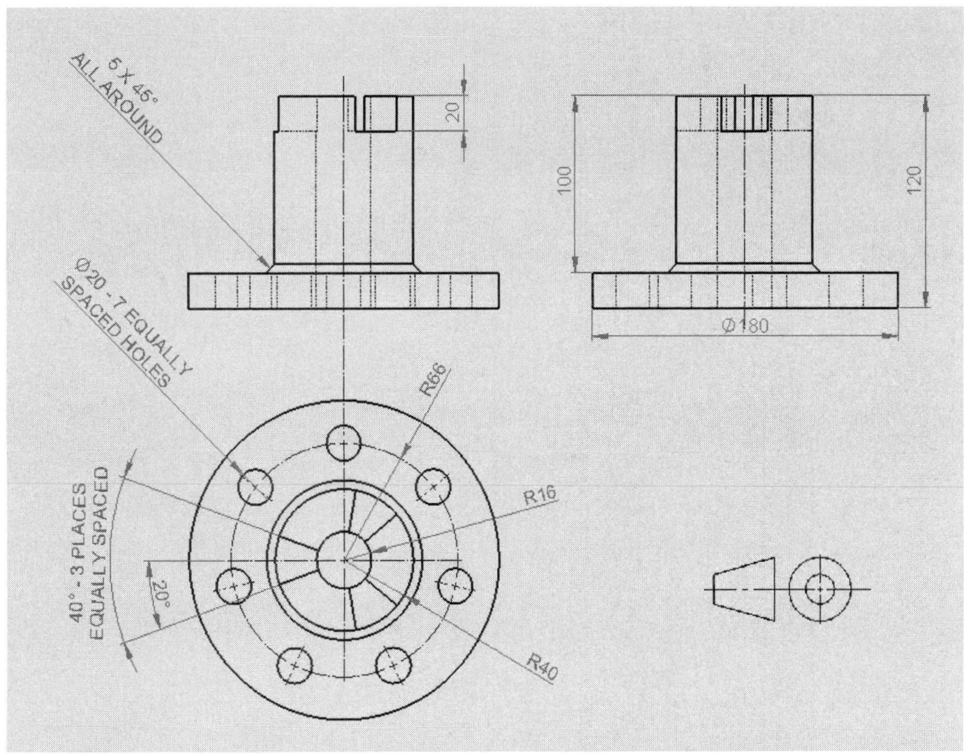

Figure P4-71
MILLIMETERS

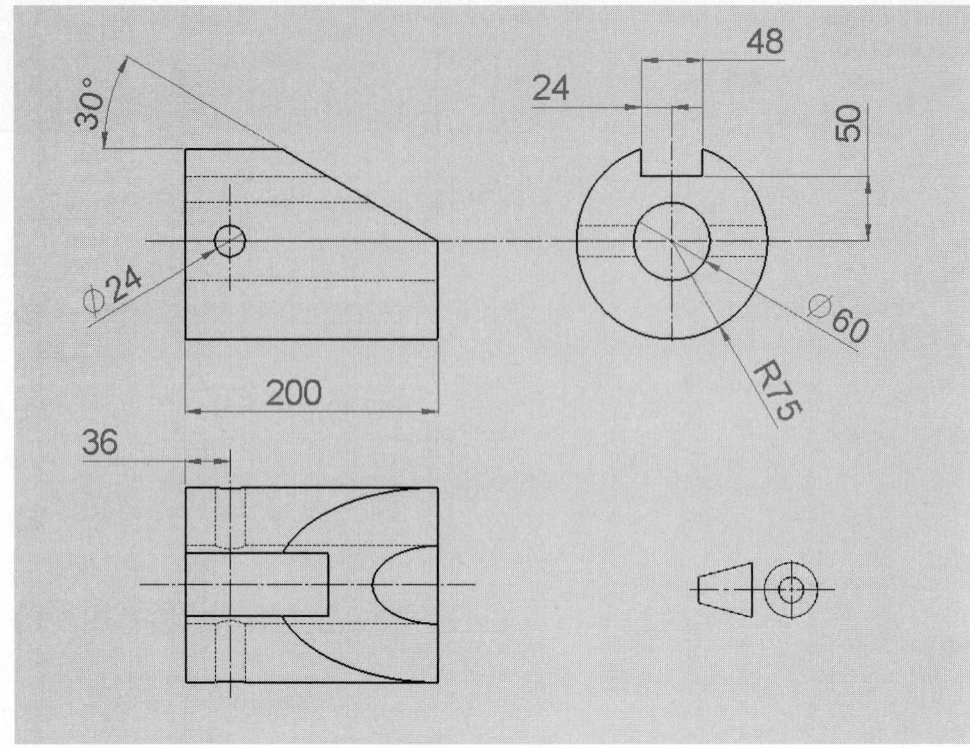

chapterfive
Assembly Drawings

CHAPTER OBJECTIVES

- Learn how to create assembly drawings
- Learn how to create a family of drawings
- Learn how to animate assembly drawings
- Learn how to edit assembly drawings

Introduction

This chapter explains how to create assembly drawings using a group of relatively simple parts to demonstrate the techniques required. The idea is to learn how to create assembly drawings and then gradually apply the knowledge to more difficult assemblies. For example, the next chapter introduces threads and fasteners and includes several exercise problems that require the use of fasteners when creating assembly drawings. Assembly drawings will be included throughout the remainder of the book.

This chapter also shows how to create bills of materials, isometric assembly drawings, title blocks, and other blocks associated with assembly drawings, as well as how to animate assembly drawings.

Bottom-Up and Top-Down Assemblies

bottom-up approach: Creating an assembly drawing by compiling files from existing drawings.

top-down approach: Creating an assembly drawing by creating model drawings on the assembly drawing.

There are three ways to create assembly drawings: bottom up, top down, or a combination of the two. A ***bottom-up*** approach uses drawings that already exist. Model drawings are pulled from files and compiled to create an assembly. The ***top-down*** approach creates model drawings from the assembly drawing. It is also possible to pull drawings from a file and then create more drawings as needed to complete the assembly.

Starting an Assembly Drawing

Assembly drawings are created using the .iam format. In this example the bottom-up approach will be used. It is assumed that a model called SQBLOCK already exists. The SQBLOCK figure was created from a 30 mm × 30 mm × 30 mm cube with a 15 mm × 15 mm × 30 mm cutout. See Figure 5-1.

Figure 5-1

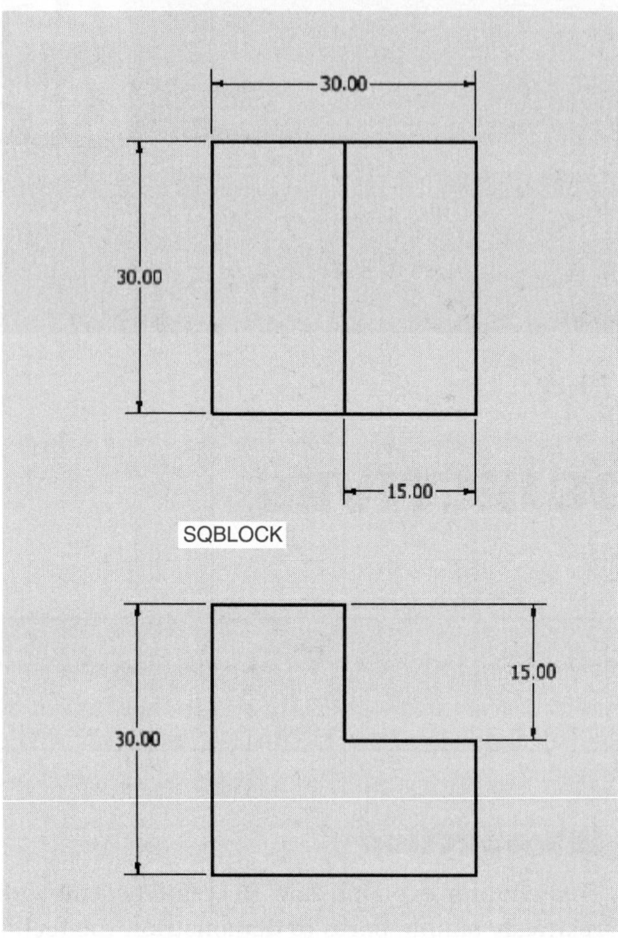

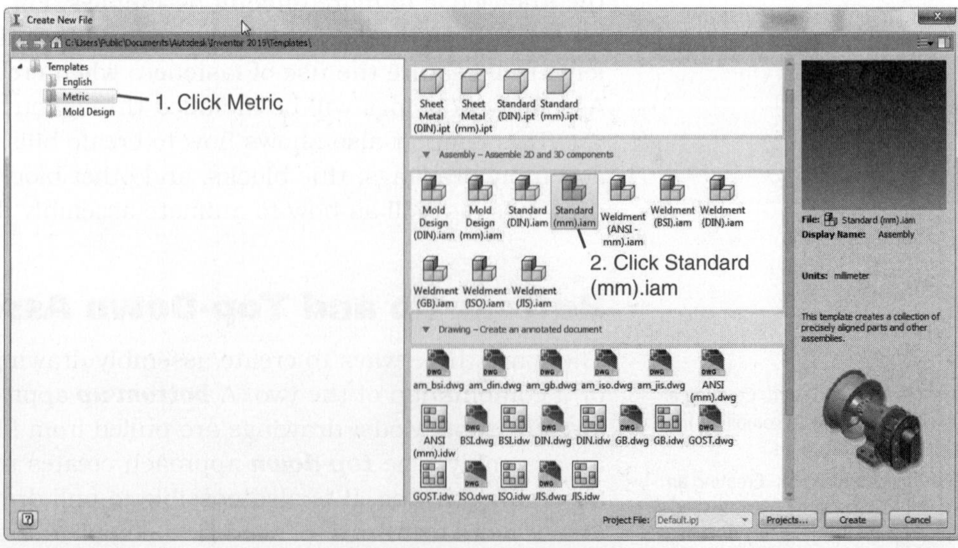

EXERCISE 5-1 Starting an Assembly Drawing

1 Click on the **New** tool (the **Create New File** dialog box appears), select the **Metric** tab, then **Standard (mm).iam.** Click **Create.**

See Figure 5-1. The **Assemble** tab will appear. See Figure 5-2.

Figure 5-2

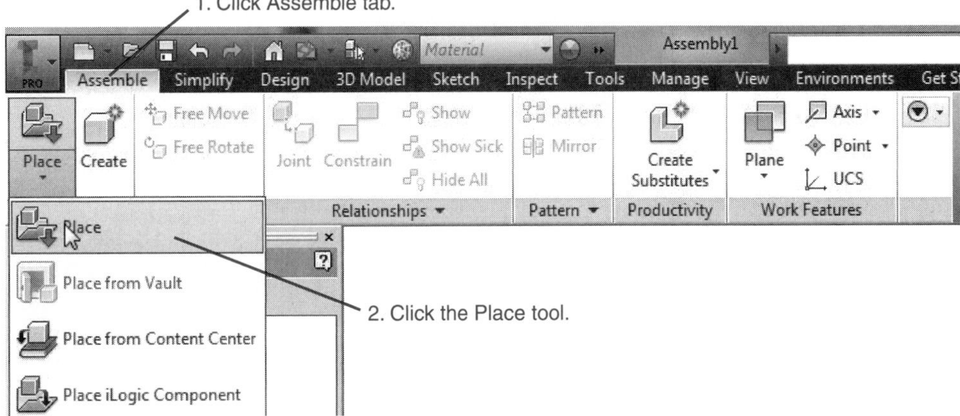

2 Click the **Place** tool located on the **Component** panel under the **Assemble** tab.

The **Place Component** dialog box will appear. See Figure 5-3.

Figure 5-3

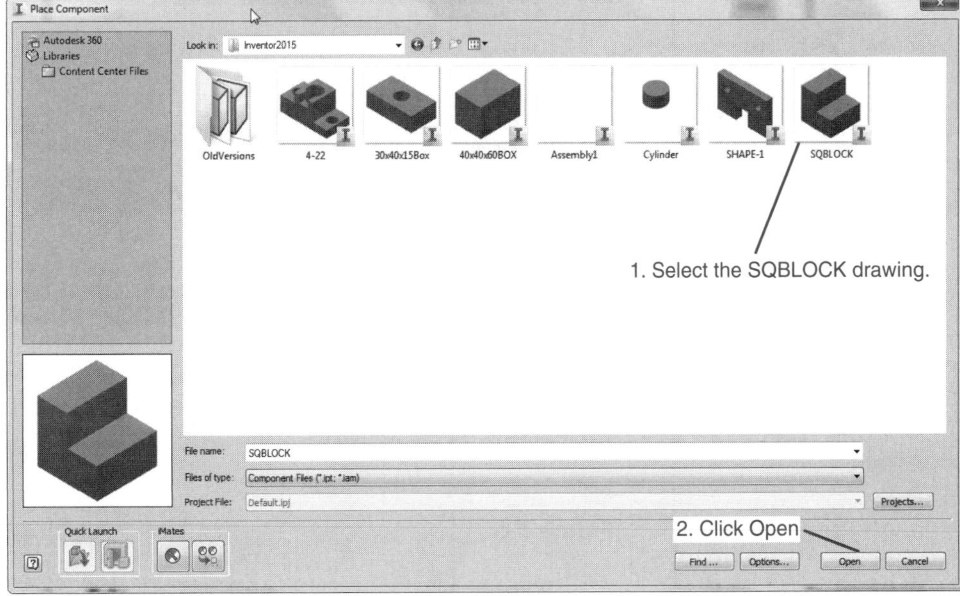

NOTE

Be sure to select the drive and file where the component is located

3 Click the desired file name, then **OPEN.**

In this example the **SQBLOCK** file was selected. The selected model (component) will appear on the screen.

4 Zoom the component to an appropriate size, then left-click the mouse to locate the component.

A second copy of the component will automatically appear.

5 Move the second component away from the first. Left-click the mouse to locate the second component, then right-click the mouse and select the **OK** option.

See Figure 5-4.

Figure 5-4

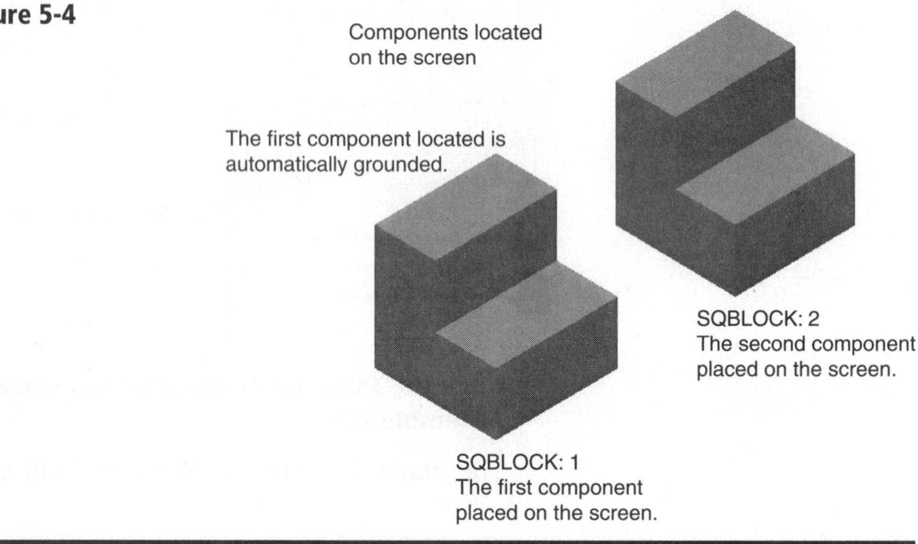
Components located on the screen
The first component located is automatically grounded.
SQBLOCK: 2
The second component placed on the screen.
SQBLOCK: 1
The first component placed on the screen.

NOTE

The mouse wheel is used to zoom the drawing. Moving the mouse while holding the wheel down will move the drawing.

Degrees of Freedom

grounded component: A component of a drawing that will not move when assembly tools are applied.

Components are either free to move or they are grounded. ***Grounded components*** will not move when assembly tools are applied. The first component will automatically be grounded. Grounded components are identified by a pushpin icon in the browser box. See Figure 5-5.

Figure 5-5

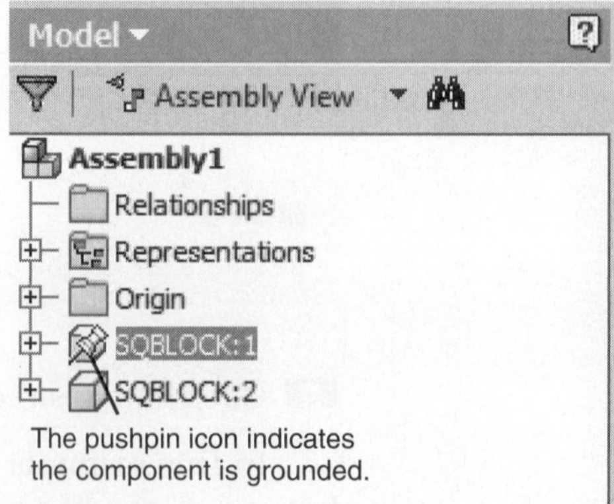
Model
Assembly View
Assembly1
Relationships
Representations
Origin
SQBLOCK:1
SQBLOCK:2
The pushpin icon indicates the component is grounded.

EXERCISE 5-2 Displaying the Degrees of Freedom

Components that are not grounded will have degrees of freedom. The available degrees of freedom for a component may be seen by using the **Degrees of Freedom** option.

1 Click the **View** tab at the top of the screen.

2 Click the **Degrees of Freedom** tool located on the **Visibility** panel.

See Figure 5-6. The available degrees of freedom will appear on the components. See Figure 5-7. Note that in Figure 5-7 the SQBLOCK:1 does not have any degrees of freedom; it is grounded. See the pushpin in the browser box.

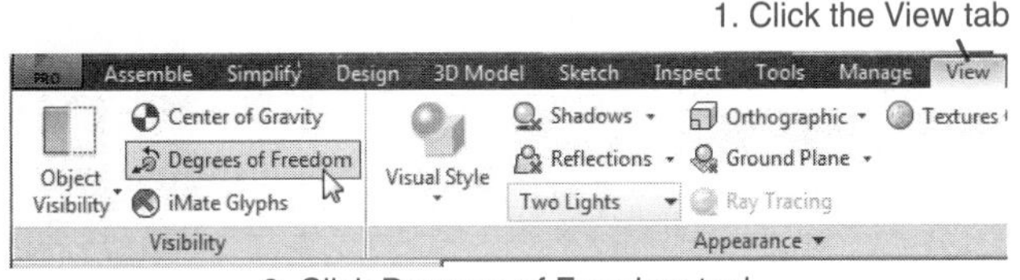

Figure 5-6

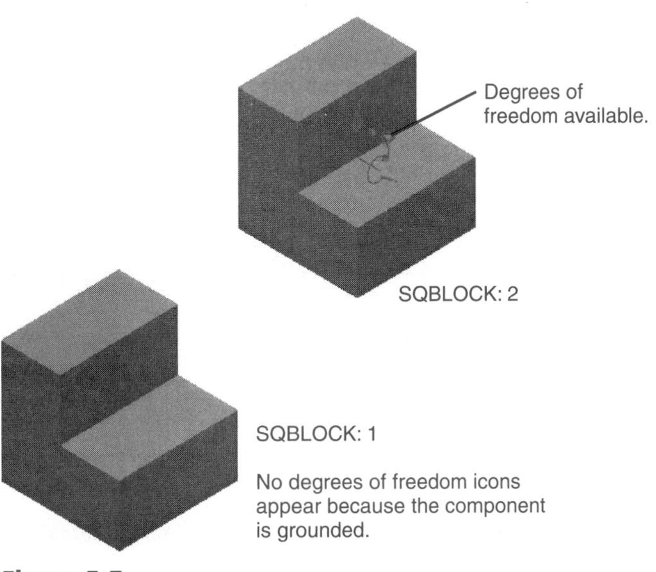

Figure 5-7

EXERCISE 5-3 Ungrounding a Component

1 Right-click on the **SQBLOCK:1** heading in the browser box.

A dialog box will appear. See Figure 5-8.

2 Click the **Grounded** option.

3 Use the **Undo** command and return the drawing to its original configuration: one block grounded, one block not grounded.

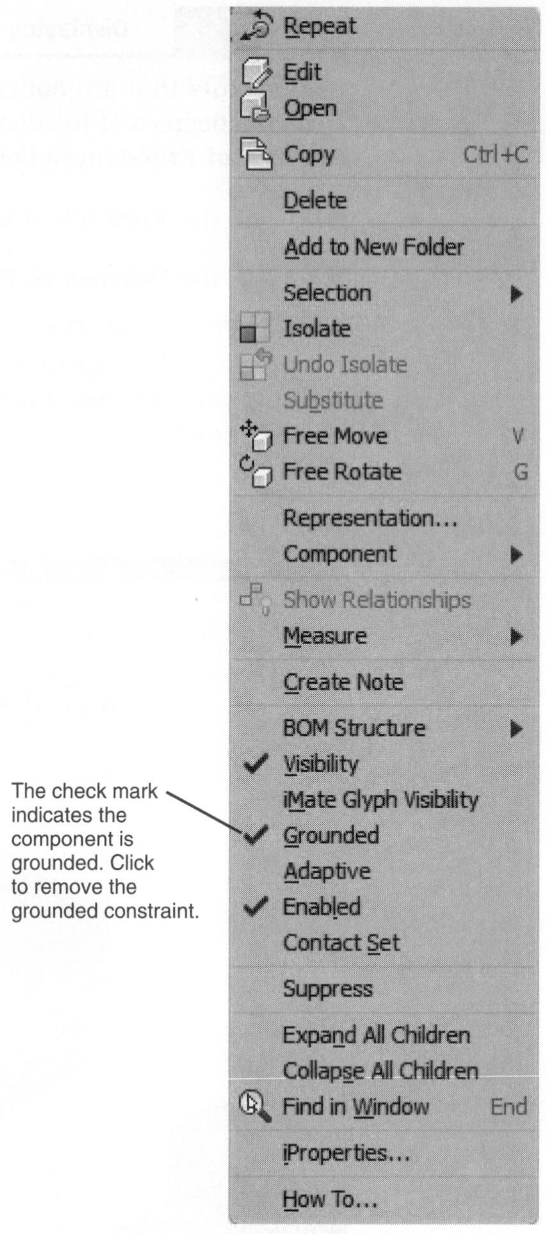

Figure 5-8

Moving Components and Rotating Components

The **Move Component** and **Rotate Component** tools are found on the **Position** panel under the **Assemble** tab. The tools are used, as their names imply, to move and rotate individual components.

> **NOTE**
>
> Hold down the mouse wheel to *move* the entire drawing, and the ViewCube to *rotate* the entire drawing.

EXERCISE 5-4 **Moving a Component**

1 Click the **Move Component** tool, and click the component **SQBLOCK:2**.

NOTE

SQBLOCK:1 is a grounded component and cannot be moved or rotated. If you want to reorient SQBLOCK:1, right-click the component callout in the browser box and remove the grounded constraint.

2 Hold the left mouse button down and move the component about the screen.

3 When the desired location is reached, release the left button.

4 Right-click the mouse and select the **OK** option.

EXERCISE 5-5 **Rotating a Component**

1 Click the **Rotate Component** tool, and click the component to rotate.

A circle will appear around the component. See Figure 5-9.

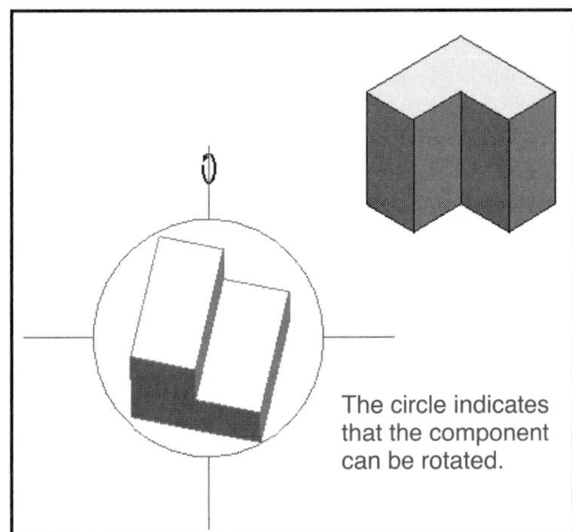

The circle indicates
that the component
can be rotated.

Figure 5-9

2 Click and hold the left mouse button outside the circle and move the cursor.

The component will rotate. Click various points outside the circle to see how the component can be rotated.

3 When the desired orientation is achieved, press the right mouse button and select the **Done** option.

Constrain

The **Constrain** tool is used to locate components relative to one another. Components may be constrained using the **Mate, Flush, Angle, Tangent,** or **Insert** option.

EXERCISE 5-6 Using the Mate Option

1. Click the **Constrain** tool located on the **Position** panel under the **Assemble** tab.

The **Place Constraint** dialog box will appear. See Figure 5-10. The **Mate** option will automatically be selected.

Figure 5-10

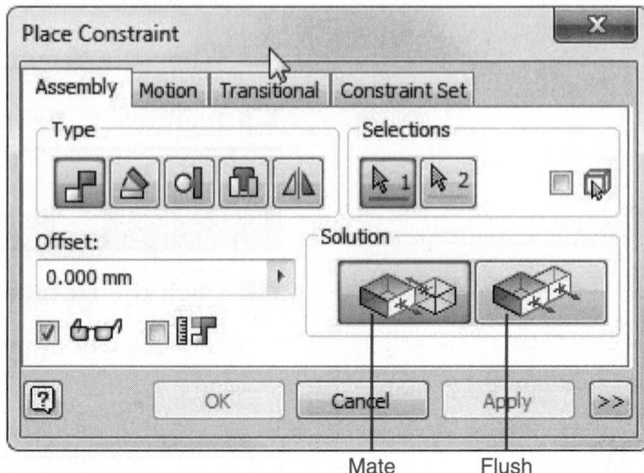

2. Click the front face of **SQBLOCK:1** as shown.
3. Click the front face of **SQBLOCK:2** as shown.
4. Click the **Apply** box on the **Place Constraint** dialog box or right-click the mouse and select the **Apply** option.

See Figure 5-11. The blocks will be joined at the selected surfaces. The blocks may not be perfectly aligned when assembled. This situation may be corrected using the **Flush** option.

Figure 5-11

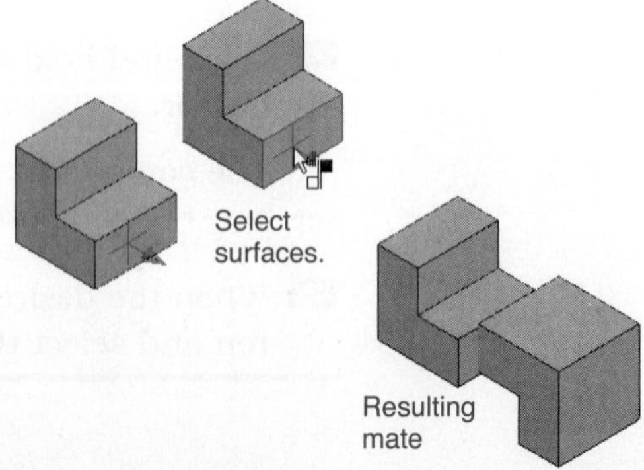

The **Mate** option may also be used to align centerlines of holes, shafts, and fasteners and the edges of models. See Figure 5-12.

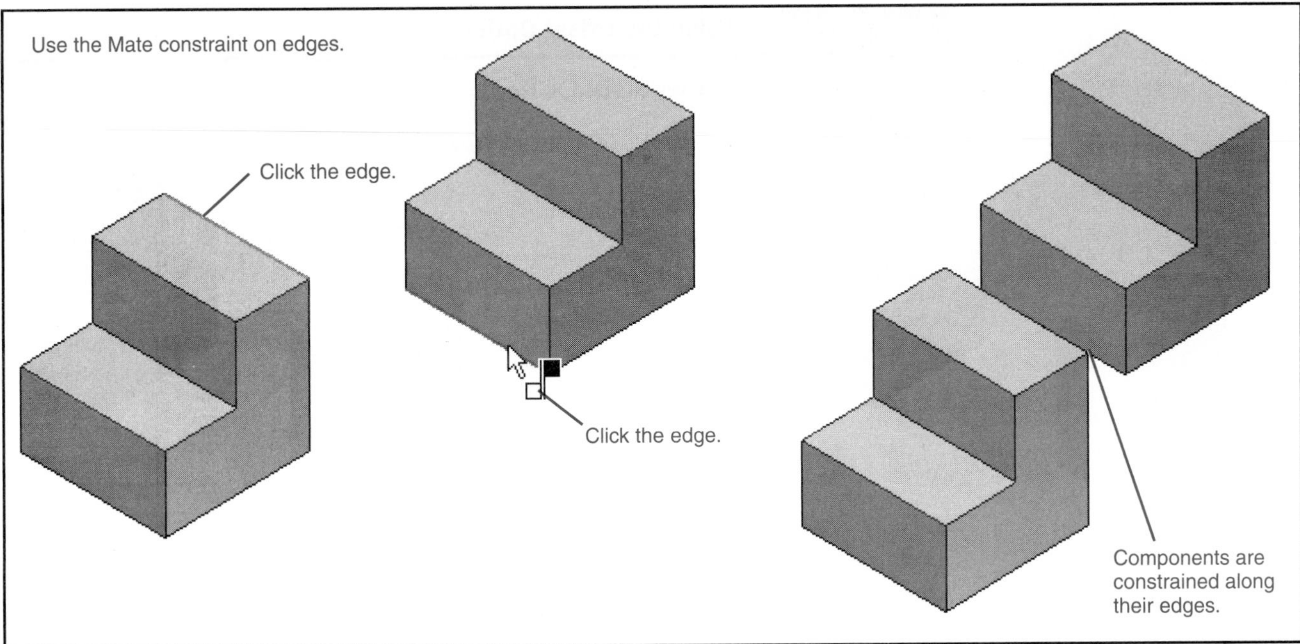

Figure 5-12

EXERCISE 5-7 Using the Flush Option

1. Move and rotate the components approximately into the position shown in Figure 5-13.

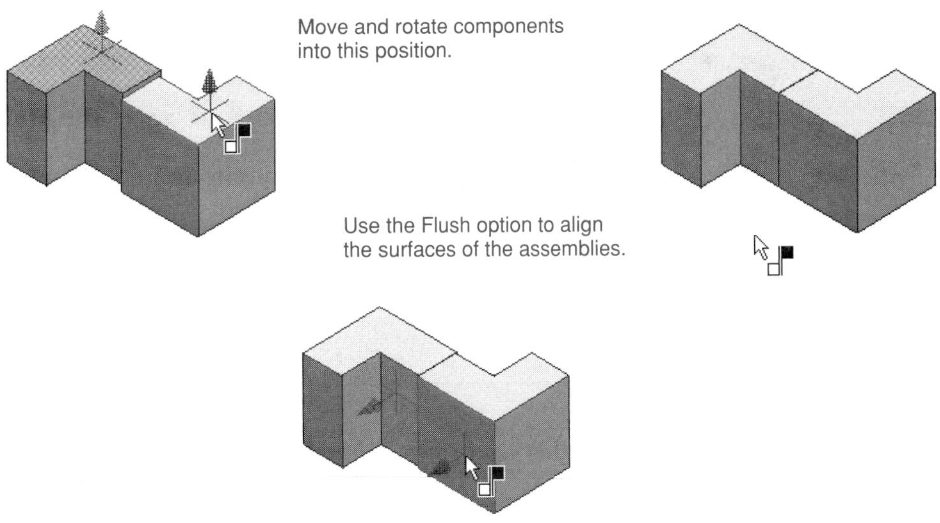

Move and rotate components into this position.
Use the Flush option to align the surfaces of the assemblies.

Figure 5-13

2. Click the **Flush** tool on the **Place Constraint** dialog box.
3. Click the top surface of each block as shown.
4. Click the **Apply** button.
5. Make other surfaces flush as needed to align the two blocks.

EXERCISE 5-8 Using the Offset Option

Figure 5-14 shows two SQBLOCKs.

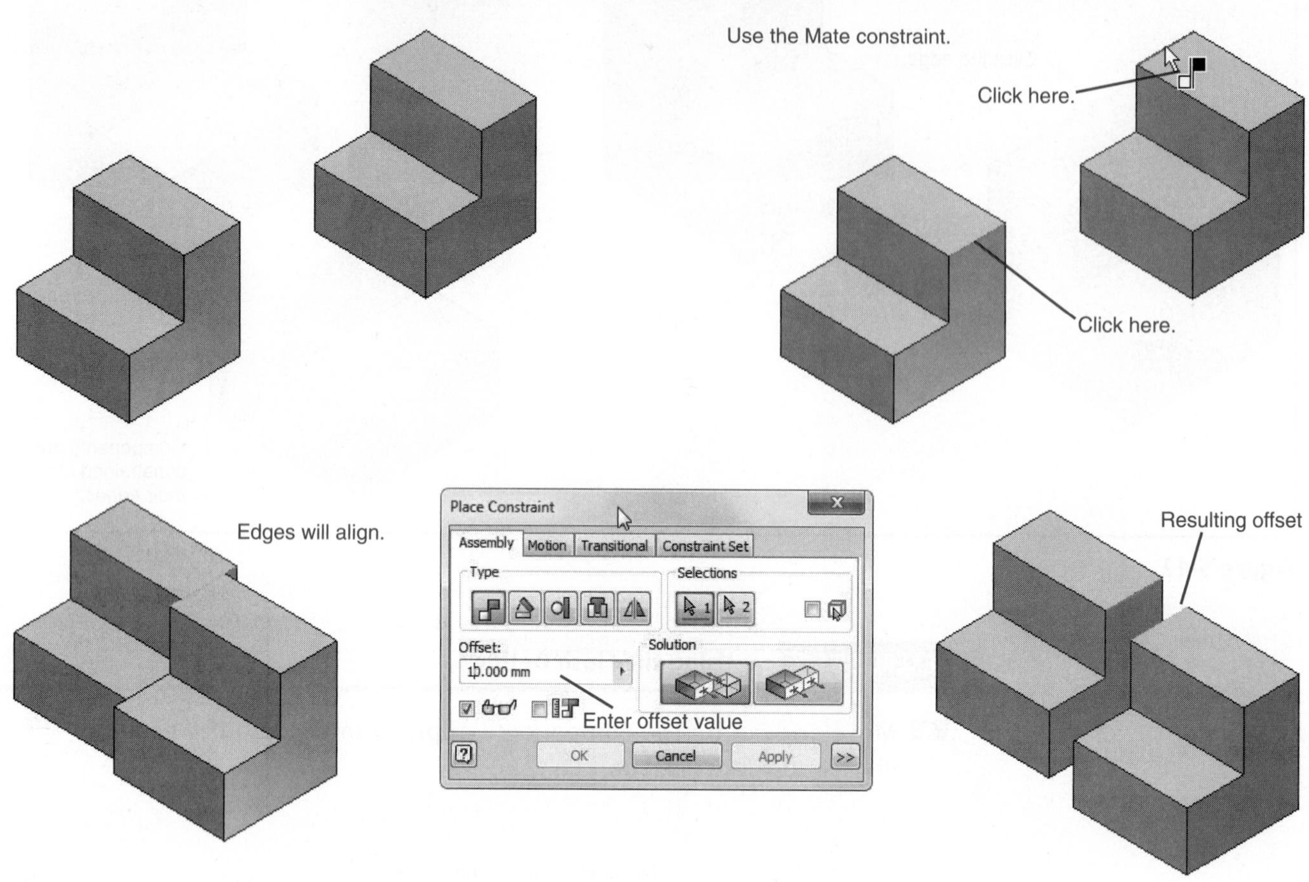

Figure 5-14

1. Click the **Constrain** tool and select the **Mate** option.
2. Use the **Mate** tool and mate the components' edges as shown.
3. Enter a value of **10 mm** into the **Offset** box.

The two mated edges will move apart 10 mm.

> **TIP**
>
> Offset values may be negative. Negative values create an offset in the direction opposite that of positive values.

EXERCISE 5-9 Positioning Objects

Sometimes components are not oriented so they can be joined as desired. In these cases first rotate or move one of the components as needed, then use the **Constrain** tools. See Figure 5-15. The left component has been rotated using the **Rotate Component** tool.

1. Click the **Constrain** tool on the **Position** panel under the **Assemble** tab.
2. Click the edge lines of the two components as shown.

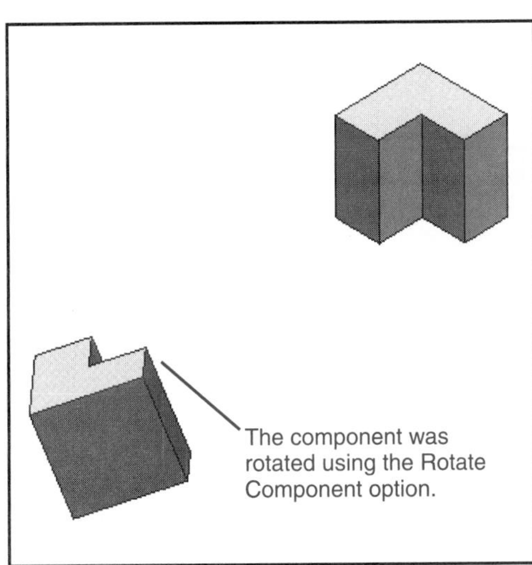
The component was
rotated using the Rotate
Component option.

Figure 5-15

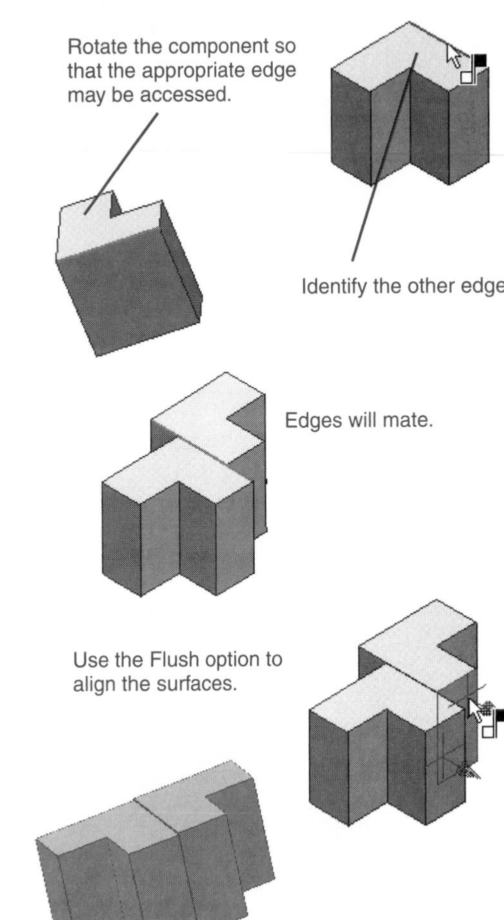
Rotate the component so
that the appropriate edge
may be accessed.
Identify the other edge.
Edges will mate.
Use the Flush option to
align the surfaces.
The final assembly

Figure 5-16

3 Click the **Apply** box.

See Figure 5-16.

4 Use the **Flush** option to align the components.

5 Click the **Apply** button.

TIP

Components cannot be moved or rotated if they are grounded. If a component does not respond to the **Rotate Component** tool, check to see whether it is grounded, that is, whether there is a pushpin to the left of its heading in the browser box.

EXERCISE 5-10 Using the Angle Option

1 Click the **Constrain** tool on the **Position** panel under the **Assemble** tab.

The **Place Constraint** dialog box will appear. See Figure 5-17.

2 Use the **Mate** constraint and align the edges of the SQBLOCKs shown in Figure 5-18.

Figure 5-17

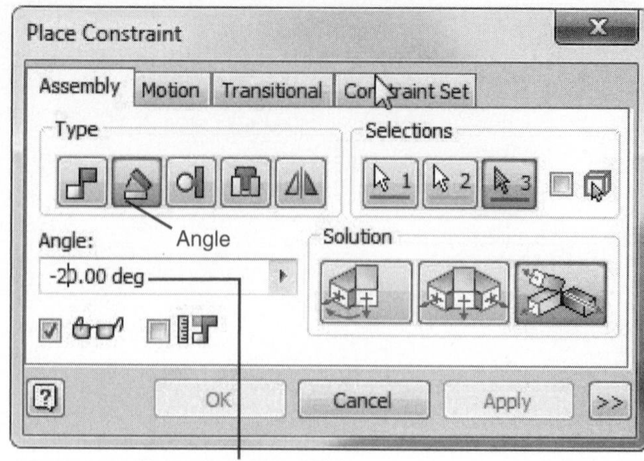

Two SQBLOCKs

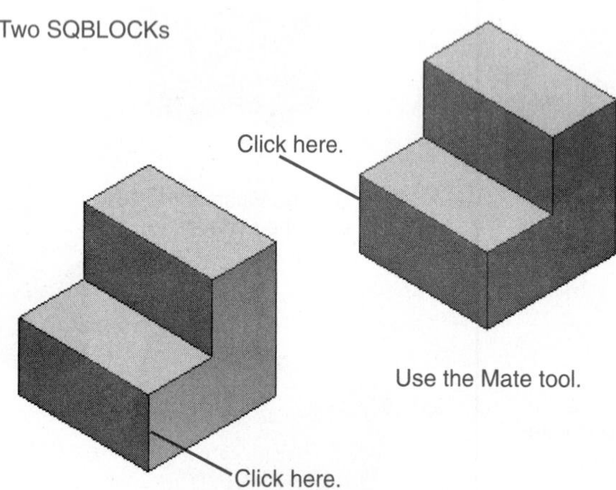

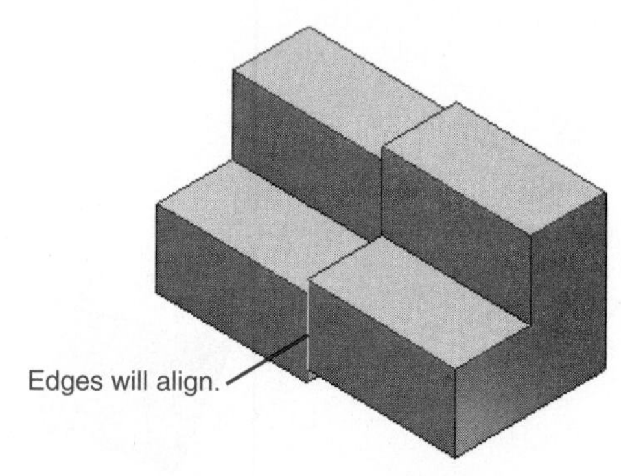

Use the Angle constraint set to –20.00°.

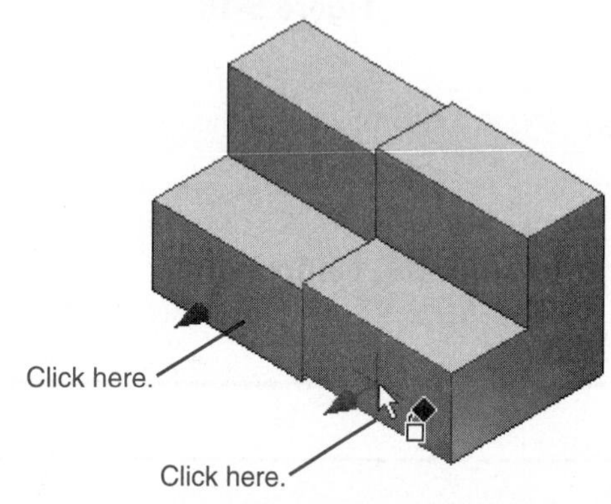

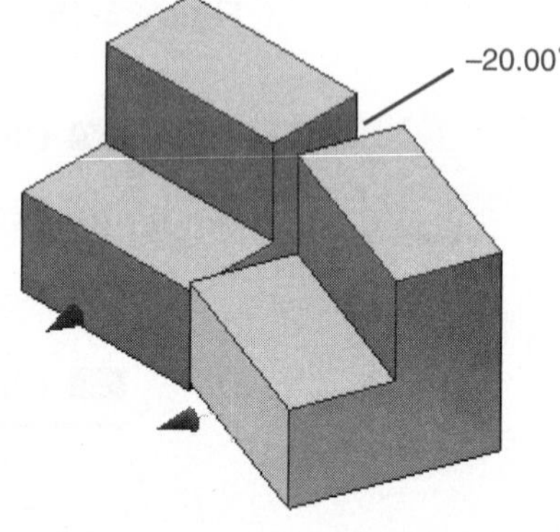

Use the Flush constraint.

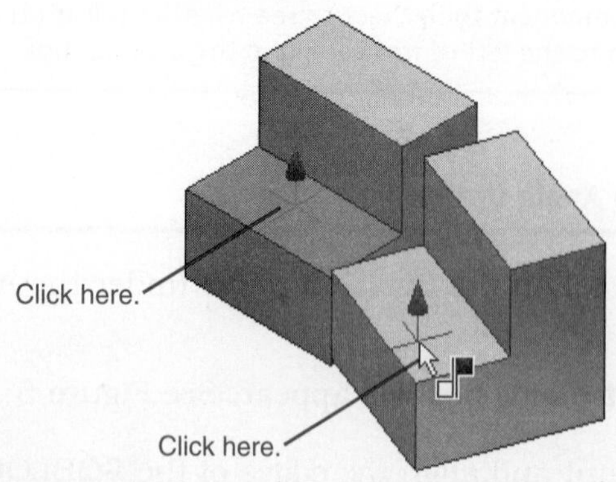

Figure 5-18

3 Access the **Angle** constraint, set the angle for **–20.00,** and click the two front surfaces of the SQBLOCKs as shown.

4 Click the **Apply** box.

5 Click the **Flush** constraint and align the two surfaces as shown.

EXERCISE 5-11 Using the Tangent Option

Figure 5-19 shows two cylinders. The smaller cylinder has dimensions of Ø10 × 20, and the larger cylinder has dimensions of Ø20 × 20 with a Ø10 centered longitudinal hole.

1 Click the **Constrain** tool.

The **Place Constraint** dialog box will appear. See Figure 5-20.

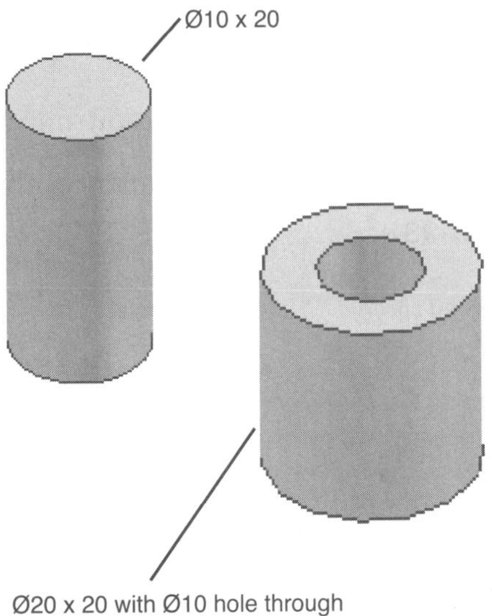

Figure 5-19

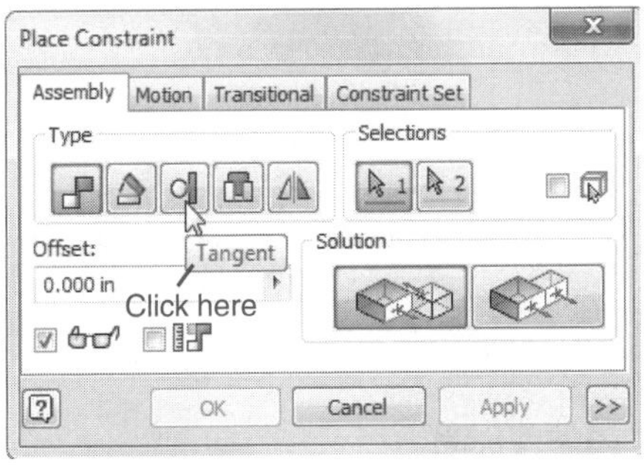

Figure 5-20

2 Click the **Tangent** box under the **Type** heading.

The **Outside** option will be selected automatically.

3 Select the outside edge of the large cylinder, then the outside edge of the smaller cylinder.

Figure 5-21 shows the resulting tangent constraint for the cylinders.

Figure 5-21

The resulting tangent cylinders

EXERCISE 5-12 Using the Insert Option

1 Click the **Constrain** tool.

The **Place Constraint** dialog box will appear. See Figure 5-22.

Figure 5-22

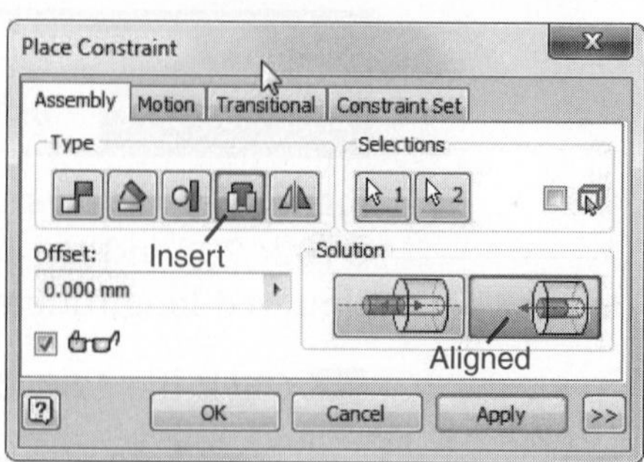

2 Click the **Insert** box under the **Type** heading, then click the **Aligned** box under the **Solution** heading.

Note that the **Aligned** box is the right-hand box.

3 Click the top surface of each cylinder as shown.

See Figure 5-23.

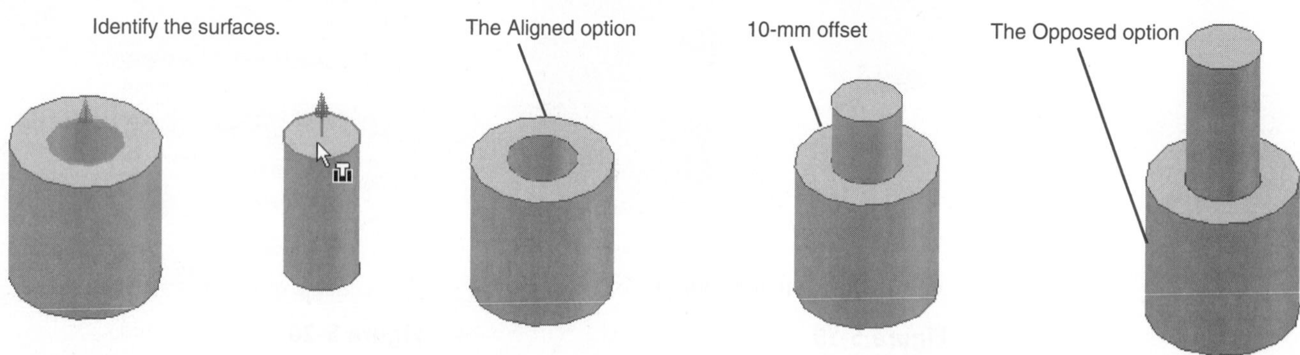

Figure 5-23

4 Click the **Apply** button.

Figure 5-23 also shows the result of using the **Opposed** option under the **Solution** heading.

Sample Assembly Problem SP5-1

Figure 5-24 shows three models that will be used to create an assembly drawing. The dimensions for the models are given in the figure.

1 Create a new drawing using the **Standard (mm).iam** format.

2 Use the **Place** tool on the **Component** panel under the **Assemble** tab and place the **Block, Bottom; Block, Top;** and **Post, Ø10 × 20** on the drawing screen.

This will be a bottom-up assembly.

3 Click the **Constrain** tool and use the **Mate** option to align the two edges as shown.

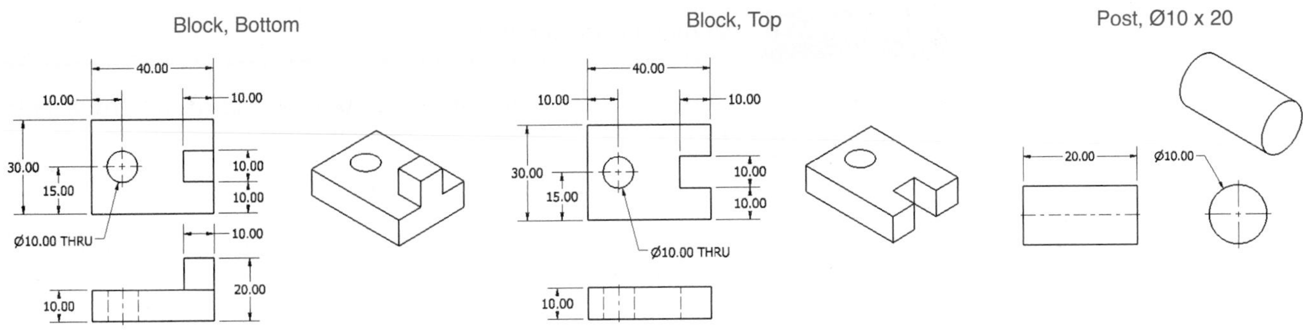

Figure 5-24

4 Use the **Flush** constraint to align the front surfaces of the two blocks.

5 Use the **Insert** tool to locate the post in the holes in the blocks. See Figure 5-25.

6 Save the assembly.

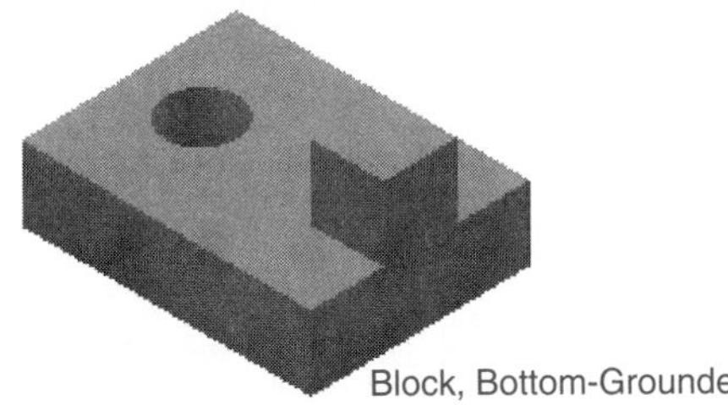

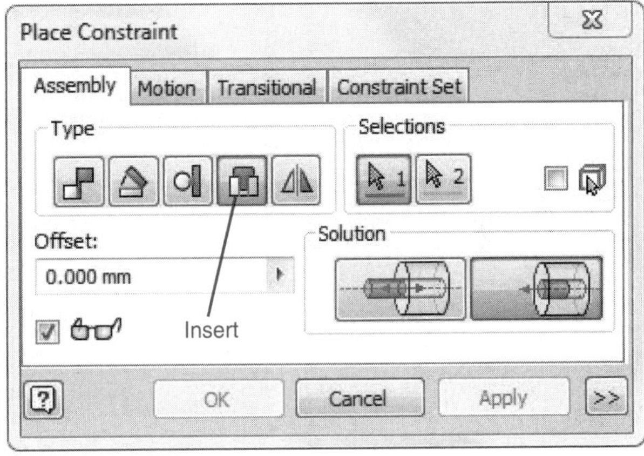

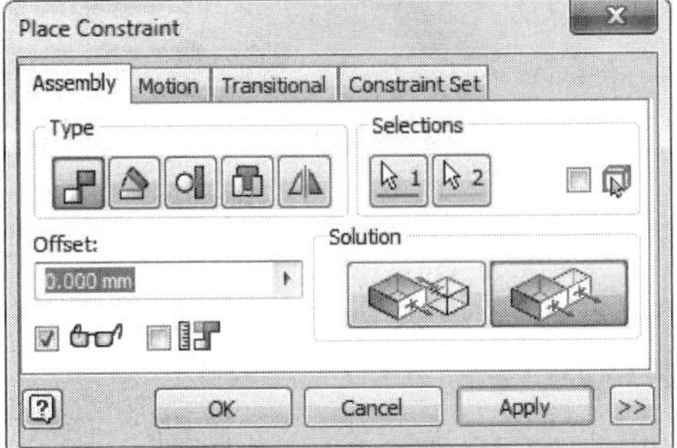

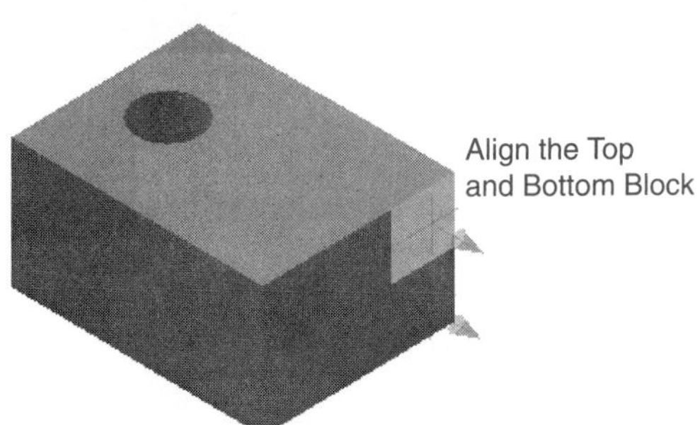

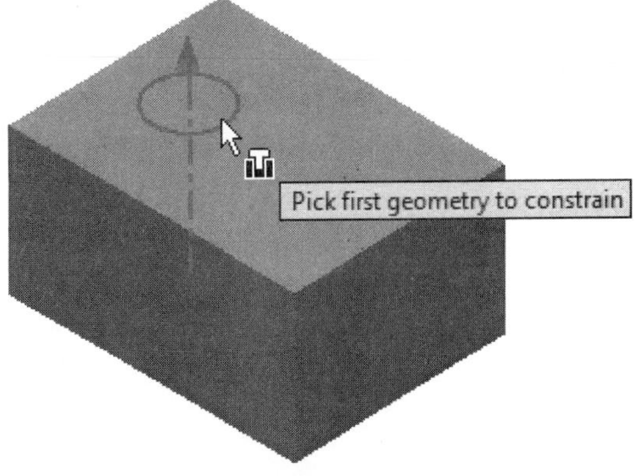

Figure 5-25

EXERCISE 5-13 Saving an Assembly

1 Create the assembly, then click the **Save As** heading under the **I** icon pull-down menu.

The file will be saved as an .iam file.

2 Save the assembly using the file name **BLOCK, ASSEMBLY;** click **Save**.

Presentation Drawings

presentation drawing: An exploded assembly drawing that can be animated to show how the assembly is to be created.

Presentation drawings are used to create exploded assembly drawings that can then be animated to show how the assembly is to be created from its components.

EXERCISE 5-14 Creating a Presentation Drawing

1 Click on the **New** tool.

The **Create New File** dialog box will appear. See Figure 5-26.

Figure 5-26

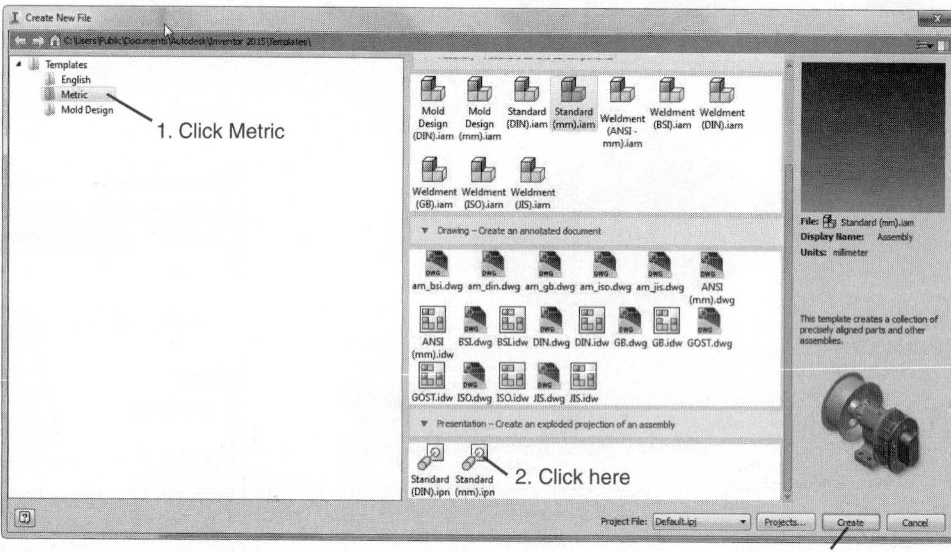

2 Click the **Standard (mm).ipn tool,** then **Create**.

The **Presentation** tab will appear. See Figure 5-27.

Figure 5-27

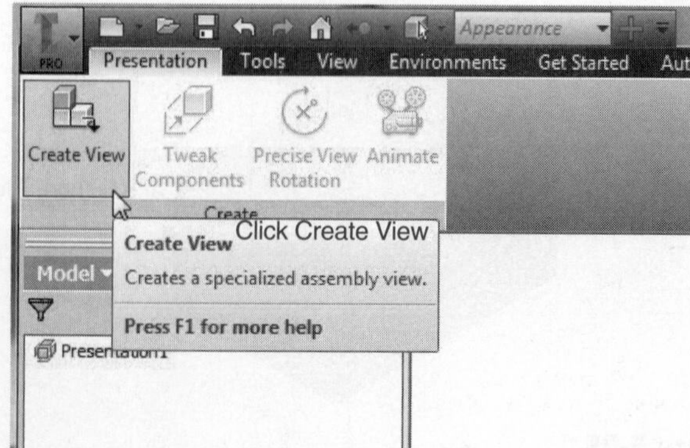

3 Click the **Create View** tool on the **Create** panel under the **Presentation** tab.

The **Select Assembly** dialog box will appear. See Figure 5-28.

Figure 5-28

4 Click the **Open an existing file** box.

The **Open** dialog box will appear, listing all the existing assembly drawings. See Figure 5-29.

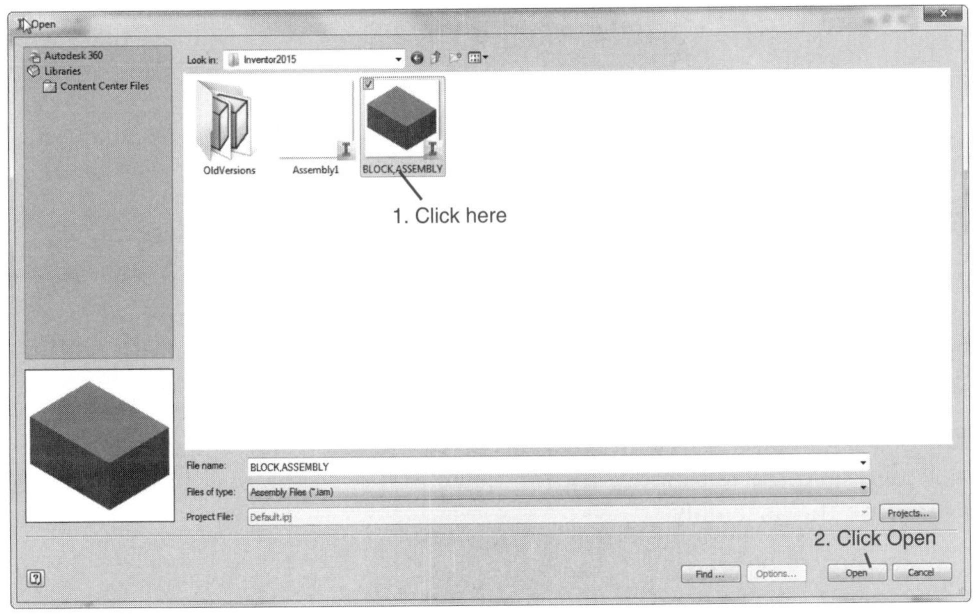

Figure 5-29

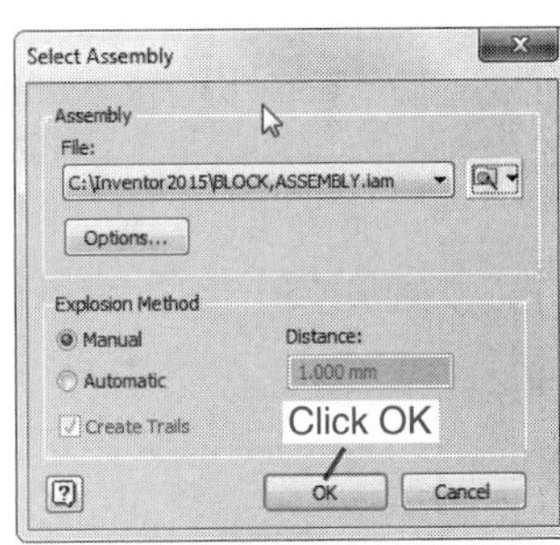

Figure 5-30

5 Select the appropriate assembly drawing, then click **Open**.

The **Select Assembly** dialog box will reappear listing the selected assembly under the **File** heading. See Figure 5-30.

6 Click **OK**.

The assembly will appear. See Figure 5-31.

A presentation drawing ready for tweaking

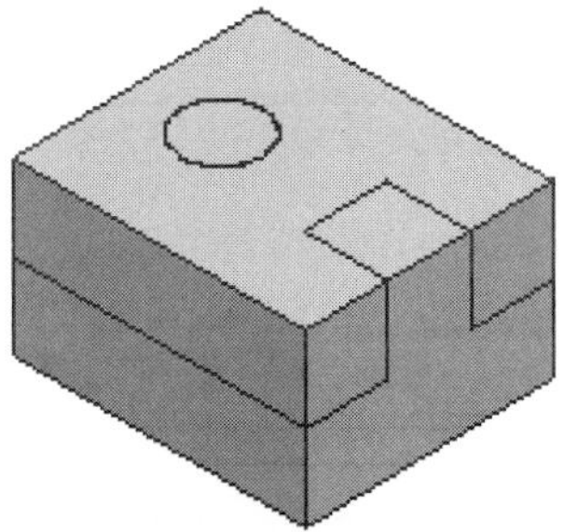

Figure 5-31

EXERCISE 5-15 Creating an Exploded Assembly Drawing

1 Click the **Tweak Components** tool located on the **Create** panel under the **Presentation** tab.

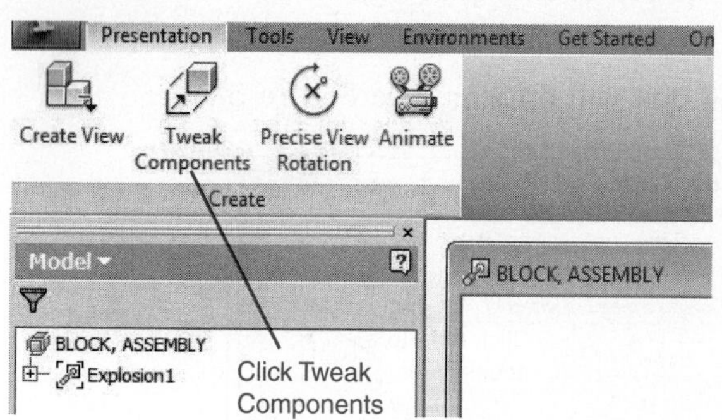

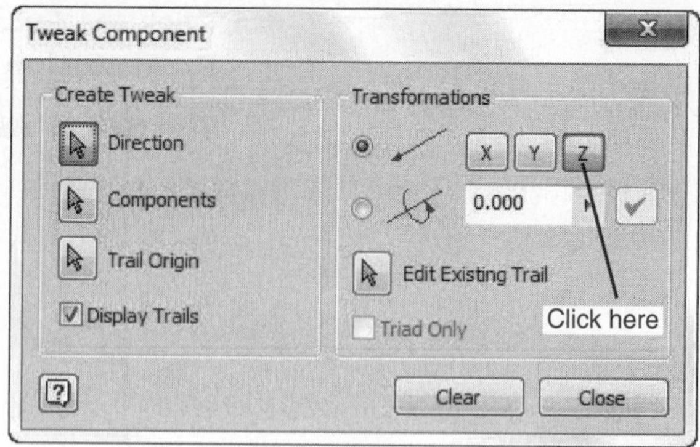

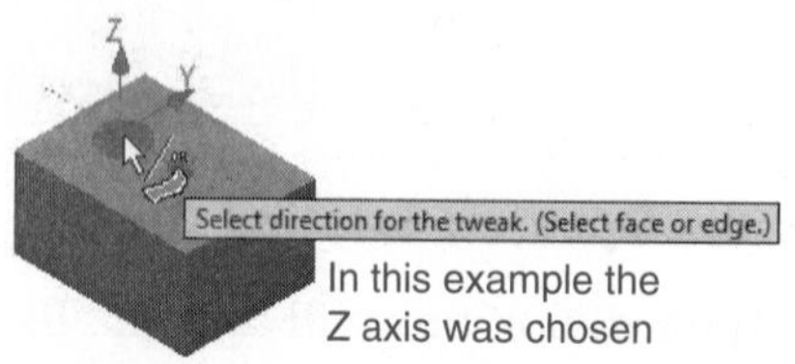

Figure 5-32

Figure 5-33

The **Tweak Component** dialog box will appear. See Figure 5-32. The **Direction** option will automatically be selected.

2 Select the direction of the tweak by selecting one of the assembly's vertical edge lines.

The **Tweak Component** dialog box will switch to the **Components** option. In this example the **Z** axis was selected, and the lower front edge of the assembly was used to define the direction.

3 Select the post (the post will change to the color blue when selected), then hold the left mouse button down and drag the post to a position above the assembly.

See Figure 5-33.

> **TIP**
>
> Do not click the **Clear** or **Close** box on the **Tweak Component** dialog box. If you do and then select the top block, the block will move independently of the post and bottom block. In the next step both the top block and the post will move.

4 Select the top block and drag it to a position above the bottom block.

See Figure 5-34.

5 Click the **Clear** box on the **Tweak Component** dialog box, then click the **Close** box.

Figure 5-34

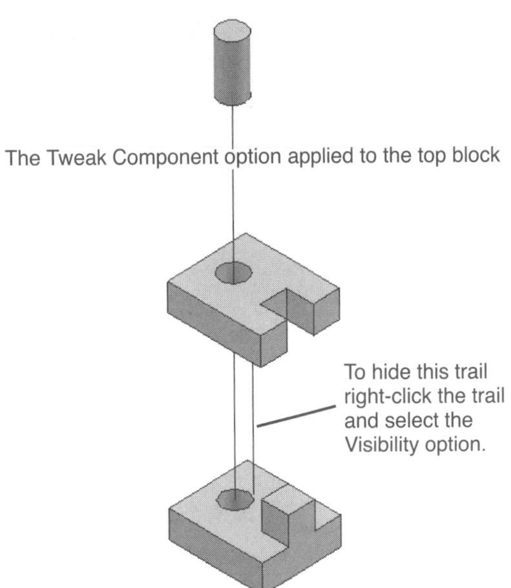
The Tweak Component option applied to the top block
To hide this trail right-click the trail and select the Visibility option.

EXERCISE 5-16 **Saving the Presentation Drawing**

1 Click the **Save As** heading on the **File** pull-down menu under the **I** icon.

The **Save As** dialog box will appear.

2 Enter the file name and click the **Save** box.

The drawing will be saved as an .ipn drawing. In this example the assembly drawing file named **BLOCK, ASSEMBLY** was used. The same name can be used because the file is saved using a different extension. In this example the presentation drawing was labeled **BLOCK, ASSEMBLY (2)**.

EXERCISE 5-17 **Hiding a Trail**

1 Right-click the trail.

2 Select the **Visibility** option.

> **TIP**
> Note that if you delete a trail, rather than using the **Visibility** option, the tweaking also will be deleted, and the models will return to their original assembled positions.

Animation

Presentation drawings can be animated using the **Animate** tool.

EXERCISE 5-18 **Animating a Presentation Drawing**

1 Click on the **Animate** tool on the **Create** panel under the **Presentation** tab.

The **Animation** dialog box will appear. See Figure 5-35. The control buttons on the **Animation** dialog box are similar to those found on CD players.

Figure 5-35

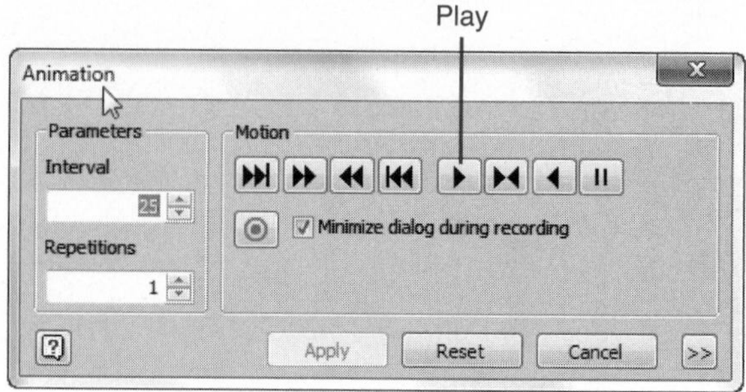

2 Click the **Play forward** button.

The assembly will slowly be reassembled in the reverse of the order used to tweak the components.

3 Click the **Reset** button to re-create the original presentation drawing.

4 Save the exploded presentation drawing. The same drawing name can be used, because presentation drawings are saved using a different file root than assembly drawings.

Isometric Drawings

Isometric drawings can be created directly from presentation drawings. Assembly numbers (balloons) can be added to the isometric drawings, and a parts list will automatically be created.

EXERCISE 5-19 **Creating an Isometric Drawing**

1 Click on the **New** tool, then the **Metric** tab.

The **Create New File** dialog box will appear. See Figure 5-36.

2 Select the **ANSI (mm).idw** tool, and click **Create**.

The **Place Views** tools will appear in the tab. See Chapter 4 for a further explanation of the **Place Views** tools.

3 Click the **Base View** tool.

The **Drawing View** dialog box will appear. See Figure 5-36.

4 Click the **Open an existing file** button.

The **Open** dialog box will appear. See Figure 5-37.

> **TIP**
>
> Set the **Files of type** box for **Inventor Files (*.ipt, *.iam, *.ipn)** to assure that all files are available.

Ensure that all files are listed.

Figure 5-36

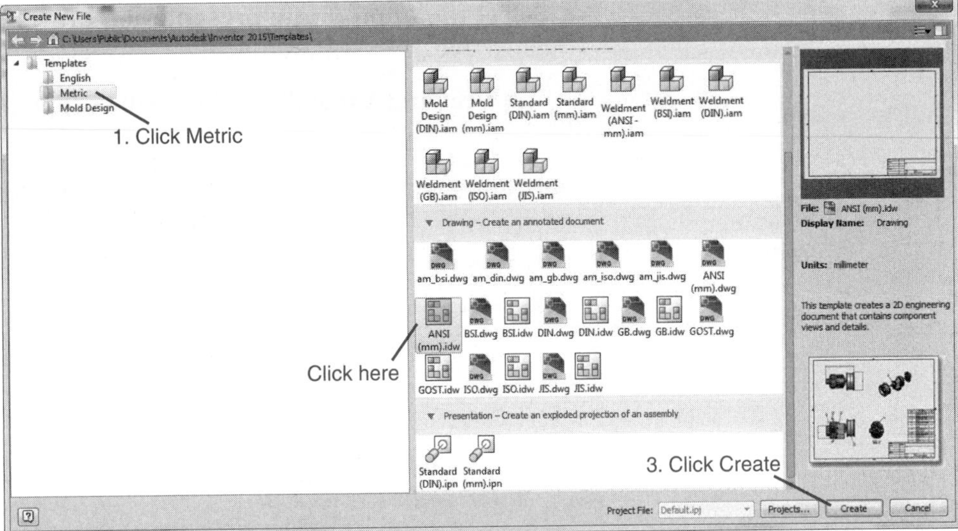

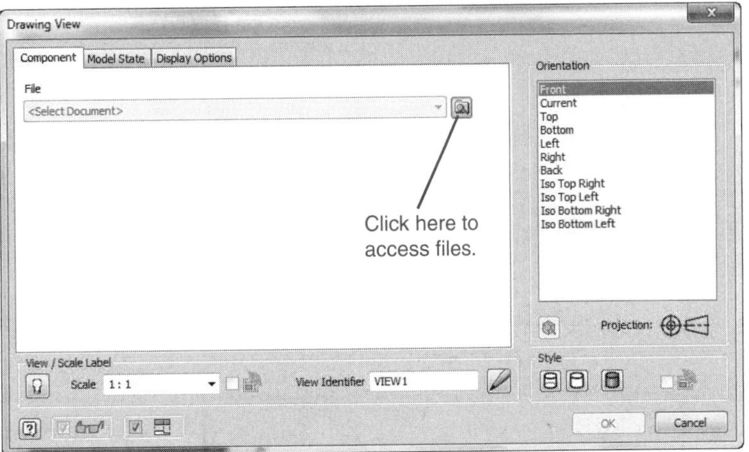

Figure 5-37

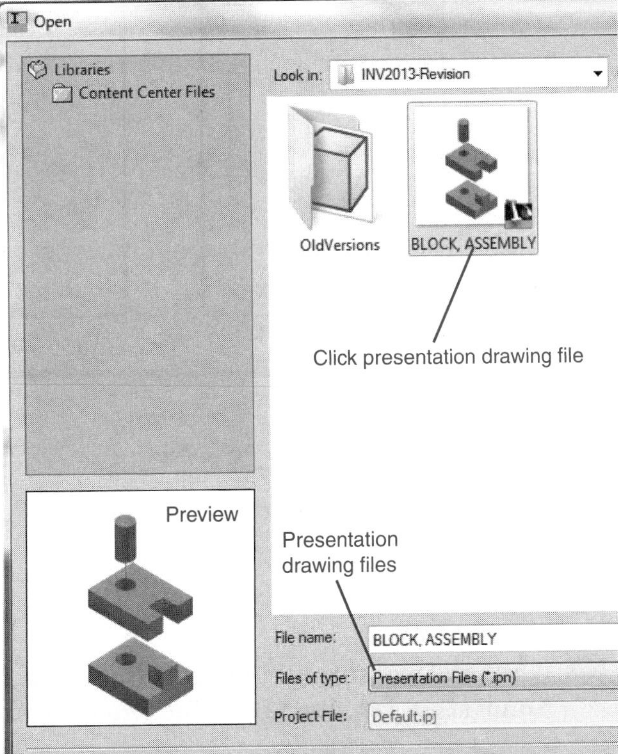

5 Select the appropriate presentation drawing (file type is .ipn), then click **Open**.

The **Drawing View** dialog box will reappear. See Figure 5-38.

Figure 5-38

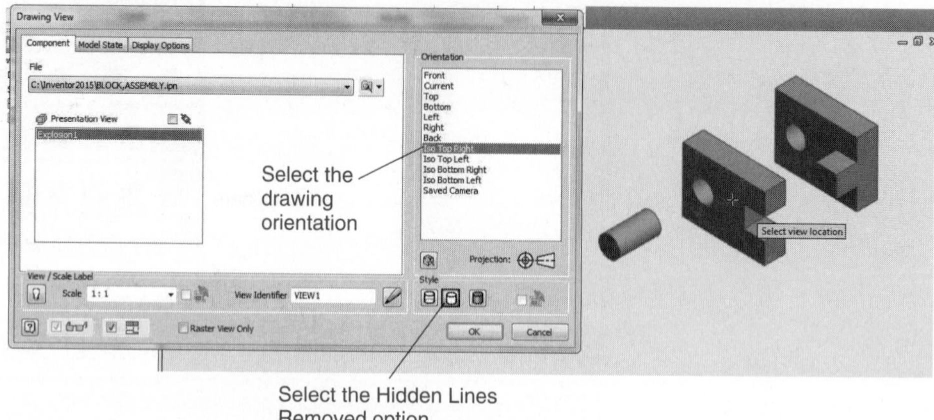
Select the drawing orientation
Select the Hidden Lines Removed option
Select view location

6 Select the **Iso Top Right** orientation and set the **Scale** as needed. Select the **Hidden Lines Removed** option under the **Style** heading.

7 Select a location for the exploded isometric drawing and click the mouse.

Figure 5-39 shows the resulting isometric view. Figure 5-40 shows the isometric drawing created using the **Shaded** option.

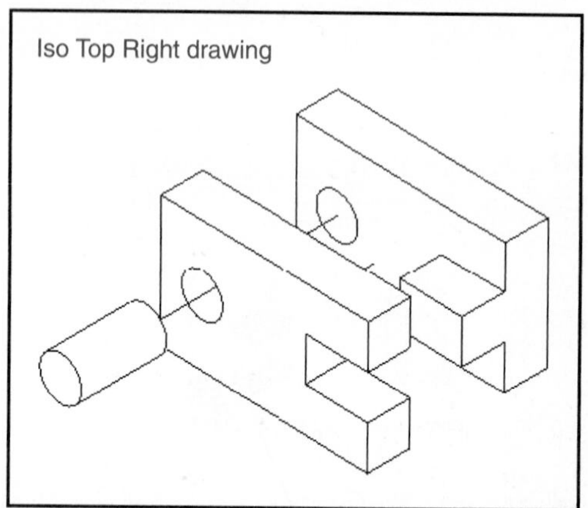
Iso Top Right drawing

Figure 5-39

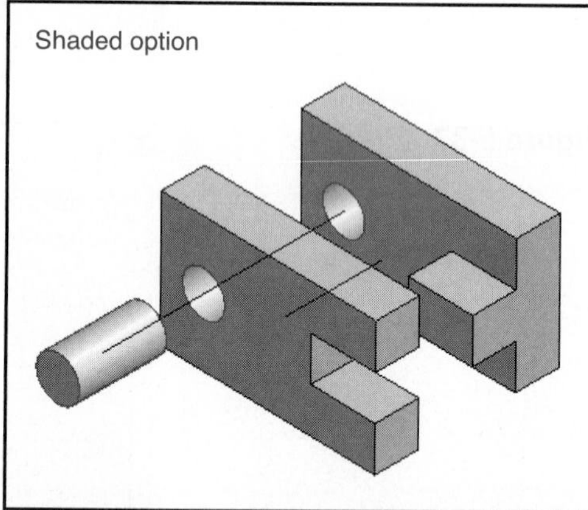
Shaded option

Figure 5-40

Assembly Numbers

Assembly numbers are added to an isometric drawing using the **Balloon** tool.

EXERCISE 5-20 Adding Balloons

1 While still working on the **ANSI (mm).idw** template, click the **Annotation** tab, then click the **Drawing Annotation** option.

The **Annotate** tab will appear. See Figure 5-41.

Figure 5-41

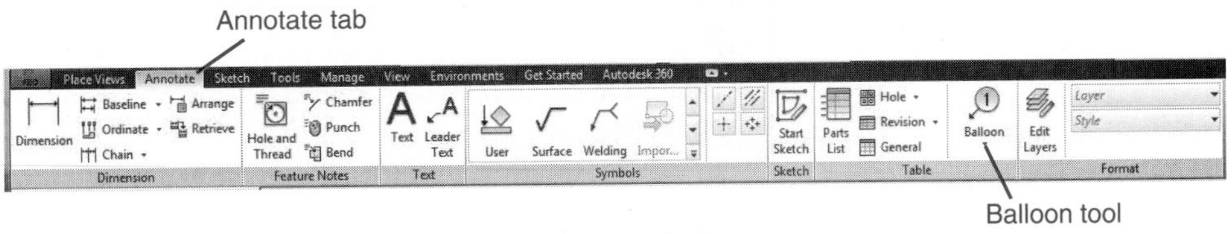

Annotate tab

Balloon tool

2 Click the **Balloon** tool on the **Table** panel, and click the exploded drawing.

The **BOM Properties** dialog box will appear. See Figure 5-42.

Figure 5-42

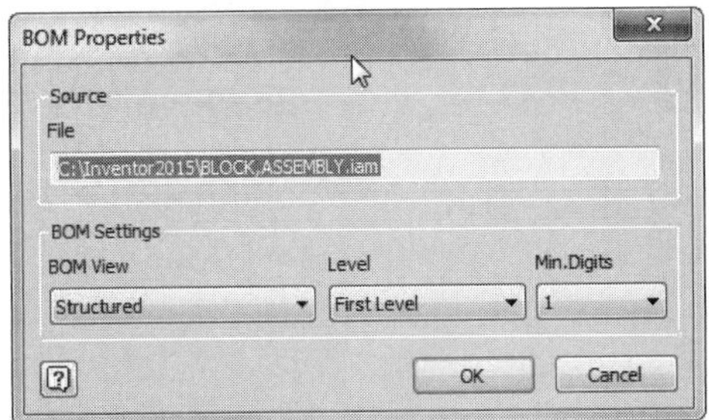

3 Click the topmost edge line of the bottom block.

4 Click **OK,** then drag the cursor away from the selected edge line.

5 Locate a position away from the component and click the left mouse button. Move the cursor in a horizontal direction and click the left mouse button again.

6 Right-click the mouse and select the **Continue** option.

7 Add balloons to the other components.

8 Press the **<Esc>** key after all balloon numbers have been assigned.

NOTE

Balloon numbers (assembly numbers) are also referred to as *bubble numbers* and are used to identify parts within an assembly. *Assembly* numbers are different from *part* numbers.

See Figure 5-43. The balloon numbers will be in the order the parts were added to the drawing.

Figure 5-43

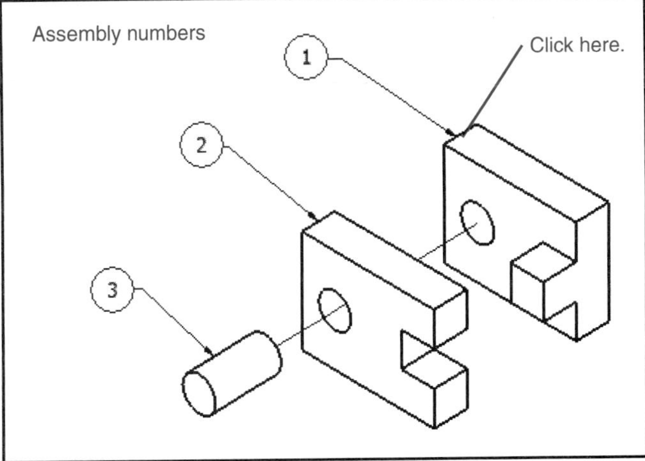

> **TIP**
> Making the balloon leaders lines the same angle will give the drawing a well-organized appearance.

EXERCISE 5-21 Editing Balloons

In general, the biggest parts have the lowest numbers. The assigned numbers can be edited.

1. Right-click the balloon to be edited and select the **Edit Balloon** option. See Figure 5-44. The **Edit Balloon** dialog box will appear. See Figure 5-45.
2. Make any needed changes in the **Edit Balloon** dialog box and click **OK**.

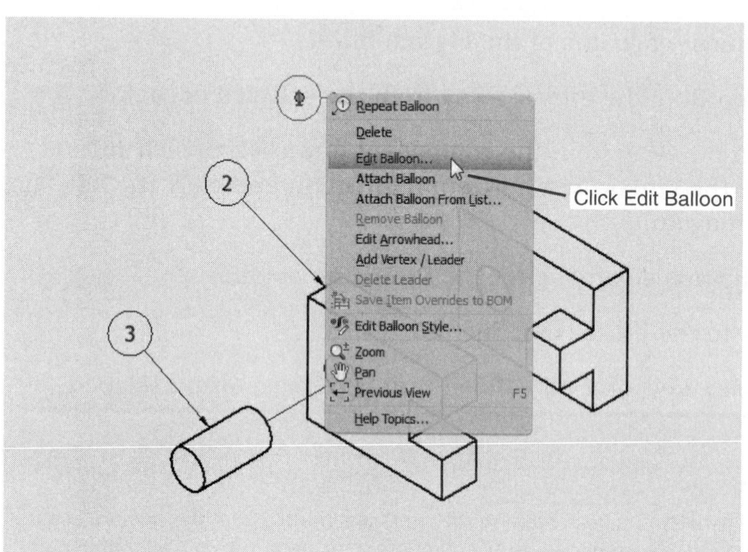

Figure 5-44

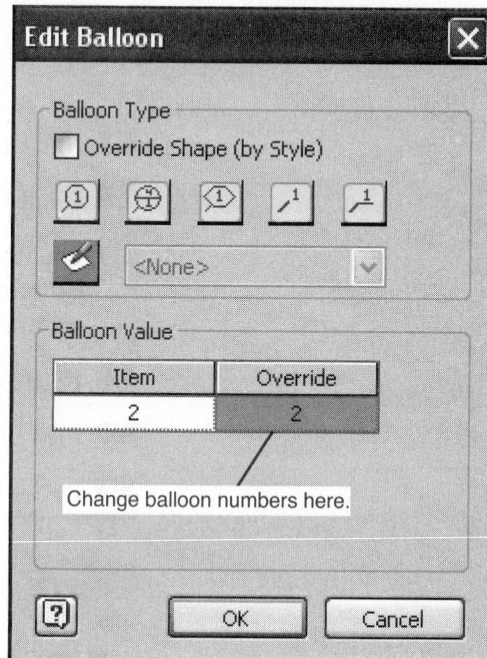

Figure 5-45

> **TIP**
> The terms *parts list* and *BOM* are interchangeable.

Parts List

A parts list can be created from an isometric drawing after the balloons have been assigned using the **Parts List** tool located on the **Table** panel under the **Annotate** tab.

EXERCISE 5-22 Creating a Parts List

1. Click the **Parts List** tool on the **Table** panel.

   The **Parts List** dialog box will appear. See Figure 5-46.

Figure 5-46

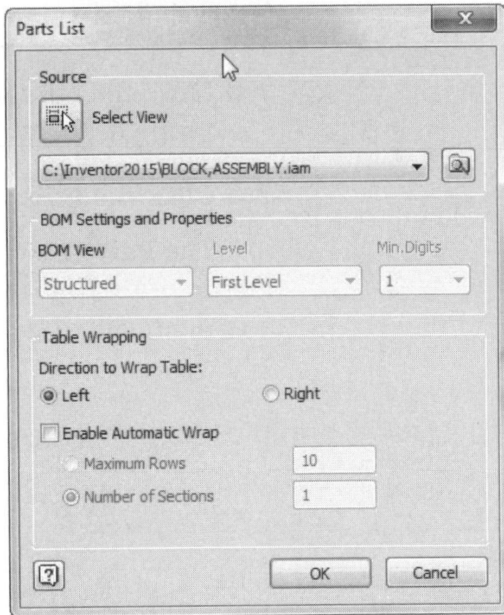

2 Move the cursor into the area around the isometric drawing.

A broken red line will appear when the cursor is in the area.

3 Click the left mouse button, then click the **OK** button on the **Parts List** dialog box. Move the cursor away from the isometric drawing area.

An outline of the parts list will appear and move with the cursor.

4 Select a location for the parts list and left-click the mouse.

Figure 5-47 shows the resulting drawing. The parts list was generated using information from the original model drawings and the presentation drawings. Your part numbers may be different. They will be the file names you assigned the part drawings.

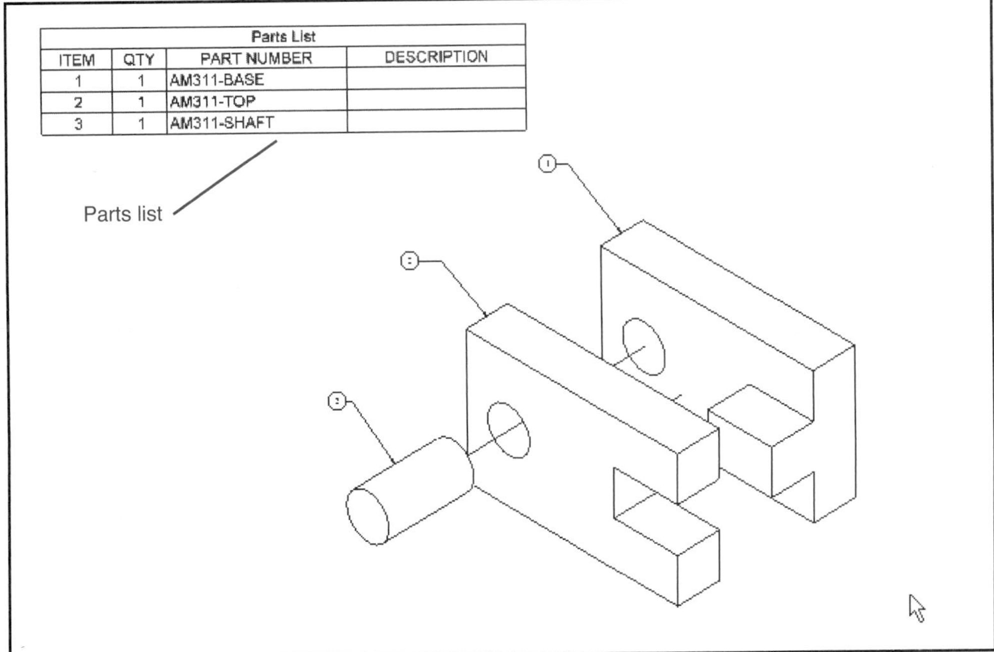

Parts List			
ITEM	QTY	PART NUMBER	DESCRIPTION
1	1	AM311-BASE	
2	1	AM311-TOP	
3	1	AM311-SHAFT	

Figure 5-47

EXERCISE 5-23 Editing a Parts List

1 Move the cursor onto the parts list and right-click the mouse. The lettering in the parts list will turn red when activated.

2 Click the **Edit Parts List** option.

The **Parts List: BLOCK, ASSEMBLY (2).iam** box will appear. See Figure 5-48. Click on a cell and either delete or add text. Figure 5-49 shows an edited parts list.

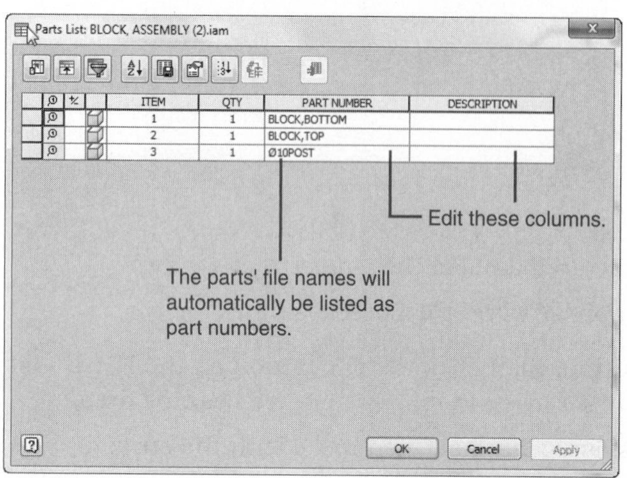

Figure 5-48

Parts List: BLOCK, ASSEMBLY (2).iam

ITEM	QTY	PART NUMBER	DESCRIPTION
1	1	BU-407-1	BLOCK, BOTTOM
2	1	BU-407-2	BLOCK, TOP
3	1	BU-407-3	Ø10 POST

The file names were removed and part numbers were inserted.

Figure 5-49

Inventor will automatically insert the file name of a component into the **PART NUMBER** column. In this example the file names were removed, and part numbers and descriptions were added. See Figure 5-48. To add or delete information from a parts list, click the appropriate box, delete existing text, and type in the new text.

Naming and Adding New Columns

Each company or organization has its own system for naming parts. In the examples in this book the noun, modifier format was used.

EXERCISE 5-24 Adding a New Column

Say two additional columns were required for the parts list shown in Figure 5-49: Material and Notes.

1 Move the cursor onto the parts list area and right-click the mouse.

2 Select the **Edit Parts List** option.

The **Parts List** dialog box will appear.

3 Select **Column Chooser**.

The **Parts List Column Chooser** dialog box will appear. See Figure 5-50.

4 Scroll down the **Available Properties** listing to see if the new column headings are listed.

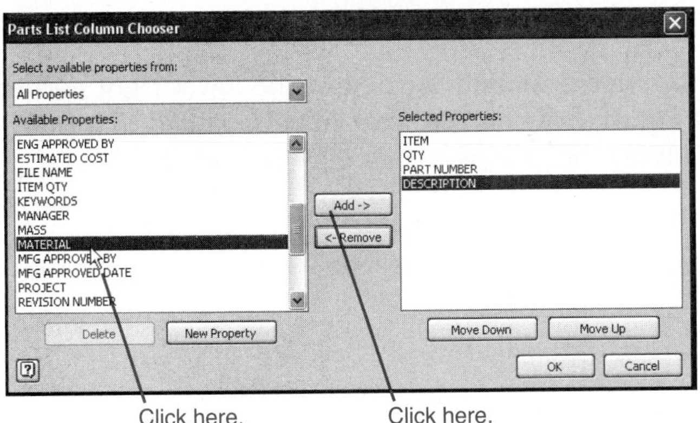

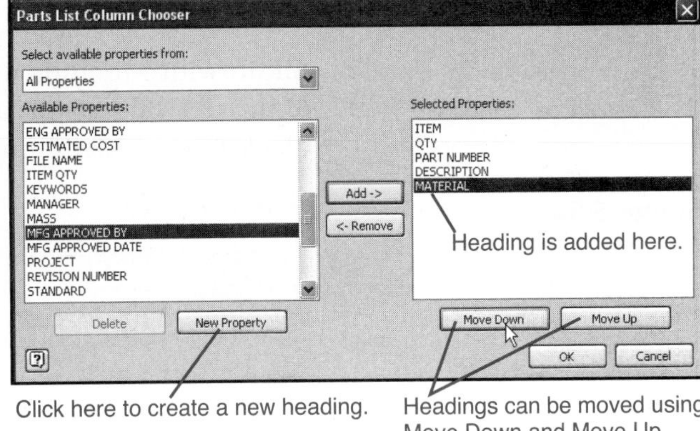

Figure 5-50

5 **MATERIAL** is listed, so click on the listing, then click the **Add** box in the middle of the screen.

The heading **MATERIAL** will appear in the **Selected Properties** area. Use the **Move Down** and **Move Up** boxes to sequence the column headings.

The heading **NOTES** is not listed, so it must be defined.

6 Click on the **New Property** box.

The **Define New Property** dialog box will appear. See Figure 5-51.

7 Type in the name of the new column, then click **OK**.

In this example a **NOTES** column was added.

Note that only uppercase letters are used to define column headings.

8 Arrange the **Selected Properties** in the desired sequence, and click **OK**.

The **Parts List** dialog box will reappear.

9 Click **Save**, then click **Done**.

10 Delete the old parts list and enter a new one. The new parts list will include the edits.

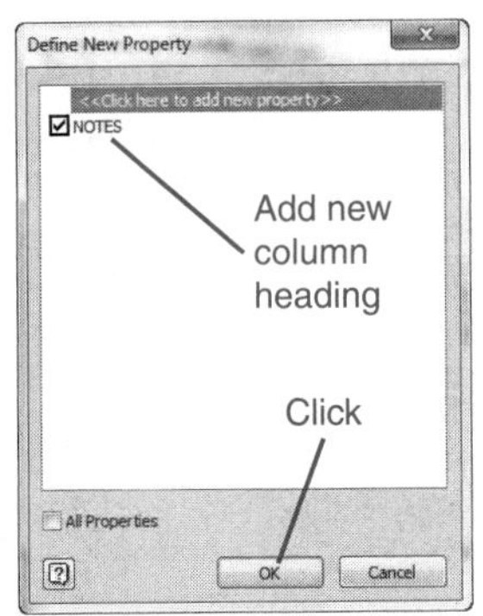

Figure 5-51

Figure 5-52 shows the revised column in the parts list. If no material is defined, the word **Default** will appear. The material for a model will be assigned to the model drawing and brought forward into the parts list. The **MATERIAL** column can be edited like the other columns.

Figure 5-53 shows the edited parts list on the drawing.

Figure 5-52

Parts List: BLOCK, ASSEMBLY (2).iam

ITEM	QTY	PART NUMBER	DESCRIPTION
1	1	AM311-1	BLOCK, BOTTOM
2	1	AM311-2	BLOCK, TOP
3	1	AM311-3	Ø10 POST

Revised part numbers

Figure 5-53

Completed parts list (BOM)

PARTS LIST					
ITEM	QTY	PART NUMBER	DESCRIPTION	MATERIAL	NOTES
1	1	AM311-1	BLOCK, BOTTOM	STEEL,MILD	
2	1	AM311-2	BLOCK, TOP	STEEL,MILD	
3	1	AM311-3	Ø10 POST	STEEL,MILD	

Title Block

All drawings include a title block, usually located in the lower right corner of the drawing sheet, as Figure 5-54 shows. Text may be added to a title block under existing headings, or new headings may be added.

Figure 5-54

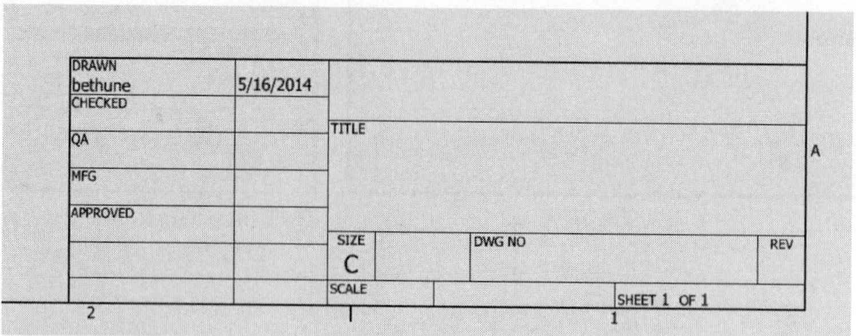

EXERCISE 5-25 **Adding Text to a Title Block**

1 Right-click the drawing name **(BLOCK, ASSEMBLY)** in the **Model** browser box, then click the **Properties** option.

See Figure 5-55. The drawing's **iProperties** dialog box will appear. Text can be typed into the **iProperties** dialog box and will appear on the title block. Figure 5-56 shows the **Summary** input.

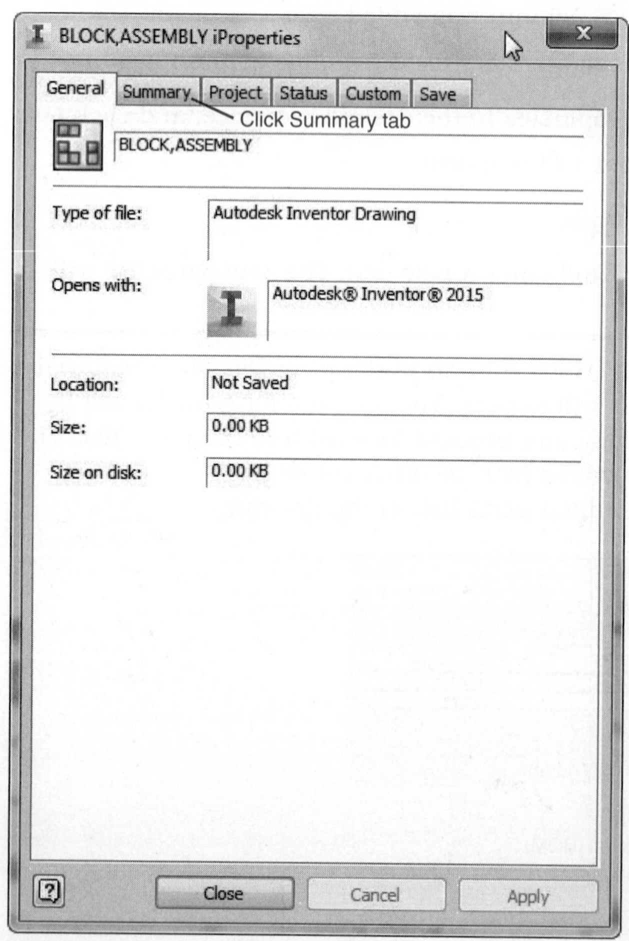

Figure 5-55

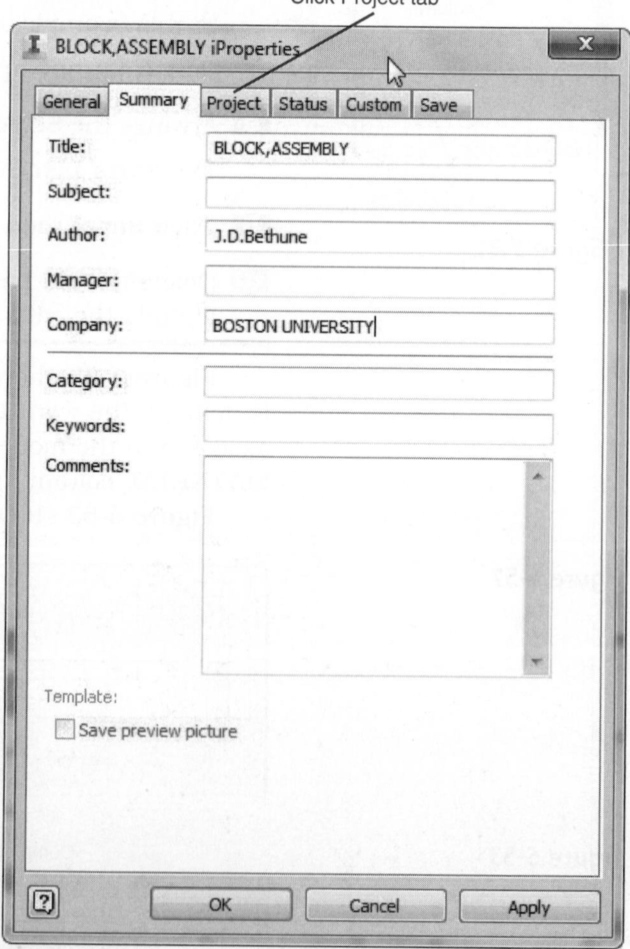

Figure 5-56

Figure 5-57

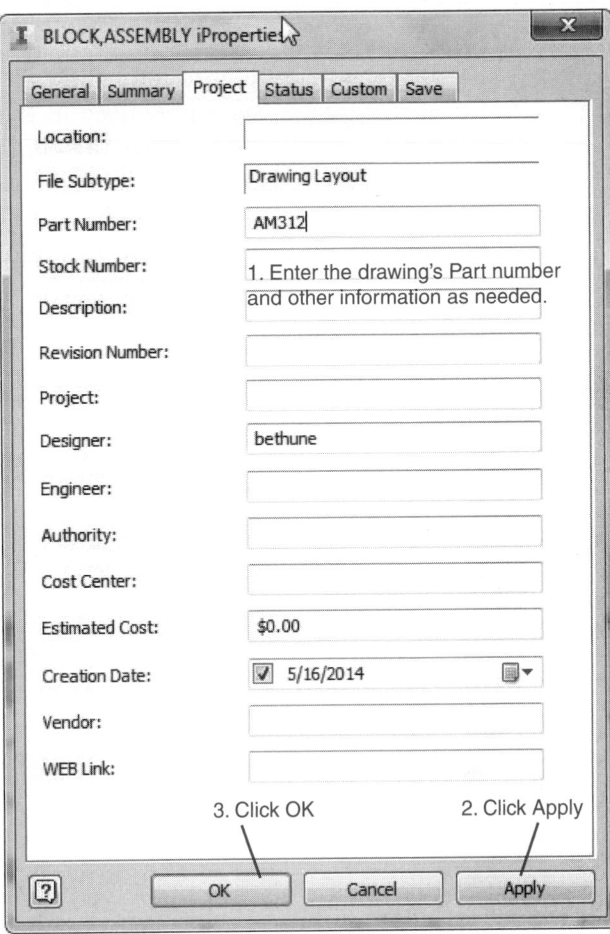

2 Click the **Summary** tab on the **iProperties** dialog box and enter the appropriate information.

In this example, the Title, Author, and Company were added.

3 Click the **Project** tab and add the **Part Number.** See Figure 5-57. In this example the number **AM312** was added. This is an assembly drawing number. Each individual part has its own number.

4 Click **Apply** and **OK**.

See Figure 5-57. Figure 5-58 shows the completed title block.

The title block included with Inventor is only one possible format. Each company and organization will have its own specifications.

Figure 5-58

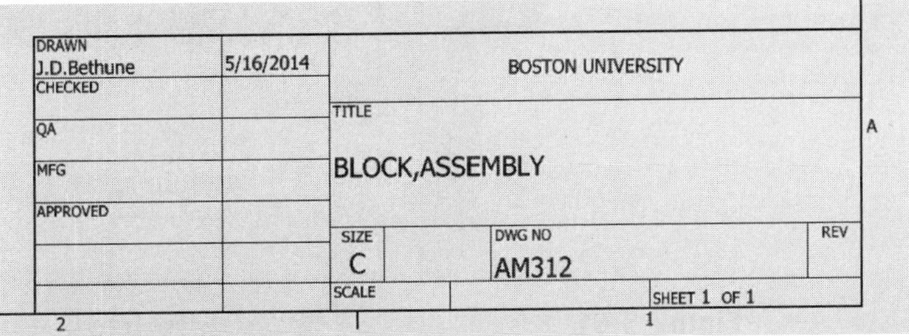

Subassemblies

Figure 5-59 shows a slightly more complicated assembly than the BLOCK assembly used in the previous sections. It is called a PIVOT assembly. Figure 5-60 shows the components needed to create the assembly. This will be a bottom-up assembly. The dimensions for the parts are found in Project 5-14.

It is sometimes difficult to control the assembly constraints. The parts seem to move randomly about the screen when constraints are added. Figure 5-61 shows an example of an incorrect application of an assembly

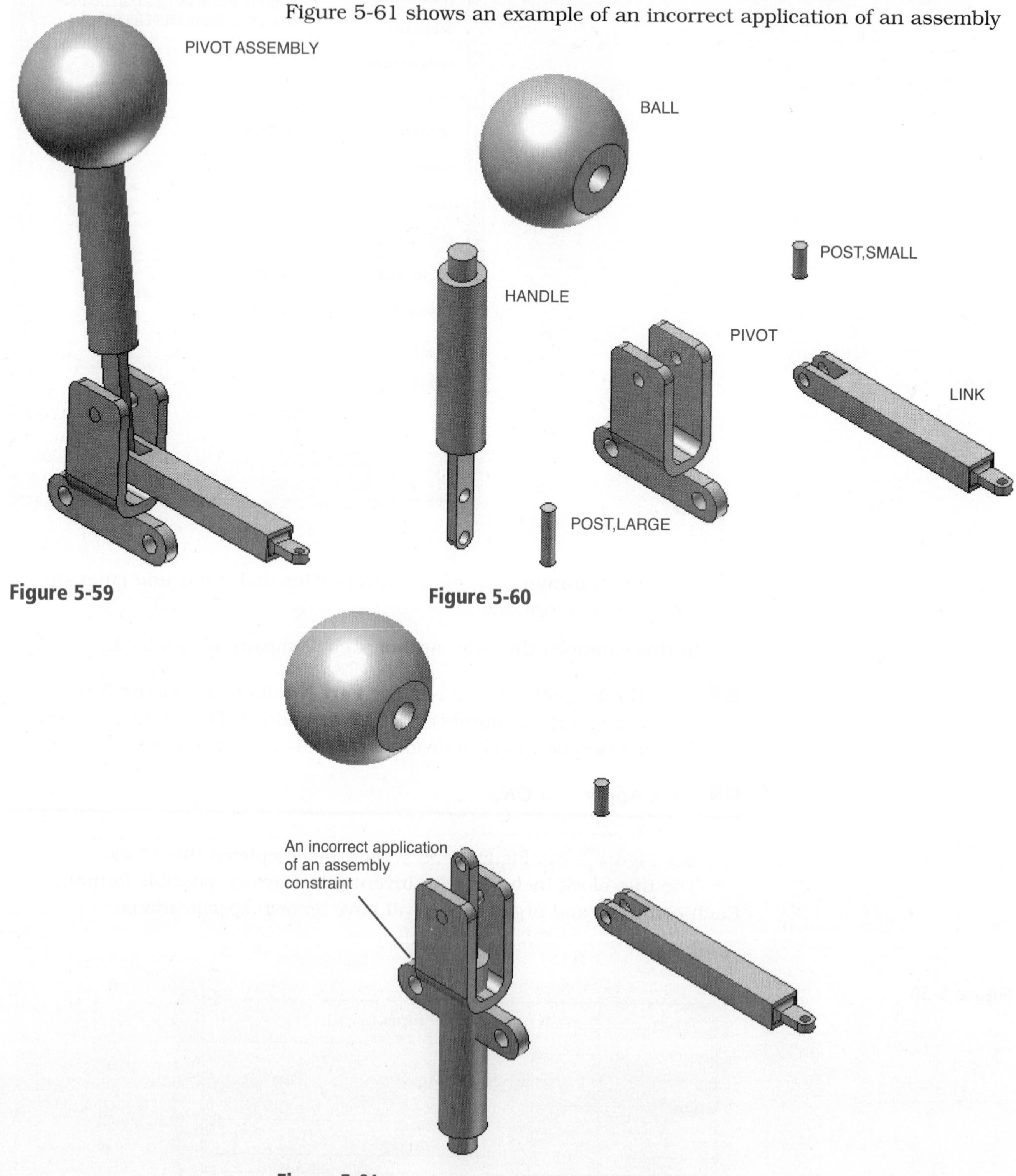

Figure 5-59

Figure 5-60

Figure 5-61

constraint. If this occurs, undo the incorrect application and consider temporarily fixing some constraints. The constraints can be deleted when no longer needed. The assembly sequence presented here is one of many different ones that could be used.

The PIVOT is grounded because it was the first component entered on the screen.

1 Right-click **HANDLE** in the browser box and ground the handle.

More than one component may be grounded at one time. See Figure 5-62.

Figure 5-62

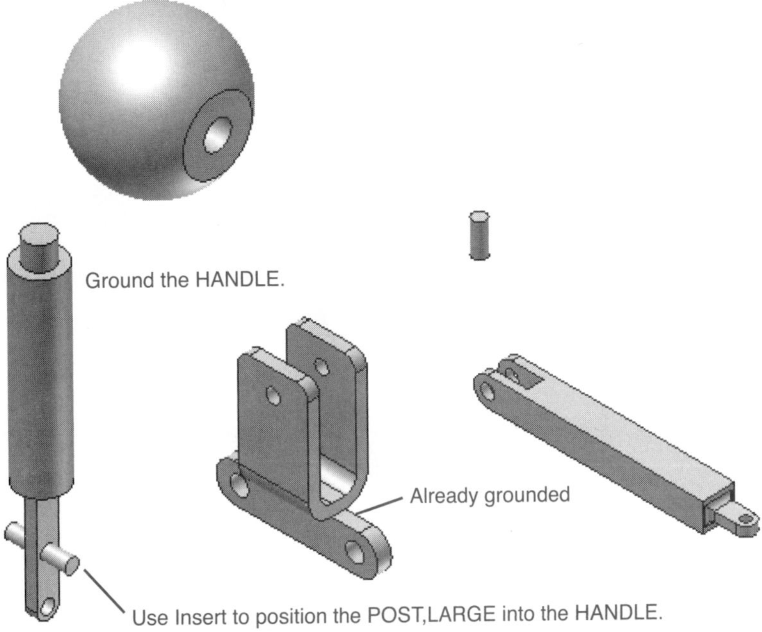
Ground the HANDLE.
Already grounded
Use Insert to position the POST,LARGE into the HANDLE.

2 Use the **Insert** constraint and insert the POST,LARGE into the top hole of the HANDLE.

Use the **Offset** option to center the post.

3 Use **Insert** and position the POST,SMALL into the HANDLE.

See Figure 5-63.

Figure 5-63

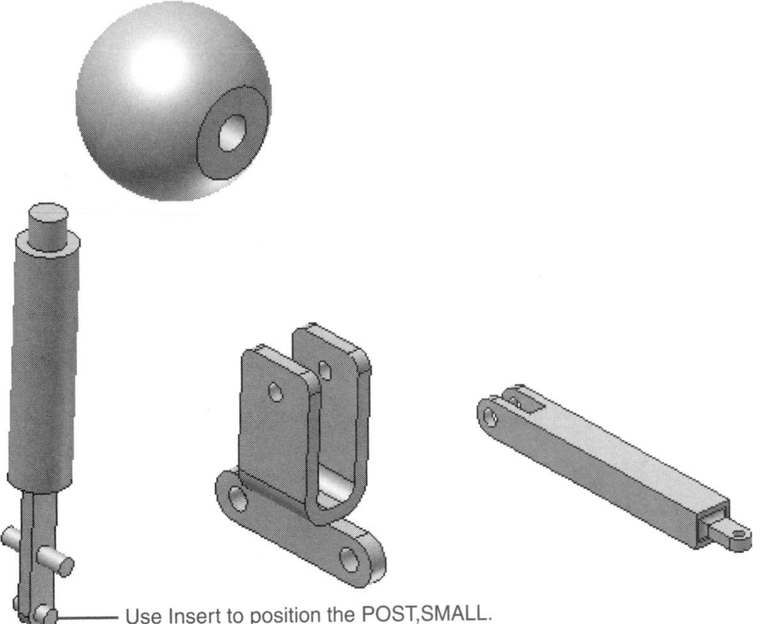
Use Insert to position the POST,SMALL.

4 Use **Insert** and position the LINK onto the POST,SMALL. See Figure 5-64.

Figure 5-64

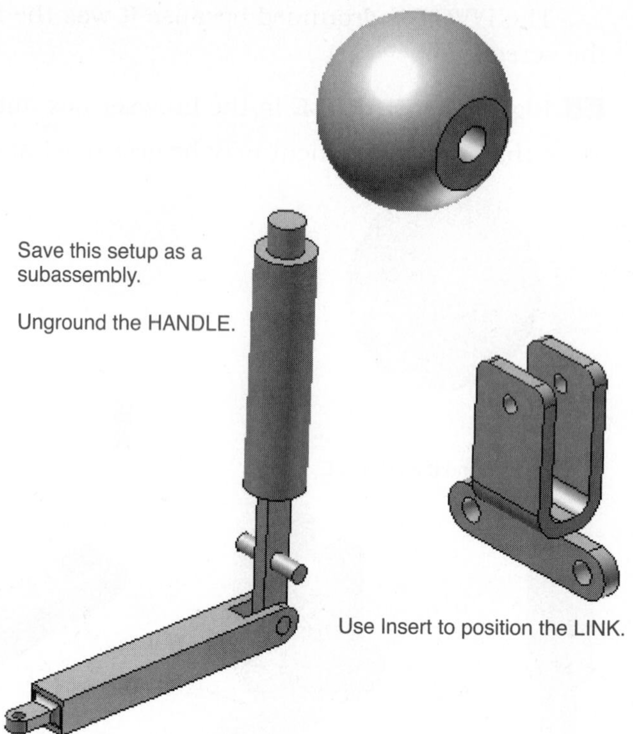

Save this setup as a subassembly.
Unground the HANDLE.
Use Insert to position the LINK.

5 Unground the HANDLE.

6 Save the assembly drawing as **PIVOT ASSEMBLY**.

7 Use **Insert** to position the subassembly into the PIVOT. See Figure 5-65.

Figure 5-65

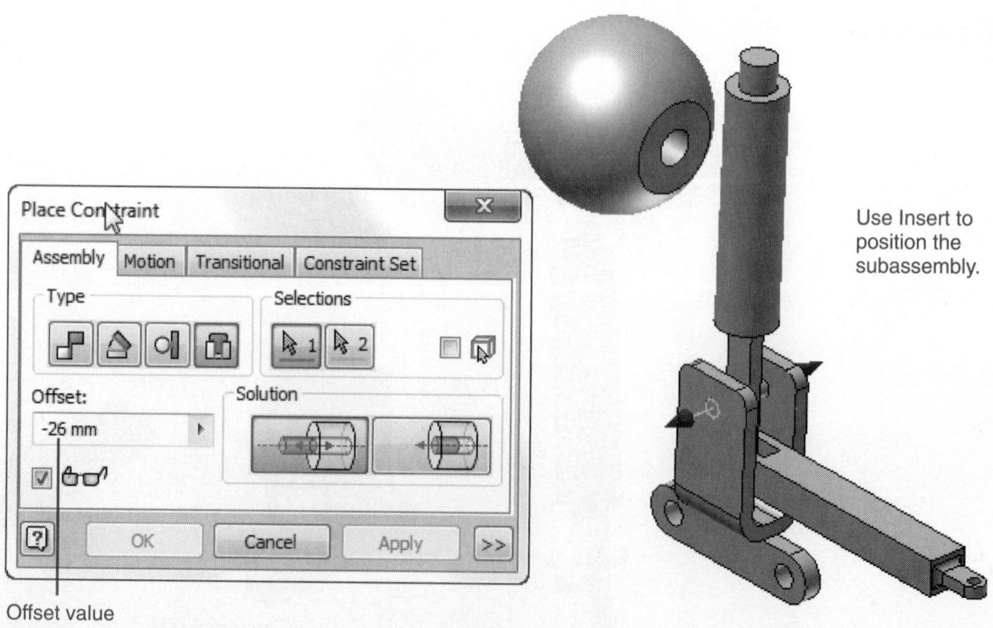

8 Use the **Angle** constraint to position the HANDLE relative to the LINK.

In this example, an angle of 100° was used. See Figure 5-66.

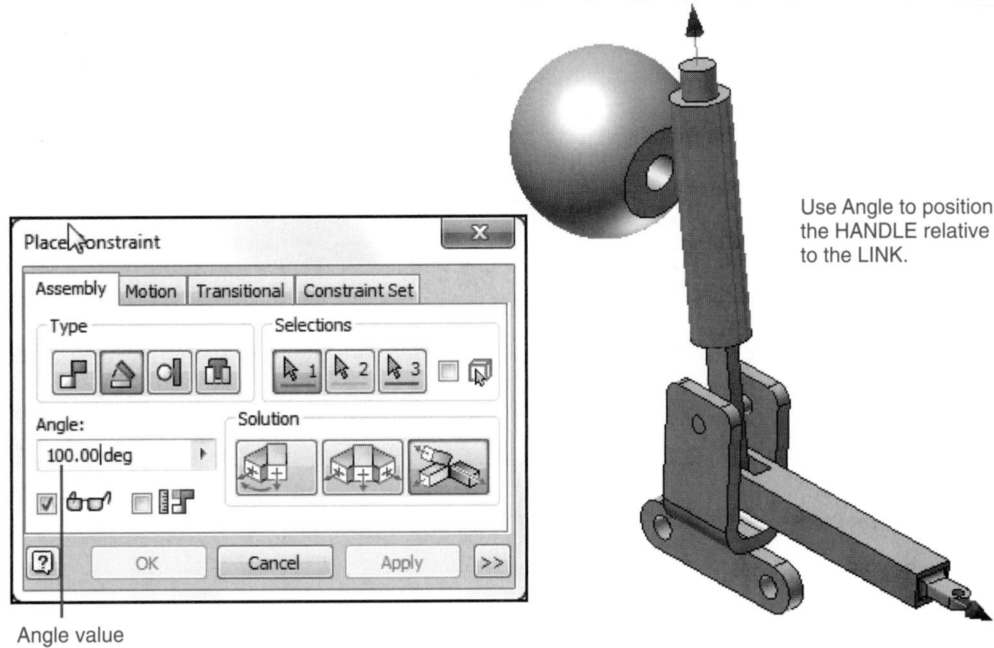

Figure 5-66

9 Use **Insert** to position the BALL on top of the HANDLE.

See Figure 5-59.

Figure 5-67 shows a presentation drawing of the PIVOT ASSEMBLY, and Figure 5-68 shows an exploded isometric drawing of the assembly and a parts list.

Figure 5-67

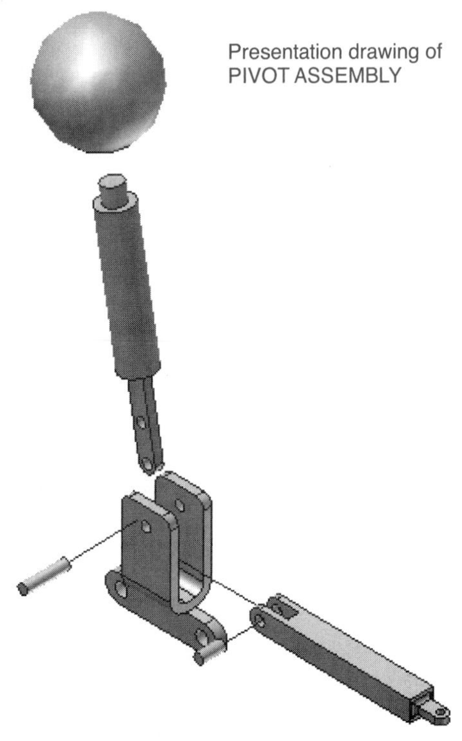

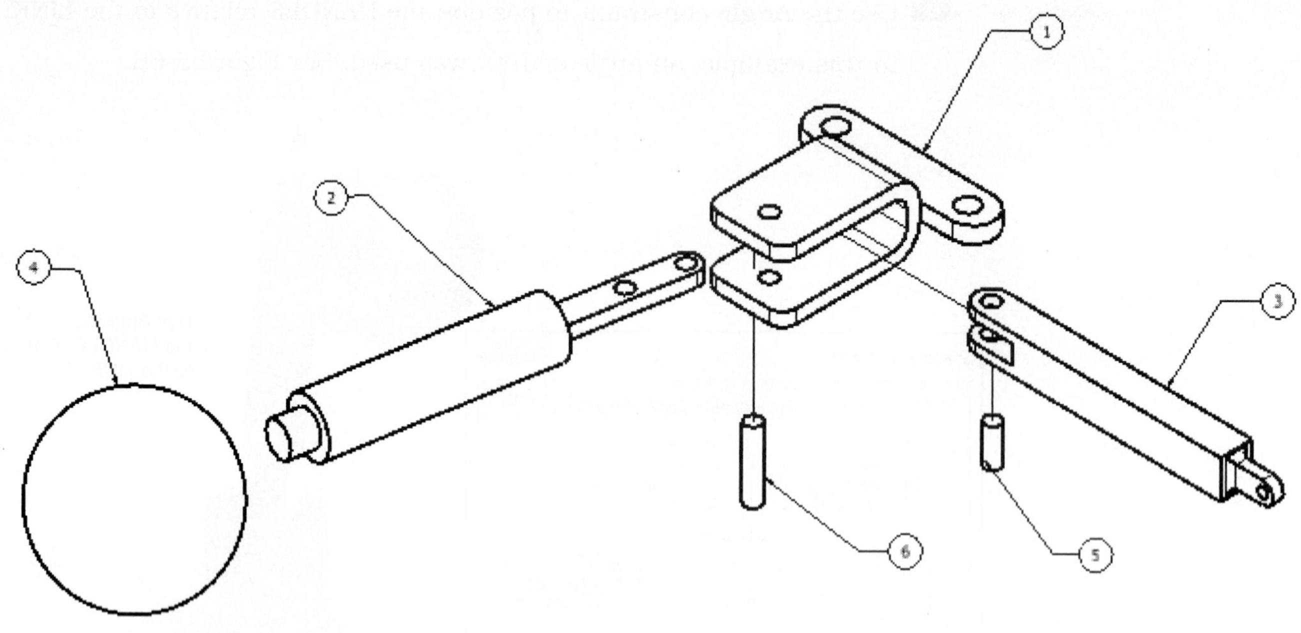

Parts List				
ITEM	PART NUMBER	DESCRIPTION	MATERIAL	QTY
1	ENG-A43	BOX,PIVOT	SAE1020	1
2	ENG-A44	POST,HANDLE	SAE1020	1
3	ENG-A45	LINK	SAE1020	1
4	AM300-1	HANDLE	STEEL	1
5	EK-132	POST-Ø6x14	STEEL	1
6	EK-131	POST-Ø6x26	STEEL	1

Figure 5-68

Drawing Sheets

Some assemblies are so large they require larger paper sheet sizes. Drawings are prepared on predefined standard-size sheets of paper. Each standard size has been assigned a letter value. Figure 5-69 shows the letter values and the sheet size assigned to each. All these sizes and more are available within Inventor.

Standard Drawing Sheet Sizes
Inches
A = 8.5 × 11
B = 11 × 17
C = 17 × 22
D = 22 × 34
E = 34 × 44

Standard Drawing Sheet Sizes
MILLIMETERS
A4 = 210 × 297
A3 = 297 × 420
A2 = 420 × 594
A1 = 594 × 841
A0 = 841 × 1189

Figure 5-69

Figure 5-70 shows a drawing done on a C-size drawing sheet. Note the letter C in the title block. The drawing is crowded on the sheet, so a larger sheet size is needed.

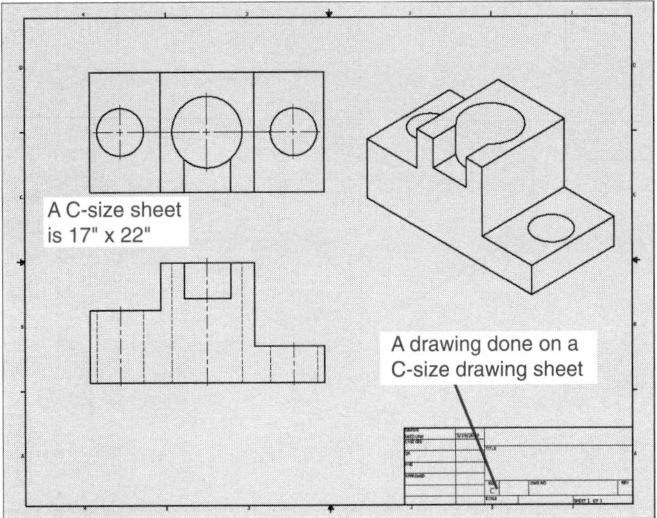

Figure 5-70

EXERCISE 5-26 Changing a Sheet Size

1 Locate the cursor on the **Sheet:1** heading in the browser box and right-click the mouse.

2 Select the **Edit Sheet** option.

The **Edit Sheet** dialog box will appear. See Figure 5-71.

3 Select the **D** option and click **OK**.

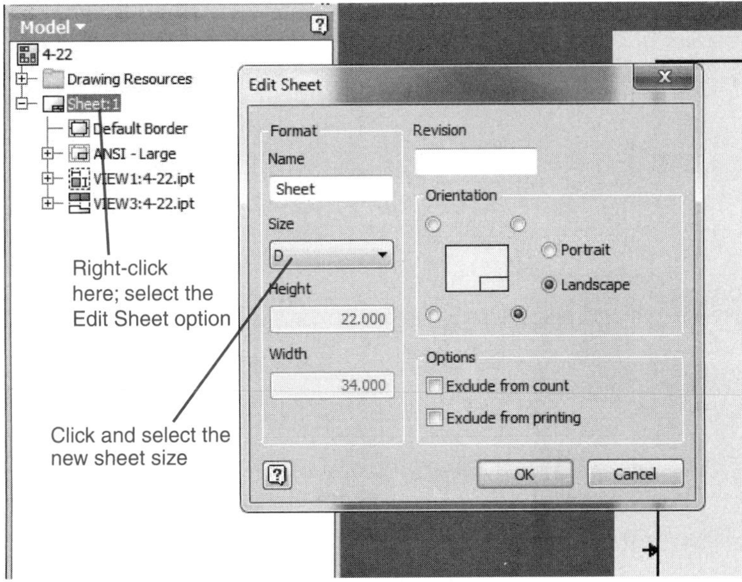

Figure 5-71

See Figure 5-72. Note that the letter C has been replaced with the letter D, and the sheet size has changed.

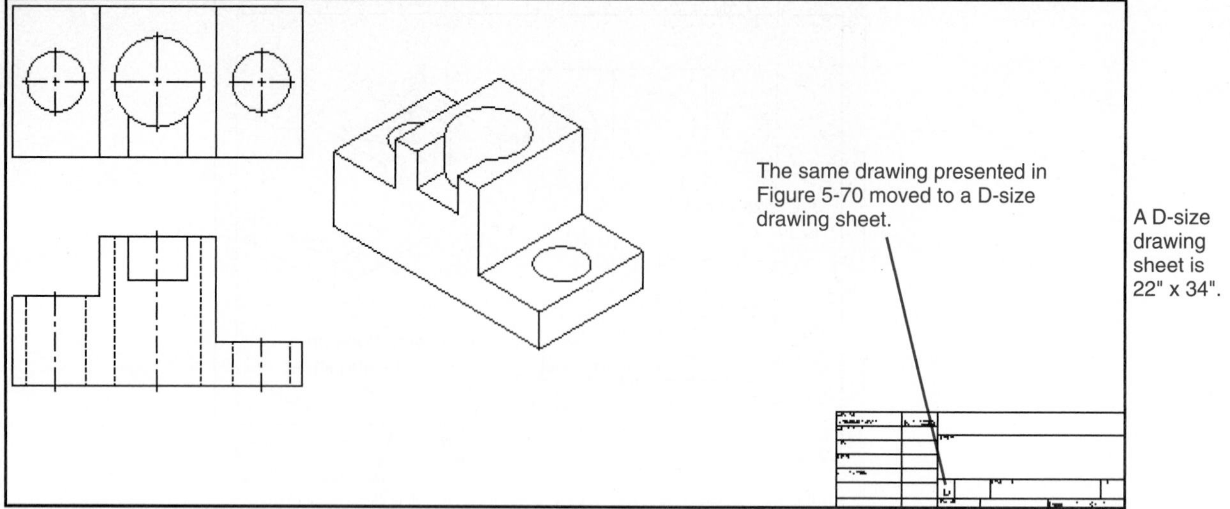

Figure 5-72

EXERCISE 5-27 Adding More Sheets to a Drawing

Sometimes assembly drawings are so large they require more than one sheet.

1. Click the **New Sheet** tool located on the **Sheets** panel under the **Place Views** tab.
2. A **Sheet:2** listing will appear in the browser box. See Figure 5-73.
3. Right-click the **Sheet:2** callout and select the **Edit Sheet** option.
4. Select the desired sheet size and click **OK**.

   A new sheet will appear on the screen. See Figure 5-74.

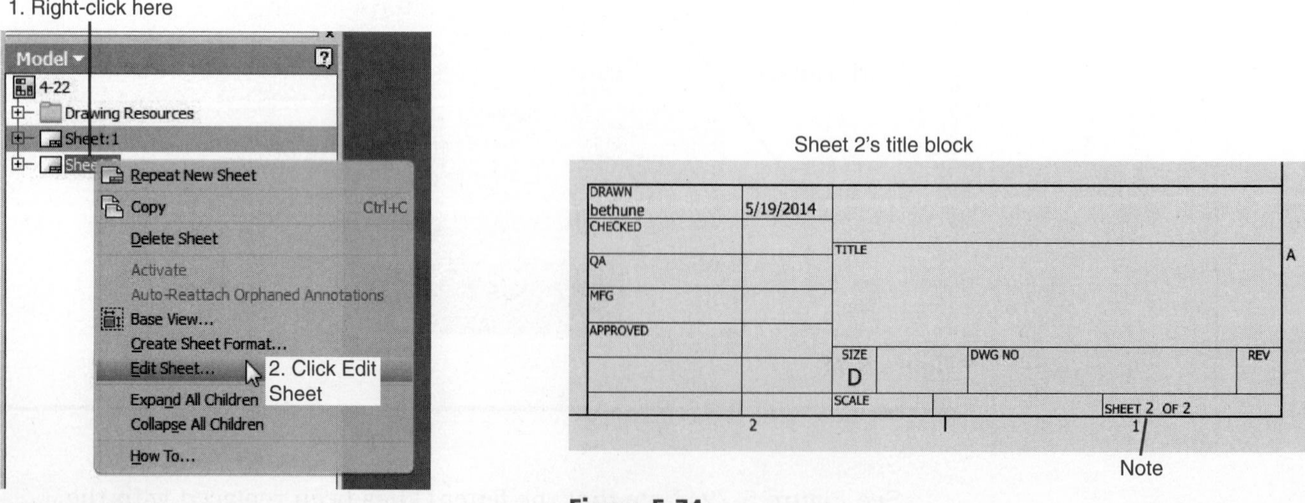

Figure 5-73

Figure 5-74

Other Types of Drawing Blocks

Release Blocks

release block: The area in a title block where required approval signatures are entered.

Figure 5-75 shows an enlarged view of the title block. The area on the left side of the block is called a ***release block***. After a drawing is completed it is first checked. If the drawing is acceptable, the checker will initial the drawing and forward it to the next approval person. Which person(s) and which department approve new drawings varies, but until a drawing is "signed off," that is, all required signatures have been entered, it is not considered a finished drawing.

Figure 5-75

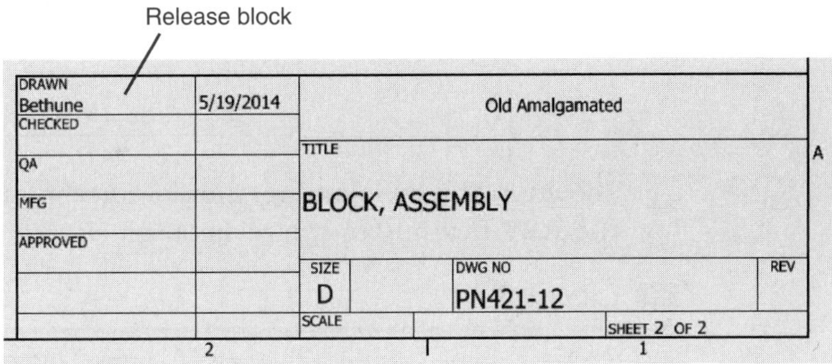

Revision Blocks

revision block: The area in a drawing where changes are listed by number with a brief description of the change.

Figure 5-76 shows a sample ***revision block***. It was created using the **Revision Table** tool located on the **Table** panel under the **Annotate** tab. Drawings used in industry are constantly being changed. Products are improved or corrected, and drawings must reflect and document these changes.

Drawing changes are listed in the revision block by number. Revision blocks are usually located in the upper right corner of the drawing.

Figure 5-76

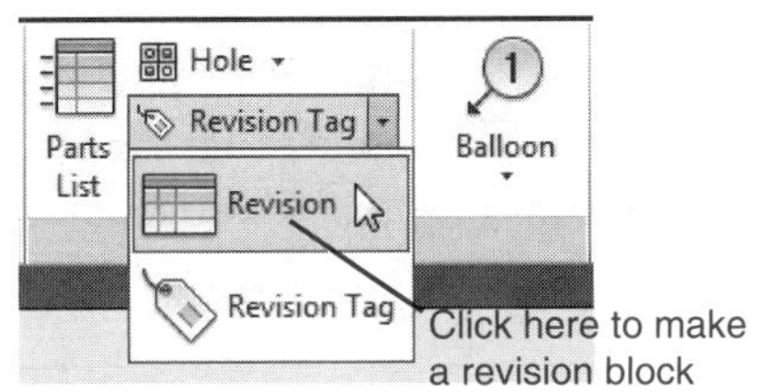

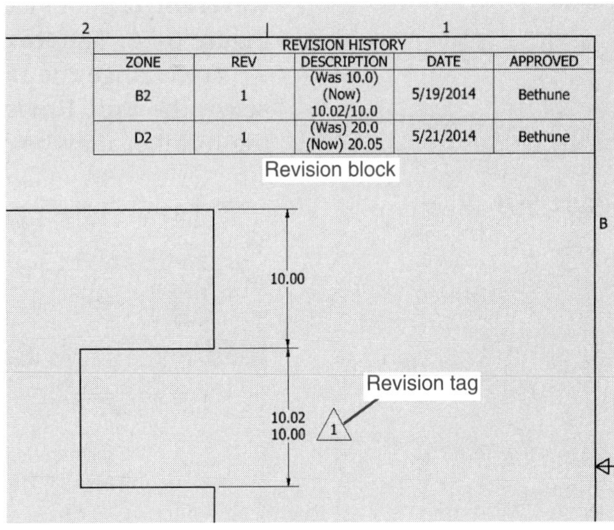

REVISION HISTORY

ZONE	REV	DESCRIPTION	DATE	APPROVED
B2	1	(Was 10.0) (Now) 10.02/10.0	5/19/2014	Bethune
D2	1	(Was) 20.0 (Now) 20.05	5/21/2014	Bethune

EXERCISE 5-28 **Creating a Revision Block**

1 Click the **Revision Table** tool on the **Table** panel.

The **Revision Table** dialog box will appear on the screen. See Figure 5-77. Revisions are usually numbered starting with 1. The default **Start Value** shown in the **Revision Table** dialog box is 1. The numbers

Figure 5-77

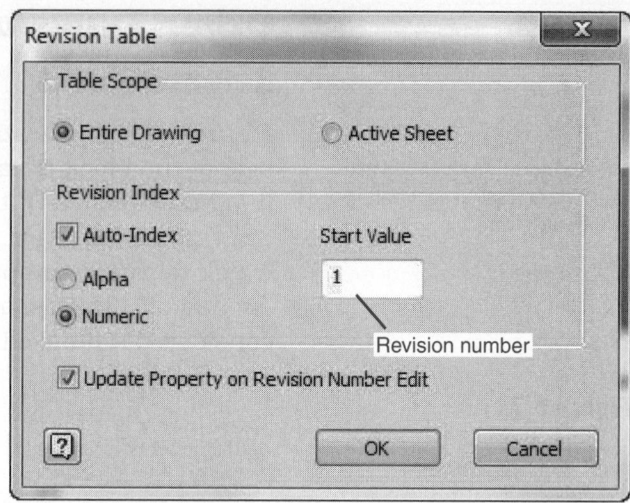

shown in the revision tag should correspond to numbers listed under the **REV** heading in the revision block.

2 Click **OK.**

An outline of the revision block will appear on the screen. Revision blocks are usually located in the upper right corner of the drawing. Move the rectangular outline to the upper right corner of the drawing and click the mouse.

Each drawing revision is listed by number in the revision block. A brief description of the change is also included. It is important that the description be as accurate and complete as possible. The zone on the drawing where the revision is located is also specified.

The revision number is added to the field of the drawing in the area where the change was made. The revision letter is located within a "flag" to distinguish it from dimensions and drawing notes. The flag is created using the **Revision Tag** tool located on the **Table** panel under the **Annotate** tab. See Figure 5-76. The **Revision Tag** tool is a flyout from the **Revision Table** tool.

To change the number within a revision tag, right-click the tag and select the **Edit Revision Tag** option. A text dialog box will appear, and the tag number may be changed. See Figure 5-78.

Figure 5-78

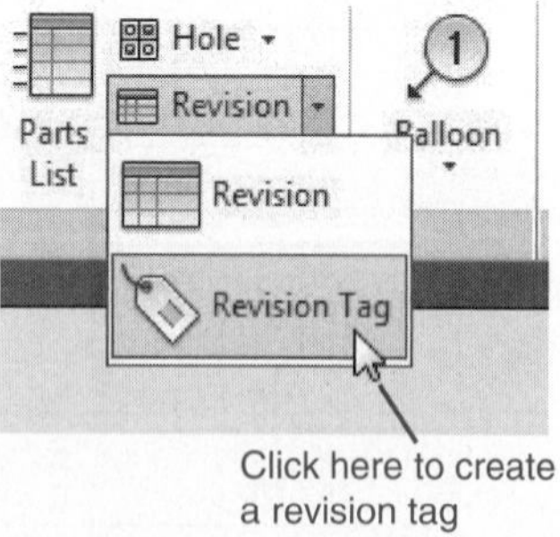

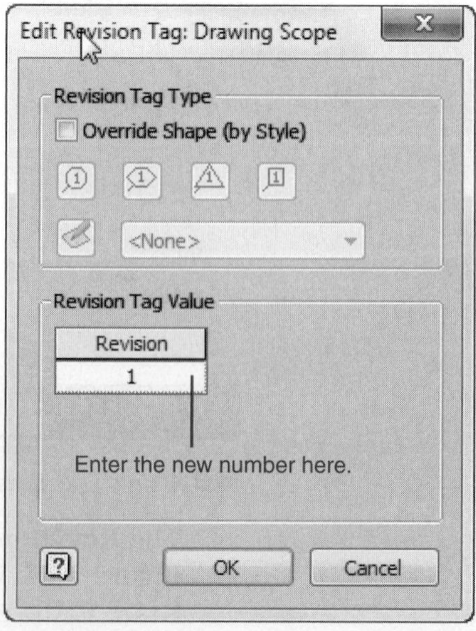

EXERCISE 5-29 Editing the Revision Block

1 Move the cursor onto the revision block.

Filled green circles will appear around the revision block. See Figure 5-79.

Figure 5-79

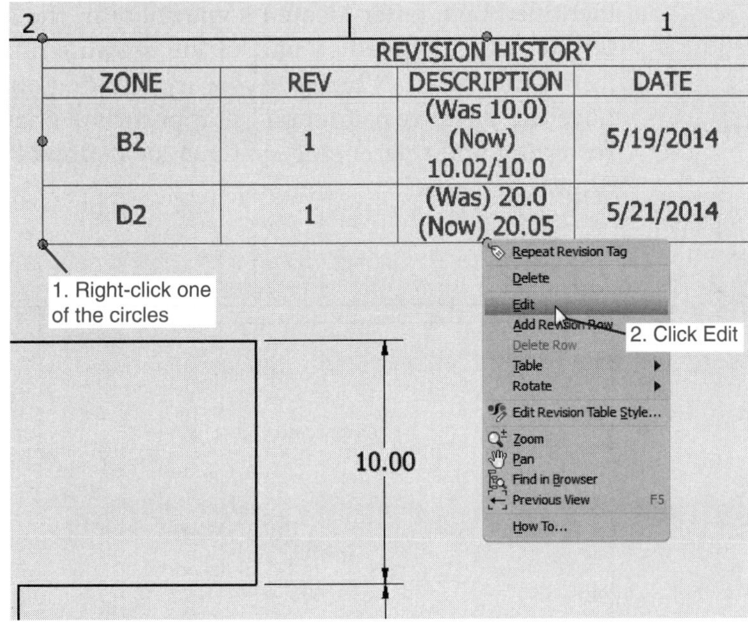

2 Right-click the mouse and select the **Edit** option.

The **Revision Table** dialog box will appear. The block's headings may be edited or rearranged as needed. See Figure 5-80.

Figure 5-80

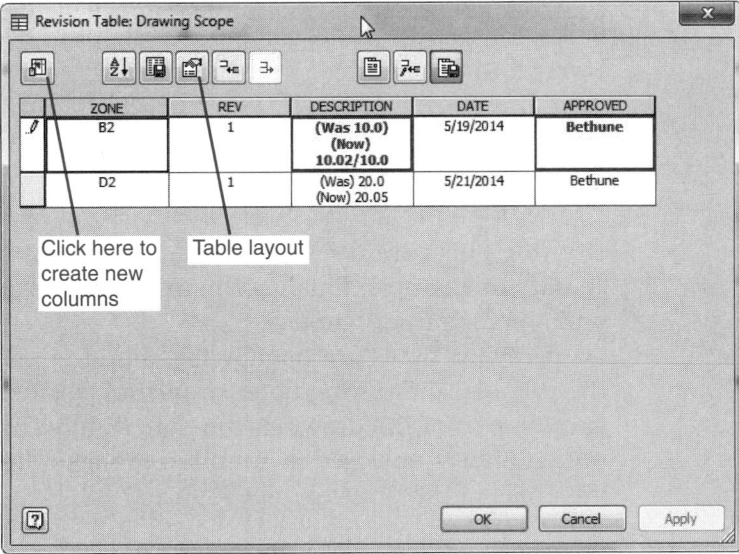

ECOs

Most companies have systems in place that allow engineers and designers to make quick changes to drawings. These change orders are called *engineering change orders* (ECOs), *engineering orders* (EOs), or *change orders* (COs), depending on the company's preference. Change orders are documented on special drawing sheets that are usually stapled to a print of the drawing. Figure 5-81 shows a sample change order attached to a drawing.

drawing revision: A version of a drawing into which change orders have been incorporated.

After a number of change orders have accumulated, they are incorporated into the drawing. This process is called a ***drawing revision***, which is different from a revision to the drawing. Drawing revisions are usually identified by a letter located somewhere in the title block. The revision letters may be included as part of the drawing number or in a separate box in the title block. Whenever you are working on a drawing make sure you have the latest revision and all appropriate change orders. Companies have recording and referencing systems for listing all drawing revisions and drawing changes.

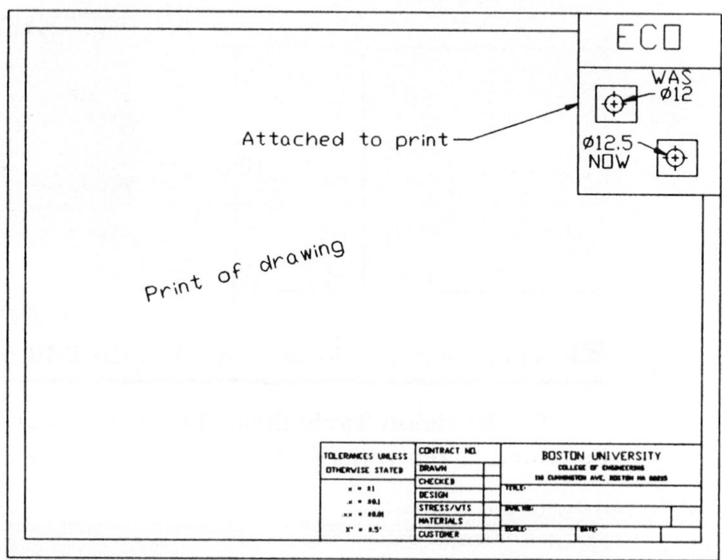

Figure 5-81

Drawing Notes

Drawing notes are used to provide manufacturing information that is not visual, for example, finishing instructions, torque requirements for bolts, and shipping instructions.

Drawing notes are usually listed on the right side of the drawing above the title block. Drawing notes are listed by number. If a note applies to a specific part of the drawing, the note number is enclosed in a triangle. The note numbers enclosed in triangles are also drawn next to the corresponding areas of the drawing. See Figure 5-82.

Top-Down Assemblies

A *top-down assembly* is an assembly that creates new parts as the assembly is created. Figure 5-83 shows a ROTATOR ASSEMBLY that was created using the top-down method. This section will explain how the assembly was created.

Figure 5-82

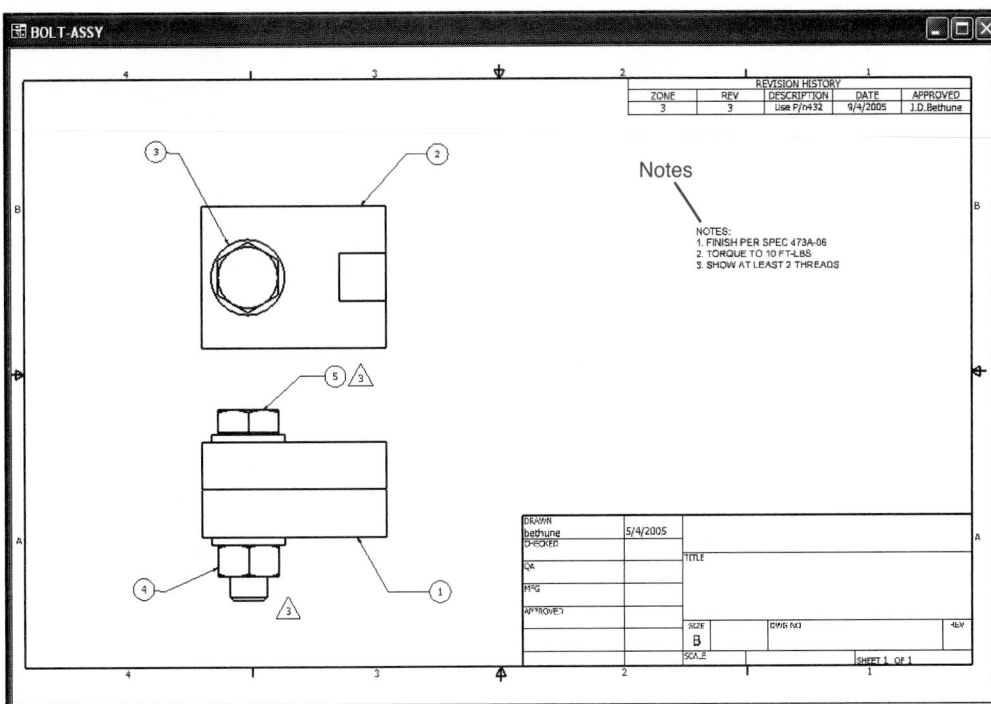

Figure 5-83

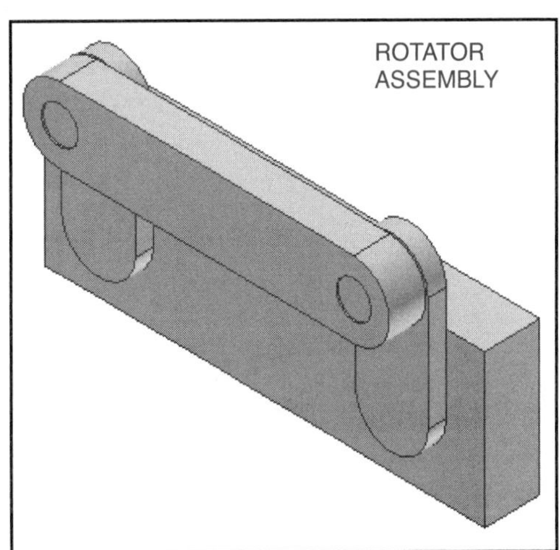

EXERCISE 5-30 Starting an Assembly

When creating a top-down assembly, start by saving the assembly drawing. Individual components can then be added to the assembly drawing.

1 Click on the **New** tool.

2 Click the **Metric** tab, and select **Standard (mm).iam**.

The **Assemble** tab will appear.

3 Left-click on the heading **Assembly1** in the browser box.

See Figure 5-84.

4 Click the arrowhead next to the large **I** icon in the upper left corner of the screen and select the **Save As** tool from the cascading menu.

The **Save As** dialog box will appear. See Figure 5-84.

Figure 5-84

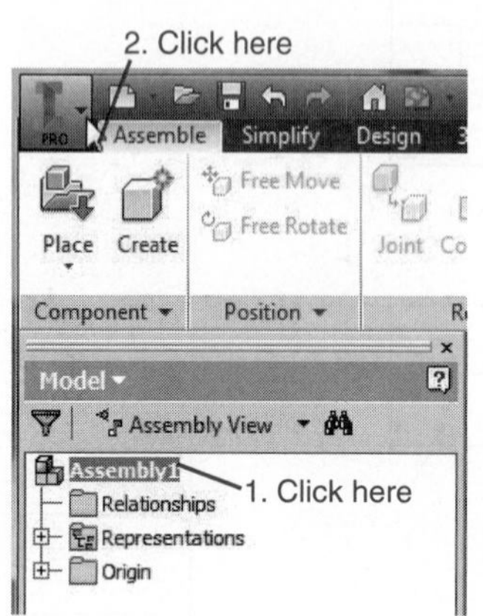

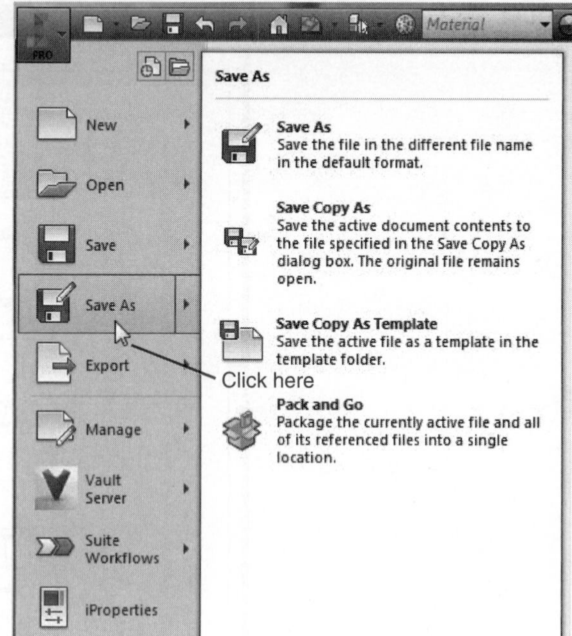

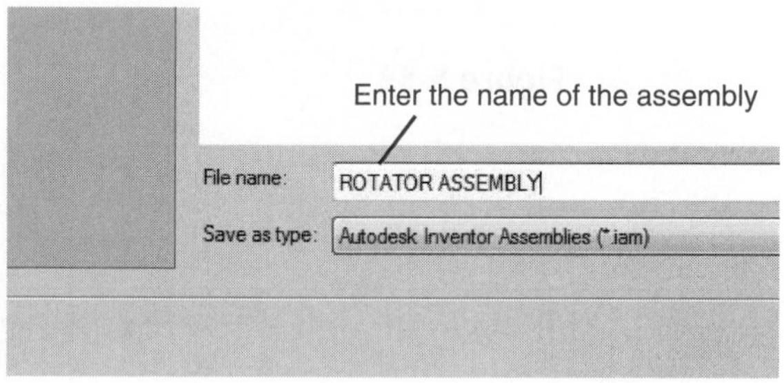

5 Name the assembly, then click **Save**.

In this example the assembly was named **ROTATOR ASSEMBLY**.

EXERCISE 5-31 Changing the Sketch Plane

The parts created for this assembly were created on the XZ plane. This gives the assembly a more realistic orientation.

1 Click the **Tools** tab at the top of the screen and select **Application Options**.

2 Click the **Part** tab.

3 Select the **Sketch on x-z plane** button, then **Apply** and **OK**.

See Figure 5-85.

EXERCISE 5-32 Creating a Part

1 Click the **Create Component** tool on the **Component** panel under the **Assemble** tab.

The **Create In-Place Component** dialog box will appear. See Figure 5-86.

Figure 5-85

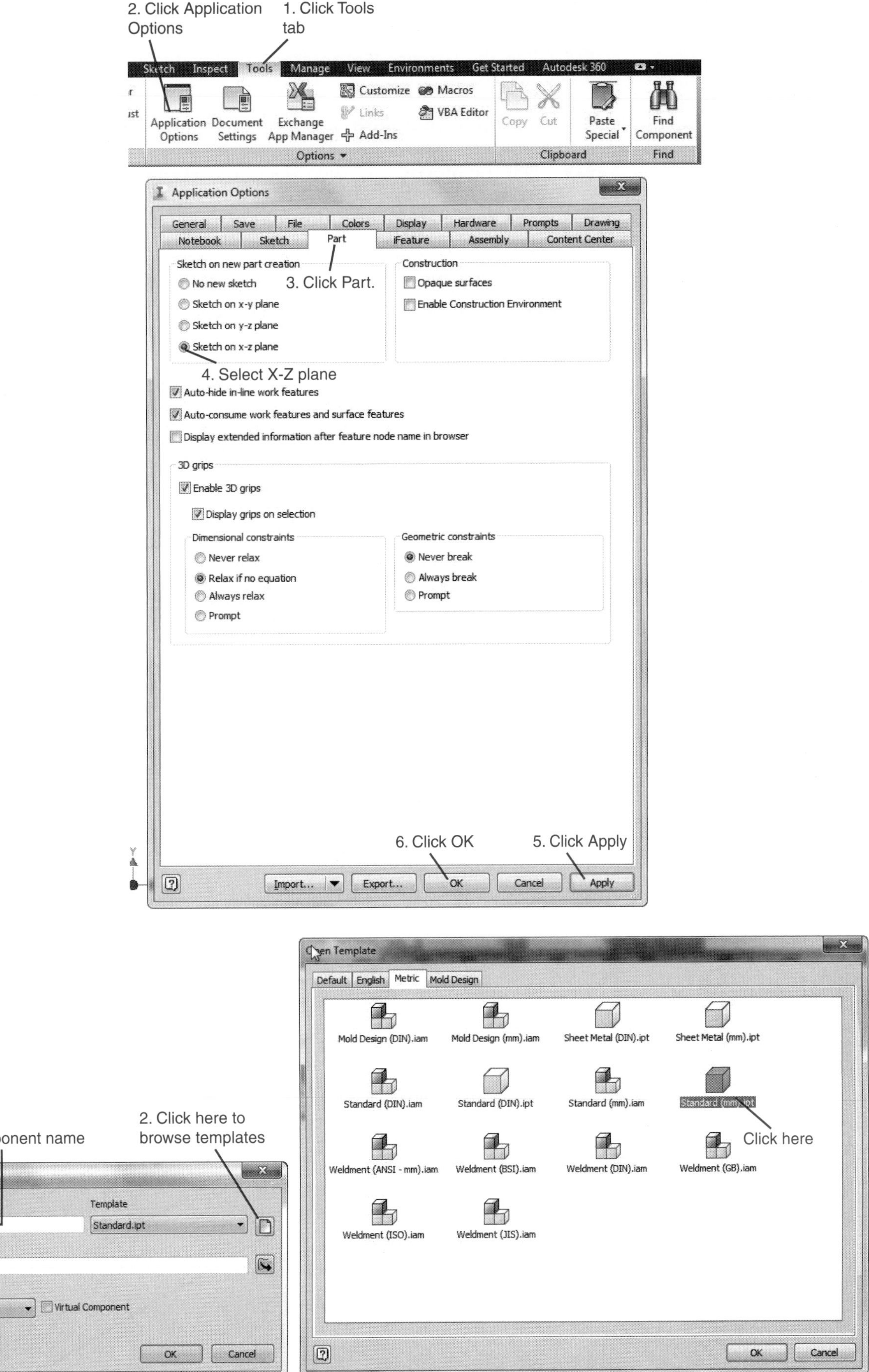

2. Click Application Options
1. Click Tools tab
3. Click Part.
4. Select X-Z plane
6. Click OK
5. Click Apply
1. Enter new component name
2. Click here to browse templates
Click here

Figure 5-86

2 Change the file name to **PLATE.** Select a file location.

3 Click the **Browse Templates** box located to the right of the **Template** box.

The **Open Template** dialog box will appear.

4 Click the **Metric** tab, then select the **Standard (mm).ipt** option, then **OK**.

5 The **Create In-Place Component** dialog box will appear. Click **OK**. Type in the **New Component Name, PLATE,** then click **OK**.

A cursor will appear with a 3D box next to it.

6 Left-click the mouse.

7 Move the cursor into the ViewCube area and click the icon that looks like a house.

This is the **Home** tool. An axis will appear on the screen.

8 Click the **Create 2D Sketch** tool.

The **Sketch** panel will appear in the panel box area.

9 Select a sketch plane.

10 Use the **Two Point Rectangle, Line,** and **Dimension** tools to create the PLATE shown in Figure 5-87.

The line labeled "Central line" is not an edge. It is a single line that will be used to help assemble the parts.

Figure 5-87

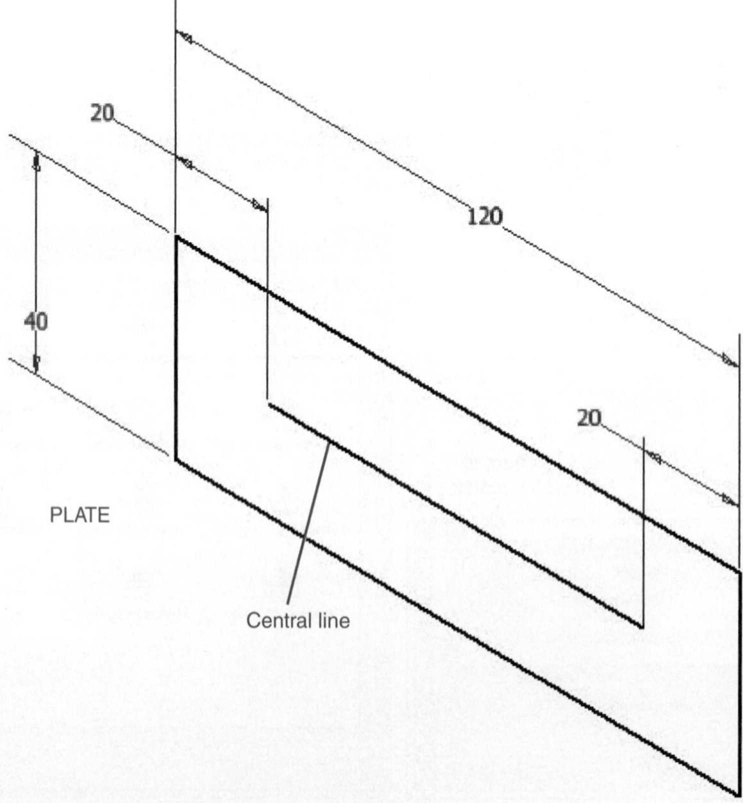

EXERCISE 5-33 Adding Work Points and a Work Axis

1 Right-click the mouse and select the **Finish 2D Sketch** option, and click the **3DModel** tab.

The **3D Model** panel will appear.

> **TIP**
> Do *not* select the **Finish Edit** option.

2 Click the **Work Point** option located on the **Work Features** panel, and locate work points at both ends of the central line.

3 Click the **Work Axis** tool and add a work axis between the two work points.

The new work axis will be through the points and parallel to the baseline of the plate. The work points and work axis will be listed in the browser area.

4 Click on **Origin** under **PLATE** in the browser area.

5 Select the **Work Axis** tool, then click the **XZ Plane** in the browser area and add a work axis through both work points.

The planes used to create this assembly are unique to this example. Other combinations of planes can be used.

6 Right-click the mouse and select the **Finish Edit** option.

See Figure 5-88.

7 Save the drawing.

> **TIP**
> The work axis perpendicular to the XZ plane must appear on both sides of the PLATE.

Figure 5-88

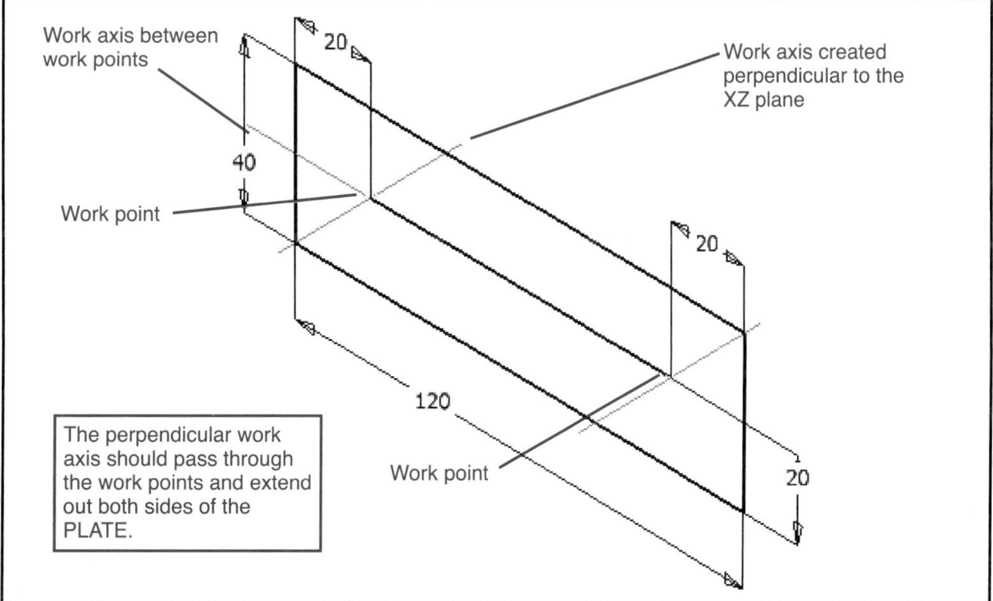

Each of the four parts that make up the assembly must be created as individual drawings. Be sure to use the **Create Component** tool for each part. Each part should be listed in the browser box.

EXERCISE 5-34 Creating LINK-L

1. Click the **Assemble** tab in the browser box, click on the **Create Component** tool, and create a new component named **LINK-L**.

   The **Create In-Place Component** dialog box will appear. See Figure 5-86. Click the **Browse Templates** box, and click the **Metric** tab in the **Open Template** dialog box that will appear. Create the LINK-L component using the **Standard (mm).ipt** format. Define the new component's file name. In this example the name **LINK-L** was assigned.

2. Click the drawing screen, click the **Create 2D Sketch** tool, select the appropriate plane, and use the **Circle**, **Line, and Dimension** tools on the **Sketch** panel to create LINK-L as shown in Figure 5-89.

Figure 5-89

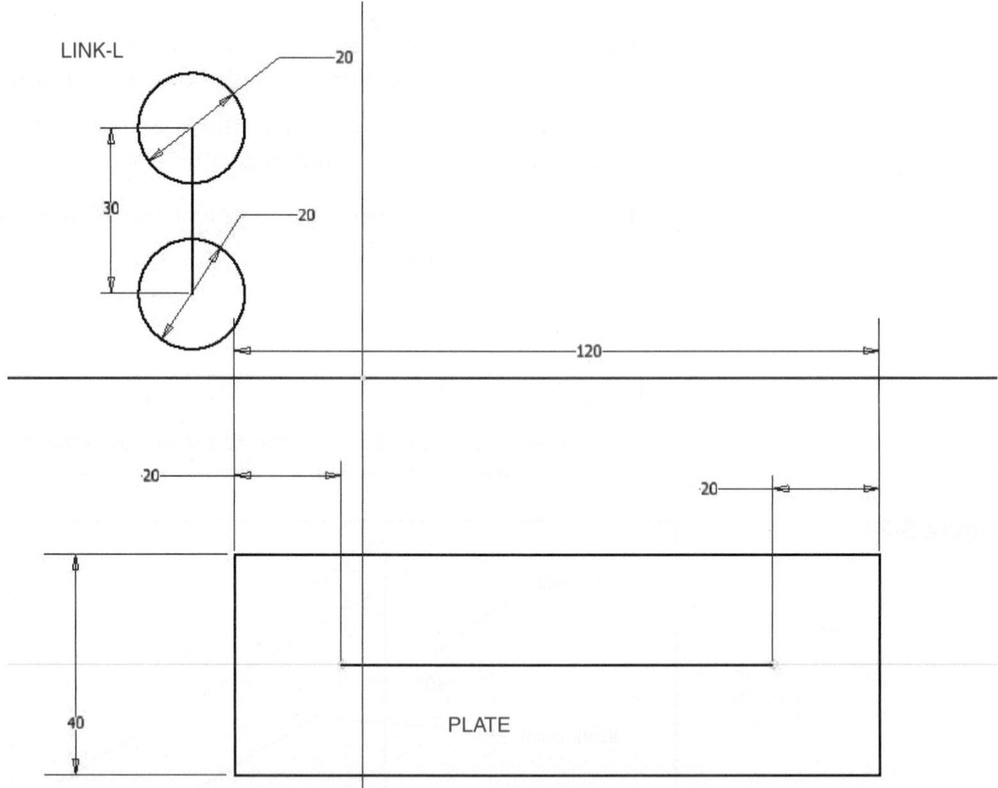

3. Right-click the mouse and select the **Finish 2D Sketch** option.

4. Click the **3D Model** tab, and use the **Work Point** tool to create work points at the center points of both circles. Use the **Work Axis** tool to create a work axis between the two holes' centers.

5. Use the **Work Axis** tool and the **XZ Plane** option listed under **Origin** under **LINK-L** in the browser area to create two work axes through the two work points, perpendicular to the XZ plane.

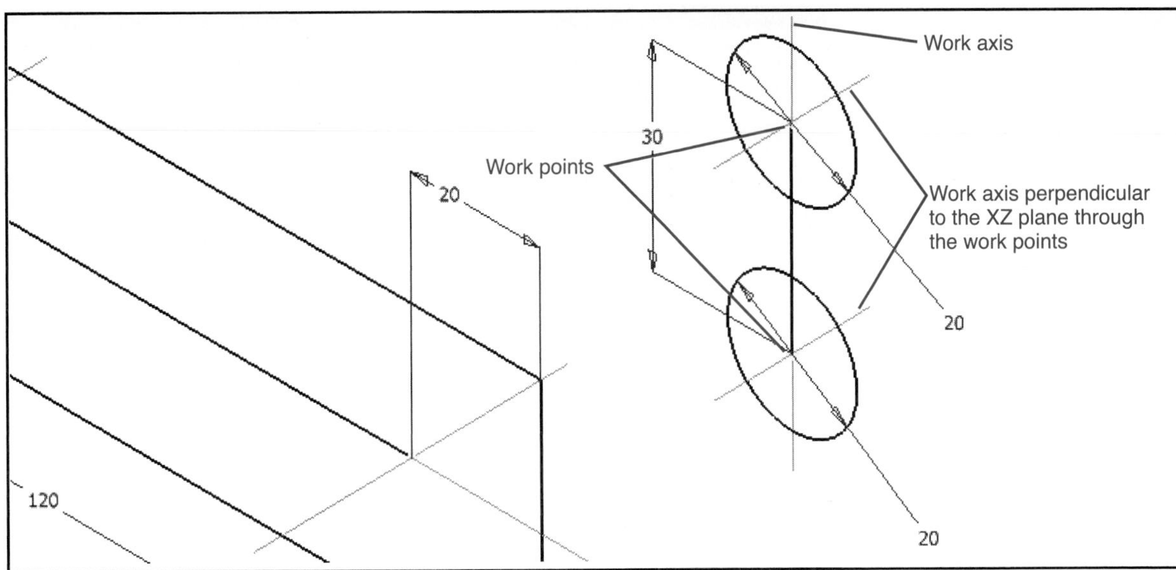

Figure 5-90

See Figure 5-90.

6 Right-click the mouse and select **Finish Edit**.

EXERCISE 5-35 Copying a Component

Figure 5-91 shows the drawing screen with the PLATE and LINK-L components. The LINK-L component will be copied and the copy's name changed to LINK-R.

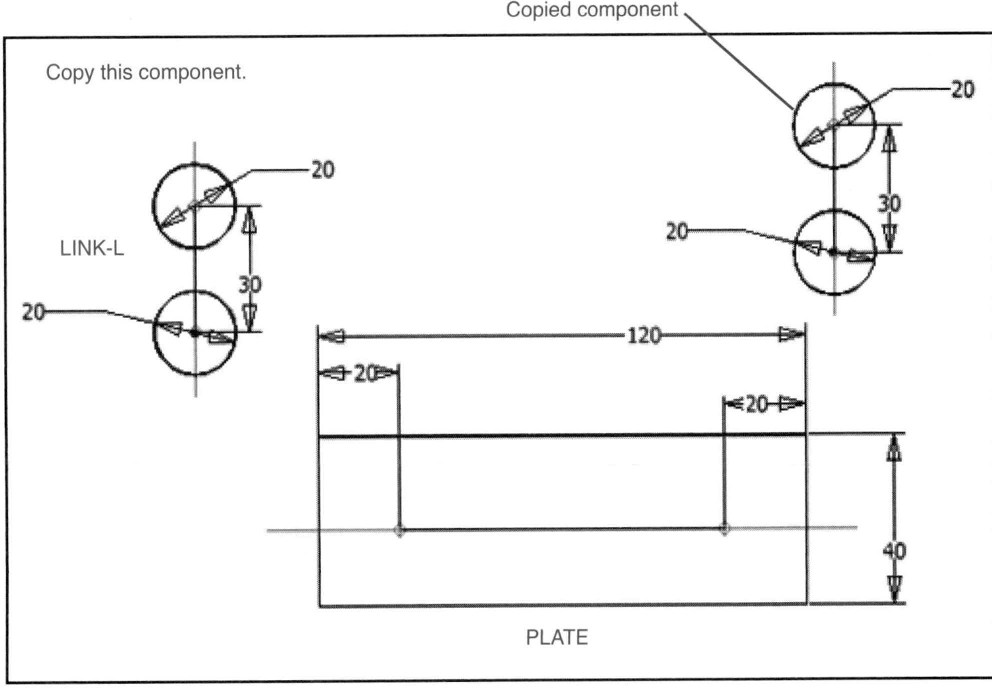

Figure 5-91

1 Click the **Tools** tab. See Figure 5-92.

2 Click the **LINK-L** heading in the browser box.

The **LINK-L** heading will appear in the dialog box.

Figure 5-92

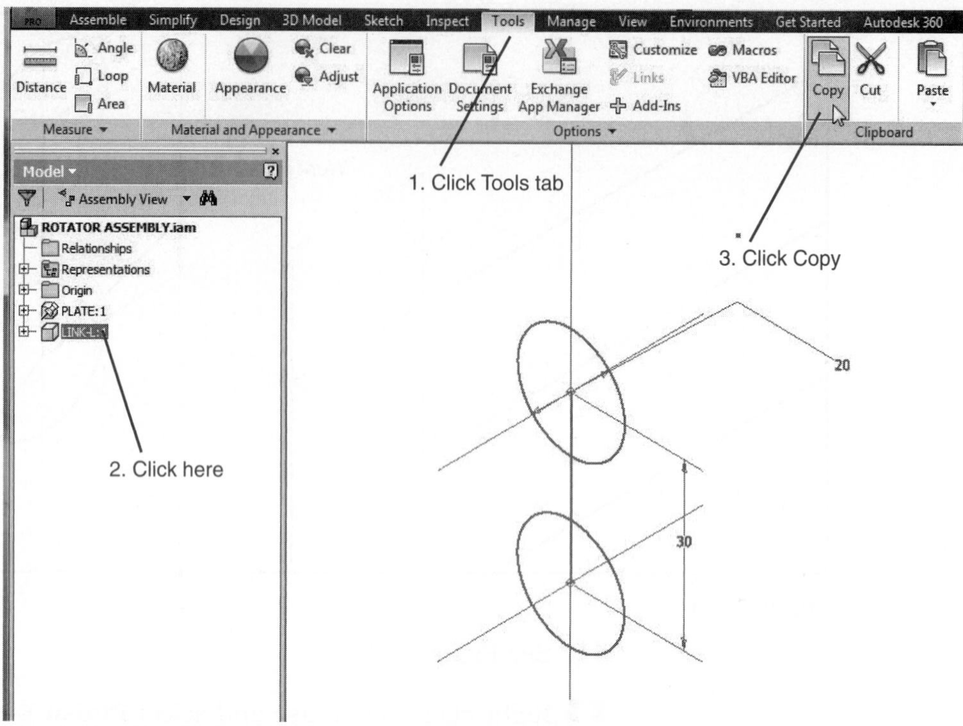

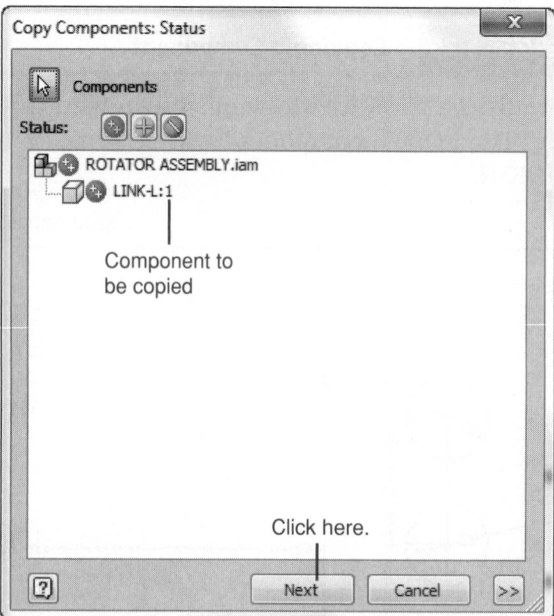

3 Click the **Copy** tool on the **Clipboard** panel under the **Tools** tab.

See Figure 5-93.

4 Right-click the mouse and select the **Paste** option.

A copy of the LINK-L will appear on the screen. LINK-L will be renamed LINK-L:1 and the copied link will be named LINK-L:2. The new names will appear in the browser box.

5 Position LINK-L:2 to the right of the PLATE. See Figure 5-94.

Check to see that the work points and axis created for LINK-L have been copied onto LINK-R in the browser box.

Figure 5-93

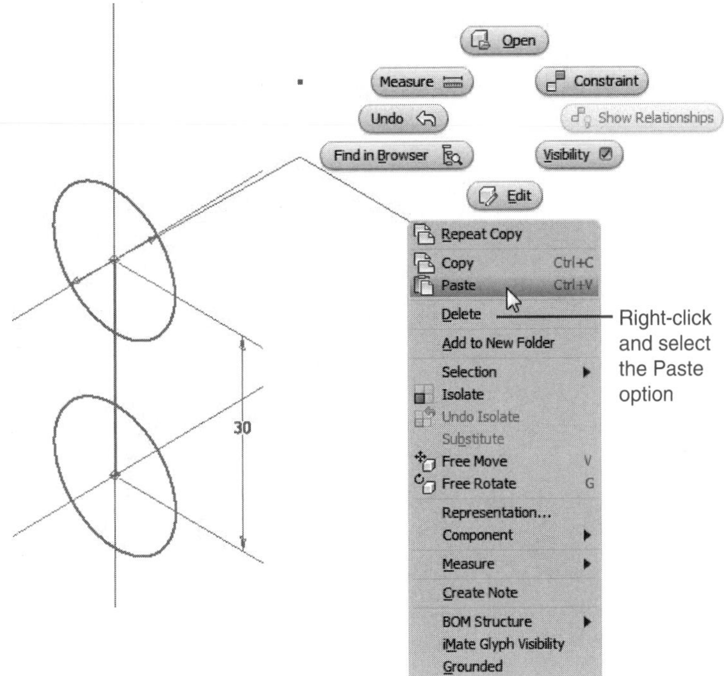

Figure 5-94

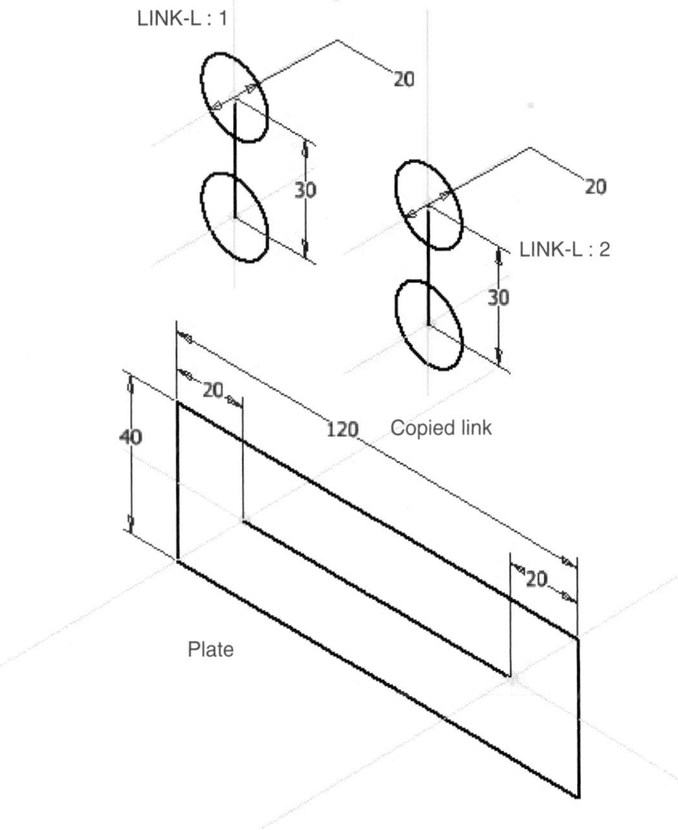

EXERCISE 5-36 Creating the CROSSLINK

1 Double-click the assembly name in the browser box, and use the **Create Component** tool to create the CROSSLINK using the dimensions presented in Figure 5-95.

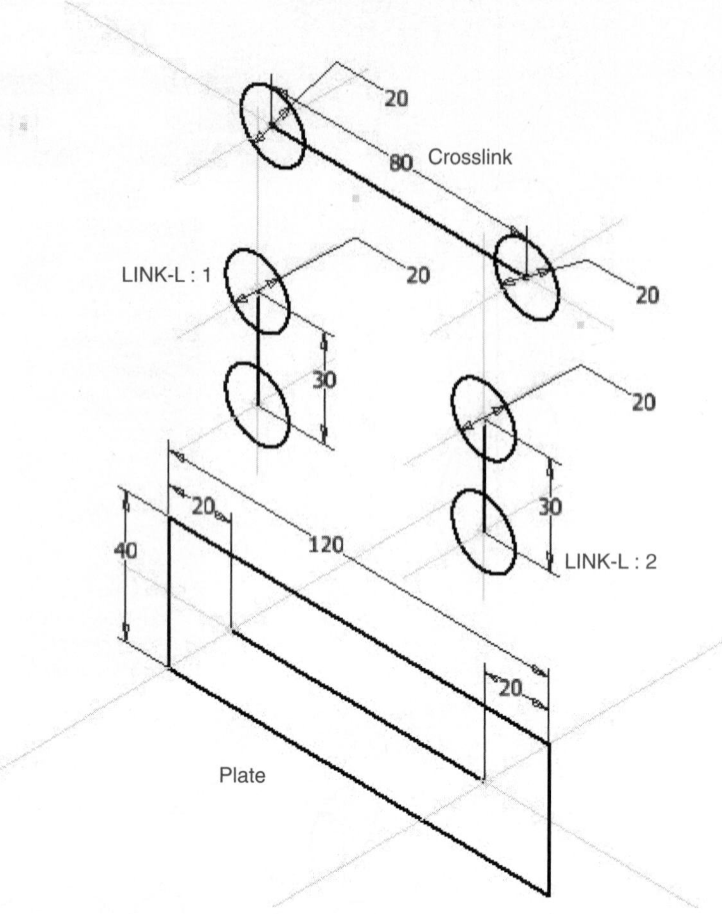

Figure 5-95

Click the **Create Component** tool. See Figure 5-86. Click the **Browse Templates** box, and click the **Metric** tab in the **Open Template** dialog box that will appear. Create the new component using the **Standard (mm).ipt** format. Define the new component's file name as CROSSLINK.

> **NOTE**
>
> All four components that make up the assembly should now be listed in the browser box.

2 Click **OK** and move the cursor into the drawing area and click the left mouse button.

3 Use the **Create 2D Sketch** tool and draw the CROSSLINK using the dimensions shown in Figures 5-95 and 5-96.

4 Right-click the mouse and select the **Finish 2D Sketch** option.

5 Select the appropriate orientation.

6 Add work points and axes as shown.

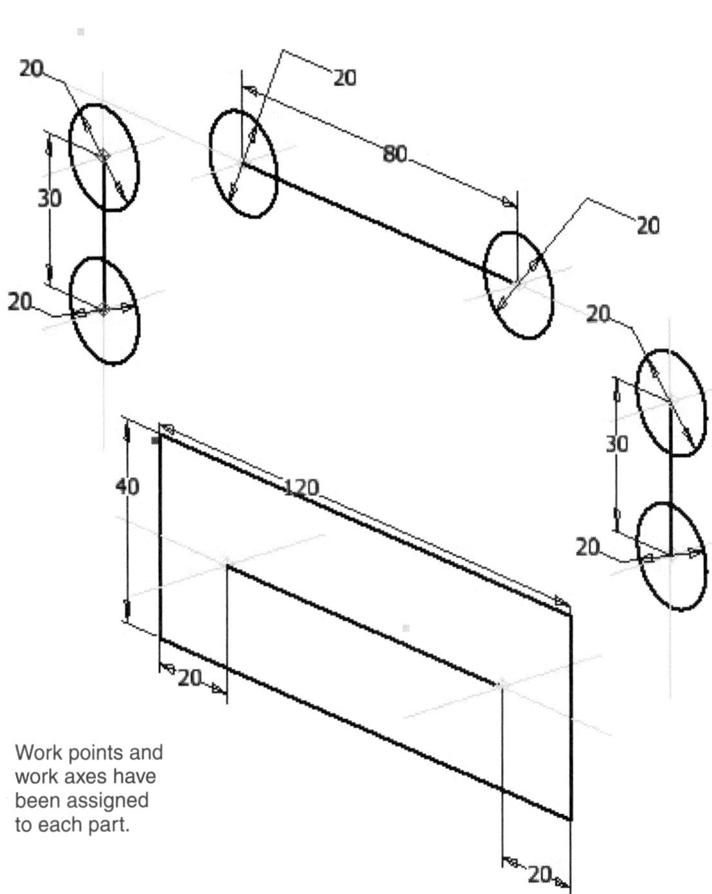

Work points and work axes have been assigned to each part.

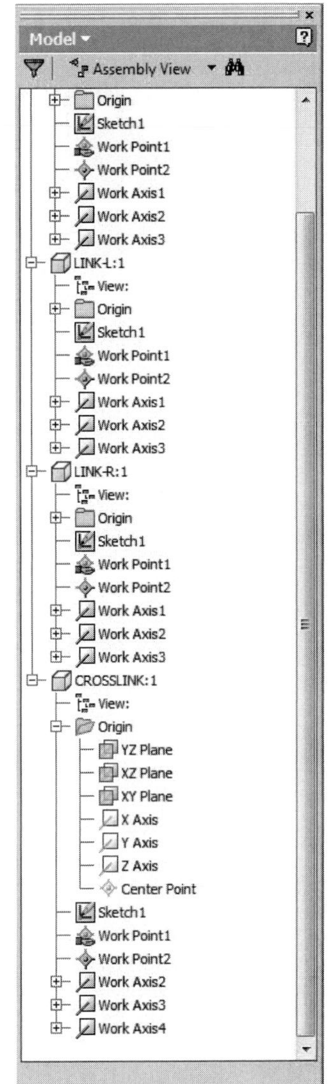

Browser box for ROTATOR ASSEMBLY.

Note the each part has 2 work points and 3 work axes.

Figure 5-96

NOTE

The work axis should appear equal on both sides of the work point. If it appears on only one side, undo and redefine the sketch.

7 Right-click the mouse and select the **Finish Edit** option.

The browser box should look like the one shown in Figure 5-96.

EXERCISE 5-37 Assembling the Parts

1 Return to the **Assemble** tab.

2 Click the **Constrain** tool.

The **Place Constraint** dialog box will appear.

3 Select the **Mate** option.

4 Select the work axes as shown in Figure 5-97.

Figure 5-97

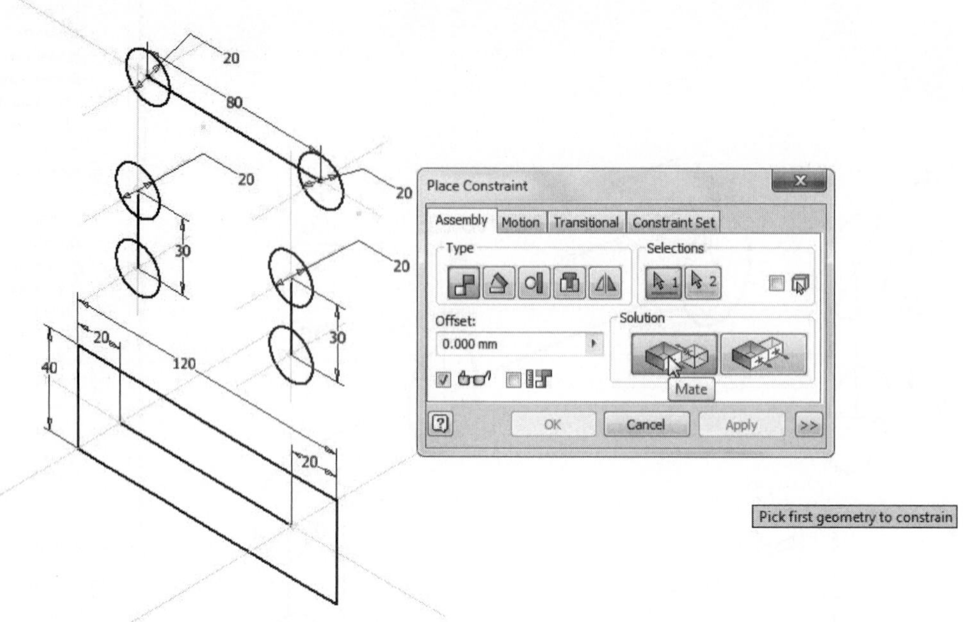

5 Click **Apply**.

6 Assure that the **Mate** tool is still active and mate the work axes shown in Figure 5-98.

Figure 5-98

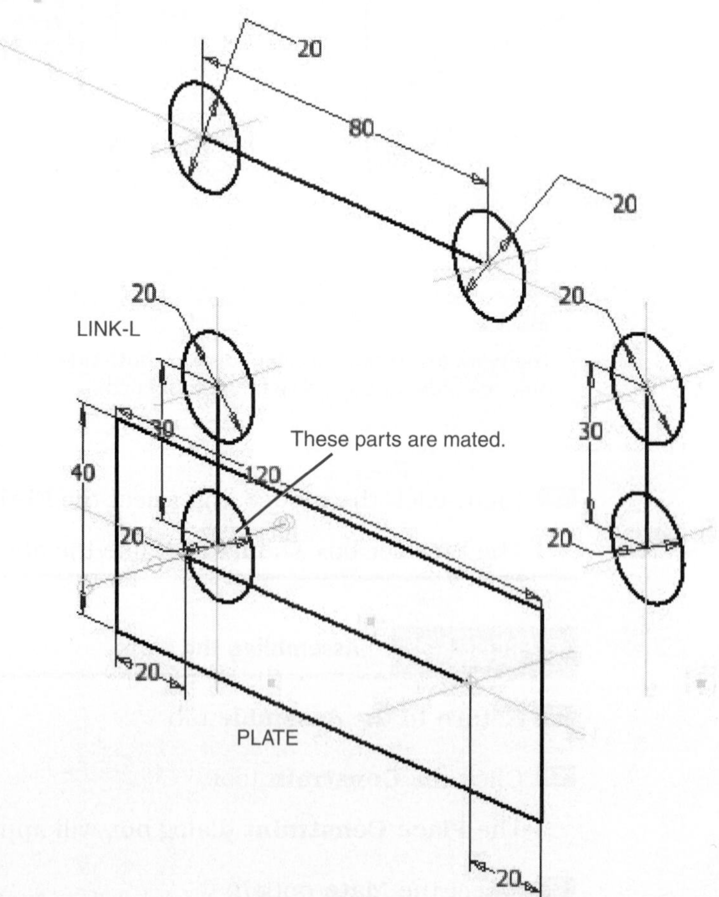

7 Assure that the **Mate** tool is still active and mate the work axes shown in Figures 5-99 and 5-100.

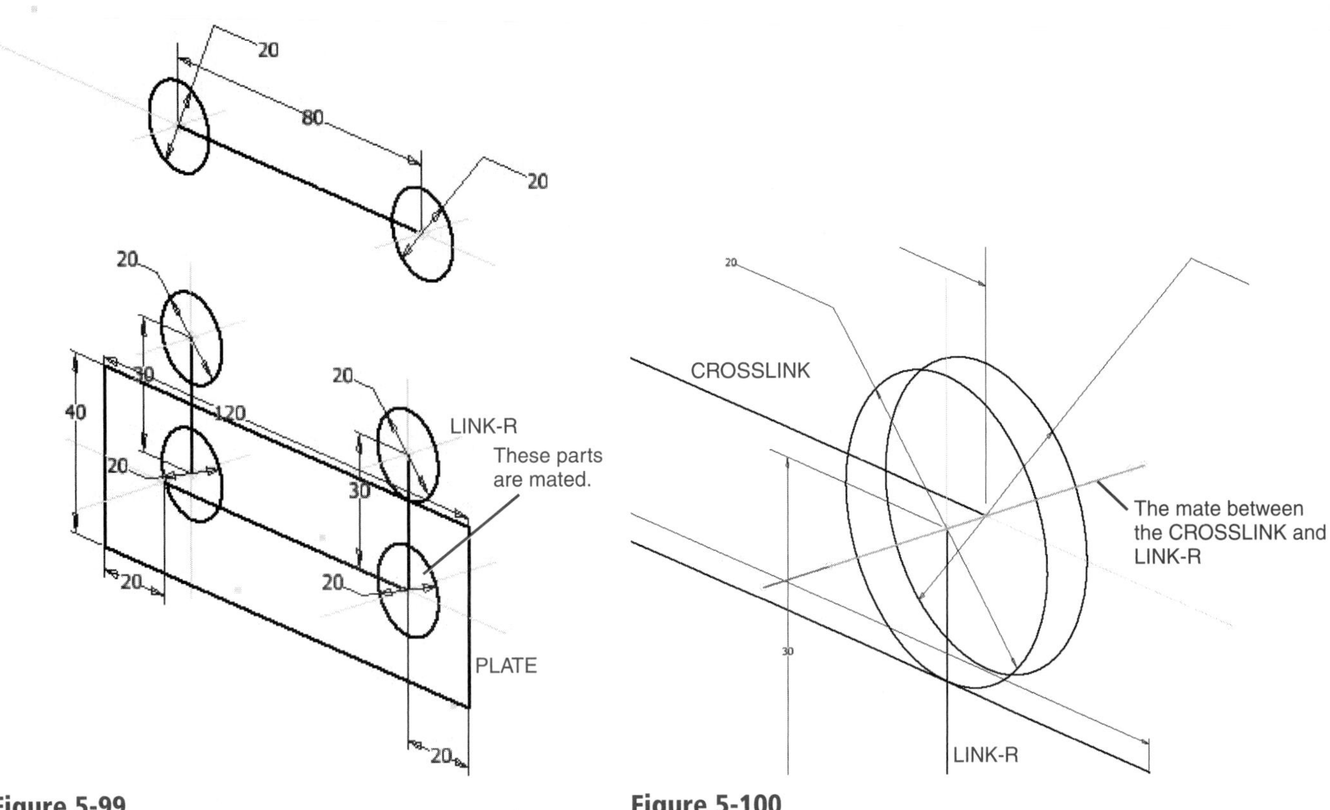

Figure 5-99

Figure 5-100

> **TIP**
>
> Select the work axes that are perpendicular to the XZ plane. Do not select the work points. Make the selections on the perpendicular work axis away from the work points.

Figure 5-101 shows the resulting assembly drawing.

Figure 5-101

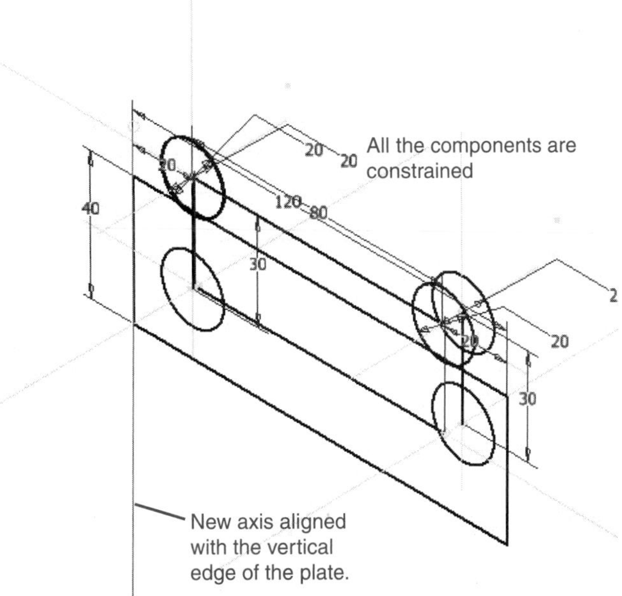

EXERCISE 5-38 Animating the LINKs

1 Access the **Assemble** tab.

2 Right-click the PLATE heading in the browser box and create a work axis on the edge of the PLATE as shown in Figure 5-101.

3 Right-click the mouse and click the **Finish Edit** box. Click the **Constrain** tool.

The **Place Constraint** dialog box will appear.

4 Select the **Angle** option.

The default angle setting is **0.00 deg**.

5 Click the vertical work axis on the PLATE created in step 2, click the vertical axis of **LINK-L:2,** and click the vertical work axis on the PLATE again. See Figure 5-102.

Figure 5-102

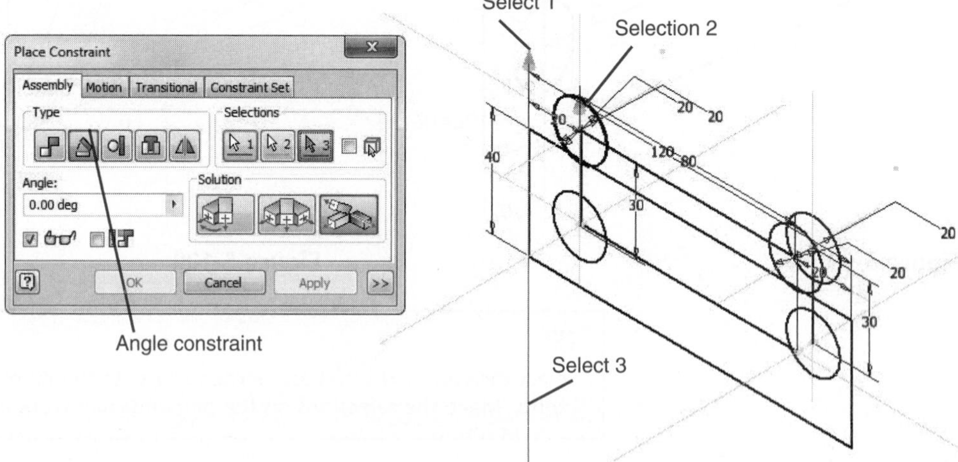

The **Apply** option will be activated when the selections have been made.

NOTE: Click the axis line located on the edge of the PLATE above the edge's centerline.

6 Click **Apply**.

EXERCISE 5-39 Setting the Assembly in Motion

1 Right-click the **Angle:1 (0.00 deg)** constraint in the browser area, and select the **Drive** option.

See Figure 5-103. The **Drive Constraint** dialog box will appear. See Figure 5-104.

2 Set the **End** angle for **720.00 deg**.

3 Click the **Forward** button.

The assembly should rotate freely. If the rotation is not correct, check the constraints.

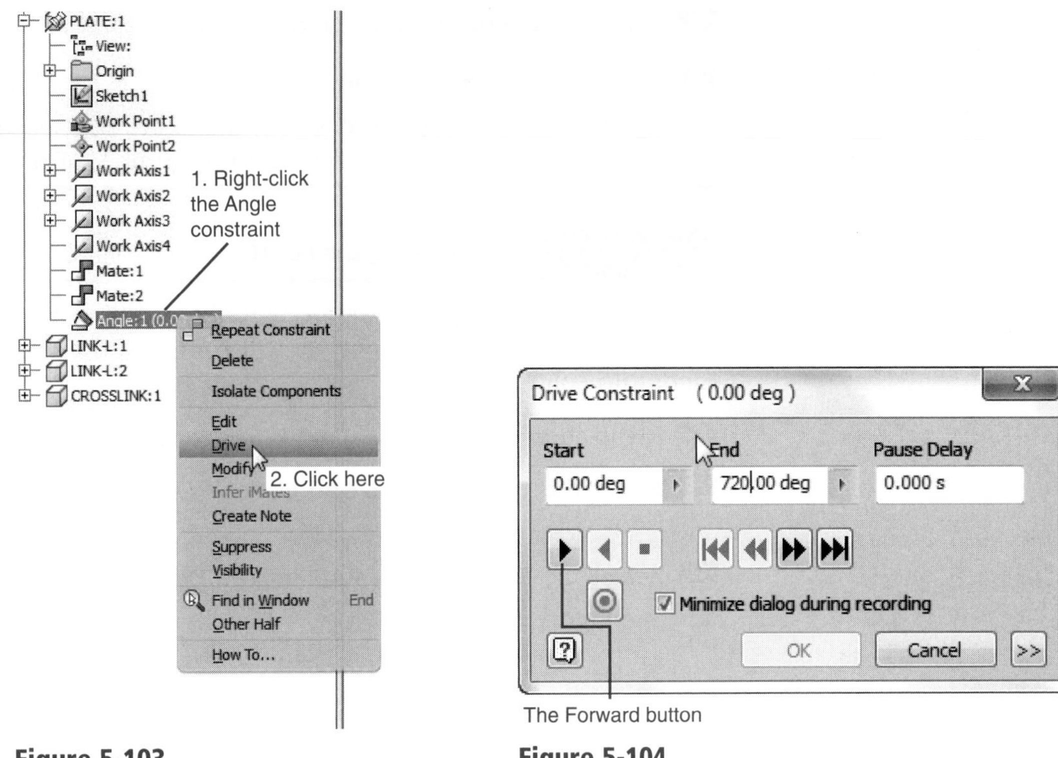

Figure 5-103

Figure 5-104

EXERCISE 5-40 Controlling the Speed of the Rotation

1 Click the **>>** button on the **Drive Constraint** dialog box.

The box will expand. See Figure 5-105.

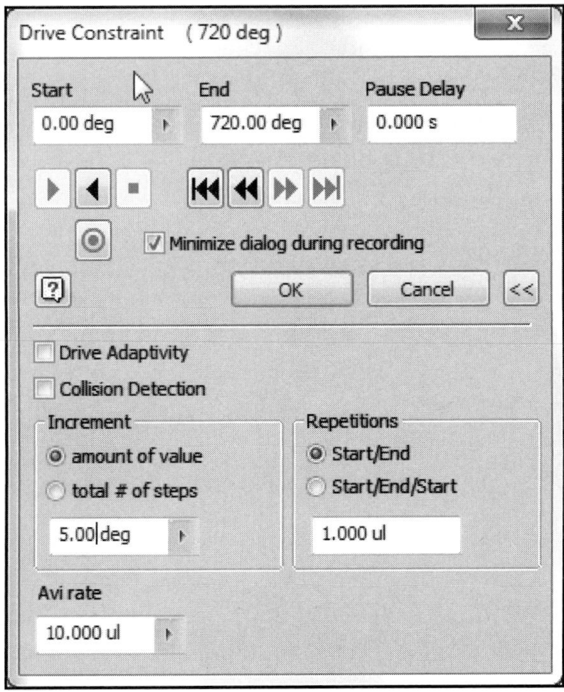

Figure 5-105

2 Set the **Increment** value to **5.00 deg**.

3 Click the **Forward** button.

The higher the **Increment** value, the faster the assembly will rotate. The **Repetitions** setting is used to control the number of revolutions generated.

EXERCISE 5-41 Completing the PLATE

So far only outlines of the components have been created. This is a good way to check the motion. Now we will complete the components, turning them into solid models.

1 Right-click the heading **PLATE** in the browser box.

2 Select the **Open** option, and click the **Home** icon to create an isometric view.

3 Select the **Extrude** tool and set the **Extents Distance** to **15 mm**.

See Figure 5-106.

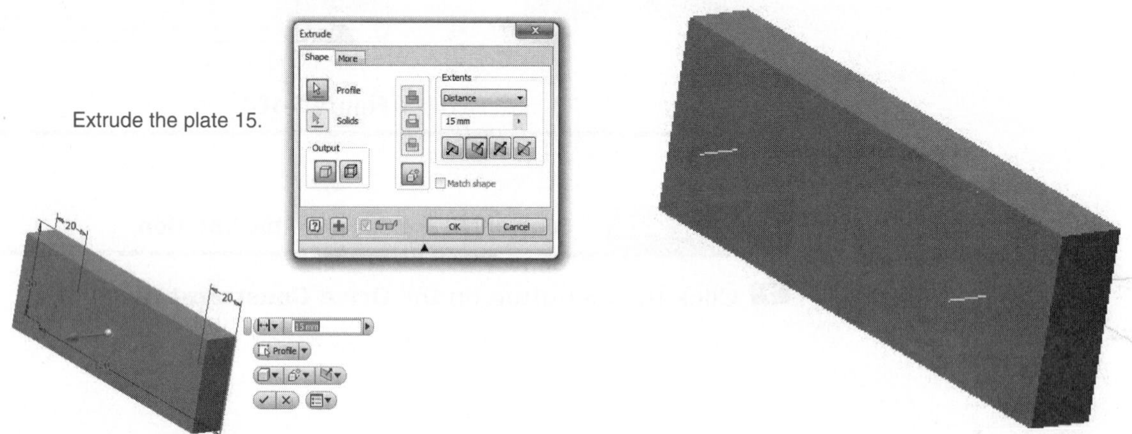

Figure 5-106

4 Click **OK**.

5 Click the **Hole** tool. Use the two existing work points as center points for two Ø10.0 holes.

See Figure 5-107.

Figure 5-107

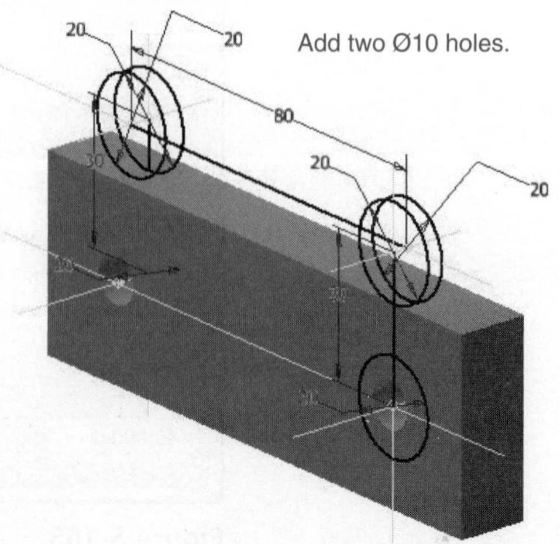

6 Save the PLATE changes.

7 Click the **X** in the upper right corner to return to the assembly drawing. **Save** the drawing.

EXERCISE 5-42 Completing LINK-L and LINK-R

1 Right-click **LINK-L:1** in the browser area and select the **Open** option.

2 Right-click the word **Sketch 1** under the **LINK-L:1** heading and select the **Edit Sketch** option.

3 Draw two tangent lines as shown in Figure 5-108.

Figure 5-108

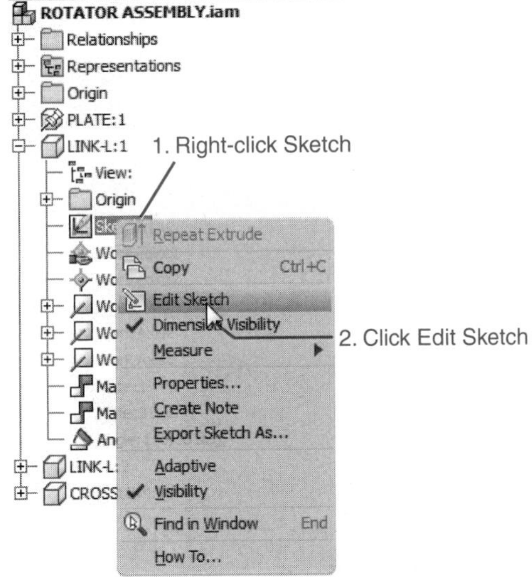

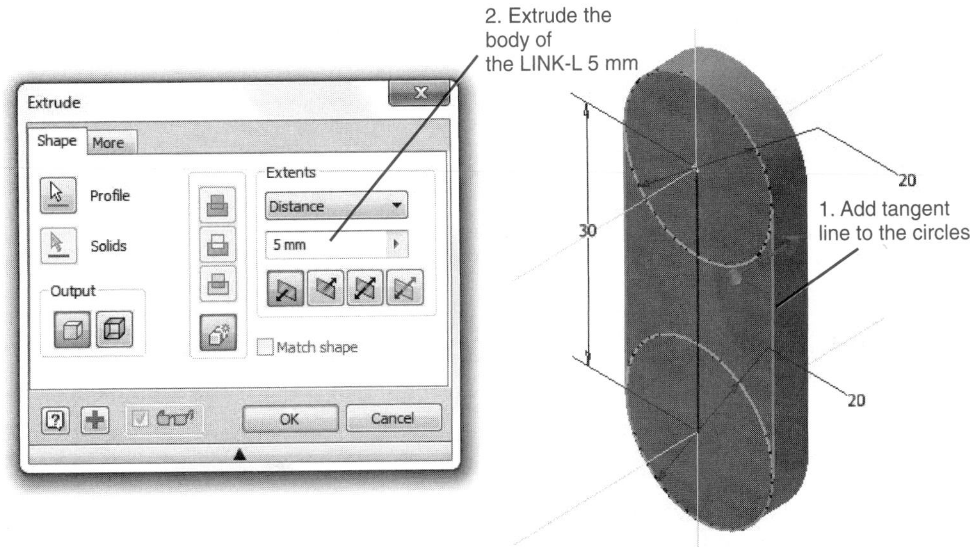

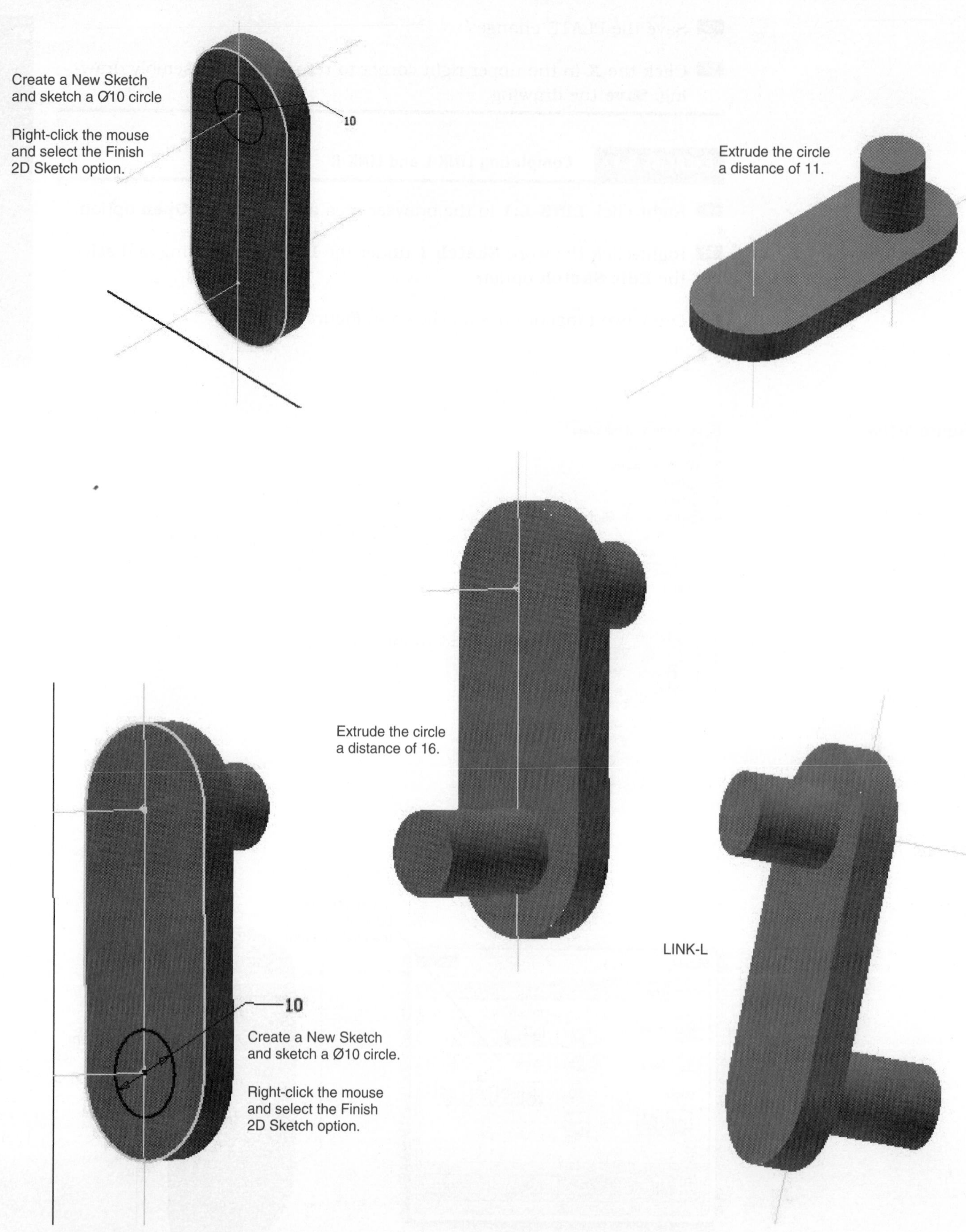
Create a New Sketch
and sketch a Ø10 circle
10
Right-click the mouse
and select the Finish
2D Sketch option.
Extrude the circle
a distance of 11.
Extrude the circle
a distance of 16.
LINK-L
10
Create a New Sketch
and sketch a Ø10 circle.
Right-click the mouse
and select the Finish
2D Sketch option.

Figure 5-108
(*Continued*)

4 Click the **Finish 2D Sketch** tool and extrude LINK-L **5 mm**.

5 Right-click the mouse and select the **New Sketch** option. Sketch a **Ø10** circle. Right-click the mouse and click the **Finish 2D Sketch** option.

6 Extrude the top circle **11** forward.

The 11 extrusions will create a 1 offset from the CROSSLINK.

7 Reorient **LINK-L:1** so that the back surface is visible, right-click the mouse, and select the **New Sketch** option.

8 Sketch a **Ø10** circle, right-click the mouse, and select the **Finish 2D Sketch** option.

9 Extrude the circle a distance of **16**.

The 16 extrusion will create a 1 offset from the PLATE.

10 Edit **LINK-L:2** to the same dimensions and features.

11 Save the LINKs.

12 Click the **X** in the upper right corner of the screen to return to the assembly drawing.

EXERCISE 5-43 Completing the CROSSLINK

1 Right-click **CROSSLINK** in the browser area and select the **Open** option.

2 Add two tangent lines and extrude the CROSSLINK **10 mm**.

See Figure 5-109.

Figure 5-109

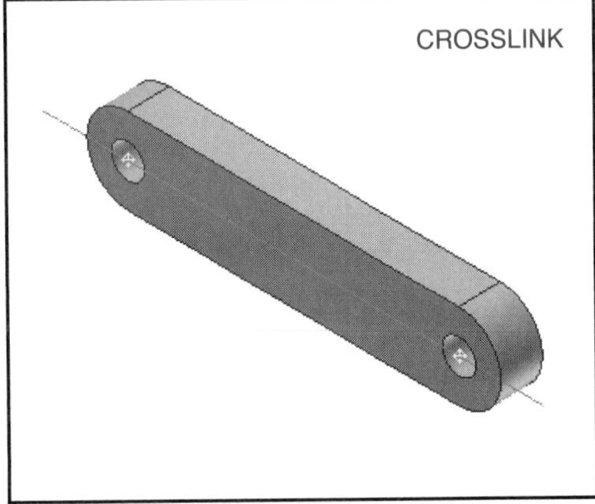

3 Create two **Ø10** holes as shown.

Save the CROSSLINK.

4 Click the **X** in the upper right corner of the screen to return to the assembly drawing.

Aligning the Assembly

Now that the components have thickness, they may not be aligned correctly; that is, they interfere with each other. The alignment could have been created as the components were constrained in the last section by defining offset values. Rotate the assembly and see if there is any interference or if the components are too far apart. See Figure 5-110. Use the **Flush** constraint and create a **1** offset between the parts if needed. The extrusion distances of 11 and 16 should help create the offset. Align the front surface of the CROSSLINK with the top of the Ø10 × 11 post on the LINKs and the back of the PLATE with the top of the Ø10 × 16 post. The finished assembly should look like Figure 5-111.

Figure 5-110

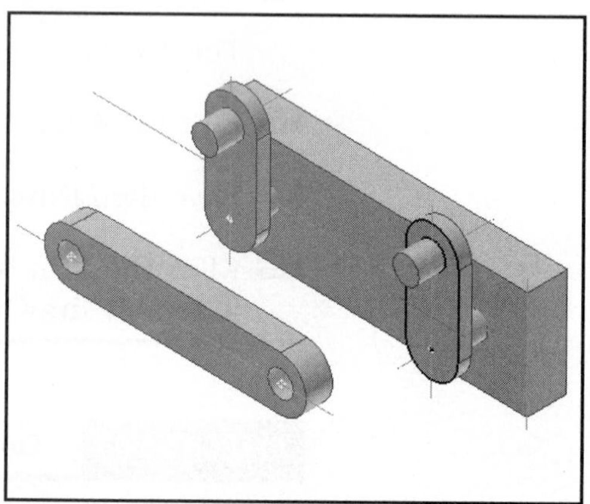

Figure 5-111

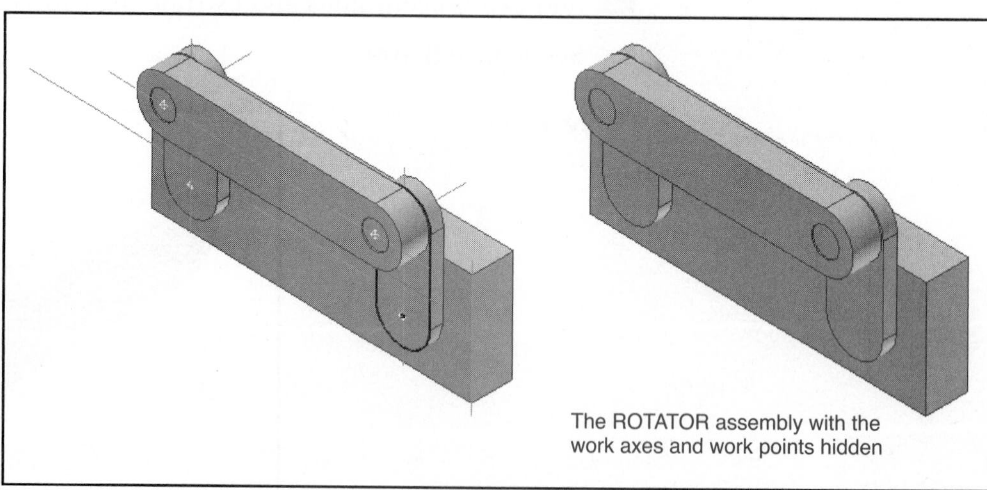

The ROTATOR assembly with the
work axes and work points hidden

Presentations

Figure 5-112 shows a presentation drawing of the ROTATOR, and Figure 5-113 shows an exploded isometric drawing created using the .idw format.

Figure 5-112

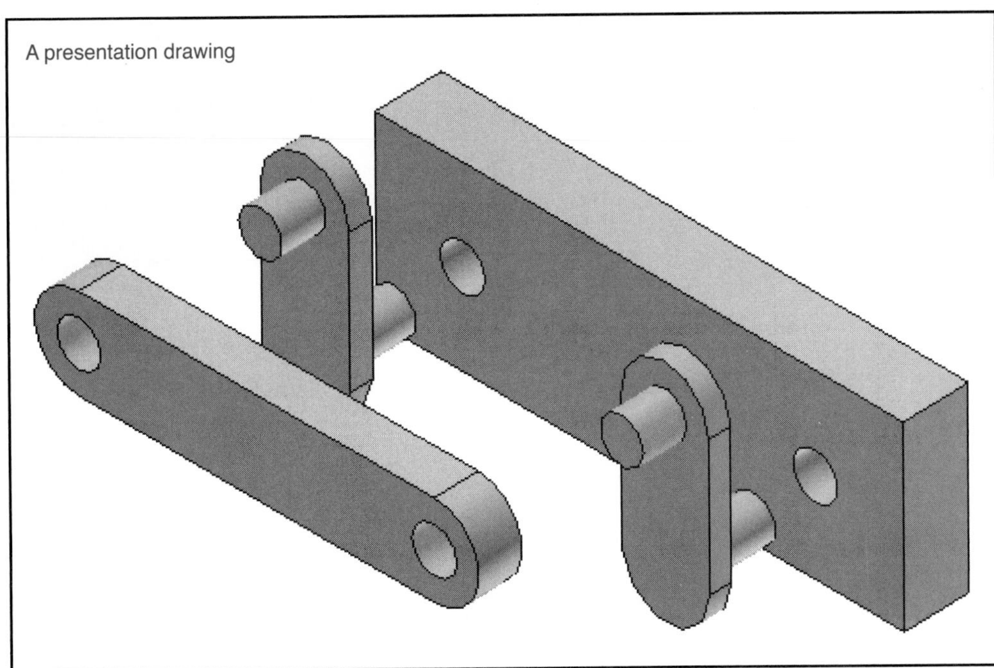

Figure 5-113

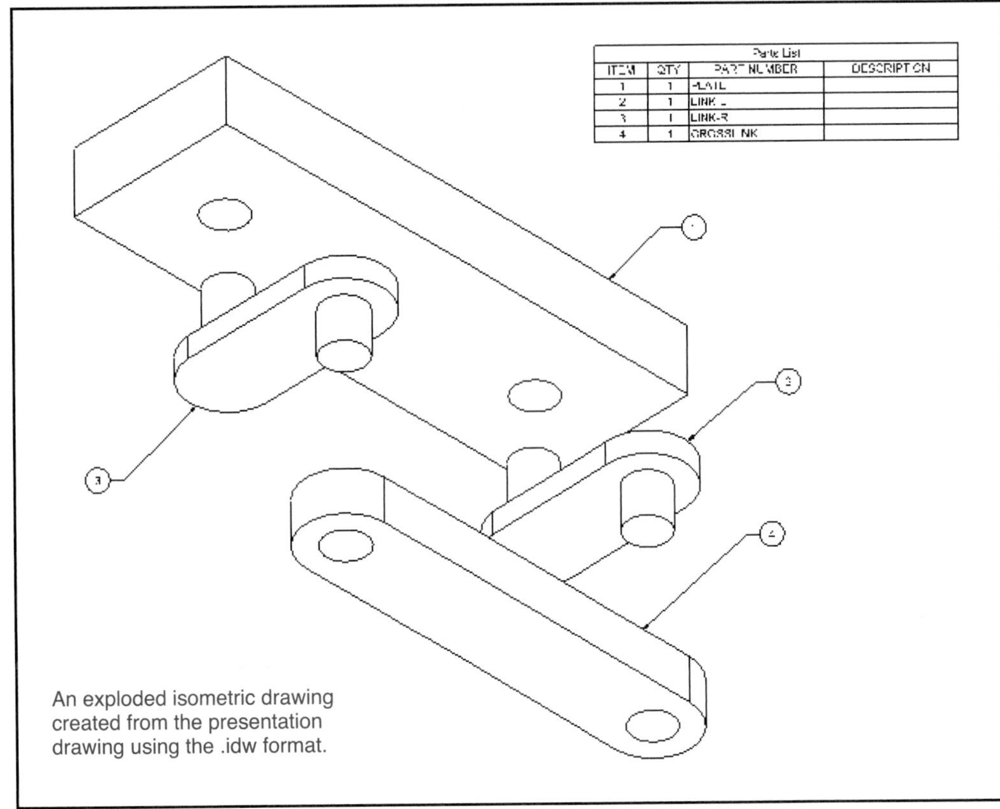

Editing a Part within an Assembly Drawing

Figure 5-114 shows an assembly drawing. The assembly is called 114-ASSEMBLY and is made from three components: 114-PLATE, 114-BRACKET, and 114-POST.

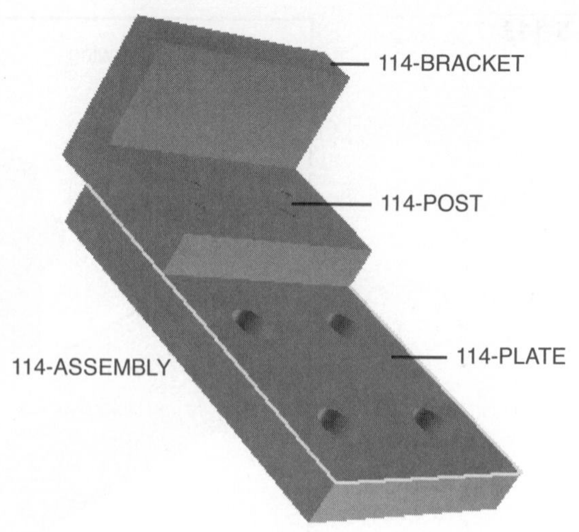

Figure 5-114

EXERCISE 5-44 **Editing a Feature**

It has been determined that the Ø10 mm holes in the bracket are too small. They are to be increased to Ø11 mm. The holes are features.

1 Right-click the **114-BRACKET** heading in the browser box.

See Figure 5-115. A list of options will appear. See Figure 5-116.

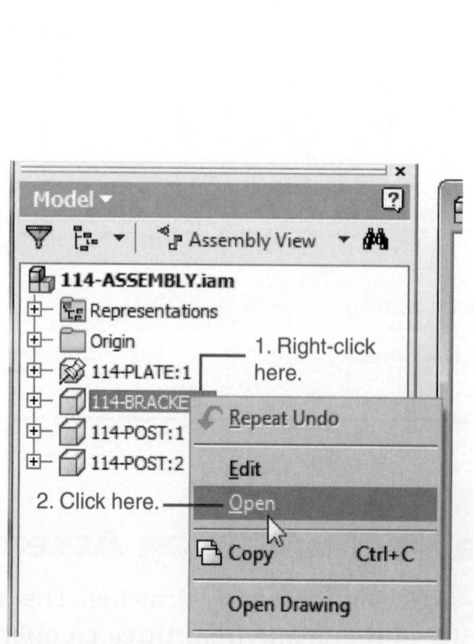

Figure 5-115

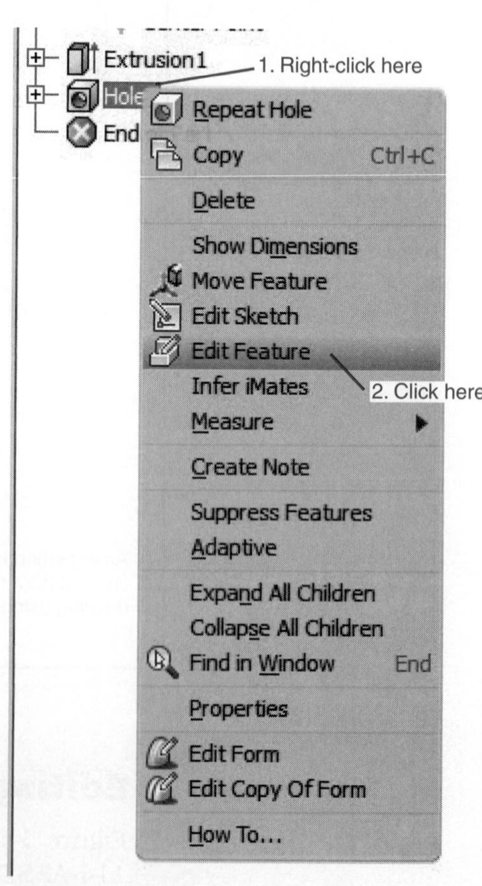

Figure 5-116

2 Click the **Open** option.

The browser box display will change.

3 Right-click the **Hole1** heading and select the **Edit Feature** option.

See Figure 5-116. The **Hole** dialog box will appear. See Figure 5-117.

4 Change the hole's diameter value from **10** to **11 mm.** Click **OK**.

Figure 5-117

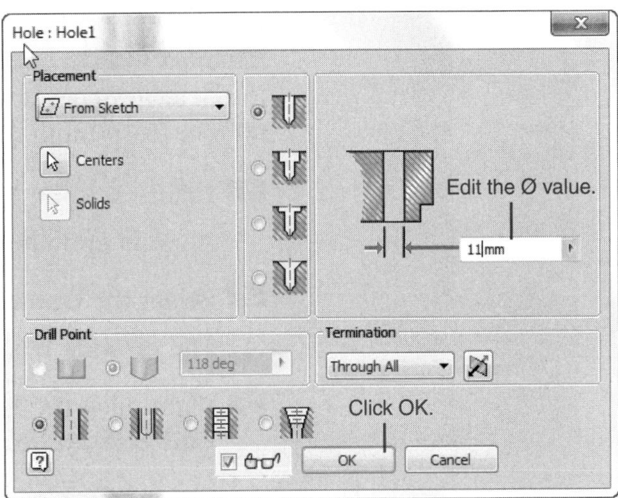

> **TIP**
>
> The hole's locational dimensions could also be edited at this time.

The bracket will appear on the screen. Click the **Close** button (the **X** in the upper right corner of the bracket's screen). A warning box will appear. See Figure 5-118.

Figure 5-118

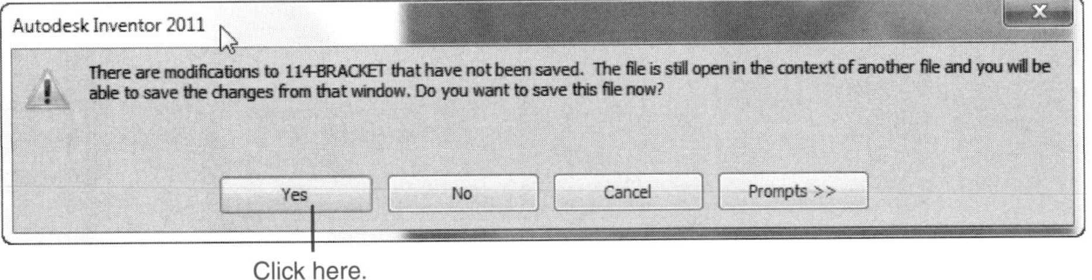

5 Click the **Yes** box.

The assembly will appear.

6 Click the bracket.

Figure 5-119 shows the edited bracket. Note that the holes appear larger than they did in Figure 5-114.

Figure 5-119

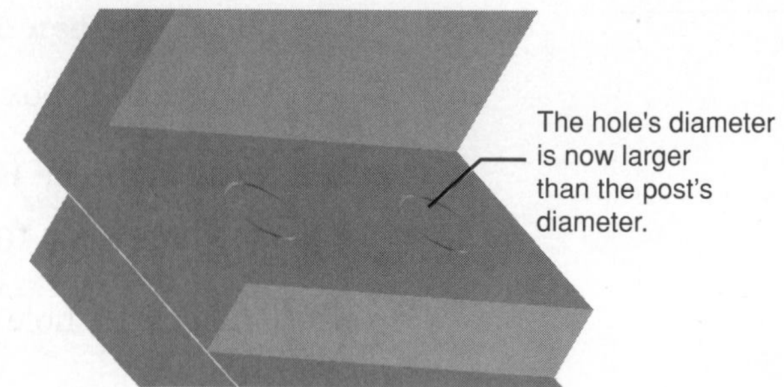

EXERCISE 5-45 Editing a Sketch

It has been determined that the length of the plate is to be increased from 90 mm to 100 mm. This change is a change to the initial sketch.

1. Right-click the **114-PLATE:1** heading in the browser box.

   A list of options will appear. See Figure 5-120.

2. Select the **Open** option.

   The browser box will change.

3. Right-click the **Extrusion1** heading in the browser box, then click the **Edit Sketch** heading.

   See Figure 5-121. The original rectangular sketch will appear. See Figure 5-122.

4. Double-click the **90** dimension value and enter a new value of **100;** click the check mark on the **Edit Dimension** box. See Figure 5-123.

5. Right-click the mouse and select the **Finish 2D Sketch** option.

   The plate will appear on the screen.

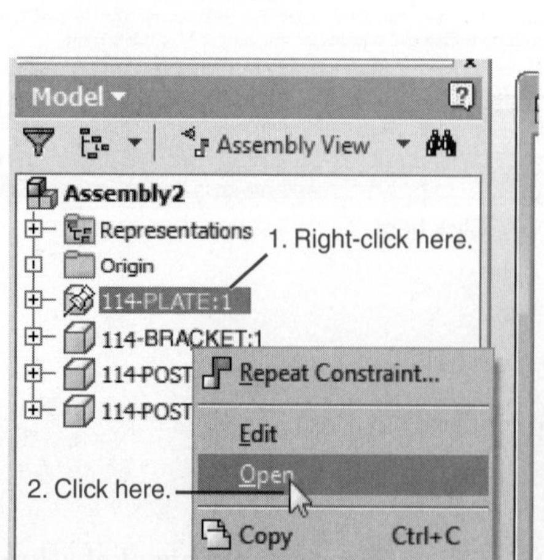

Figure 5-120

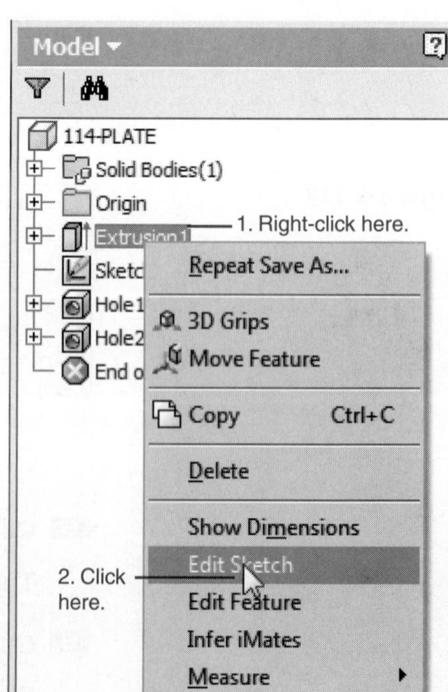

Figure 5-121

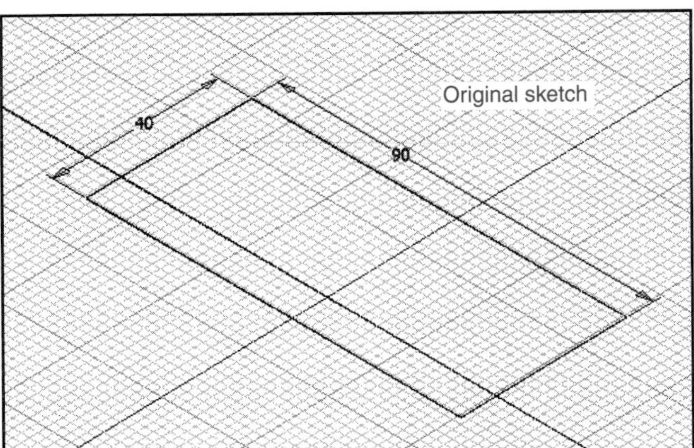

Figure 5-122

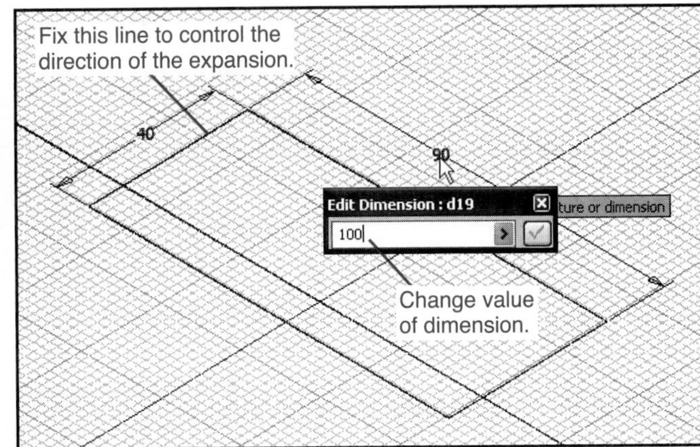

Figure 5-123

6 Click the **Close** box in the upper right corner of the plate's screen.

A warning box will appear. See Figure 5-124.

Figure 5-124

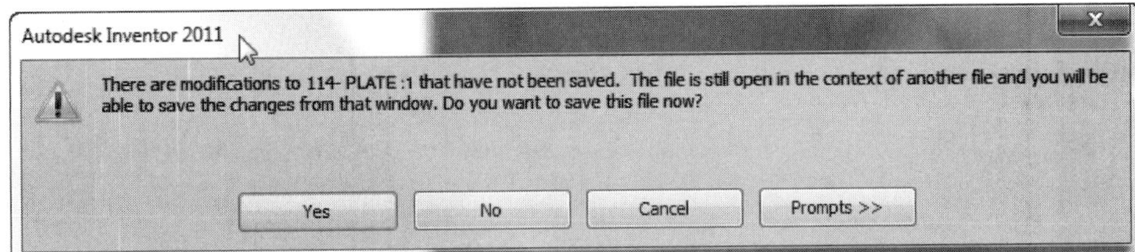

7 Click the **Yes** box.

The assembly will appear on the screen. See Figure 5-125.

Figure 5-125

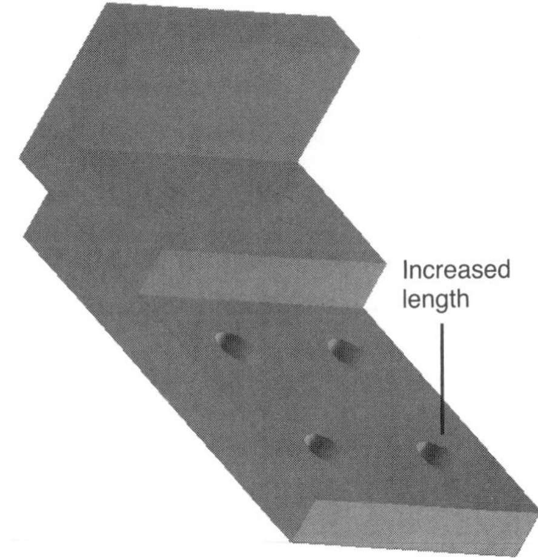

8 Click the plate.

Note the increase in the plate's length.

Patterning Components

Figure 5-126 shows an assembly in which a post is inserted into a plate. Posts are to be inserted into all 16 holes in the plate. The plate is 80 × 80 × 10. There are 16 Ø10 holes located 10 from each edge and 20 apart. The post is Ø10 × 20.

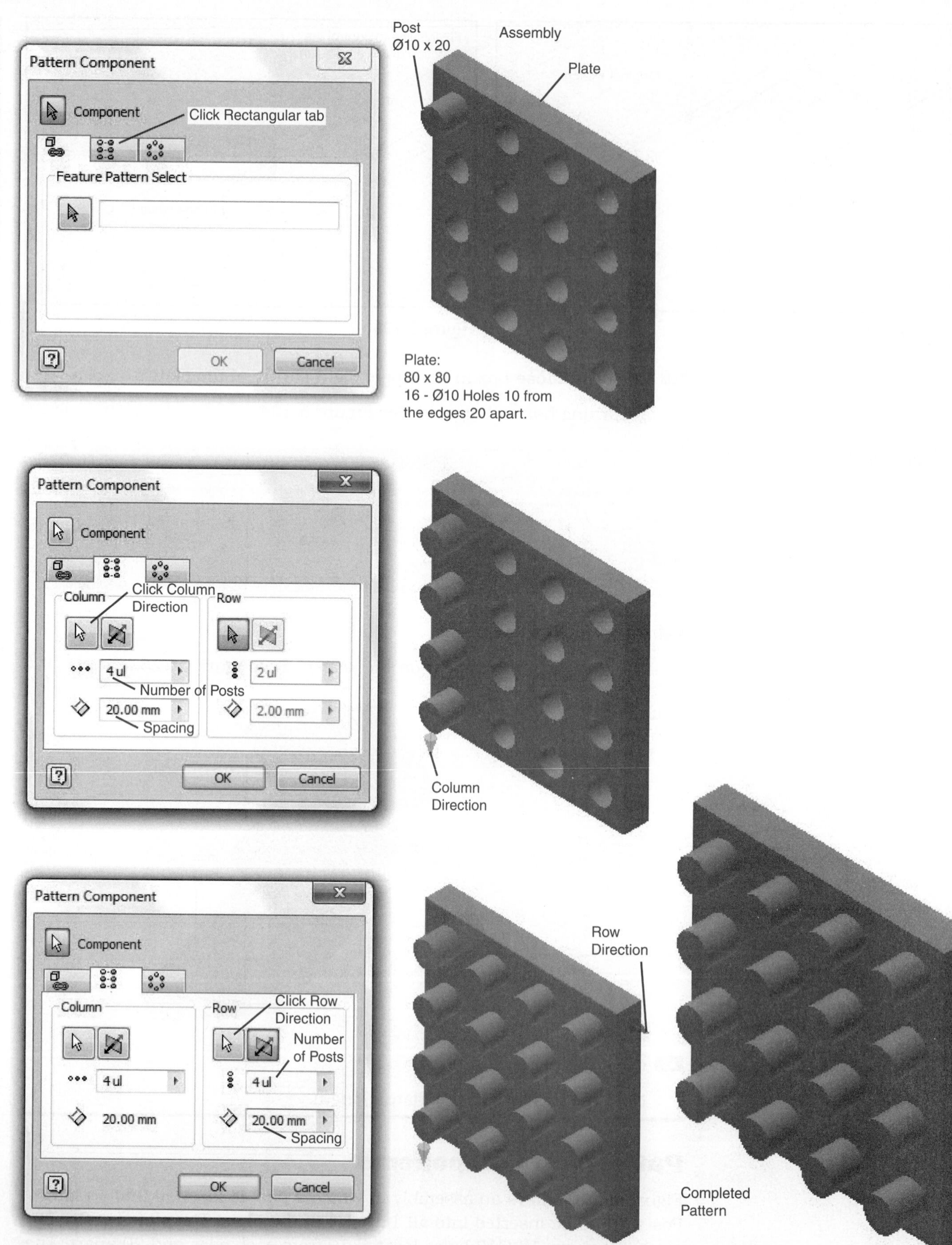
Pattern Component
Component
Click Rectangular tab
Feature Pattern Select
OK
Cancel
Post
Ø10 x 20
Assembly
Plate
Plate:
80 x 80
16 - Ø10 Holes 10 from
the edges 20 apart.
Pattern Component
Component
Column
Click Column
Direction
Row
4 ul
Number of Posts
20.00 mm
Spacing
2 ul
2.00 mm
OK
Cancel
Column
Direction
Pattern Component
Component
Column
Row
Click Row
Direction
Number
of Posts
4 ul
20.00 mm
4 ul
20.00 mm
Spacing
OK
Cancel
Row
Direction
Completed
Pattern

Figure 5-126

EXERCISE 5-46 **Creating a Pattern of Components**

1. Select the **Pattern Component** tool located on the **Component** panel under the **Assemble** tab.

   The **Pattern Component** dialog box will appear. See Figure 5-126.

2. Select the **Post** as the component.

3. Select the **Rectangular** tab.

4. Define the **Column** direction, the number of components, and the spacing between components.

   In this example the holes in the plate are **20 mm** apart.

5. Define the **Row** direction and spacing.

6. Click **OK**.

Mirroring Components

Components within an assembly may be mirrored and added to the assembly. See Figure 5-127.

Figure 5-127

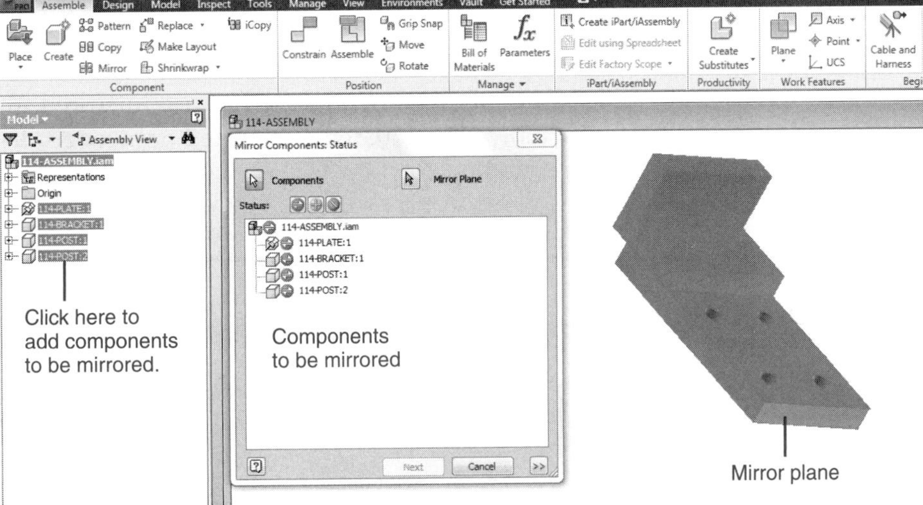

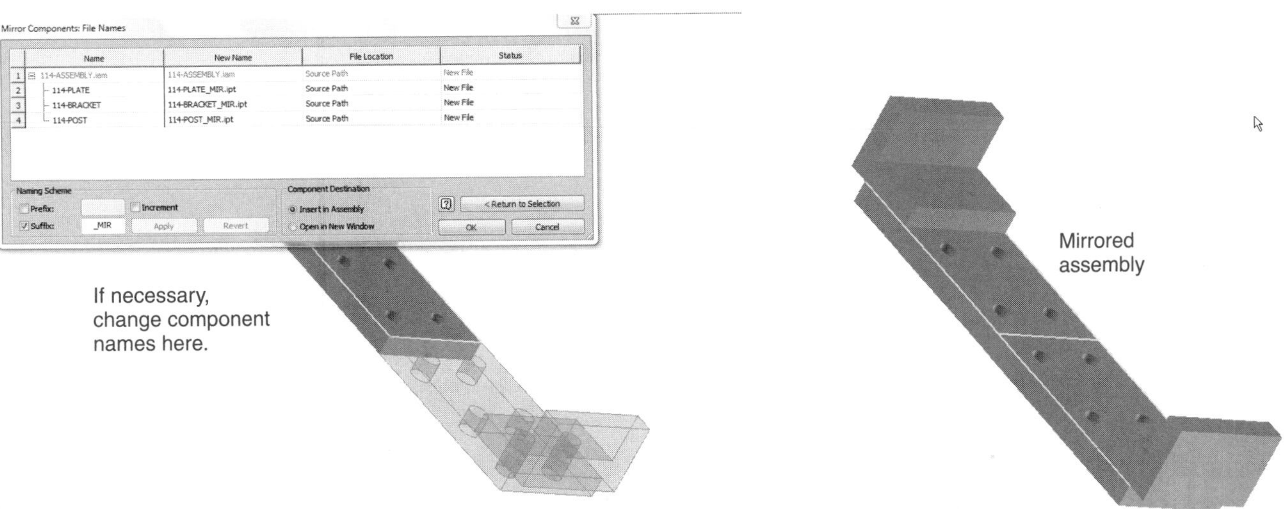

	Name	New Name	File Location	Status
1	114-ASSEMBLY.iam	114-ASSEMBLY.iam	Source Path	New File
2	114-PLATE	114-PLATE_MIR.ipt	Source Path	New File
3	114-BRACKET	114-BRACKET_MIR.ipt	Source Path	New File
4	114-POST	114-POST_MIR.ipt	Source Path	New File

EXERCISE 5-47 **Mirroring an Assembly**

1 Click the **Mirror Component** tool located on the **Component** panel under the **Assemble** tab.

The **Mirror Components: Status** dialog box will appear.

> **TIP**
>
> The assembly shown in Figure 5-128 is the 114-ASSEMBLY drawing used on page 278. Note how the dimension change from 90 mm to 100 mm affects the material under the copied bracket.

2 Click the components to be mirrored.

In this example all the components were selected. A listing of the selected components will appear in the **Mirror Components: Status** dialog box.

3 Click the **Mirror Plane** box.

4 Select the right front vertical plane as the mirror plane.

5 Click **Next**.

The **Mirror Components: File Names** dialog box will appear. This is a listing of all components that are mirrored and have been added to the assembly. The box may be edited as needed. In this example no changes were made.

6 Click **OK**.

Copying Components

Components already on an assembly may be copied and added to the assembly. See Figure 5-128. In this example the bracket and posts will be copied.

Figure 5-128

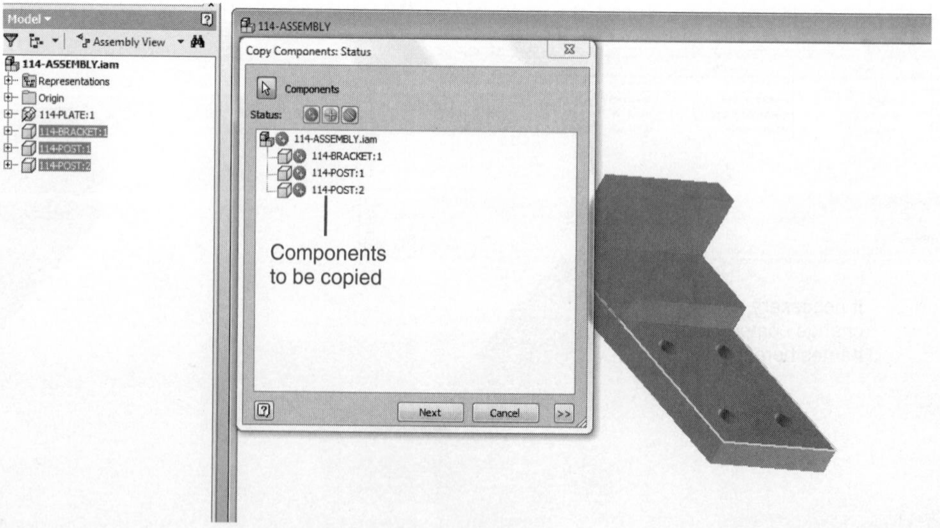

Figure 5-128
(*Continued*)

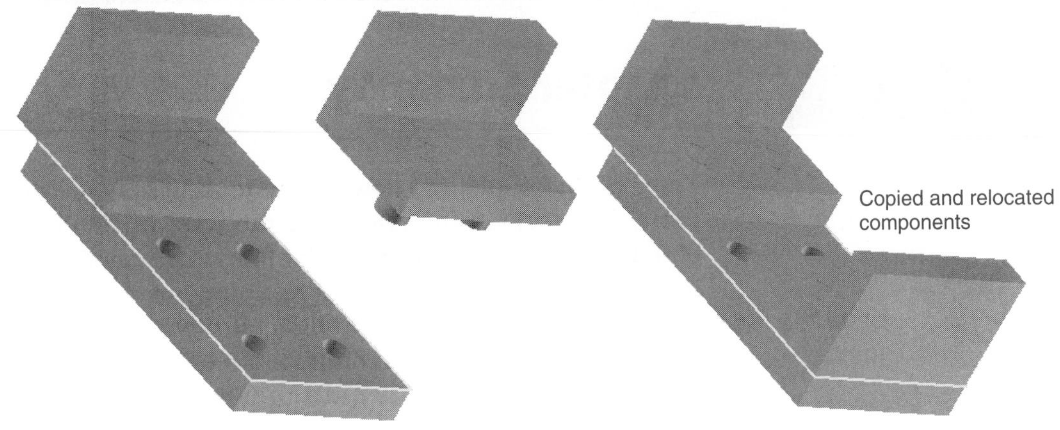

EXERCISE 5-48 **Copying a Component**

1 Click the **Copy Component** tool located on the **Pattern** panel under the **Assembly** tab.

The **Copy** tool is a flyout on the **Pattern** panel. The **Copy Components: Status** dialog box will appear.

2 Select the bracket and two posts.

3 Select **OK**.

The **Copy Components: File Names** dialog box will appear. This is a listing of all components that have been copied and added to the assembly.

4 Move the copied component away from the assembly.

5 Use the **Constrain** tool to position the copied component into the assembly.

Chapter Summary

This chapter explained how to create assembly drawings from individual parts. The tools in the **Assemble** tab were introduced and used in a bottom-up approach to create the first assembly drawing from an existing model. The method for grounding components was demonstrated. The **Move** and **Rotate** tools as well as all the options of the **Constrain** tool were used to manipulate components of the drawing.

The tools in the **Presentation** tab were used to explode an assembly drawing to show how the assembly is created from its components. The presentation drawing was then animated. Assembly numbers were added and edited to isometric views of the presentation drawing.

The other elements of a presentation drawing—the parts list or bill of materials, the title block, release blocks, and revision blocks—were also demonstrated.

The top-down approach to creating an assembly drawing was also illustrated, and various techniques for editing assembly drawings were explained, including patterning, mirroring, and copying components.

Chapter Test Questions

Multiple Choice

Circle the correct answer.

1. Which of the following is *not* an assembly constraint tool?
   a. **Flush** c. **Bisect**
   b. **Insert** d. **Angle**
2. Which tool is used to locate a shaft into a hole?
   a. **Flush** c. **Bisect**
   b. **Insert** d. **Angle**
3. Why is the **Tweak** tool used in presentation drawings?
   a. To pull components apart
   b. To edit components
   c. To rotate components
4. What is the function of the **Column Chooser** option of the **Edit Parts List** dialog box?
   a. To change the size of the columns
   b. To arrange the columns in numerical order
   c. To add, subtract, and edit columns
5. What is a subassembly?
   a. Another name for an assembly drawing
   b. A group of parts assembled together and then inserted as a group into an assembly
   c. An assembly that is located below other assemblies

6. What are the dimensions of a C-size drawing sheet?
   a. 8.5 × 11
   b. 11 × 17
   c. 17 × 22
7. What is the purpose of drawing revisions?
   a. To document recent changes to a drawing
   b. To indicate future changes to a drawing
   c. To specify the drawing's tolerances
8. Why are drawing notes added to a drawing?
   a. To explain the functions of the drawing's parts
   b. To list the comments of the drawing's engineer
   c. To define manufacturing information that cannot be shown visually
9. Which tool is used to animate an assembly drawing?
   a. **Drive Constraint**
   b. **Increment Constraint**
   c. **Repetition Constraint**
10. Which tool may be used to change the dimensions of an individual part?
   a. **Change Drawing**
   b. **Edit Component**
   c. **Modify Entity**

Matching

Write the number of the correct answer on the line.

Column A	Column B
a. A4 _____	1. 8.5″ × 11″
b. C _____	2. 11″ × 17″
c. A _____	3. 17″ × 22″
d. A1 _____	4. 22″ × 34″
e. A2 _____	5. 210 mm × 297 mm
f. B _____	6. 297 mm × 420 mm
g. A3 _____	7. 420 mm × 594 mm
h. D _____	8. 594 mm × 841 mm

True or False

Circle the correct answer.

1. **True or False:** A bottom-up assembly drawing is created from existing drawings.
2. **True or False:** A top-down assembly drawing is created from existing drawings.
3. **True or False:** A grounded component has no degrees of freedom.

4. **True or False:** A presentation drawing is used to pull assembled components apart.
5. **True or False:** A presentation drawing can be animated.
6. **True or False:** An isometric assembly drawing is created using the **ANSI.idw** format.
7. **True or False:** Assembly numbers are the same as part numbers.
8. **True or False:** Assembly numbers are added to an assembly drawing using the **Balloon** tool.
9. **True or False:** A BOM is the same as a parts list.
10. **True or False:** ECOs are used to make quick changes to an existing drawing.

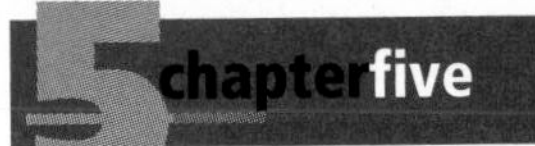

Chapter Projects

Project 5-1

A dimensioned block is shown in Figure P5-1. Redraw this block and save it as **SQBLOCK.** See page 218. Use the SQBLOCK to create assemblies as shown.

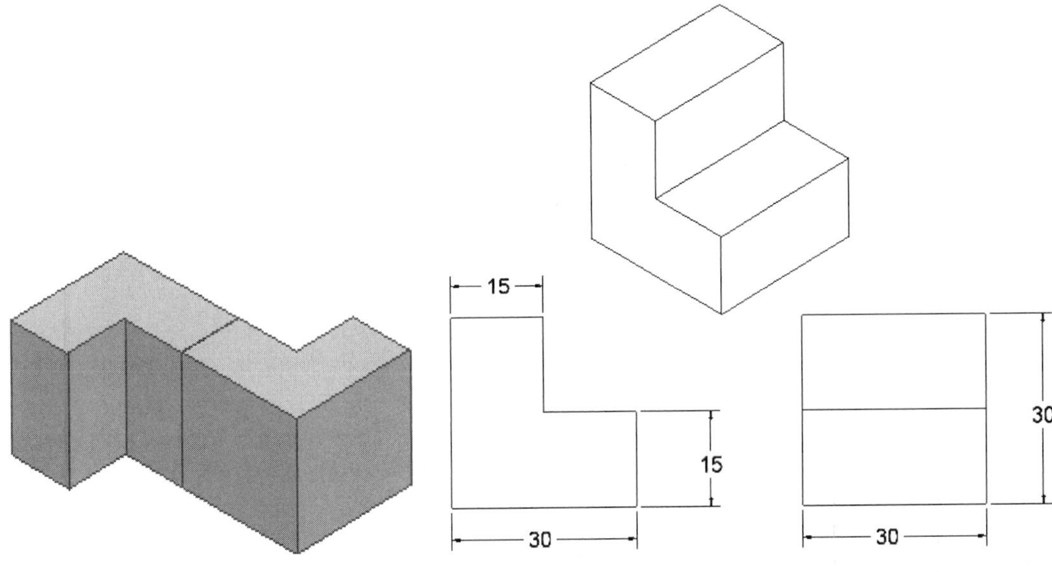

Figure P5-1A
MILLIMETERS

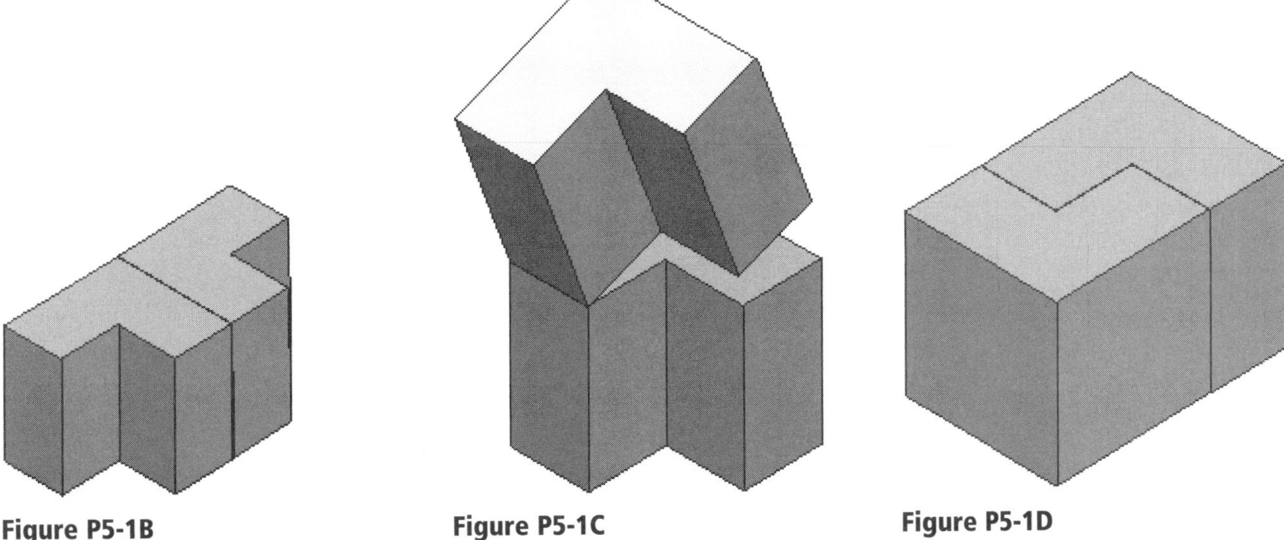

Figure P5-1B **Figure P5-1C** **Figure P5-1D**

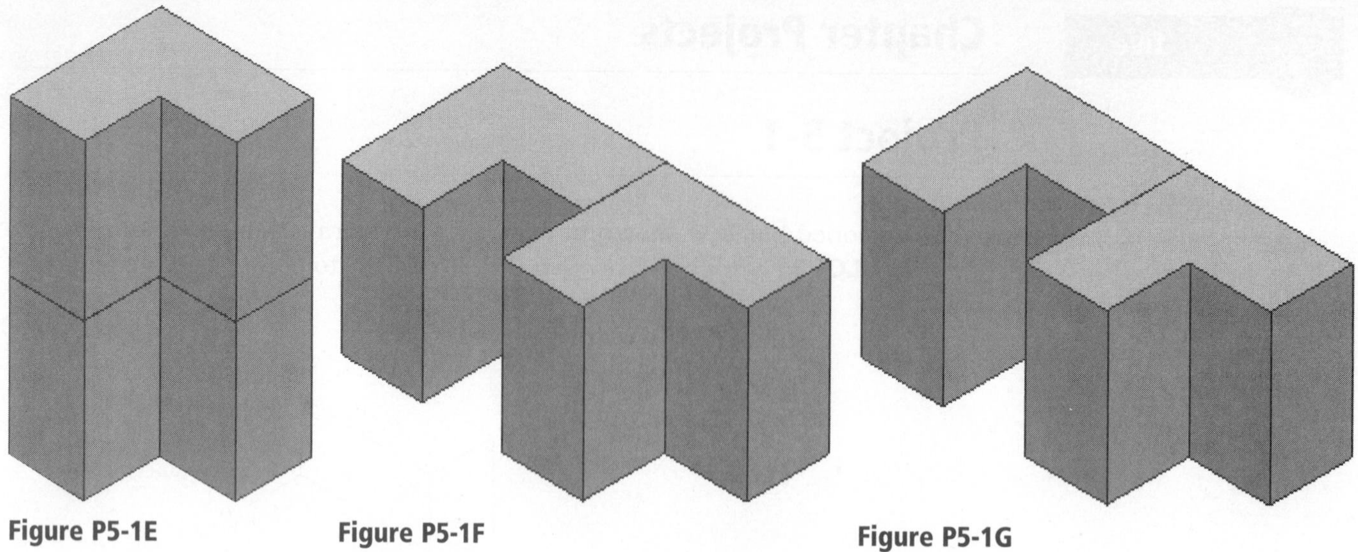

Figure P5-1E

Figure P5-1F

Figure P5-1G

Pages 289–290 show a group of parts. These parts are used to create the assemblies presented as problems in this section. Use the given descriptions, part numbers, and materials when creating BOMs for the assemblies.

Project 5-2

Redraw the following models and save them as **Standard (mm).ipn** files. All dimensions are in millimeters.

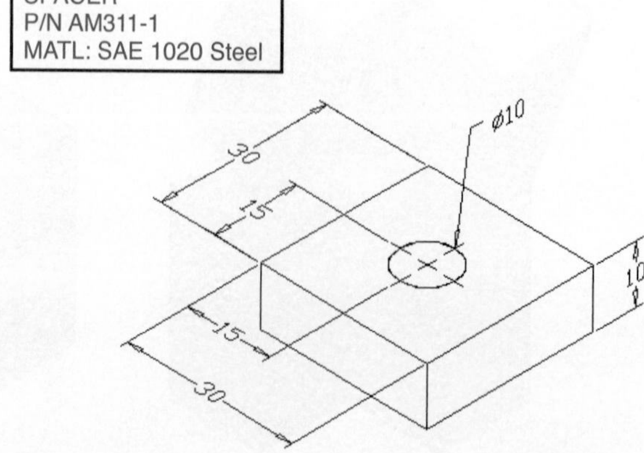

Figure P5-2A

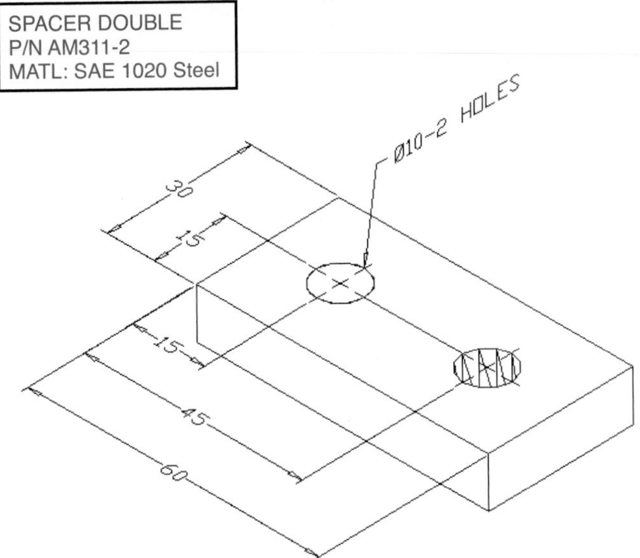

Figure P5-2B

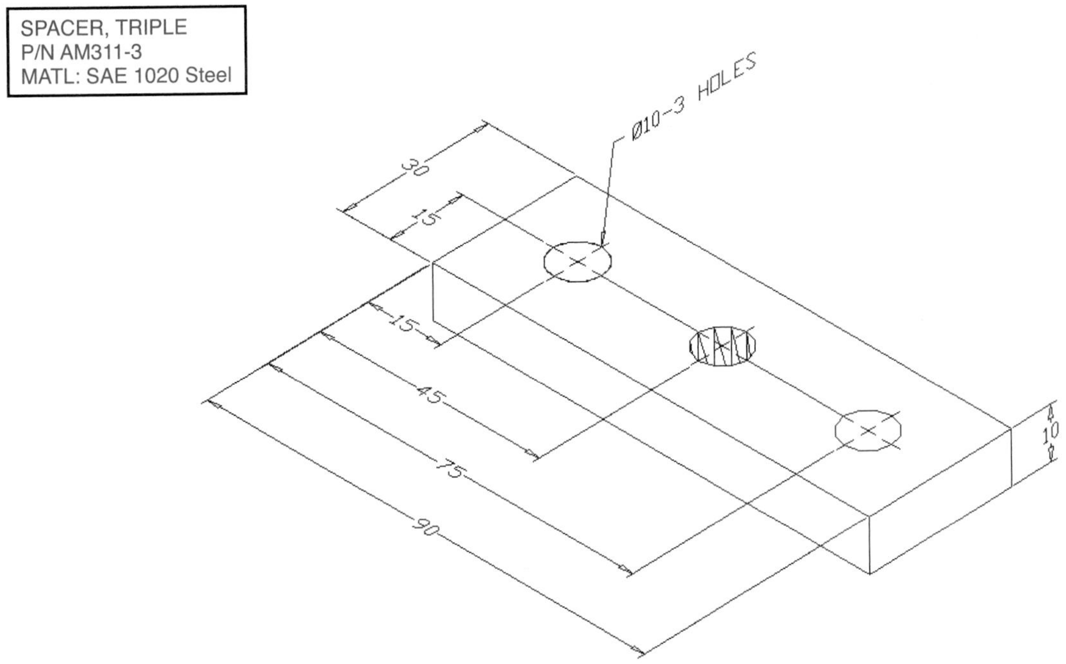

Figure P5-2C

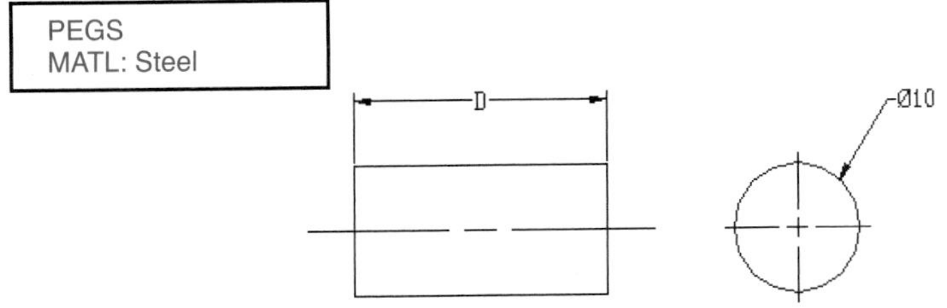

DESCRIPTION	PART NO.	D
PEG, SHORT	PG20-1	20
PEG	PG30-1	30
PEG, LONG	PG40-1	40

ALL DISTANCES IN MILLIMETERS

Figure P5-2D

Figure P5-2E

L-BRACKET
P/N BK20-1
MATL: SAE 1040 Steel

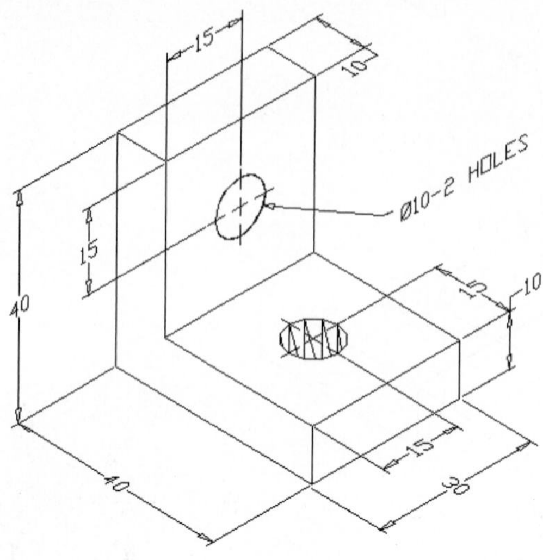

Figure P5-2F

Z-BRACKET
P/N BK20-2
MATL: SAE 1040 Steel

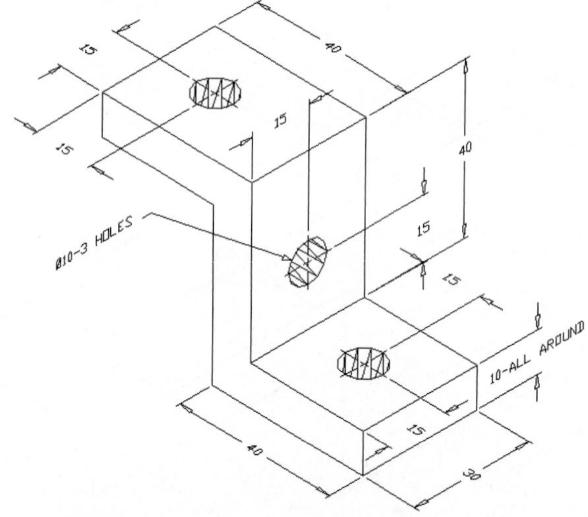

Figure P5-2H

C-BRACKET
P/N BK20-3
SAE 1040 Steel

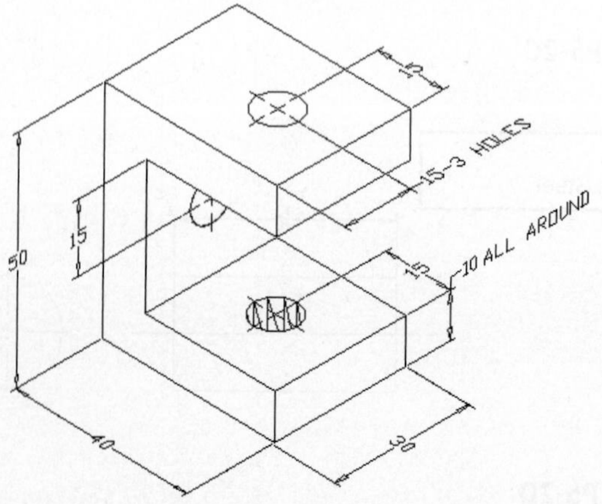

Figure P5-2G

PLATE, BASE
SAE 1020 Steel

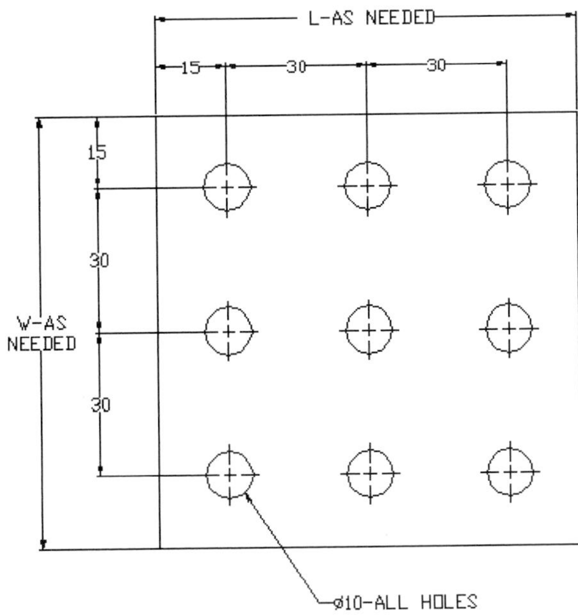

PART NO.	TOTAL NO OF HOLES	L	W	HOLE PATTERN
PL110-9	9	90	90	3x3
PL110-16	16	120	120	4x4
PL110-6	6	60	90	2x3
PL110-8	8	60	120	2x4
PL80-4	4	60	60	2x2

Project 5-3

Draw an exploded isometric assembly drawing of Assembly 1. Create a BOM.

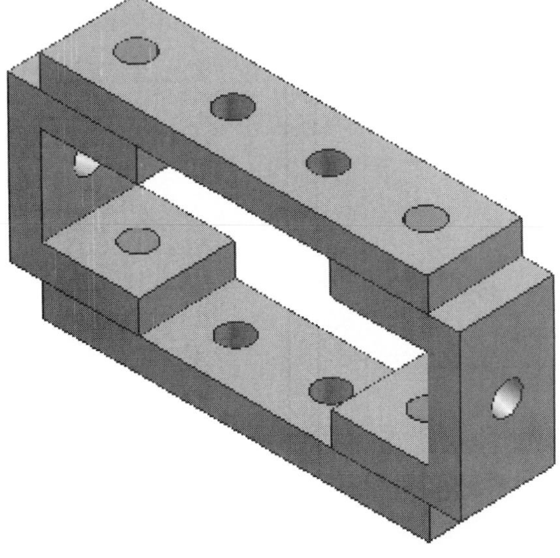

Figure P5-3A
MILLIMETERS

ASSEMBLY 1

Figure P5-3B

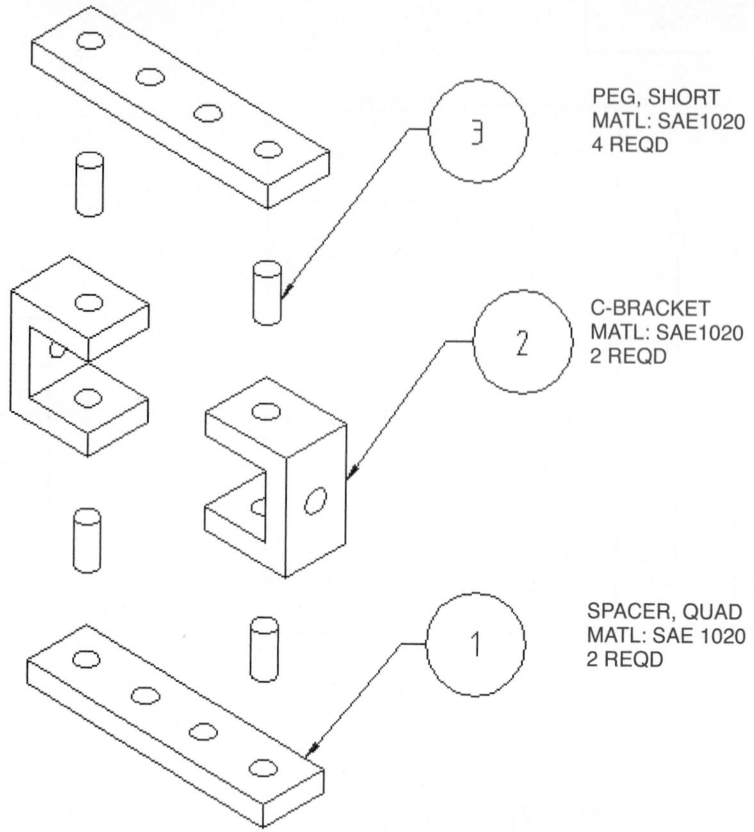

Project 5-4

Draw an exploded isometric assembly drawing of Assembly 2. Create a BOM.

Figure P5-4A
MILLIMETERS

ASSEMBLY 2

Figure P5-4B

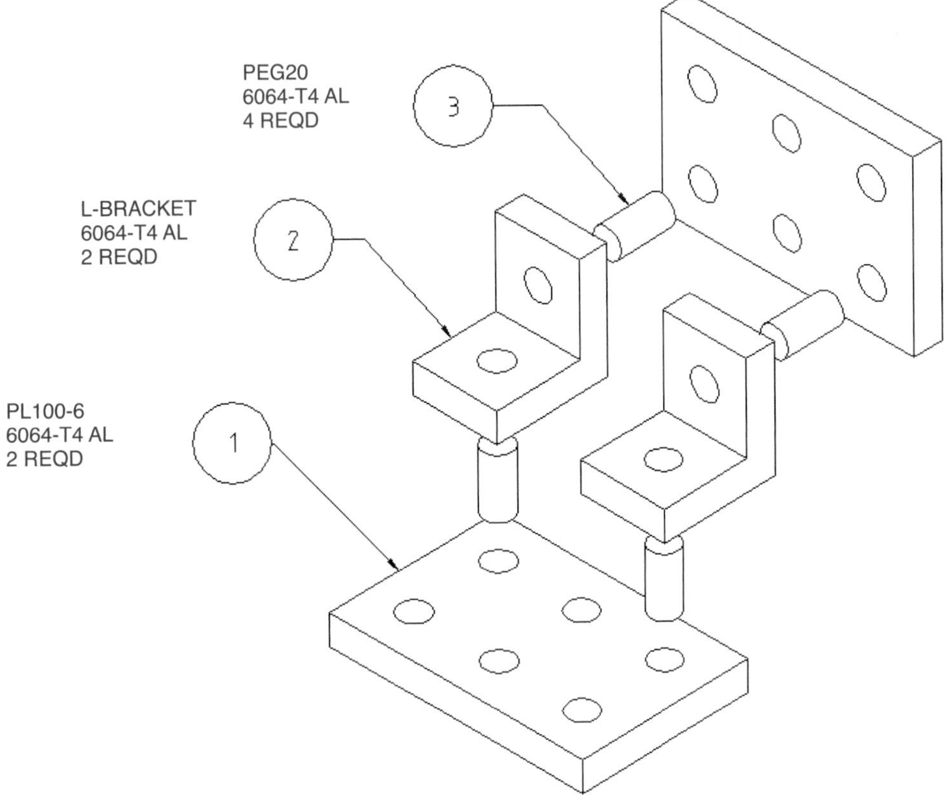

Project 5-5

Draw an exploded isometric assembly drawing of Assembly 3. Create a BOM.

Figure P5-5A
MILLIMETERS

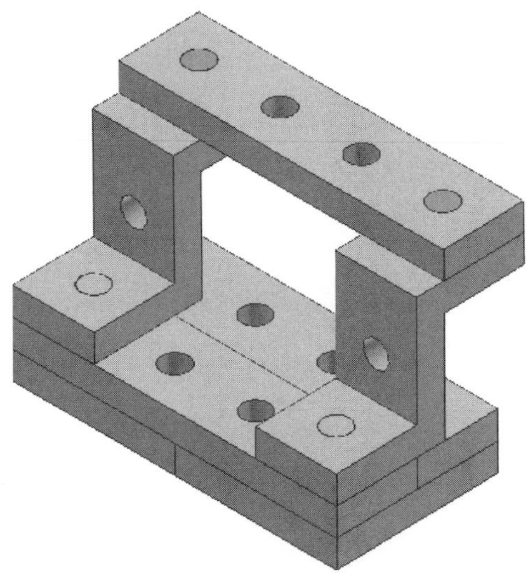

Figure P5-5B

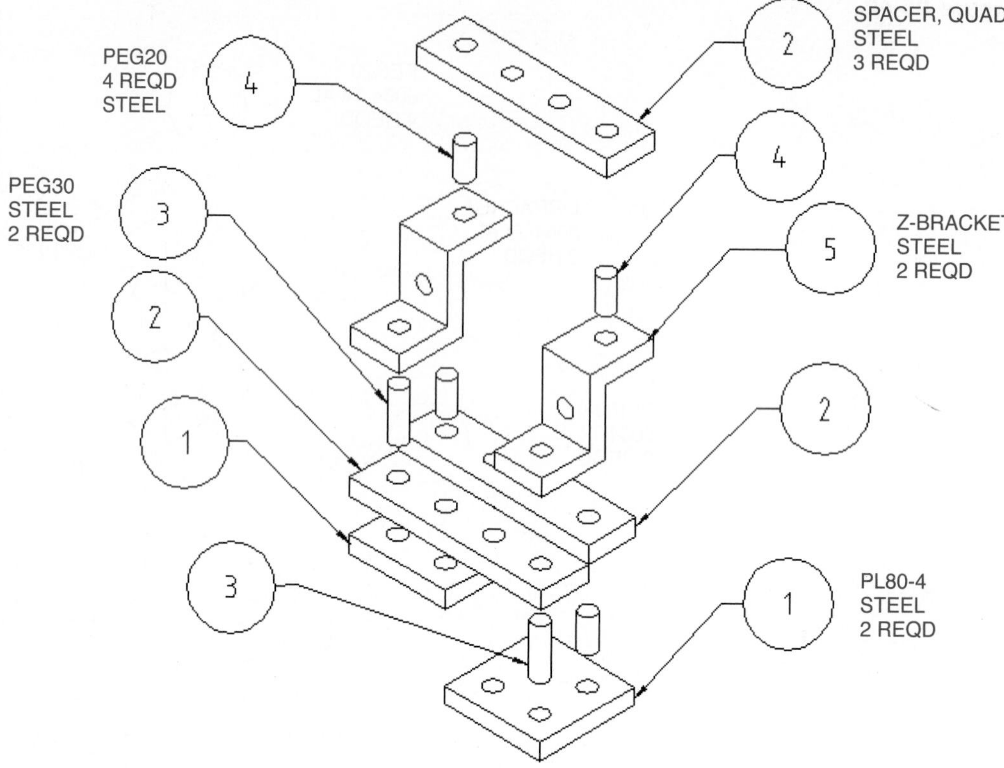

Project 5-6

Draw an exploded isometric assembly drawing of Assembly 4. Create a BOM.

Figure P5-6A
MILLIMETERS

ASSEMBLY 4

Figure P5-6B

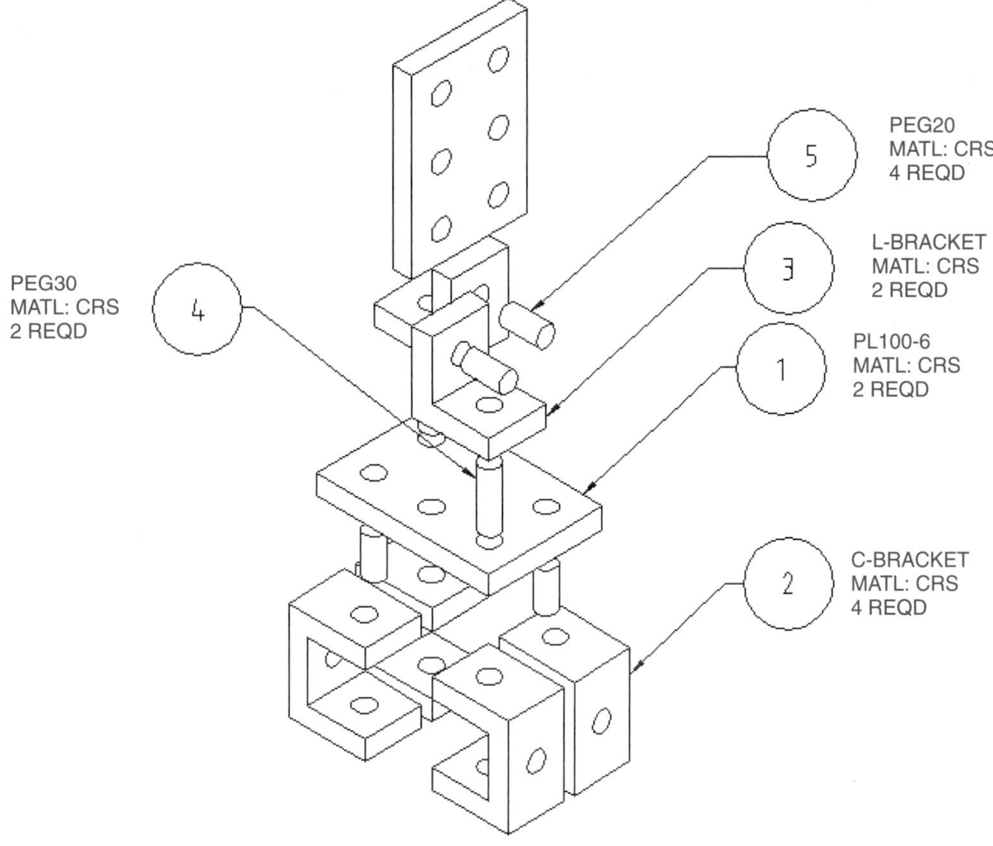

Project 5-7

Draw an exploded isometric assembly drawing of Assembly 5. Create a BOM.

Figure P5-7A
MILLIMETERS

ASSEMBLY 5

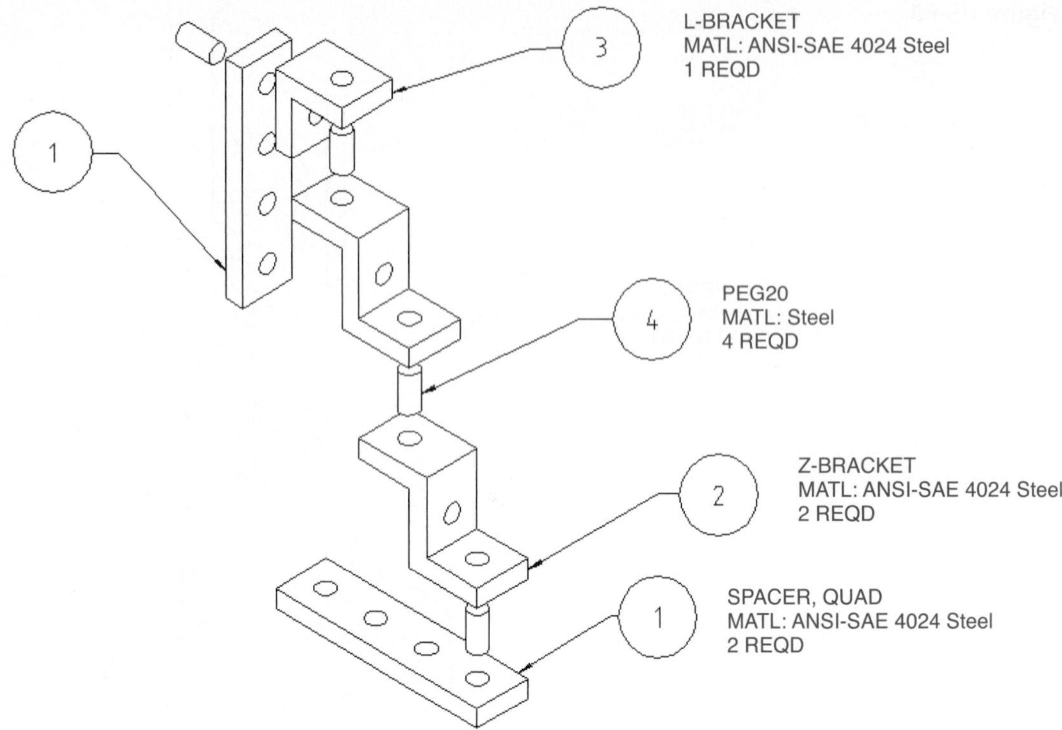

Figure P5-7B

Project 5-8

Draw an exploded isometric assembly drawing of Assembly 6. Create a BOM.

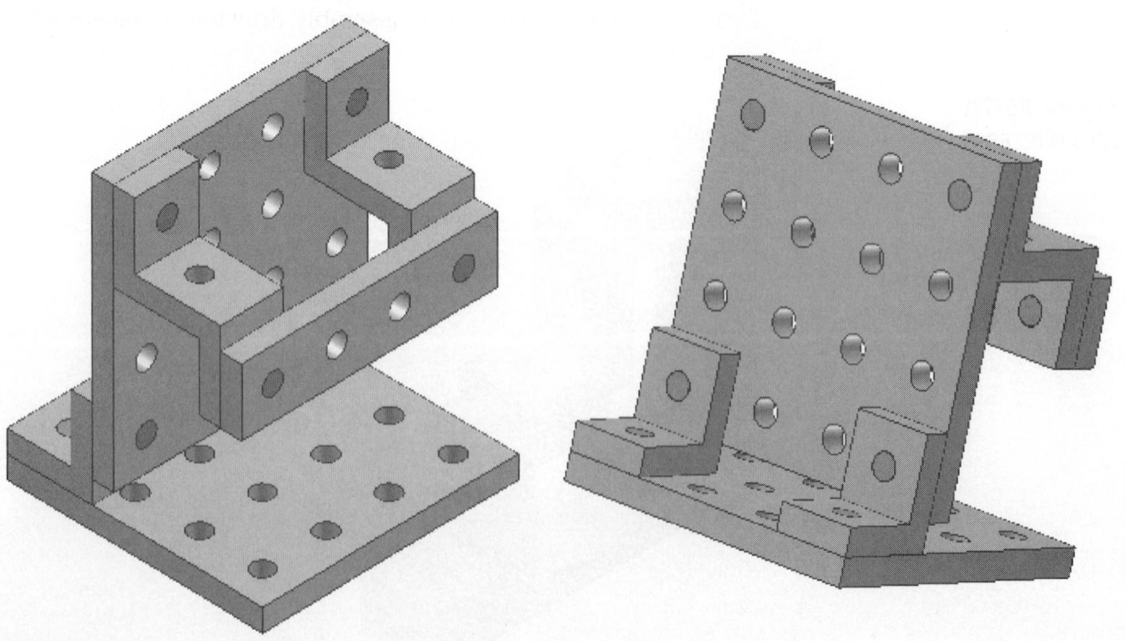

Figure P5-8A

ASSEMBLY 6

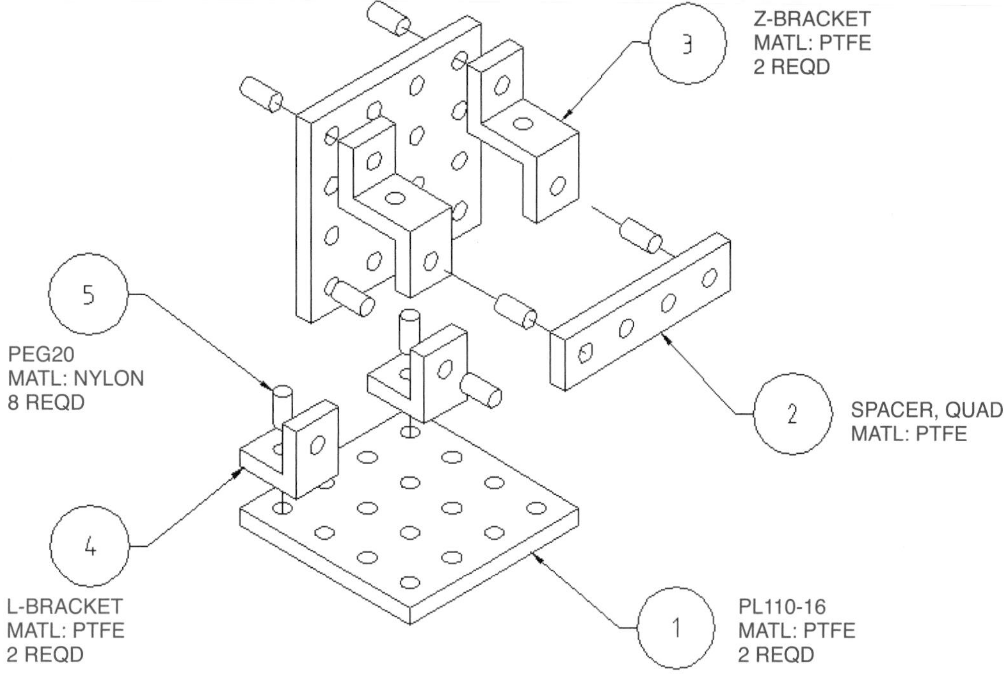

Figure P5-8B

Project 5-9

Create an original assembly based on the parts shown on pages 290–293. Include a scene, an exploded isometric drawing with assembly numbers, and a BOM. Use at least 12 parts.

Project 5-10

Draw the ROTATOR ASSEMBLY shown. Include the following:

A. An assembly drawing.
B. An exploded presentation drawing.
C. An isometric drawing with assembly numbers.
D. A parts list.
E. An animated assembly drawing; the LINKs should rotate relative to the PLATE. The LINKs should carry the CROSSLINK. The CROSSLINK should remain parallel during the rotation.

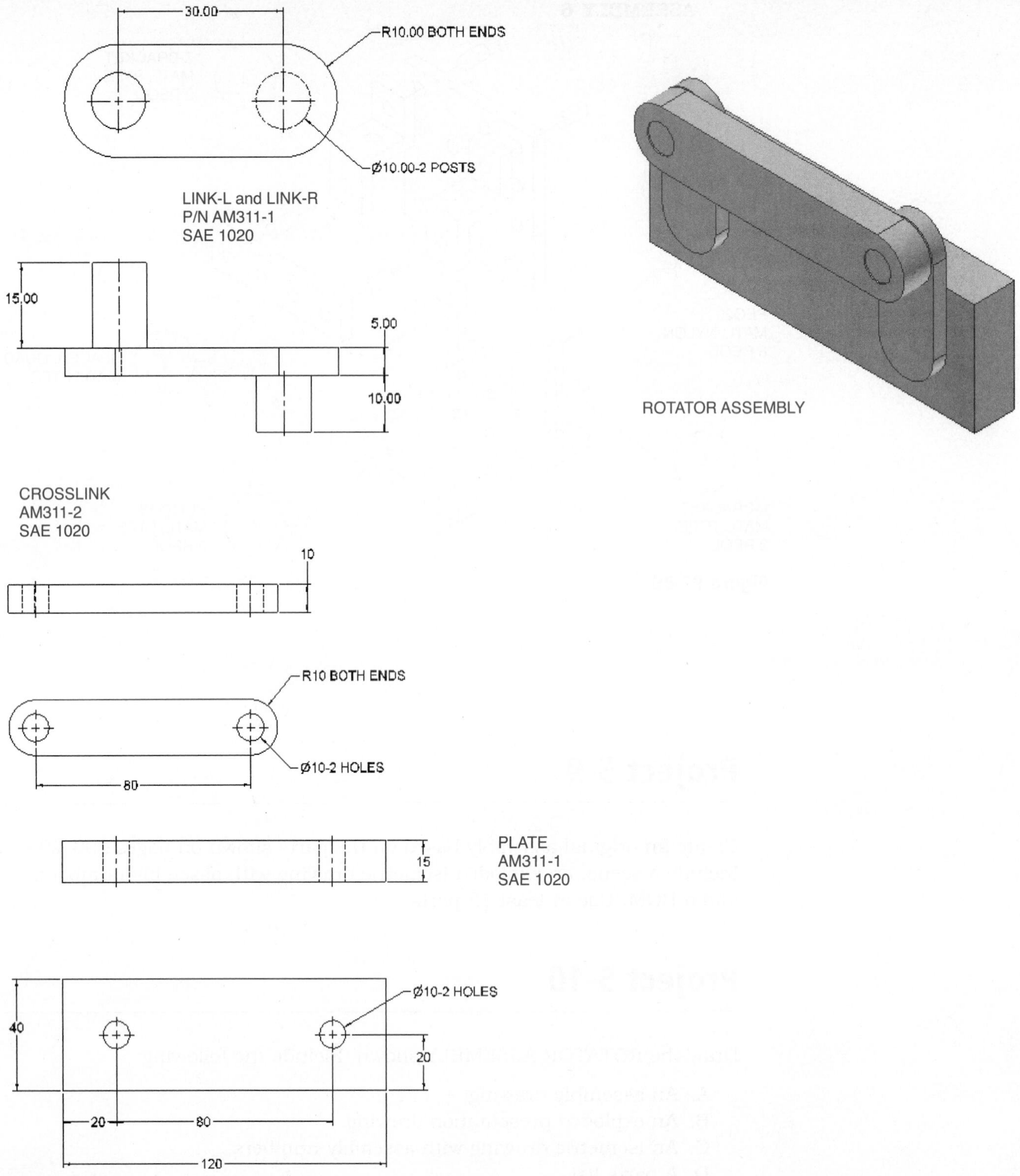

30.00
R10.00 BOTH ENDS
Ø10.00-2 POSTS
LINK-L and LINK-R
P/N AM311-1
SAE 1020
15.00
5.00
10.00
ROTATOR ASSEMBLY
CROSSLINK
AM311-2
SAE 1020
10
R10 BOTH ENDS
Ø10-2 HOLES
80
15
PLATE
AM311-1
SAE 1020
Ø10-2 HOLES
40
20
20
80
120

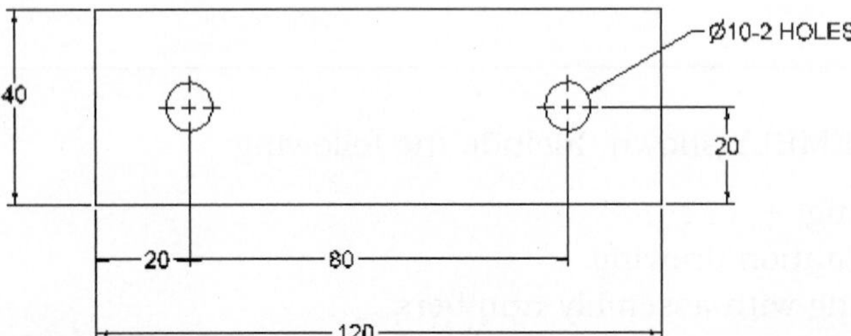

Figure P5-10

NOTE

This assembly was used in the section on top-down assemblies. See page 217.

Project 5-11

Draw the FLY ASSEMBLY shown. Include the following:

A. An assembly drawing.
B. An exploded presentation drawing.
C. An isometric drawing with assembly numbers.
D. A parts list.
E. An animated assembly drawing; the FLYLINK should rotate around the SUPPORT base.

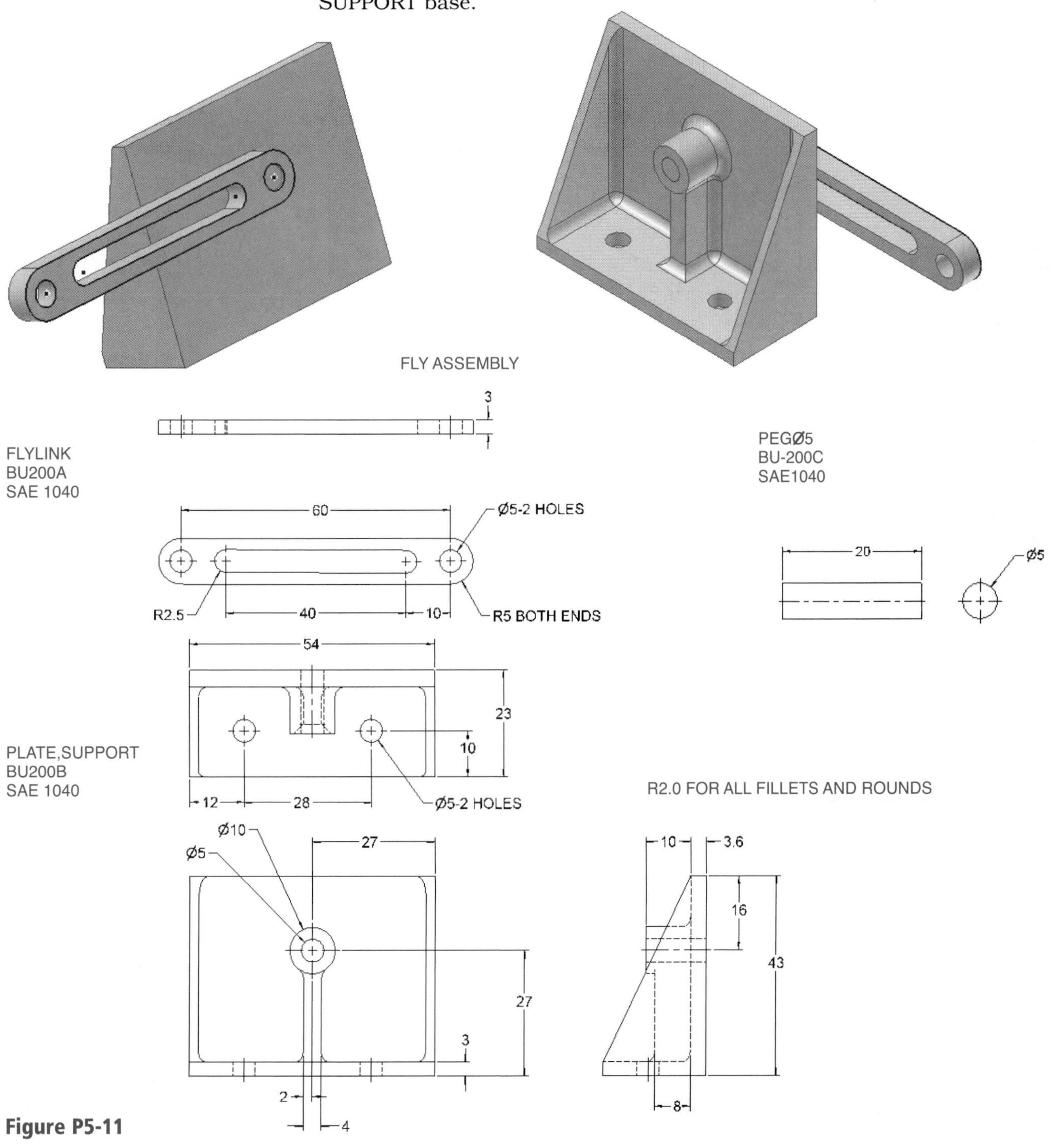

Figure P5-11

Project 5-12

Draw the ROCKER ASSEMBLY shown. Include the following:

A. An assembly drawing
B. An exploded presentation drawing
C. An isometric drawing with assembly numbers
D. A parts list
E. An animated assembly drawing

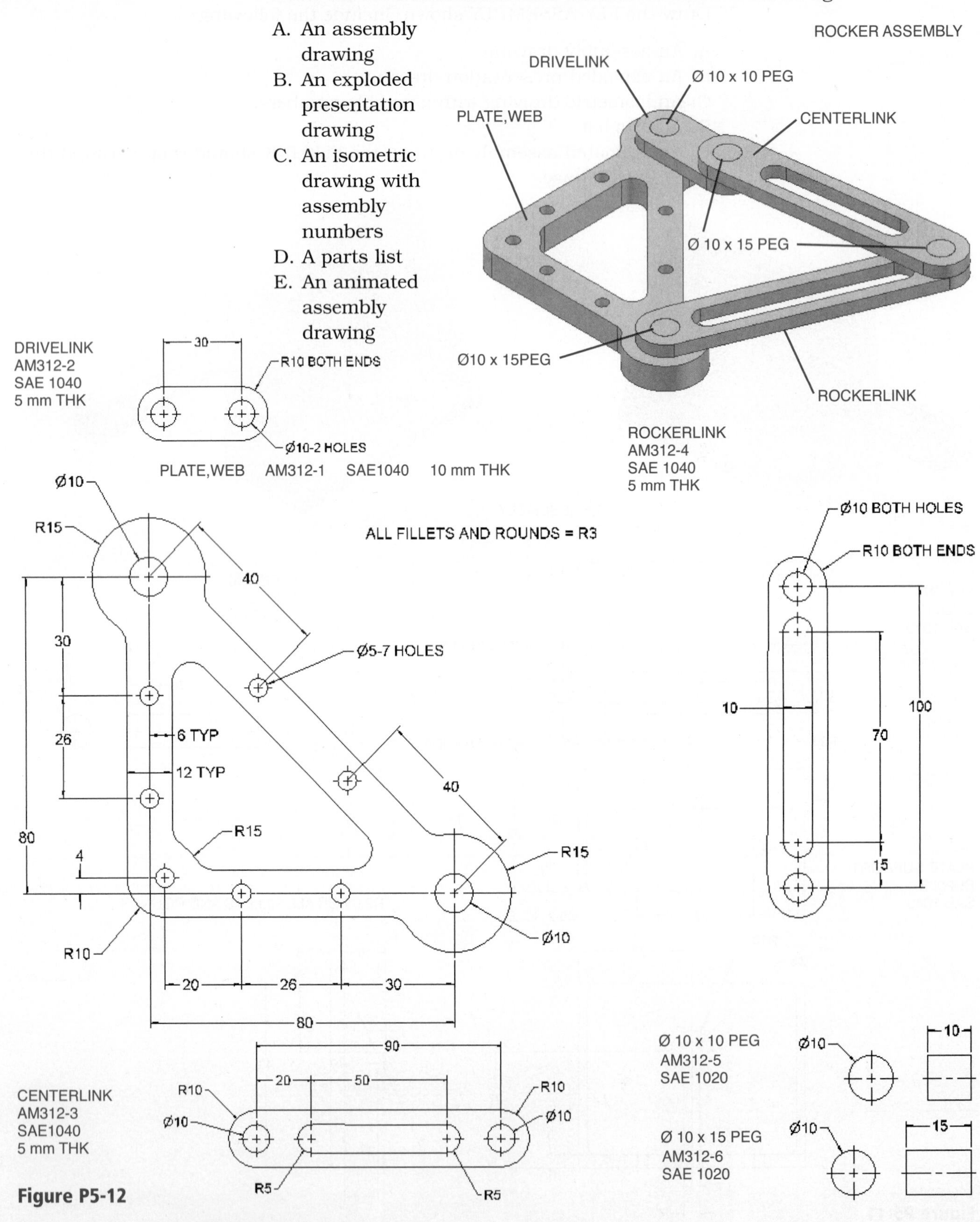

Figure P5-12

Project 5-13

Draw the LINK ASSEMBLY shown. Include the following:

A. An assembly drawing.
B. An exploded presentation drawing.
C. An isometric drawing with assembly numbers.
D. A parts list.
E. An animated assembly drawing; the HOLDER ARM should rotate between –30° and +30°.

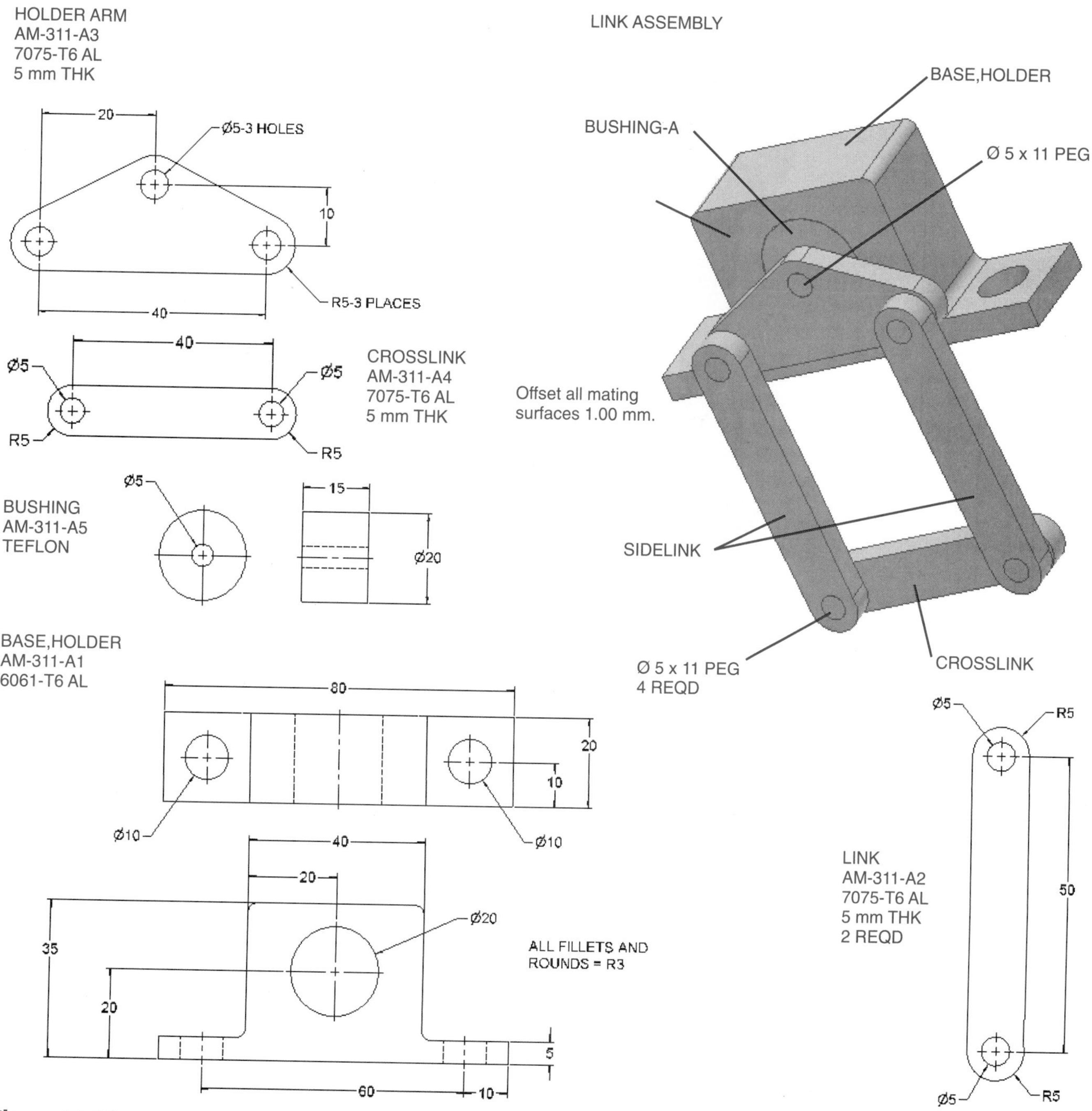

Figure P5-13

Project 5-14

Draw the PIVOT ASSEMBLY shown using the dimensioned components given. Include the following:

A. A presentation drawing
B. A 3D exploded isometric drawing
C. A parts list

TIP

This assembly was used in the section on subassemblies. See page 246.

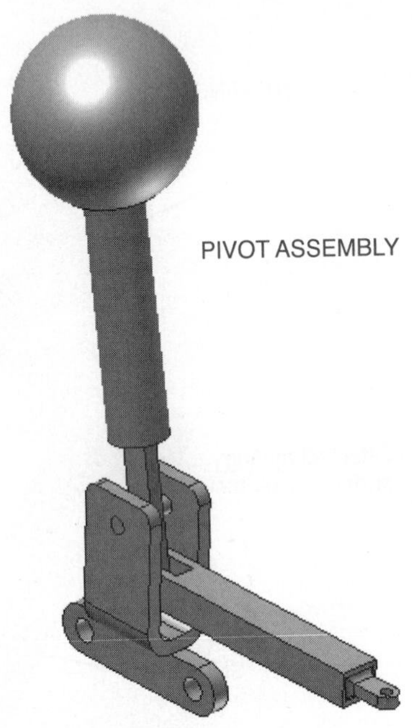

Figure P5-14A
MILLIMETERS

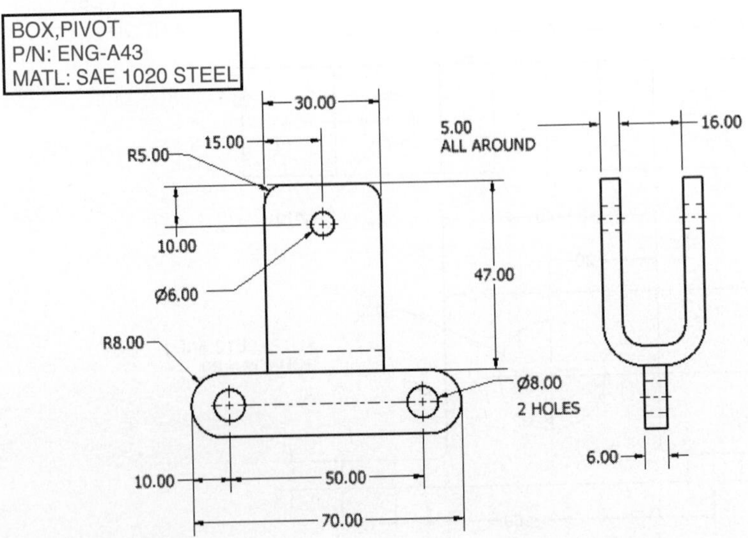

Figure P5-14B

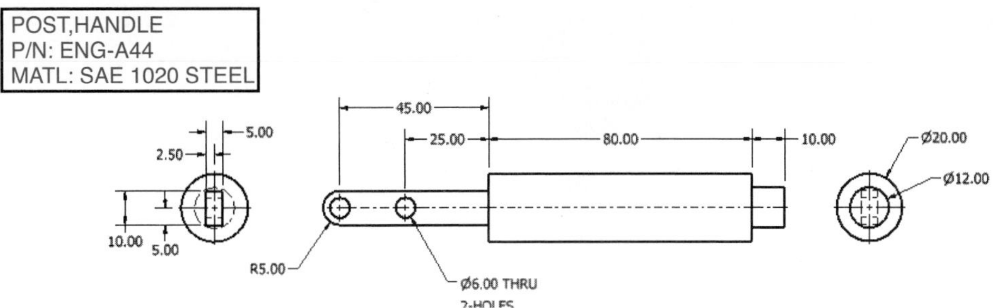

Figure P5-14C

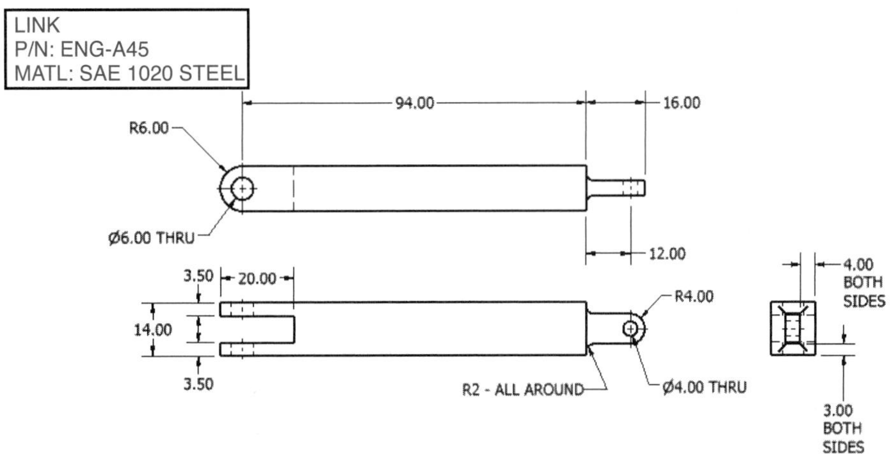

Figure P5-14D

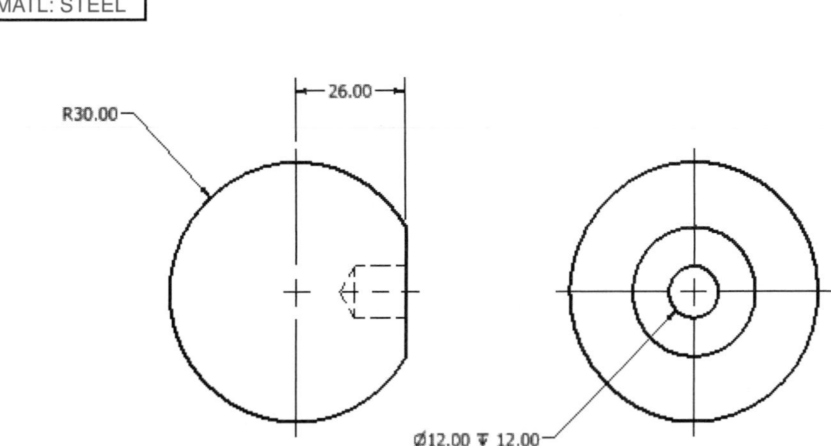

Figure P5-14E

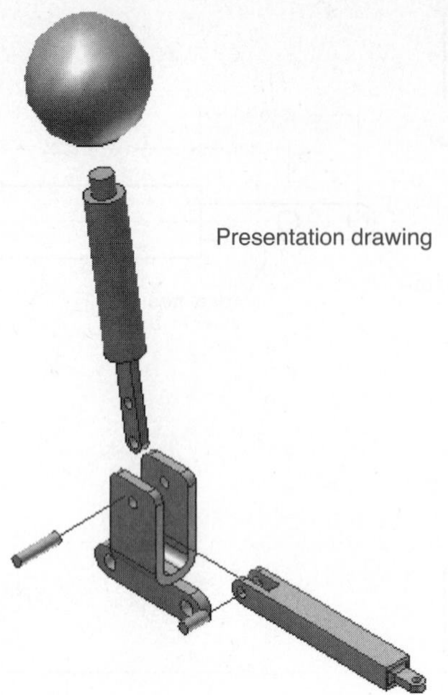

Figure P5-14F

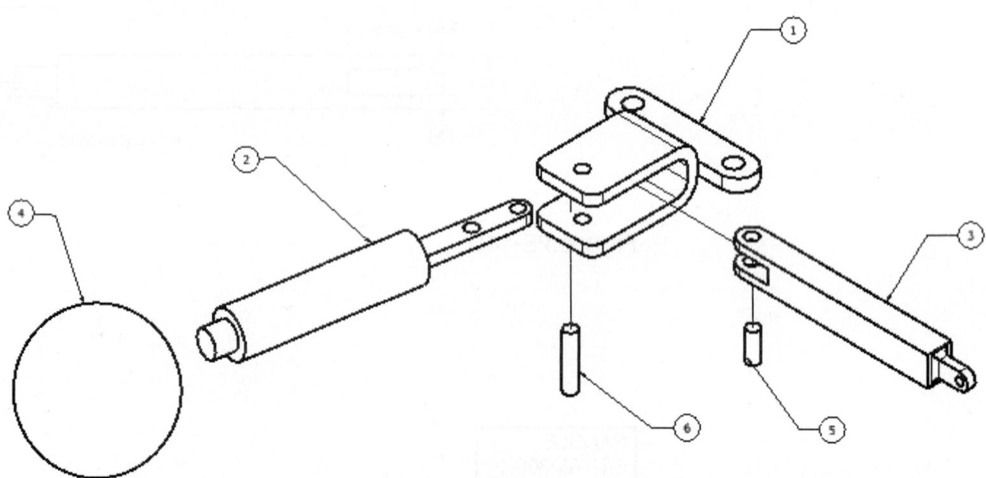

Parts List				
ITEM	PART NUMBER	DESCRIPTION	MATERIAL	QTY
1	ENG-A43	BOX,PIVOT	SAE1020	1
2	ENG-A44	POST,HANDLE	SAE1020	1
3	ENG-A45	LINK	SAE1020	1
4	AM300-1	HANDLE	STEEL	1
5	EK-132	POST-Ø6x14	STEEL	1
6	EK-131	POST-Ø6x26	STEEL	1

Figure P5-14G

6 chaptersix Threads and Fasteners

CHAPTER OBJECTIVES

- Learn thread terminology and conventions
- Learn how to draw threads
- Learn how to size both internal and external threads
- Learn how to use standard-sized threads
- Learn how to use and size washers, nuts, and setscrews

Introduction

This chapter explains how to draw threads and washers. It also explains how to select fasteners and how to design using fasteners, washers, and keys.

Threads are created in Inventor using either the **Hole** or the **Thread** tool located on the **Modify** panel under the **3D Model** tab. See Figure 6-1. Predrawn fasteners may be accessed using the **Content Center** tool. The **Content Center** library is explained later in the chapter.

Figure 6-1

Figure 6-2

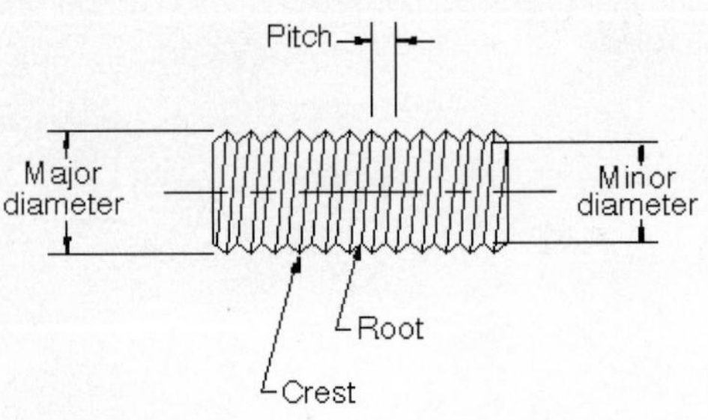

Thread Terminology

crest: The peak of a thread.

root: The valley of a thread.

major diameter: The distance across a thread from crest to crest.

minor diameter: The distance across a thread from root to root.

Figure 6-2 shows a thread. The peak of a thread is called the ***crest,*** and the valley portion is called the ***root.*** The ***major diameter*** of a thread is the distance across the thread from crest to crest. The ***minor diameter*** is the distance across the thread from root to root.

Thread Callouts—Metric Units

Threads are specified on a drawing using drawing callouts. See Figure 6-3. The M at the beginning of a drawing callout specifies that the callout is for a metric thread. Holes that are not threaded use the ∅ symbol.

Figure 6-3

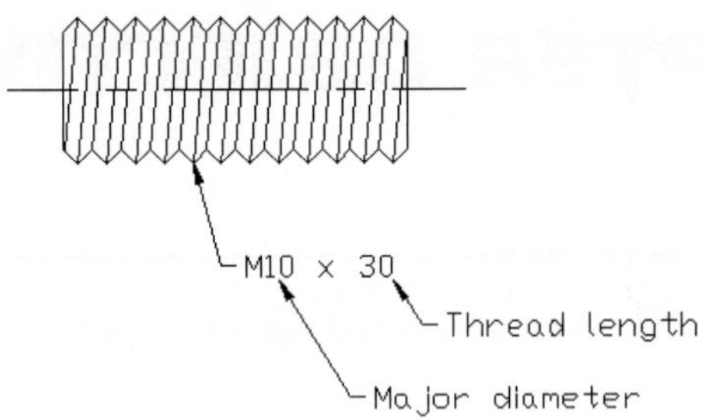

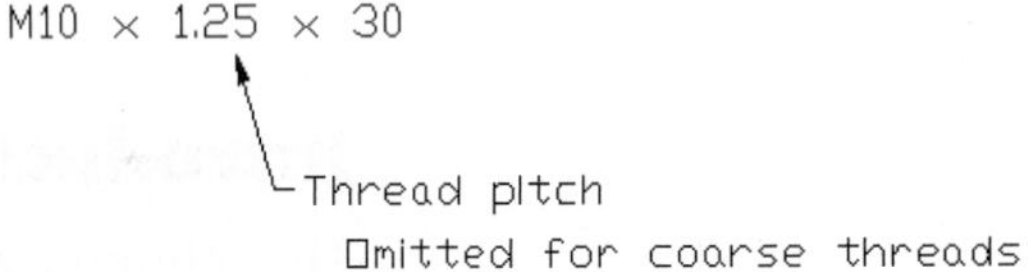

The number following the M is the major diameter of the thread. An M10 thread has a major diameter of 10 mm. The pitch of a metric thread is assumed to be a coarse thread unless otherwise stated. The callout M10 × 30 assumes a coarse thread, or a thread length of 1.5 mm per thread. The number 30 is the thread length in millimeters. The "×" is read as "by," so the thread is called a "ten by thirty."

The callout M10 × 1.25 × 30 specifies a pitch of 1.25 mm per thread. This is not a standard coarse thread size, so the pitch must be specified.

Figure 6-4 shows a listing of standard metric thread sizes. Other sizes may be located by scrolling through the given nominal sizes. Inventor lists metric threads according to ANSI (American National Standards Institute) Metric M Profile standards.

Whenever possible use preferred thread sizes for designing. Preferred thread sizes are readily available and are usually cheaper than nonstandard sizes. In addition, tooling such as wrenches is also readily available for preferred sizes.

Figure 6-4

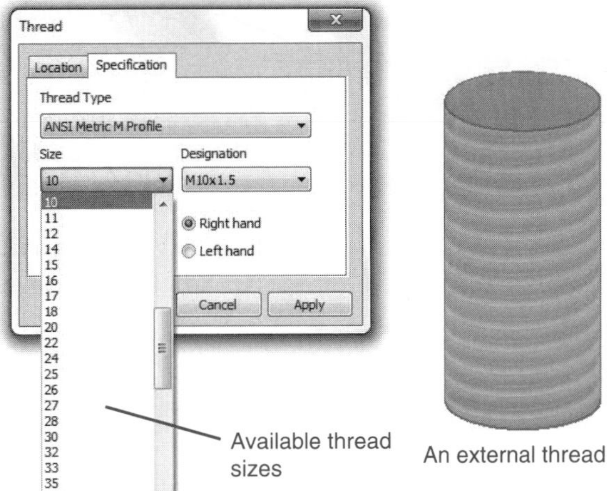

Thread Callouts—ANSI Unified Screw Threads

ANSI Unified screw threads (English units) always include a thread form specification. Thread form specifications are designated by capital letters, as shown in Figure 6-5, and are defined as follows.

UNC—Unified National Coarse

UNF—Unified National Fine

UNEF—Unified National Extra Fine

UN—Unified National, or constant-pitch threads

An ANSI (English units) thread callout starts by defining the major diameter of the thread followed by the pitch specification. The callout .500-13 UNC means a thread whose major diameter is .500 in. with 13 threads per inch. The thread is manufactured to the UNC standards.

There are three possible classes of fit for a thread: 1, 2, and 3. The different class specifications specify a set of manufacturing tolerances. A class 1 thread is the loosest, and a class 3 the most exact. A class 2 fit is the most common.

The letter A designates an external thread, B an internal thread. The symbol × means "by" as in 2 × 4, "two by four." The thread length (3.00) may be followed by the word LONG to prevent confusion about which value represents the length.

Pitch is defined as 1/number of threads per inch. ANSI English thread callouts specify the number of threads per inch, so, for example, a .500-13 UNC thread would have a pitch equal to 1/13, or .077 inch. ANSI metric thread designations specify the pitch directly.

Drawing callouts for ANSI (English unit) threads are sometimes shortened, such as in Figure 6-5. The callout .500-13 UNC-2A × 3.00 LONG is shortened to .500-13 × 3.00. Only a coarse thread has 13 threads per inch, and it should be obvious whether a thread is internal or external, so these specifications may be dropped. Most threads are class 2, so it is tacitly accepted that all threads are class 2 unless otherwise specified. The shortened callout form is not universally accepted. When in doubt, use a complete thread callout.

A listing of standard ANSI (English unit) threads, as presented in Inventor, is shown in Figure 6-6. Some of the drill sizes listed use numbers and letters. The decimal equivalents to the numbers are listed in Figure 6-6.

Figure 6-5

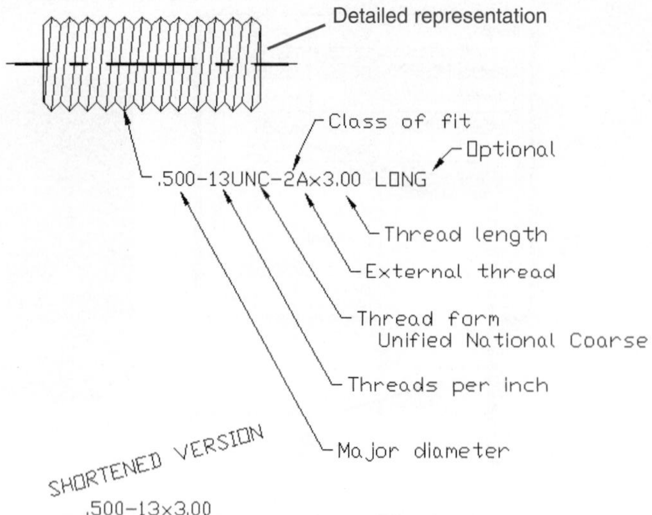

Figure 6-6

The decimal equivalents for threads specifed by numbers

#1	Ø.073
#2	Ø.086
#3	Ø.090
#4	Ø.112
#5	Ø.126
#6	Ø.138
#8	Ø.164
#10	Ø.190
#12	Ø.216

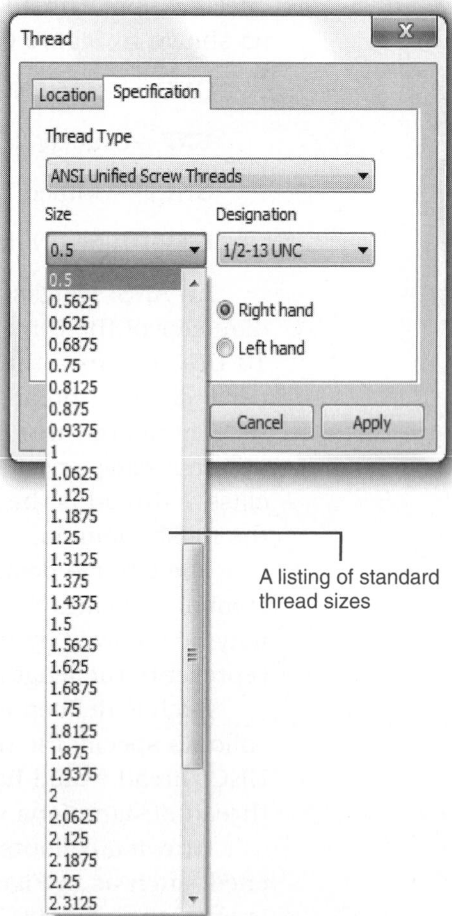

Thread Representations

There are three ways to graphically represent threads on a technical drawing: detailed, schematic, and simplified. Figure 6-5 shows an external detailed representation, and Figure 6-7 shows both the external and internal simplified and schematic representations.

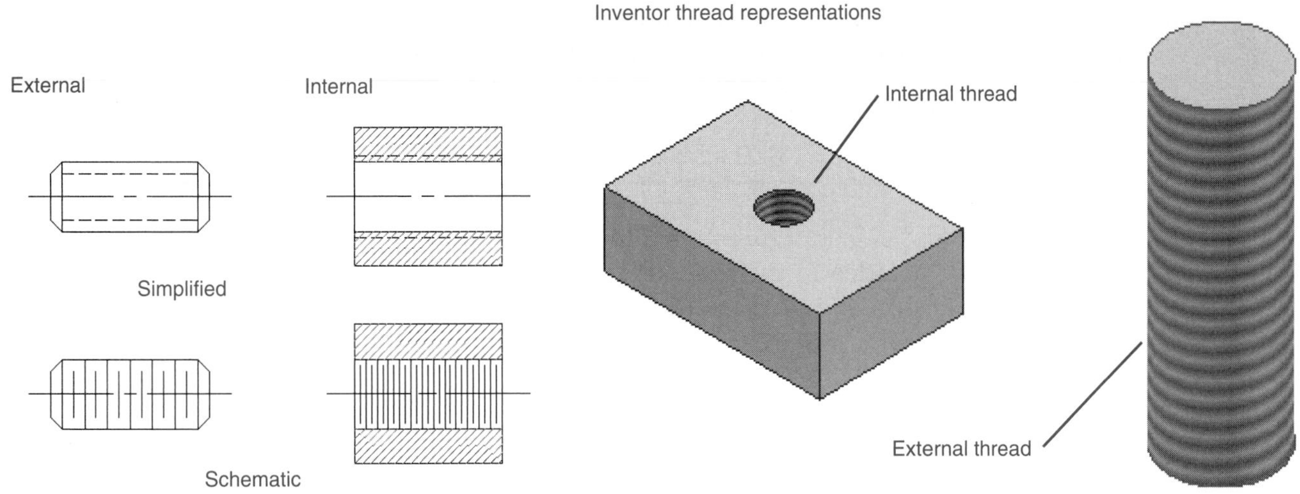

Figure 6-7

Figure 6-8

Figure 6-8 shows an internal and an external thread created using Inventor. The threads will automatically be sized to the existing hole. Threads may be created only around existing holes and cylinders.

Internal Threads

Figure 6-9 shows a 20 × 30 × 10 box with a ∅6.0 hole drilled through its center. The hole was created using the **Hole** tool.

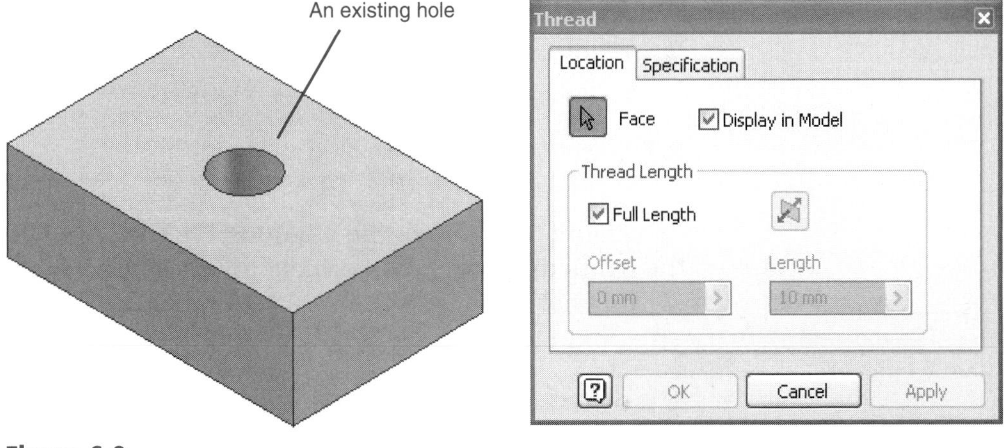

Figure 6-9

EXERCISE 6-1 **Adding Threads to an Existing Hole**

1 Click on the **Thread** tool located on the **Modify** panel under the **3D Model** tab.

The **Thread** dialog box will appear. See Figures 6-4, 6-6, and 6-9.

2 Click on the existing hole (the internal surface).

The threads will automatically be created to match the hole's diameter. Because the hole's diameter is 6.0, an M6 × 1 thread is created.

3 Click **OK.**

Figure 6-10 shows the resulting threaded hole.

When a thread is added to an existing hole, a thread listing will be included in the browser box. See Figure 6-11. The listing will confirm that a thread has been added, but it will not define the size or type of thread.

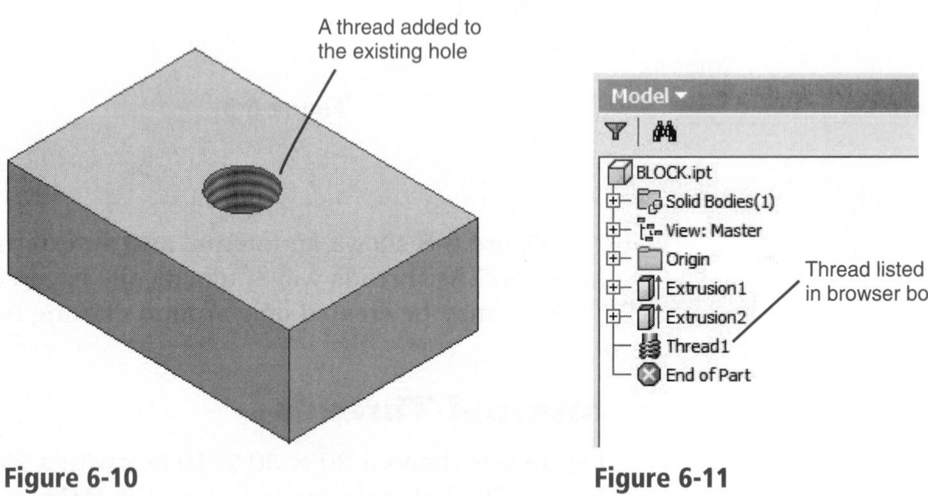

Figure 6-10

Figure 6-11

EXERCISE 6-2 **Determining the Thread Size**

1 Right-click on the **Hole1** listing in the browser box.

A dialog box will appear. See Figure 6-12.

2 Select the **Show Dimensions** option.

Figure 6-13 shows the resulting dimensions. The dimensions define the hole's diameter, and because Inventor will match the thread size to the existing hole diameter, the thread is an M6.

Threaded Blind Holes

The internally threaded holes presented in the last section passed completely through the material. This section shows how to draw holes that do not pass completely through the object but have a defined depth.

Figure 6-14 shows a tapped hole. It is drawn using the simplified representation. Note that there are three separate portions to the hole's representation: the threaded portion, the unthreaded portion, and the 120° conical point. Only the threaded portion of the tapped hole is used to define the hole's depth. The unthreaded portion and the conical point are shown but are not included in the depth calculation.

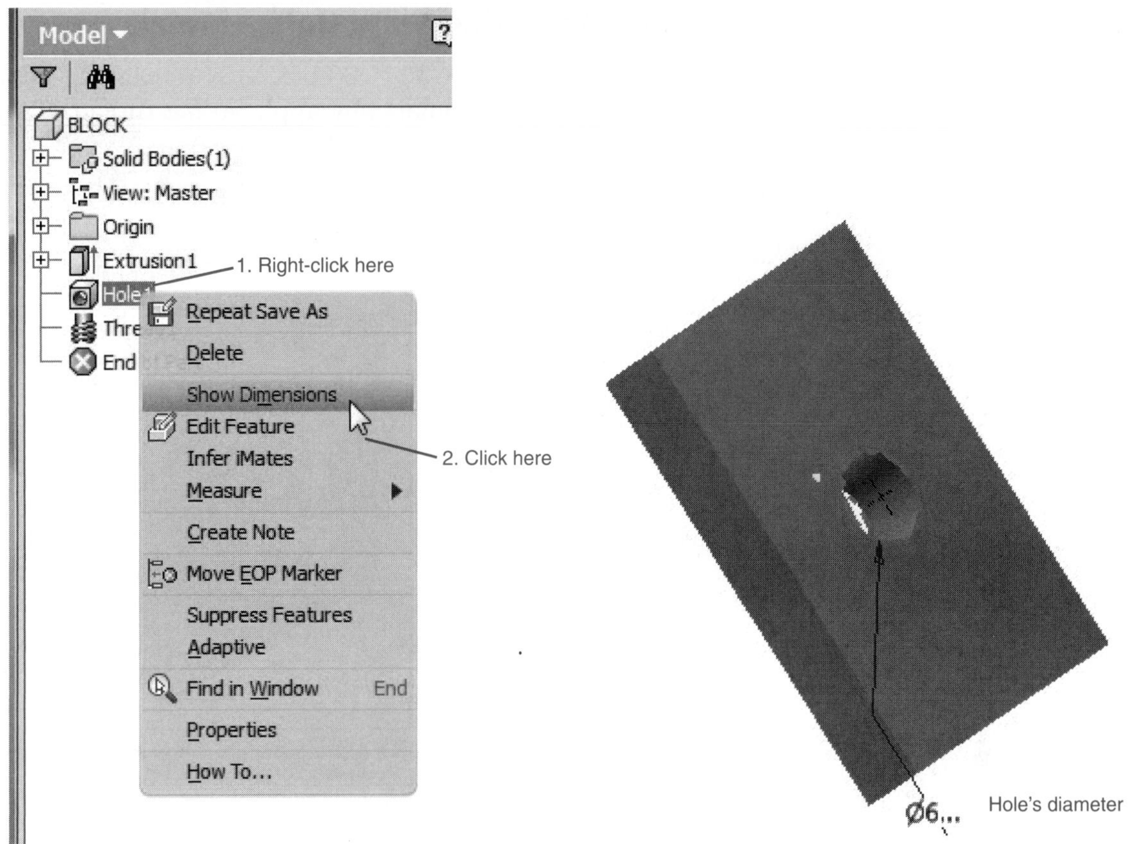

Figure 6-12

Figure 6-13

Figure 6-14

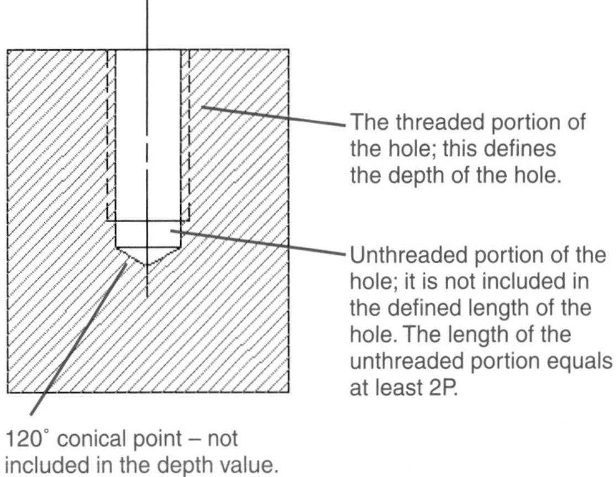

A tapped hole is manufactured by first drilling a pilot hole that is slightly smaller than the major diameter of the threads. The threads are then cut into the side of the pilot hole using a tapping bit. The tapping bit has no cutting edges on its bottom surface, so if it strikes the bottom of the hole, the bit can be damaged. Convention calls for the unthreaded portion of the pilot hole to extend about the equivalent of two thread pitches (2P) beyond the end of the threaded portion. The conical portion is added to the bottom surface of the pilot hole.

Drawing a Threaded Blind Hole—Metric

Figure 6-15 shows a 20 × 30 × 30 box with an existing Ø8 × 16 deep hole. The **Thread** tool will automatically apply an M8 thread to the Ø8 hole.

Figure 6-15

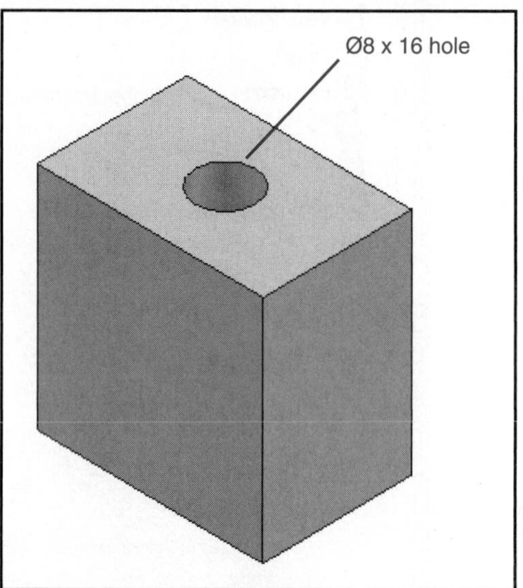

A distance equal to two pitches (2P) is recommended between the bottom of the hole and the end of the threads. One pitch equals 1.25 mm, so 2P = 2.50 mm. The existing hole is 16 mm, so the thread depth is 13.5 mm.

> **NOTE**
> The 2P recommendation is a minimal value. Distances of 3P, 4P, or greater may be used depending on the application.

EXERCISE 6-3 Creating a Threaded Blind Hole

1. Click on the **Thread** tool on the **Modify** panel located under the **3D Model** tab.

   The **Thread** dialog box will appear. See Figure 6-16.

   The **Face** tool will automatically be activated.
2. Select **Face** by clicking the internal surface of the hole.
3. Remove the check mark from the **Full Length** box.
4. Enter a **Length** value of **13.5.**
5. Click the **Specification** tab and set the **Designation** for **M8 × 1.25.**
6. Click **OK.**

> **NOTE**
> Inventor will automatically create a coarse thread with a pitch of 1.25. There are other pitch sizes available: M8 × 1, M8 × 0.8, M8 × 0.75, and M8 × 0.5. These pitch sizes are accessed using the arrow on the right side of the **Designation** box on the **Specification** tab of the **Thread** dialog box.

Figure 6-16

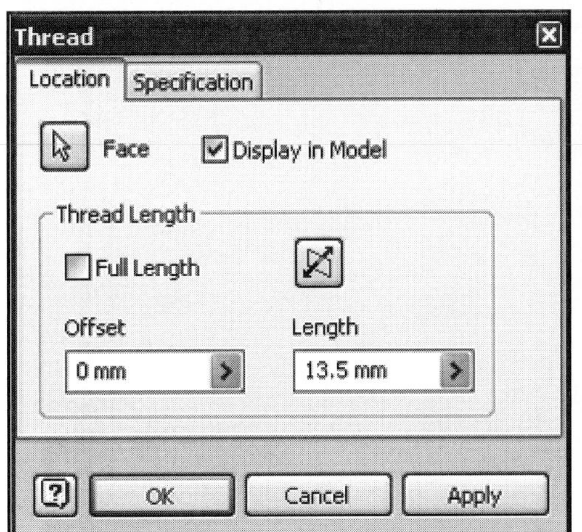

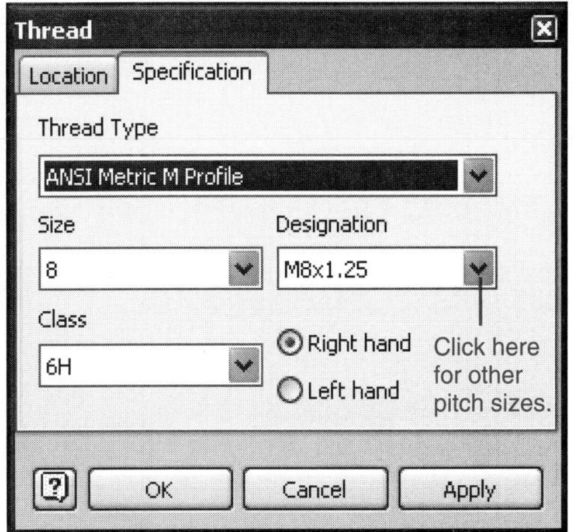

Chapter 6

Figure 6-17 shows a section view of the threaded blind hole. Note how hidden lines are used to represent threads in both the top and the section views.

Figure 6-17

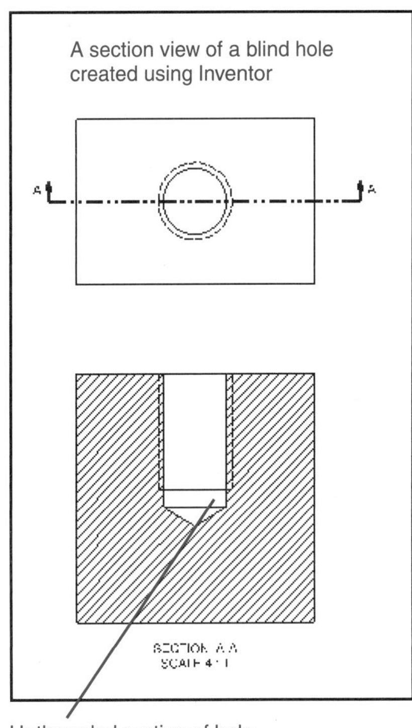

Drawing a Blind Hole—ANSI Threads

Pitch for ANSI threads is defined as threads per inch. A ¼-20 UNC thread has a pitch of 20 threads per inch. The length in inches of one thread is 1⁄20, or 0.05 in. Therefore, 2P = 2(0.05) = 0.10 in.

If a block includes a hole 1.50 deep, then the appropriate thread length is 1.50 – 0.10 = 1.40 in.

Creating Threaded Holes Using the Hole Tool

Threaded holes may be created directly using the **Hole** tool. Figure 6-18 shows a 30 × 40 × 30 mm block with a hole center defined using the **Point, Center Point** tool in the center of the top surface.

Figure 6-18

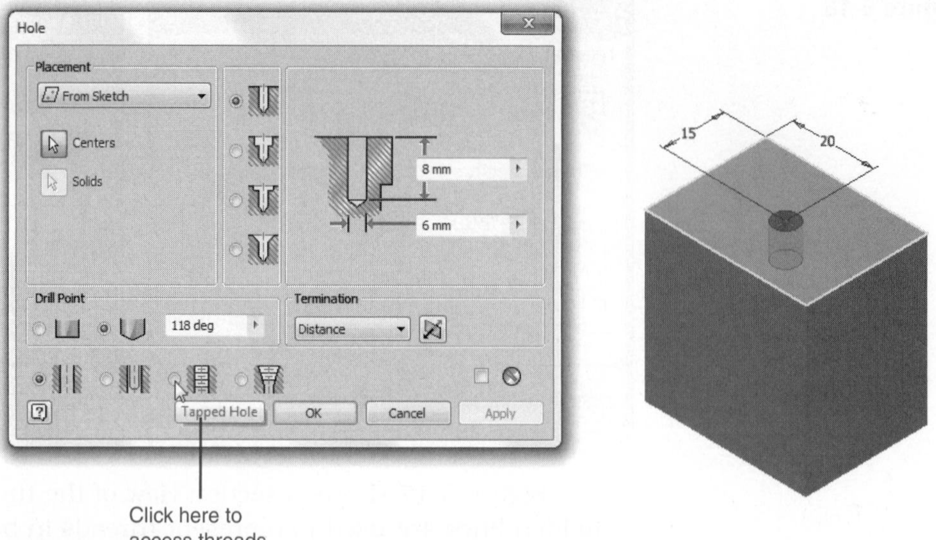

EXERCISE 6-4 Creating a Threaded Through Hole

1. Create a new sketch plane on the top surface of the block and use the **Point** tool to locate a point at the center of the surface.
2. Right-click the mouse and select the **Finish 2D Sketch** option.
3. Click on the **Hole** tool on the **Modify** panel under the **3D Model** tab. The **Hole** dialog box will appear. See Figure 6-18.
4. Click the **Termination** box and select the **Through All** option.
5. Click the **Tapped Hole** button.
6. Click on **Thread Type.** Select the **ANSI Metric M Profile** option. See Figure 6-19.
7. Set the thread **Size** for **10.** Select the **M10 × 1.5** pitch in the **Designation** box.

Figure 6-19

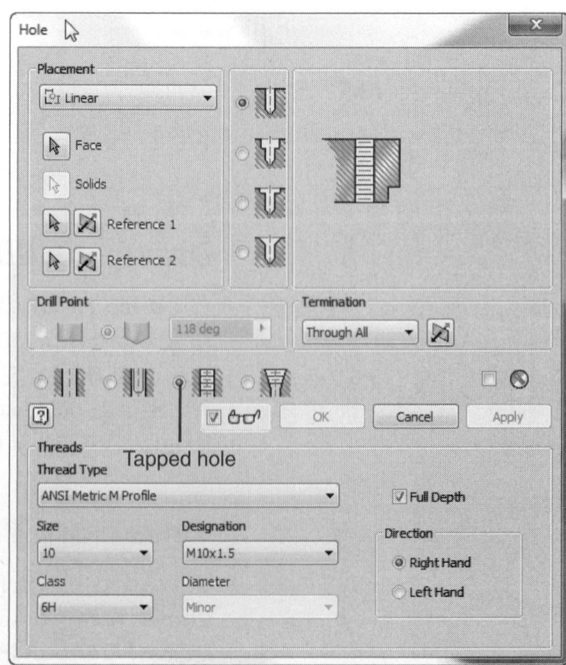

More than one pitch size is available. In addition to the 1.50 pitch a 1.25 or 0.75 option is also available. These are Fine and Extra Fine designations. See Figure 6-20.

8 Click **OK.**

Figure 6-20

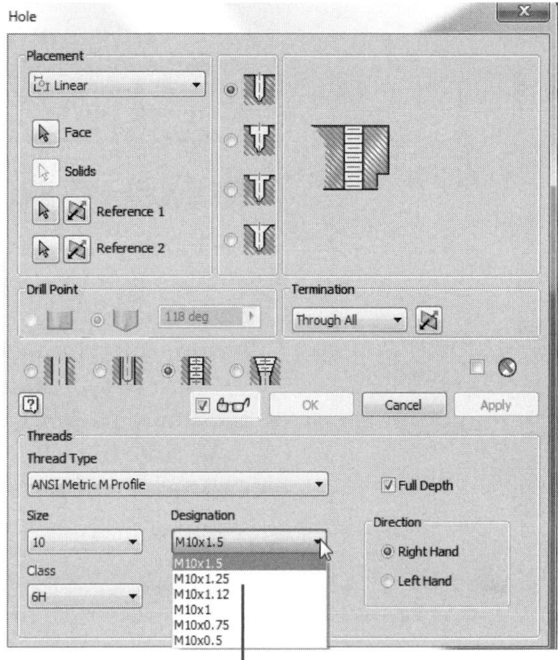

EXERCISE 6-5 Creating a Threaded Blind Hole

Create a 30 × 40 × 30-mm block with a hole's center defined in the center of the top surface.

1 Click on the **Hole** tool.

The **Hole** dialog box will appear. See Figure 6-21.

Figure 6-21

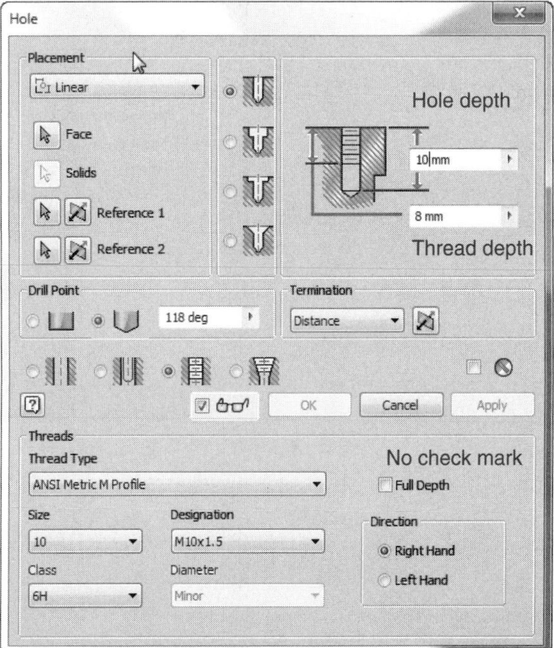

2 Set the **Termination** for **Distance,** and click the **Tapped Hole** button.

3 Set the **Size** for **10** and the **Designation** for **M10 × 1.5.**

4 Click the **Full Depth** box so that no check appears.

5 Set the thread depth for **8** and the hole depth for **10.**

The hole depth should, with rare exceptions, be greater than the thread depth.

6 Click **OK.**

Standard Fasteners

Fasteners, such as screws and bolts, and their associated hardware, such as nuts and washers, are manufactured to standard specifications. Using standard-sized fasteners in designs saves production costs and helps assure interchangeability.

Inventor includes a library of standard parts that may be accessed using the **Place from Content Center** tool on the **Component** panel under the **Assemble** tab. Clicking on the **Place from Content Center** tool accesses the **Content Center** dialog box. The **Content Center** may also be accessed by right-clicking the mouse and selecting the **Place from Content Center** option.

> **NOTE**
>
> The drawing must be in the **Standard.iam** format, either mm or inch designation.

See Figure 6-22. Figure 6-23 shows the **Place from Content Center** dialog box. Click the + sign to the left of the **Fasteners** heading under

Figure 6-22

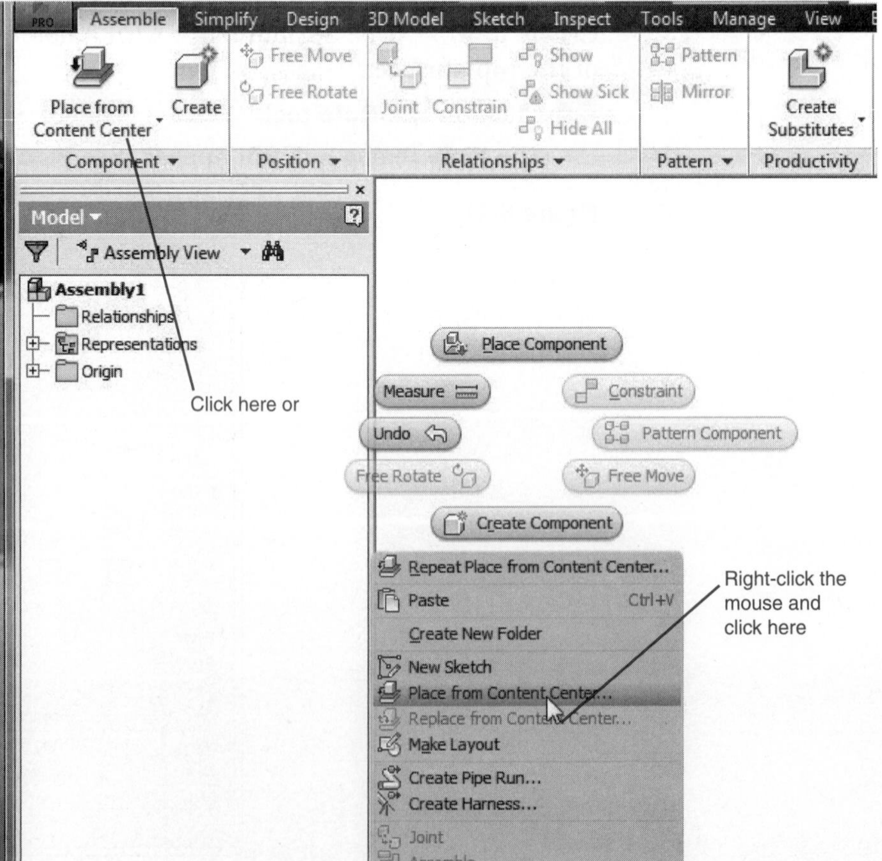

Figure 6-23

Category View. Click **Fasteners, Bolts, Hex Head,** and select the **Hex Bolt-Metric** bolt. A listing of available diameters and lengths will appear. See Figure 6-24. Click the **Table View** tab to access a table of sizes and dimensions that apply to the selected bolt. See Figure 6-25.

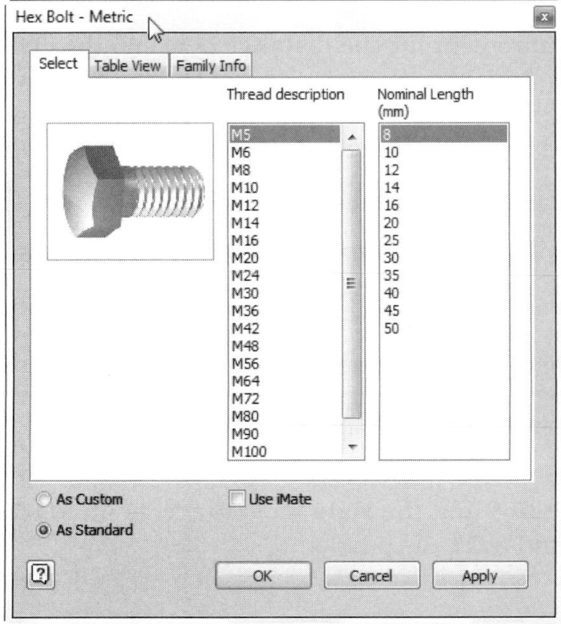

Figure 6-24

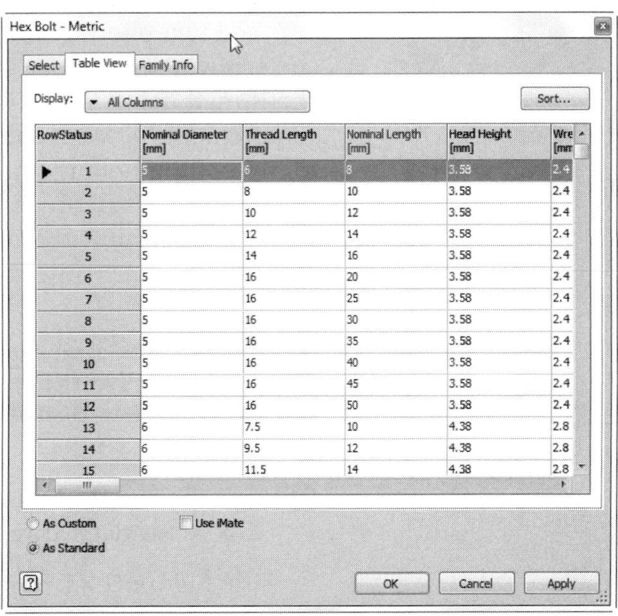

RowStatus	Nominal Diameter [mm]	Thread Length [mm]	Nominal Length [mm]	Head Height [mm]	Wre [mm
1	5	6	8	3.58	2.4
2	5	8	10	3.58	2.4
3	5	10	12	3.58	2.4
4	5	12	14	3.58	2.4
5	5	14	16	3.58	2.4
6	5	16	20	3.58	2.4
7	5	16	25	3.58	2.4
8	5	16	30	3.58	2.4
9	5	16	35	3.58	2.4
10	5	16	40	3.58	2.4
11	5	16	45	3.58	2.4
12	5	16	50	3.58	2.4
13	6	7.5	10	4.38	2.8
14	6	9.5	12	4.38	2.8
15	6	11.5	14	4.38	2.8

Figure 6-25

Sizing a Threaded Hole to Accept a Screw

Say we wish to determine the length of thread and the depth of hole needed to accept an M10 × 25 hex head screw and to create a drawing of the screw in the threaded hole. The length of the threaded hole must extend a minimum of two pitches (2P) beyond the end of the screw, and the untapped portion of the hole must extend a minimum of two pitches (2P) beyond the threaded portion of the hole. This requirement ensures that all the fastener threads will be in contact with the hole's threads and that fasteners will not bottom out. In this example the thread pitch equals 1.50 mm. Therefore, 2P = 2(1.50) = 3.0 mm. This is the minimum distance and can be increased but never decreased. See Figure 6-26.

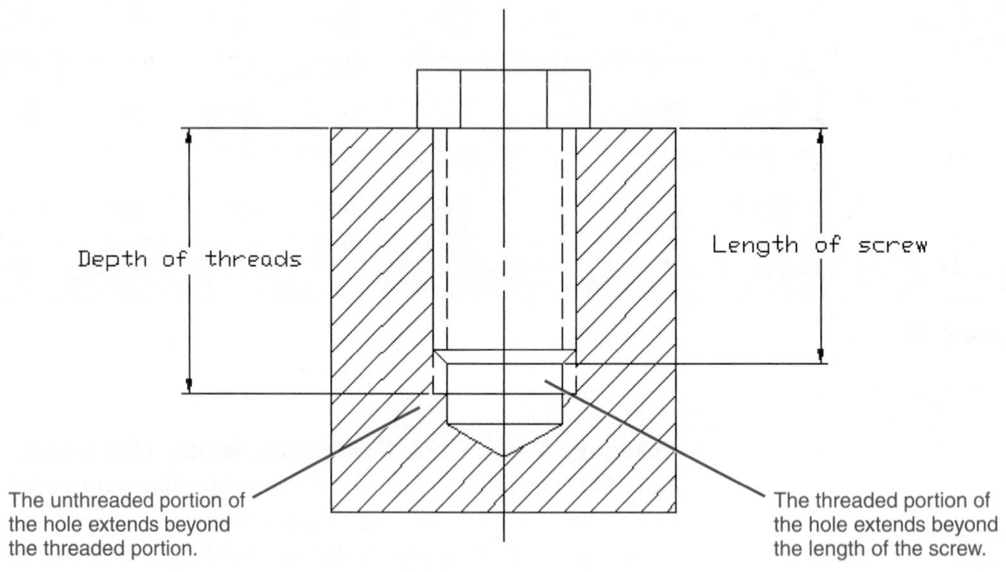
Depth of threads
Length of screw
The unthreaded portion of the hole extends beyond the threaded portion.
The threaded portion of the hole extends beyond the length of the screw.

Figure 6-26

The two-pitch length requirement for the distance between the end of the screw and the end of the threaded portion of the hole determines that the thread depth should be 25.0 + 3.0 = 28.0 mm. The two-pitch length requirement between the end of the threaded portion of the hole and the bottom of the hole requires that the hole must have a depth of 28.0 + 3.0, or 31.0 mm.

It is important that complete hole depths be specified, as they will serve to show any interference with other holes or surfaces.

EXERCISE 6-6 **Drawing a Blind Threaded Hole**

1. Create a new **Standard (mm).ipt** drawing and create a 40 × 40 × 60 block.
2. Use the **Point, Center Point** and the **Hole** tools to create an **M10 × 1.5 × 28** deep thread and a **31** deep hole.

   See Figure 6-27.
3. Save the threaded block as **BLOCK, THREADED.**

Figure 6-27

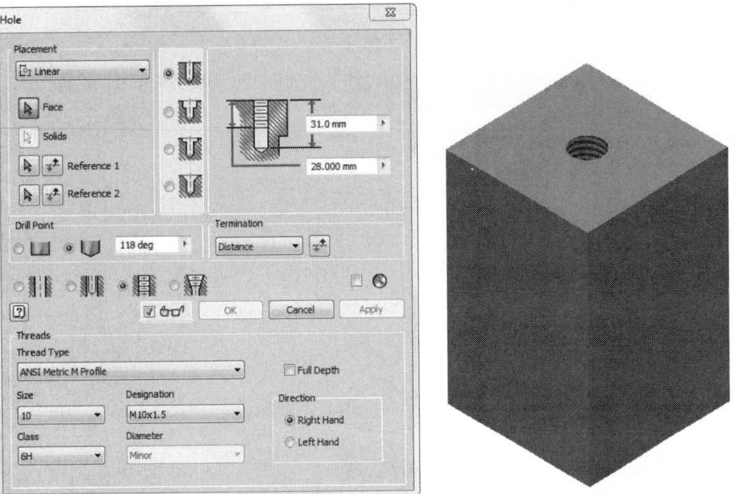

EXERCISE 6-7 **Creating an Assembly with an M10 × 1.5 × 25 Hex Head Bolt**

1. Create a new **Standard (mm).iam** drawing called **M10ASSEMBLY.**
2. Use the **Place Component** tool and locate one copy of the **BLOCK, THREADED** block on the drawing screen.
3. Click the **Place from Content Center** tool located on the **Component** panel under the **Assemble** tab. The **Place from Content Center** tool is a flyout from the **Place Component** tool.

   The **Place from Content Center** dialog box will appear. See Figure 6-28.

Figure 6-28

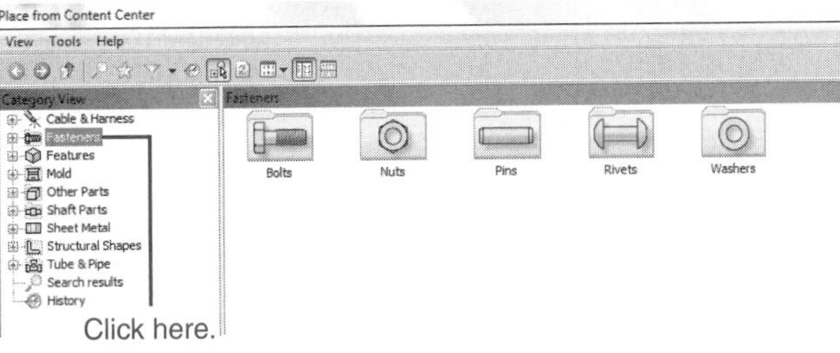

4. Click the + sign to the left of the **Fasteners** option.

   See Figure 6-29.
5. Click the **Bolts** option, then click the **Hex Head** option.
6. Scroll through the options and select the **Hex Bolt-Metric** option, and click **OK.**

   A small icon will appear on the screen attached to the cursor.
7. Click the screen.

   A picture of the Hex Bolt-Metric fastener will appear in the dialog box along with a listing of available thread sizes and lengths. See Figure 6-30.
8. Set the nominal diameter for **10,** the pitch for **1.5,** and the nominal length for **25.**
9. Set the **Thread description** to **M10** and the **Nominal Length** to **25.**
10. Click the **Apply** box.

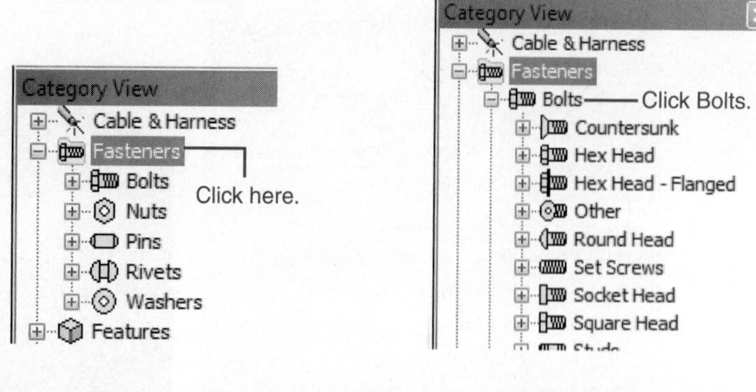

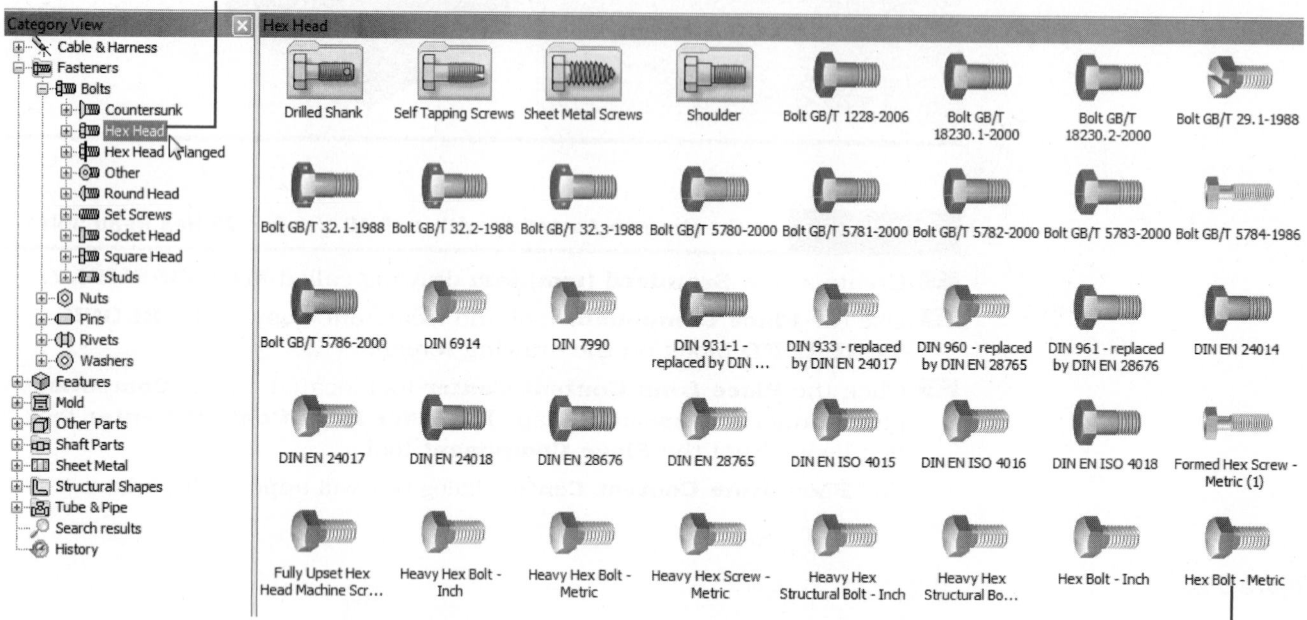

Figure 6-29

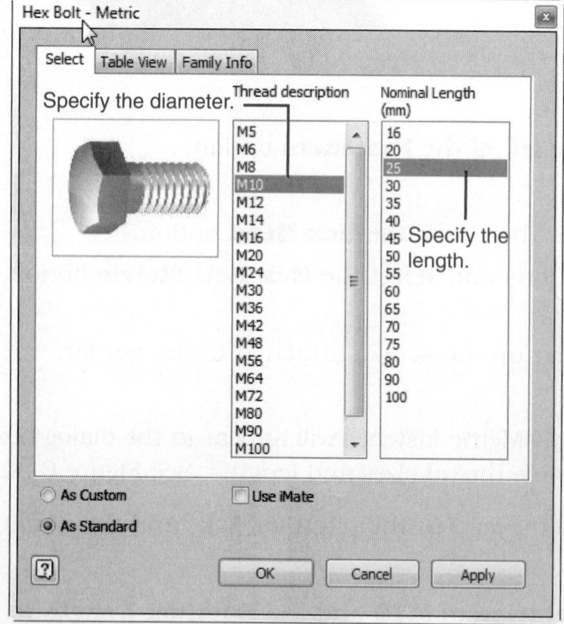

Hex Bolt - Metric

Select | Table View | Family Info

Display: All Columns — Sort...

RowStatus	Designation	Cone Point Diameter [mm]	Width Across Corner [mm]	Thread Diameter [mm]	St [n
36	M8 x 1.25 x 60	6.46637	15.015	8	8
37	M8 x 1.25 x 65	6.46637	15.015	8	8
38	M8 x 1.25 x 70	6.46637	15.015	8	8
39	M8 x 1.25 x 75	6.46637	15.015	8	8
40	M8 x 1.25 x 80	6.46637	15.015	8	8
41	M10 x 1.5 x 16	8.15965	18.48	10	1(
42	M10 x 1.5 x 20	8.15965	18.48	10	1(
43	M10 x 1.5 x 25	8.15965	18.48	10	1(
44	M10 x 1.5 x 30	8.15965	18.48	10	1(
45	M10 x 1.5 x 35	8.15965	18.48	10	1(
46	M10 x 1.5 x 40	8.15965	18.48	10	1(
47	M10 x 1.5 x 45	8.15965	18.48	10	1(
48	M10 x 1.5 x 50	8.15965	18.48	10	1(
49	M10 x 1.5 x 55	8.15965	18.48	10	1(
50	M10 x 1.5 x 60	8.15965	18.48	10	1(

As Custom — Use iMate — As Standard

Thread designation

OK — Cancel — Apply

Figure 6-30

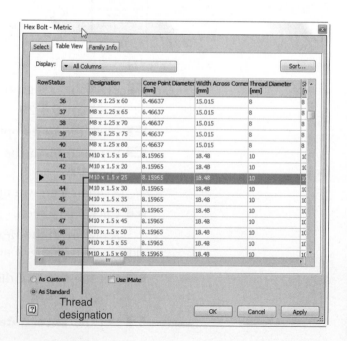

The M10 bolt will appear on the drawing screen with the BLOCK, THREADED. See Figure 6-31. If the bolt interferes with the BLOCK, THREADED, use the **Move Component** option to position the bolt away from the block. See Figure 6-31. Both the **Move** and **Rotate** tools may be used to locate the bolt above the block.

Figure 6-31

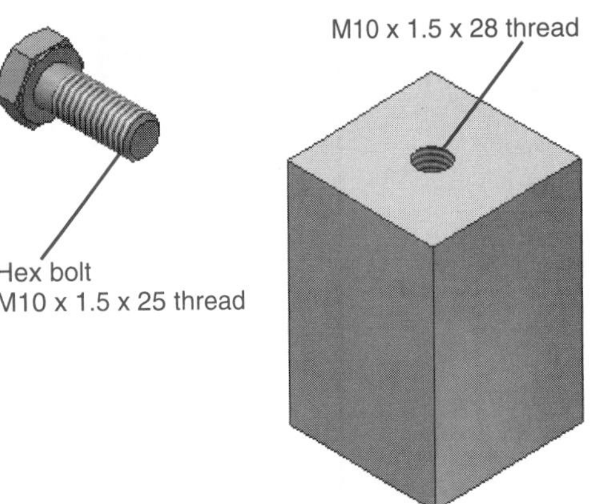

Inventor will automatically create a second fastener for insertion.

11 Press the **<Esc>** key to remove the second fastener.

EXERCISE 6-8 Inserting the M10 Bolt

1 Click the **Constrain** tool on the **Position** panel under the **Assemble** tab.

The **Place Constraint** dialog box will appear. See Figure 6-32.

2 Click the **Insert** option.

3 Click the bottom surface of the bolt's hex head, then click the threaded hole in the BLOCK, THREADED block.

Figure 6-33 shows the resulting assembly.

4 Save the assembly.

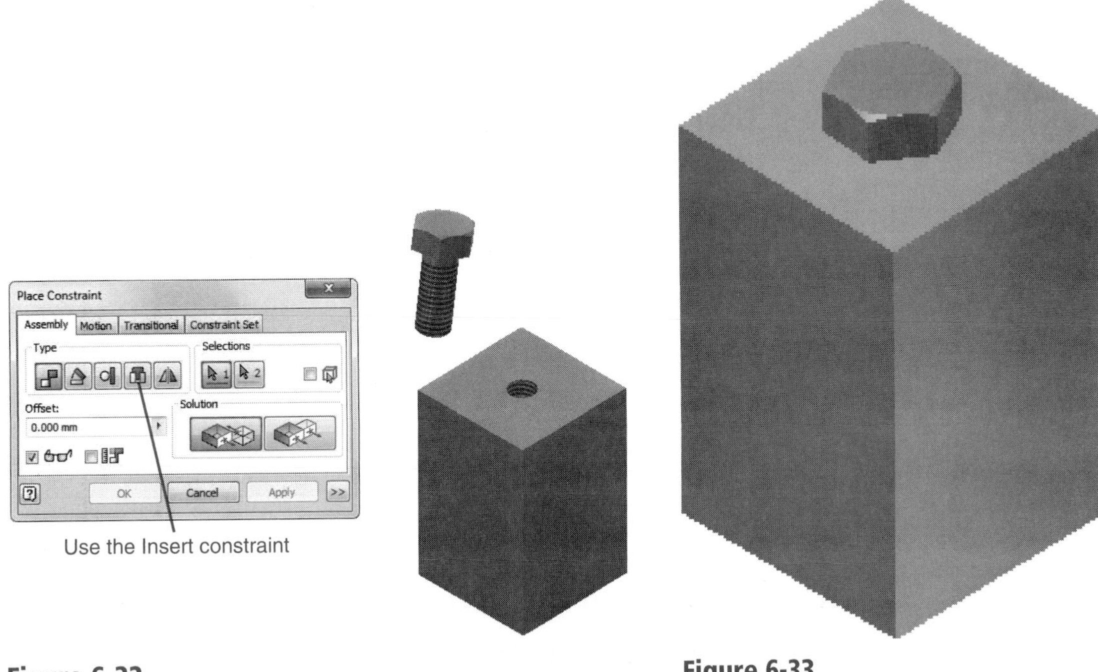

Figure 6-32

Figure 6-33

Figure 6-34 shows a top view and a section view of the M10ASSEMBLY created using the **ISO.idw** option. Note the open area between the bottom of the M10 bolt and the bottom of the hole.

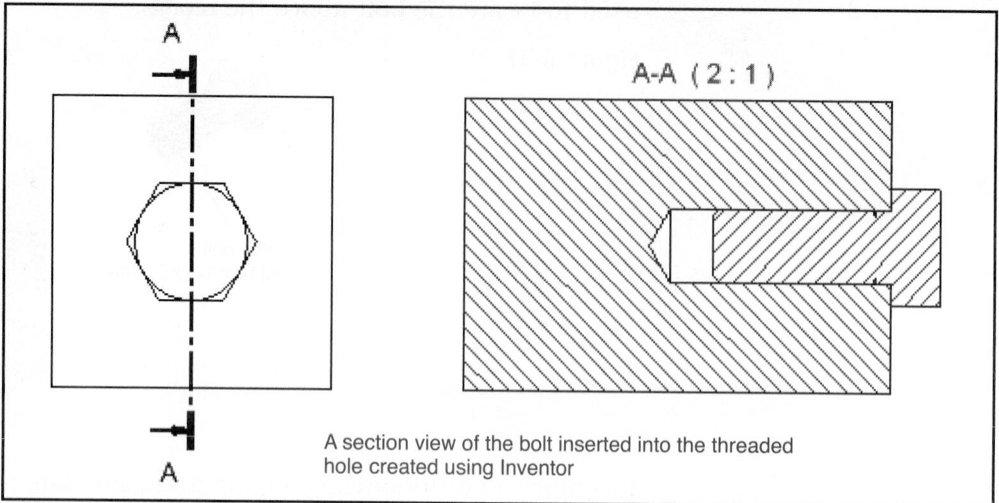

Figure 6-34

Screws and Nuts

Screws often pass through an object or group of objects and are secured using nuts. The threads of a nut must match the threads of the screw.

Nuts are manufactured at a variety of heights depending on their intended application. Nuts made for heavy loads are thicker than those intended for light loads. In general, nut thickness can be estimated as 0.88 of the major diameter of the thread. For example, if the nut has an M10 thread, the thickness will be about 0.88(10) = 8.8 mm.

It is good practice to specify a screw length that allows for at least two threads beyond the nut. This will ensure that all threads of the nut are in contact with the screw threads. Strength calculations for screw threads are based on 100% contact with the nut. See Figure 6-35.

Figure 6-35

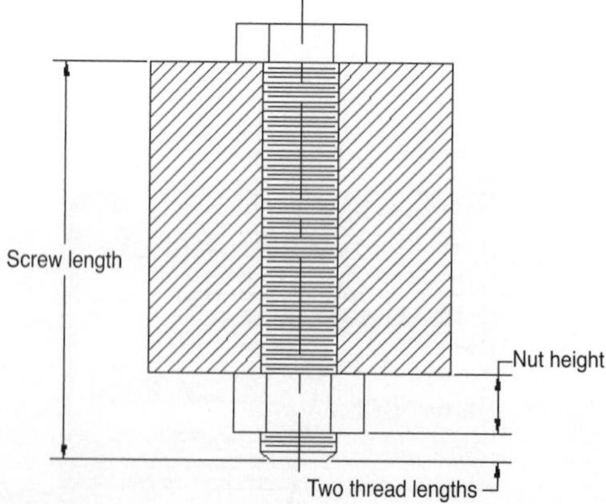

Calculating the Screw Thread Length

In the following example an M10 bolt will pass through a box that has a height of 30 mm and be held in place using an M10 hex nut. The required length of the screw is calculated by adding the height of the box to the height of the nut, then adding at least two thread lengths (2P).

For an M10 coarse thread 2P = 1.60 mm. The height of the nut = 0.88(10) = 8.80 mm, and the height of the box = 30 mm:

30.00 + 8.80 + 1.60 = 40.4 mm

Refer to Figure 6-30 and find the nearest M10 standard thread length that is greater than 40.4. The table shows the next available standard length that is greater than 40.4 is 45 mm.

EXERCISE 6-9 Adding an M10 × 45 Hex Head Screw to a Drawing

1 Draw a **40 × 40 × 30** box.

2 Locate a **Ø11** hole in the center of the top surface of the box.

The hole does not have threads. It is a clearance hole and so should be slightly larger than the M10 thread. See Figure 6-36.

3 Save the block as **Ø11BLOCK.**

Figure 6-36

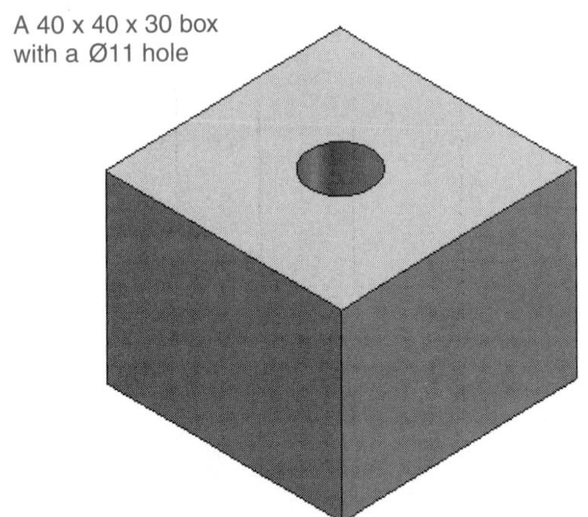

EXERCISE 6-10 Adding a Bolt and Nut to the Drawing

1 Start a new assembly drawing called **M10NUT.** Use the **Standard (mm).iam** format.

2 Use the **Place Component** tool and place a copy of the Ø11BLOCK on the drawing screen.

3 Use the **Save As** command on the **File** pull-down menu to save and name the new assembly as **M10NUT.iam.**

4 Click the **Place from Content Center** tool.

5 Click the **Fasteners** option, the **Bolts** option, the **Hex Head** option, the **Hex Bolt-Metric** bolt, and **OK.**

A small icon will appear on the screen attached to the cursor.

6 Click the screen.

7 Set the **Thread description** for **M10** and the **Nominal Length** for **45.**

8 Click the **Apply** box.

See Figure 6-37.

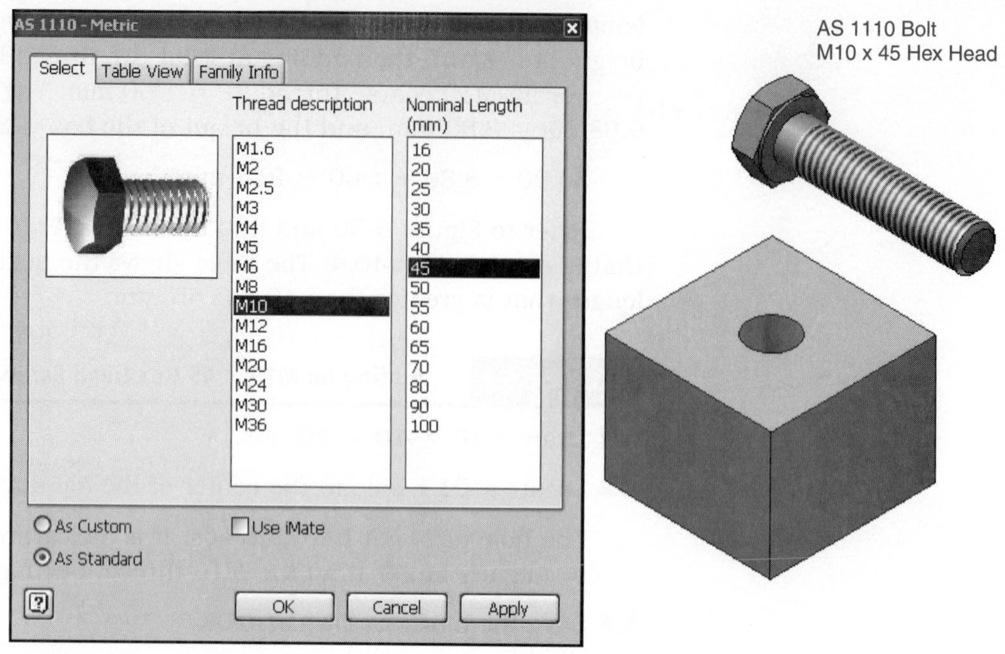

Figure 6-37

9 Click the **Place from Content Center** tool, click the **Nuts** heading in the **Category View** listing, click the **Hex** option, and select the **Hex Nut-Metric** nut. Select the **M10 Thread description** and **Style 1.** See Figure 6-38.

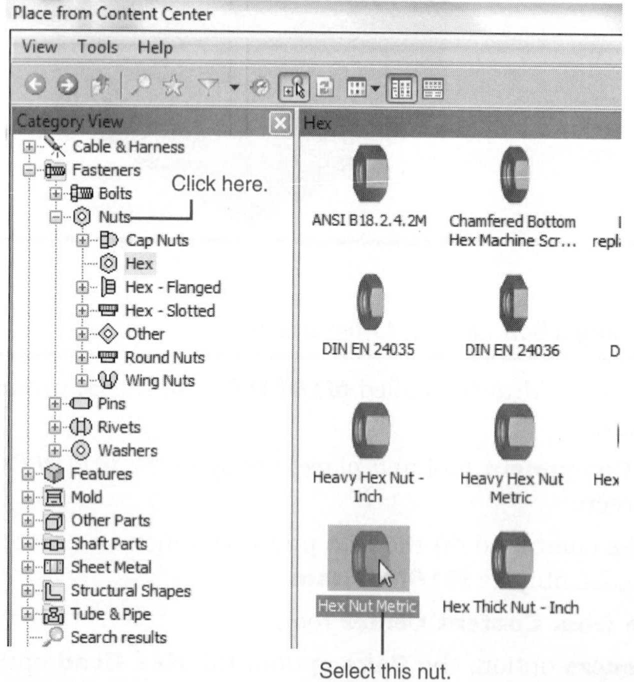

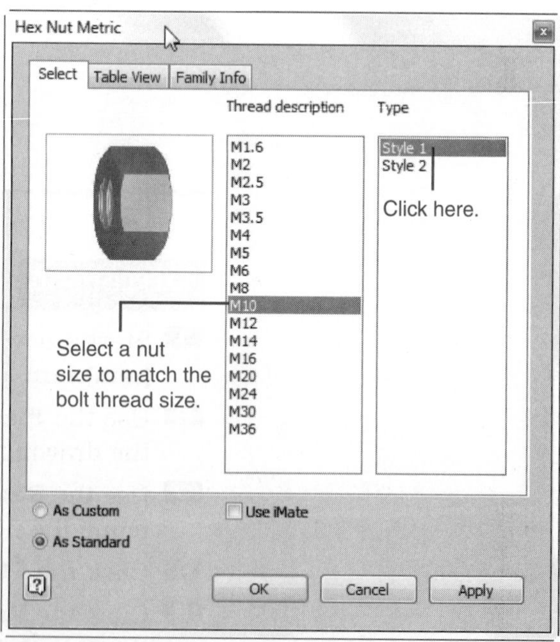

Figure 6-38

10 Click the **Apply** box.

See Figure 6-39.

Inventor will automatically create a second nut for insertion.

11 Press the **<Esc>** key to remove the second nut.

Figure 6-39

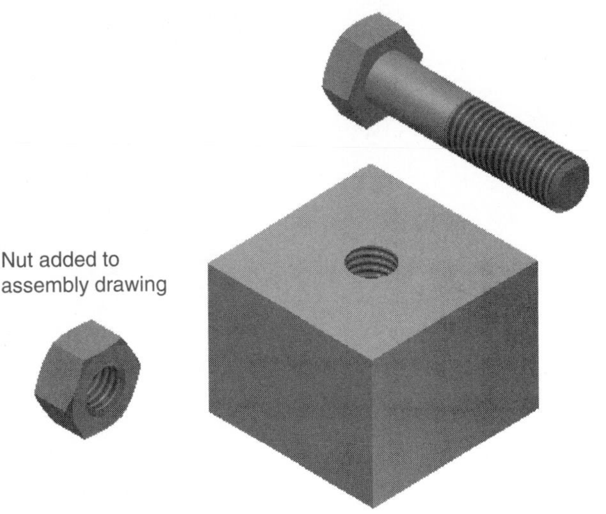
Nut added to
assembly drawing

EXERCISE 6-11 Assembling the Components

1. Click the **Constrain** tool.
2. Insert the bolt into the hole as described previously.

The nut is to be located on the bottom surface of the block, which is presently not visible.

3. Click the **ViewCube** and rotate the block so that the bottom surface is visible.

   See Figure 6-40.

Figure 6-40

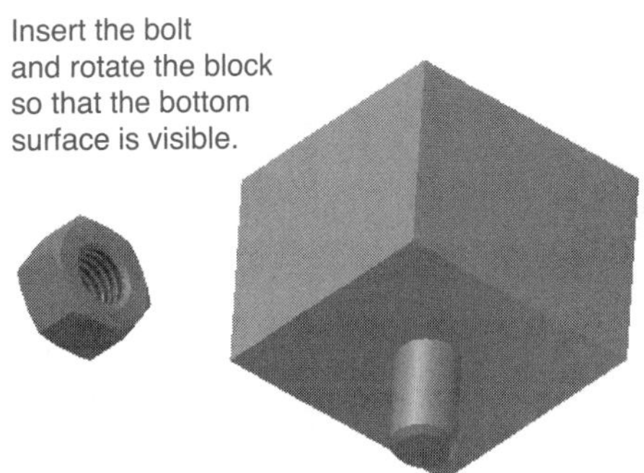
Insert the bolt
and rotate the block
so that the bottom
surface is visible.

4. Click the **Constrain** tool and select the **Insert** option.
5. Click the bottom of the nut and the edge line of the ∅11 hole.

   See Figure 6-41.

Figure 6-41

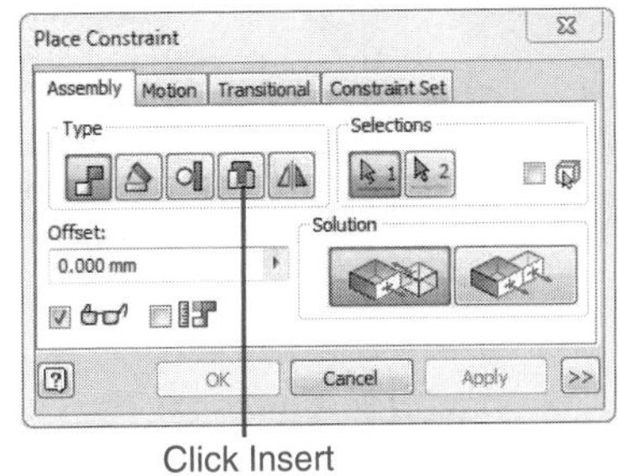
Place Constraint
Assembly
Motion
Transitional
Constraint Set
Type
Selections
Offset:
0.000 mm
Solution
OK
Cancel
Apply
>>
Click Insert

Use the Move and
Rotate tools to
position the nut for
easy assembly

6 Apply the constraints.

Figure 6-42 shows the resulting assembly.

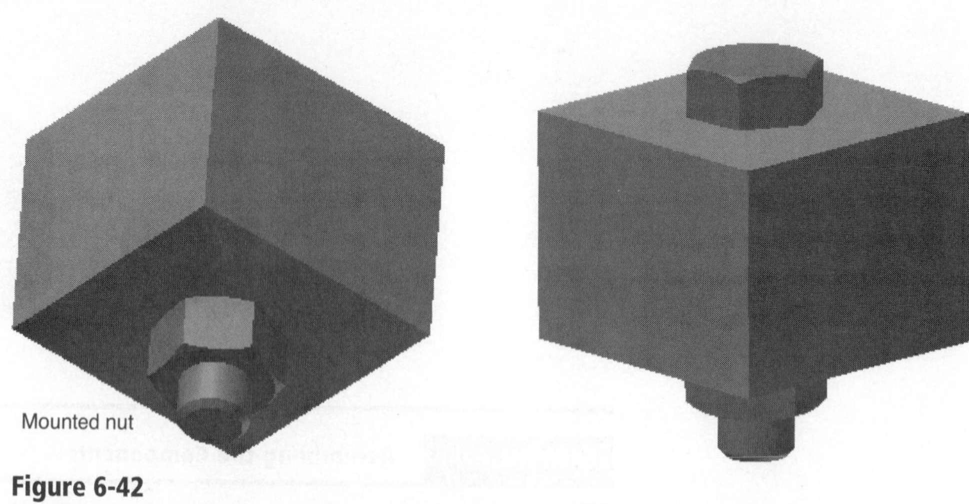

Figure 6-42

Types of Threaded Fasteners

There are many different types of screws generally classified by their head types. Figure 6-43 shows six of the most commonly used types.

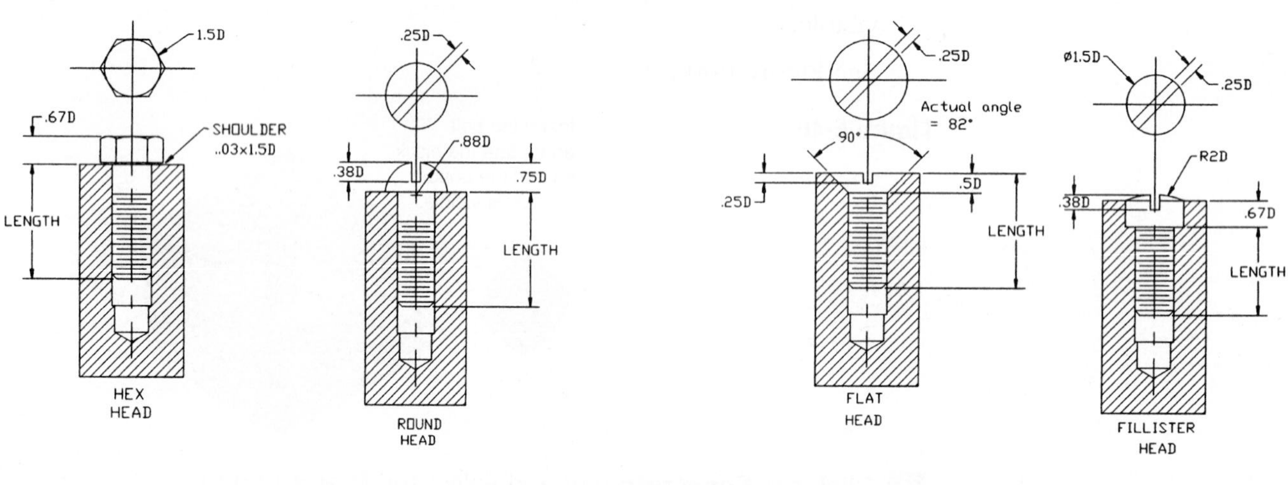

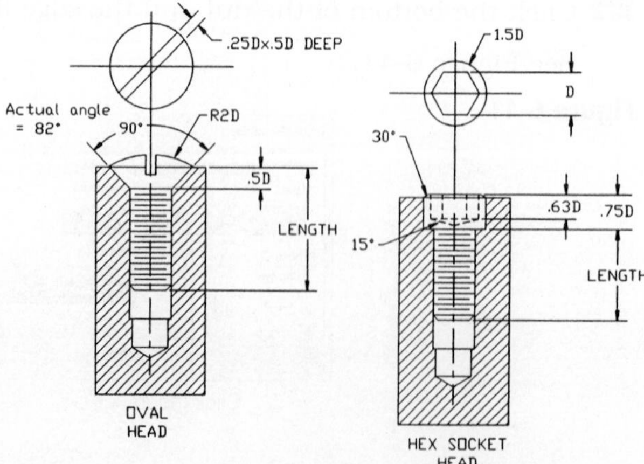

Note:
The dimensions listed are for reference only. See manufacturers' specifications for the actual sizes.

Figure 6-43

The choice of head type depends on the screw's application and design function. A product design for home use would probably use screws that had slotted heads, as most homes possess a blade screwdriver. Hex head screws can be torqued to higher levels than slotted pan heads but require socket wrenches. Flat head screws are used when the screw is located in a surface that must be flat and flush.

Sometimes a screw's head shape is selected to prevent access. For example, the head of the screw used to open most fire hydrants is pentagon-shaped and requires a special wrench to open it. This is to prevent unauthorized access that could affect a district's water pressure.

Screw connections for oxygen lines in hospitals have left-handed threads. They are the only lines that have left-handed threads, to ensure that no patient needing oxygen is connected to anything but oxygen.

Inventor's **Content Center** lists many different types of fasteners. See Figure 6-44. There are many subfiles to each of the fastener headings.

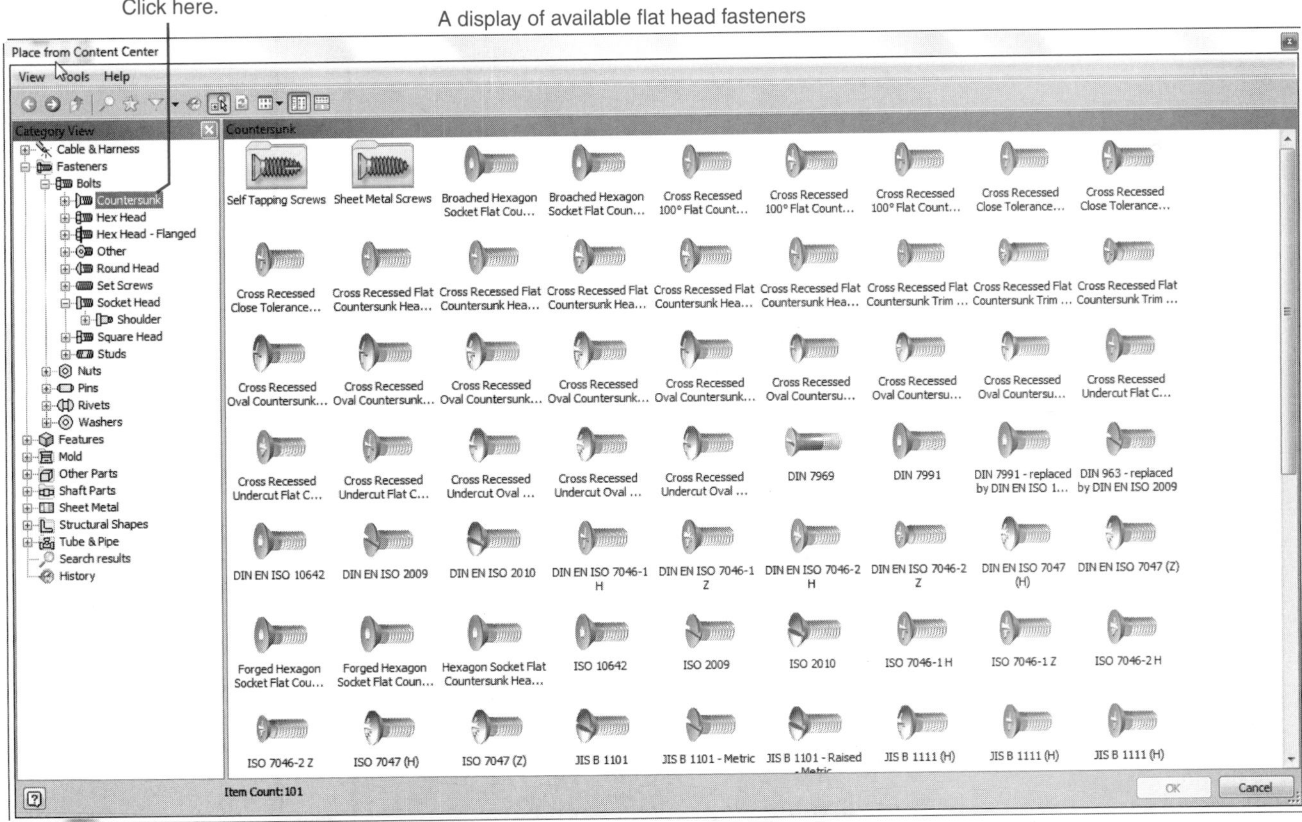

Figure 6-44

Flat Head Screws—Countersunk Holes

Flat head screws are inserted into countersunk holes. The procedure is first to create a countersunk hole on a component, then to create an assembly using the component along with the appropriate screw listed in the **Content Center.**

The following example uses an M8 × 50 Forged Hexagon Socket Flat Countersunk Head Cap Screw-Metric.

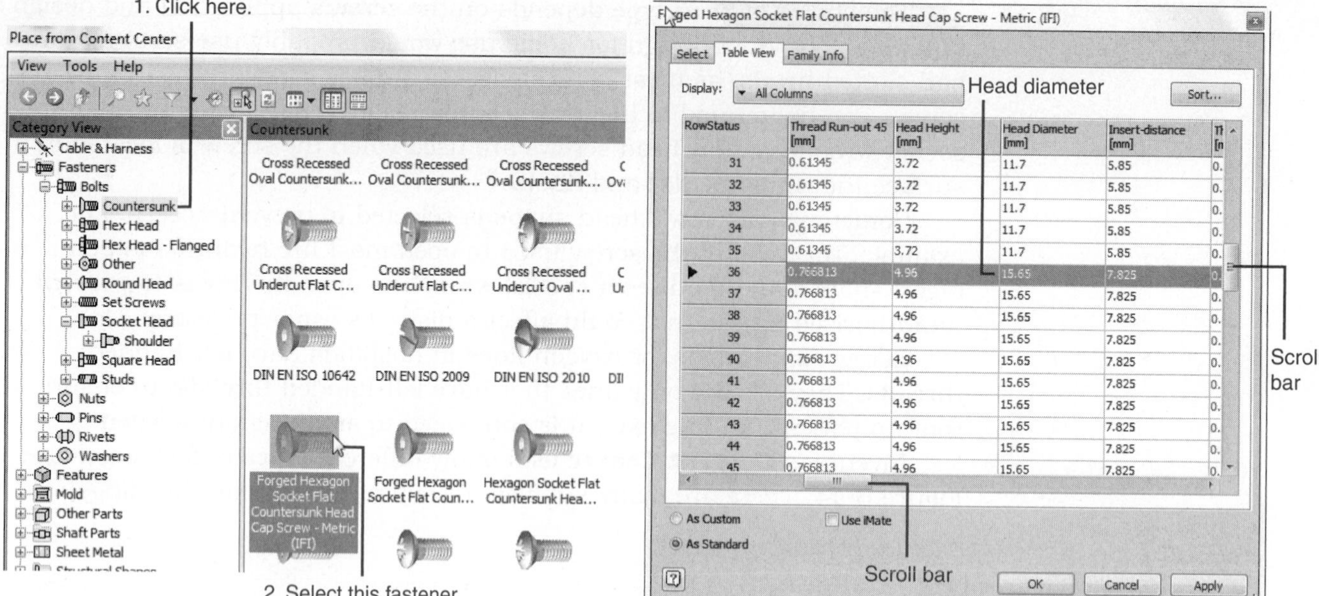

Figure 6-45

The hole in the block must be sized to accept the M8 × 50 screw. This information is available from the **Table View** option of the **Place from Content Center** dialog box. See Figure 6-45. For example, the table defines the screw's head diameter. This value will be used to size the countersunk hole in the 40 × 40 × 80 block. In this example the head diameter is 15.65 mm.

EXERCISE 6-12 Creating a Countersunk Hole

1. Create a **40 × 40 × 80** block.
2. Locate a hole's center point in the center (20 × 20) of the top surface of the block using the **Point, Center Point** tool.
3. Go to the **3D Model** tab and click the **Hole** tool.
4. Click the **Countersink** and **Tapped** boxes.

   See Figure 6-46.

   The pitch length of an M8 thread is 1.25, so two thread lengths (2P) equals 2.50.

   The hole's threads must be at least 50.00 + 2.50 = 52.50, and the pilot hole must be at least 52.50 + 2.50 = 55.00 deep.
5. Click the **Full Depth** box (remove the check mark), set the **Termination** option to **Distance**, set the **Thread Type** option for **ANSI Metric M Profile,** the thread depth for **52.5,** the hole's depth for **55,** and the head diameter for **15.65.**

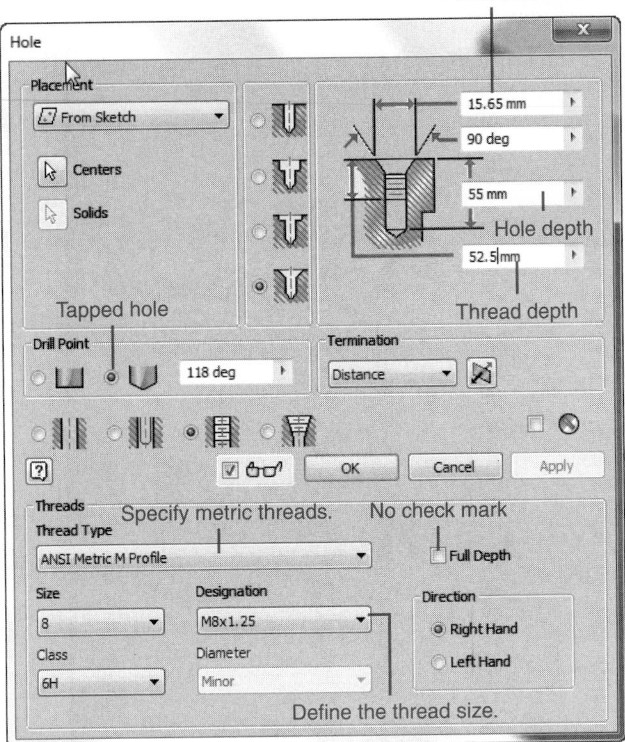

Figure 6-46

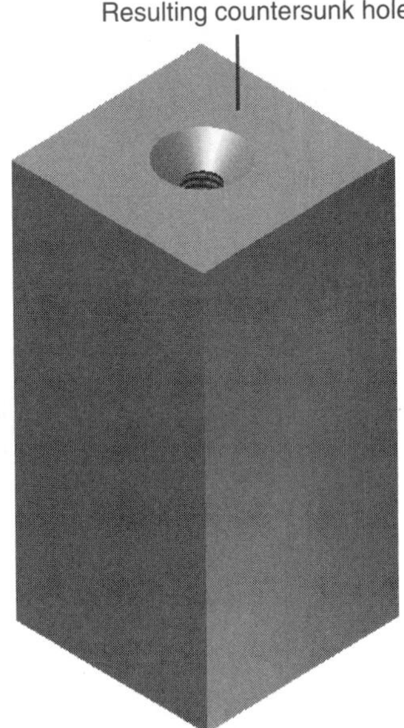

Figure 6-47

The value 15.65 came from the **Table View** portion of the **Place from Content Center** dialog box. Figure 6-47 shows the countersunk hole located in the 40 × 40 × 80 block.

6 Save the block as **BLOCK, COUNTERSINK.**

EXERCISE 6-13 Creating and Inserting a Flat Head Screw

1 Create an assembly drawing using the **Standard (mm).iam** format.

2 Use the **Place Component** tool and locate one copy of the BLOCK, COUNTERSINK on the drawing screen.

3 Use the **Save As** command to save and name the assembly.

In this example the name **COUNTERSINK ASSEMBLY** was used.

4 Click the **Place from Content Center** tool.

5 Select the **Fasteners, Bolts** option, then **Countersunk.**

6 Select the **Forged Hexagon Socket Flat Countersunk Head Cap Screw—Metric** and set the **Thread description** for **M8** and the **Nominal Length** for **50.**

See Figure 6-48.

7 Click the **Insert** box.

Figure 6-48

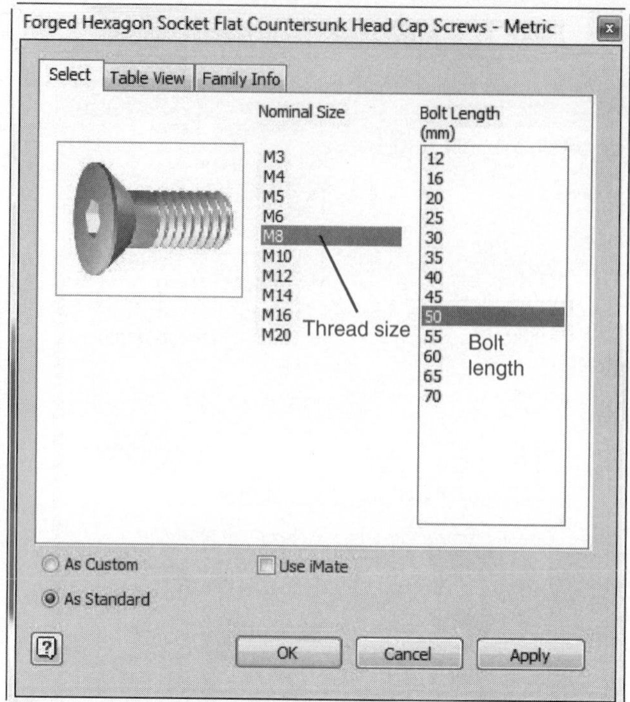

8 Click the **Constrain** tool, then select the **Insert** option on the **Place Constraint** dialog box.

9 Insert the screw into the block.

Figure 6-49 shows the resulting assembly. Figure 6-50 shows a top and a section view of the countersunk screw inserted into the block. Note that the portion of the hole below the bottom of the M8 × 50 screw is

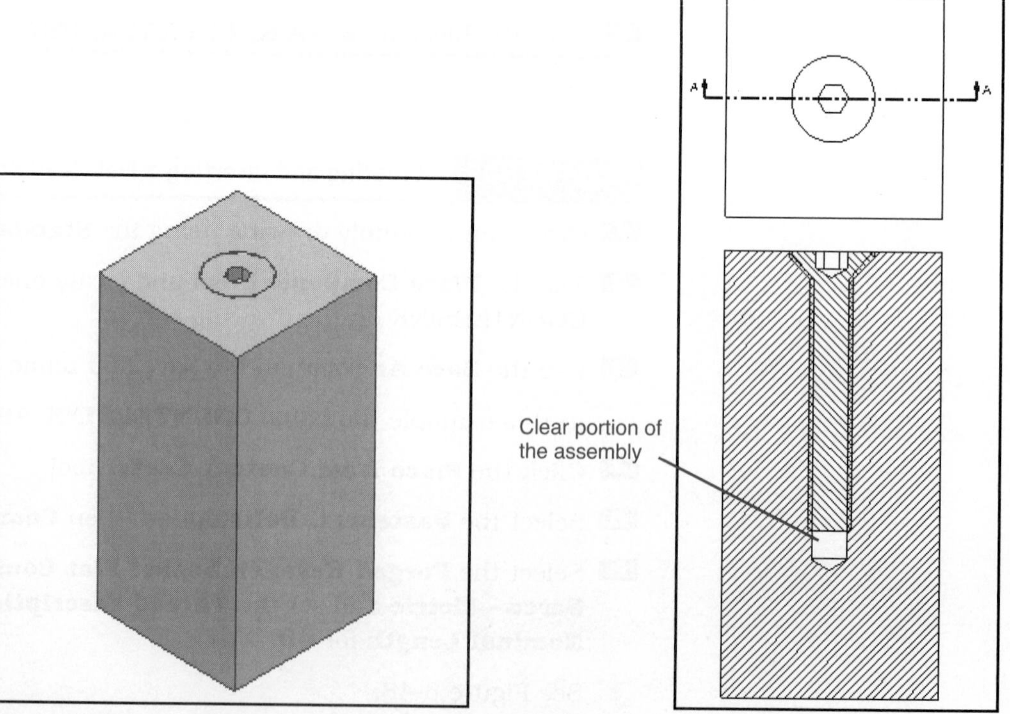

Figure 6-49

Figure 6-50

clear; that is, it does not show the unused threads. This is a drawing convention that is intended to add clarity to the drawing. Inventor will automatically omit the unused threads.

Counterbores

A counterbored hole is created by first drilling a hole, then drilling a second larger hole aligned with the first. Counterbored holes are often used to recess the heads of fasteners.

Say we wish to fit a 3/8-16 UNC × 1.50 LONG hex head screw into a block that includes a counterbored hole, and that after assembly the head of the screw is to be below the surface of the block.

Determining the Counterbore Depth

It is necessary to access the **Content Center** to find the head height of the bolt to be able to determine the depth of the **counterbore**.

EXERCISE 6-14 Determining the Head Height of a Bolt

1. Create a new assembly drawing and click the **Place from Content Center** tool.
2. Click **Fasteners, Bolts,** and **Hex Head** to access the hex head bolt listing.
3. Select the **Hex Bolt - Inch** bolt.

   The **Hex Bolt - Inch** dialog box will appear. See Figure 6-51.

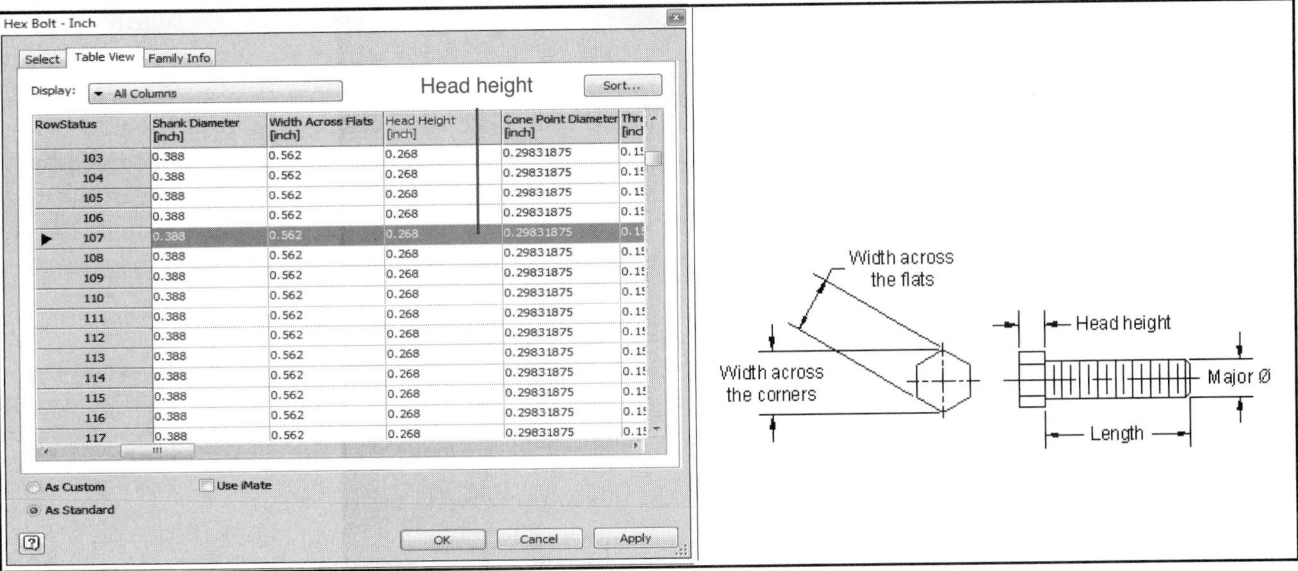

RowStatus	Shank Diameter [inch]	Width Across Flats [inch]	Head Height [inch]	Cone Point Diameter [inch]	Thr [ind
103	0.388	0.562	0.268	0.29831875	0.1
104	0.388	0.562	0.268	0.29831875	0.1
105	0.388	0.562	0.268	0.29831875	0.1
106	0.388	0.562	0.268	0.29831875	0.1
▶ 107	0.388	0.562	0.268	0.29831875	0.1
108	0.388	0.562	0.268	0.29831875	0.1
109	0.388	0.562	0.268	0.29831875	0.1
110	0.388	0.562	0.268	0.29831875	0.1
111	0.388	0.562	0.268	0.29831875	0.1
112	0.388	0.562	0.268	0.29831875	0.1
113	0.388	0.562	0.268	0.29831875	0.1
114	0.388	0.562	0.268	0.29831875	0.1
115	0.388	0.562	0.268	0.29831875	0.1
116	0.388	0.562	0.268	0.29831875	0.1
117	0.388	0.562	0.268	0.29831875	0.1

Figure 6-51

4. Select the **3/8-16 UNC × 1.50 LONG** bolt.
5. Click the **Table View** tab.

See Figure 6-43 for reference.

In this example the head height is 0.268 in. (See Figure 6-51.) The **counterbore** must have a depth greater than the head height. A distance of 0.313 (5/16) was selected.

Determining the Thread Length

The screw is 1.50 in. long and has a pitch of 16 threads per inch. Each thread is therefore 1/16, or 0.0625 in. It is recommended that there be at least two threads beyond the end of the screw. Two thread pitches would be 2(0.0625) = 0.125 in. This is a minimum recommendation. A larger value could also be used.

The thread depth is 1.50 + 0.125, or 1.625 in. minimum; however, the thread is created below the counterbore, so in Inventor the value must include the depth of the counterbore. The thread depth is 1.625 + 0.313 = 1.938 in.

Determining the Depth of the Hole

The hole should extend at least two pitch lengths beyond the threaded portion of the hole, plus the depth of the counterbore, so 1.938 + 0.125 = 2.063 in.

Determining the Counterbore's Diameter

The distance across the corners of the screw is listed in the **Table View** portion of the **Place from Content Center** dialog box as 0.650 in. The counterbored hole must be at least this large plus have an allowance for the tool (socket wrench) needed to assemble and disassemble the screw. In general, the diameter is increased 0.125 in., or 1/8 in. [3.2 mm], all around to allow for tooling needs.

If space is a concern, then designers will change the head type of a screw so that the tooling will not add to the required diameter. For example, a socket head type screw may be used if design requirements permit.

The minimum counterbore diameter is 0.650 + 0.125 = 0.775. The 0.650 value was determined using the .649 distance across the corners value that was found in the **Table View** in the **Hex Bolt-Inch** dialog box. The 0.125 value is a tool allowance added so a socket wrench can access the bolt head. For this example 0.8125, or 13/16 in., was selected. We can now create the component.

EXERCISE 6-15 **Drawing a Counterbored Hole**

1. Draw a **3.00** × **3.00** × **5.00** block using the **Standard (in).ipt** format.
2. Locate a hole's center point in the center of the 3.00 × 3.00 surface.
3. Click the **3D Model** tab and click the **Hole** tool.

   The **Hole** dialog box will appear.
4. Click the **Counterbore, Tapped** tools, and turn off the **Full Depth** option.
5. Set the hole depth for **2.063,** the thread depth for **1.938,** the **counterbore** diameter for **.8125,** and the **counterbore** depth for **.313.**
6. Select **3/8-16 UNC** threads.

   See Figure 6-52.
7. Click **OK,** then save the block as **BLOCK, COUNTERBORE.**

Figure 6-52

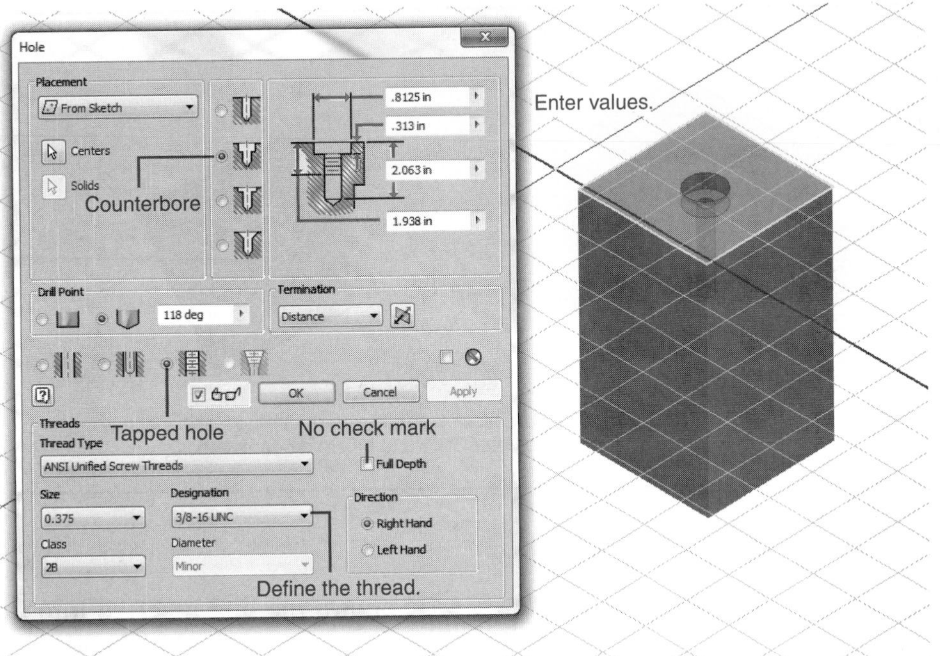

EXERCISE 6-16 Assembling the Screw

1. Create an assembly drawing using the **Standard (in).iam** format.
2. Use the **Place Component** tool to place a copy of the **counterbored** block on the screen.
3. Use the **Save As** command to save and name the assembly **COUNTERBORE ASSEMBLY.**
4. Click the **Place from Content Center** tool and select a **3/8-16 UNC** × **1.50** hex head bolt and insert it onto the drawing.

   See Figure 6-53.
5. Use the **Place Constraint** tool to insert the bolt into the hole.

Figure 6-53

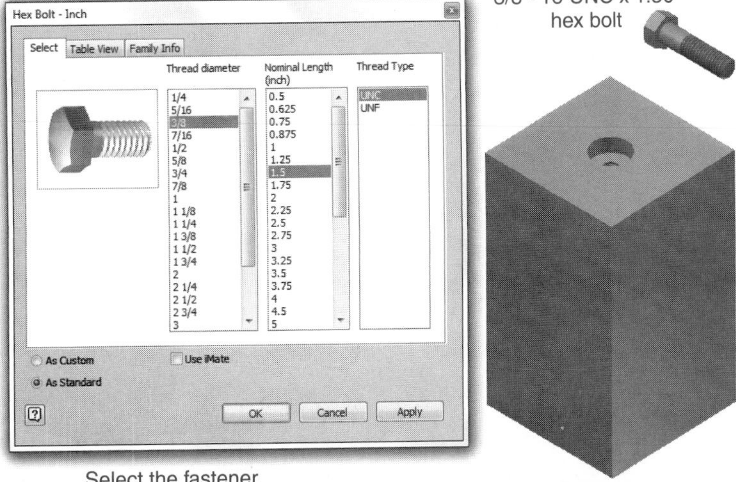

Select the fastener.

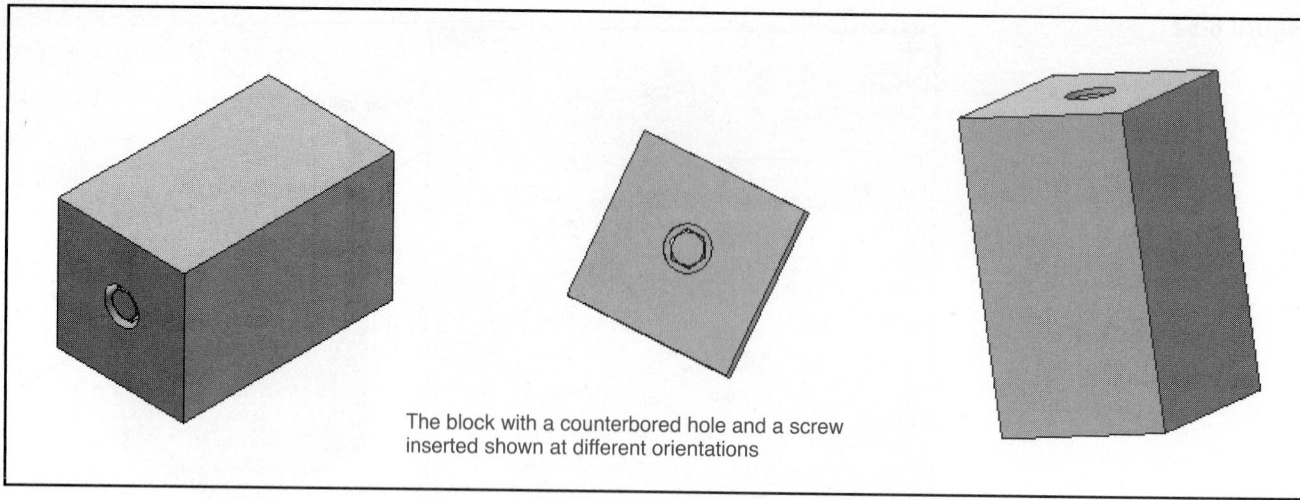

The block with a counterbored hole and a screw inserted shown at different orientations

Figure 6-54

Figure 6-54 shows the bolt inserted into the counterbored hole. Note the tooling clearance around the hex head and the clearance between the top of the bolt and the top surface of the block.

Drawing Fasteners Not Included in the Content Center

The **Content Center** contains a partial listing of fasteners. There are many other sizes and styles available. Inventor can be used to draw specific fasteners that can then be saved and used on assemblies.

Say we wish to draw an M8 × 25 hex head screw and that this size is not available in the **Content Center.**

EXERCISE 6-17 Drawing an M8 × 25 Hex Head Screw

1. Create a **Standard (mm).ipt** drawing.
2. Draw a **Ø8 × 25** cylinder. Draw the cylinder with its top surface on the XY plane so that it extends in the negative Z direction.

   See Figure 6-55.

Figure 6-55

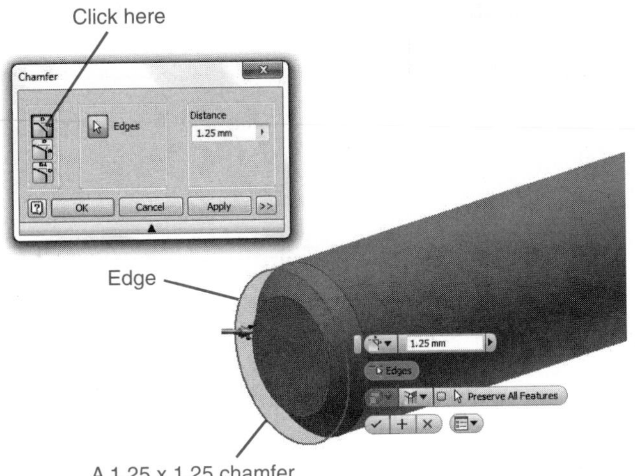

Figure 6-56

3 Add a **1.25 × 1.25** chamfer to the bottom of the cylinder.

See Figure 6-56. In general, the chamfer will approximately equal one pitch length. The **Chamfer** tool is located on the **Modify** panel under the **3D Model** tab.

4 Create a new sketch plane on the top surface of the cylinder.

5 Use the **Polygon** tool and draw a hexagon centered on the top surface of the cylinder. Sketch the hexagon **12 mm** across the flats (see Chapter 2) and extrude it to a height of **5.4 mm** above the XY plane.

See Figure 6-57.

The head height and distance across the flats were determined using the general values defined in Figure 6-43. Specific values may be obtained from manufacturers, many of whom list their products on the Web, or from reference books such as *Machinery's Handbook.*

6 Add threads to the cylinder, using the **Thread** tool. Click the cylinder, and Inventor will automatically create threads to match the cylinder's diameter.

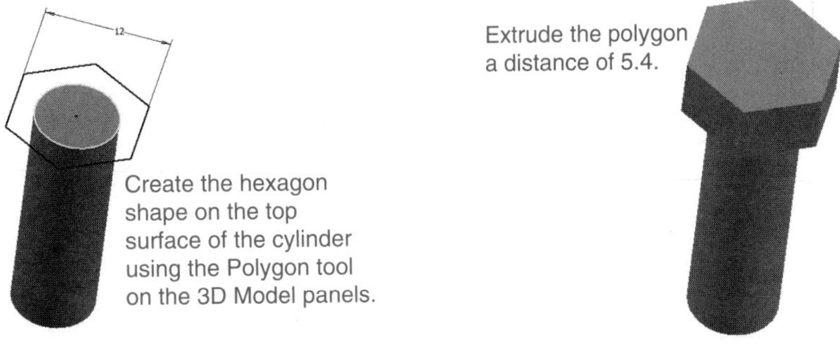

12
Create the hexagon
shape on the top
surface of the cylinder
using the Polygon tool
on the 3D Model panels.
Extrude the polygon
a distance of 5.4.

Figure 6-57

See Figure 6-58. Check the thread specification to assure that correct threads were created. Note that the coarse pitch of 1.25 was automatically selected, but other pitch values are also available.

7 Save the drawing as **M8 × 1.25 × 25 HEXHEADSCREW.**

Figure 6-58

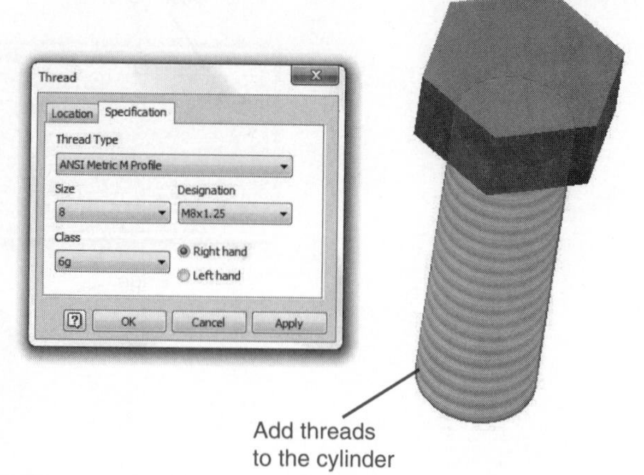

Sample Problem SP6-1

Nuts are used with externally threaded objects to hold parts together. There are many different styles of nuts. The **Content Center** library includes listings for hex, slotted hex, and wing nuts, among others. Figure 6-59 shows the **Table View** portion of the **Content Center** dialog box that includes nut heights.

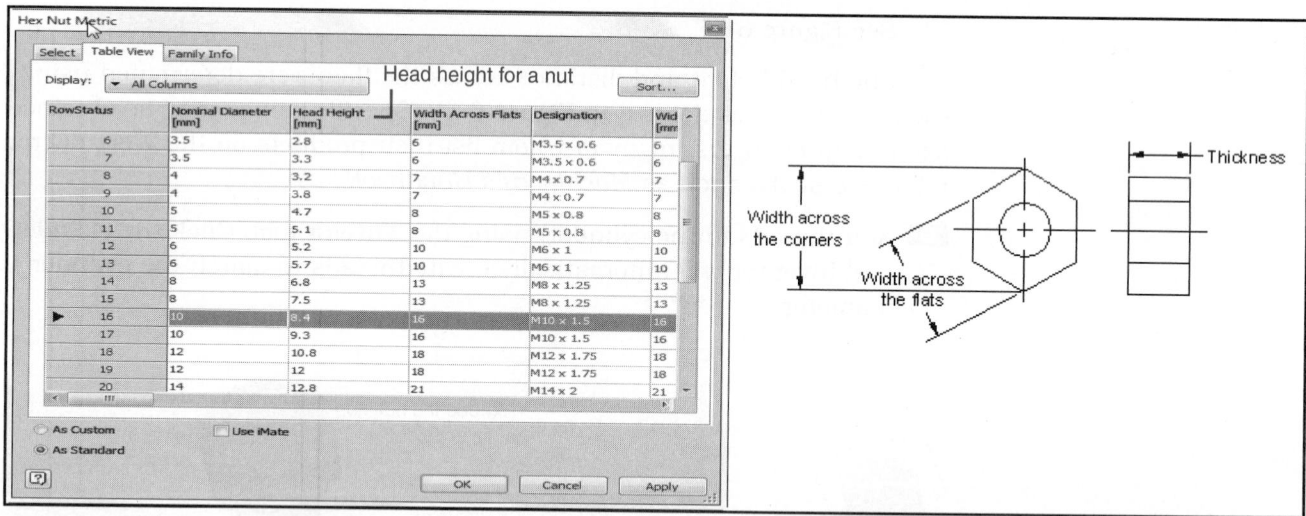

Figure 6-59

The threads of a nut must be exactly the same as the external threads inserted into them. For example, if a screw with an M8 × 1.25 thread is selected, an M8 × 1.25 nut thread must be selected.

The head height of the nut must be considered when determining the length of a bolt. It is good practice to have a minimum of two threads extend beyond the nut to help ensure that the nut is fully secured. Strength calculations are based on all nut threads' being 100% engaged, so having threads extend beyond a nut is critical.

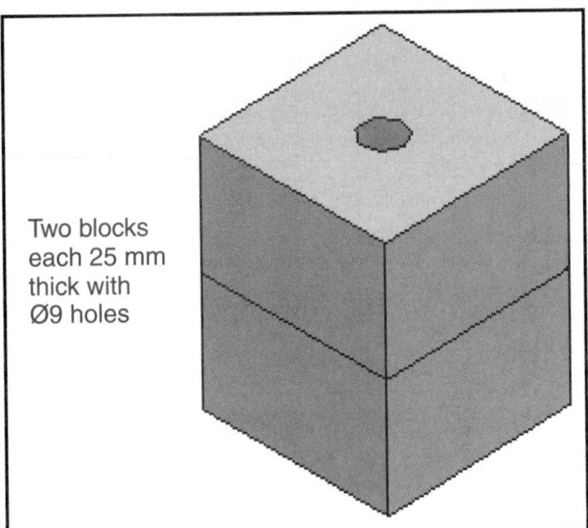

Figure 6-60

Figure 6-60 shows two blocks, each 25 mm thick with a center hole of ∅9.00 mm. The holes are clearance holes and do not include threads. The blocks are to be held together using an M8 hex head screw and a compatible nut.

Determining the Minimum Thread Length Required

Each block is 25 mm thick, for a total of 50 mm. The nut height, from Figure 6-59, for an M8 hex nut is 6.80, so the minimum thread length that will pass through both parts and the nut is 56.80 mm. Two threads must extend beyond the nut to ensure that it is fully secured. From Figure 6-59 the length of an M8 thread is given as 1.25 mm, so two threads equal 2.50 mm. Therefore, the minimum thread length must be 50.00 + 6.80 + 2.50 = 59.3 mm.

Bolts are manufactured in standard lengths, some of which are listed in the **Content Center** library. If the required thread length was not available from the library, manufacturers' catalogs would have to be searched and a new screw drawing created. In this example the next available standard length is 60 mm.

EXERCISE 6-18 Selecting a Screw

1. Click the **Place from Content Center** tool.
2. Select the **Fasteners** option, then **Bolts,** then **Hex Head.** Select the **Hex Bolt - Metric.**

   See Figure 6-61. The standard thread length that is closest to, but still greater than, 59.3 is 60 mm. The 60 mm length is selected and applied to the drawing.
3. Define the values for an **M8** × **60** hex head screw and insert it into the drawing.

   See Figure 6-62.
4. Insert the screw into the two assembled parts.

   See Figure 6-63. Note that the screw extends beyond the bottom of the two assembled parts.

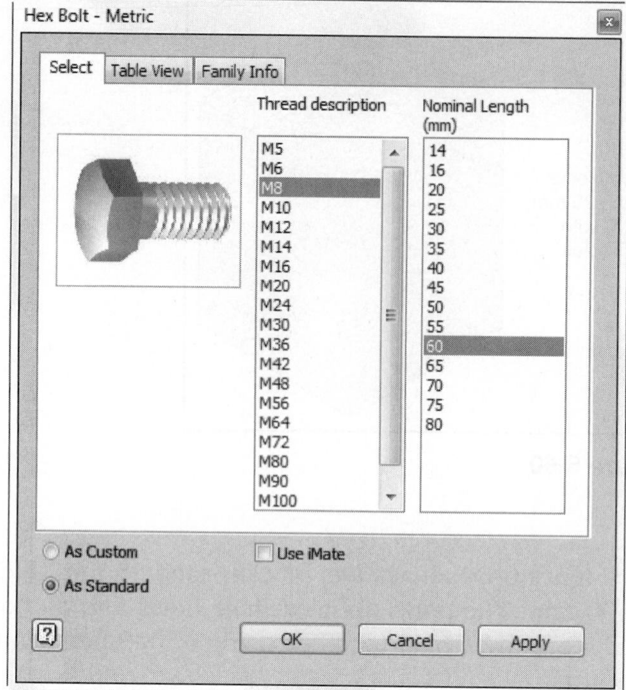

Figure 6-61

An M8 x 60 hex head screw

Figure 6-62

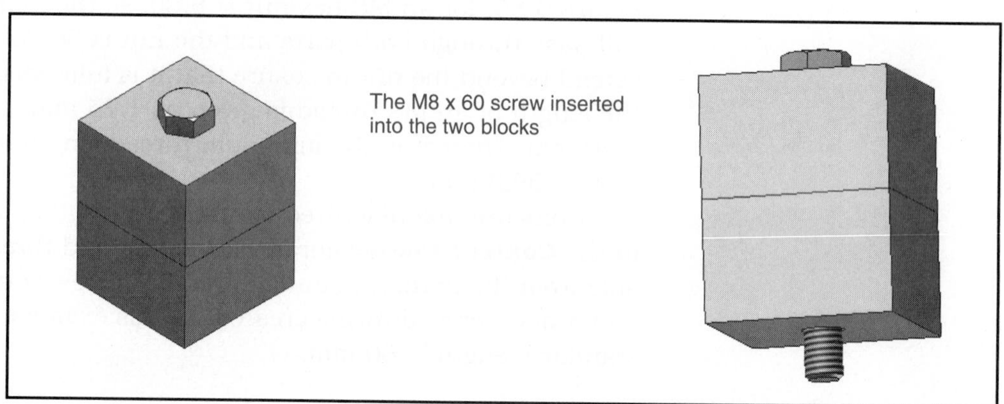

The M8 x 60 screw inserted into the two blocks

Figure 6-63

EXERCISE 6-19 Selecting a Nut

1. Click the **Place from Content Center** and select the **Fasteners** option, then **Nuts,** then **Hex,** then the **Hex Nut- Metric** listing.

   See Figure 6-64.

2. Insert a copy of the nut into the drawing area.

   Figure 6-65 shows the nut added to the drawing screen.

3. Use the **Place Constraint** tool and insert the nut onto the screw so that it is flush with the bottom surface of the blocks.

   Figure 6-66 shows the nut inserted onto the screw.

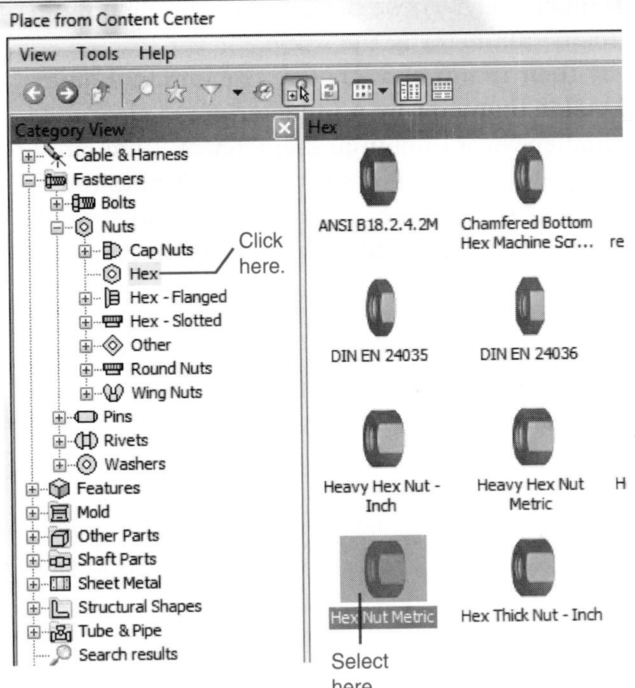

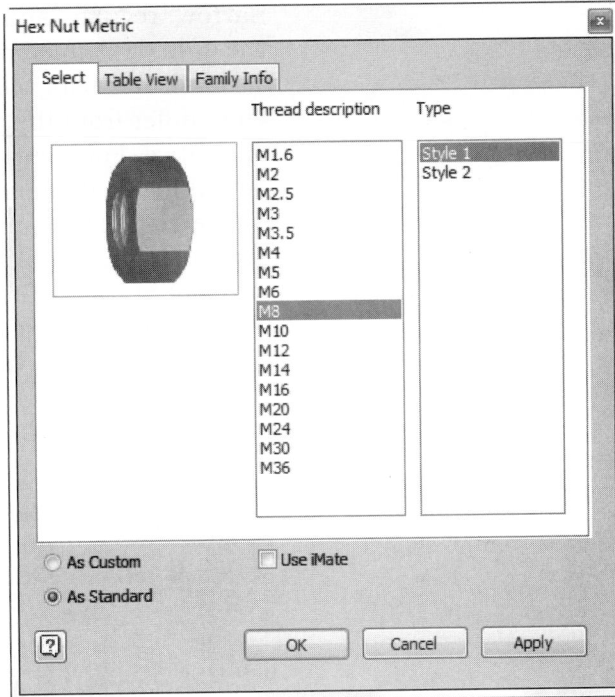

Figure 6-64

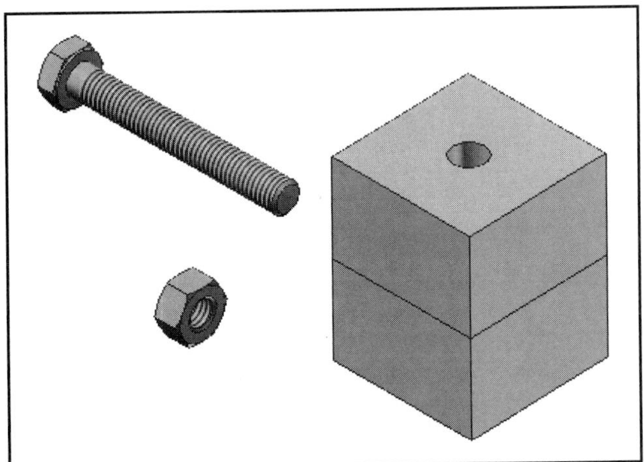

Figure 6-65

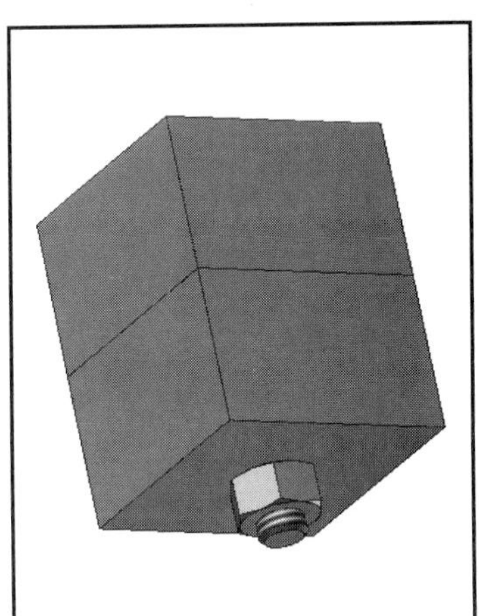

Figure 6-66

The nuts listed in the **Content Center** library represent only a partial listing of the sizes and styles of nuts available. If a design calls for a nut size or type not listed in the **Content Center** library, refer to manufacturers' specifications, then draw the nut and save it as an individual drawing. It can then be added to the design drawings as needed.

Washers

washer: A flat thin ring used to increase the bearing area under a fastener or as a spacer.

Washers are used to increase the bearing area under fasteners or as spacers. Washers are identified by their inside diameter, outside diameter, and thickness. In addition, washers can be designated N, R, or W for narrow, regular, and wide, respectively. These designations apply only to the outside diameters; the inside diameter is the same.

Inventor lists washers by their nominal diameter. The nominal diameters differ from the actual inside diameter by a predetermined clearance allowance. For example, a washer with a nominal diameter of 8 has an actual inside diameter of 8.40, or 0.40 mm greater than the 8 nominal diameter. This means that washer sizes can easily be matched to thread sizes using nominal sizes. A washer with a nominal diameter designation of 8 will fit over a thread designated M8.

There are different types of washers including, among others, plain and tapered. The **Place from Content Center** dialog box includes a listing of both plain and tapered washers. Figure 6-67 shows the **Table View** portion of the **Place from Content Center** dialog box for a plain washer ISO 7089.

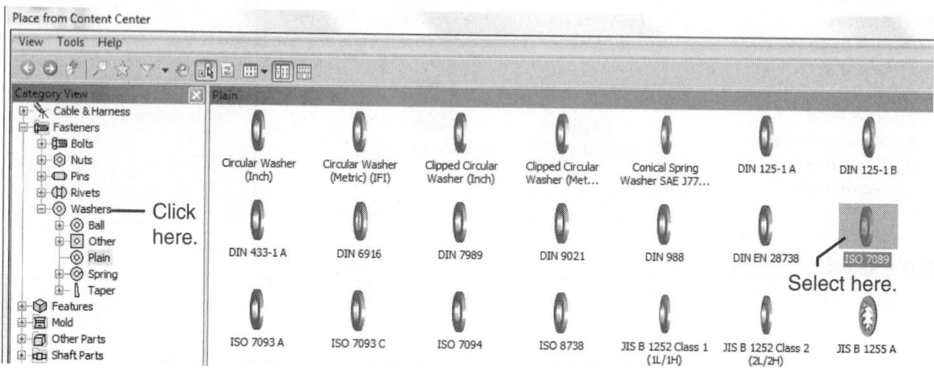

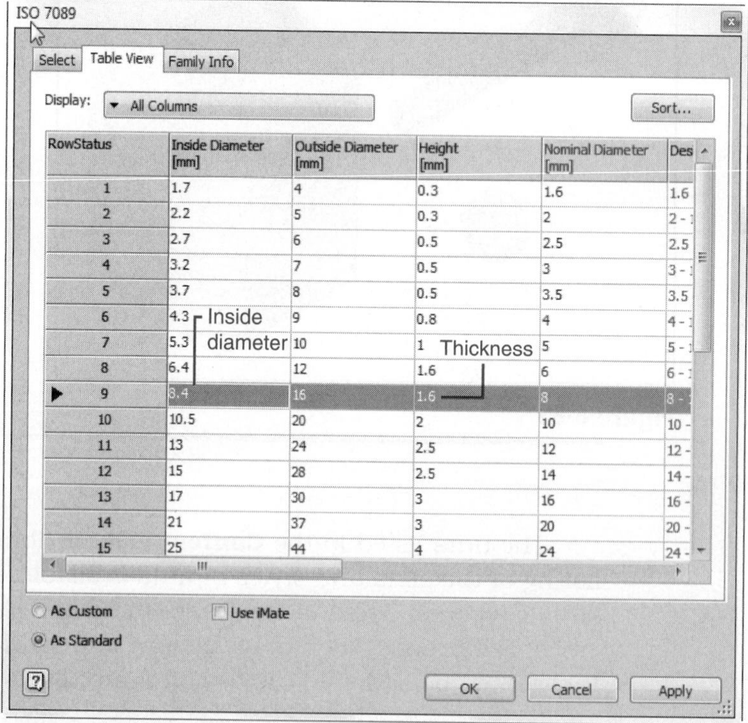

RowStatus	Inside Diameter [mm]	Outside Diameter [mm]	Height [mm]	Nominal Diameter [mm]	Des
1	1.7	4	0.3	1.6	1.6
2	2.2	5	0.3	2	2 -
3	2.7	6	0.5	2.5	2.5
4	3.2	7	0.5	3	3 -
5	3.7	8	0.5	3.5	3.5
6	4.3	9	0.8	4	4 -
7	5.3	10	1	5	5 -
8	6.4	12	1.6	6	6 -
9	8.4	16	1.6	8	8 -
10	10.5	20	2	10	10 -
11	13	24	2.5	12	12 -
12	15	28	2.5	14	14 -
13	17	30	3	16	16 -
14	21	37	3	20	20 -
15	25	44	4	24	24 -

Figure 6-67

Inserting Washers onto a Fastener

We again start with the two blocks shown in Figure 6-60. Each has a height of 25 mm. We know from the previous section that the nut height is

6.80 mm and that the requirement that at least two threads extend beyond the end of the nut adds 2.50 mm, yielding a total thread length requirement of 59.3. We now have to add the thickness of the two washers and recalculate the minimum required bolt length.

Say we selected a plain regular washer number ISO 7089 with a nominal size of 8. From Figure 6-67 the **Table View** on the **1SO 7089** dialog box the thickness is found to be 1.6 mm, or a total of 3.20 mm for the two washers. This extends the minimum bolt length requirement to 59.3 + 3.20 = 62.5. The nearest standard thread bolt length listed in the **Place from Content Center** dialog box that is greater than the 62.5 requirement is 65.

EXERCISE 6-20 Adding Washers to an Assembly

1 Use the **Place Component** tool and locate two copies of the block on the drawing, then align the blocks.

2 Use the **Save As** command to save and name the assembly.

3 Click the **Place from Content Center** tool and select an **M8 × 65 Hex Head** screw and an **M8 Hex Nut** and insert them into the drawing.

See Figure 6-68.

4 Access the **Place from Content Center** dialog box and select the **Washers** option, then **Plain,** then an **ISO 7089** washer.

See Figure 6-69. Figure 6-67 shows the **Table View** values for the washer.

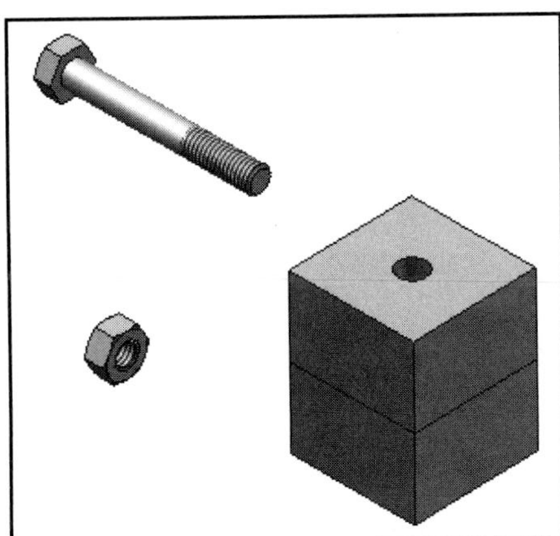

Figure 6-68

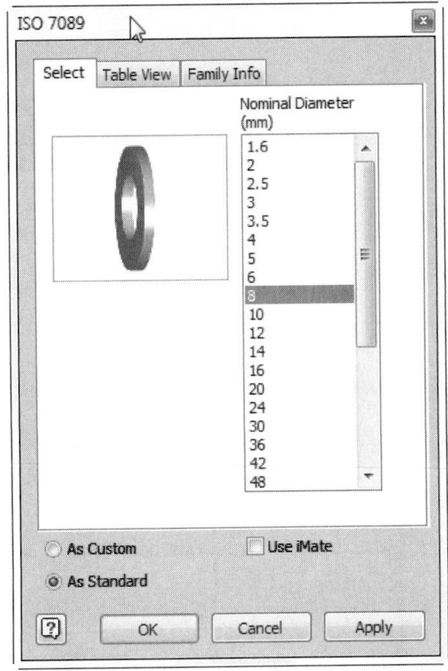

Figure 6-69

5 Insert two copies into the drawing.

See Figure 6-70.

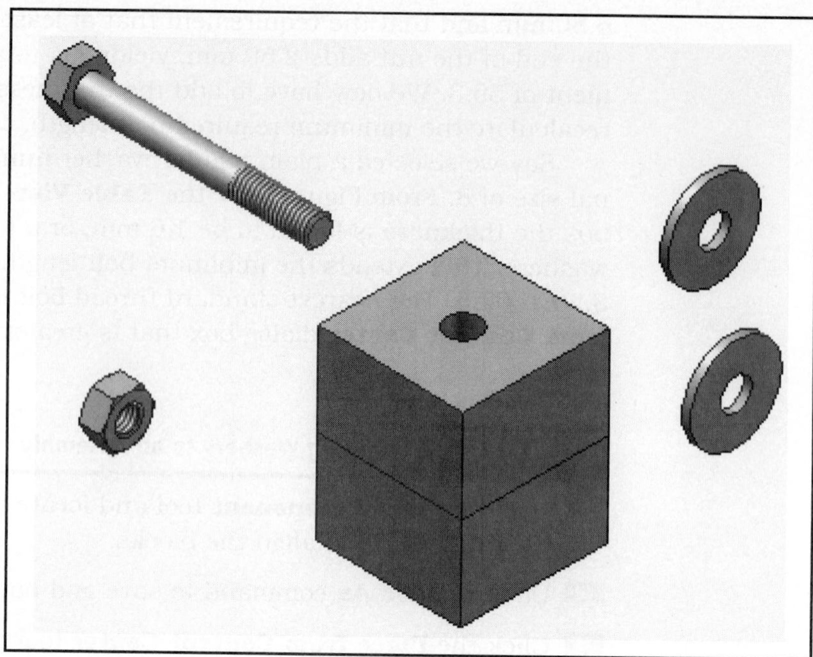

Figure 6-70

6 Use the **Place Constraint, Insert** option tool and align the washers with the holes in the blocks.
See Figure 6-71.

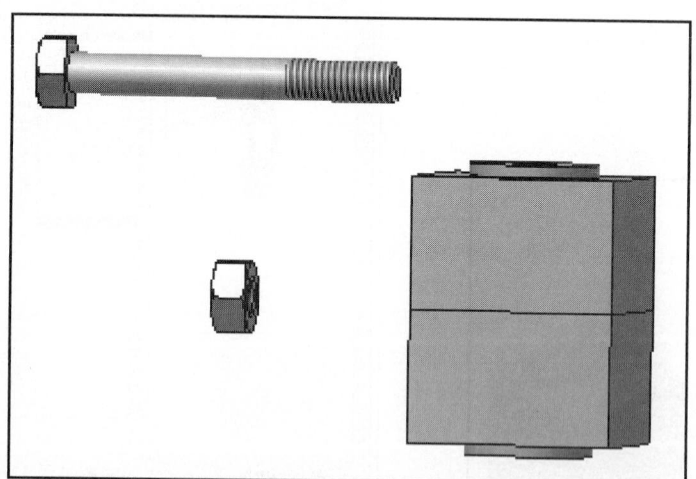

Figure 6-71

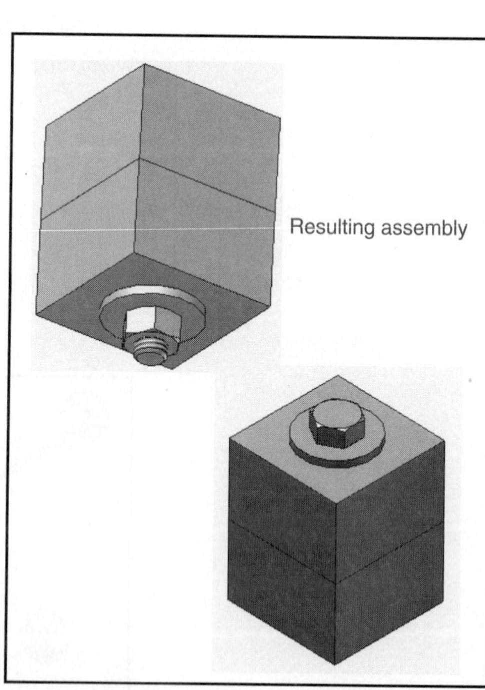

Figure 6-72

7 Use the **Place Constraint** tool and insert the M8 × 65 screw and the nut.
Figure 6-72 shows the resulting assembly.

The washers listed in the **Content Center** dialog box library represent only a partial listing of the washers available. If a design calls for a washer size or type not listed in the library, refer to manufacturers' specifications, then draw the washer and save it as an individual drawing. It can then be added to the design drawings as needed.

Setscrews

setscrew: A fastener used to hold parts to rotating shafts or other moving objects to prevent slippage between the two objects.

Setscrews are fasteners used to hold parts like gears and pulleys to rotating shafts or other moving objects to prevent slippage between the two objects. See Figure 6-73.

Most setscrews have recessed heads to help prevent interference with other parts.

Many different head styles and point styles are available. See Figure 6-74. The dimensions shown in Figure 6-74 are general sizes for use in this book. For actual sizes, see the manufacturer's specifications.

Figure 6-73

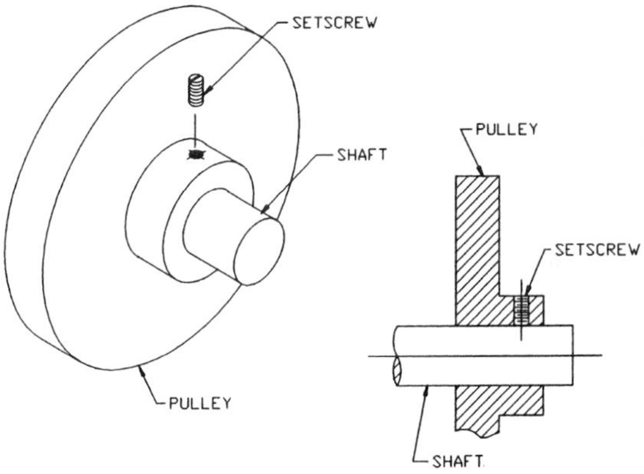

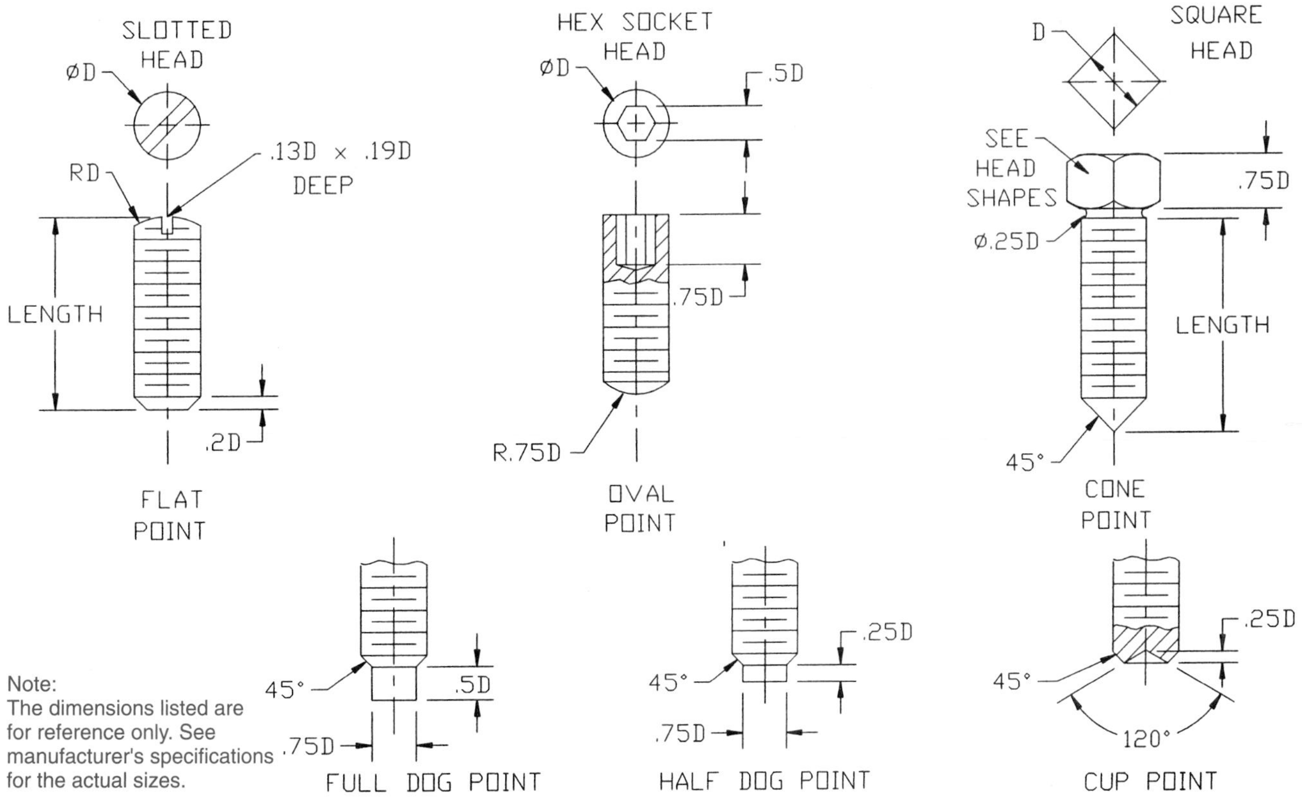

Note:
The dimensions listed are for reference only. See manufacturer's specifications for the actual sizes.

Figure 6-74

A collar with two #10(.19)-32UNF threaded holes

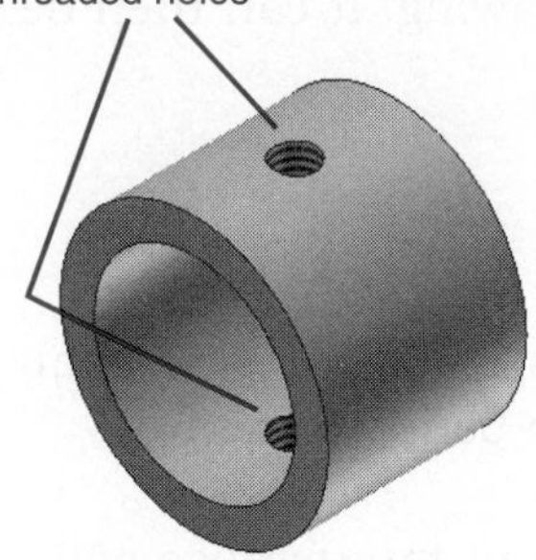

Collar I.D. = .750
O.D. = 1.000
Length = .750

Figure 6-75

Figure 6-75 shows a collar with two 10(.19)-32UNF threaded holes.

EXERCISE 6-21 Adding a Setscrew

1. Create an assembly drawing using the **Standard (in).iam** format.
2. Use the **Place Component** tool and place one copy of the collar on the drawing.
3. Use the **Save As** tool to save and name the assembly.
4. Access the **Place from Content Center** dialog box and select the **Fasteners** option, then **Bolts,** then **Set Screws.**
5. Select the **Type E - Spline Socket Set Screw - Cup Point - Inch** option and set the (nominal diameter) **Thread description** for **#10,** the **Nominal Length** for **0.3125,** and the **Thread Type** for **UNF.**

See Figure 6-76.

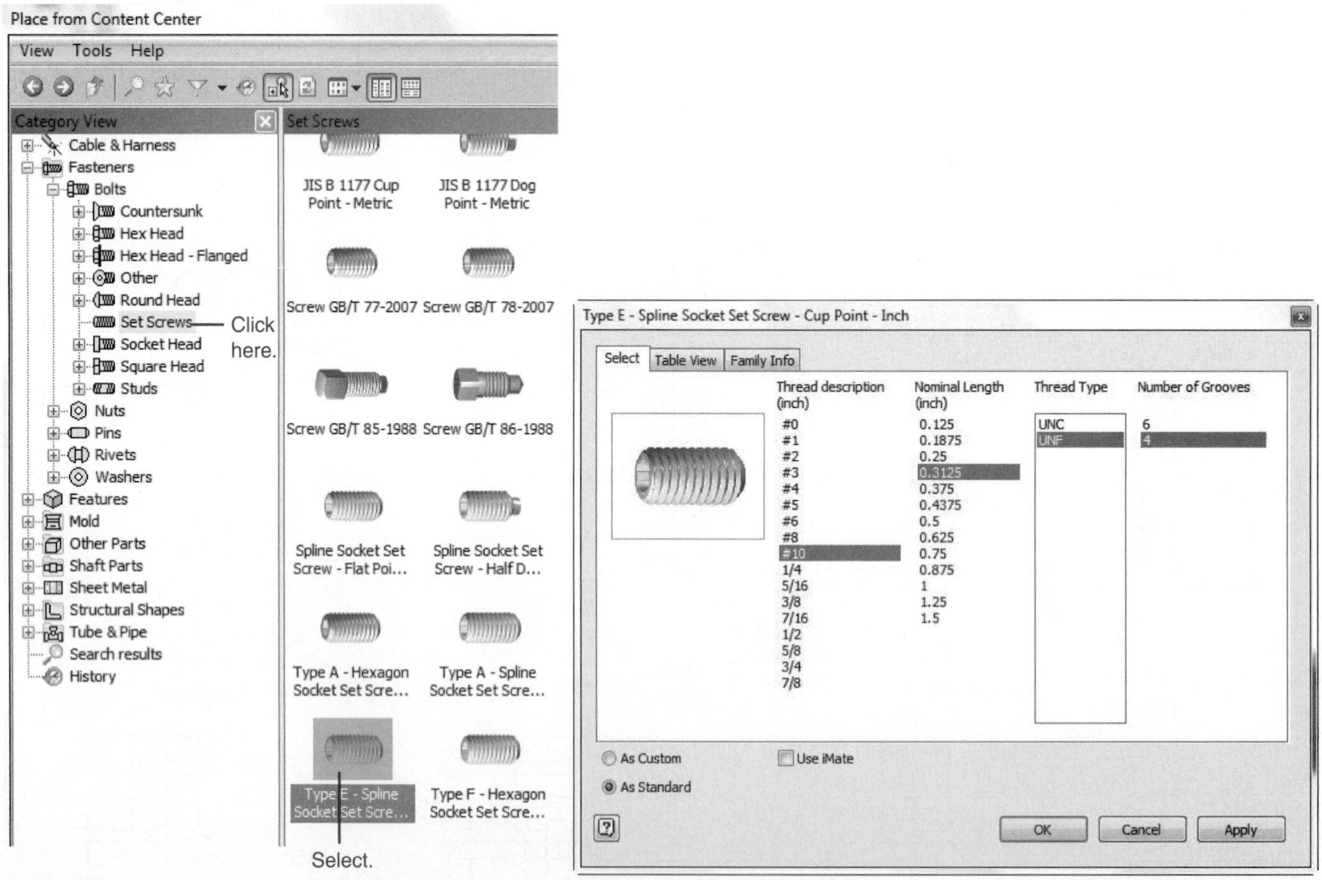

Click here.

Select.

Figure 6-76

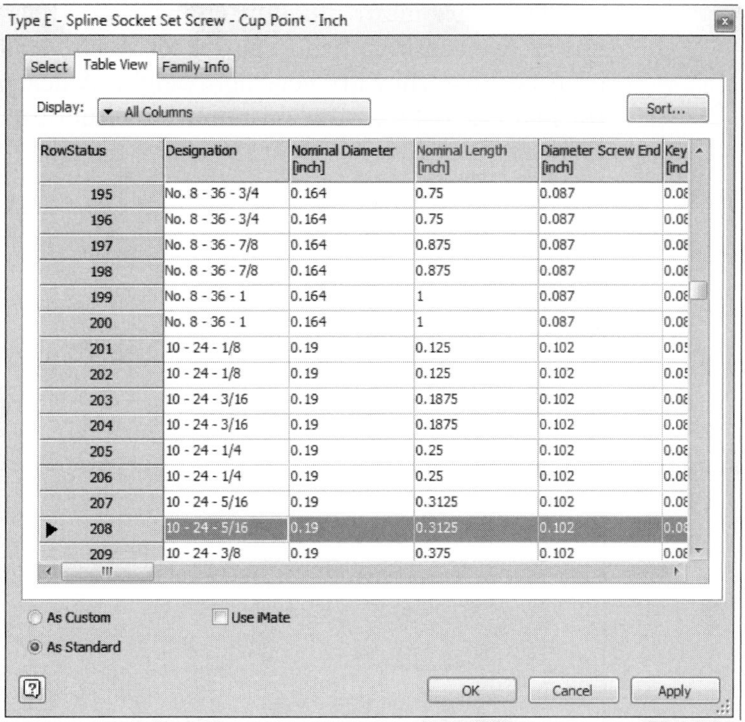

RowStatus	Designation	Nominal Diameter [inch]	Nominal Length [inch]	Diameter Screw End [inch]	Key [ind
195	No. 8 - 36 - 3/4	0.164	0.75	0.087	0.08
196	No. 8 - 36 - 3/4	0.164	0.75	0.087	0.08
197	No. 8 - 36 - 7/8	0.164	0.875	0.087	0.08
198	No. 8 - 36 - 7/8	0.164	0.875	0.087	0.08
199	No. 8 - 36 - 1	0.164	1	0.087	0.08
200	No. 8 - 36 - 1	0.164	1	0.087	0.08
201	10 - 24 - 1/8	0.19	0.125	0.102	0.05
202	10 - 24 - 1/8	0.19	0.125	0.102	0.05
203	10 - 24 - 3/16	0.19	0.1875	0.102	0.08
204	10 - 24 - 3/16	0.19	0.1875	0.102	0.08
205	10 - 24 - 1/4	0.19	0.25	0.102	0.08
206	10 - 24 - 1/4	0.19	0.25	0.102	0.08
207	10 - 24 - 5/16	0.19	0.3125	0.102	0.08
▶ 208	10 - 24 - 5/16	0.19	0.3125	0.102	0.08
209	10 - 24 - 3/8	0.19	0.375	0.102	0.08

Figure 6-76

(Continued)

6 Insert a copy of the setscrew on the drawing.

See Figure 6-77.

7 Use the **Place Constraint** tool and insert the setscrew into one of the threaded holes.

Figure 6-78 shows the resulting assembly.

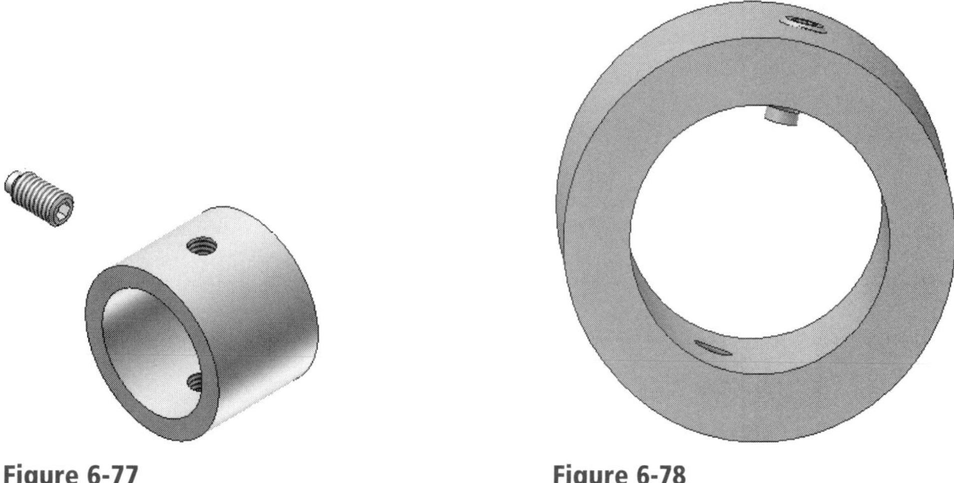

Figure 6-77 **Figure 6-78**

Rivets

rivet: A metal fastener with a head and a straight shaft for holding together overlapping and adjoining objects.

Rivets are fasteners that hold together adjoining or overlapping objects. A rivet starts with a head at one end and a straight shaft at the other end. The rivet is then inserted into the object, and the headless end is "bucked" or otherwise forced into place. A force is applied to the headless end that changes its shape so that another head is formed holding the objects together.

There are many different shapes and styles of rivets. Figure 6-79 shows five common head shapes for rivets. Hollow rivets are used on aircraft because they are very lightweight. A design advantage of rivets is that they can be drilled out and removed without damage to the objects they hold together.

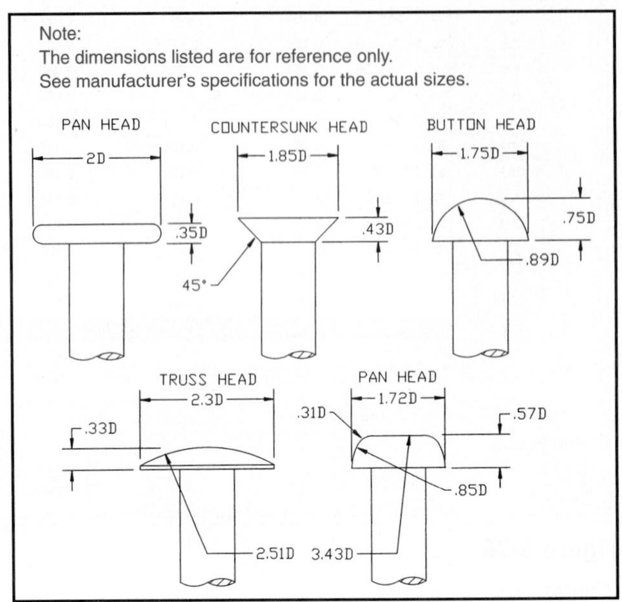

Figure 6-79

Rivet types are represented on technical drawings using a coding system. See Figure 6-80. Since rivets are sometimes so small and the material they hold together so thin that it is difficult to clearly draw the rivets, some companies draw only the rivet's centerline in the side view and identify the rivets using a drawing callout.

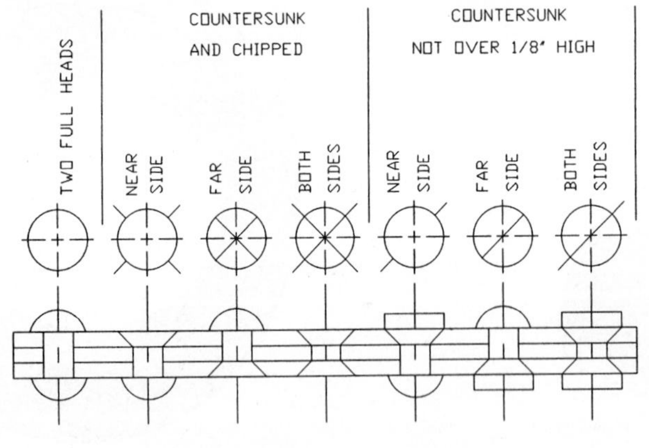

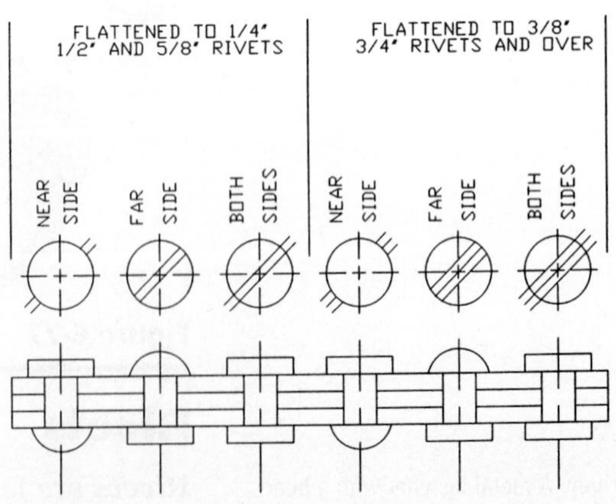

Figure 6-80

Figure 6-81 shows the **Place from Content Center** dialog box for plain rivets.

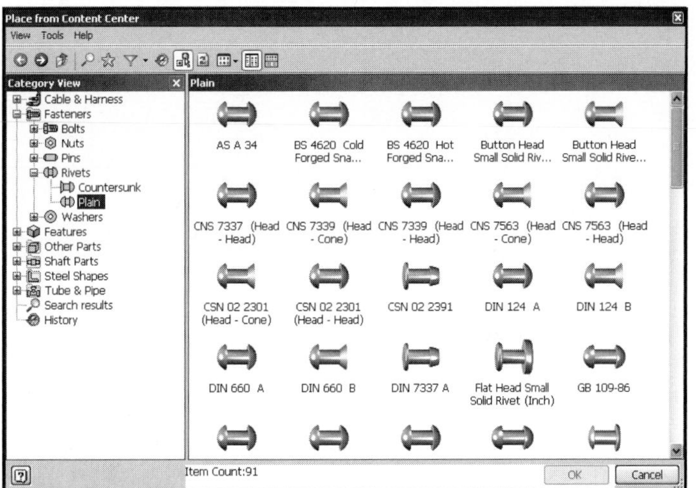

Figure 6-81

Sample Problem SP6-2

Figure 6-82 shows an assembly drawing. The assembly is to be held together using M8 hex head screws. It was created as follows.

1 Use the **Standard (mm).ipt** format and create the TOP BRACKET and the SOLID BASE.

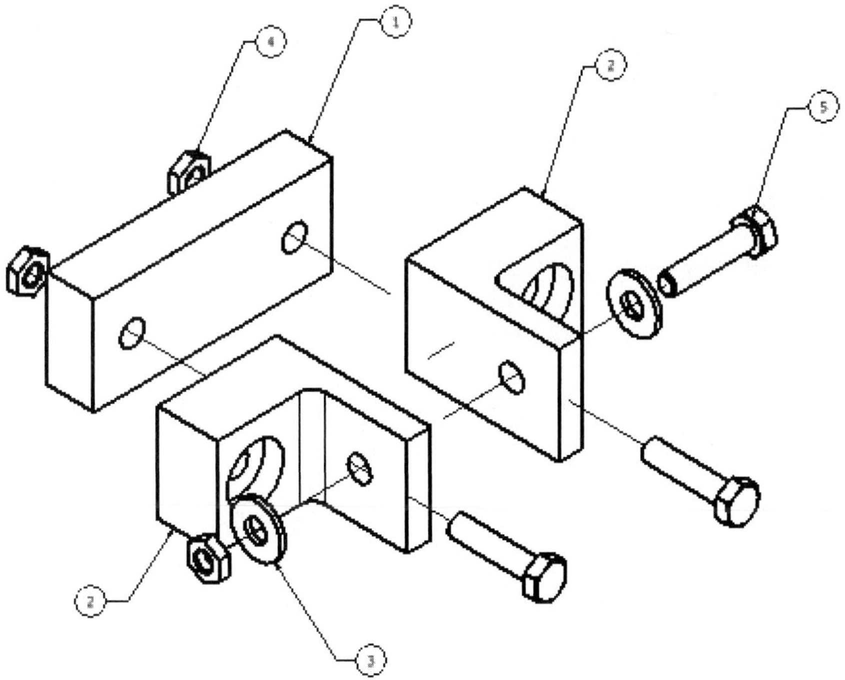

Figure 6-82

This assembly was created using the bottom-up approach. It could also have been created using the top-down approach. Figure 6-83 shows dimensioned drawings of the top bracket and solid base.

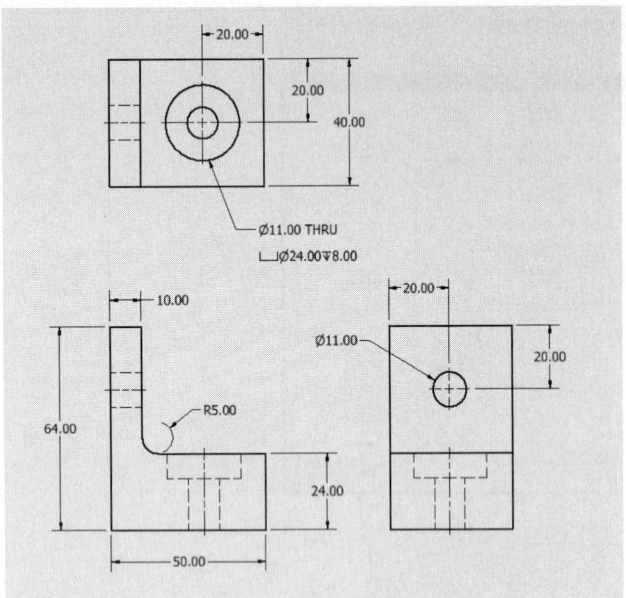

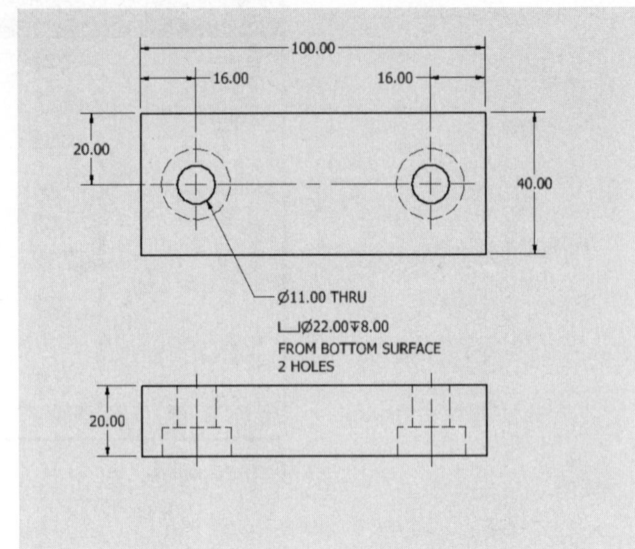

Figure 6-83

2. Assemble two brackets and one base as shown. See Figure 6-84.

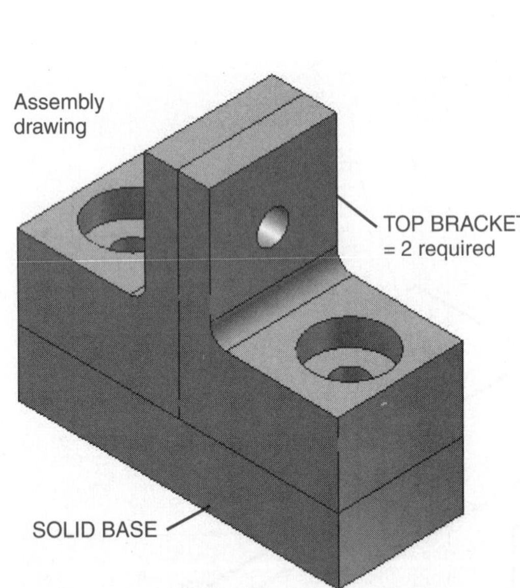

Figure 6-84

Figure 6-85

3. Access the **Place from Content Center** dialog box.
4. Click the **Washers** heading and select the **Plain Washer (Metric).** See Figure 6-85.
5. Select a **Nominal Diameter** of **10.** See Figure 6-86.
6. Click the **Table View** tab on the **Plain Washer (Metric)** dialog box. See Figure 6-87.
7. Note that the height of the washer is **2.3,** the inside diameter is **10.85,** and the outside diameter is **20.** The washer specification is

Figure 6-86

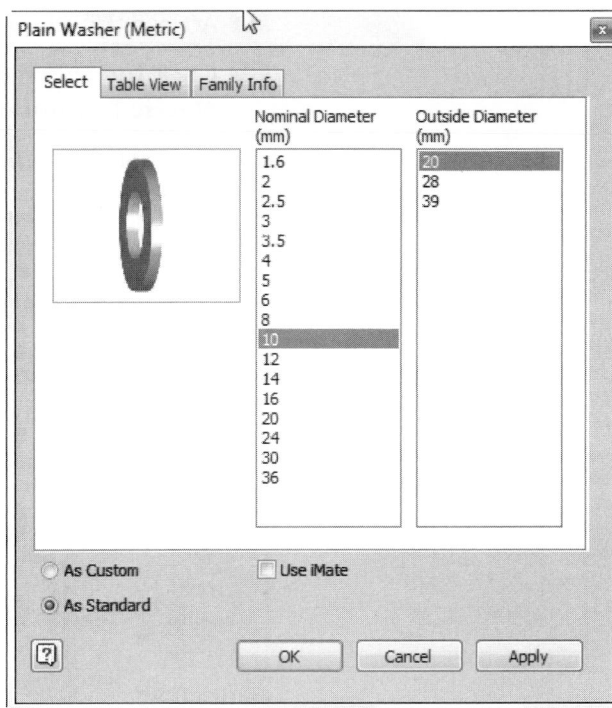

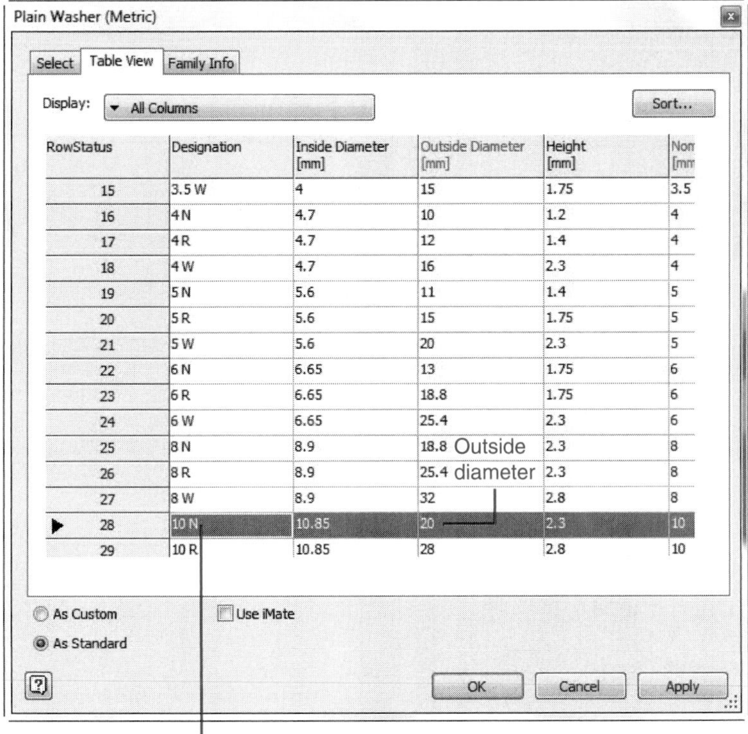

RowStatus	Designation	Inside Diameter [mm]	Outside Diameter [mm]	Height [mm]	Nom [mm
15	3.5 W	4	15	1.75	3.5
16	4 N	4.7	10	1.2	4
17	4 R	4.7	12	1.4	4
18	4 W	4.7	16	2.3	4
19	5 N	5.6	11	1.4	5
20	5 R	5.6	15	1.75	5
21	5 W	5.6	20	2.3	5
22	6 N	6.65	13	1.75	6
23	6 R	6.65	18.8	1.75	6
24	6 W	6.65	25.4	2.3	6
25	8 N	8.9	18.8	2.3	8
26	8 R	8.9	25.4	2.3	8
27	8 W	8.9	32	2.8	8
28	10 N	10.85	20	2.3	10
29	10 R	10.85	28	2.8	10

Figure 6-87

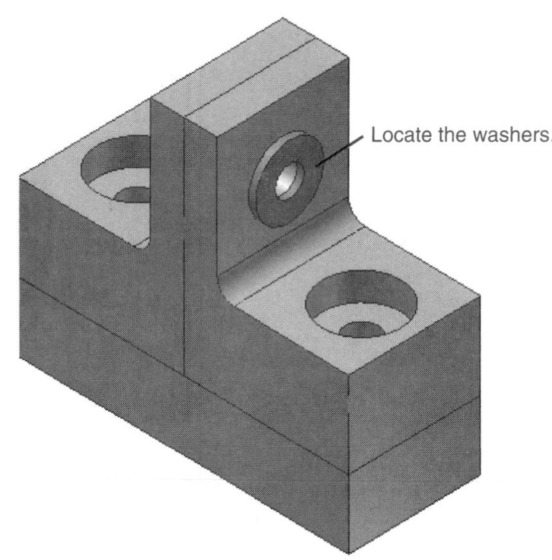

Figure 6-88

10 × 20 × 2.3. The inside nominal diameter is used in the washer specification.

8 Click **OK.**

9 Add two washers to the assembly drawing.

10 Use the **Insert** constraint to locate the washers around the holes on the two top brackets.

See Figure 6-88.

11 Access the **Place from Content Center** dialog box.

12 Click the **Hex** heading under **Nuts** and select the **Hex Jam Nut - Metric** hex nut option.

See Figure 6-89.

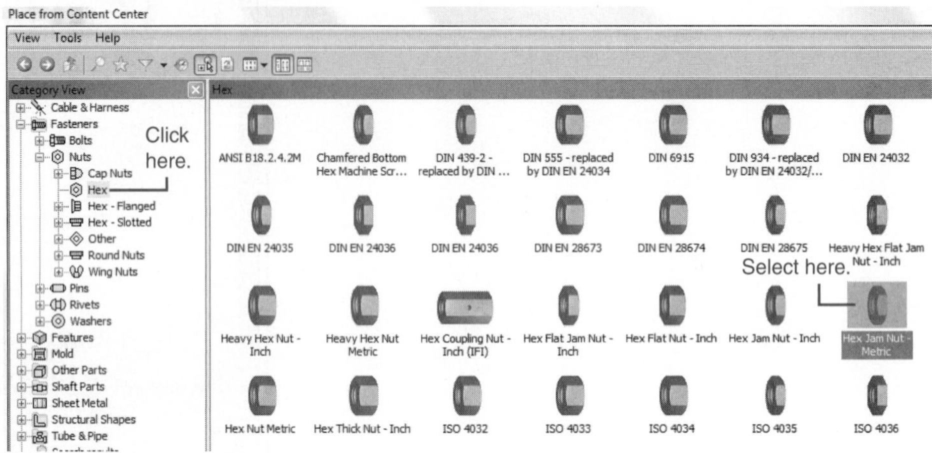

Figure 6-89

13 Select the **M10** thread option.

See Figure 6-90.

Figure 6-90

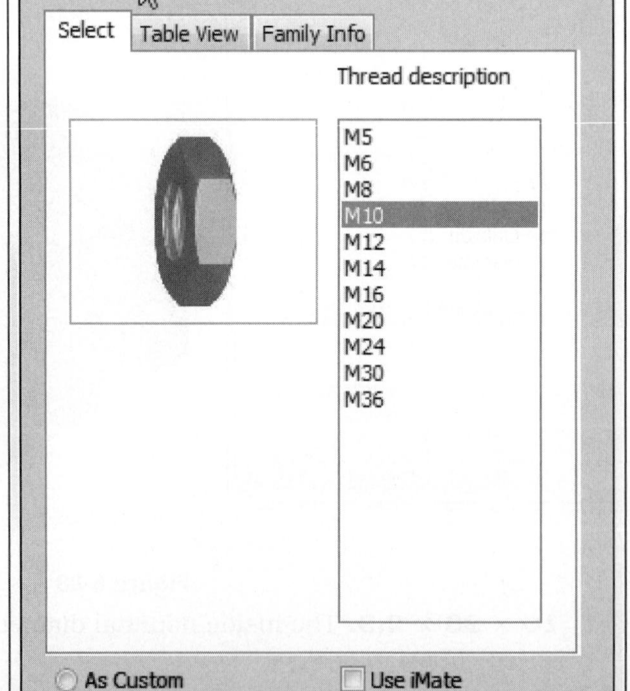

14 Click the **Table View** tab on the **Hex Jam Nut - Metric** dialog box.

See Figure 6-91.

Figure 6-91

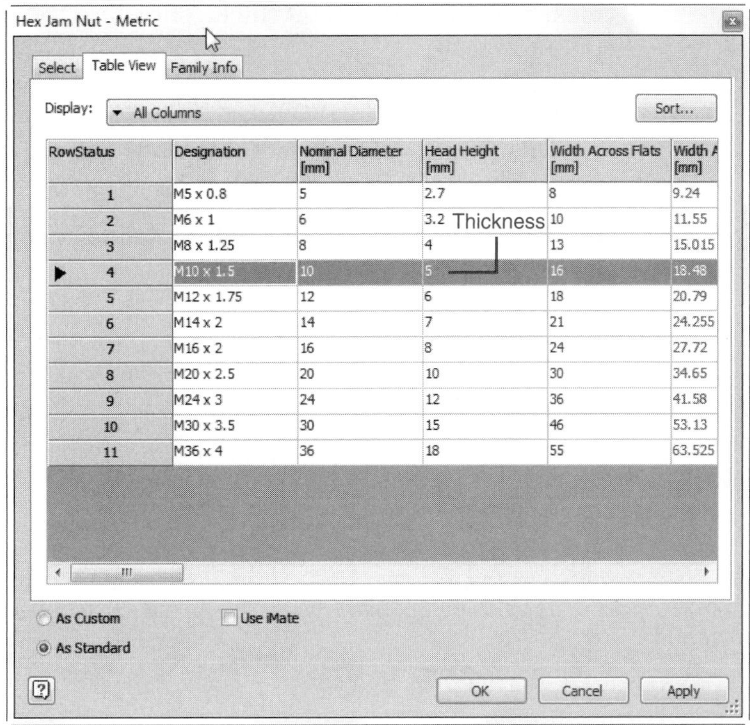

RowStatus	Designation	Nominal Diameter [mm]	Head Height [mm]	Width Across Flats [mm]	Width A [mm]
1	M5 x 0.8	5	2.7	8	9.24
2	M6 x 1	6	3.2	10	11.55
3	M8 x 1.25	8	4	13	15.015
4	M10 x 1.5	10	5	16	18.48
5	M12 x 1.75	12	6	18	20.79
6	M14 x 2	14	7	21	24.255
7	M16 x 2	16	8	24	27.72
8	M20 x 2.5	20	10	30	34.65
9	M24 x 3	24	12	36	41.58
10	M30 x 3.5	30	15	46	53.13
11	M36 x 4	36	18	55	63.525

15 Note that the height of the nut is **5.0.** Scroll the screen to the right and note that the pitch is **1.50.**

16 Click **OK.**

17 Add three nuts to the assembly drawing.

18 Use the **Insert** constraint and locate the nuts aligned with the matching holes in the top brackets and with the holes between the top brackets and the solid base as shown.

See Figure 6-92.

Figure 6-92

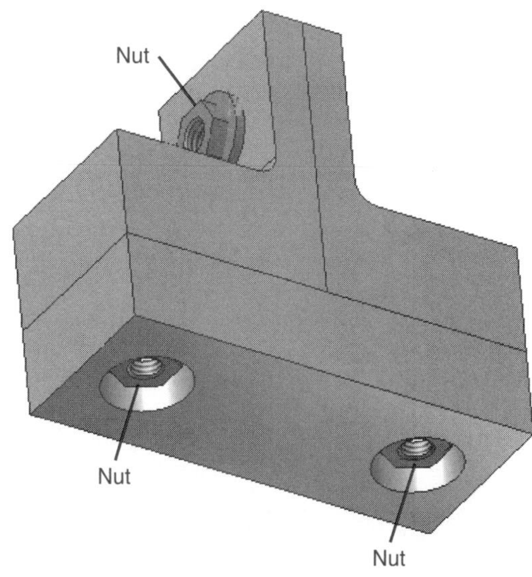

19 Calculate the required screw length for the screw that mounts between the back-to-back top brackets.

The flanges on the two top brackets are 10 each, the washers are 2.30 thick, and the nut is 5.00 thick. Therefore the minimum screw length is equal to

$$10 + 10 + 2.30 + 2.30 + 5.00 = 29.60$$

This length value does not include the required two pitches that must extend beyond the nut.

20 Access the **Place from Content Center** and click the **Fasteners** heading.

21 Select the **Hex Head** option.

See Figure 6-93.

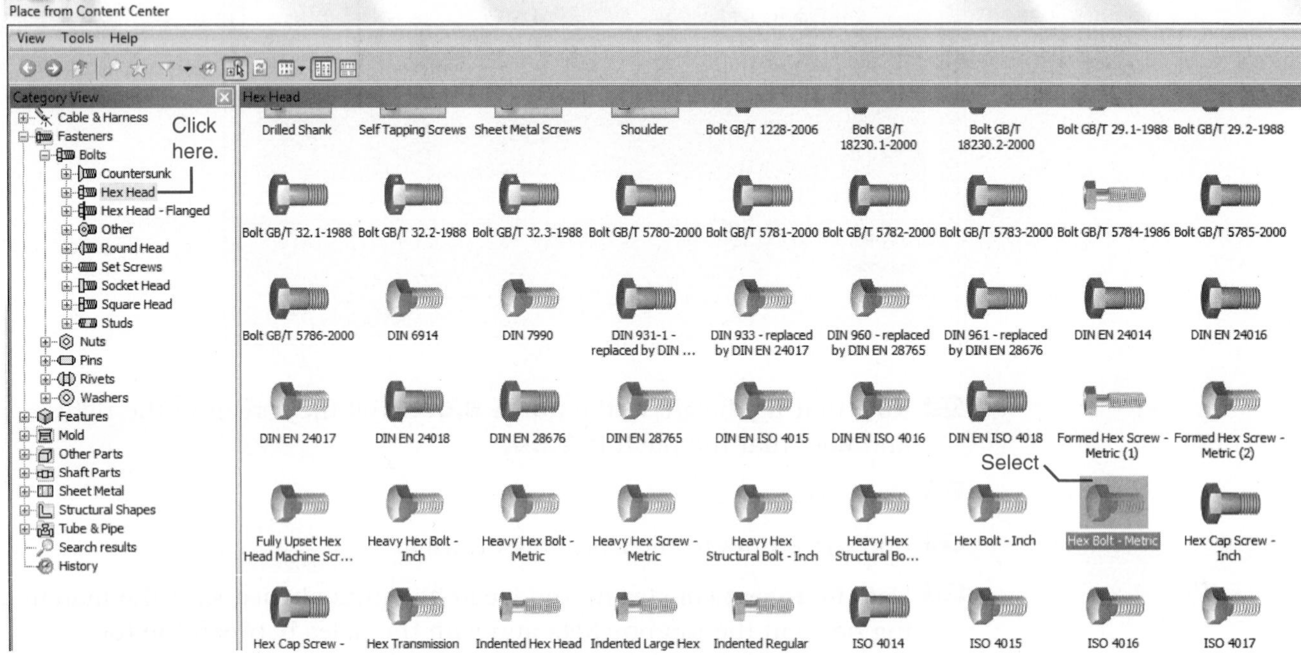

Figure 6-93

22 Select the **Hex Bolt - Metric** option.

See Figure 6-94.

23 Click the **Table View** tab on the **Hex Bolt - Metric** dialog box and note that the pitch of the M10 thread is **1.50.**

24 Click the **Select** tab on the **Hex Bolt - Metric** dialog box.

The value for 2P (two pitches) equals 2(1.50) = 3.00. Add this value to the 29.60 length requirement calculated previously:

$$29.60 + 3.00 = 32.60$$

The next standard length value is **35.** See Figure 6-95.

25 Select the **35** nominal length; click **OK.**

26 Use the **Insert** constraint tool and insert the screw into the assembly.

See Figure 6-96.

27 Calculate the minimum thread length for the two screws inserted between the top bracket and the solid base.

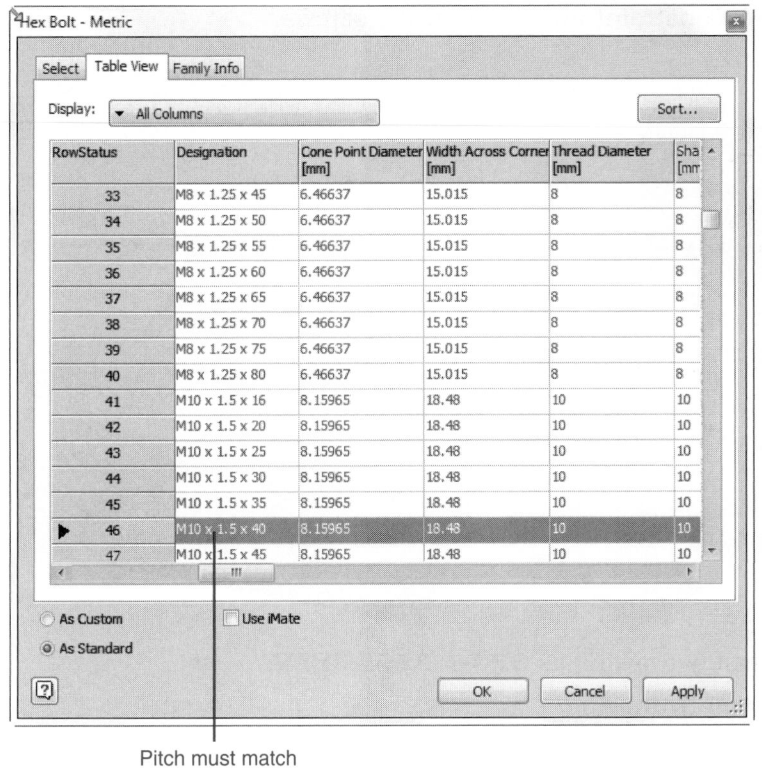

Figure 6-94

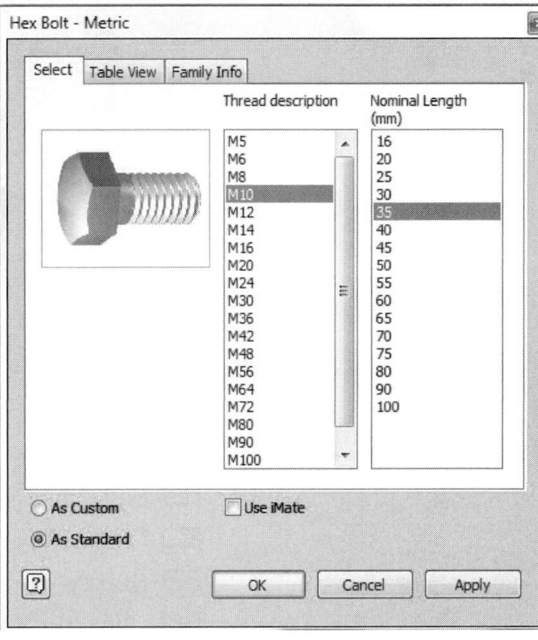

Figure 6-95

Figure 6-96

From the dimensioned drawings shown in Figure 6-83, the depth of the counterbore on the top bracket is 8.00. The total height of the flange is 24.00, so the threaded portion of each screw must pass through a difference of 16.00. The counterbore on the solid base is 8.00 deep, and the flange is 20 thick, so the thread must pass through 12.00. The total distance is 16.00 + 12.00 = 28.00.

If the same nut as selected previously is used, the nut height is 5.0. The thread pitch for an M10 screw is 1.50 (2P = 3.00). Therefore, the minimum required thread length is as follows:

$$16.00 + 12.00 + 5.00 + 3.00 = 36.00$$

This value indicates that an M10 × 40 screw is required.

28 Use the **Insert** constraint and position the screws.

See Figure 6-97.

Figure 6-97

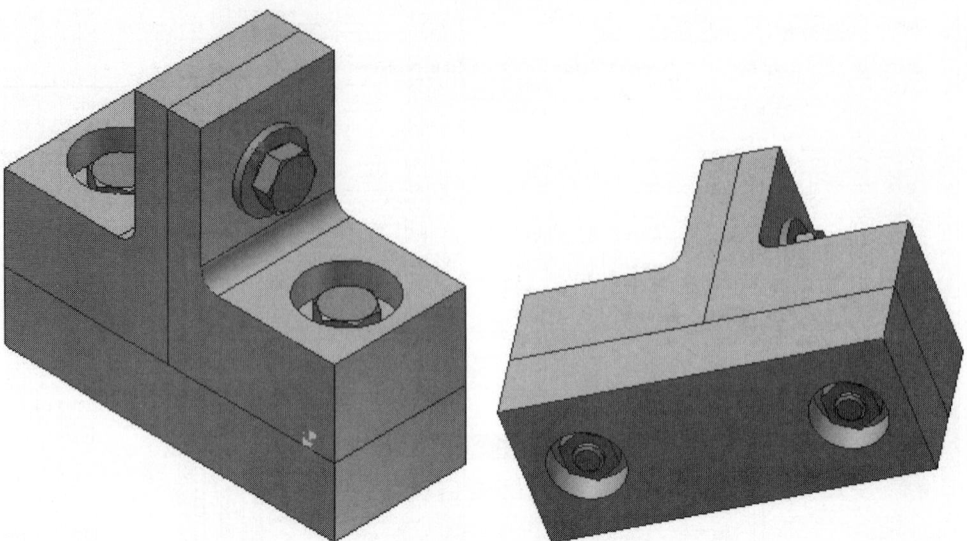

29 Save the assembly drawing as **SP6-1 ASSEMBLY.**

30 Close the assembly drawing.

31 Start a new drawing using the **Standard (mm).ipn** format.

32 Click the **Create View** tool and access the **SP6-1 ASSEMBLY.**

33 Use the **Tweak Components** tool and create an exploded isometric drawing showing all the parts of the assembly.

See Figure 6-98.

Figure 6-98

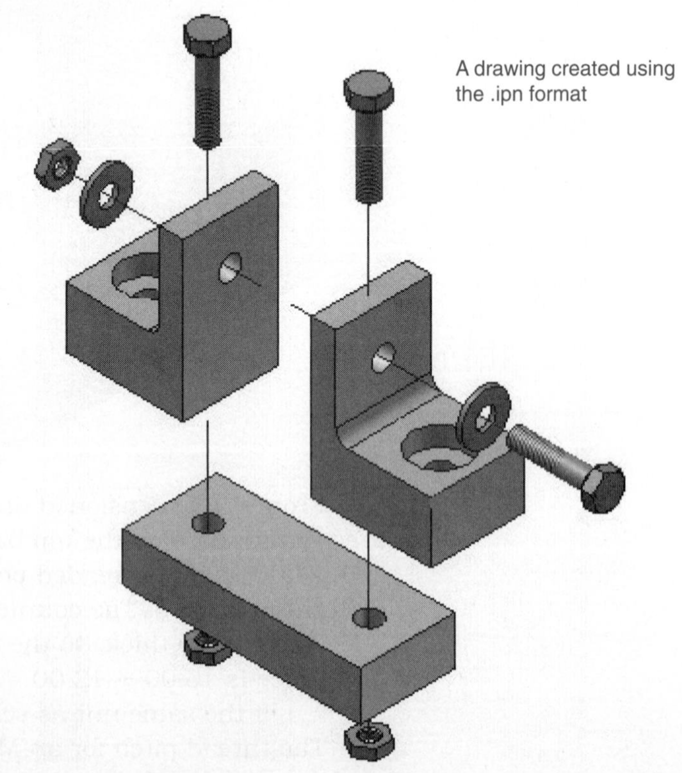

TIP

Try the animation options.

34 Save the exploded assembly drawing as **SP6-1 ASSEMBLY.**

35 Close the assembly drawing.

36 Start a new drawing using the **ANSI (mm).idw** format.

37 Create an **Iso Top Left** exploded assembly drawing based on the .ipn bracket assembly drawing.

See Figure 6-99.

Figure 6-99

An exploded isometric drawing created using the ANSI (mm).idw format

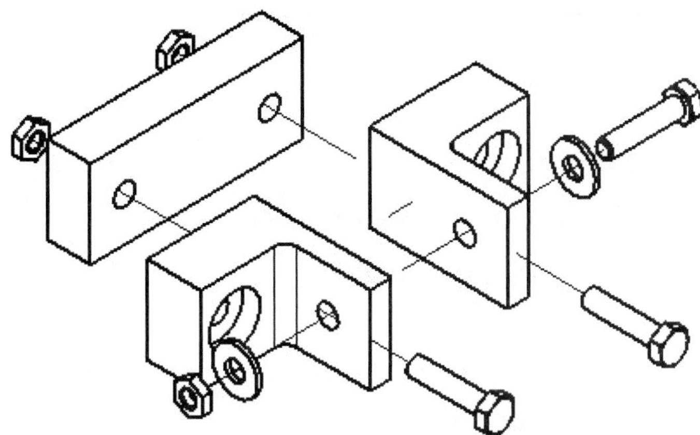

38 Add assembly numbers (balloon numbers) to the assembly drawing.

Start the assembly numbers with the biggest parts. The solid base should be part 1, the top bracket part 2, and so on.

See Figure 6-100.

Figure 6-100

Add assembly numbers.

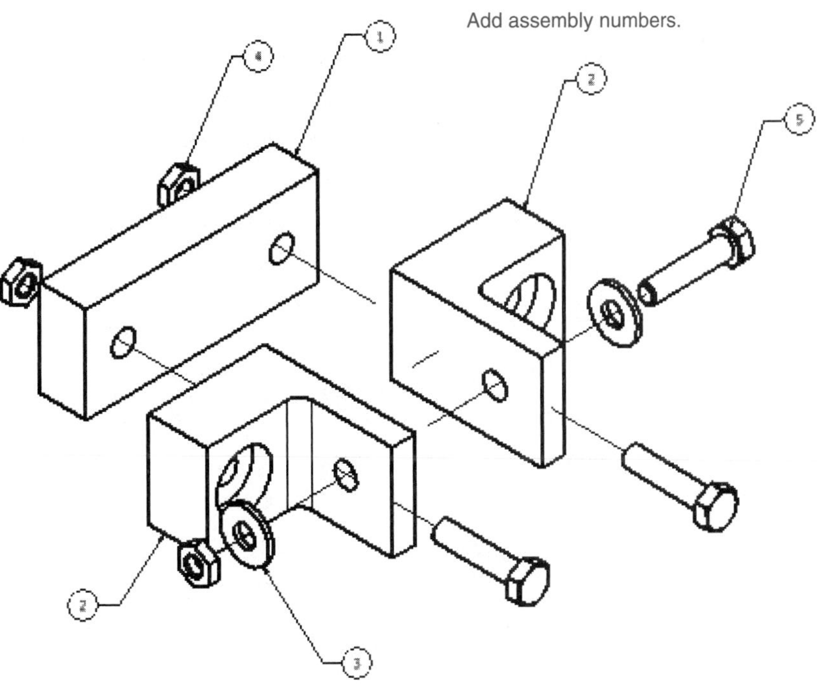

39 Click the **Parts List** tool on the **Annotate** panel and locate the parts list on the assembly drawing.

See Figure 6-101. The part numbers that are displayed on the parts list are the file numbers for the parts. These numbers may be edited as needed.

Figure 6-101

PARTS LIST			
ITEM	QTY	PART NUMBER	DESCRIPTION
1	1	SOLID BASE	
2	2	TOP BRACKET	
3	2	ANSI B18.22M - 10 N	Metric Plain Washers
4	3	ANSI B18.2.4.5M - M10 x 1.5	Hex Jam Nut
5	1	ANSI B18.2.3.5M - M10 x 1.5 x 35	Hex Bolt
6	2	ANSI B18.2.3.5M - M10 x 1.5 x 40	Hex Bolt

TIP

The parts list may be located on a separate sheet if needed.

40 Right-click the parts list and select the **Edit Parts List** option.

41 Click the **Column Chooser** option.

See Figure 6-102.

Figure 6-102

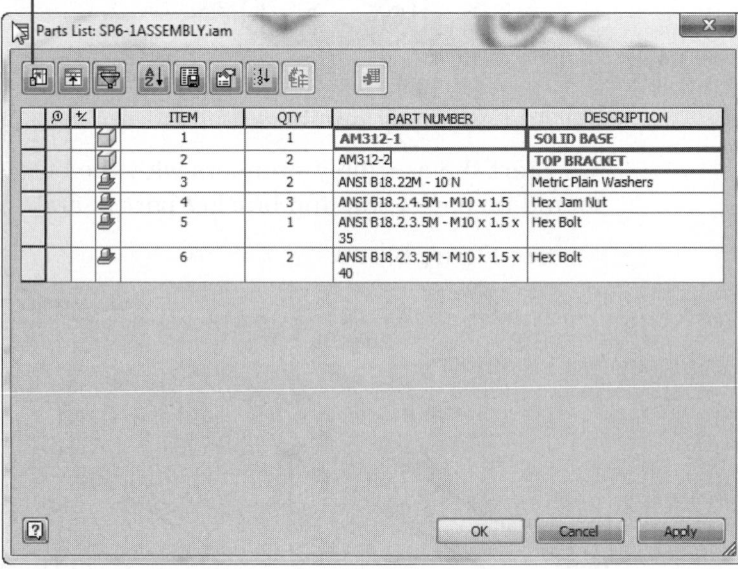

42 Scroll down the **Available Properties** options and add **MATERIAL** to the parts list.

See Figure 6-103.

43 Use the **Move Down** option on the **Parts List Column Chooser** dialog box and move the **QTY** heading to the bottom of the list.

44 Click **OK.**

See Figure 6-104.

45 Add the description and material requirements for the solid base and top brackets.

See Figure 6-105.

46 Click **OK.**

Figure 6-105 shows the finished isometric exploded assembly drawing.

Figure 6-103

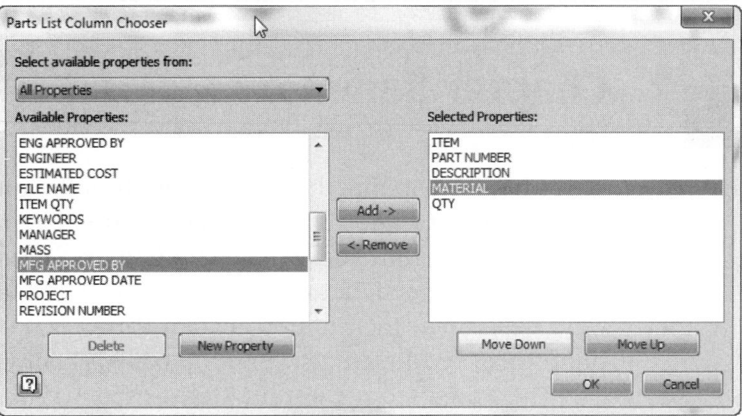

Figure 6-104

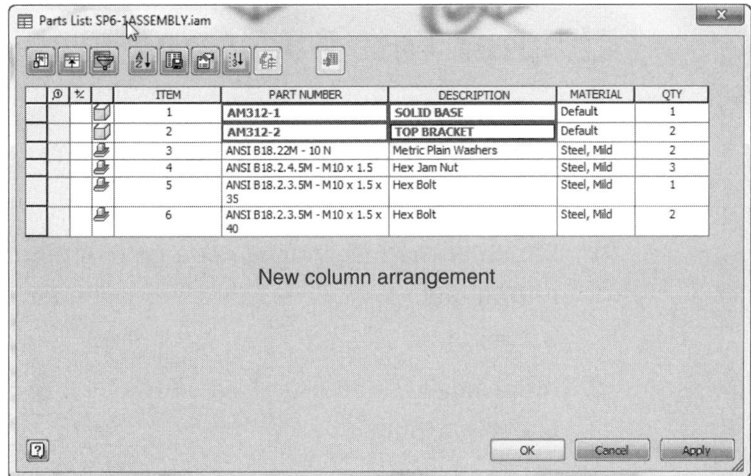

Figure 6-105

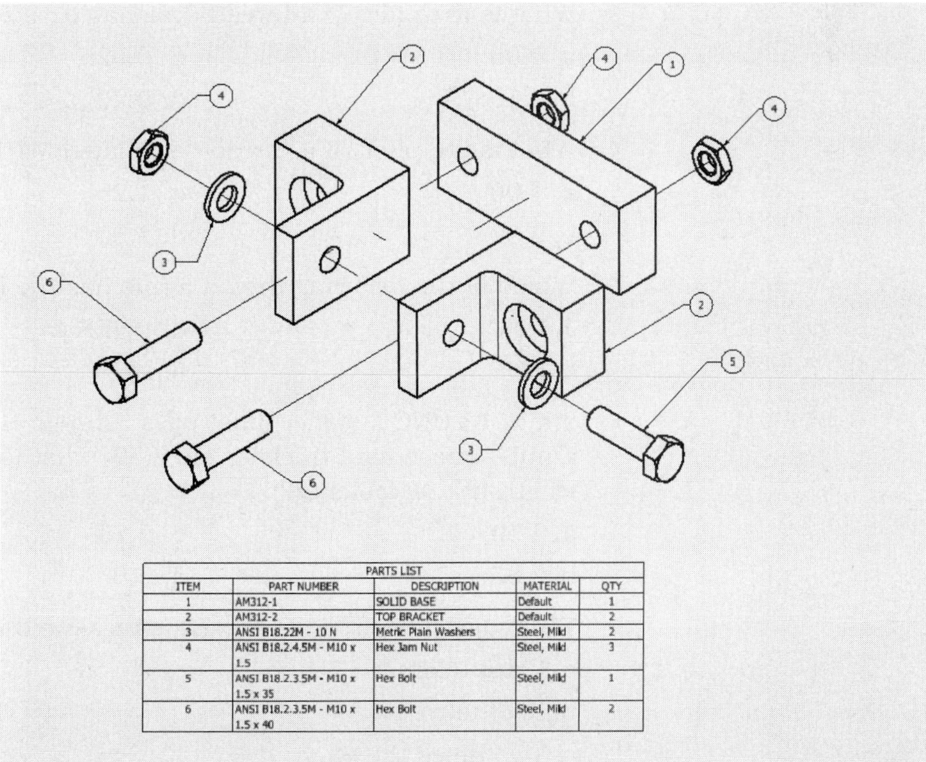

PARTS LIST

ITEM	PART NUMBER	DESCRIPTION	MATERIAL	QTY
1	AM312-1	SOLID BASE	Default	1
2	AM312-2	TOP BRACKET	Default	2
3	ANSI B18.22M - 10 N	Metric Plain Washers	Steel, Mild	2
4	ANSI B18.2.4.5M - M10 x 1.5	Hex Jam Nut	Steel, Mild	3
5	ANSI B18.2.3.5M - M10 x 1.5 x 35	Hex Bolt	Steel, Mild	1
6	ANSI B18.2.3.5M - M10 x 1.5 x 40	Hex Bolt	Steel, Mild	2

Chapter Summary

This chapter explained how to draw threads and washers as well as how to select fasteners and how to design using fasteners, washers, and keys.

Thread terminology was explained and illustrated, and the different thread form specifications and ways of graphically representing threads were described, including threaded through holes, internal threads, and blind threaded holes. ANSI standards and conventions were followed.

The **Place from Content Center** tool was used to specify different types of fasteners in drawings, including bolts and screws coupled with nuts. Countersunk screws and counterbored holes were described and illustrated.

Chapter Test Questions

Multiple Choice

Circle the correct answer.

1. Which of the following is *not* a type of point for a setscrew?

a. Half dog
b. Cone
c. Cylinder
d. Flat

2. A flat head fastener is used with which of the following hole shapes?

a. Counterbore
b. Countersink
c. Spotface

3. What is the pitch of a thread designated ¼-20 UNC?

a. 0.05
b. 0.10
c. 0.025
d. 0.20

4. What is the pitch of a thread designated M10 × 1.25 × 20?

a. 1.00
b. 2.00
c. 1.25
d. 1.50

5. Which of the following thread forms has the most threads per inch?

a. UNC
b. UNF
c. UNEF
d. UN

6. If a ½-13 UNC fastener must pass through two parts—a washer and a nut—whose total thickness is 2.50, what is the minimum possible length for the fastener?

a. 2.50
b. 2.65
c. 2.75
d. 2.88

7. The major diameter of a thread is measured from

a. Crest to crest
b. Root to root
c. Two times the pitch

8. Which tool is used to edit an existing threaded hole?

   a. **Edit Sketch**

   b. **Edit Feature**

   c. **Edit Text**

9. If an M8 × 1 × 16 fastener is to be inserted into a threaded hole, the thread depth must be at least

   a. 16

   b. 17

   c. 18

   d. 19

10. Which type of thread advances when turned in the clockwise direction?

    a. Right-hand

    b. Left-hand

    c. Center-thread

Matching

Given the following thread callout, identify the meaning of each term.
¼-20 UNC-2A × 1.625

Column A	Column B
a. 20________	1. Length
b. UNC________	2. Class of fit
c. ¼________	3. External
d. 1.625________	4. Major diameter
e. A________	5. Pitch
f. 2________	6. Thread form

Given the following thread callout, identify the meaning of each term.
M10 × 1.25 × 30

Column A	Column B
g. M10________	7. Length
h. 1.25________	8. Thread designation
i. 30________	9. Pitch

True or False

Circle the correct answer.

1. **True or False:** The notation ∅10 designates either a hole or a cylindrical-shaped object, whereas the notation M10 designates either an internal or external thread.

2. **True or False:** The pitch of a thread is the linear distance along the thread from crest to crest.

3. **True or False:** The thread designation M10 × 1.25 × 30 indicates that the thread's pitch is 30.

4. **True or False:** A coarse thread pitch is not included in a metric thread **callout**.
5. **True or False:** The thread designation .500 13 UNC × 3.00 LONG indicates that the thread's pitch is 13.
6. **True or False:** UNC, UNF, UNEF, and UN are all standard thread forms.
7. **True or False:** Threads can be drawn in Inventor using the **Hole** tool on the **3D Model** tab.
8. **True or False:** The threaded portion of an internal hole must extend at least two pitches beyond the end of the fastener inserted into the hole.
9. **True or False:** The Inventor **Content Center** includes a listing of standard threaded fasteners.
10. **True or False:** A washer is defined using the inside diameter × outside diameter × thickness.

Chapter Projects

Project 6-1: Millimeters

Figure P6-1 shows three blocks. Assume that the blocks are each 30 × 30 × 10 and that the hole is ∅9. Assemble the three blocks so that their holes are aligned and they are held together by a hex head bolt secured by an appropriate hex nut. Locate a washer between the bolt head and the top block and between the nut and the bottom block. Create all drawings using either an A4 or A3 drawing sheet, as needed. Include a title block on all drawing sheets.

A. Define the bolt.
B. Define the nut.
C. Define the washers.
D. Draw an assembly drawing including all components.
E. Create a BOM for the assembly.
F. Create a presentation drawing of the assembly.
G. Create an isometric exploded drawing of the assembly.
H. Create an animation drawing of the assembly.

Figure P6-1

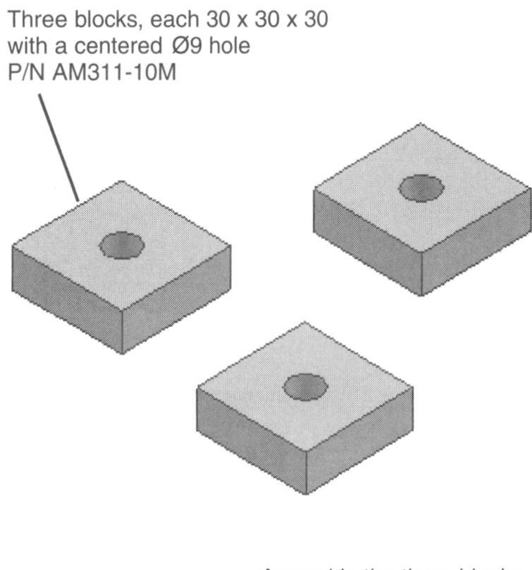

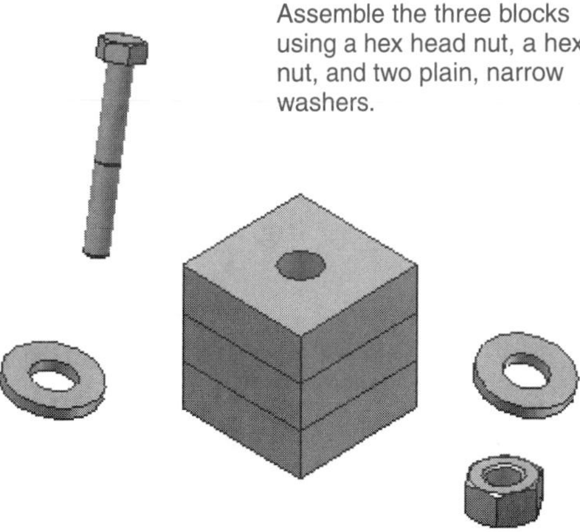

Assemble the three blocks
using a hex head nut, a hex
nut, and two plain, narrow
washers.

Project 6-2: Millimeters

Figure P6-2 shows three blocks, one 30 × 30 × 50 with a centered M8 threaded hole, and two 30 × 30 × 10 blocks with centered ∅9 holes. Join the two 30 × 30 × 10 blocks to the 30 × 30 × 50 block using an M8 hex head bolt. Locate a regular plain washer under the bolt head.

A. Define the bolt.
B. Define the thread depth.
C. Define the hole depth.
D. Define the washer.
E. Draw an assembly drawing including all components.
F. Create a BOM for the assembly.
G. Create a presentation drawing of the assembly.
H. Create an isometric exploded drawing of the assembly.
I. Create an animation drawing of the assembly.

Figure P6-2

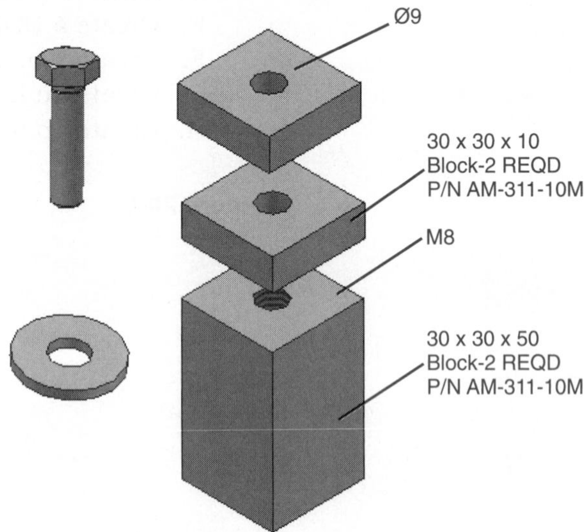

Project 6-3: Inches

Figure P6-3 shows three blocks. Assume that each block is 1.00 × 1.00 × .375 and that the hole is ∅.375. Assemble the three blocks so that their holes are aligned and that they are held together by a 5/16-18 UNC indented regular hex head bolt secured by an appropriate hex nut. Locate a washer between the bolt head and the top block and between the nut and the bottom block. Create all drawings using either an A4 or A3 drawing sheet, as needed. Include a title block on all drawing sheets.

A. Define the bolt.
B. Define the nut.
C. Define the washers.
D. Draw an assembly drawing including all components.
E. Create a BOM for the assembly.
F. Create a presentation drawing of the assembly.
G. Create an isometric exploded drawing of the assembly.
H. Create an animation drawing of the assembly.

Figure P6-3

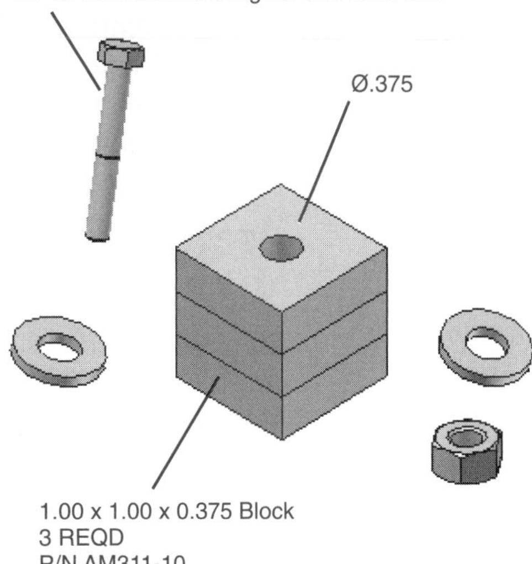

Project 6-4: Inches

Figure P6-4 shows three blocks, one 1.00 × 1.00 × 2.00 with a centered threaded hole, and two 1.00 × 1.00 × .375 blocks with centered ∅.375 holes. Join the two 1.00 × 1.00 × .375 blocks to the 1.00 × 1.00 × 2.00 block using a 5/16-18 UNC hex head bolt. Locate a regular plain washer under the bolt head.

A. Define the bolt.
B. Define the thread depth.
C. Define the hole depth.
D. Define the washer.

Figure P6-4

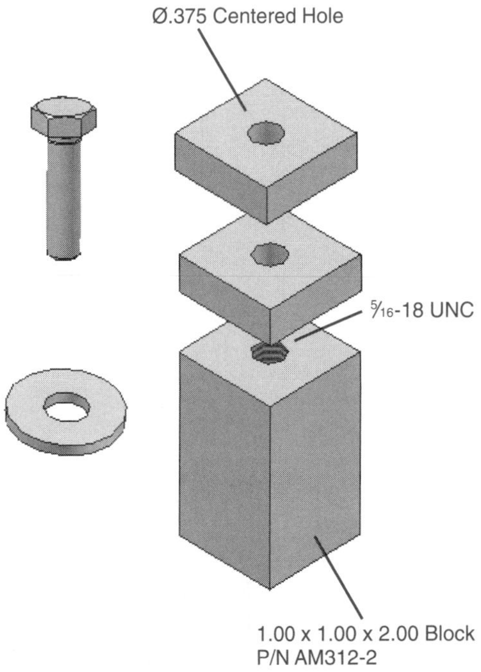

E. Draw an assembly drawing including all components.
F. Create a BOM for the assembly.
G. Create a presentation drawing of the assembly.
H. Create an isometric exploded drawing of the assembly.
I. Create an animation drawing of the assembly.

Project 6-5: Inches or Millimeters

Figure P6-5 shows a centering block. Create an assembly drawing of the block and insert three setscrews into the three threaded holes so that they extend at least .25 in. or 6 mm into the center hole.

A. Use the inch dimensions.
B. Use the millimeter dimensions.
C. Define the setscrews.
D. Draw an assembly drawing including all components.
E. Create a BOM for the assembly.
F. Create a presentation drawing of the assembly.
G. Create an isometric exploded drawing of the assembly.
H. Create an animation drawing of the assembly.

Centering Block
P/N BU2004-5
SAE 1020 Steel

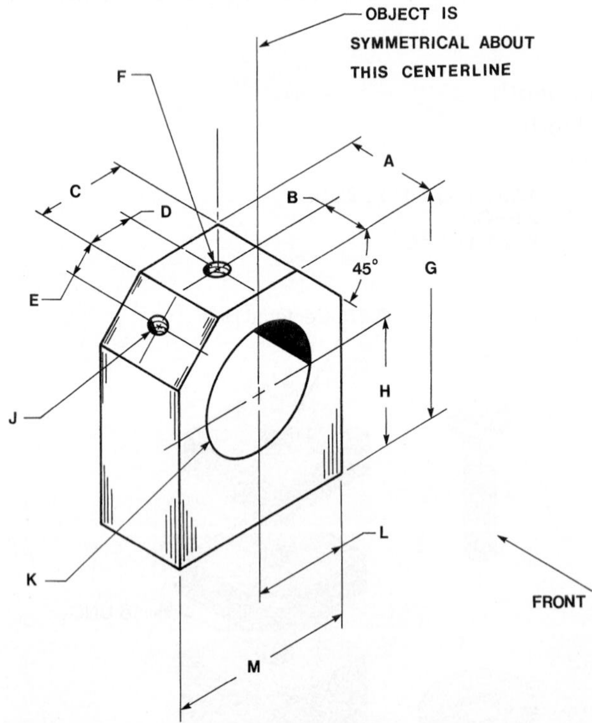

DIMENSION	INCHES	mm
A	1.00	26
B	.50	13
C	1.00	26
D	.50	13
E	.38	10
F	.190-32 UNF	M8X1
G	2.38	60
H	1.38	34
J	.164-36 UNF	M6
K	Ø1.25	Ø30
L	1.00	26
M	2.00	52

Figure P6-5

Project 6-6: Millimeters

Figure P6-6 shows two parts: a head cylinder and a base cylinder. The head cylinder has outside dimensions of ∅40 × 20, and the base cylinder has outside dimensions of ∅40 × 50. The holes in both parts are located on a ∅24 bolt circle. Assemble the two parts using hex head bolts.

A. Define the bolt.
B. Define the holes in the head cylinder, the counterbore diameter and depth, and the clearance hole diameter.
C. Define the thread depth in the base cylinder.
D. Define the hole depth in the base cylinder.
E. Draw an assembly drawing including all components.
F. Create a BOM for the assembly.
G. Create a presentation drawing of the assembly.
H. Create an isometric exploded drawing of the assembly.
I. Create an animation drawing of the assembly.

Figure P6-6

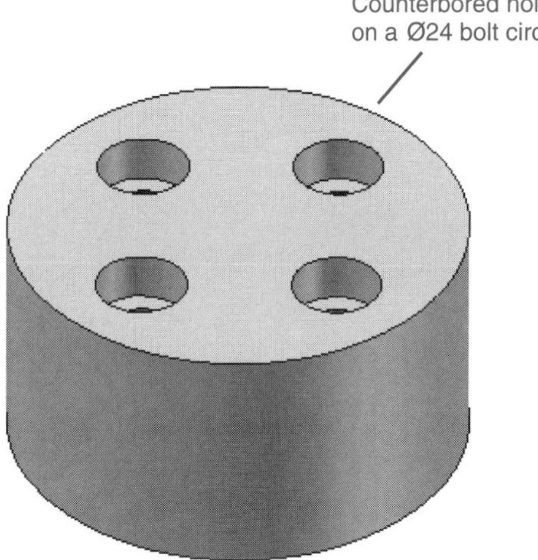

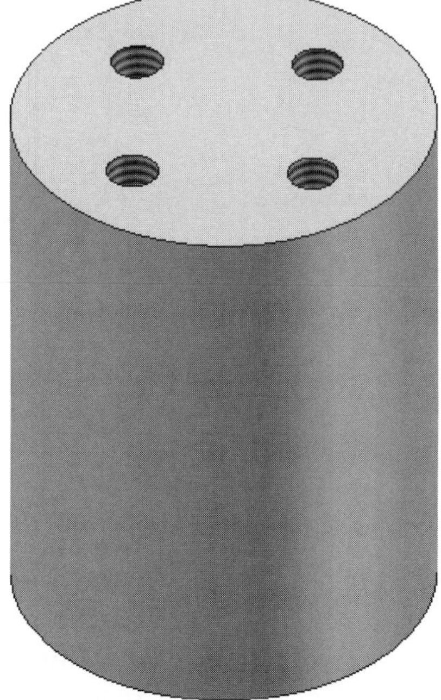

Project 6-7: Millimeters

Figure P6-7 shows a pressure cylinder assembly.

A. Draw an assembly drawing including all components.
B. Create a BOM for the assembly.
C. Create a presentation drawing of the assembly.
D. Create an isometric exploded drawing of the assembly.
E. Create an animation drawing of the assembly.

Figure P6-7

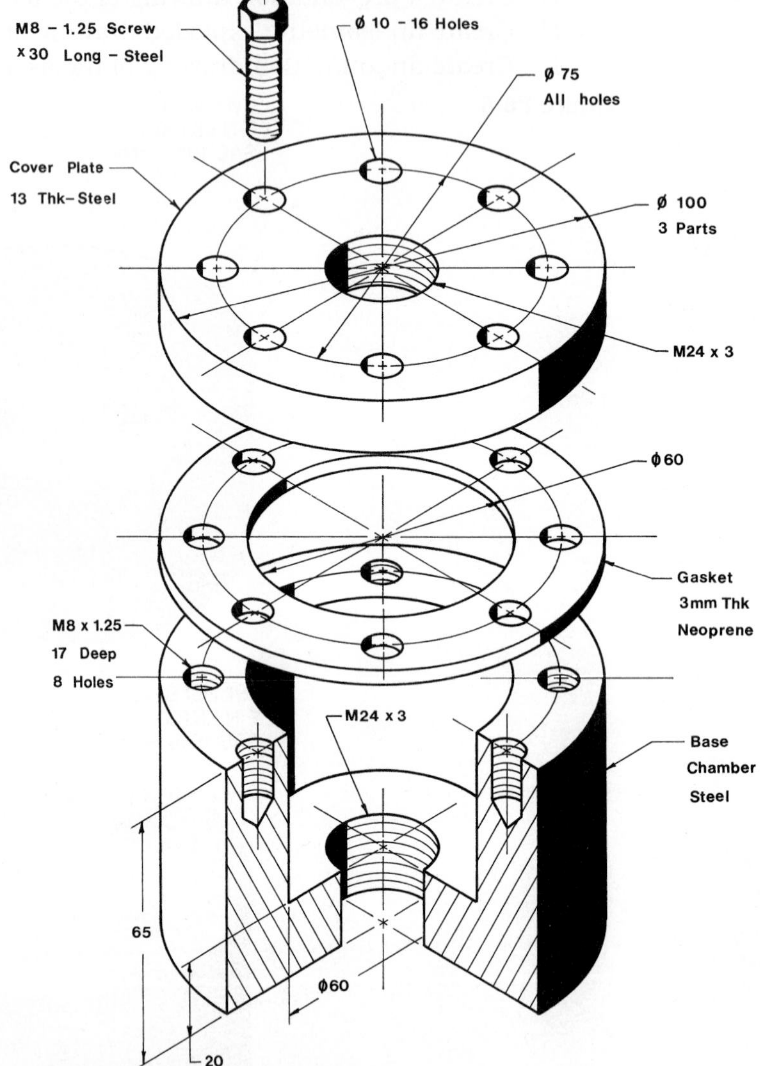

Project 6-8: Millimeters

Figure P6-7 shows a pressure cylinder assembly.

A. Revise the assembly so that it uses M10 × 35 hex head bolts.
B. Draw an assembly drawing including all components.
C. Create a BOM for the assembly.
D. Create a presentation drawing of the assembly.
E. Create an isometric exploded drawing of the assembly.
F. Create an animation drawing of the assembly.

Figure P6-9

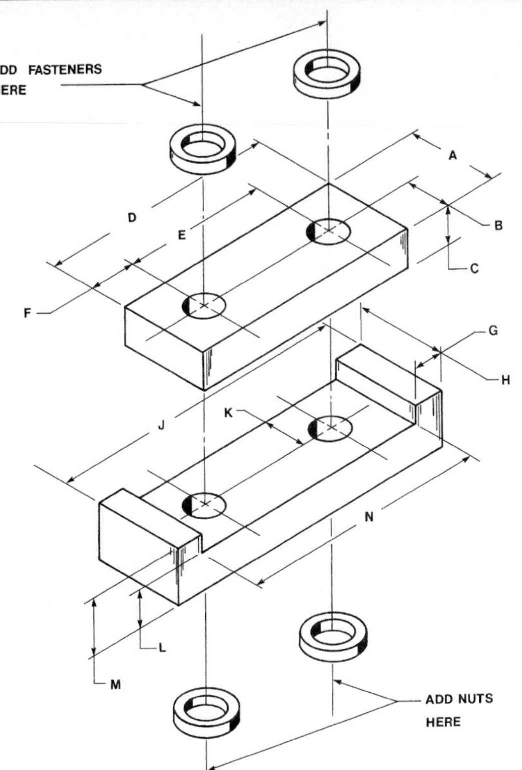

DIMENSION	INCHES	mm
A	1.25	32
B	.63	16
C	.50	13
D	3.25	82
E	2.00	50
F	.63	16
G	.38	10
H	1.25	32
J	4.13	106
K	.63	16
L	.50	13
M	.75	10
N	3.38	86

Project 6-9: Inches and Millimeters

Figure P6-9 shows a C-block assembly.

Use one of the following fasteners assigned by your instructor.

1. M12 hex head
2. M10 square head
3. ¼-20 UNC hex head
4. ⅜-16 UNC square head
5. M10 socket head
6. M8 slotted head
7. ¼-20 UNC slotted head
8. ⅜-16 UNC socket head

A. Define the bolts.
B. Define the nuts.
C. Define the washers.
D. Draw an assembly drawing including all components.
E. Create a BOM for the assembly.
F. Create a presentation drawing of the assembly.
G. Create an isometric exploded drawing of the assembly.
H. Create an animation drawing of the assembly.

Project 6-10: Millimeters

Figure P6-10 shows an exploded assembly drawing. There are no standard parts, so each part must be drawn individually.

A. Draw an assembly drawing including all components.
B. Create a BOM for the assembly.

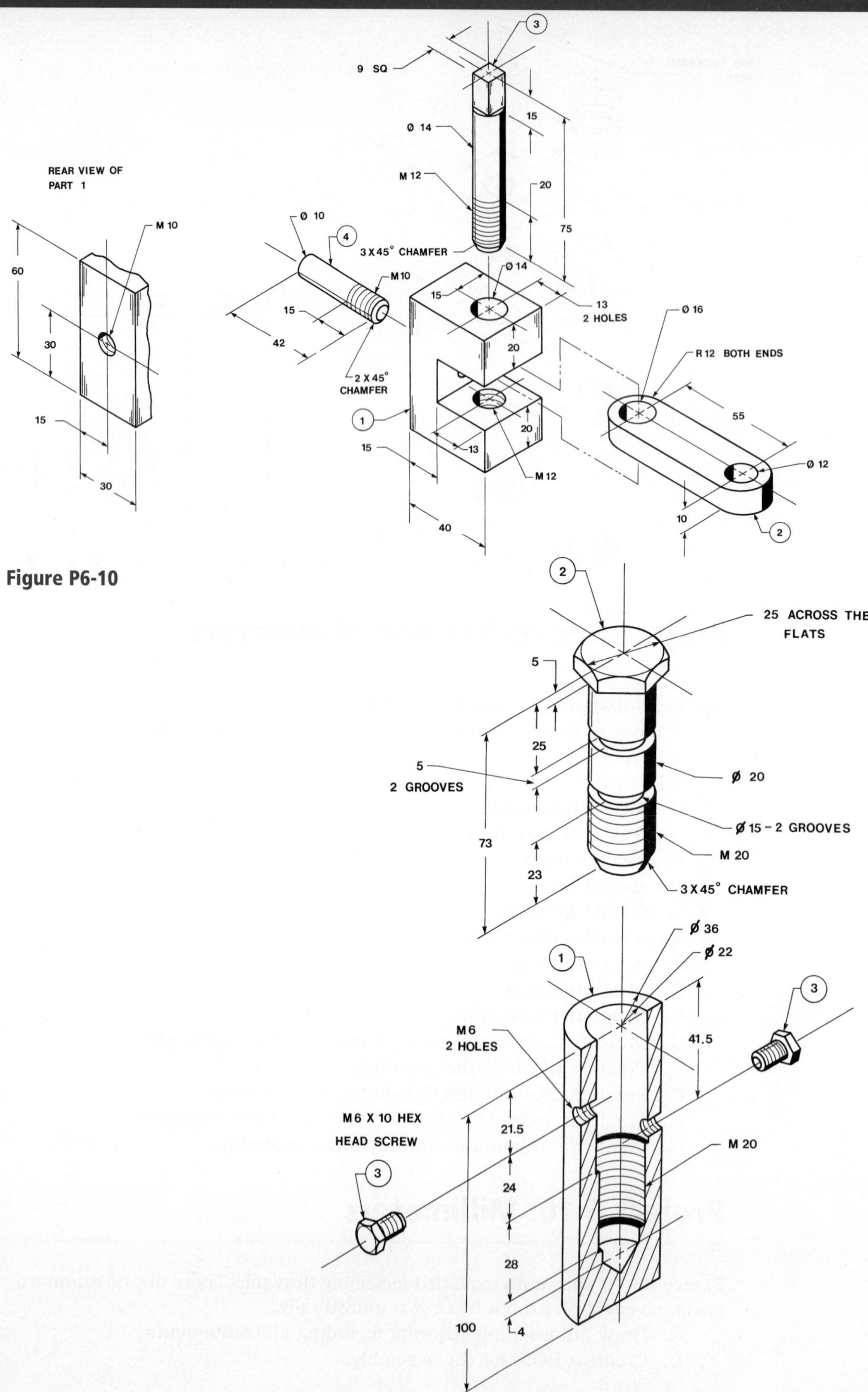

Figure P6-10

Figure P6-11

C. Create a presentation drawing of the assembly.
D. Create an isometric exploded drawing of the assembly.
E. Create an animation drawing of the assembly.

Project 6-11: Millimeters

Figure P6-11 shows an exploded assembly drawing.

A. Draw an assembly drawing including all components.
B. Create a BOM for the assembly.
C. Create a presentation drawing of the assembly.
D. Create an isometric exploded drawing of the assembly.
E. Create an animation drawing of the assembly.

Project 6-12: Inches or Millimeters

Figure P6-12 shows an exploded assembly drawing. No dimensions are given. If parts 3 and 5 have either M10 or 3/8-16 UNC threads, size parts 1 and 2. Based on these values estimate and create the remaining sizes and dimensions.

Figure P6-12

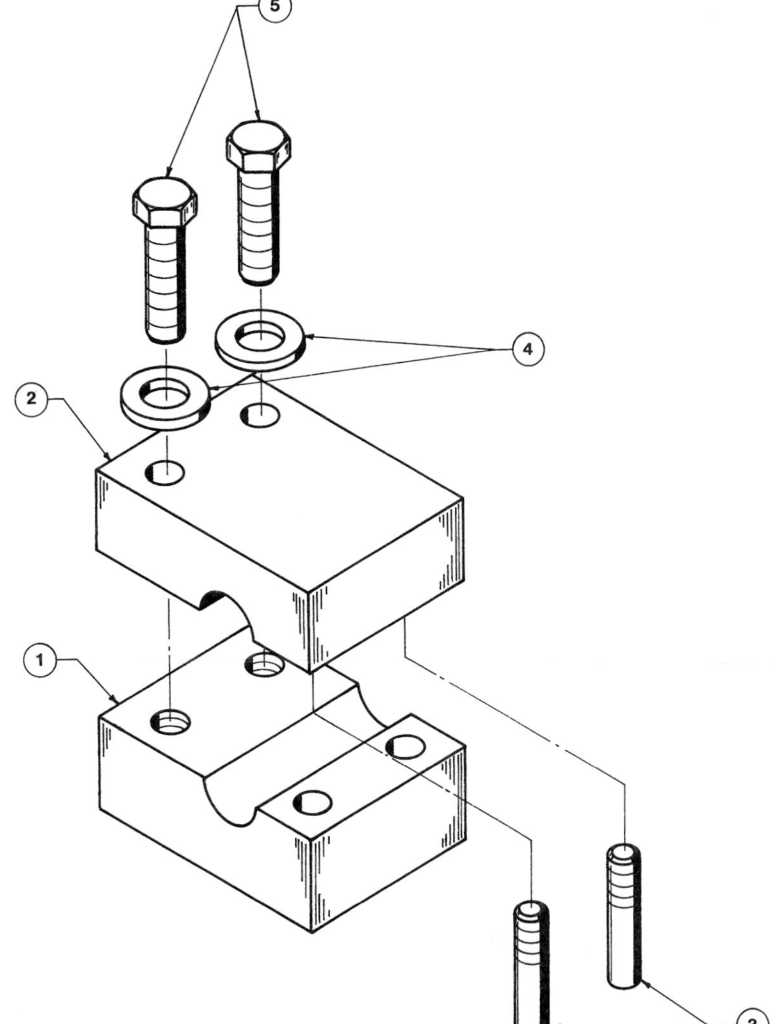

A. Draw an assembly drawing including all components.
B. Create a BOM for the assembly.
C. Create a presentation drawing of the assembly.
D. Create an isometric exploded drawing of the assembly.
E. Create an animation drawing of the assembly.

SIMPLIFIED SURFACE GAGE

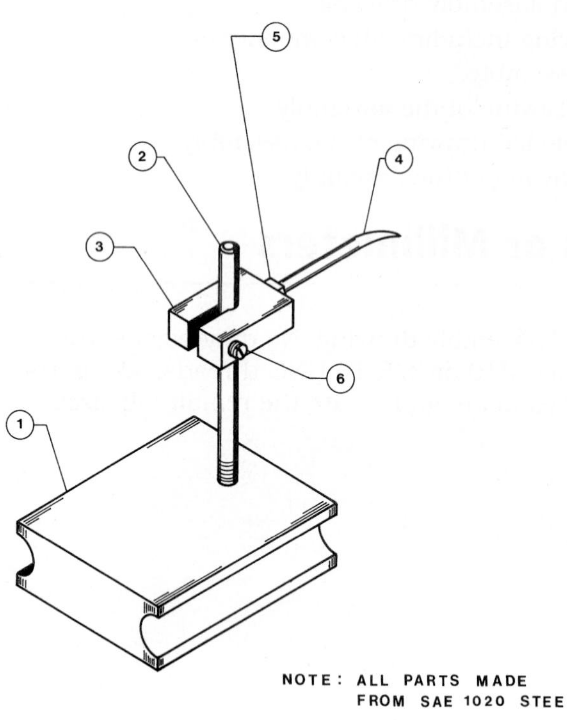

NOTE: ALL PARTS MADE FROM SAE 1020 STEEL

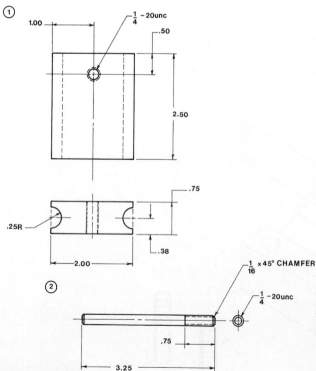

1
1.00
$\frac{1}{4}$ - 20unc
.50
2.50
.75
.25R
.38
2.00

2
$\frac{1}{16}$ x 45° CHAMFER
$\frac{1}{4}$ -20unc
.75
3.25

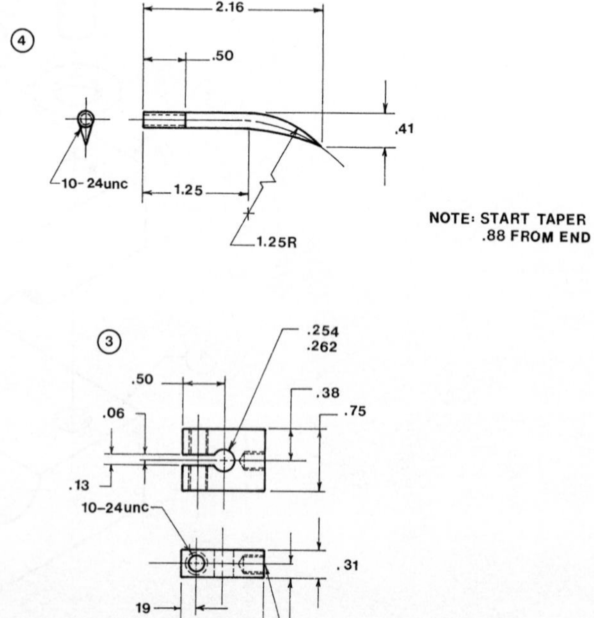

Figure P6-13

Project 6-13: Inches

Figure P6-13 shows an assembly drawing and detail drawings of a surface gauge.

A. Draw an assembly drawing including all components.
B. Create a BOM for the assembly.
C. Create a presentation drawing of the assembly.
D. Create an isometric exploded drawing of the assembly.
E. Create an animation drawing of the assembly.

Project 6-14: Millimeters

Figure P6-14 shows an assembly made from parts defined on pages 290–293. Assemble the parts using M10 threaded fasteners.

A. Define the bolt.
B. Define the nut.
C. Draw an assembly drawing including all components.
D. Create a BOM for the assembly.
E. Create a presentation drawing of the assembly.
F. Create an isometric exploded drawing of the assembly.
G. Create an animation drawing of the assembly.
H. Consider possible interference between the nuts and ends of the fasteners both during and after assembly. Recommend an assembly sequence.

Project 6-15: Millimeters

Figure P6-15 shows an assembly made from parts defined on pages 290–293. Assemble the parts using M10 threaded fasteners.

A. Define the bolt.
B. Define the nut.
C. Draw an assembly drawing including all components.

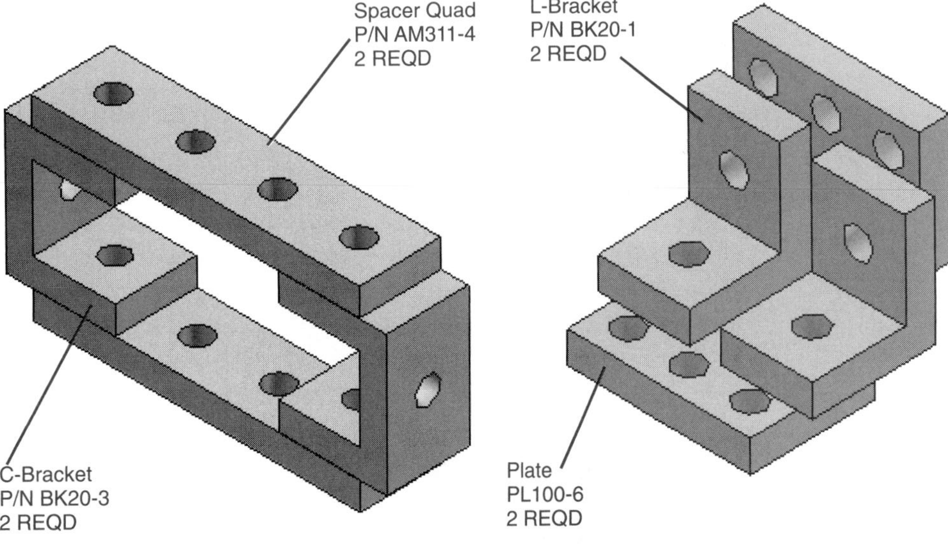

Figure P6-14

Figure P6-15

D. Create a BOM for the assembly.
E. Create a presentation drawing of the assembly.
F. Create an isometric exploded drawing of the assembly.
G. Create an animation drawing of the assembly.
H. Consider possible interference between the nuts and ends of the fasteners both during and after assembly. Recommend an assembly sequence.

Project 6-16: Millimeters

Figure P6-16 shows an assembly made from parts defined on pages 290–293. Assemble the parts using M10 threaded fasteners.

A. Define the bolt.
B. Define the nut.
C. Draw an assembly drawing including all components.
D. Create a BOM for the assembly.
E. Create a presentation drawing of the assembly.
F. Create an isometric exploded drawing of the assembly.
G. Create an animation drawing of the assembly.
H. Consider possible interference between the nuts and ends of the fasteners both during and after assembly. Recommend an assembly sequence.

Project 6-17: Access Controller

Design an access controller based on the information given in Figure P6-17. The controller works by moving an internal cylinder up and down within

Figure P6-16

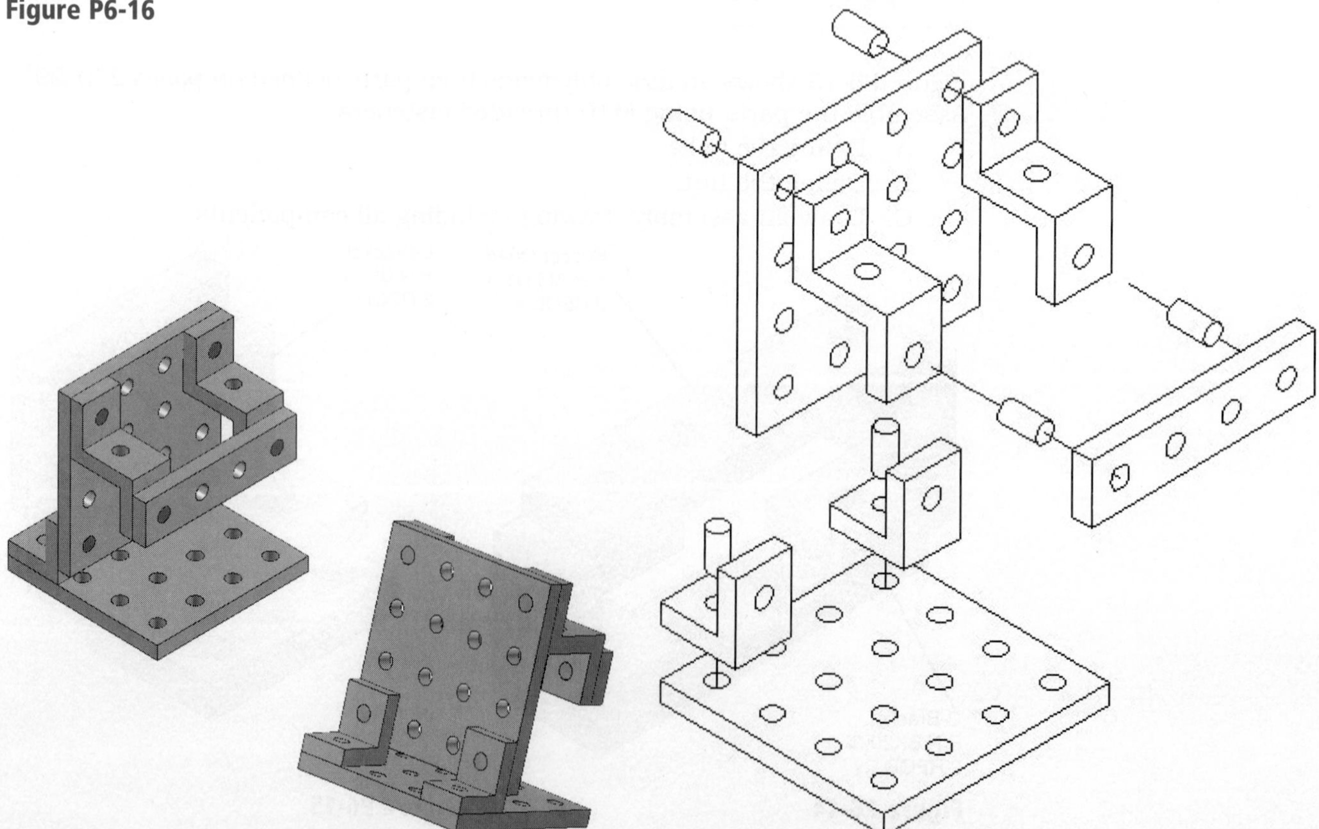

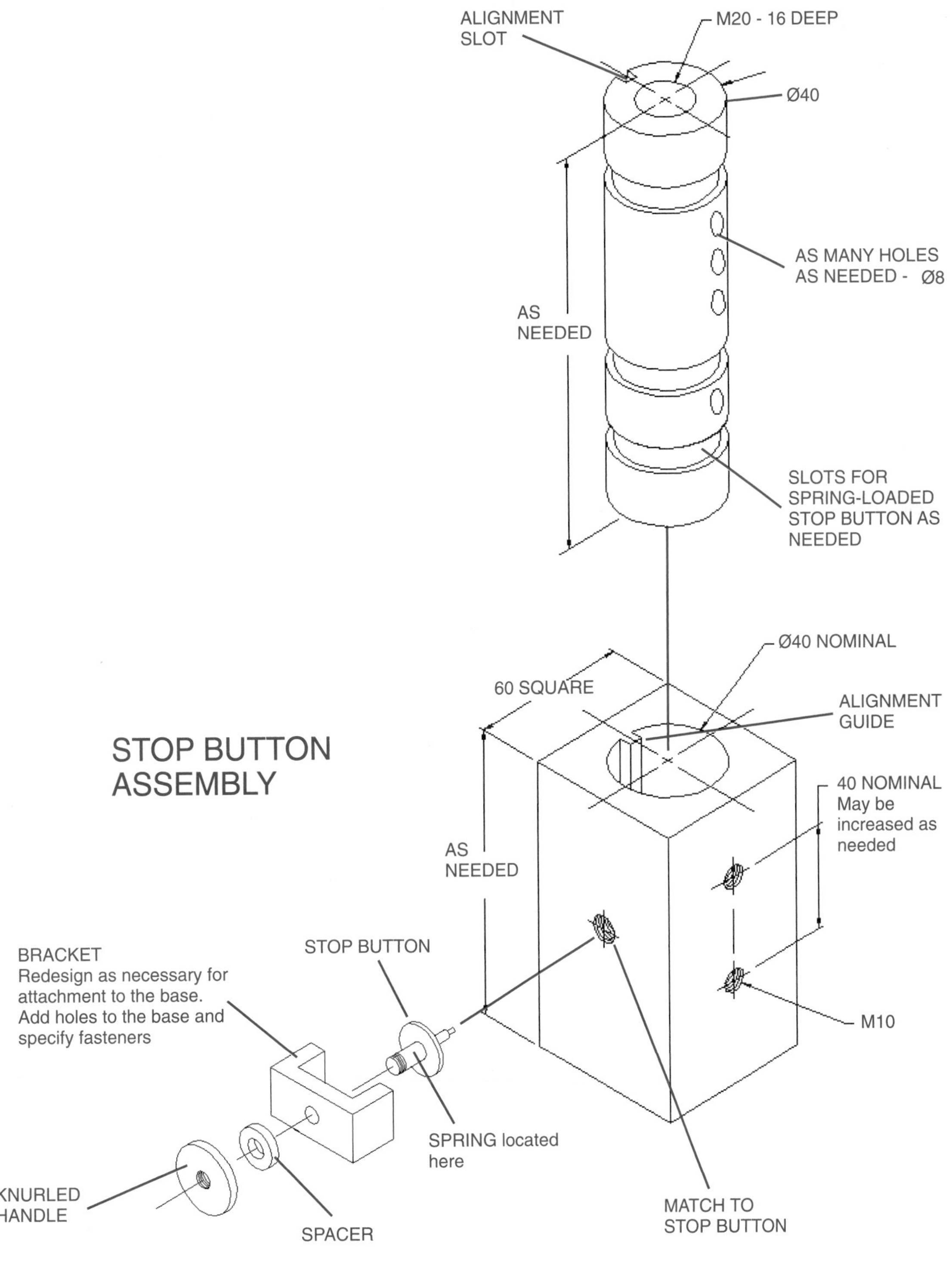
INTERNAL CYLINDER
ALIGNMENT SLOT
M20 - 16 DEEP
Ø40
AS MANY HOLES AS NEEDED - Ø8
AS NEEDED
SLOTS FOR SPRING-LOADED STOP BUTTON AS NEEDED
Ø40 NOMINAL
60 SQUARE
ALIGNMENT GUIDE
STOP BUTTON ASSEMBLY
40 NOMINAL May be increased as needed
AS NEEDED
BRACKET Redesign as necessary for attachment to the base. Add holes to the base and specify fasteners
STOP BUTTON
M10
SPRING located here
KNURLED HANDLE
SPACER
MATCH TO STOP BUTTON
BASE

Figure P6-17

the base so the cylinder aligns with output holes A and B. Liquids will enter the internal cylinder from the top, then exit the base through holes A and B. Include as many holes in the internal cylinder as necessary to create the following liquid exit combinations.

1. A open, B closed
2. A open, B open
3. A closed, B open

The internal cylinder is to be held in place by an alignment key and a stop button. The stop button is to be spring-loaded so that it will always be held in place. The internal cylinder will be moved by pulling out the stop button, repositioning the cylinder, then reinserting the stop button.

Prepare the following drawings.

A. Draw an assembly drawing.
B. Draw detail drawings of each nonstandard part. Include positional tolerances for all holes.
C. Prepare a parts list.

Project 6-18: Grinding Wheel

Design a hand-operated grinding wheel as shown in Figure P6-18 specifically for sharpening a chisel. The chisel is to be located on an adjustable rest while it is being sharpened. The mechanism should be able to be clamped to a table during operation using two thumbscrews. A standard grinding wheel is ∅6.000 in. and 1/2 in. thick, and has an internal mounting hole with a 50.00 ± .03 bore.

Prepare the following drawings.

A. Draw an assembly drawing.
B. Draw detail drawings of each nonstandard part. Include positional tolerances for all holes.
C. Prepare a parts list.

Project 6-19: Millimeters

Given the assembly shown in Figure P6-19 on page 372, add the following fasteners.

1. Create an assembly drawing.
2. Create a parts list including assembly numbers.
3. Create a dimensioned drawing of the support block and specify a dimension for each hole including the thread size and the depth required.

Fasteners:

A.

1. M10 × 35 HEX HEAD BOLT
2. M10 × 35 HEX HEAD BOLT
3. M10 × 30 HEX HEAD BOLT
4. M10 × 25 HEX HEAD BOLT

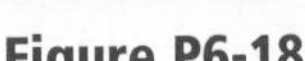

Figure P6-18

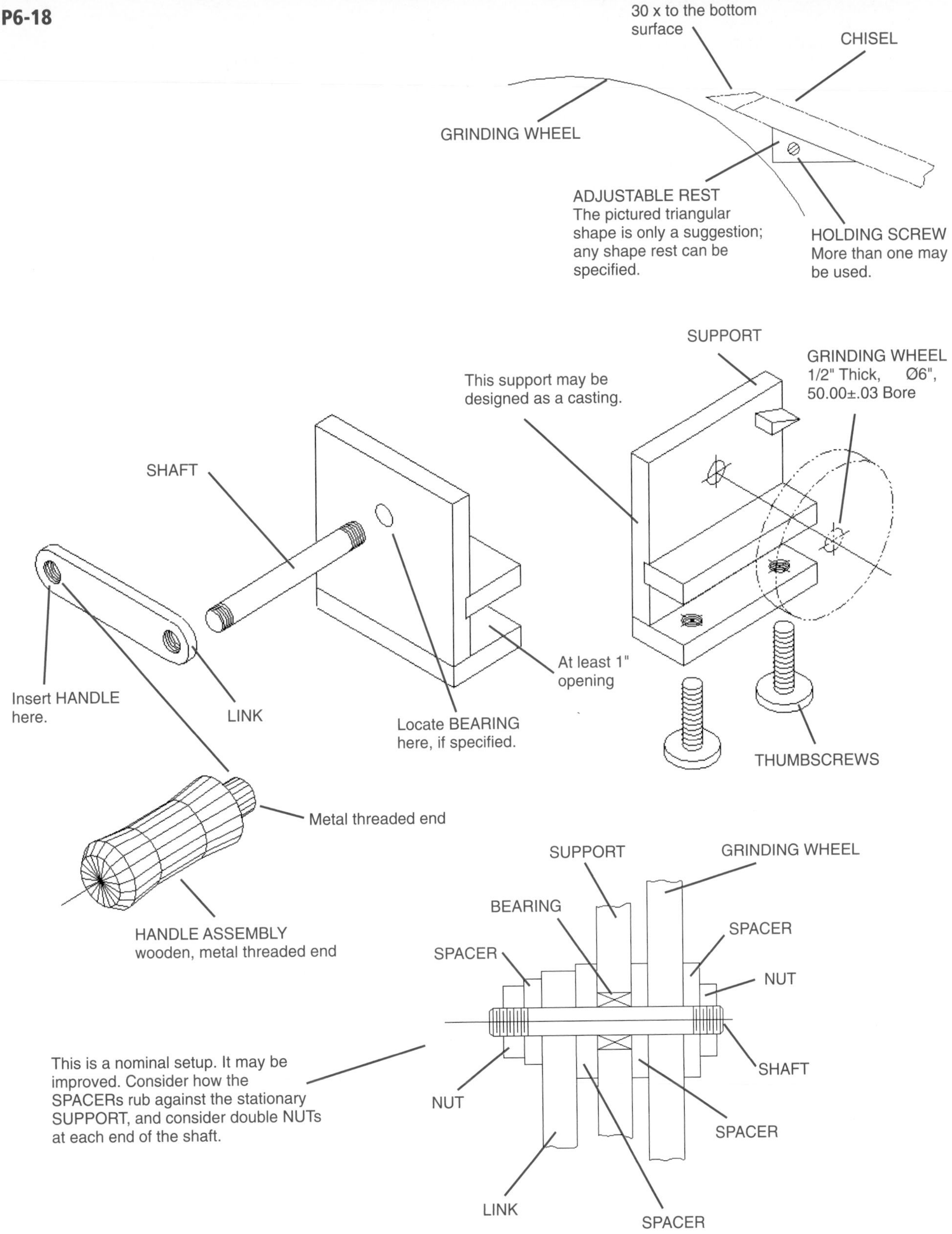

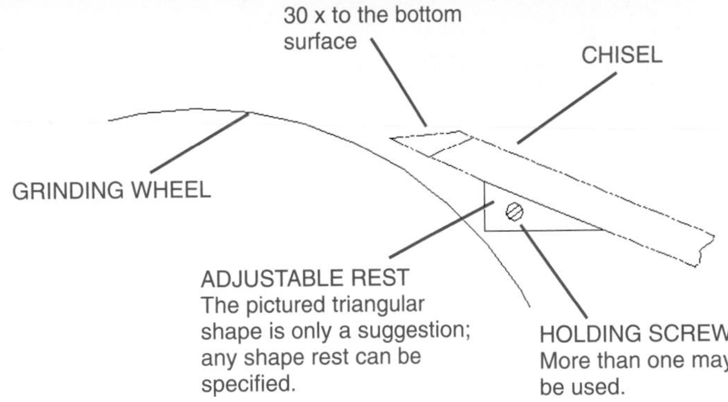

30 x to the bottom surface
CHISEL
GRINDING WHEEL
ADJUSTABLE REST
The pictured triangular shape is only a suggestion; any shape rest can be specified.
HOLDING SCREW
More than one may be used.
SUPPORT
GRINDING WHEEL
1/2" Thick, Ø6", 50.00±.03 Bore
This support may be designed as a casting.
SHAFT
Insert HANDLE here.
LINK
Locate BEARING here, if specified.
At least 1" opening
THUMBSCREWS
Metal threaded end
HANDLE ASSEMBLY
wooden, metal threaded end
SUPPORT
GRINDING WHEEL
BEARING
SPACER
SPACER
NUT
SHAFT
This is a nominal setup. It may be improved. Consider how the SPACERs rub against the stationary SUPPORT, and consider double NUTs at each end of the shaft.
NUT
SPACER
LINK
SPACER

B.

1. M10 × 1.5 × 35 HEX HEAD BOLT
2. M8 × 35 ROUND HEAD BOLT
3. M10 × 30 HEXAGON SOCKET HEAD CAP SCREW
4. M6 × 30 SQUARE BOLT

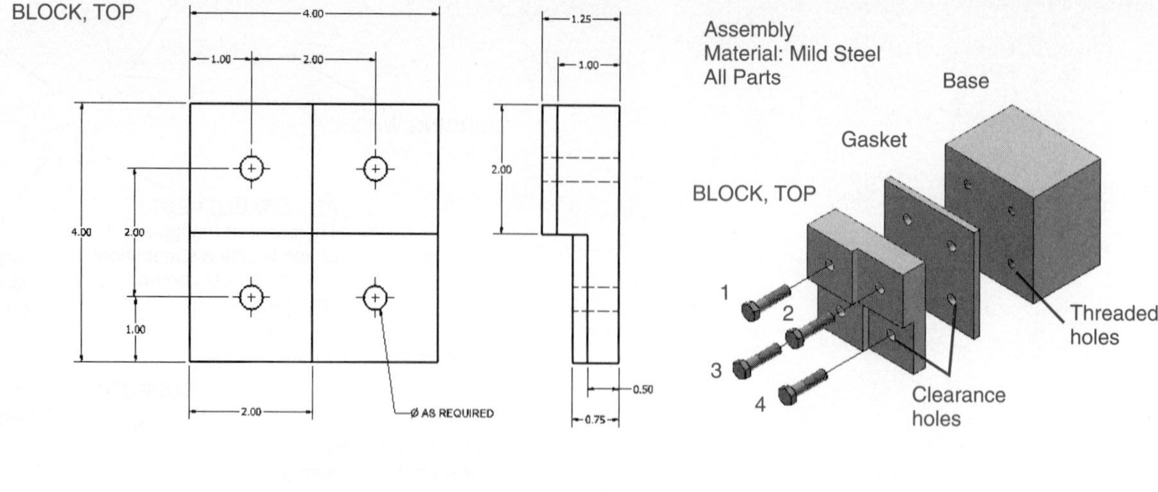

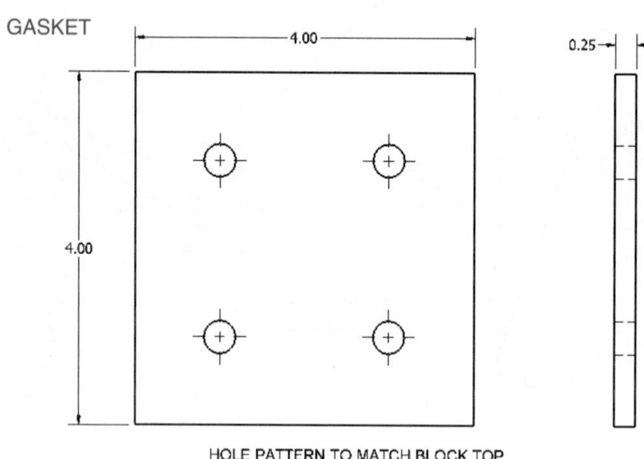

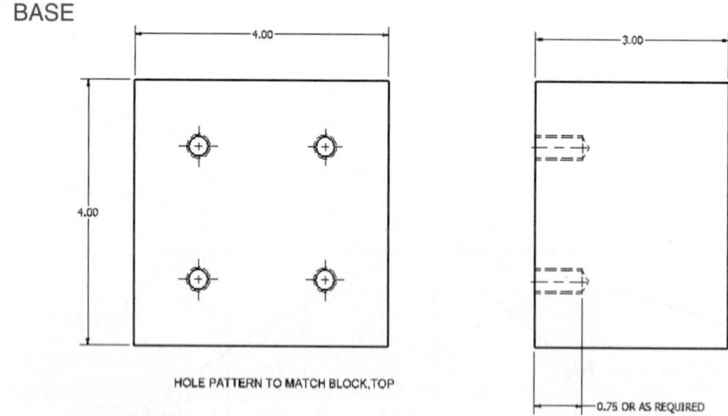

Figure P6-19

Project 6-20: Inches

1. Create an assembly drawing.
2. Create a parts list including assembly numbers.
3. Create a dimensioned drawing of the base and specify a dimension for each hole including the thread size and the depth required.

Fasteners:

A.

1. 3/8-16 UNC × 2.50 HEX HEAD SCREW
2. 1/4-20 UNC × 2.00 HEX HEAD SCREW
3. 7/16-14 UNC × 1.75 HEX HEAD SCREW
4. 5/16-18 UNC × 2.25 HEX HEAD SCREW

B.

1. 1/4-28 UNF × 2.00 HEX HEAD SCREW
2. #8 (.164)-32 UNC × 2.00 INDENTED LARGE HEX HEAD SCREW
3. 3/8-16 UNC × 1.75 CROSS RECESSED PAN HEAD MACHINE SCREW
4. 5/16-18 UNC × 1.75 HEXAGON SOCKET HEAD CAP SCREW

Project 6-21: Millimeters

Given the collar shown in Figure P6-21, add the following setscrews.

1. Create an assembly drawing.
2. Create a parts list.
3. Create a dimensioned drawing of the collar. Specify a thread specification for each hole as required by the designated setscrew.

Holes:

A.

1. M4 × 6-AS1421 DOG POINT-METRIC
2. M3 × 3 BROACHED HEXAGON SOCKET SET SCREW - FLAT POINT - METRIC
3. M2.5 × 4 BS4168: PART 3 HEXAGON SOCKET SET SCREW - CONE POINT - METRIC
4. M4 × 5 FORGED HEXAGON SOCKET SET SCREW - HALF DOG POINT - METRIC

B.

1. M2 × 4 JIS B 1117 TRUNCATED CONE POINT - METRIC
2. M3 × 6 JIS B 1117 LONG DOG POINT - METRIC

Figure P6-21

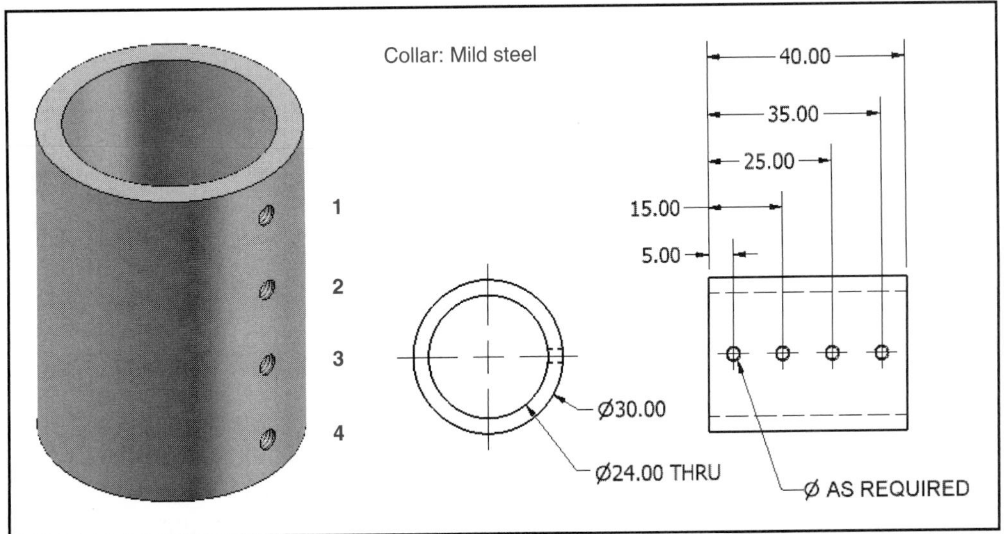

3. M4 × 5 SS-ISO 4766 SLOTTED HEADLESS SET SCREW
4. M1.6 × 4 JIS B 1117 FLAT POINT HEXAGON SOCKET SET SCREW

Project 6-22: Inches

Given the collar shown in Figure P6-22, add the following setscrews.

1. Create an assembly drawing.
2. Create a parts list.
3. Create a dimensioned drawing of the collar. Specify a thread specification for each hole as required by the designated setscrew.

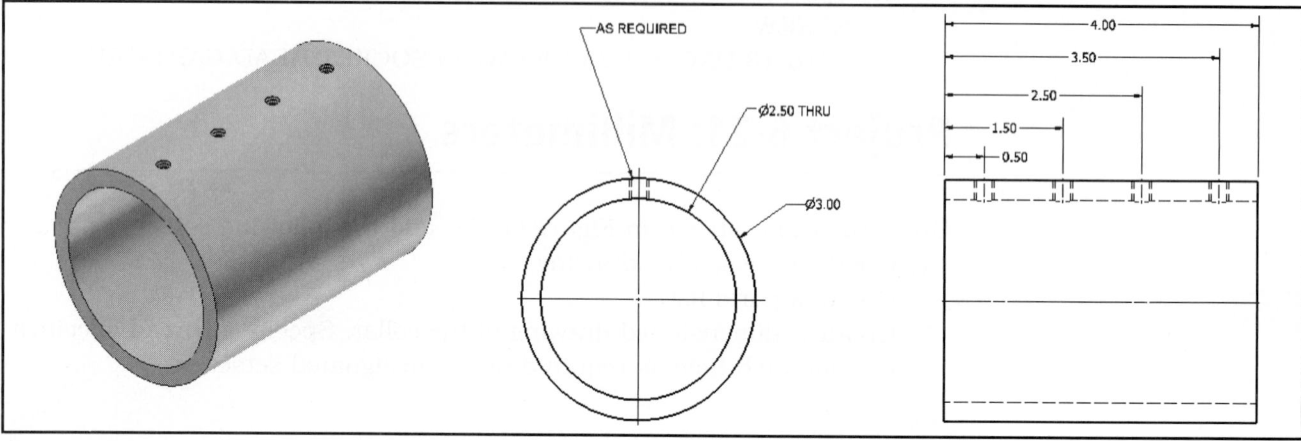

Figure P6-22

Holes:

A.

1. #10 (0.190) × .375 SQUARE HEAD SET SCREW - DOG POINT - INCH
2. #6 (0.138) × .125 SLOTTED HEADLESS SET SCREW - FLAT POINT - INCH
3. #8 (0.164) × .375 TYPE C - SPLINE SOCKET SET SCREW - CUP POINT - INCH
4. #5 (0.126) × .45 HEXAGON SOCKET SET SCREW - UNBRAKO CONE POINT - INCH

B.

1. #6 (0.138) × .25 TYPE D - SPLINE SOCKET SET SCREW - CUP POINT - INCH
2. #8 (0.164) × .1875 SLOTTED HEADLESS SET SCREW - DOG POINT - INCH
3. #10 (0.190) × .58 HEXAGON SOCKET SET SCREW - FLAT POINT - INCH
4. #6 (0.138) × .3125 SPLINE SOCKET SET SCREW - HALF DOG POINT - INCH

Project 6-23: Millimeters

Given the components shown in Figure P6-23:

1. Create an assembly drawing.
2. Create a presentation drawing.

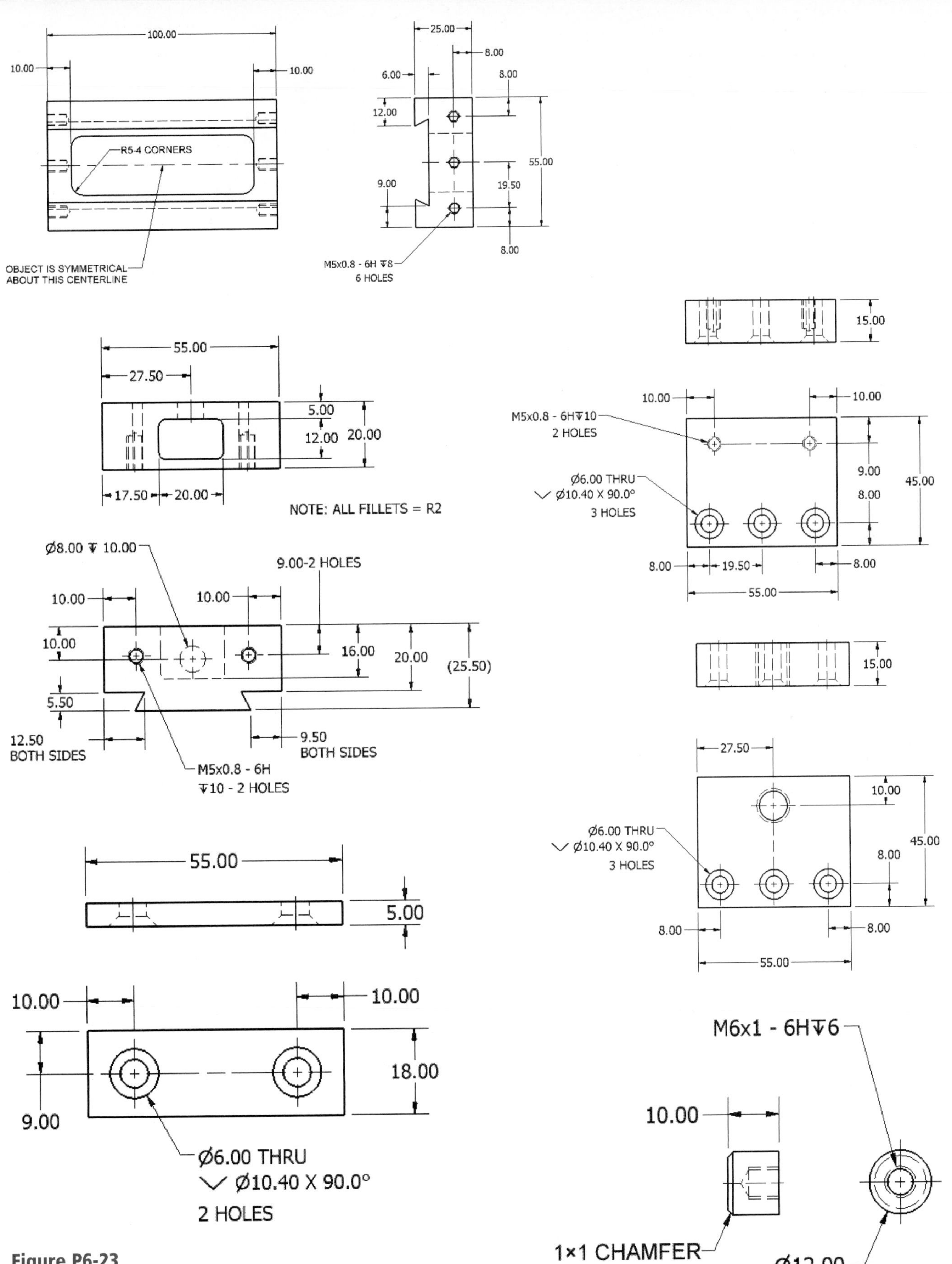
100.00
10.00
10.00
R5-4 CORNERS
OBJECT IS SYMMETRICAL ABOUT THIS CENTERLINE
25.00
8.00
6.00
8.00
12.00
55.00
9.00
19.50
8.00
M5x0.8 - 6H ↧8
6 HOLES
55.00
27.50
5.00
12.00
20.00
17.50
20.00
NOTE: ALL FILLETS = R2
15.00
10.00
10.00
M5x0.8 - 6H↧10
2 HOLES
Ø6.00 THRU
⌵ Ø10.40 X 90.0°
3 HOLES
9.00
8.00
45.00
8.00
19.50
8.00
55.00
Ø8.00 ↧ 10.00
9.00-2 HOLES
10.00
10.00
10.00
16.00
20.00
(25.50)
5.50
12.50
BOTH SIDES
9.50
BOTH SIDES
M5x0.8 - 6H
↧10 - 2 HOLES
15.00
27.50
10.00
Ø6.00 THRU
⌵ Ø10.40 X 90.0°
3 HOLES
45.00
8.00
8.00
8.00
55.00
55.00
5.00
10.00
10.00
18.00
9.00
Ø6.00 THRU
⌵ Ø10.40 X 90.0°
2 HOLES
M6x1 - 6H↧6
10.00
1×1 CHAMFER
Ø12.00

Figure P6-23

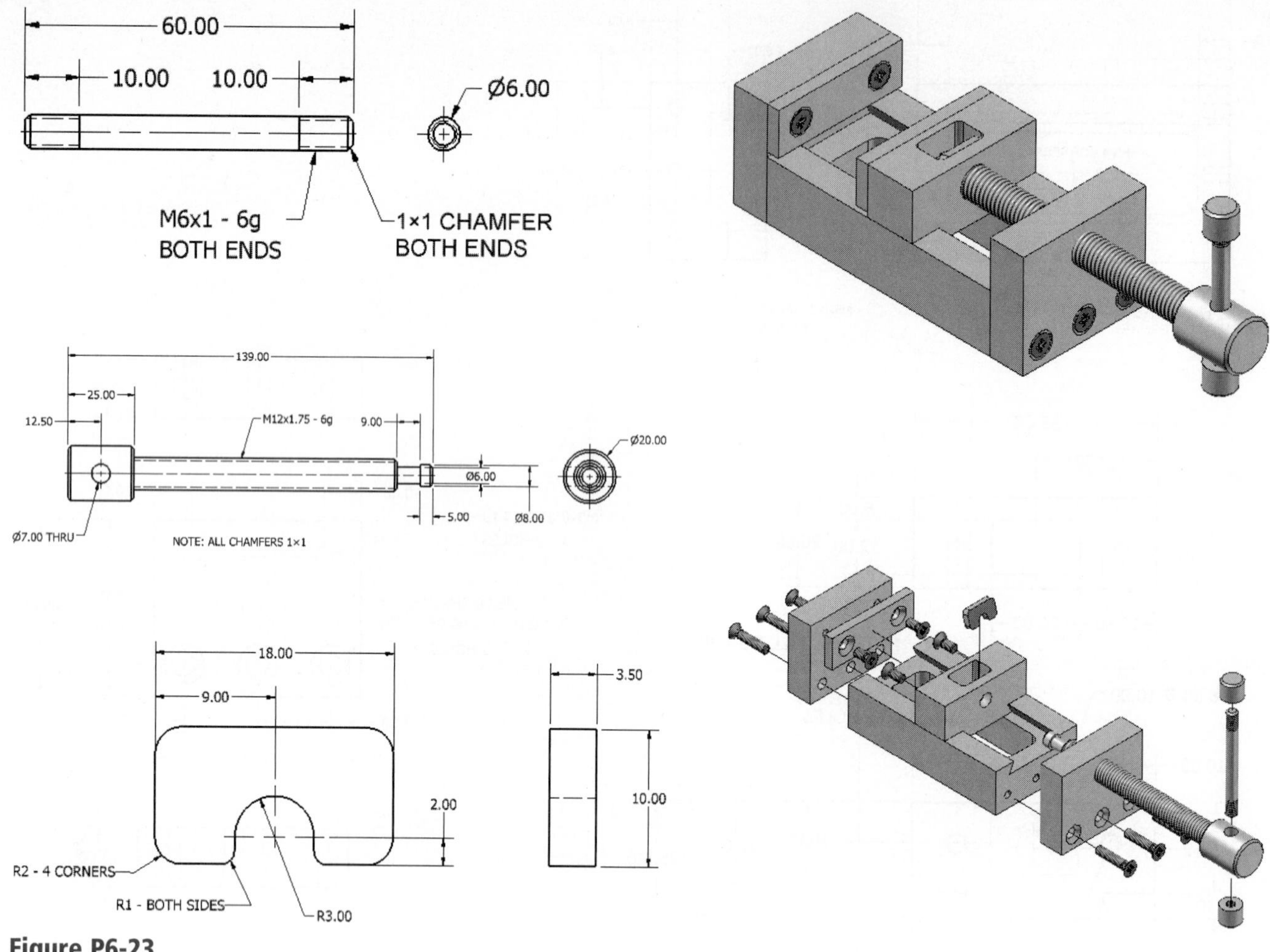

Figure P6-23
(Continued)

3. Animate the presentation drawing.
4. Create an exploded isometric drawing.
5. Create a parts list.

Project 6-24: Inches

Given the assembly drawing shown in Figure P6-24:

1. Create an assembly drawing.
2. Create a presentation drawing.
3. Animate the presentation drawing.
4. Create an exploded isometric drawing.
5. Create a parts list.

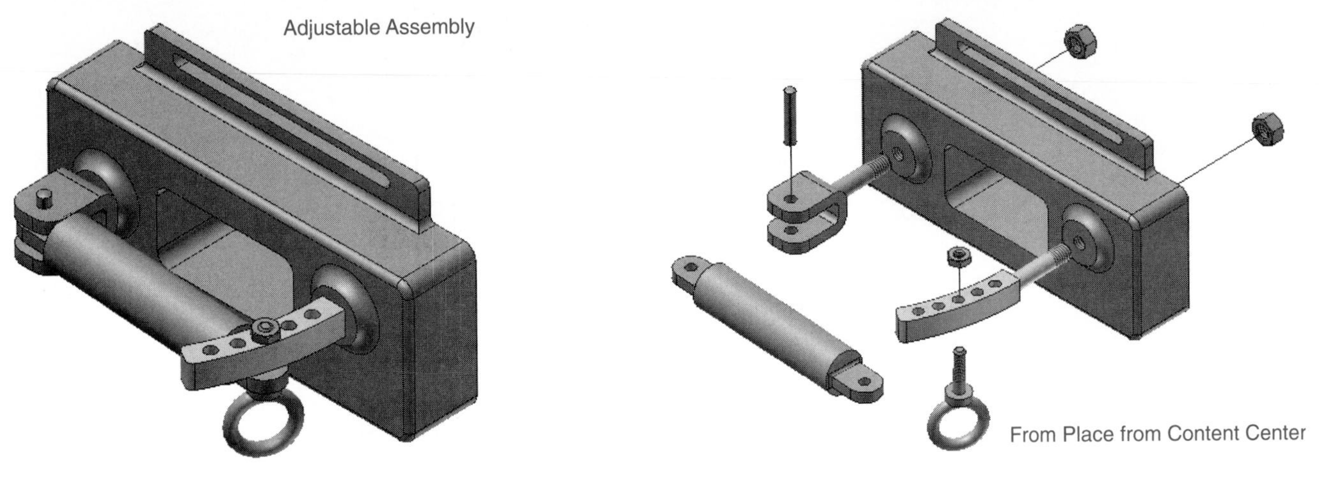
Adjustable Assembly
From Place from Content Center

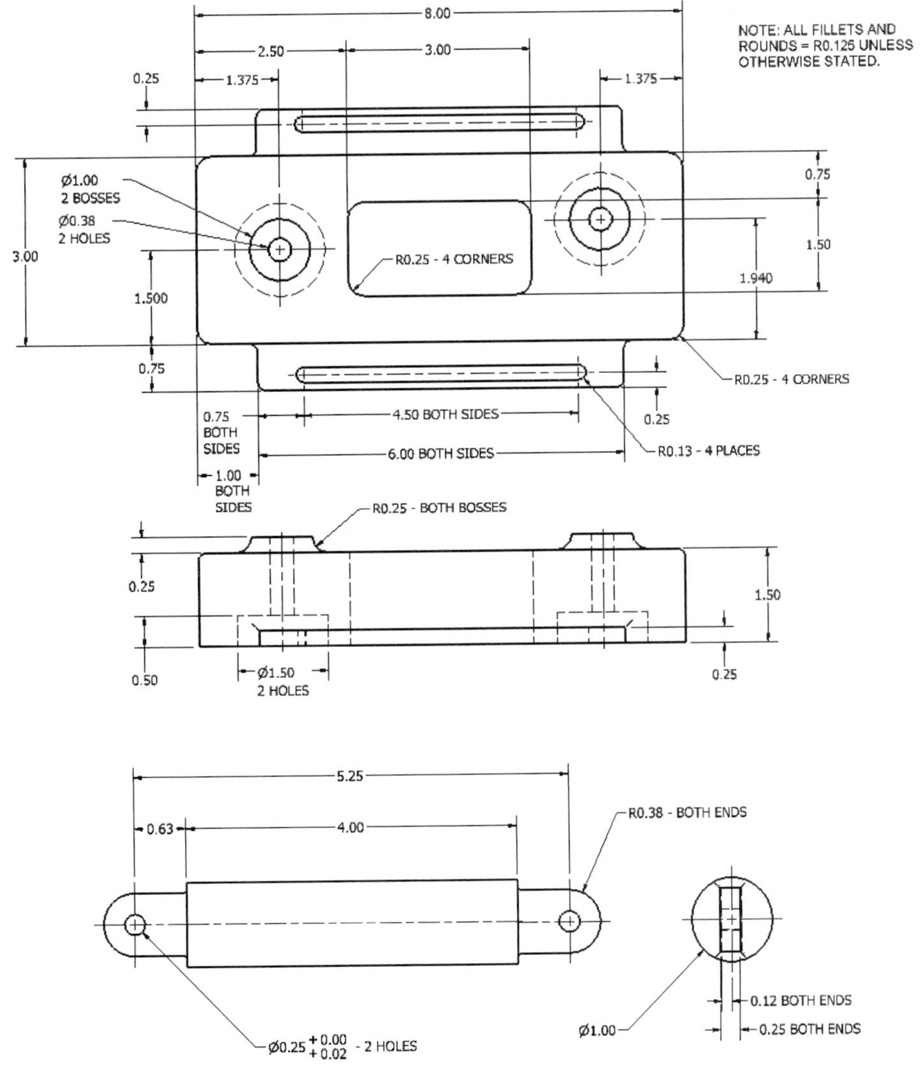
8.00
2.50
3.00
1.375
1.375
0.25
NOTE: ALL FILLETS AND ROUNDS = R0.125 UNLESS OTHERWISE STATED.
Ø1.00
2 BOSSES
Ø0.38
2 HOLES
3.00
R0.25 - 4 CORNERS
0.75
1.50
1.940
1.500
0.75
R0.25 - 4 CORNERS
0.75
BOTH
SIDES
4.50 BOTH SIDES
0.25
6.00 BOTH SIDES
R0.13 - 4 PLACES
1.00
BOTH
SIDES
R0.25 - BOTH BOSSES
0.25
1.50
0.50
Ø1.50
2 HOLES
0.25
5.25
0.63
4.00
R0.38 - BOTH ENDS
Ø0.25 +0.00 +0.02 - 2 HOLES
Ø1.00
0.12 BOTH ENDS
0.25 BOTH ENDS
NOTE: ALL FILLETS = R 0.125

Figure P6-24

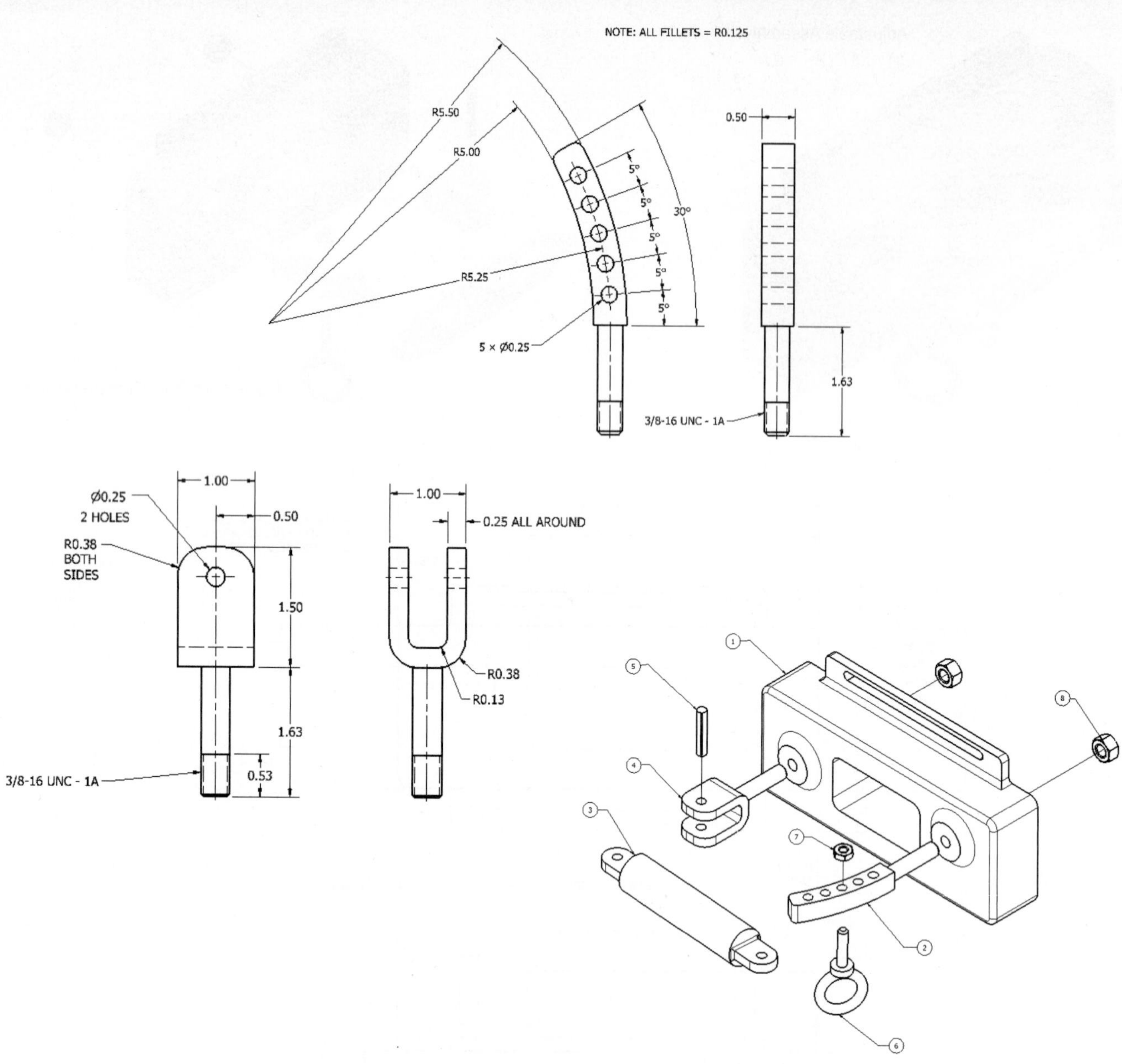

Parts List				
ITEM	PART NUMBER	DESCRIPTION	MATERIAL	QTY
1	ENG-311	BASE#4, CAST	Cast Iron	1
2	ENG-312	SUPPORT, ROUNDED	SAE 1040 STEEL	1
3	ENG-404	POST, ADJUSTABLE	Steel, Mild	1
4	BU-1964	YOKE	Cast Iron	1
5	ANSI B18.8.2 1/4x1.3120	Grooved pin, Type C - 1/4x1.312 ANSI B18.8.2	Steel, Mild	1
6	ANSI B18.15 - 1/4 - 20. Shoulder Pattern Type 2 - Style A	Forged Eyebolt	Steel, Mild	1
7	ANSI B18.6.3 - 1/4 - 20	Hex Machine Screw Nut	Steel, Mild	1
8	ANSI B18.2.2 - 3/8 - 16	Hex Nut	Steel, Mild	2

Figure P6-24
(Continued)

7 chapterseven Dimensioning a Drawing

CHAPTER OBJECTIVES

- Learn how to dimension objects
- Learn about ANSI standards and conventions
- Learn how to dimension different shapes and features
- Begin 3D dimensioning

Introduction

model dimension: A dimension created as a model is being constructed; it may be edited to change the shape of a model.

drawing dimension: A dimension attached to a specified distance on a drawing that can be edited without changing the shape of the model.

Inventor uses two types of dimensions: model and drawing. ***Model dimensions*** are created as the model is being constructed and may be edited to change the shape of a model. ***Drawing dimensions*** can be edited, but the changes will not change the shape of the model. If the shape of a model is changed, the drawing dimensions associated with the revised surfaces will change to reflect the new values.

Dimensions are usually applied to a drawing using either American National Standards Institute (ANSI) or International Organization for Standardization (ISO) standards. If English units are selected when a new drawing is started, the ANSI inch standards (ANSI (in).idw) will be invoked. If metric units are selected, the ISO standards (ISO.idw) may be invoked. This book uses ANSI standards for both inch and metric units. Metric units will use the ANSI-Metric standards.

Figure 7-1 shows a drawing that includes only the drawing dimensions. The **Metric** option was selected before the model was drawn. The model dimensions were created automatically as the model was created. The **Dimension** tool was used to edit the sketch dimensions.

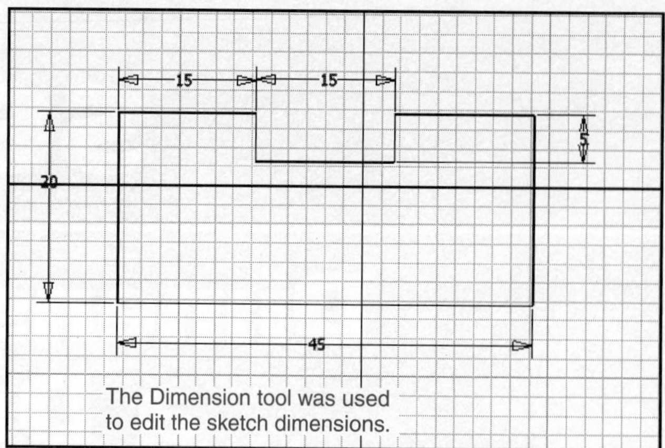

Figure 7-1

Terminology and Conventions—ANSI

Some Common Terms

Figure 7-2 shows both ANSI and ISO style dimensions. The terms apply to both styles.

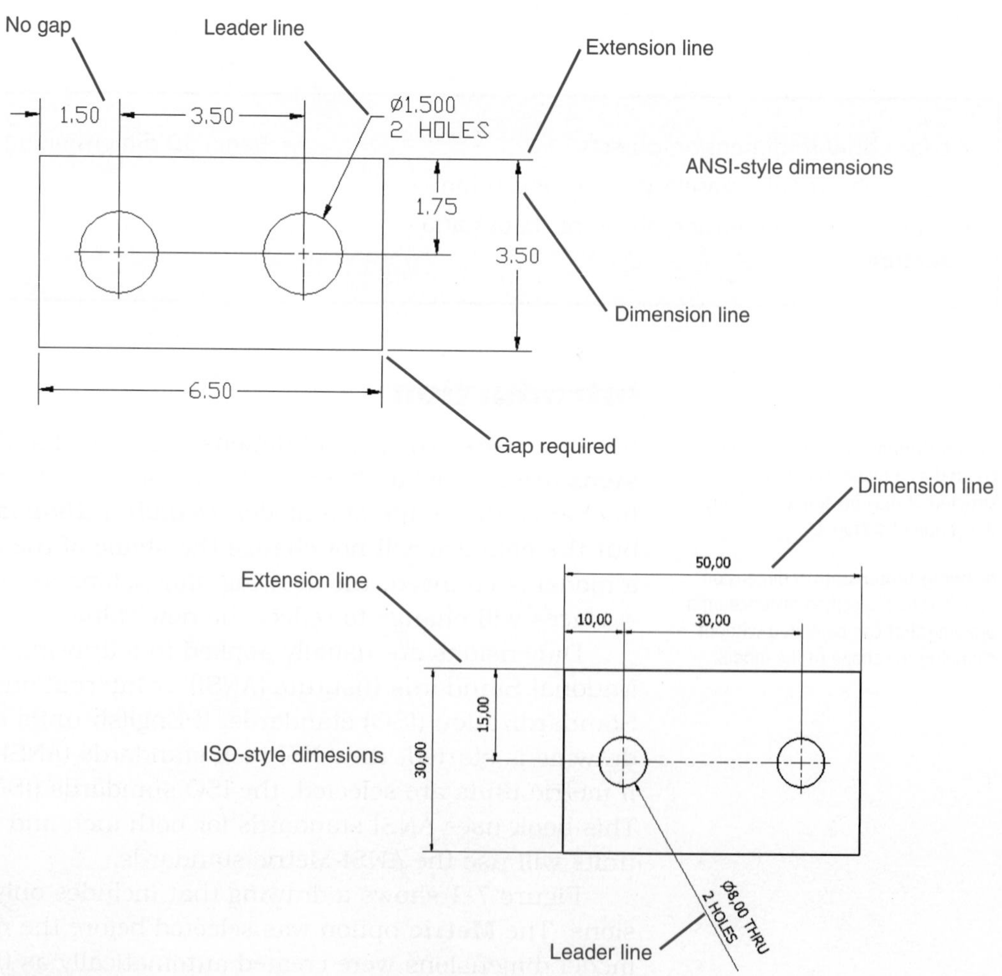

Figure 7-2

Dimension Lines: In mechanical drawings, lines between extension lines that end with an arrowhead and include a numerical dimensional value located within the line.

Extension Lines: Lines that extend away from an object and allow dimensions to be located off the surface of an object.

Leader Lines: Lines drawn at an angle, not horizontal or vertical, that are used to dimension specific shapes such as holes. The start point of a leader line includes an arrowhead. Numerical values are drawn at the end opposite the arrowhead.

Linear Dimensions: Dimensions that define the straight-line distance between two points.

Angular Dimensions: Dimensions that define the angular value, measured in degrees, between two straight lines.

Some Dimensioning Conventions

See Figure 7-3.

1 Dimension lines should be drawn evenly spaced; that is, the distance between dimension lines should be uniform. A general rule of thumb is to locate dimension lines about ½ in. or 15 mm apart.

2 There should a noticeable gap between the edge of a part and the beginning of an extension line. This serves as a visual break between the object and the extension line. The visual difference between the line types can be enhanced by using different colors for the two types of lines.

3 Leader lines are used to define the size of holes and should be positioned so that the arrowhead points toward the center of the hole.

4 Centerlines may be used as extension lines. No gap is used when a centerline is extended beyond the edge lines of an object.

5 Align dimension lines whenever possible to give the drawing a neat, organized appearance.

Figure 7-3

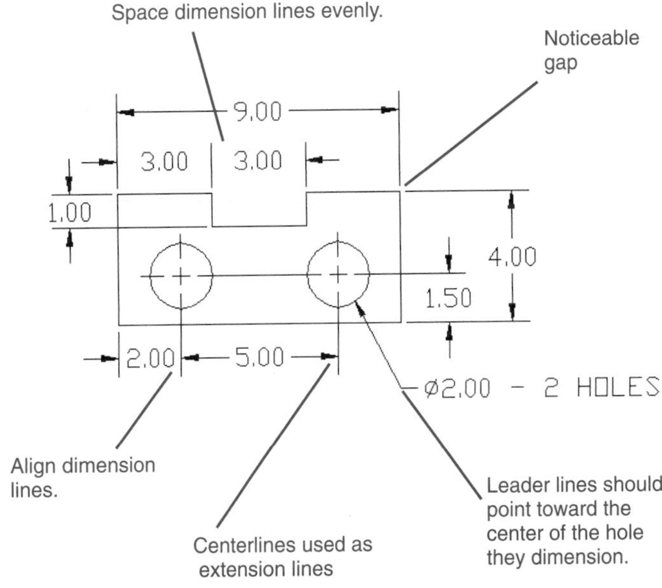

Space dimension lines evenly.
Noticeable gap
9.00
3.00
3.00
1.00
4.00
1.50
2.00
5.00
Ø2.00 - 2 HOLES
Align dimension lines.
Centerlines used as extension lines
Leader lines should point toward the center of the hole they dimension.

Some Common Errors to Avoid

See Figure 7-4.

1 Avoid crossing extension lines. Place longer dimensions farther away from the object than shorter dimensions.

2 Do not locate dimensions within cutouts; always use extension lines.

3 Do not locate any dimension close to the object. Dimension lines should be at least 1/2 in. or 15 mm from the edge of the object.

4 Avoid long extension lines. Locate dimensions in the same general area as the feature being defined.

Figure 7-4

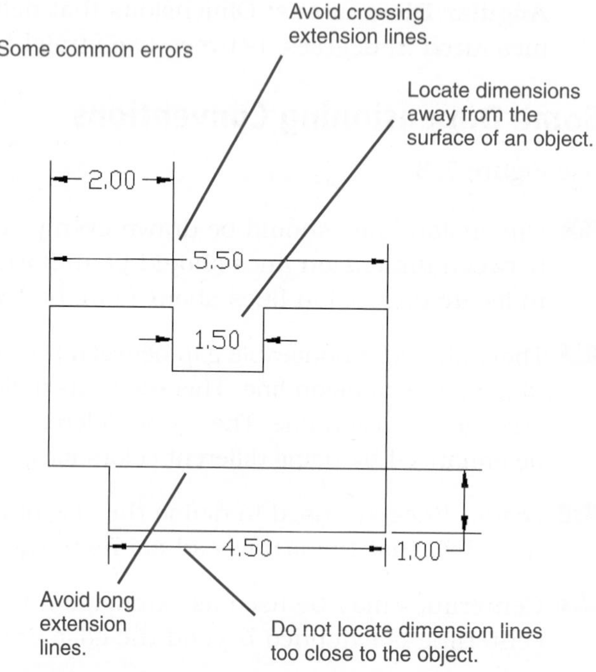

Creating Drawing Dimensions

Drawing dimensions are added to a drawing using the **Dimension** tool. The **Dimension** tool is located on the **Dimension** panel under the **Annotate** tab.

This section will add drawing dimensions to the model view shown in Figure 7-5. The dimensions will be in compliance with ANSI Metric standards. The model was drawn using the **Standard (mm).ipt** format. All values are in millimeters.

Figure 7-5

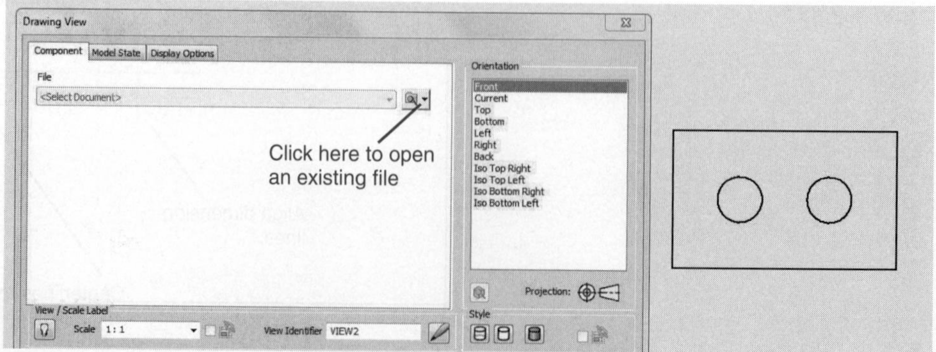

A dimensioned view of the model used in this example is available in Figure 7-11. The holes have a diameter of 10. Use the **Standard (mm).ipt** format and create a solid model. Extrude the model to a thickness of **10.** Save the model as **BLOCK, 2 HOLES.**

Drawing dimensions are different from model dimensions. *Model dimensions* are created as the model is created and can be used to edit (change the shape of) the model. *Drawing dimensions* are attached to a specified distance. Changing a drawing dimension will not change the shape of the model.

> **TIP**
>
> Changing the scale of a drawing will change the size of the drawing. The size of the dimension text will not change.

EXERCISE 7-1 Creating an Orthographic View for Dimensioning

1. Click on the **New** tool, then the **Metric** tab. Select the **ANSI (mm).idw** format and click **OK.**
2. Select the **Base** tool from the **Create** panel under the **Place Views** tab.

   The **Drawing View** dialog box will appear. See Figure 7-5.
3. Select the **Open an existing file** box, select the **BLOCK, 2 HOLES** file, click **Open,** locate the view on the drawing screen, and click the mouse.
4. If needed, change the scale of the drawing.
5. Click **OK.**

EXERCISE 7-2 Accessing the Annotate panel

1. Click the **Annotate** tab.

   The panel bar will change to the **Annotate** panel bar shown in Figure 7-6. The **Dimension** tool is the first tool listed.

Figure 7-6

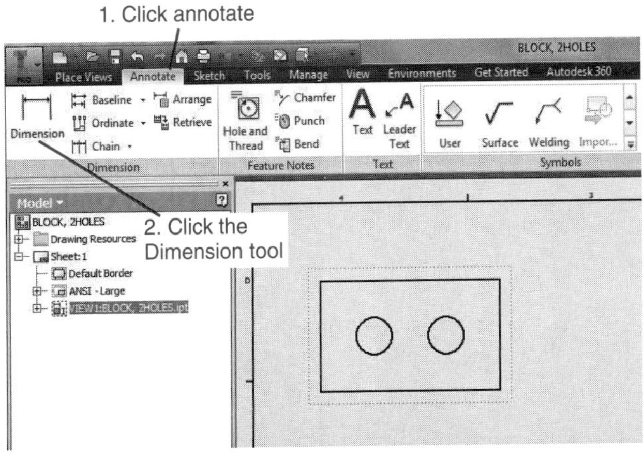
1. Click annotate
2. Click the Dimension tool

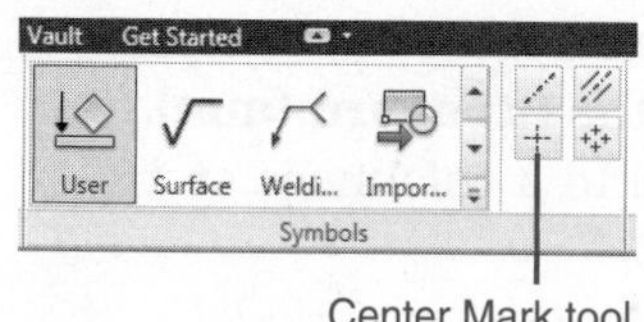

Figure 7-7

Adding Centerlines to Holes

The tools for adding centerlines to a drawing are located on the **Symbols** panel under the **Annotate** tab. See Figure 7-7.

Figure 7-8 shows a front orthographic view of the model to be dimensioned. Centerlines were added to the model using the **Center Mark** tool located on the **Symbols** panel.

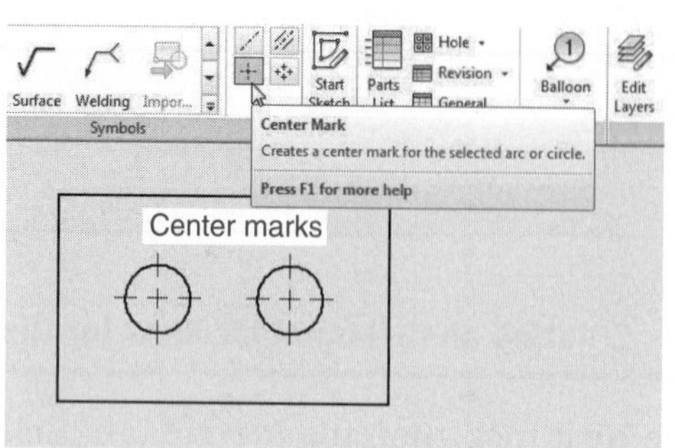

Figure 7-8

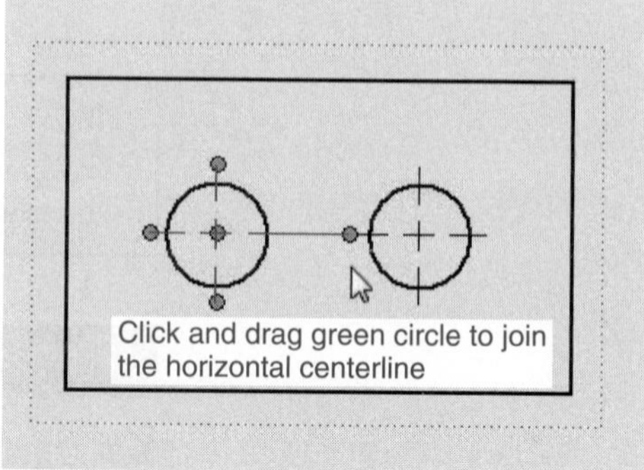

Figure 7-9

1 Click the **Center Mark** tool.

2 Click the edge of the circles.

Center marks will appear on the holes.

3 Right-click the mouse and click the **OK** option.

4 Move the cursor over the horizontal centreline of the left circle. Green circles will appear.

5 Click and drag the horizontal centreline so that it intersects the horizontal centreline of the right circle.

See Figure 7-9. This common centerline will be used to indicate that the holes are horizontally aligned.

EXERCISE 7-3 Adding Horizontal Dimensions

1 Click the **Dimension** tool.

2 Move the cursor into the drawing area and first click the upper left corner of the model.

A green circle will appear on the corner, indicating that it has been selected.

3 Click the top end of the left hole's vertical centerline.

4 Move the cursor, locating the dimension **(15),** then press the left mouse button.

The **Edit Dimension** dialog box will appear; click **OK**.

5 Still using the **Dimension** tool, click the top end of the two vertical centerlines and locate a second horizontal dimension between the two vertical hole centerlines **(20)**.

See Figure 7-10.

Figure 7-10

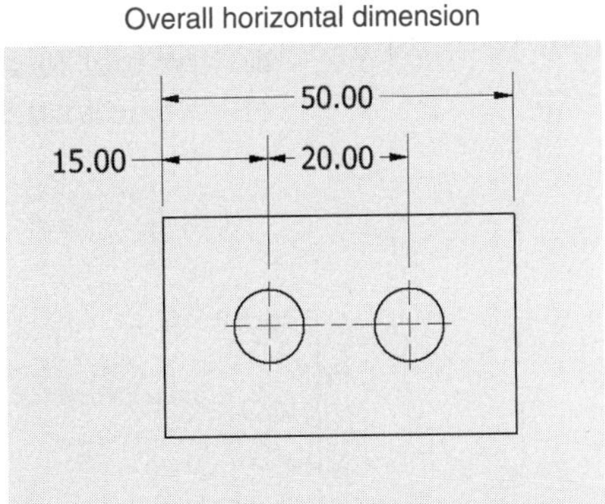

Overall Dimensions

overall dimensions: Dimensions that define the outside sizes of a model: the maximum length, width, and height.

Overall dimensions define the outside sizes of a model, the maximum length, width, and height. It is important that overall dimensions be easy to find and read, as they are often used to determine the stock sizes needed to produce the model.

Convention calls for overall dimensions to be located farther away from the model than any other dimensions. In Figure 7-10 the 50 overall dimension was located above the other two horizontal dimensions, that is, farther away from the model. The dimension could also have been located below the model.

Note that the spacing between the model's edge and the two horizontal dimensions is approximately equal to the distance between the overall dimension and the two horizontal dimensions.

Vertical Dimensions

ANSI standards call for the text of vertical dimensions to be written *unidirectionally.* This means that the text should be written horizontally and be read from left to right. Figure 7-11 shows two vertical dimensions added to

Figure 7-11

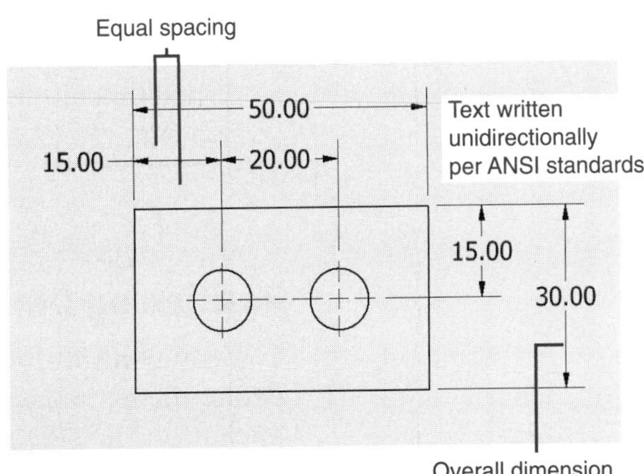

the model. Both use unidirectional text. Note also that the overall height dimension is located the farthest away from the model's edge.

EXERCISE 7-4 **Creating Unidirectional Text**

1 Click the **Manage** tab at the top of the screen, and select the **Styles Editor** tool located on the **Styles and Standards** panel.
The **Style and Standard Editor** dialog box will appear. See Figure 7-12.

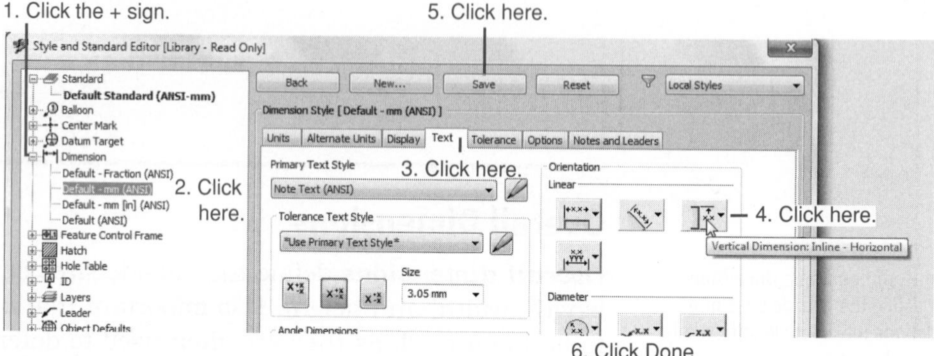

Figure 7-12

2 Click the + **sign** to the left of the **Dimension** listing.

3 Select the **Default - mm (ANSI)** option.

The **Style and Standard Editor** dialog box will change.

4 Select the **Text** tab under the **Dimension Style** heading.

5 Go to the **Orientation** box and select the **Vertical Dimension: Inline - Horizontal** option as shown.

6 Click the **Save** box, then click **Done.**

All vertical dimensions will now be written using unidirectional text.

7 Use the **Dimension** tool and add the vertical dimensions to the block. The vertical dimensions, 15.00 and 30.00, should be unidirectional as shown.

Positioning Dimension Text

You can control the location of dimension text by the sequence used to define the distance to be dimensioned. See Figure 7-13. The text will appear on the side of the first selected edge or point. For example, the text appears on the left when the left vertical line is selected first and on the right when the right vertical line is selected first.

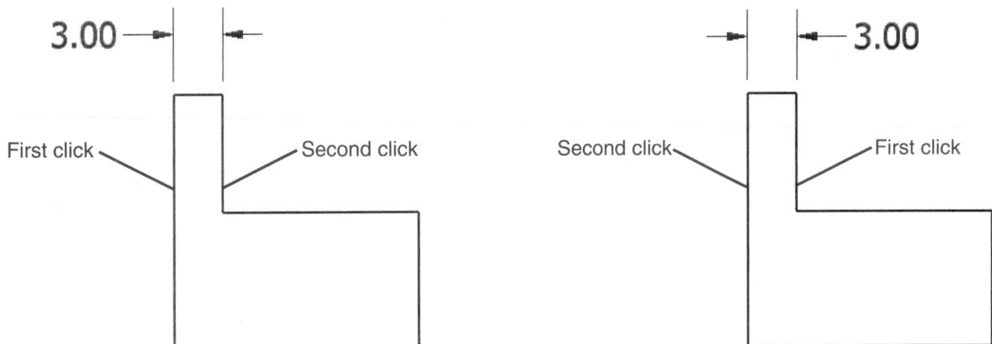

Figure 7-13

Dimensioning Holes

There are two Ø10 holes in the BLOCK, 2 HOLES model. Two hole dimensions could be applied, or one dimension could be used with the additional note Ø10 - 2 HOLES. In general, it is desirable to use as few dimensions as possible to clearly and completely define the model's size. This helps prevent a cluttered and confusing drawing.

1. Click the **Hole and Thread** tool on the **Feature Notes** panel under the **Annotate** tab, then the edge of one of the holes.
2. Drag the dimension away from the model.
3. Locate the hole dimension and left-click the mouse.
4. Right-click the mouse and select the **OK** option.

   See Figure 7-14.

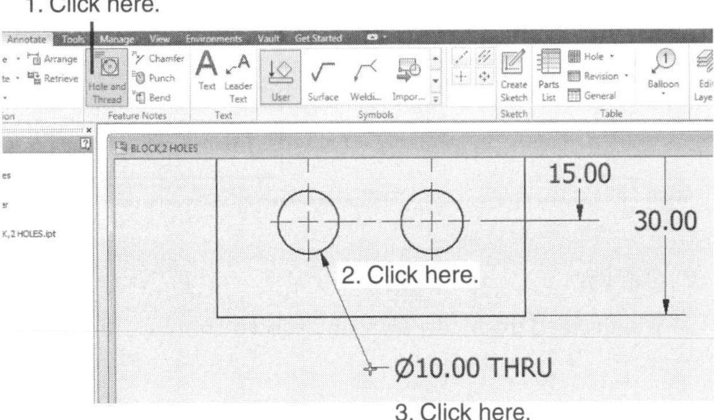

Figure 7-14

EXERCISE 7-5 **Adding Text to the Hole Dimension**

1. Move the cursor onto the hole dimension and right-click the mouse. See Figure 7-15.
2. Click the **Text** option.

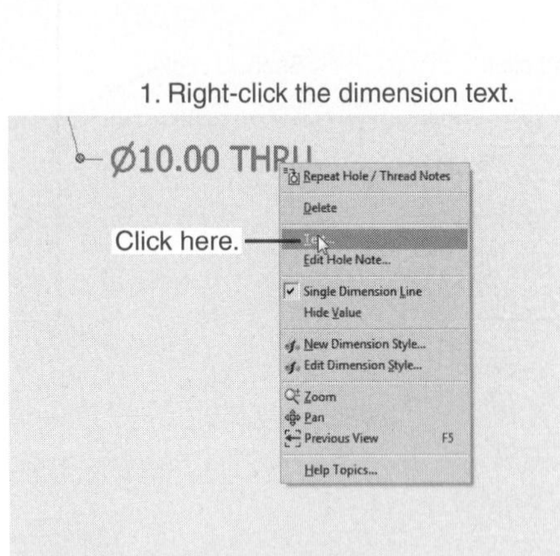

Figure 7-15

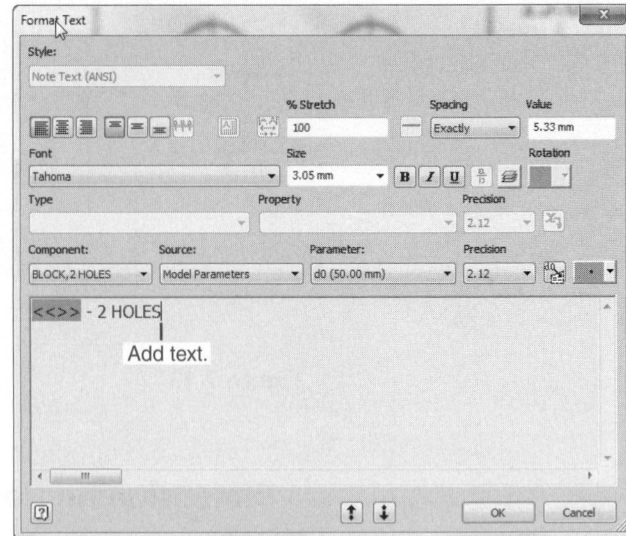

Figure 7-16

The **Format Text** dialog box will appear. See Figure 7-16. The ≪ ≫ symbol represents the existing text.

3 Locate the cursor to the right of the ≪ ≫ symbol and type - **2 HOLES.**

4 Click the **OK** box.

Figure 7-17 shows the resulting dimension.

Figure 7-17

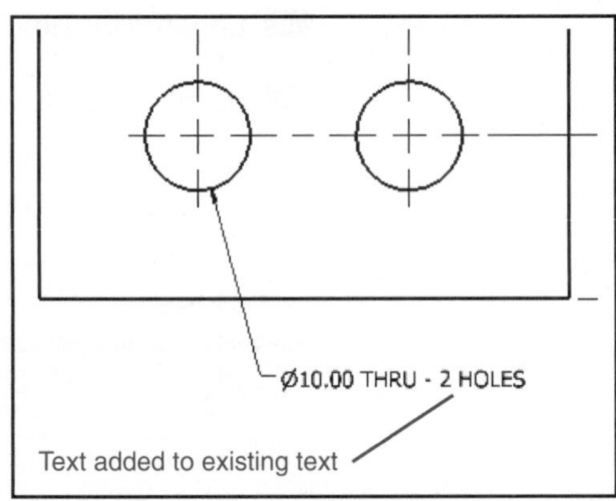

NOTE

The **Styles Editor** can also be used to change the height of the text.

TIP

Drawing convention calls for all drawing text to use uppercase letters.

EXERCISE 7-6 Editing a Hole Dimension

The hole dimension shown in Figure 7-17 includes the word **THRU** after the dimension value. Including the word THRU after a hole dimension is an optional practice, and not all companies include it. The word may be removed using the **Edit Hole Note** option.

1. Move the cursor to the hole dimension and right-click the mouse.
2. Select the **Edit Hole Note** option.

   The **Edit Hole Note** dialog box will appear. See Figure 7-18.
3. Backspace out the word **THRU.**
4. Click the **OK** box.

Figure 7-19 shows the resulting hole dimension. If the 2 HOLES note is not added using the **Text** option, it can be added using the **Edit Hole Note** option.

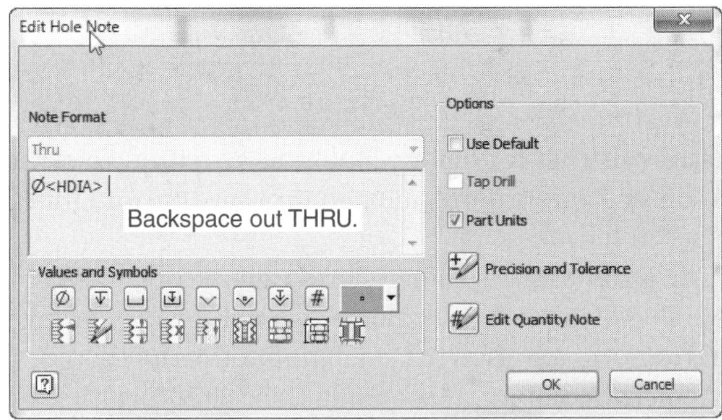

Figure 7-18

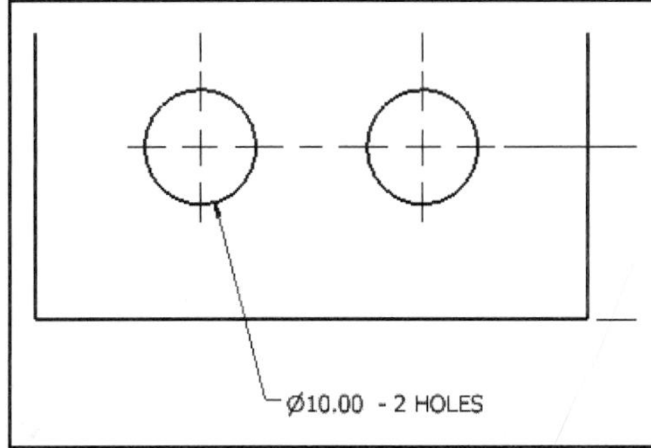

Figure 7-19

Drawing Scale

Drawings are often drawn "to scale" because the actual part is either too big to fit on a sheet of drawing paper or too small to be seen. For example, a microchip circuit must be drawn at several thousand times its actual size to be seen.

Drawing scales are written using the following formats:

SCALE: 1 = 1

SCALE: FULL

SCALE: 1000 = 1

SCALE: .25 = 1

In each example the value on the left indicates the scale factor. A value greater than 1 indicates that the drawing is larger than actual size. A value smaller than 1 indicates that the drawing is smaller than actual size.

Regardless of the drawing scale selected the dimension values must be true size. Figure 7-20 shows the same rectangle drawn at two different scales. The top rectangle is drawn at a scale of 1 = 1, or its true size. The bottom rectangle is drawn at a scale of 2 = 1, or twice its true size. In both examples the 3.00 dimension remains the same.

Units

It is important to understand that dimensional values are not the same as mathematical units. Dimensional values are manufacturing instructions and always include a tolerance, even if the tolerance value is not stated. Manufacturers use a predefined set of standard dimensions that are applied

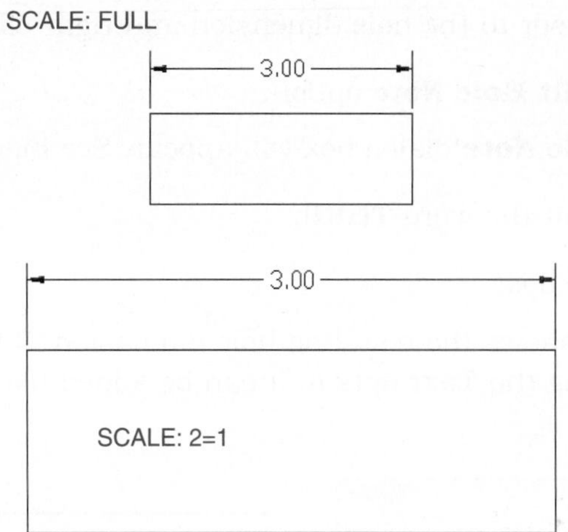

Figure 7-20

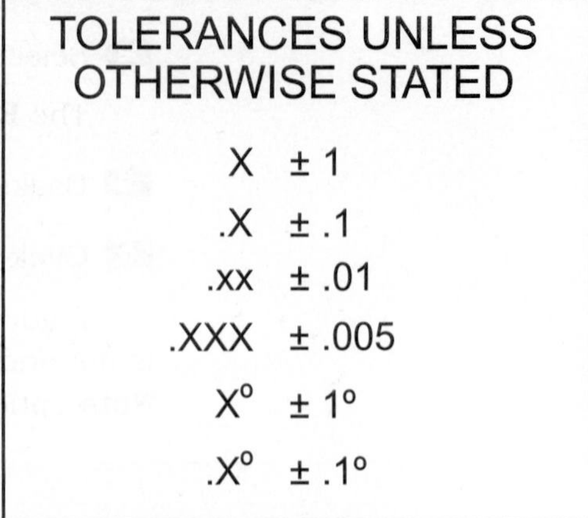

TOLERANCES UNLESS OTHERWISE STATED

X	± 1
.X	± .1
.XX	± .01
.XXX	± .005
X°	± 1°
.X°	± .1°

Figure 7-21

to any dimensional value that does not include a written tolerance. Standard tolerance values differ from organization to organization. Figure 7-21 shows a chart of standard tolerances.

In Figure 7-22 a distance is dimensioned twice: once as 5.50 and a second time as 5.5000. Mathematically these two values are equal, but they are not the same manufacturing instruction. The 5.50 value could, for example, have a standard tolerance of ±.01, whereas the 5.5000 value could have a standard tolerance of ±.0005. A tolerance of ±.0005 is more difficult and, therefore, more expensive to manufacture than a tolerance of ±.01.

These dimensions are not the same. They have different tolerance requirements.

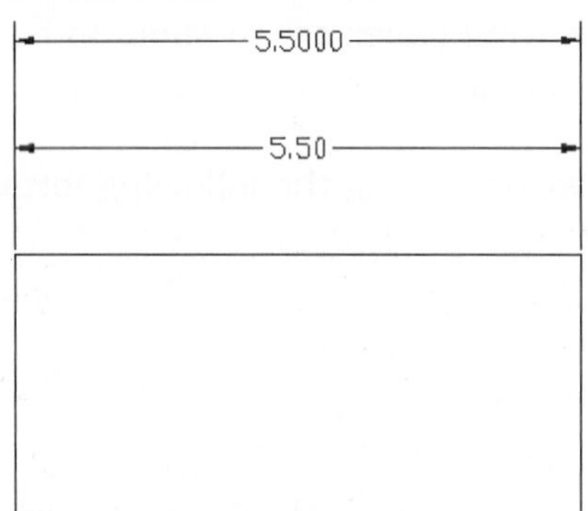

Figure 7-22

Millimeters

0.25	0.5	0.033
32	14.5	3

Zero required

Inches

No zero required

.25	.05	.033
32.00	14.50	3.000

Figure 7-23

Figure 7-23 shows examples of units expressed in millimeters and in decimal inches. A zero is not required to the left of the decimal point for decimal inch values less than one. Whole-number millimeter values do not require zeros to the right of the decimal point unless needed to specify a tolerance. Millimeter and decimal inch values never include symbols; the units will be defined in the title block of the drawing.

EXERCISE 7-7 **Preventing a 0 from Appearing to the Left of the Decimal Point**

1 Click the **Manage** tab at the top of the screen, then select the **Style and Standard Editor** option. See Figure 7-24.

The **Style and Standard Editor** dialog box will appear.

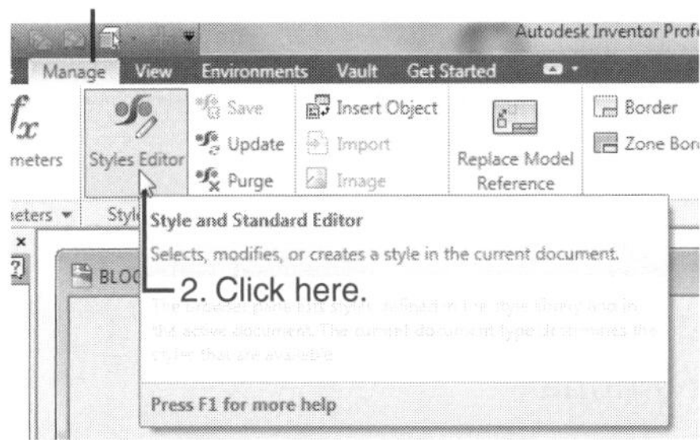

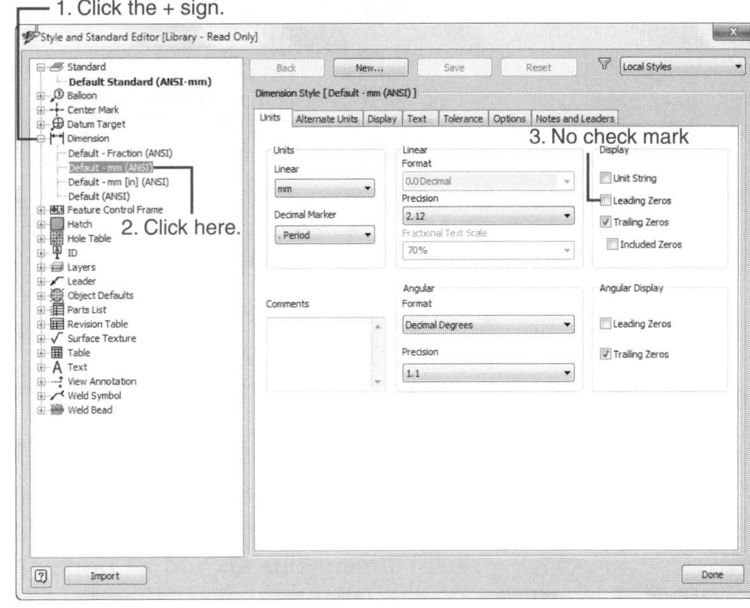

Figure 7-24

2 Click the + **sign** to the left of **Dimension,** and select **Default - mm (ANSI).**

3 Select the **Units** tab.

4 Remove the check mark from the **Leading Zero** box in the **Display** area.

This procedure creates a new dimension style that suppresses all leading zeros.

5 Click the **Save** and **Done** boxes. The **Default - mm (ANSI)** standard may already have the leading zero suppressed.

EXERCISE 7-8 Changing the Number of Decimal Places in a Dimension Value

1 Click on the **Manage** tab at the top of the screen, then select the **Style and Standard Editor** option.
The **Style and Standard Editor** dialog box will appear. See Figure 7-25.

Figure 7-25

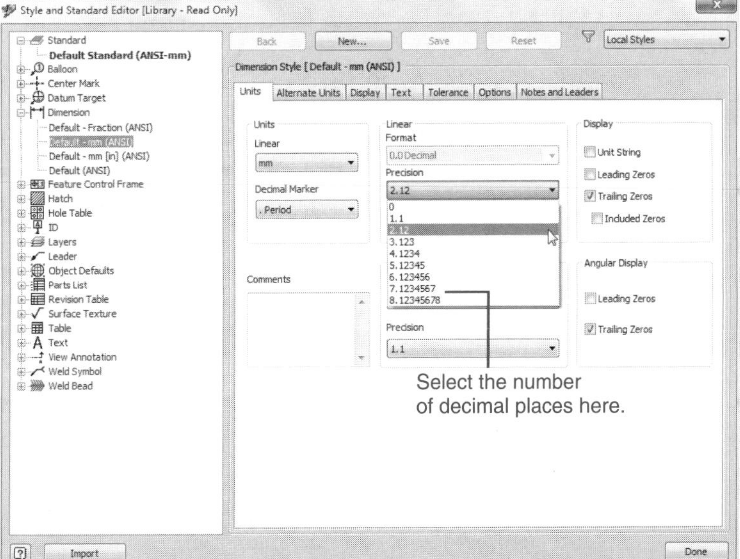

2. Click the + **sign** to the left of **Dimension,** and select **Default - mm (ANSI).**
3. Select the **Units** tab.
4. Click the scroll arrow on the right side of the **Precision** box.

   A listing of available precision settings will cascade down.
5. Select the desired precision value.

   In this example, 2 decimal places was selected.
6. Click **Save** then **Done.**

Aligned Dimensions

Aligned dimensions are dimensions that are parallel to a slanted edge or surface. They are not horizontal or vertical. The unit values for aligned dimensions should be horizontal or unidirectional.

EXERCISE 7-9 Creating an Aligned Dimension

Figure 7-26 shows an orthographic view of a model that includes a slanted surface. The model was created using the **Standard (mm).ipt** format, and the orthographic view was created using the **ANSI (mm).idw** format. The **Dimension** tool was used to add the aligned dimension.

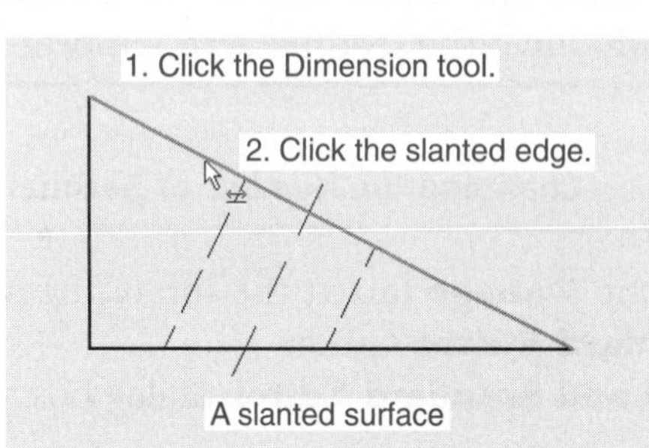

Figure 7-26

1. Click the **Annotate** tab.
2. Click the **Dimension** tool, located on the **Dimension** panel, then move the cursor onto the slanted edge to be dimensioned.

   The edge will change color, indicating that it has been selected.
3. Left-click the mouse.
4. Move the mouse as necessary to locate the aligned dimension.

   See Figure 7-26.

TIP

If an aligned dimension is not created, right-click the mouse and select **Dimension Type, Aligned** before locating the dimension.

EXERCISE 7-10 **Creating a Unidirectional Aligned Dimension**

Aligned dimensions can be made unidirectional by using the **Orientation** option on the **Style and Standard Editor** dialog box.

1 Click on the **Manage** tab at the top of the screen, then select the **Style and Standard Editor** tool.

The **Style and Standard Editor** dialog box will appear. See Figure 7-27.

Figure 7-27

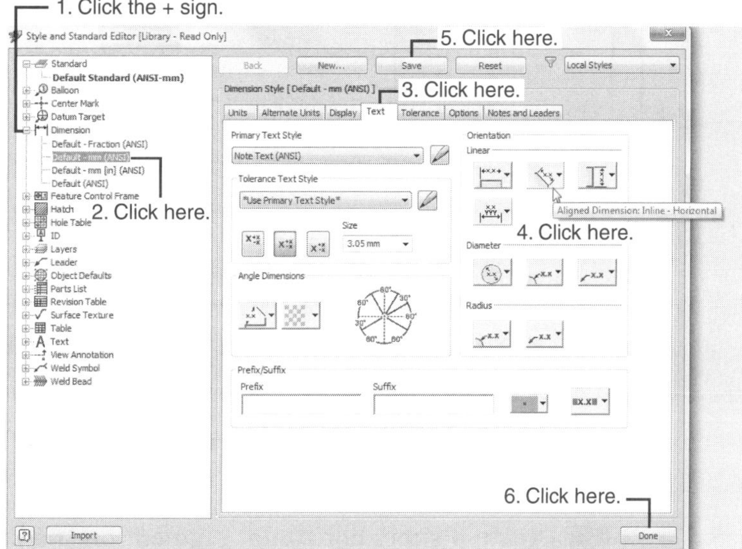

2 Click the + **sign** to the left of the **Dimension** listing.

3 Select **Default - mm (ANSI).**

The **Style and Standard Editor** dialog box will change.

4 Select the **Text** tab.

5 Go to the **Orientation** box and select the **Align Dimension: Inline - Horizontal** option as shown.

6 Click the **Save** box, then click **Done.**

The aligned text will become unidirectional.

Radius and Diameter Dimensions

Figure 7-28 shows an object that includes both fillets and circles. The general rule is to dimension arcs using a radius dimension, and circles using diameter dimensions. This convention is consistent with the tooling required to produce the feature shape. Any arc greater than 180° is considered a circle and is dimensioned using a diameter. In this example, the hole was omitted. The horizontal and vertical dimensions were added.

Figure 7-28

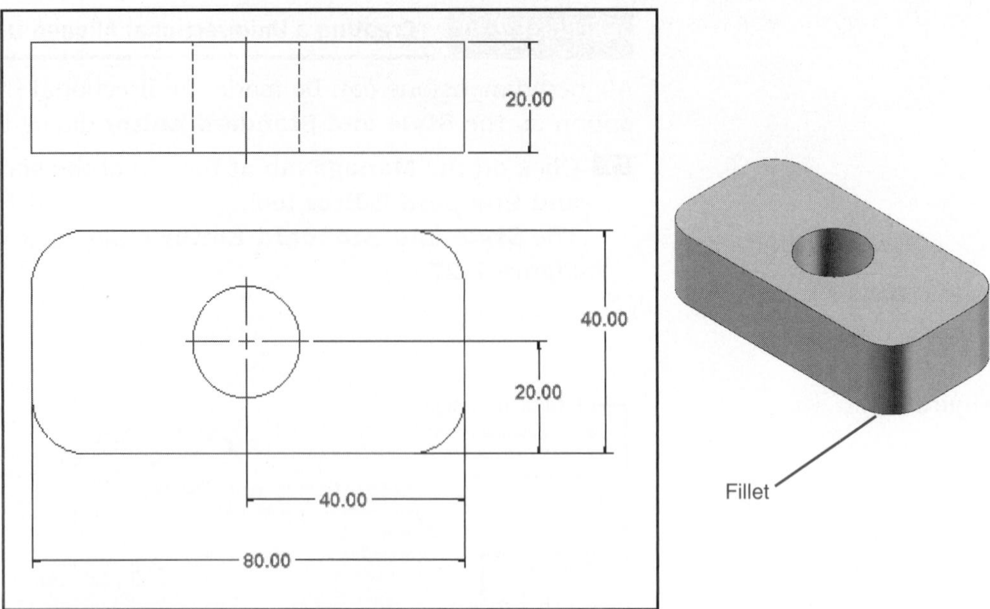

EXERCISE 7-11 Creating a Radius Dimension

1. Click the **Leader Text** tool located on the **Text** panel under the **Annotate** tab.
2. Click one of the filleted corners and move the cursor away from the corner.
3. Create a short horizontal segment on the leader line.

   See Figure 7-29.

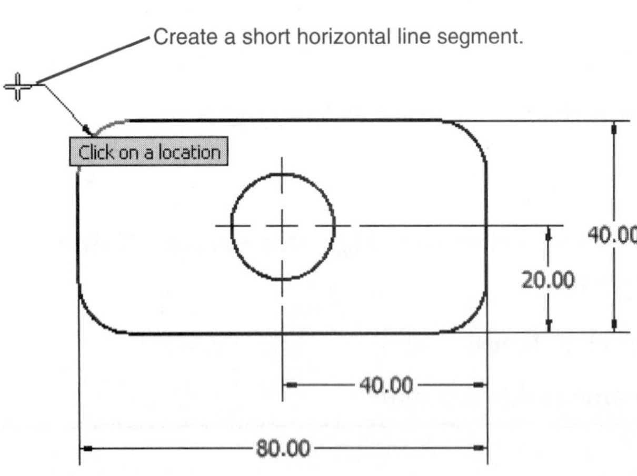

Figure 7-29

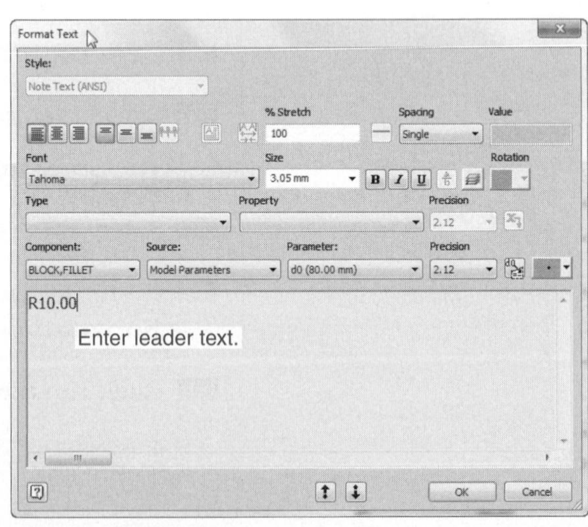

Figure 7-30

4. Right-click the mouse and select the **Continue** option.

   The **Format Text** dialog box will appear. See Figure 7-30.
5. Enter **R10.00**, and click **OK.**
6. Press the **<Esc>** key to end the dimensioning sequence.

Figure 7-31 shows the resulting dimension. There are four equal arcs on the object, and they all must be dimensioned. It would be better to add the words 4 CORNERS to the radius dimension than to include four radius dimensions.

Figure 7-31

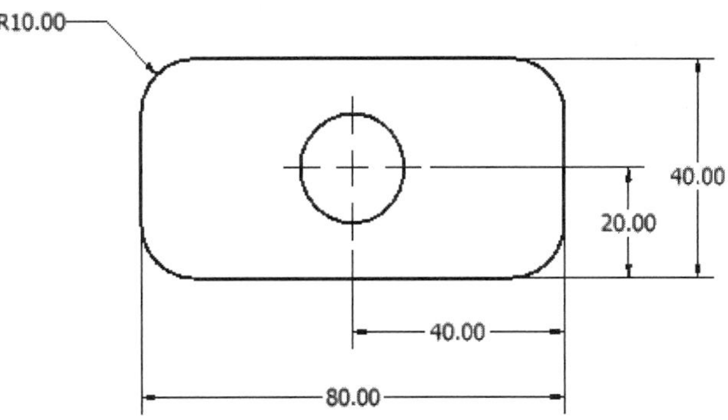

EXERCISE 7-12 Adding Text to an Existing Drawing Dimension

1 Move the cursor to the fillet dimension.

Filled colored circles will appear, indicating that the dimension has been selected.

2 Right-click the mouse and select the **Edit Leader Text** option.

The **Format Text** dialog box will appear with the existing text. See Figure 7-32.

3 Type **4 CORNERS** under the existing text and click the **OK** box.

Figure 7-33 shows the resulting dimension.

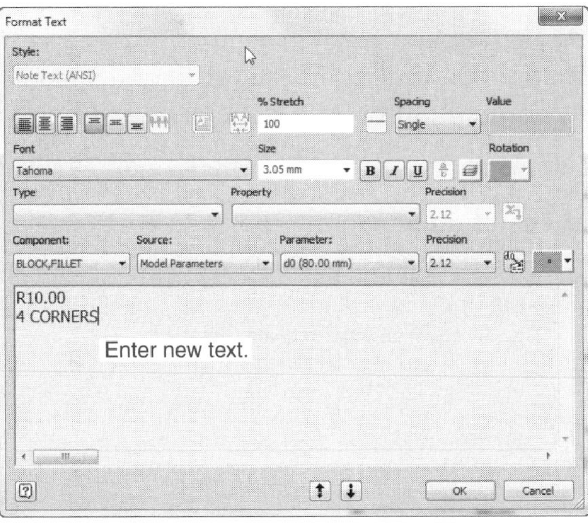

Figure 7-32

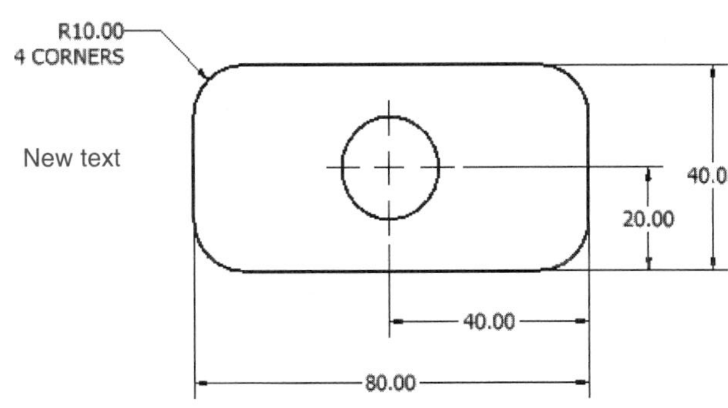

Figure 7-33

Dimensioning Holes

Holes are dimensioned by stating their diameter and depth, if any. Holes that go completely through an object are defined using only a diameter dimension. See Figure 7-34. The word **THRU** may be added if desired.

Figure 7-34

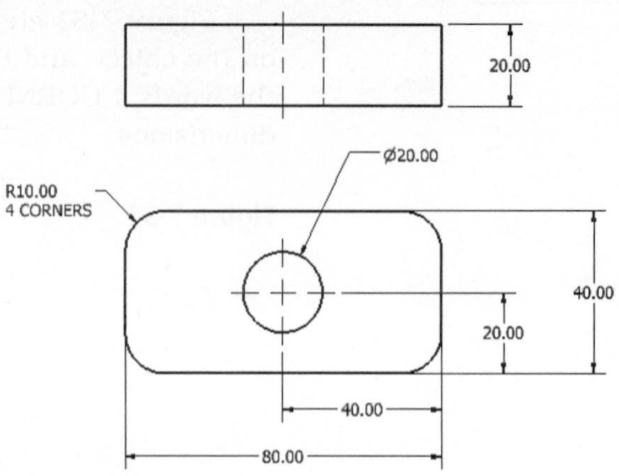

EXERCISE 7-13 Dimensioning a Through Hole

1. Click the **Hole and Thread Notes** tool on the **Feature Notes** panel under the **Annotate** tab.
2. Click the edge of the hole and move the cursor away from the hole, off the surface of the part.
3. Locate and enter the dimension by checking the left mouse button.
4. Right-click the mouse and select the **OK** option.
5. Move the cursor over the hole dimension, right-click the mouse, and select the **Edit Hole Note** option.
6. Backspace out the word **THRU,** then click **OK.**

TIP

The **Dimension** tool may also be used.

Dimensioning Individual Holes

Figure 7-35 shows three different methods that can be used to dimension a hole that does not go completely through an object. Dimension the hole using the **Dimension** tool. The depth value will be included automatically. The symbol indicates depth.

Figure 7-36 shows three methods of dimensioning holes in section views. The single-line-note version is the preferred method.

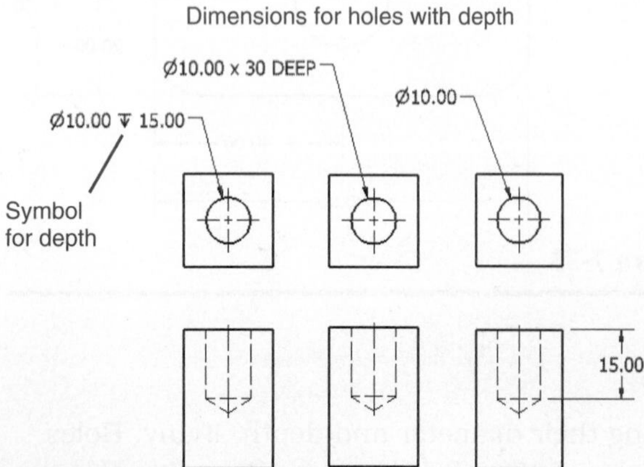

Figure 7-35

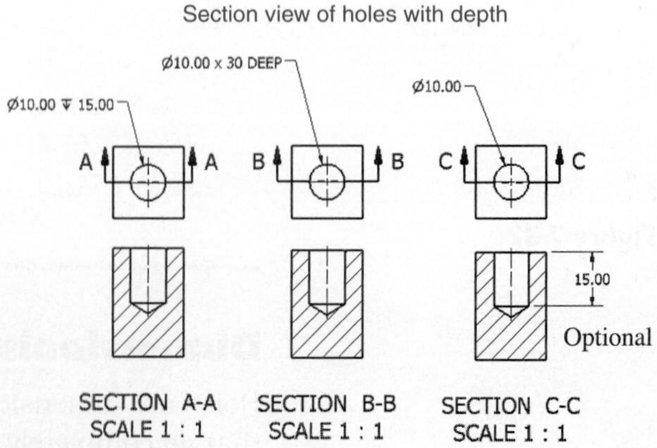

Section view of holes with depth

Ø10.00 x 30 DEEP

Ø10.00

Ø10.00 ⧩ 15.00

A A B B C C

15.00

Optional

SECTION A-A SCALE 1 : 1

SECTION B-B SCALE 1 : 1

SECTION C-C SCALE 1 : 1

Figure 7-36

Sample Problem SP7-1

Figure 7-37 shows orthographic views of a model that includes two holes that do not pass through the model. The dimensions other than the dimensions for the holes are applied using the **Dimension** tool located under the **Annotate** tab.

To dimension the holes:

1. Click the **Hole and Thread Notes** tool on the **Feature Notes** panel under the **Annotate** tab.
2. Click the edge of the left hole and move the cursor away from the hole.
3. Locate a position for the hole's dimension and click the left mouse button.
4. Right-click the mouse and select the **OK** option.
5. Repeat the same procedure for the right hole.

Remember that a hole's depth dimension does not include the conical point at the bottom of the hole.

See Figure 7-38. Note that the depth symbol is used automatically. The symbol can be edited if desired.

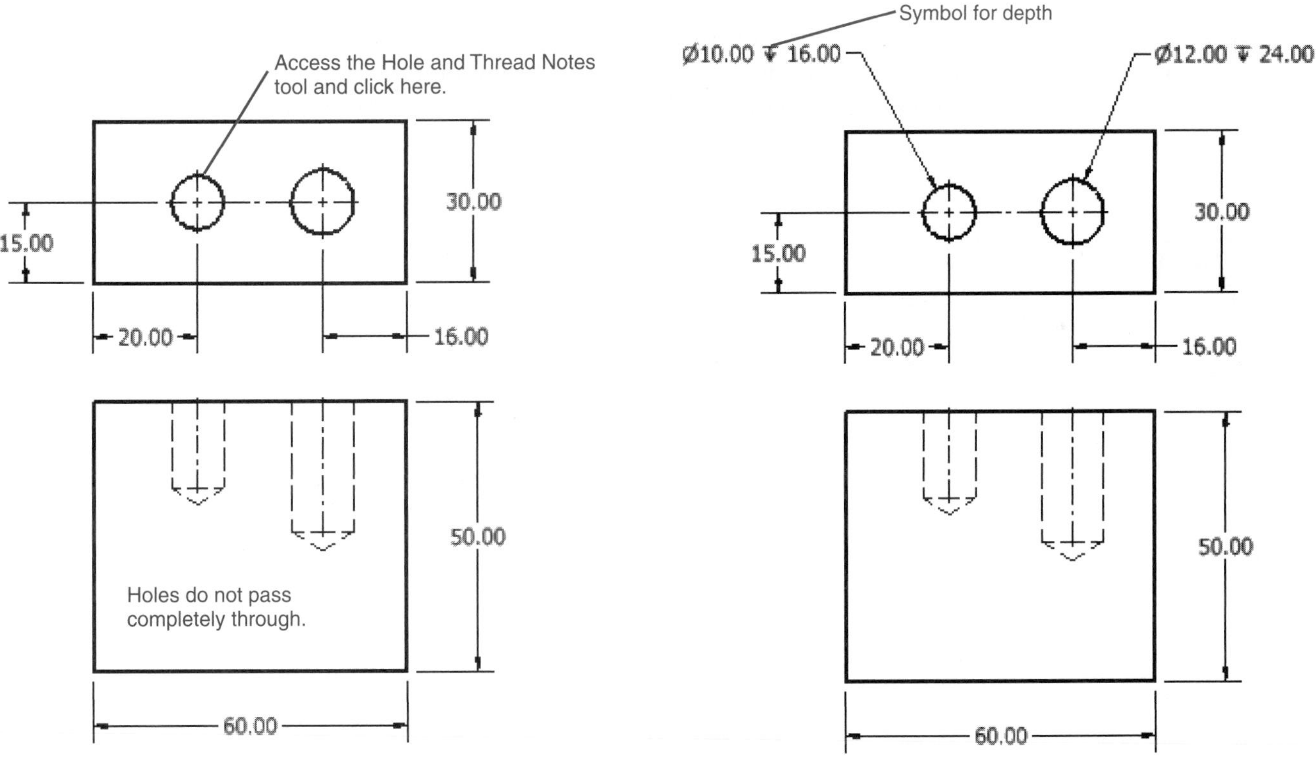

Figure 7-37

Figure 7-38

Dimensioning Hole Patterns

Figure 7-39 shows two different hole patterns dimensioned. The circular pattern includes the note Ø10 - 4 HOLES. This note serves to define all four holes within the object.

Figure 7-39 also shows a rectangular object that contains five holes of equal diameter, equally spaced from one another. The notation 5 × Ø10

Figure 7-39 Dimensions for hole patterns

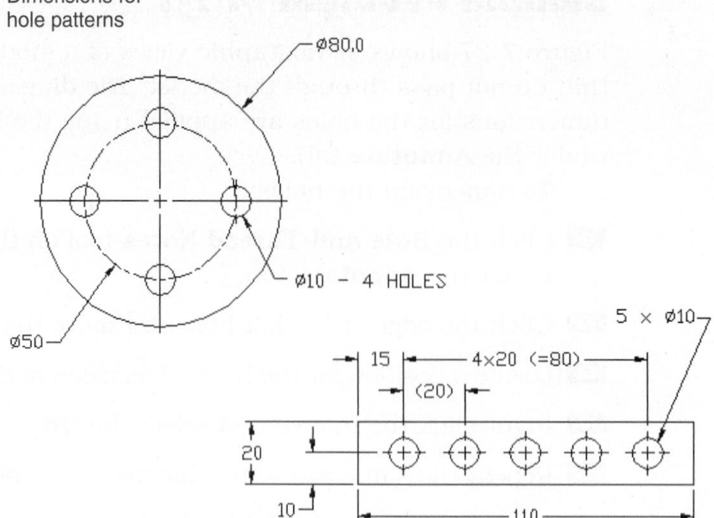

specifies 5 holes of 10 diameter. The notation 4 × 20 (=80) means 4 equal spaces of 20. The notation (=80) is a *reference dimension* and is included for convenience. Reference dimensions are explained in Chapter 8.

Figure 7-40 shows two additional methods for dimensioning repeating hole patterns. Figure 7-41 shows a circular hole pattern that includes two different hole diameters. The hole diameters are not noticeably different and could be confused. One group is defined by indicating letter (A); the other is dimensioned in a normal manner.

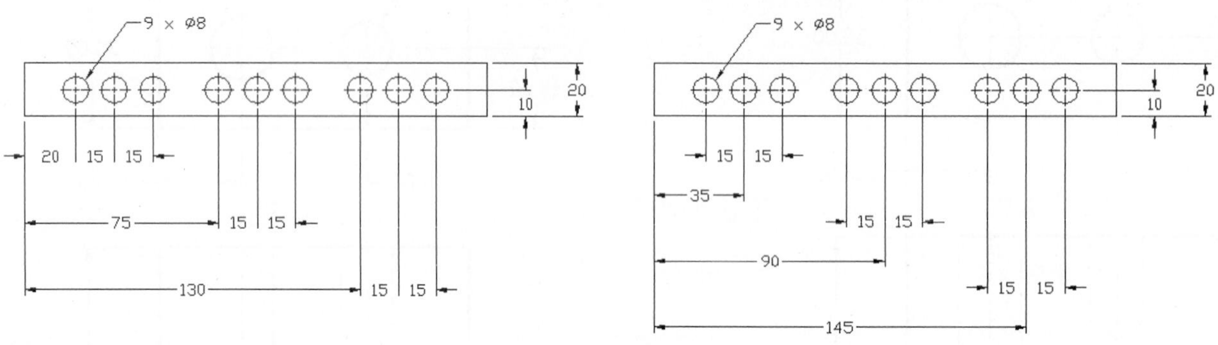

Figure 7-40

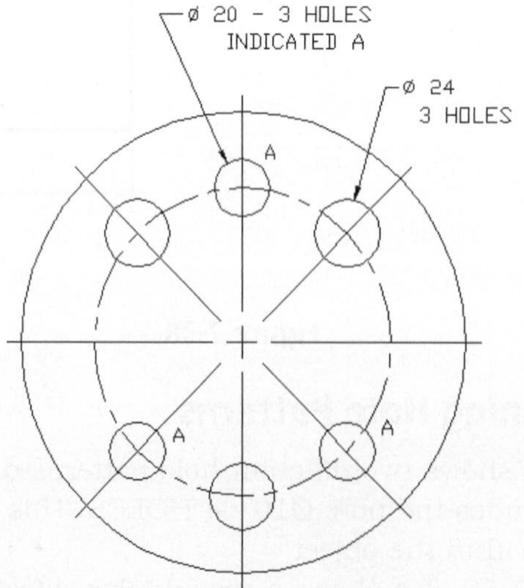

Figure 7-41

Using Symbols with Dimensions

In an attempt to eliminate language restrictions from drawings, ANSI standards permit the use of certain symbols for dimensions. Like musical notation that can be universally read by people who speak different languages, symbolic dimensions can be read by different people regardless of which language they speak. For example, the symbol Ø replaces the notation DIA.

Figure 7-42 shows a Ø10 hole that has a depth of 15. Inventor will automatically apply symbolic dimensions. The shown dimension is created as follows.

Figure 7-42

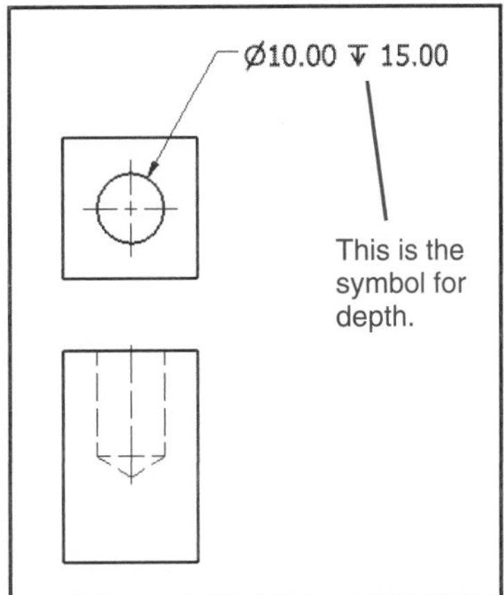

1 Access the **Hole and Thread Notes** tool under the **Annotate** tab.

2 Click the edge of the circular view of the hole.

3 Move the cursor away from the hole's edge and locate the dimension.

4 Right-click the mouse and select the **OK** option.

EXERCISE 7-14 Adding Symbols to Existing Dimension Text

Figure 7-43 shows an object with a Ø20.00 dimension. Add a depth requirement of 5.00.

Figure 7-43

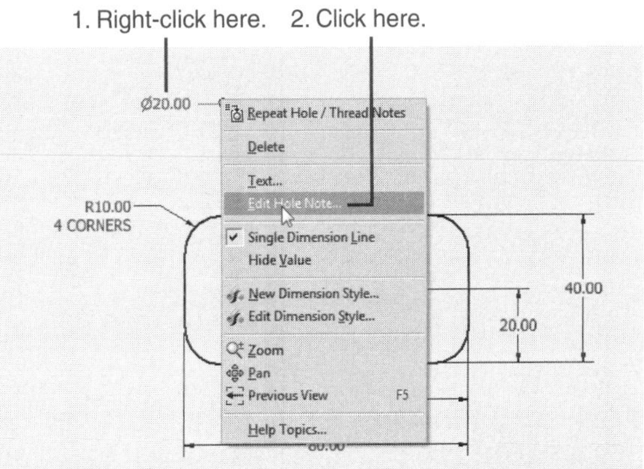

1 Right-click the dimension and select the **Edit Hole Note** option.

The **Edit Hole Note** dialog box will appear. See Figure 7-44.

2 Click the **Insert Symbol** box and click the depth symbol.

The depth symbol will appear next to the existing dimension in the **Note Format** box.

3 Type **5.00,** then click **OK.**

Figure 7-45 shows the resulting dimension.

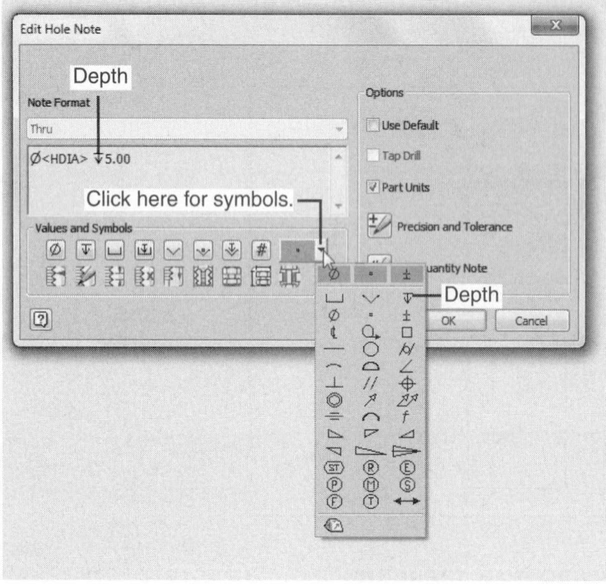

Figure 7-44

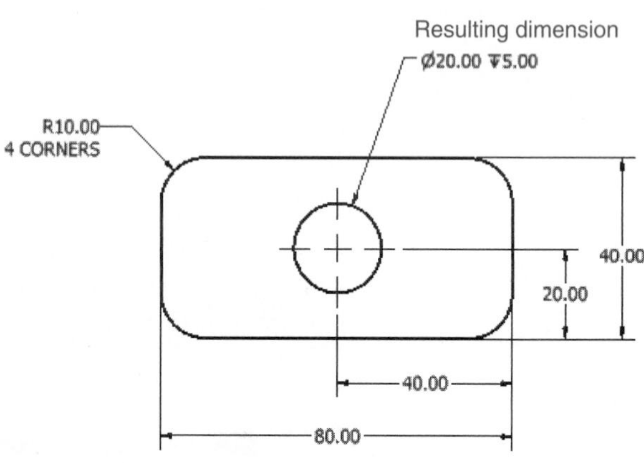

Figure 7-45

> **TIP**
>
> Drawing symbols can also be accessed using the **Style and Standard Editor** option. Click the **Manage** tab at the top of the screen, click the **Styles Editor** option, click **Dimension,** click **Default - mm (ANSI),** click the **Text** tab, and click the **Insert Symbol** box in the lower right corner of the dialog box. See Figure 7-46.

Figure 7-46

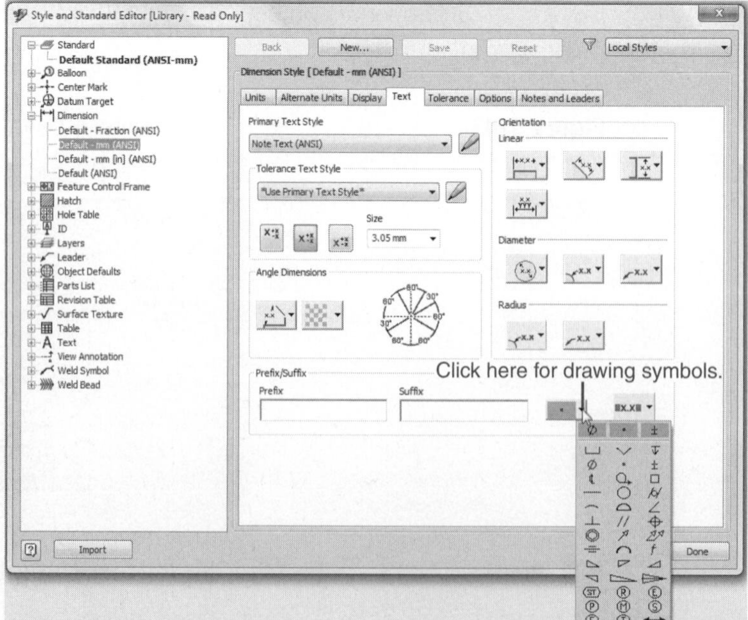

Dimensioning Counterbored, Countersunk Holes

Chapter 6 explained how to draw counterbored and countersunk holes. This section shows how to dimension countersunk and counterbored holes.

Figure 7-48 shows a 70 × 30 × 20 block that includes both a counterbored and a countersunk hole. The sizes for the holes were determined based on the fastener information given in Chapter 6. The **Hole** dialog boxes used to create the holes are also shown.

The information used to create the holes will automatically be used to create the hole's dimension. Prefixes and suffixes may be added to the dimensions, and the dimension value may be changed using the **Edit Hole Note** option. Hole dimension values may also be changed by editing the hole's feature value on the original model.

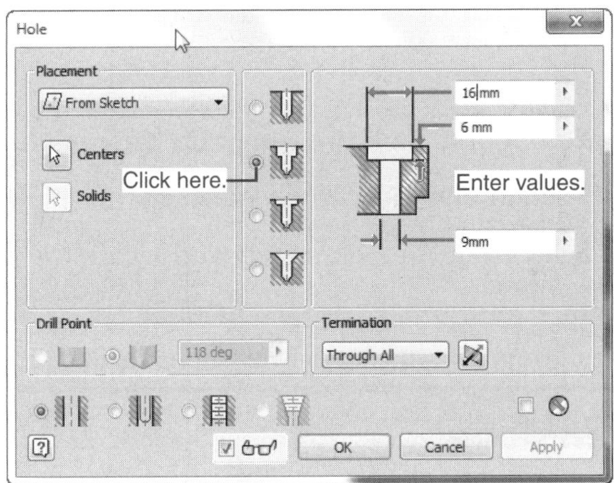

Figure 7-47

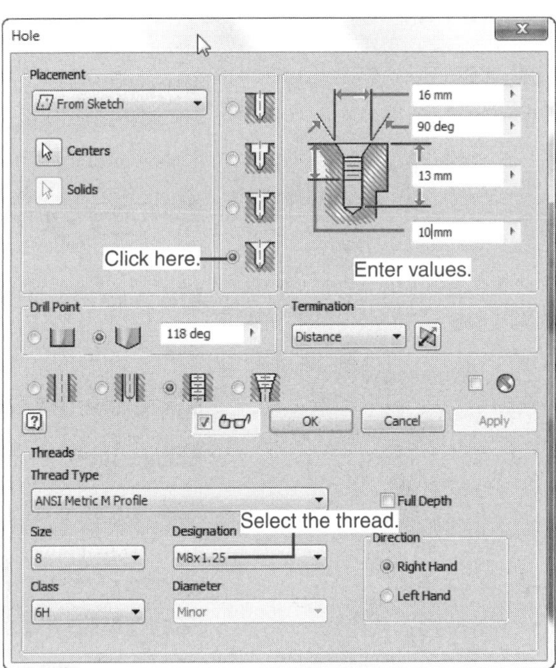

EXERCISE 7-15 Dimensioning a Counterbored Hole

The counterbored hole has a clear hole that goes completely through the block.

1. Click the **Annotate** tab and click the **Hole and Thread Notes** tool.
2. Click the edge of the circular view of the hole.
3. Move the cursor away from the hole's edge and locate the dimension.
4. Right-click the mouse and select the **OK** option.

Note how the dimensions match exactly the values used to create the hole. See Figures 7-47 and 7-48.

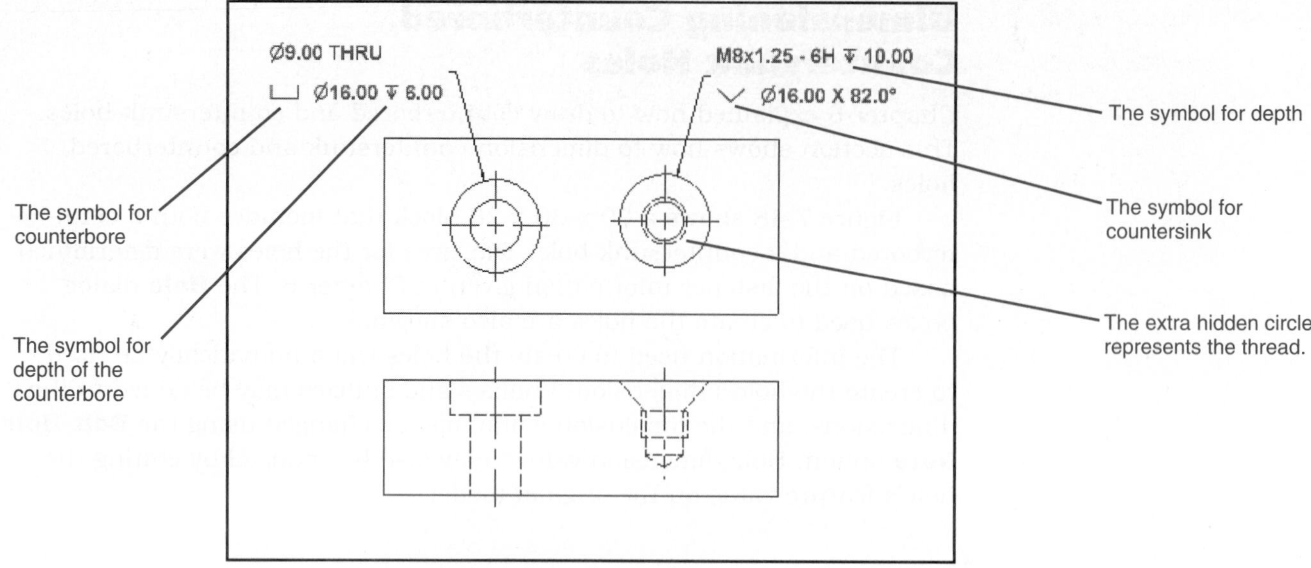
Ø9.00 THRU
⌴ Ø16.00 ↧ 6.00
M8x1.25 - 6H ↧ 10.00
⌵ Ø16.00 X 82.0°
The symbol for depth
The symbol for counterbore
The symbol for countersink
The extra hidden circle represents the thread.
The symbol for depth of the counterbore

Figure 7-48

EXERCISE 7-16 **Dimensioning a Countersunk Hole**

The countersunk hole has M8 threads and a thread depth of 10 mm.

1. Create an orthographic view of the model using the **ANSI (mm).idw** format.
2. Add centerlines.
3. Click the **Annotate** tab and click the **Hole and Thread Notes** tool.
4. Click the edge of the circular view of the hole.
5. Move the cursor away from the hole's edge and locate the dimension.
6. Right-click the mouse and select the **OK** option.

   Note how the dimensions match exactly the values used to create the hole.

Angular Dimensions

Figure 7-49 shows a model that includes a slanted surface. The dimension value is located beyond the model between two extension lines. Locating dimensions between extension lines is preferred to locating the value between an extension line and the edge of the model.

Figure 7-49

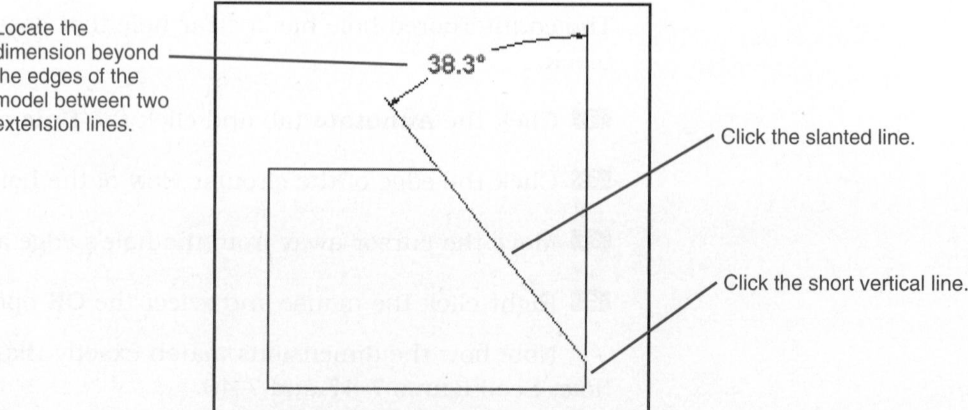
Locate the dimension beyond the edges of the model between two extension lines.
38.3°
Click the slanted line.
Click the short vertical line.

EXERCISE 7-17 **Creating an Angular Dimension**

1 Create an orthographic view of the model using the **ANSI (mm).idw** format.

2 Click the **Annotate** tab and click the **Dimension** tool.

3 Click the slanted line, then click the short vertical line on the right side of the model.

4 Move the cursor away from the hole's edge and locate the dimension.

5 Right-click the mouse and select the **OK** option.

Avoiding Overdimensioning

Figure 7-50 shows a shape dimensioned using an angular dimension. The shape is completely defined. Any additional dimension would be an error. It is tempting, in an effort to make sure a shape is completely defined, to add more dimensions, such as a horizontal dimension for the short horizontal edge at the top of the shape. This dimension is not needed and is considered *double dimensioning.*

Figure 7-51 also shows the same front view dimensioned using only linear dimensions. The choice of whether to use angular or linear dimensions depends on the function of the model and which distances are more critical.

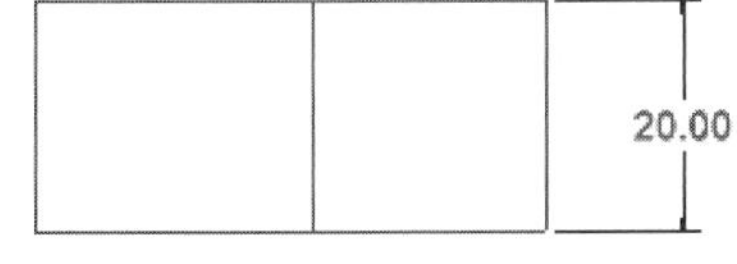

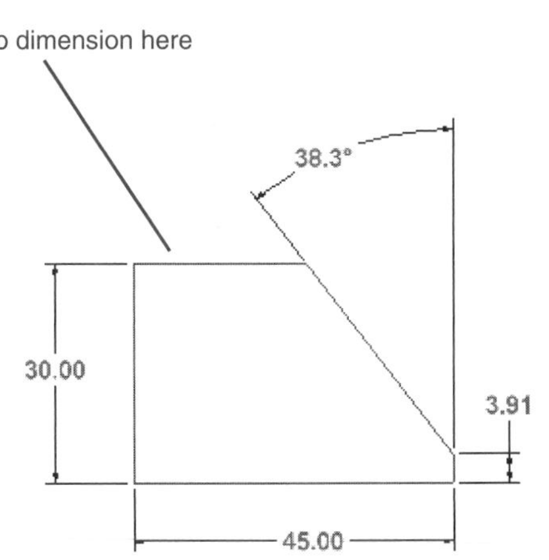

Figure 7-50

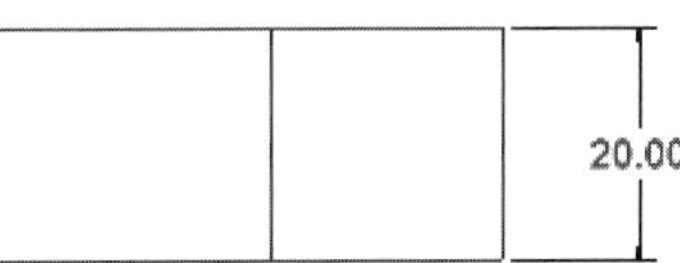

There are different ways to dimension the same model. Do not include more dimensions than are needed.

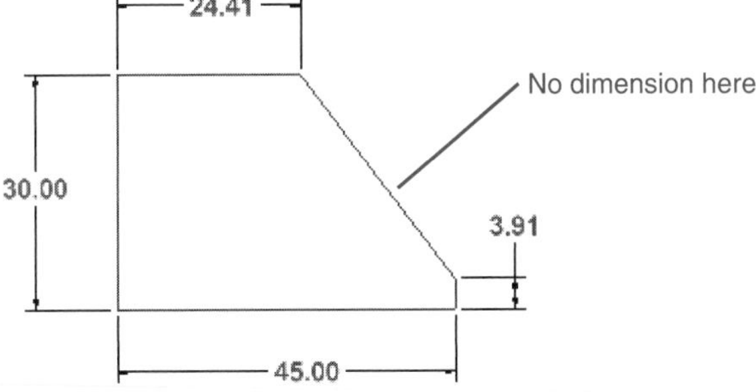

Figure 7-51

Figure 7-52 shows an object dimensioned two different ways. The dimensions used in the top example do not include a dimension for the width of the slot. This dimension is allowed to *float,* that is, allowed to accept any tolerance buildup. The dimensions used in the bottom example dimension the width of the slot but not the upper right edge. In this example the upper right edge is allowed to float or accept any tolerance buildup. The choice of which edge to float depends on the function of the part. If the slot were to interface with a tab on another part, then it would be imperative that it be dimensioned and toleranced to match the interfacing part.

Figure 7-52

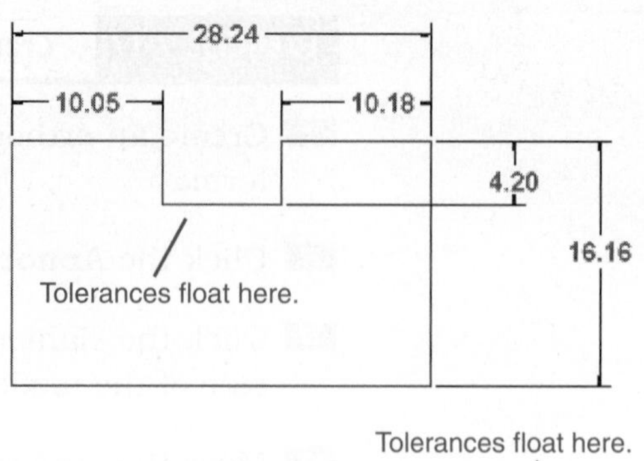

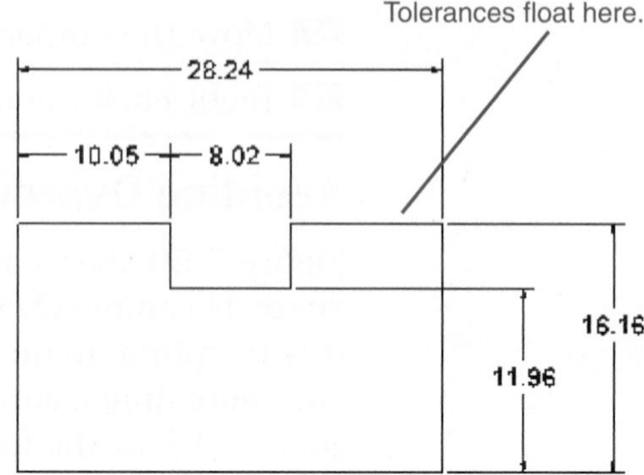

Ordinate Dimensions

ordinate dimension: A dimension based on an X, Y coordinate system that consists simply of horizontal and vertical leader lines.

Ordinate dimensions are dimensions based on an X,Y coordinate system. Ordinate dimensions do not include extension lines, dimension lines, or arrowheads, but simply horizontal and vertical leader lines drawn directly from the features of the object. Ordinate dimensions are particularly useful when dimensioning an object that includes many small holes.

Figure 7-53 shows a model that is to be dimensioned using ordinate dimensions. Ordinate dimensions values are calculated from the X,Y origin, which, in this example, is the lower left corner of the front view of the model.

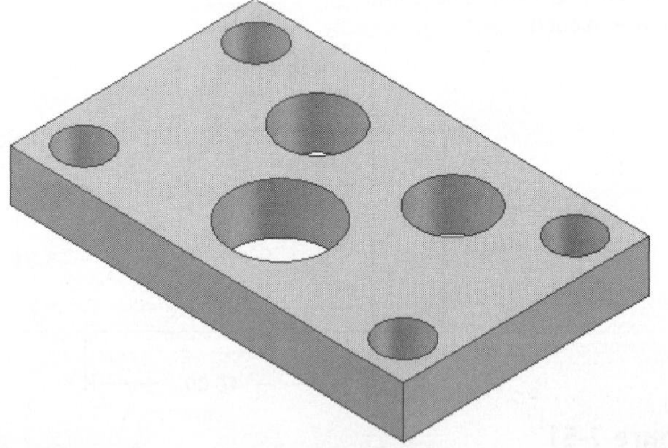

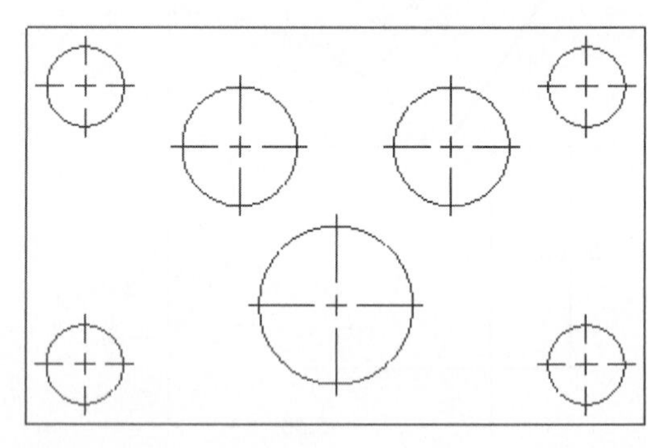

Figure 7-53

EXERCISE 7-18 Creating Ordinate Dimensions

1 Create an orthographic view of the model using the **ANSI (mm).idw** format.

2 Click the **Annotate** tab, click on the **Ordinate** tool, and click the **Ordinate Set** tool.

See Figure 7-54.

3 Move the cursor into the drawing area and click the lower left corner of the view.

The horizontal edge line will change color.

4 Move the cursor away from the corner, right-click the mouse, and select the **Continue** option.

The number **0.00** will appear. If the number is not unidirectional, use the **Style and Standard Editor** tool to reorient the dimension. See Figure 7-55.

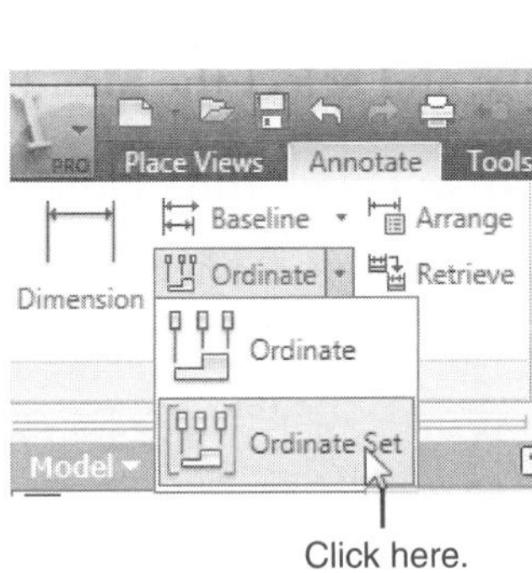

Figure 7-54

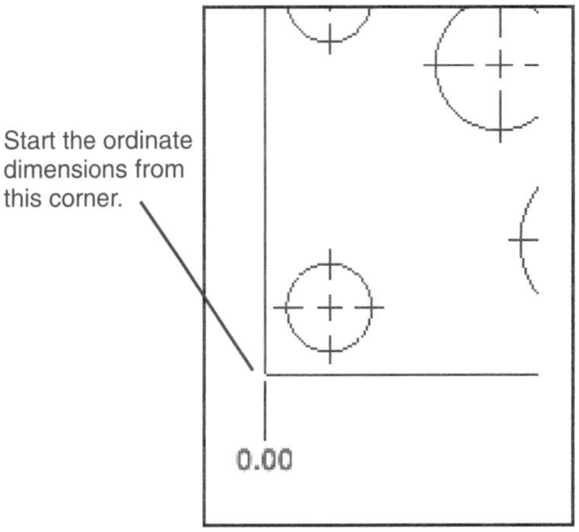

Start the ordinate dimensions from this corner.

0.00

Figure 7-55

5 Right-click the mouse again and select the **Make Origin** option.

6 Locate the number and left-click the mouse.

7 Click the lower end of the vertical centerline of the first hole and position the dimension so that it is in line with the first dimension.

See Figure 7-56.

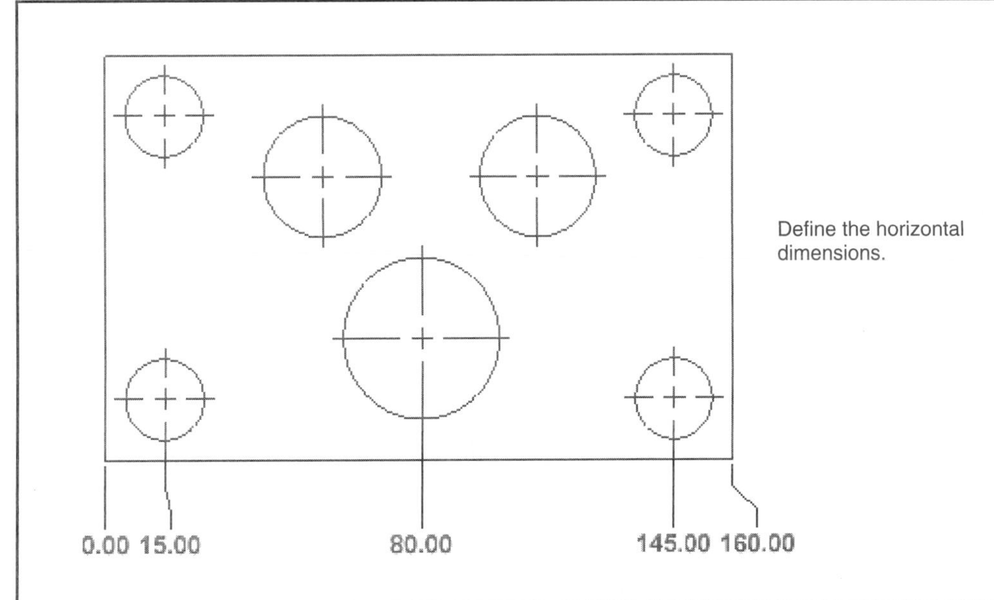

Figure 7-56

Note how the extension line from the first hole's centerline curves so that the dimension value may be located in line with the first dimension. Inventor will automatically align ordinate dimensions.

8 Right-click the mouse and click the **Create** option.

9 Right-click the mouse and select the **Repeat Ordinate Dimension Set** option.

10 Click the left corner so that the vertical edge is highlighted.

11 Move the cursor to the left, away from the corners, and right-click the mouse.

12 Select the **Continue** option.

13 Right-click the mouse and select the **Make Origin** option.

14 Locate the dimension and click the mouse.

Again start with the lower left corner of the model, then click the appropriate horizontal centerlines. See Figure 7-57.

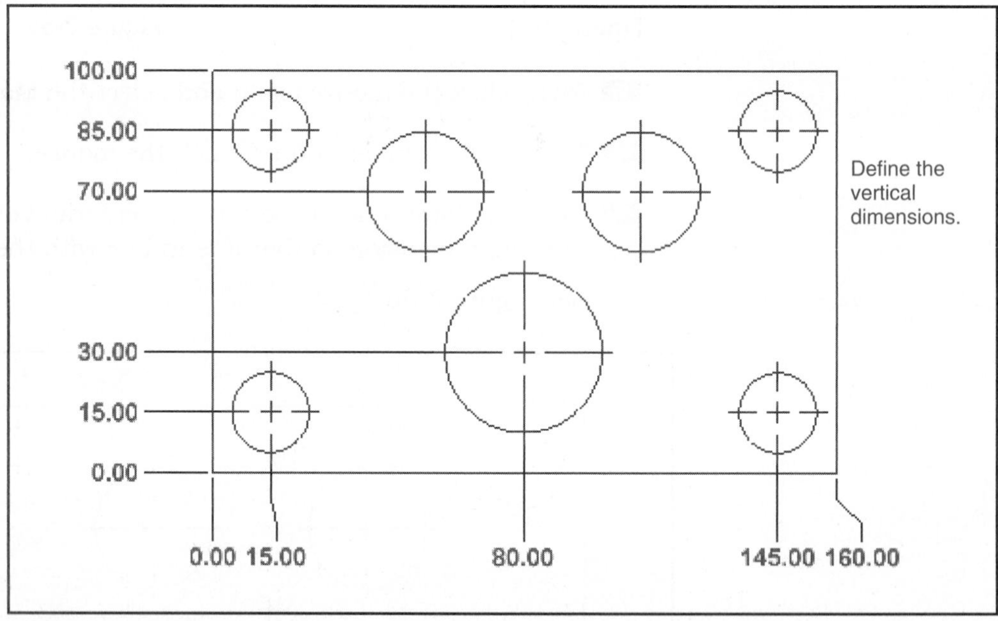

Figure 7-57

EXERCISE 7-19 Adding Hole Dimensions

1 Click the **Hole and Thread Notes** option under the **Annotate** tab.

2 Add the appropriate hole dimensions.

See Figure 7-58.

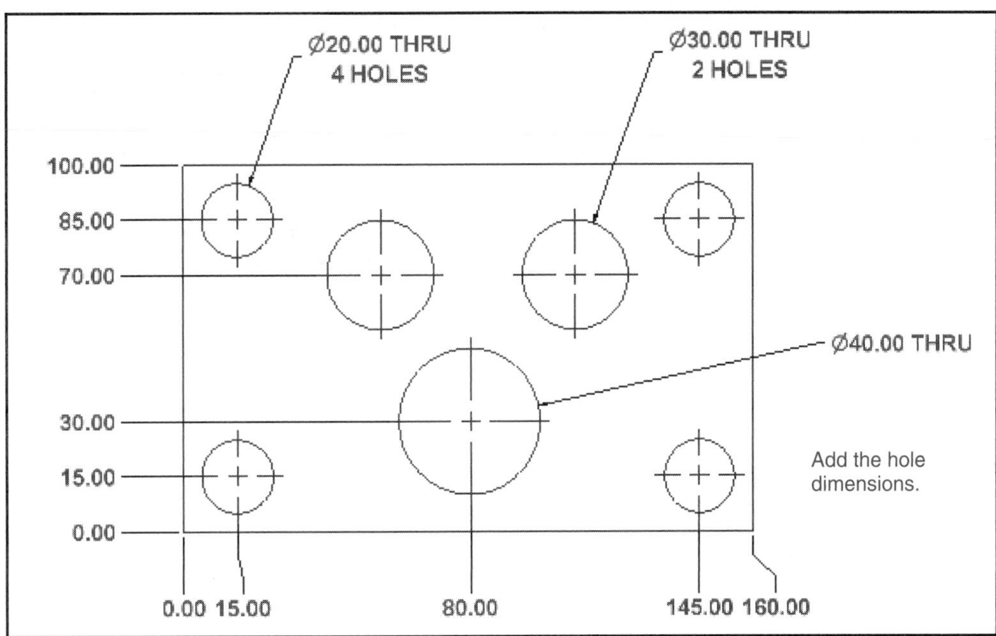

Figure 7-58

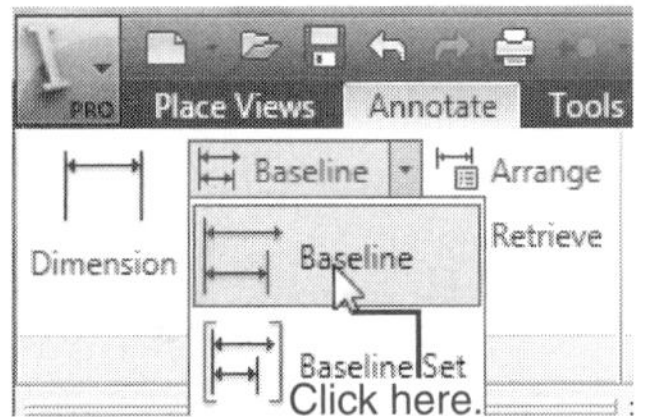

Figure 7-59

baseline dimensions: A series of dimensions that originate from a common baseline or datum line.

Baseline Dimensions

Baseline dimensions are a series of dimensions that originate from a common baseline or datum line. Baseline dimensions are very useful because they help eliminate the tolerance buildup that is associated with chain-type dimensions. The **Baseline Dimension** tool is found on the **Dimension** panel under the **Annotate** tab. See Figure 7-59.

EXERCISE 7-20 Using the Baseline Dimension Tool

See Figure 7-60.

1. Click the **Baseline Dimension** tool under the **Annotate** tab.
2. Click the left vertical edge of the model.
3. Click the lower end on each vertical centerline and the right vertical edge line.
4. Right-click the mouse and select the **Continue** option.
5. Move the cursor away from the surface of the part and left-click the mouse to locate the dimensions.
6. Left-click the mouse again, then right-click the mouse and select the **Create** option.

> **TIP**
>
> If the dimension line spacing is not acceptable, left-click and drag the dimension value to a new location.

7. Click the **Baseline Dimension** tool again.

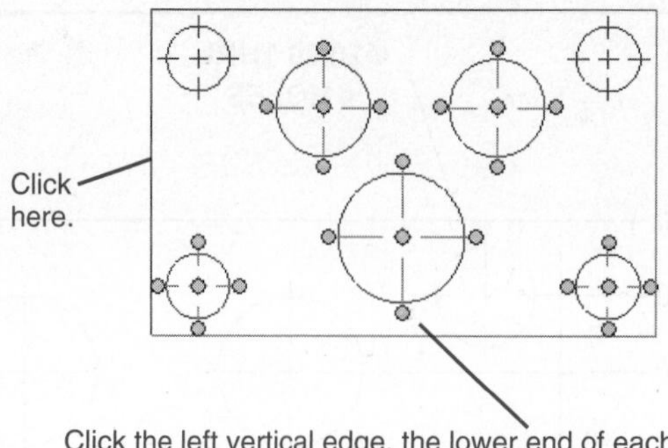

Press the right mouse button, select the Continue option, locate the dimensions, left-click the mouse, right-click again, and select the Create option.

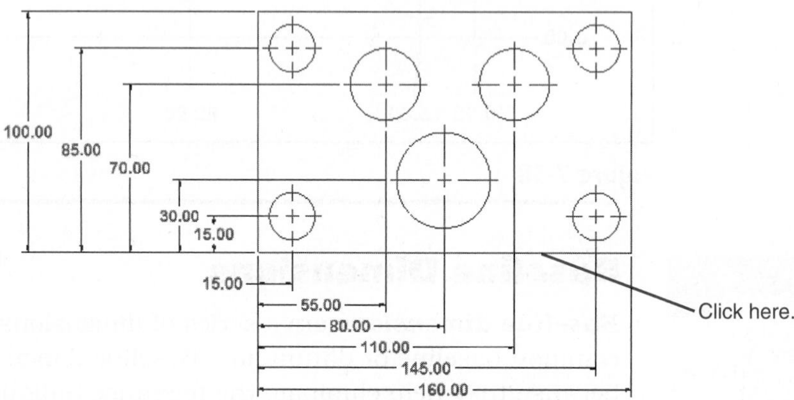

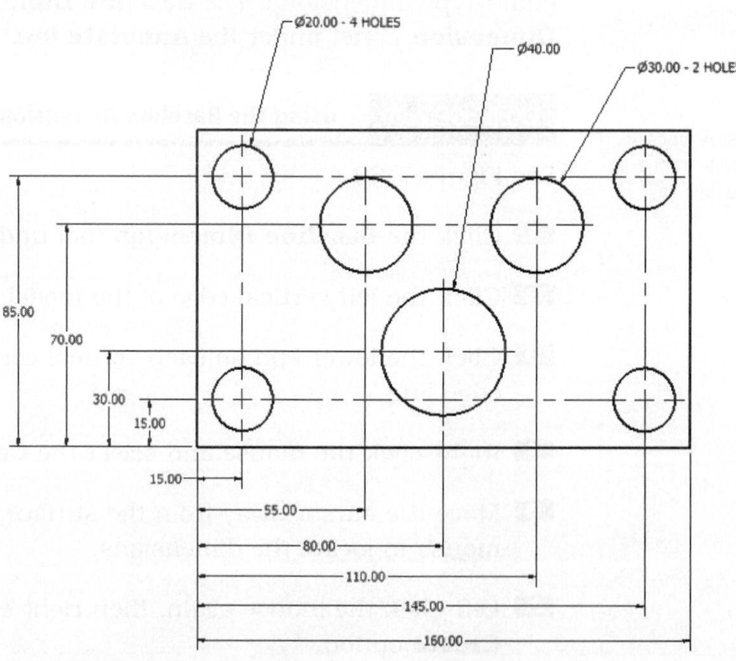

Figure 7-60

8. Click the bottom horizontal edge line then the left end of the horizontal centerlines.

9. Click the left end of the holes' horizontal centerline.

10. Move the cursor away from the surface of the part and left-click the mouse to locate the dimensions.

11 Left-click the mouse again, then right-click the mouse and select the **Create** option.

12 Add the appropriate centerline.

13 Add the hole diameter dimensions using the **Hole and Thread Notes** tool.

Hole Tables

Inventor will generate hole tables that list holes' diameters and locations. There are two options: list all the holes, or list selected holes.

EXERCISE 7-21 Listing All Holes in a Table

1 Click the **Hole Table - Selection** tool located on the **Table** panel under the **Annotate** tab.

2 Select the **Hole View** option.

See Figure 7-61.

Figure 7-61

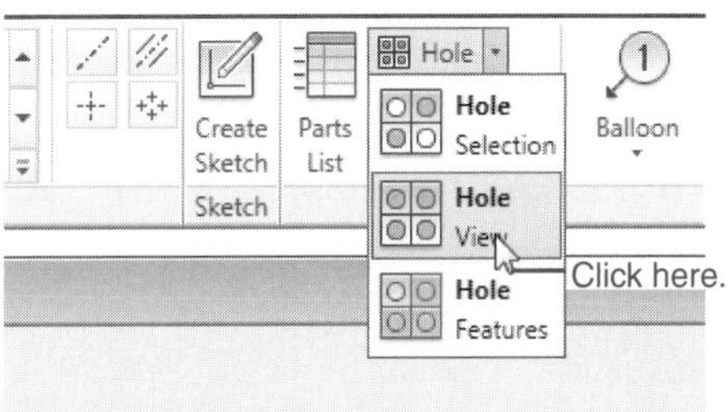

3 Move the mouse into the part area and left-click the mouse.

The origin cursor symbol will appear.

4 Click on the lower left corner of the model.

This will define the origin for the table's X and Y values.

5 Move the cursor away from the part area.

A rectangular shape will appear. This rectangular shape represents the size of the dimension table that will be created.

6 Locate the table and left-click the mouse.

See Figure 7-62.

Figure 7-62

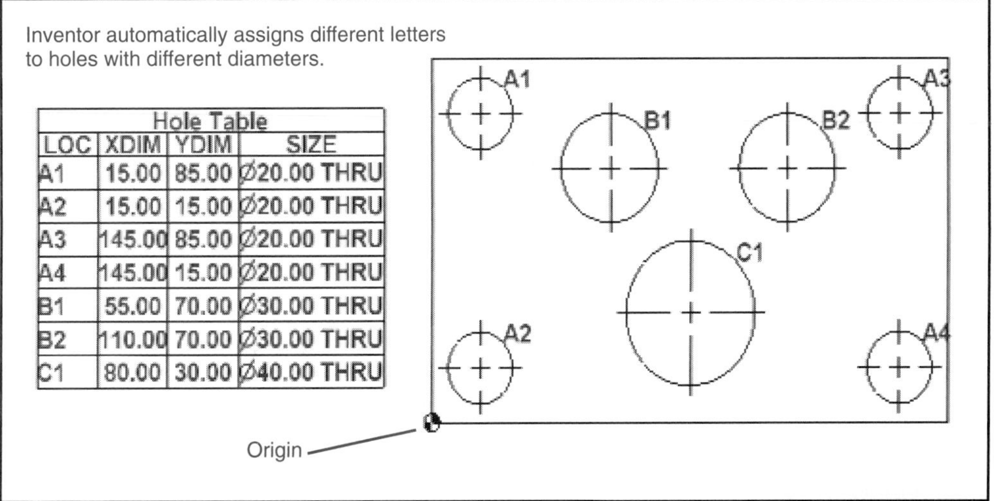

Hole Table

LOC	XDIM	YDIM	SIZE
A1	15.00	85.00	Ø20.00 THRU
A2	15.00	15.00	Ø20.00 THRU
A3	145.00	85.00	Ø20.00 THRU
A4	145.00	15.00	Ø20.00 THRU
B1	55.00	70.00	Ø30.00 THRU
B2	110.00	70.00	Ø30.00 THRU
C1	80.00	30.00	Ø40.00 THRU

Locating Dimensions

There are eight general rules concerning the location of dimensions. See Figure 7-63.

1 Locate dimensions near the features they are defining.

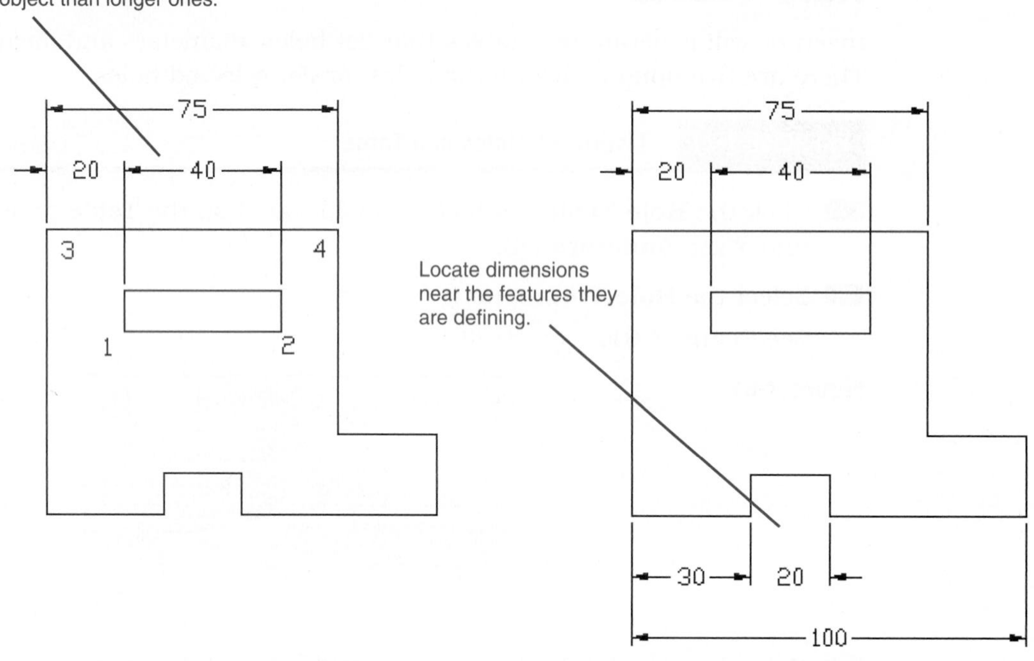
Locate shorter dimensions closer to the object than longer ones.
75
20
40
3
4
1
2
Locate dimensions near the features they are defining.
75
20
40
30
20
100

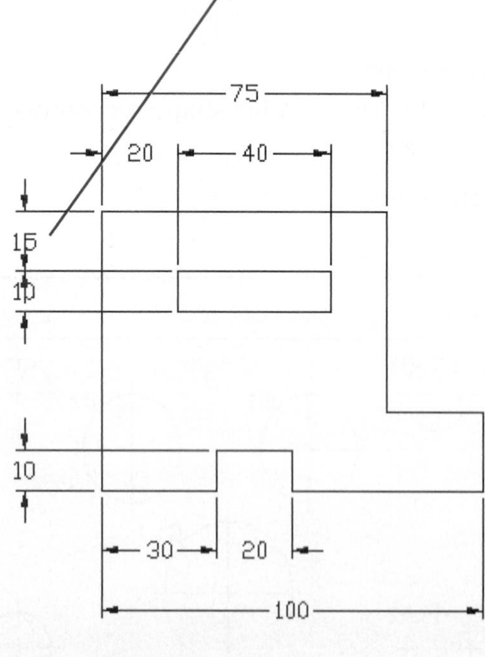
Use the Explode, Erase, and Move tools to reconstruct and relocate inappropriate dimensions.
75
20
40
15
10
10
30
20
100

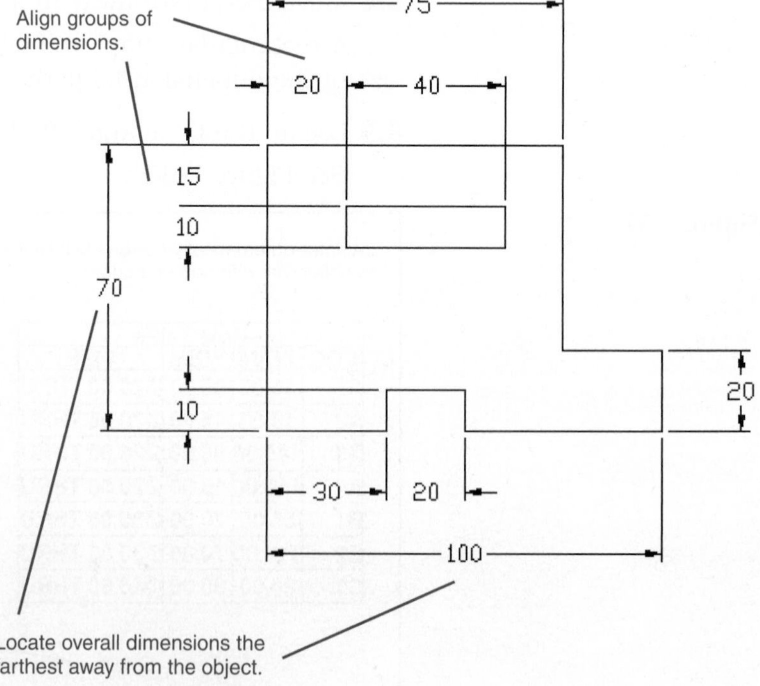
DO NOT LOCATE DIMENSIONS ON THE SURFACE OF THE OBJECT.
Align groups of dimensions.
75
20
40
15
10
70
10
20
30
20
100
Locate overall dimensions the farthest away from the object.

Figure 7-63

2 Do not locate dimensions on the surface of the object.

3 Align and group dimensions so that they are neat and easy to understand.

4 Avoid crossing extension lines.

Sometimes it is impossible not to cross extension lines because of the complex shape of the object, but whenever possible, avoid crossing extension lines.

5 Do not cross dimension lines.

6 Locate shorter dimensions closer to the object than longer ones.

7 Always locate overall dimensions the farthest away from the object.

8 Do not dimension the same distance twice. This is called double dimensioning and will be discussed in Chapter 8.

Fillets and Rounds

Fillets and rounds may be dimensioned individually or by a note. In many design situations all the fillets and rounds are the same size, so a note as shown in Figure 7-64 is used. Any fillets or rounds that have a different radius from that specified by the note are dimensioned individually.

Figure 7-64

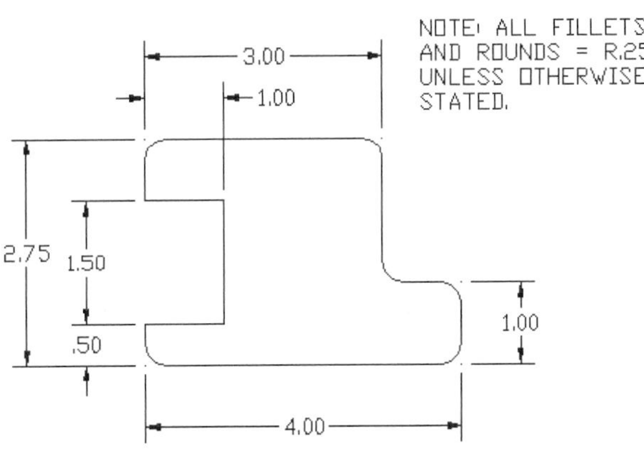

Rounded Shapes—Internal

slot: An internal rounded shape.

Internal rounded shapes are called ***slots***. Figure 7-65 shows three different methods for dimensioning slots. The end radii are indicated by the note R - 2 PLACES, but no numerical value is given. The width of the slot is

Figure 7-65

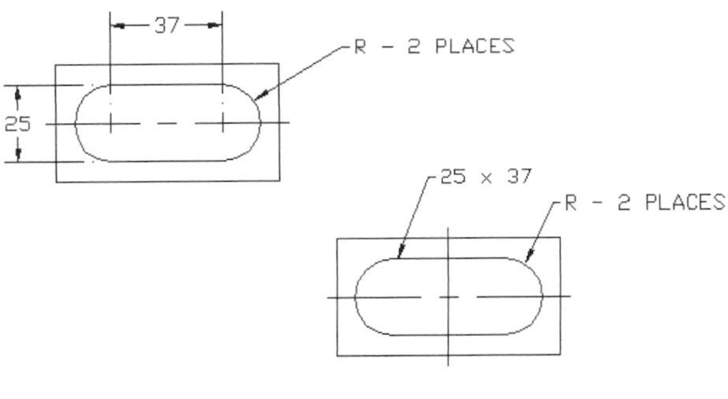

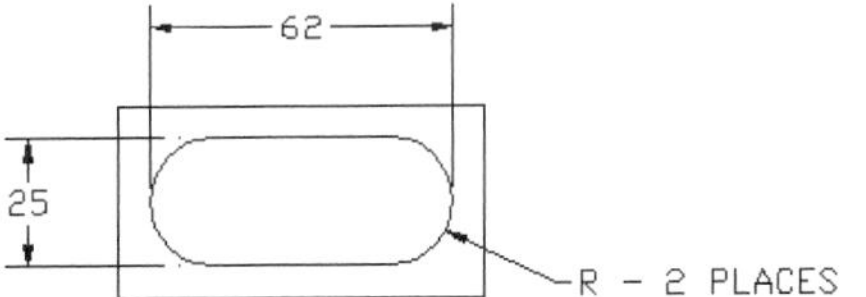

dimensioned, and it is assumed that the radius of the rounded ends is exactly half of the stated width.

Rounded Shapes—External

Figure 7-66 shows two shapes with external rounded ends. As with internal rounded shapes, the end radii are indicated but no value is given. The width of the object is given, and the radius of the rounded end is assumed to be exactly half of the stated width.

The second example shown in Figure 7-66 shows an object dimensioned using the object's centerline. This type of dimensioning is done when the distance between the holes is more important than the overall length of the object; that is, the tolerance for the distance between the holes is more exact than the tolerance for the overall length of the object.

The overall length of the object is given as a reference dimension (100). This means the object will be manufactured based on the other dimensions, and the 100 value will be used only for reference.

Objects with partially rounded edges should be dimensioned as shown in Figure 7-66. The radii of the end features are dimensioned. The center point of the radii is implied to be on the object centerline. The overall dimension is given; it is not referenced unless specific radii values are included.

Figure 7-66

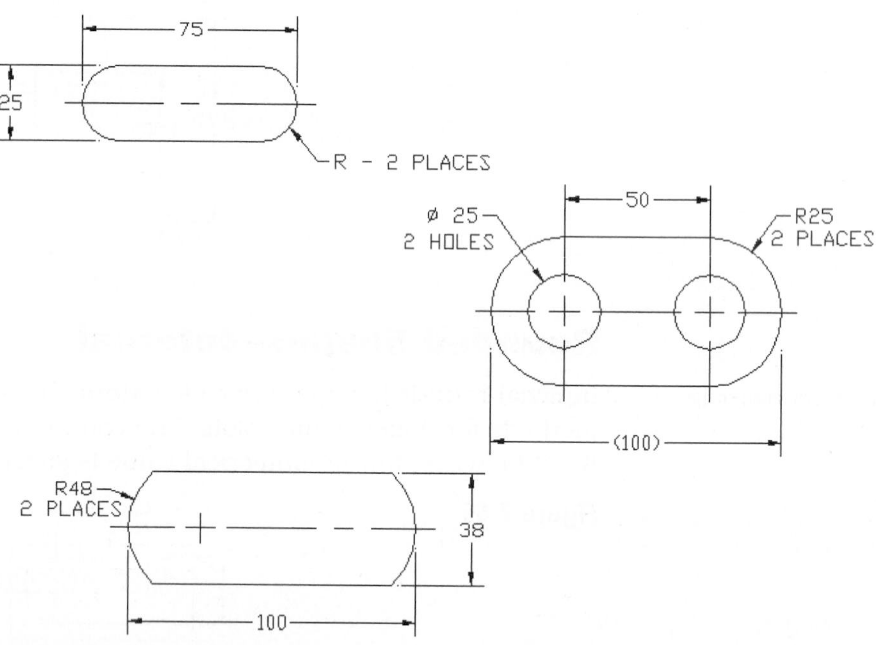

Irregular Surfaces

There are three different methods for dimensioning irregular surfaces: tabular, baseline, and baseline with oblique extension lines. Figure 7-67 shows an irregular surface dimensioned using the tabular method. An XY axis is defined using the edges of the object. Points are then defined relative to the XY axis. The points are assigned reference numbers, and the reference numbers and XY coordinate values are listed in chart form as shown.

Station	1	2	3	4	5	6
X	0	20	40	55	62	70
Y	40	38	30	16	10	0

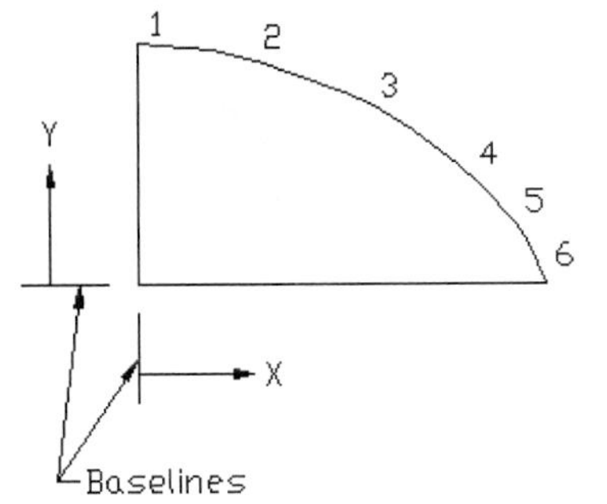

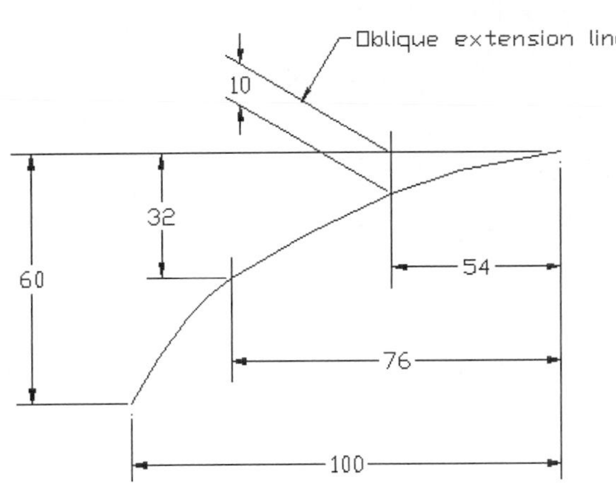

Figure 7-67

Figure 7-68 shows an irregular curve dimensioned using baseline dimensions. The baseline method references all dimensions to specified baselines. Usually there are two baselines, one horizontal and one vertical.

It is considered poor practice to use a centerline as a baseline. Centerlines are imaginary lines that do not exist on the object and would make it more difficult to manufacture and inspect the finished objects.

Figure 7-68

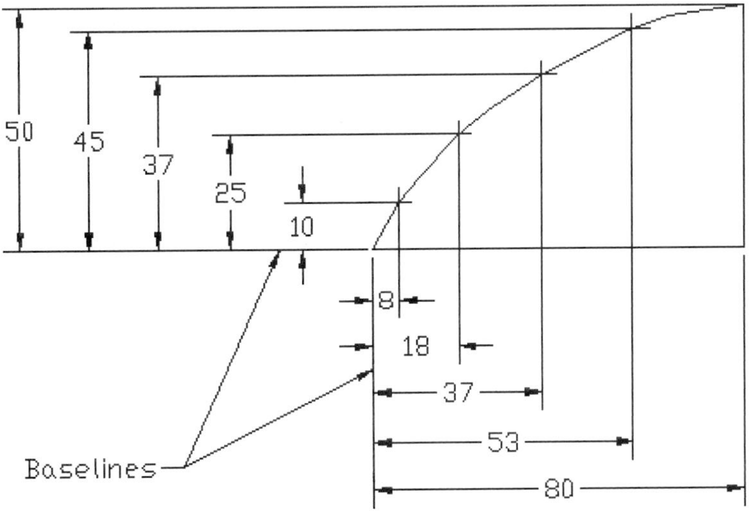

Baseline dimensioning is very common because it helps eliminate tolerance buildup and is easily adaptable to many manufacturing processes. Inventor has a special **Baseline Dimension** tool for use in creating baseline dimensions.

Polar Dimensions

polar dimension: A dimension defined by a radius and an angle.

Polar dimensions are similar to polar coordinates. A location is defined by a radius (distance) and an angle. Figure 7-69 shows an object that includes polar dimensions. The holes are located on a circular centerline, and their positions from the vertical centerline are specified using angles.

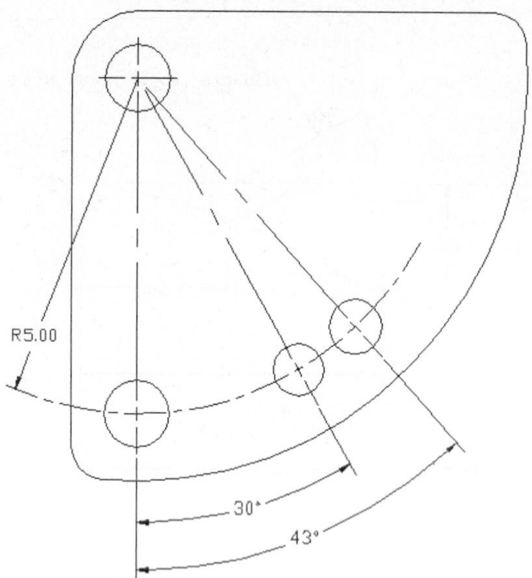

Figure 7-69

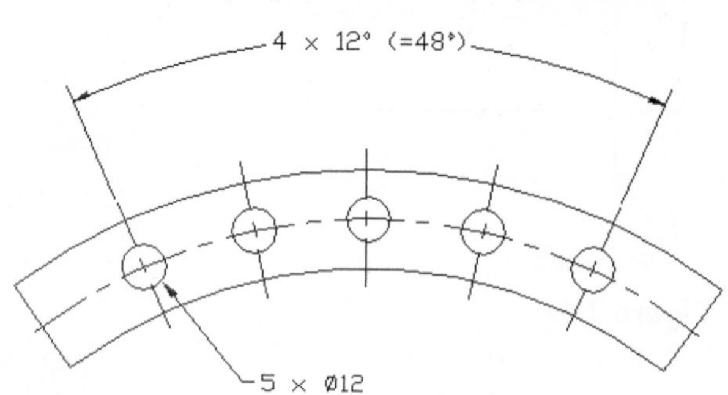

Figure 7-70

Figure 7-70 shows an example of a hole pattern dimensioned using polar dimensions.

Chamfers

chamfer: An angular cut made on the edge of an object.

Chamfers are angular cuts made on the edges of objects. They are usually used to make it easier to fit two parts together. They are most often made at 45° angles but may be made at any angle. Figure 7-71 shows two objects with chamfers between surfaces 90° apart and two examples between surfaces that are not 90° apart. Either of the two types of dimensions shown for the 45° dimension may be used. If an angle other than 45° is used, the angle and setback distance must be specified.

Figure 7-72 shows two examples of internal chamfers. Both define the chamfer using an angle and diameter. Internal chamfers are very similar to countersunk holes.

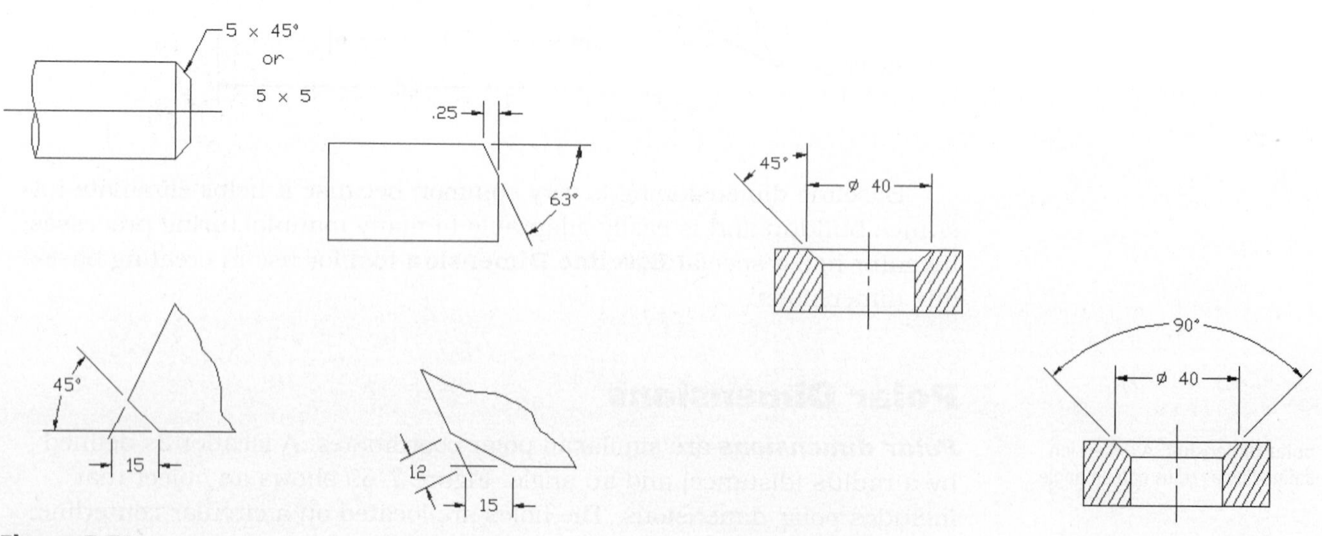

Figure 7-71

Figure 7-72

Knurling

knurl: A series of small ridges on a metal surface used to make it easier to grip a shaft, or to roughen a surface before it is used in a press fit; may be diamond or straight.

There are two types of ***knurls***: diamond and straight. Knurls are used to make it easier to grip a shaft, or to roughen a surface before it is used in a press fit.

Knurls are defined by their pitch and diameter. See Figure 7-73. The pitch of a knurl is the ratio of the number of grooves on the circumference to the diameter. Standard knurling tools sized to a variety of pitch sizes are used to manufacture knurls for both English and metric units.

Figure 7-73

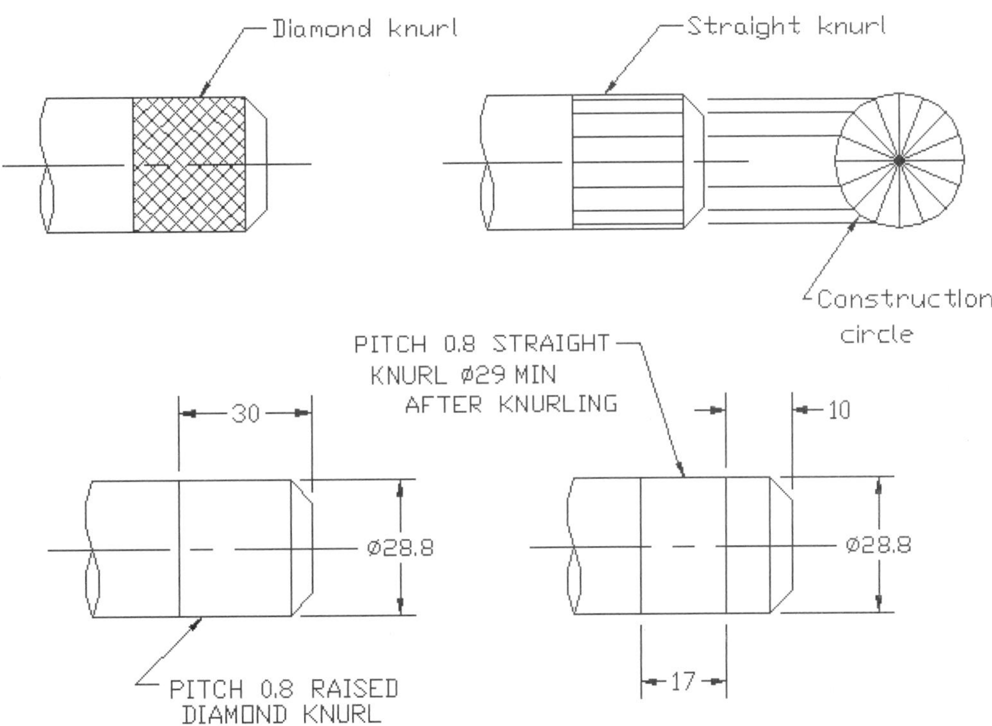

Diamond knurls may be represented by a double hatched pattern or by an open area with notes. The **Hatch** tool is used to draw the double hatched lines. Straight knurls may be represented by straight lines in the pattern shown or by an open area with notes. The straight-line pattern is created by projecting lines from a construction circle. The construction points are evenly spaced on the circle.

Keys and Keyseats

key: A small piece of material used to transmit power, such as rotary motion from a shaft to a gear.

Keys are small pieces of material used to transmit power. For example, Figure 7-74 shows how a key can be fitted between a shaft and a gear so that the rotary motion of the shaft can be transmitted to the gear.

There are many different styles of keys. See the keys listed on the **Content Center** under **Shaft Parts.** The key shown in Figure 7-74 has a rectangular cross section and is called a *square key*. Keys fit into grooves called ***keyseats*** or *keyways*.

keyseat: A groove into which a key fits; also called *keyway*.

Keyways are dimensioned from the bottom of the shaft or hole as shown.

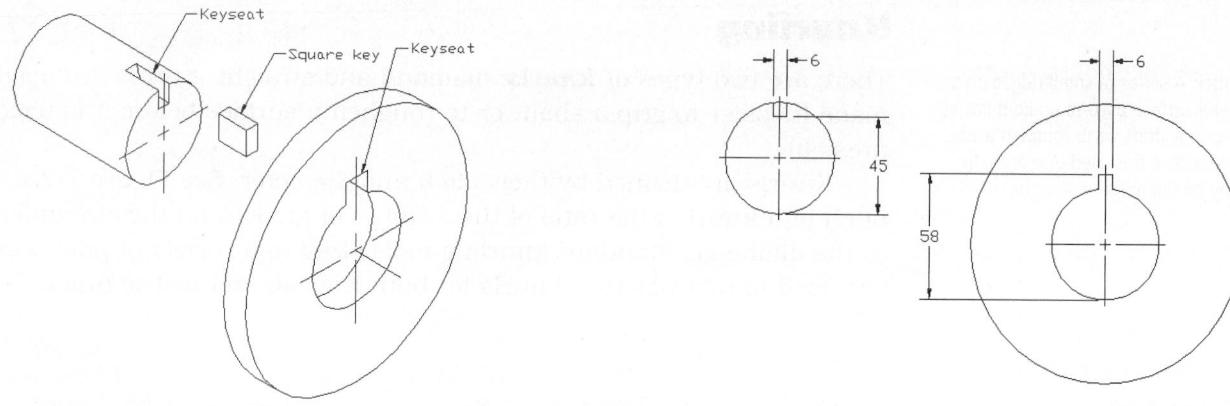

Figure 7-74

Symbols and Abbreviations

Symbols are used in dimensioning to help accurately display the meaning of the dimension. Symbols also help eliminate language barriers when reading drawings. Figure 7-75 shows a list of dimensioning symbols available on the **Style and Standard Editor** dialog box accessed by clicking the **Styles Editor** under the **Manage** tab, and clicking the **Notes and Leaders** tab. How to apply symbols to dimensions was explained earlier in the chapter.

Figure 7-75

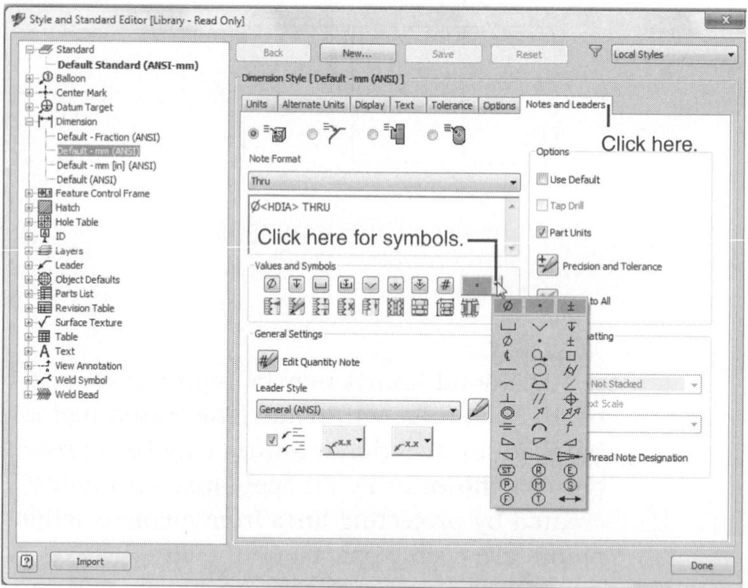

Abbreviations should be used very carefully on drawings. Whenever possible, write out the full word including correct punctuation. Figure 7-76 shows several standard abbreviations used on technical drawings.

Symmetrical and Centerline Symbols

An object is symmetrical about an axis when one side is an exact mirror image of the other. Figure 7-77 shows a symmetrical object. The two short parallel lines symbol or the note OBJECT IS SYMMETRICAL ABOUT THIS AXIS (centerline) may be used to designate symmetry.

AL	= Aluminum
C'BORE	= Counterbore
CRS	= Cold Rolled Steel
CSK	= Countersink
DIA	= Diameter
EQ	= Equal
HEX	= Hexagon
MAT'L	= Material
R	= Radius
SAE	= Society of Automotive Engineers
SFACE	= Spotface
ST	= Steel
SQ	= Square
REQD	= Required

Figure 7-76

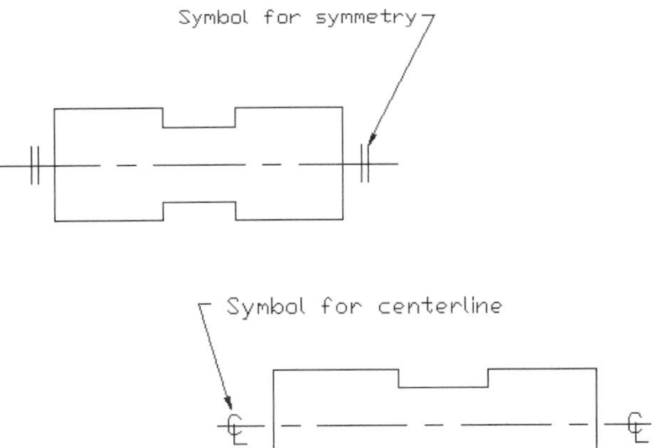

Figure 7-77

If an object is symmetrical, only half the object need be dimensioned. The other dimensions are implied by the symmetry note or symbol.

Centerlines are slightly different from the axis of symmetry. An object may or may not be symmetrical about its centerline. See Figure 7-77. Centerlines are used to define the center of both individual features and entire objects. Use the centerline symbol when a line is a centerline, but do not use it in place of the symmetry symbol.

Dimensioning to a Point

Curved surfaces can be dimensioned using theoretical points. See Figure 7-78. There should be a small gap between the surface of the object and the lines used to define the theoretical point. The point should be defined by the intersection of at least two lines.

Figure 7-78

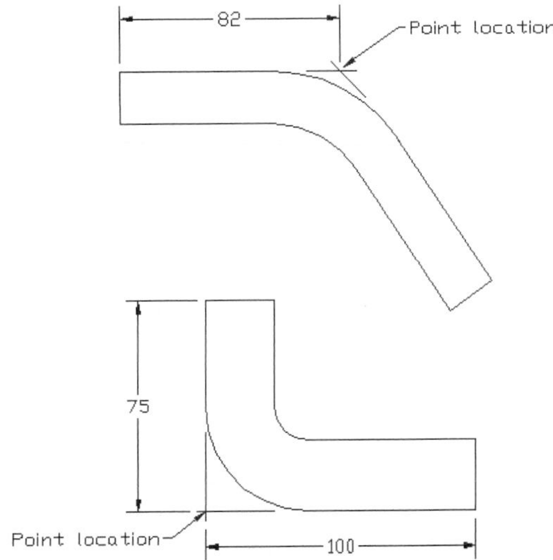

There should also be a small gap between the extension lines and the theoretical point used to locate the point.

Dimensioning Section Views

Section views are dimensioned, as are orthographic views. See Figure 7-79. The section lines should be drawn at an angle that allows the viewer to clearly distinguish between the section lines and the extension lines.

Figure 7-79

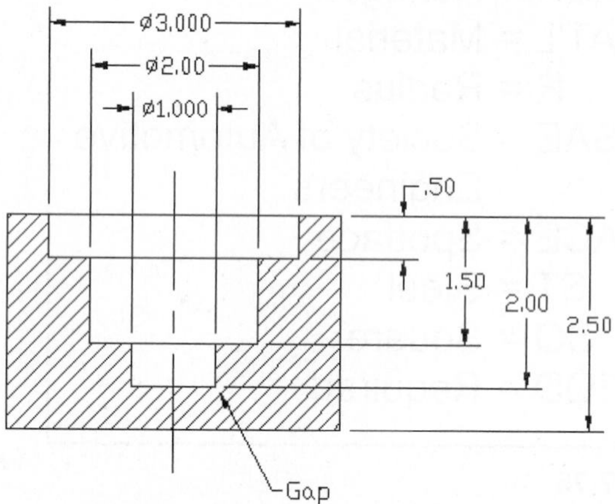

Dimensioning Orthographic Views

Dimensions should be added to orthographic views where the features appear in contour. Holes should be dimensioned in their circular views. Figure 7-80 shows three views of an object that has been dimensioned.

Figure 7-80

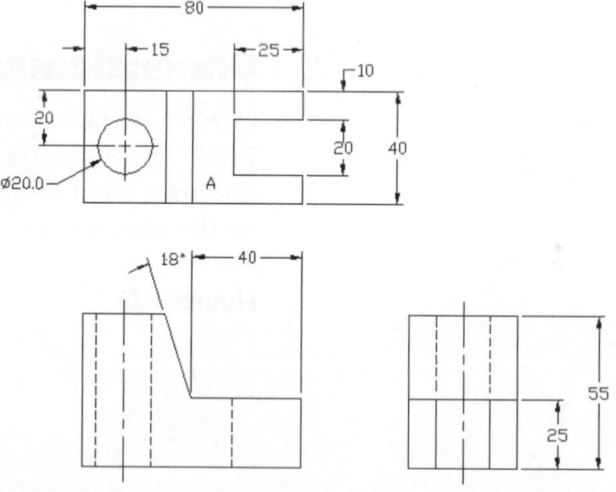

The hole dimensions are added to the top view, where the hole appears circular. The slot is also dimensioned in the top view because it appears in contour. The slanted surface is dimensioned in the front view.

The height of surface A is given in the side view rather than run along extension lines across the front view. The length of surface A is given in the front view. This is a contour view of the surface.

It is considered good practice to keep dimensions in groups. This makes it easier for the viewer to find dimensions.

Be careful not to double-dimension a distance. A distance should be dimensioned only once. If a 30 dimension were added above the 25 dimension on the right-side view, it would be an error. The distance would be double dimensioned: once with the 25 + 30 dimension and again with the 55 overall dimension. The 25 + 30 dimensions are mathematically equal to the 55 overall dimension, but there is a distinct difference in how they affect the manufacturing tolerances. Double dimensions are explained more fully in Chapter 8.

Dimensions Using Centerlines

Figure 7-81 shows an object dimensioned from its centerline. This type of dimensioning is used when the distance between the holes relative to each other is critical.

Figure 7-81

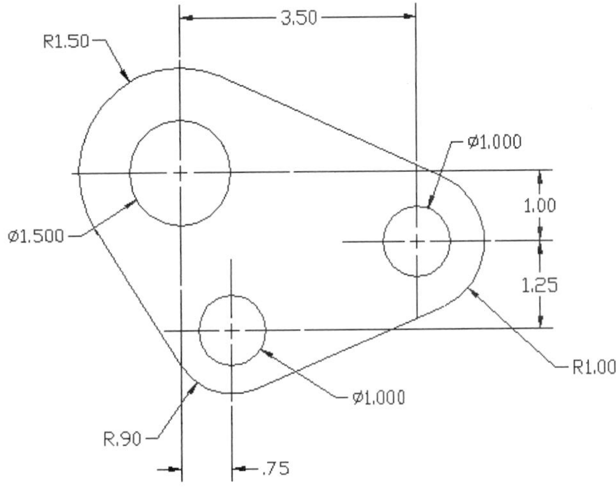

3D Dimensions

Inventor can be used to create dimensions on a three-dimensional object. Figure 7-82 shows an object drawn using the **ANSI (mm).idw** format. The drawing shows a front orthographic view of the object and an isometric view projected from the front view. Figure 7-83 shows the isometric view dimensioned. The **Dimension** tool was used to create the 3D dimensions. The procedure for creating 3D dimensions is the same as for 2D dimensioning.

Figure 7-82

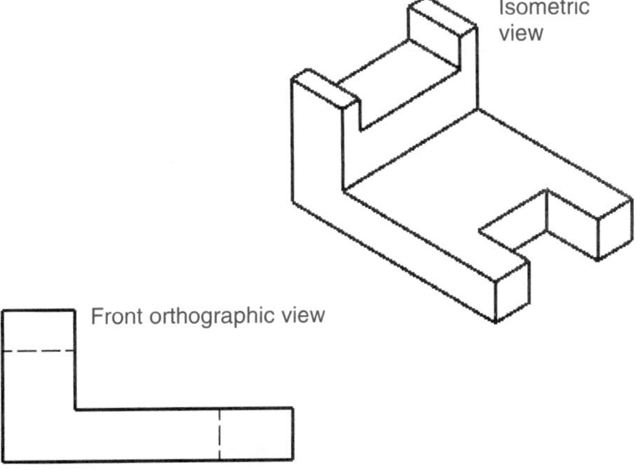

Figure 7-83

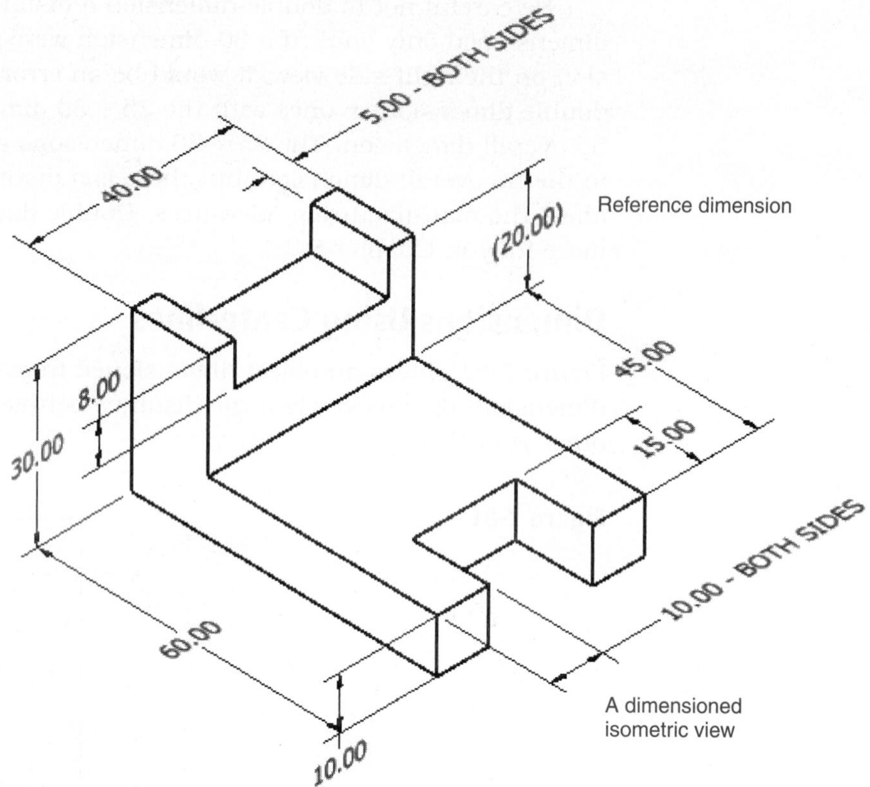

Sample Problem SP7-2

Figure 7-84 shows a shape that is to be dimensioned. All dimensions are in inches. The object is .50 thick.

The general procedure for applying dimensions is to start from the center of the shape and work out.

For example, in this example the hole and cutout are dimensioned first. The leader line for the hole is placed last, as it can be located in many different positions to avoid existing extension and dimension lines.

1 Dimension the rectangular cutout.

Click the left edge line first to ensure that the .50 dimension is on the outside left of the extension lines.

> **TIP**
> Every hole requires three dimensions: two for location and one for diameter.

2 Dimension the hole's location.

3 Dimension the circular cutout.

Use the **Leader Text** tool.

4 Add the overall dimensions.

Locate the overall dimensions so that there is an open area for the chamfer note and the hole's diameter.

5 Use the **Chamfer Note** tool and dimension the chamfers.

6 Edit the chamfer note to include **2 PLACES.**

7 Use the **Hole and Thread Notes** tool to dimension the hole's diameter.

Figure 7-84

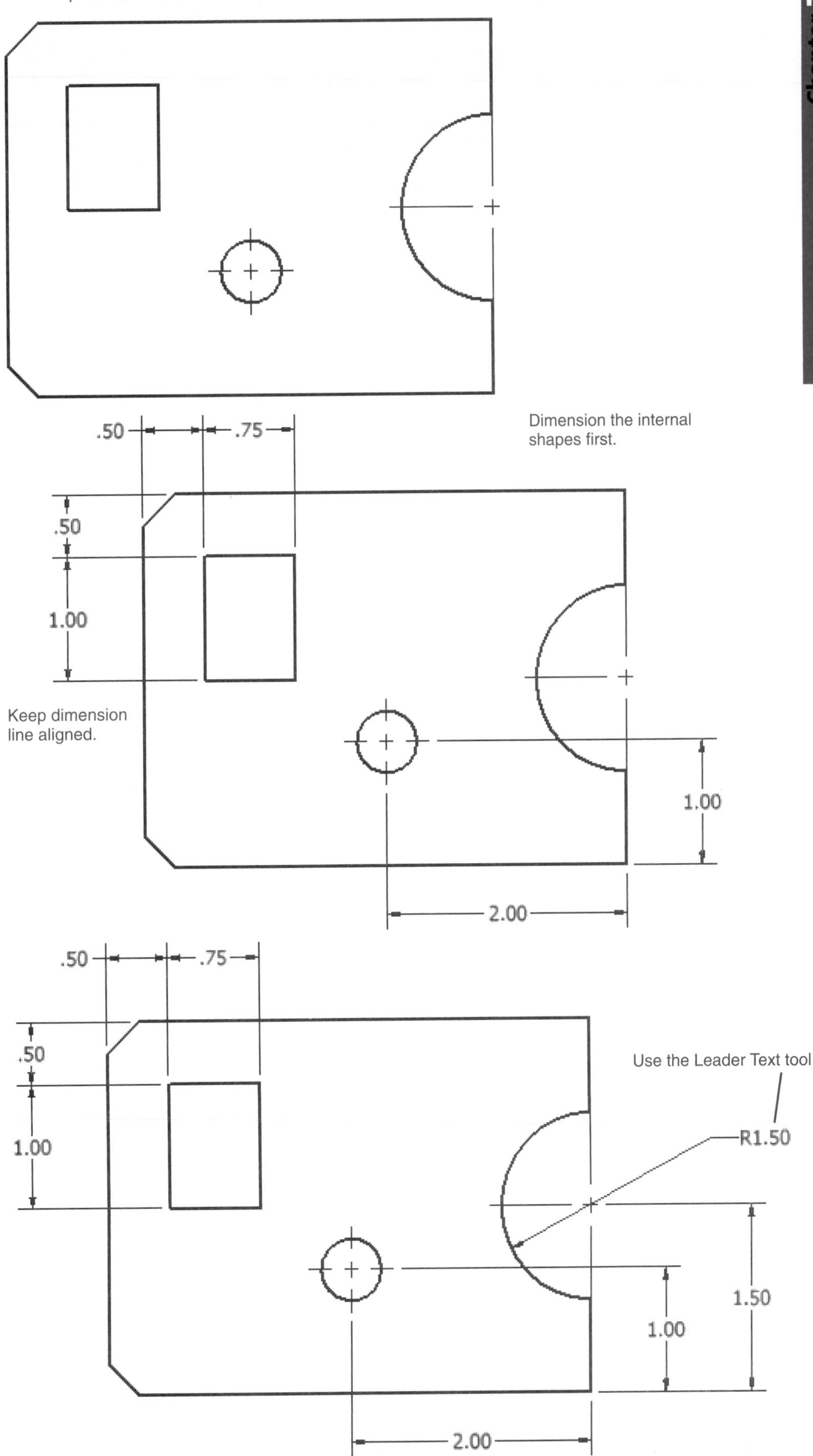

Shape to be dimensioned
.50
.75
Dimension the internal shapes first.
.50
1.00
Keep dimension line aligned.
1.00
2.00
.50
.75
.50
1.00
Use the Leader Text tool.
R1.50
1.50
1.00
2.00

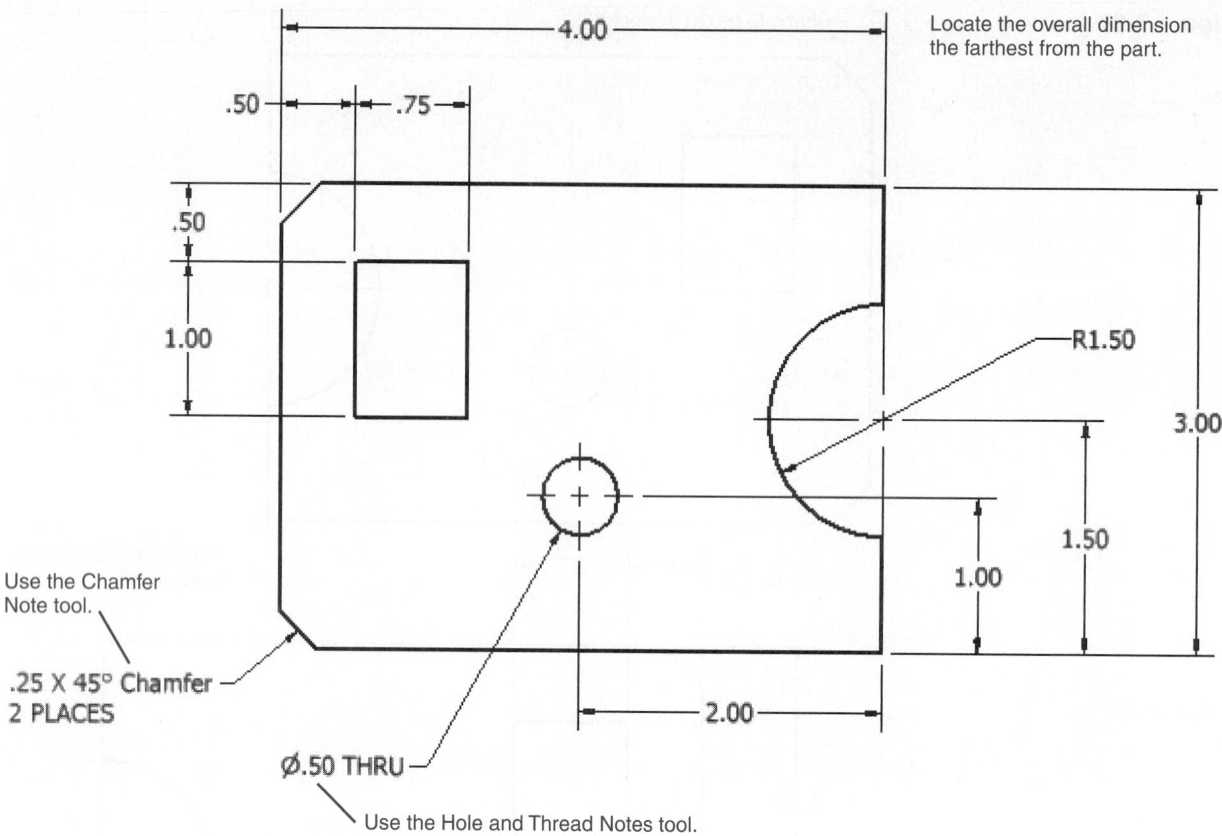

Figure 7-84
(Continued)

Sample Problem SP7-3

Figure 7-85 shows a shape to be dimensioned. All dimensions are in millimeters. The object is 10 thick.

The shape is unusual in that one end is rounded. It is difficult to create an overall dimension to a rounded surface. There are tolerances in both the shape of the circular end (how round it is) and in the location of the arc's center point. It this example a reference dimension will be used.

Note that the shape is symmetrical about its horizontal centerline. A symmetrical shape is one that is *exactly* the same on both sides of a centerline.

1 Define the object as symmetrical about its horizontal centerline by adding the symbol for symmetry.

The symbol for symmetry is created using the **Text** tool.

2 Use the **Baseline Dimension** tool and dimension the cutouts along the top edge of the part.

Note that the 5.00 depth was located along the bottom edge of the part. This was done to allow space for the internal slot's dimensions.

3 Dimension the internal slot.

Position the dimension values so that they do not interfere with other dimensions.

4 Use the **Hole and Thread Notes** tool and dimension the hole.

5 Use the **Leader** tool and add the radius for the rounded end.

The R30 dimension defines the width of the part, so no other dimension is needed.

6 Add overall reference dimensions for the length and width.

Reference dimensions are indicated by parentheses. Reference dimensions are not to be used for manufacture or inspection. These are included on a drawing for convenience.

7 Use the **Text** tool and add a note defining the fillet size for the internal slot.

There are several different ways to dimension the shape shown in Figure 7-85. Consider chain dimensions or ordinate dimensions as two other possible methods.

Figure 7-85

Dimension this shape.

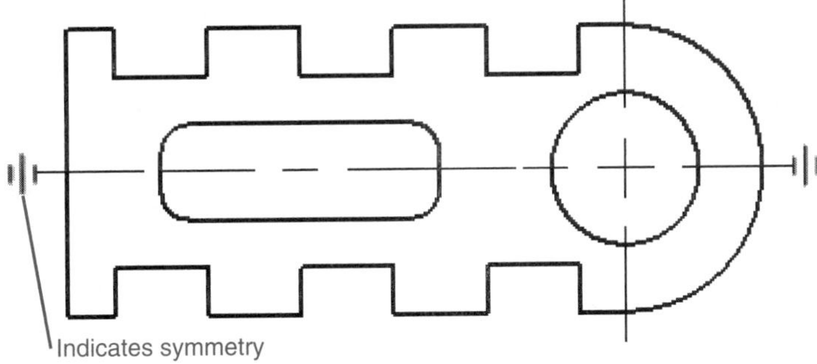

Use the Baseline Dimension tool.

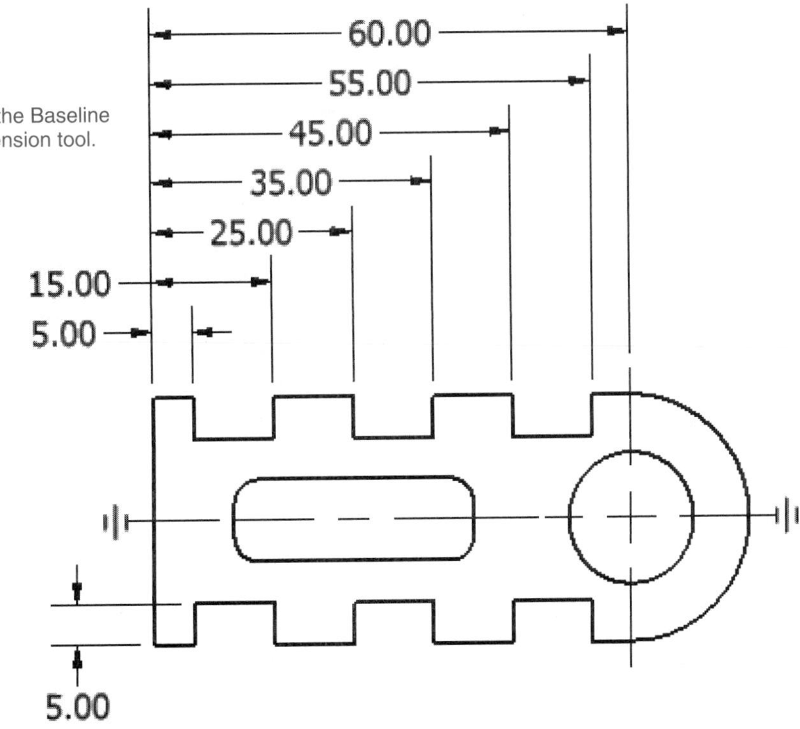

Figure 7-85
(Continued)

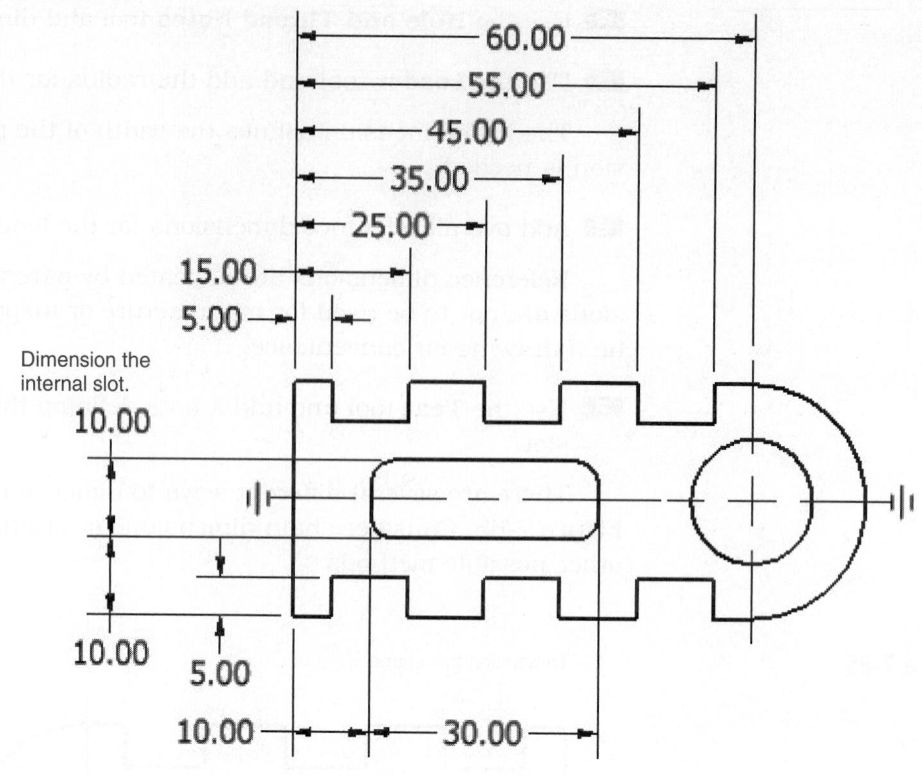

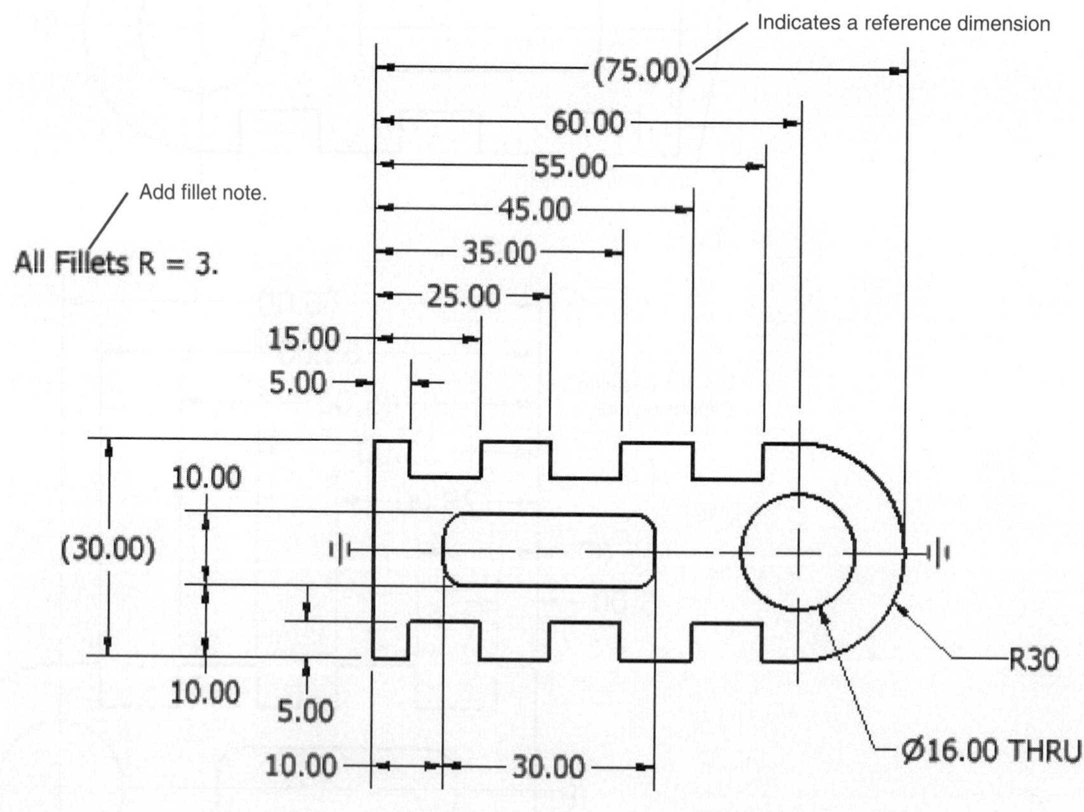

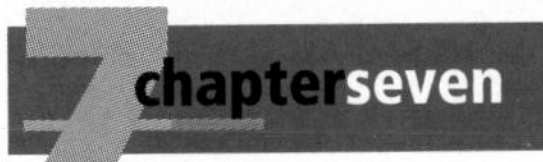

Chapter Summary

This chapter explained how to dimension objects with different shapes and features. It presented the ANSI standards and conventions and showed how to dimension different shapes and features, including various types of holes, slanted surfaces, fillets, and rounds. Drawing scales and dimensional values were discussed, including the use of leading zeros with decimal notation and the appropriate use of numbers of decimal places in a dimensional value. Dimensioning of sectional and orthographic views was also illustrated.

Chapter Test Questions

Multiple Choice

Circle the correct answer.

1. When dimensions are drawn, lines that have gaps for numerical values that end with arrowheads are called
   a. Centerlines c. Dimension lines
   b. Extension lines d. Leader lines
2. When dimensions are drawn, lines that project away from the surface of the part are called
   a. Centerlines c. Dimension lines
   b. Extension lines d. Leader lines
3. When dimensions are drawn, lines used to define hole diameters are called
   a. Centerlines c. Dimension lines
   b. Extension lines d. Leader lines
4. Polar dimensions are created using
   a. Two linear dimensions
   b. A linear and an angular dimension
   c. Two angular dimensions
5. Which of the following is *not* a type of knurl?
   a. Crosshatch c. Straight
   b. Diamond d. Raised diamond
6. A drawing scale designation of 2 = 1 indicates that the shape on the drawing is related how to the actual part?
   a. The shape is twice as big.
   b. The shape is the same size.
   c. The shape is half the size.
7. To edit a center mark's size
   a. Use the **Feature Control frame** option on the **Style and Standard Editor** dialog box

b. Use the **Dimension** option on the **Style and Standard Editor** dialog box

c. Use the **Center Mark** option on the **Style and Standard Editor** dia-log box

8. Dimensions created using the ANSI standards have decimal points (5.25). Dimensions created using the ISO standards use

a. Decimal points
c. Semicolons
b. Commas
d. En dashes

9. Dimensions that define straight distances are called

a. Linear dimensions
b. Angular dimensions
c. Horizontal dimensions

10. The rule for crossing extension lines states that it

a. Is recommended
b. Should be avoided
c. Is never allowed

Matching

Figure 7-86 shows a dimensioned object. Match the numbers with the type of dimension.

Column A **Column B**

a. The symbol for countersink ________
b. Indicates that the hole goes completely through the part ________
c. The inclusive angle for the countersink ________
d. An overall dimension ________
e. Indicates the number of Ø9 holes on the part ________
f. A chain dimension ________

True or False

Circle the correct answer.

1. **True or False:** Dimensions are located on a drawing in compliance with ANSI standards.

2. **True or False:** Model dimensions are used to construct a shape but do not appear on the **ANSI.idw** drawing.

3. **True or False:** Dimensions may be located on the surface of a part.

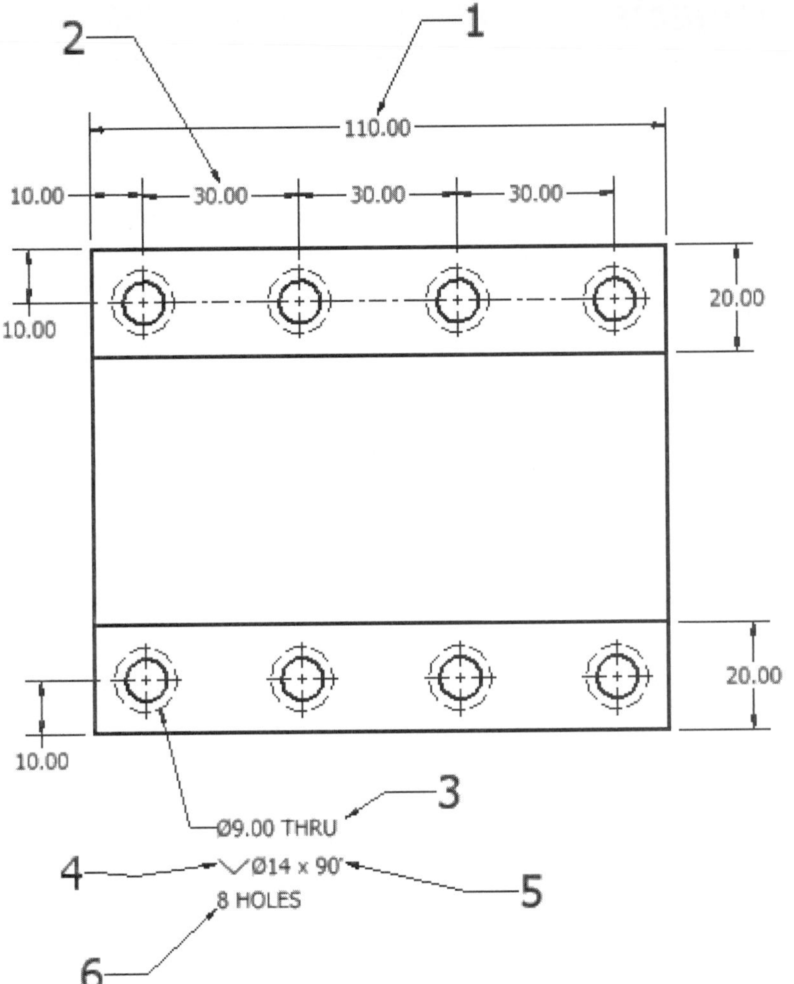

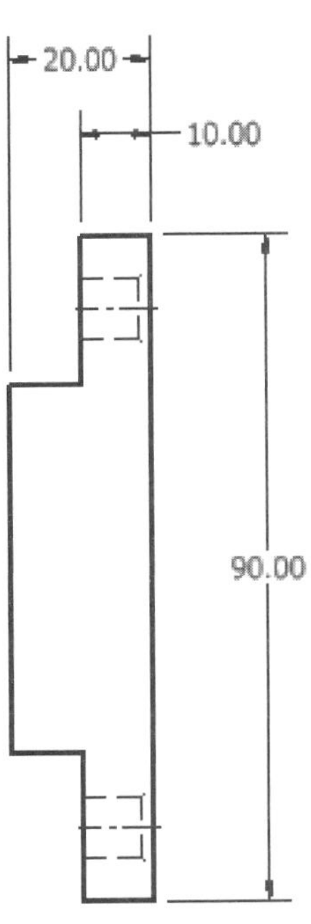

Figure 7-86

4. **True or False:** Dimensions are added to an **ANSI.idw** drawing using the **Dimension** tool located on the **Annotate** tab.

5. **True or False:** All dimensional values must be located using the unidirectional format.

6. **True or False:** To edit an existing hole dimension, right-click the dimension and select the **Edit Hole Note** option.

7. **True or False:** Inch dimensions that are less than 1.00 require a 0 to the left of the decimal point.

8. **True or False:** Metric dimensions that are less than 1.00 require a 0 to the left of the decimal point.

9. **True or False:** Aligned dimensions must be written using the unidirectional format.

10. **True or False:** A counterbored hole is best dimensioned using a note.

Chapter Projects

Project 7-1

Measure and redraw the shapes in Figures P7-1 through P7-18. The dotted grid background has either .50-in. or 10-mm spacing. All holes are through holes. Specify the units and scale of the drawing. Create a model by using the **Extrude** tool. Create a set of multiviews (front, top, side, and isometric views) using the .idw format and add the appropriate dimensions.

A. Measure using millimeters.

B. Measure using inches.

All dimensions are within either .25 in. or 5 mm. All fillets and rounds are R.50 in., R.25 in. or R10 mm, R5 mm.

THICKNESS:
40 mm
1.50 in.

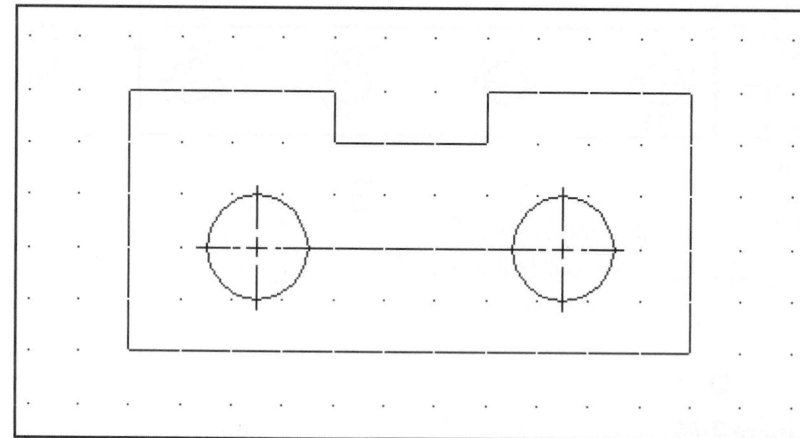

Figure P7-1

THICKNESS:
20 mm
.75 in.

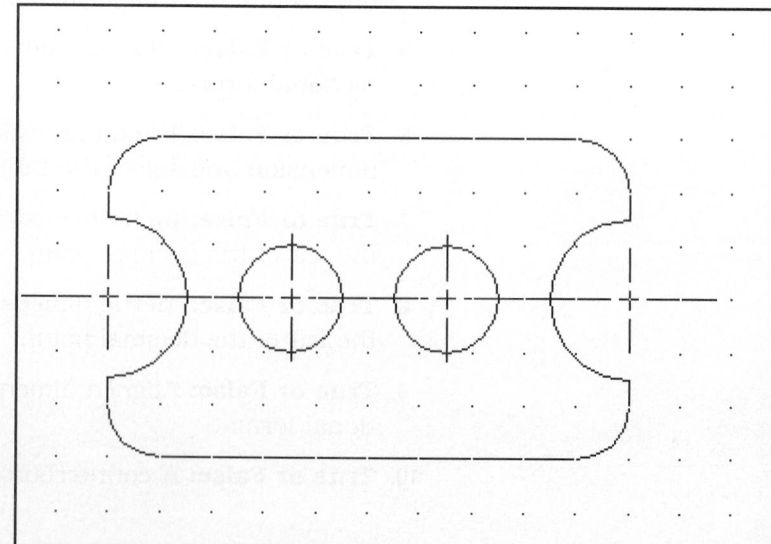

Figure P7-2

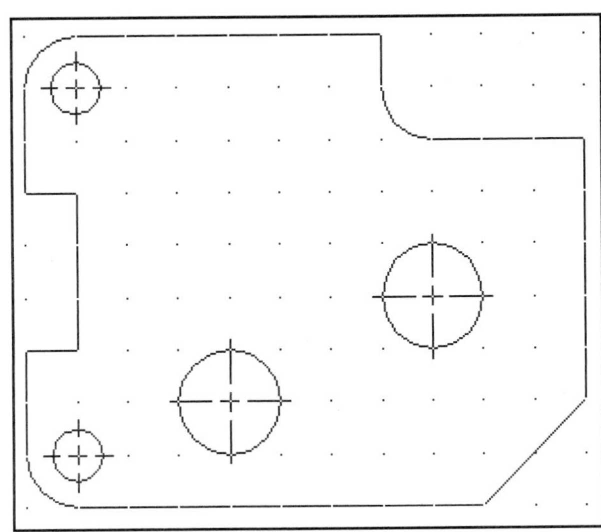

THICKNESS:
35 mm
1.25 in.

Figure P7-3

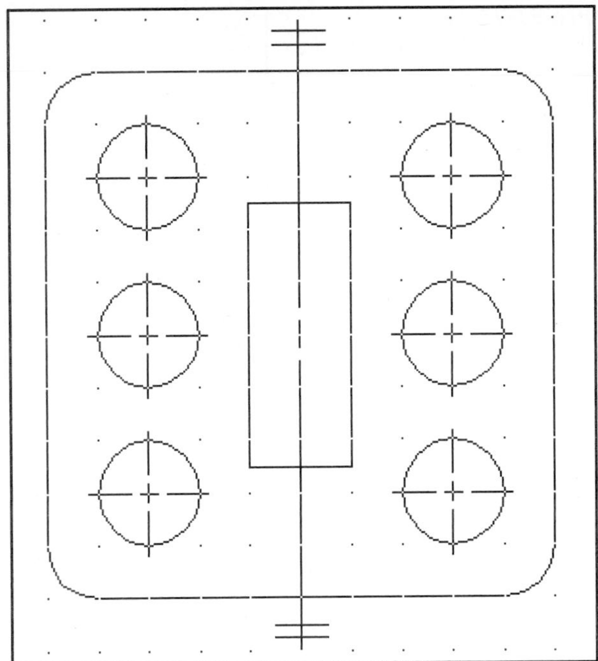

THICKNESS:
15 mm
.50 in.

Figure P7-4

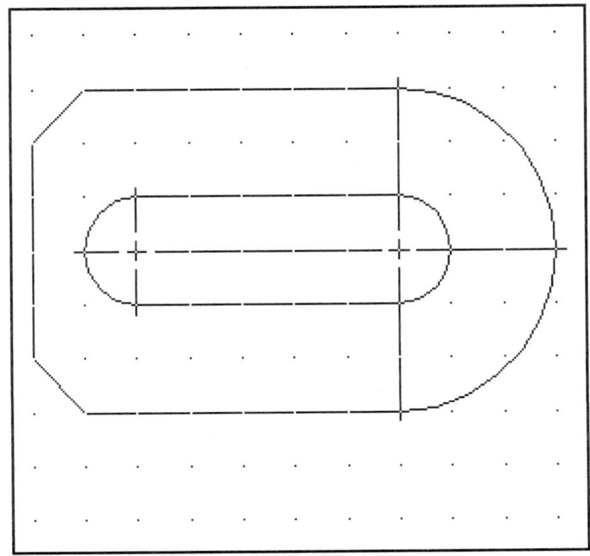

THICKNESS:
10 mm
.50 in.

Figure P7-5

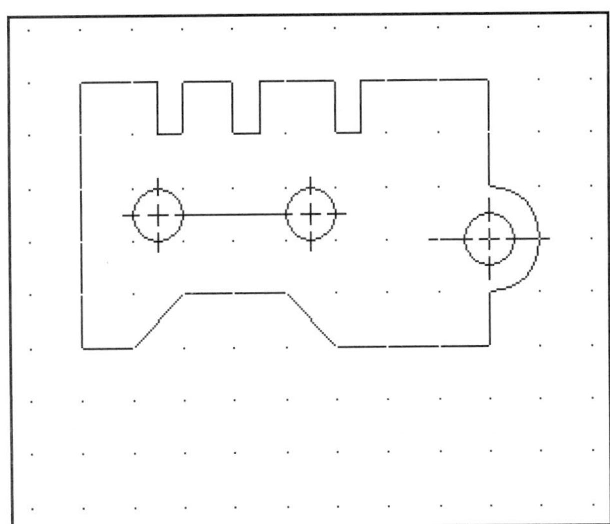

THICKNESS:
5 mm
.25 in.

Figure P7-6

Figure P7-7

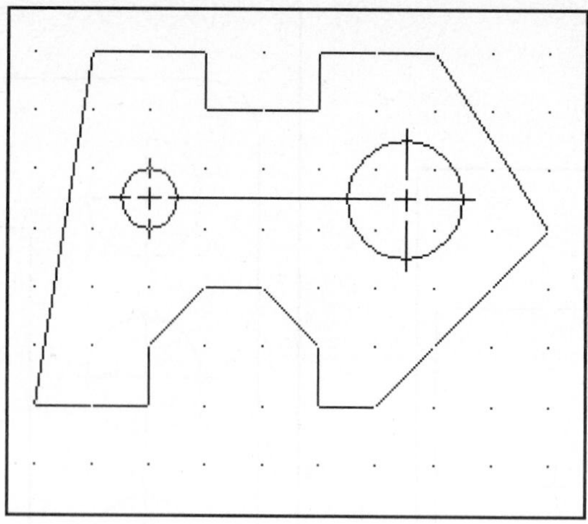

THICKNESS:
10 mm
.25 in.

THICKNESS:
8 mm
.25 in.

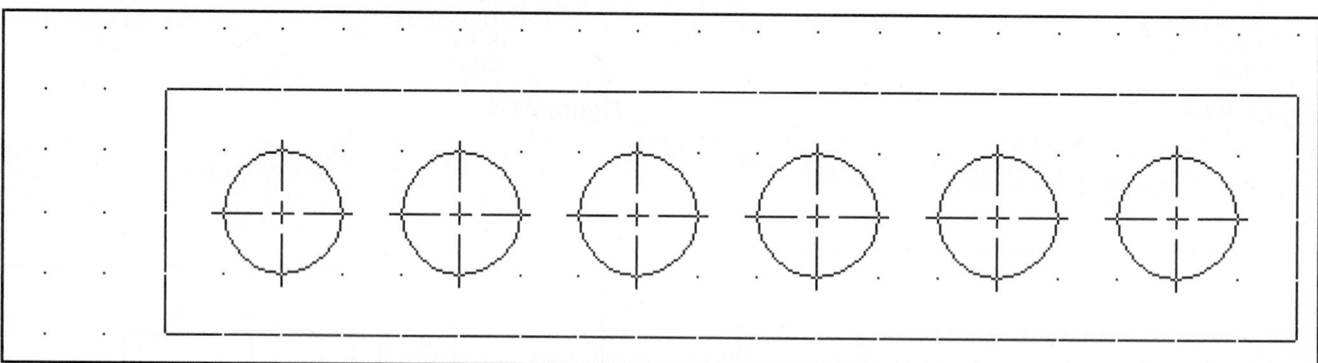

Figure P7-8

Figure P7-9

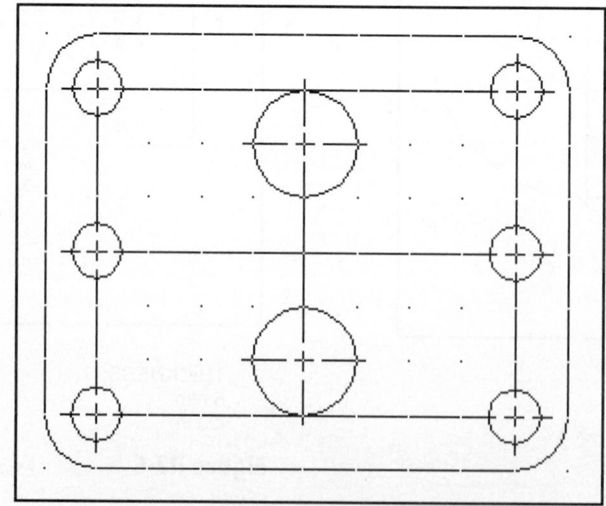

THICKNESS:
20 mm
.75 in.

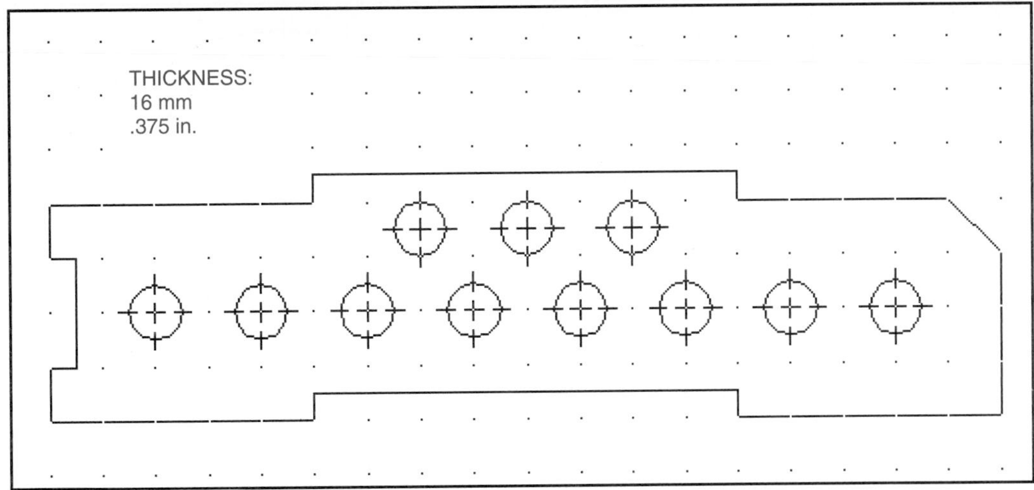

Figure P7-10

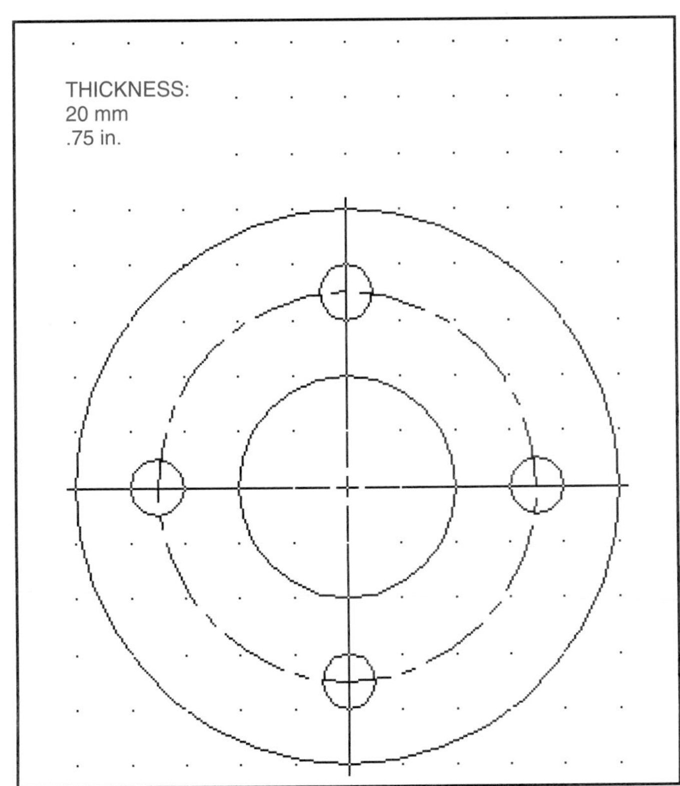

Figure P7-11

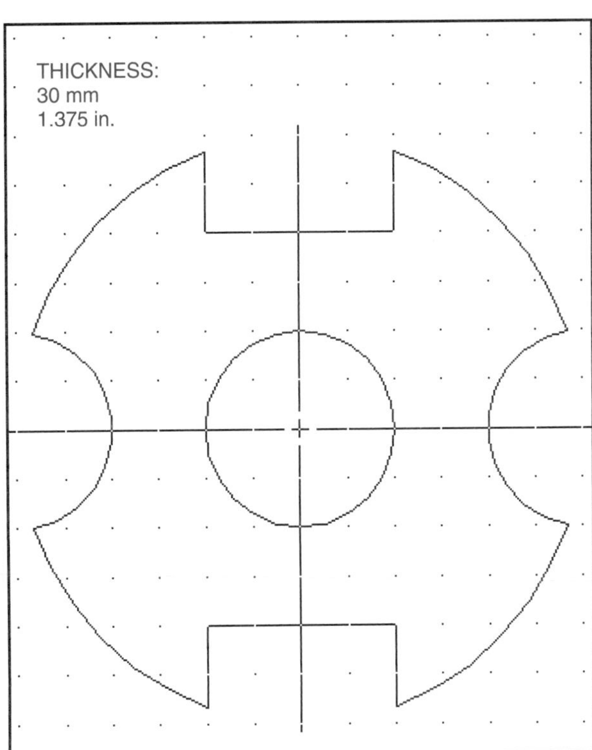

Figure P7-12

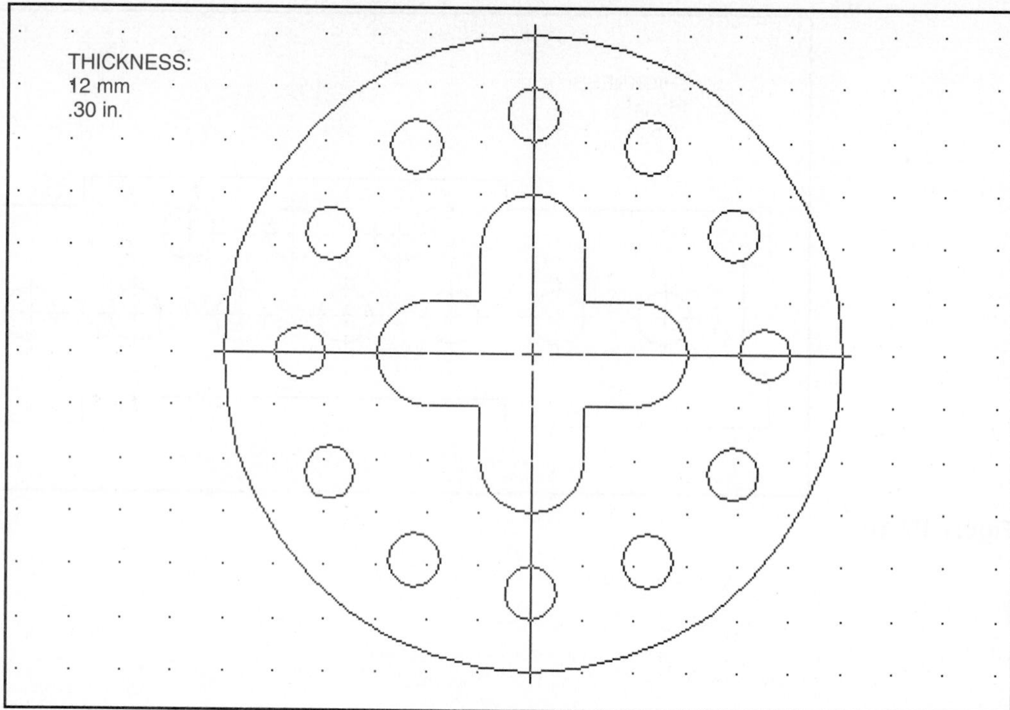

Figure P7-13

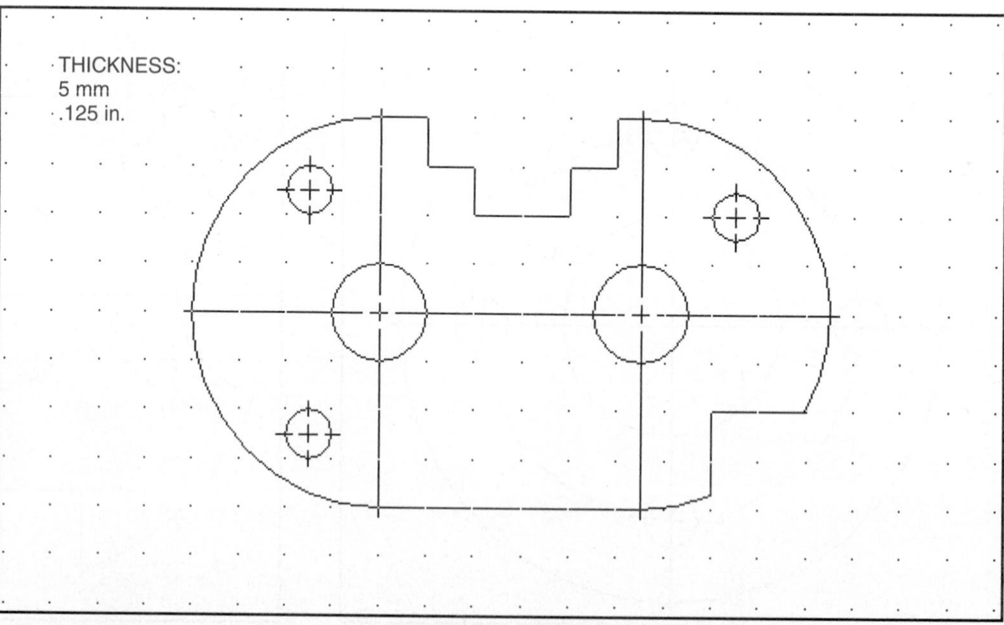

Figure P7-14

THICKNESS:
10 mm
.25 in.

Dimension using baseline dimensions.

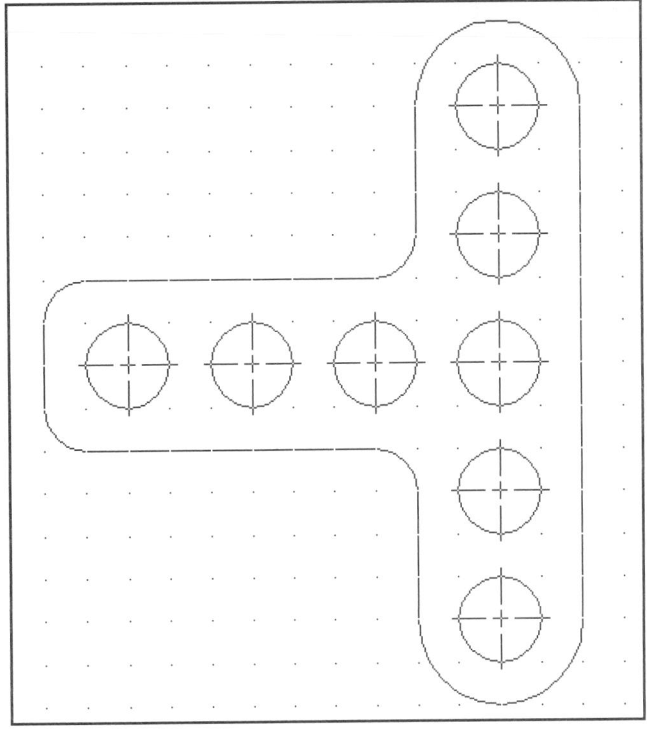

Figure P7-15

THICKNESS:
15 mm
.50 in.

Dimension using
A. Baseline dimensions.
B. Ordinate dimensions.
C. Chain dimensions.
D. Hole table.

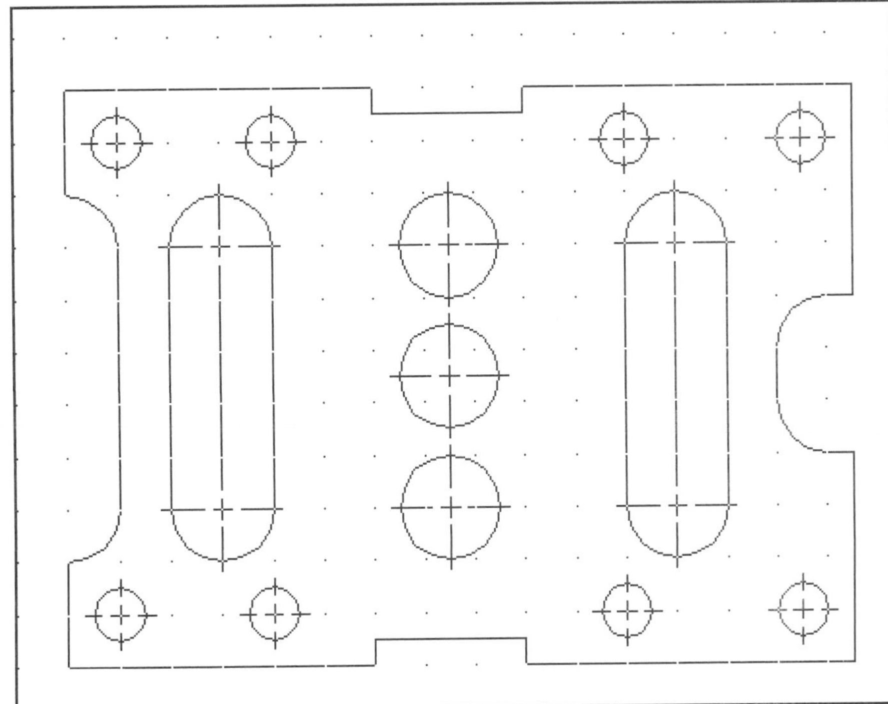

Figure P7-16

THICKNESS:
5 mm
.19 in.

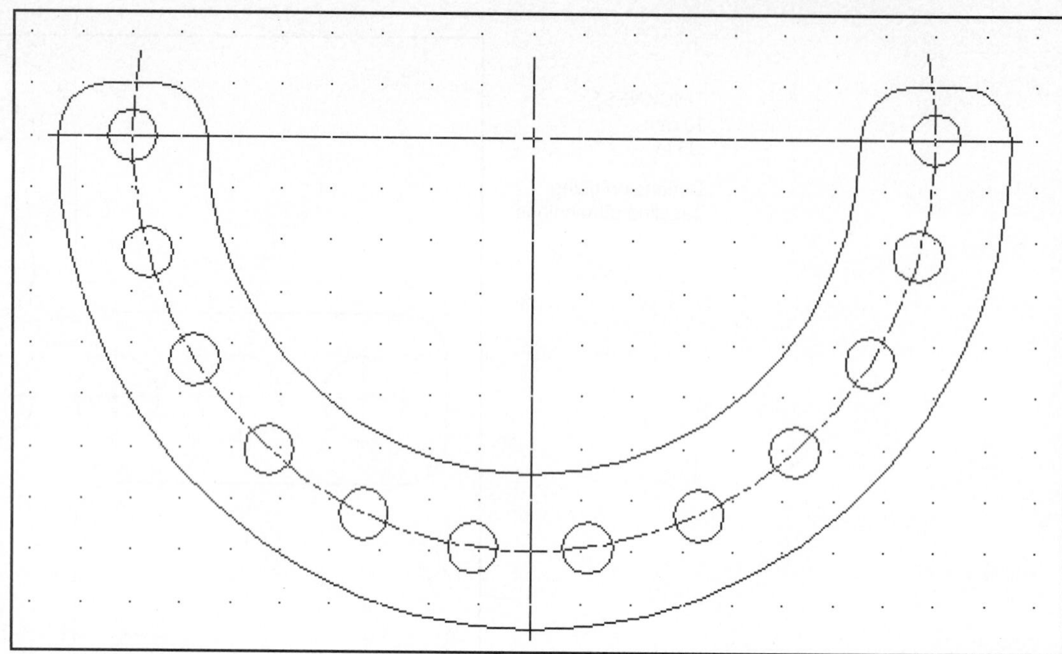

Figure P7-17

THICKNESS:
15 mm
.625 in.

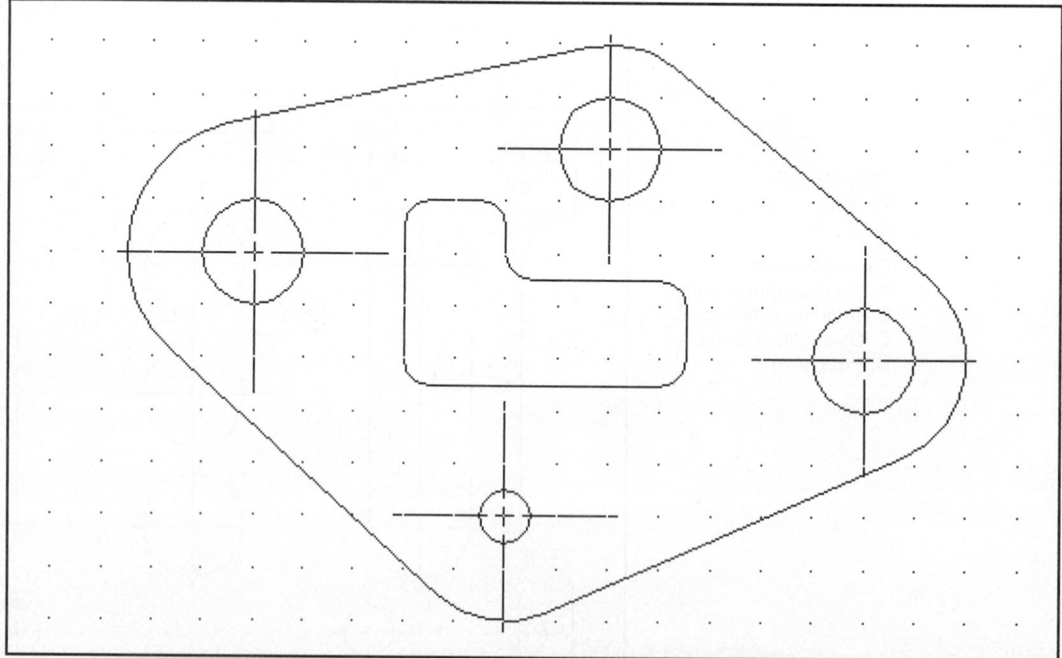

Figure P7-18

Project 7-2

Draw models of the objects shown in Figures P7-19 through P7-36.

1. Create orthographic views of the objects. Dimension the orthographic views.
2. Create 3D models of the objects. Dimension the 3D models.

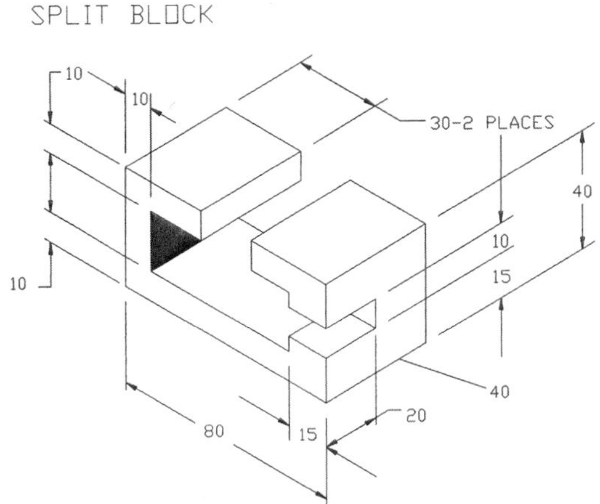

Figure P7-19
MILLIMETERS

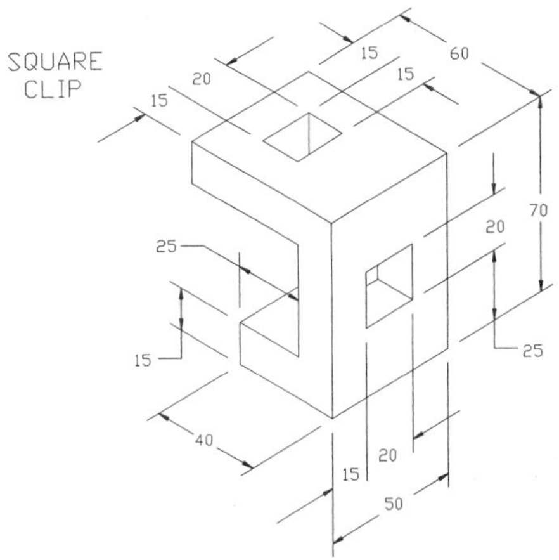

Figure P7-20
MILLIMETERS

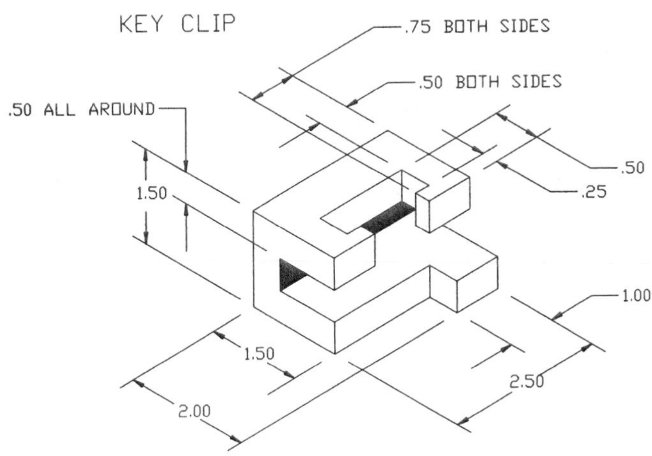

Figure P7-21
INCHES

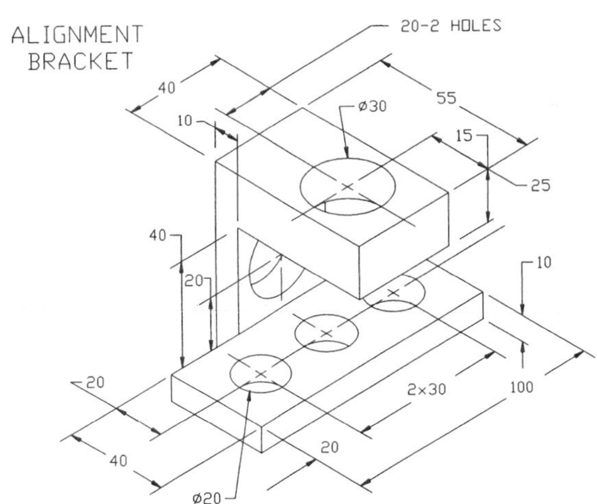

Figure P7-22
MILLIMETERS

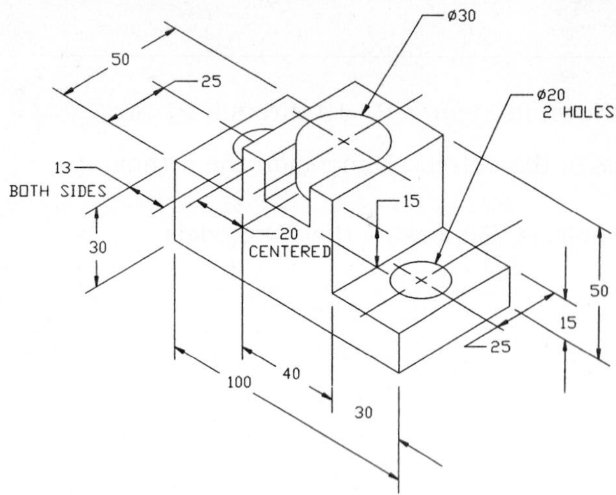

Figure P7-23
MILLIMETERS

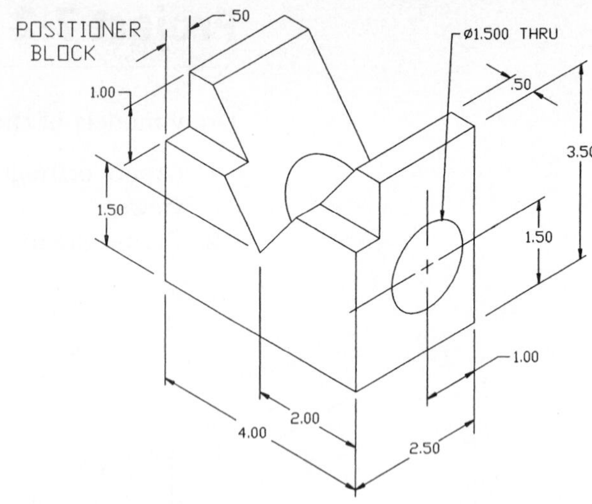

Figure P7-24
INCHES

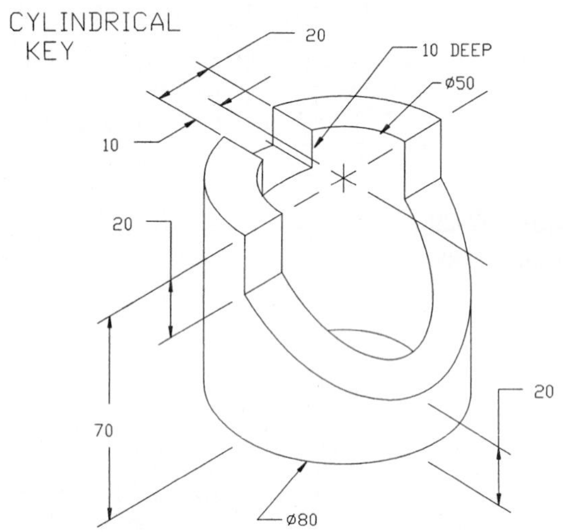

Figure P7-25
MILLIMETERS

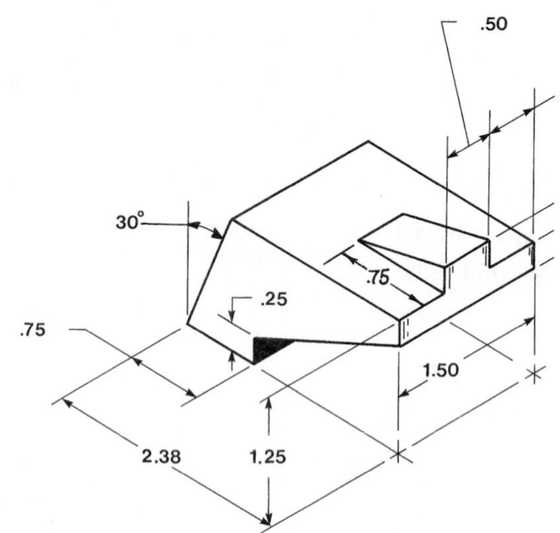

Figure P7-26
INCHES

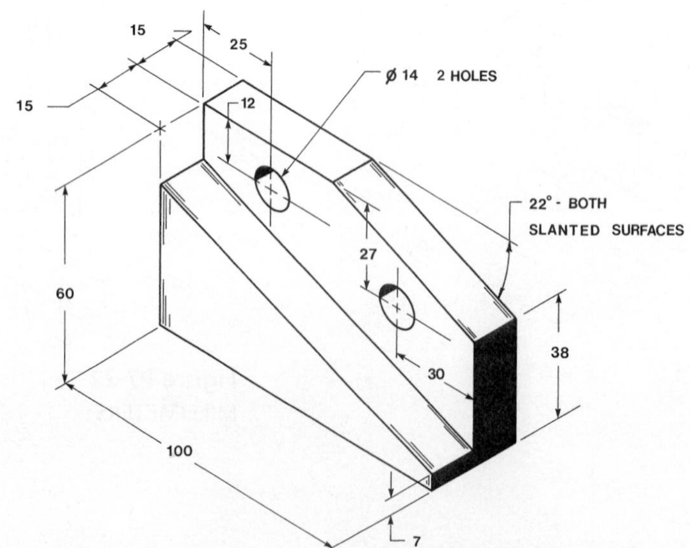

Figure P7-27
MILLIMETERS

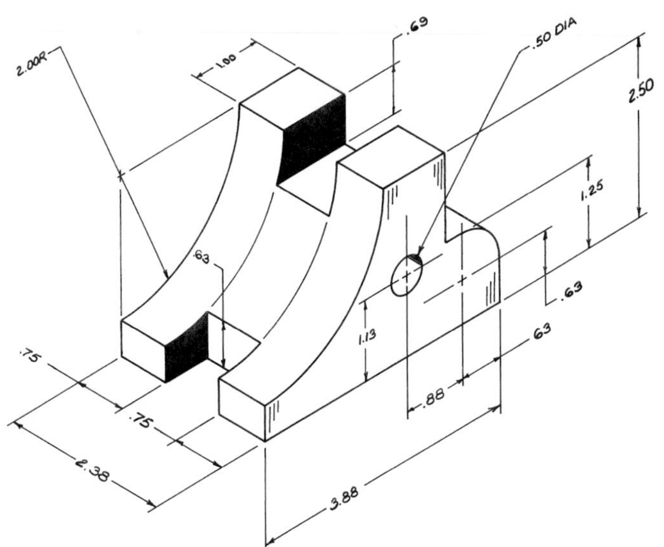

Figure P7-28
INCHES

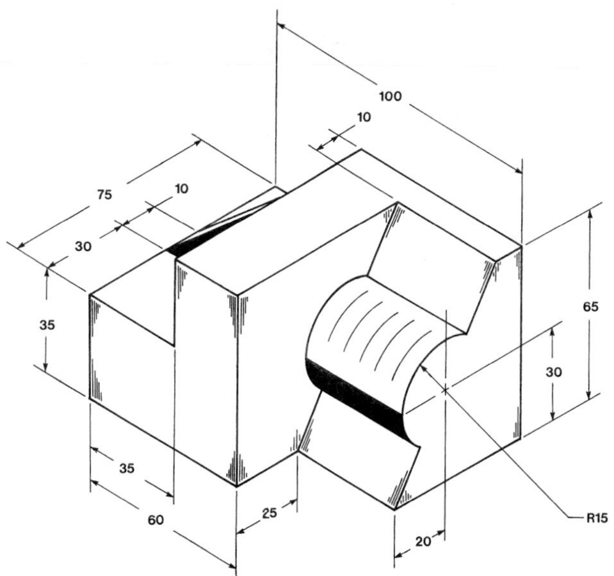

Figure P7-29
MILLIMETERS

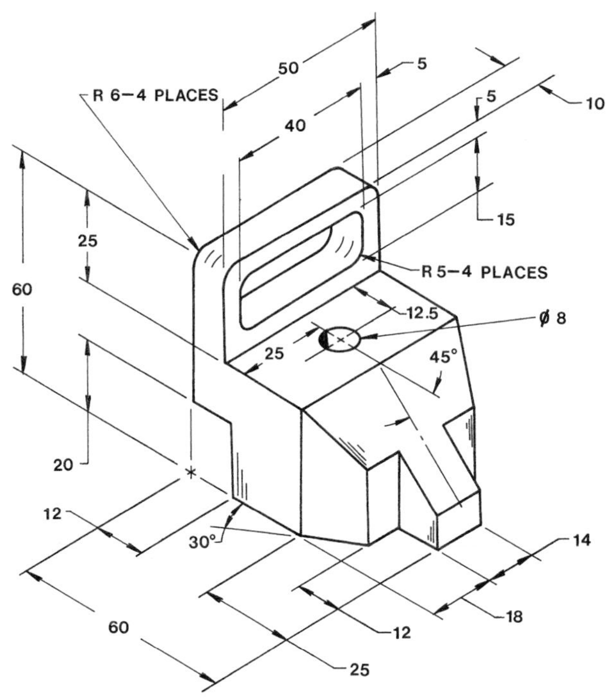

Figure P7-30
MILLIMETERS

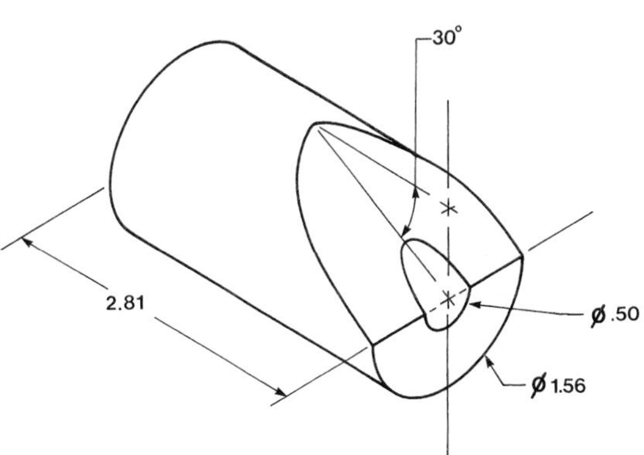

Figure P7-31
INCHES

NOTE: ALL FILLET AND ROUNDS=R3

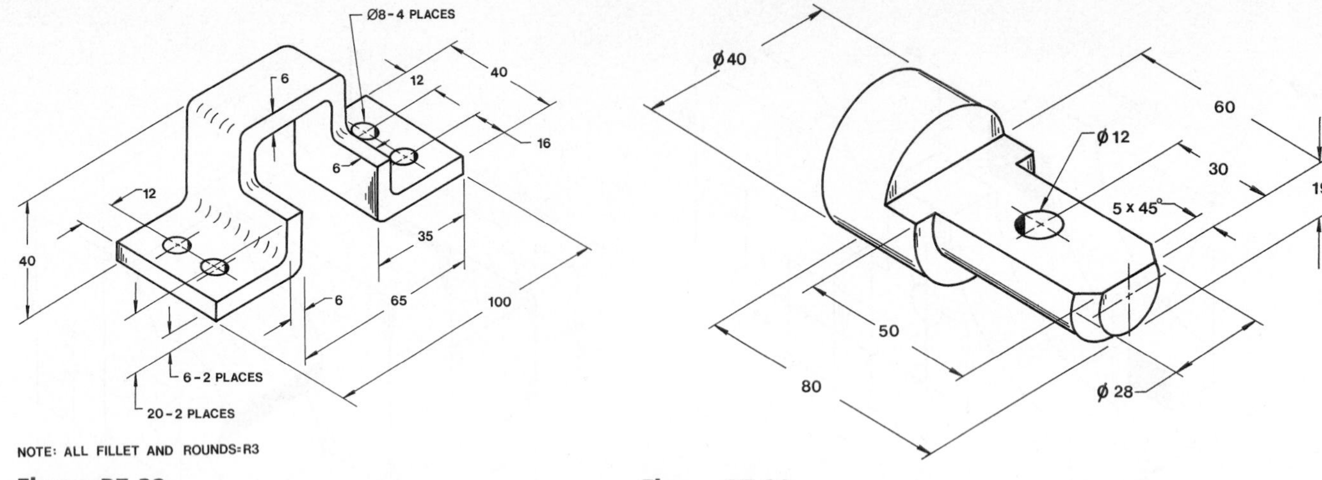

Figure P7-32
MILLIMETERS

Figure P7-33
MILLIMETERS

ALL FILLETS AND ROUNDS=R5
MATL 5 THK

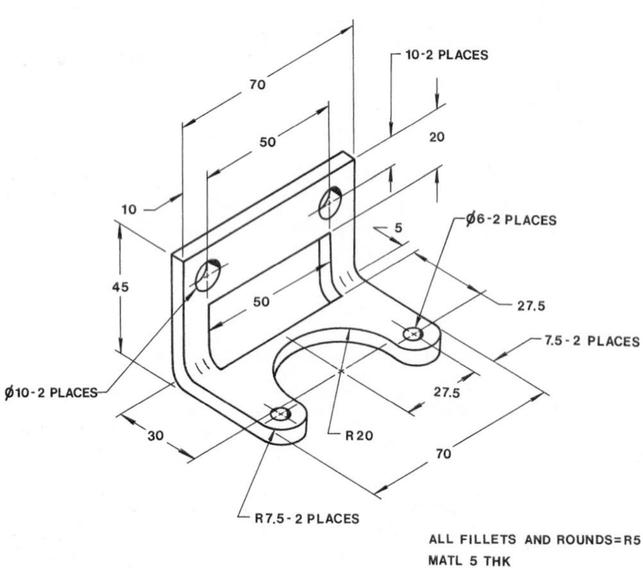

Figure P7-34
MILLIMETERS

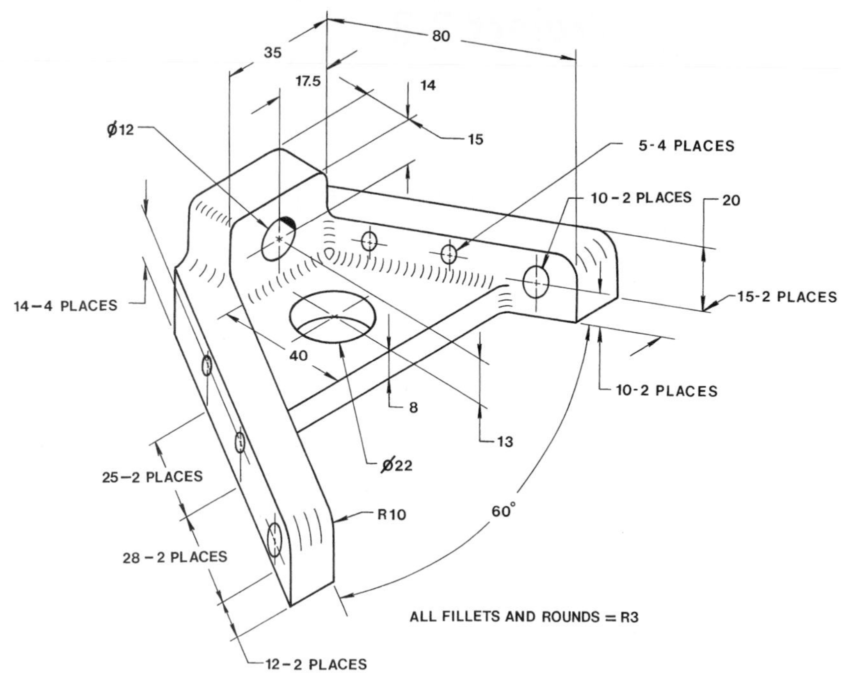

Figure P7-35
MILLIMETERS

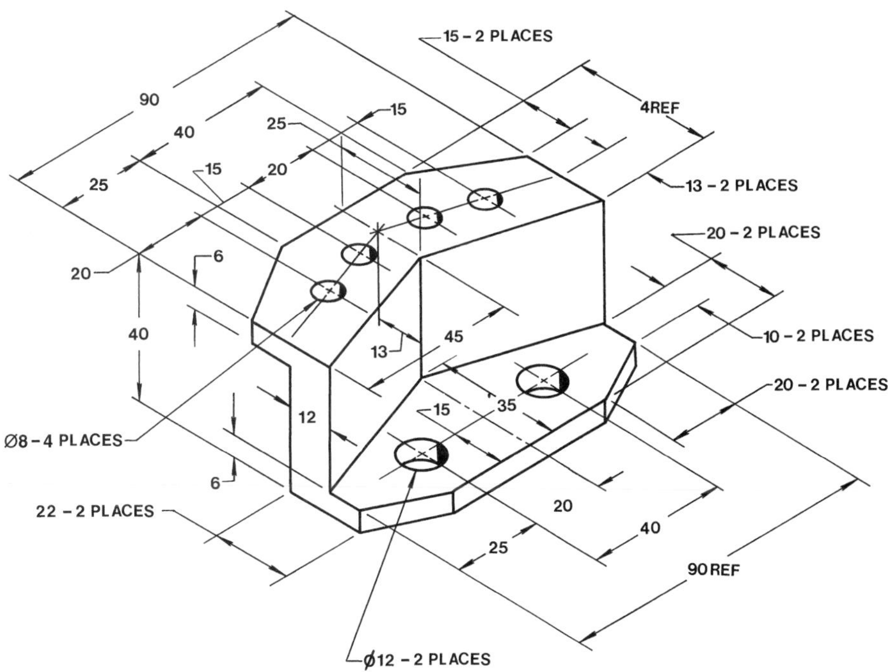

Figure P7-36
MILLIMETERS

Project 7-3

1. Draw a 3D model from the given top orthographic and sectional views in Figure P7-37.
2. Draw a top orthographic view and a section view of the object and add dimensions.

Project 7-4

1. Draw a 3D model from the given top orthographic and sectional views in Figure P7-38.
2. Draw a top orthographic view and a section view of the object and add dimensions.

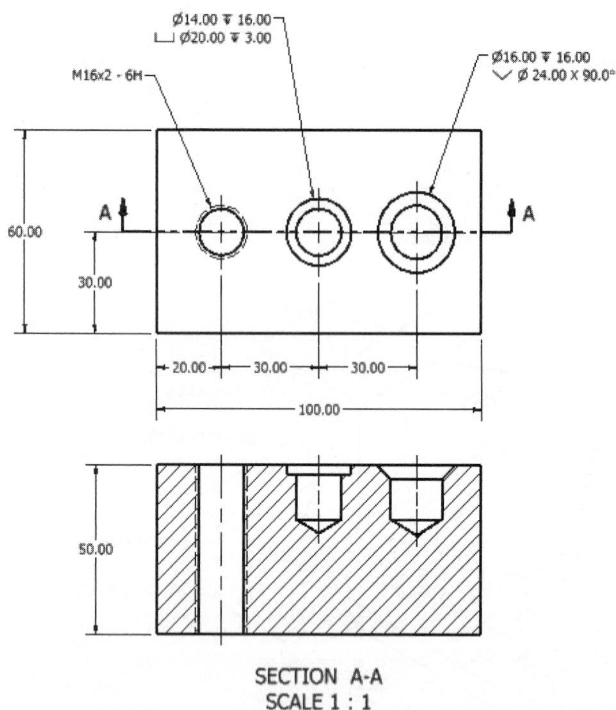

Figure P7-37
MILLIMETERS

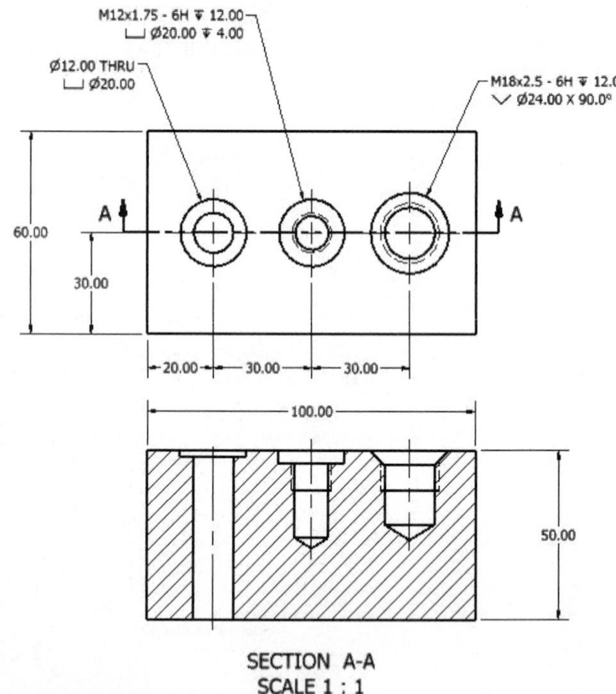

Figure P7-38
MILLIMETERS

Project 7-5

1. Draw a 3D model from the given top orthographic and section views in Figure P7-39.
2. Draw a top orthographic view and a section view of the object and add dimensions.

Project 7-6

1. Draw a 3D model from the given top orthographic and section views in Figure P7-40.
2. Draw a top orthographic view and a section view of the object and add dimensions.

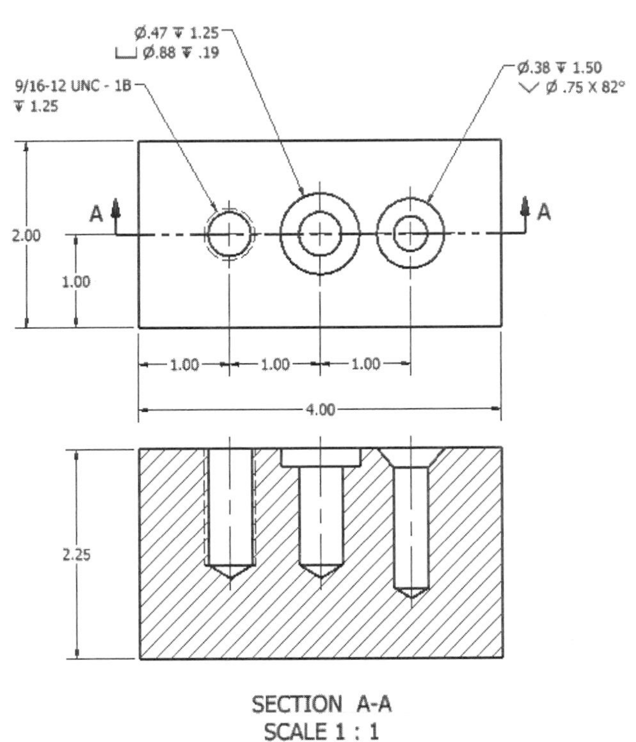

Figure P7-39
INCHES

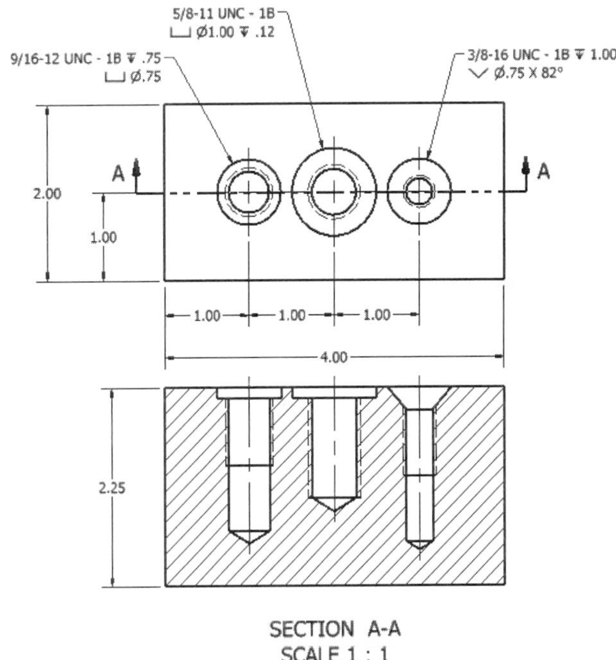

Figure P7-40
INCHES

8 chapter eight
Tolerancing

CHAPTER OBJECTIVES

- Understand tolerance conventions
- Understand the meaning of tolerances
- Learn how to apply tolerances
- Understand geometric tolerances
- Understand positional tolerances

Introduction

tolerance: The manufacturing limits for dimensions.

Tolerances define the manufacturing limits for dimensions. All dimensions have tolerances either written directly on the drawing as part of the dimension or implied by a predefined set of standard tolerances that apply to any dimension that does not have a stated tolerance.

This chapter explains general tolerance conventions and how they are applied using Inventor. It includes a sample tolerance study and an explanation of standard fits and surface finishes.

Direct Tolerance Methods

There are two methods used to include tolerances as part of a dimension: *plus and minus*, and *limits*. Plus and minus tolerances can be expressed in either bilateral (deviation) or unilateral (symmetric) form.

bilateral tolerance: A tolerance with both a plus and a minus value.

unilateral tolerance: A tolerance with either the plus or the minus value equal to zero.

A ***bilateral tolerance*** has both a plus and a minus value, whereas a ***unilateral tolerance*** has either the plus or the minus value equal to zero. Figure 8-1 shows a horizontal dimension of 60 mm that includes a bilateral tolerance of plus or minus 1 and another dimension of 60.00 mm that

Figure 8-1

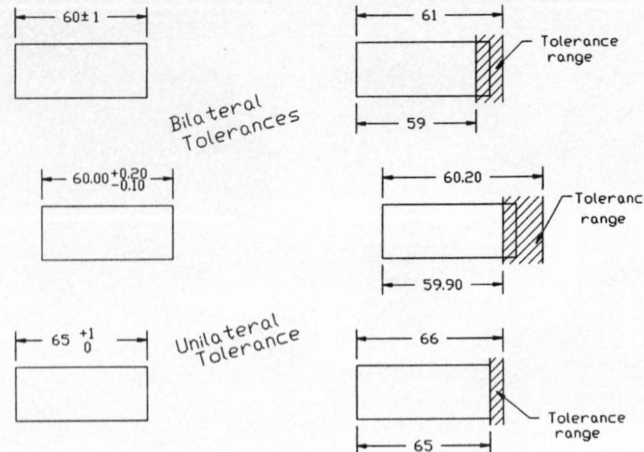

includes a bilateral tolerance of plus 0.20 or minus 0.10. Figure 8-1 also shows a dimension of 65 mm that includes a unilateral tolerance of plus 1 or minus 0.

> **NOTE**
>
> Bilateral tolerances are called *symmetric* in Inventor. Unilateral tolerances are called *deviation.*

Plus or minus tolerances define a range for manufacturing. If inspection shows that all dimensioned distances on an object fall within their specified tolerance range, the object is considered acceptable; that is, it has been manufactured correctly.

The dimension and tolerance of 60 ± 0.1 means that the part must be manufactured within a range no greater than 60.1 nor less than 59.9. The dimension and tolerance 65 + 1/−0 defines the tolerance range as 65.0 to 66.0.

Figure 8-2 shows some bilateral and unilateral tolerances applied using decimal inch values. Inch dimensions and tolerances are written using a slightly different format than millimeter dimensions and tolerances, but they also define manufacturing ranges for dimension values. The horizontal bilateral dimension and tolerance 2.50 ± .02 defines the longest acceptable distance as 2.52 in. and the shortest as 2.48. The unilateral dimension 2.50 + .02/−.00 defines the longest acceptable distance as 2.52 and the shortest as 2.50.

Figure 8-2

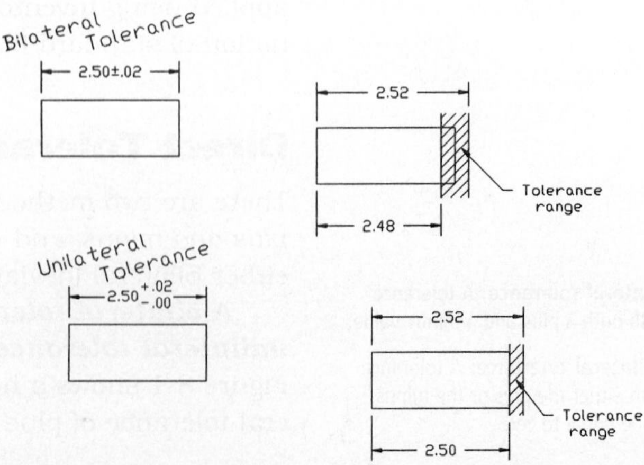

Tolerance Expressions

Dimension and tolerance values are written differently for inch and millimeter values. See Figure 8-3. Unilateral dimensions for millimeter values specify a zero limit with a single 0. A zero limit for inch values must include the same number of decimal places given for the dimension value. In the example shown in Figure 8-3, the dimension value .500 has a unilateral tolerance with minus zero tolerance. The zero limit is written as .000, with three decimal places for both the dimension and the tolerance.

Figure 8-3

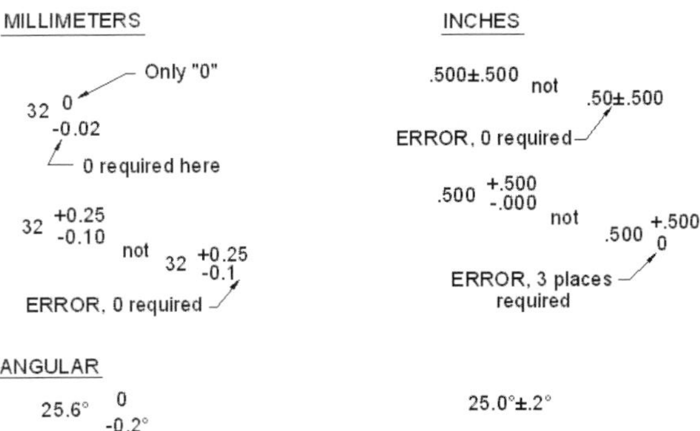

Both values in a bilateral tolerance for inch values must contain the same number of decimal places; for millimeter values the tolerance values need not include the same number of decimal places as the dimension value. In Figure 8-3 the dimension value 32 is accompanied by tolerances of +0.25 and −0.10. This form is not acceptable for inch dimensions and tolerances. An equivalent inch dimension and tolerance would be written 32.00 + .25/−.10.

Degree values must include the same number of decimal places in both the dimension value and the tolerance values for bilateral tolerances. A single 0 may be used for unilateral tolerances.

Understanding Plus and Minus Tolerances

A millimeter dimension and tolerance of 12.0 + 0.2/−0.1 means the longest acceptable distance is 12.2000 . . . 0, and the shortest is 11.9000 . . . 0. The total range is 0.3000 . . . 0.

After an object is manufactured, it is inspected to ensure that the object has been manufactured correctly. Each dimensioned distance is measured and, if it is within the specified tolerance, is accepted. If the measured distance is not within the specified tolerance, the part is rejected. Some rejected objects may be reworked to bring them into the specified tolerance range, whereas others are simply scrapped.

Figure 8-4 shows a dimension with a tolerance. Assume that five objects were manufactured using the same 12.0 + 0.2/−0.1 dimension and

Figure 8-4

GIVEN (mm)

12 +0.2 −0.1

MEANS

TOL MAX = 12.2
TOL MIN = 11.9
TOTAL TOL = 0.3

OBJECT	AS MEASURED	ACCEPTABLE?
1	12.160	OK
2	12.020	OK
3	12.203	TOO LONG
4	11.920	OK
5	11.895	TOO SHORT

tolerance. The objects were then inspected and the results were as listed. Inspected measurements are usually expressed to at least one more decimal place than that specified in the tolerance. Which objects are acceptable and which are not? Object 3 is too long and object 5 is too short because their measured distances are not within the specified tolerances.

Figure 8-5 shows a dimension and tolerance of 3.50 ± .02 in. Object 3 is not acceptable because it is too short, and object 4 is too long.

Figure 8-5

GIVEN (inches)
3.50±.02

MEANS
TOL MAX = 3.52
TOL MIN = 3.48
TOTAL TOL = .04

OBJECT	AS MEASURED	ACCEPTABLE?
1	3.520	OK
2	3.486	OK
3	3.470	TOO SHORT
4	3.521	TOO LONG
5	3.515	OK

Creating Plus and Minus Tolerances

Plus and minus tolerances may be created in Inventor using the **Tolerance** option associated with existing dimensions, or by setting the plus and minus values using the **Dimension Styles** tool.

Create a drawing using the **Metric** tab and the **ANSI (mm).ipt** format. See Figure 8-6 for the model's dimensions. Dimension the drawing.

Figure 8-6

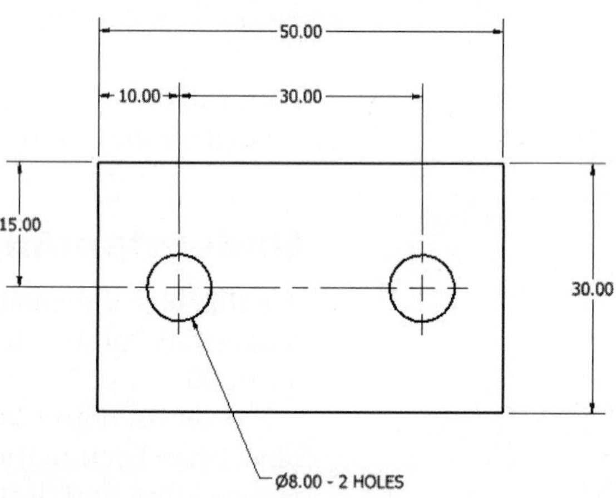

EXERCISE 8-1 **Creating Plus and Minus Tolerances**

1 Create a new drawing using the **ANSI (mm).dwg** format.

2 Use the model created from Figure 8-6 and add the dimensions as shown.

3 Click the **30.00** dimension.

The dimension will change color.

4 Right-click the mouse and select the **Edit** option.

See Figure 8-7. The **Edit Dimension** dialog box will appear. See Figure 8-8.

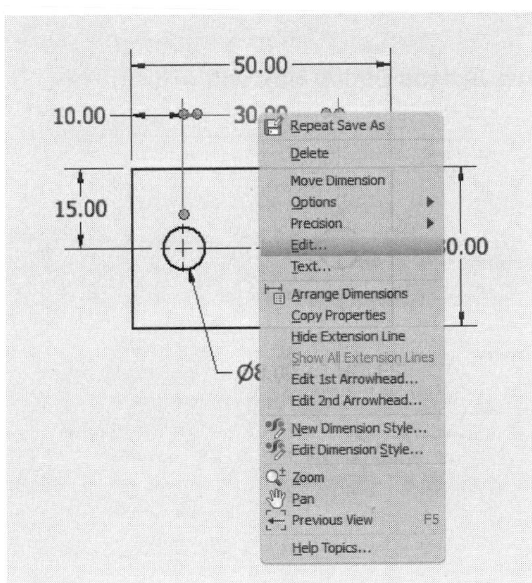

Figure 8-7

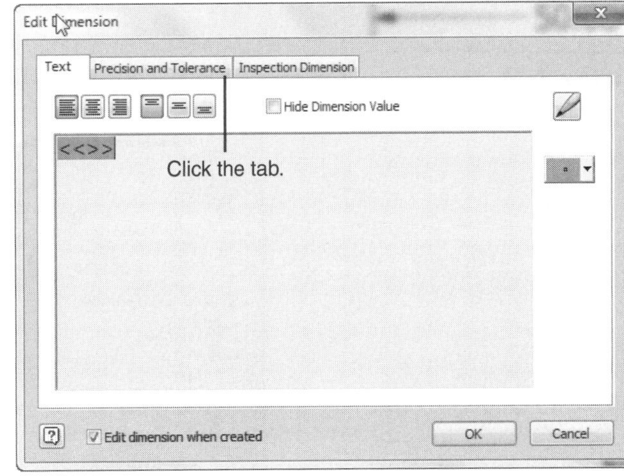

Figure 8-8

5 Click the **Precision and Tolerance** tab.

The **Edit Dimension** dialog box will change. See Figure 8-9.

6 Select the **Symmetric** option and enter an upper value of **0.20.** Click **OK.**

Figure 8-10 shows the resulting symmetric tolerances.

Figure 8-9

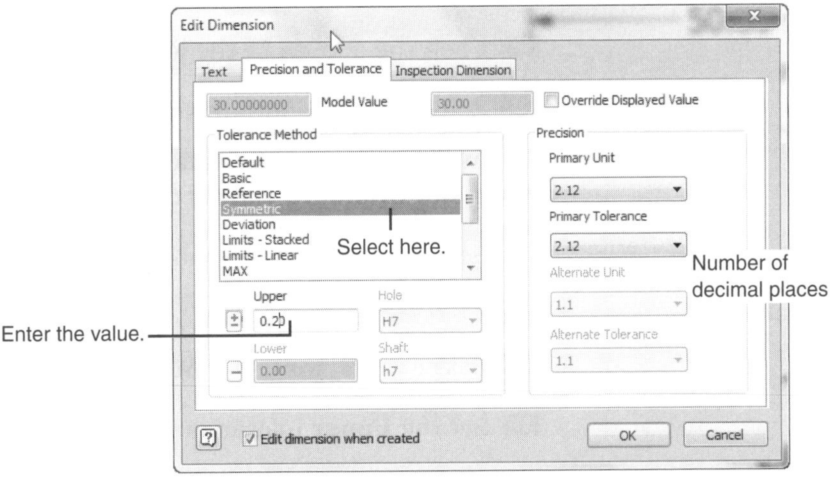

Figure 8-10

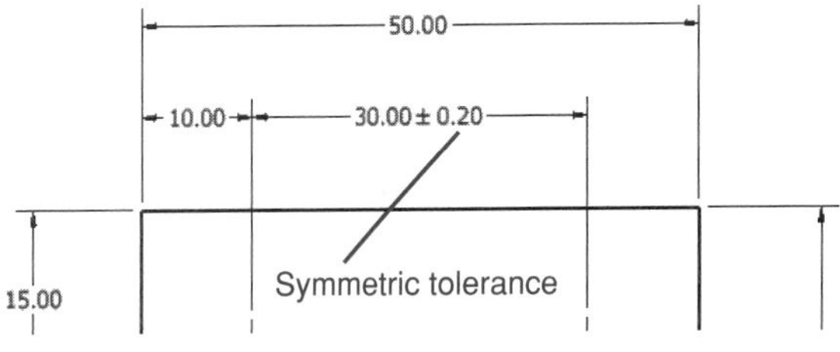

EXERCISE 8-2 Creating Plus and Minus Tolerances Using Styles Editor

1 Click the **Manage** tab at the top of the screen and select the **Styles Editor** option.

The **Style and Standard Editor** dialog box will appear. See Figure 8-11.

Figure 8-11

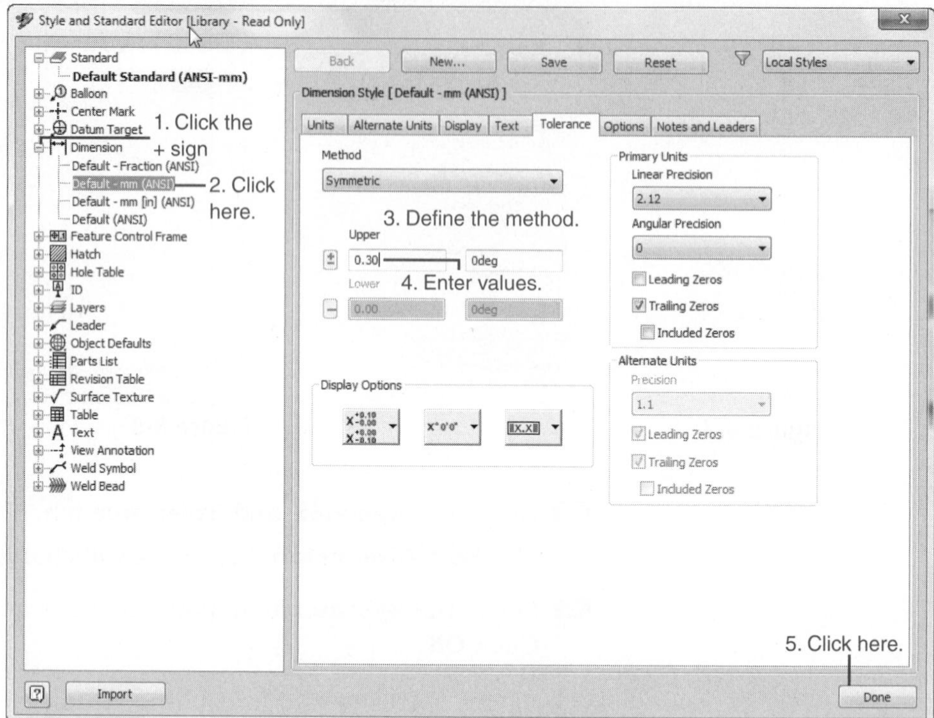

1. Click the + sign
2. Click here.
3. Define the method.
4. Enter values.
5. Click here.

2 Click the + **sign** to the left of the **Dimension** listing.

3 Select the appropriate standard.

In this example the **Default.mm (ANSI)** standard was selected. The **Style and Standard Editor** dialog box will change.

4 Select the **Tolerance** tab.

See Figure 8-11.

5 Select the **Symmetric Method.**

6 Set the **Upper** tolerance value for **0.30.**

Inventor will automatically make this a ± tolerance.

7 Click **Save,** then **Done.**

Figure 8-12 shows the resulting tolerances. Note that all dimensions, except the hole value, have been changed to include a ±0.03 tolerance. The **Style and Standard Editor** is used to create tolerances for an entire drawing, that is, all the dimensions. The **Edit Dimension** dialog box is used to apply tolerances to individual dimension values.

Standard tolerances will be discussed later in the chapter.

Figure 8-12

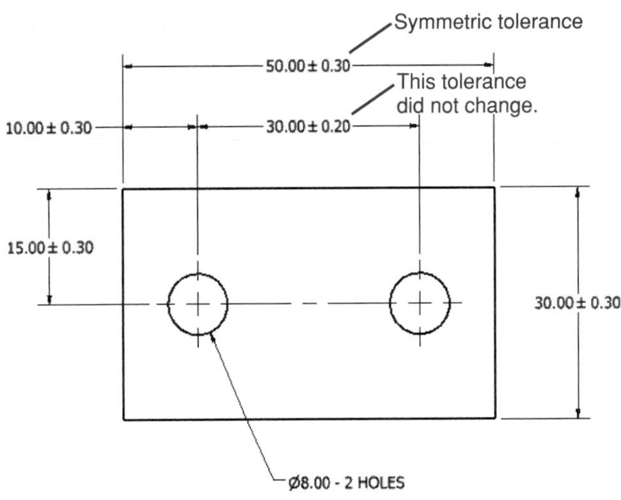

> **NOTE**
> Note that all dimensions are toleranced.

EXERCISE 8-3 Creating Unequal Plus and Minus Tolerances

1 Click on the dimension to be toleranced.

Circles will appear on the dimension, indicating that it has been selected.

2 Right-click the mouse and select the **Edit** option.

The **Edit Dimension** dialog box will appear. See Figure 8-13.

Figure 8-13

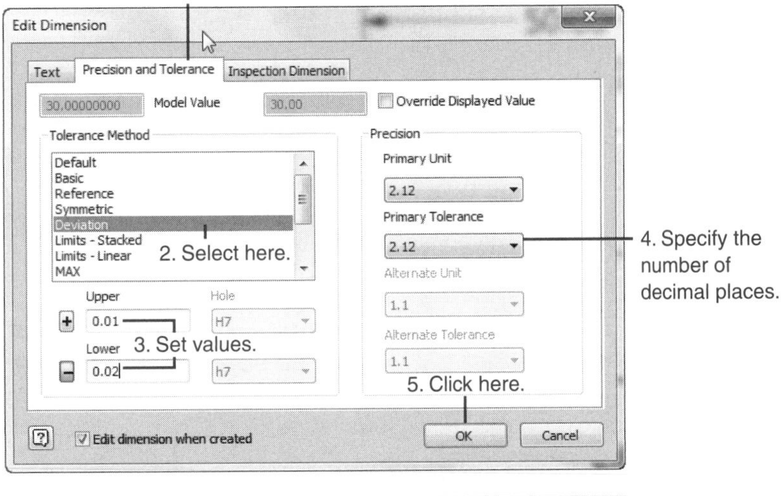

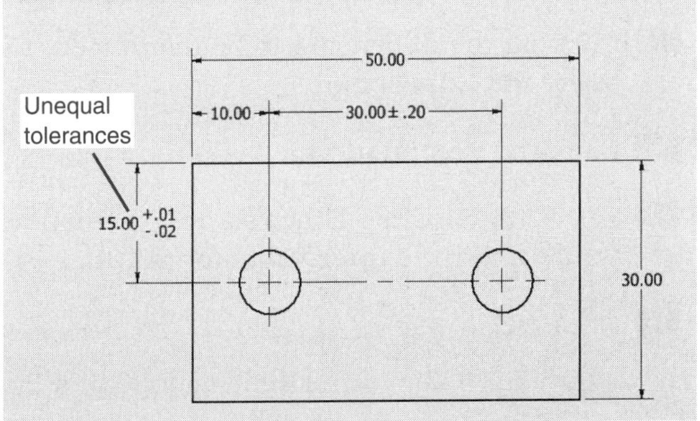

3 Click the **Precision and Tolerance** tab and select the **Deviation** option, then set the **Upper** value for **0.01** and the **Lower** value for **0.02.**

4 Click **OK.**

Figure 8-13 shows the dimension with the unequal tolerances assigned.

Limit Tolerances

Figure 8-14 shows examples of limit tolerances. Limit tolerances replace dimension values. Two values are given: the upper and lower limits for the dimension value. The limit tolerance 62.1 and 61.9 is mathematically equal to 62 ± 0.1, but the stated limit tolerance is considered easier to read and understand.

Figure 8-14

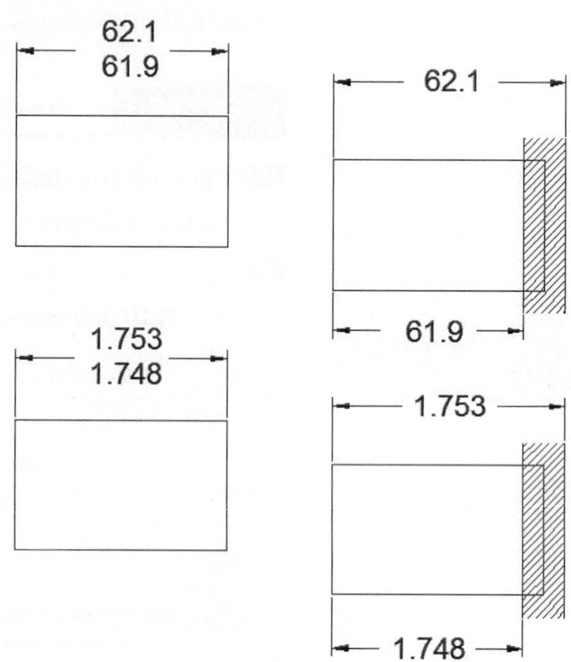

Limit tolerances define a range for manufacture. Final distances on an object must fall within the specified range to be acceptable.

EXERCISE 8-4 **Creating Limit Tolerances**

1 Click on the dimension to be toleranced, right-click the mouse, and select the **Edit** option.

2 Click the **Precision and Tolerance** tab.

3 Select the **Limits - Stacked** option, then set the **Upper** value for **30.03** and the **Lower** value for **29.96.**

4 Click **OK.**

Figure 8-15 shows the dimension with a limit tolerance assigned.

Figure 8-15

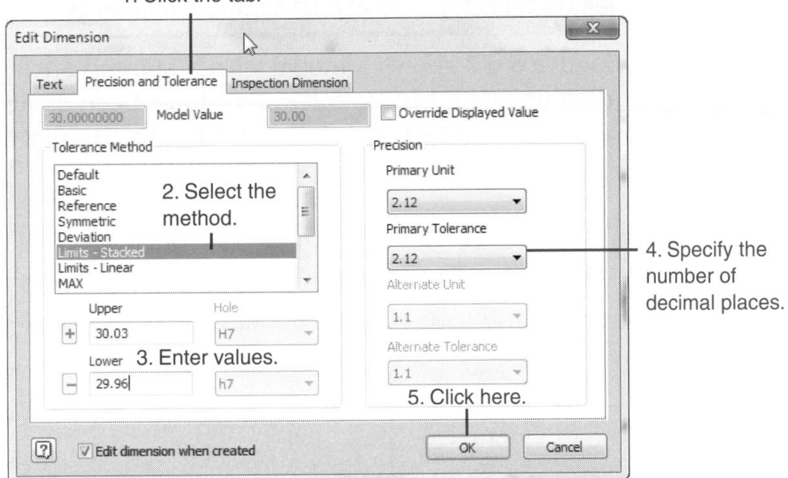

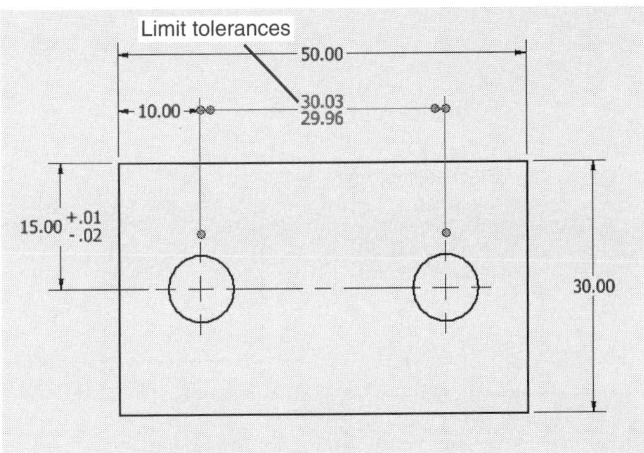

Angular Tolerances

Figure 8-16 shows an example of an angular dimension with a symmetric tolerance. The procedures explained for applying different types of tolerances to linear dimensions also apply to angular dimensions.

Figure 8-16

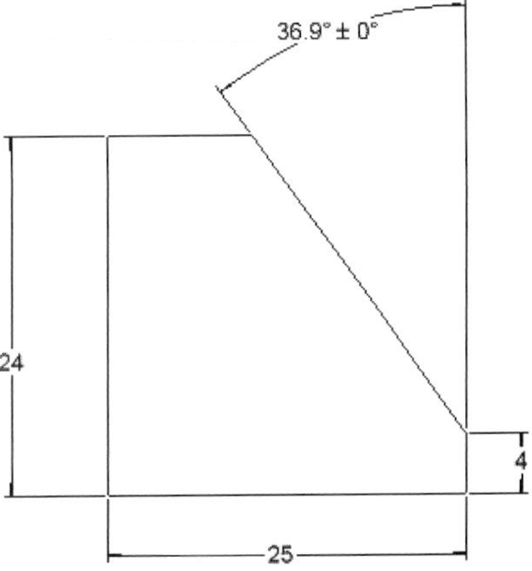

EXERCISE 8-5 Creating Angular Tolerances

Figure 8-17 shows a model with a slanted surface. The model has been dimensioned, but no tolerances have been assigned. This example will assign a stacked limit tolerance to the angular dimension.

Figure 8-17

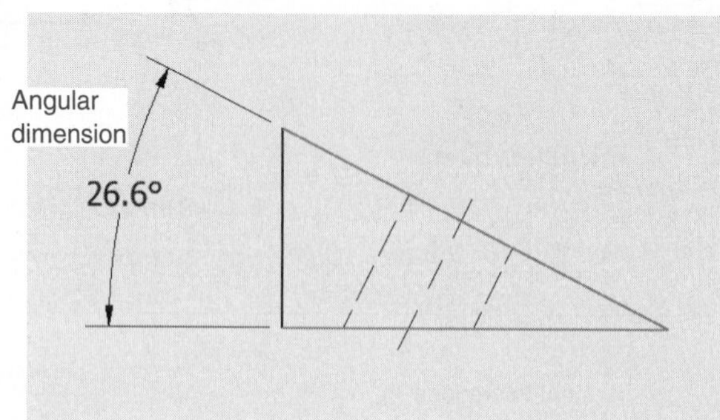

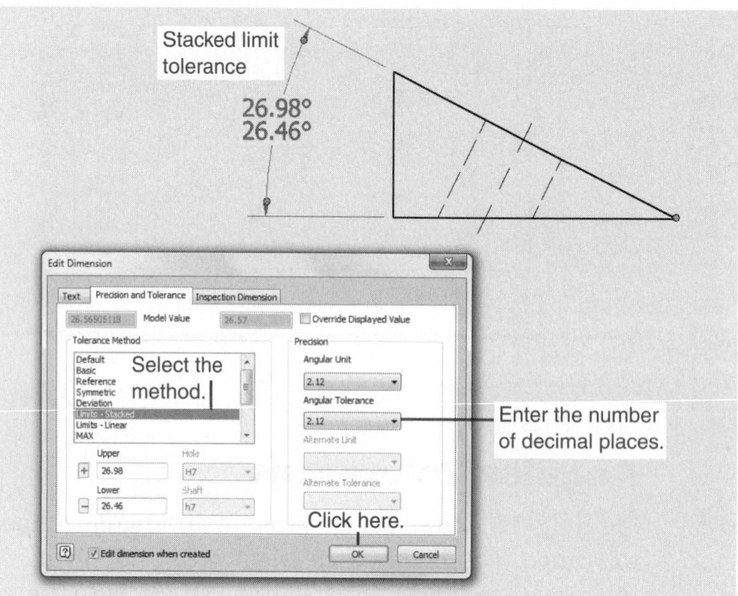

1. Click the angular dimension, right-click the mouse, and select the **Edit** option.

   The **Edit Dimension** dialog box will appear.

2. Click the **Precision and Tolerance** tab.

3. Change the precision of both the **Unit** and **Tolerance** to **2** places (0.00).

4. Select the **Limits - Stacked** option and set the upper and lower values for the tolerance.

5. Click **OK.**

Angular tolerances also can be assigned using the **Dimension Styles** tool. Figure 8-18 shows the **Style and Standard Editor** dialog box. Click the **Format** heading at the top of the screen, then click the **Style and Standard Editor** option.

Figure 8-18

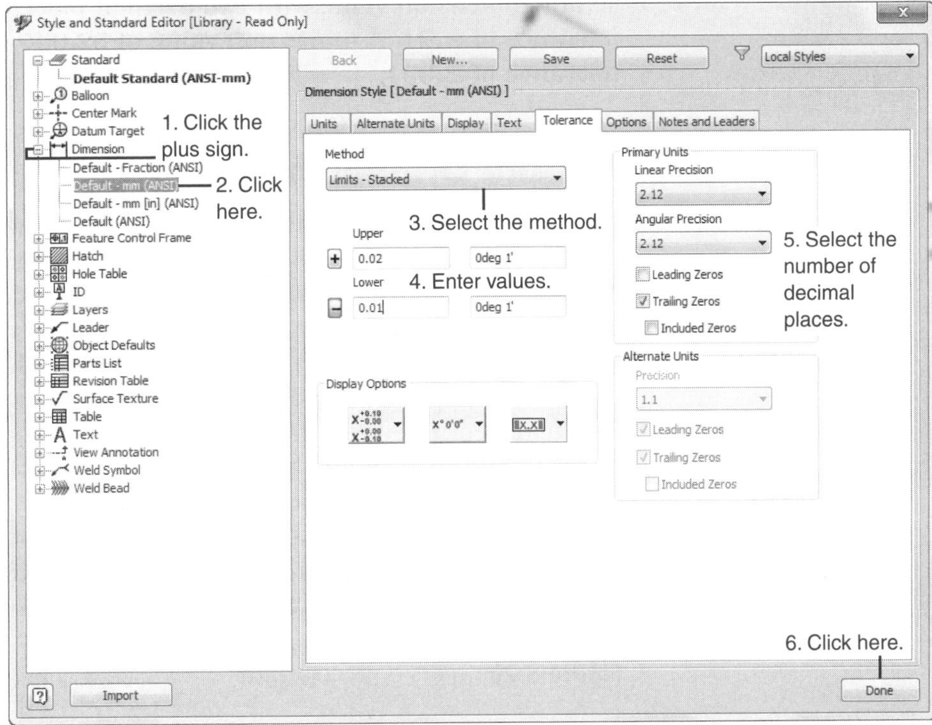

TIP

Remember that if values are defined using the **Style and Standard Editor,** all angular dimensions will have the assigned tolerance.

Standard Tolerances

Most manufacturers establish a set of standard tolerances that are applied to any dimension that does not include a specific tolerance. Figure 8-19 shows some possible standard tolerances. Standard tolerances vary from

Figure 8-19

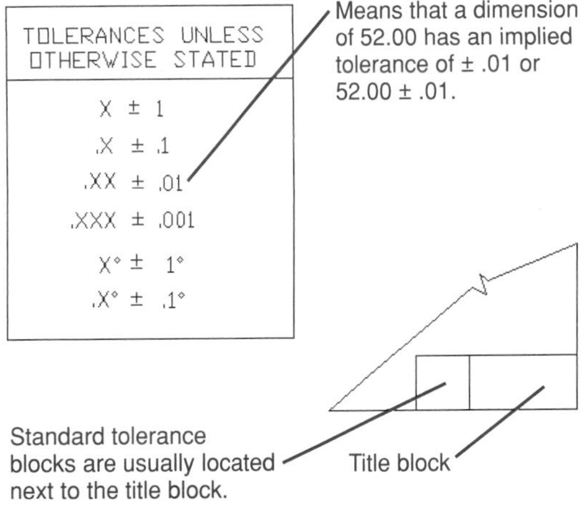

company to company. Standard tolerances are usually listed on the first page of a drawing to the left of the title block, but this location may vary.

The X value used when specifying standard tolerances means any X stated in that format. A dimension value of 52.00 would have an implied tolerance of ±.01 because the stated standard tolerance is .XX ± .01, so any dimension value with two decimal places has a standard implied tolerance of ±.01. A dimension value of 52.000 would have an implied tolerance of ±.001.

Double Dimensioning

double dimensioning: An error in which the same distance is dimensioned twice.

It is an error to dimension the same distance twice. This mistake is called ***double dimensioning***. Double dimensioning is an error because it does not allow for tolerance buildup across a distance.

Figure 8-20 shows an object that has been dimensioned twice across its horizontal length, once using three 30-mm dimensions and a second time using the 90-mm overall dimension. The two dimensions are mathematically equal but are not equal when tolerances are considered. Assume that each dimension has a standard tolerance of ±1 mm. The three 30-mm dimensions could create an acceptable distance of 90 ± 3 mm, or a maximum distance of 93 and a minimum distance of 87. The overall dimension of 90 mm allows a maximum distance of 91 and a minimum distance of 89. The two dimensions yield different results when tolerances are considered.

Figure 8-20

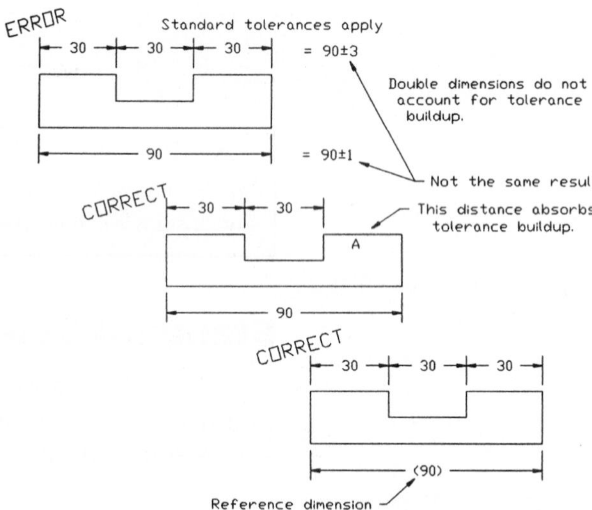

The size and location of a tolerance depends on the design objectives of the object, how it will be manufactured, and how it will be inspected. Even objects that have similar shapes may be dimensioned and toleranced very differently.

NOTE

Never dimension the same distance twice.

One possible solution to the double dimensioning shown in Figure 8-20 is to remove one of the 30-mm dimensions and allow that distance to "float," that is, absorb the accumulated tolerances. The choice of which 30-mm dimension to eliminate depends on the design objectives of the part. For this example the far-right dimension was eliminated to remove the double-dimensioning error.

Another possible solution to the double-dimensioning error is to retain the three 30-mm dimensions and to change the 90-mm overall dimension to a reference dimension. A reference dimension is used only for mathematical convenience. It is not used during the manufacturing or inspection process. A reference dimension is designated on a drawing using parentheses: (90).

If the 90-mm dimension was referenced, then only the three 30-mm dimensions would be used to manufacture and inspect the object. This would eliminate the double-dimensioning error.

Chain Dimensions and Baseline Dimensions

chain dimension: A dimension in which each feature is dimensioned to the feature next to it.

baseline dimension: A dimension in which all features are dimensioned from a single baseline or datum.

There are two systems for applying dimensions and tolerances to a drawing: chain and baseline. Figure 8-21 shows examples of both systems. ***Chain dimensions*** dimension each feature to the feature next to it. ***Baseline dimensions*** dimension all features from a single baseline or datum.

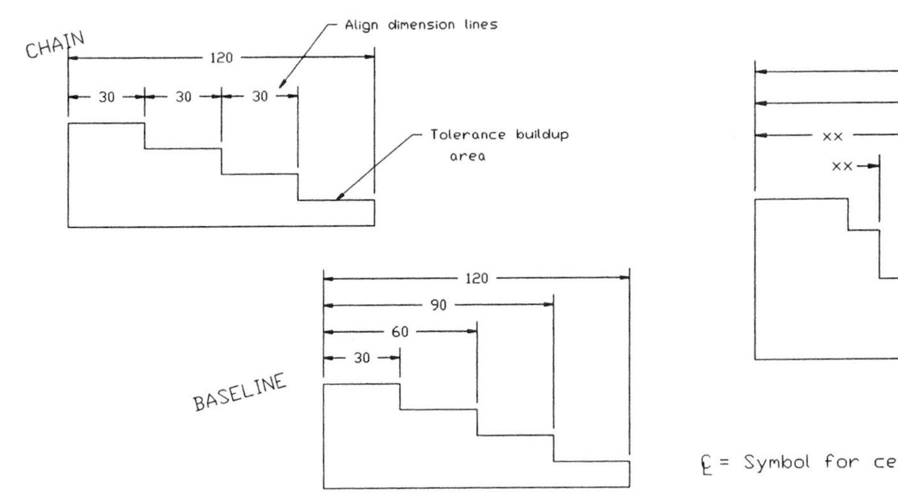

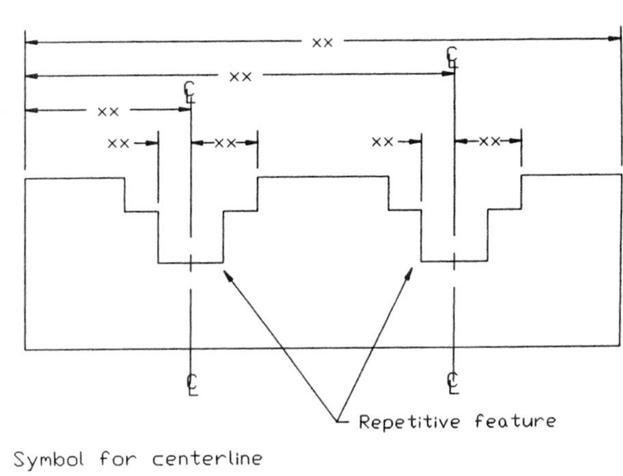

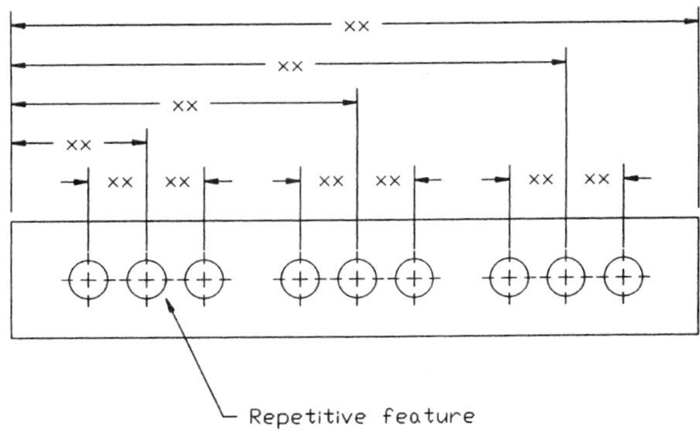

Figure 8-21

Chain and baseline dimensions may be used together. Figure 8-21 also shows two objects with repetitive features; one object includes two slots, and the other, three sets of three holes. In each example, the center of the repetitive feature is dimensioned to the left side of the object, which serves as a baseline. The sizes of the individual features are dimensioned using chain dimensions referenced to centerlines.

Baseline dimensions eliminate tolerance buildup and can be related directly to the reference axis of many machines. They tend to take up much more area on a drawing than do chain dimensions.

Chain dimensions are useful in relating one feature to another, such as the repetitive hole pattern shown in Figure 8-21. In this example the distance between the holes is more important than the individual hole's distance from the baseline.

Figure 8-22 shows the same object dimensioned twice, once using chain dimensions and once using baseline dimensions. All distances are assigned a tolerance range of 2 mm, stated using limit tolerances. The maximum distance for surface A is 28 mm using the chain system and 27 mm using the baseline system. The 1-mm difference comes from the elimination of the first 26–24 limit dimension found on the chain example but not on the baseline.

Figure 8-22

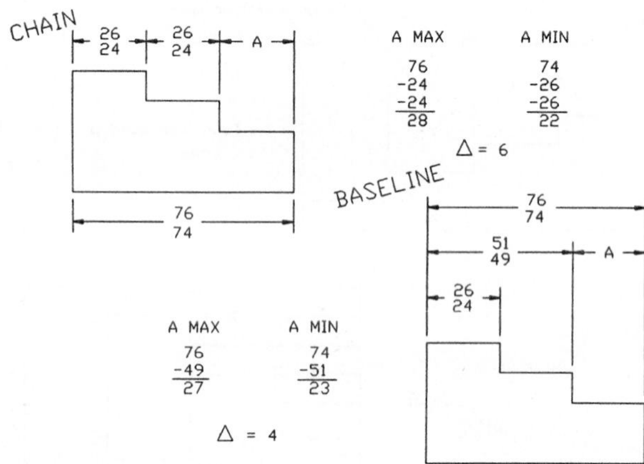

The total tolerance difference is 6 mm for the chain and 4 mm for the baseline. The baseline reduces the tolerance variations for the object simply because it applies the tolerances and dimensions differently. So why not always use baseline dimensions? For most applications, the baseline system is probably better, but if the distance between the individual features is more critical than the distance from the feature to the baseline, use the chain system.

Baseline Dimensions Created Using Inventor

See Figure 8-23. See also Chapter 7.

Note in the example of baseline dimensioning shown in Figure 8-23 that each dimension is independent of the other. This means that if one of the dimensions is manufactured incorrectly, it will not affect the other dimensions.

Figure 8-23

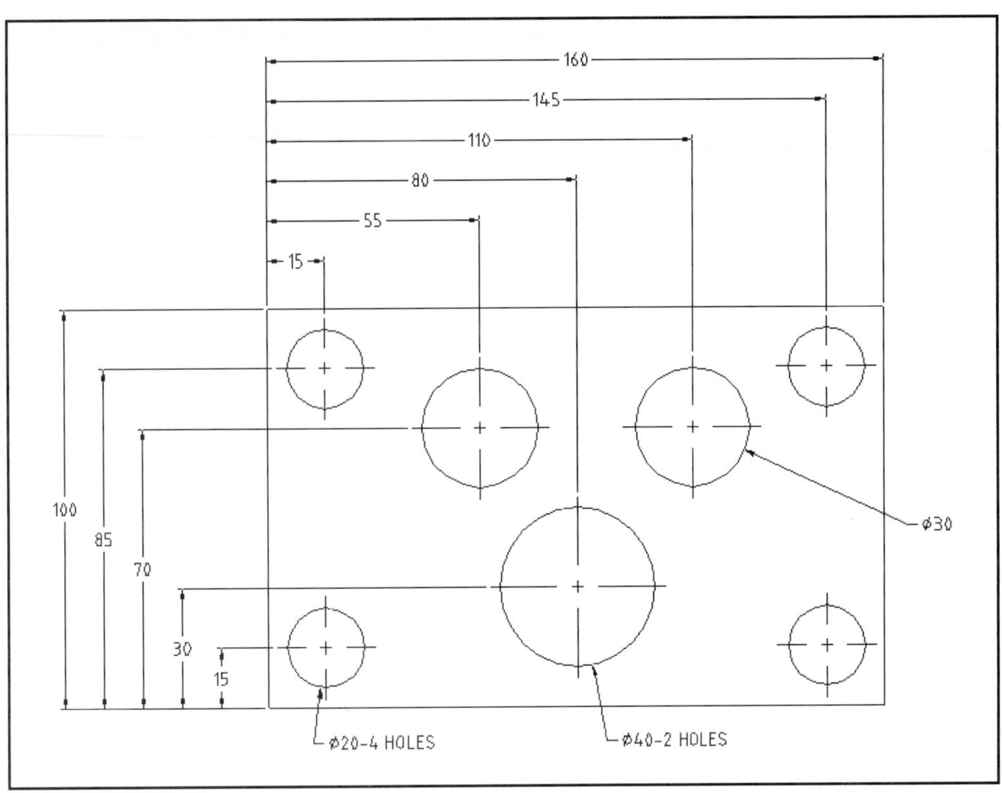

Tolerance Studies

tolerance study: An analysis of the effects of a group of tolerances on one another and on an object.

The term ***tolerance study*** is used when analyzing the effects of a group of tolerances on one another and on an object. Figure 8-24 shows an object with two horizontal dimensions. The horizontal distance A is not dimensioned. Its length depends on the tolerances of the two horizontal dimensions.

Figure 8-24

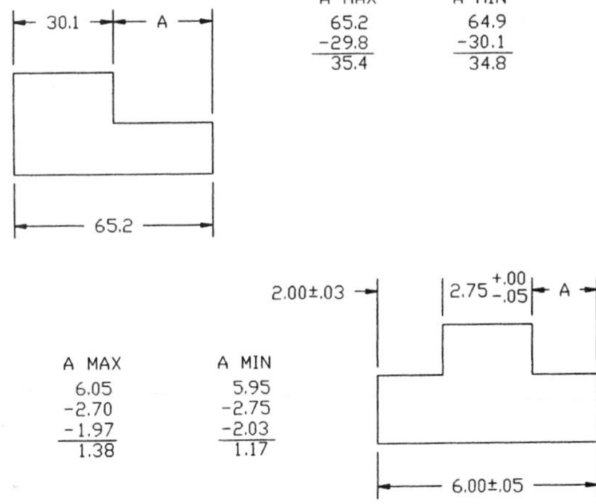

Calculating the Maximum Length of A

Distance A will be longest when the overall distance is at its longest and the other distance is at its shortest.

$$\begin{array}{r} 65.2 \\ -29.8 \\ \hline 35.4 \end{array}$$

Calculating the Minimum Length of A

Distance A will be shortest when the overall length is at its shortest and the other length is at its longest.

$$\begin{array}{r} 64.9 \\ -30.1 \\ \hline 34.8 \end{array}$$

Rectangular Dimensions

Figure 8-25 shows an example of rectangular dimensions referenced to baselines. Figure 8-26 shows a circular object on which dimensions are referenced to a circle's centerlines. Dimensioning to a circle's centerline is critical to accurate hole location.

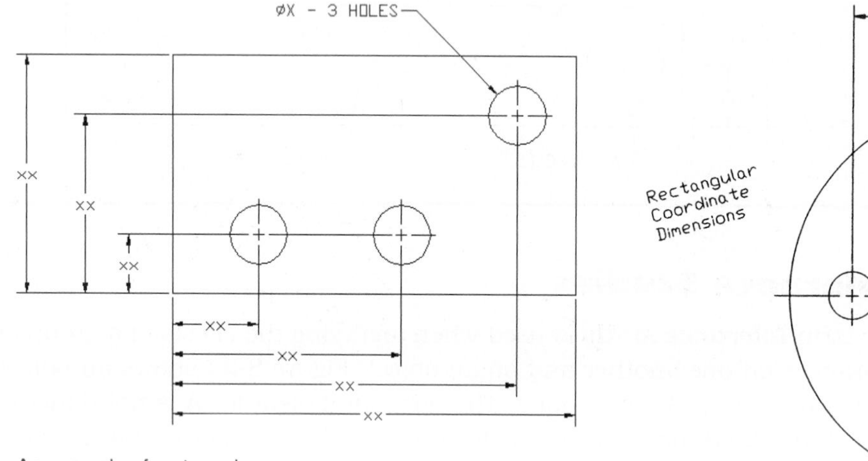

An example of rectangular coordinate dimensions.

Figure 8-25

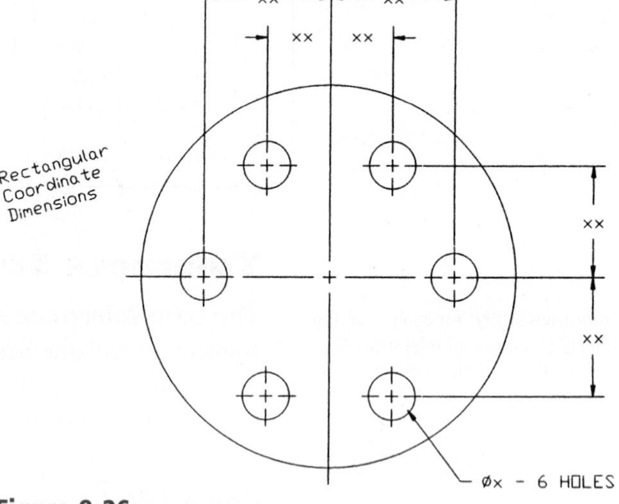

Figure 8-26

> **NOTE**
> The hole locations can also be defined using polar dimensions.

Hole Locations

When rectangular dimensions are used, the location of a hole's center point is defined by two linear dimensions. The result is a rectangular tolerance zone whose size is based on the linear dimension's tolerances. The shape of the center point's tolerance zone may be changed to circular using positioning tolerancing as described later in the chapter.

Figure 8-27 shows the location and size dimensions for a hole. Also shown are the resulting tolerance zone and the overall possible hole shape. The center point's tolerance is .2 by .3 based on the given linear locating tolerances.

The hole diameter has a tolerance of ±.05. This value must be added to the center point location tolerances to define the maximum overall possible shape of the hole. The maximum possible hole shape is determined

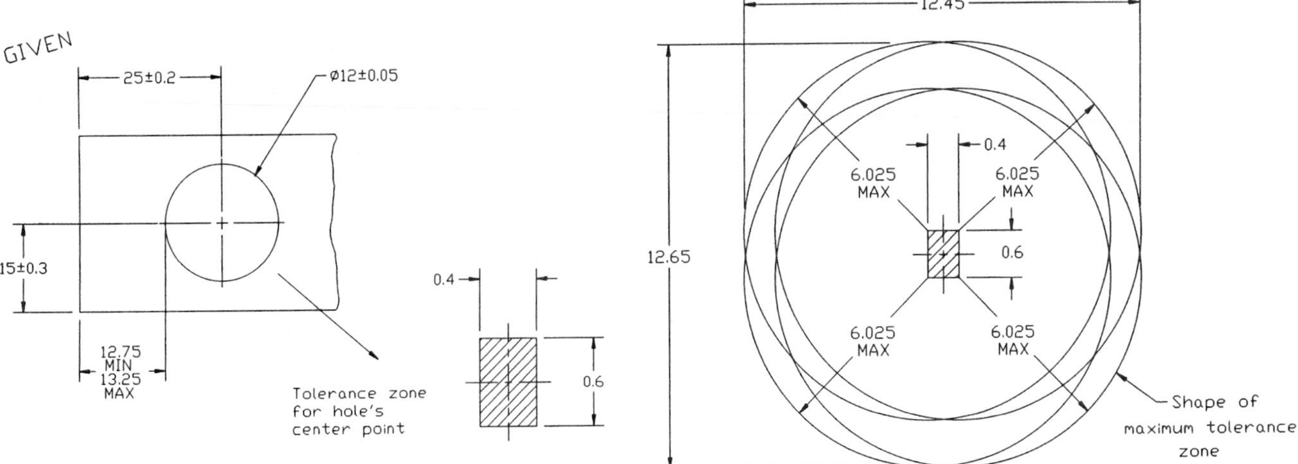

Figure 8-27

by drawing the maximum radius from the four corner points of the tolerance zone.

This means that the left edge of the hole could be as close to the vertical baseline as 12.75 or as far as 13.25. The 12.75 value was derived by subtracting the maximum hole diameter value 12.05 from the minimum linear distance 24.80 (24.80 − 12.05 = 12.75). The 13.25 value was derived by subtracting the minimum hole diameter 11.95 from the maximum linear distance 25.20 (25.20 − 11.95 = 13.25).

Figure 8-28 shows a hole's tolerance zone based on polar dimensions. The zone has a sector shape, and the possible hole shape is determined by locating the maximum radius at the four corner points of the tolerance zone.

Figure 8-28

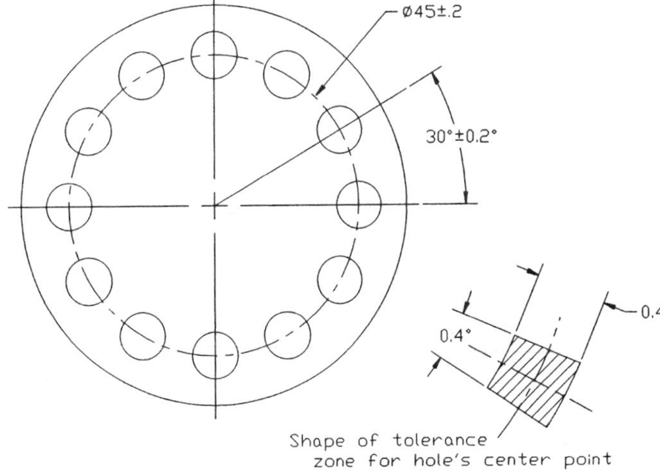

Choosing a Shaft for a Toleranced Hole

Given the hole location and size shown in Figure 8-27, what is the largest diameter shaft that will always fit into the hole?

Figure 8-29 shows the hole's center point tolerance zone based on the given linear locating tolerances. Four circles have been drawn centered at

Figure 8-29

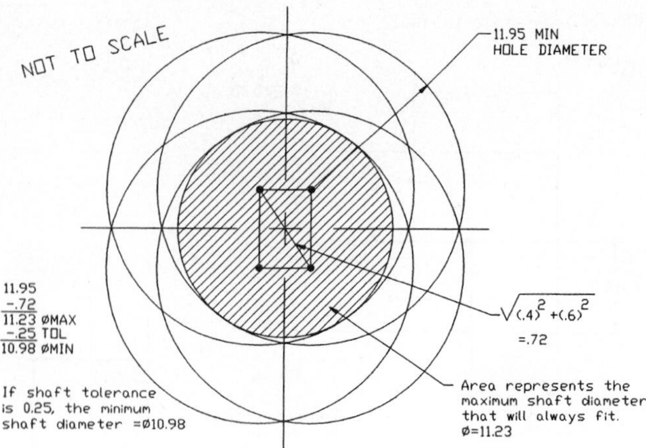

the four corners on the linear tolerance zone that represent the smallest possible hole diameter. The circles define an area that represents the maximum shaft size that will always fit into the hole, regardless of how the given dimensions are applied.

The diameter size of this circular area can be calculated by subtracting the maximum diagonal distance across the linear tolerance zone (corner to corner) from the minimum hole diameter.

The results can be expressed as a formula.

For Linear Dimensions and Tolerances

$S_{max} = H_{min} - DTZ$

where

S_{max} = maximum shaft diameter

H_{min} = minimum hole diameter

DTZ = diagonal distance across the tolerance zone

In the example shown the diagonal distance is determined using the Pythagorean theorem:

$$DTZ = \sqrt{(.4)^2 + (.6)^2}$$

$$= \sqrt{.16 + .36}$$

$$DTZ = .72$$

This means that the maximum shaft diameter that will always fit into the given hole is 11.23.

$$S_{max} = H_{min} - DTZ$$

$$= 11.95 - .72$$

$$S_{max} = 11.23$$

This procedure represents a restricted application of the general formula presented later in the chapter for positioning tolerances.

Once the maximum shaft size has been established, a tolerance can be applied to the shaft. If the shaft had a total tolerance of .25, the minimum shaft diameter would be 11.23 − .25, or 10.98. Figure 8-29 shows a shaft dimensioned and toleranced using these values.

The formula presented is based on the assumption that the shaft is perfectly placed on the hole's center point. This assumption is reasonable if two objects are joined by a fastener and both objects are free to move. When both objects are free to move about a common fastener, they are called ***floating objects***.

floating objects: Objects that are free to move about a common fastener.

NOTE

Linear tolerances generate a square or rectangular tolerance zone.

Sample Problem SP8-1

Parts A and B in Figure 8-30 are to be joined by a common shaft. The total tolerance for the shaft is to be .05. What are the maximum and minimum shaft diameters?

Figure 8-30

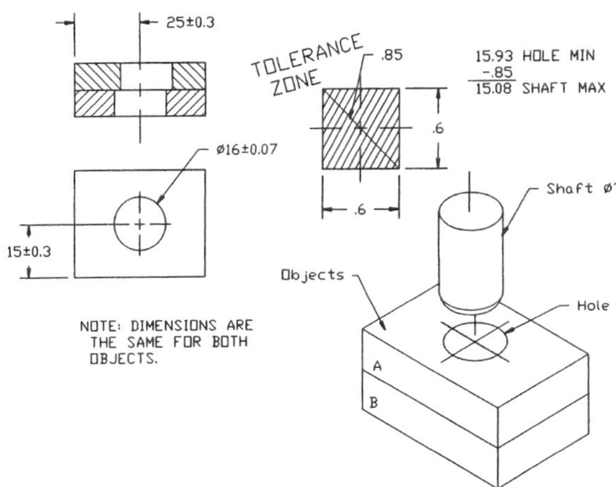

Both objects have the same dimensions and tolerances and are floating relative to each other.

$$S_{max} = H_{min} - DTZ$$
$$= 15.93 - .85$$
$$S_{max} = 15.08$$

The shaft's minimum diameter is found by subtracting the total tolerance requirement from the calculated maximum diameter:

$$15.08 - .05 = 15.03$$

Therefore,

Shaft max = 15.08
Shaft min = 15.03

Sample Problem SP8-2

The procedure presented in Sample Problem SP8-1 can be worked in reverse to determine the maximum and minimum hole size based on a given shaft size.

Objects AA and BB as shown in Figure 8-31 are to be joined using a bolt whose maximum diameter is .248. What is the minimum hole size for

Figure 8-31

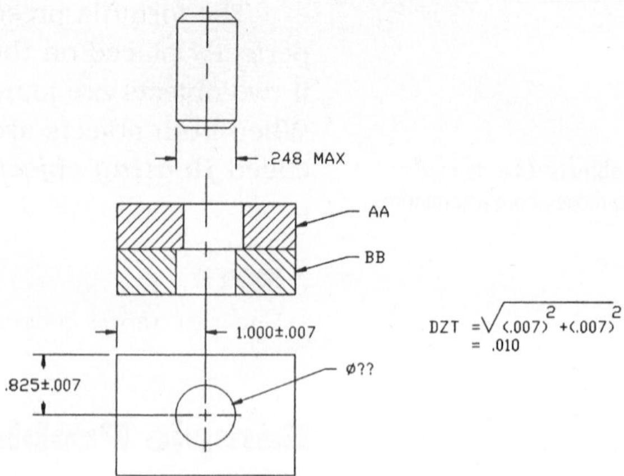

objects that will always accept the bolt? What is the maximum hole size if the total hole tolerance is .005?

$$S_{max} = H_{min} - DZT$$

In this example H_{min} is the unknown factor, so the equation is rewritten as

$$H_{min} = S_{max} + DZT$$
$$= .248 + .010$$
$$H_{min} = 2.58$$

This is the minimum hole diameter, so the total tolerance requirement is added to this value:

$$.258 + .005 = .263$$

Therefore,

Hole max = .263

Hole min = .258

Nominal Sizes

nominal size: The approximate size of an object that matches a common fraction or whole number.

The term ***nominal*** refers to the approximate size of an object that matches a common fraction or whole number. A shaft with a dimension of 1.500 ± .003 is said to have a nominal size of "one and a half inches." A dimension of 1.500 +.000/−.005 is still said to have a nominal size of one and a half inches. In both examples 1.5 is the closest common fraction.

Standard Fits (Metric Values)

Calculating tolerances between holes and shafts that fit together is so common in engineering design that a group of standard values and notations has been established. These values may be calculated using the **Limits/Fits** tool.

There are three possible types of fits between a shaft and a hole: clearance, transition, and interference. There are several subclassifications within each of these categories.

clearance fit: A fit that always defines the maximum shaft diameter as smaller than the minimum hole diameter.

A ***clearance fit*** always defines the maximum shaft diameter as smaller than the minimum hole diameter. The difference between the two diameters is the amount of clearance. It is possible for a clearance fit to be defined with zero clearance; that is, the maximum shaft diameter is equal to the minimum hole diameter.

interference fit: A fit that always defines the minimum shaft diameter as larger than the maximum hole diameter.

An ***interference fit*** always defines the minimum shaft diameter as larger than the maximum hole diameter; that is, the shaft is always bigger than the hole. This definition means that an interference fit is the converse of a clearance fit. The difference between the diameter of the shaft and the hole is the amount of interference.

An interference fit is primarily used to assemble objects together. Interference fits eliminate the need for threads, welds, or other joining methods. Using an interference fit for joining two objects is generally limited to light load applications.

It is sometimes difficult to visualize how a shaft can be assembled into a hole with a diameter smaller than that of the shaft. It is sometimes done using a hydraulic press that slowly forces the two parts together. The joining process can be augmented by the use of lubricants or heat. The hole is heated, causing it to expand, the shaft is inserted, and the hole is allowed to cool and shrink around the shaft.

transition fit: A fit that may be either a clearance or an interference fit.

A ***transition fit*** may be either a clearance or an interference fit. It may have a clearance between the shaft and the hole or an interference.

The notations are based on Standard International Tolerance values. A specific description for each category of fit follows.

Clearance Fits

H11/c11 or C11/h11 = loose running fit

H8/d8 or D8/h8 = free running fit

H8/f7 or F8/h7 = close running fit

H7/g6 or G7/h6 = sliding fit

H7/h6 = locational clearance fit

Transition Fits

H7/k6 or K7/h6 = locational transition fit

H7/n6 or N7/h6 = locational transition fit

Interference Fits

H7/p6 or P7/h6 = locational transition fit

H7/s6 or S7/h6 = medium drive fit

H7/u6 or U7/h6 = force fit

Using Inventor's Limits and Fits Calculator

Limits and fits can be derived using Inventor's **Limits/Fits Calculator.** The limit values are also presented in visual form so you can better visualize the type of fit being created.

EXERCISE 8-6 Accessing the Limits/Fits Calculator

1 Start a new drawing using the **Standard (mm).iam** format.

2 Click the **Design** tab.

3. Click the arrowhead to the right of the **Power Transmission** heading.

See Figure 8-32

4. Click the arrowhead to the right of the **Limits/Fits Calculator** option.

See Figure 8-33.

5. Click the **Limits/Fits Calculator** option.

See Figure 8-34.

The **Limits and Fits Mechanical Calculator** dialog box will appear. See Figure 8-35.

Figure 8-32

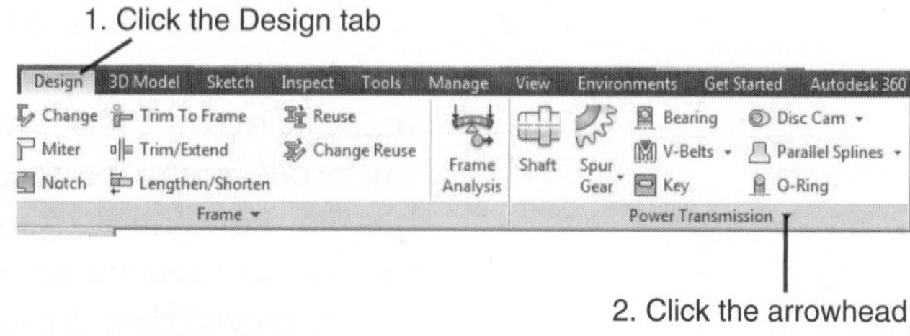

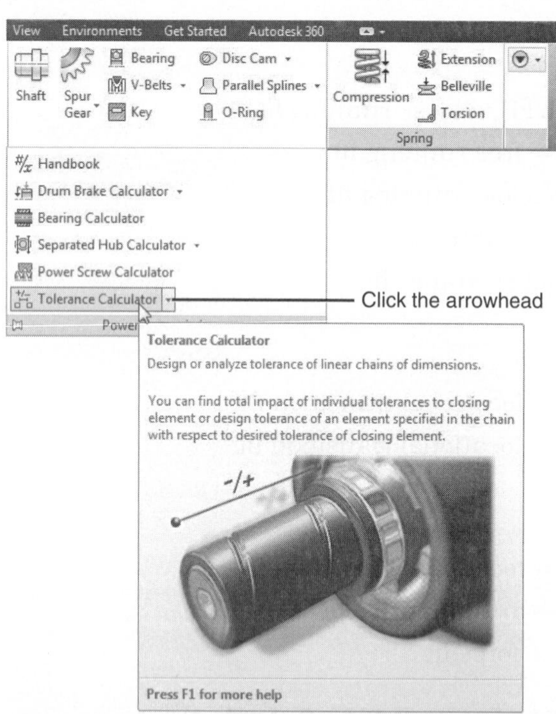

Figure 8-33

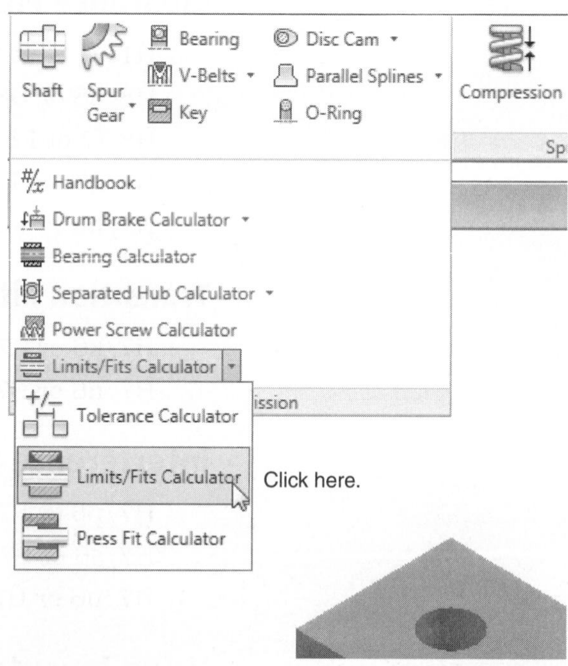

Figure 8-34

Hole and Shaft Basis

The **Limits and Fits Mechanical Calculator** shown in Figure 8-35 applies tolerances starting with the nominal hole sizes, called *hole-basis tolerances;* the other option applies tolerances starting with the shaft nominal sizes, called *shaft-basis tolerances.* The choice of which set of values to use depends on the design application. In general, hole-basis numbers are used more often

Figure 8-35

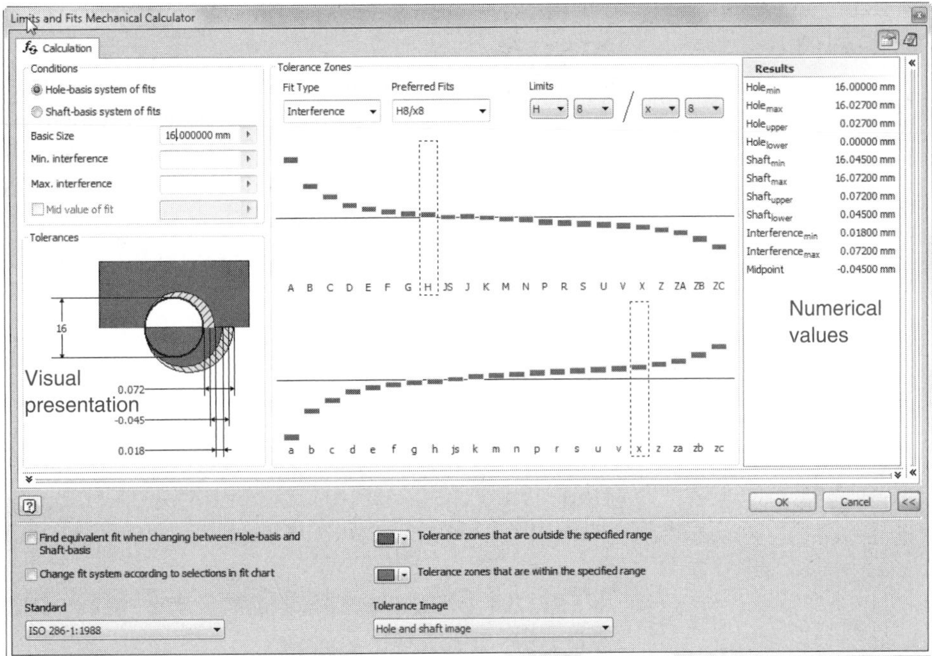

because it is more difficult to vary hole diameters manufactured using specific drill sizes than shaft sizes manufactured using a lathe. Shaft sizes may be used when a specific fastener diameter is used to assemble several objects.

Figure 8-36 shows calculations done for a nominal diameter of 16 mm using the hole-basis system. An H7/g6 fit was selected.

Figure 8-36

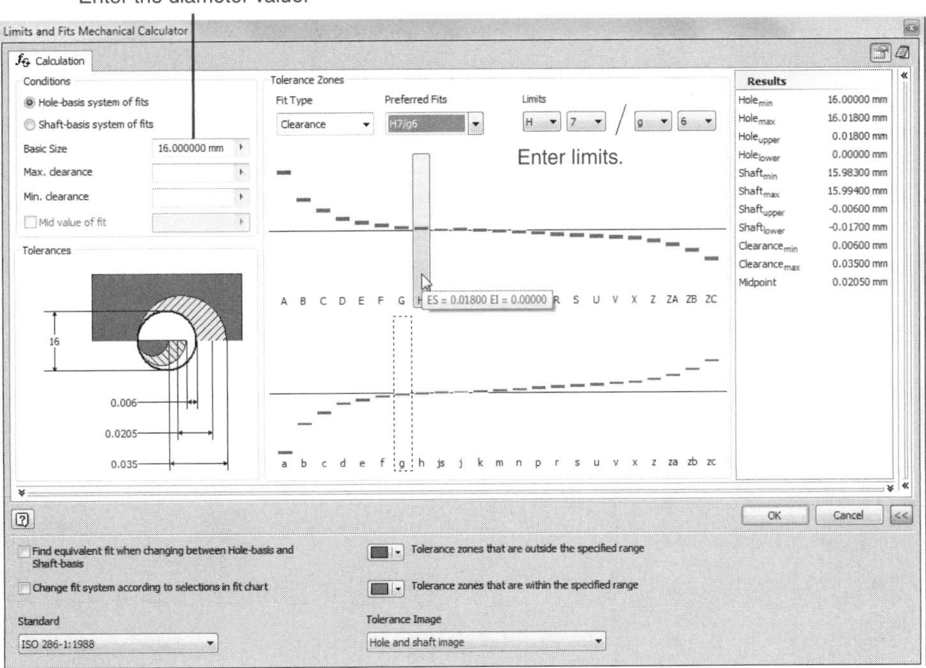

EXERCISE 8-7 Determining the Hole/Shaft Sizes

1 Access the **Limits and Fits Mechanical Calculator.**

2 Define the **Basic Size** and select the **Condition.**

In this example Ø**16.00** was entered and **Hole-basis system of fits** was chosen.

The limit designation H7/g6 is based on an international standard for hole and shaft tolerances. The hole's tolerance is always first and uses an uppercase letter designation. The shaft's tolerance is second and uses a lowercase letter designation. The numerical values for the H7/g6 tolerance can be found in one of two ways: by reading the values in the **Results** box or by moving the cursor into the rectangular area defined by a broken line. The hole's limits will appear. See Figure 8-36.

Note that tolerances listed in the **Results** box and in the **Tolerance Zones** area are the same. The $Hole_{min}$ is 16.0000 mm in the **Results** box and EI = 0.0000 in the **Tolerance Zones** area. The 16.0000 value or 0.0000 value comes from the **Hole-basis** designation. The nominal hole value of 16.00 becomes the starting point for all other tolerance designations. Had the **Shaft-basis** condition been selected for the Ø16.00 nominal size, the hole and shaft tolerances would be different.

Visual Presentations of the Hole and Shaft Tolerances

Figure 8-37 shows the visual presentation of the H7/g6 tolerances along with the **Results** box values. The clearance max value is **0.03500,** as stated in the **Results** box. Note how the 0.035 clearance max value is presented visually in the **Tolerances** box.

Figure 8-38 shows the tolerances applied to a hole and a shaft.

Figure 8-37

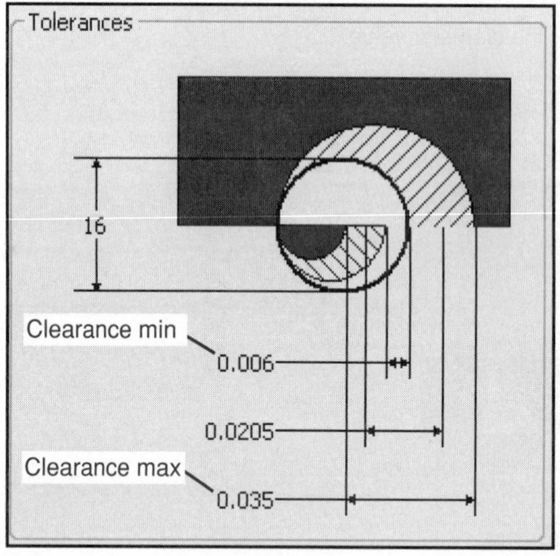

Results	
$Hole_{min}$	16.00000 mm
$Hole_{max}$	16.01800 mm
$Hole_{upper}$	0.01800 mm
$Hole_{lower}$	0.00000 mm
$Shaft_{min}$	15.98300 mm
$Shaft_{max}$	15.99400 mm
$Shaft_{upper}$	-0.00600 mm
$Shaft_{lower}$	-0.01700 mm
$Clearance_{min}$	0.00600 mm
$Clearance_{max}$	0.03500 mm
Midpoint	0.02050 mm

Same values

Figure 8-38

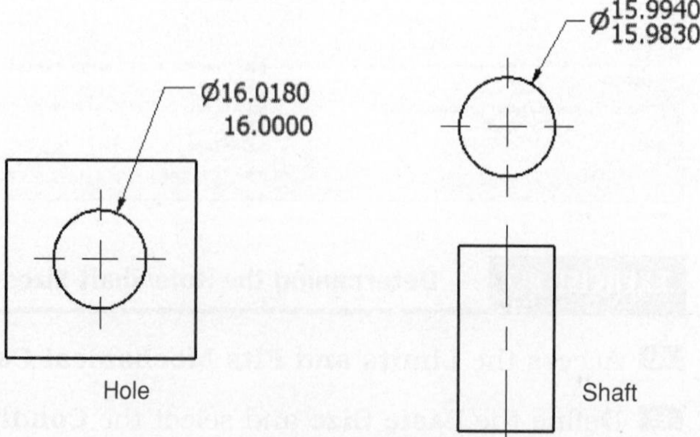

Standard Fits (Inch Values)

Inch values are accessed in the **Limits/Fits Calculator** by selecting the ANSI standards.

Fits defined using inch values are classified as follows:

RC = running and sliding fits

LC = clearance locational fits

LT = transitional locational fits

LN = interference fits

FN = force fits

Each of these general categories has several subclassifications within it defined by a number, for example, Class RC1, Class RC2, through Class RC8. The letter designations are based on International Tolerance Standards, as are metric designations.

Charts of tolerance values can be found in the appendix.

Preferred and Standard Sizes

It is important that designers always consider preferred and standard sizes when selecting sizes for designs. Most tooling is set up to match these sizes, so manufacturing is greatly simplified when preferred and standard sizes are specified. Figure 8-39 shows a listing of preferred sizes for metric values.

Consider the case of design calculations that call for a 42-mm-diameter hole. A 42-mm-diameter hole is not a preferred size. A diameter of 40 mm is the closest preferred size, and a 45-mm diameter is a second choice. A 42-mm hole could be manufactured but would require an unusual drill size that may not be available. It would be wise to reconsider the design to see if a 40-mm-diameter hole could be used, and if not, possibly a 45-mm-diameter hole.

A production run of a very large quantity could possibly justify the cost of special tooling, but for smaller runs it is probably better to use preferred sizes. Machinists will have the required drills, and maintenance personnel will have the appropriate tools for these sizes.

Figure 8-39

PREFERRED SIZES			
First Choice	Second Choice	First Choice	Second Choice
1	1.1	12	14
1.2	1.4	16	18
1.6	1.8	20	22
2	2.2	25	28
2.5	2.8	30	35
3	3.5	40	45
4	4.5	50	55
5	5.5	60	70
6	7	80	90
8	9	100	110
10	11	120	140

Figure 8-40

Fraction	Decimal Equivalent	Fraction	Decimal Equivalent	Fraction	Decimal Equivalent
7/64	.1094	21/64	.3281	11/16	.6875
1/8	.1250	11/32	.3438	3/4	.7500
9/64	.1406	23/64	.3594	13/16	.8125
5/32	.1562	3/8	.3750	7/8	.8750
11/64	.1719	25/64	.3906	15/16	.9375
3/16	.1875	13/32	.4062	1	1.0000
13/64	.2031	27/64	.4219	Partial List of standard Twist Drill Sizes (Fractional sizes)	
7/32	.2188	7/16	.4375		
1/4	.2500	29/64	.4531		
17/64	.2656	15/32	.4688		
9/32	.2812	1/2	.5000		
19/64	.2969	9/16	.5625		
5/16	.3125	5/8	.6250		

Figure 8-40 shows a listing of standard fractional drill sizes. Most companies now specify metric units or decimal inches; however, many standard items are still available in fractional sizes, and many older objects may still require fractional-sized tools and replacement parts. A more complete listing is available in the appendix.

Surface Finishes

surface finish: The accuracy (flatness) of a surface.

The term ***surface finish*** refers to the accuracy (flatness) of a surface. Metric values are measured using micrometers (μm), and inch values are measured in microinches (μin.).

The accuracy of a surface depends on the manufacturing process used to produce the surface. Figure 8-41 shows a listing of manufacturing processes and the quality of the surface finish they can be expected to produce.

datum surface: The surface used for baseline dimensioning.

Surface finishes have several design applications. ***Datum surfaces***, or surfaces used for baseline dimensioning, should have fairly accurate surface finishes to help assure accurate measurements. Bearing surfaces should have good-quality surface finishes for better load distribution, and parts that operate at high speeds should have smooth finishes to help reduce friction. Figure 8-42 shows a screw head sitting on a very wavy surface. Note that the head of the screw is actually in contact with only two wave peaks, meaning the entire bearing load is concentrated on the two peaks. This situation could cause stress cracks and greatly weaken the surface. A better-quality surface finish would increase the bearing contact area.

Figure 8-42 also shows two very rough surfaces moving in contact with each other. The result will be excess wear to both surfaces because the surfaces touch only on the peaks, and these peaks will tend to wear faster than flatter areas. Excess vibration can also result when interfacing surfaces are too rough.

surface texture: The overall quality and accuracy of a surface.

Surface finishes are classified into three categories: surface texture, roughness, and lay. ***Surface texture*** is a general term that refers to the overall quality and accuracy of a surface.

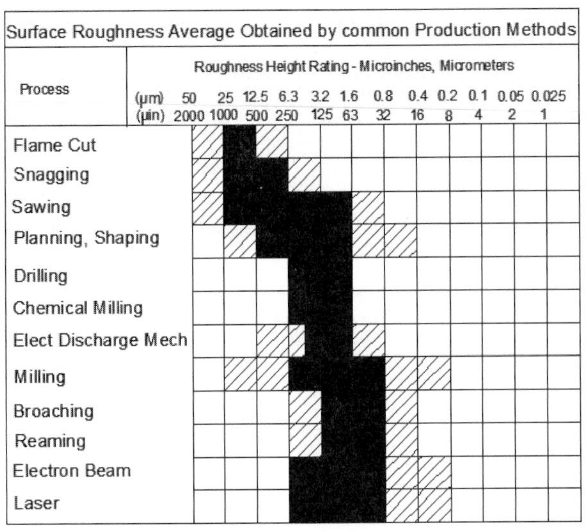

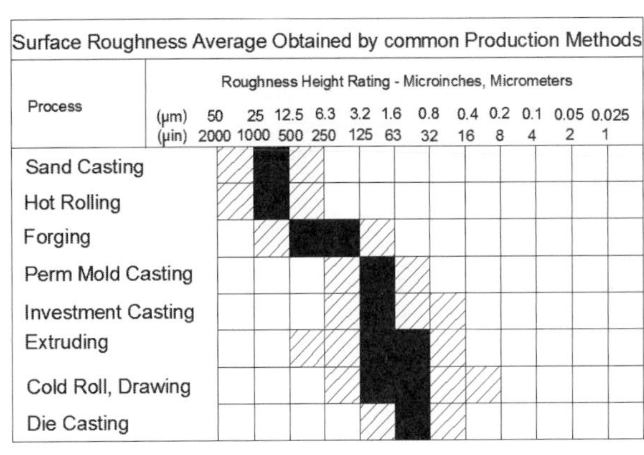

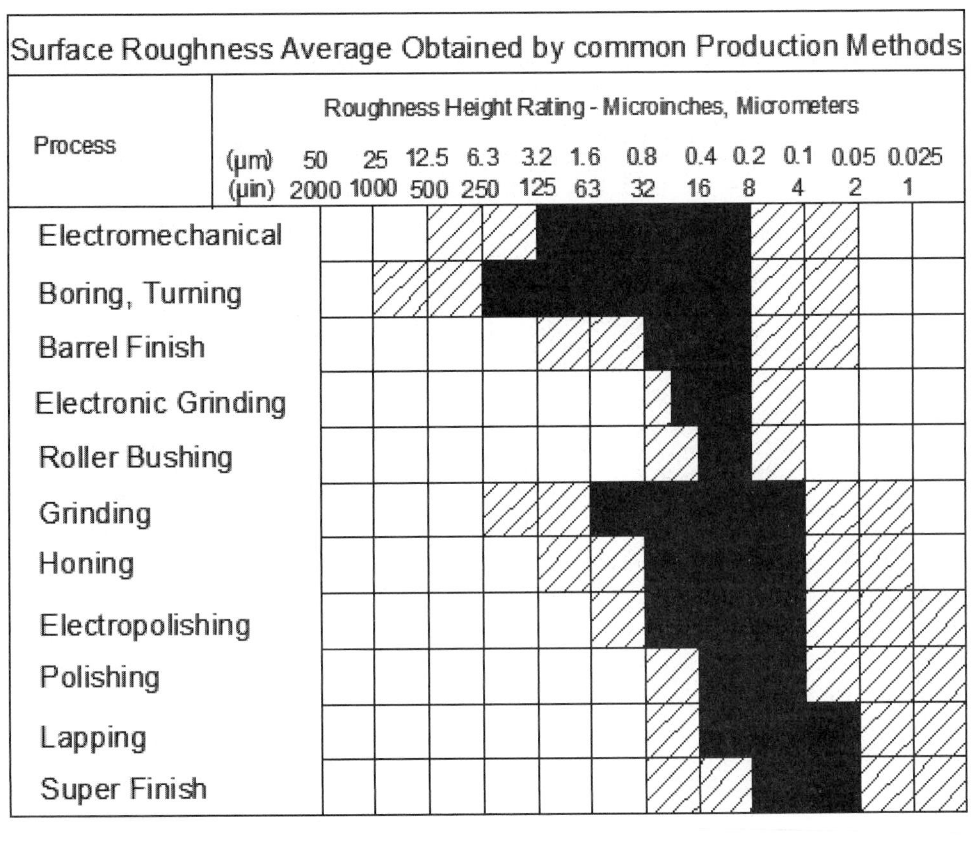

Average Application   Less Frequent Application

Figure 8-41

roughness: A measure of the average deviation of a surface's peaks and valleys.

lay: The direction of machine marks on a surface.

Roughness is a measure of the average deviation of a surface's peaks and valleys. See Figure 8-43.

Lay refers to the direction of machine marks on a surface. See Figure 8-43. The lay of a surface is particularly important when two moving objects are in contact with each other, especially at high speeds.

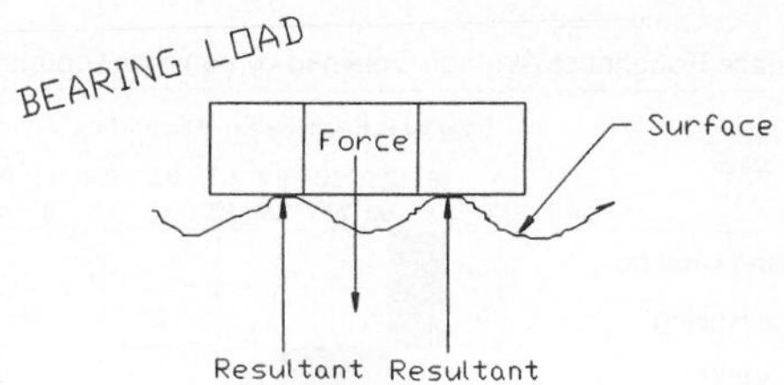

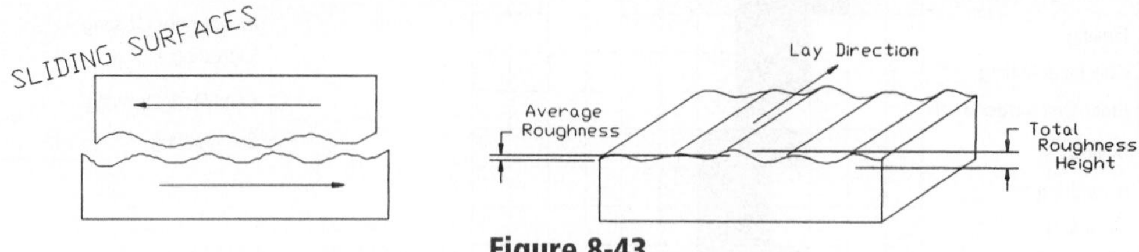

Figure 8-42

Figure 8-43

Surface Control Symbols

Surface finishes are indicated on a drawing using surface control symbols. See Figure 8-44. The general surface control symbol looks like a check mark. Roughness values may be included with the symbol to specify the

Figure 8-44

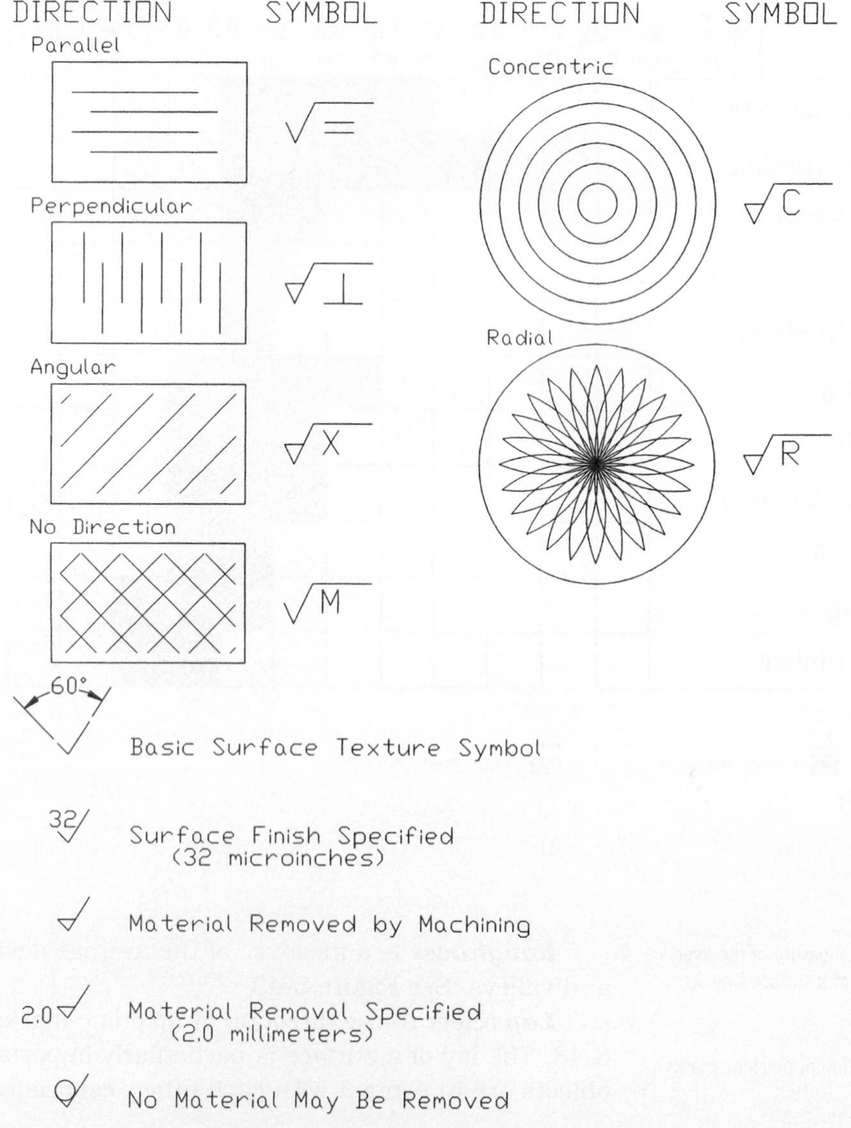

Figure 8-45

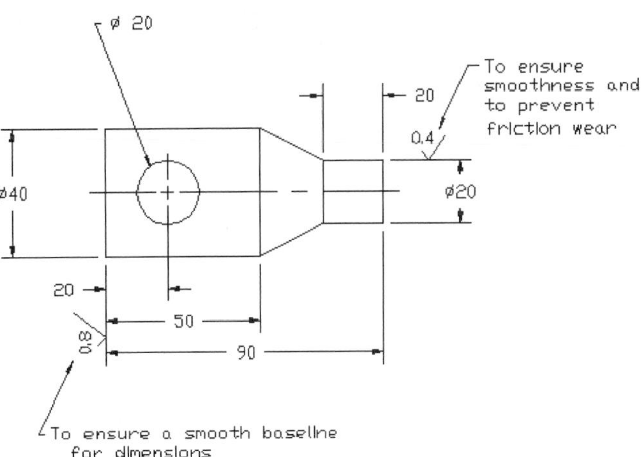

required accuracy. Surface control symbols can also be used to specify the manufacturing process that may or may not be used to produce a surface.

Figure 8-45 shows two applications of surface control symbols. In the first example, a 0.8-μm (32-μin.) surface finish is specified on the surface that serves as a datum for several horizontal dimensions. A 0.8-μm surface finish is generally considered the minimum acceptable finish for datums.

A second finish mark with a value of 0.4 μm is located on an extension line that refers to a surface that will be in contact with a moving object. The extra flatness will help prevent wear between the two surfaces.

EXERCISE 8-8 Applying Surface Control Symbols Using Inventor

Figure 8-47 shows a dimensioned view of a model. Surface symbols are to be added to the sides of the slot. This example requires that you be working with a drawing created using the **ANSI (mm).idw** format.

1 Click the **Annotate** tab and click the **Surface** tool.

See Figure 8-46.

Figure 8-46

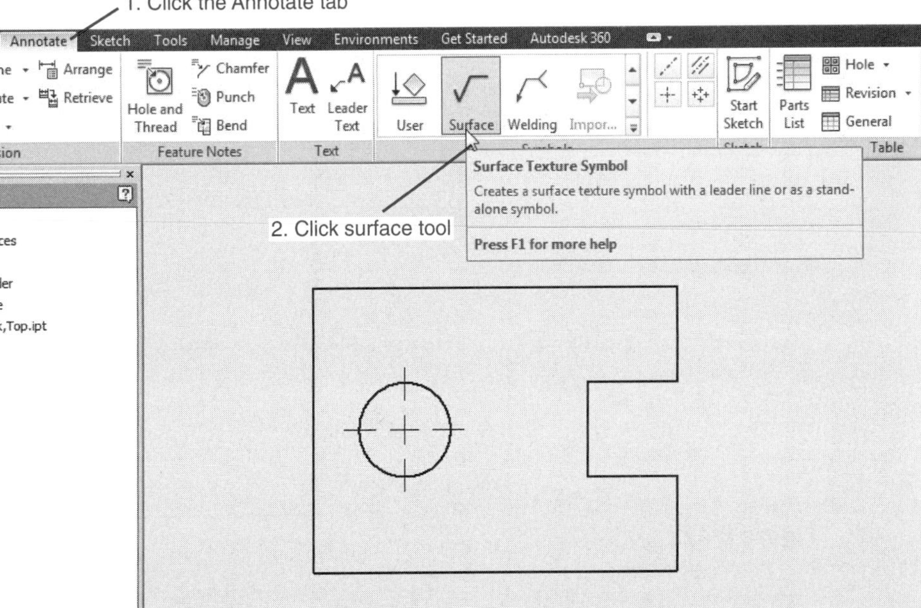

2 Move the cursor to the drawing area and click the lower horizontal edge of the slot.

This step locates the surface texture symbol on the drawing. See Figure 8-47.

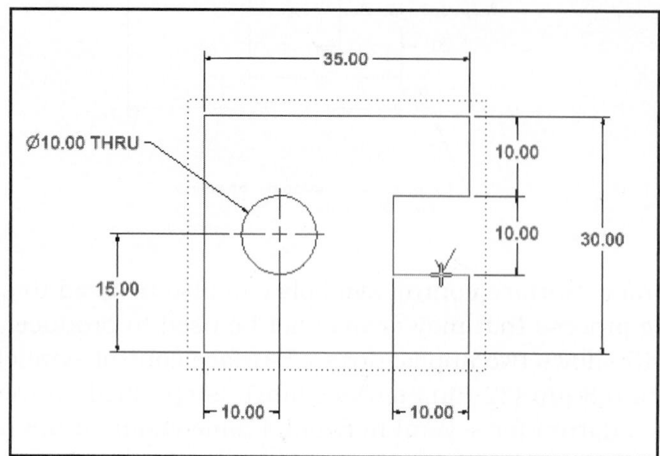

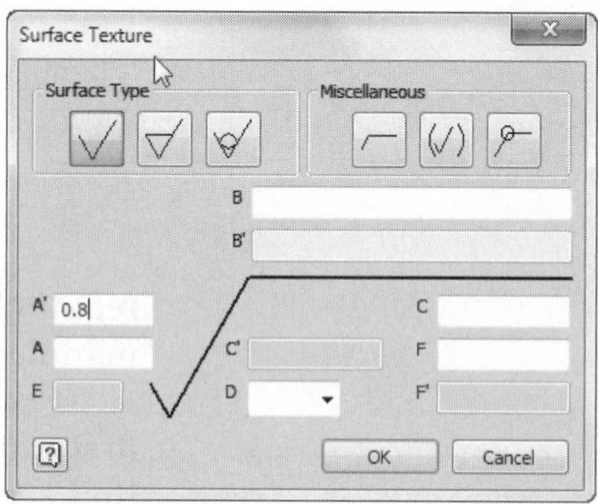

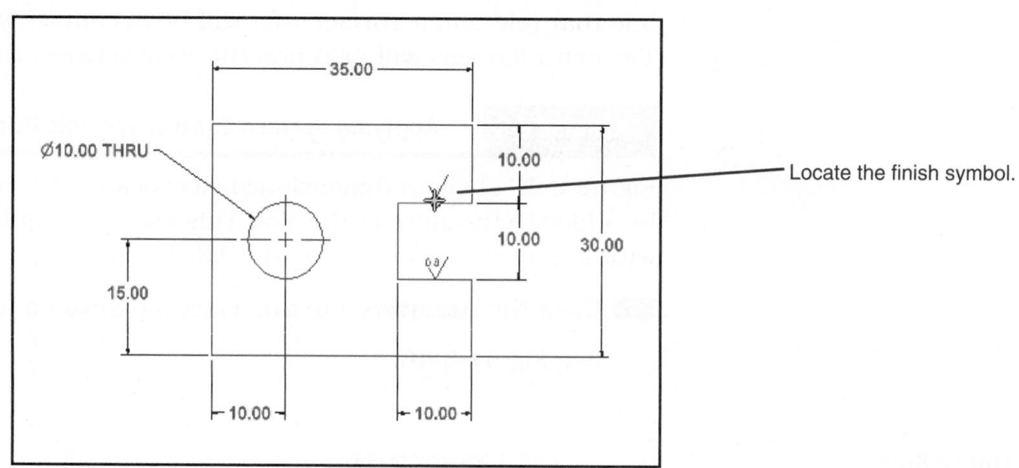

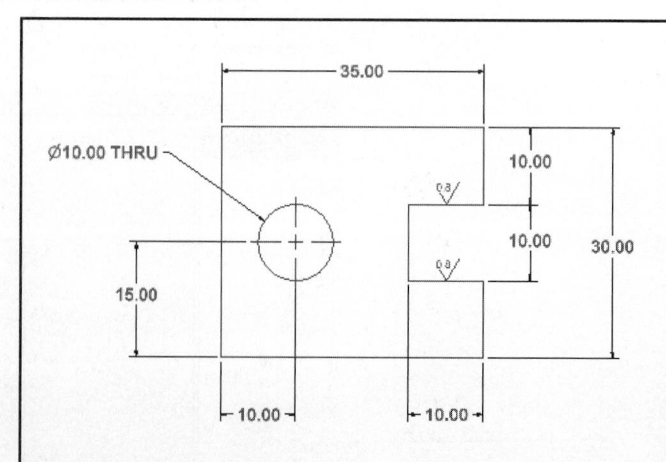

Figure 8-47

3 Right-click the mouse, and select the **Continue** option.

The **Surface Texture** dialog box will appear.

4 Enter a surface texture value of **0.8.**

5 Click **OK.**

The surface symbol will be added to the drawing.

The cursor will remain in **Surface Texture** mode so that other symbols may be applied.

6 Add a second symbol to the upper edge of the slot.

7 Right-click the mouse and select the **Done** option.

Available Lay Symbols

Inventor includes a group of lay symbols that can be added to the drawing using the **Surface** tool. The definition of the symbols is located on the **Style and Standard Editor** dialog box.

1 Click the **Manage** tab at the top of the screen, then the **Styles Editor** option.

See Figure 8-48.

2 Select the **Surface Texture** option, then select the lay symbol.

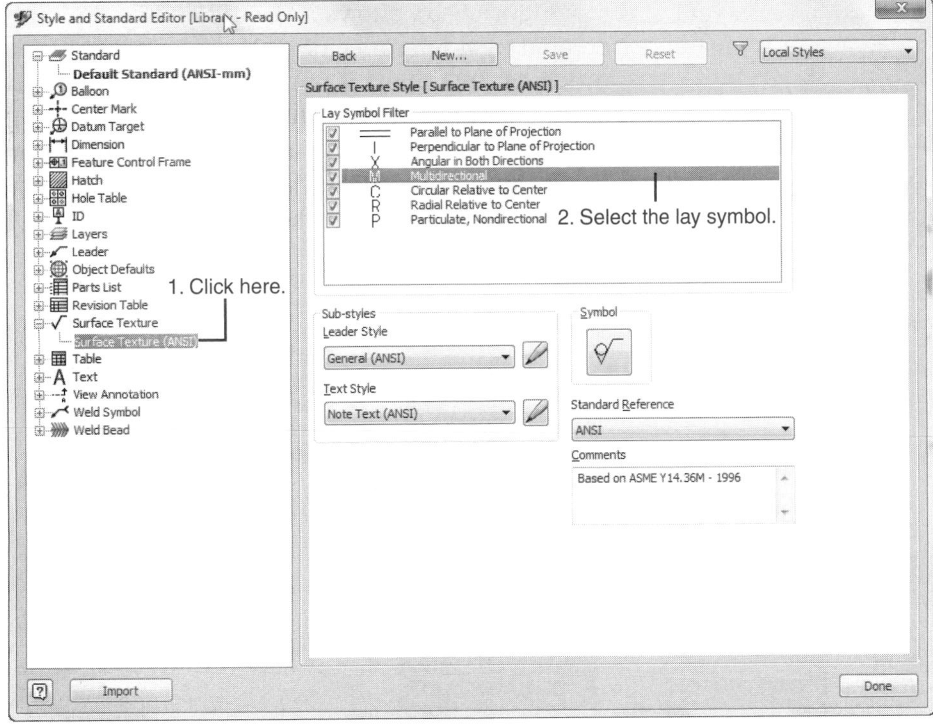

Figure 8-48

EXERCISE 8-9 Adding Lay Symbols

1. Click on the **Surface** tool, locate a symbol, then right-click the mouse. A dialog box will appear. See Figure 8-49.
2. Scroll down the available symbols in box **D** and select an appropriate symbol. In this example the symbol **M** for **Multidirectional** was selected.
3. Click **OK.**

Design Problems

Figure 8-50 shows two objects that are to be fitted together using a fastener such as a screw-and-nut combination. For this example a cylinder will be used to represent a fastener. Only two nominal dimensions are given. The dimensions and tolerances were derived as follows.

The distance between the centers of the holes is given as 50 nominal. The term *nominal* means that the stated value is only a starting point. The final dimensions will be close to the given value but do not have to equal it.

Assigning tolerances is an iteration process; that is, a tolerance is selected and other tolerance values are calculated from the selected initial values. If the results are not satisfactory, the initial value is modified, and the other values are recalculated. As your experience grows you will become better at selecting realistic initial values.

In the example shown in Figure 8-50, start by assigning a tolerance of ±.01 to both the top and bottom parts for both the horizontal and vertical dimensions used to locate the holes. This means that there is a possible center point variation of .02 for both parts. The parts must always fit together, so tolerances must be assigned based on the worst-case condition, or when the parts are made at the extreme ends of the assigned tolerances.

Figure 8-51 shows a greatly enlarged picture of the worst-case condition created by a tolerance of ±.01. The center points of the holes could be as much as .028 apart if the two center points were located at opposite corners of the tolerance zones. This means that the minimum hole diameter must always be at least .028 larger than the maximum stud diameter. In addition, there should be a clearance tolerance assigned so that the hole and stud are never exactly the same size. Figure 8-52 shows the resulting tolerances.

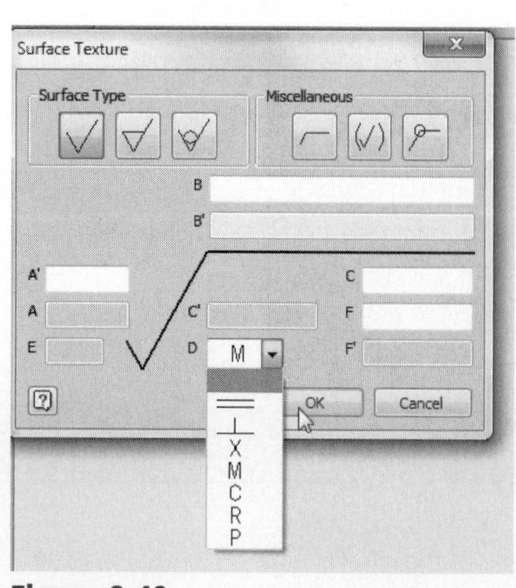

Figure 8-49

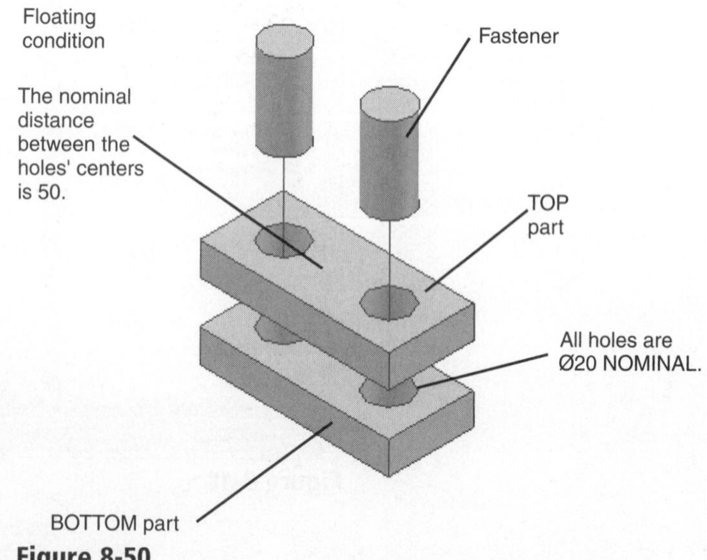

Figure 8-50

Tolerance zones for the center points of
the holes using linear tolerances 50±.01

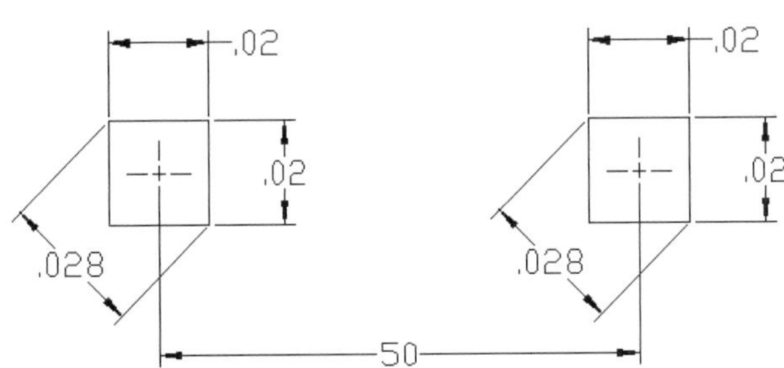

Figure 8-51

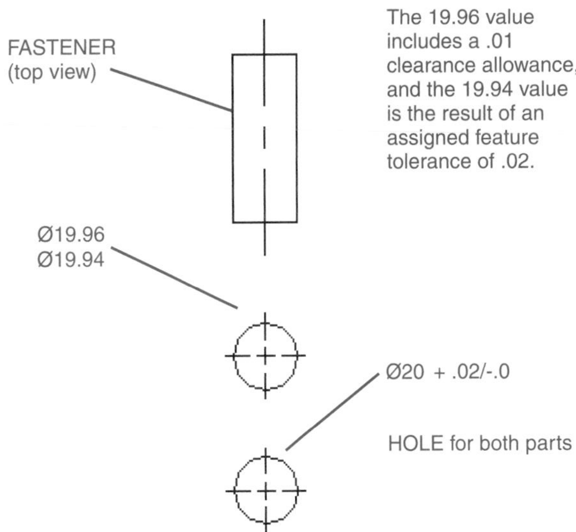

Figure 8-52

> **TIP**
> The tolerance zones in this section are created by the ±.01 tolerance on the linear dimensions that generate square tolerance zones.

Floating Condition

The top and bottom parts shown in Figure 8-50 are to be joined by two independent fasteners; that is, the location of one fastener does not depend on the location of the other. This situation is called a ***floating condition***.

floating condition: A situation in which the location of one fastener does not depend on the location of another.

This means that the tolerance zones for both the top and bottom parts can be assigned the same values and that a fastener diameter selected to fit one part will also fit the other part.

The final tolerances were developed by first defining a minimum hole size of 20.00. An arbitrary tolerance of .02 was assigned to the hole and was expressed as 20.00 + .02/−0, so that the hole can never be any smaller than 20.00.

The 20.00 minimum hole diameter dictates that the maximum fastener diameter can be no greater than 19.97, or .03 (the rounded-off diagonal distance across the tolerance zone: .028) less than the minimum hole diameter. A .01 clearance was assigned. The clearance ensures that the hole and fastener are never exactly the same diameter. The resulting maximum allowable diameter for the fastener is 19.96. Again, an arbitrary tolerance of .02 was assigned to the fastener. The final fastener dimensions are therefore 19.96 to 19.94.

The assigned tolerances ensure that there will always be at least .01 clearance between the fastener and the hole. The other extreme condition occurs when the hole is at its largest possible size (20.02) and the fastener is at its smallest (19.94). This means that there could be as much as .08 clearance between the parts. If this much clearance is not acceptable, then the assigned tolerances will have to be reevaluated.

Figure 8-53 shows the top and bottom parts dimensioned and toleranced. Any dimensions that do not have assigned tolerances are assumed to have standard tolerances.

Note, in Figure 8-53, that the top edge of each part has been assigned a surface finish. This was done to help ensure the accuracy of the

20 ± .01 dimension. If this edge surface was rough, it could affect the tolerance measurements.

This example will be done later in the chapter using geometric tolerances. Geometric tolerance zones are circular rather than rectangular.

Fixed Condition

fixed condition: A situation in which the location of fasteners is fixed to a part.

Figure 8-54 shows the same nominal conditions presented in Figure 8-50, but the fasteners are now fixed to the top part. This situation is called the ***fixed condition***. In analyzing the tolerance zones for the fixed condition, two positional tolerances must be considered: the positional tolerances for the holes in the bottom part, and the positional tolerances for the fixed fasteners in the top part. This relationship may be expressed in an equation as follows:

$$S_{max} + DTSZ = H_{min} - DTZ$$

where:

S_{max} = maximum shaft (fastener) diameter

H_{min} = minimum hole diameter

DTSZ = diagonal distance across the shaft's center point tolerance zone

DTZ = diagonal distance across the hole's center point tolerance zone

If a dimension and tolerance of 50 ± .01 is assigned to both the center distance between the holes and the center distance between the fixed fasteners, the values for DTSZ and DTZ will be equal. The formula can then be simplified as follows:

$$S_{max} = H_{min} - 2(DTZ)$$

where DTZ equals the diagonal distance across the tolerance zone. If a hole tolerance of 20.00 +.02/−0 is also defined, the resulting maximum shaft size can be determined, assuming that the calculated distance of .028 is rounded off to .03. See Figure 8-55.

$$S_{max} = 20.00 - 2(0.03)$$
$$= 19.94$$

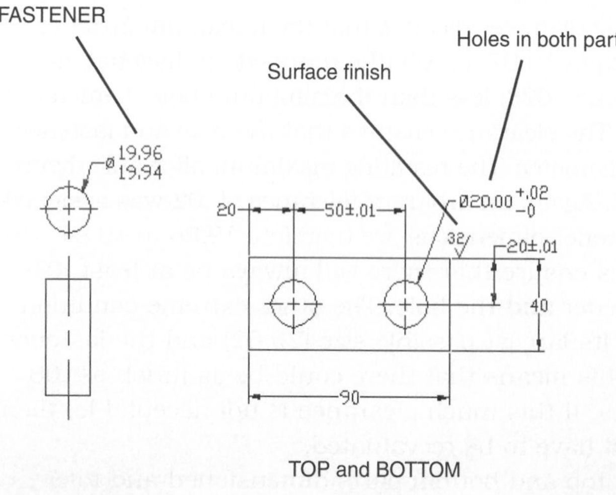

Figure 8-53

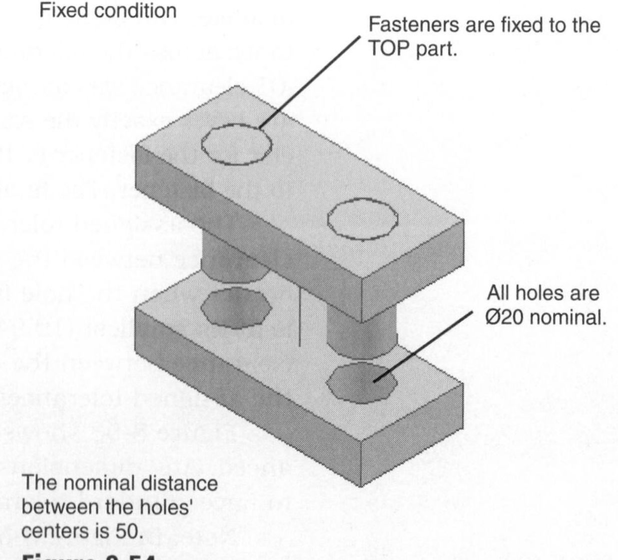

Figure 8-54

Figure 8-55

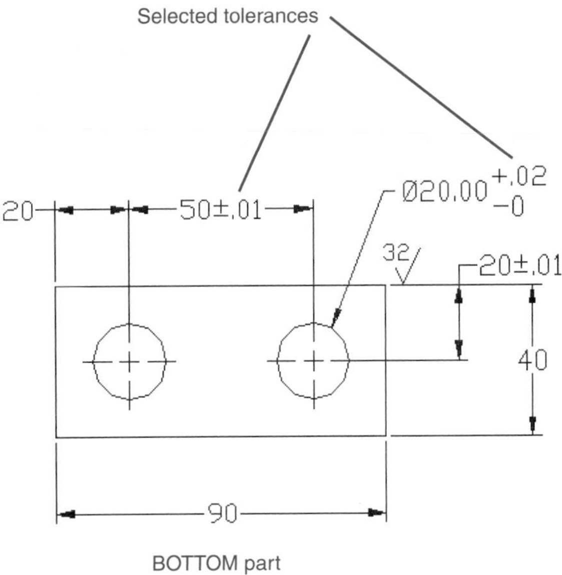

BOTTOM part

This means that 19.94 is the largest possible shaft diameter that will just fit. If a clearance tolerance of .01 is assumed to ensure that the shaft and hole are never exactly the same size, the maximum shaft diameter becomes 19.93. See Figure 8-56.

Figure 8-56

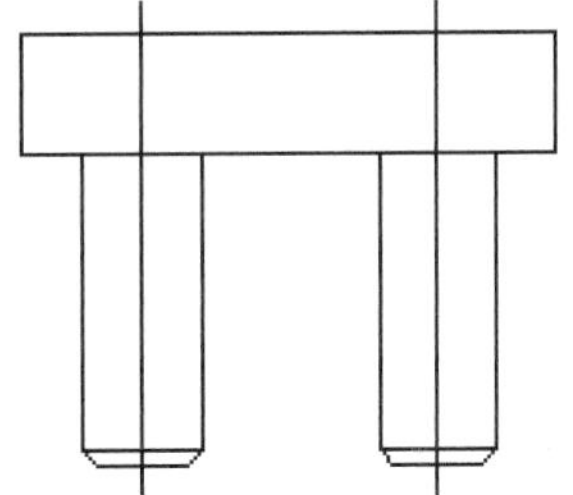

The shaft values were derived as follows:

20.00	The selected value for the minimum hole diameter.
−.03	The rounded-off value for the hole positional tolerance
−.03	The rounded-off value for the shaft positional tolerance
−.01	The selected clearance value
19.93	The maximum shaft value
−.02	The selected tolerance value
19.91	The minimum shaft diameter

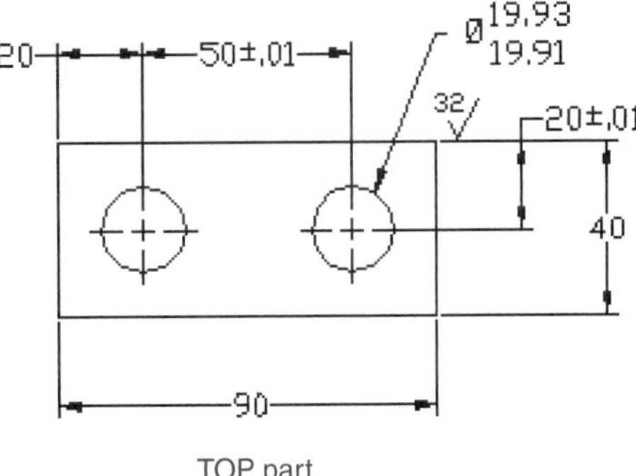

TOP part

A feature tolerance of .02 on the shaft will result in a minimum shaft diameter of 19.91. Note that the .01 clearance tolerance and the .02 feature tolerance were arbitrarily chosen. Other values could have been used.

Designing a Hole Given a Fastener Size

The previous two examples started with the selection of a minimum hole diameter and then calculation of the resulting fastener size. Figure 8-57 shows a situation in which the fastener size is defined, and the problem is to determine the appropriate hole sizes. Figure 8-58 shows the dimensions and tolerances for both top and bottom parts.

Figure 8-57

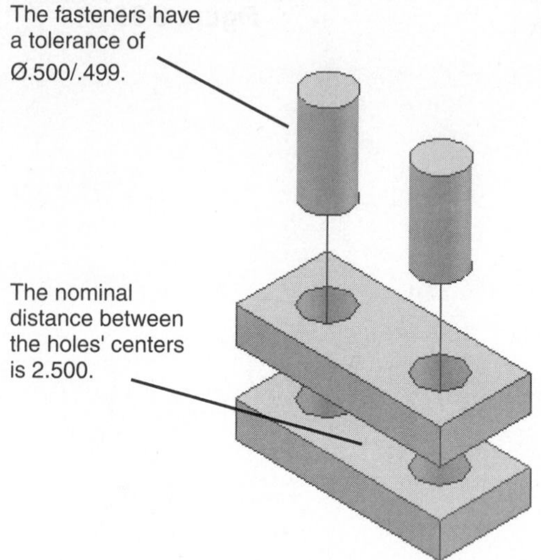

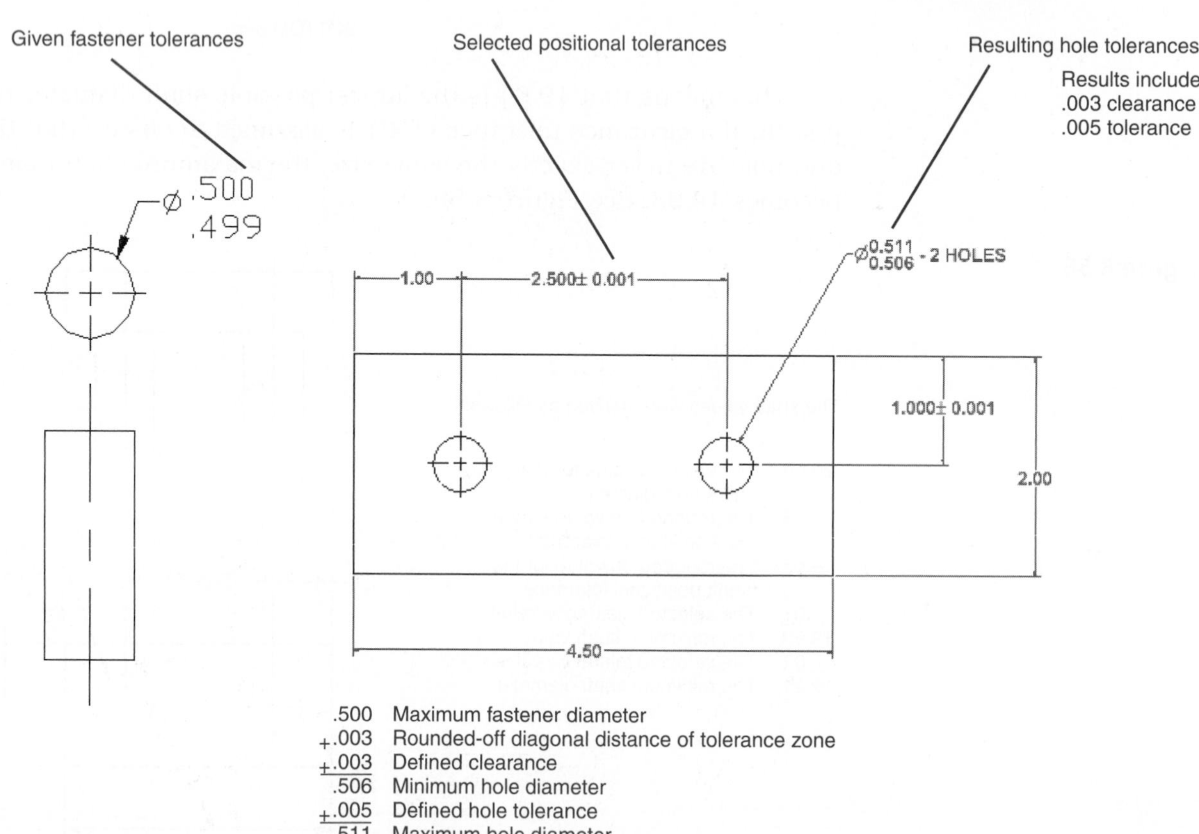

Figure 8-58

Requirements:

Clearance, minimum = .003

Hole tolerances = .005

Positional tolerance = .002

Geometric Tolerances

geometric tolerancing: A dimensioning and tolerancing system based on the geometric shape of an object.

Geometric tolerancing is a dimensioning and tolerancing system based on the geometric shape of an object. Surfaces may be defined in terms of their flatness or roundness, or in terms of how perpendicular or parallel they are to other surfaces.

Geometric tolerances allow a more exact definition of the shape of an object than do conventional coordinate-type tolerances. Objects can be toleranced in a manner more closely related to their design function or so that their features and surfaces are more directly related to each other.

Tolerances of Form

tolerance of form: A tolerance used to define the shape of a surface relative to itself.

Tolerances of form are used to define the shape of a surface relative to itself. There are four classifications: flatness, straightness, circularity, and cylindricity. Tolerances of form are not related to other surfaces but apply only to an individual surface.

Flatness

flatness tolerance: A tolerance used to define the amount of variation permitted on an individual surface.

Flatness tolerances are used to define the amount of variation permitted in an individual surface. The surface is thought of as a plane not related to the rest of the object.

Figure 8-59 shows a rectangular object. How flat is the top surface? The given plus or minus tolerances allow a variation of (±0.5) across the surface. Without additional tolerances the surface could look like a series of waves varying between 30.5 and 29.5.

If the example in Figure 8-59 was assigned a flatness tolerance of 0.3, the height of the object—the feature tolerance—could continue to vary based on the 30 ± 0.5 tolerance, but the surface itself could not vary by more than 0.3. In the most extreme condition, one end of the surface could be 30.5 above the bottom surface and the other end 29.5, but the surface would still be limited to within two parallel planes 0.3 apart as shown.

To better understand the meaning of flatness, consider how the surface would be inspected. The surface would be acceptable if a gauge could be moved all around the surface and never vary by more than 0.3. See Figure 8-60. Every point in the plane must be within the specified tolerance.

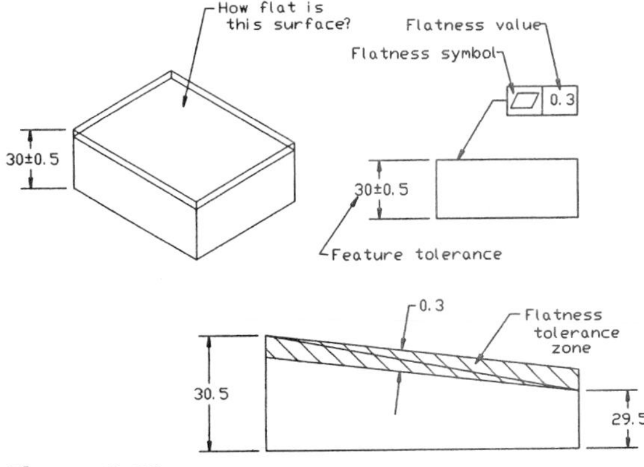

Figure 8-59

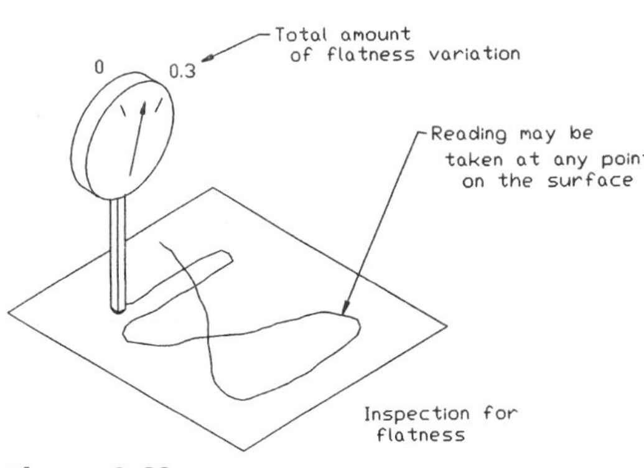

Figure 8-60

Straightness

straightness tolerance: A tolerance used to measure the variation of an individual feature along a straight line in a specified direction.

Straightness tolerances are used to measure the variation of an individual feature along a straight line in a specified direction. Figure 8-61 shows an object with a straightness tolerance applied to its top surface. Straightness differs from flatness because straightness measurements are checked by

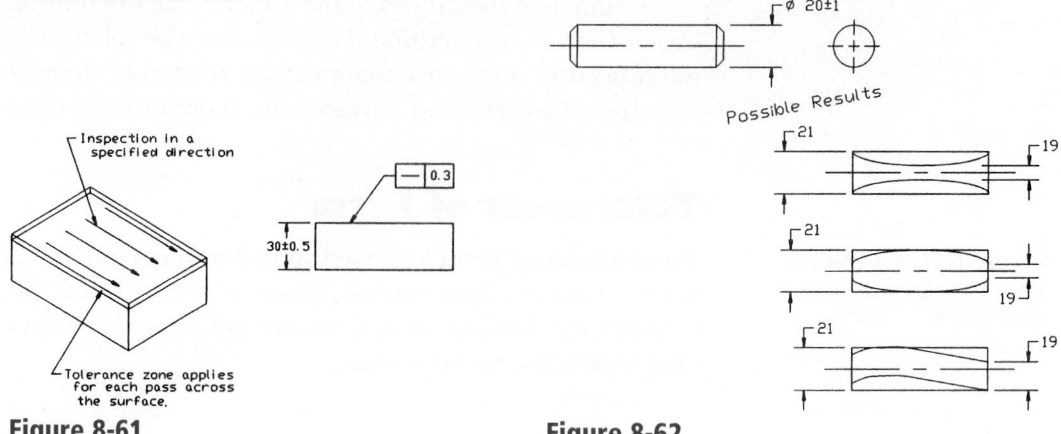

Figure 8-61

Figure 8-62

moving a gauge directly across the surface in a single direction. The gauge is not moved randomly about the surface, as is required by flatness.

Straightness tolerances are most often applied to circular or matching objects to help ensure that the parts are not barreled or warped within the given feature tolerance range and, therefore, do not fit together well. Figure 8-62 shows a cylindrical object dimensioned and toleranced using a standard feature tolerance. The surface of the cylinder may vary within the specified tolerance range as shown.

Figure 8-63 shows the same object shown in Figure 8-62 dimensioned and toleranced using the same feature tolerance but also including a 0.05 straightness tolerance. The straightness tolerance limits the surface variation to 0.05 as shown.

Figure 8-63

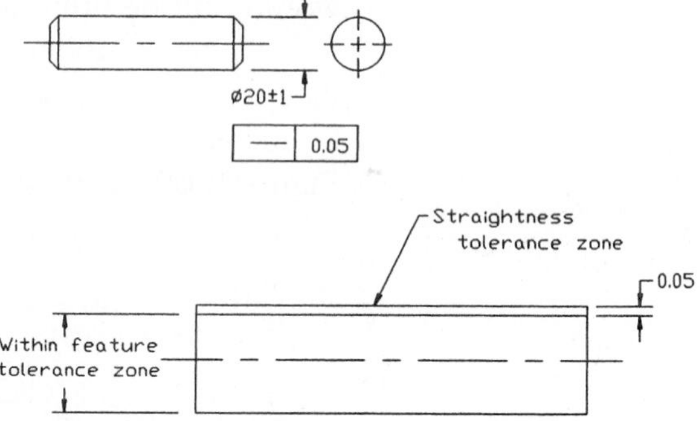

Straightness (RFS and MMC)

Figure 8-64 again shows the same cylinder shown in Figures 8-62 and 8-63. This time the straightness tolerance is applied about the cylinder's centerline. This type of tolerance permits the feature tolerance and geometric

Figure 8-64

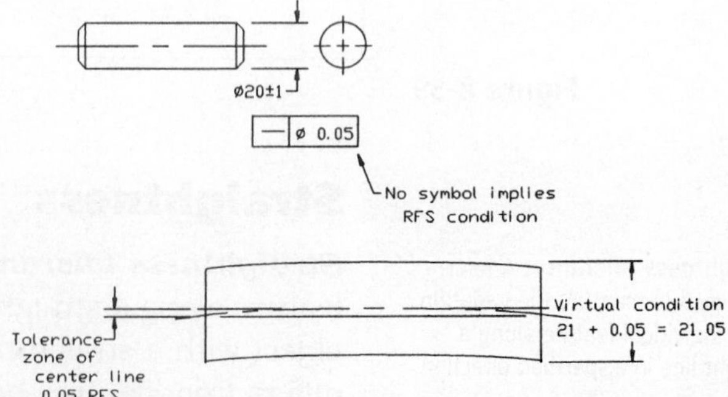

tolerance to be used together to define a *virtual condition.* A virtual condition is used to determine the maximum possible size variation of the cylinder or the smallest diameter hole that will always accept the cylinder.

The geometric tolerance specified in Figure 8-64 is applied to any circular segment along the cylinder, regardless of the cylinder's diameter. This means that the 0.05 tolerance is applied equally when the cylinder's diameter measures 19 or when it measures 21. This application is called RFS, *regardless of feature size.* RFS conditions are specified in a tolerance either by an S with a circle around it or implied tacitly when no other symbol is used. In Figure 8-59 no symbol is listed after the 0.05 value, so it is assumed to be applied RFS.

Figure 8-65 shows the cylinder dimensioned with an MMC condition applied to the straightness tolerance. MMC stands for *maximum material condition* and means that the specified straightness tolerance (0.05) is applied only at the MMC condition or when the cylinder is at its maximum diameter size (21).

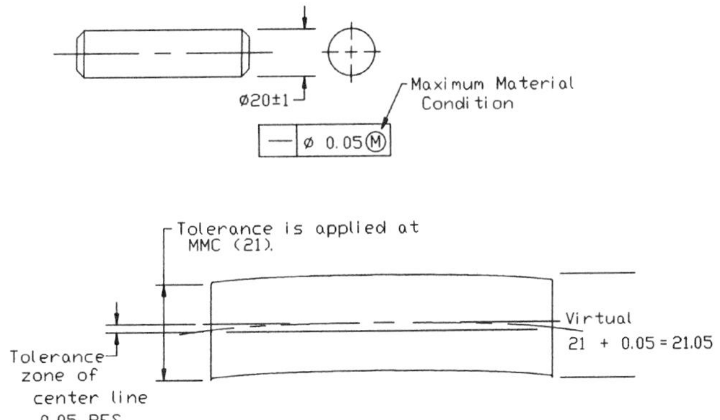

Measured Size	Allowable Tolerance Zone	Virtual Condition
21.0	0.05	21.05
20.9	0.15	21.15
20.8	0.25	21.25
⋮	⋮	⋮
20.0	1.05	22.05
⋮	⋮	⋮
19.0	2.05	23.05

Figure 8-65

A shaft is an external feature, so its largest possible size or MMC occurs when it is at its maximum diameter. A hole is an internal feature. A hole's MMC condition occurs when it is at its smallest diameter. The MMC condition for holes will be discussed later in the chapter along with positional tolerances.

Applying a straightness tolerance at MMC allows for a variation in the resulting tolerance zone. Because the 0.05 flatness tolerance is applied at MMC, the virtual condition is still 21.05, the same as with the RFS condition; however, the tolerance is applied only at MMC. As the cylinder's diameter varies within the specified feature tolerance range the acceptable tolerance zone may vary to maintain the same virtual condition.

The table in Figure 8-65 shows how the tolerance zone varies as the cylinder's diameter varies. When the cylinder is at its largest size or MMC, the tolerance zone equals 0.05, or the specified flatness variation. When the cylinder is at its smallest diameter, the tolerance zone equals 2.05, or the total feature size plus the total flatness size. In all variations the virtual size remains the same, so at any given cylinder diameter value, the size of the tolerance zone can be determined by subtracting the cylinder's diameter value from the virtual condition.

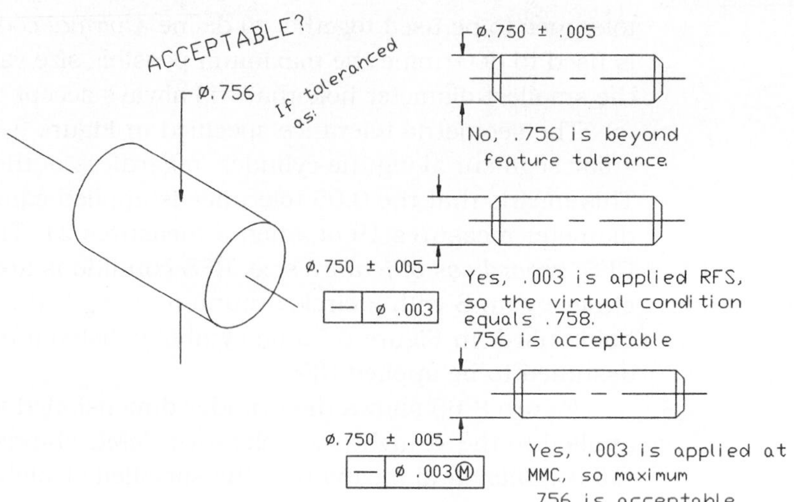

Figure 8-66

The RFS condition does not allow the tolerance zone to grow, as does the same tolerance applied at MMC.

Figure 8-66 shows a comparison between different methods used to dimension and tolerance a .750 shaft. The first example uses only a feature tolerance. This tolerance sets an upper limit of .755 and a lower limit of .745. Any variations within that range are acceptable.

NOTE

Geometric tolerances applied at MMC allow the tolerance zone to grow.

The second example in Figure 8-66 sets a straightness tolerance of .003 about the cylinder's centerline. No conditions are defined, so the tolerance is applied RFS. This limits the variations in straightness to .003 at all feature sizes. For example, when the shaft is at its smallest possible feature size of .745, the .003 still applies. This means that a shaft measuring .745 that had a straightness variation greater than .003 would be rejected. If the tolerance had been applied at MMC, the part would have been accepted. This does not mean that straightness tolerances should always be applied at MMC. If straightness is critical to the design integrity or function of the part, then straightness should be applied in the RFS condition.

The third example in Figure 8-66 applies the straightness tolerance about the centerline at MMC. This tolerance creates a virtual condition of .758. The MMC condition allows the tolerance to vary as the feature tolerance varies, so when the shaft is at its smallest feature size, .745, a straightness tolerance of .003 is acceptable (.005 feature tolerance + .003 straightness tolerance).

If the tolerance specification for the cylinder shown in Figure 8-66 was .000 applied at MMC, it would mean that the shaft would have to be perfectly straight at MMC or when the shaft was at its maximum value (.755); however, the straightness tolerance can vary as the feature size varies, as discussed for the other tolerance conditions. A .000 tolerance means that the MMC and the virtual conditions are equal.

Figure 8-67 shows a very long .750 diameter shaft. Its straightness tolerance includes a length qualifier that serves to limit the straightness variations over each inch of the shaft length and to prevent excess waviness over the full length. The tolerance .002/1.000 means that the total straightness may vary over the entire length of the shaft by .003 but that the variation is limited to .002 per 1.000 of shaft length.

Figure 8-67

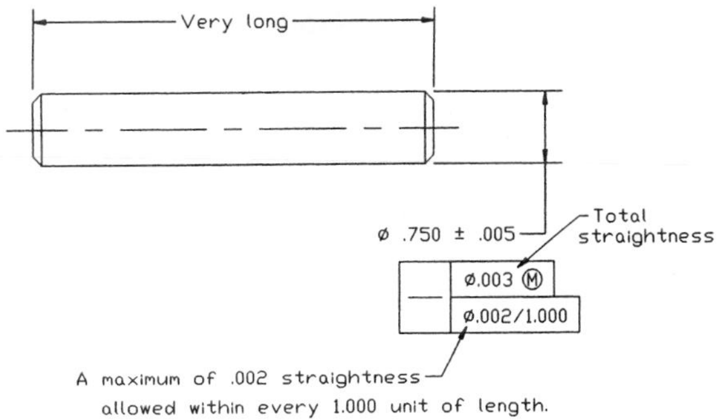

Chapter 8

Circularity

circularity tolerance: A tolerance used to limit the amount of variation in the roundness of a surface of revolution.

A ***circularity tolerance*** is used to limit the amount of variation in the roundness of a surface of revolution. It is measured at individual cross sections along the length of the object. The measurements are limited to the individual cross sections and are not related to other cross sections. This means that in extreme conditions the shaft shown in Figure 8-68 could actually taper from a diameter of 21 to a diameter of 19 and never violate the circularity requirement. It also means that qualifications such as MMC cannot be applied.

Figure 8-68 shows a shaft that includes a feature tolerance and a circularity tolerance of 0.07. To understand circularity tolerances, consider an individual cross section or slice of the cylinder. The actual shape of the outside edge of the slice varies around the slice. The difference between the maximum diameter and the minimum diameter of the slice can never exceed the stated circularity tolerance.

Circularity tolerances can be applied to tapered sections and spheres, as shown in Figure 8-69. In both applications, circularity is measured around individual cross sections, as it was for the shaft shown in Figure 8-68.

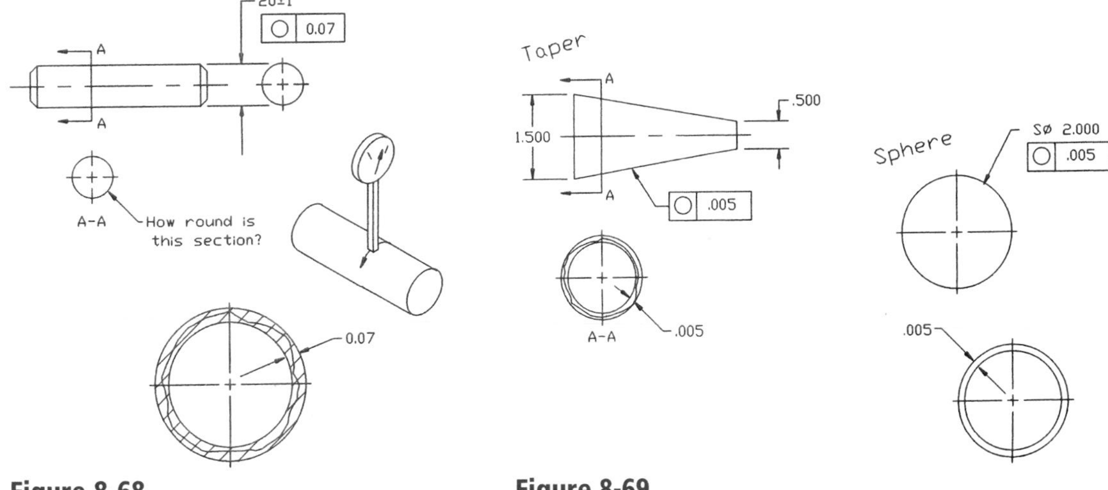

Figure 8-68

Figure 8-69

Cylindricity

cylindricity tolerance: A tolerance used to define a tolerance zone both around individual circular cross sections of an object and also along its length.

Cylindricity tolerances are used to define a tolerance zone both around individual circular cross sections of an object and also along its length. The resulting tolerance zone looks like two concentric cylinders.

Figure 8-70

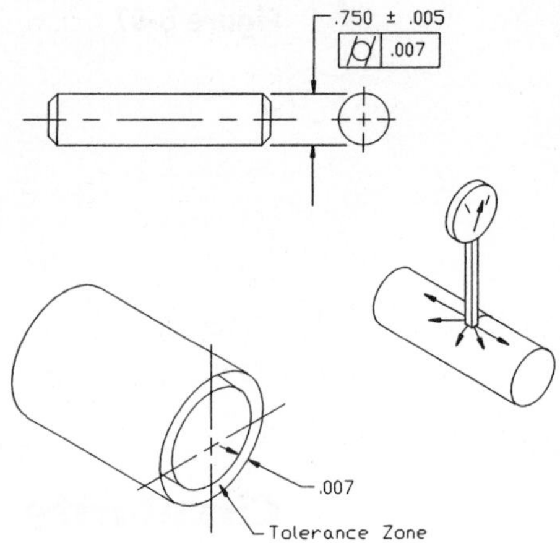

Figure 8-70 shows a shaft that includes a cylindricity tolerance that establishes a tolerance zone of .007. This means that if the maximum measured diameter is determined to be .755, the minimum diameter cannot be less than .748 anywhere on the cylindrical surface. Cylindricity and circularity are somewhat analogous to flatness and straightness. Flatness and cylindricity are concerned with variations across an entire surface or plane. In the case of cylindricity, the plane is shaped like a cylinder. Straightness and circularity are concerned with variations of a single element of a surface: a straight line across the plane in a specified direction for straightness, and a path around a single cross section for circularity.

Geometric Tolerances Using Inventor

geometric tolerance: A tolerance that limits dimensional variations based on the geometric properties of an object.

Geometric tolerances are tolerances that limit dimensional variations based on the geometric properties. Figure 8-71 shows three different ways geometric tolerance boxes can be added to a drawing.

Figure 8-71

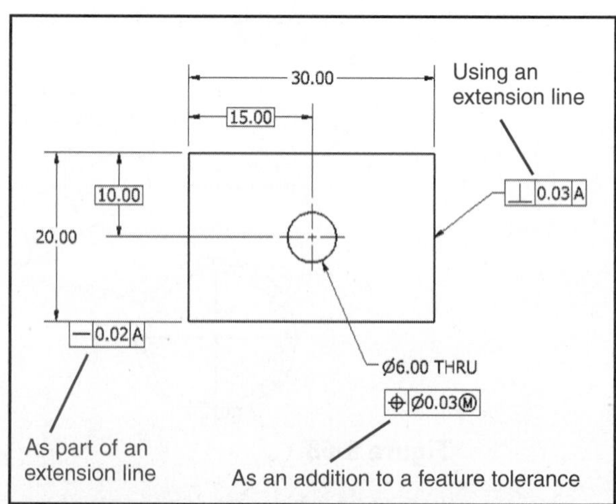

Figure 8-72 shows lists of geometric tolerance symbols. Figure 8-73 shows an object dimensioned using geometric tolerances. The geometric tolerances were created as follows.

	TYPE OF TOLERANCE	CHARACTERISTIC	SYMBOL
FOR INDIVIDUAL FEATURES	FORM	STRAIGHTNESS	—
		FLATNESS	▱
		CIRCULARITY	○
		CYLINDRICITY	⌭
INDIVIDUAL OR RELATED FEATURES	PROFILE	PROFILE OF A LINE	⌒
		PROFILE OF A SURFACE	⌓
RELATED FEATURES	ORIENTATION	ANGULARITY	∠
		PERPENDICULARITY	⊥
		PARALLELISM	//
	LOCATION	POSITION	⌖
		CONCENTRICITY	◎
	RUNOUT	CIRCULAR RUNOUT	↗
		TOTAL RUNOUT	⌰

TERM	SYMBOL
AT MAXIMUM MATERIAL CONDITION	Ⓜ
REGARDLESS OF FEATURE SIZE	Ⓢ
AT LEAST MATERIAL CONDITION	Ⓛ
PROJECTED TOLERANCE ZONE	Ⓟ
DIAMETER	Ø
SPHERICAL DIAMETER	SØ
RADIUS	R
SPHERICAL RADIUS	SR
REFERENCE	( )
ARC LENGTH	⌒

Figure 8-72

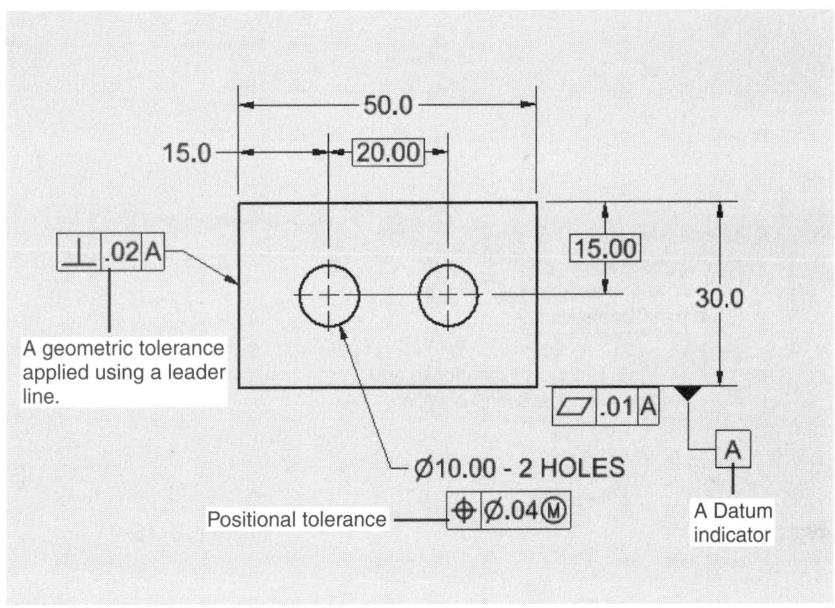

Figure 8-73

EXERCISE 8-10 Defining a Datum—Tolerance Tool

1 Click the **Annotate** tab and click the icon in the lower right corner of the **Symbols** panel.

See Figure 8-74.

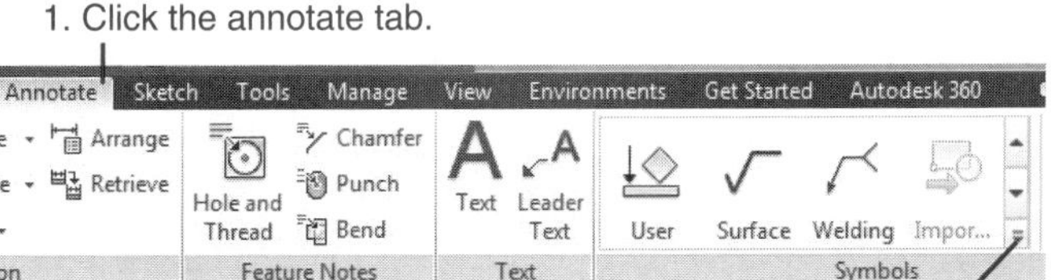

Figure 8-74

2 Click the **Datum Identifier Symbol** tool and move the cursor into the drawing area.

A datum box will appear on the cursor. See Figure 8-75.

3 Position the datum identifier directly over an extension line and press the left mouse button.

4 Move the cursor away from the extension line, and right-click the mouse.

5 Click the **Continue** option.

The **Format Text** dialog box will appear. See Figure 8-76. The letter **A** will automatically be selected. If another letter is required, backspace out the existing letter and type in a new one.

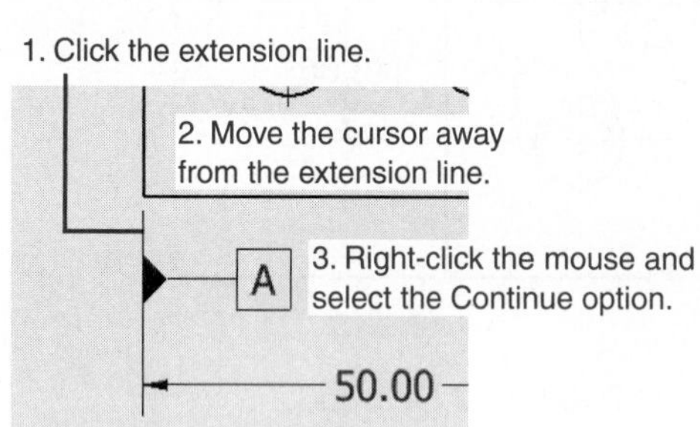

Figure 8-75

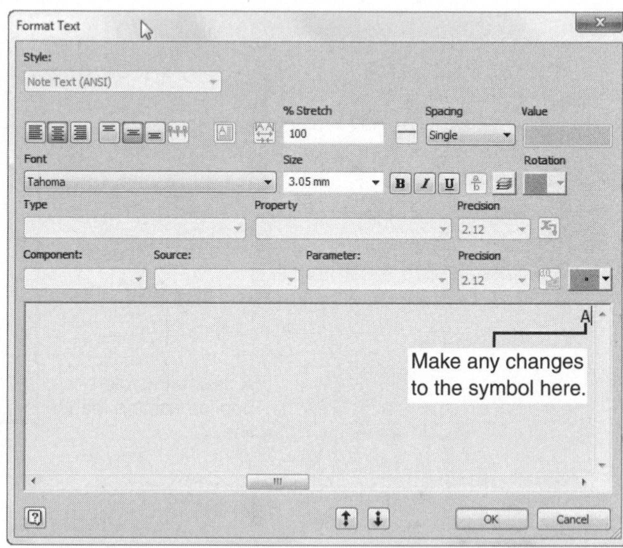

Figure 8-76

6 Click **OK,** then right-click the mouse and select the **Done** option.

The symbol may be moved after it is created if needed.

EXERCISE 8-11 Defining a Perpendicular Tolerance

1 Click the **Annotate** tab and access the **Feature Control** option on the **Symbols** panel.

2 Move the cursor into the drawing area, and select a location for the control frame.

3 Left-click the mouse, then right-click it. Select the **Continue** option.

The **Feature Control Frame** dialog box will appear. See Figure 8-77.

4 Click the **Sym** box and select the perpendicularity symbol.

5 Set the **Tolerance 1** value for **.02** and the **Datum 1** value for **A.**

Figure 8-78 shows the resulting feature control frame.

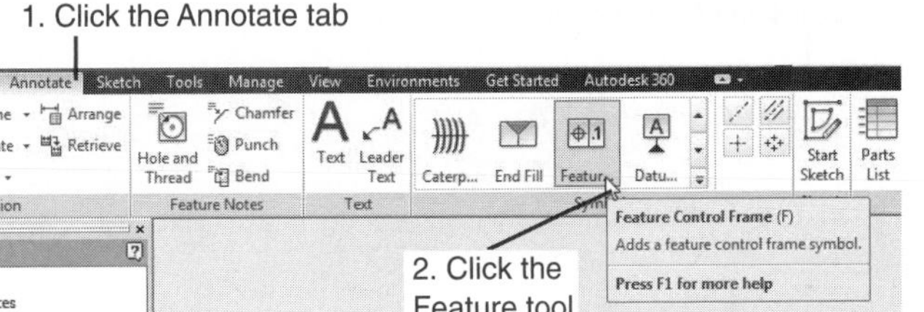

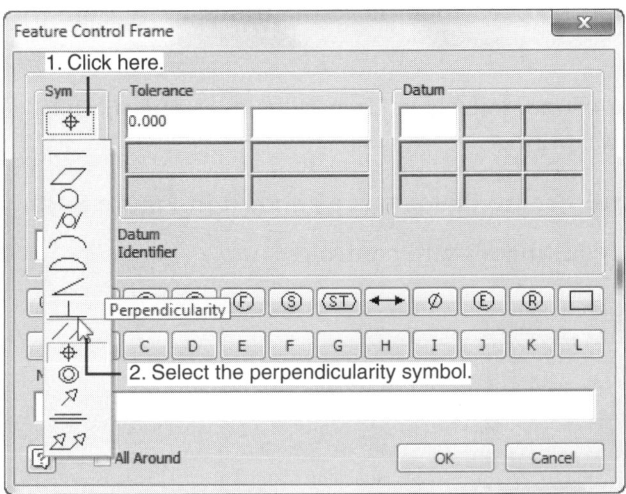

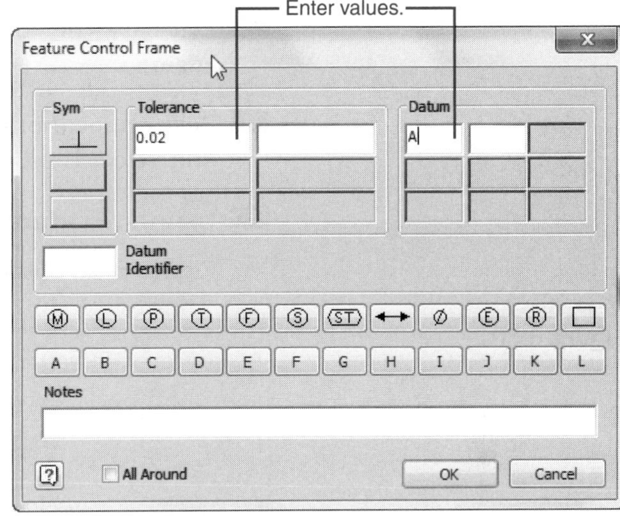

Figure 8-77

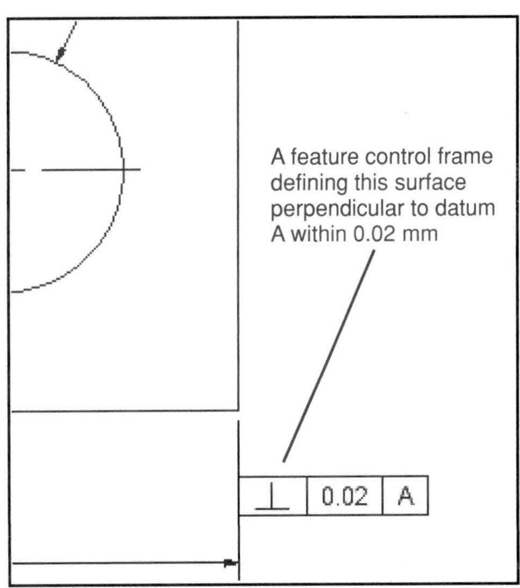

Figure 8-78

EXERCISE 8-12 Defining a Straightness Tolerance with a Leader Line

1 Click the **Annotate** tab and access the **Feature Control** option on the **Symbols** panel.

See Figure 8-79.

2 Move the cursor into the drawing area.

A feature control box will appear.

Figure 8-79

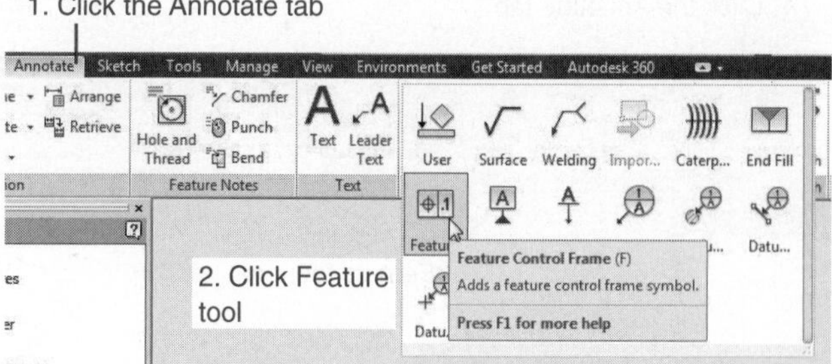

3 Left-click the edge line of the part, then move the frame away from the edge.

4 Select a location for the frame, left-click, then right-click the mouse and select the **Continue** option.

5 Edit the **Feature Control Frame** dialog box as shown in Figure 8-80.

Figure 8-80 shows the resulting feature control.

Figure 8-80

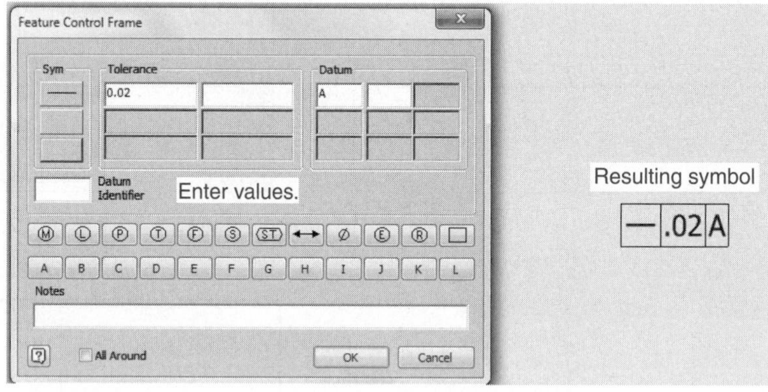

Positional Tolerance

positional tolerance: A tolerance used to locate and tolerance a hole in an object.

A ***positional tolerance*** is used to locate and tolerance a hole in an object. Positional tolerances require basic locating dimensions for the hole's center point. Positional tolerances also require a feature tolerance to define the diameter tolerances of the hole, and a geometric tolerance to define the positional tolerance for the hole's center point.

NOTE

A geometric positional tolerance creates a circular tolerance zone.

EXERCISE 8-13 Creating a Basic Dimension

basic dimension: A dimension enclosed in a rectangle.

The 15 and 20 dimensions in Figure 8-73 used to locate the center position of the hole are basic dimensions. ***Basic dimensions*** are dimensions enclosed in rectangles.

1 Create dimensions using the **Dimension** tool.

2 Right-click the existing dimension and select the **Edit** option.

The **Edit Dimension** dialog box will appear. See Figure 8-81.

Figure 8-81

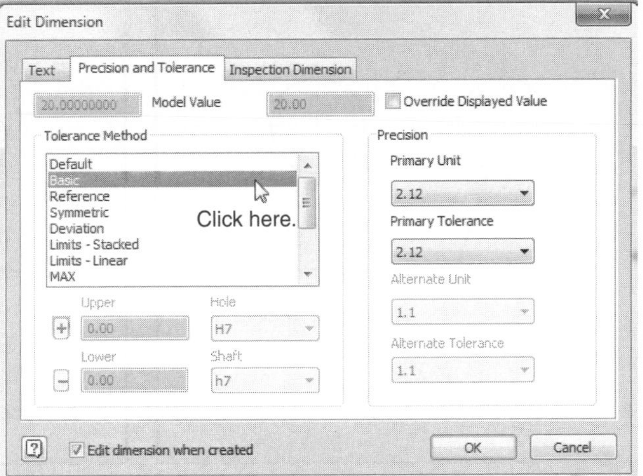

3 Click the **Precision and Tolerance** tab.

4 Select the **Basic** option, then click **OK.**

The selected dimension will be enclosed in a rectangle. This is a basic dimension. See Figure 8-82.

Figure 8-82

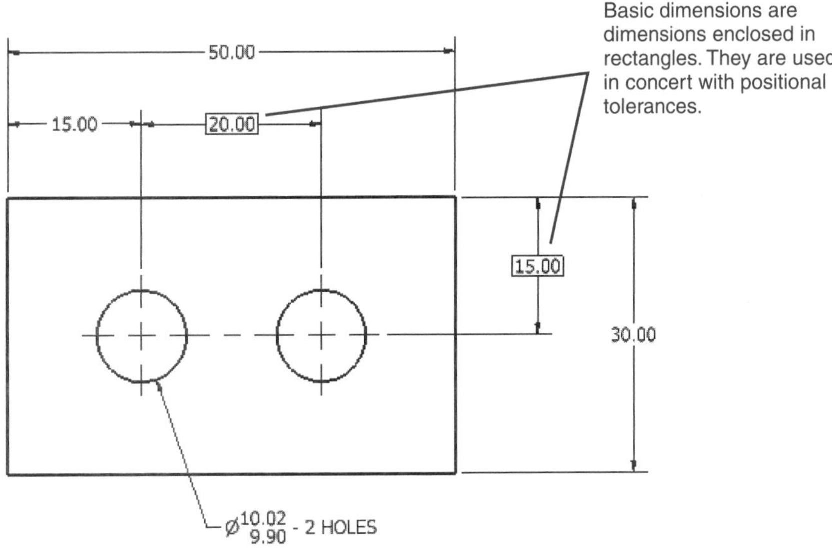

EXERCISE 8-14 Adding a Positional Tolerance to a Hole's Feature Tolerance

Figure 8-83 shows a feature tolerance for a hole. A feature tolerance defines the hole's size limits. In the example shown in Figure 8-83 the hole is defined as 10.2 to 9.9. These values define the tolerance of the hole. In addition, the location of the hole's center point must be defined and toleranced.

1 Click the **Annotate** tab and access the **Feature Control** option on the **Symbols** panel.

2 Move the cursor and locate the feature control frame below the hole's feature control dimensions.

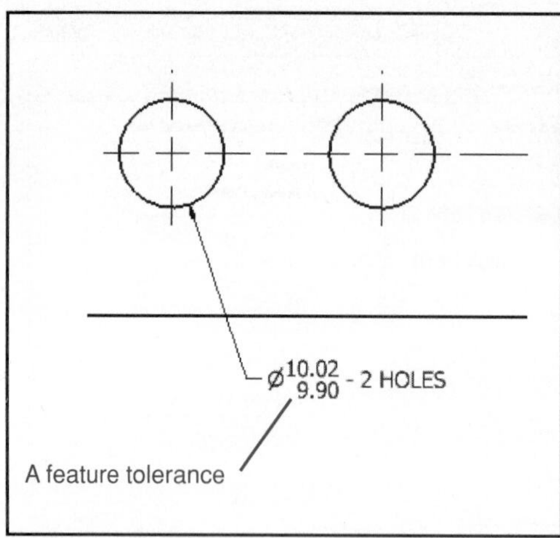

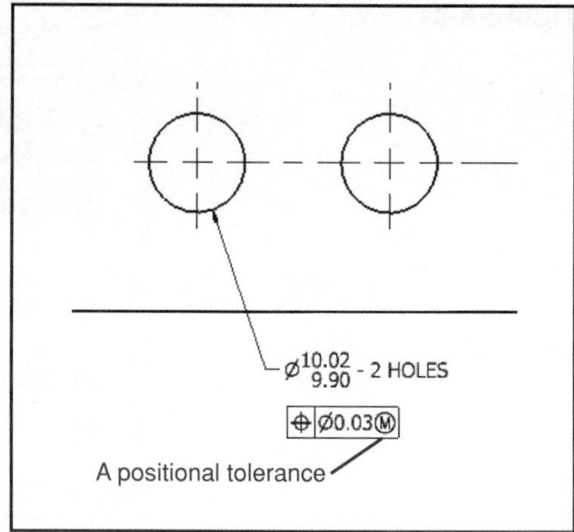

Figure 8-83

3. Click the left mouse button, then the right mouse button. Select the **Continue** option.

   The **Feature Control Frame** dialog box will appear. See Figure 8-84.

Figure 8-84

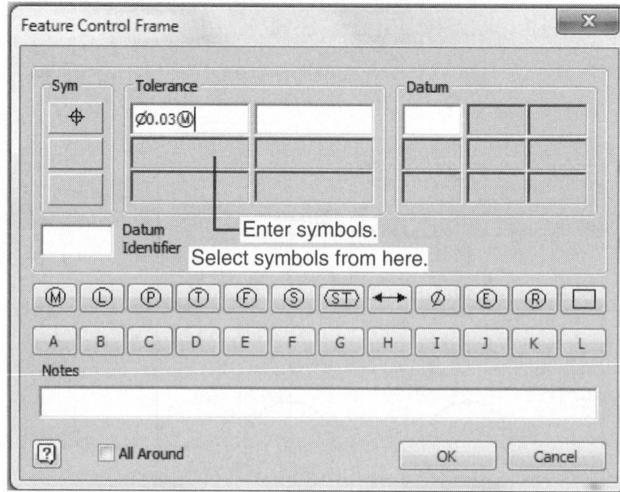

4. Select the positional symbol, then move to the **Tolerance** box and locate the cursor to the left of the **0.000** default tolerance value.

5. Select the centerline symbol, **Ø.**

6. Create a tolerance value of **0.03,** then click the maximum material condition symbol, an M with a circle around it.

7. Click **OK.**

8. Right-click the mouse and select the **Done** option.

Creating More Complex Geometric Tolerance Drawing Callouts

Figure 8-85 shows a model that has several, more complex, geometric tolerance drawing callouts. These callouts are created using the procedure and the same **Feature Control Frame** dialog box as shown in Figure 8-84,

Figure 8-85

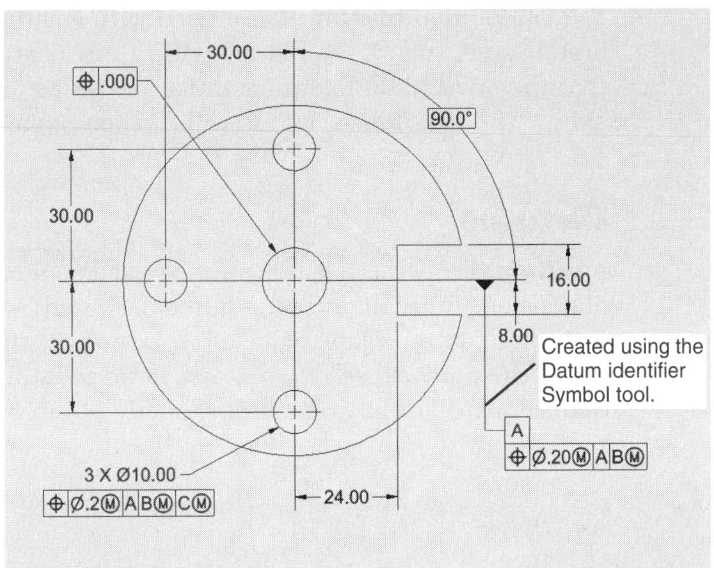

but data must be entered. The individual datum callouts are created using the **Datum Identifier Symbol** tool.

Figure 8-86 shows the **Feature Control Frame** dialog box used to create the slot dimension in Figure 8-85. Figure 8-86 also shows the **Feature Control Frame** dialog box for a multilined tolerance callout.

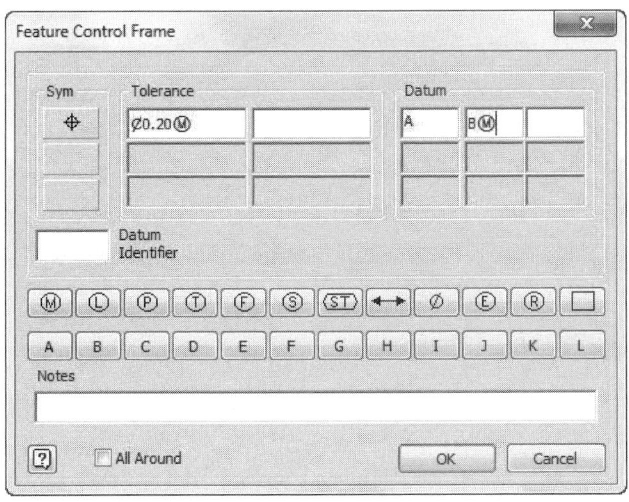

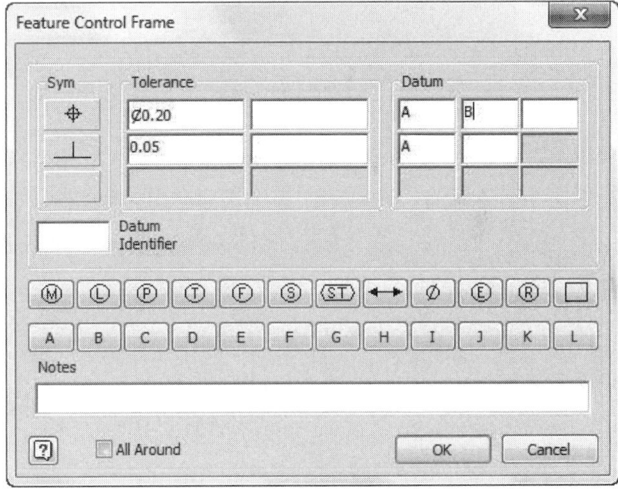

Figure 8-86

Tolerances of Orientation

tolerance of orientation: A tolerance used to relate a feature or surface to another feature or surface.

Tolerances of orientation are used to relate a feature or surface to another feature or surface. Tolerances of orientation include perpendicularity, parallelism, and angularity. They may be applied using RFS or MMC conditions, but they cannot be applied to individual features by themselves. To define a surface as parallel to another surface is very much like assigning a flatness value to the surface. The difference is that flatness applies only within the surface; every point on the surface is related to a defined set of limiting parallel planes. Parallelism defines every point in the surface relative to another surface. The two surfaces are therefore directly related to each other, and the condition of one affects the other.

Orientation tolerances are used with locational tolerances. A feature is first located, then it is oriented within the locational tolerances. This means that the orientation tolerance must always be less than the locational tolerances. The next four sections will further explain this requirement.

Datums

datum: A point, axis, or surface used as a starting reference point for dimensions and tolerances.

A ***datum*** is a point, axis, or surface used as a starting reference point for dimensions and tolerances. Figures 8-87 and 8-88 show a rectangular object with three datum planes labeled A, B, and C. The three datum planes are called the *primary*, *secondary*, and *tertiary* datums, respectively. The three datum planes are, by definition, exactly 90° to one another.

Figure 8-87

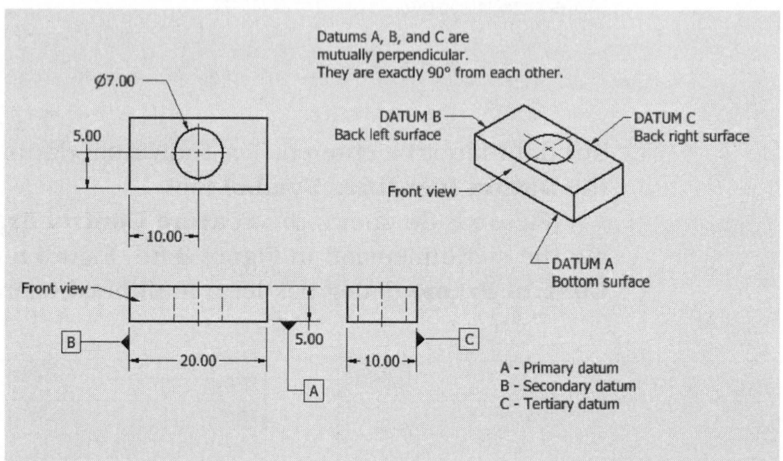

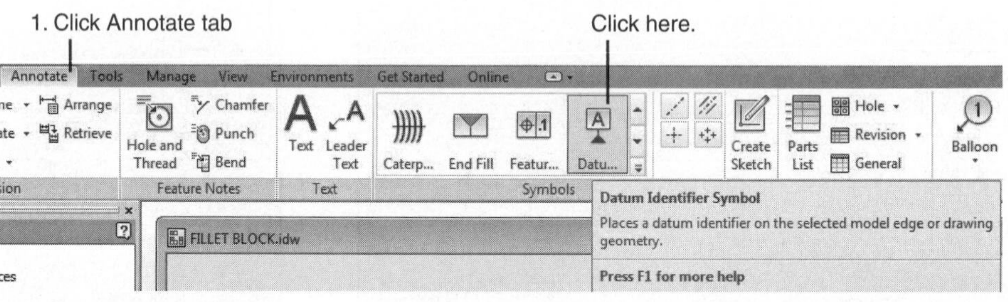

Figure 8-88

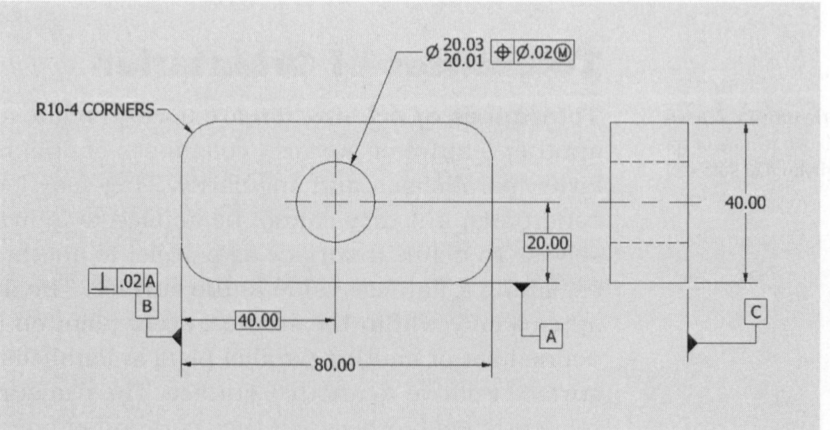

Datum symbols are created using the **Datum Identifier Symbol** tool located on the **Symbols** panel under the **Annotate** tab.

Creating a Datum Symbol

1. Click the **Datum Identifier Symbol** tool.

   The cursor will change to a +-like icon.

2. Click on the appropriate extension line, move the cursor away from the extension line, right-click the mouse, and click the **Continue** option.

   The **Format Text** dialog box will appear.

3. Enter the appropriate datum letter.

4. Click **OK.**

The +-like icon will remain active so you can identify another datum. When all datums have been identified, press the <**Esc**> key.

Datum symbols can be moved after they are created.

1. Click the datum symbol.

   Filled circles will appear on the symbol.

2. Click and drag the symbol to a new location.

3. Release the mouse button.

Figure 8-89 shows a cylindrical datum frame that includes three planes. The YZ and XY planes are perpendicular to each other, and the base XZ plane is perpendicular to the datum axis between the YZ and XY planes. The plane definitions will vary based on the drawing's original orientation. The planes shown in Figure 8-89 are defined as datums A, B, and C.

Datum planes are assumed to be perfectly flat. When assigning a datum status to a surface, be sure that the surface is reasonably flat. This means that datum surfaces should be toleranced using surface finishes, or created using machine techniques that produce flat surfaces.

Perpendicularity

perpendicularity tolerance: A tolerance used to limit the amount of variation for a surface or feature with two planes perpendicular to a specified datum.

Perpendicularity tolerances are used to limit the amount of variation for a surface or feature within two planes perpendicular to a specified datum. Figure 8-90 shows a rectangular object. The bottom surface is assigned as datum A, and the right vertical edge is toleranced so that it must be perpendicular within a limit of 0.05 to datum A. The perpendicularity tolerance defines a tolerance zone 0.05 wide between two parallel planes that are perpendicular to datum A.

The object also includes a horizontal dimension and tolerance of 40 ± 1. This tolerance is called a *locational tolerance* because it serves to locate the right edge of the object. As with rectangular coordinate tolerances, discussed earlier in the chapter, the 40 ± 1 controls the location of the edge—how far away or how close it can be to the left edge—but does not directly control the shape of the edge. Any shape that falls within the specified tolerance range is acceptable. This may in fact be sufficient for a given design, but if a more controlled shape is required, a perpendicularity tolerance must be added. The perpendicularity tolerance works within the locational tolerance to ensure that the edge is not only within the locational tolerance but is also perpendicular to datum A.

Figure 8-89

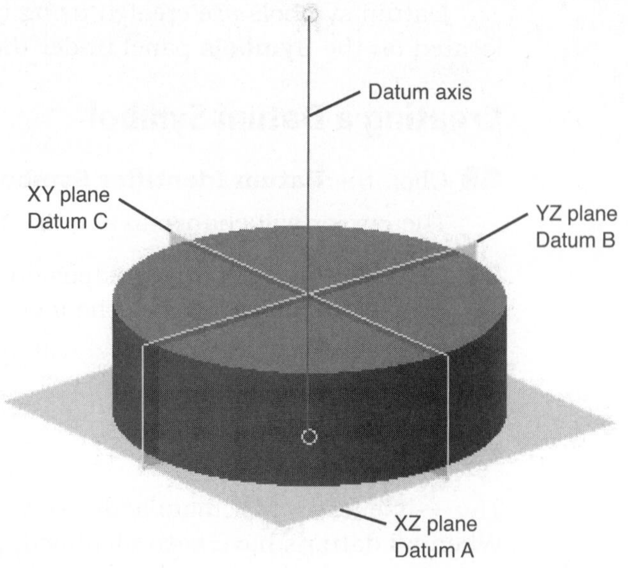

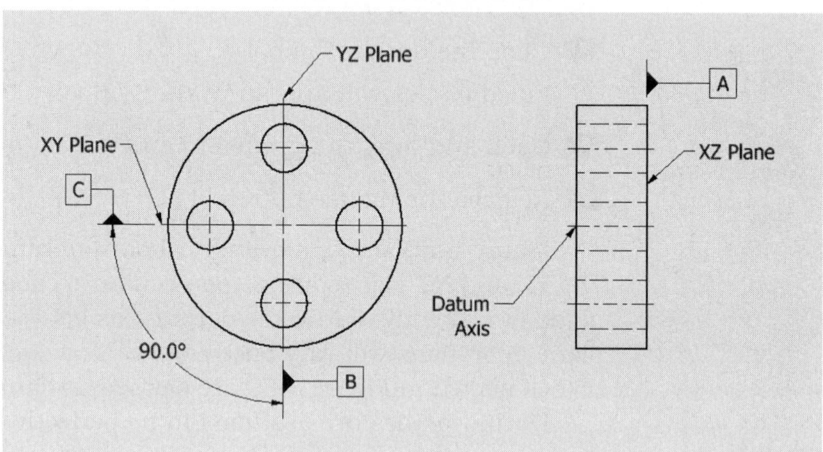

Figure 8-90 shows the two extreme conditions for the 40 ± 1 locational tolerance. The perpendicularity tolerance is applied by first measuring the surface and determining its maximum and minimum lengths. The difference between these two measurements must be less than 0.05. So if the measured maximum distance is 41, then no other part of the surface may be less than 41 − 0.05 = 40.95.

Tolerances of perpendicularity serve to complement locational tolerances, to make the shape more exact, so tolerances of perpendicularity

Figure 8-90

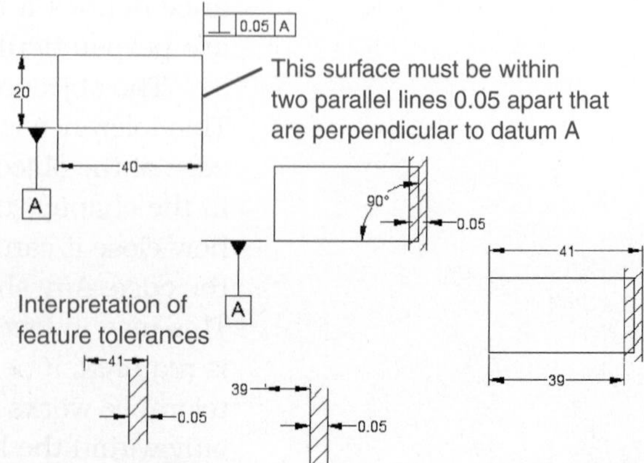

must always be smaller than tolerances of location. It would be of little use, for example, to assign a perpendicularity tolerance of 1.5 for the object shown in Figure 8-91. The locational tolerance would prevent the variation from ever reaching the limits specified by such a large perpendicularity tolerance.

Figure 8-92 shows a perpendicularity tolerance applied to cylindrical features: a shaft and a hole. The figure includes examples of both RFS and MMC applications. As with straightness tolerances applied at MMC, perpendicularity tolerances applied about a hole or shaft's centerline allow the tolerance zone to vary as the feature size varies.

The inclusion of the Ø symbol in a geometric tolerance is critical to its interpretation. See Figure 8-93. If the Ø symbol is not included, the tolerance applies only to the view in which it is written. This means that the tolerance zone is shaped like a rectangular slice, not a cylinder, as would be the case if the Ø symbol were included. In general it is better to always include the Ø symbol for cylindrical features because it generates a tolerance zone more like that used in positional tolerancing.

Figure 8-93 shows a perpendicularity tolerance applied to a slot, a noncylindrical feature. Again, the MMC specification is always for variations in the tolerance zone.

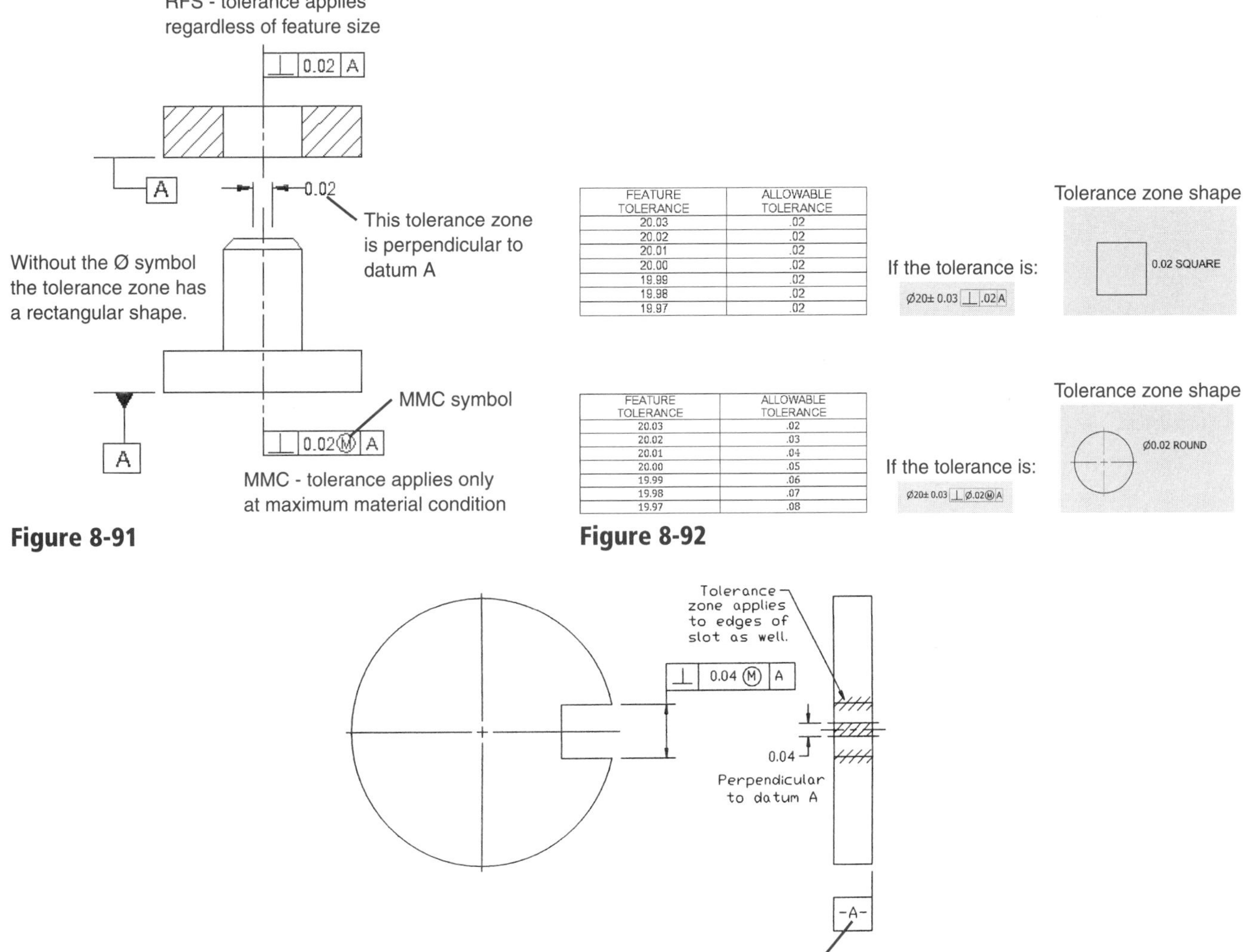

FEATURE TOLERANCE	ALLOWABLE TOLERANCE
20.03	.02
20.02	.02
20.01	.02
20.00	.02
19.99	.02
19.98	.02
19.97	.02

FEATURE TOLERANCE	ALLOWABLE TOLERANCE
20.03	.02
20.02	.03
20.01	.04
20.00	.05
19.99	.06
19.98	.07
19.97	.08

Figure 8-91

Figure 8-92

Figure 8-93

Parallelism

parallelism tolerance: A tolerance used to ensure that all points within a plane are within two parallel planes that are parallel to a referenced datum plane.

Parallelism tolerances are used to ensure that all points within a plane are within two parallel planes that are parallel to a referenced datum plane. Figure 8-94 shows a rectangular object that is toleranced so that its top surface is parallel to the bottom surface within 0.02. This means that every point on the top surface must be within a set of parallel planes 0.02 apart. These parallel tolerancing planes are located by determining the maximum and minimum distances from the datum surface. The difference between the maximum and minimum values may not exceed the stated 0.02 tolerance.

In the extreme condition of maximum feature size, the top surface is located 40.5 above the datum plane. The parallelism tolerance is then applied, meaning that no point on the surface may be closer than 40.3 to the datum. This is an RFS condition. The MMC condition may also be applied, thereby allowing the tolerance zone to vary as the feature size varies.

Figure 8-94

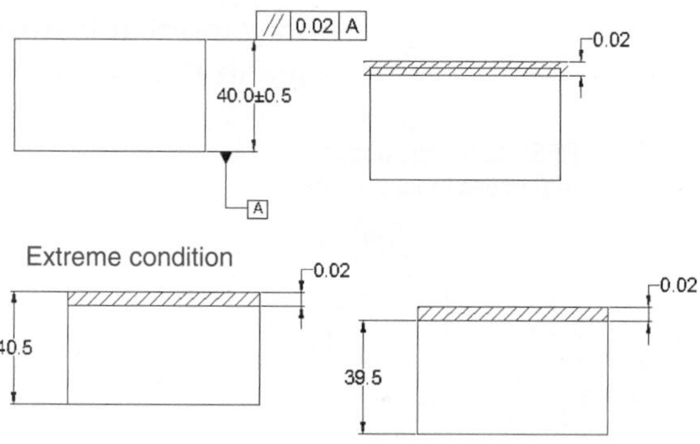

Angularity

angularity tolerance: A tolerance used to limit the variance of surfaces and axes that are at an angle relative to a datum.

Angularity tolerances are used to limit the variance of surfaces and axes that are at an angle relative to a datum. Angularity tolerances are applied like perpendicularity and parallelism tolerances as a way to better control the shape of locational tolerances.

Figure 8-95 shows an angularity tolerance and several ways it is interpreted at extreme conditions.

Figure 8-95

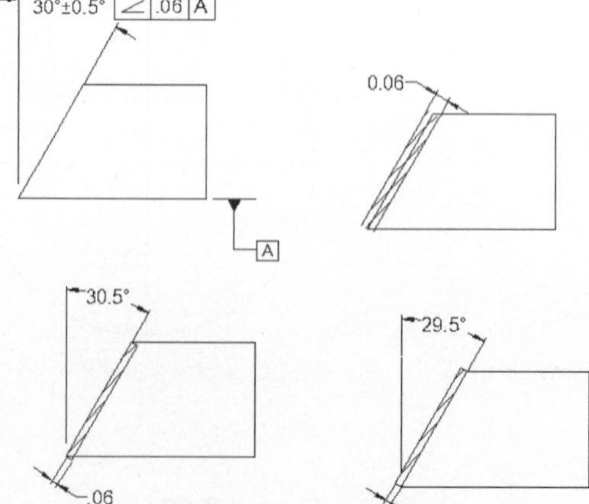

Profiles

profile tolerance: A tolerance used to limit the variations of irregular surfaces.

Profile tolerances are used to limit the variations of irregular surfaces. They may be assigned as either bilateral or unilateral tolerances. There are two types of profile tolerances: surface and line. *Surface* profile tolerances limit the variation of an entire surface, whereas a *line* profile tolerance limits the variations along a single line across a surface.

Figure 8-96 shows an object that includes a surface profile tolerance referenced to an irregular surface. The tolerance is considered a bilateral tolerance because no other specification is given. This means that all points on the surface must be located between two parallel planes 0.08 apart that are centered about the irregular surface. The measurements are taken perpendicular to the surface.

Figure 8-96

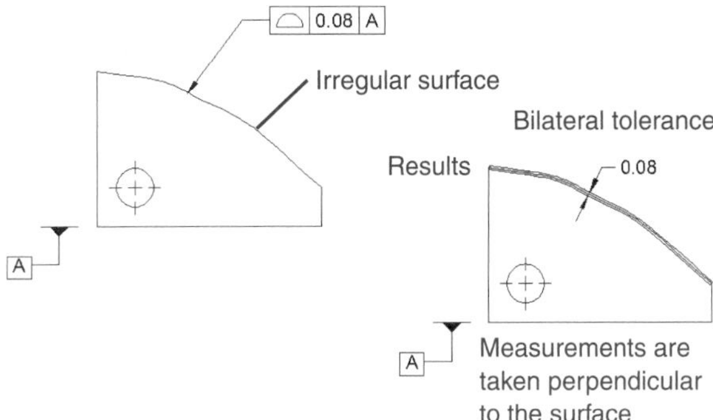

Unilateral applications of surface profile tolerances must be indicated on the drawing using phantom lines. The phantom line indicates the side of the true profile line of the irregular surface on which the tolerance is to be applied. A phantom line above the irregular surface indicates that the tolerance is to be applied using the true profile line as 0, and then the specified tolerance range is to be added above that line. See Figures 8-97 and 8-98.

Profiles of line tolerances are applied to irregular surfaces, as shown in Figure 8-98. Profiles of line tolerances are particularly helpful when tolerancing an irregular surface that is constantly changing, such as the surface of an airplane wing.

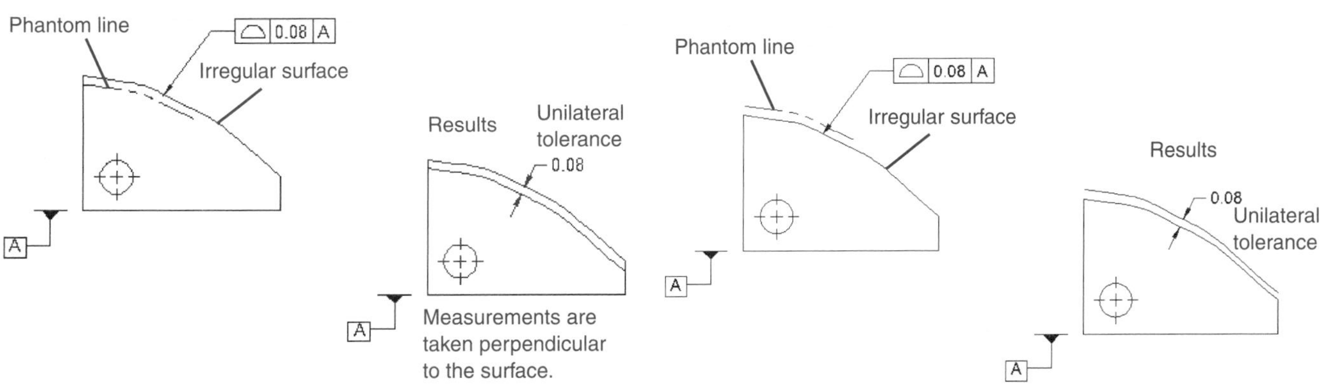

Figure 8-97

Figure 8-98

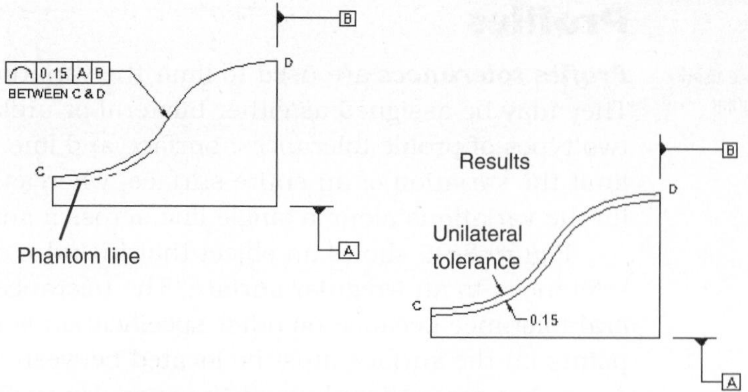

Surface and line profile tolerances are somewhat analogous to flatness and straightness tolerances. Flatness and surface profile tolerances are applied across an entire surface, whereas straightness and line profile tolerances are applied only along a single line across the surface.

Runouts

runout tolerance: A tolerance used to limit the variations between features of an object and a datum.

A ***runout tolerance*** is used to limit the variations between features of an object and a datum. More specifically they are applied to surfaces around a datum axis such as a cylinder or to a surface constructed perpendicular to a datum axis. There are two types of runout tolerances: circular and total.

Figure 8-99 shows a cylinder that includes a circular runout tolerance. The runout requirements are checked by rotating the object about its longitudinal axis or datum axis while holding an indicator gauge in a fixed position on the object's surface.

Figure 8-99

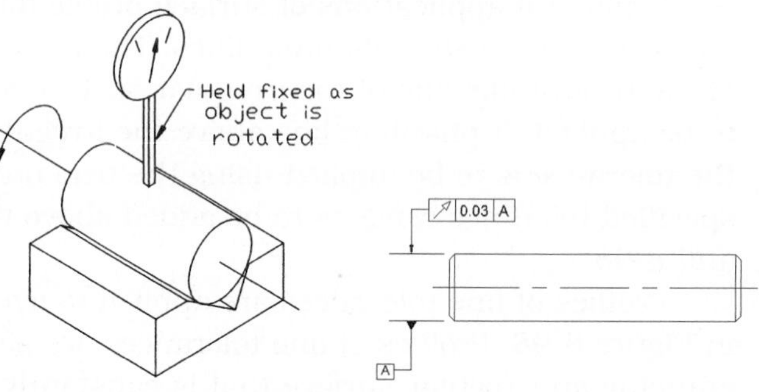

Runout tolerances may be either bilateral or unilateral. A runout tolerance is assumed to be bilateral unless otherwise indicated. If a runout tolerance is to be unilateral, a phantom line is used to indicate the side of the object's true surface to which the tolerance is to be applied. See Figure 8-100.

Figure 8-100

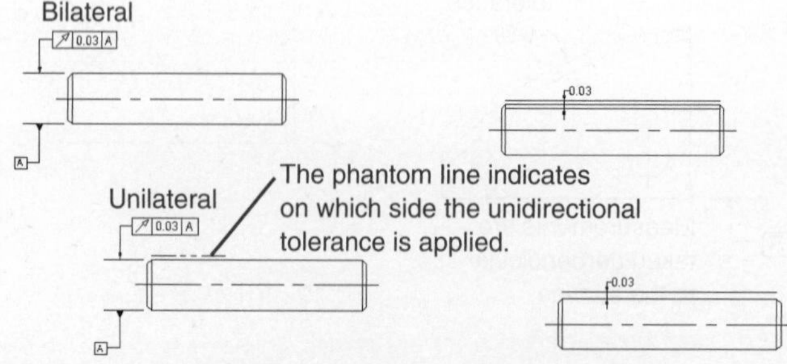

Runout tolerances may be applied to tapered areas of cylindrical objects, as shown in Figure 8-101. The tolerance is checked by rotating the object about a datum axis while holding an indicator gauge in place.

A total runout tolerance limits the variation across an entire surface. See Figure 8-102. An indicator gauge is not held in place while the object is rotated, as it is for circular runout tolerances, but is moved about the rotating surface.

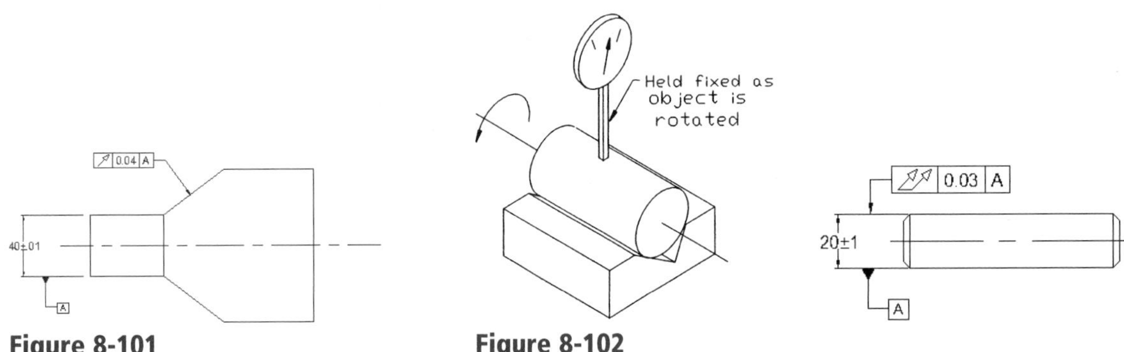

Figure 8-101

Figure 8-102

Figure 8-103 shows a circular runout tolerance that references two datums. The two datums serve as one datum. The object can then be rotated about both datums simultaneously as the runout tolerances are checked.

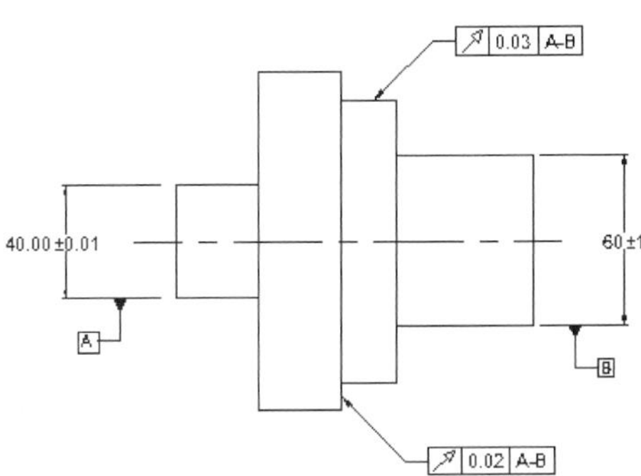

Figure 8-103

Positional Tolerances

As defined earlier, *positional tolerances* are used to locate and tolerance holes. Positional tolerances create a circular tolerance zone for hole center point locations, in contrast with the rectangular tolerance zone created by linear coordinate dimensions. See Figure 8-104. The circular tolerance zone allows for an increase in acceptable tolerance variation without compromising the design integrity of the object. Note that some of the possible hole center points fall in an area outside the rectangular tolerance zone but are still within the circular tolerance zone. If the hole had been located using linear coordinate dimensions, center points located beyond the rectangular tolerance zone would have been rejected as beyond tolerance, and yet holes produced using these locations would function correctly from a design standpoint. The center point locations would be acceptable if positional tolerances had been specified. The finished hole is round, so a round

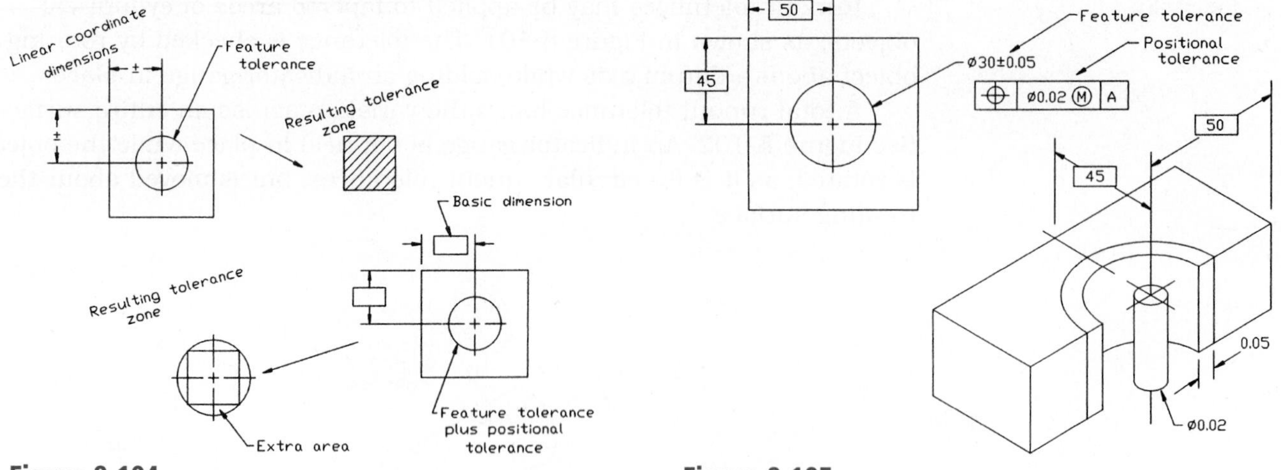

Figure 8-104

Figure 8-105

tolerance zone is appropriate. The rectangular tolerance zone rejects some holes unnecessarily.

Holes are dimensioned and toleranced using geometric tolerances by a combination of locating dimensions, feature dimensions and tolerances, and positional tolerances. See Figure 8-105. The locating dimensions are enclosed in rectangular boxes and are called *basic dimensions.* Basic dimensions are assumed to be exact.

The feature tolerances for the hole are as presented earlier in the chapter. They can be presented using plus and minus or limit-type tolerances. In the example shown in Figure 8-105 the diameter of the hole is toleranced using a ±0.05 tolerance.

The basic locating dimensions of 45 and 50 are assumed to be exact. The tolerances that would normally accompany linear locational dimensions are replaced by the positional tolerance. The positional tolerance also specifies that the tolerance be applied at the centerline at maximum material condition. The resulting tolerance zones are as shown in Figure 8-105.

Figure 8-106 shows an object containing two holes that are dimensioned and toleranced using positional tolerances. There are two consecutive horizontal basic dimensions. Because basic dimensions are exact, they

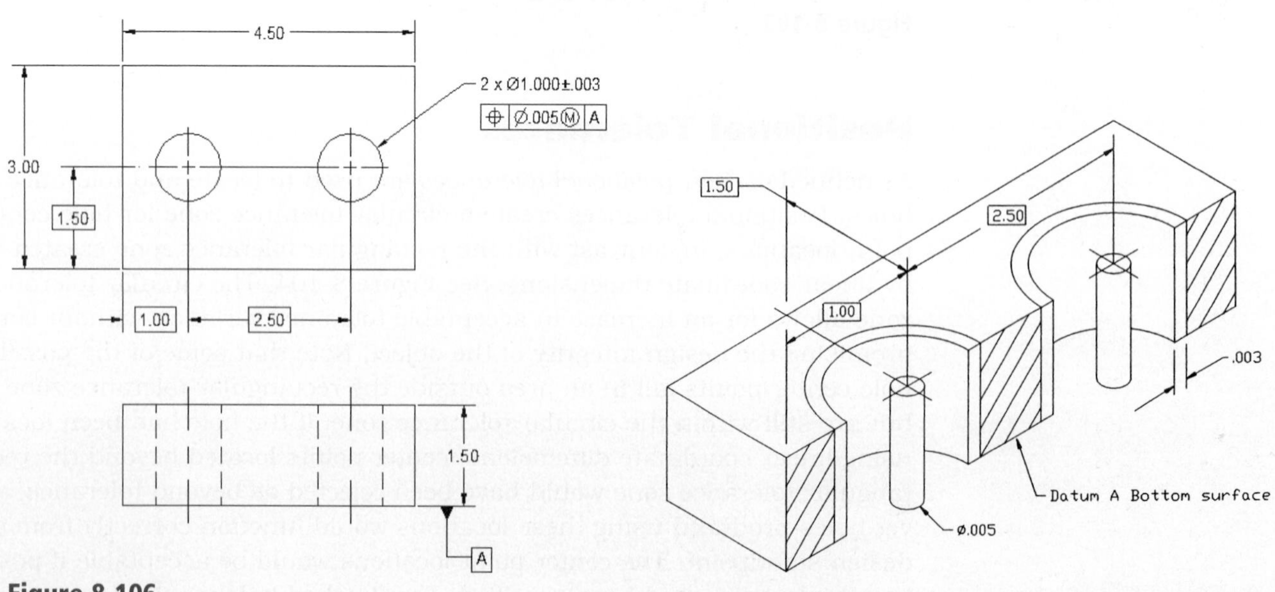

Figure 8-106

do not have tolerances that accumulate; that is, there is no tolerance buildup.

> **TIP**
> Geometric positional tolerances must include basic dimensions.

Virtual Condition

virtual condition: The combination of a feature's maximum material condition and its geometric tolerance.

Virtual condition is a combination of a feature's MMC and its geometric tolerance. For external features (shafts) it is the MMC plus the geometric tolerance; for internal features (holes) it is the MMC minus the geometric tolerance.

The following calculations are based on the dimensions shown in Figure 8-107.

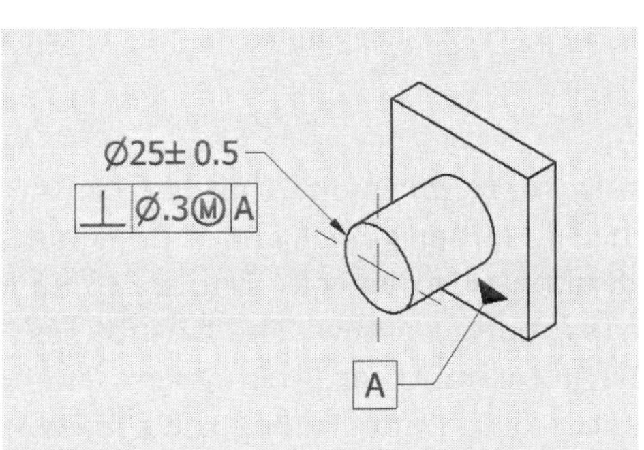

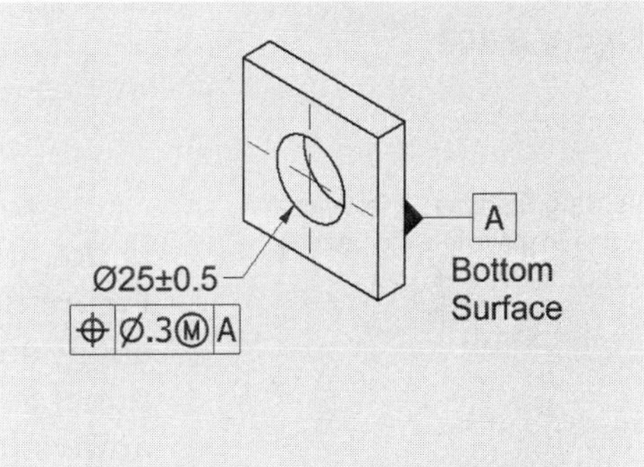

Figure 8-107

Calculating the Virtual Condition for a Shaft

25.5 MMC for shaft—maximum diameter
+0.3 Geometric tolerance
25.8 Virtual condition

Calculating the Virtual Condition for a Hole

24.5 MMC for hole—minimum diameter
−0.3 Geometric tolerance
24.2 Virtual condition

Floating Fasteners

Positional tolerances are particularly helpful when dimensioning matching parts. Because basic locating dimensions are considered exact, the sizing of mating parts is dependent only on the hole and shaft's MMC and the geometric tolerance between them.

The relationship for floating fasteners and holes in objects may be expressed as a formula:

$$H - T = F$$

where:

H = hole at MMC
T = geometric tolerance
F = shaft at MMC

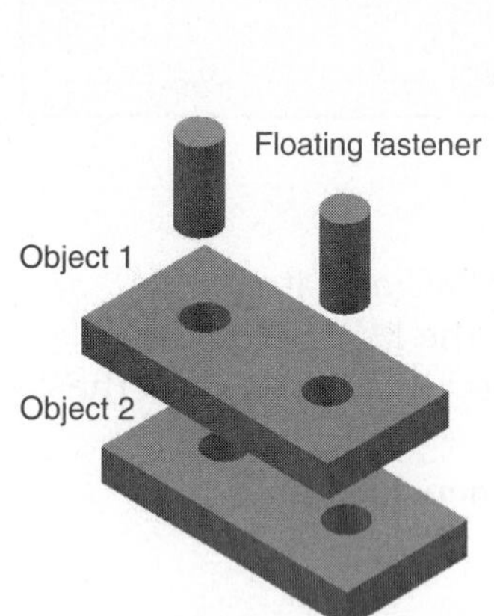

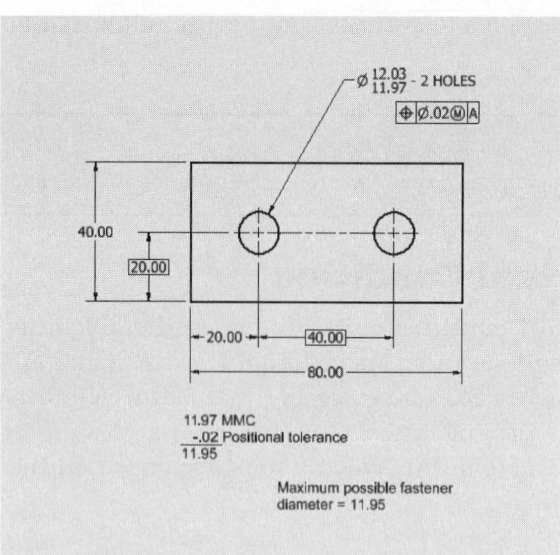

SIZE	TOLERANCE ZONE
11.97 MMC	.02
11.98	.03.
11.99	.04
12.00	.05
12.01	.06
12.02	.07
12.03 LMC	.08

Figure 8-108

floating fastener: A fastener that is free to move in either mating object.

A ***floating fastener*** is one that is free to move in either object. It is not attached to either object and it does not screw into either object. Figure 8-108 shows two objects that are to be joined by a common floating shaft, such as a bolt or screw. The feature size and tolerance and the positional geometric tolerance are both given. The minimum size hole that will always just fit is determined using the preceding formula.

$$H - T = F$$
$$11.97 - .02 = 11.95$$

Therefore, the shaft's diameter at MMC, the shaft's maximum diameter, equals 11.95. Any required tolerance would have to be subtracted from this shaft size.

The .02 geometric tolerance is applied at the hole's MMC, so as the hole's size expands within its feature tolerance, the tolerance zone for the acceptable matching parts also expands.

Sample Problem SP8-3

The situation presented in Figure 8-108 can be worked in reverse; that is, hole sizes can be derived from given shaft sizes.

The two objects shown in Figure 8-109 are to be joined by a .250-in. bolt. The parts are floating; that is, they are both free to move, and the fastener is not joined to either object. What is the MMC of the holes if the positional tolerance is to be .030?

A manufacturer's catalog specifies that the tolerance for .250 bolts is .2500 to .2600.

Rewriting the formula

$$H - T = F$$

to isolate the H yields

$$\begin{aligned} H &= F + T \\ &= .260 + .030 \\ &= .290 \end{aligned}$$

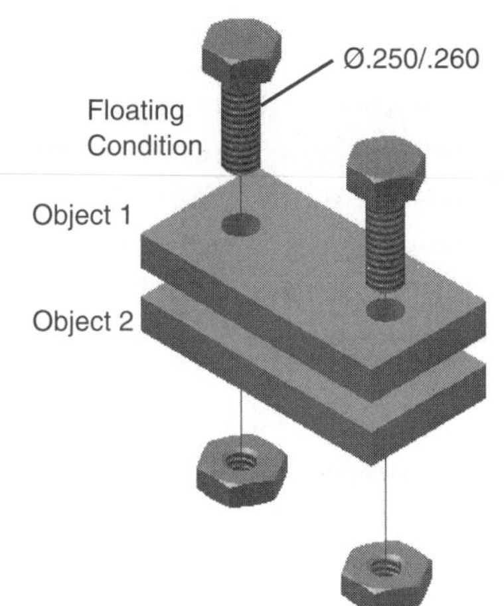

Figure 8-109

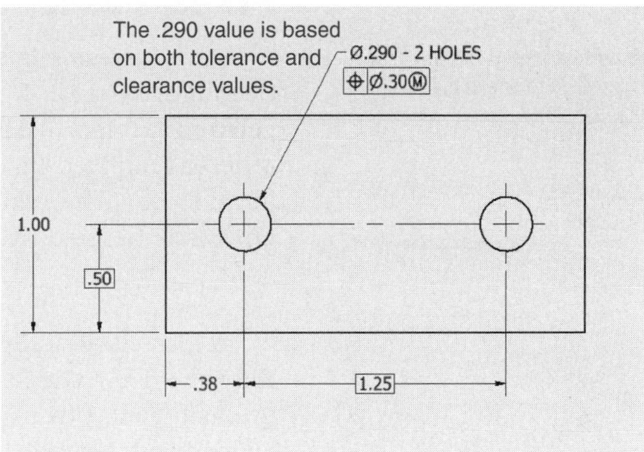

Figure 8-110

The .290 value represents the minimum hole diameter, MMC, for all four holes that will always accept the .250 bolt. Figure 8-110 shows the resulting drawing callout.

Any clearance requirements or tolerances for the hole would have to be added to the .290 value.

Sample Problem SP8-4

Repeat the problem presented in sample problem SP8-3 but be sure that there is always a minimum clearance of .002 between the hole and the shaft, and assign a hole tolerance of .008.

Sample problem SP8-3 determined that the maximum hole diameter that will always accept the .250 bolt is .290 based on the .030 positioning tolerance. If the minimum clearance is to be .002, the maximum hole diameter is found as follows:

.290 Minimum hole diameter that will always accept the bolt (0 clearance at MMC)
+.002 Minimum clearance
.292 Minimum hole diameter including clearance

Now, assign the tolerance to the hole:

.292 Minimum hole diameter
+.001 Tolerance
.293 Maximum hole diameter

See Figure 8-111 for the appropriate drawing callout. The choice of clearance size and hole tolerance varies with the design requirements for the objects.

Figure 8-111

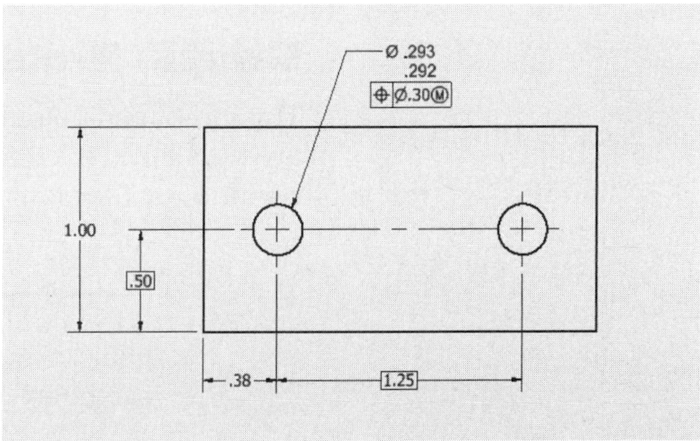

Fixed Fasteners

fixed fastener: A fastener that is attached to one of the mating objects.

A ***fixed fastener*** is one that is attached to one of the mating objects. See Figure 8-112. Because the fastener is fixed to one of the objects, the geometric tolerance zone must be smaller than that used for floating fasteners. The fixed fastener cannot move without moving the object it is attached to. The relationship between fixed fasteners and holes in mating objects is defined by the formula

$$H - 2T = F$$

This relationship can be demonstrated by the objects shown in Figure 8-113. The same feature sizes that were used in Figure 8-113 are assigned, but in this example the fasteners are fixed. Solving for the geometric tolerance yields a value as follows:

$$\begin{aligned} H - F &= 2T \\ 11.97 - 11.95 &= 2T \\ .02 &= 2T \\ .01 &= T \end{aligned}$$

The resulting positional tolerance is half that obtained for floating fasteners.

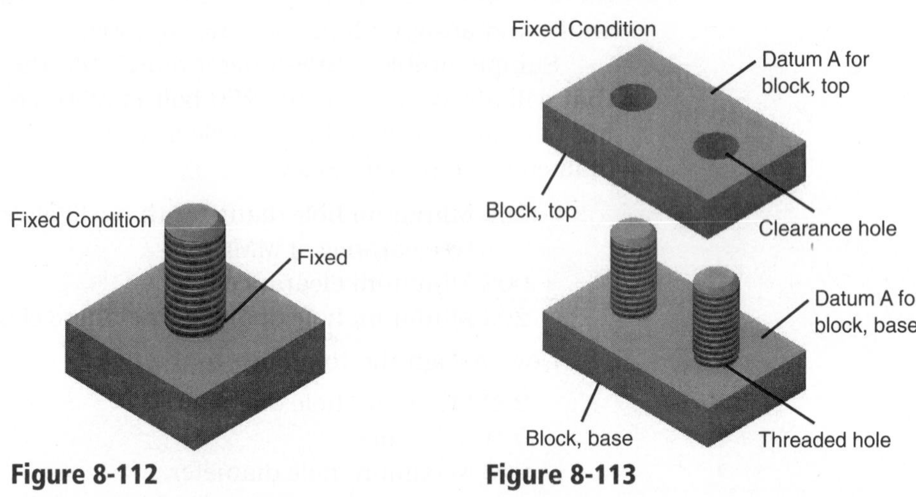

Figure 8-112 **Figure 8-113**

Sample Problem SP8-5

This problem is similar to sample problem SP8-3, but the given conditions are applied to fixed fasteners rather than floating fasteners. Compare the resulting shaft diameters for the two problems. See Figure 8-114.

A. What is the minimum diameter hole that will always accept the fixed fasteners?

B. If the minimum clearance is .005 and the hole is to have a tolerance of .002, what are the maximum and minimum diameters of the hole?

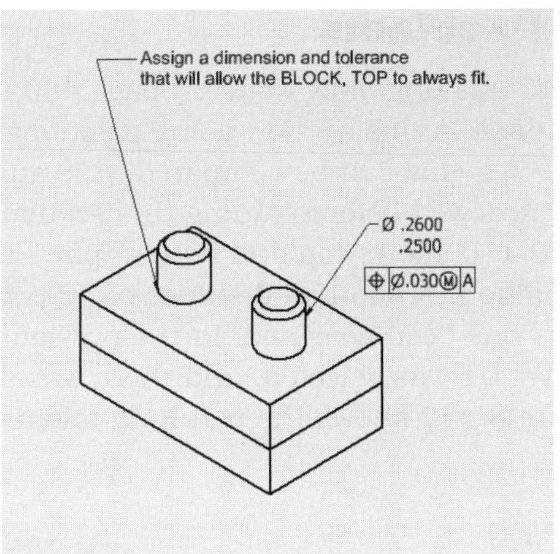

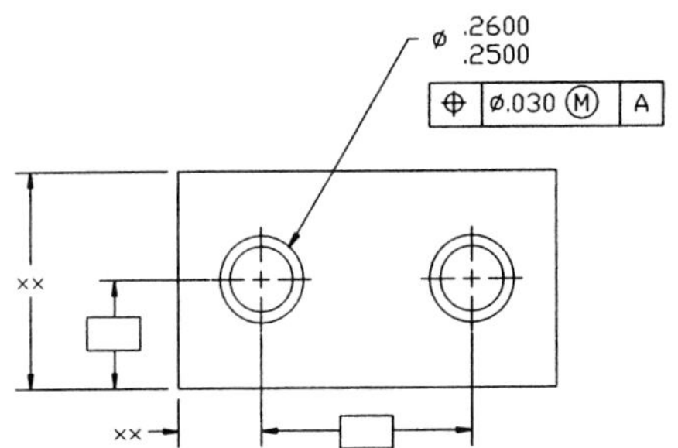

Figure 8-114

$$H - 2T = F$$
$$H = F + 2T$$
$$= .260 + 2(.030)$$
$$= .260 + .060$$
$$= .320 \text{ Minimum diameter that will always accept the fastener}$$

If the minimum clearance is .005 and the hole tolerance is .002,

.320 Virtual condition
+.005 Clearance
.325 Minimum hole diameter

.325 Minimum hole diameter
+.002 Tolerance
.327 Maximum hole diameter

The maximum and minimum values for the hole's diameter can then be added to the drawing of the object that fits over the fixed fasteners. See Figure 8-115.

Figure 8-115

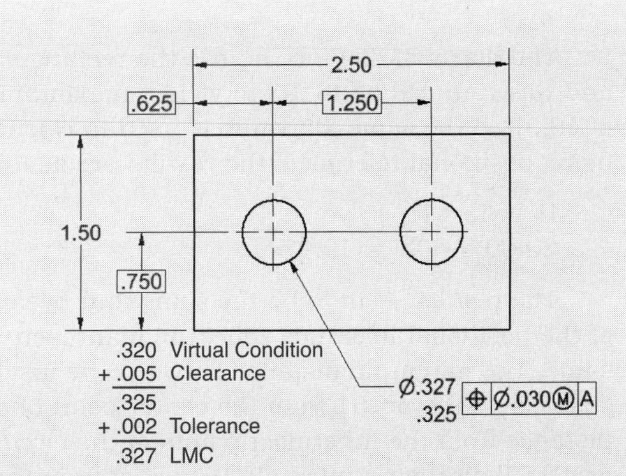

Design Problems

This problem was originally done on page 480 using rectangular tolerances. It is done in this section using positional geometric tolerances so that the two systems can be compared. It is suggested that the previous problem be reviewed before reading this section.

Figure 8-116 shows top and bottom parts that are to be joined in the floating condition. A nominal distance of 50 between hole centers and 20 for the holes has been assigned. In the previous solution a rectangular tolerance of ±.01 was selected, and there was a minimum hole diameter of 20.00. Figure 8-117 shows the resulting tolerance zones.

Figure 8-116

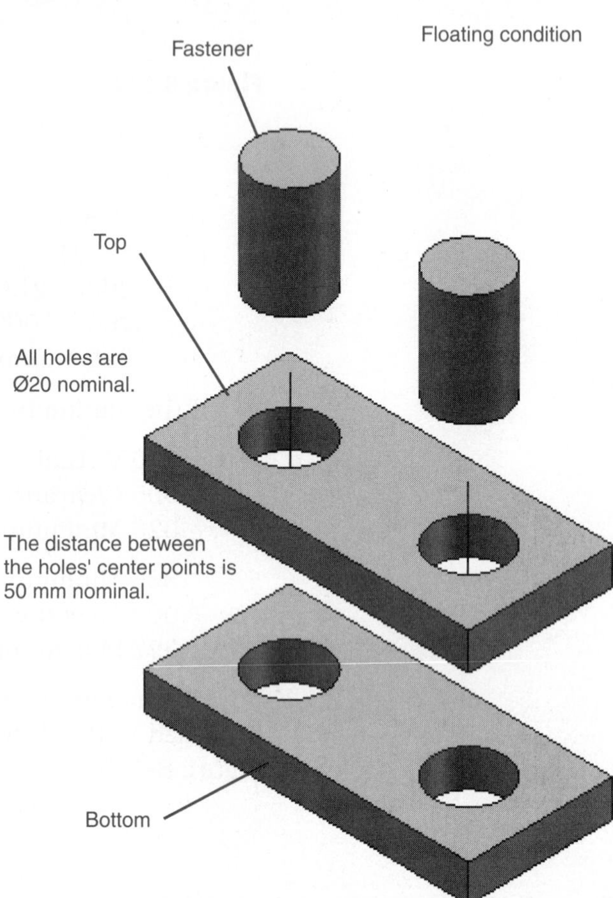

Floating condition
Fastener
Top
All holes are
Ø20 nominal.
The distance between
the holes' center points is
50 mm nominal.
Bottom

The diagonal distance across the rectangular tolerance zone is .028 and was rounded off to .03 to yield a maximum possible fastener diameter of 19.97. If the same .03 value is used to calculate the fastener diameter using positional tolerance, the results are as follows:

$$H - T = F$$
$$20.00 - .03 = 19.97$$

The results seem to be the same, but because of the circular shape of the positional tolerance zone, the manufactured results are not the same. The minimum distance between the inside edges of the rectangular zones is 49.98, or .01 from the center point of each hole. The minimum distance from the innermost points of the circular tolerance zones is 49.97, or .015 (half the rounded-off .03 value) from the center point of each hole. The same value difference also occurs for the maximum distance between

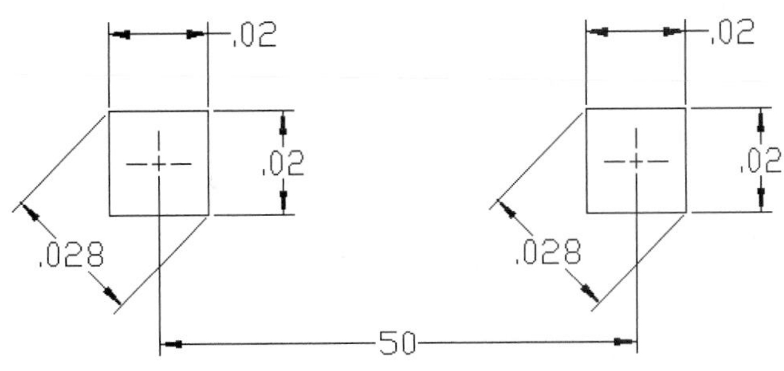

Rectangular range: 49.98 to 50.02
Circular range: 49.97 to 50.03

These crescent-shaped areas account for the increased tolerance range of the circular tolerances.

This increased area of acceptability is the result of assigning the positional tolerance at MMC.

.02
.02
.028

Figure 8-117

center points, where 50.02 is the maximum distance for the rectangular tolerances, and 50.03 is the maximum distance for the circular tolerances. The size of the circular tolerance zone is larger because the hole tolerances are assigned at MMC. Figure 8-117 shows a comparison between the tolerance zones, and Figure 8-118 shows how the positional tolerances would be presented on a drawing of either the top or bottom part.

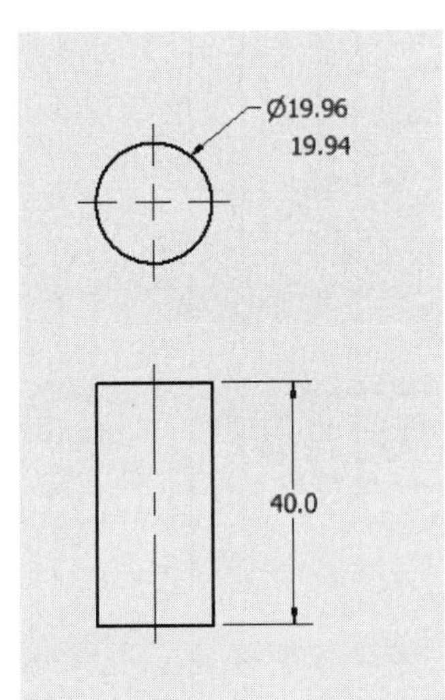

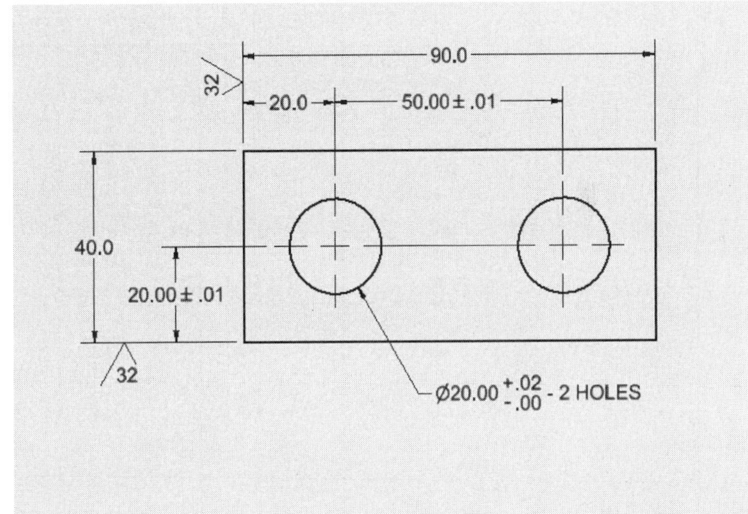

Figure 8-118

Figure 8-119 shows the same top and bottom parts joined together in the fixed condition. The initial nominal values are the same. If the same .03 diagonal value is assigned as a positional tolerance, the results are as follows:

$$H - 2T = F$$
$$20.00 - .06 = 19.94$$

These results appear to be the same as those generated by the rectangular tolerance zone, but the circular tolerance zone allows a greater variance in acceptable manufactured parts. Figure 8-120 shows how the positional tolerance would be presented on a drawing.

Figure 8-119

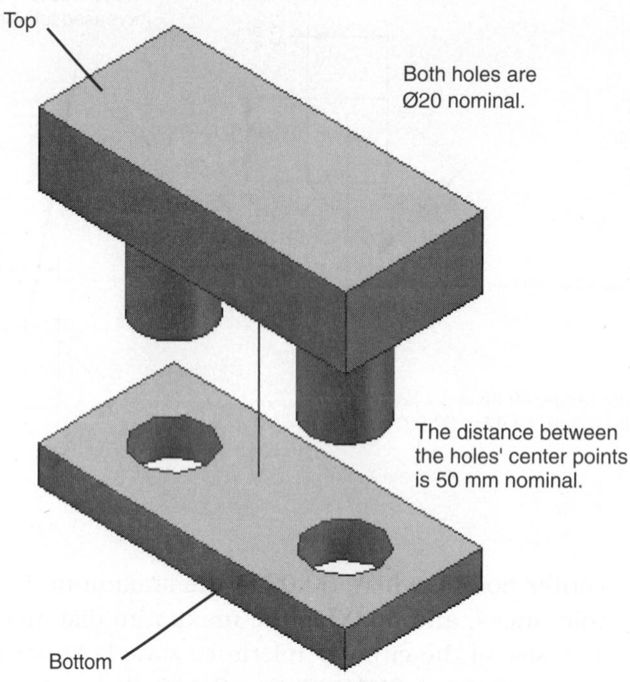

Fixed condition
Top
Both holes are
Ø20 nominal.
The distance between
the holes' center points
is 50 mm nominal.
Bottom

Figure 8-120

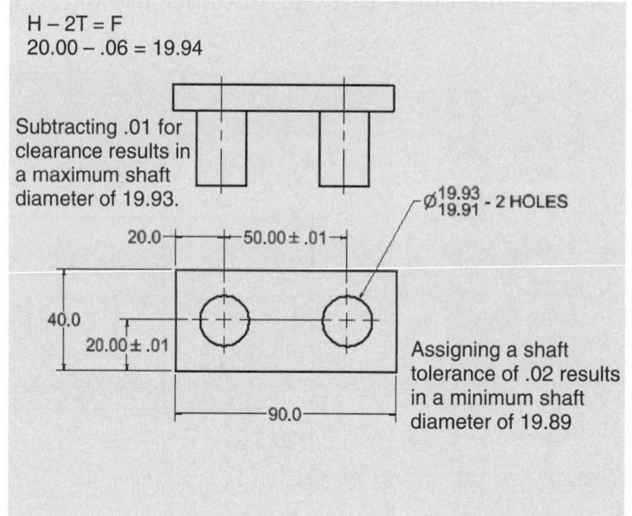

H – 2T = F
20.00 – .06 = 19.94
Subtracting .01 for
clearance results in
a maximum shaft
diameter of 19.93.
Ø19.93 19.91 - 2 HOLES
20.0
50.00 ± .01
40.0
20.00 ± .01
90.0
Assigning a shaft
tolerance of .02 results
in a minimum shaft
diameter of 19.89

Chapter Summary

Tolerances define the manufacturing limits for dimensions. This chapter defined the various tolerance conventions and illustrated how to create plus and minus tolerances and limit tolerances, including angular tolerances. The two systems for applying dimensions and tolerances to a drawing—chain and baseline—were illustrated. Sample problems explained the application of positional tolerances to holes and shafts in determining fits.

Surface finishes and the use of surface control symbols were also discussed and illustrated.

Examples were given of the various forms of geometric tolerancing—defining surfaces in terms of their flatness or roundness or in terms of how perpendicular or parallel they are to other surfaces. The four classifications of tolerances of form—flatness, straightness, roundness, and cylindricity—were also illustrated. Sample problems involving both fixed and floating fasteners demonstrated the use of tolerances.

Chapter Test Questions

Multiple Choice

Circle the correct answer.

1. In what type of fit is the shaft always smaller than the hole?

a. Interference fit

b. Clearance fit

c. Transition fit

2. Which of the following is *not* a type of tolerance?

a. Limits

b. Symmetric

c. Deviation

d. Reference

3. An RC fit is which of the following?

a. Running and sliding fit

b. Clearance fit

c. Force fit

d. Interference fit

4. In which type of condition are fasteners assembled into a part and then inserted through (clearance) a second part?

a. Floating condition

b. Fixed condition

c. Absolute condition

5. Flatness and straightness are considered tolerances of

a. Profile

b. Runout

c. Form

d. MMC

6. What is a datum?
   a. A reference point or surface
   b. The product of a series of tolerances
   c. A listing of data points
7. Positional tolerances are used to
   a. Locate a part
   b. Locate and tolerance a hole
   c. Define the angle of a part's feature
8. The equation H − T = F is used to define the relationship between holes, fasteners, and tolerances in which condition?
   a. Floating
   b. Fixed
   c. Absolute
   d. Loose running
9. Tolerances of orientation are used to
   a. Relate a feature to the XY axis
   b. Relate a surface to a machine mark
   c. Relate a feature or surface to another feature or surface
10. How are a primary datum, a secondary datum, and a tertiary datum generally referenced on a drawing?
   a. 1, 2, and 3
   b. A, B, and C
   c. X, Y, and Z
   d. Front, top, and side

Matching

Figure 8-121 shows an object dimensioned and toleranced using geometric tolerances. Match the numbers with the appropriate definitions.

Figure 8-121

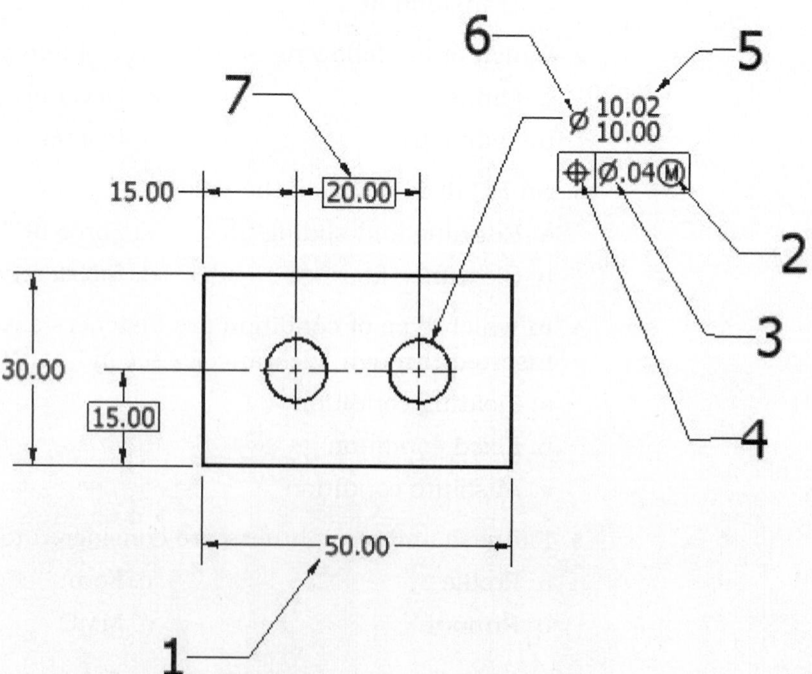

Column A

1. ______
2. ______
3. ______
4. ______
5. ______
6. ______
7. ______

Column B

a. Symbol for diameter
b. Stacked limit tolerances
c. Overall dimension
d. Basic dimension
e. Symbol for positional tolerance
f. Symbol for centerline
g. Symbol for maximum material condition

True or False

Circle the correct answer.

1. **True or False:** There are three types of fits: clearance, location, and interference.
2. **True or False:** A tolerance written using the ± symbol is called either symmetric or deviation.
3. **True or False:** Positional tolerances require the use of basic dimensions.
4. **True or False:** For an interference fit the shaft is smaller than the hole.
5. **True or False:** Fit tolerances are listed on the **Design** panel.
6. **True or False:** English units specify surface finishes using microinches.
7. **True or False:** MMC is an abbreviation for maximum material condition.
8. **True or False:** Every dimension value on a drawing always has a tolerance.
9. **True or False:** Chain dimensions are the same as baseline dimensions.
10. **True or False:** In the floating condition, fasteners pass through clearance holes.

Chapter Projects

Project 8-1

Draw a model of the objects shown in Figures P8-1A through P8-1D using the given dimensions and tolerances. Create a drawing layout with a view of the model as shown. Add the specified dimensions and tolerances.

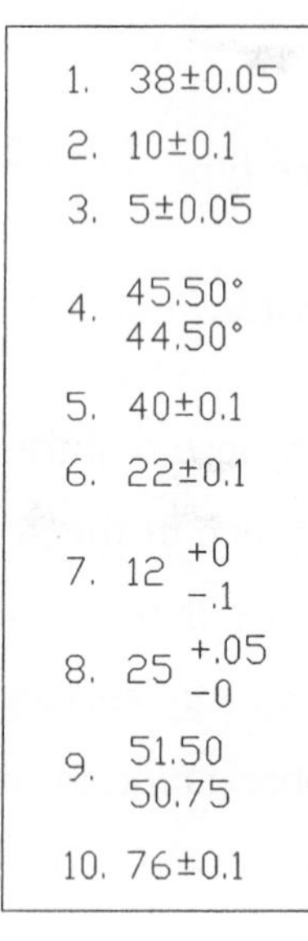

1. 38±0.05
2. 10±0.1
3. 5±0.05
4. 45.50° / 44.50°
5. 40±0.1
6. 22±0.1
7. 12 +0 / −.1
8. 25 +.05 / −0
9. 51.50 / 50.75
10. 76±0.1

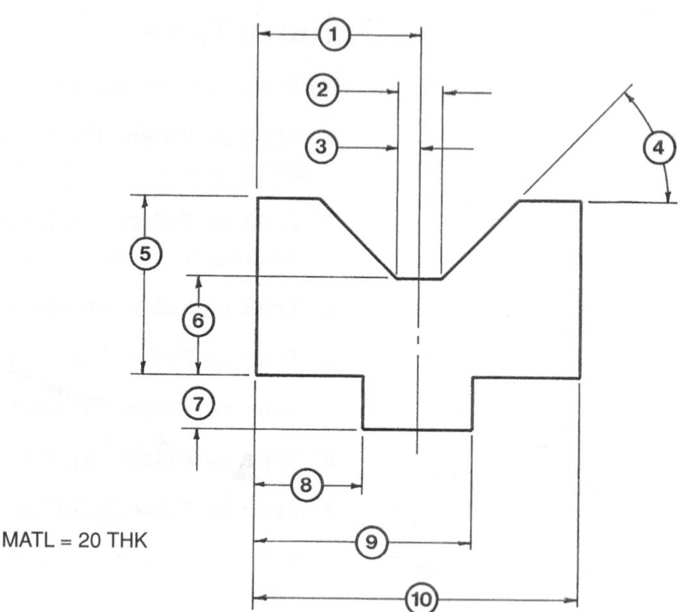

Figure P8-1A
MILLIMETERS

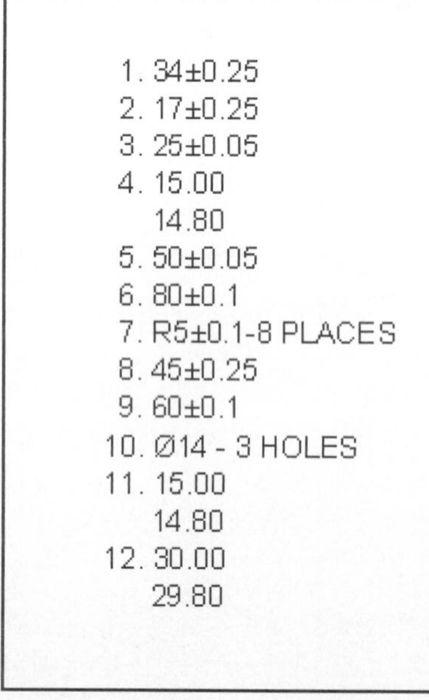

1. 34±0.25
2. 17±0.25
3. 25±0.05
4. 15.00 / 14.80
5. 50±0.05
6. 80±0.1
7. R5±0.1-8 PLACES
8. 45±0.25
9. 60±0.1
10. Ø14 - 3 HOLES
11. 15.00 / 14.80
12. 30.00 / 29.80

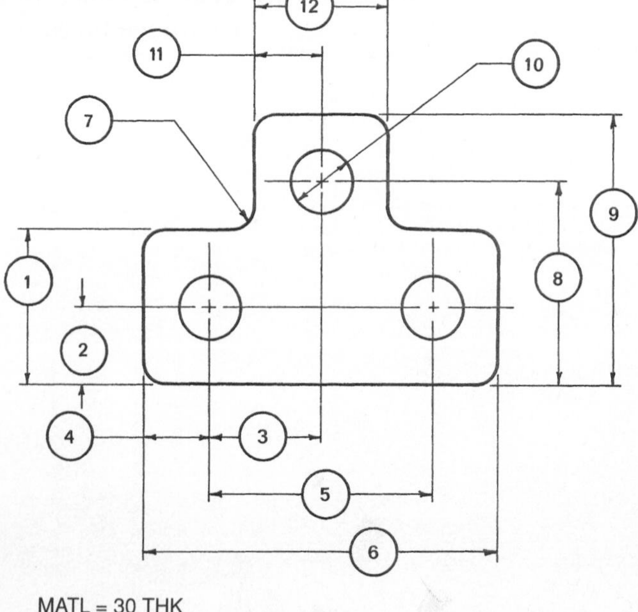

Figure P8-1B
MILLIMETERS

1. 3.00±.01
2. 1.56±.01
3. 46.50° / 45.50°
4. .750±.005
5. 2.75 / 2.70
6. 3.625±.010
7. 45°±.5°
8. 2.250±.005

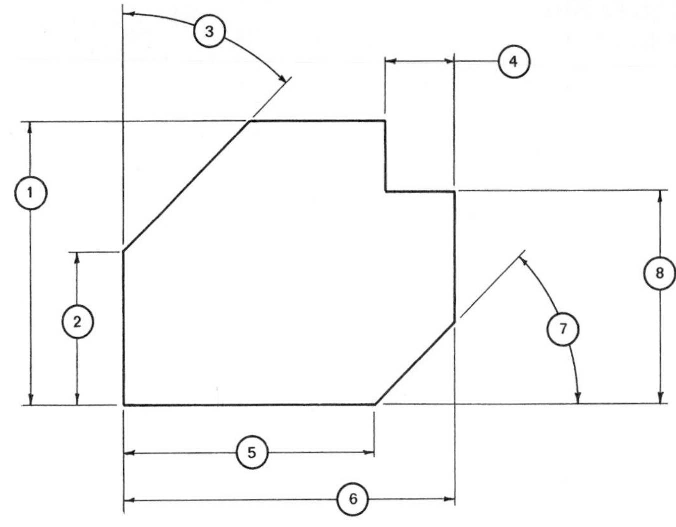

MATL = .75 THK

Figure P8-1C
INCHES

1. 50 $^{+.2}_{0}$
2. R45±.1 - 2 PLACES
3. 63.5 $^{0}_{-.2}$
4. 76±.1
5. 38±.1
6. Ø12.00 $^{+.05}_{0}$ - 3 HOLES
7. 30±.03
8. 30±.03
9. 100 $^{+.4}_{0}$

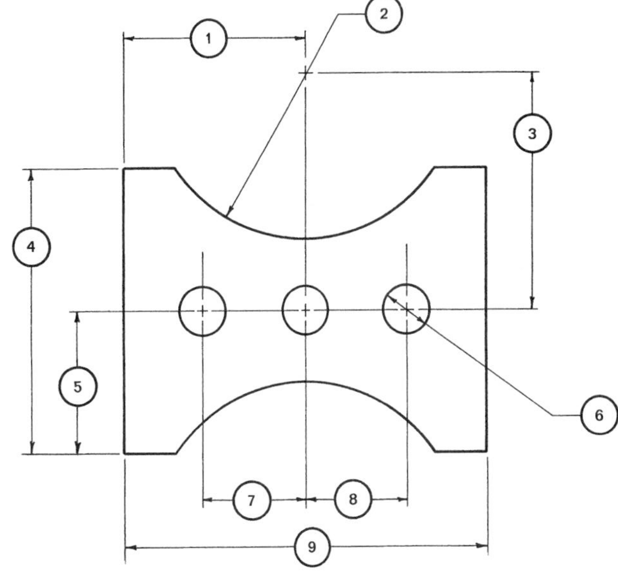

MATL = 10 THK

Figure P8-1D
MILLIMETERS

Project 8-2

Redraw the following object, including the given dimensions and tolerances. Calculate and list the maximum and minimum distances for surface A.

Figure P8-2
MILLIMETERS

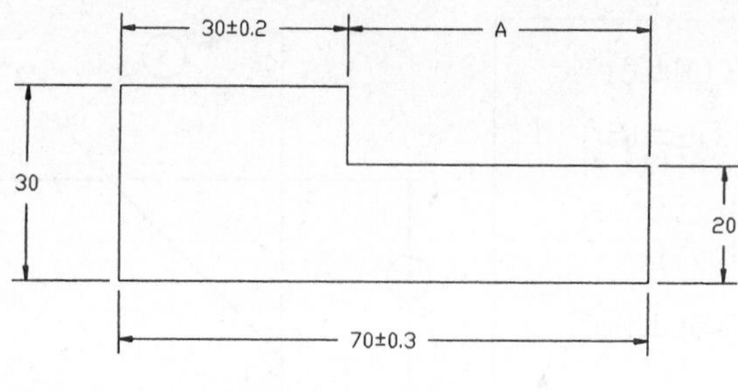

Project 8-3

A. Redraw the following object, including the dimensions and tolerances. Calculate and list the maximum and minimum distances for surface A.
B. Redraw the given object and dimension it using baseline dimensions. Calculate and list the maximum and minimum distances for surface A.

Figure P8-3
INCHES

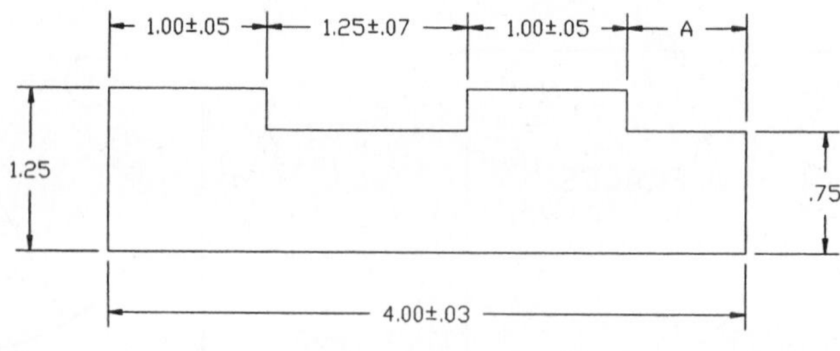

Project 8-4

Redraw the following object, including the dimensions and tolerances. Calculate and list the maximum and minimum distances for surfaces D and E.

Figure P8-4
MILLIMETERS

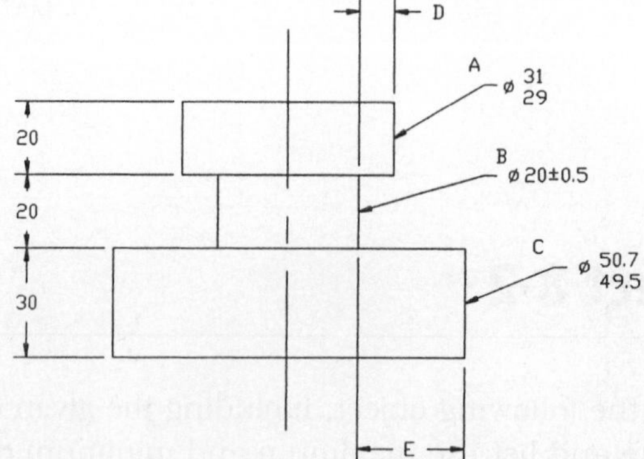

Project 8-5

Dimension the following object twice, once using chain dimensions and once using baseline dimensions. Calculate and list the maximum and minimum distances for surface D for both chain and baseline dimensions. Compare the results.

Figure P8-5
MILLIMETERS

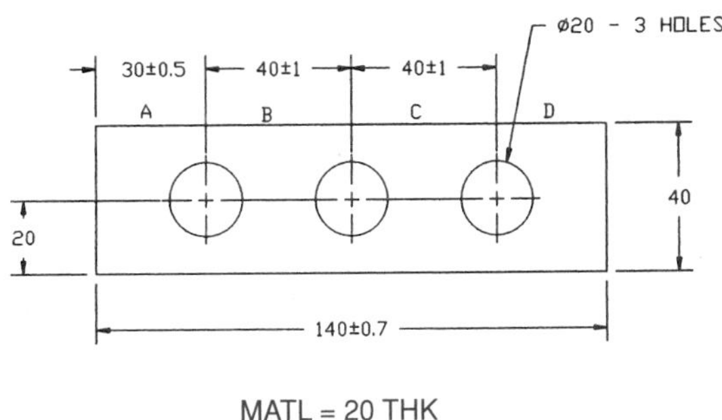

Project 8-6

Redraw the following shapes, including the dimensions and tolerances. Also list the required minimum and maximum values for the specified distances.

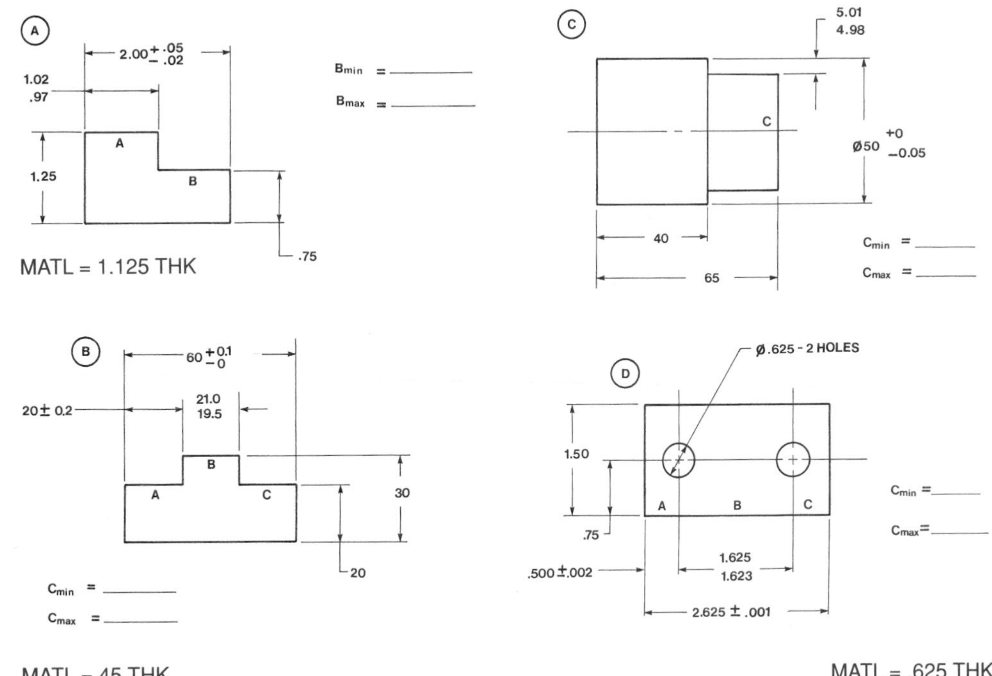

Figure P8-6
INCHES

Project 8-7

Redraw and complete the following inspection report. Under the RESULTS column classify each "AS MEASURED" value as OK if the value is within the stated tolerances, REWORK if the value indicates that the measured value is beyond the stated tolerance but can be reworked to bring it into the acceptable range, or SCRAP if the value is not within the tolerance range and cannot be reworked to make it acceptable.

Figure P8-7
MILLIMETERS

INSPECTION REPORT

PART NAME AND NO: 10755002

INSPECTOR:

DATE:

BASE DIMENSION	TOLERANCES MAX	TOLERANCES MIN	AS MEASURED	RESULTS
(1) 100 ± 0.5			99.8	
(2) $\phi\, ^{57}_{56}$			57.01	
(3) 22 ± 0.3			21.72	
(4) $^{40.05}_{39.95}$			39.98	
(5) 22 ± 0.3			21.68	
(6) $R52^{+0}_{-0.2}$			51.99	
(7) $35^{+0.2}_{-0.3}$			35.20	
(8) $30^{+0.4}_{0}$			30.27	
(9) $6.0^{+.1}_{-.2}$			5.85	
(10) 12.0 ± 0.2			11.90	

1.00 3 PLACES

.50 — 10 PLACES

Project 8-8

Redraw the following charts and complete them based on the following information. All values are in millimeters.

A. Nominal = 16, Fit = H8/d8
B. Nominal = 30, Fit = H11/c11
C. Nominal = 22, Fit = H7/g6
D. Nominal = 10, Fit = C11/h11
E. Nominal = 25, Fit = F8/h7
F. Nominal = 12, Fit = H7/k6
G. Nominal = 3, Fit = H7/p6
H. Nominal = 18, Fit = H7/s6
I. Nominal = 27, Fit = H7/u6
J. Nominal = 30, Fit = N7/h6

Figure P8-8
MILLIMETERS

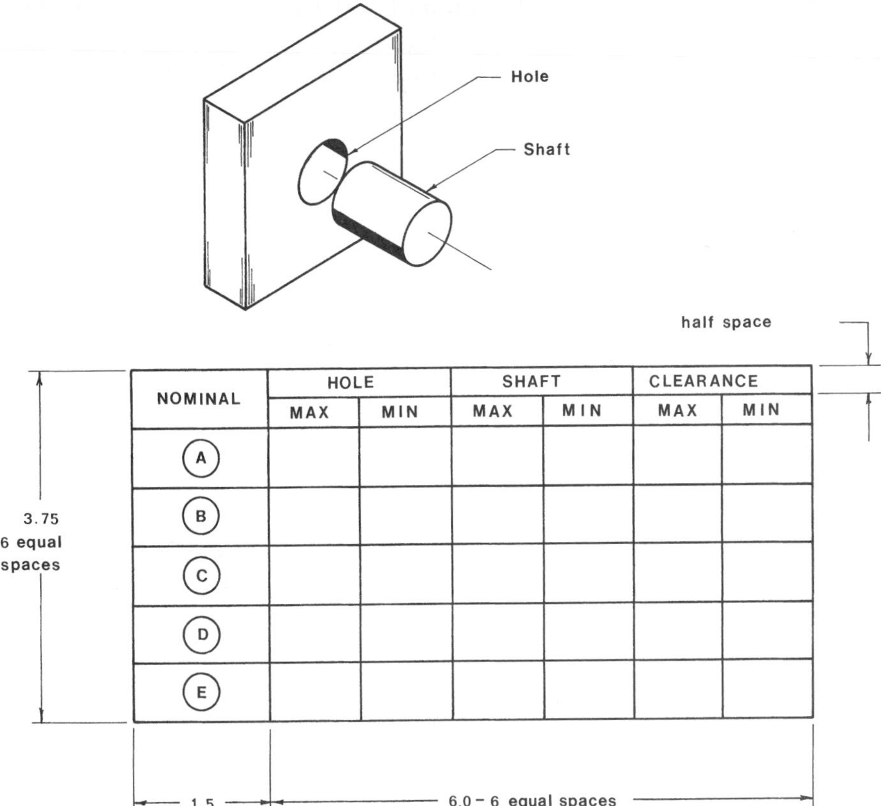

NOMINAL	HOLE		SHAFT		CLEARANCE	
	MAX	MIN	MAX	MIN	MAX	MIN
A						
B						
C						
D						
E						

NOMINAL	HOLE		SHAFT		INTERFERENCE	
	MAX	MIN	MAX	MIN	MAX	MIN
F						
G						
H						
I						
J						

Use the same dimensions given above

Project 8-9

Redraw the following charts and complete them based on the following information. All values are in inches.

A. Nominal = 0.25, Fit = Class LC5, H7/g6
B. Nominal = 1.00, Fit = Class LC7, H10/e9
C. Nominal = 1.50, Fit = Class LC9, F11/h11

D. Nominal = 0.75, Fit = Class RC3, H7/f6
E. Nominal = 1.75, Fit = Class RC6, H9/e8
F. Nominal = .500, Fit = Class LT2, H8/js7
G. Nominal = 1.25, Fit = Class LT5, H7/n6
H. Nominal = 1.38, Fit = Class LN3, J7/h6
I. Nominal = 1.625, Fit = Class FN, H7/s6
J. Nominal = 2.00, Fit = Class FN4, H7/u6

Figure P8-9
INCHES

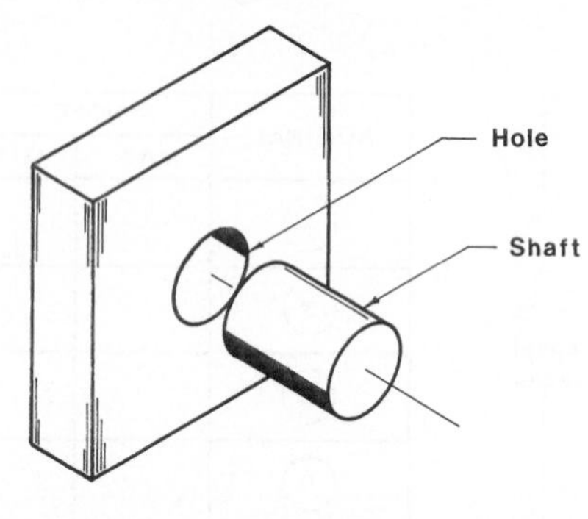

half space

NOMINAL	HOLE		SHAFT		CLEARANCE	
	MAX	MIN	MAX	MIN	MAX	MIN
(A)						
(B)						
(C)						
(D)						
(E)						

3.75
6 equal
spaces

1.5

6.0 – 6 equal spaces

NOMINAL	HOLE		SHAFT		INTERFERENCE	
	MAX	MIN	MAX	MIN	MAX	MIN
(F)						
(G)						
(H)						
(I)						
(J)						

Use the same dimensions given above

Project 8-10

Draw the chart shown and add the appropriate values based on the dimensions and tolerances given in Figures P8-10A through P8-10D.

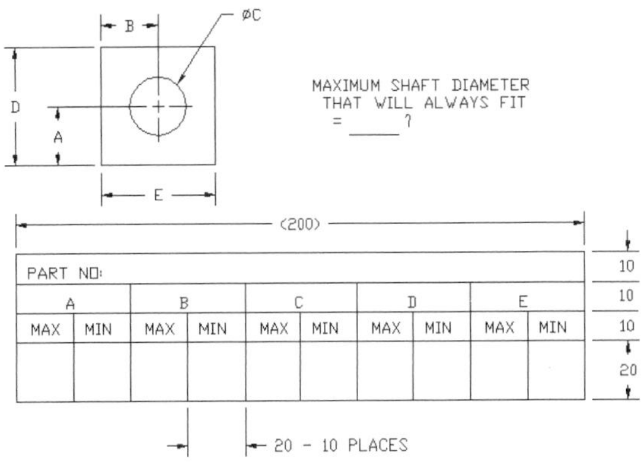

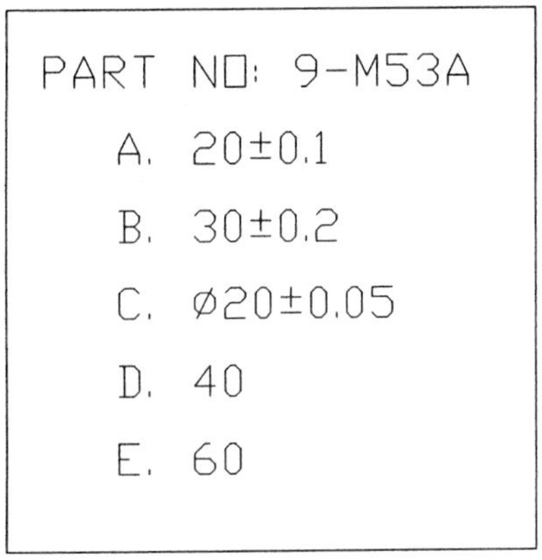

MAXIMUM SHAFT DIAMETER THAT WILL ALWAYS FIT = ____ ?

PART NO:									
A		B		C		D		E	
MAX	MIN	MAX	MIN	MAX	MIN	MAX	MIN	MAX	MIN

(200) — 20 – 10 PLACES

PART NO: 9-M53A

A. 20±0.1

B. 30±0.2

C. Ø20±0.05

D. 40

E. 60

Figure P8-10A
MILLIMETERS

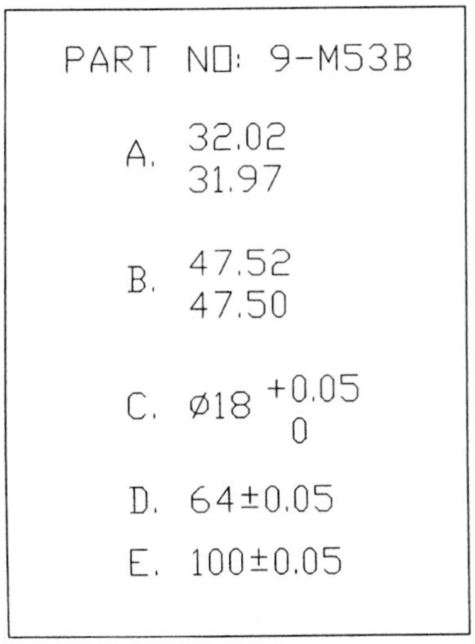

PART NO: 9-M53B

A. 32.02 / 31.97

B. 47.52 / 47.50

C. Ø18 $^{+0.05}_{0}$

D. 64±0.05

E. 100±0.05

Figure P8-10B
MILLIMETERS

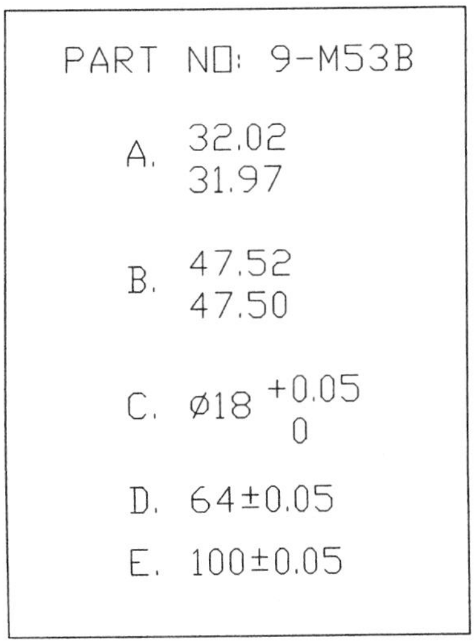

PART NO: 9-M53B

A. 32.02 / 31.97

B. 47.52 / 47.50

C. Ø18 $^{+0.05}_{0}$

D. 64±0.05

E. 100±0.05

Figure P8-10C
MILLIMETERS

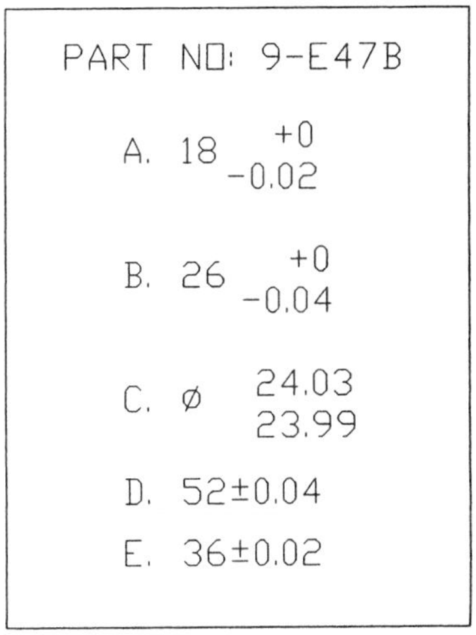

PART NO: 9-E47B

A. 18 $^{+0}_{-0.02}$

B. 26 $^{+0}_{-0.04}$

C. Ø 24.03 / 23.99

D. 52±0.04

E. 36±0.02

Figure P8-10D
MILLIMETERS

Project 8-11

Prepare front and top views of parts 4A and 4B based on the given dimensions. Add tolerances to produce the stated clearances.

Figure P8-11
MILLIMETERS

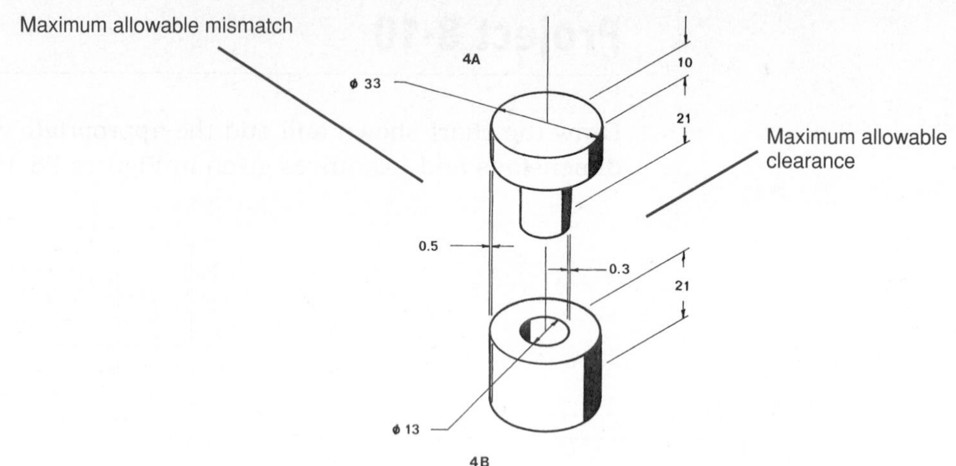

Project 8-12

Redraw parts A and B and dimensions and tolerances to meet the "UPON ASSEMBLY" requirements.

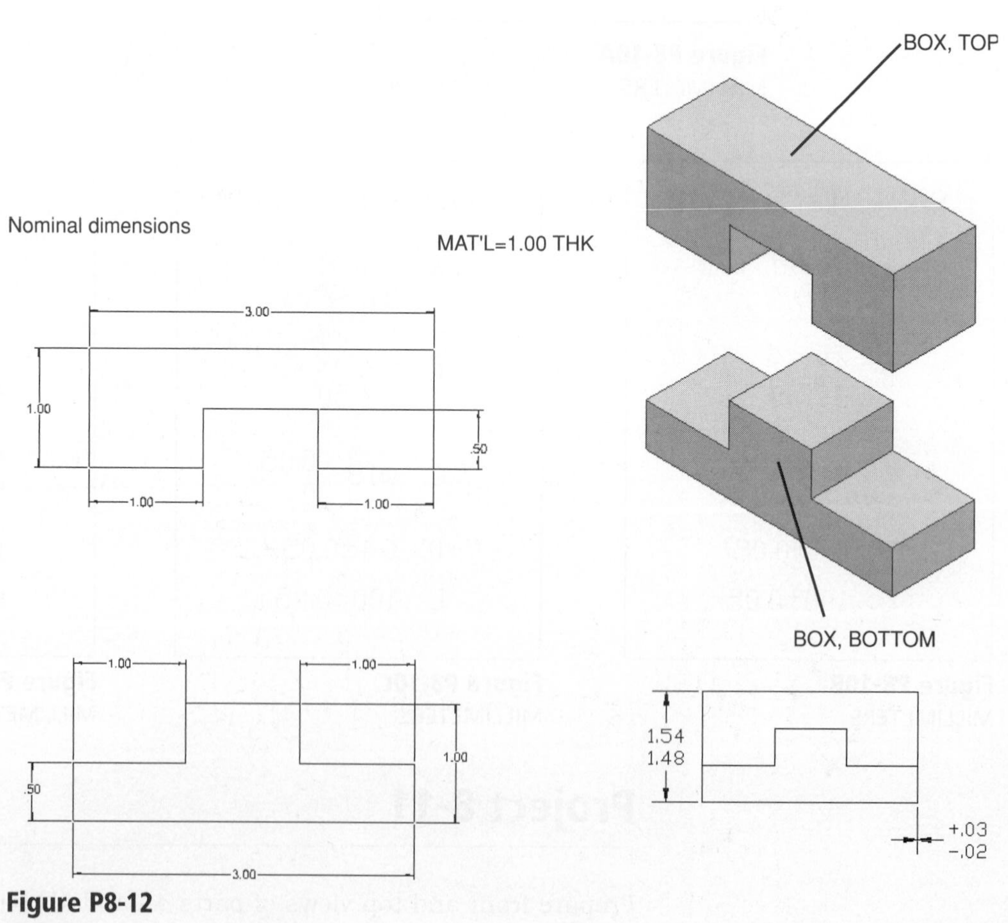

Figure P8-12
INCHES

Project 8-13

Draw a front and top view of both given objects. Add dimensions and tolerances to meet the “FINAL CONDITION” requirements.

Figure P8-13
MILLIMETERS

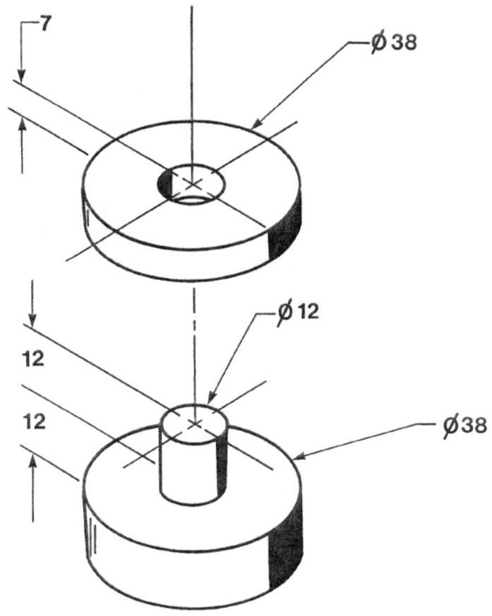

FINAL CONDITION

MAX =0.03
MIN =0.01

MIN =0.00
MAX =0.04

Project 8-14

Given the following nominal sizes, dimension and tolerance parts AM311 and AM312 so that they always fit together regardless of orientation. Further, dimension the overall lengths of each part so that in the assembled condition they will always pass through a clearance gauge with an opening of 80.00 ± 0.02.

Figure P8-14
MILLIMETERS

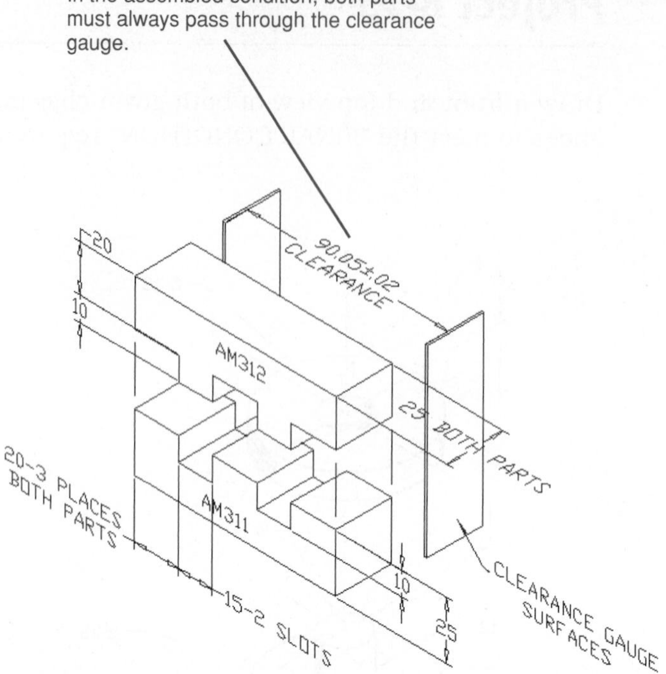

Project 8-15

Given the following rail assembly, add dimensions and tolerances so that the parts always fit together as shown in the assembled position.

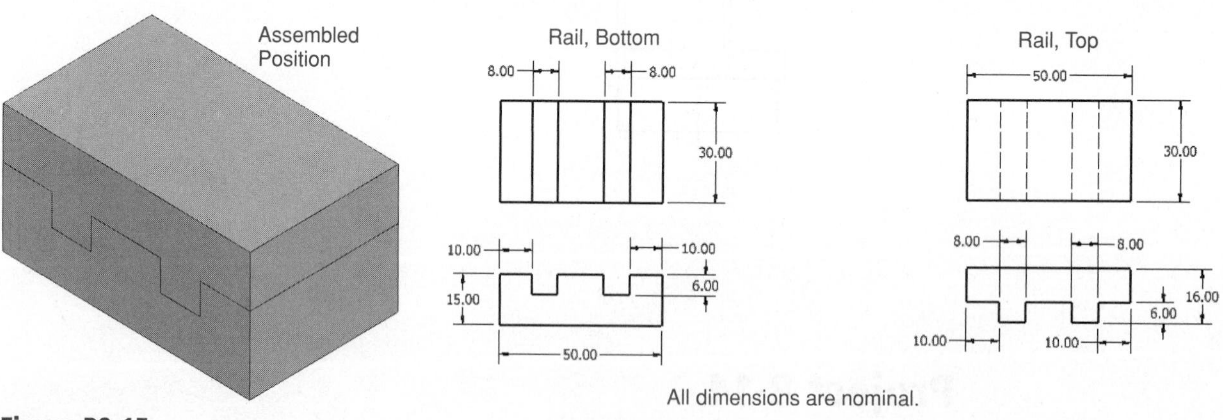

Figure P8-15
MILLIMETERS

Project 8-16

Given the following peg assembly, add dimensions and tolerances so that the parts always fit together as shown in the assembled position.

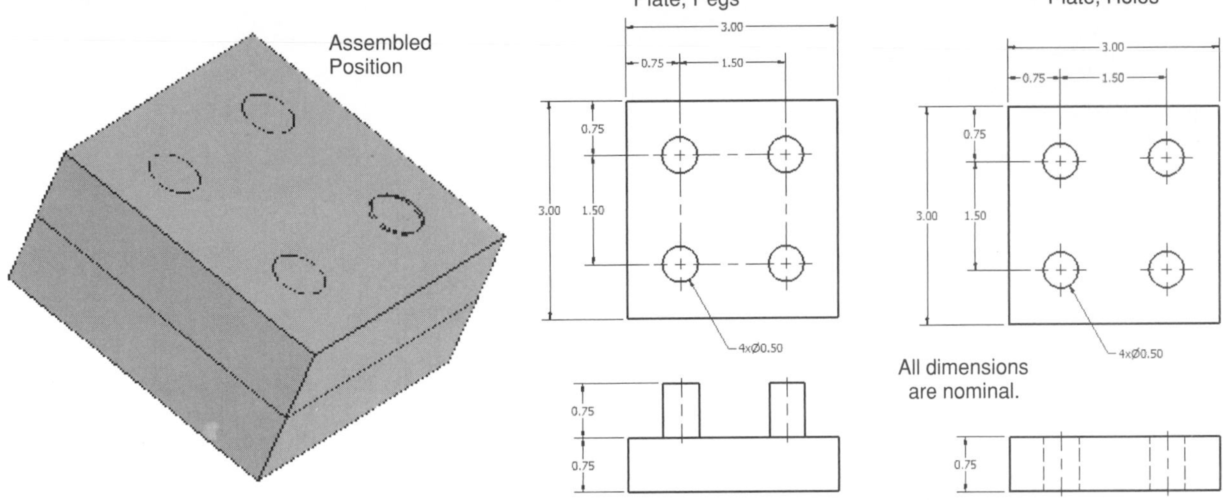

Figure P8-16
INCHES

Project 8-17

Given the following collar assembly, add dimensions and tolerances so that the parts always fit together as shown in the assembled position.

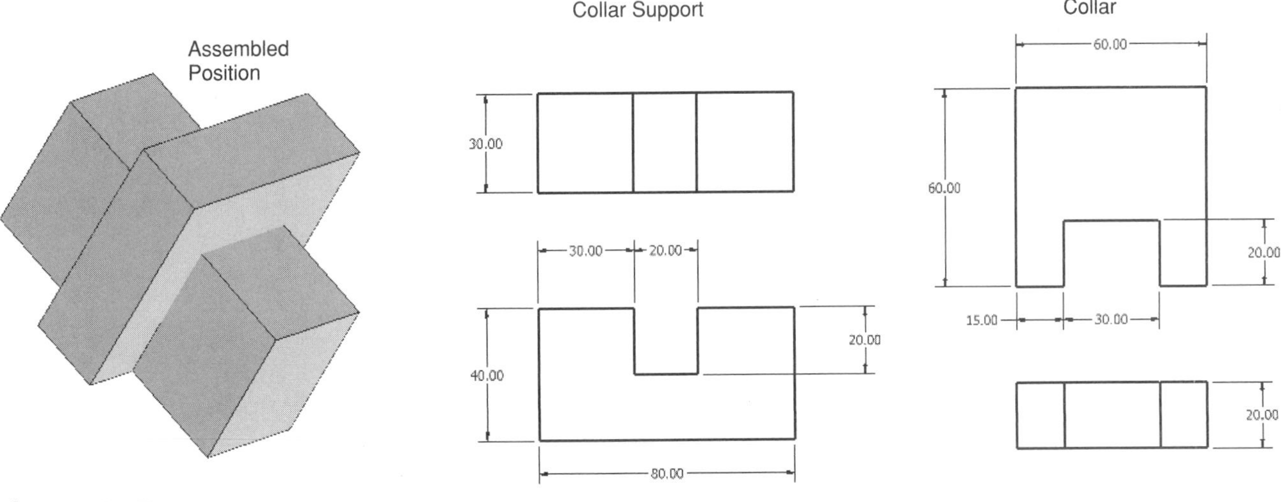

Figure P8-17
MILLIMETERS

Project 8-18

Given the following vee-block assembly, add dimensions and tolerances so that the parts always fit together as shown in the assembled position. The total height of the assembled blocks must be between 4.45 and 4.55 in.

Assembled
Position
Block, Vee
6.00
2.50
1.50
1.50
60° BOTH SIDES
1.50
3.00
Vee, Top
1.50
(3.00)
1.50
BOTH
SIDES
120° BOTH SIDES
1.50
All dimensions are nominal.
2.50
6.00

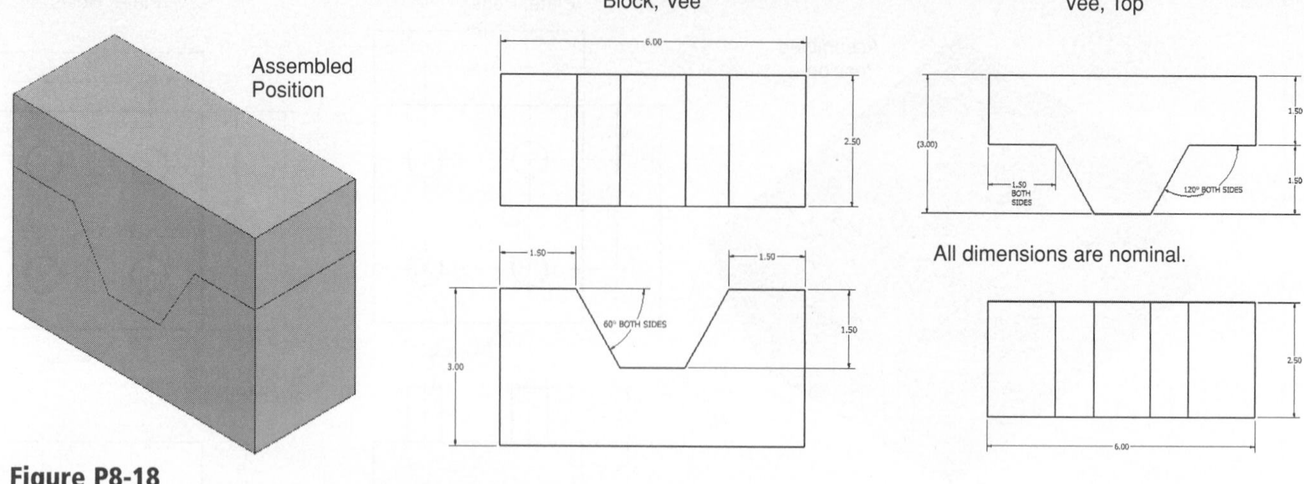

Figure P8-18
INCHES

Project 8-19

Design a bracket that will support the three Ø100 wheels shown. The wheels will utilize three Ø5.00 ± 0.01 shafts attached to the bracket. The bottom of the bracket must have a minimum clearance of 10 mm from the ground. The wall thickness of the bracket must always be at least 5 mm, and the minimum bracket opening must be at least 15 mm.

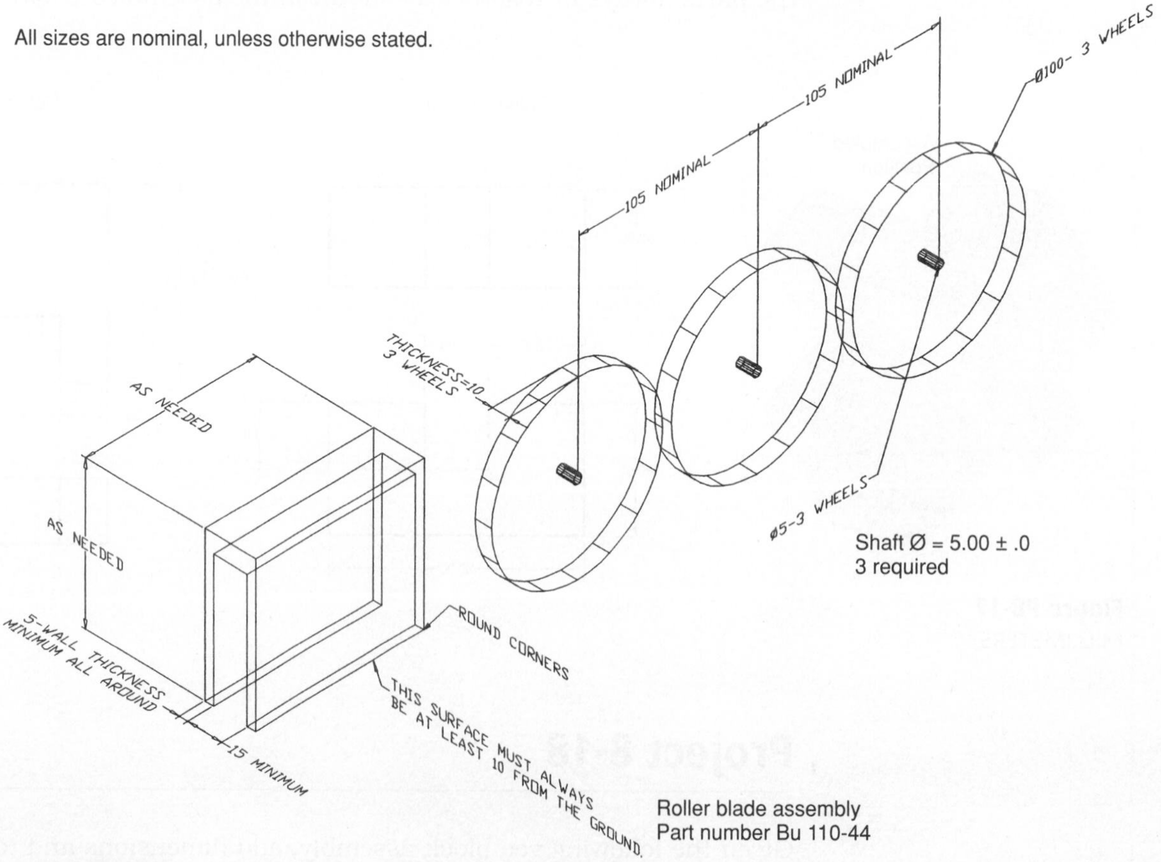

All sizes are nominal, unless otherwise stated.
105 NOMINAL
105 NOMINAL
Ø100- 3 WHEELS
THICKNESS=10
3 WHEELS
AS NEEDED
AS NEEDED
Ø5-3 WHEELS
Shaft Ø = 5.00 ± .0
3 required
5-WALL THICKNESS
MINIMUM ALL AROUND
ROUND CORNERS
THIS SURFACE MUST ALWAYS
BE AT LEAST 10 FROM THE GROUND.
15 MINIMUM
Roller blade assembly
Part number Bu 110-44

Figure P8-19A
MILLIMETERS

1. Prepare a front and a side view of the bracket.
2. Draw the wheels in their relative positions using phantom lines.
3. Add all appropriate dimensions and tolerances.

Given a TOP and a BOTTOM part in the floating condition as shown in Figure P8-19B, satisfy the requirements given in projects P8-20 through P8-23 so that the parts always fit together regardless of orientation. Prepare drawings of each part including dimensions and tolerances.

A. Use linear tolerances.
B. Use positional tolerances.

Figure P8-19B

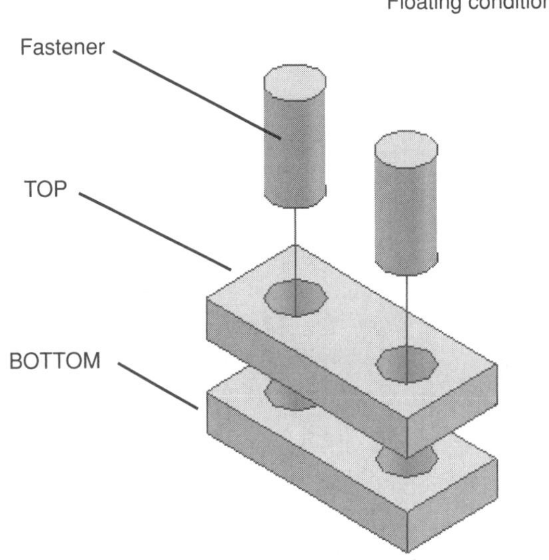

> **NOTE**
>
> The term *tolerance range* as used in the following projects indicates the total tolerance value. A tolerance range of .002 could be defined as
>
> ±.001
>
> +.002
> −.000
>
> .000
> −.002
>
> or by using limit tolerances

Project 8-20: Inches

A. The distance between the holes' center points is 2.00 nominal.
B. The holes are Ø.375 nominal.
C. The fasteners have a tolerance range of .001.
D. The holes have a tolerance range of .002.
E. The minimum allowable clearance between the fasteners and the holes is .003.
F. The positional tolerance range is .002.

Project 8-21: Millimeters

A. The distance between the holes' center points is 80 nominal.
B. The holes are Ø12 nominal.

C. The fasteners have a tolerance range of 0.05.
D. The holes have a tolerance range of 0.03.
E. The minimum allowable clearance between the fasteners and the holes is 0.02.
F. The positional tolerance range is 0.02.

Project 8-22: Inches

A. The distance between the holes' center points is 3.50 nominal.
B. The holes are Ø.625 nominal.
C. The fasteners have a tolerance range of .005.
D. The holes have a tolerance range of .003.
E. The minimum allowable clearance between the fasteners and the holes is .002.
F. The positional tolerance range is .004.

Project 8-23: Millimeters

A. The distance between the holes' center points is 65 nominal.
B. The holes are Ø16 nominal.
C. The fasteners have a tolerance range of 0.03.
D. The holes have a tolerance range of 0.04.
E. The minimum allowable clearance between the fasteners and the holes is 0.03.
F. The positional tolerance range is 0.02.

Given a top and a bottom part in the fixed condition as shown in Figure P8-23, satisfy the requirements given in projects P8-24 through P8-27 so that the parts fit together regardless of orientation. Prepare drawings of each part including dimensions and tolerances.

A. Use linear tolerances.
B. Use positional tolerances.

Figure P8-23

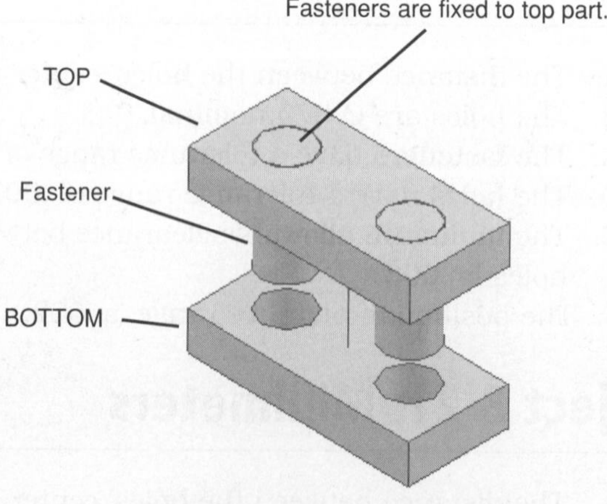

Project 8-24: Millimeters

A. The distance between the holes' center points is 60 nominal.
B. The holes are Ø10 nominal.
C. The fasteners have a tolerance range of 0.04.
D. The holes have a tolerance range of 0.02.
E. The minimum allowable clearance between the fasteners and the holes is 0.02.
F. The positional tolerance range is 0.02.

Project 8-25: Inches

A. The distance between the holes' center points is 3.50 nominal.
B. The holes are Ø.563 nominal.
C. The fasteners have a tolerance range of .005.
D. The holes have a tolerance range of .003.
E. The minimum allowable clearance between the fasteners and the holes is .002.
F. The positional tolerance range is .002.

Project 8-26: Millimeters

A. The distance between the holes' center points is 100 nominal.
B. The holes are Ø18 nominal.
C. The fasteners have a tolerance range of 0.02.
D. The holes have a tolerance range of 0.01.
E. The minimum allowable clearance between the fasteners and the holes is 0.03
F. The positional tolerance range is 0.02.

Project 8-27: Inches

A. The distance between the holes' center points is 1.75 nominal.
B. The holes are Ø.250 nominal.
C. The fasteners have a tolerance of .002.
D. The holes have a tolerance of .003.
E. The minimum allowable clearance between the fasteners and the holes is .001.

Project 8-28: Millimeters

Dimension and tolerance the rotator assembly shown in Figure P8-28. Use the given dimensions as nominal and add sleeve bearings between the links and both the cross-link and the plate. Create drawings of each part. Modify the dimensions as needed and add the appropriate tolerances. Specify the selected sleeve bearing.

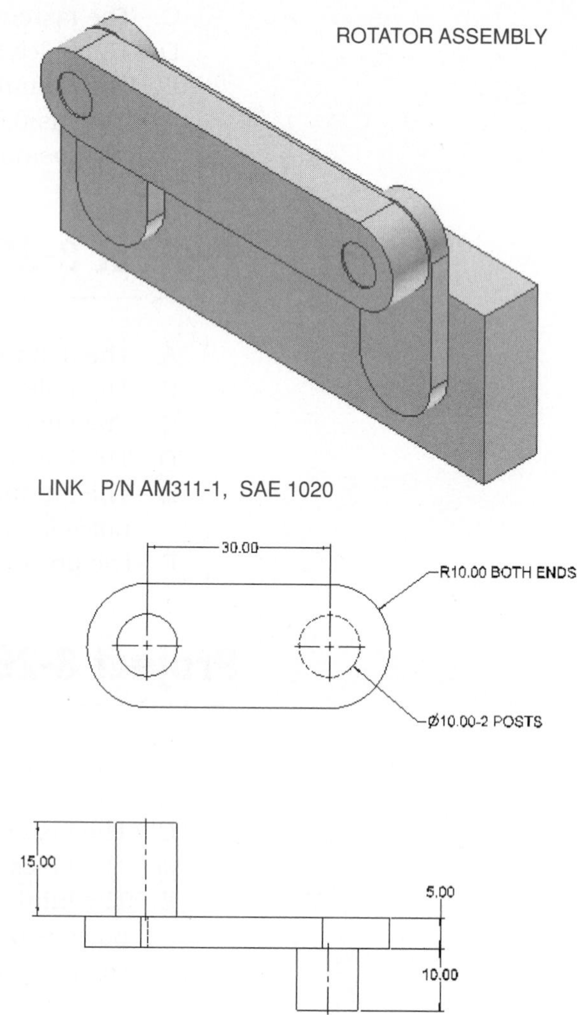

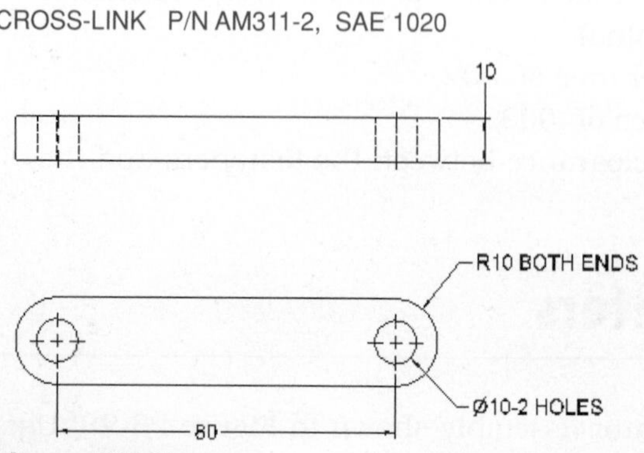

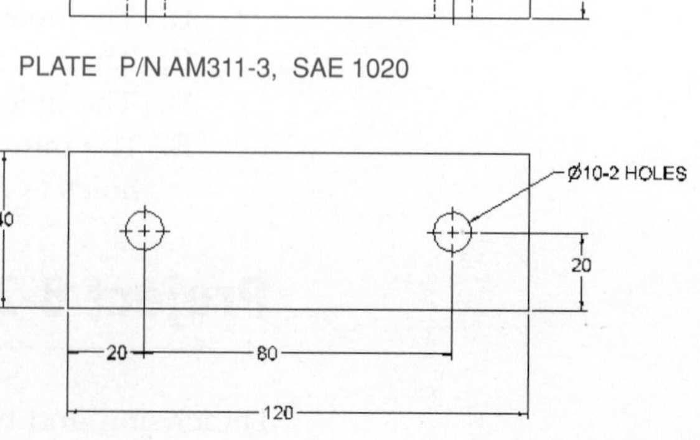

Figure P8-28

Project 8-29

Dimension and tolerance the rocker assembly shown in Figure P8-29. Use the given dimensions as nominal, and add sleeve bearings between all moving parts. Create drawings of each part. Modify the dimensions as needed and add the appropriate tolerances. Specify the selected sleeve bearing.

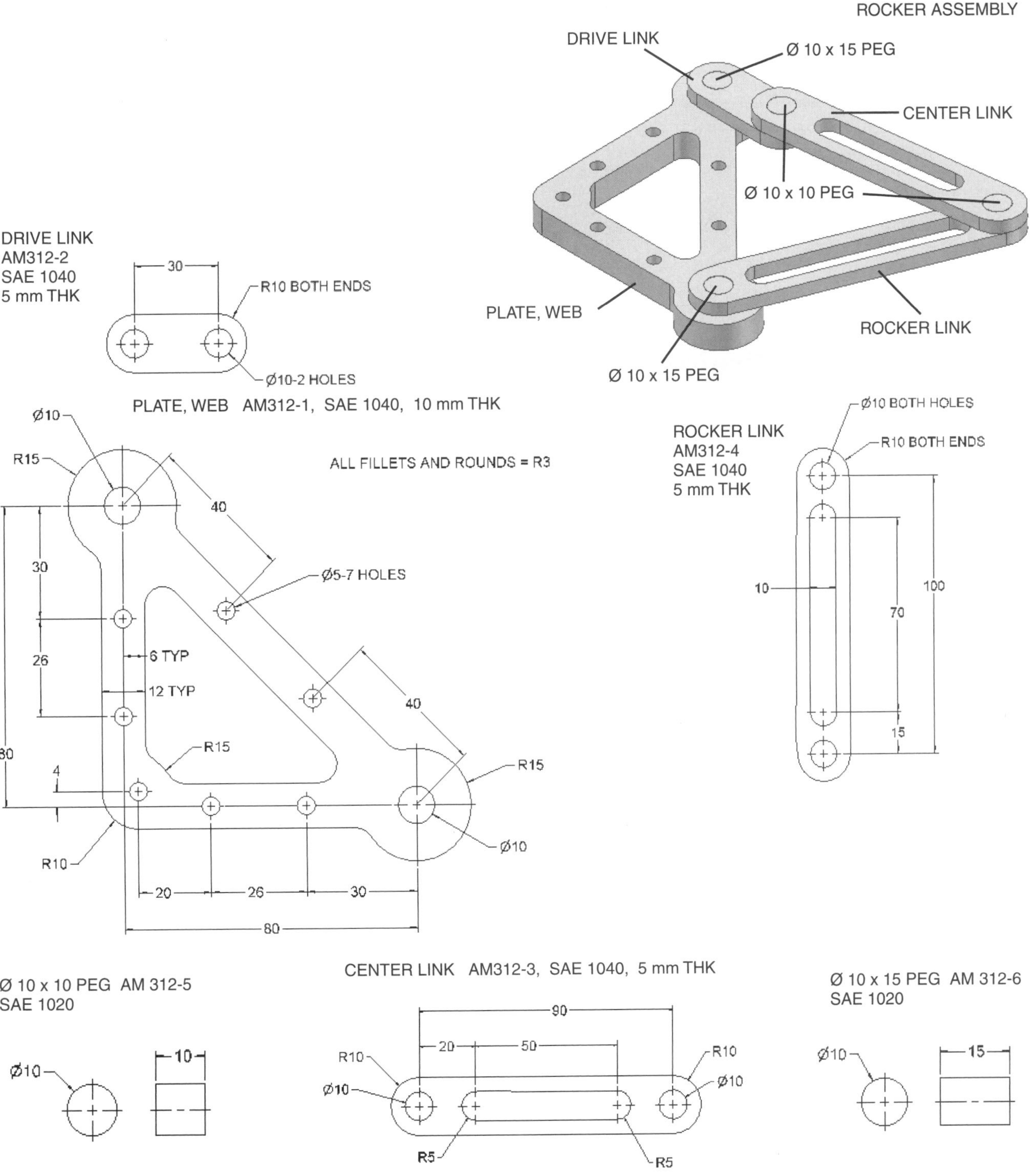

Figure P8-29

Project 8-30

Draw the model shown in Figure P8-30, create a drawing layout with the appropriate views, and add the specified dimensions and tolerances.

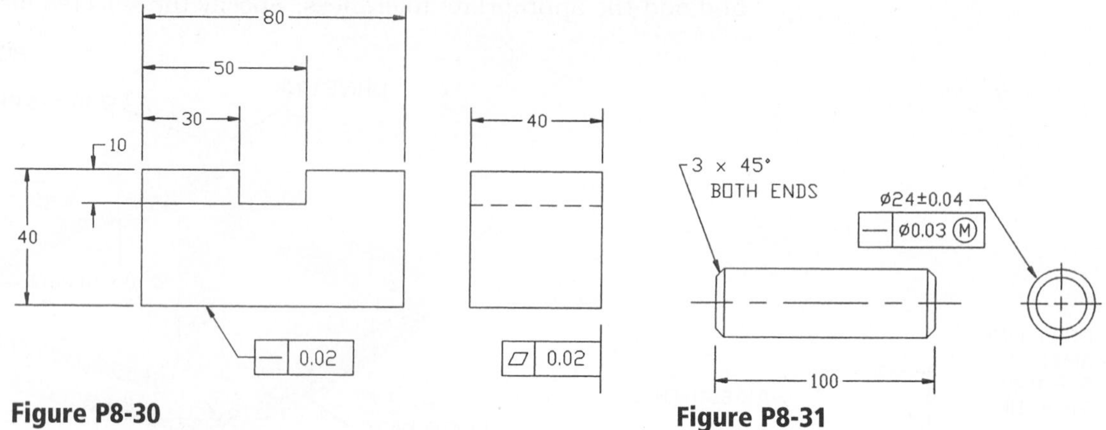

Figure P8-30

Figure P8-31

Project 8-31

Redraw the shaft shown in Figure P8-31, create a drawing layout with the appropriate views, and add a feature dimension and tolerance of 36 ± 0.1 and a straightness tolerance of 0.07 about the centerline at MMC.

Project 8-32

A. Given the shaft shown in Figure P8-32, what is the minimum hole diameter that will always accept the shaft?
B. If the minimum clearance between the shaft and a hole is equal to 0.02, and the tolerance on the hole is to be 0.6, what are the maximum and minimum diameters for the hole?

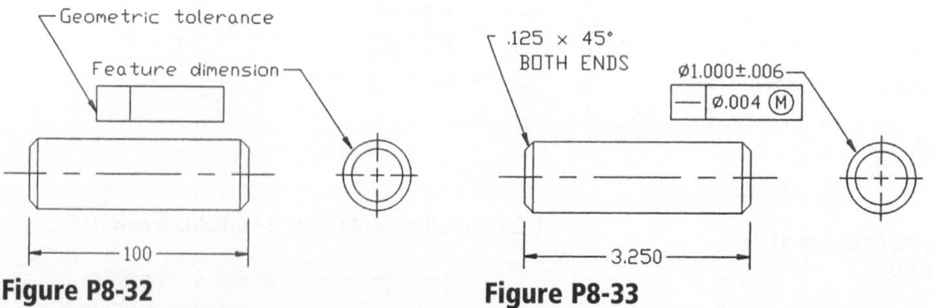

Figure P8-32

Figure P8-33

Project 8-33

A. Given the shaft shown in Figure P8-33, what is the minimum hole diameter that will always accept the shaft?

B. If the minimum clearance between the shaft and a hole is equal to .005, and the tolerance on the hole is to be .007, what are the maximum and minimum diameters for the hole?

Project 8-34

Draw a front and a right-side view of the object shown in Figure P8-34 and add the appropriate dimensions and tolerances based on the following information. Numbers located next to an edge line indicate the length of the edge.

A. Define surfaces A, B, and C as primary, secondary, and tertiary datums, respectively.
B. Assign a tolerance of ±0.5 to all linear dimensions.
C. Assign a feature tolerance of 12.07 - 12.00 to the protruding shaft.
D. Assign a flatness tolerance of 0.01 to surface A.
E. Assign a straightness tolerance of 0.03 to the protruding shaft.
F. Assign a perpendicularity tolerance to the centerline of the protruding shaft of 0.02 at MMC relative to datum A.

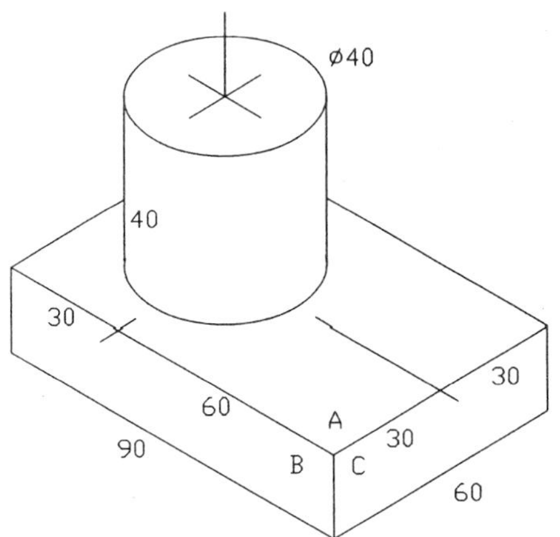

Figure P8-34

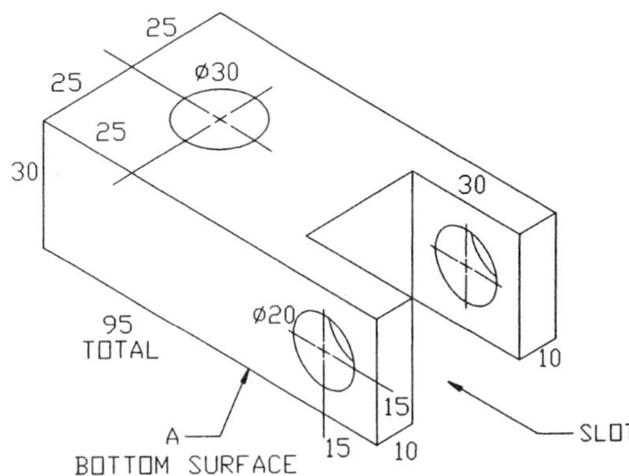

Figure P8-35

Project 8-35

Draw a front and a right-side view of the object shown in Figure P8-35 and add the following dimensions and tolerances.

A. Define the bottom surface as datum A.
B. Assign a perpendicularity tolerance of 0.4 to both sides of the slot relative to datum A.
C. Assign a perpendicularity tolerance of 0.2 to the centerline of the 30 diameter hole at MMC relative to datum A.
D. Assign a feature tolerance of ±0.8 to all three holes.
E. Assign a parallelism tolerance of 0.2 to the common centerline between the two 20 diameter holes relative to datum A.
F. Assign a tolerance of ±0.5 to all linear dimensions.

Project 8-36

Draw a circular front and the appropriate right-side view of the object shown in Figure P8-36 and add the following dimensions and tolerances.

A. Assign datum A as indicated.
B. Assign the object's longitudinal axis as datum B.
C. Assign the object's centerline through the slot as datum C.
D. Assign a tolerance of ±0.5 to all linear tolerances.
E. Assign a tolerance of ±0.5 to all circular features.
F. Assign a parallelism tolerance of 0.01 to both edges of the slot.
G. Assign a perpendicularity tolerance of 0.01 to the outside edge of the protruding shaft.

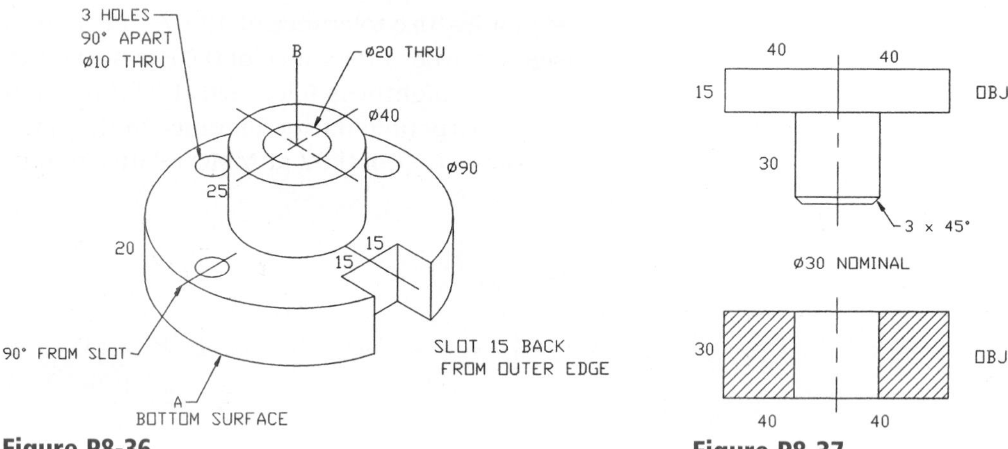

Figure P8-36

Figure P8-37

Project 8-37

Given the two objects shown in Figure P8-37, draw a front and a side view of each. Assign a tolerance of ±0.5 to all linear dimensions. Assign a feature tolerance of ±0.4 to the shaft, and also assign a straightness tolerance of 0.2 to the shaft's centerline at MMC.

Tolerance the hole so that it will always accept the shaft with a minimum clearance of 0.1 and a feature tolerance of 0.2. Assign a perpendicularity tolerance of 0.05 to the centerline of the hole at MMC.

Project 8-38

Given the two objects shown in Figure P8-38, draw a front and a side view of each. Assign a tolerance of ±0.005 to all linear dimensions. Assign a feature tolerance of ±0.004 to the shaft, and also assign a straightness tolerance of 0.002 to the shaft's centerline at MMC.

Tolerance the hole so that it will always accept the shaft with a minimum clearance of 0.001 and a feature tolerance of 0.002.

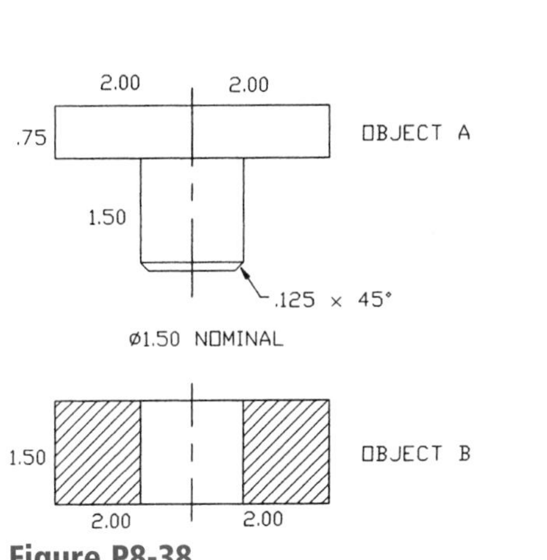

Figure P8-38

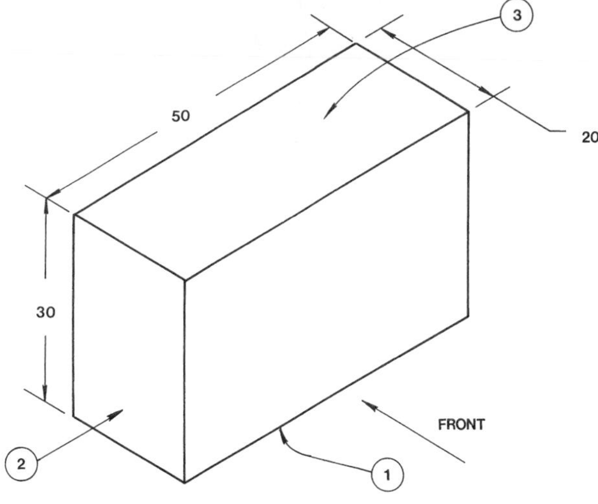

Figure P8-39

Project 8-39

Draw a model of the object shown in Figure P8-39, then create a drawing layout including the specified dimensions. Add the following tolerances and specifications to the drawing.

A. Surface 1 is datum A.
B. Surface 2 is datum B and is perpendicular to datum A within 0.1 mm.
C. Surface 3 is datum C and is parallel to datum A within 0.3 mm.
D. Locate a 16-mm diameter hole in the center of the front surface that goes completely through the object. Use positional tolerances to locate the hole. Assign a positional tolerance of 0.02 at MMC perpendicular to datum A.

Project 8-40

Draw a model of the object shown in Figure P8-40, then create a drawing layout including the specified dimensions. Add the following tolerances and specifications to the drawing.

A. Surface 1 is datum A.
B. Surface 2 is datum B and is perpendicular to datum A within .003 in.
C. Surface 3 is parallel to datum A within .005 in.
D. The cylinder's longitudinal centerline is to be straight within .001 in. at MMC.
E. Surface 2 is to have circular accuracy within .002 in.

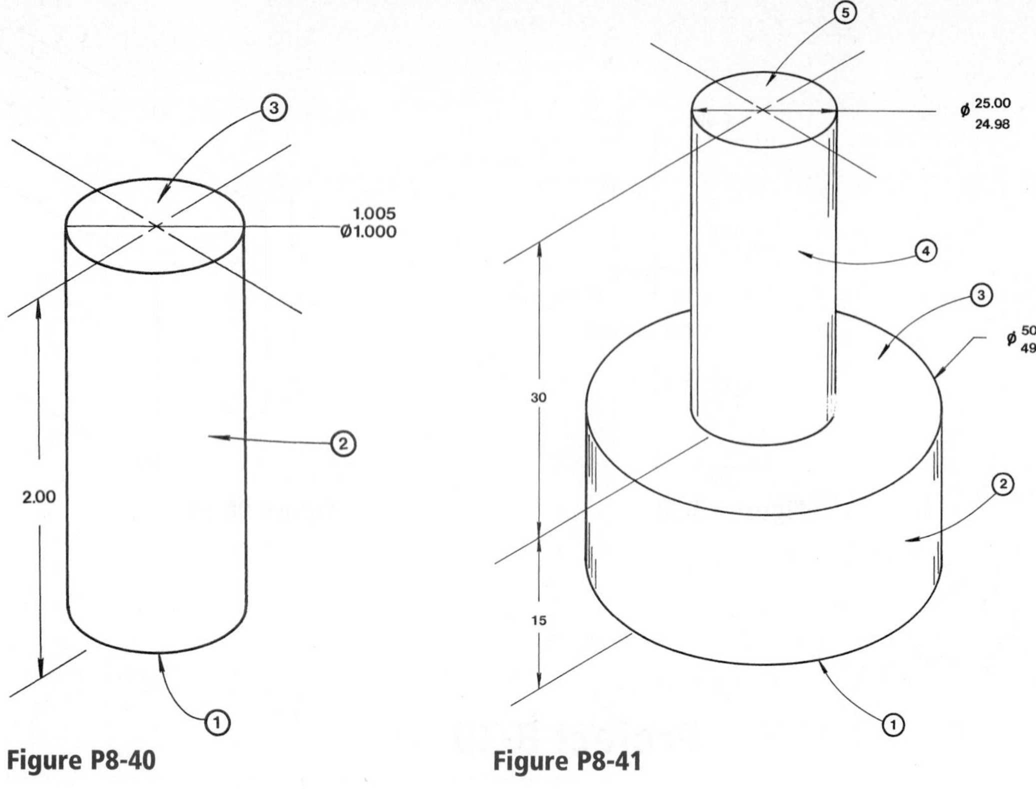

Figure P8-40

Figure P8-41

Project 8-41

Draw a model of the object shown in Figure P8-41, then create a drawing layout including the specified dimensions. Add the following tolerances and specifications to the drawing.

A. Surface 1 is datum A.
B. Surface 4 is datum B and is perpendicular to datum A within 0.08 mm.
C. Surface 3 is flat within 0.03 mm.
D. Surface 5 is parallel to datum A within 0.01 mm.
E. Surface 2 has a runout tolerance of 0.2 mm relative to surface 4.
F. Surface 1 is flat within 0.02 mm.
G. The longitudinal centerline is to be straight within 0.02 at MMC and perpendicular to datum A.

Project 8-42

Draw a model of the object shown in Figure P8-42, then create a drawing layout including the specified dimensions. Add the following tolerances and specifications to the drawing.

A. Surface 2 is datum A.
B. Surface 6 is perpendicular to datum A with .000 allowable variance at MMC but with a .002 in. MAX variance limit beyond MMC.

C. Surface 1 is parallel to datum A within .005.
D. Surface 4 is perpendicular to datum A within .004 in.

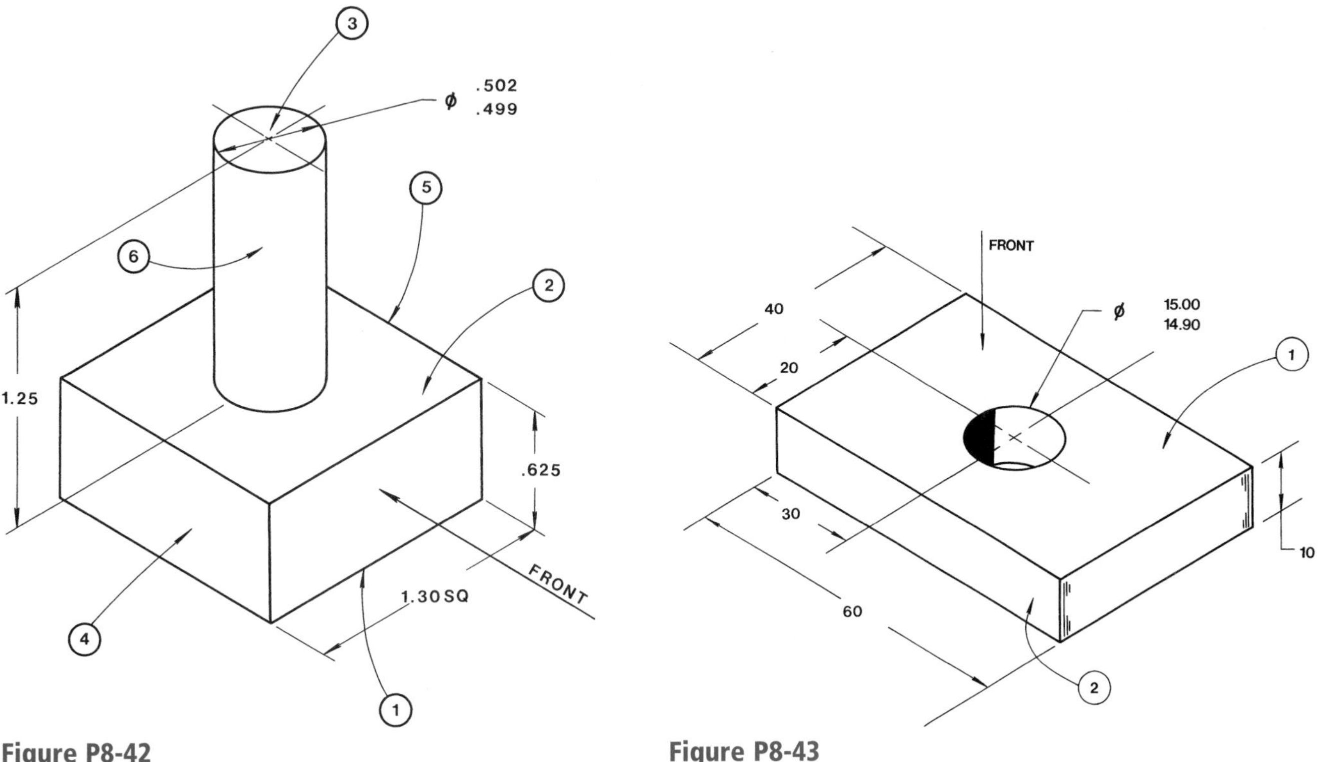

Figure P8-42

Figure P8-43

Project 8-43

Draw a model of the object shown in Figure P8-43, then create a drawing layout including the specified dimensions. Add the following tolerances and specifications to the drawing.

A. Surface 1 is datum A.
B. Surface 2 is datum B.
C. The hole is located using a true position tolerance value of 0.13 mm at MMC. The true position tolerance is referenced to datums A and B.
D. Surface 1 is to be straight within 0.02 mm.
E. The bottom surface is to be parallel to datum A within 0.03 mm.

Project 8-44

Draw a model of the object shown in Figure P8-44, then create a drawing layout including the specified dimensions. Add the following tolerances and specifications to the drawing.

A. Surface 1 is datum A.
B. Surface 2 is datum B.
C. Surface 3 is perpendicular to surface 2 within 0.02 mm.

D. The four holes are to be located using a positional tolerance of 0.07 mm at MMC referenced to datums A and B.
E. The centerlines of the holes are to be straight within 0.01 mm at MMC.

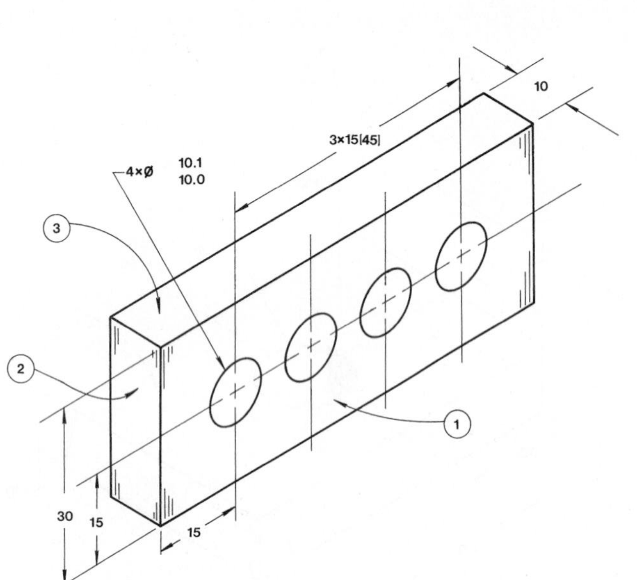

Figure P8-44

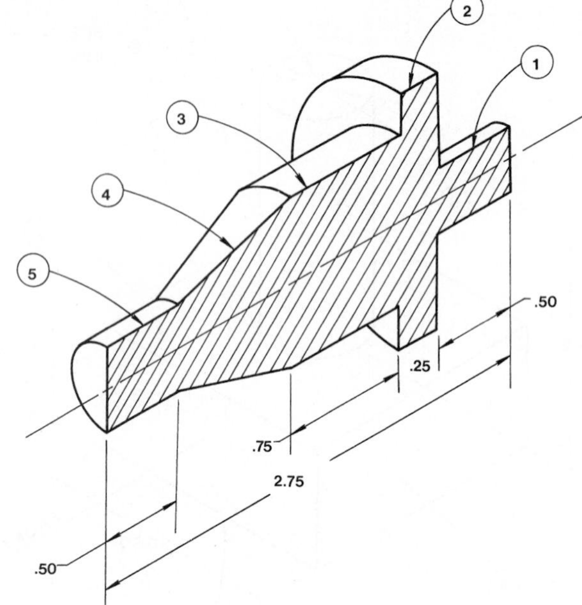

Figure P8-45

Project 8-45

Draw a model of the object shown in Figure P8-45, then create a drawing layout including the specified dimensions. Add the following tolerances and specifications to the drawing.

A. Surface 1 has a dimension of .378–.375 in. and is datum A. The surface has a dual primary runout with datum B to within .005 in. The runout is total.
B. Surface 2 has a dimension of 1.505–1.495 in. Its runout relative to the dual primary datums A and B is .008 in. The runout is total.
C. Surface 3 has a dimension of 1.000 ± .005 and has no geometric tolerance.
D. Surface 4 has no circular dimension but has a total runout tolerance of .006 in. relative to the dual datums A and B.
E. Surface 5 has a dimension of .500–.495 in. and is datum B. It has a dual primary runout with datum A within .005 in. The runout is total.

Project 8-46

Draw a model of the object shown in Figure P8-46, then create a drawing layout including the specified dimensions. Add the following tolerances and specifications to the drawing.

A. Hole 1 is datum A.
B. Hole 2 is to have its circular centerline parallel to datum A within 0.2 mm at MMC when datum A is at MMC.
C. Assign a positional tolerance of 0.01 to each hole's centerline at MMC.

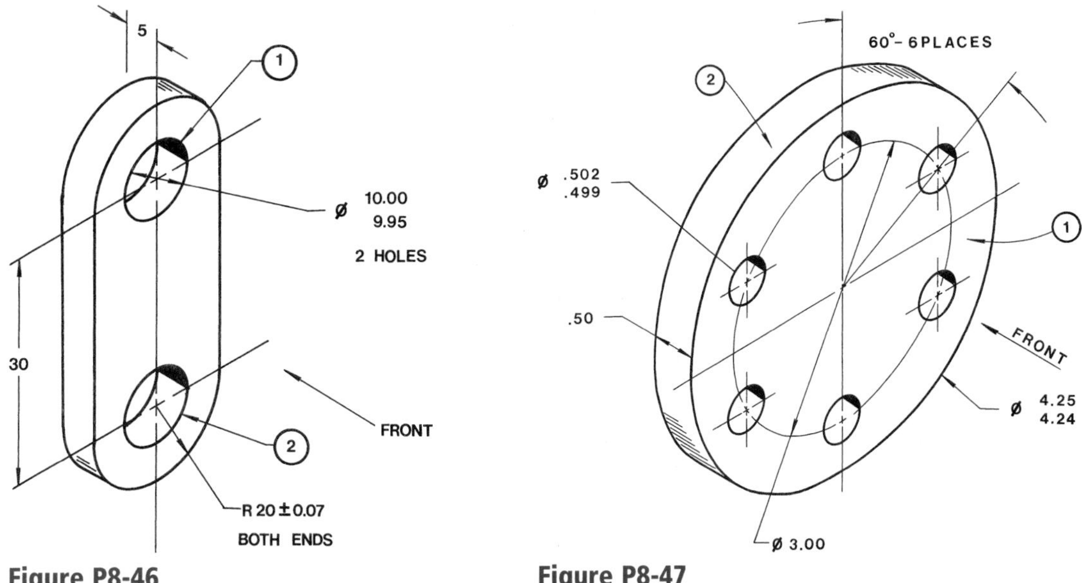

Figure P8-46

Figure P8-47

Project 8-47

Draw a model of the object shown in Figure P8-47, then create a drawing layout including the specified dimensions. Add the following tolerances and specifications to the drawing.

A. Surface 1 is datum A.
B. Surface 2 is datum B.
C. The six holes have a diameter range of .502–.499 in. and are to be located using positional tolerances so that their centerlines are within .005 in. at MMC relative to datums A and B.
D. The back surface is to be parallel to datum A within .002 in.

Project 8-48

Draw a model of the object shown in Figure P8-48, then create a drawing layout including the specified dimensions. Add the following tolerances and specifications to the drawing.

A. Surface 1 is datum A.
B. Hole 2 is datum B.
C. The eight holes labeled 3 have diameters of 8.4–8.3 mm with a positional tolerance of 0.15 mm at MMC relative to datums A and B. Also, the eight holes are to be counterbored to a diameter of 14.6–14.4 mm and to a depth of 5.0 mm.
D. The large center hole is to have a straightness tolerance of 0.2 at MMC about its centerline.

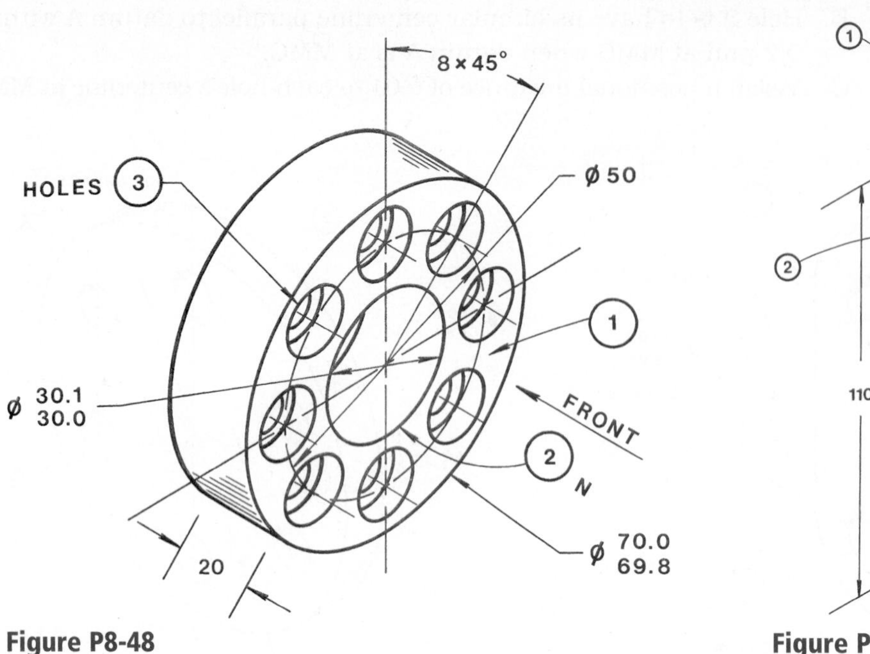

Figure P8-48

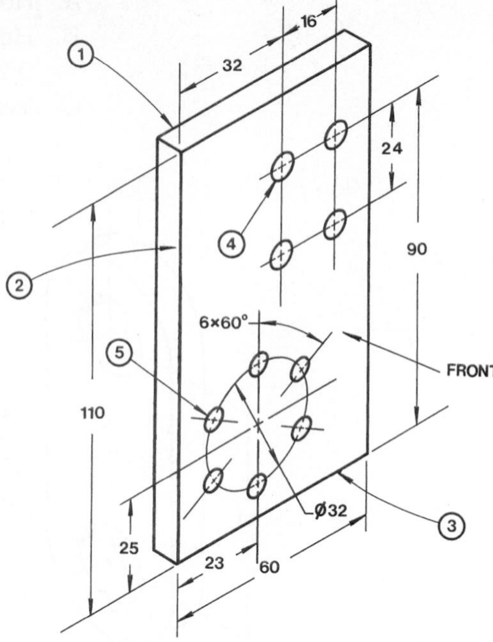

Figure P8-49

Project 8-49

Draw a model of the object shown in Figure P8-49, then create a drawing layout including the specified dimensions. Add the following tolerances and specifications to the drawing.

A. Surface 1 is datum A.
B. Surface 2 is datum B.
C. Surface 3 is datum C.
D. The four holes labeled 4 have a dimension and tolerance of 8 + 0.3, −0 mm. The holes are to be located using a positional tolerance of 0.05 mm at MMC relative to datums A, B, and C.
E. The six holes labeled 5 have a dimension and tolerance of 6 + 0.2, −0 mm. The holes are to be located using a positional tolerance of 0.01 mm at MMC relative to datums A, B, and C.

Project 8-50

The objects in Figure P8-50B labeled A and B are to be toleranced using four different tolerances as shown. Redraw the charts shown in Figure P8-50A and list the appropriate allowable tolerance for "as measured" increments of 0.1 mm or .001 in. Also include the appropriate geometric tolerance drawing called out above each chart.

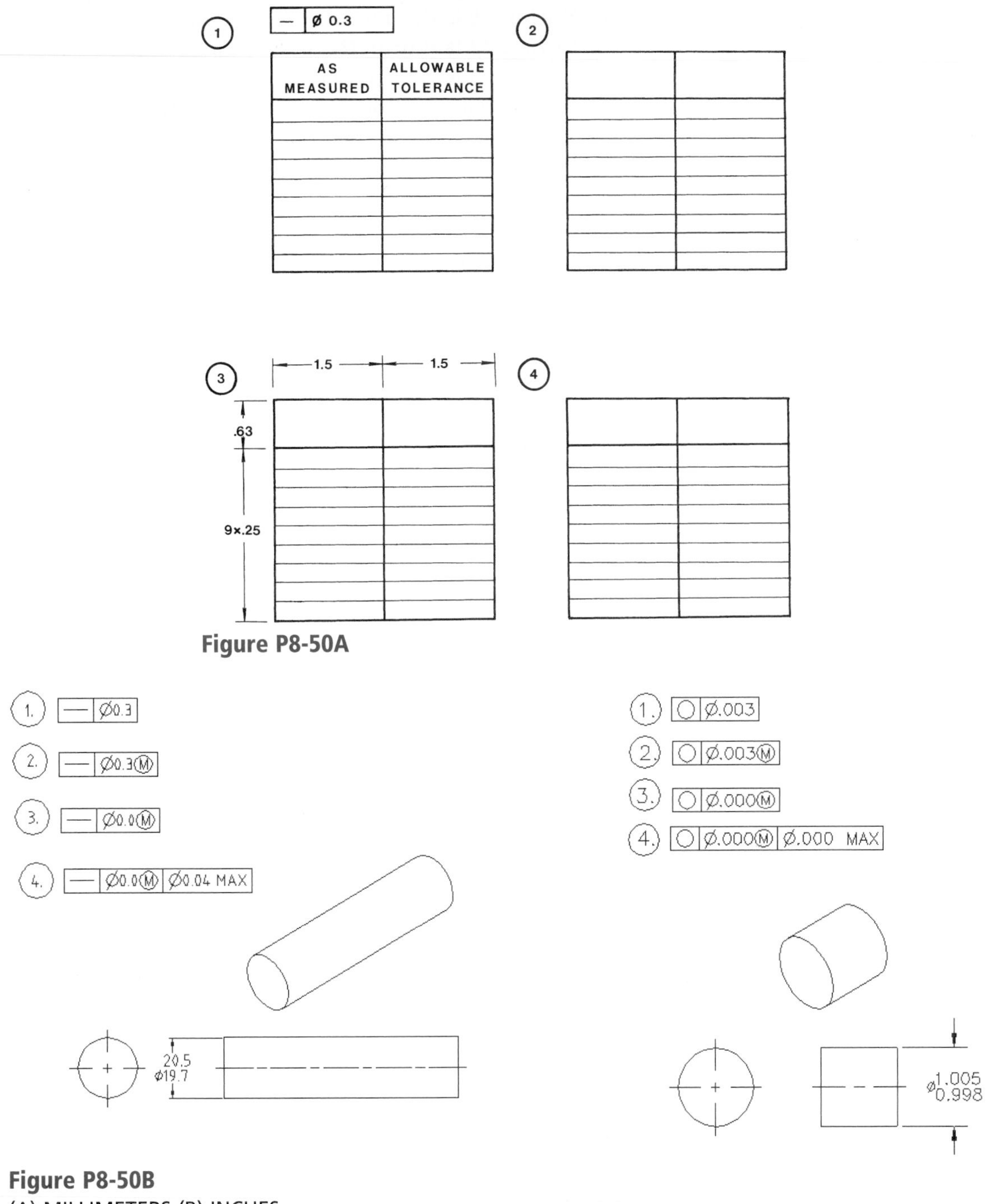

Figure P8-50A

Figure P8-50B
(A) MILLIMETERS (B) INCHES

Project 8-51

Assume that there are two copies of the part in Figure P8-51 and that these parts are to be joined together using four fasteners in the floating condition. Draw front and top views of the object, including dimensions and tolerances. Add the following tolerances and specifications to the drawing, then draw front and top views of a shaft that can be used to join the two objects. The shaft should be able to fit into any of the four holes.

A. Surface 1 is datum A.
B. Surface 2 is datum B.
C. Surface 3 is perpendicular to surface 2 within 0.02 mm.
D. Specify the positional tolerance for the four holes applied at MMC.
E. The centerlines of the holes are to be straight within 0.01 mm at MMC.
F. The clearance between the shafts and the holes is to be 0.05 minimum and 0.10 maximum.

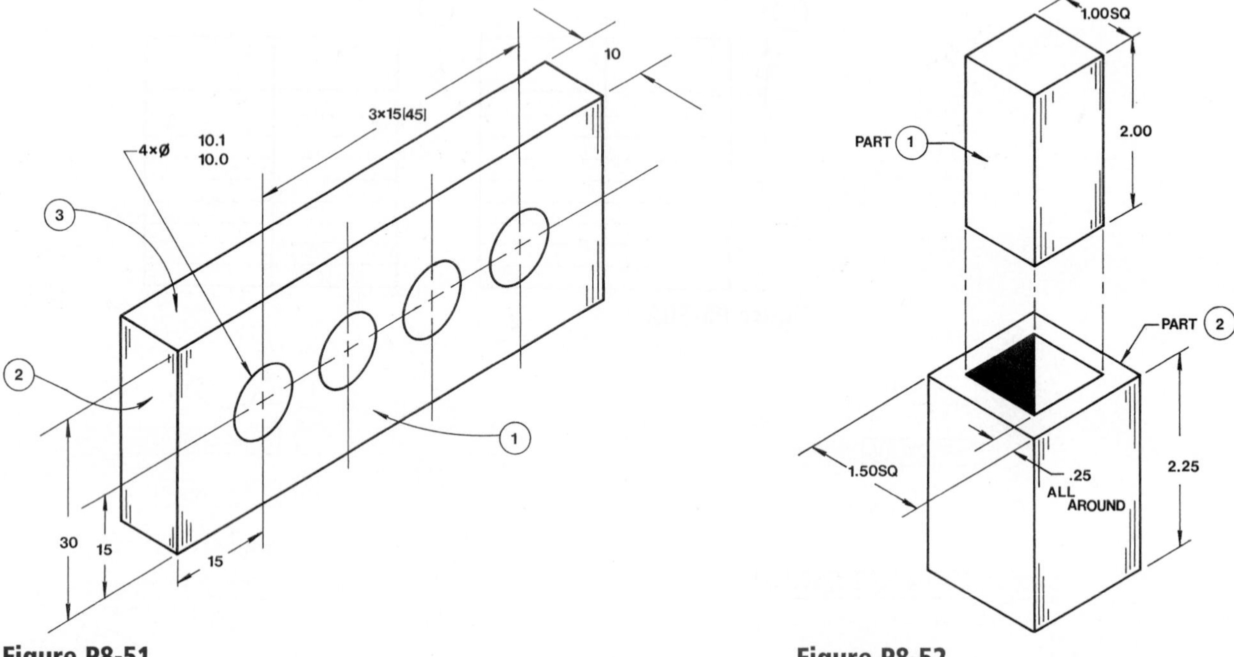

Figure P8-51
MILLIMETERS

Figure P8-52
INCHES

Project 8-52

Dimension and tolerance parts 1 and 2 of Figure P8-52 so that part 1 always fits into part 2 with a minimum clearance of .005 in. The tolerance for part 1's outer matching surface is .006 in.

Project 8-53

Dimension and tolerance parts 1 and 2 of Figure P8-53 so that part 1 always fits into part 2 with a minimum clearance of 0.03 mm. The tolerance for part 1's diameter is 0.05 mm. Take into account the fact that the interface is long relative to the diameters.

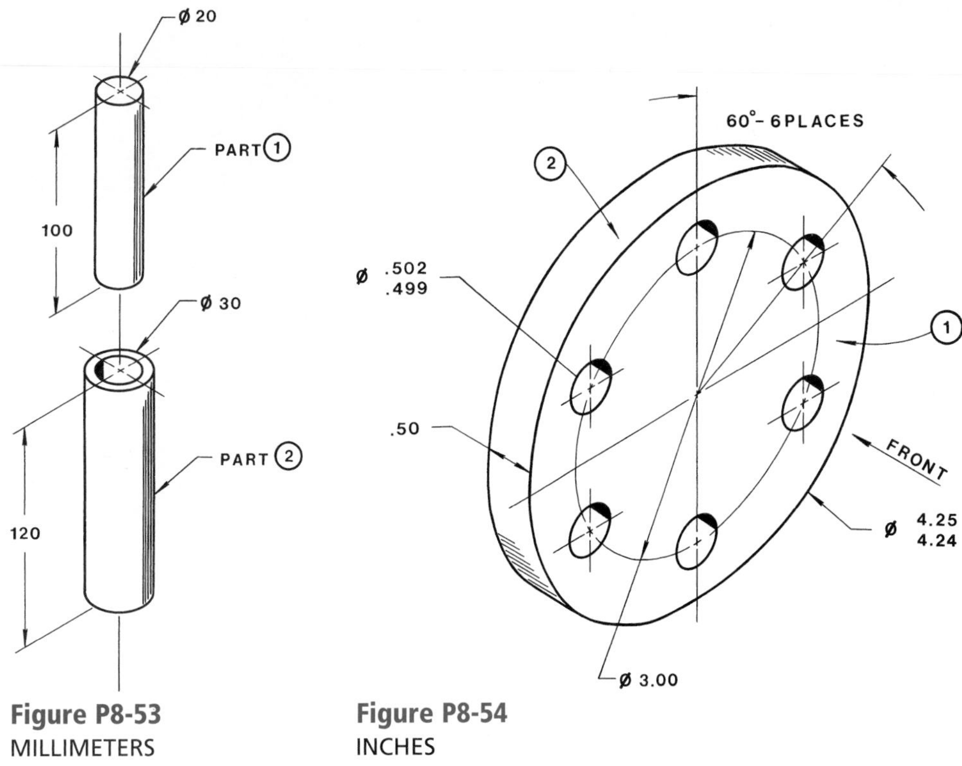

Figure P8-53
MILLIMETERS

Figure P8-54
INCHES

Project 8-54

Assume that there are two copies of the part in Figure P8-54 and that these parts are to be joined together using six fasteners in the floating condition. Draw front and top views of the object, including dimensions and tolerances. Add the following tolerances and specifications to the drawing, then draw front and top views of a shaft that can be used to join the two objects. The shaft should be able to fit into any of the six holes.

A. Surface 1 is datum A.
B. Surface 2 is round within .003.
C. Specify the positional tolerance for the six holes applied at MMC.
D. The clearance between the shafts and the holes is to be .001 minimum and .003 maximum.

Project 8-55

The assembly shown in Figure P8-55 is made from parts defined in Chapter 5.

1. Draw an exploded assembly drawing.
2. Draw a BOM.
3. Use the drawing layout mode and draw orthographic views of each part. Include dimensions and geometric tolerances. The pegs should have a minimum clearance of 0.02. Select appropriate tolerances and define them for each hole using positional tolerance.

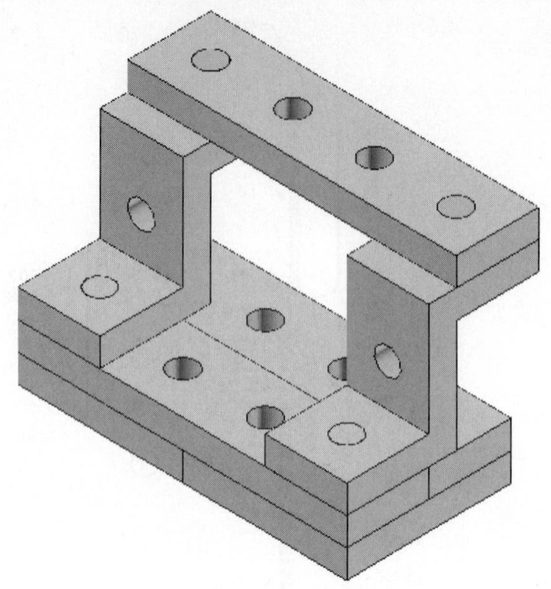

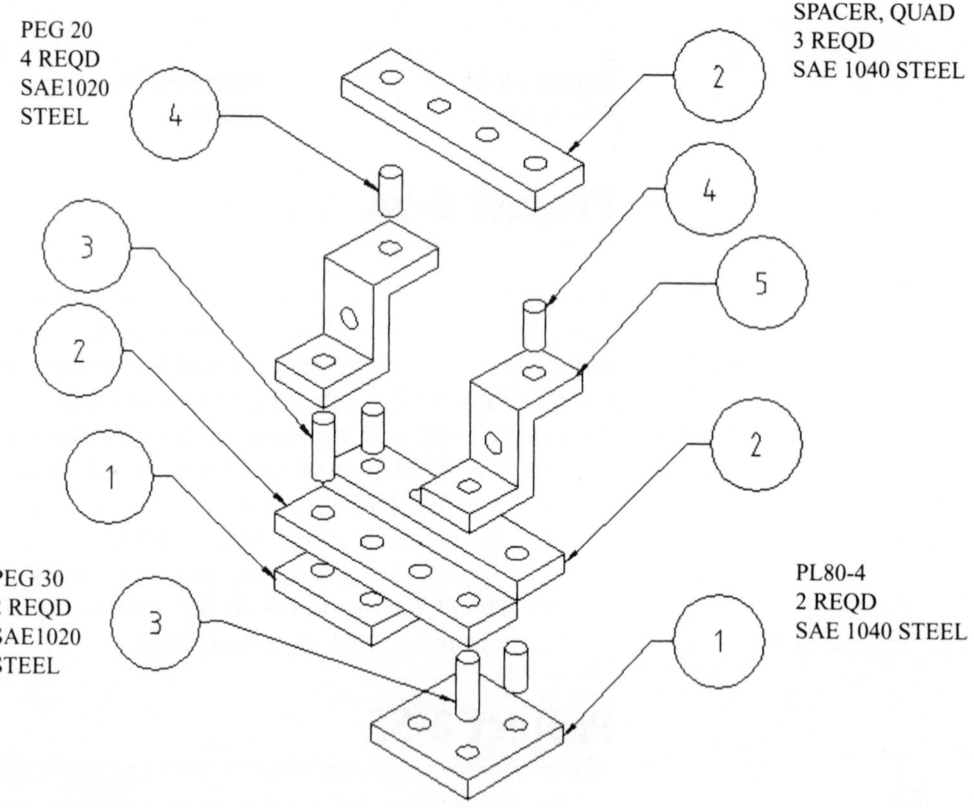

Figure P8-55
MILLIMETERS

CHAPTER OBJECTIVES

- Learn how to draw springs using the **Coil** tool and the tools on the **Spring** panel under the **Design** tab
- Learn how to draw compression springs
- Learn how to draw extension springs
- Learn how to draw torsion springs
- Learn how to draw Belleville springs

Introduction

This chapter shows how to draw springs. Both the **Coil** tool and the tools from the **Spring** panel are used to draw springs. Compression, extension, torsion, and Belleville springs are introduced.

Compression Springs

EXERCISE 9-1 Drawing a Compression Spring Using the Coil Tool

1 Start a new metric drawing using the **Standard (mm).ipt** format, select the **Create 2D Sketch** option and select the XY plane.

2 Create an isometric view (use the **Home** tool next to the ViewCube) and draw a line and a circle as shown in Figure 9-1.

The circle diameter is the wire's diameter (5), and the distance between the line and the center point of the circle equals the spring's mean diameter (Ø20, R = 10).

Figure 9-1

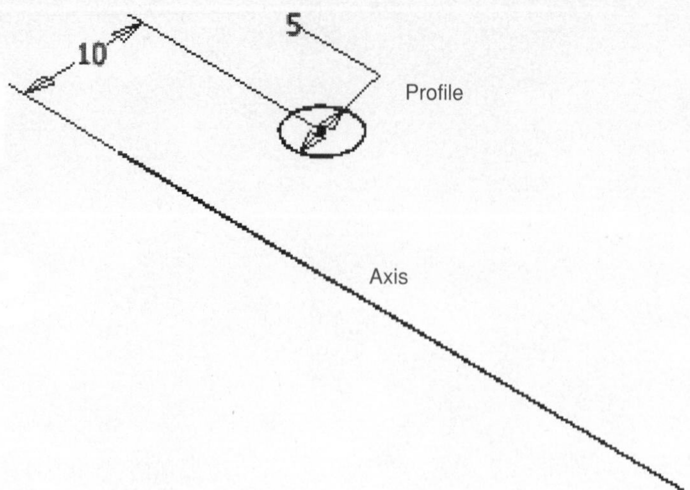

3 Right-click the mouse and select the **Finish 2D Sketch** option.

4 Select the **Coil** tool on the **Create** panel under the **3D Model** tab. The **Coil** dialog box will appear. See Figure 9-2.

Figure 9-2

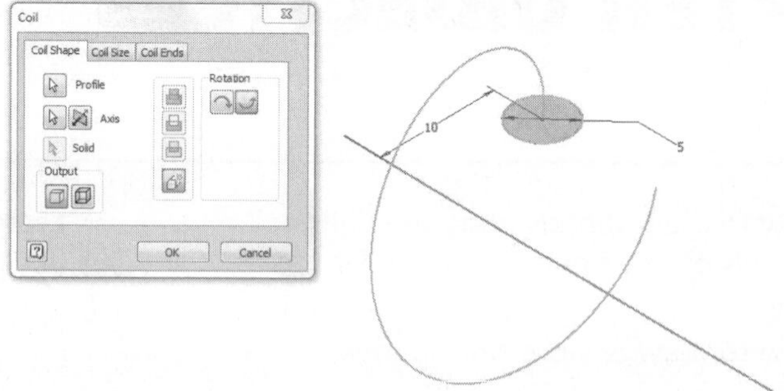

5 Select the circle as the **Profile** and the line as the **Axis.**

6 Click on the **Coil Size** tab on the **Coil** dialog box. See Figure 9-3.

Figure 9-3

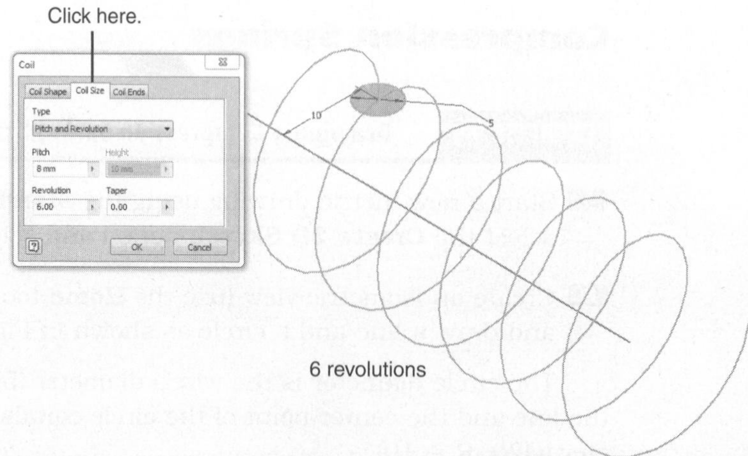

7 Set the **Pitch** value for **8** and the **Revolution** for **6.**

8 Click **OK.**

See Figure 9-4.

Figure 9-4

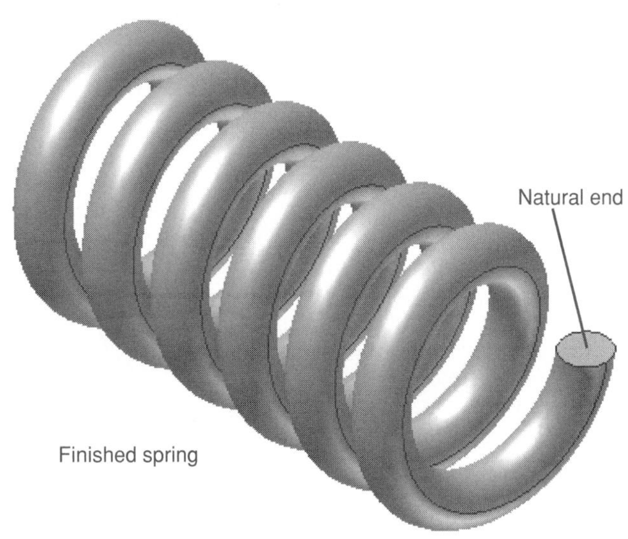

Coil Ends

The ends of a spring created using the **Coil** tool can be drawn in one of two ways: natural or flat. Figure 9-5 shows three different possible coil end configurations.

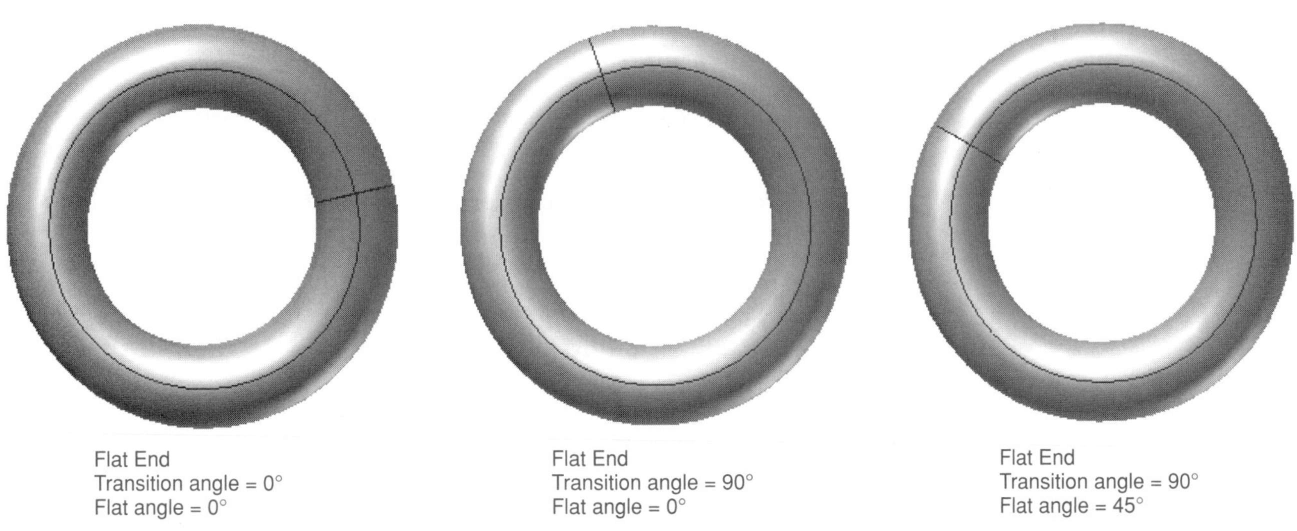

Figure 9-5

EXERCISE 9-2 **Changing the End of a Spring**

1 Click the **Coil Ends** tab on the **Coil** dialog box.

See Figure 9-6. The spring shown in Figure 9-4 has natural ends. Springs with flat ends must have their transition and flat angles defined, as was done in Figure 9-5.

Figure 9-6

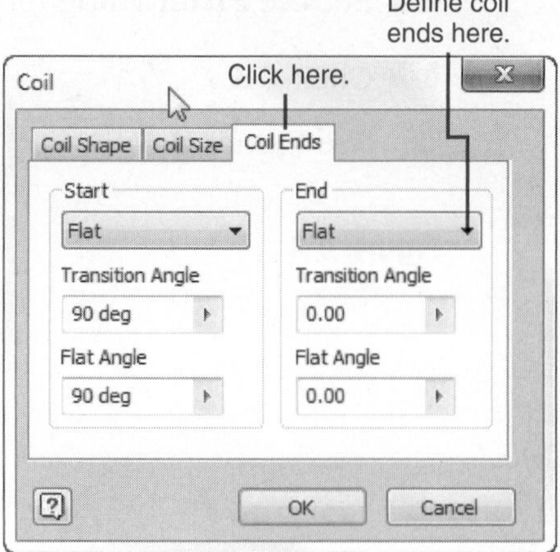

EXERCISE 9-3 Drawing a Compression Spring Using Design Accelerator Options

1 Start a new drawing using the **English** tab.

2 Select the **Standard (in).iam** format.

The **Assemble** panels will appear. See Figure 9-7.

Figure 9-7

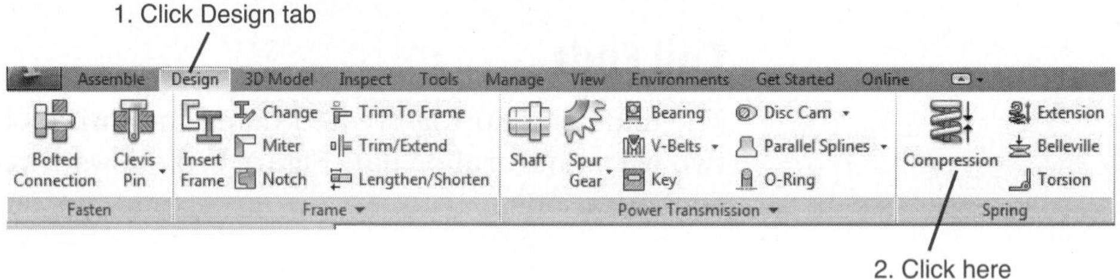

3 Click the **Design** tab.

4 Select the **Compression** tool located on the **Spring** panel under the **Design** tab.

The **Compression Spring Component Generator** dialog box will appear. See Figure 9-8.

5 Enter values for both the **Spring Start** and **Spring End** as follows:

Closed End Coils = 2.000

Transition Coils = 1.000

Ground Coils = 0.750

6 Click the **Calculation** tab.

The dialog box will change. See Figure 9-9.

7 Set the **Spring Strength Calculation** for **Work Forces Calculation.**

8 Set the **Dimensions** values as follows:

Wire Diameter = 0.125

Outside Diameter = 0.875

Loose Spring Length = 2.00

Figure 9-8

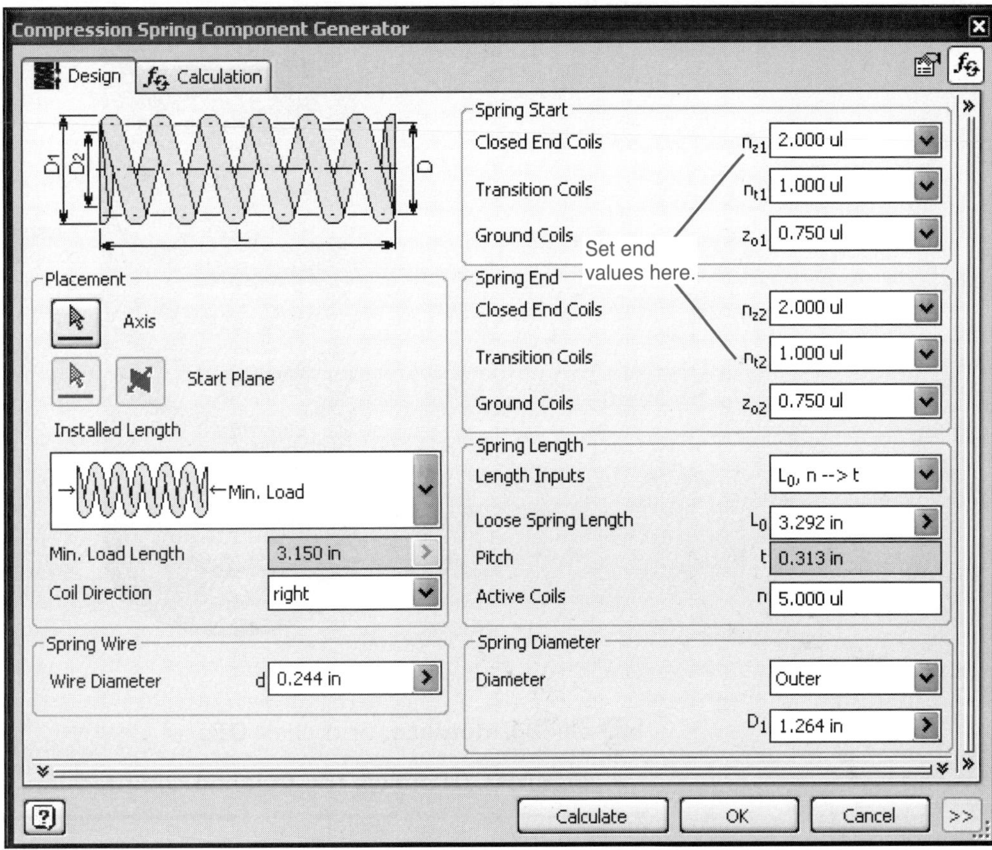

Figure 9-9

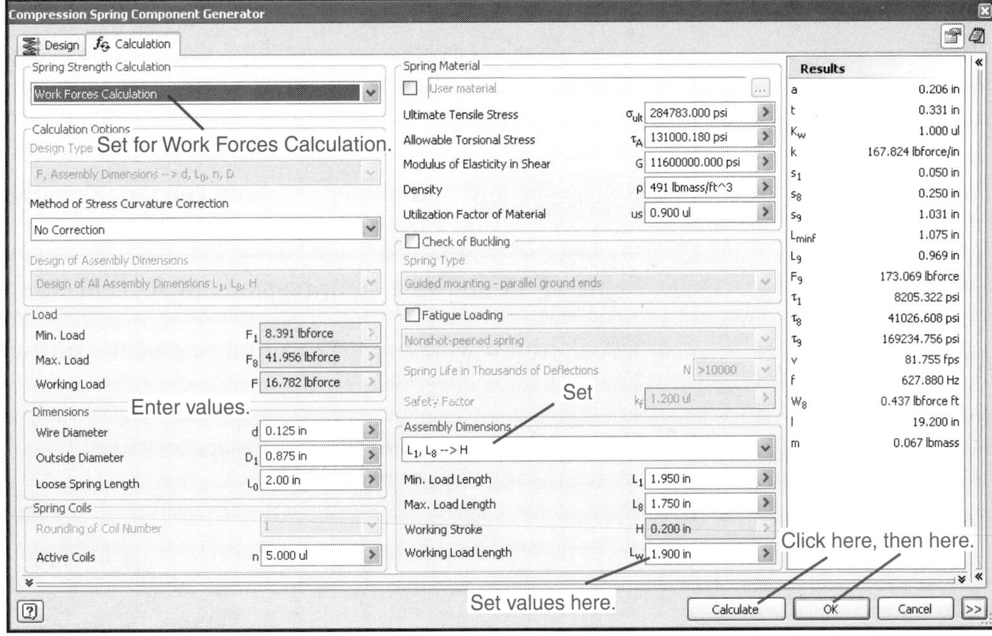

9 Set the **Assembly Dimensions** values for the following.

L_1,L_8--> H

Min. Load Length = 1.950

Max. Load Length = 1.750

Working Load Length = 1.900

Figure 9-10

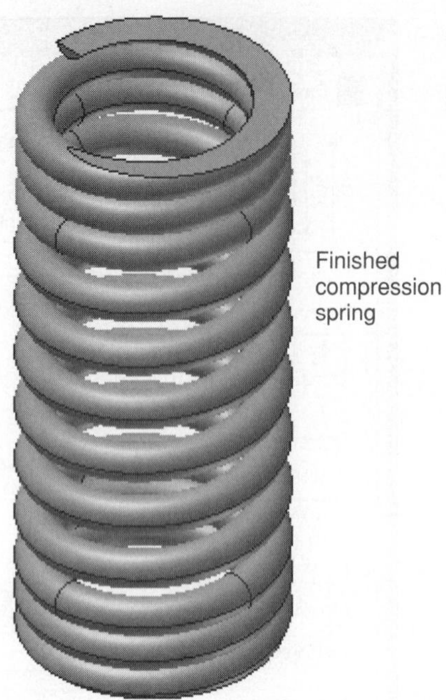

10 Click **Calculate,** and click **OK.**

Figure 9-10 shows the finished compression spring.

TIP

The spring's ends are square ground. Square-ground ends sit better on flat surfaces and tend not to buckle when put under a load.

EXERCISE 9-4 Drawing a Ground End on a Compression Spring Drawn Using the Coil Tool

Figure 9-11 shows a compression spring that was drawn using the **Coil** tool.

1 Create a work plane through one of the end coils.

Figure 9-11

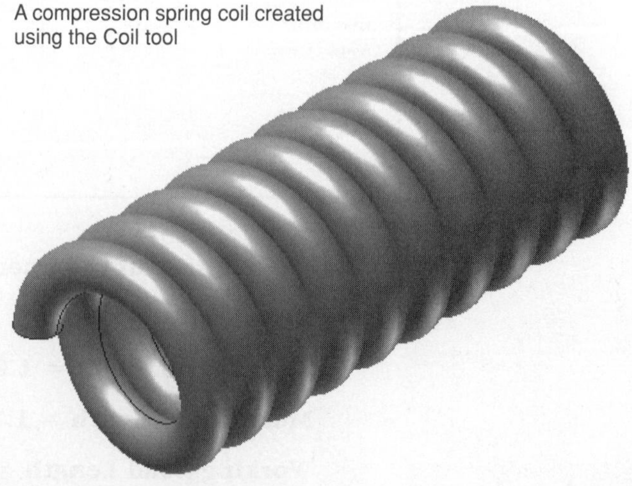

See Figure 9-12. In this example an XZ work plane was created offset 4.00 from the end of the spring.

2 Create a new sketch plane on the work plane. Draw a rectangle on the new sketch plane. Size the rectangle so that it is larger than the outside diameter of the spring.

See Figure 9-13.

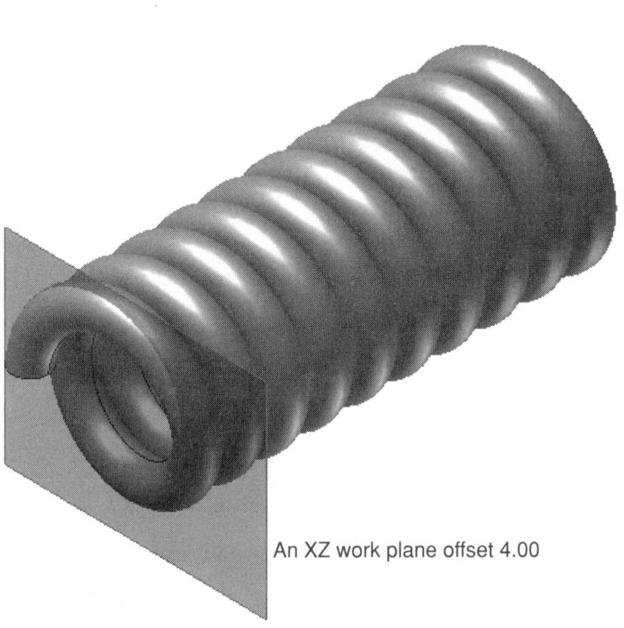

Figure 9-12

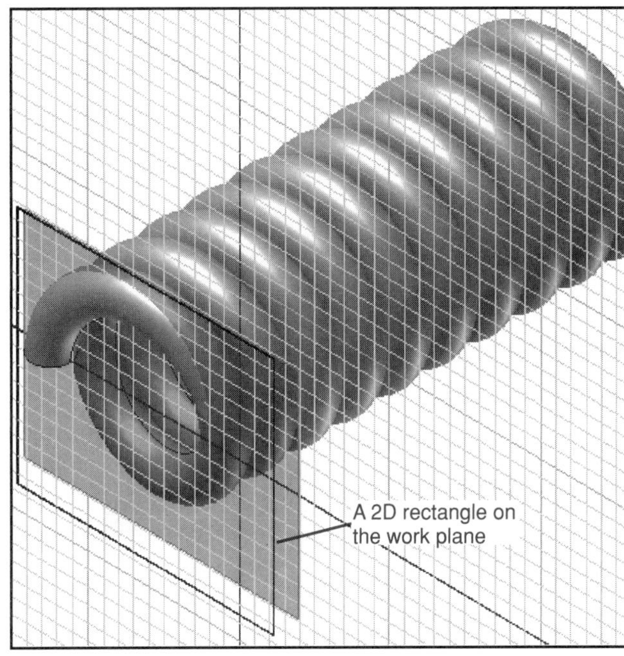

Figure 9-13

3 Right-click the mouse and select the **Extrude** tool. Extrude the rectangle so that it extends beyond the end of the spring.

In this example an extrusion value of **10** was used. See Figure 9-14.

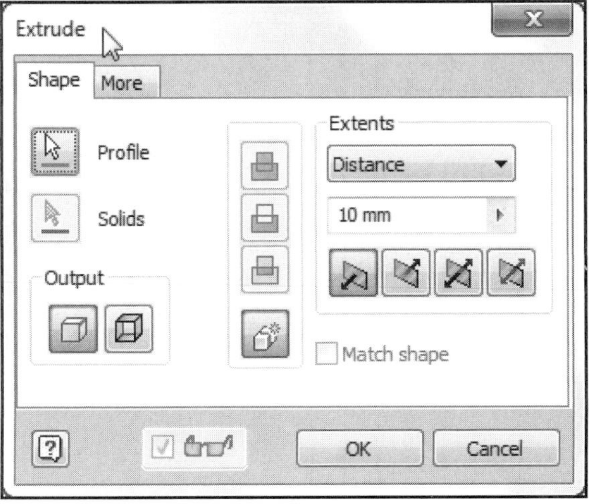

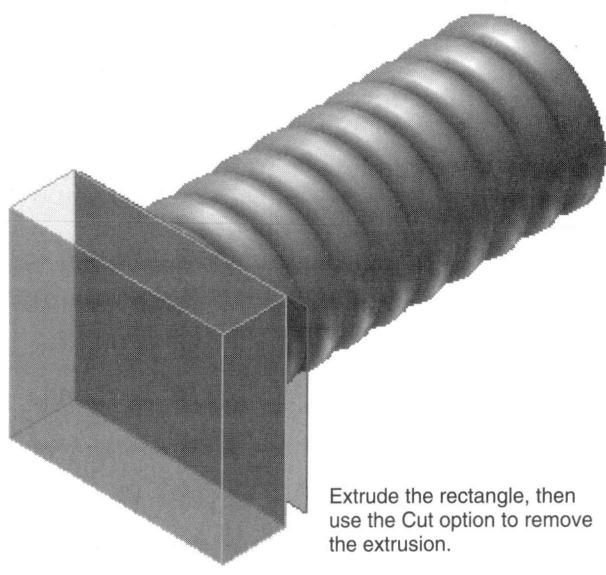

Extrude the rectangle, then use the Cut option to remove the extrusion.

Figure 9-14

4 Use the **Cut** option of the **Extrude** tool and cut out the extruded rectangle.

Figure 9-15 shows the finished spring.

Figure 9-15

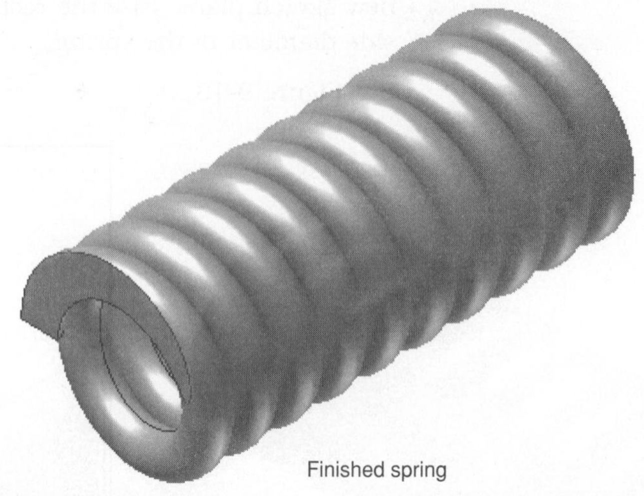

Extension Springs

EXERCISE 9-5 **Drawing an Extension Spring Using Tools on the Spring Panel**

1 Start a new drawing using **English** units and the **Standard (in).iam** format.

2 Click the **Design** tab.

See Figure 9-16.

Figure 9-16

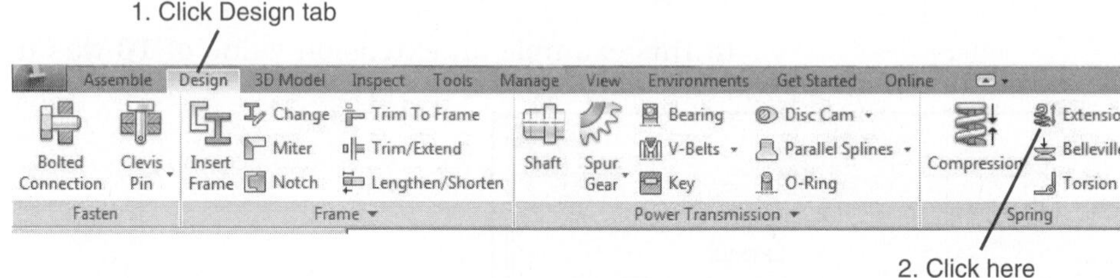

3 Click the **Extension** tool on the **Spring** panel.

The **Extension Spring Component Generator** dialog box will appear. See Figure 9-17.

4 Set the following values and inputs:

Coil Direction = **right**

Wire Diameter = **0.125**

Diameter, Outer = **1.000**

Start Hook Type = **Full Loop**

Start Hook Length = **0.800**

End Hook Type = **Full Loop**

Figure 9-17

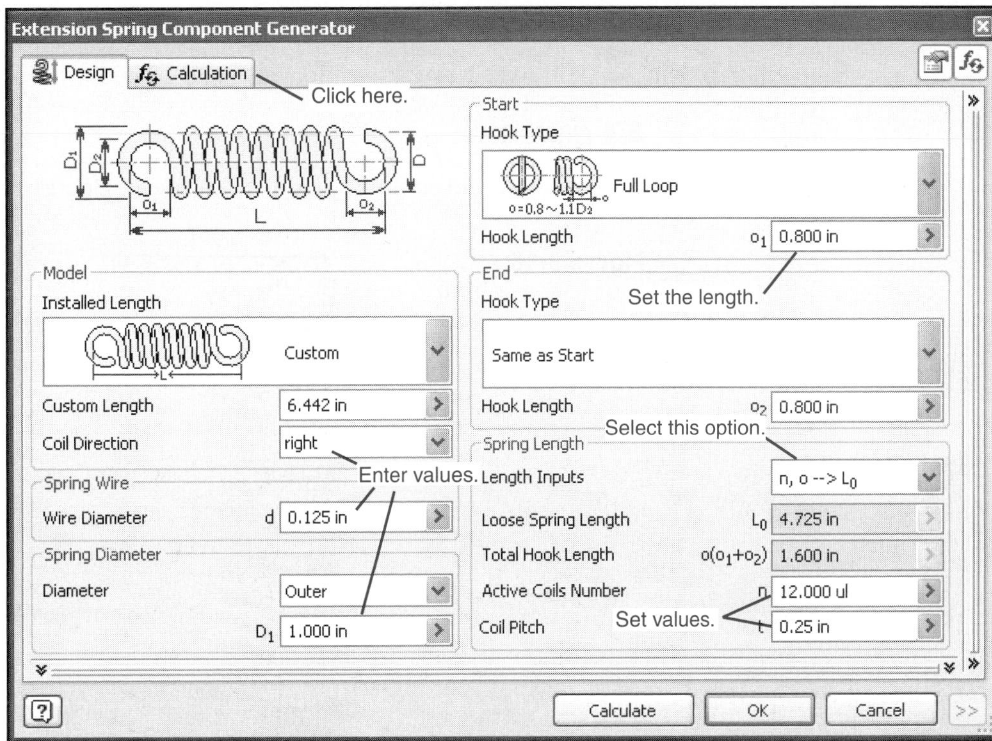

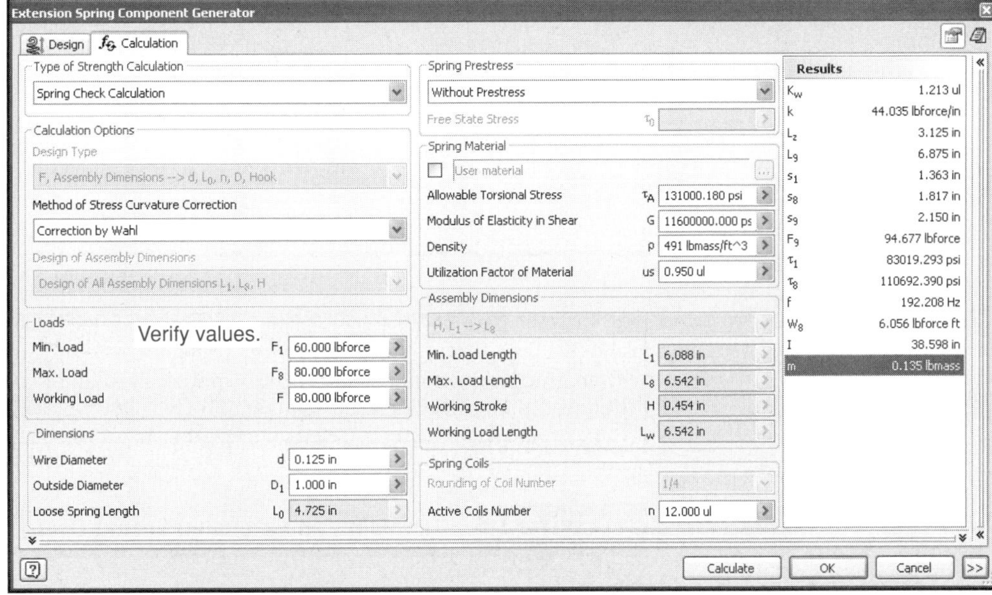

End Hook Length = 0.800

Length Inputs = n, o --> L_0

n = 12.000

t = 0.25

Click the **Calculation** tab and verify that the values are as shown in Figure 9-17.

5 Click **Calculate.**

6 Click **OK.**

The **File Naming** dialog box will appear.

7 Click **OK.**

The finished extension spring will appear. See Figure 9-18.

Figure 9-18

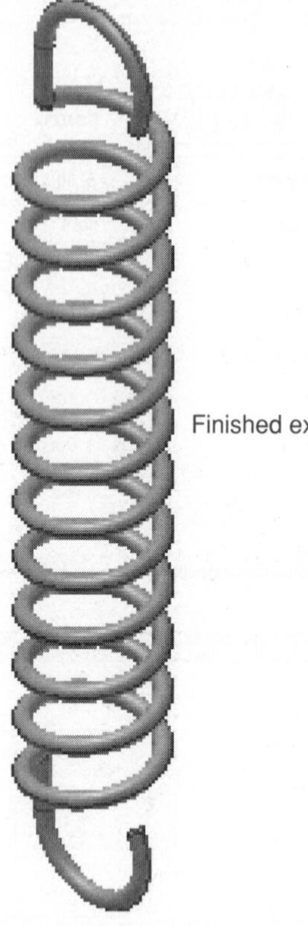

> **TIP**
>
> If an input error is made, when **OK** is clicked a red line will appear across the bottom of the screen, and the incorrect value will be highlighted.

EXERCISE 9-6 Drawing an Extension Spring Using the Coil Tool

1 Start a new drawing using the **Standard (mm).ipt** format, select the **Create 2D Sketch** option, and select the XY plane.

2 Use the **Home** tool to orient the drawing to an isometric view and draw a Ø**5**-mm circle **15** mm from a line.

See Figure 9-19.

3 Right-click the mouse and select **Finish 2D Sketch** option. Select the **Coil** tool.

The **Coil** dialog box will appear. See Figure 9-20.

4 Select the circle as the **Profile** and the line as the **Axis.** Click the **Coil Size** tab.

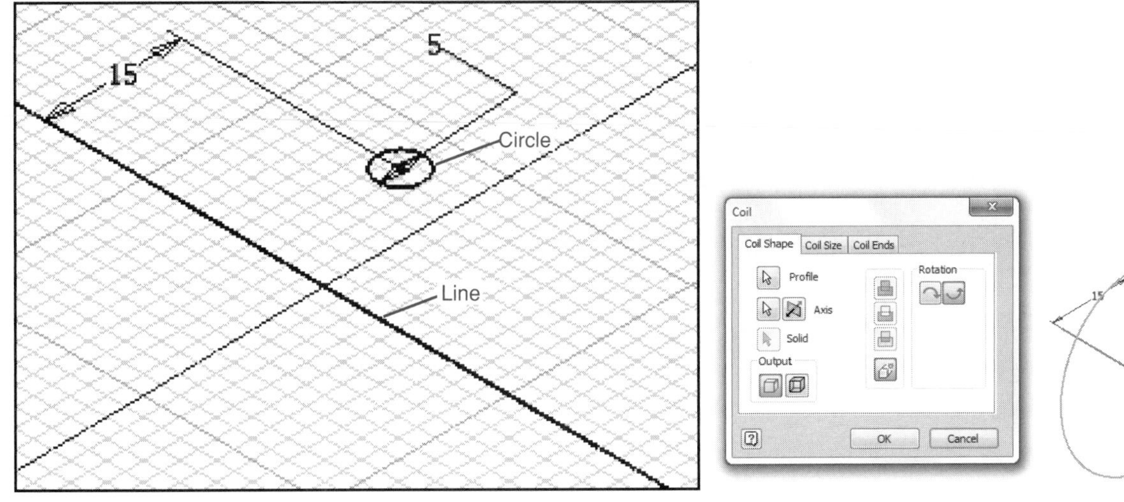

Figure 9-19

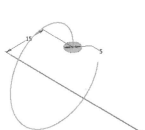

Figure 9-20

5 Set the **Pitch** for **15,** the **Revolution** for **6.00,** and the **Taper** for **0.00.** Click the **Coil Ends** tab.

See Figure 9-21.

6 Set the **Start** for **Flat** and both the **Transition Angle** and **Flat Angle** for **90°.** Set the **End** for **Flat** and the **Transition Angle** and **Flat Angle** for **0.00°.** Click **OK.**

See Figure 9-22. The finished spring will appear. See Figure 9-23.

Figure 9-21

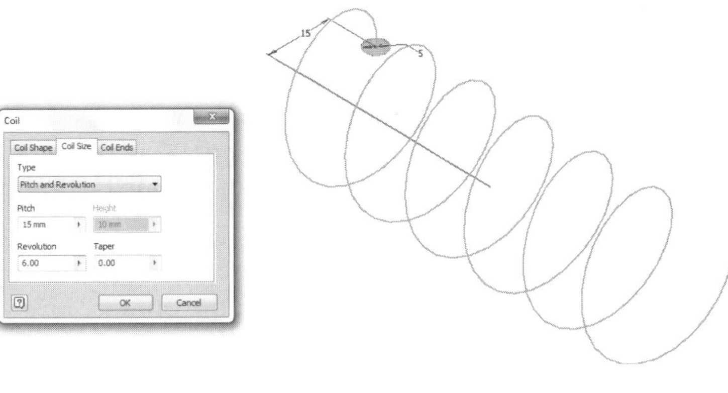

Figure 9-22

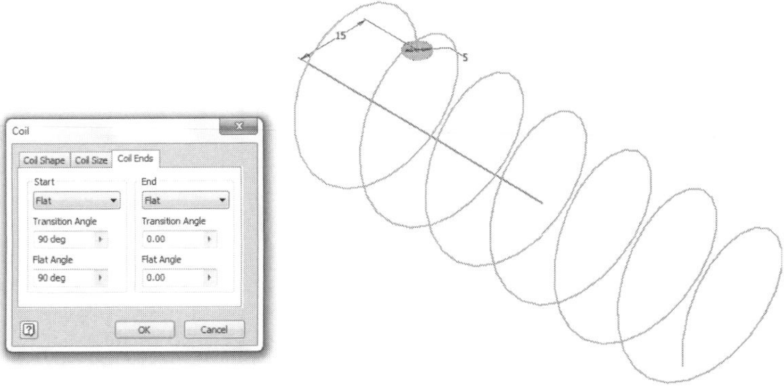

7 Use the **Free Orbit** tool located on the right edge of the screen and rotate the spring so that the start end is completely visible.

See Figure 9-24.

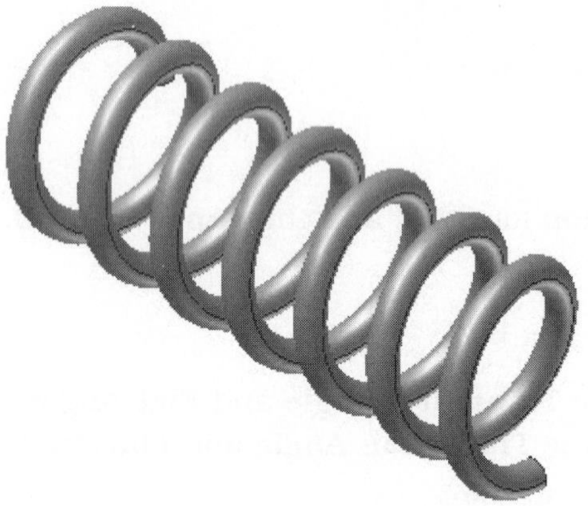

Start end

Figure 9-23

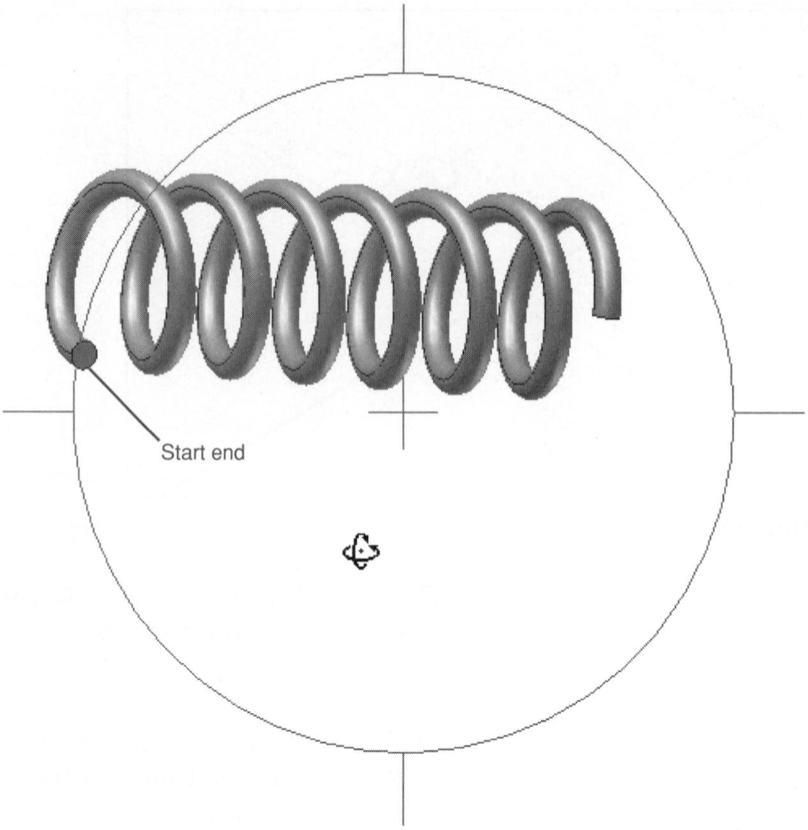

Figure 9-24

8 Create a new sketch plane aligned with the start end of the spring.

See Figure 9-25.

9 Right-click the mouse and select the **Finish 2D Sketch** option. Create an XY work plane aligned with the start end of the spring.

To create an XY work plane, click on the **Plane** tool on the **Work Features** panel, then on the **XY Plane** tool located under the **Origin** heading under **Part name** in the browser box, and click the start end's center point.

See Figure 9-26.

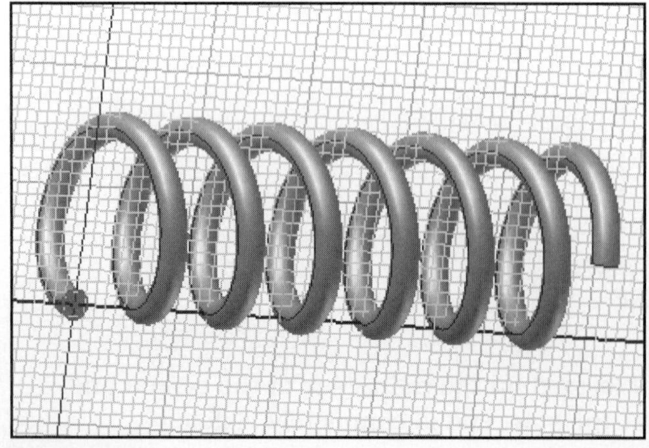

Figure 9-25

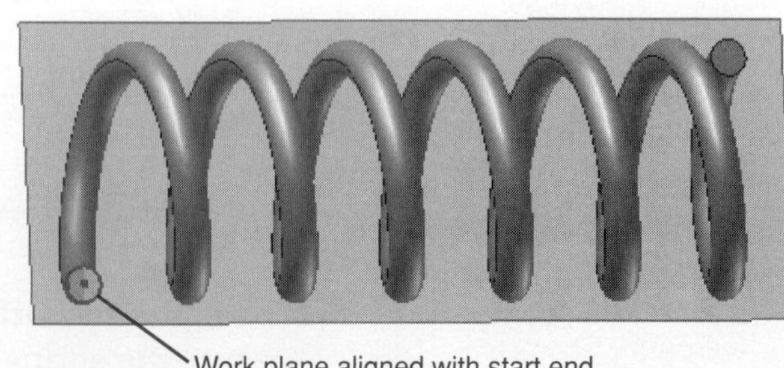

Figure 9-26

10 Add a YZ work plane through the center point of the spring's start end.

The YZ work plane is created by clicking the **Work Plane** tool, clicking the **YZ Plane** tool located under the **Origin** heading under the **Part Name** in the browser box, and clicking the starting end's center point.

See Figure 9-27.

11 Create a new sketch plane on the YZ work plane and draw a Ø**5** circle centered on the spring's start end.

See Figure 9-28. Use the **Free Orbit** and **Zoom** tools to orient and enlarge the end planes so they are clearly visible.

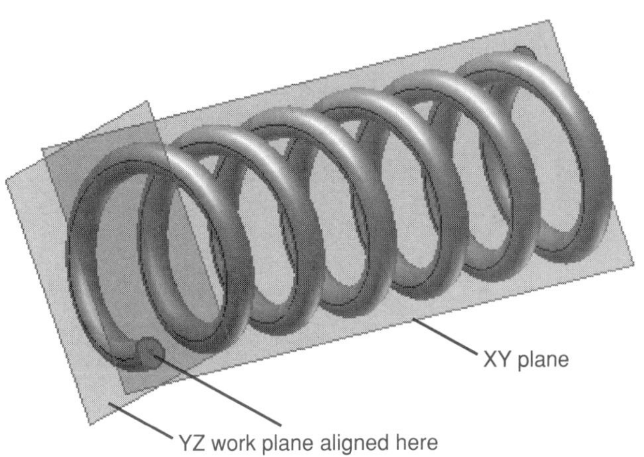

Figure 9-27

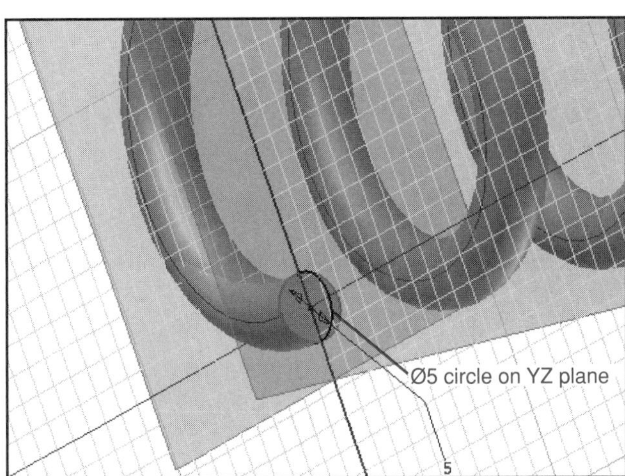

Figure 9-28

12 Right-click the mouse and select the **Finish 2D Sketch** option. Extrude the Ø5 circle **12 mm** away from the spring.

See Figure 9-29.

13 Rotate the spring and use the **Loft** tool to fill in the area between the extruded cylinder and the spring's start end.

See Figure 9-30.

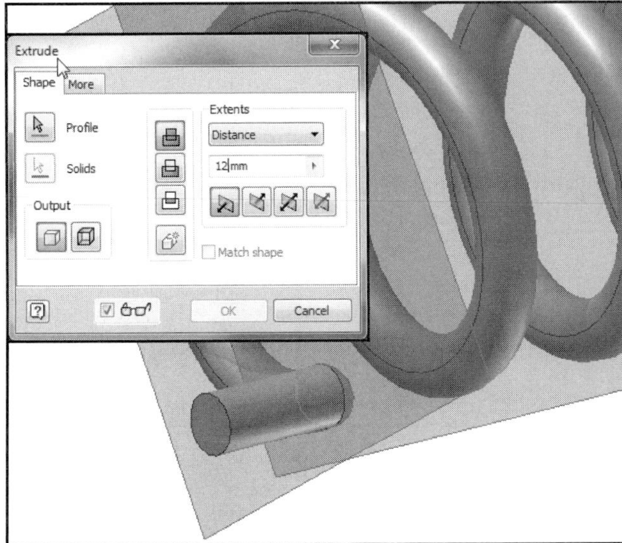

Figure 9-29

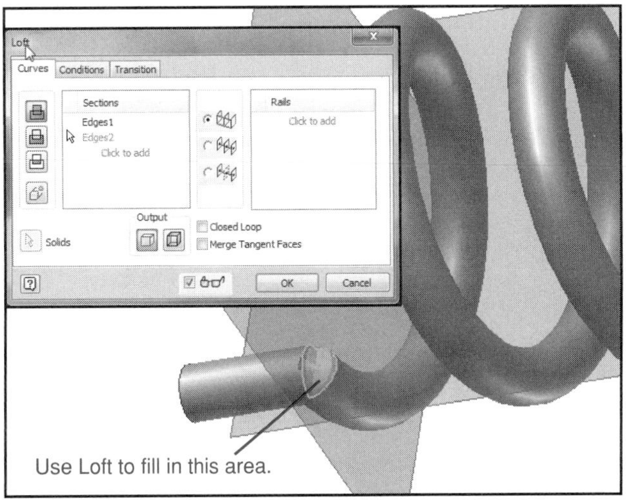

Figure 9-30

14 Create a new sketch plane on the end of the 12-mm extrusion created in step 12. Draw a **Ø5** circle centered about the center point of the extrusion's end.

See Figure 9-31.

15 Create a fixed work point on the extrusion's center point as shown. Create a new sketch plane on the XY work plane and draw a **15-mm, 180°** arc using the **Center Point Arc** tool, which is located on the **Draw** panel under the **Sketch** tab.

See Figure 9-32. Use the background grid and visually align the arc's center point with the extrusion's center point.

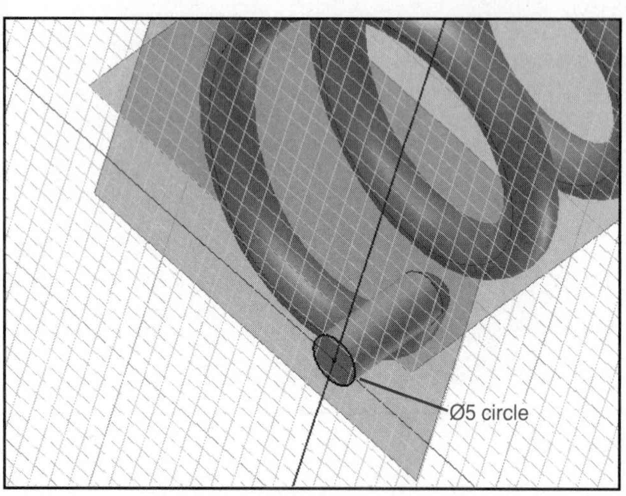

Figure 9-31

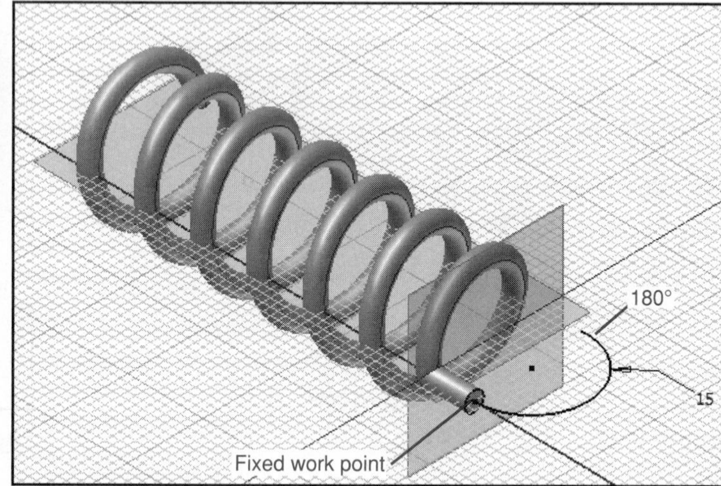

Figure 9-32

16 Use the **Sweep** tool to draw the spring's hooked end.

See Figures 9-33 and 9-34.

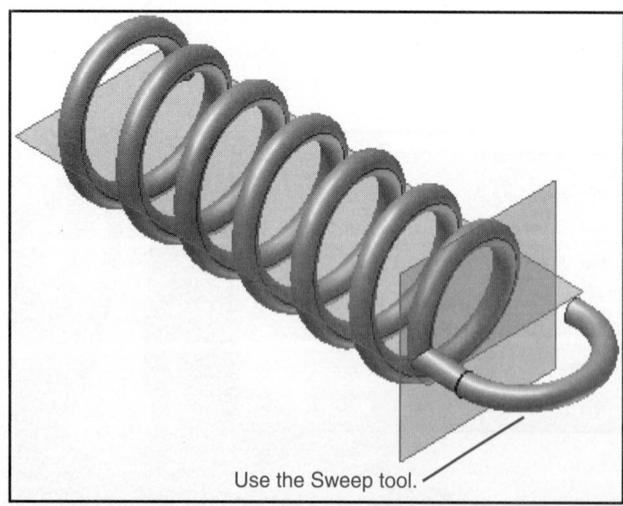

Figure 9-33

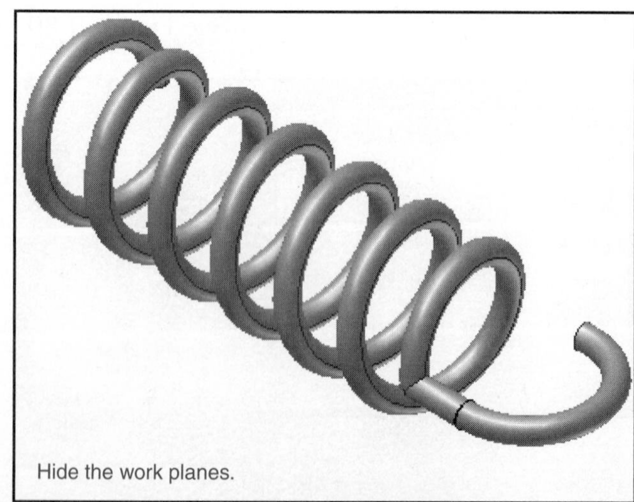

Figure 9-34

17 Apply the same procedure to the other end of the spring.

See Figures 9-35 and 9-36.

Figure 9-35

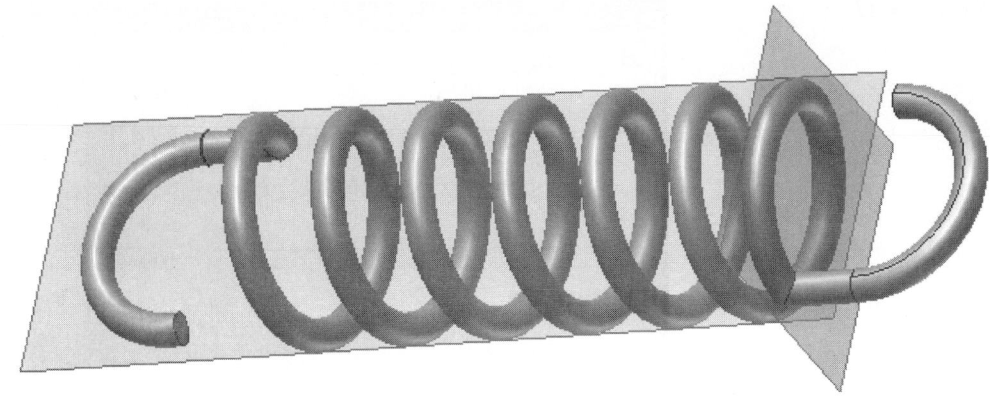

Figure 9-36

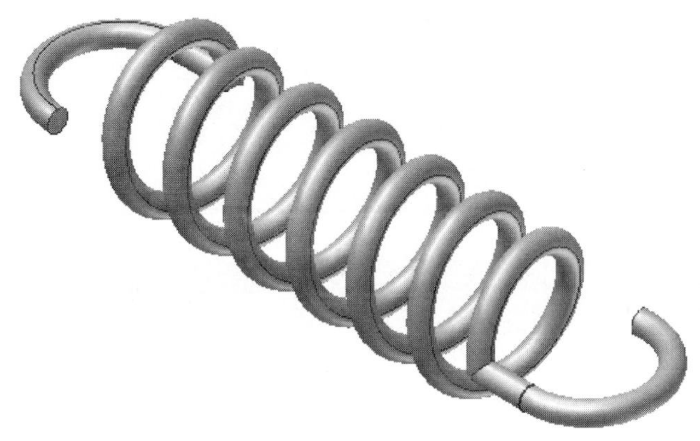

Torsion Springs

EXERCISE 9-7 **Drawing a Torsion Spring Using Tools on the Spring Panel**

1 Start a new drawing using the **Standard (in).iam** format.

The **Assemble** panels will appear.

2 Click the **Design** tab and select the **Torsion** tool located on the **Spring** panel.

The **Torsion Spring Component Generator** dialog box will appear. See Figure 9-37.

3 Enter the following values:

Coil Direction = **right**

Wire Diameter = **0.125**

Diameter, Outer = **1.2883**

Start Arm Length = **1.25**

End Arm Length = **1.00**

Length Inputs = **t, n --> L_0**

Active Coils Number = **14**

Figure 9-37

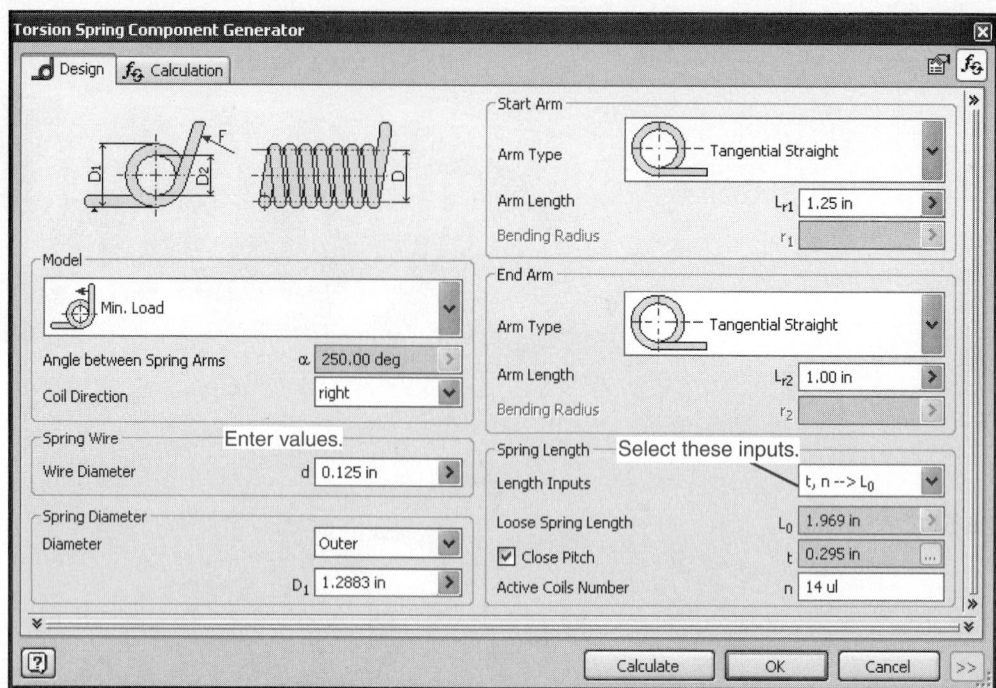

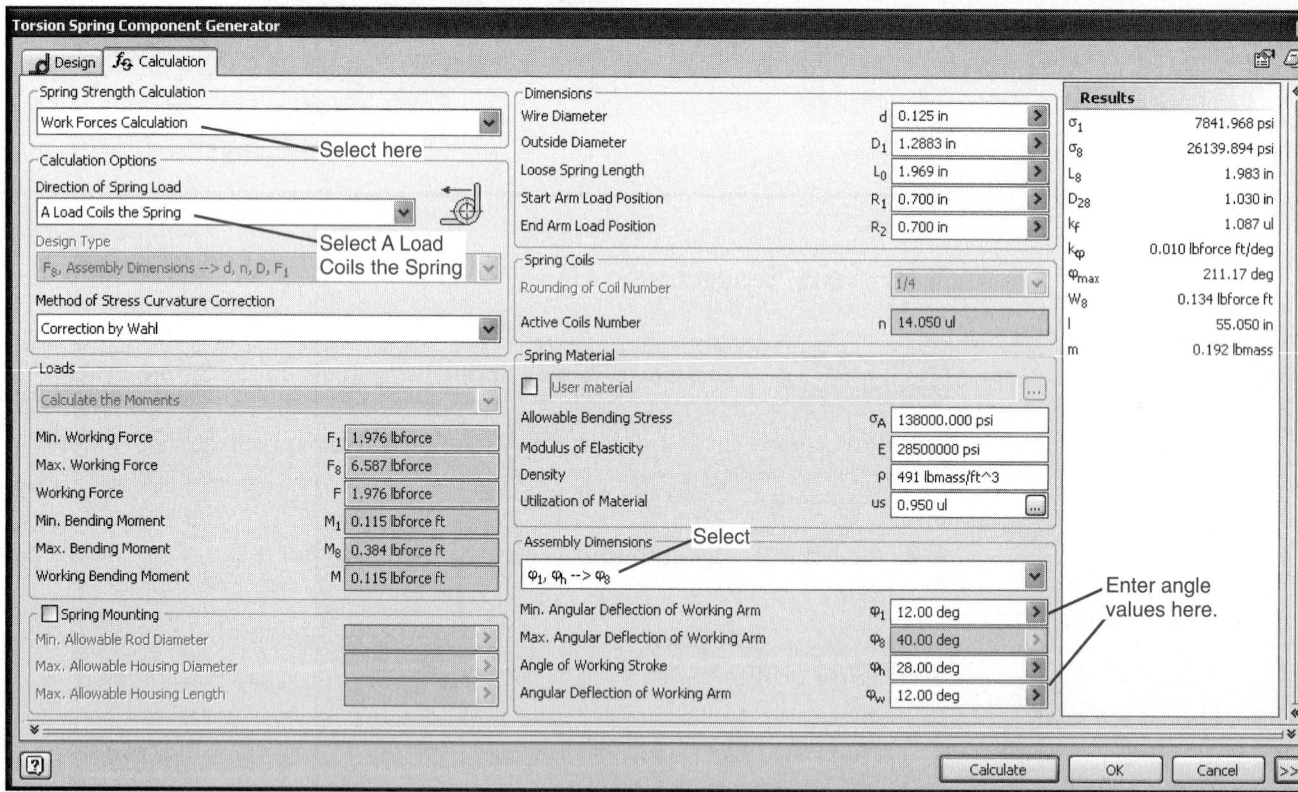

Figure 9-38

4 Click the **Calculation** tab and enter the following values and options. See Figure 9-38.

Spring Strength Calculation = Work Forces Calculation
Direction of Spring Load = A Load Coils the Spring
Min. Angular Deflection of Working Arm = 12.00
Angle of Working Stroke = 28
Angular Deflection of Working Arm = 12.00

5 Click **OK.**

The **File Naming** dialog box will appear.

6 Click **OK.**

Figure 9-39 shows the finished torsion spring.

Figure 9-39

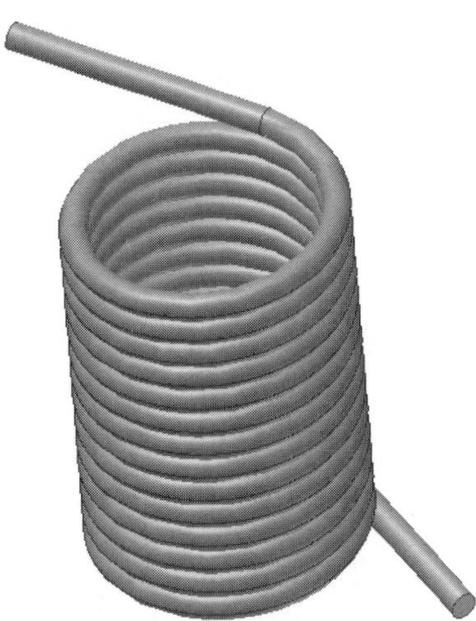

EXERCISE 9-8 **Drawing a Torsion Spring Using the Coil Tool**

1 Start a new drawing using the **Standard (mm).ipt** format, select the **Create 2D Sketch** option, and select the XY plane.

2 Use the **Home** tool and create an isometric orientation. Define a wire diameter of **6 mm** and a mean diameter of **32 mm.**

3 Right-click the mouse and select the **Finish 2D Sketch** option.

See Figure 9-40.

4 Access the **Coil** tool, define the **Profile** and **Axis,** click the **Coil Size** tab, and set the **Pitch** for **10** and the **Revolution** for **8.**

See Figure 9-41.

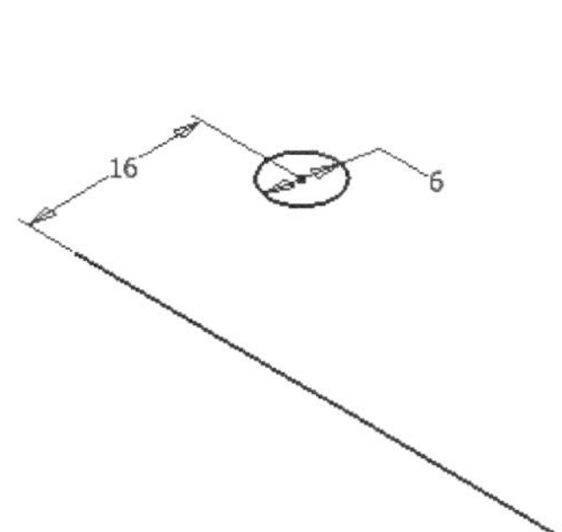

Figure 9-40

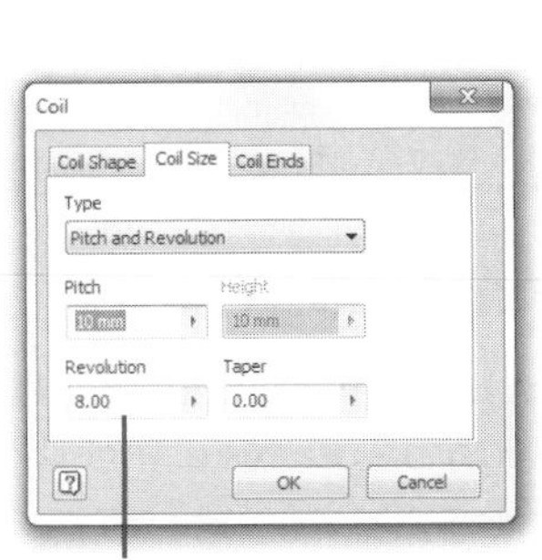

Figure 9-41

5 Set the **Start** for **Flat** and a **Transition Angle** and **Flat Angle** of **90°.** Set the **End** for **Flat** and the **Transition Angle** and **Flat Angle** for **45°.**

See Figure 9-42.

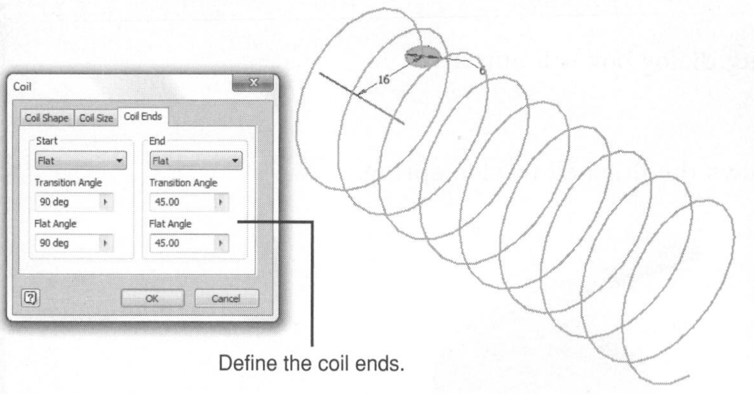

Figure 9-42

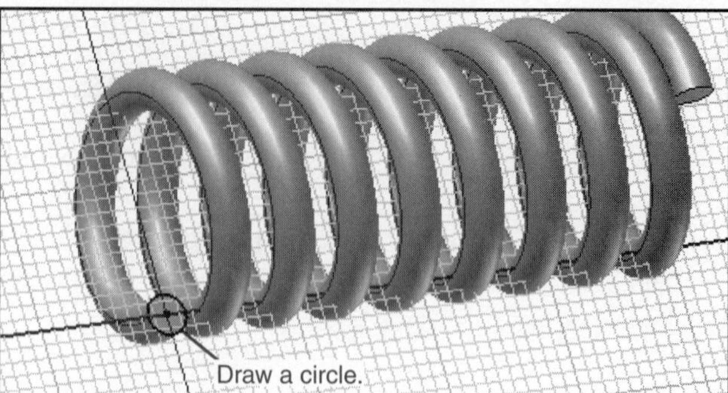

Figure 9-43

6 Use the **Free Orbit** tool and rotate the spring to access the start end. Create a new sketch plane on the end. Draw a Ø**6**-mm circle on the end centered on the end's center point.

See Figure 9-43.

7 Right-click the mouse and select the **Finish 2D Sketch** option. Extrude the start end **50** mm.

See Figure 9-44.

8 Extrude the other spring end **50 mm.**

Figure 9-45 shows the finished spring.

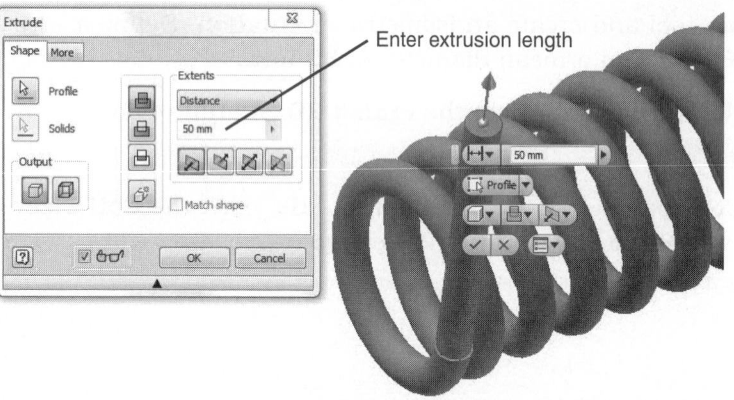

Figure 9-44

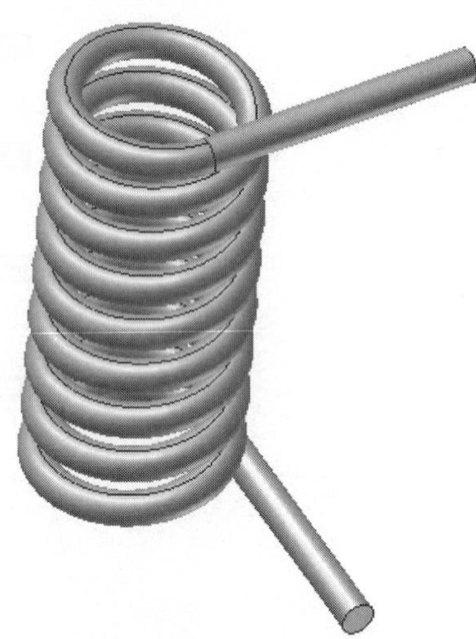

Figure 9-45

Belleville Springs

Belleville spring: A disklike device that resists bending; springs can be stacked bending in the same direction or in opposition.

A ***Belleville spring*** is a disklike device that resists bending. Belleville springs can be stacked together, all bending in the same direction, or stacked in opposition. The tools on the **Spring** panel can be used to generate either condition.

EXERCISE 9-9 Drawing a Belleville Spring Using Tools on the Spring Panel

1 Start a new drawing using the **Standard (in).iam** format.

2 Click the **Design** tab and select the **Belleville** tool on the **Spring** panel.

Figure 9-46

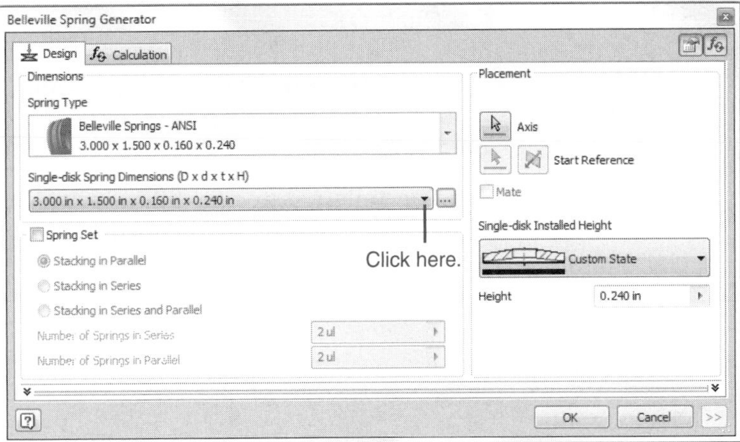

The **Belleville Spring Generator** dialog box will appear. See Figure 9-46.

3 Click on the arrow to the right of the **Single-disk Spring Dimensions** heading.

A list of options will cascade down.

4 Select the **3.000 in x 1.500 in x 0.100 in x 0.175 in** option.

See Figure 9-47.

Figure 9-47

Belleville Spring - ANSI

Name	D [in]	d [in]	t [in]	H [in]
2.25 x 1.125 x 0.056 x 0.126	2.25	1.125	0.056	0.126
2.5 x 1.25 x 0.132 x 0.198	2.5	1.25	0.132	0.198
2.5 x 1.25 x 0.085 x 0.149	2.5	1.25	0.085	0.149
2.5 x 1.25 x 0.063 x 0.142	2.5	1.25	0.063	0.142
2.75 x 1.375 x 0.15 x 0.225	2.75	1.375	0.15	0.225
2.75 x 1.375 x 0.095 x 0.166	2.75	1.375	0.095	0.166
2.75 x 1.375 x 0.071 x 0.16	2.75	1.375	0.071	0.16
3 x 1.5 x 0.16 x 0.24	3	1.5	0.16	0.24
3 x 1.5 x 0.1 x 0.175	3	1.5	0.1	0.175
3 x 1.5 x 0.075 x 0.169	3	1.5	0.075	0.169
3.5 x 1.75 x 0.18 x 0.27	3.5	1.75	0.18	0.27
3.5 x 1.75 x 0.118 x 0.207	3.5	1.75	0.118	0.207
3.5 x 1.75 x 0.09 x 0.203	3.5	1.75	0.09	0.203

Select here.

OK Cancel

5 Enter the **Height** value.

In this example a value of **0.16** was used. See Figure 9-48.

Figure 9-48

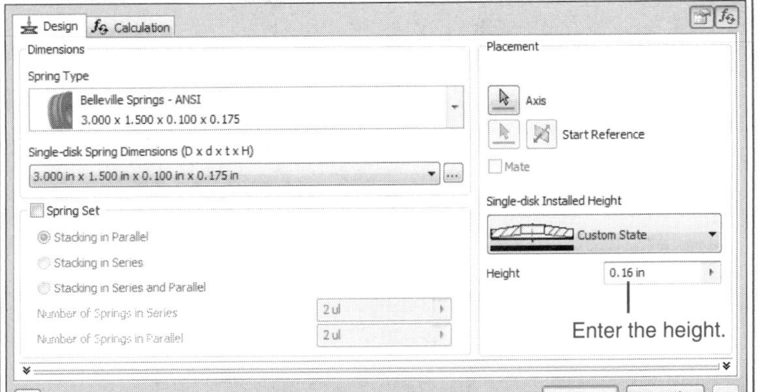

6 Click **OK.**

The **File Naming** dialog box will appear.

7 Click **OK.**

The finished spring will appear. See Figure 9-49. Figure 9-50 shows orthographic views of the spring.

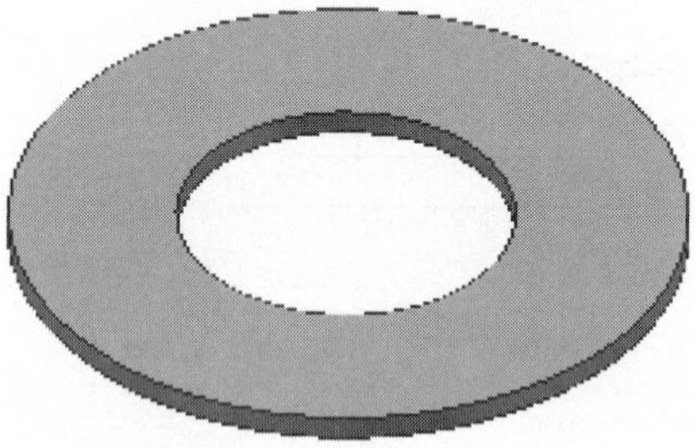

Finished Belleville spring

Figure 9-49

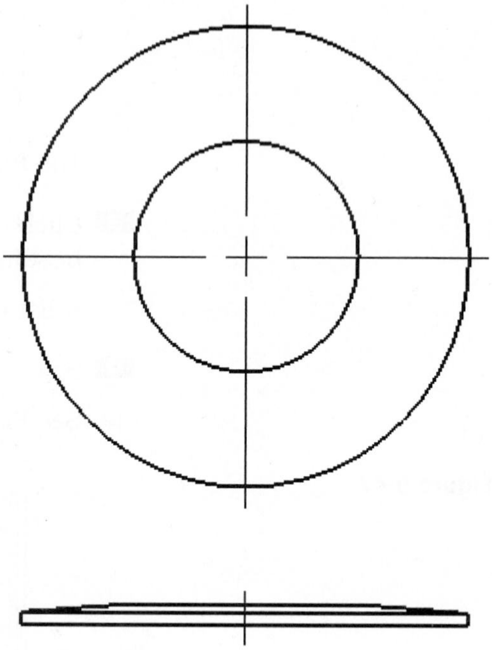

Orthographic views of a Belleville spring

Figure 9-50

Springs in Assembly Drawings

Figure 9-51 shows five components that are to be assembled together. The drawing was created using the **Standard (in).iam** format, and the springs were created using tools from the **Spring** panel.

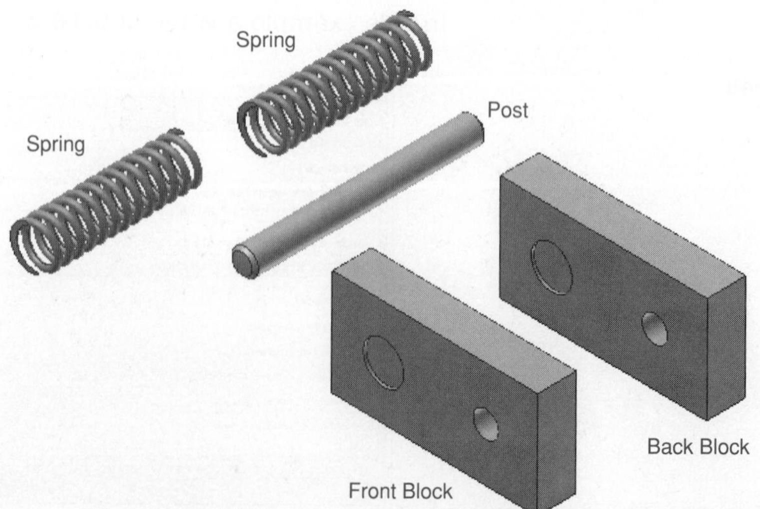

Figure 9-51

Figure 9-52 shows dimensioned drawings of the Blocks and the Post. Figure 9-53 shows the values used with the **Compression Spring Component Generator** dialog box.

The assembly procedure presented here represents one of several possible procedures.

Figure 9-52

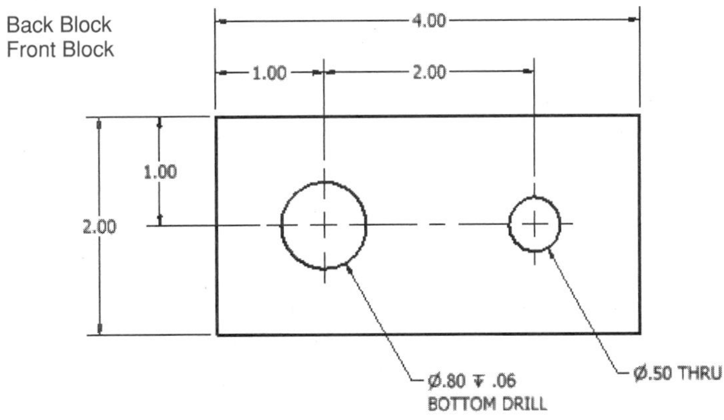

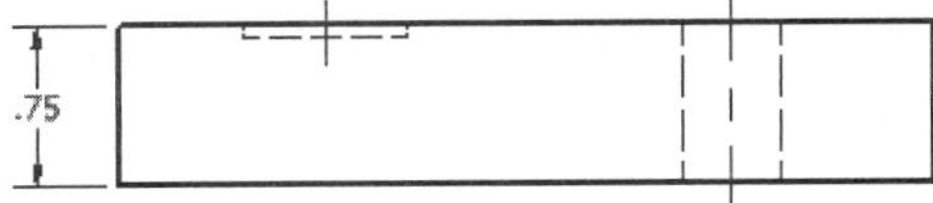

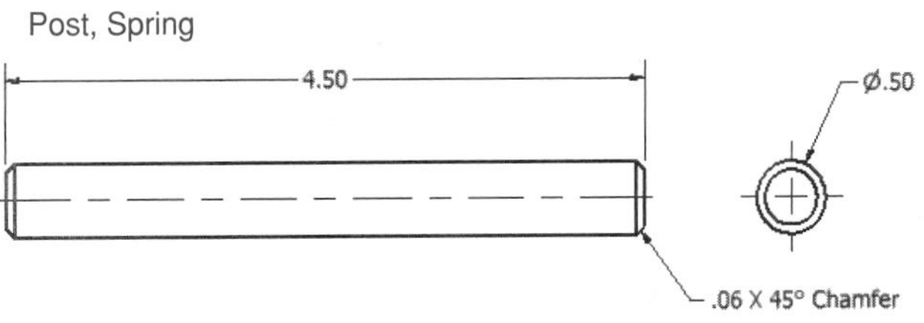

Figure 9-53

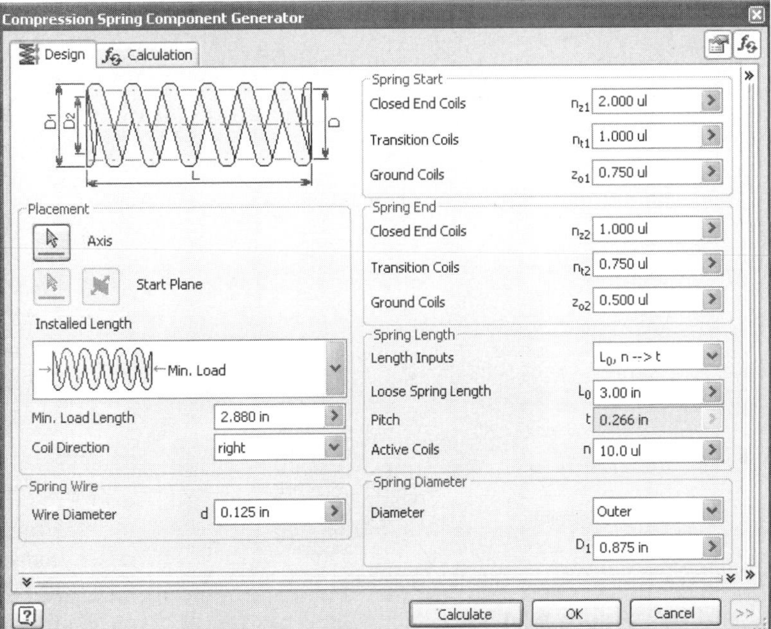

Figure 9-53
(Continued)

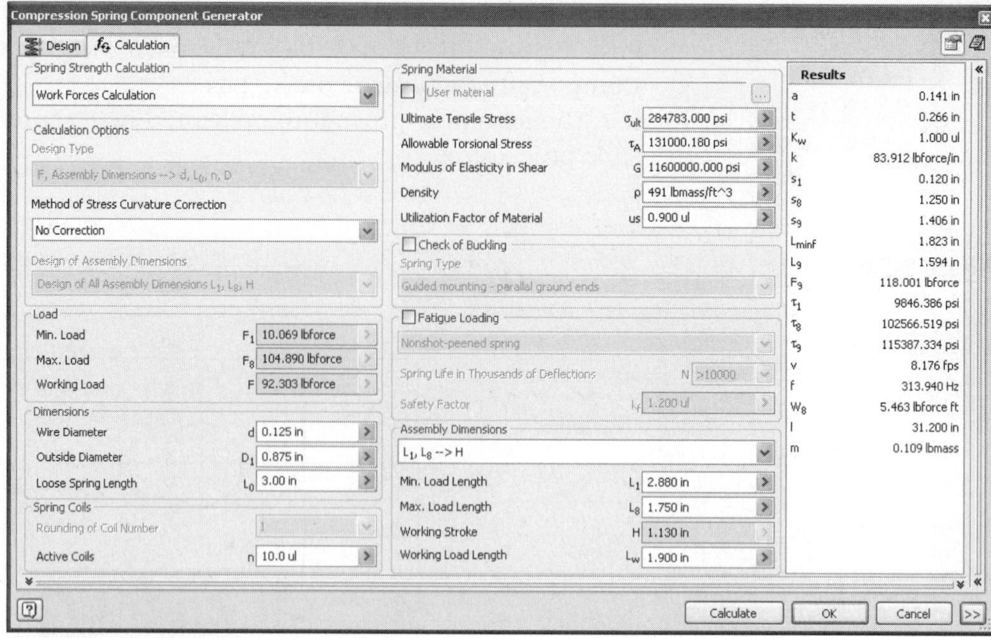

EXERCISE 9-10 Creating a Work Axis on the Front Block

1 Right-click on the Front Block and select the **Edit** option.

The other components will fade to a lighter color.

2 Create a work axis through the center point of the through hole.

3 Create a new sketch plane aligned with the bottom surface of the shallow hole. Click the right mouse button and select the **Finish 2D Sketch** (*not* **Finish Edit**) option. Create a work point on the center point of the shallow hole.

See Figure 9-54.

4 Create an XY work plane aligned with the work point.

5 Create a work axis perpendicular to the work plane through the work point.

See Figure 9-55.

6 Right-click the mouse and select **Finish Edit.**

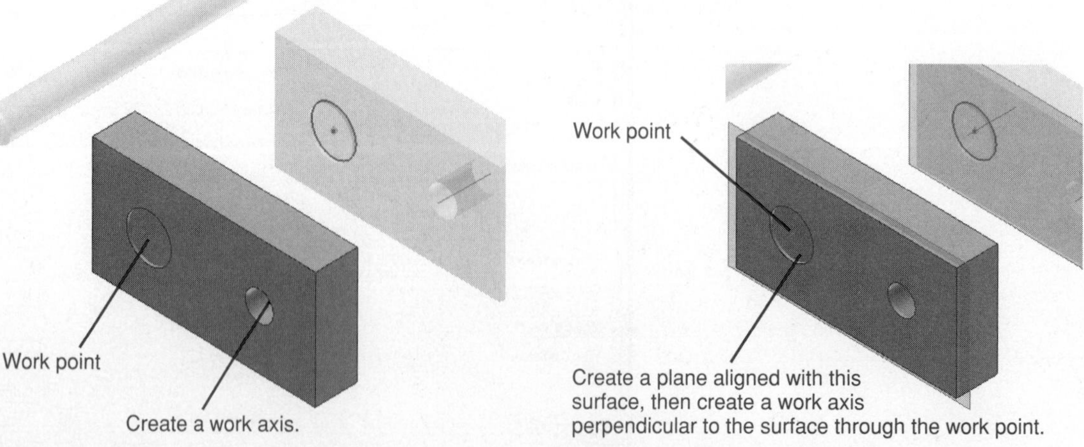

Figure 9-54

Figure 9-55

EXERCISE 9-11 Adding a Work Axis to the Springs

1 In the browser box, click the **Compression Spring** heading, and click the **WorkAxis1** heading.

See Figure 9-56.

2 Right-click on **WorkAxis1** and select the **Visibility** option.

The work axis will appear on the spring. See Figure 9-57.

3 Repeat the procedure for the second spring.

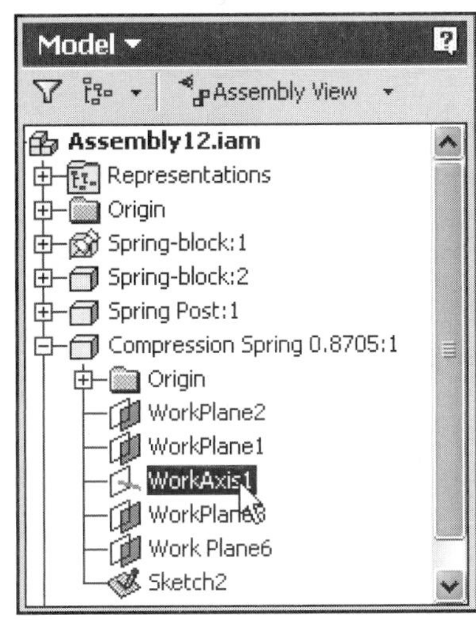

Figure 9-56

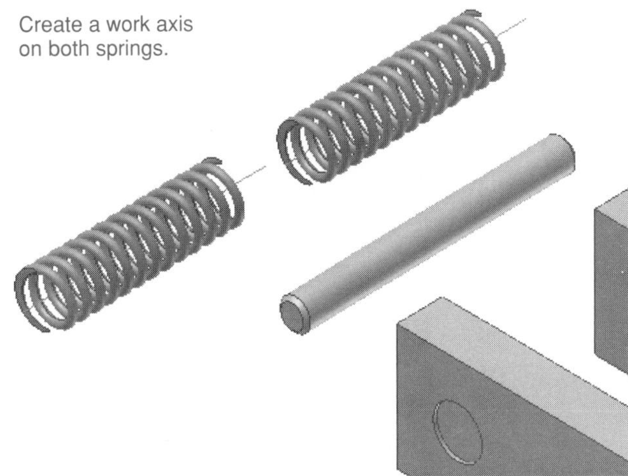

Figure 9-57

EXERCISE 9-12 Assembling the Components

1 Insert the Post into the Front Block.

See Figure 9-58. If the front surface of the Front Block is used, offset the post **–0.75** so that it aligns with the back surface of the Front Block.

2 Rotate one of the springs so that the ground end is accessible. Click the **Constrain** tool on the **Position** panel under the **Assemble** tab and select the **Mate** option, and align the spring's ground surface with the front surface of the Front Block.

See Figure 9-59.

Figure 9-58

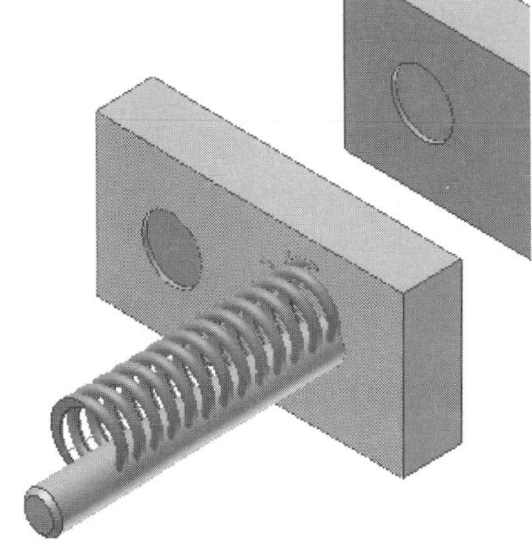

Figure 9-59

3 Use the **Mate** constraint to align the spring's work axis with the work axis through the through hole.

See Figure 9-60.

4 Repeat the procedure to position the second spring.

See Figure 9-61.

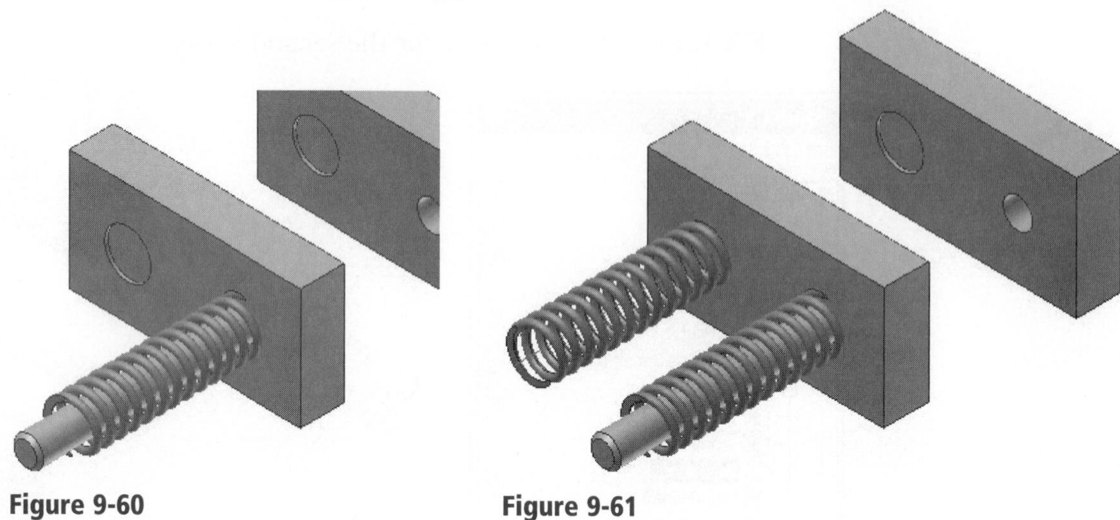

Figure 9-60

Figure 9-61

5 Add the Rear Block to the assembly.

See Figure 9-62.

Figure 9-62

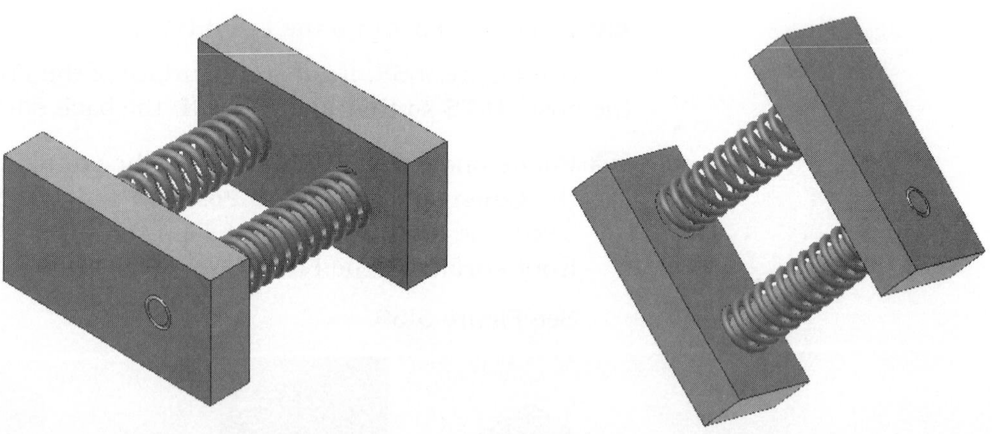

Chapter Summary

This chapter showed how to draw four types of springs—compression, extension, torsion, and Belleville—using both the **Coil** and the **Spring** tools. Coil features such as natural ends and flat ends with different transition and flat angles were illustrated.

Chapter Test Questions

Multiple Choice

Circle the correct answer.

1. A compression spring is one in which the forces
   a. Twist the spring
   b. Pull the spring apart lengthwise
   c. Push the spring together lengthwise
2. An extension spring is one in which the forces
   a. Twist the spring
   b. Pull the spring apart lengthwise
   c. Push the spring together lengthwise
3. A torsion spring is one in which the forces
   a. Twist the spring
   b. Pull the spring apart lengthwise
   c. Push the spring together lengthwise
4. A compression spring is defined using the **Coil** tool by defining
   a. An axis and a work plane
   b. An axis and a profile
   c. A work plane and a work axis
5. Which of the following is not a type of hook for an extension spring?
   a. Full loop
   b. Raised hook
   c. Back loop
   d. Double twisted full loop

Matching

Match the following notations with their definition for creating a spring using the tools from the ***Spring*** *panel.*

Column A	Column B
a. n ________	1. Wire diameter
b. D_1 ________	2. Outer spring diameter
c. d ________	3. Number of active coils
d. L_0 ________	4. Spring start ground coils
e. z_{01} ________	5. Loose spring length

True or False

Circle the correct answer.

1. **True or False:** Springs can be drawn using either the **Coil** tool or the tools on the **Spring** panel.
2. **True or False:** The ends of compression springs are often ground flat.
3. **True or False:** Extension springs have hooked ends.
4. **True or False:** Belleville springs have coils.
5. **True or False:** The pitch of a spring is the distance from the center of one coil to the center of the next coil.
6. **True or False:** The coil direction of a spring can be either left or right.

Chapter Projects

Project 9-1: Inches

A. Draw the following springs using the **Coil** tool.
   Mean Diameter = 2.00
   Wire Diameter = 0.125
   Pitch = 0.375
   Revolution = 16
   Ends = Natural

B. Draw a second spring from the same data that has ground ends and a preloaded length of 5.00.

Figure P9-1

Project 9-2: Inches

A. Draw the following springs using the **Coil** tool.
   Mean Diameter = 0.50
   Wire Diameter = 0.06
   Pitch = 0.12
   Revolution = 10
   Ends = Flat, Transition angle = 90°, Flat angle = 0°

B. Draw a second spring from the same data that has ground ends and a preloaded length of 1.00.

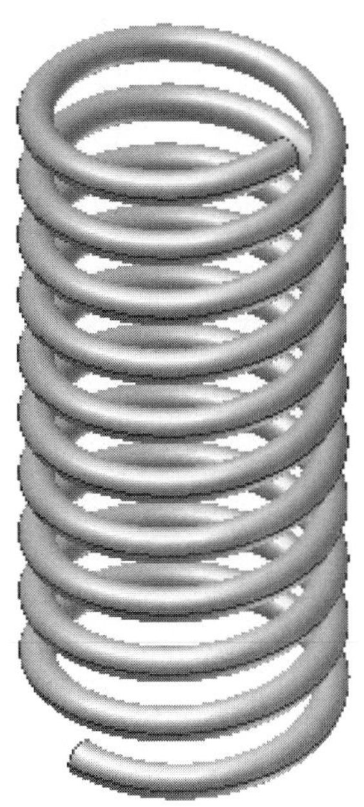

Figure P9-2

Project 9-3: Millimeters

A. Draw the following springs using the **Coil** tool.
   Mean Diameter = 8
   Wire Diameter = 3
   Pitch = 4
   Revolution = 20
   Ends = Natural

B. Draw a second spring from the same data that has ground ends and a preloaded length of 65.00.

Project 9-4: Millimeters

A. Draw the following springs using the **Coil** tool.
   Mean Diameter = 24
   Wire Diameter = 12
   Pitch = 14
   Revolution = 6
   Ends = Flat, Transition Angle = 90°, Flat Angle = 45°

B. Draw a second spring from the same data that has ground ends and a preloaded length of 75.00.

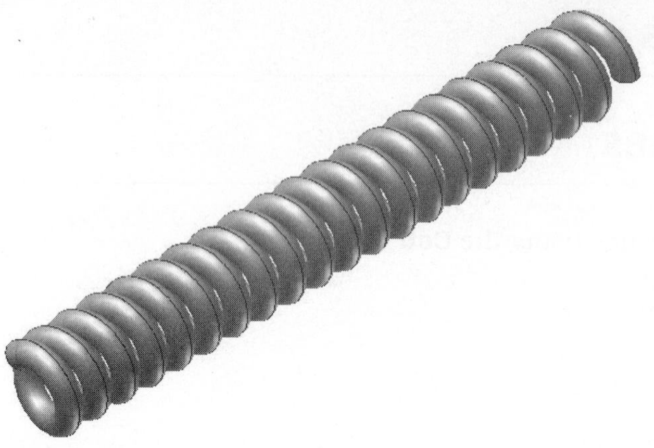

Figure P9-3

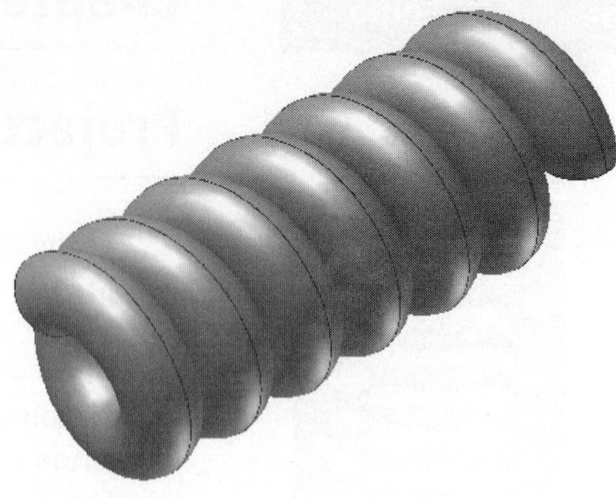

Figure P9-4

Project 9-5: Millimeters

Draw the following springs using the **Coil** tool.

Mean Diameter = 25
Wire = 5 x 5 square
Pitch = 6
Revolution = 8
Ends = Natural

Draw the following compression springs using the tools on the **Spring** panel.

Figure P9-5

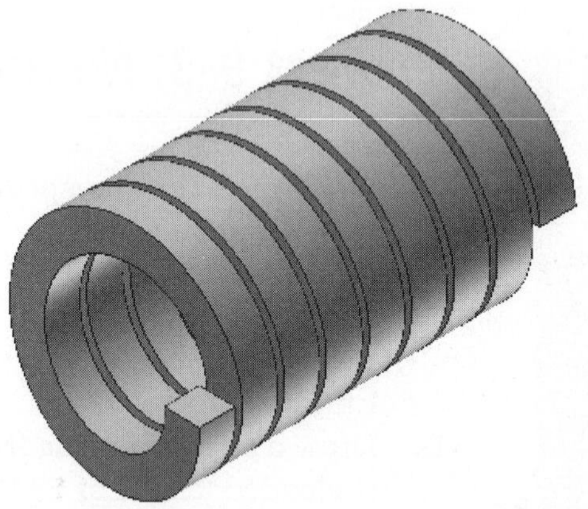

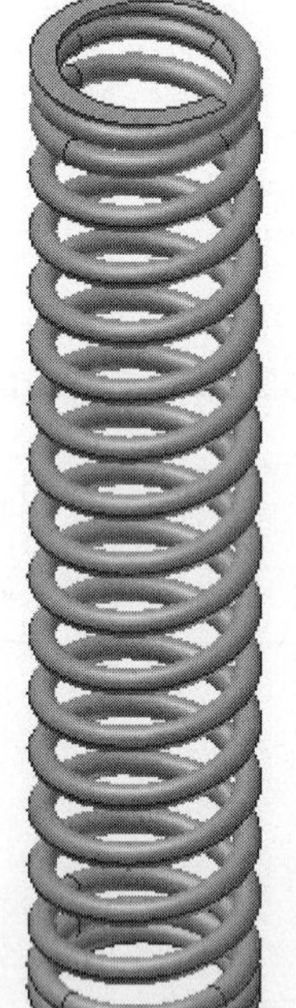

Figure P9-6

Project 9-6: Inches

Wire Diameter = 0.072
Outside Diameter = 0.625
Loose Spring Length = 3.15
Min. Load Length = 3.00
Working Stroke = 0.15
Max. Load Length = 2.95
Coil Direction = right

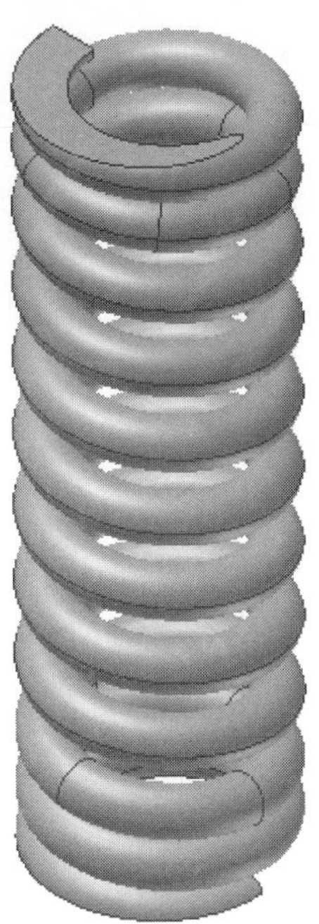
Figure P9-7

Active Coils = 16
Both coils
- Closed End Coils = 1.5
- Transition Coils = 1.0
- Ground Coils = 0.75

Project 9-7: Inches

Wire Diameter = 0.2437
Outside Diameter = 1.250
Loose Spring Length = 4.1
Min. Load Length = 3.825
Working Stroke = 0.20
Max. Load Length = 3.82
Coil Direction = left
Active Coils = 9
Both coils
- Closed End Coils = 1.5
- Transition Coils = 1.0
- Ground Coils = 0.75

Project 9-8: Millimeters

Wire Diameter = 3.00
Outside Diameter = 20.0
Loose Spring Length = 81.0
Min. Load Length = 80.0
Working Stroke = 4.0
Max. Load Length = 80.0
Coil Direction = right
Active Coils = 10
Both coils
- End Coils = 1.5
- Transition Coils = 1.0
- Ground Coils = 0.75

Project 9-9: Millimeters

Wire Diameter = 5.0
Outer Diameter = 30.0
Loose Spring Length = 101.0
Min. Load Length = 100.0
Working Stroke = 6.0
Working Load Length = 94.0
Coil Direction = right

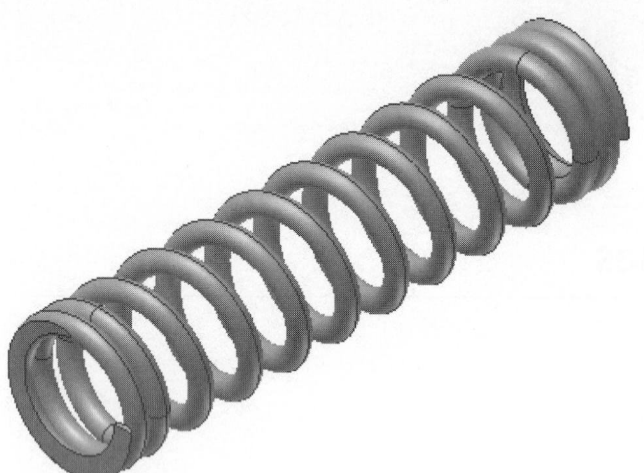

Figure P9-8

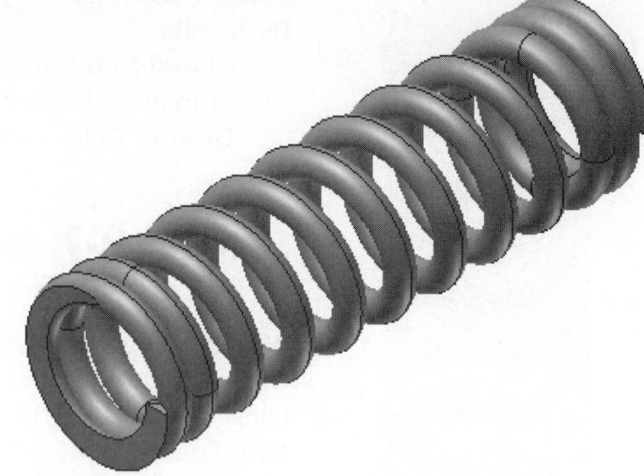

Figure P9-9

Active Coils = 9
Both coils
    End Coils = 1.5
    Transition Coils = 1.0
    Ground Coils = 0.75

Draw the following extension springs using the tools on the **Spring** panel.

Note: $L_0 < L_c < L_8$. The **Custom Length** of the spring must be greater than the **Loose Spring Length** and less than the **Max. Load Length**.

Project 9-10: Inches

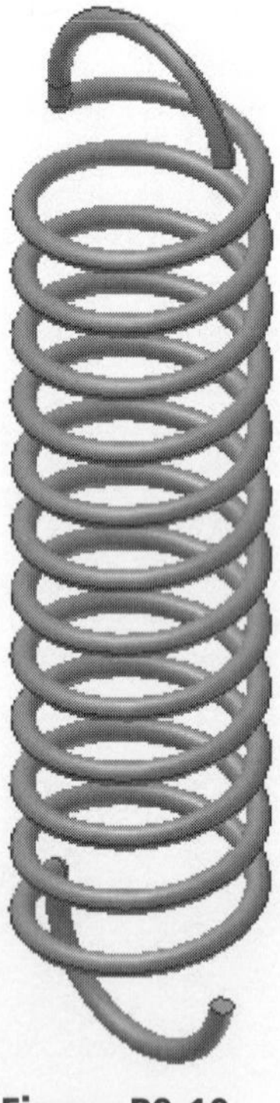

Spring Prestress = With Prestress
Coil Direction = right
Wire Diameter = 0.0938
Outer Diameter = 1.000
Min. Load Length = 6.00
Active Coils Number = 12
Both hooks
    Spring Hook = Half Hook
    Spring Hook Height = 0.50
    Working Stroke = 0.25
    Working Load Length = 6.1

Figure P9-10

Project 9-11: Inches

Spring Prestress = With Prestress
Coil Direction = right
Wire Diameter = 0.1000

Figure P9-11

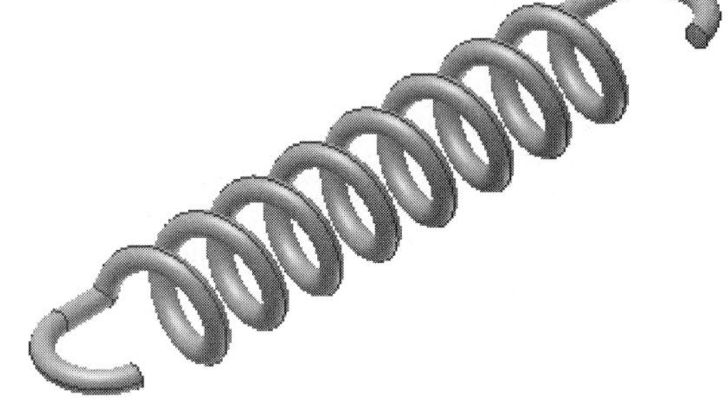

Figure P9-12

Outer Diameter = 1.500
Loose Spring Length = 6.00
Active Coils Number = 16
Both hooks

- Spring Hook = Non-specified Hook Type
- Max. Load Length = 6.25
- Working Stroke = 0.300
- Working Load Length = 6.00

Project 9-12: Millimeters

Spring Prestress = Without Prestress
Coil Direction = right
Wire Diameter = 4
Outer Diameter = 22
Loose Spring Length = 130
Active Coils Number = 8
Both hooks

- Spring Hook = Raised Hook
- Hook Length = 16.936
- Max. Load Spring Length = 138
- Working Stroke = 4
- Working Load Length = 136

Project 9-13: Millimeters

Spring Prestress = With Prestress
Coil Direction = right
Wire Diameter = 3.75
Outer Diameter = 40
Loose Spring Length = 64.923
Active Coils Number = 4.18
Both hooks
    Spring Hook = Half Hook
    Hook Length = 22.514
    Max. Load Spring Length = 80
    Working Stroke = 6
    Working Load Length = 74

Draw the following extension springs using the **Coil** tool.

Figure P9-13

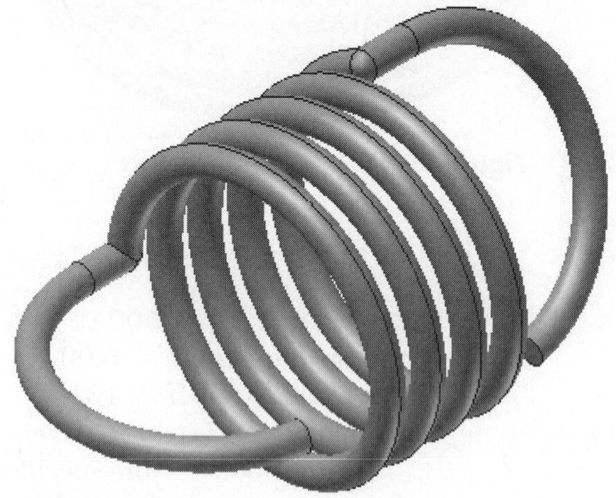

Project 9-14: Inches

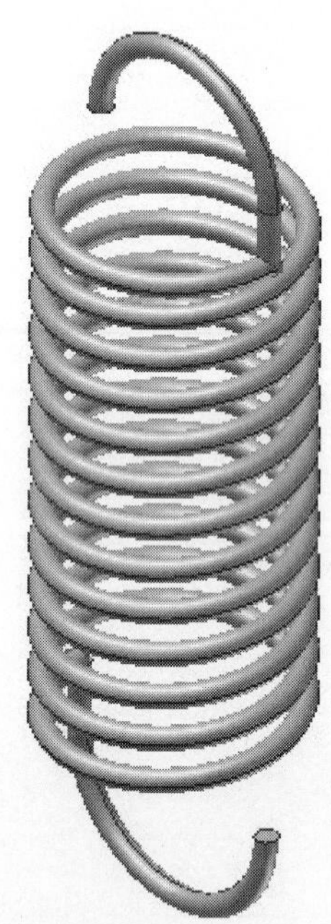

Figure P9-14

Wire Diameter = 0.19
Mean Diameter = 1.00
Pitch = .375
Revolution = 12
Taper = 0
Start Coil End = Flat, Transition Angle = 90.0°, Flat Angle = 90.0°
End Coil End = Flat, Transition Angle = 0.0°, Flat Angle = 0.0°
Extension Distance = 0.50
Arc Radius = 1.00

Project 9-15: Inches

Wire Diameter = 0.375
Mean Diameter = 1.50
Pitch = .625

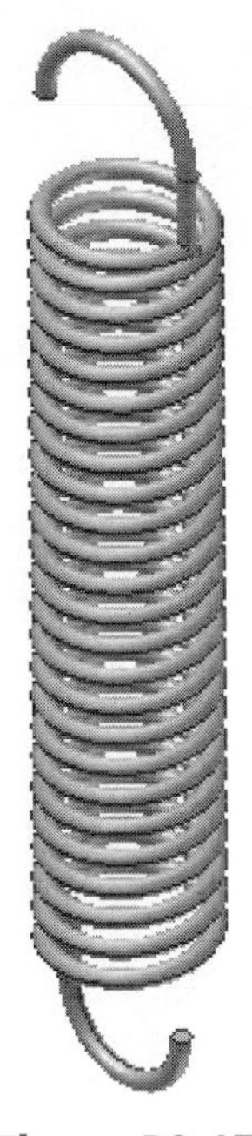

Figure P9-15

Revolution = 24
Taper = 0
Start Coil End = Flat, Transition Angle = 90.0°, Flat Angle = 90.0°
End Coil End = Flat, Transition Angle = 0.0°, Flat Angle = 0.0°
Extension Distance = 1.50
Arc Radius = 1.50

Project 9-16: Millimeters

Wire Diameter = 16
Mean Diameter = 30
Pitch = 30
Revolution = 10
Taper = 0
Start Coil End = Flat, Transition Angle = 90.0°, Flat Angle = 90.0°
End Coil End = Flat, Transition Angle = 0.0°, Flat Angle = 0.0°
Extension Distance = 18
Arc Radius = 30

Figure P9-16

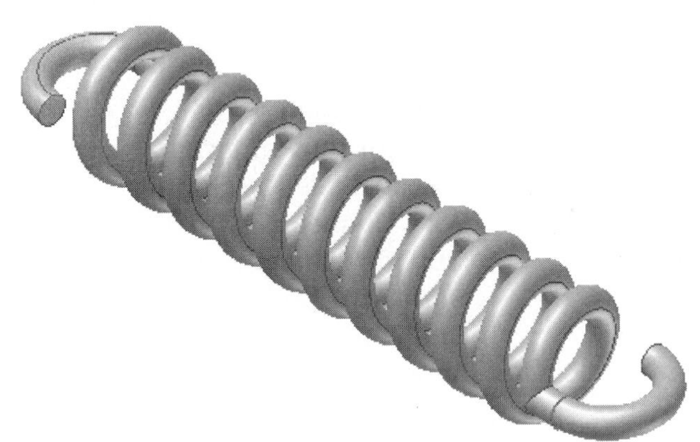

Project 9-17: Millimeters

Wire Diameter = 8
Mean Diameter = 24
Pitch = 14
Revolution = 16
Taper = 0
Start Coil End = Flat, Transition Angle = 90.0°, Flat Angle = 90.0°
End Coil End = Flat, Transition Angle = 0.0°, Flat Angle = 0.0°
Extension Distance = 26
Arc Radius = 24

Draw the following torsion springs using the tools on the **Spring** panel. Use the default values for any values not specified.

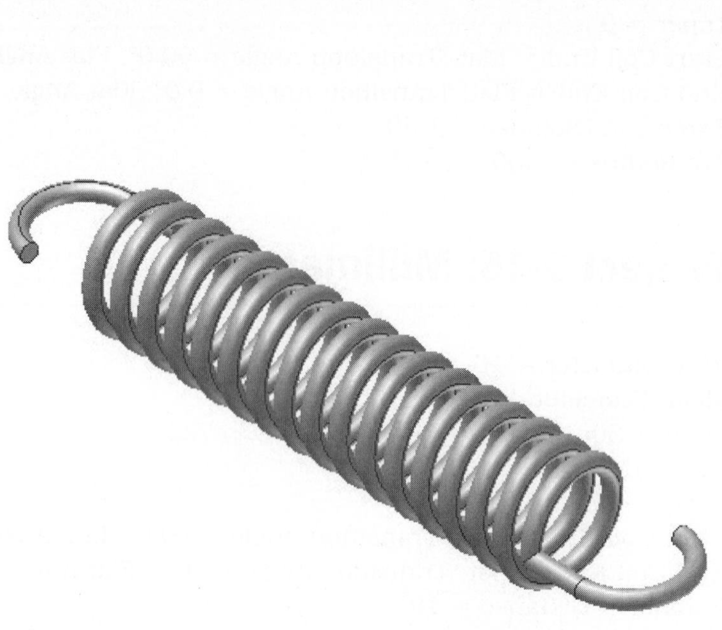

Figure P9-17

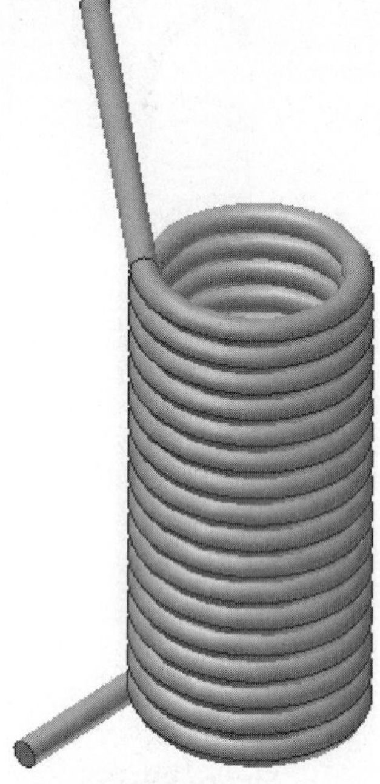

Figure P9-18

Project 9-18: Inches

Coil Direction = right
Straight torsion arms
A Load Coils the Spring
Wire Diameter = 0.1055
Outside Diameter = 0.875
Start Arm Length = 1.50
End Arm Length = 0.75
Active Coils Number = 12
Min. Angular Deflection of Working Arm = 12
Angle of Working Stroke = 36
Angular Deflection of Working Arm = 15

Project 9-19: Inches

Coil Direction = right
Straight torsion arms
A Load Uncoils the Spring
Wire Diameter = 0.072
Outside Diameter = 0.500
Start Arm Length = 0.50
End Arm Length = 0.50
Active Coils Number = 18
Min. Angular Deflection of Working Arm = 20
Angle of Working Stroke = 10
Angular Deflection of Working Arm = 30

Project 9-20: Millimeters

Coil Direction = left
Straight torsion arms
A Load Coils the Spring
Wire Diameter = 2.00
Outside Diameter = 16
Arm of Working Force = 20.0
Arm of Support = 20.0
Active Coils Number = 10
Min. Angular Deflection of Working Arm = 10
Angle of Working Stroke = 40
Angular Deflection of Working Arm = 10

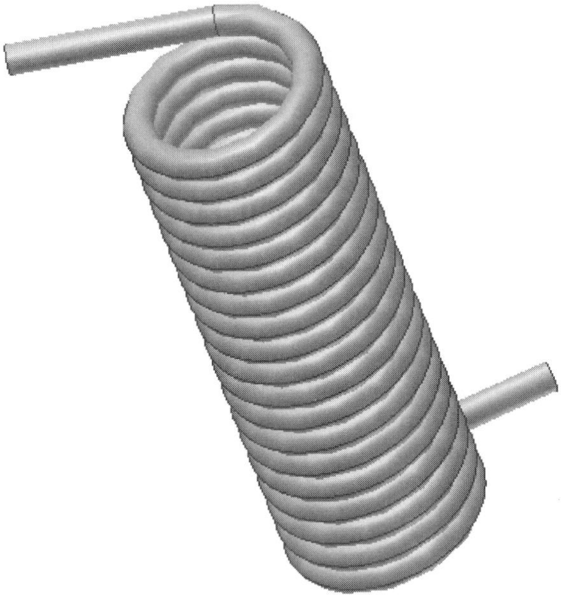

Figure P9-19

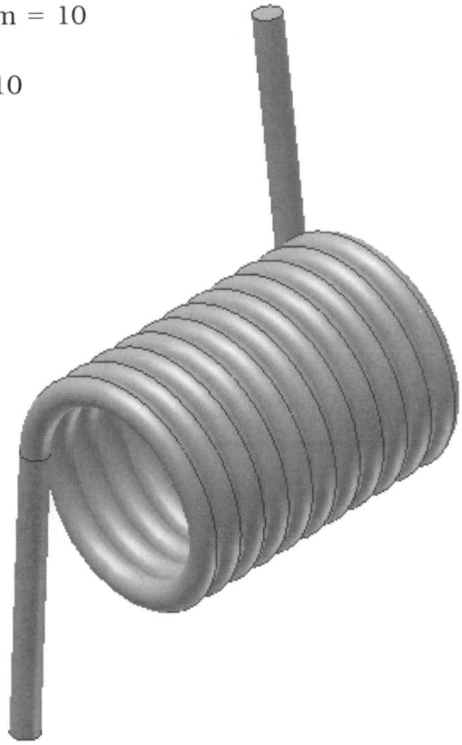

Figure P9-20

Project 9-21: Millimeters

Coil Direction = right
Straight torsion arms
A Load Uncoils the Spring
Wire Diameter = 3.0
Mean Diameter = 40
Arm of Working Force = 35
Arm of Support = 30
Active Coils Number = 20
Min. Angular Deflection of Working Arm = 15
Angle of Working Stroke = 50
Angular Deflection of Working Arm = 15

Draw the following torsion springs using the **Coil** tool.

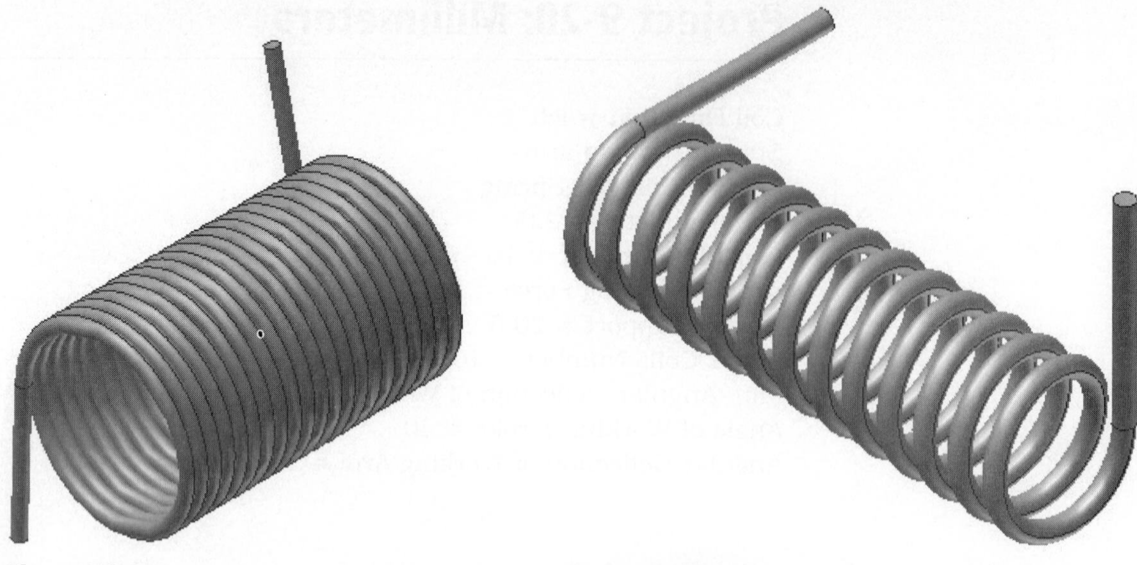

Figure P9-21

Figure P9-22

Project 9-22: Inches

Create a drawing using the **Standard (in).ipt** format.
Wire Diameter = 0.19
Mean Diameter = 1.25
Pitch = 0.375
Revolution = 12
Start Coil = Flat, Transition Angle = 90°, Flat Angle = 90°
End Coil = Flat, Transition Angle = 45°, Flat Angle = 45°
Arm Length (both arms) = 2.00

Project 9-23: Inches

Create a drawing using the **Standard (in).ipt** format.
Wire Diameter = 0.06
Mean Diameter = 0.50
Pitch = 0.125
Revolution = 20
Start Coil = Flat, Transition Angle = 90°, Flat Angle = 90°
End Coil = Flat, Transition Angle = 45°, Flat Angle = 45°
Arm Length (both arms) = 1.00

Project 9-24: Millimeters

Create a drawing using the **Standard (mm).ipt** format.
Wire Diameter = 4.0
Mean Diameter = 12.0
Pitch = 6
Revolution = 18

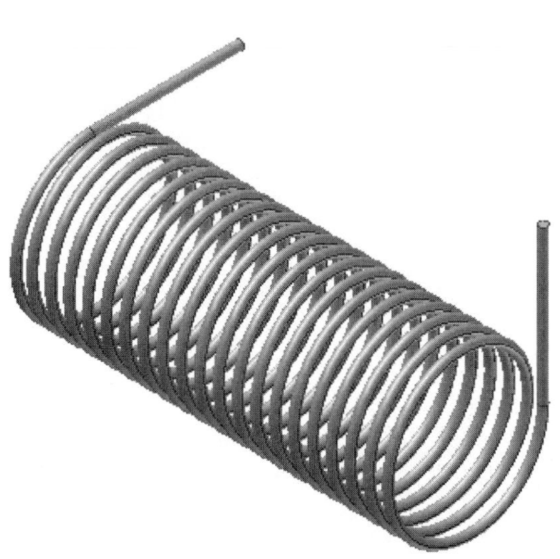

Figure P9-23

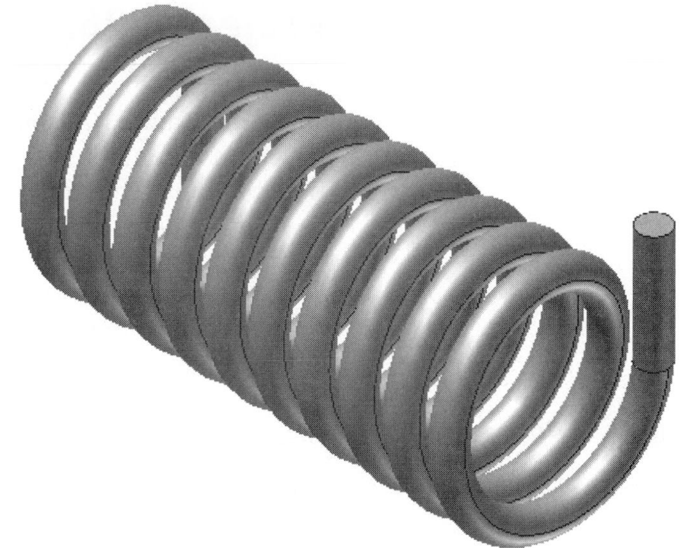

Figure P9-24

Start Coil = Flat, Transition Angle = 90°, Flat Angle = 90°
End Coil = Flat, Transition Angle = 45°, Flat Angle = 45°
Arm Length (both arms) = 15

Project 9-25: Millimeters

Create a drawing using the **Standard (mm).ipt** format.
Wire Diameter = 5
Mean Diameter = 40.0
Pitch = 10
Revolution = 18
Start Coil = Flat, Transition Angle = 90°, Flat Angle = 90°
End Coil = Flat, Transition Angle = 45°, Flat Angle = 45°
Arm Length (both arms) = 60

Figure P9-25

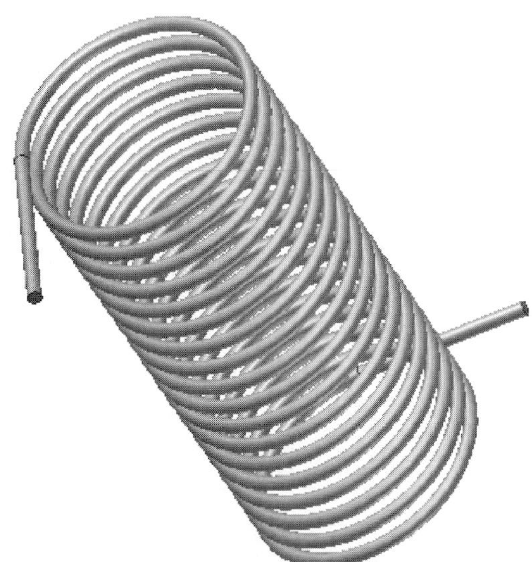

Project 9-26: Inches

Draw the Damper Assembly shown in Figure P9-26. This exercise is loosely based on a damper system placed under the seat of helicopter pilots to help minimize the amount of vibration they experience.

Spring data:
Spring Strength Calculation = Work Forces Calculation
Wire Diameter = 0.125
Outside Diameter = 1.000
Min. Load Length = 2.10

Figure P9-26

Damper Assembly

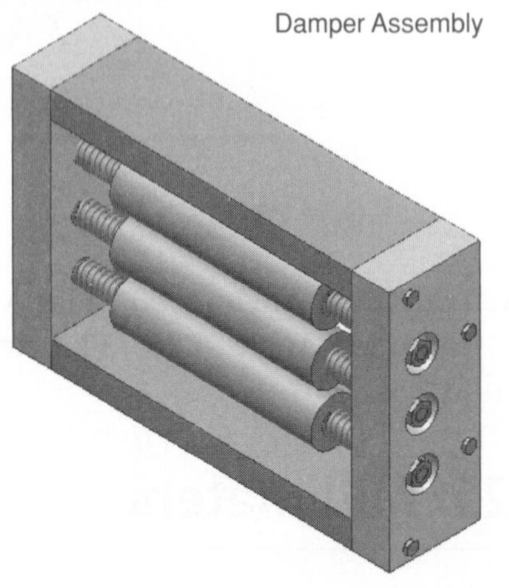

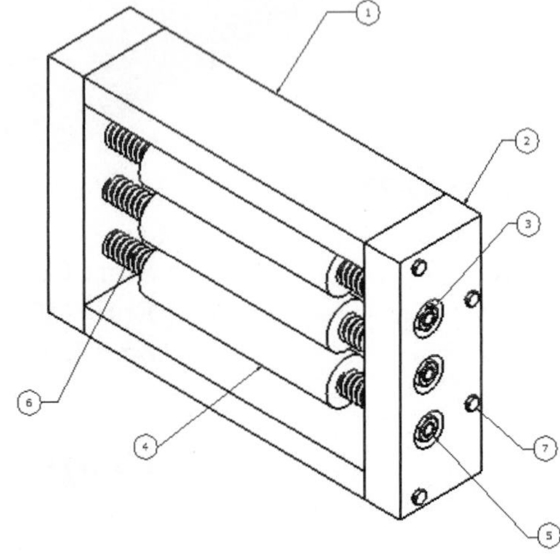

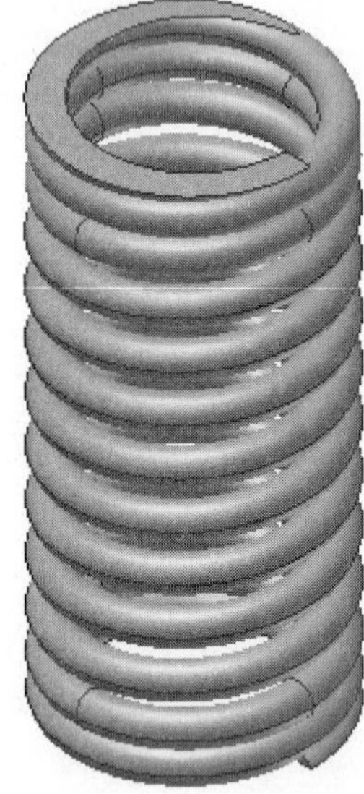

Parts List				
ITEM	PART NUMBER	DESCRIPTION	MATERIAL	QTY
1	AM311-1	BASE	Steel, Mild	1
2	AM311-2	PLATE,END	Steel, Mild	2
3	AM311-3	POST, GUIDE	Steel, Mild	3
4	EK-152	WEIGHT	Steel, Mild	3
5	AS 2465 - 1/2 UNC	HEX NUT	Steel, Mild	6
6		COMPRESSION SPRING	Steel, Mild	6
7	AS 2465 - 1/4 x 2 1/2 UNC	HEX BOLT	Steel, Mild	8

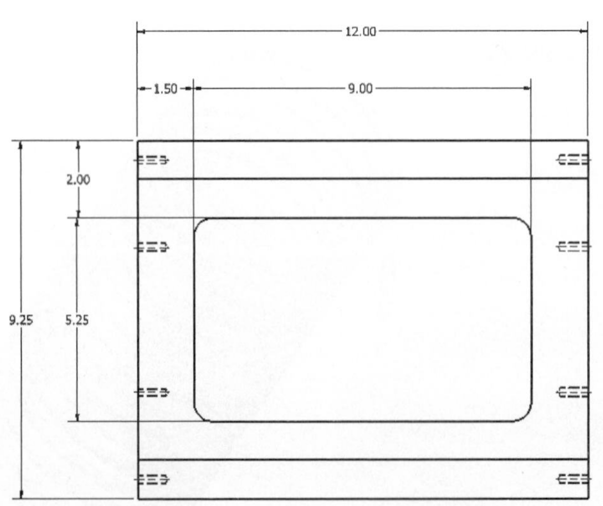

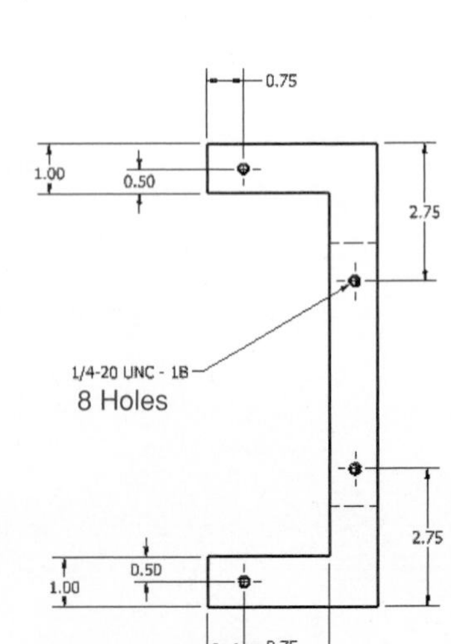

Figure P9-26
(Continued)

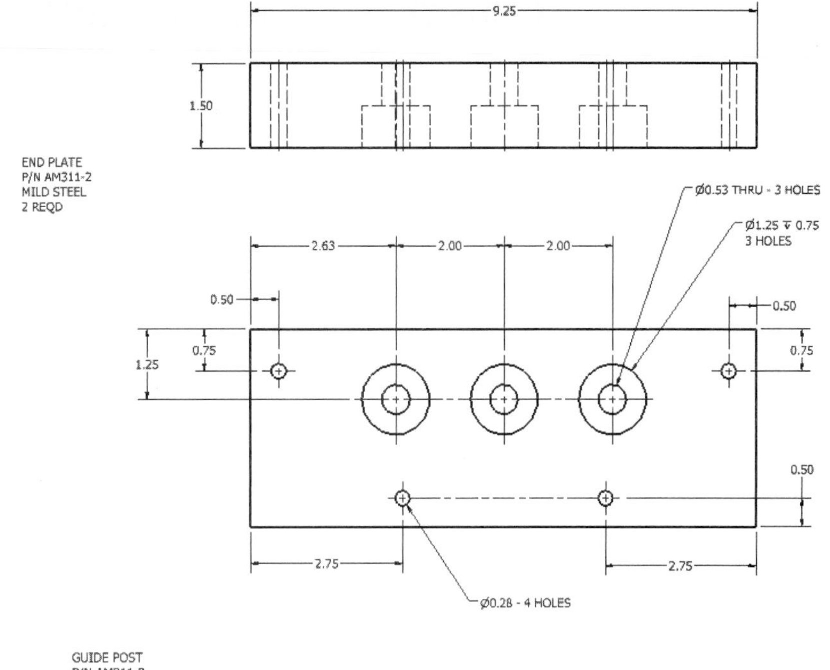

GUIDE POST
P/N AM311-3
MILD STEEL
3 REQD

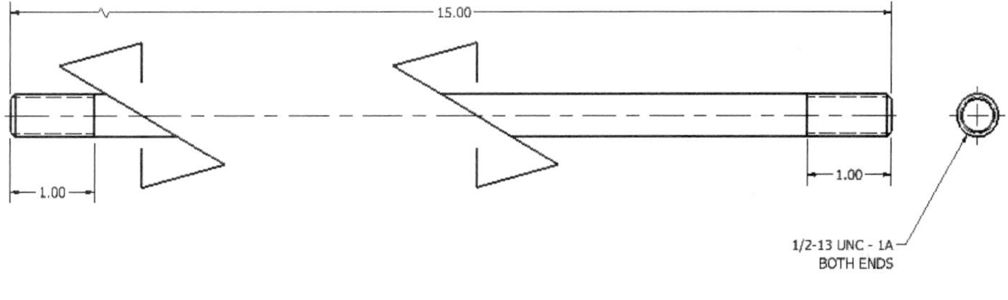

WEIGHT
P/N EK-152
MILD STEEL
3 REQD

8.00

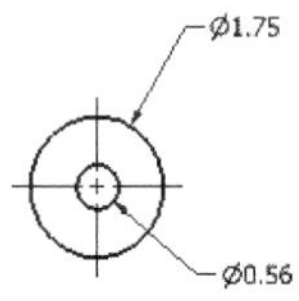

Working Stroke = 0.25
Working Load Length = 2.00
Both coils
- Closed End Coils = 1.50
- Transition Coils = 1.00
- Ground Coils = 0.75
- Active Coils = 10

Draw the following:
A. An assembly drawing
B. A presentation drawing
C. An exploded isometric drawing
D. A parts list

Project 9-27: Millimeters

Draw the Circular Damper Assembly shown in Figure P9-27.
The Threaded Post is M18 × 530 mm.
Spring data:

Wire Diameter = 3.0
Outer Diameter = 28
Loose Spring Length = 81
Min. Load Length = 80
Working Stroke = 5
Working Load Length = 78
Coil Direction = right
Closed End Coils = 1.5

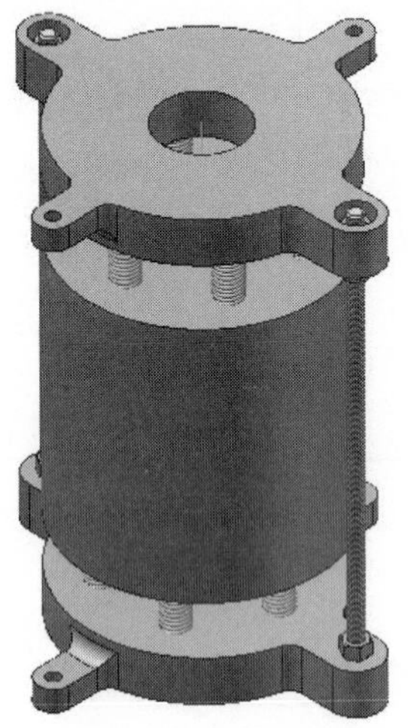

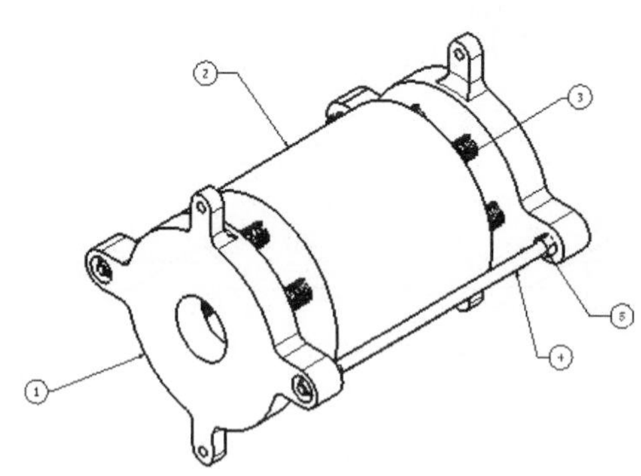

Parts List				
ITEM	PART NUMBER	DESCRIPTION	MATERIAL	QTY
1	BU2008-1	BASE, HOLDER	Steel	2
3	BU2008-2	SPRING, COMPRESSION	Steel, Mild	12
2	BU2008-3	COUNTERWEIGHT	Steel	1
4	AM312-12	POST, THREADED	Steel	2
5	AS 1112 - M18 Type	HEX NUT	Steel, Mild	8

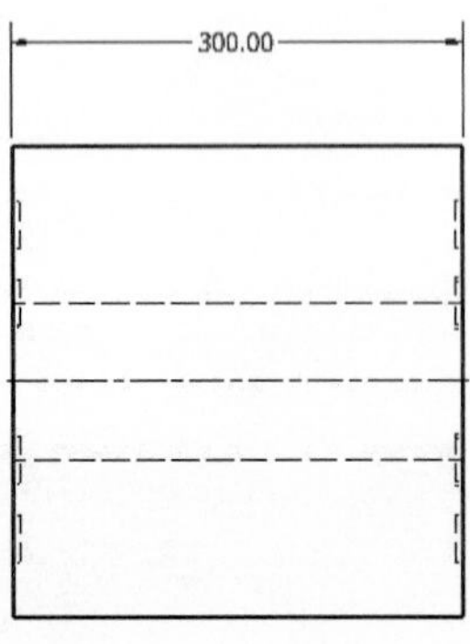

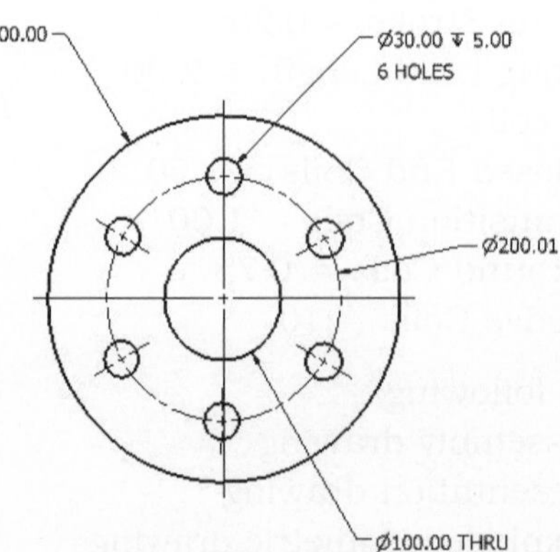

Figure P9-27

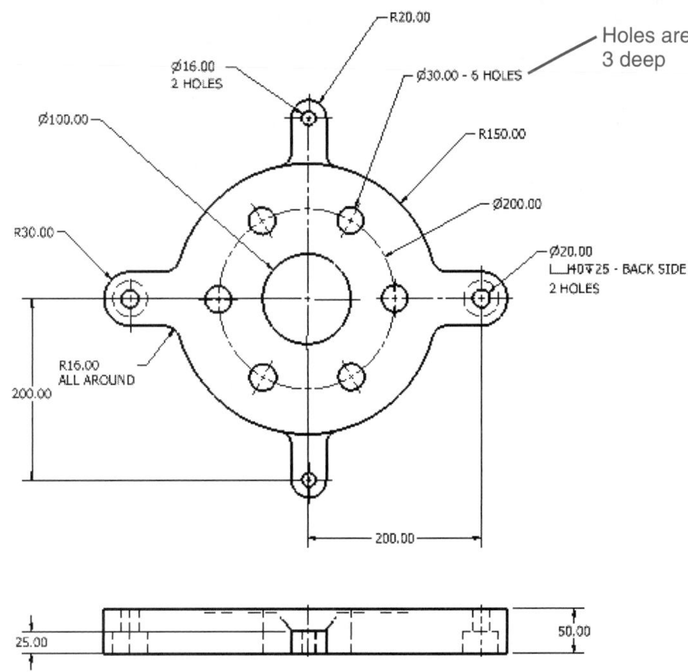

Figure P9-27
(Continued)

Transition Coils = 1.00
Ground Coils = 0.75
Active Coils = 16

Draw the following:

A. An assembly drawing
B. A presentation drawing
C. An exploded isometric drawing
D. A parts list

chapterten
Shafts

CHAPTER OBJECTIVES

- Learn how to use the **Design** panel to draw shafts
- Learn how to add retaining ring grooves to shafts
- Learn how to add keyways to shafts
- Learn how to add O-ring grooves to a shaft
- Learn how to add pin holes to a shaft
- Learn how to use the **Content Center** to add retaining rings, keys, O-rings, and pins to shafts

Introduction

The tools on the **Design** panel can be used to draw many different styles of shafts. Figure 10-1 shows a uniform shaft with chamfered ends that was created using the **Shaft** tool located on the **Power Transmission** panel under the **Design** tab. Shafts may also be created by extruding a circle,

Figure 10-1

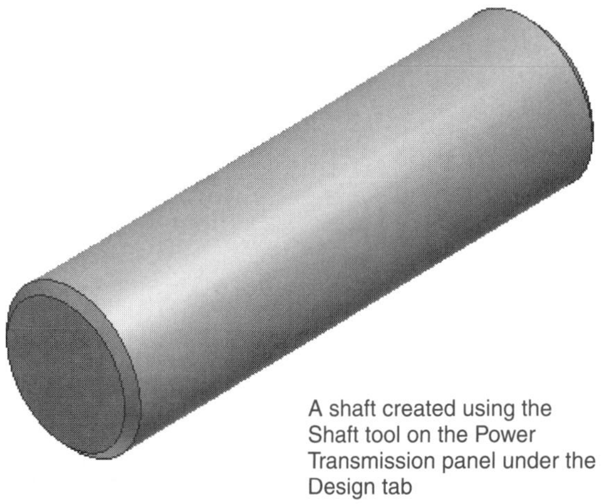

A shaft created using the Shaft tool on the Power Transmission panel under the Design tab

but the **Shaft** tool allows features such as keyways and retaining ring grooves to be added to the shaft.

Uniform Shafts and Chamfers

EXERCISE 10-1 Drawing a Uniform Shaft with Chamfered Ends

The shaft is to be Ø1.00 × 4.00 in.

1. Create a new drawing using the **Standard (in).iam** format.
2. Click the **Design** tab and select the **Shaft** tool located on the **Power Transmission** panel.

   See Figure 10-2.

The **Shaft Component Generator** dialog box will appear. See Figure 10-3.

Figure 10-2

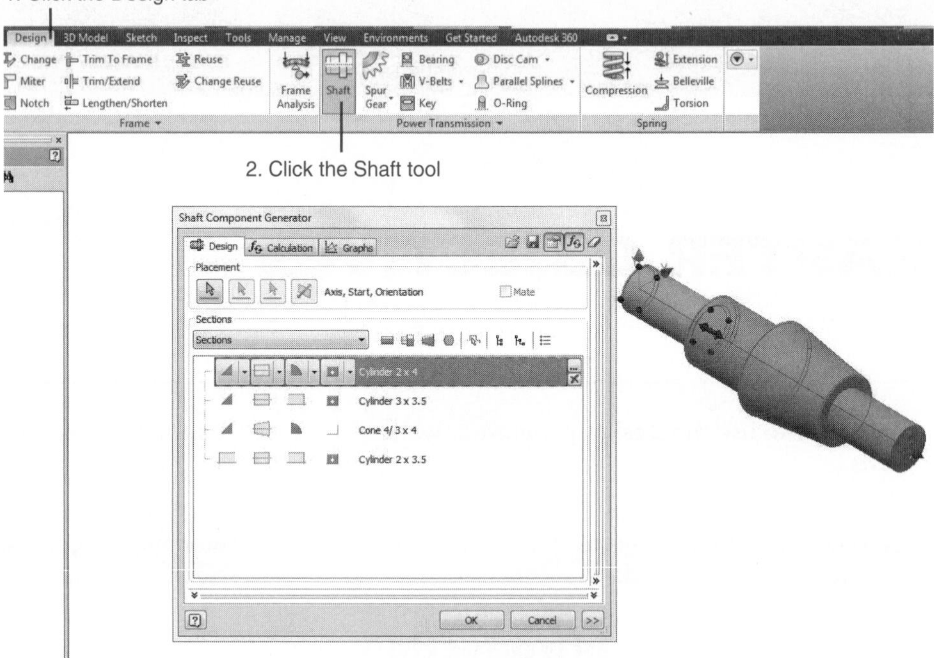

Figure 10-3

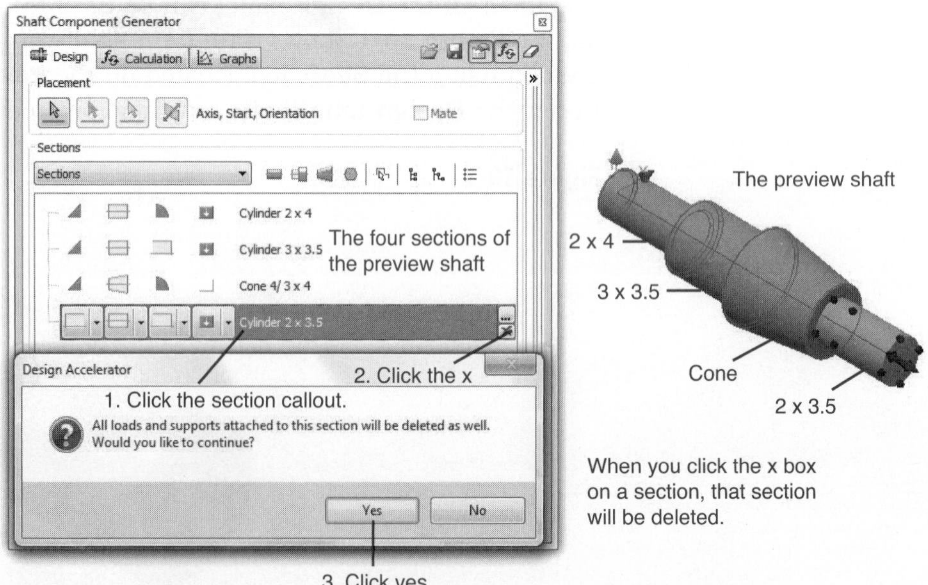

The **Shaft Component Generator** dialog box shows a shaft made of four sections. These sections can be manipulated or deleted as needed, making it easier to create a shaft than if you had to start from scratch.

EXERCISE 10-2 Deleting a Shaft Section

1. Move the cursor onto the **Shaft Component Generator** dialog box and click the shaft section callout to be deleted.
   The section's heading will be highlighted.
2. Click the **X** box in the lower right corner of the section's name.
   A check box will appear.
3. Click **Yes.**

The selected section will be deleted. See Figure 10-3. Repeat the procedure deleting both the **Cone** and **3** × **3.5 Cylinder** sections. Retain the **2** × **4 Cylinder** section. See Figure 10-4.

Figure 10-4

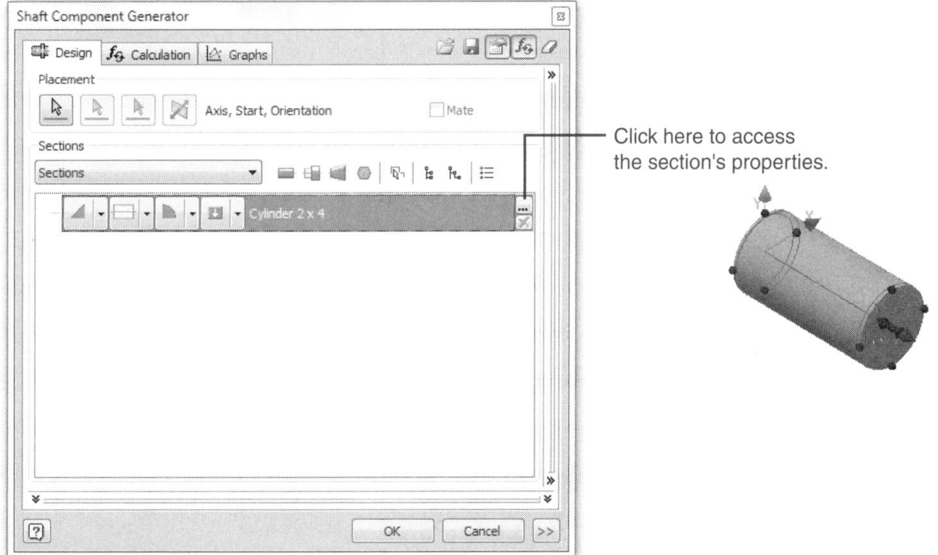

EXERCISE 10-3 Defining the Cylinder's Diameter and Length

1. Click the **Section Properties** box.
   The **Cylinder** dialog box will appear. See Figure 10-5.
2. Click to the right of the default diameter value.
   An arrowhead will appear.
3. Click the arrowhead.
   A list of options will appear.
4. Click the **Values** option. A listing of standard shaft diameter values will cascade down.
   See Figure 10-5.

NOTE
The existing value could be deleted from the cylinder box and new values typed in.

Figure 10-5

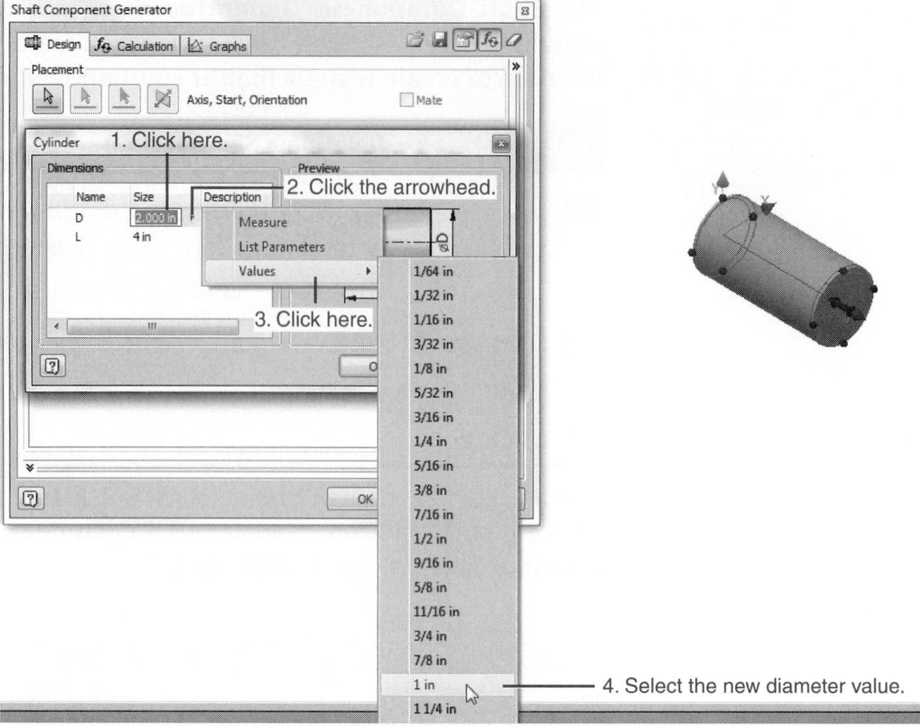

5 Select the appropriate diameter.

In this example a value of **1 in** was selected.

6 Repeat the procedure for the shaft length.

In this example a length of **3 in** was selected.

See Figure 10-6.

7 Click the **OK** button on the **Cylinder** dialog box.

Figure 10-6

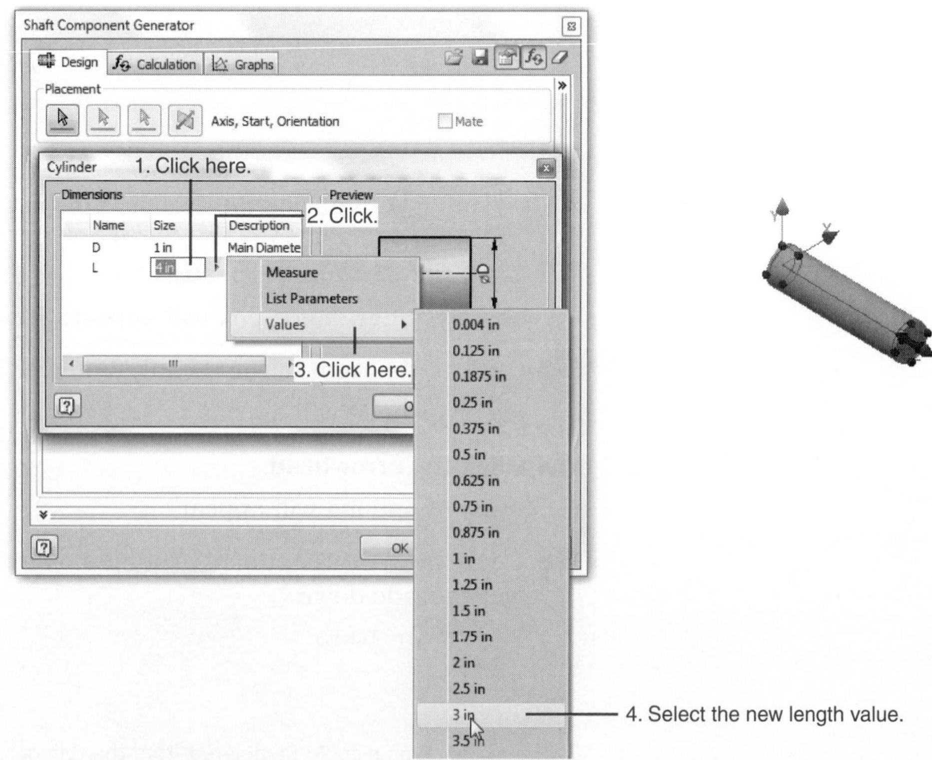

NOTE

Do not click the **OK** button on the **Shaft Component Generator** dialog box until the cylinder has been completely defined. If you do, Inventor will create a second shaft adding the chamfers. If this happens, delete the incomplete shaft.

EXERCISE 10-4 Creating Chamfers

1. Click the **First Edge Feature** box.

   See Figure 10-7.

2. Select the **Chamfer** option.

   The **Chamfer** dialog box will appear. See Figure 10-8.

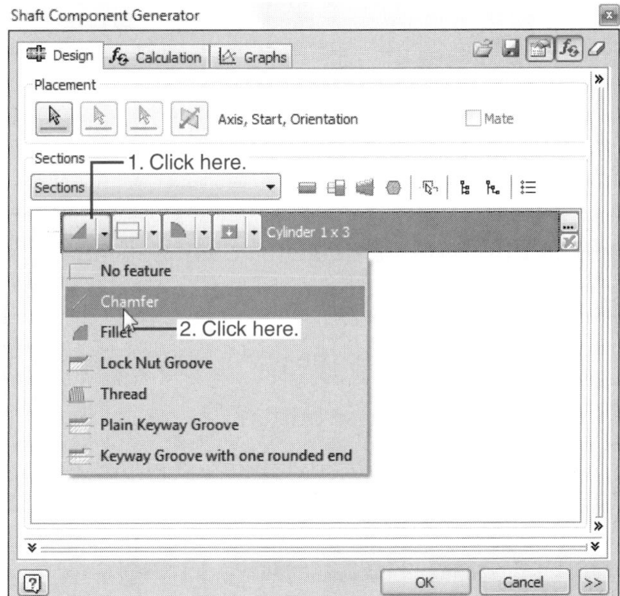

Figure 10-7

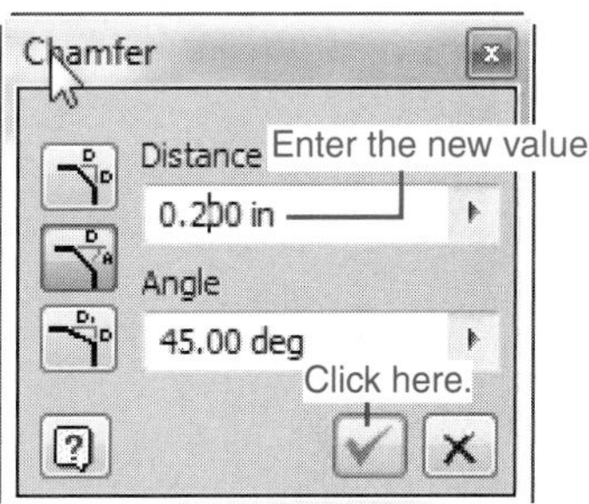

Figure 10-8

3. Enter the chamfer values.

   In this example a value of **.200 in** × **45°** was entered.

4. Click the check mark.

   A preview of the chamfers will appear.

5. Click the **Second Edge Feature** box and create a chamfer of the same size on the right edge of the shaft.

   See Figure 10-9.

NOTE

The preview of the shaft that appears on the screen will show the size and location of the chamfers.

6. Click the check mark.

7. Click **OK.**

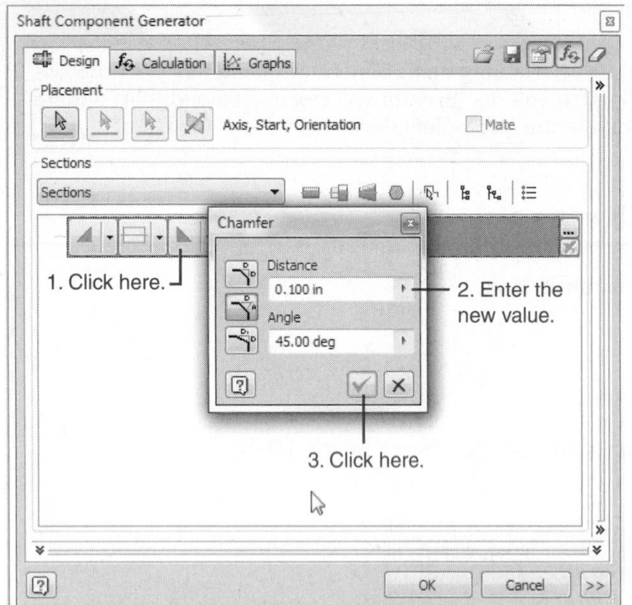

Figure 10-9

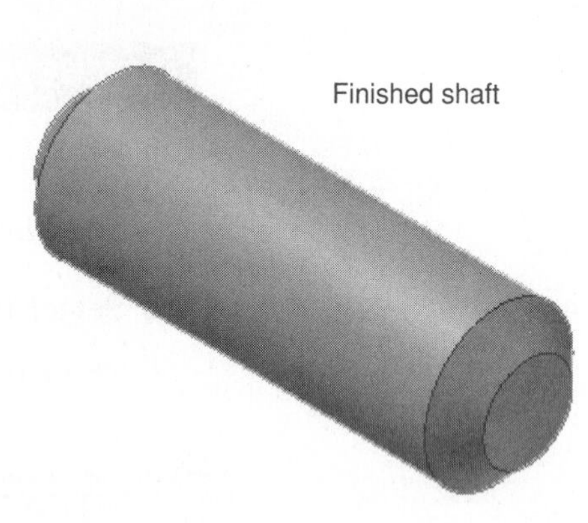

Figure 10-10

A preview of the chamfers will appear on the shaft.

The **File Naming** dialog box will appear.

8 Click **OK.**

A grayed picture of the shaft will appear on the screen.

9 Click the left mouse button.

Figure 10-10 shows the finished shaft.

Shafts and Retaining Rings

retaining ring: Used to prevent longitudinal movement of a shaft. The ring fits into a groove cut into the shaft.

Retaining rings are used to prevent longitudinal movement of shafts. Grooves are cut into the shafts and the rings fitted into the groves. Both internal and external rings are available.

For this section, grooves will be cut into a Ø20 shaft to match the requirements of an ANSI B 27.7M external retaining ring.

The **Content Center** is used to determine the required diameter and width of the groove required by the ANSI B 27.7M retaining ring.

1 Create a new drawing using the **Standard (mm).iam** format.

2 Click the **Place from Content Center** tool.

3 Click the + **sign** next to the **Shaft Parts** heading.

4 Click the **Circlips** option, click the **External** option, click the **ANSI B 27.7M** ring, then click **OK.**

See Figure 10-11. The **ANSI B 27.7M** dialog box will appear. See Figure 10-12.

5 Select a shaft diameter of **20.**

6 Click the **Table View** tab.

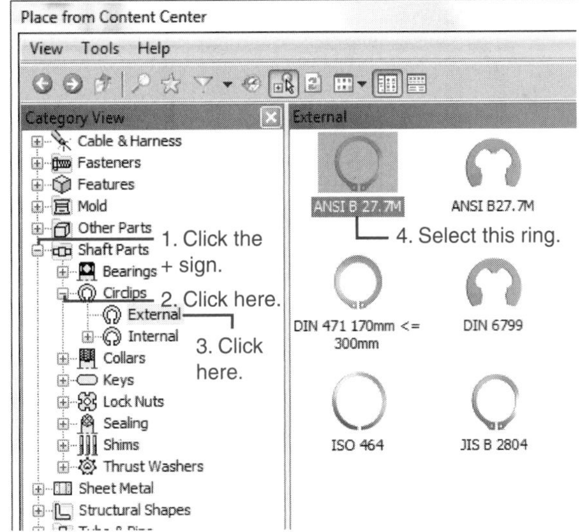

Figure 10-11

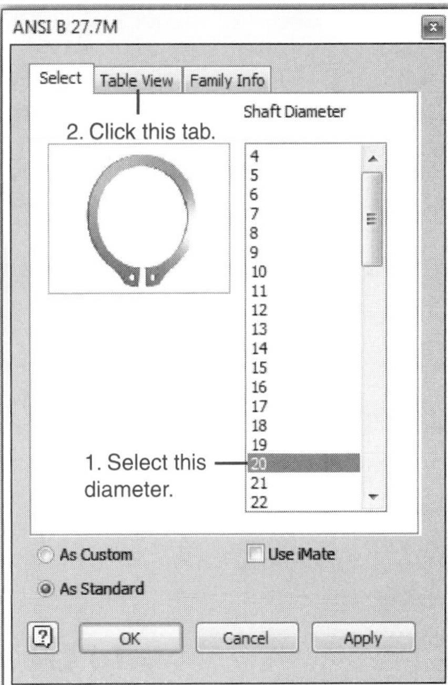

Figure 10-12

7 Scroll across the table until the **Groove Diameter** and **Groove Width** values are revealed.

See Figure 10-13. Note that the **Groove Diameter** value is **18.85** and the **Groove Width** is **1.2.**

8 Click **OK.**

A retaining ring will appear on the screen.

Figure 10-13

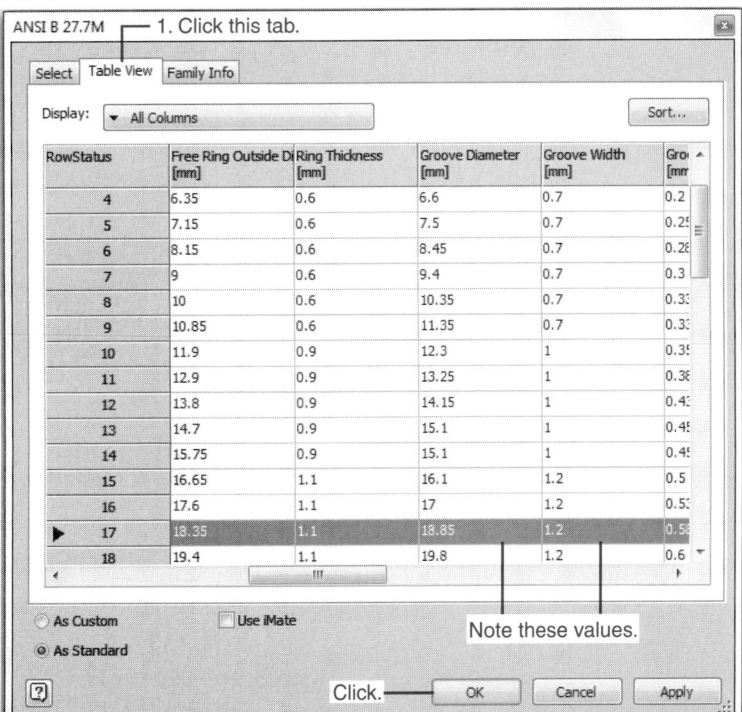

RowStatus	Free Ring Outside Di [mm]	Ring Thickness [mm]	Groove Diameter [mm]	Groove Width [mm]
4	6.35	0.6	6.6	0.7
5	7.15	0.6	7.5	0.7
6	8.15	0.6	8.45	0.7
7	9	0.6	9.4	0.7
8	10	0.6	10.35	0.7
9	10.85	0.6	11.35	0.7
10	11.9	0.9	12.3	1
11	12.9	0.9	13.25	1
12	13.8	0.9	14.15	1
13	14.7	0.9	15.1	1
14	15.75	0.9	15.1	1
15	16.65	1.1	16.1	1.2
16	17.6	1.1	17	1.2
17	18.35	1.1	18.85	1.2
18	19.4	1.1	19.8	1.2

9 Click the mouse to create a second ring, and press the **<Esc>** key.

Park these rings at the side of the drawing screen for now. They will be added to the shaft once it is created. See Figure 10-14.

Figure 10-14

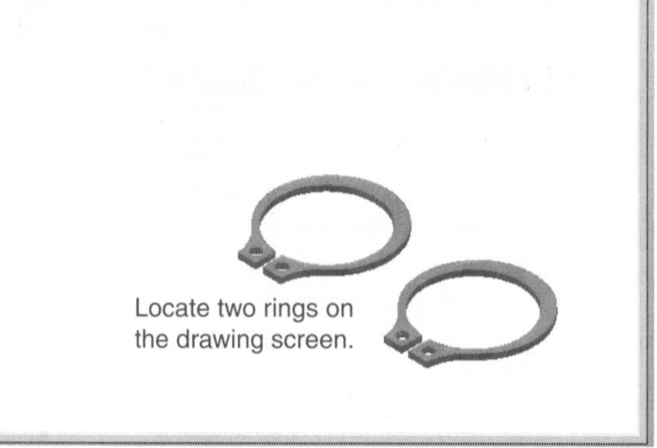

To Create the Shaft

1 Click the **Design** tab, and select the **Shaft** tool. The **Shaft Component Generator** dialog box will appear. Create a Ø20 × 40 shaft.

2 Click the **Section Features** box and select the **Add Retaining Ring** option.

See Figure 10-15. The selected **20 ANSI B 27.7M** retaining ring reference will automatically be selected. See Figure 10-16.

3 Click the **Retaining ring 20 ANSI B 17.7M** callout and click the **Feature Properties** box.

4 The **Retaining Ring Groove** dialog box will appear. See Figure 10-17.

5 Set the **Position** for **Measure from second edge** and the **x** value to **3.000.**

A preview of the groove will appear. The 3.000 value was selected to locate the ring 3.000 mm from the end of the shaft.

Figure 10-15

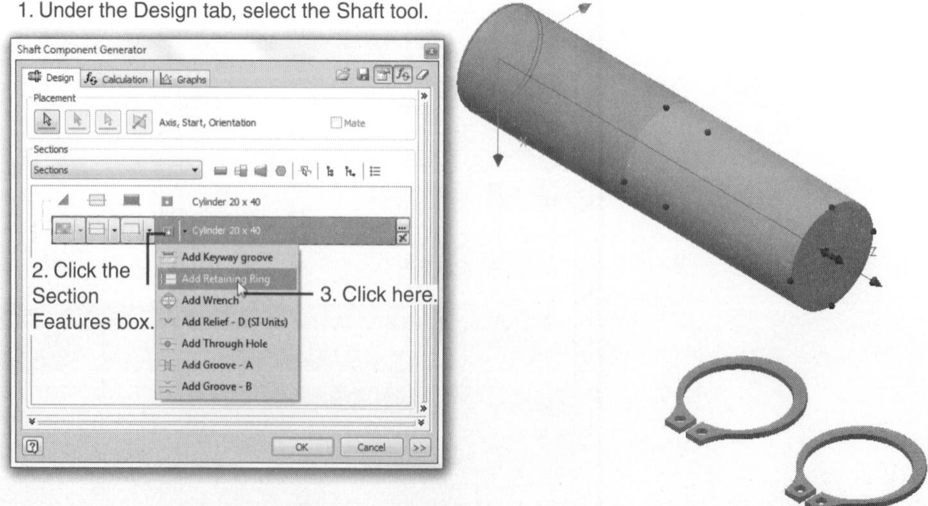

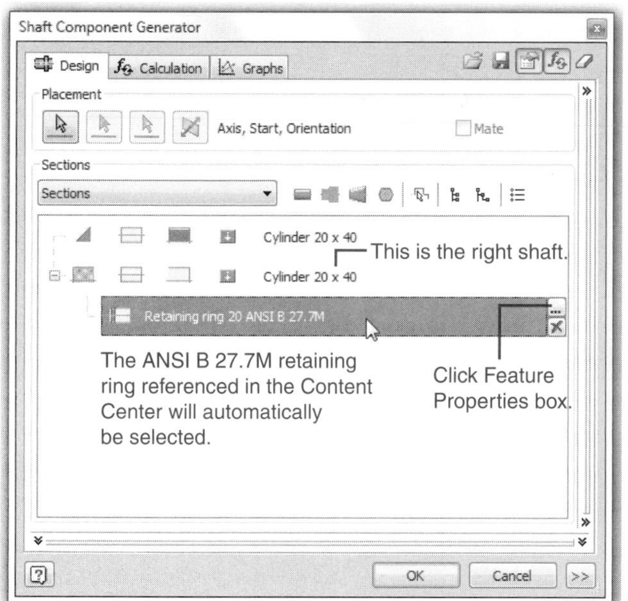

Figure 10-16

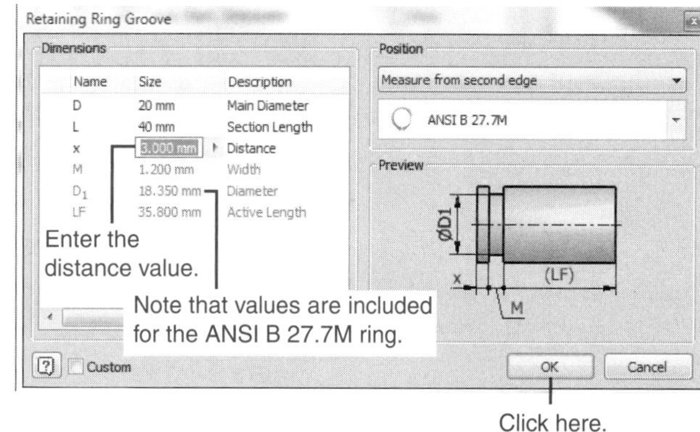

Figure 10-17

6 Click **OK.**

7 Click the **Feature Properties** box for the retaining ring.

See Figure 10-18.

Figure 10-18

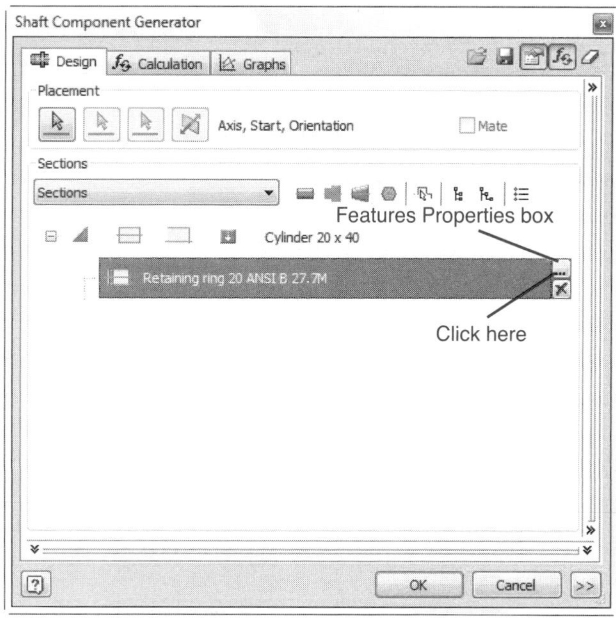

The **Retaining Ring Groove** dialog box will appear. See Figure 10-19.

8 Click the **Feature Properties** box and set the **Position** for **Measure from second edge** and the **x** value to **3.000.**

A preview of the groove will appear. The **3.000** value was selected to locate the ring 3.000 mm from the end of the shaft.

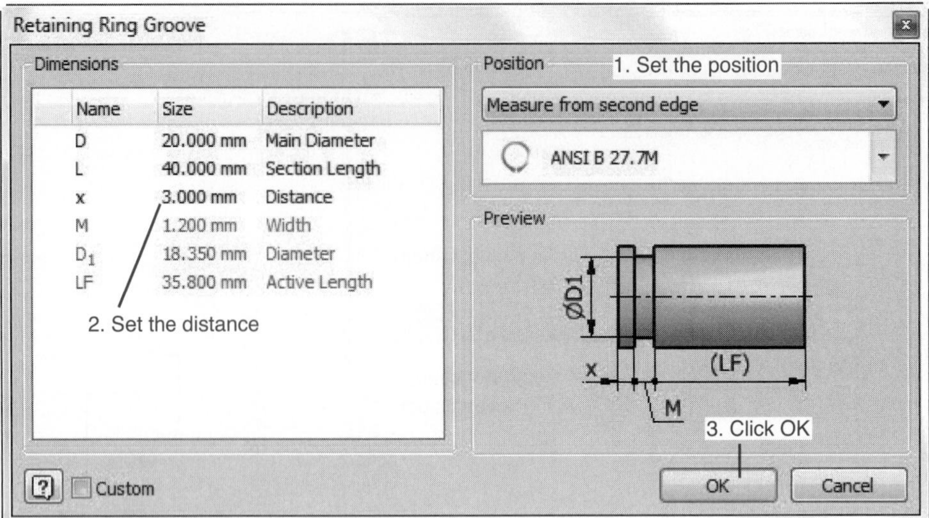

Figure 10-19

9 Click **OK.**

The **File Naming** dialog box will appear.

10 Click **OK.**

The newly created shaft will appear with the previously drawn retaining rings. See Figure 10-20.

Figure 10-20

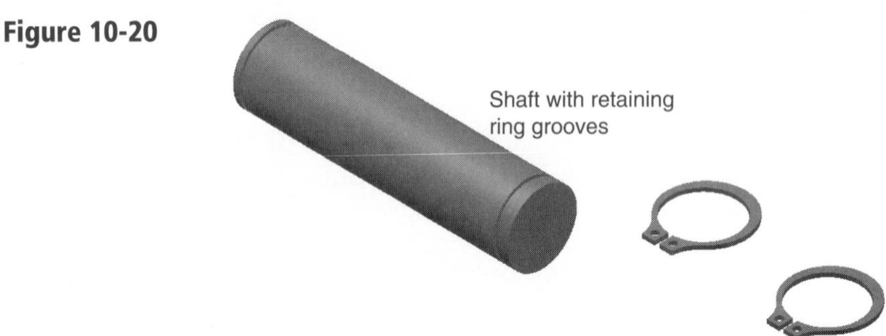

The first retaining ring created will be grounded, as indicated by a pushpin next to the ring's callout in the browser box. Inventor always grounds the first component entered in a new assembly. We wish to remove the grounding from the retaining ring and ground the shaft.

1 Access the browser box area and right-click the grounded retaining ring.

See Figure 10-21.

2 Click the check mark next to the **Grounded** option.

The pushpin icon will disappear. The ring is no longer grounded.

3 Right-click the shaft callout and click the **Grounded** option.

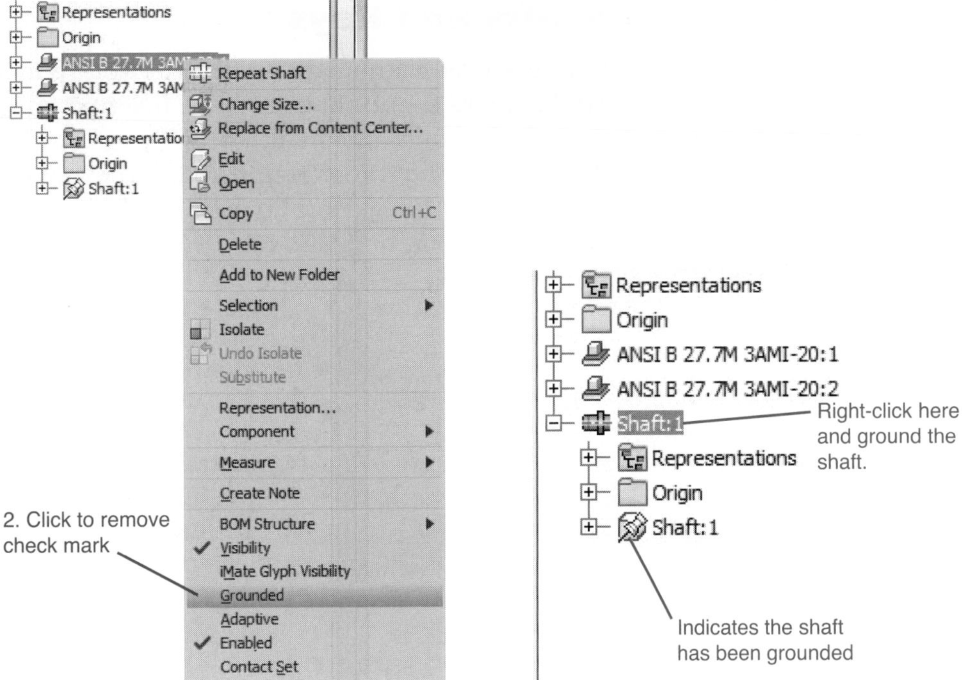

Figure 10-21

Figure 10-22

The shaft is now grounded. This can be verified by clicking the + **sign** to the left of the shaft's callout. There should be a pushpin next to the **Shaft:1** callout. See Figure 10-22.

Add the rings to the shaft.

4 Click the **Assemble** tab and use the **Constrain** tool to locate the retaining rings on the shaft.

Hint: Use the **Insert** tool and align the edge of the retaining rings with the edge of the grooves.

Figure 10-23 shows the retaining ring mounted on the shaft.

Figure 10-23

Shafts and Keys

Keys are used with shafts to transfer rotary motion and torque. Figure 10-24 shows a hub that has been inserted onto a shaft with a square key between the hub and shaft. As the shaft turns, the motion and torque of the shaft will be transferred through the key into the hub.

There are five general types of keys: Pratt and Whitney, square, rectangular, Woodruff, and Gib. See Figure 10-25.

Figure 10-24

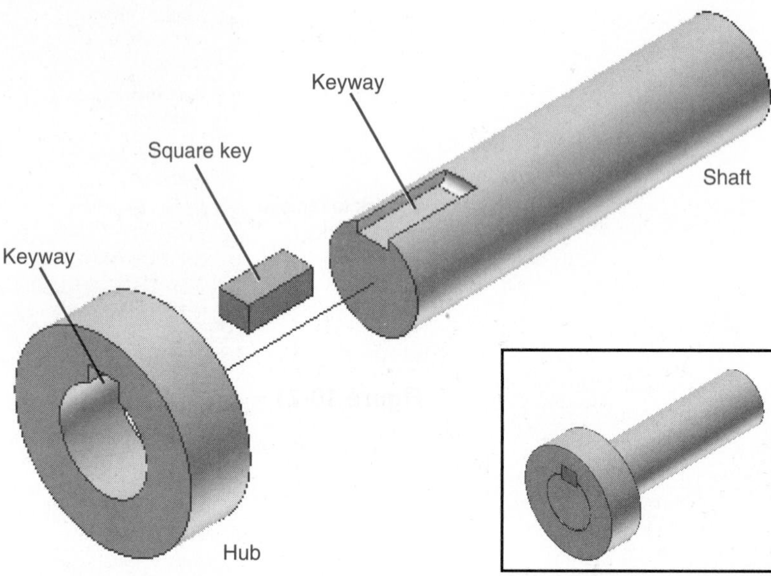

Figure 10-25

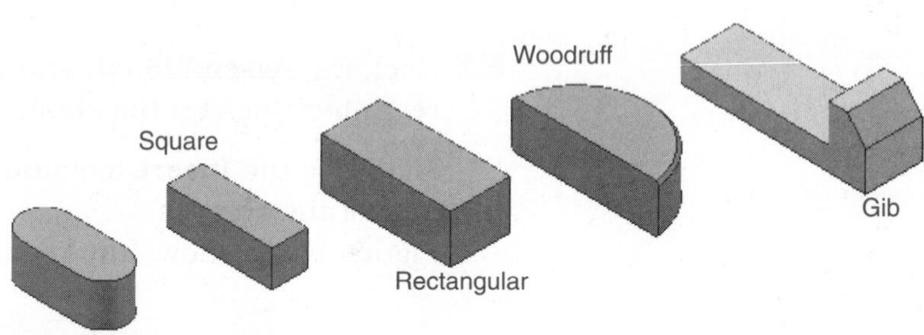

Square Keys

The following exercise will place a keyway in a Ø30 × 60 shaft. A rectangular DIN 6885 B key will be used. In an actual design situation the key selected would be based on load shaft speed and shaft size considerations.

EXERCISE 10-5 Drawing a Keyway on a Shaft

1. Start a new drawing using the **Standard (mm).iam** format.
2. Create a **Ø30** × **60** shaft with no end features.
3. Right-click the mouse and click the **Place from Content Center** option. See Figure 10-26.

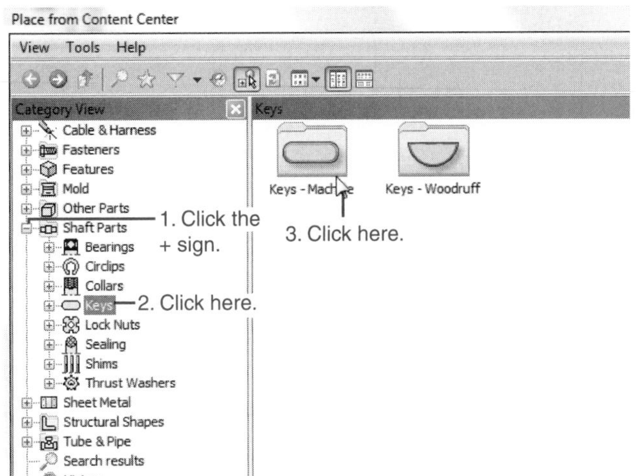

Figure 10-26

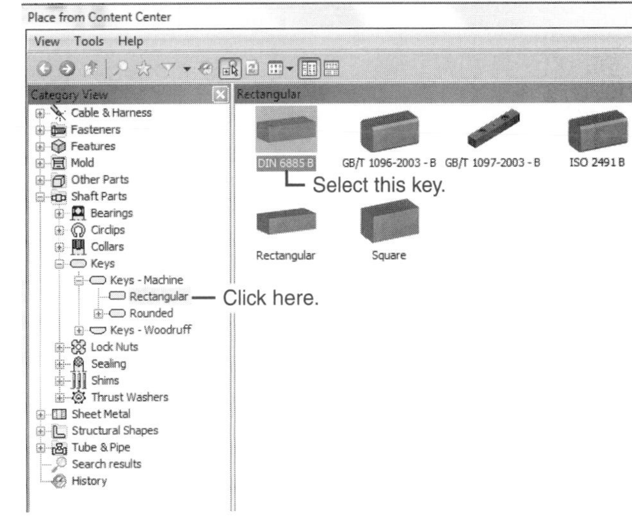

Figure 10-27

4 Click **Shaft Parts, Keys, Keys - Machine,** and select the **Rectangular** option.

The **Rectangular** key options will be shown. See Figure 10-27.

5 Select the **DIN 6885 B** key.

The **DIN 6885 B** dialog box will appear. See Figure 10-28.

Figure 10-28

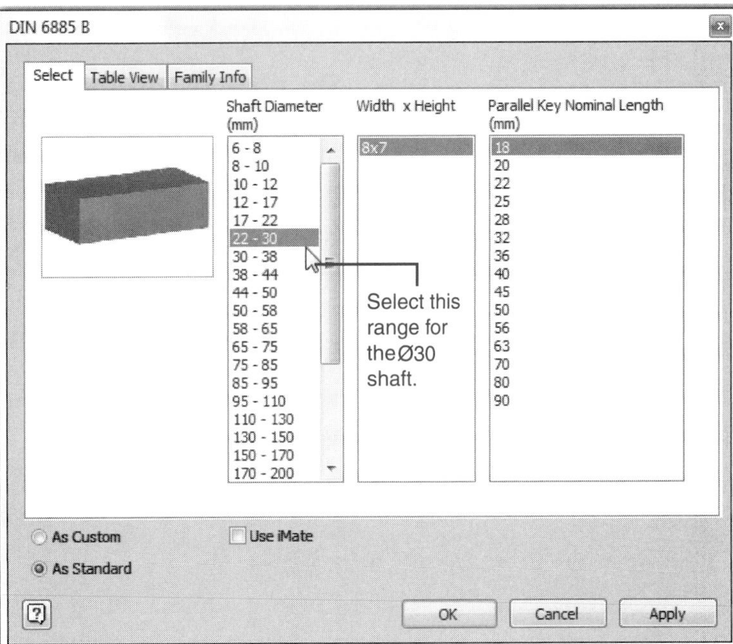

6 Select the **Shaft Diameter** range **22 - 30.**

Note that the width × height of the key is listed as 8 × 7 and that the key's nominal length is 18.

7 Click the **Table View** tab and record the specified keyway values.

The required keyway depth is **4,** and the keyway groove length is **22.** See Figure 10-29.

8 Click **OK** and add a rectangular key to the drawing.

Figure 10-29

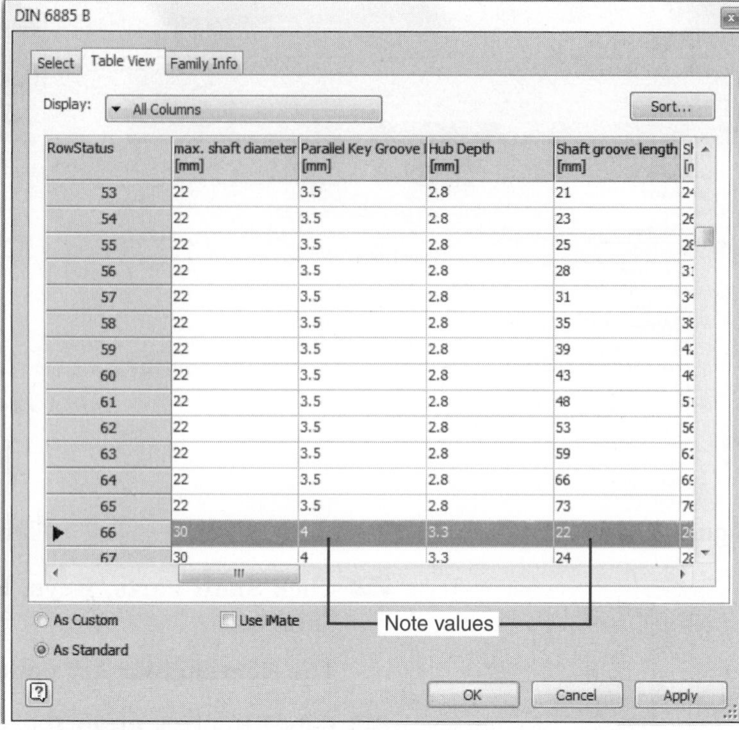

RowStatus	max. shaft diameter [mm]	Parallel Key Groove [mm]	Hub Depth [mm]	Shaft groove length [mm]
53	22	3.5	2.8	21
54	22	3.5	2.8	23
55	22	3.5	2.8	25
56	22	3.5	2.8	28
57	22	3.5	2.8	31
58	22	3.5	2.8	35
59	22	3.5	2.8	39
60	22	3.5	2.8	43
61	22	3.5	2.8	48
62	22	3.5	2.8	53
63	22	3.5	2.8	59
64	22	3.5	2.8	66
65	22	3.5	2.8	73
66	30	4	3.3	22
67	30	4	3.3	24

Now, create a **Ø30** × **60** shaft.

1 Access the **Shaft** tool from the **Design** tab.

The previous shaft data will appear on the screen. These data could be erased and a new shaft created, but Inventor allows you to easily modify an existing shaft using dynamic inputs.

2 Delete the retaining ring values and one of the two Ø20 × 40 shafts by clicking the **X** boxes on the right side of the callouts.

See Figure 10-30. Note the brown filled circles at each end of the cylinder. These are the dynamic inputs for changing the diameter and length of the shaft.

Figure 10-30

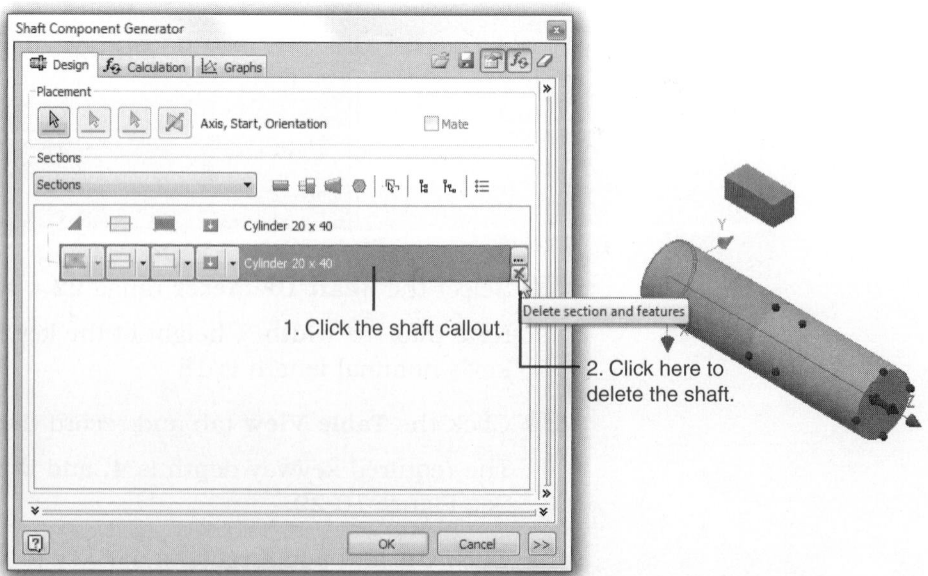

> **TIP**
> Ensure that no chamfers or end rounds have been included on the regenerated shaft. If they have been, enter a **No feature** option for both ends of the shaft.

3 Click one of the dynamic arrows on the YZ plane and drag the shaft to a new diameter of **30.**

4 Use the dynamic arrows to change the length to **60.**

See Figure 10-31.

Figure 10-31

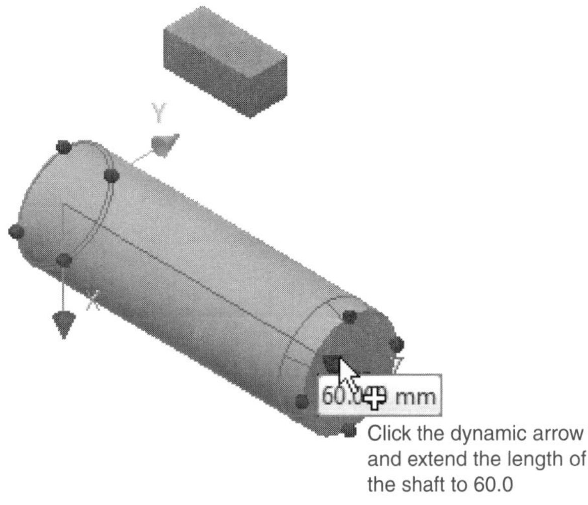

> **TIP**
> The dynamic inputs are preset to specific increments. If you need a shaft size that is not included in the preset list, click the **Cylinder** box and enter the desired values.

5 Click the **Second Edge Features** box and select a **Plain Keyway Groove** option.

See Figure 10-32. A default keyway will be added to the shaft. This keyway will be modified to fit the selected 8 × 7 × 18 rectangular key.

Figure 10-32

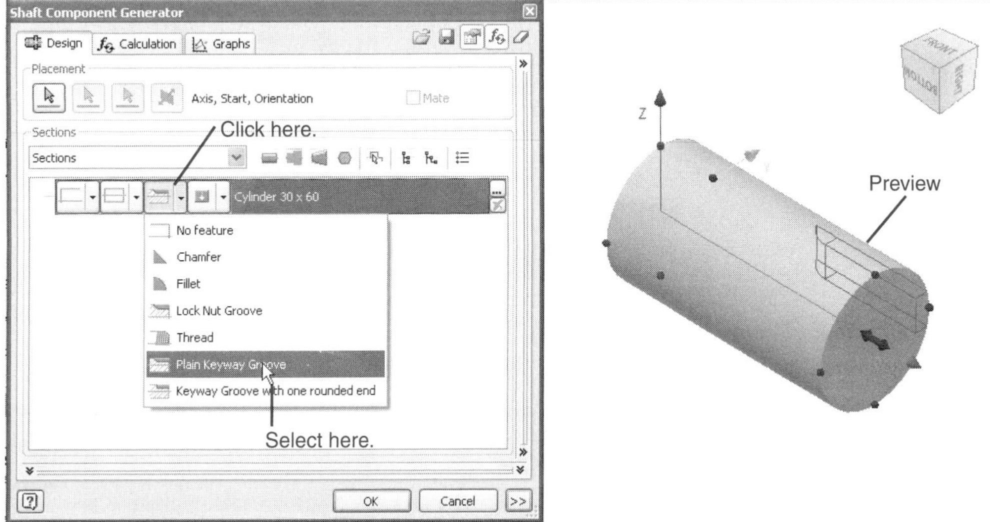

6 Click the **Second Edge Features** box again.

The **Plain Keyway Groove** dialog box will appear. See Figure 10-33.

Figure 10-33

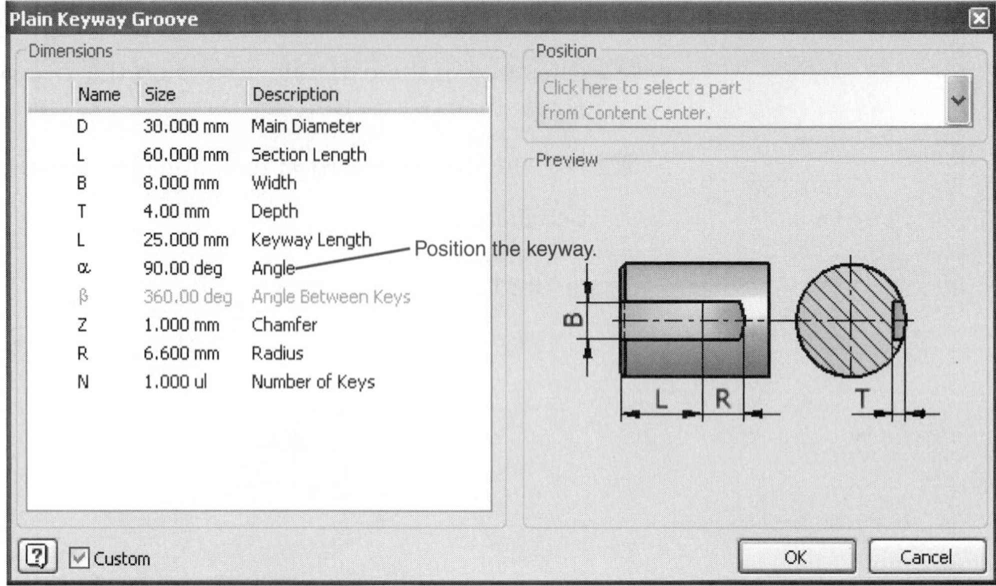

7 Enter the appropriate values.

In this example **T** = **4.00, L** = **25.000,** and **B** = **8.000.** The length of the keyway could be any number greater than 18. The angle value is used to position the keyway. In this example a value of **90**° was entered.

> **NOTE**
>
> A **1** × **45**° chamfer will automatically be added unless you enter a **0** value for Z.

8 Click OK.

The **Shaft Component Generator** dialog box will appear.

9 Click OK.

The **File Naming** dialog box will appear.

10 Click OK.

Figure 10-34 shows the shaft with a keyway and the shaft assembled with the 8 × 7 × 18 key.

Figure 10-34

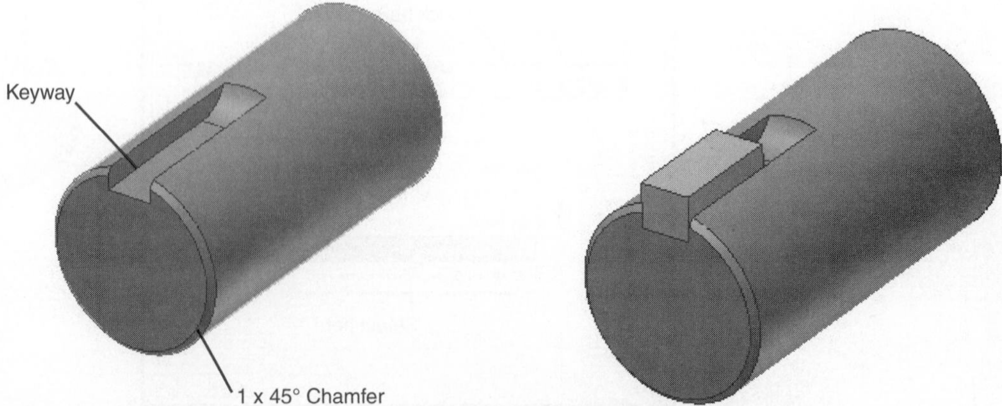

This section shows how to draw shafts and keyways using the tools on the **2D Sketch Panel** and the **Part Features** panel bar.

EXERCISE 10-6 Drawing a Keyway Using the Sketch and Extrude Panel Tools

Figure 10-35 shows a Ø30 × 60 shaft drawn by first drawing a **Ø30** circle using the **Circle** tool from the **Draw** panel under the **Sketch** tab, and then extruding to a height of **60** using the **Extrude** tool from the **Create** panel under the **3D Model** tab. Draw a keyway that is **8** wide, **4** deep, and **18** long with end radius equal to **4.**

1 Create a work plane tangent to the edge of the shaft.

2 Create a new sketch plane on the work plane and draw an **8** × **22** rectangle. The 22 value includes the 4 needed to create the radius at the end of the keyway.

See Figure 10-36.

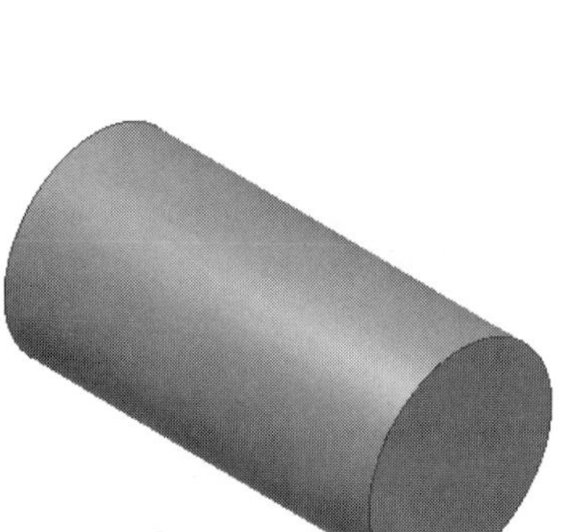

Figure 10-35

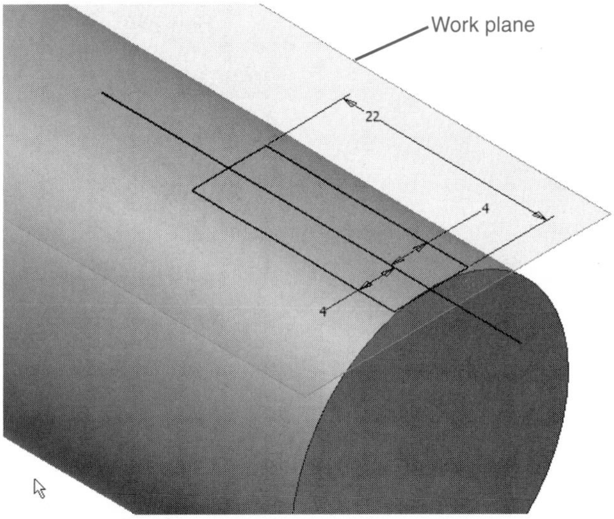

Work plane
22
4
4

Figure 10-36

3 Use the **Cut** option on the **Extrude** tool to cut out the keyway.

4 Draw an **R4** radius at the end of the keyway.

See Figure 10-37. Figure 10-38 shows the finished keyway.

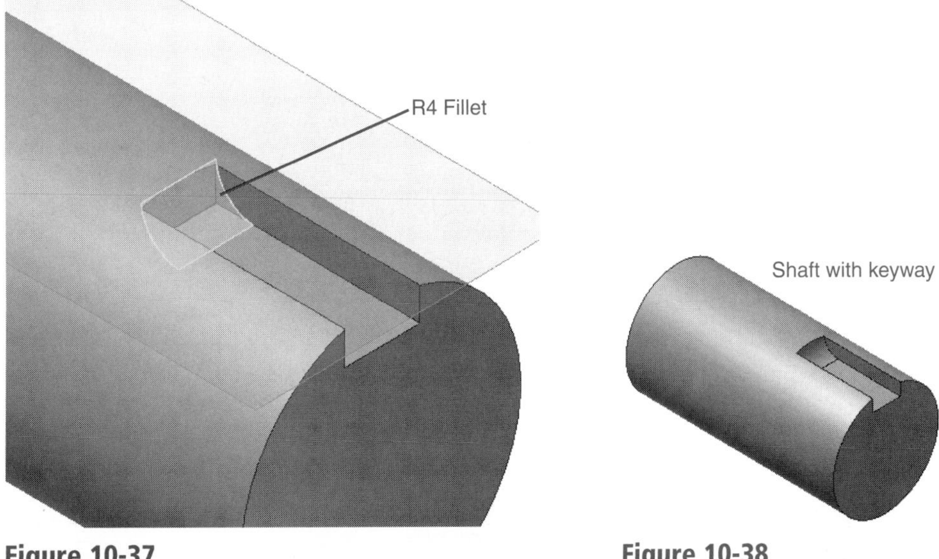

Figure 10-37

Figure 10-38

Pratt and Whitney Keys

Pratt and Whitney key: A key similar to a square key but with rounded ends.

Pratt and Whitney keys are similar to square keys but have rounded ends. See Figure 10-25.

EXERCISE 10-7 Drawing a Shaft with a Pratt and Whitney Keyway

1. Start a new drawing using the **Standard (mm).iam** format.
2. Click the **Design** tab and select the **Shaft** option.
3. Draw a **Ø30** × **65** shaft with **3** × **45**° chamfers at each end; click **OK.** See Figure 10-39.
4. Select the **Add Keyway groove** located under the **Select Features** tool. See Figure 10-40. A default preview of a keyway will appear.

Figure 10-39

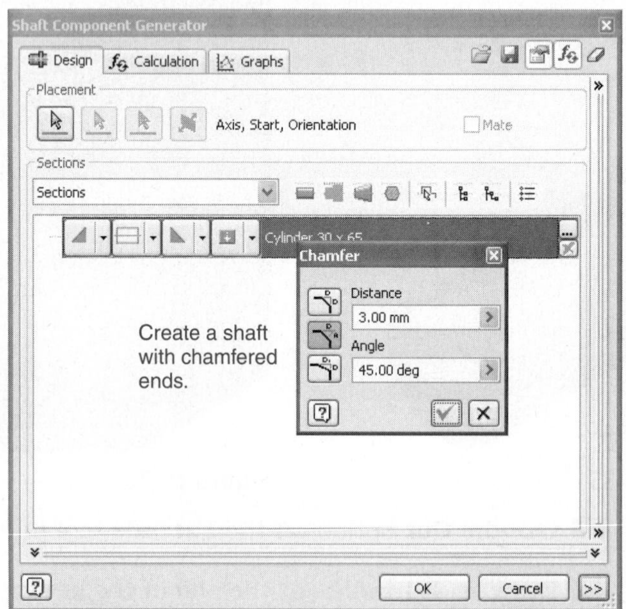

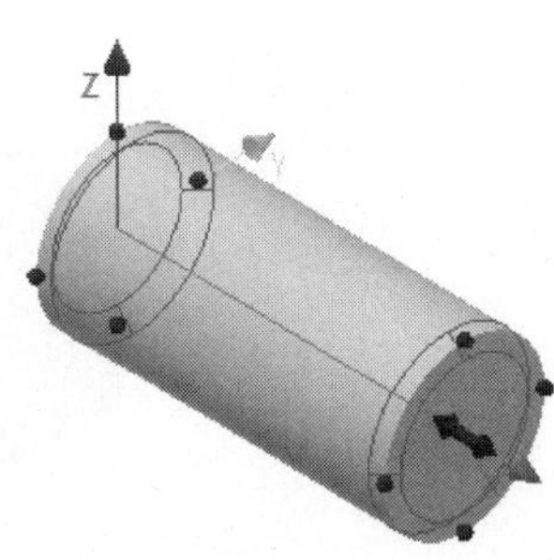

Figure 10-40

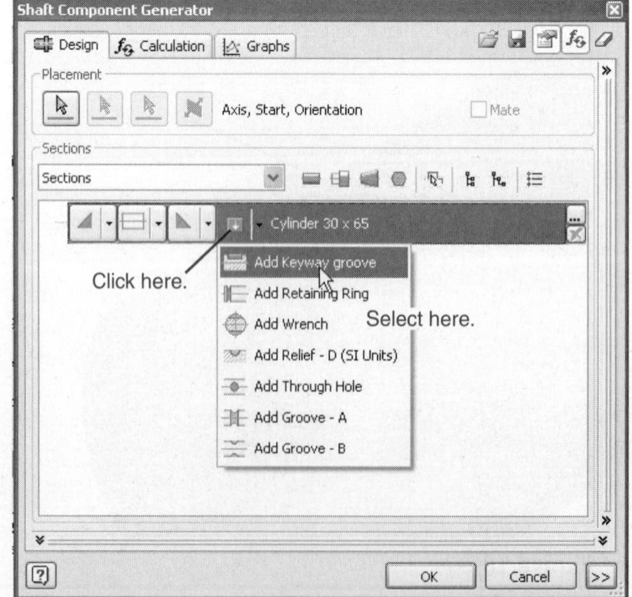

Figure 10-41

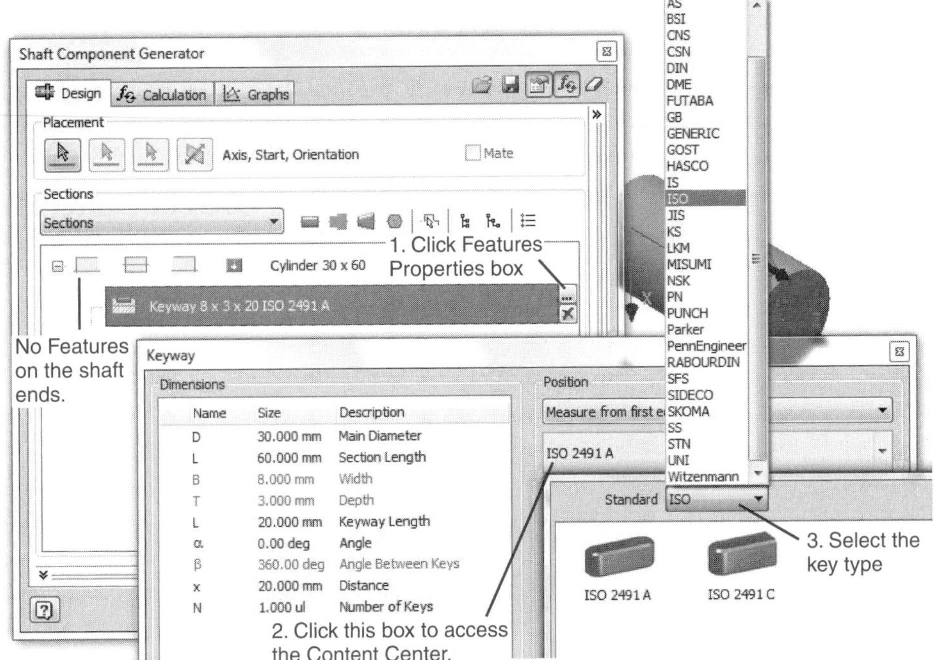

5 Click the **Feature Properties** box on the keyway callout line. The **Keyway** dialog box will appear. See Figure 10-41.

6 Access the **Content Center** and select an **ISO 2491 A** key. See Figure 10-42

7 Return to the **Keyway** dialog box; click **OK**. See Figure 10-43.

Figure 10-42

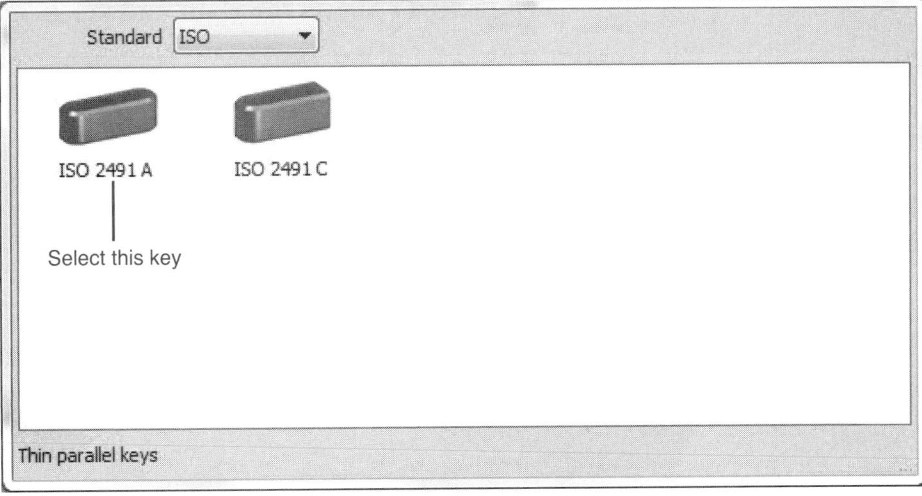

Figure 10-43

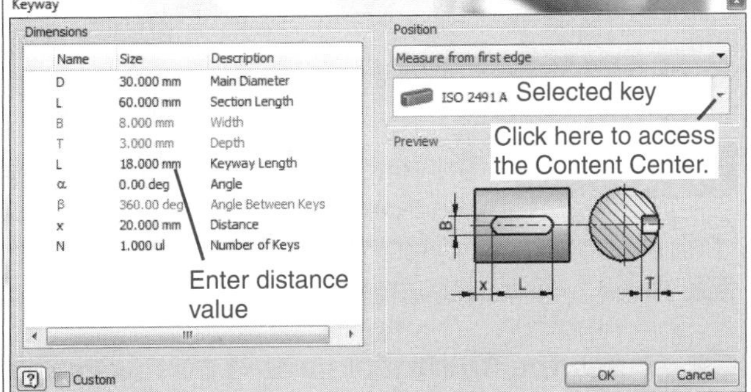

The grayed numbers cannot be changed; they were generated when the **ISO 2491 A** key was selected. The values in black can be changed.

8 Change the distance value to **18;** click **OK.**

Figure 10-44 shows the finished keyway.

9 Access the **Place from Content Center** dialog box and add the key.

See Figure 10-45.

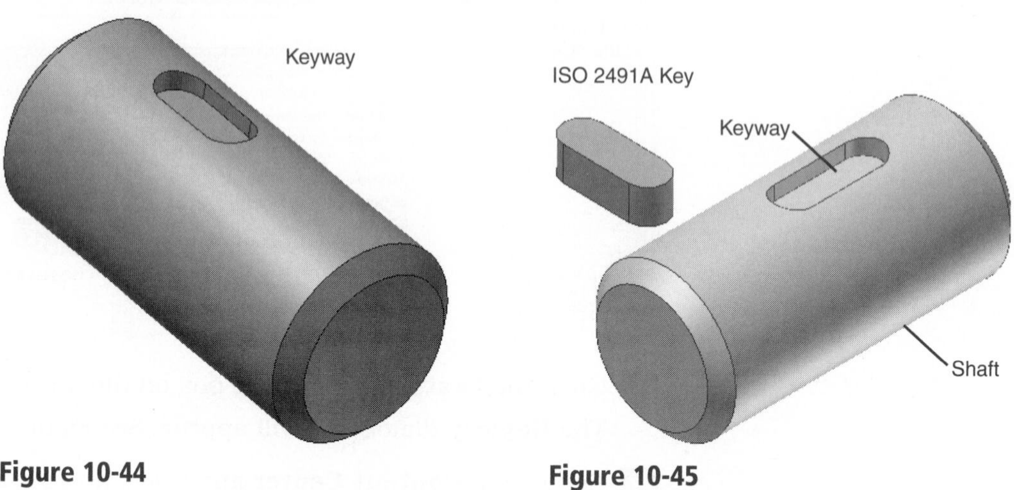

Figure 10-44

Figure 10-45

10 Use the **Constrain** tool and place the key into the keyway.

See Figure 10-46.

Figure 10-47 shows the shaft and key mounted into a hub.

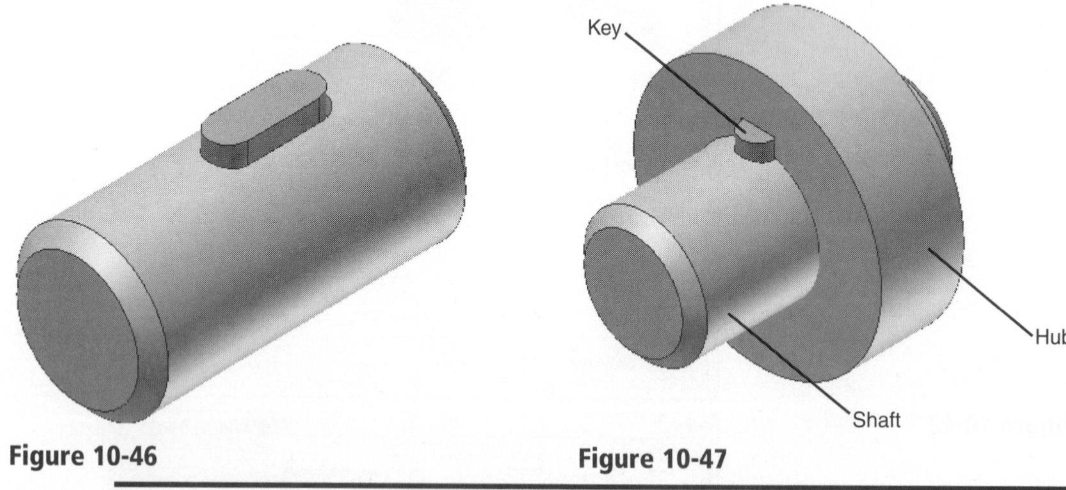

Figure 10-46

Figure 10-47

EXERCISE 10-8 **Drawing a Pratt and Whitney Keyway Using the Key Option on the Power Transmission Panel Located under the Design Tab**

1 Start a new drawing using the **Standard (mm).iam** format.

2 Access the **Shaft** tool located on the **Power Transmission** panel.

3 Draw a **Ø30 × 65** shaft.

See Figure 10-48.

4 Click the **Key** tool on the **Power Transmission** panel under the **Design** tab.

See Figure 10-49. The **Parallel Key Connection Generator** box will appear. See Figure 10-50.

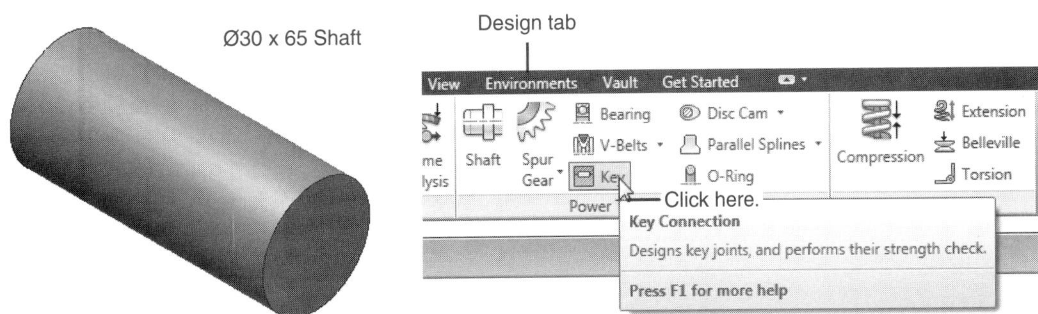

Figure 10-48

Figure 10-49

Figure 10-50

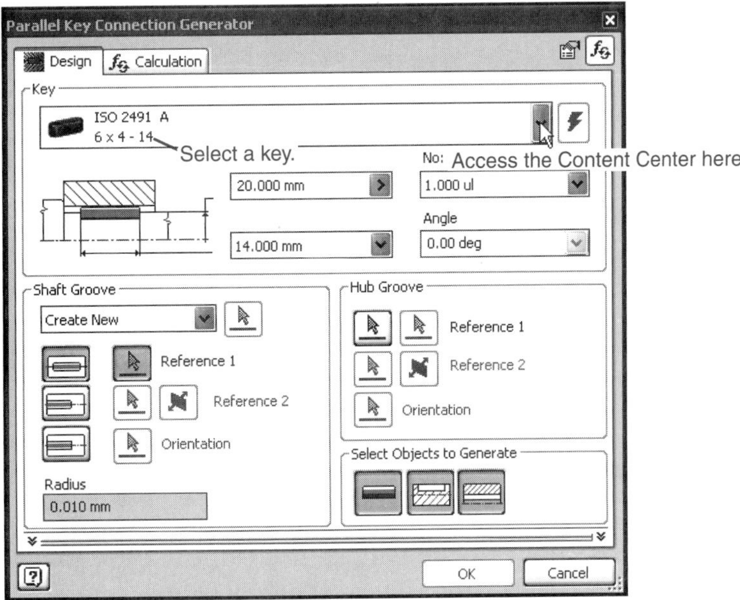

5 Access the **Content Center** by clicking the arrowhead as shown, and select a key.

6 Click the **Groove with Rounded Edges** box, then click the rounded side of the shaft. After the **Groove with Rounded Edges** box is clicked, the **Reference 1** box will activate automatically, followed by the **Reference 2** box.

See Figure 10-51.

The **Planar Face of Work Plane** box will automatically be highlighted. A preview of the keyway will appear on the shaft. See Figure 10-52. The arrows on the preview can be used to control the position, length, and orientation of the keyway.

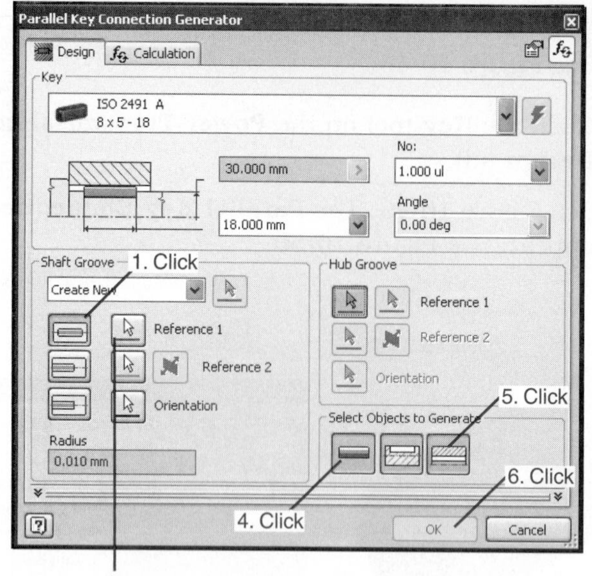

Figure 10-51

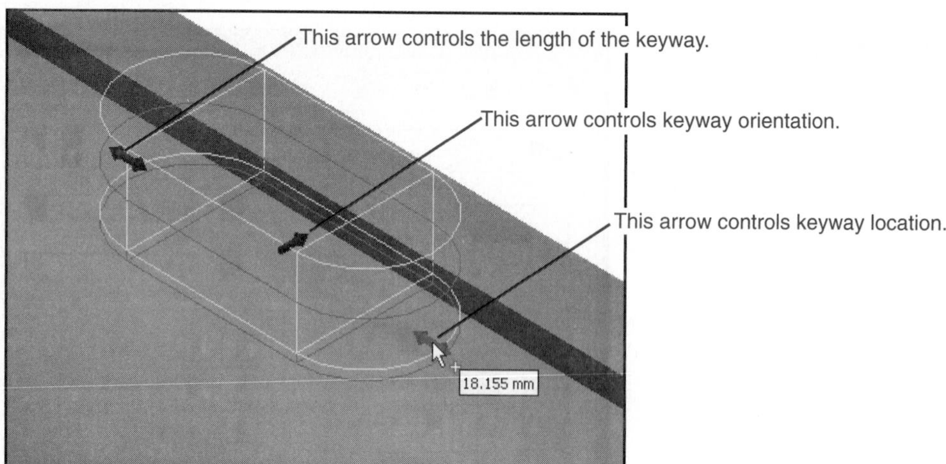

Figure 10-52

7 Edit the keyway as needed, click the **Insert Key** box, click the **Insert Hub Groove** box, then click **OK.**

Figure 10-53 shows the finished keyway.

Figure 10-53

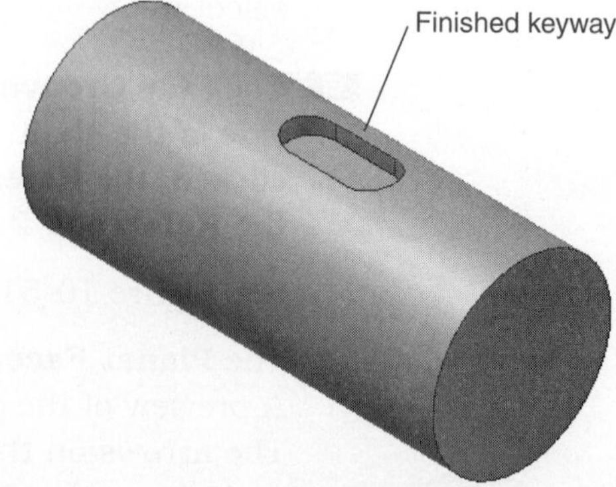

Chapter 10

EXERCISE 10-9 Creating a Plain Groove Keyway at the End of a Shaft

Plain grooves are used with square keys.

1. Start a new drawing using the **Standard (mm).iam** format.
2. Click the **Shaft** tool located on the **Power Transmission** panel under the **Design** tab.
3. Draw a **Ø30** × **65** shaft.
4. Access the **Key** tool on the **Power Transmission** panel.
5. Select a key.

   In this example a **DIN 6885 B** key was selected.
6. Click the **Plain Groove** box.

   See Figure 10-54.
7. Click the **Reference 1** box, and click the rounded side of the shaft.

   The **Reference 2** box will automatically be highlighted.
8. Click the end of the shaft, click the **Insert Key** box, click the **Insert Hub Groove** box, then click **OK.**

   Figure 10-55 shows the resulting plain groove.

Figure 10-54

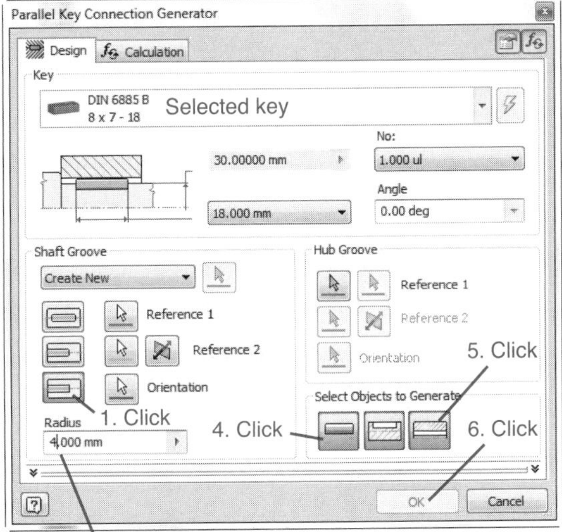

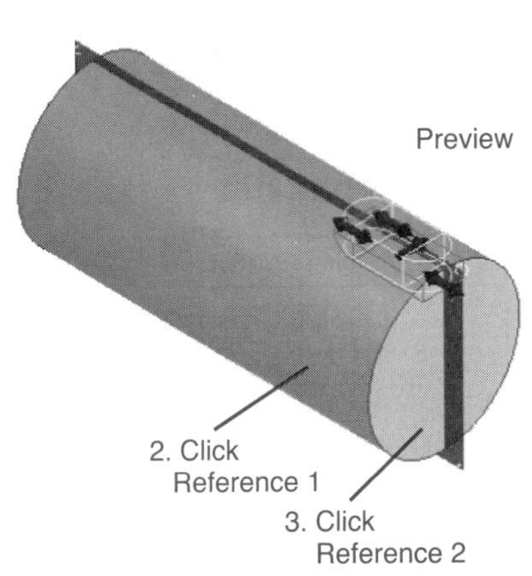

Figure 10-55

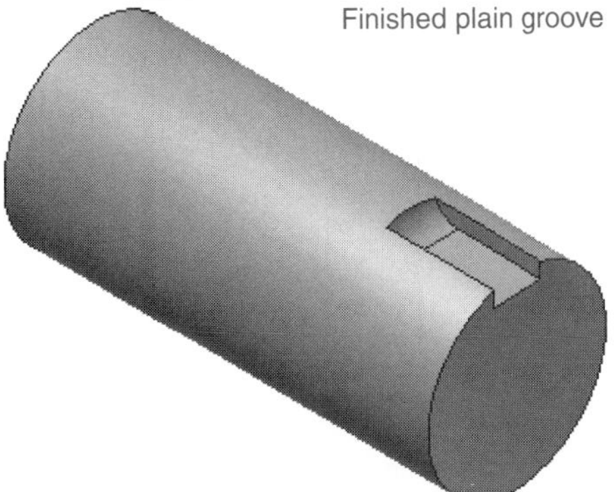

Woodruff Keys

Woodruff key: A crescent-shaped key that allows for some slight rotation between the shaft and hub.

Woodruff keys are crescent shaped. The crescent shape allows for some slight rotation between the shaft and hub.

EXERCISE 10-10 Drawing a Shaft with a Woodruff Key

1 Create a new drawing using the **Standard (mm).iam** format.

2 Access the **Place from Content Center** tool, click **Shaft Parts,** click **Keys,** then click **Keys - Woodruff.**

See Figure 10-56.

3 Select an **ISO 3912 (I)** key and select the **32–38** shaft diameter option.

See Figure 10-57. Click the **Table View** tab. The dimensions listed will be used to create the shaft keyway.

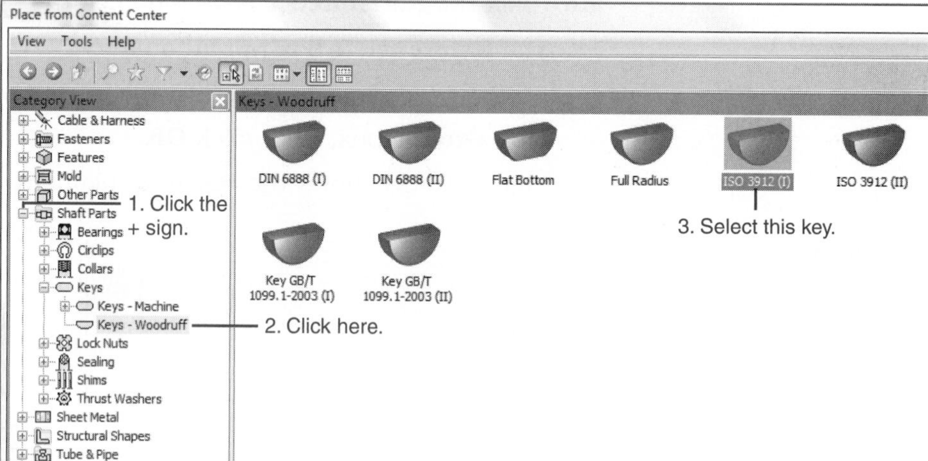

Figure 10-56

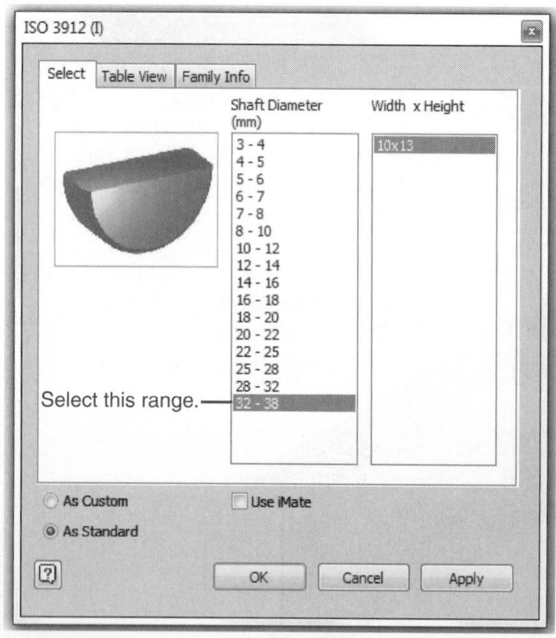

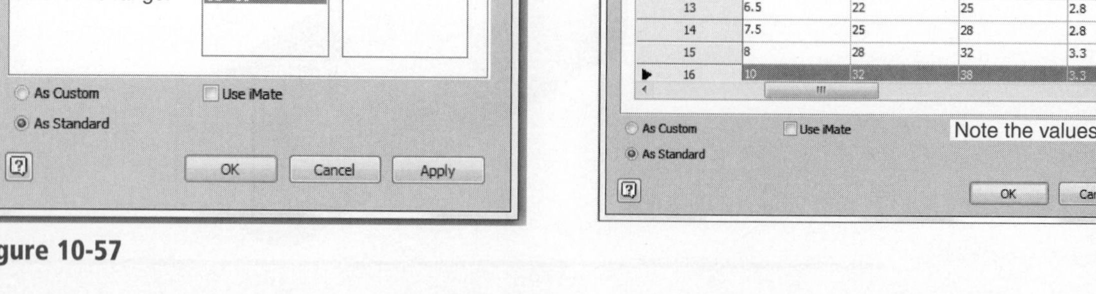

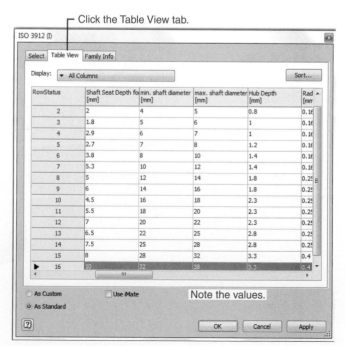

RowStatus	Shaft Seat Depth fo [mm]	min. shaft diameter [mm]	max. shaft diameter [mm]	Hub Depth [mm]	Rad [mm
2	2	4	5	0.8	0.16
3	1.8	5	6	1	0.16
4	2.9	6	7	1	0.16
5	2.7	7	8	1.2	0.16
6	3.8	8	10	1.4	0.16
7	5.3	10	12	1.4	0.16
8	5	12	14	1.8	0.25
9	6	14	16	1.8	0.25
10	4.5	16	18	2.3	0.25
11	5.5	18	20	2.3	0.25
12	7	20	22	2.3	0.25
13	6.5	22	25	2.8	0.25
14	7.5	25	28	2.8	0.25
15	8	28	32	3.3	0.4
16	10	32	38	3.3	0.4

Figure 10-57

4 Click **Apply.**

The key will appear on the screen. We must determine the size of the key to create an appropriate keyway in the shaft. Refer to the table values in Figure 10-57. The Woodruff key diameter is **32,** and the shaft seat depth is **10.** The key itself has a height of **13.**

> **NOTE**
>
> The depth of the keyway is 10, and the key diameter is 32.

5 Access the **Shaft** tool on the **Power Transmission** panel and draw a **Ø32** × **65** shaft. Create an offset YZ work plane through the center of the shaft and create a new sketch plane on the work plane.

See Figure 10-58. The shaft has a diameter of 32, so the work plane will be offset 16 from the edge of the shaft.

6 Rotate the shaft into a 2D view and add lines and a circle as shown in Figure 10-59.

The dimensions used came from the **Table View** values for the key.

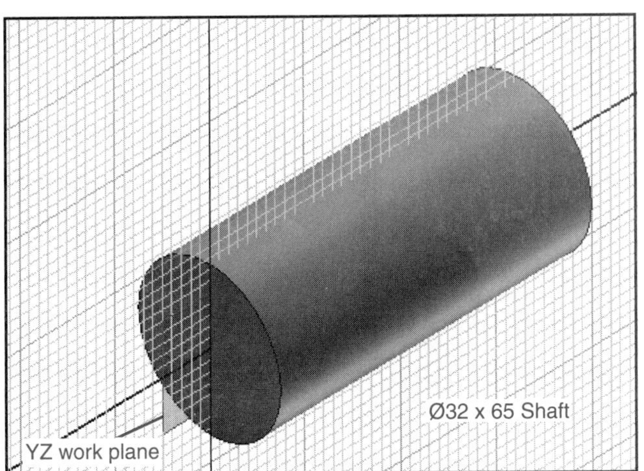

Figure 10-58

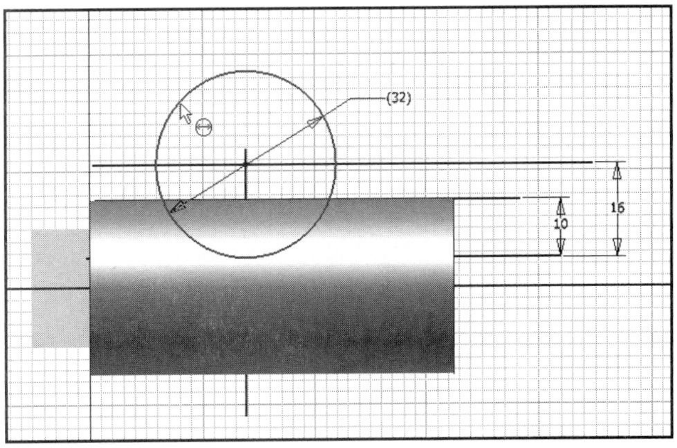

Figure 10-59

7 Return to an isometric view and select the **Extrude** tool. Use the **Cut** and **Intersection** options to create a cylinder from the circle drawn in step 6.

See Figures 10-60 and 10-61.

8 Use the **Constrain** tool to locate the key into the shaft's keyway.

See Figure 10-62.

> **TIP**
>
> You may have to use work planes to locate the key into the shaft.

Figure 10-63 shows a hub design to fit over the shaft and Woodruff key shown in Figure 10-62. The 19.3 dimension value was derived from adding the bore hole radius of 16.0 to a hub depth value of 3.3 listed in the **Table View** for the key.

Figure 10-60

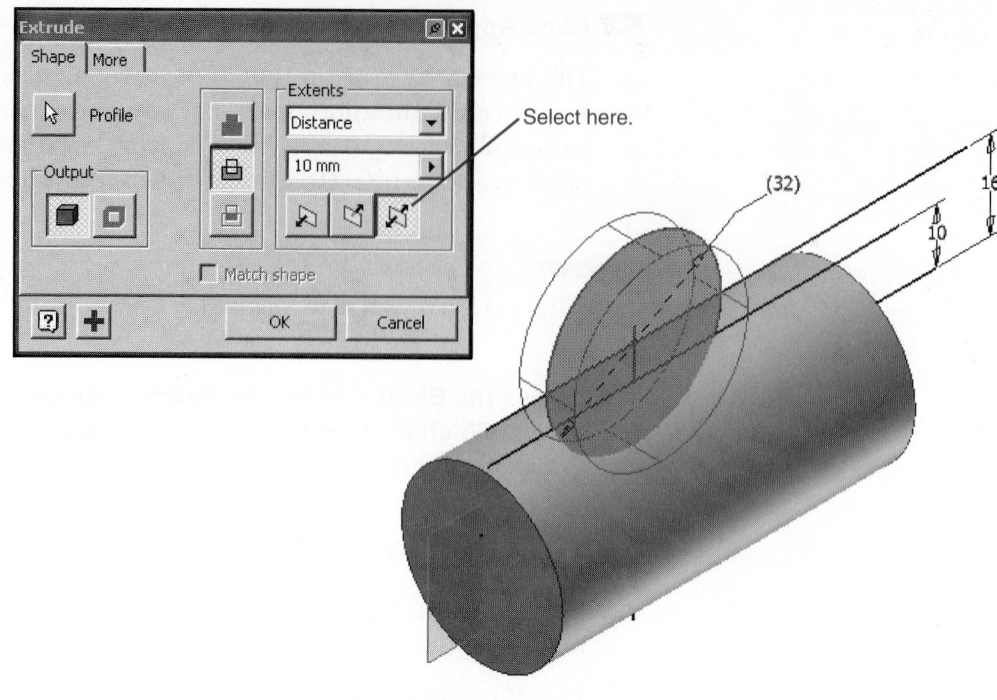

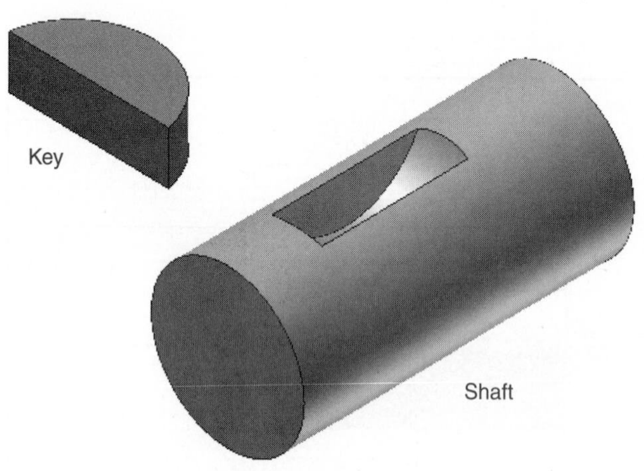

Figure 10-61

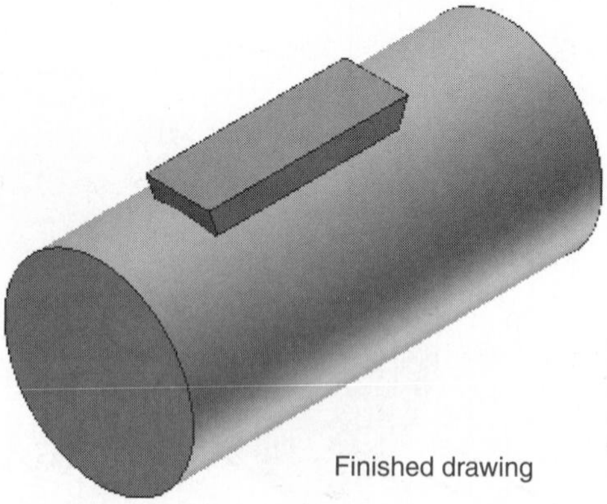

Figure 10-62

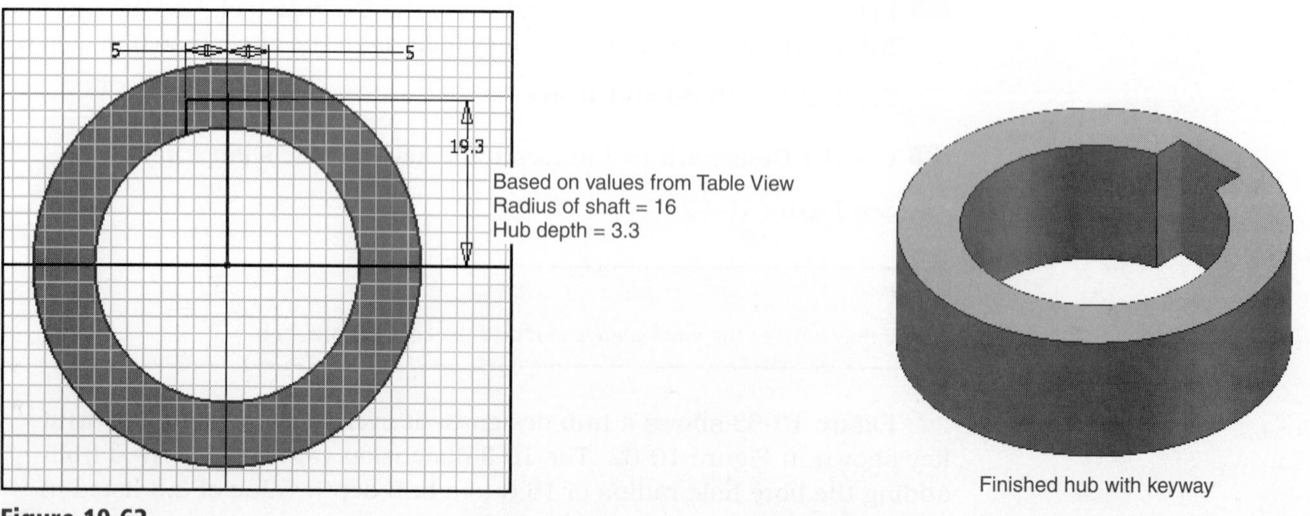

Figure 10-63

Shafts with Splines

splines: A series of cutouts in a shaft that are sized to match a corresponding set of cutouts in a hub; they are used to transmit torque.

Splines are a series of cutouts in a shaft that are sized to match a corresponding set of cutouts in a hub. Splines are used to transmit torque.

TIP

Splines are generally used on larger shafts. Smaller shafts use setscrews or pins.

EXERCISE 10-11 Drawing a Spline

1 Create a new drawing using the **Standard (mm).iam** format.

2 Use the **Shaft** tool on the **Power Transmission** panel under the **Design** tab to create a **Ø32** × **65** shaft.

See Figure 10-64.

NOTE

The shaft could have been drawn directly on the .iam drawing using the **Create Component** tool.

3 Click the **Parallel Splines** tool on the **Power Transmission** panel under the **Design** tab.

See Figure 10-65.

Figure 10-64

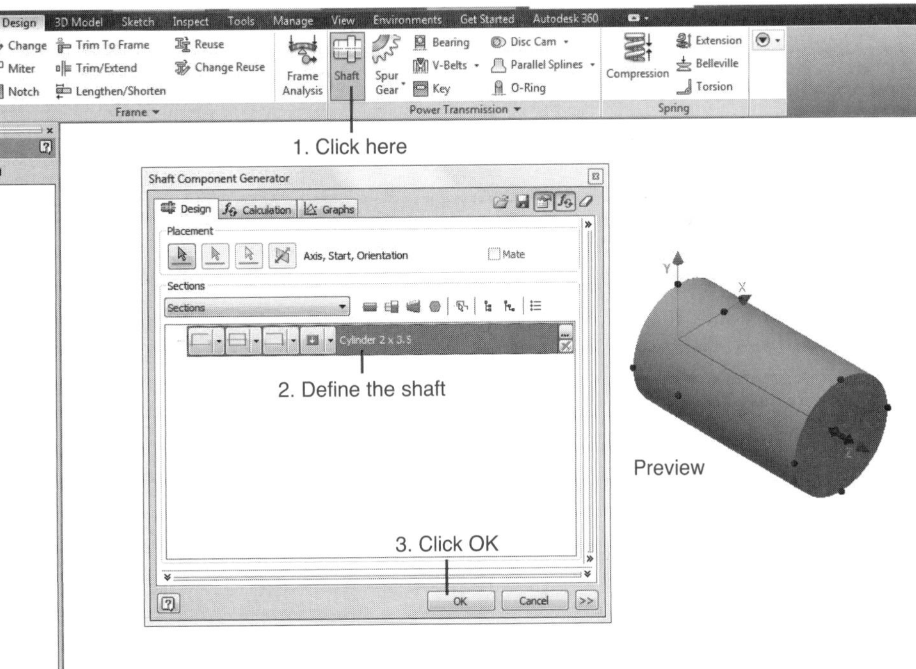

Figure 10-65

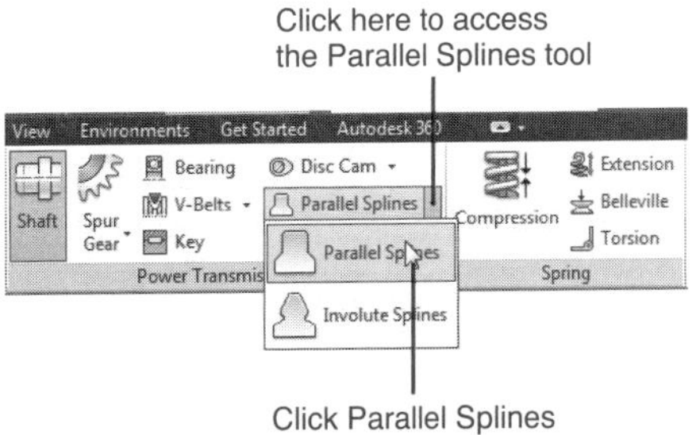

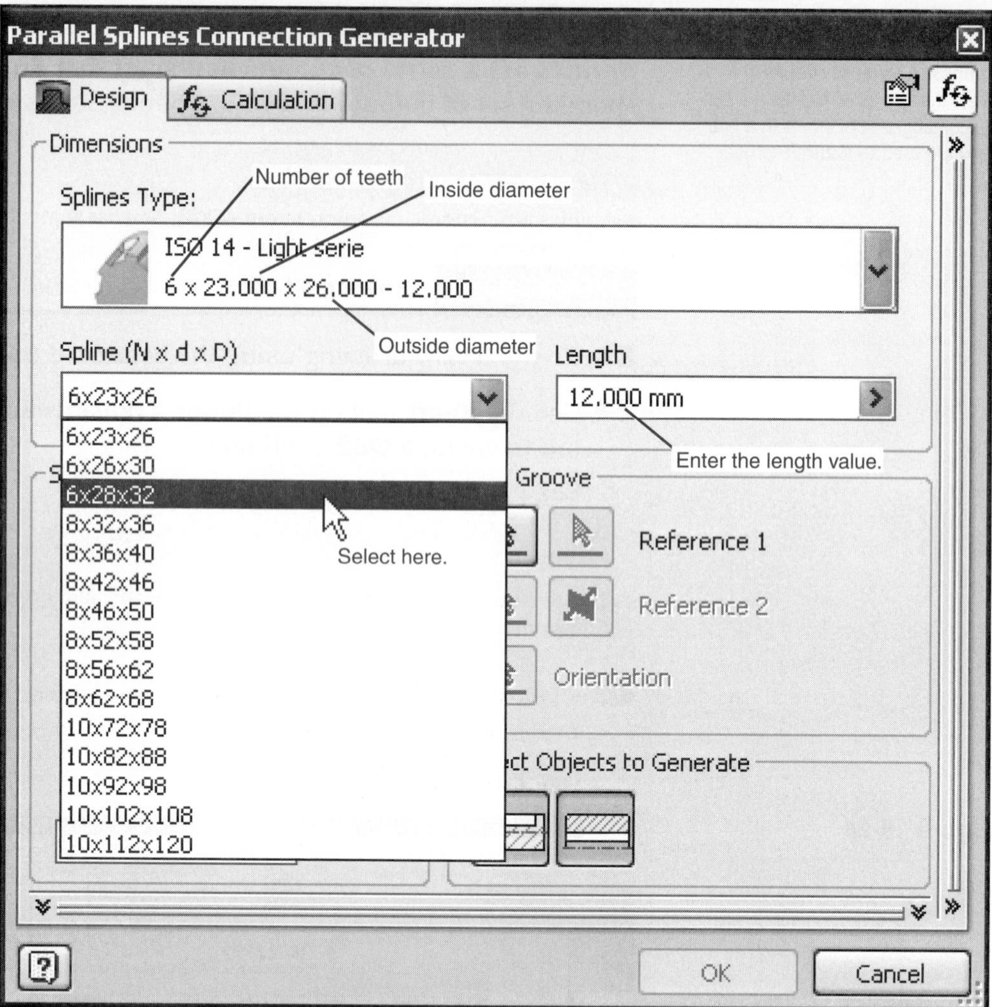

Figure 10-66

The **Parallel Splines Connection Generator** dialog box will appear. See Figure 10-66.

NOTE

The sequence of clicks used to create a spline is critical.

4 Set the **Splines Type** for **light,** the **Spline (N × d × D)** dimensions for **6 × 28 × 32,** the spline **Length** for 12, and the **Radius** for 4.

See Figure 10-67.

TIP

The outside diameter of the spline must equal the outside diameter of the shaft.

5 Click the **Reference 1** box, click the rounded sides of the shaft (the **Reference 2** box will automatically become active), click the flat end of the shaft, click the right-hand box in the **Select Objects to Generate** box, then click **OK.**

The **File Naming** dialog box will appear.

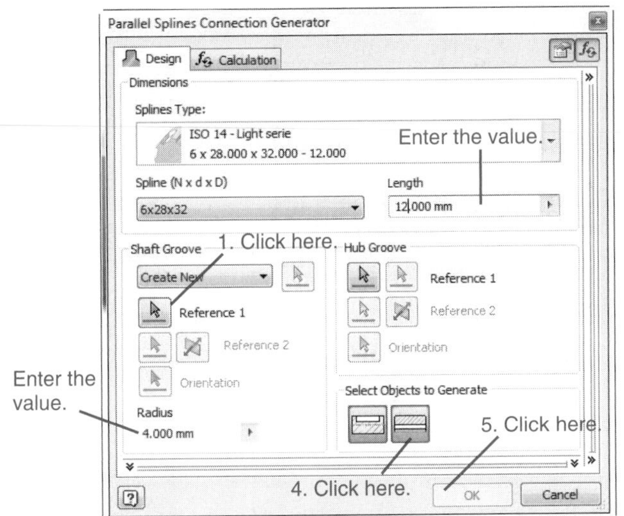

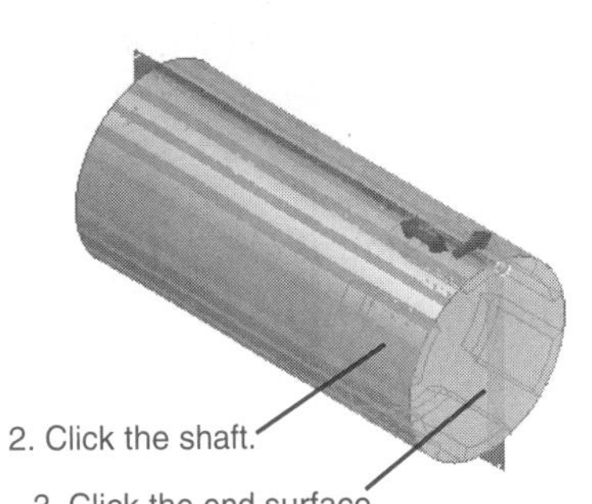

Figure 10-67

6 Click **OK.**

Figure 10-68 shows the finished spline.

Figure 10-68

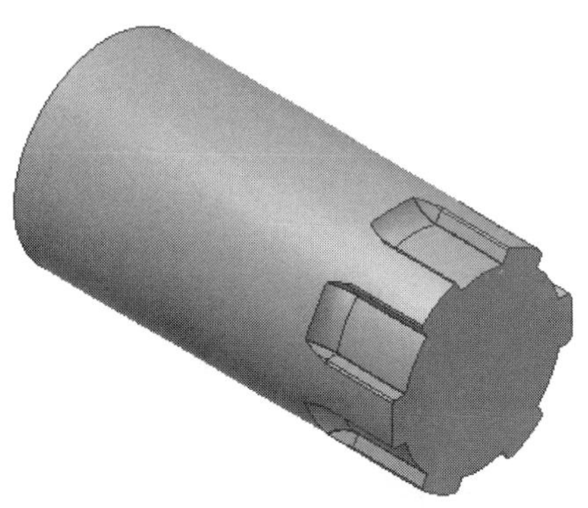

Retain the Ø32 × 65 shaft with the splined end and continue to work on the same assembly drawing for the next exercise. A Ø90 × 10 hub will be added to the assembly. A spline will be cut into the hub and inserted onto the Ø32 shaft.

> **TIP**
>
> Splines are defined by the dimensions N×d×D, where N is the number of teeth on the spline, d is the inside diameter, and D is the outside diameter.

EXERCISE 10-12 Drawing a Hub

1 Click the **Shaft** tool on the **Power Transmission** panel and create a **Ø90** × **10** shaft.

See Figures 10-69 and 10-70. The new shaft will be listed in the browser box as **Shaft:2.**

2 Click the **Shaft:2** heading in the browser box and select the **Open** option.

See Figure 10-71.

3 Right-click the hub **(Shaft:2)** and select the **Edit** option.

See Figure 10-72.

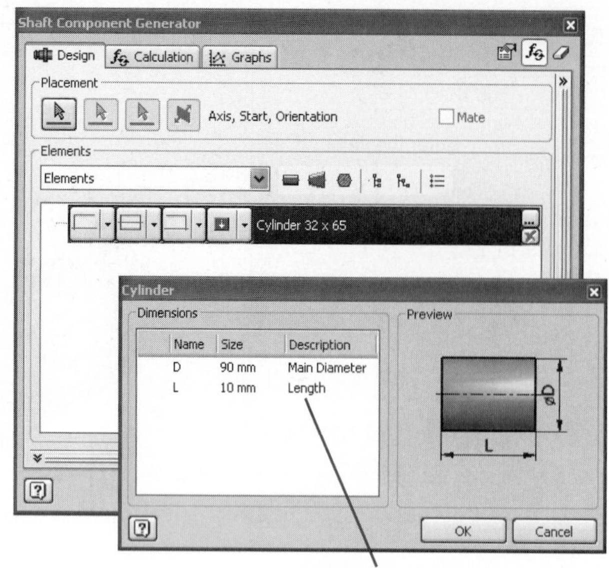

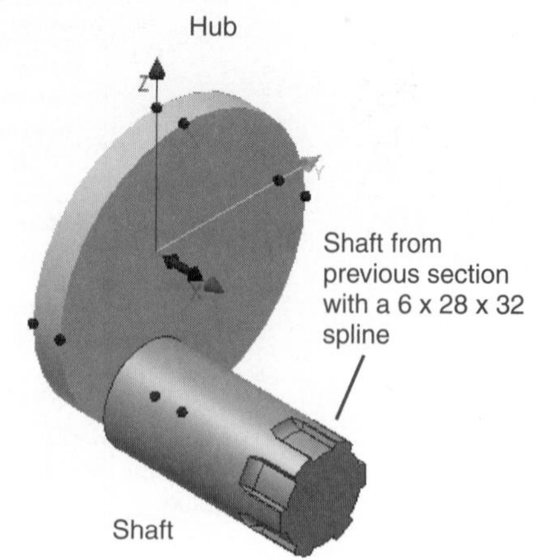

Figure 10-69

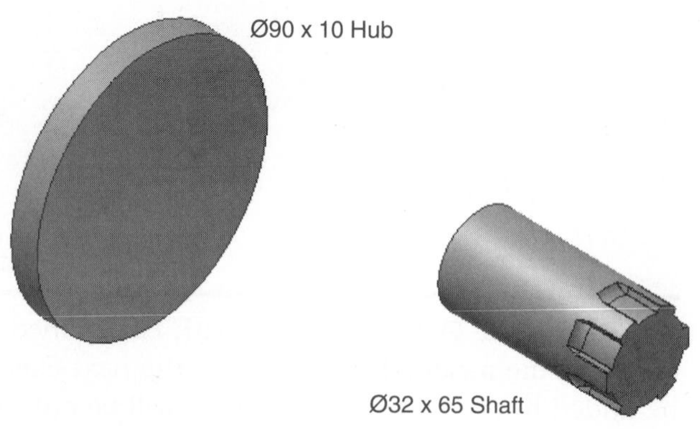

Figure 10-70

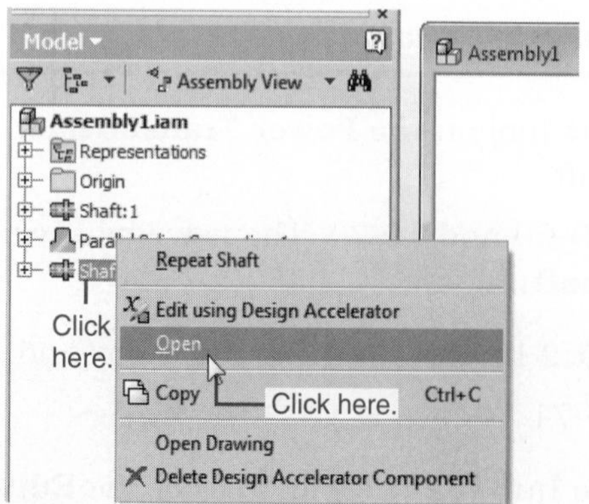

Figure 10-71

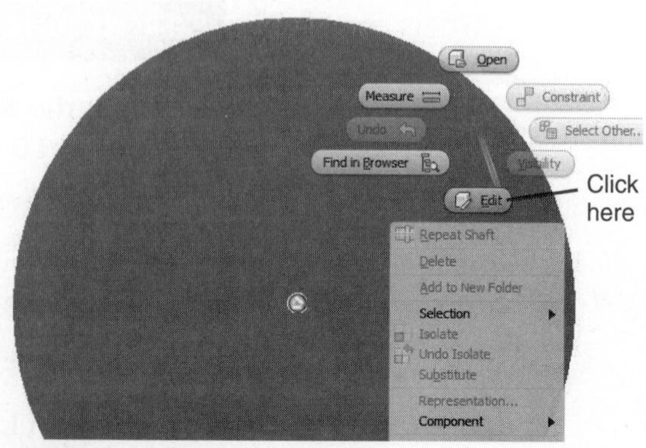

Figure 10-72

> **TIP**
> The hole diameter must equal the inside diameter of the spline.

4 Right-click the front surface of the Ø90 × 10 shaft and create a new sketch plane on the front surface of the hub as shown and add a **Point, Center Point** at the origin of the shaft.

See Figure 10-73.

5 Right-click and select **Finish 2D Sketch.**

See Figure 10-74.

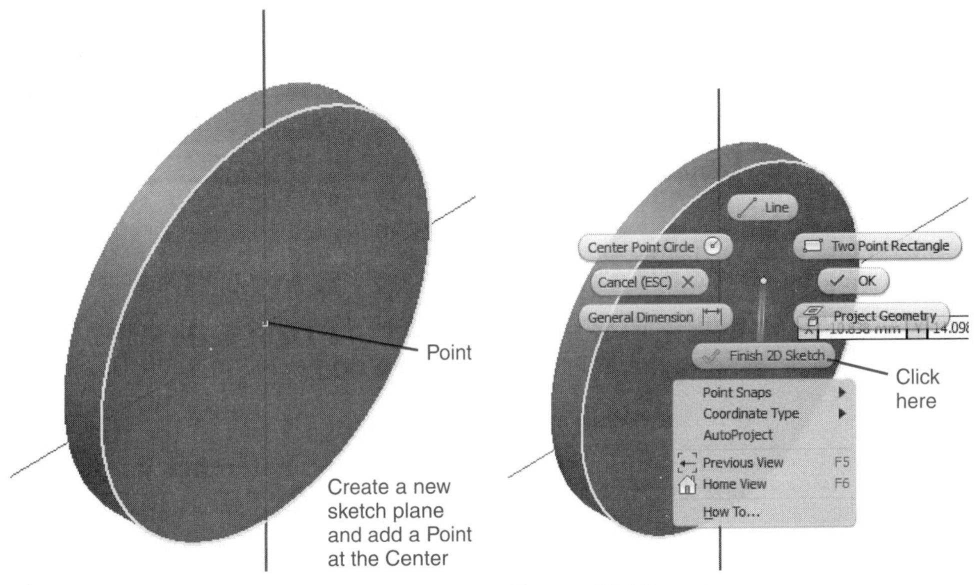

Figure 10-73

Figure 10-74

6 Access the **Hole** tool on the **Modify** panel and create a Ø**28** hole through the **Ø90** shaft (hub).

7 Right-click the mouse and select the **Finish Edit** option.

8 Right-click the mouse and click the **X** box in the upper right corner of the **Shaft** drawing screen.

A warning box will appear.

9 Click the **Yes** box.

The **Save** dialog box will appear.

10 Click **OK.**

The drawing screen will return to the assembly screen showing both the Ø32 shaft and the Ø90 hub.

11 Click the hub.

The Ø32 hole will appear in the hub. See Figure 10-75.

12 Click the **Parallel Splines** tool on the **Power Transmission** panel.

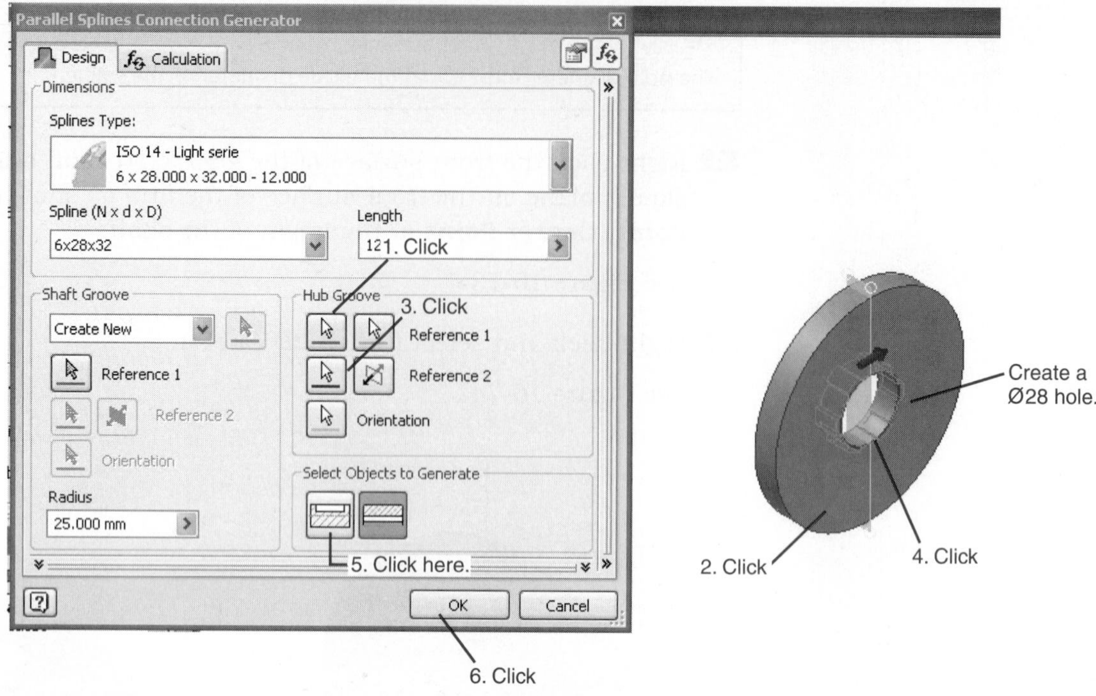

Figure 10-75

13 Select an **ISO - 14 Light serie** spline with dimensions of **6** × **28.000** × **32.000 - 12.000.**

14 Click the **Reference 1** box in the **Hub Groove** box, click the surface of the hub, click the **Reference 2** box, click the edge of the Ø28 hole, click the left box in the **Select Objects to Generate** box, then click **OK.**

The **File Naming** dialog box will appear.

15 Click **OK.**

Figure 10-76 shows the hub with a splined hole.

16 Return to the **Assemble** panel bar and use the **Insert** tool on the **Constrain** panel and mount the hub onto the shaft.

See Figure 10-77.

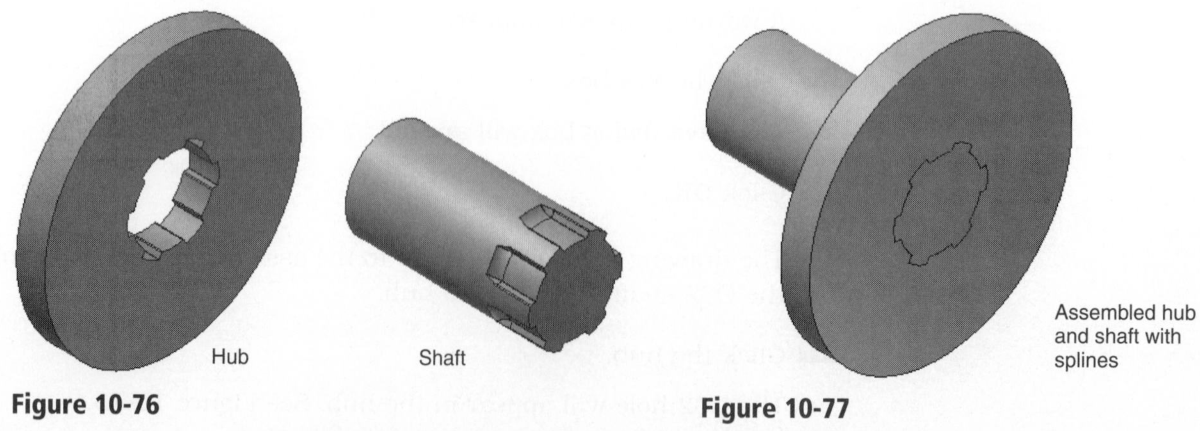

Figure 10-76

Figure 10-77

Collars

collar: Used to hold a shaft in place as it rotates; may be mounted on the inside or outside of the support structure.

Collars are used to hold shafts in position as they rotate. Collars may be mounted inside or outside the support structure. See Figure 10-78.

Figure 10-78

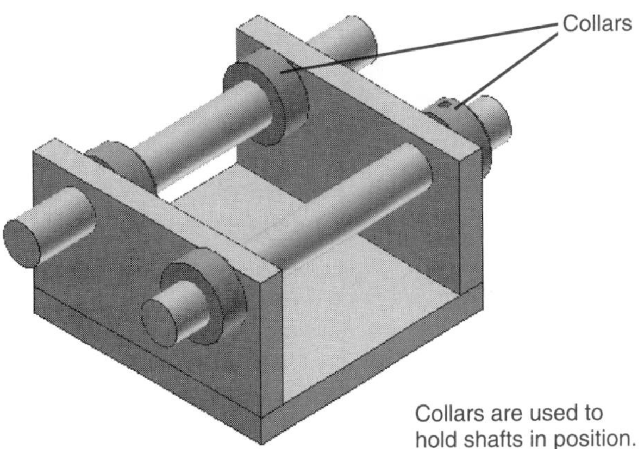

EXERCISE 10-13 Adding Collars to a Shaft Assembly—Setscrews

1 Start a new drawing using the **Standard (mm).iam** format.

2 Click the **Place from Content Center** tool, click **Shaft Parts,** then **Collars.**

See Figure 10-79. In this example a **DIN 705 B** collar with a **Ø16** was selected.

Figure 10-79

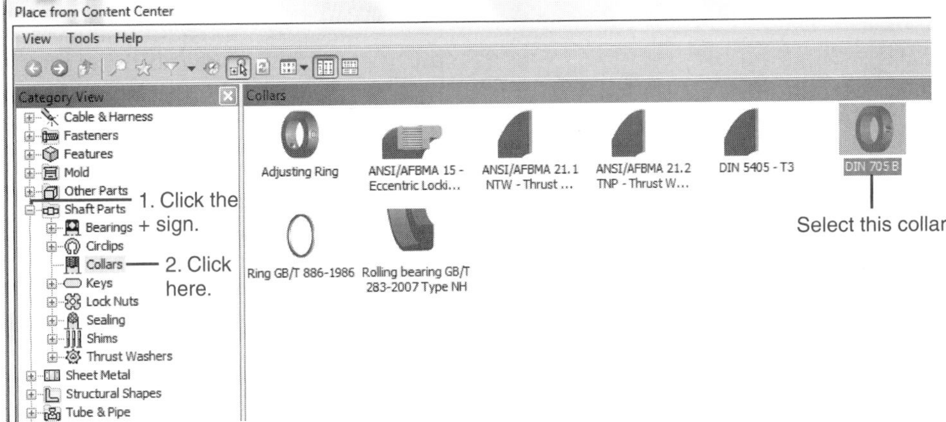

3 Click the **DIN 705 B** icon, and click **OK.**

The **DIN 705 B** dialog box will appear.

4 Select an **Inside Diameter** of **16** and click the **Table View** tab, then **All Columns.**

See Figure 10-80. Note the size specifications for the collar. In this example the **Nominal Diameter** is the diameter of the small hole in the collar. The diameter of the hole is **4.**

This hole can be threaded and a setscrew added to hold the collar on the shaft.

See Exercise 3-15 for instructions on how to create an offset work plane. Once a work plane is established, a new sketch may be created and used to create a threaded hole.

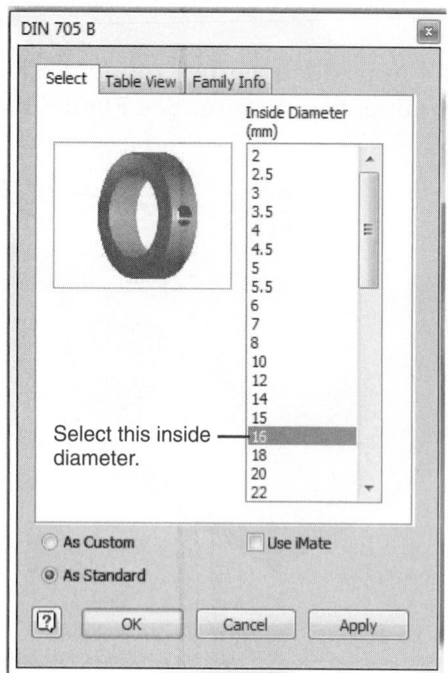

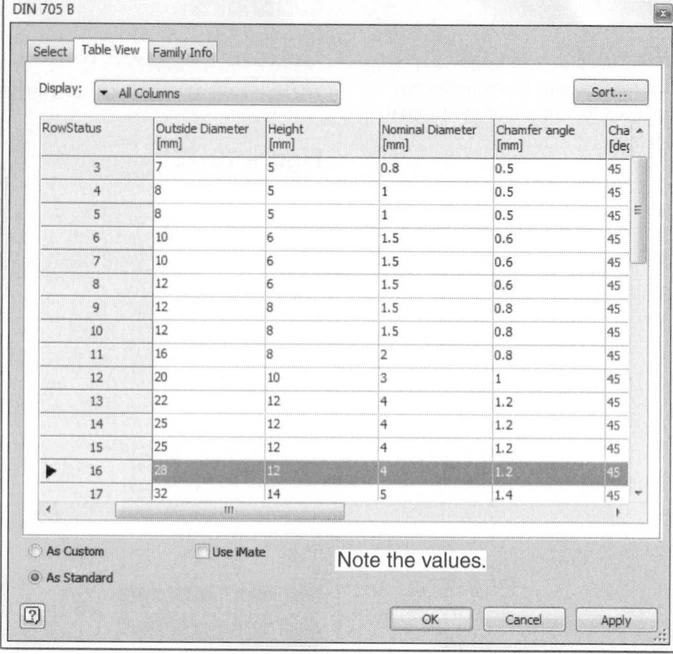

RowStatus	Outside Diameter [mm]	Height [mm]	Nominal Diameter [mm]	Chamfer angle [mm]	Cha [deg
3	7	5	0.8	0.5	45
4	8	5	1	0.5	45
5	8	5	1	0.5	45
6	10	6	1.5	0.6	45
7	10	6	1.5	0.6	45
8	12	6	1.5	0.6	45
9	12	8	1.5	0.8	45
10	12	8	1.5	0.8	45
11	16	8	2	0.8	45
12	20	10	3	1	45
13	22	12	4	1.2	45
14	25	12	4	1.2	45
15	25	12	4	1.2	45
16	28	12	4	1.2	45
17	32	14	5	1.4	45

Figure 10-80

5 Access the **Place from Content Center** dialog box and select a setscrew. See Exercise 6-21 for information about setscrews.

See Figure 10-81. In this example a **DIN EN 27434 Set Screw** with an **M4** thread, a **6** length, and a conical point was selected.

6 Use the **Constrain** tool and insert the setscrew into the hole.

The assembly sequence presented here is one of several possibilities.

7 Insert the setscrew into the drawing and position it above the hole.

See Figure 10-81.

8 Draw a work axis through the hole and use the **Mate** tool on the **Constrain** panel to align the hole's work axis with the setscrew's axis.

9 Draw a work plane tangent to the collar and hole's edge. Use the **Flush** tool on the **Constrain** panel to make the setscrew flush with the work plane.

10 Hide the work plane and work axis.

Adding Collars to a Shaft Assembly—Pins

A collar may be held in place using pins. Figure 10-82 shows a shaft and a DIN 705 B collar, both with Ø4.00 holes. A pin will be inserted through the hole in the collar into the hole in the shaft.

Figure 10-83 shows some of the different types of pins available in the Inventor **Content Center.** For this example a spring-type cylindrical pin will be used. These type pins are squeezed to a smaller diameter, inserted into the hole, and then released back to their original diameter.

1 Insert the collar onto the shaft with the holes aligned.

See Figure 10-84.

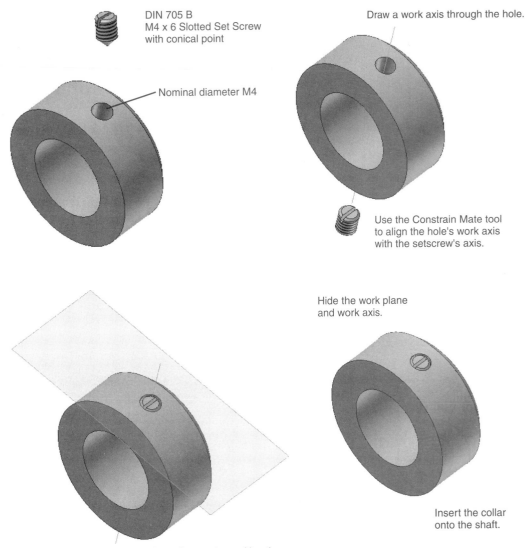

DIN 705 B
M4 x 6 Slotted Set Screw
with conical point

Nominal diameter M4

Draw a work axis through the hole.

Use the Constrain Mate tool
to align the hole's work axis
with the setscrew's axis.

Hide the work plane
and work axis.

Insert the collar
onto the shaft.

Draw a work plane tangent to the collar as shown. Use the Constrain Flush tool to make the top surface of the setscrew flush with the work plane.

Figure 10-81

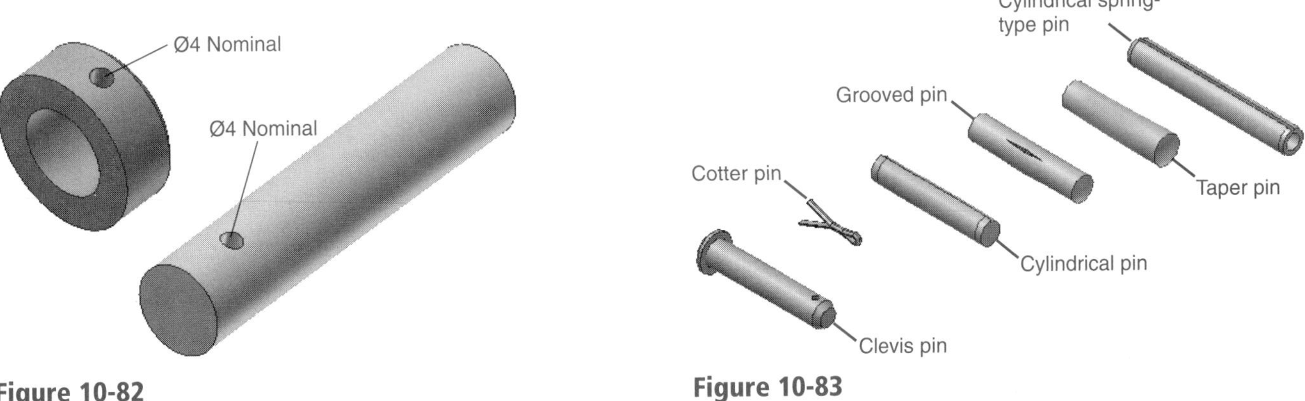

Figure 10-82

Figure 10-83

NOTE

Consider defining a work axis through the holes to help with the alignment.

2 Click the **Place from Content Center** tool, and select the **Fasteners, Pins,** and **Cylindrical** options.

Figure 10-84

See Figures 10-85 and 10-86. In this example a **DIN EN ISO 13337** spring-type pin with a **Nominal Diameter** of **4** and a **Nominal Length** of **20** was selected. The length 20 was selected because the collar thickness is 6, and the shaft diameter is 16, for a total of 22. There is only one hole in the collar, so the pin length cannot exceed 22 or it will extend beyond the collar. A length value of 20 allows for a clearance of 2. Figure 10-87 shows the selected pin with the collar and shaft.

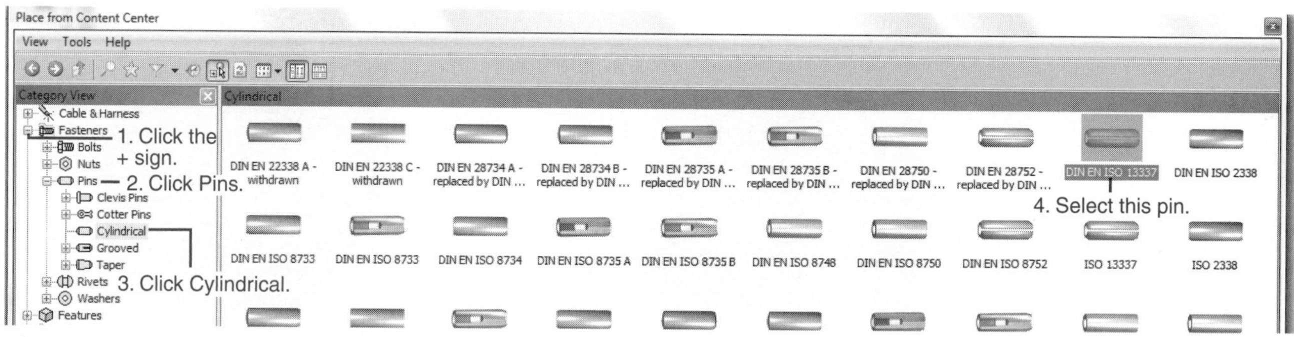

Figure 10-85

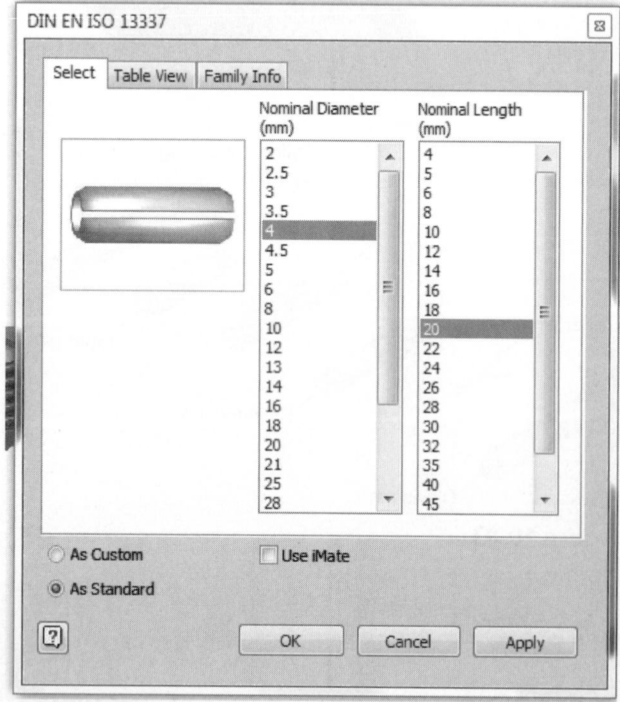

Figure 10-86

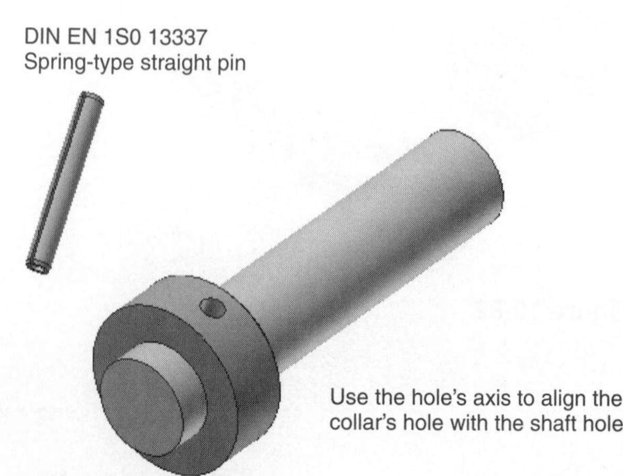

Figure 10-87

> **NOTE**
> Consider using work planes and a work axis when constraining the pin and collar.

3 Use the **Constrain** tool to assemble the pin into the collar and shaft.

See Figure 10-88.

Insert the pin into the collar and shaft.

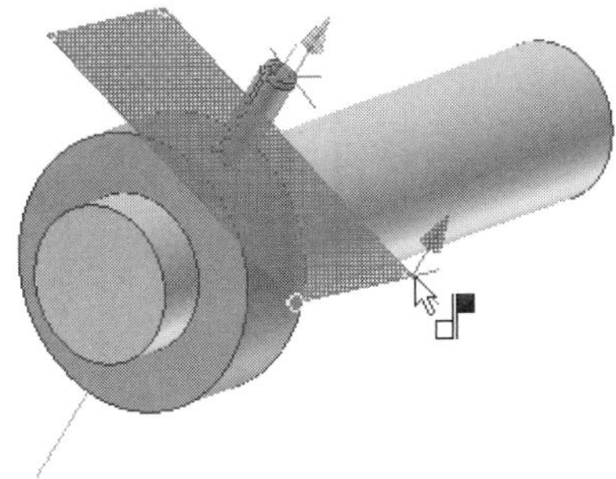

Use a work plane and the Constrain Flush tool to align the pin with the collar.

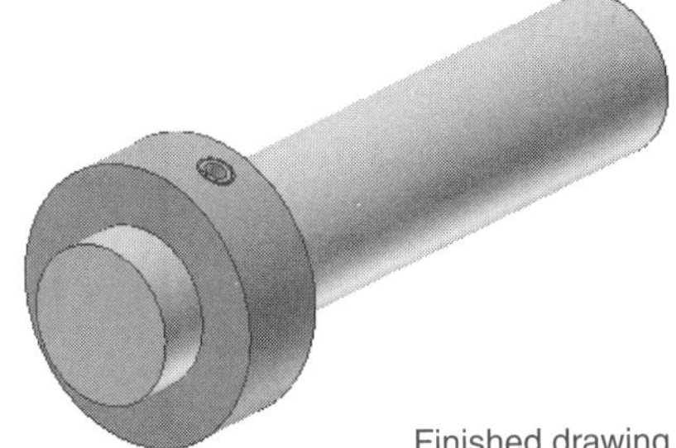

Figure 10-88

O-Rings

O-ring: A ring that is forced between two objects to create a seal.

O-rings are used to create seals. The rings are forced between two different objects, which distorts the rings and creates a seal.

EXERCISE 10-14 Drawing an O-Ring and a Shaft

Assume the nominal diameter of the shaft is 0.500.

1 Create a new drawing using the **Standard (in).iam** format.

2 Click the **Design** tab and use the **Shaft** tool on the **Power Transmission** panel to draw a **Ø0.500** × **2.50** shaft.

3 Click the **Assemble** tab and use the **Place from Content Center** dialog box to click **Shaft Parts, Sealing, O-Rings,** and select an **AS 568 O-Ring**.

See Figure 10-89.

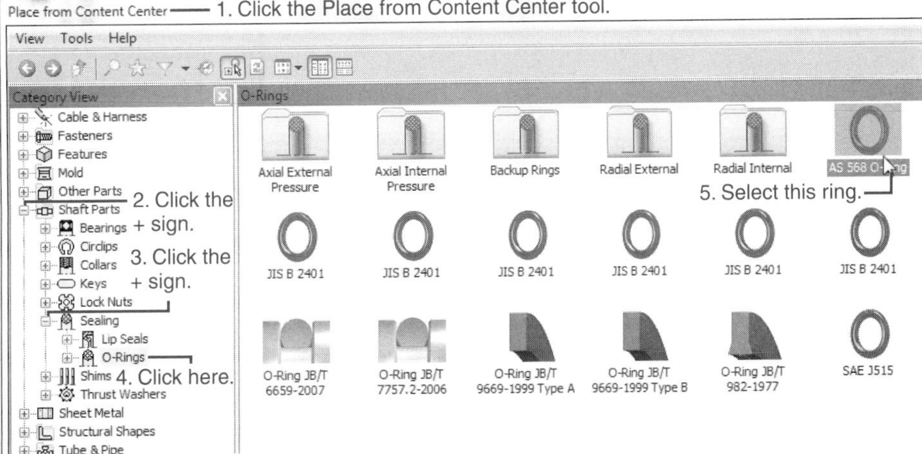

Figure 10-89

4 Click the **Table View** tab, then **All Columns.** Select a **DASH # 014 O-Ring** because it has a 0.489 inside diameter.

See Figure 10-90.

> **NOTE**
>
> The cross section for this O-ring is 0.07.

Figure 10-90

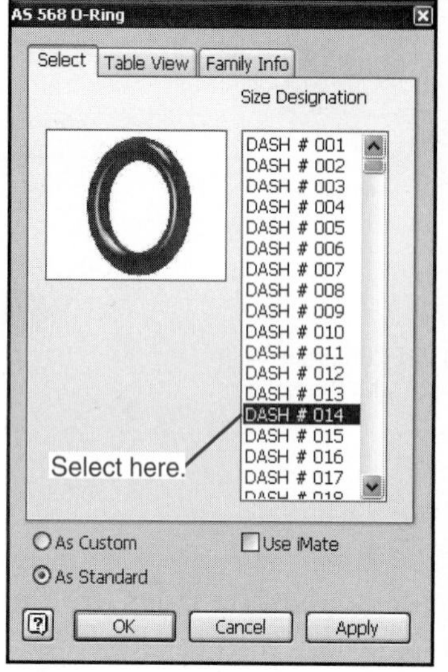

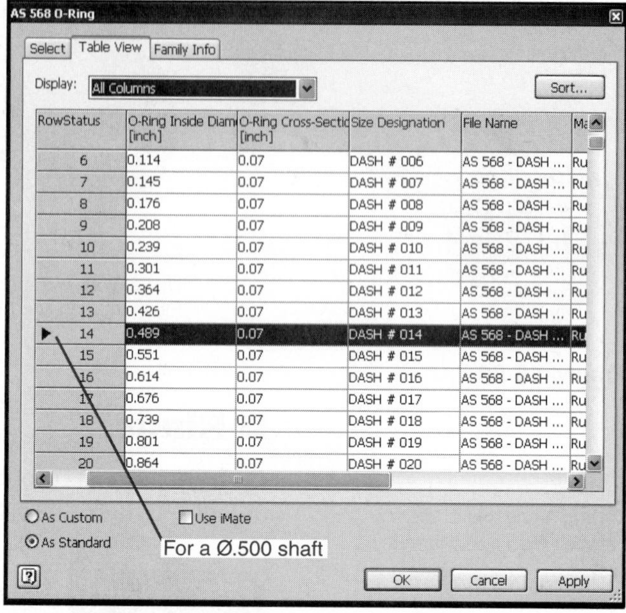

5 Insert the ring into the drawing.

6 Right-click the **Shaft:1** heading in the browser box and select the **Edit using Design Accelerator** option.

See Figure 10-91. The **Shaft Component Generator** dialog box will appear. See Figure 10-92.

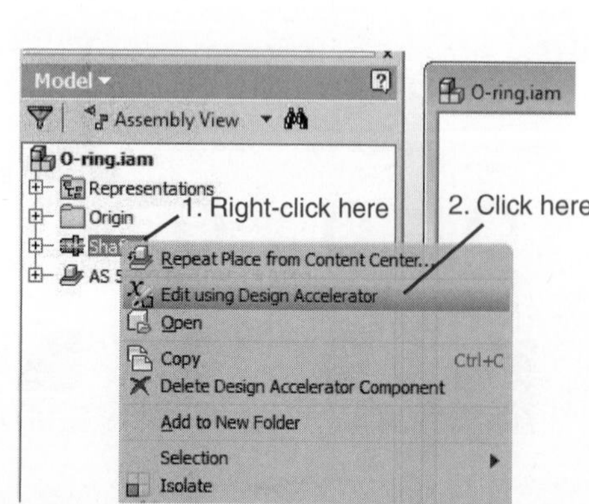

Figure 10-91

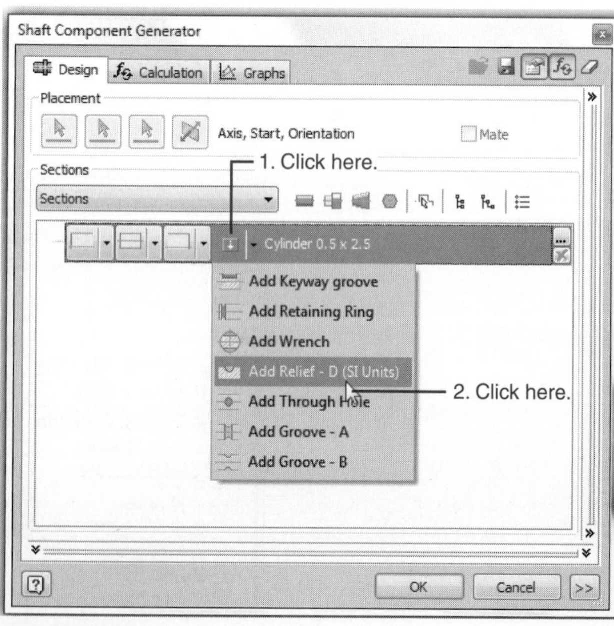

Figure 10-92

7 Select the **Add Relief–D (SI Units)** option.

The **Relief-D** option will appear on the screen.

See Figure 10-93.

8 Click the **Element Properties** box.

The **Relief–D (SI Units)** dialog box will appear. See Figure 10-94.

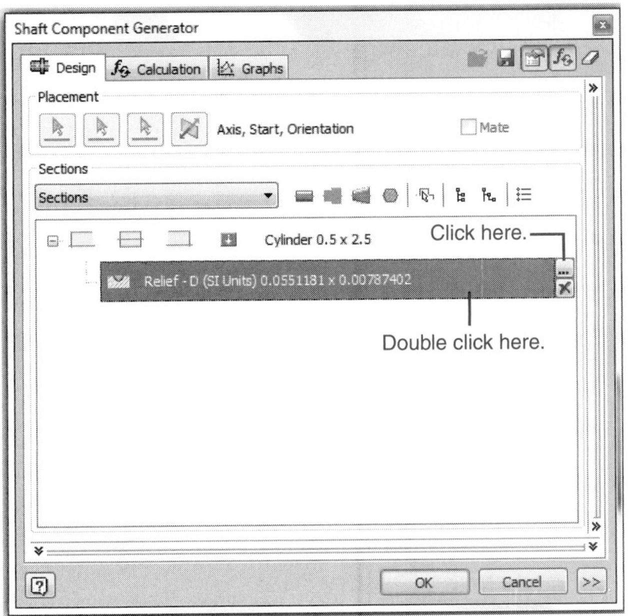

Figure 10-93

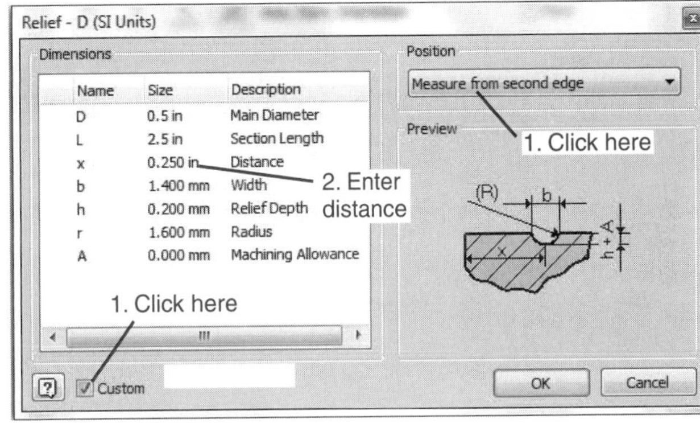

Figure 10-94

9 Enter the distance from the second edge. Click the **Custom** box and enter the values as shown in Figure 10-94. (These values were derived from the **Content Center.**) Unclick the **Custom** box. Click **OK.**

The **Shaft Component Generator** dialog box will appear.

10 Click **OK.**

Figure 10-95 shows the O-ring and modified shaft.

Figure 10-95

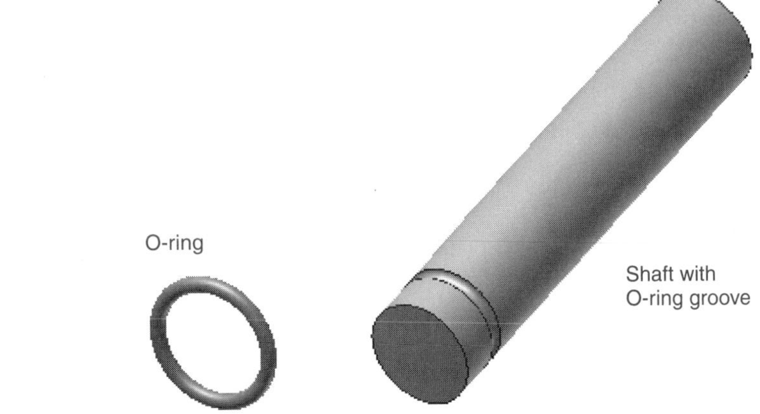

11 Use the **Constrain** tool to position the O-ring onto the shaft's groove.

The procedure presented here is one possibility.

12 Draw an XY work plane through the O-ring.

See Figure 10-96.

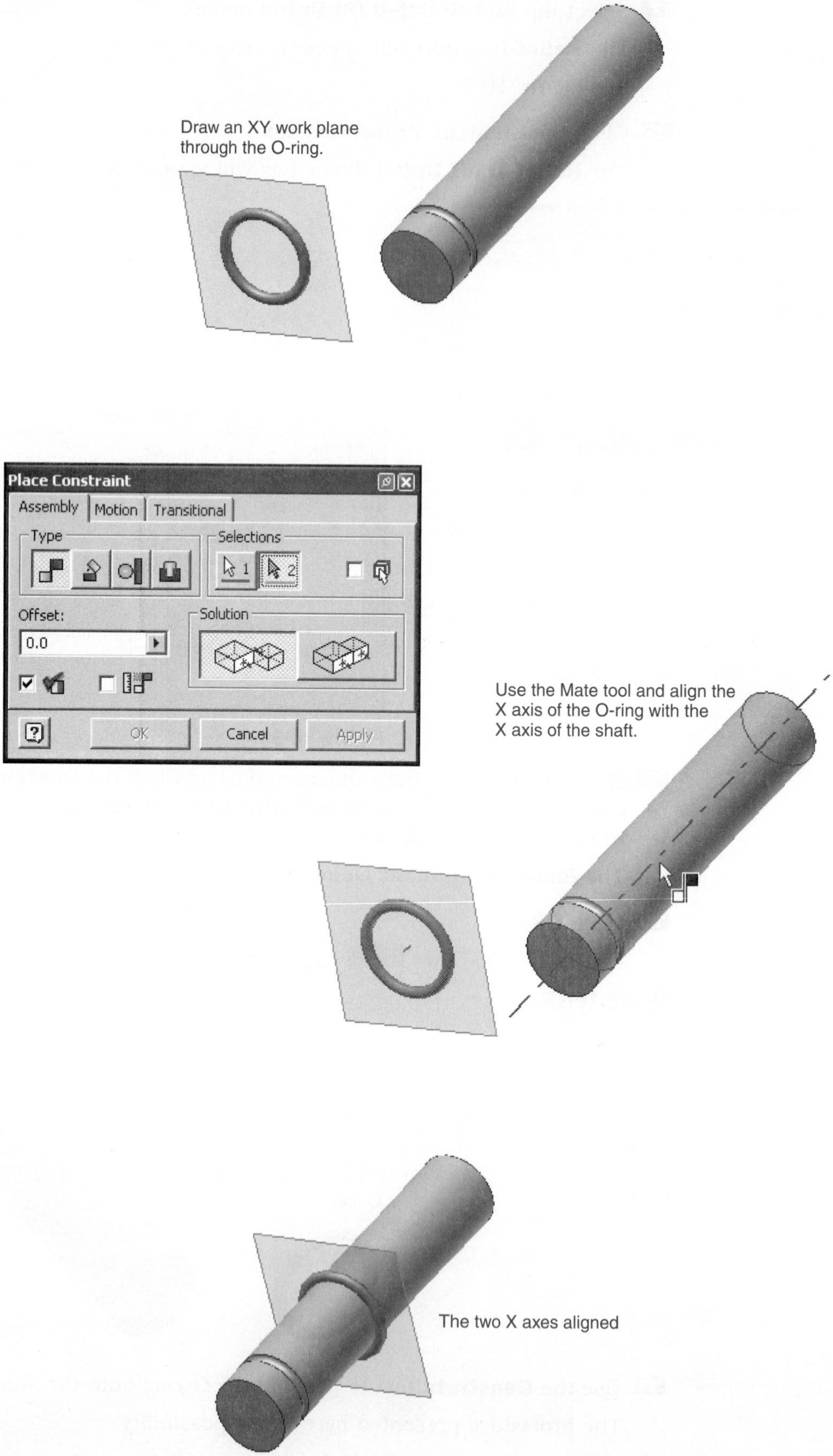

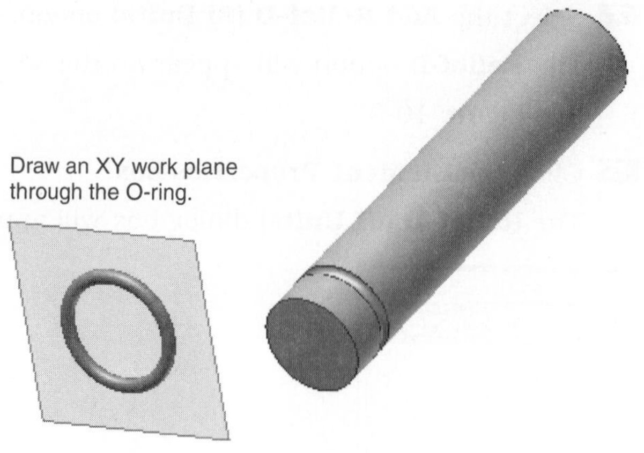

Draw an XY work plane
through the O-ring.
Place Constraint
Assembly
Motion
Transitional
Type
Selections
Offset:
0.0
Solution
OK
Cancel
Apply
Use the Mate tool and align the
X axis of the O-ring with the
X axis of the shaft.
The two X axes aligned

Figure 10-96

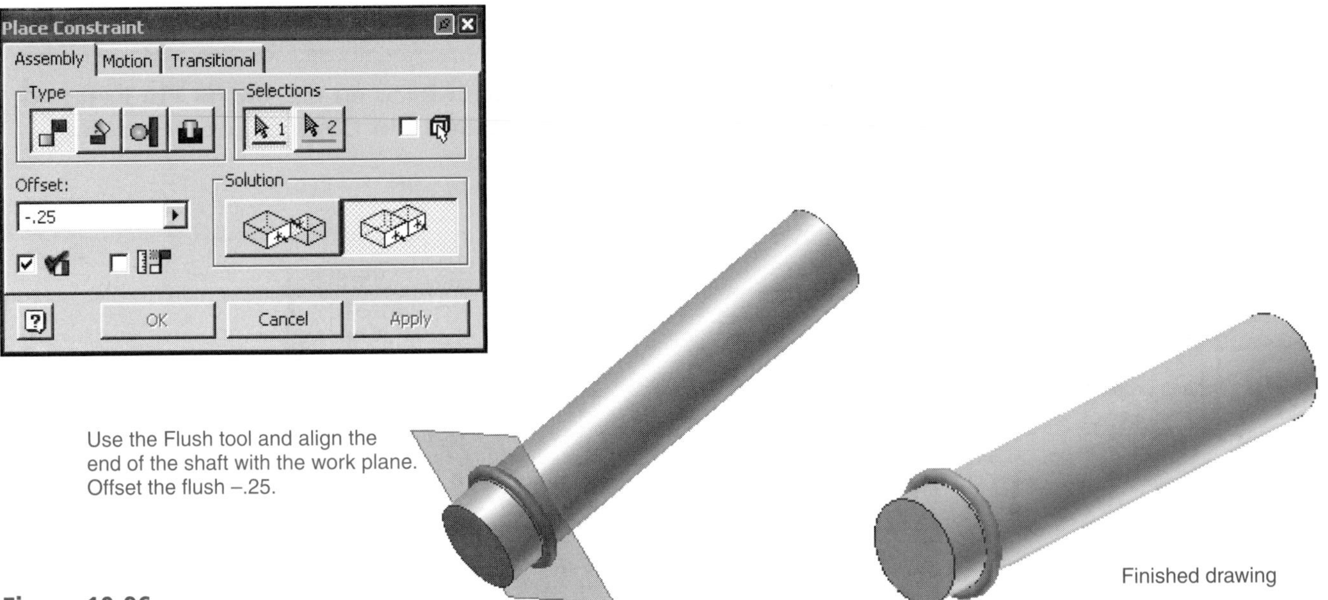

Figure 10-96
(Continued)

13 Use the **Mate** tool on the **Constrain** panel and align the X axis of the O-ring with the X axis of the shaft.

14 Use the **Flush** tool to align the front surface of the shaft with the work plane on the O-ring, then define a **–0.25** offset.

Figure 10-96 shows the finished O-ring and shaft.

Drawing Shafts and Pins Using the Tools Under the Design Tab

Pins can also be drawn using the tools located under the **Design** tab. Figure 10-97 shows a hub with a 20 inside diameter, a 40 outside diameter, a thickness of 16, and a through hole of Ø6. A work plane has been created tangent to the hub. Figure 10-97 also shows a Ø20 × 30 shaft with a Ø6 hole and a 1 × 45° chamfer at each end. The shaft was created using the **Shaft** tool located on the **Power Transmission** panel under the **Design** tab. The hub and the shaft are to be assembled together and held together using a pin. The drawing was created using the **Standard (mm).iam** format.

Figure 10-97

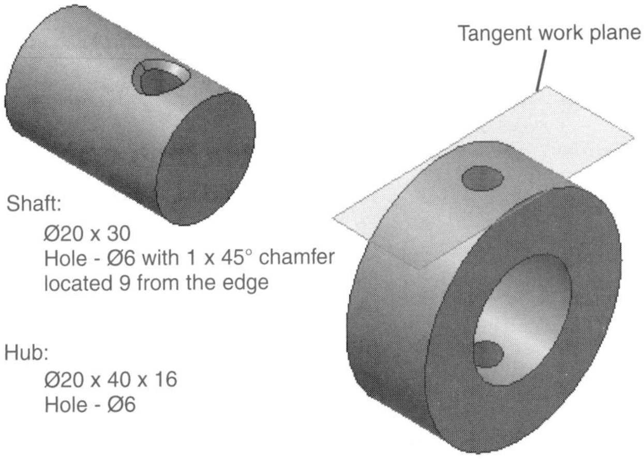

Tangent work plane

Shaft:
Ø20 x 30
Hole - Ø6 with 1 x 45° chamfer located 9 from the edge

Hub:
Ø20 x 40 x 16
Hole - Ø6

EXERCISE 10-15 Drawing a Pin Using Design Tab

1. Click the **Design** tab, click the arrowhead on the **Clevis Pin** icon and select the **Radial Pin** tool. The **Radial Pin Component Generator** dialog box will appear.

   See Figure 10-98.

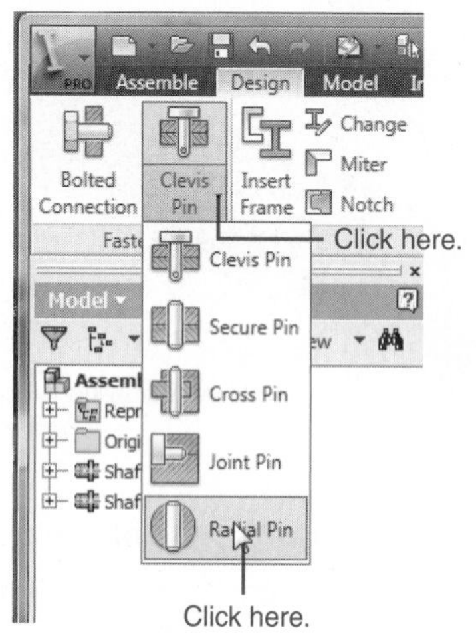

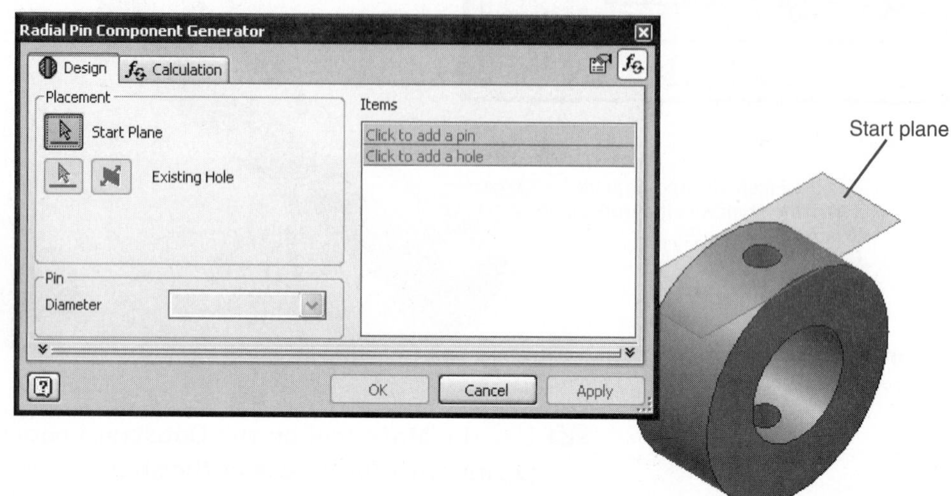

Figure 10-98

2. Select the **Start Plane** box, then the work plane on the hub. Select the **Existing Hole** box, then the **Ø6** hole on the hub. Click the **Click to add a pin** box to access the **Content Center.**

   See Figure 10-99.

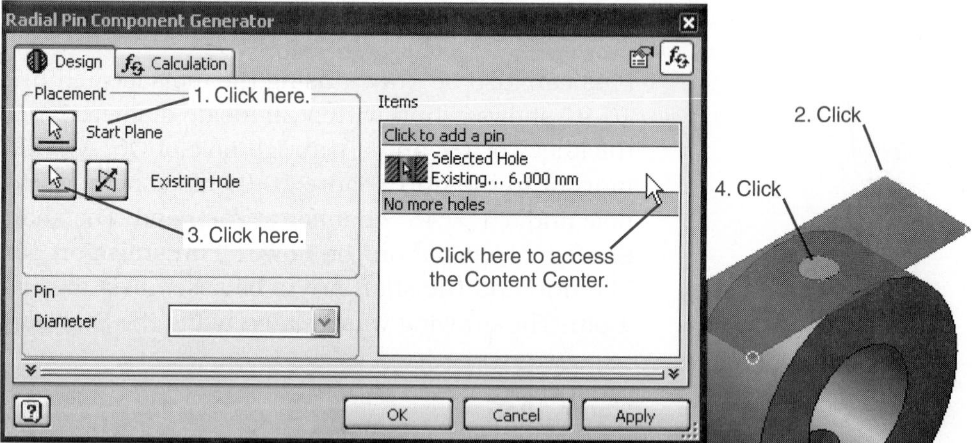

Figure 10-99

3. In the **Place from Content Center** dialog box select the **Cylindrical** pin option and a **DIN EN ISO 8752** pin. Select a **Ø6** × **40** pin.

   See Figures 10-100 and 10-101. Figure 10-102 and Figure 10-103 shows the pin in the assembly drawing.

> **TIP**
>
> Create work axes for each of the three parts to help align the assembly.

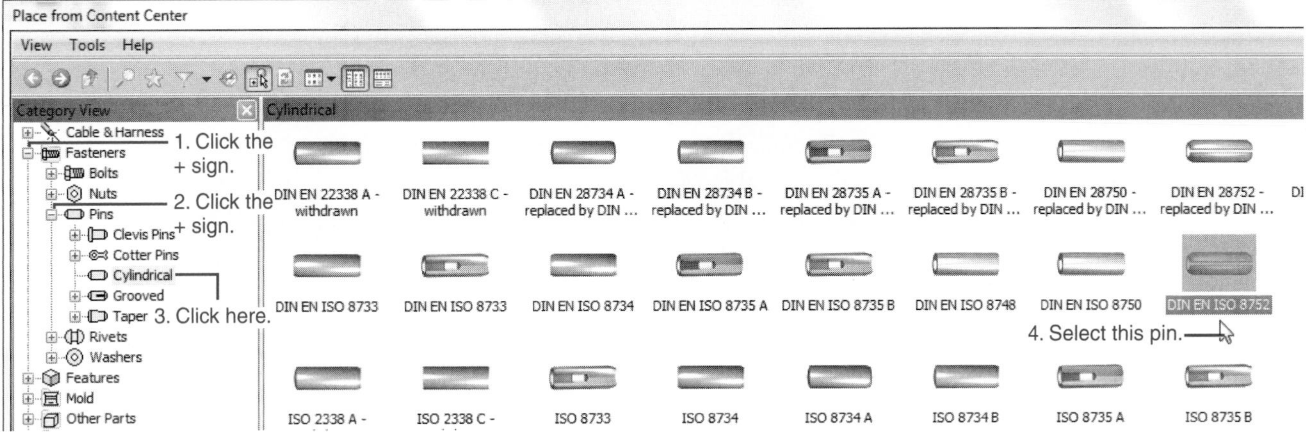

Figure 10-100

Figure 10-101

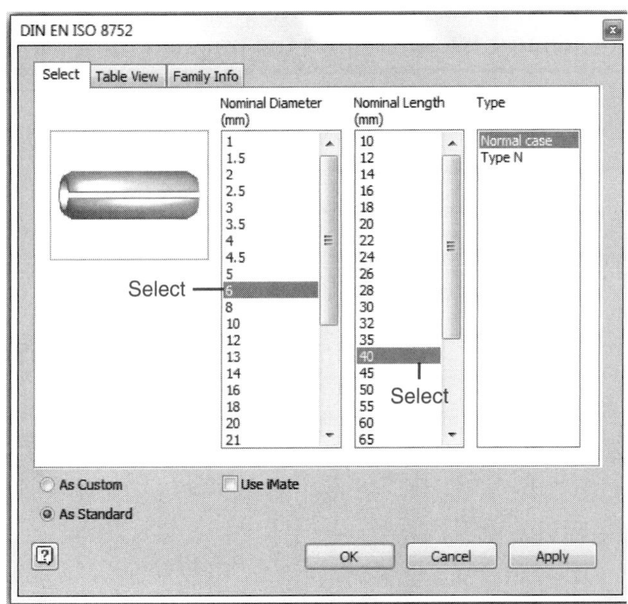

Figure 10-102

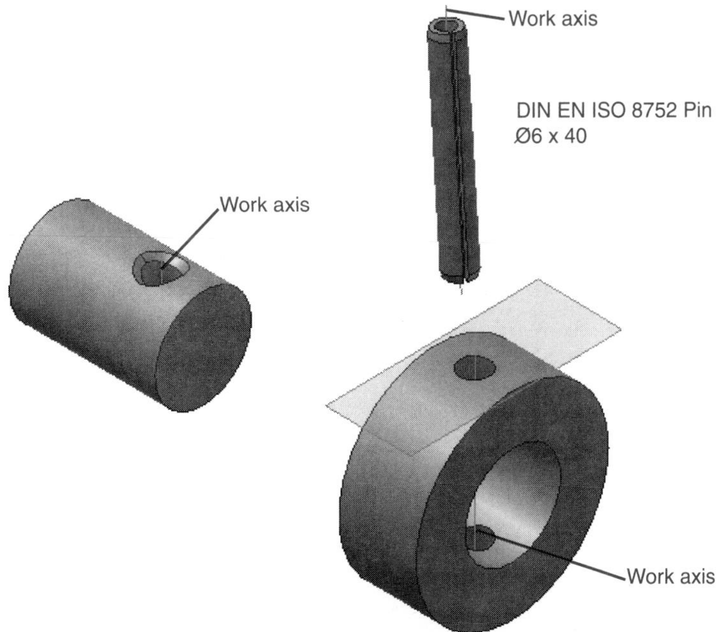

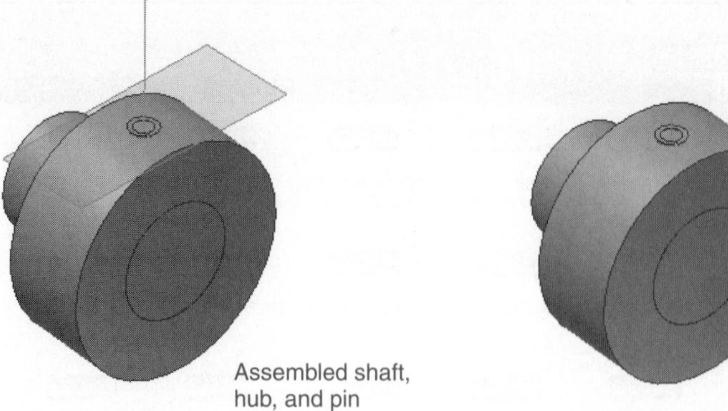

Figure 10-103

4 Use the **Constrain** tool and assemble the parts.

Figure 10-103 shows the finished assembly.

Chapter Summary

This chapter illustrated how to use the tools located under the **Design** tab to draw shafts, how to add retaining ring grooves to shafts, how to add keyways to shafts, how to add O-ring grooves to shafts, how to add pin holes to shafts, and how to use the **Content Center** to add retaining rings, keys, O-rings, and pins to shafts. The five general types of keys were described and illustrated, namely, Pratt and Whitney, square, rectangular, Woodruff, and Gib.

Splines on a shaft, for transmitting torque, were also introduced and used in a drawing.

Chapter Test Questions

Multiple Choice

Circle the correct answer.

1. Which of the following is *not* a type of key?
   a. Pratt and Whitney c. Russell
   b. Square d. Woodruff
2. Which term best describes the shape of a Woodruff key?
   a. Round c. Square
   b. Rectangular d. Crescent
3. Which of the following is *not* a type of pin?
   a. Round c. Taper
   b. Cotter d. Clevis
4. The dimensions needed to define a groove in a shaft for a retaining ring are found in the
   a. **Design** tab panel tools
   b. **Content Center**
   c. **Drawing Annotation** panel
5. Which type of key is crescent-shaped?
   a. Pratt and Whitney c. Rounded
   b. Woodruff d. Machine
6. What is a slot cut through a hub sized to accept a key called?
   a. Keyslot c. Keyway
   b. Hubslot d. Slot
7. Spring pins are fitted into holes by
   a. Squeezing them to a smaller diameter, inserting them into a hole, then releasing them back to original size
   b. Fitting them with compression springs that are released once the pins are inserted into a hole
   c. Press fitting them into a hole with spring pressure

8. Threaded holes in collars are usually fitted with
   a. Spring pins
   b. Setscrews
   c. Rivets
   d. Cotter pins
9. Splines are used to transfer which type of forces?
   a. Linear
   b. Rotary
   c. Impact
   d. Compression
10. When a shaft is a simple cylinder, the ends are said to have
   a. Chamfers
   b. Fillets
   c. No features
   d. Orthographic ends

Matching

Match the key and pin names with their pictures.

1. Woodruff ________
2. Pratt and Whitney ________
3. Rectangular ________
4. Gib ________
5. Square ________
6. Cotter ________
7. Cylindrical spring type ________
8. Clevis ________
9. Cylindrical ________
10. Taper ________
11. Grooved ________

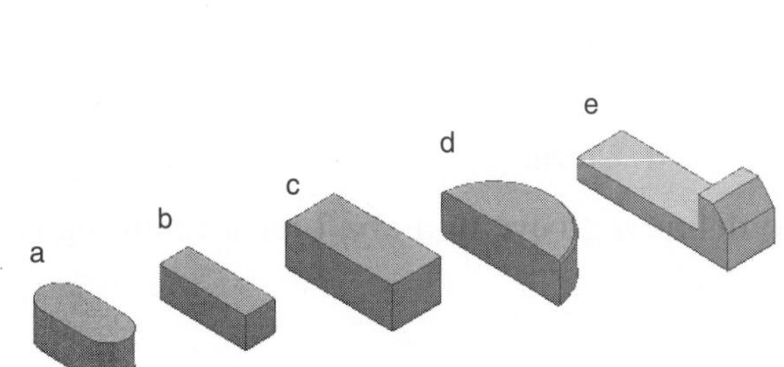

Figure 10-104

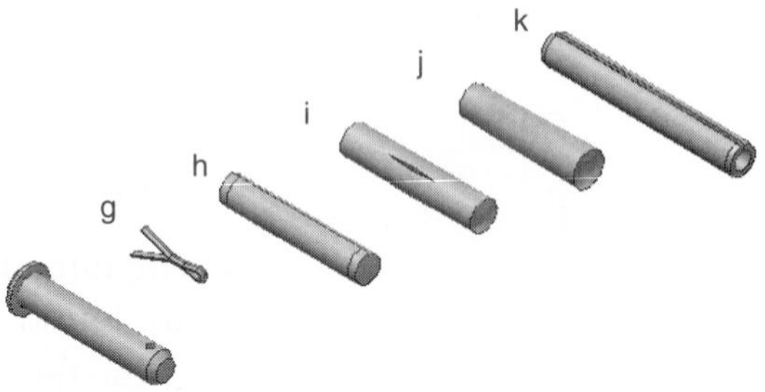

Figure 10-105

True or False

Circle the correct answer.

1. **True or False:** Chamfers can be added to shafts using **Design** tab panel tools.
2. **True or False:** Shafts must always have chamfers on their ends.
3. **True or False:** Retaining rings are used to prevent longitudinal movement of shafts.
4. **True or False:** Retaining rings can be either external or internal.

5. **True or False:** Keys are used with shafts to transfer rotary motion and torque.
6. **True or False:** Pratt and Whitney keys have rounded ends.
7. **True or False:** Woodruff keys are rectangular.
8. **True or False:** Splines are a series of cutouts in a shaft that are sized to match a corresponding set of cutouts in a hub.
9. **True or False:** Collars are used to hold shafts in position as they rotate.
10. **True or False:** O-rings are used to create seals.

Chapter Projects

Draw the following uniform shafts.

Project 10-1: Inches

Ø.50 × 3.50
0.10 × 45° chamfer, both ends

Project 10-2: Inches

Ø2.00 × 10.00
0.125 × 45° chamfer, both ends

Figure P10-1

Figure P10-2

Project 10-3: Millimeters

Ø20 × 110
3 × 45° chamfer, both ends

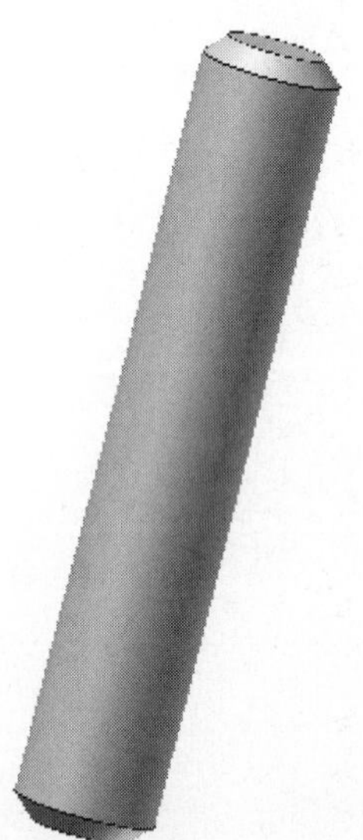

Figure P10-3

Project 10-4: Millimeters

Ø60 × 100
5 × 45° chamfer, both ends
Draw the following shafts and retaining rings. Create the appropriate grooves on the shaft, and position the retaining rings onto the shaft.

Project 10-5: Inches

Draw a Ø1.00 × 5.00 shaft.
Mount two BS 3673: Part 1 Inch retaining rings located 0.25 from each end.
Draw a 0.06 × 45° chamfer on each end.

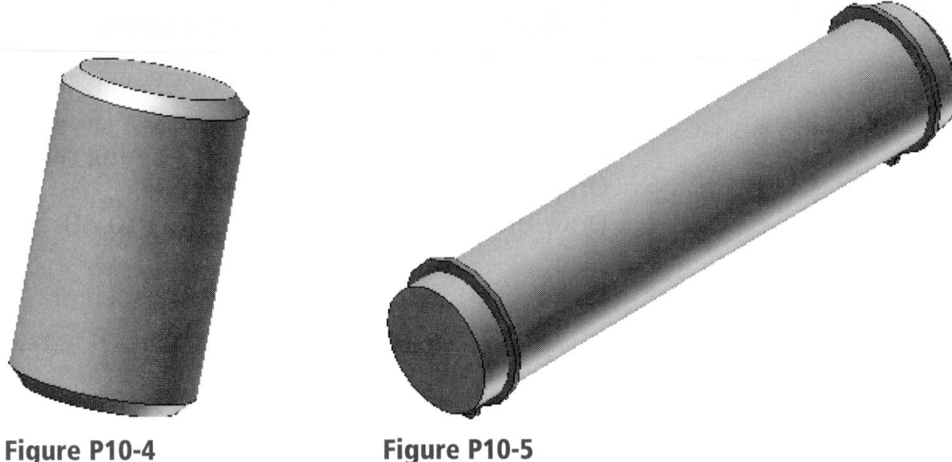

Figure P10-4 Figure P10-5

Project 10-6: Inches

Draw a Ø.25 × 6.00 shaft.
Mount two BS 3673: Part 1 Inch retaining rings located 0.375 from each end.
Draw a 0.03 × 45° chamfer on each end.

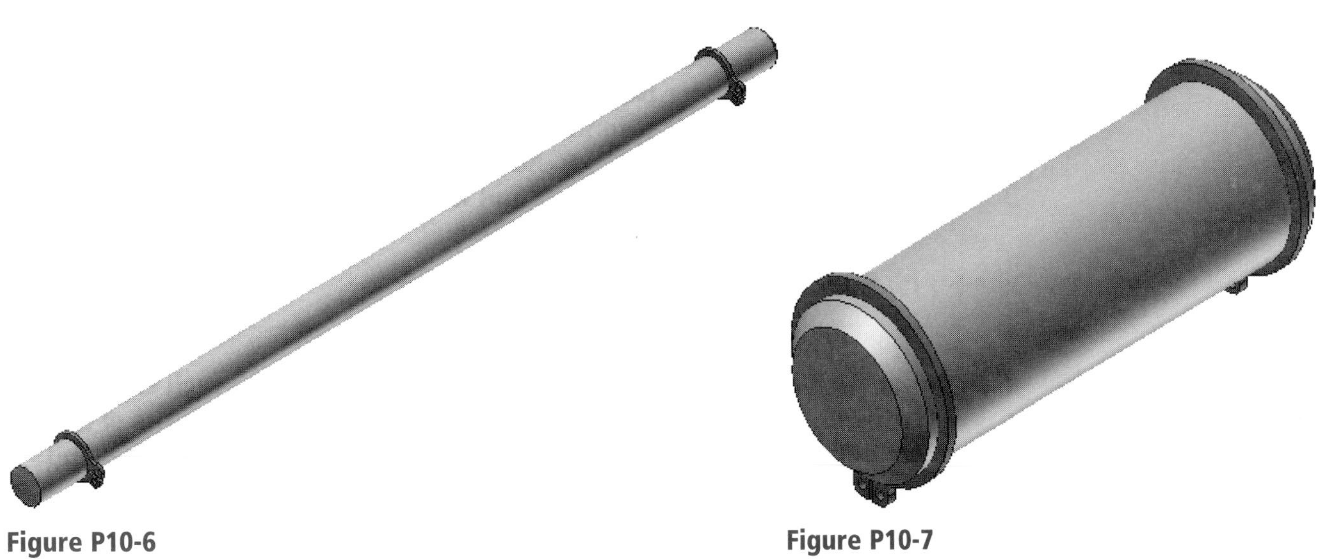

Figure P10-6 Figure P10-7

Project 10-7: Millimeters

Draw a Ø20 × 50 shaft.
Mount two CSN 02 2930 Metric retaining rings located 2 from each end.
Draw a 3 × 45° chamfer on each end.

Project 10-8: Millimeters

Draw a Ø16 × 64 shaft.
Mount two CSN 02 2930 Metric retaining rings located 10 from each end.
Draw a 4 × 45° chamfer on each end.
Draw the shafts and keys in Figure P10-8B.

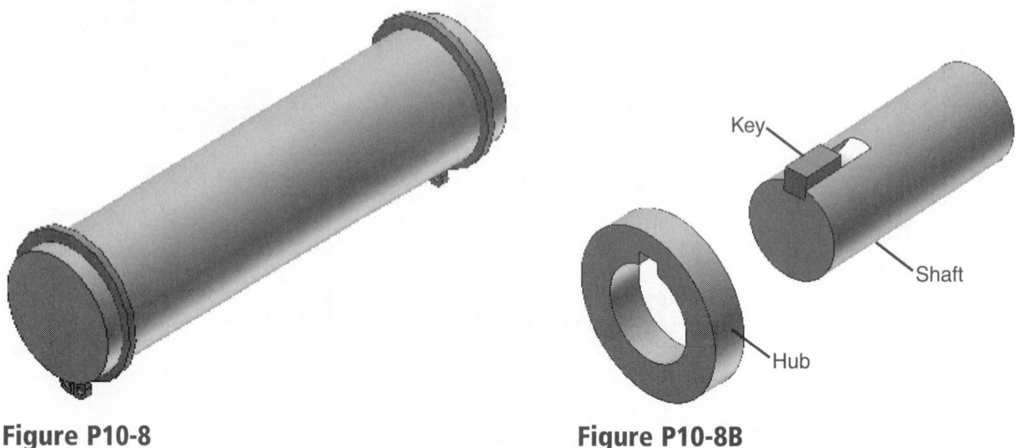

Figure P10-8

Figure P10-8B

Project 10-9: Inches

Draw a Ø2.00 × 5.25 shaft with 0.125 × 45° chamfers at each end. Draw a keyway on one end on the shaft to match the size requirements of a square key (see the **Content Center**) that is 1/2 × 1/2 × 1.25 long and has a nominal diameter range of 1.75 - 2.25. Insert the key into the shaft's keyway. Draw a hub with an inside diameter of 2.00, an outside diameter of 3.50, and a thickness of 0.75. Add the appropriate keyway to the hub, then insert the hub over the shaft and key.

Project 10-10: Inches

Draw a Ø3.50 × 10.00 shaft with 0.25 × 45° chamfers at each end. Draw a keyway on one end on the shaft to match the size requirements of a square key (see the **Content Center**) that is 7/8 × 7/8 × 2.50 long and has a nominal diameter range of 3.25 - 3.75. Insert the key into the shaft's keyway. Draw a hub with an inside diameter of 3.50, an outside diameter of 5.00, and a thickness of 1.00. Add the appropriate keyway to the hub, then insert the hub over the shaft and key.

Project 10-11: Millimeters

Draw a Ø26 × 72 shaft with 3 × 45° chamfers at each end. Draw a keyway on one end on the shaft to match the size requirements of an IS 2048 B key (see the **Content Center**) that is 2 × 2 × 22 long and has a nominal diameter range of 22 - 30. Insert the key into the shaft's keyway. Draw a hub with an inside diameter of 26, an outside diameter of 40, and a thickness of 14. Add the appropriate keyway to the hub, then insert the hub over the shaft and key.

Project 10-12: Millimeters

Draw a Ø60 × 240 shaft with 5 × 45° chamfers at each end. Draw a keyway on one end on the shaft to match the size requirements of an IS 2048 B key (see the **Content Center**) that is 18 × 11 × 70 long and has a nominal diameter range of 58 - 65. Insert the key into the shaft's keyway. Draw a hub with an inside diameter of 60, an outside diameter of 80, and a thickness of 20. Add the appropriate keyway to the hub, then insert the hub over the shaft and key. Draw the shafts and keys in Figure P10-12.

Figure P10-12

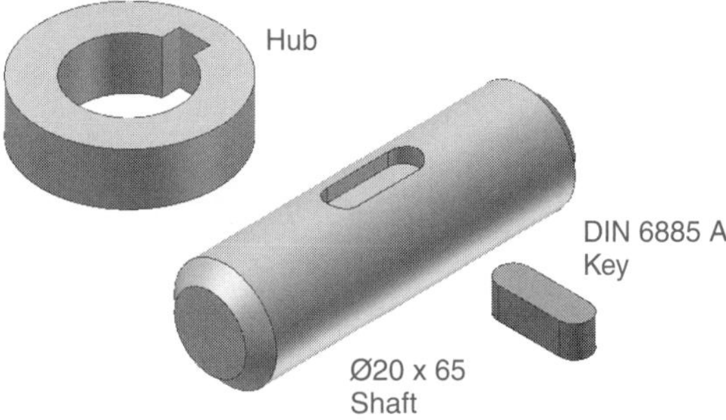

Project 10-13: Millimeters

Draw a Ø20 × 65 shaft with 3 × 45° chamfers at each end. Draw a keyway on the shaft to match the size requirements of a DIN 6885 A key (see the **Content Center**). Locate the shaft's keyway 24 from the end of the shaft. Insert the key into the shaft's keyway. Draw a hub with an inside diameter of 20, an outside diameter of 35, and a thickness of 10. Add the appropriate keyway to the hub, then insert the hub over the shaft and key.

Project 10-14: Millimeters

Draw a Ø52 × 100 shaft with 6 × 45° chamfers at each end. Draw a keyway on the shaft to match the size requirements of a DIN 6885 A key (see the **Content Center**). Locate the shaft's keyway 34 from the edge of the shaft. Insert the key into the shaft's keyway. Draw a hub with an inside diameter of 52, an outside diameter of 70, and a thickness of 16. Add the appropriate keyway to the hub, then insert the hub over the shaft and key.

Project 10-15: Inches

Draw a Ø0.750 × 4.25 shaft with 0.125 × 45° chamfers at each end. Draw a keyway 0.500 from the end of the shaft to match the size requirements of a 0.1875 × 0.1875 × 0.74 rectangular or square parallel key that has a nominal diameter range of 0.57 - 0.88 (see the **Content Center**). Insert the key into the shaft's keyway. Draw a hub with an inside diameter of 0.75, an outside diameter of 2.50, and a thickness of 0.625. Add the appropriate keyway to the hub, then insert the hub over the shaft and key.

Project 10-16: Inches

Draw a Ø1.50 × 7.50 shaft with 0.19 × 45° chamfers at each end. Draw a keyway 2.50 from the end of the shaft to match the size requirements of a 0.375 × 0.375 × 0.875 rectangular or square parallel key that has a nominal diameter range of 1.38 - 1.75. Insert the key into the shaft's keyway. Draw a hub with an inside diameter of 1.50, an outside diameter of 2.75, and a thickness of 0.750. Add the appropriate keyway to the hub, then insert the hub over the shaft and key.

Project 10-17: Millimeters

Draw a Ø163 × 40 shaft with 2 × 45° chamfers at each end. Draw a keyway 10 from the end of the shaft to match the size requirements of a CSN 30 1385 Woodruff key. Insert the key into the shaft's keyway. Draw a hub with an inside diameter of 16, an outside diameter of 40, and a thickness of 8. Add the appropriate keyway to the hub, then insert the hub over the shaft and key.

Project 10-18: Millimeters

Draw a Ø40 × 105 shaft with 4 × 45° chamfers at each end. Draw a keyway 27 from the end of the shaft to match the size requirements of a CSN 30 1385 Woodruff key. Insert the key into the shaft's keyway. Draw a hub with an inside diameter of 40, an outside diameter of 75, and a thickness of 12. Add the appropriate keyway to the hub, then insert the hub over the shaft and key.

Project 10-19: Inches

Draw a Ø0.750 × 4.25 shaft with 0.125 × 45° chamfers at each end. Draw a keyway 0.500 from the end of the shaft to match the size requirements of a Full Radius No. 606 3/16 × 0.75 Woodruff key (see the **Content Center**). Insert the key into the shaft's keyway. Draw a hub with an inside diameter of 0.75, an outside diameter of 2.50, and a thickness of 0.625. Add the appropriate keyway to the hub, then insert the hub over the shaft and key.

Project 10-20: Inches

Draw a Ø1.50 × 7.50 shaft with 0.19 × 45° chamfers at each end. Draw a keyway 2.50 from the end of the shaft to match the size requirements of a Full Radius No. 1422 I 7/16 × 2 Woodruff key (see the **Content Center**). Insert the key into the shaft's keyway. Draw a hub with an inside diameter of 1.50, an outside diameter of 2.75, and a thickness of 0.750. Add the appropriate keyway to the hub, then insert the hub over the shaft and key.

Draw the shafts and hubs in Figures P10-20A through P10-20D.

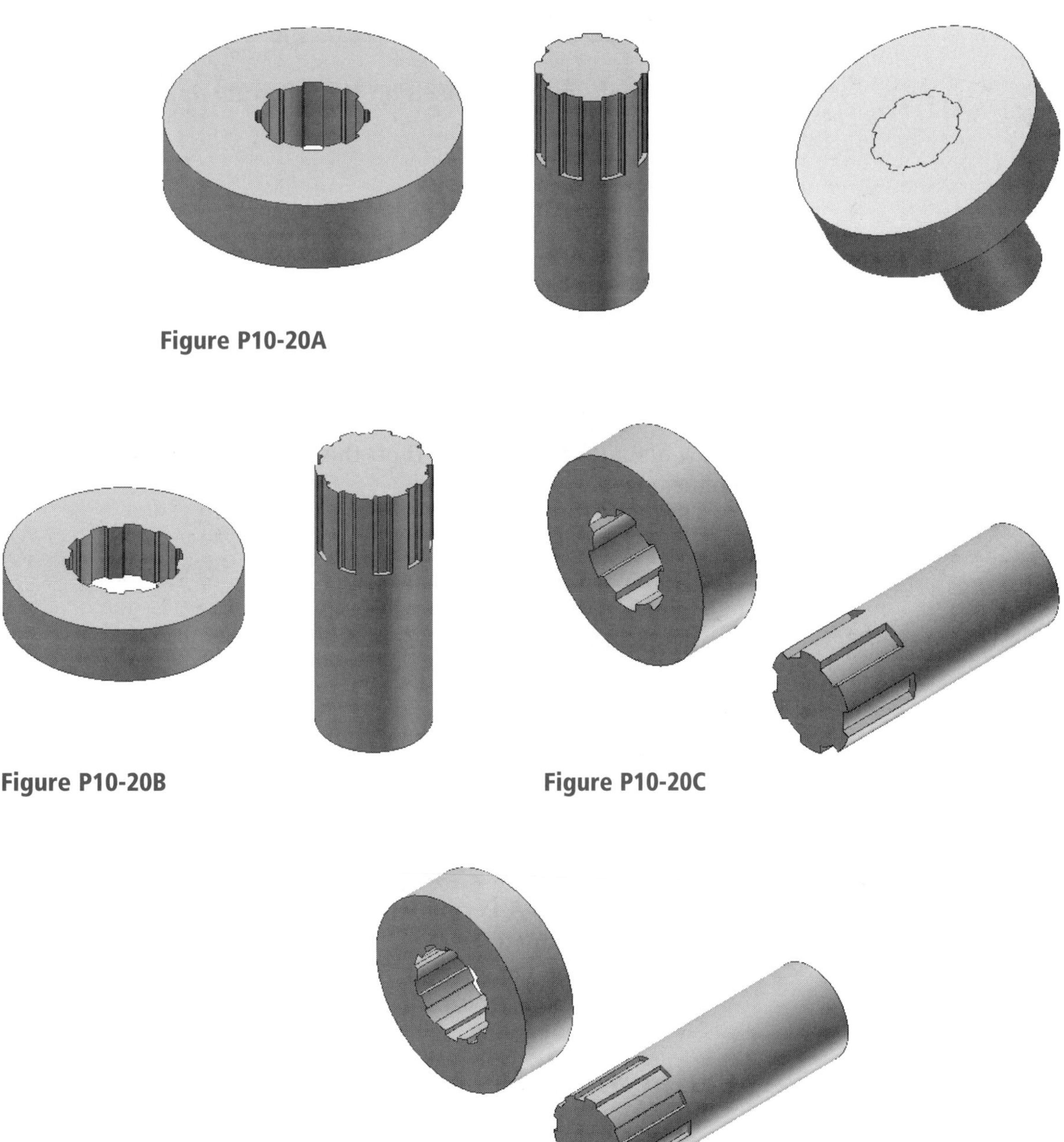

Figure P10-20A

Figure P10-20B

Figure P10-20C

Figure P10-20D

Project 10-21: Millimeters

Draw a shaft and hub based on the following data and join them using a spline.

Shaft: Ø120 × 30
Spline Type = Light
Spline Dimensions = 8 × 42 × 46
Active Spline Length = 38
Hub: Ø46 × 100

Project 10-22: Millimeters

Draw a shaft and hub based on the following data and join them using a spline.

Shaft: Ø78 × 200
Spline Type = Light
Spline Dimensions = 10 × 72 × 78
Active Spline Length = 60
Hub: Ø160 × 40

Project 10-23: Inches

Draw a shaft and hub based on the following data and join them using a spline.

Shaft: Ø1.50 × 4
Fit Type = 6B To Side - No Load
Nominal Diameter = 1.50
Active Spline Length = 1.25
Hub: Ø3 × 1

Project 10-24: Inches

Draw a shaft and hub based on the following data and join them using a spline.

Shaft: Ø2 × 6
Fit Type = 10 A Permanent Fit
Nominal Diameter = 2
Active Spline Length = 1.75
Hub: Ø4.5 × 1.25

Project 10-25: Millimeters

Create a 3D drawing of the Shaft Support Assembly in Figure P10-25 for each of the following sets of parameters. Also include an exploded isometric drawing with assembly numbers and a parts list.

A. A straight uniform shaft, Ø16 × 320 with 2 × 45° chamfers at each end. The shaft is to protrude 20 from the face of the holders.
B. A straight uniform shaft, Ø16 × 320 with 2 × 45° chamfers at each end. The shaft is to protrude 20 from the face of the holders. Add the appropriate grooves and insert two CNS 9074 external retaining rings located 17 from each end of the shaft.
C. A straight uniform shaft, Ø16 × 320 with 2 × 45° chamfers at each end. The shaft is to protrude 20 from the face of the holders. Add the appropriate grooves and insert two E-ring - Type - 3CM external retaining rings located 18 from each end of the shaft.
D. A straight uniform shaft, Ø16 × 320 with 2 × 45° chamfers at each end. The shaft is to protrude 20 from the face of the holders. Add two CNS - 122 collars with Ø4 set screws.
E. A straight uniform shaft, Ø16 × 320 with 2 × 45° chamfers at each end. The shaft is to protrude 20 from the face of the support. Add two CNS - 122 collars with Ø4 spring-type straight pins.

Figure P10-25A

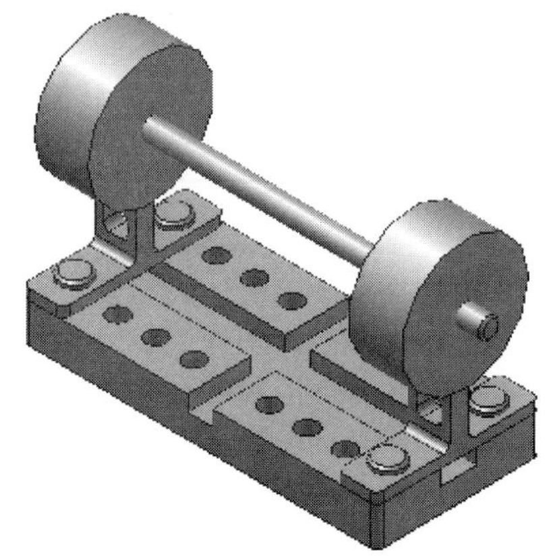

Figure P10-25B

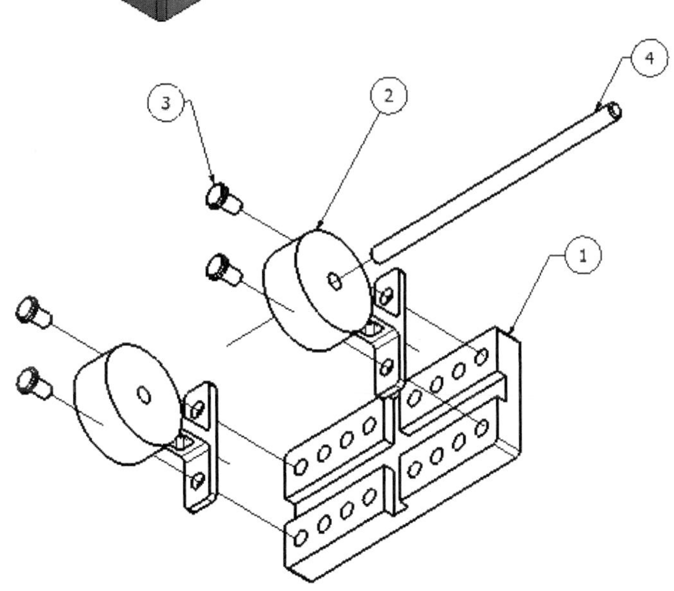

Parts List				
ITEM	PART NUMBER	DESCRIPTION	MATERIAL	QTY
1	ENG-311-1	BASE,CAST	CAST IRON	1
2	ENG-312	HOLDERS	CAST IRON	2
3	ENG-132A	PINS	2040 STEEL	4
4	ENG-131	Shaft	STEEL	1

BASE, CAST
MAT'L = CAST IRON
P/N = ENG-311-1

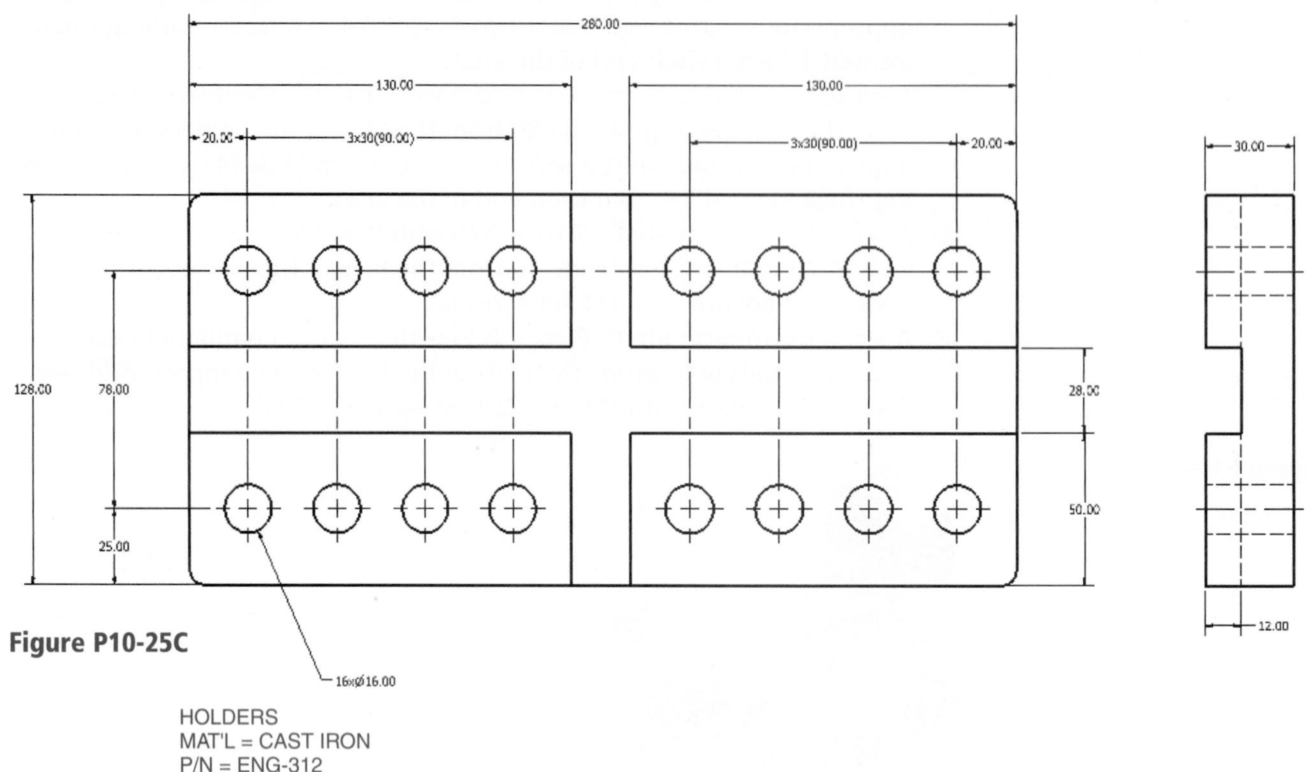

Figure P10-25C

HOLDERS
MAT'L = CAST IRON
P/N = ENG-312

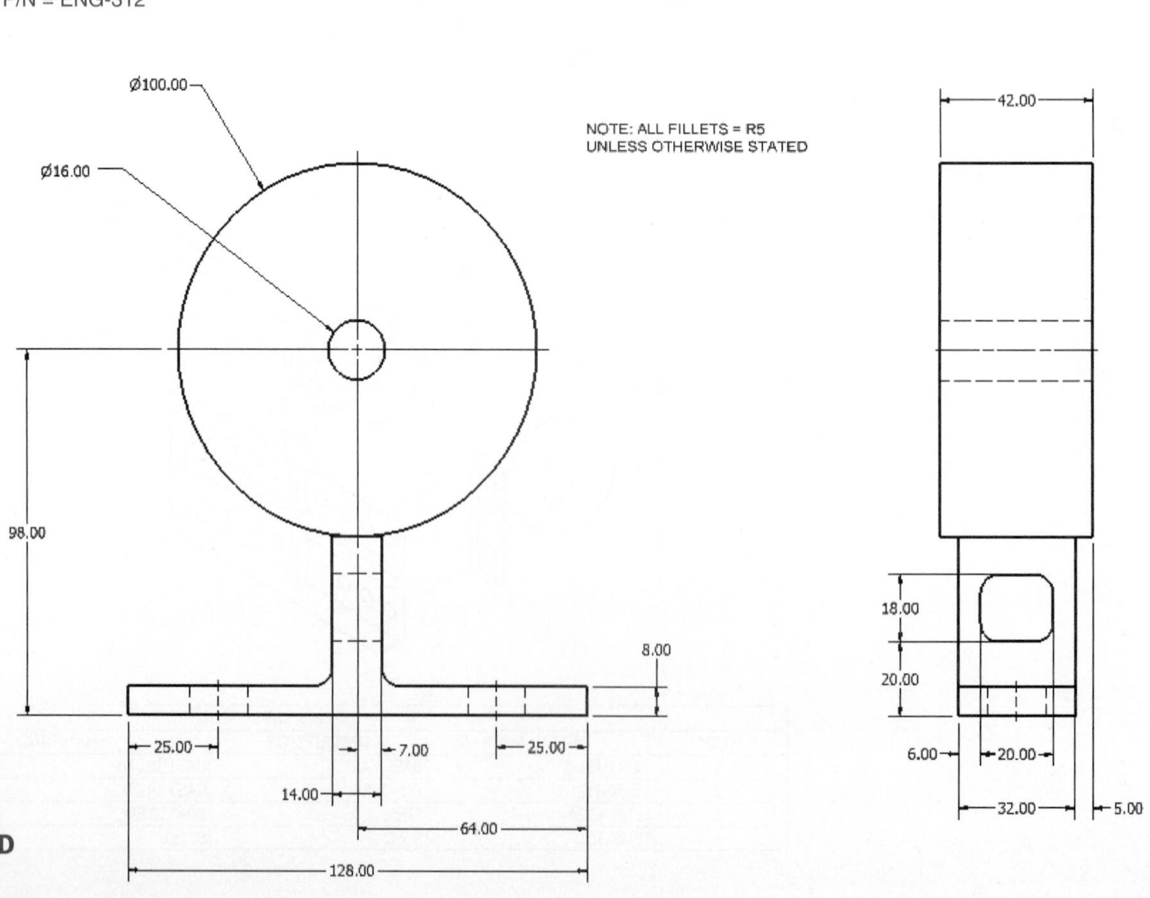

Figure P10-25D

F. A straight uniform shaft, Ø16 × 280 with 2 × 45° chamfers at each end. On one end insert a square key between the shaft and the holders.
G. A straight uniform shaft, Ø16 × 320 with 2 × 45° chamfers at each end. On one end insert a UNI 7510A key between the shaft and the support. The shaft is to protrude 20 from the surface of the holders.
H. A straight uniform shaft, Ø16 × 320 with 2 × 45° chamfers at each end. On one end insert a JIS B 1302 - Type B Woodruff key between the shaft and the support. The shaft is to protrude 20 from the surface of the holders.

Project 10-26: Inches

Create the drawings of the Adjustable Assembly shown in Figure P10-26.

The grooved pin was created using the tools under the **Design** tab. The forged eyebolt, hex machine screw nut, and hex nut were created using the **Content Center.**

A. A 3D assembly drawing
B. An exploded 3D assembly drawing
C. An exploded isometric drawing with assembly numbers
D. A parts list

PINS
MAT'L = 2040 STEEL
P/N = ENG-132A

Note: Consider both the pins found in the Content Center and those available using the tools under the Design tab.

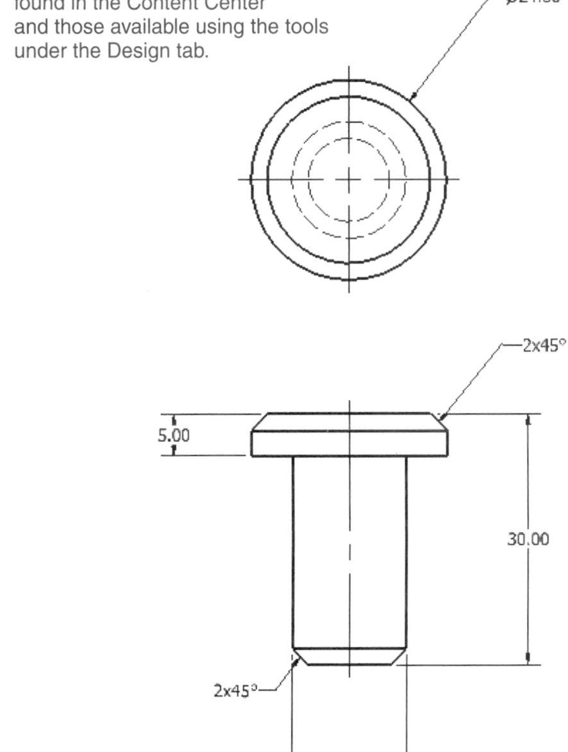

Figure P10-25E

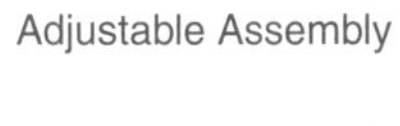

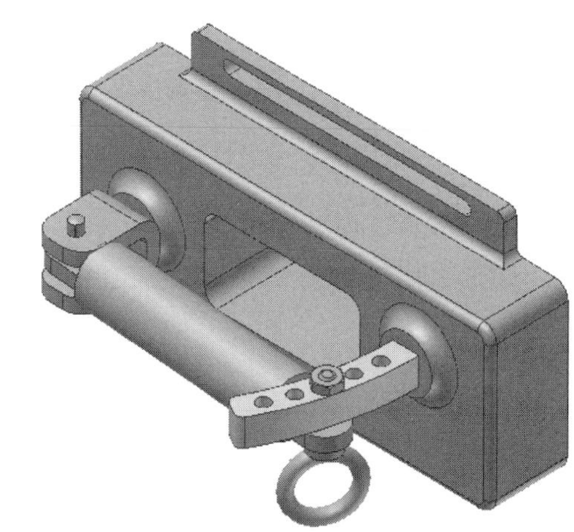

Figure P10-26A

POST, ADJUSTABLE
P/N = ENG 404
MAT'L = MILD STEEL

5.25
0.63
4.00
R0.38 - BOTH ENDS
Ø1.00
0.12 BOTH ENDS
0.25 BOTH ENDS
Ø0.25 +0.00 +0.02 - 2 HOLES

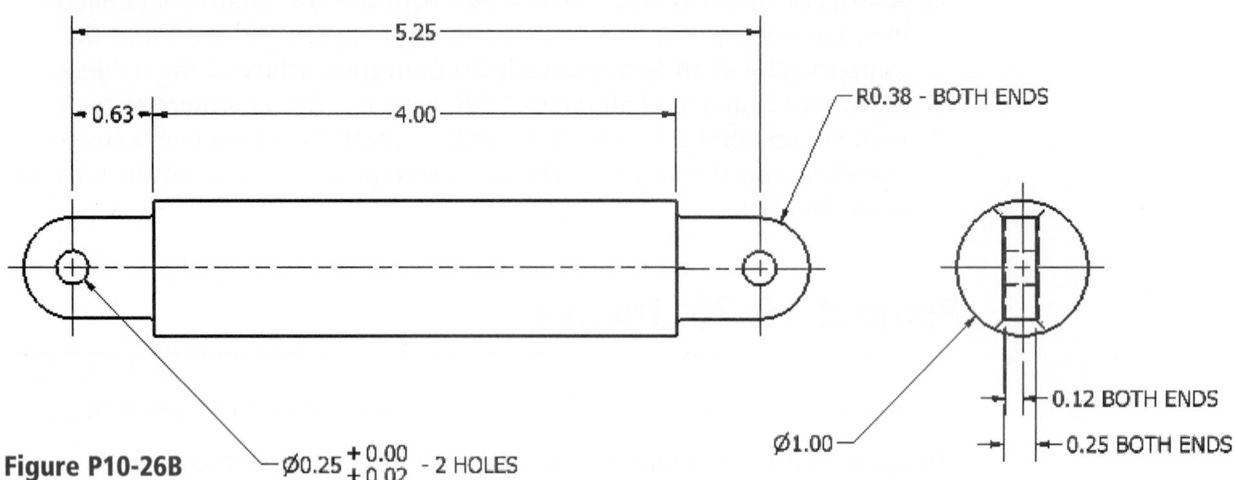

Figure P10-26B

BASE #4, CAST
P/N = ENG 311
MAT'L = CAST IRON

NOTE: ALL FILLETS = R 0.125

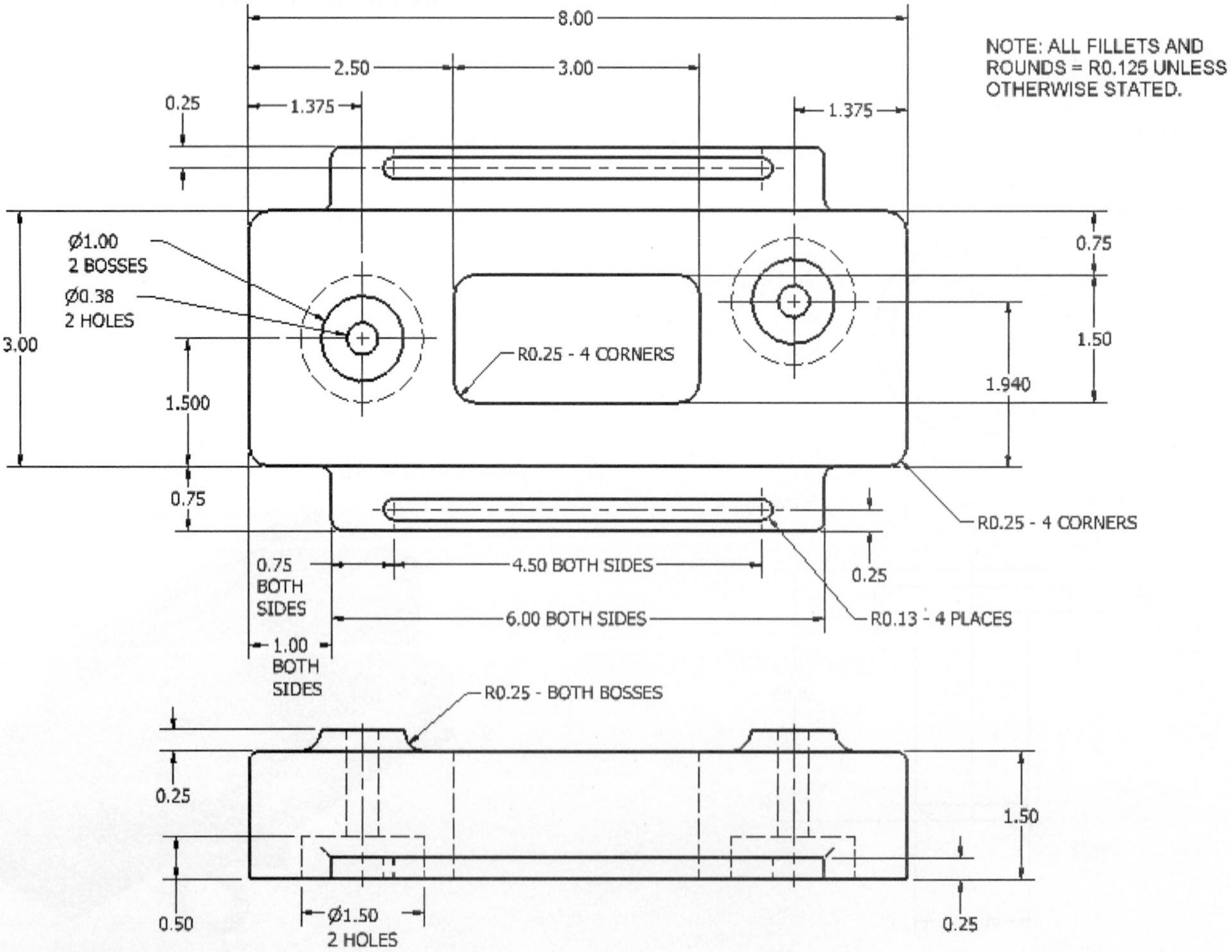

8.00
2.50
3.00
0.25
1.375
1.375
NOTE: ALL FILLETS AND ROUNDS = R0.125 UNLESS OTHERWISE STATED.
Ø1.00
2 BOSSES
Ø0.38
2 HOLES
3.00
R0.25 - 4 CORNERS
0.75
1.50
1.940
1.500
0.75
R0.25 - 4 CORNERS
0.75 BOTH SIDES
4.50 BOTH SIDES
0.25
6.00 BOTH SIDES
R0.13 - 4 PLACES
1.00 BOTH SIDES
R0.25 - BOTH BOSSES
0.25
1.50
0.50
Ø1.50
2 HOLES
0.25

Figure P10-26C

Figure P10-26D

YOKE
P/N = BU 1964
MAT'L = CAST IRON

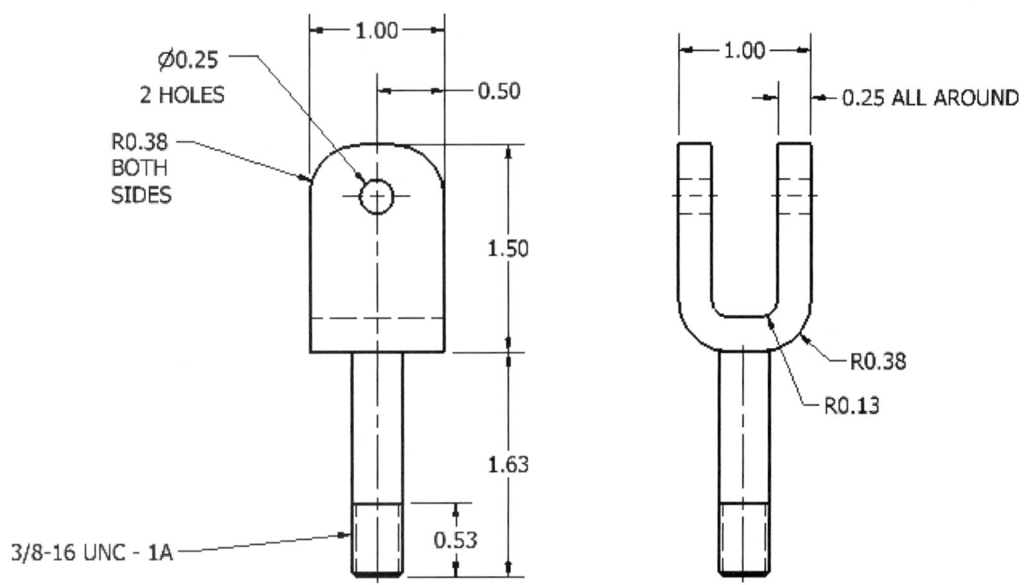

SUPPORT, ROUNDED
P/N = ENG-312
MAT'L = SAE 1040 STEEL

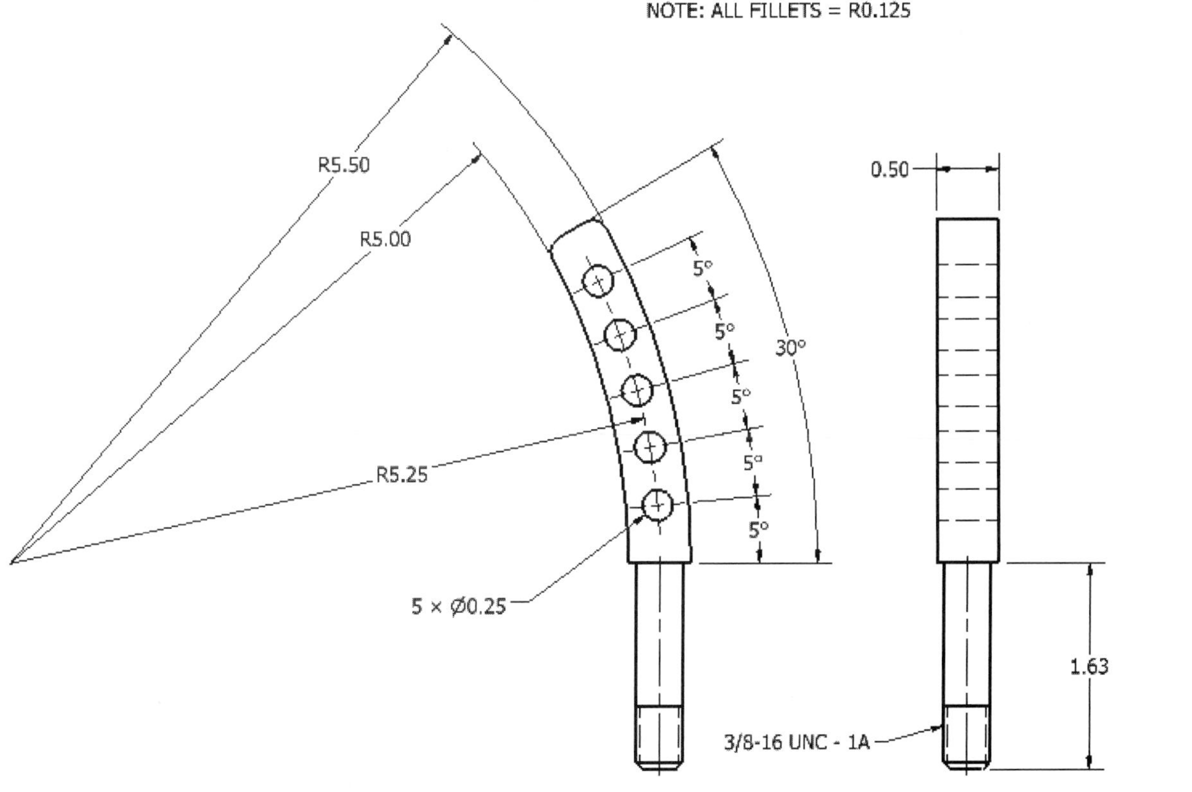

Figure P10-26E

Figure P10-26F

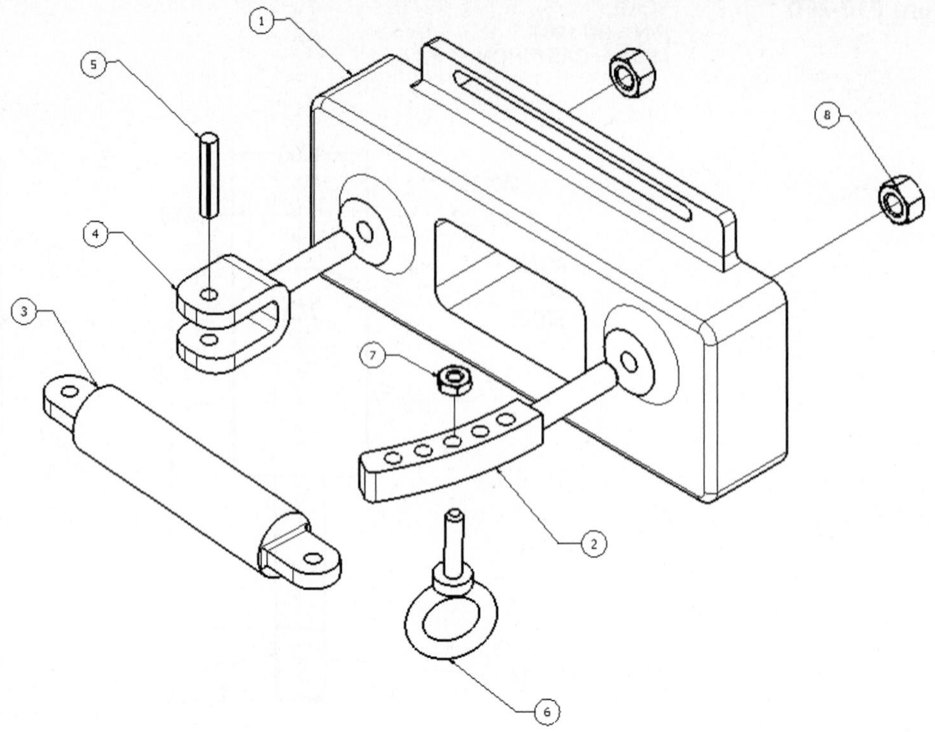

Parts List				
ITEM	PART NUMBER	DESCRIPTION	MATERIAL	QTY
1	ENG-311	BASE#4, CAST	Cast Iron	1
2	ENG-312	SUPPORT, ROUNDED	SAE 1040 STEEL	1
3	ENG-404	POST, ADJUSTABLE	Steel, Mild	1
4	BU-1964	YOKE	Cast Iron	1
5	ANSI B18.8.2 1/4x1.3120	Grooved pin, Type C - 1/4x1.312 ANSI B18.8.2	Steel, Mild	1
6	ANSI B18.15 - 1/4 - 20. Shoulder Pattern Type 2 - Style A	Forged Eyebolt	Steel, Mild	1
7	ANSI B18.6.3 - 1/4 - 20	Hex Machine Screw Nut	Steel, Mild	1
8	ANSI B18.2.2 - 3/8 - 16	Hex Nut	Steel, Mild	2

chapter eleven
Bearings

CHAPTER OBJECTIVES

- Understand the different types of bearings: plain, ball, and thrust
- Learn how to select bearings from the **Content Center**
- Understand how tolerances are used with bearings
- Learn how to use bearings in assemblies

Introduction

plain bearing: A hollow cylinder that may have a flange at one end and may or may not be lubricated; also called a *sleeve bearing or bushing.*

ball bearing: A bearing made from two cylindrical rings separated by a row of balls or rollers.

thrust bearing: A bearing used to absorb a load along the axial direction of a shaft.

This chapter discusses three types of bearings: plain, ball, and thrust. See Figure 11-1. ***Plain bearings*** are hollow cylinders that may have flanges at one end. Plain bearings are made from nylon or Teflon for dry (no lubrication) applications, and impregnated bronze or other materials when lubrications are used.

Plain bearings require less space than other types of bearings and are less expensive but have higher friction properties.

Ball bearings are made from two cylindrical rings separated by a row of balls. Shapes other than spheres may be used (rollers), and more than one row of balls may be included. Ball bearings usually take more space than plain bearings and are more expensive. They have better friction properties than plain bearings.

Thrust bearings are used to absorb loads along the axial direction (the length of the shaft). They are similar in size and cost to ball bearings.

Figure 11-1 Types of Bearings

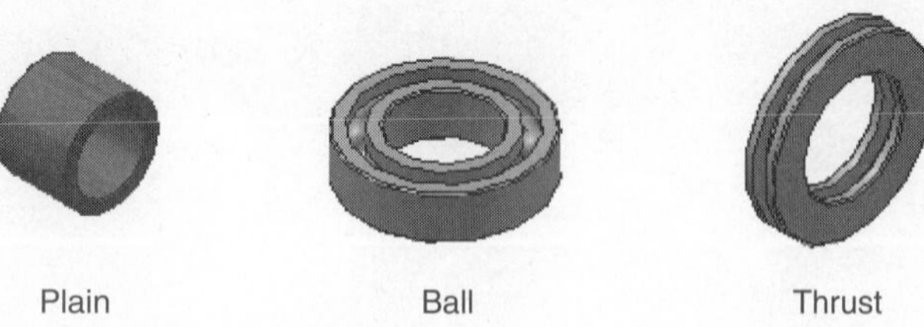

Plain Bearings

Figure 11-2 shows a U-bracket that will be used to support a rotating shaft. Plain bearings will be inserted between the shaft and the U-bracket. The bearings will be obtained from the **Content Center.**

The shaft has a nominal diameter of 6.00 mm, and the shaft will be inserted into the bearing and the bearing into the U-bracket.

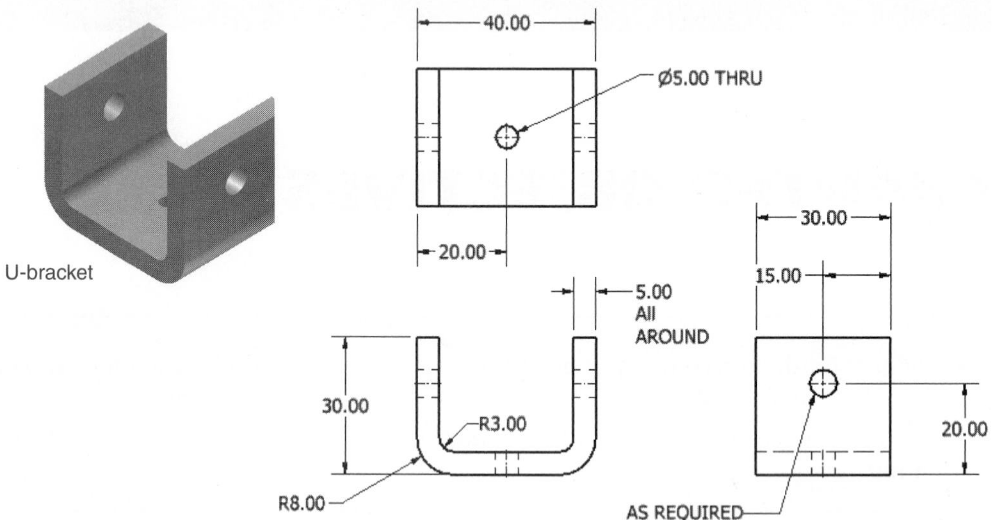

Figure 11-2

Nomenclature

Plain bearings are also called *sleeve bearings* or *bushings*. The terms are interchangeable

NOTE

When drawing a shaft use the nominal dimensions.

Shaft Tolerances

For this example it is assumed that the shaft will be purchased from a vendor. Tolerances for purchased shafts can be found in the manufacturer's catalogs or from listings posted on the Web. The shaft selected for this example has a tolerance of +0.00/−0.02.

Figure 11-3 shows a Ø6.0000 shaft 50.00 long. Each end has a 0.50 × 45° chamfer. It was created using the **Shaft** tool under the **Design** tab. See Chapter 10. The orthographic views of the shaft include the shaft's diametric tolerance.

Figure 11-3

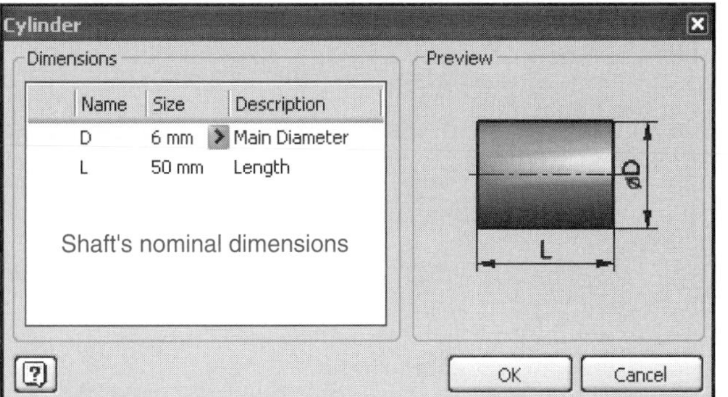

Created using the Shaft tool under the Design tab; see Chapter 10.

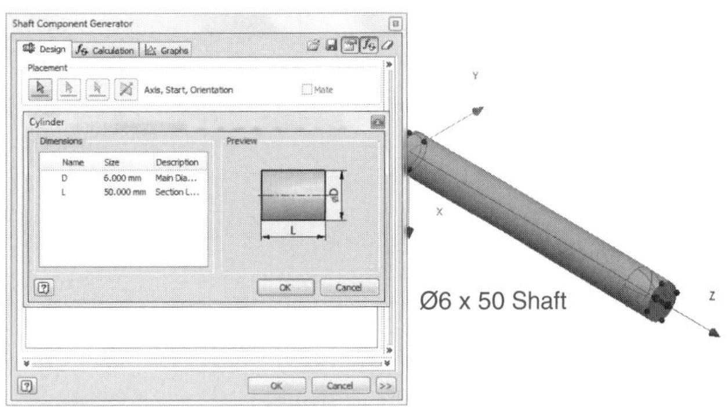

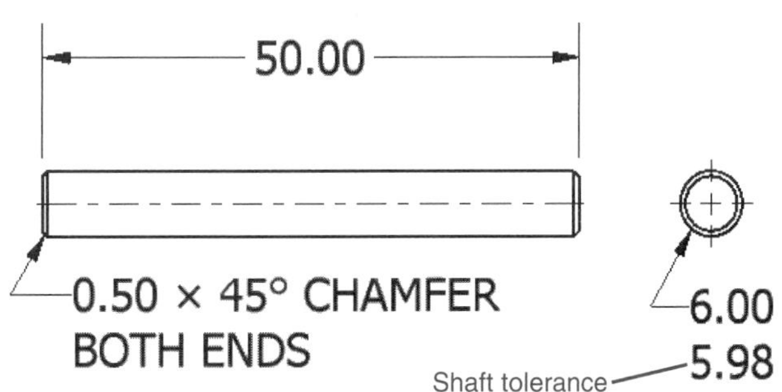

Created using the ANSI (mm).idw format; see Chapter 4.

Chapter 11

> **TIP**
>
> Purchased parts do not require individual drawings. They are listed in the parts list by manufacturer and manufacturer's part number or catalog number.

EXERCISE 11-1 Selecting a Plain Bearing

1. Draw the U-bracket shown in Figure 11-2 and save the drawing.
2. Start a new drawing using the **Standard (mm).iam** format.
3. Use the **Place Component** tool and place the U-BRACKET on the drawing.
4. Click the **Place from Content Center** tool or right-click the mouse and click the **Place from Content Center** option, click **Shaft Parts, Bearings-Plain,** and select an **ISO 2795 (Cylindrical) plain bearing.**

   See Figure 11-4.
5. Select the **6** × **10** × **4** nominal-sized bearing.

   See Figure 11-5.

Figure 11-4

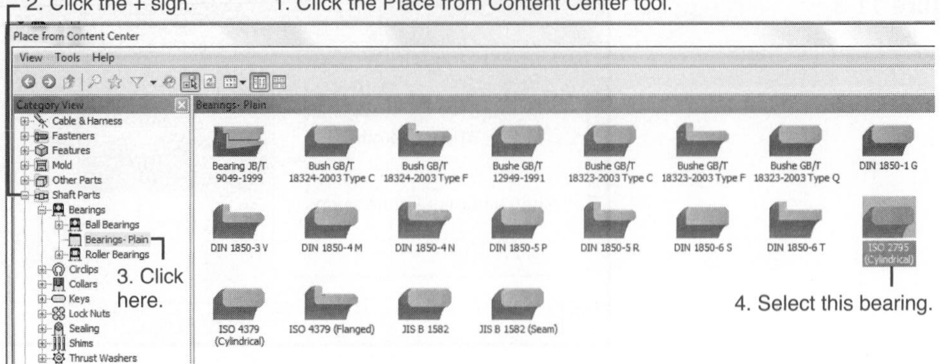

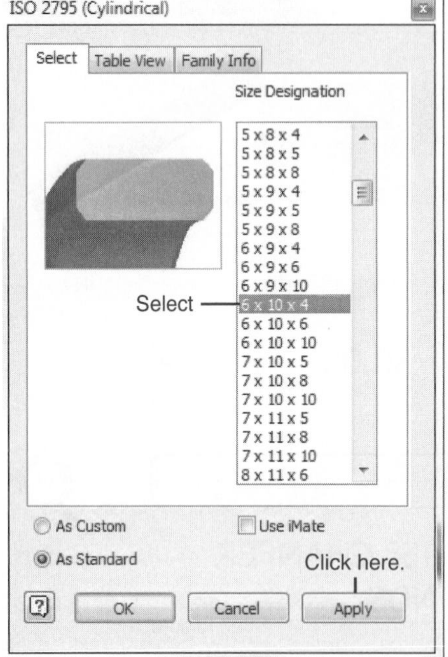

Figure 11-5

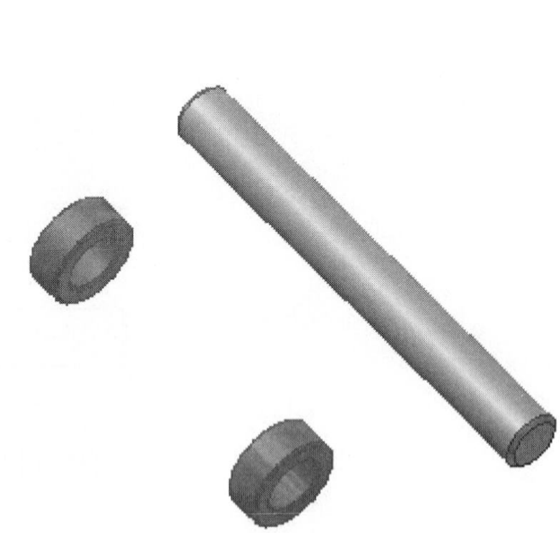

Figure 11-6

6 Click **OK** and add two bearings to the drawing.

See Figure 11-6.

Shaft/Bearing Interface

The shaft and inside diameter of the bearing (the *bore*) are generally fitted together using a clearance fit. The shaft has a tolerance of +0.00/−0.02. From manufacturer's specifications we know that the bearing's inside diameter is 6.00 nominal with a tolerance of +0.02/0.00. Therefore, the maximum clearance is 0.04, and the minimum is 0.00.

NOTE

If the clearance is too large, excessive vibrations could result at high speeds.

The range of clearance tolerance depends on the load and speed of the application.

The Hole in the U-Bracket

The bearing is to be inserted into the hole in the U-bracket using a force fit; that is, the outside diameter of the bearing will be larger than the diameter of the hole. Fits are often defined using a standard notation such as F7/h6, where the uppercase letter defines the hole tolerance and the lowercase letter defines the shaft tolerance. In this example the outside diameter of the bearing uses the shaft values. See the tables in the appendix. For a Ø10.00 the tolerances for an F7/h6 combination are

Hole: 9.991	Shaft: 10.000	Fit: 0.000
9.976	9.991	–0.024

This means that the hole in the U-bracket should be 9.991/9.976. However, a problem can arise if the manufacturer's specifications for the outside diameter of the bearing do not match the stated h6 value of 10.000/9.991.

Say, for example, the given outside diameter for a bearing is 10.009/10.000. Although the tolerance range is the same, 0.009 (10.000 – 9.991 = 0.009), the absolute values are different. We use the fit values to determine the new hole values. The fit values are given as 0.000/–0.024. These values are the maximum and minimum interference. Applying these values to the hole we get 10.000/9.985 (10.009 – 0.024 = 9.985). For this bearing the hole in the U-bracket would be 10.000/9.985.

Draw the hole using the nominal value of 10.00, and dimension the hole using the derived values. See Figure 11-7.

Figure 11-7

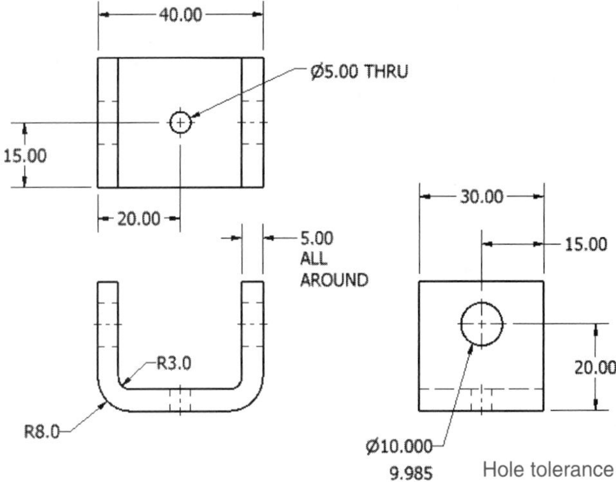

Figure 11-8 shows the finished assembly. The components were assembled using the **Constrain** tool. The shaft was offset 5 from the edge of the bearing.

Figure 11-9 shows the bearing assembly along with its parts list. Note that the bearing does not have an assigned part number but uses the manufacturer's number assigned by the **Content Center.** The U-bracket and the shaft both have assigned part numbers and would require detail drawings including dimensions and tolerances. Because the bearing is a purchased part, no drawing is required.

Figure 11-10 shows two plain bearings that include shoulders.

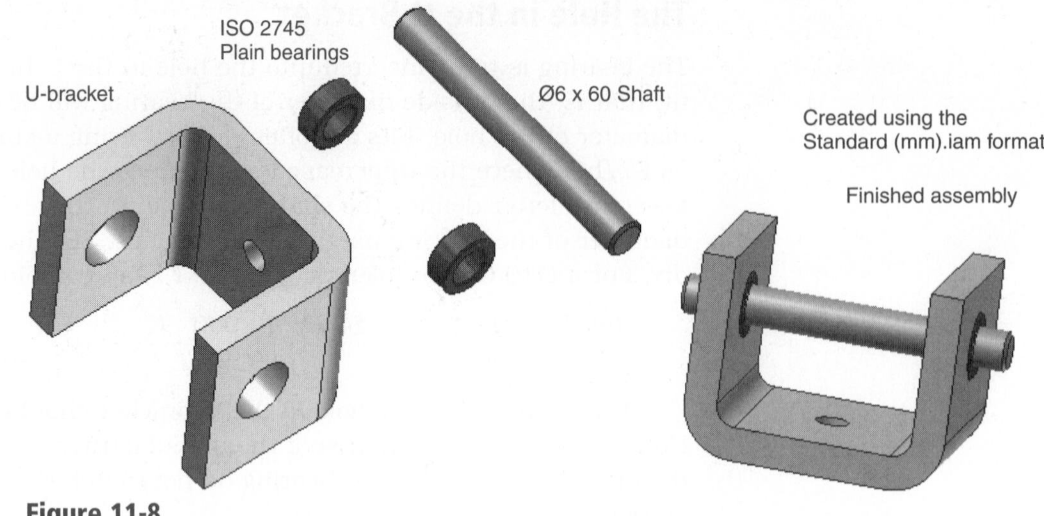

Figure 11-8

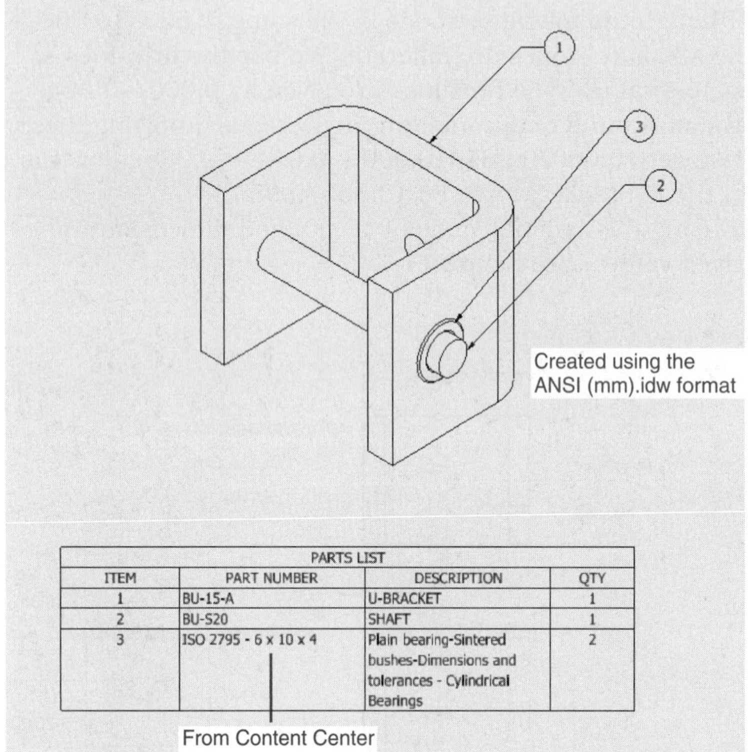

PARTS LIST			
ITEM	PART NUMBER	DESCRIPTION	QTY
1	BU-15-A	U-BRACKET	1
2	BU-S20	SHAFT	1
3	ISO 2795 - 6 x 10 x 4	Plain bearing-Sintered bushes-Dimensions and tolerances - Cylindrical Bearings	2

Figure 11-9

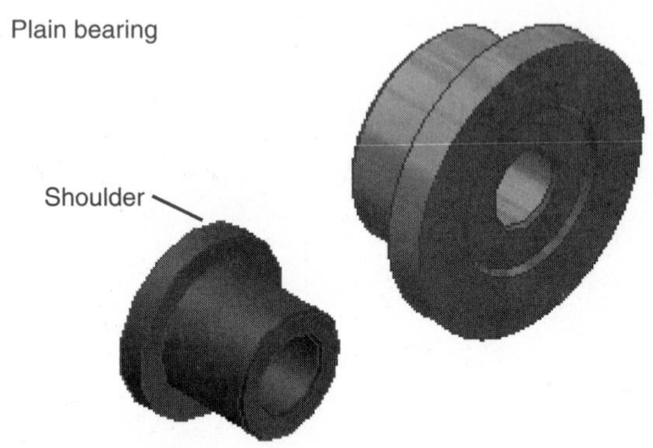

Figure 11-10

Ball Bearings

A bearing is to be selected for a Ø6.00 shaft. The shaft is to be mounted into the bearing and the bearing inserted into a U-bracket.

EXERCISE 11-2 Selecting a Ball Bearing

1 Click the **Place from Content Center** tool and select **Shaft**, click **Shaft Parts, Ball Bearings, Deep Groove Ball Bearings,** and select a **DIN 625** bearing.

See Figure 11-11. The **DIN 625** dialog box shows the bearings are defined by coded numbers.

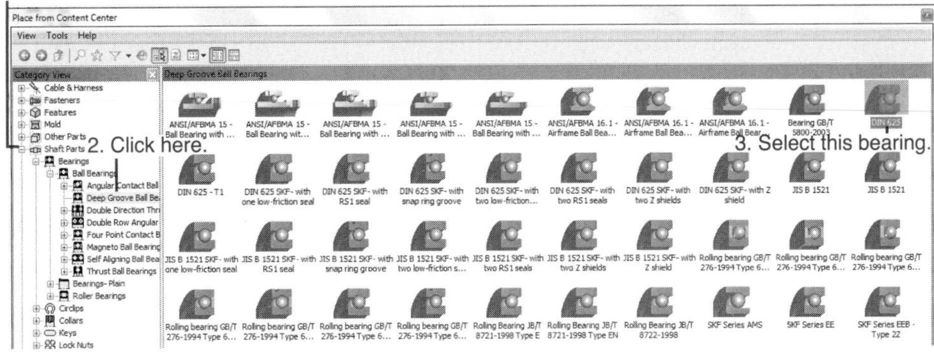

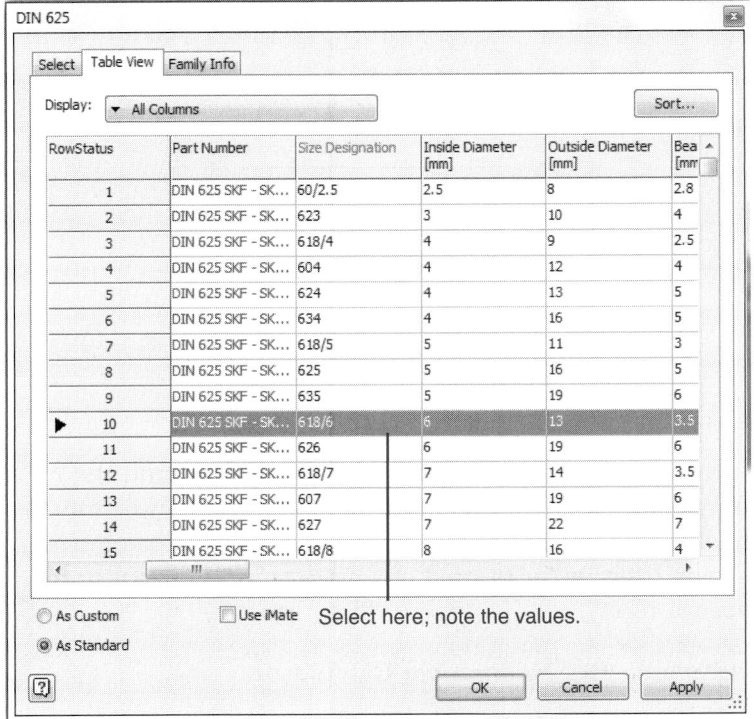

RowStatus	Part Number	Size Designation	Inside Diameter [mm]	Outside Diameter [mm]	Bea [mm
1	DIN 625 SKF - SK...	60/2.5	2.5	8	2.8
2	DIN 625 SKF - SK...	623	3	10	4
3	DIN 625 SKF - SK...	618/4	4	9	2.5
4	DIN 625 SKF - SK...	604	4	12	4
5	DIN 625 SKF - SK...	624	4	13	5
6	DIN 625 SKF - SK...	634	4	16	5
7	DIN 625 SKF - SK...	618/5	5	11	3
8	DIN 625 SKF - SK...	625	5	16	5
9	DIN 625 SKF - SK...	635	5	19	6
▶ 10	DIN 625 SKF - SK...	618/6	6	13	3.5
11	DIN 625 SKF - SK...	626	6	19	6
12	DIN 625 SKF - SK...	618/7	7	14	3.5
13	DIN 625 SKF - SK...	607	7	19	6
14	DIN 625 SKF - SK...	627	7	22	7
15	DIN 625 SKF - SK...	618/8	8	16	4

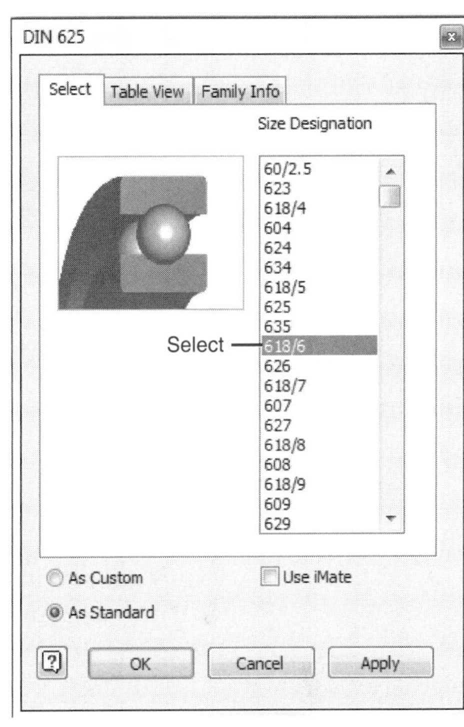

Figure 11-11

2 Click the **Table View** tab to see the dimensions for the different DIN 625 bearings.

In this example, a **618/6** bearing was selected. This bearing has a nominal bore of Ø6.00 and an outside nominal diameter of 13.00. Assume the bearing has the tolerances shown in Figure 11-12.

Figure 11-12

DIN 625 Ball bearing

Assume the following tolerances:

Ø13.039
13.028

H7/s6 FIT

3.50±0.02

6.02
6.00

3 Calculate the shaft's diameter.

Say the clearance requirements for the shaft are

Clearance = minimum = 0.00

Clearance = maximum = 0.03

This means that the shaft diameter tolerances are 6.00/5.99.

4 Use the **Shaft** tool under the **Design** tab to draw the shaft.

Use the nominal diameter of **6.00** and a length of **50.** Add a **0.50** × **45°** chamfer to each end. Figure 11-13 shows a solid model drawing of the shaft and a dimensioned orthographic drawing. The model was drawn using the nominal Ø6, and the orthographic views include the required tolerances.

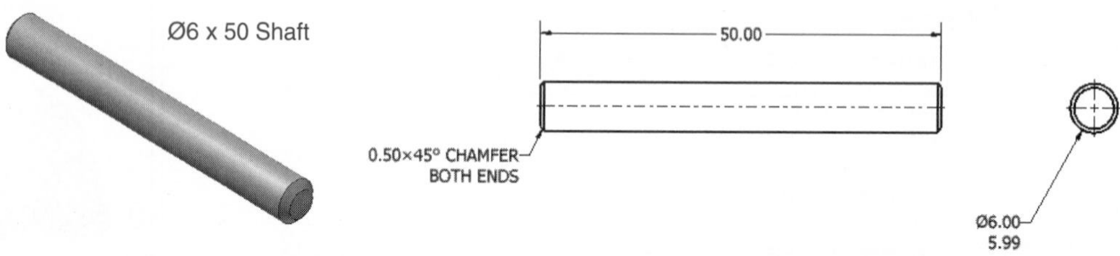

Figure 11-13

5 Determine the size for the hole in the U-bracket.

The interface between the outside diameter of the bearing and the U-bracket is defined as H7/s6. The values for this tolerance can be found in the tables in the appendix or by using the **Limits and Fits** dialog boxes found in the **Tolerance Calculator** located on the **Power Transmission** panel under the **Design** tab.

6 Access the **Limits and Fits Mechanical Calculator.**

See Figure 11-14. The **Limits and Fits Calculator** is located by clicking the arrowhead next to the **Power Transmission** panel heading, clicking the arrowhead next to the **Limits/Fits Calculator,** and clicking the **Limits/Fits Calculator** option.

7 Select the **Hole-basis system of fits,** enter a **Basic Size** of **13.00,** and select an **Interference** fit.

See Figure 11-15. The tolerance values will appear under the **Results** heading. The hole's values can also be obtained by moving the cursor to the **H** tolerance zone.

> **TIP**
>
> The terms *Hole-basis* and *Shaft-basis* refer to how the tolerance was calculated. In this example, note in the **Results** box that the minimum hole diameter is 13.000. This is the starting point for the tolerance calculations, so the calculation is hole-basis. If the minimum shaft diameter had been defined as 13.000, the tolerance calculations would have started from this value. The results would have been shaft-basis.

From Figure 11-15 we see that the hole has a tolerance of 13.018/13.000.

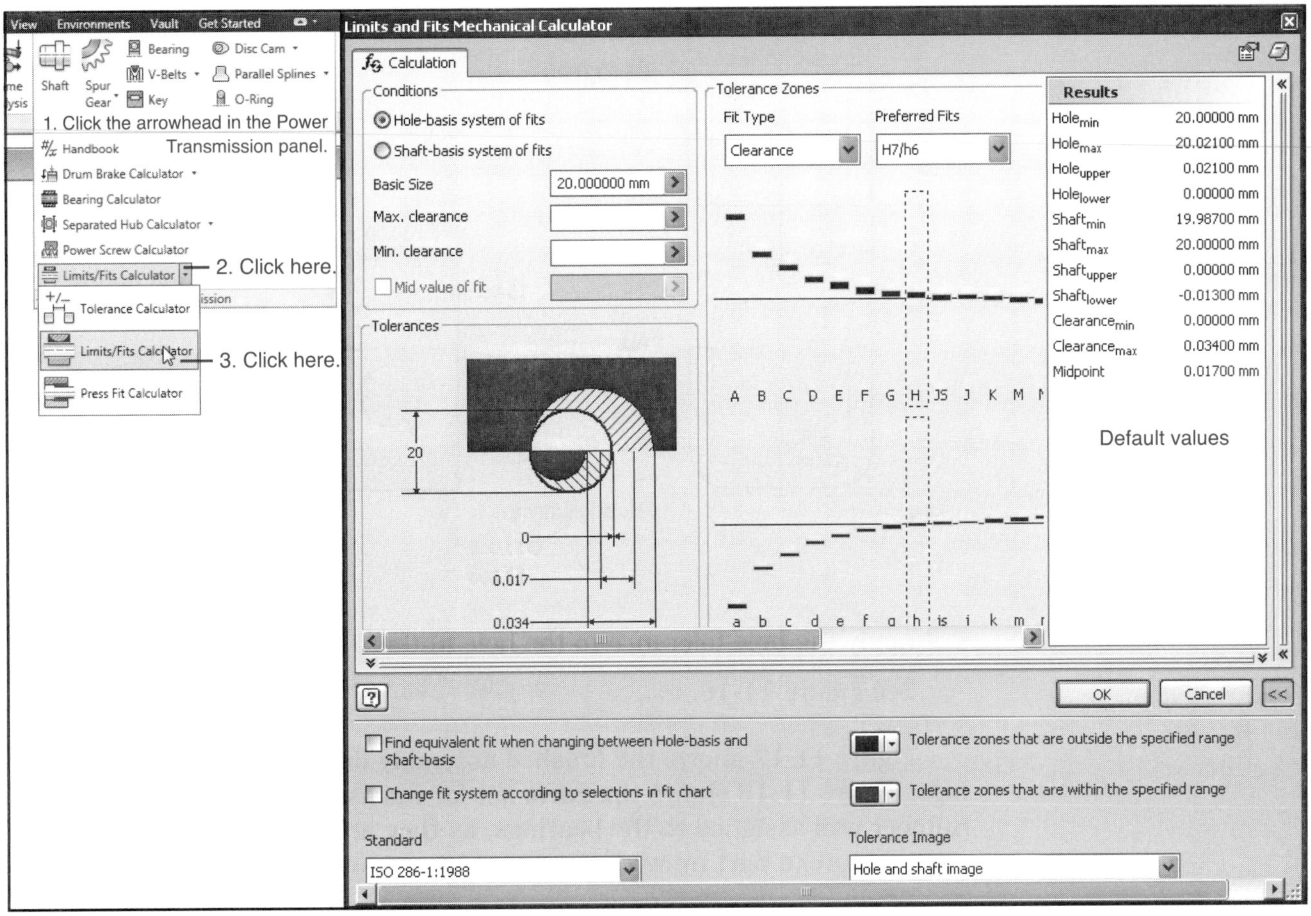

Figure 11-14

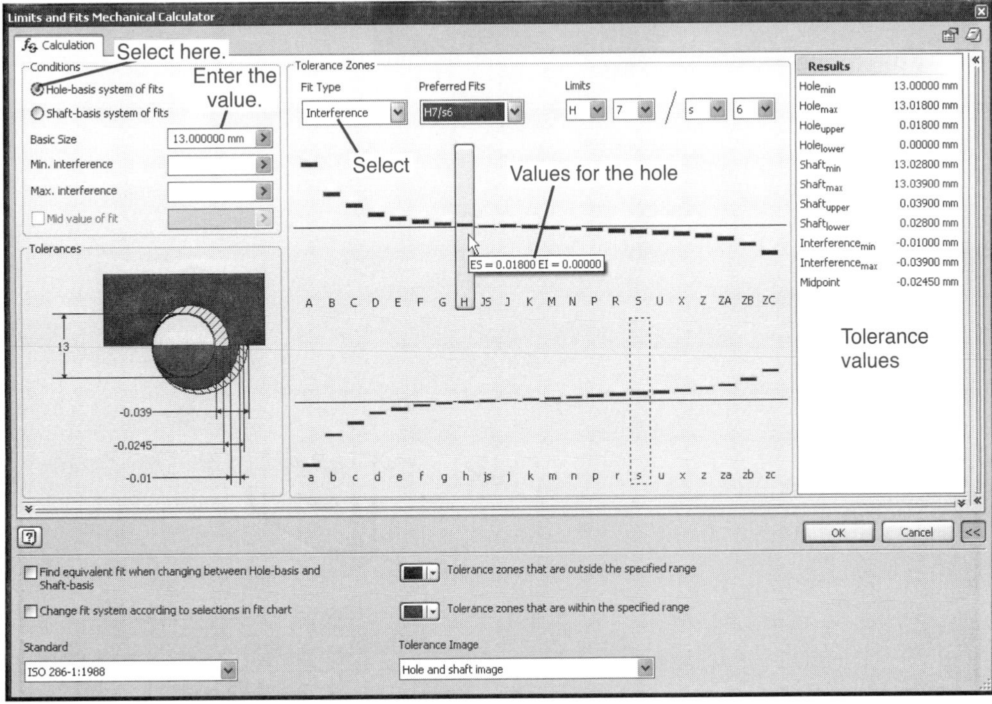

Figure 11-15

Figure 11-16

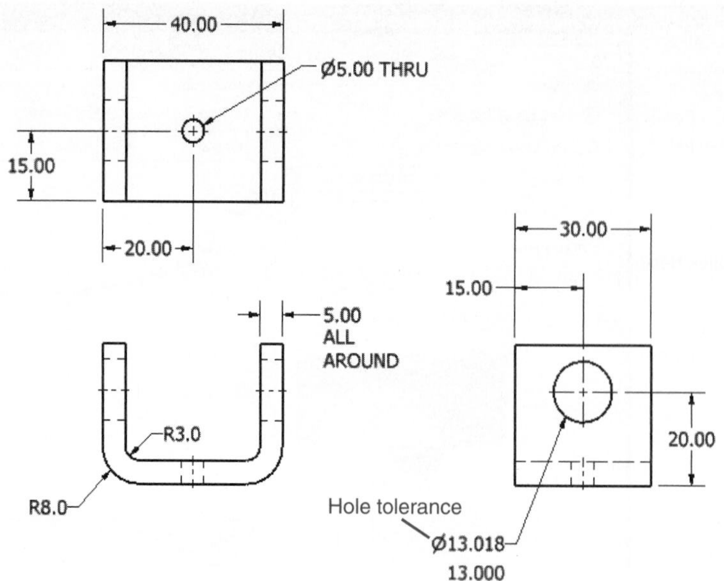

8 Apply the hole tolerance to the hole in the U-bracket.

See Figure 11-16.

Figure 11-17 shows the finished assembly drawing with the bearings. Figure 11-18 shows the parts list for the assembly. Note that no part number was assigned to the bearings, as they are purchased parts. The manufacturer's part numbers were used.

Figure 11-17

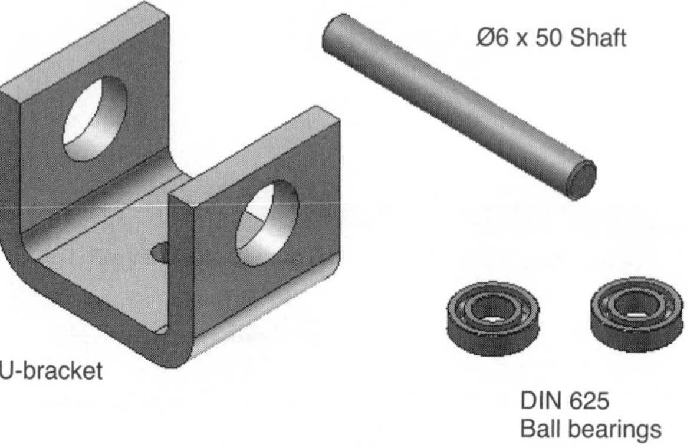

Figure 11-18

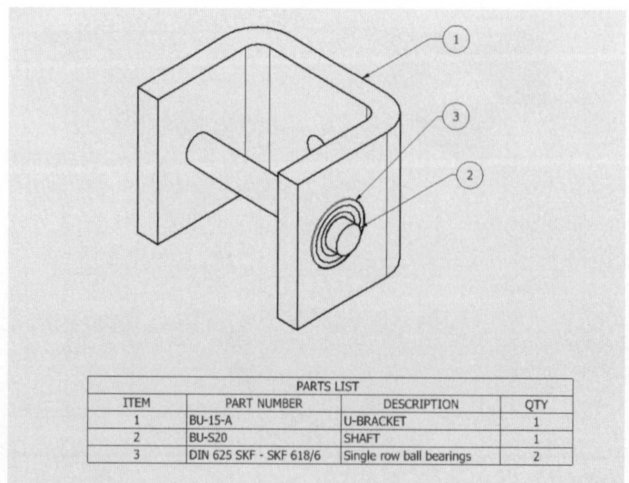

PARTS LIST			
ITEM	PART NUMBER	DESCRIPTION	QTY
1	BU-15-A	U-BRACKET	1
2	BU-S20	SHAFT	1
3	DIN 625 SKF - SKF 618/6	Single row ball bearings	2

Thrust Bearings

Tolerances for thrust bearings are similar to ball bearings. In general, a clearance fit is used between the shaft and the bearing's bore (inside diameter), and an interference fit is used between the housing and the bearing's outside diameter.

EXERCISE 11-3 Selecting Thrust Bearings

1 Create a drawing using the **Standard (mm).iam** format and access the **Place from Content Center** dialog box.

See Figure 11-19.

2 Click on **Thrust Ball Bearings** and select an **ANSI/AFBMA 24.1 TA - Thrust Ball Bearing.**

The dialog box will appear showing the **Size Designation** using coded numbers. See Figure 11-20.

Figure 11-19

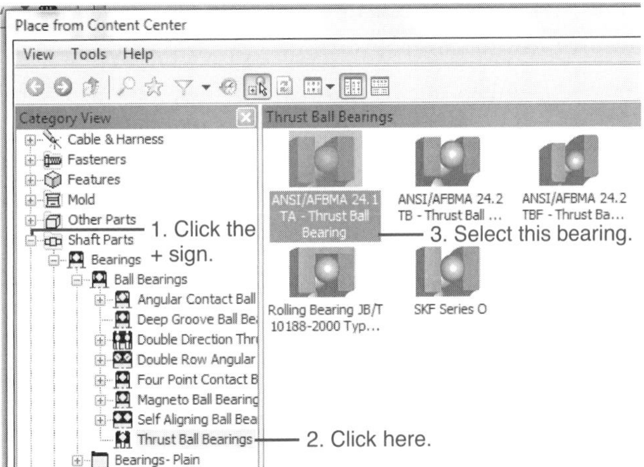

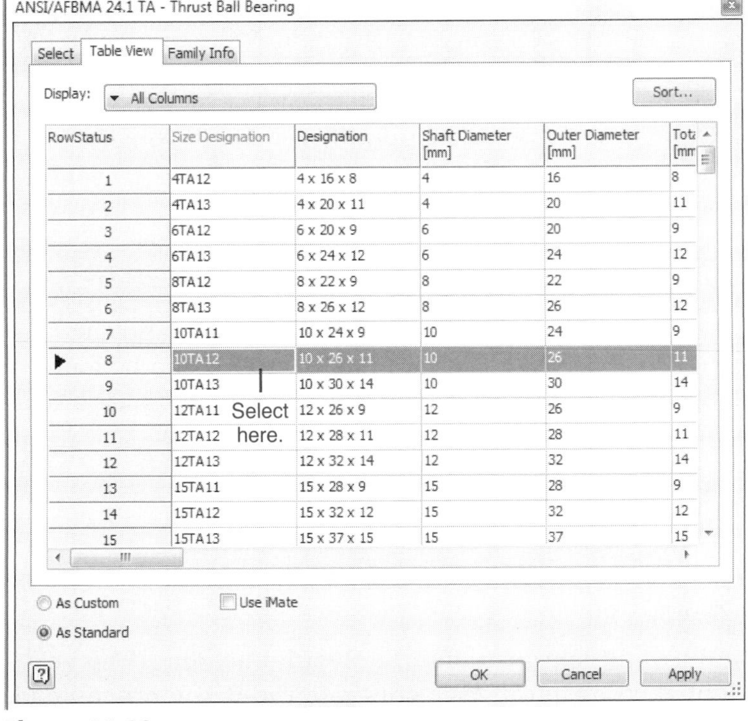

RowStatus	Size Designation	Designation	Shaft Diameter [mm]	Outer Diameter [mm]	Tota [mm
1	4TA12	4 x 16 x 8	4	16	8
2	4TA13	4 x 20 x 11	4	20	11
3	6TA12	6 x 20 x 9	6	20	9
4	6TA13	6 x 24 x 12	6	24	12
5	8TA12	8 x 22 x 9	8	22	9
6	8TA13	8 x 26 x 12	8	26	12
7	10TA11	10 x 24 x 9	10	24	9
▶ 8	10TA12	10 x 26 x 11	10	26	11
9	10TA13	10 x 30 x 14	10	30	14
10	12TA11	12 x 26 x 9	12	26	9
11	12TA12	12 x 28 x 11	12	28	11
12	12TA13	12 x 32 x 14	12	32	14
13	15TA11	15 x 28 x 9	15	28	9
14	15TA12	15 x 32 x 12	15	32	12
15	15TA13	15 x 37 x 15	15	37	15

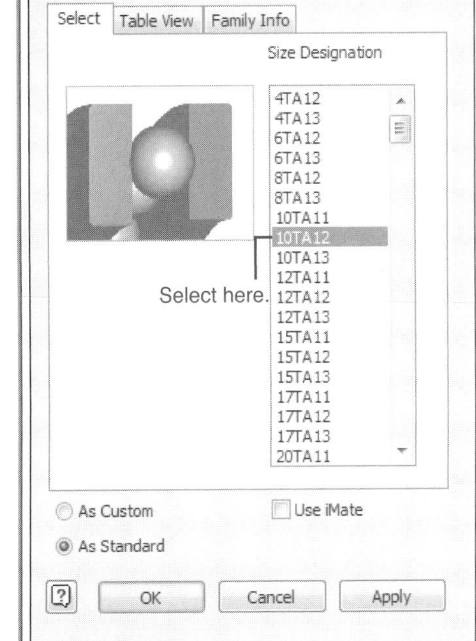

Figure 11-20

3 Click the **Table View** tab to access the dimensions for the listed bearings.

4 Select the **10TA12 Size Designation.**

Figure 11-21 shows the bearing with dimensions and tolerances.

Figure 11-21

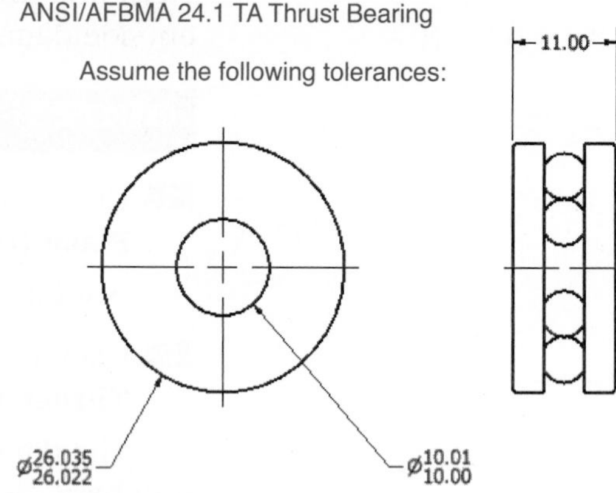

5 Calculate the shaft's diameter.

Say the clearance requirements are

Clearance = minimum = 0.00

Clearance = maximum = 0.02

This means the shaft diameter tolerance is 10.00/9.99.

6 Draw the shaft using the **Shaft** tool on the **Design** tab.

Use the nominal diameter of **10.00** and a length of **30.** Add a **0.50** × **45**° chamfer to each end. Figure 11-22 shows a solid model drawing of the shaft and a dimensioned orthographic drawing. The model was drawn using the nominal Ø10, and the orthographic views include the required tolerances.

Figure 11-22

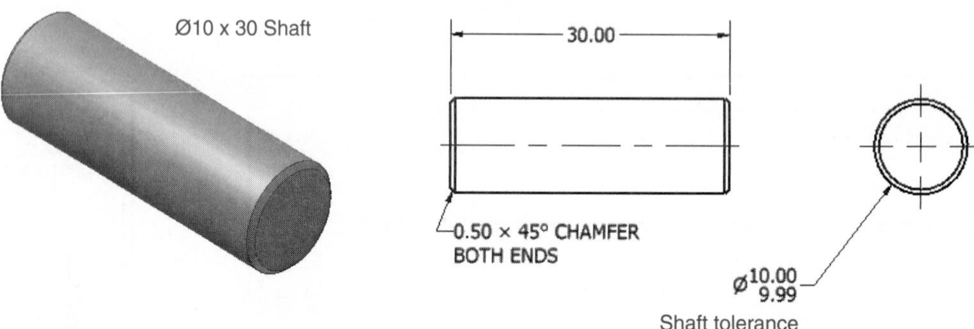

7 Determine the size for the hole in the T-bracket.

The interface between the outside diameter of the bearing and the T-bracket is defined as H7/p6. The values for this tolerance can be found in the tables in the appendix or by using the **Limits and Fits** dialog boxes found on the **Power Transmission** panel under the **Design** tab.

8 Access the **Limits/Fits Calculator** tool.

See Figure 11-23.

9 Select the **Hole-basis system of fits,** enter a **Basic Size** of **26.00,** and select an **Interference** fit.

The tolerance values will appear under the **Results** heading. The hole's values can also be obtained by moving the cursor to the **H tolerance zone.** The hole's tolerance is Ø26.021/26.000.

Figure 11-23

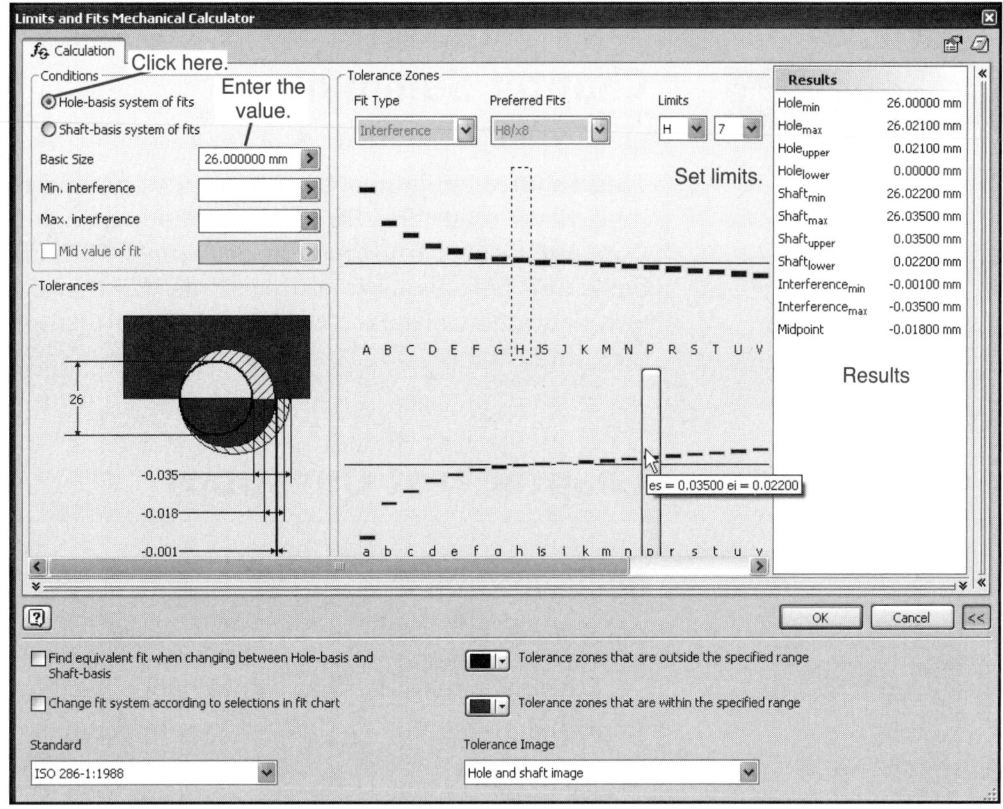

In this example the hole in the T-bracket will use a counterbored hole. The bearing has a thickness of 11.00, so the counterbored hole will have a depth of 11 and a diameter of 26.021/26.000.

10 Apply the hole tolerances to the T-bracket.

See Figure 11-24.

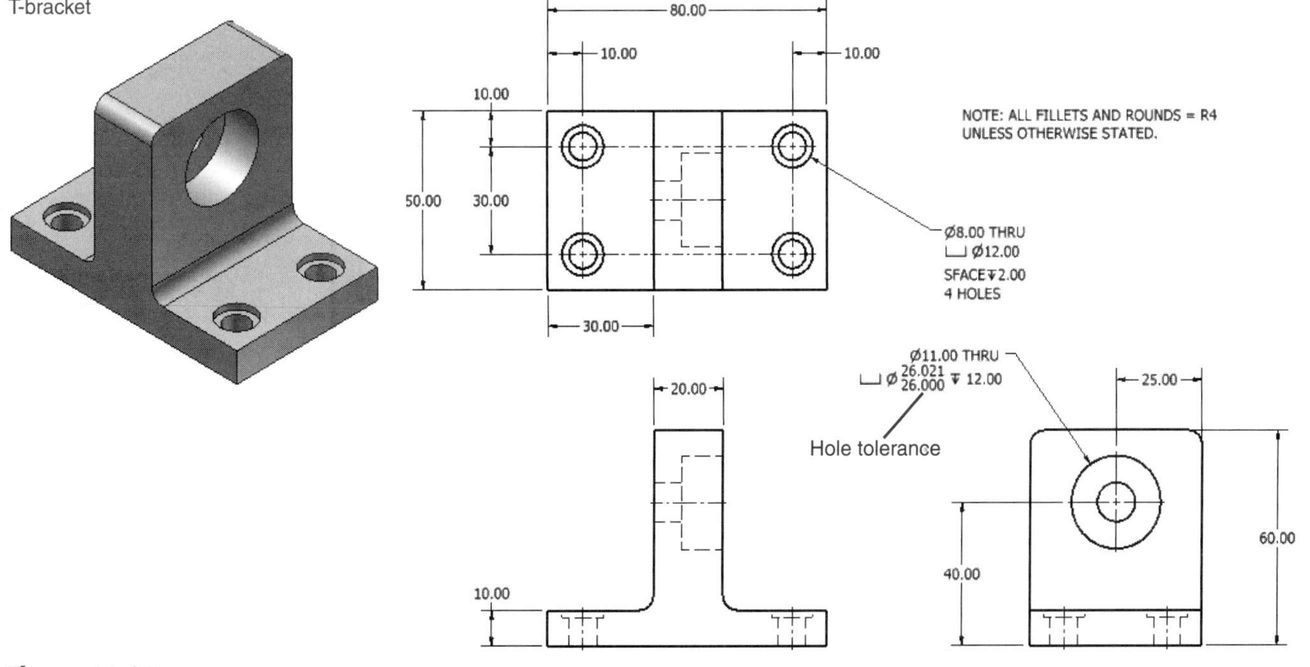

Figure 11-24

Chapter Summary

This chapter explained the differences among three types of bearings: plain, ball, and thrust. Various bearings were selected from the **Content Center** and used in drawings. The difference between shaft-basis and hole-basis tolerances was discussed and illustrated. Shafts and bearings were used in assemblies utilizing both clearance and force fits.

Chapter Test Questions

Multiple Choice

Circle the correct answer.

1. Which is a material that would *not* be used to make a sleeve bearing?
   a. Teflon
   b. Cast iron
   c. Impregnated bronze
   d. Nylon
2. In the notation H7/n6 the H7 specifies a standard tolerance for the
   a. Hole
   b. Shaft
   c. Bearing diameter
   d. Interference
3. In the notation H7/n6 the n6 specifies a standard tolerance for the
   a. Hole
   b. Shaft
   c. Bearing diameter
   d. Interference
4. Hole/shaft tolerance calculations that start with a hole's diameter are called
   a. First hole consideration
   b. Datum hole calculations
   c. Hole basis
5. In general, what type of fit is the tolerance between a shaft and sleeve bushing?
   a. Clearance
   b. Interference
   c. Transitional

Matching

Column A

a. Sleeve _________
b. Ball _________
c. Thrust _________

Column B

1. A bearing that absorbs loads along an axial direction
2. A bearing that looks like a hollow cylinder
3. A bearing that includes a row of spheres

True or False

Circle the correct answer.

1. **True or False:** A sleeve bearing is also called a bushing.
2. **True or False:** A listing of bearings is found in the **Content Center.**
3. **True or False:** In general, ball bearings are less expensive than sleeve bearings.
4. **True or False:** Values for limit and fit tolerances are located under the **Design** tab.
5. **True or False:** The designation H7/h6 specifies a standard hole/shaft fit tolerance.

Chapter Projects

Figure P11-1 shows a Bearing Assembly drawing, a parts list, and detail drawings of all manufactured parts. Use the Bearing Assembly with Projects 11-1 through 11-8.

Project 11-1: Millimeters

A. Insert two Ø12 × 90 shafts with 0.50 × 45° chamfers at each end. Insert each shaft into a CNS 02 3481 A plain bearing. Insert the two bearing shaft subassemblies into the Bearing Assembly shown in Figure P11-1. See Figure P11-2.

B. Assume the CNS 02 3481 A plain bearing has an inside diameter of 12.01/12.00 and that the required clearance with the shaft is 0.00 minimum and 0.02 maximum. What is the diametric tolerance of the shafts? Create an orthographic drawing of one of the shafts with dimensions and tolerances.

C. Assume the CNS 02 3481 A plain bearing has an outside diameter of 16.039/16.028 and that the bearing will fit into the Side Part of the Bearing Assembly using an H7/s6 interference fit. What is the dimension and tolerance for the holes in the Side Part? Create an orthographic drawing of the Side Part and include all dimensions and tolerances.

D. Create a presentation drawing of the completed Bearing Assembly.

E. Create an exploded isometric drawing of the completed Bearing Assembly. Include assembly numbers and a parts list. The part number for the shaft is SH07-12.

Figure P11-1 Bearing Assembly

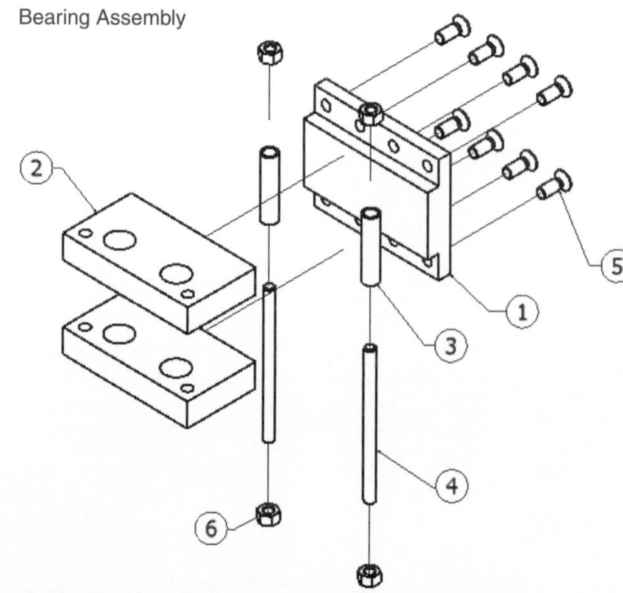

Figure P11-1
(Continued)

Bearing Assembly
Parts list

Parts List				
ITEM	PART NUMBER	DESCRIPTION	MATERIAL	QTY
1	BU2007-1	BASE	SAE 1020	1
2	BU2007-2	PART, SIDE	SAE 1020	2
3	BUENG-A	POST, GUIDE	SAE 1040	2
4	BUENG-B	POST, THREADED	SAE 1040	2
5	AS 1427 - M8 x 20	Pozidriv ISO metric machine screws	Steel, Mild	8
6	ANSI B18.2.4.2M - M8x1.25	Metric Hex Nuts Styles 2	Steel, Mild	4

Base
P/N BU2007-1
SAE 1020

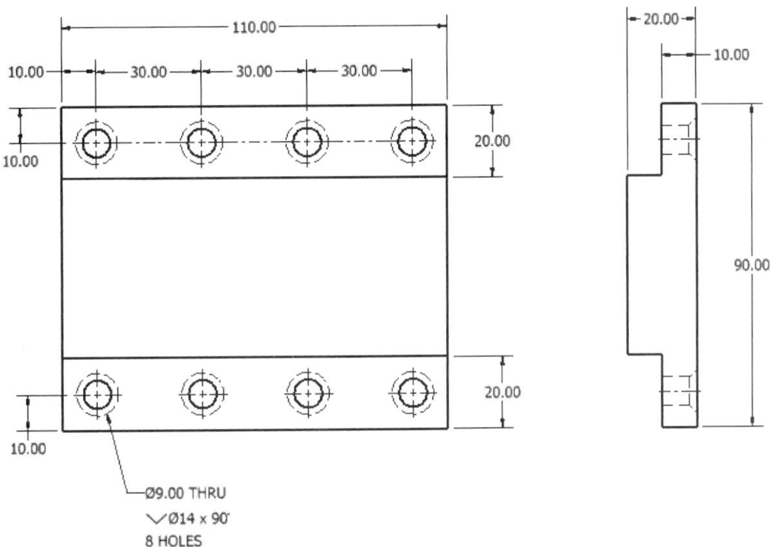

Part, Side
P/N BU2007-2
SAE 1020

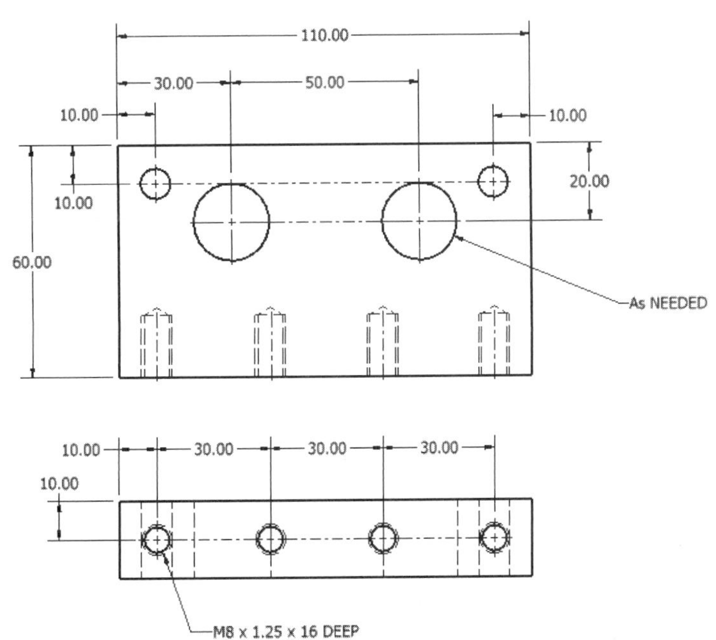

Post, Guide
BUENG-A
SAE 1040

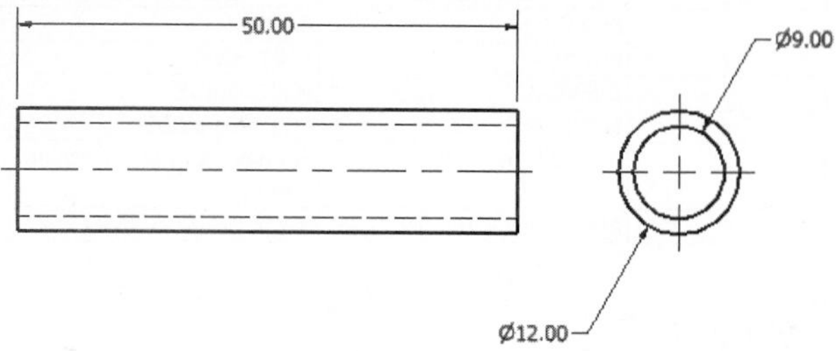

Post, Threaded
P/N BUENG-B
SAE 1040

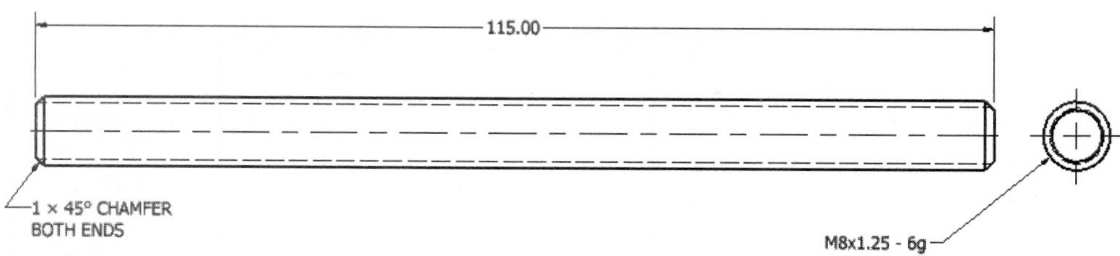

Figure P11-1
(Continued)

Project 11-2: Millimeters

A. Insert two Ø16 × 90 shafts with 0.50 × 45° chamfers at each end. Insert each shaft into a CSN 9352 plain bearing. Insert the two bearing shaft subassemblies into the Bearing Assembly shown in Figure P11-1.

B. Assume the CSN 9352 A plain bearing has an inside diameter of 16.02/12.01 and that the required clearance with the shaft is 0.01 minimum and 0.04 maximum. What is the diametric tolerance of the shafts? Create an orthographic drawing of one of the shafts with dimensions and tolerances.

C. Assume the CSN 9352 A plain bearing has an outside diameter of 22.000/21.987 and that the bearing will fit into the Side Part of the Bearing Assembly using an S7/h6 interference fit (shaft basis). What is the dimension and tolerance for the holes in the Side Part? Create an orthographic drawing of the Side Part and include all dimensions and tolerances.

D. Create a presentation drawing of the completed Bearing Assembly.

E. Create an exploded isometric drawing of the completed Bearing Assembly. Include assembly numbers and a parts list. The part number for the shaft is SH07-16.

Figure P11-2

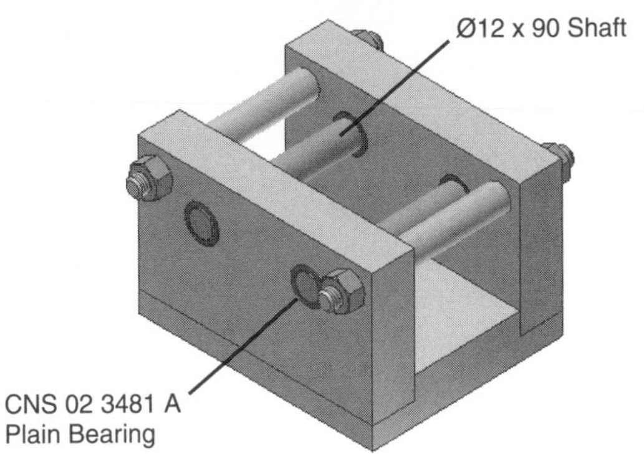

Project 11-3: Millimeters

A. Insert two Ø10 × 130 shafts with 0.50 × 45° chamfers at each end. Insert each shaft into a DIN 1850-5 P plain bearing. Insert the two bearing shaft subassemblies into the Bearing Assembly shown in Figure P11-1.

B. Assume the DIN 1850-5 P plain bearing has an inside diameter of 10.02/12.00 and that the required clearance with the shaft is 0.00 minimum and 0.04 maximum. What is the diametric tolerance of the shafts? Create an orthographic drawing of one of the shafts with dimensions and tolerances.

C. Assume the DIN 1850-5 P plain bearing has an outside diameter of 16.000/15.989 and that the bearing will fit into the Side Part of the Bearing Assembly using an S7/h6 interference fit. What is the dimension and tolerance for the holes in the Side Part? Create an orthographic drawing of the Side Part and include all dimensions and tolerances.

D. Add CNS 122 collars to both ends of both shafts.

E. Create a presentation drawing of the completed Bearing Assembly.

F. Create an exploded isometric drawing of the completed Bearing Assembly. Include assembly numbers and a parts list. The part number for the shaft is SH08-16.

Project 11-4: Millimeters

A. Insert two Ø20 × 110 shafts with 1.00 × 45° chamfers at each end. Insert each shaft into a JIS B 1582 plain bearing. Insert the two bearing shaft subassemblies into the Bearing Assembly shown in Figure P11-1.

B. Assume the JIS B 1582 plain bearing has an inside diameter of 20.02/20.00 and that the required clearance with the shaft is 0.01 minimum and 0.05 maximum. What is the diametric tolerance of the shaft? Create an orthographic drawing of the shaft with dimensions and tolerances.

C. Assume the JIS B 1582 plain bearing has an outside diameter of 28.039/28.028 and that the bearing will fit into the Side Part of the Bearing Assembly using an H7/s6 interference fit. What is the dimension and tolerance for the holes in the Side Part? Create an orthographic drawing of the Side Part and include all dimensions and tolerances.

D. Add CNS 9074 retaining rings 5 mm from each end of both shafts.

E. Create a presentation drawing of the completed Bearing Assembly.

F. Create an exploded isometric drawing of the completed Bearing Assembly. Include assembly numbers and a parts list. The part number for the shaft is SH07-12.

Project 11-5: Millimeters

A. Insert two Ø10 × 90 shafts with 0.50 × 45° chamfers at each end. Insert each shaft into a BS 290–61800 ball bearing. Insert the two bearing shaft subassemblies into the Bearing Assembly shown in Figure P11-1.

B. Assume the BS 290–61800 ball bearing has an inside diameter of 10.02/10.00 and that the required clearance with the shaft is 0.00 minimum and 0.04 maximum. What is the diametric tolerance of the shafts? Create an orthographic drawing of one of the shafts with dimensions and tolerances.

C. Assume the BS 290–61800 ball bearing has an outside diameter of 16.000/15.989 and that the bearing will fit into the Side Part of the Bearing Assembly using a U7/h6 interference fit. What is the dimension and tolerance for the holes in the Side Part? Create an orthographic drawing of the Side Part and include all dimensions and tolerances.

D. Create a presentation drawing of the completed Bearing Assembly.

E. Create an exploded isometric drawing of the completed Bearing Assembly. Include assembly numbers and a parts list. The part number for the shaft is SH07-12.

Project 11-6: Millimeters

A. Insert two Ø15 × 130 Shafts with 0.50 × 45° chamfers at each end. Insert each shaft into a CSN 02 4630 – 61802 ball bearing. Insert the two bearing shaft subassemblies into the Bearing Assembly shown in Figure P11-1.

B. Assume the CSN 02 4630–61802 ball bearing has an inside diameter of 15.01/12.00 and that the required clearance with the shaft is 0.00 minimum and 0.02 maximum. What is the diametric tolerance of the shafts? Create an orthographic drawing of one of the shafts with dimensions and tolerances.

C. Assume the CSN 02 4630–61802 ball bearing has an outside diameter of 24.000/24.987 and that the bearing will fit into the Side Part of the Bearing Assembly using an F7/h6 interference fit. What is the dimension and tolerance for the holes in the Side Part? Create an orthographic drawing of the Side Part and include all dimensions and tolerances.

D. Add a DIN 705 B collar to each end of both shafts.

E. Create a presentation drawing of the completed Bearing Assembly.

F. Create an exploded isometric drawing of the completed Bearing Assembly. Include assembly numbers and a parts list. The part number for the shaft is SH07-12.

Project 11-7: Millimeters

A. Insert two Ø12 × 110 shafts with 1.00 × 45° chamfers at each end. Insert each shaft into a CSN 02 4630–16103 ball bearing. Insert the two bearing shaft subassemblies into the Bearing Assembly shown in Figure P11-1. For this exercise replace the through hole with a counterbored hole with a nominal size Ø13.00 THRU, Ø30 CBORE 8.00 DEEP.

B. Assume the CSN 02 4630–16103 ball bearing has an inside diameter of 12.01/12.00 and that the required clearance with the shaft is 0.00 minimum and 0.02 maximum. What is the diametric tolerance of the shafts? Create an orthographic drawing of one of the shafts with dimensions and tolerances.

C. Assume the CSN 02 4630–16103 ball bearing has an outside diameter of 30.000/29.987 and that the bearing will fit into the Side Part of the Bearing Assembly using an S7/h6 interference fit. What is the dimension and tolerance for the holes in the Side Part? Create an orthographic drawing of the Side Part and include all dimensions and tolerances.

D. Add a DIN 471 5.00 mm from each end of both shafts.

E. Create a presentation drawing of the completed Bearing Assembly.

F. Create an exploded isometric drawing of the completed Bearing Assembly. Include assembly numbers and a parts list. The part number for the shaft is SH07-12.

Project 11-8: Millimeters

A. Insert two Ø16 × 100 shafts with 0.75 × 45° chamfers at each end. Insert each shaft into a GB 273.2-87–7/70 16 × 26 × 5 thrust bearing. Insert the two bearing shaft subassemblies into the Bearing Assembly shown in Figure P11-1. For this exercise replace the through hole with a counterbored hole with a nominal size Ø17.00 THRU, Ø26 CBORE 5.00 DEEP.

B. Assume the GB 273.2-87–7/70 16 × 26 × 5 thrust bearing has an inside diameter of 16.02/16.00 and that the required clearance with the shaft is 0.00 minimum and 0.04 maximum. What is the diametric tolerance of the shafts? Create an orthographic drawing of one of the shafts with dimensions and tolerances.

C. Assume the GB 273.2-87–7/70 16 × 26 × 5 thrust bearing has an outside diameter of 26.039/26.028 and that the bearing will fit into the Side Part of the Bearing Assembly using an H7/s6 interference fit. What is the dimension and tolerance for the holes in the Side Part? Create an orthographic drawing of the Side Part and include all dimensions and tolerances.

D. Create a presentation drawing of the completed Bearing Assembly.

E. Create an exploded isometric drawing of the completed Bearing Assembly. Include assembly numbers and a parts list. The part number for the shaft is SH07-16.

Project 11-9: Inches

Figure P11-9 shows a Handle Assembly.

A. Draw an assembly drawing. The Link is offset .75 from the support.

B. Create a presentation drawing.

C. Create an exploded isometric drawing with a parts list.

D. Animate the drawing so that the Ball, Threaded Post, Link, Rectangular Key, and Shaft rotate within the bearing.

Figure P11-9

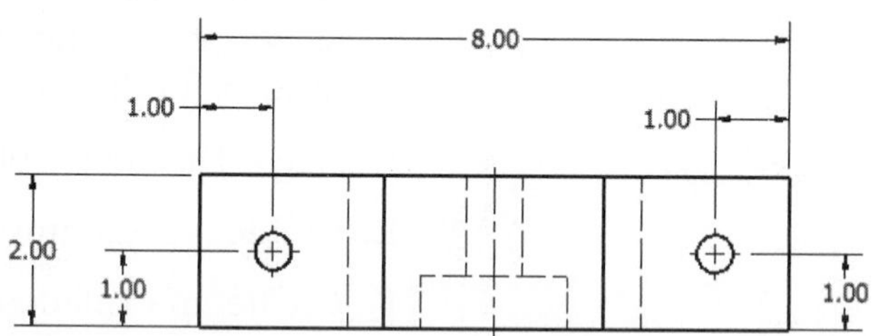

NOTE: ALL FILLETS AND ROUNDS R=0.250
UNLESS OTHERWISE STATED.

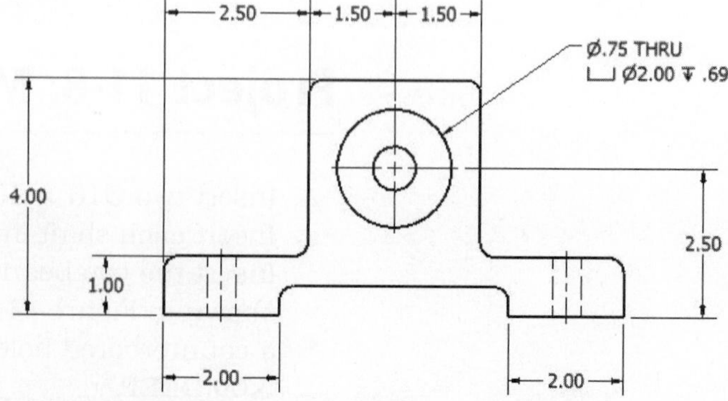

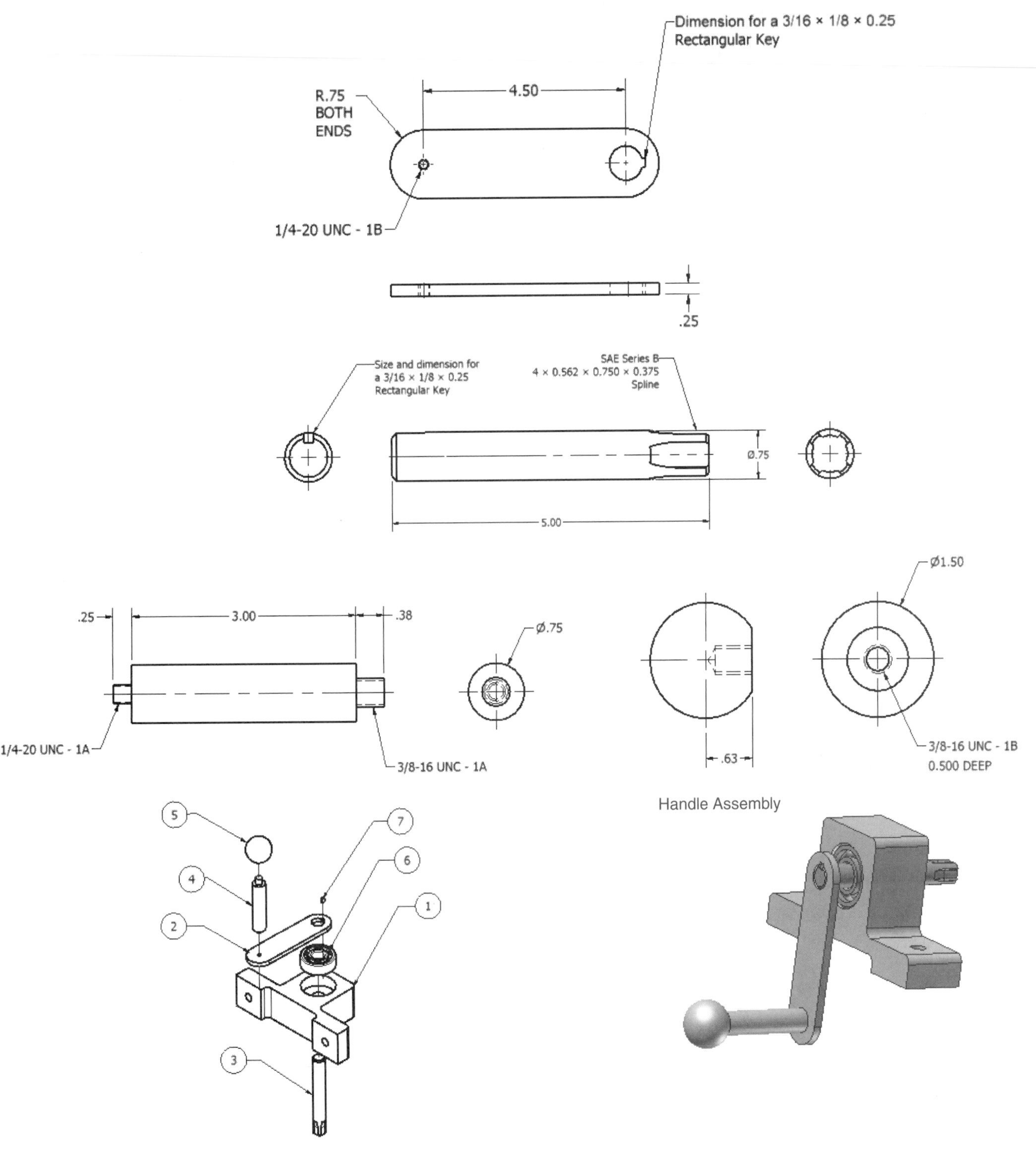

Parts List				
ITEM	PART NUMBER	DESCRIPTION	MATERIAL	QTY
1	EK131-1	SUPPORT	STEEL	1
2	EK131-2	LINK	STEEL	1
3	EK131-3	SHAFT,DRIVE	STEEL	1
4	EK131-4	POST, THREADED	STEEL	1
5	EK131-5	BALL	STEEL	1
6	BS 292 - BRM 3/4	Deep Groove Ball Bearings	STEEL,MILD	1
7	3/16x1/8x1/4	RECTANGULAR KEY	STEEL	1

Figure P11-9
(Continued)

chapter twelve
12 Gears

CHAPTER OBJECTIVES

- Understand the different types of gears: spur, bevel, and worm
- Understand gear terminology
- Learn how to select gears from the **Content Center**
- Learn how to draw hubs on gears and to add setscrews
- Learn how to draw keyways on gears and to add keys
- Learn how to use gears in assemblies

Introduction

This chapter explains how to draw gears using the **Gear** tool located on the **Power Transmission** panel under the **Design** tab. Three types of gears are covered: spur, bevel, and worm. See Figure 12-1. The chapter shows how to add hubs and splines to the gears and how to combine gears to create gear trains.

Gear Terminology

Pitch Diameter (*D*): The diameter used to define the spacing of gears. Ideally, gears are exactly tangent to each other along their pitch diameters.

Diametral Pitch (*P*): The number of teeth per inch. **Meshing gears must have the same diametral pitch.** Manufacturers' gear charts list gears with the same diametral pitch.

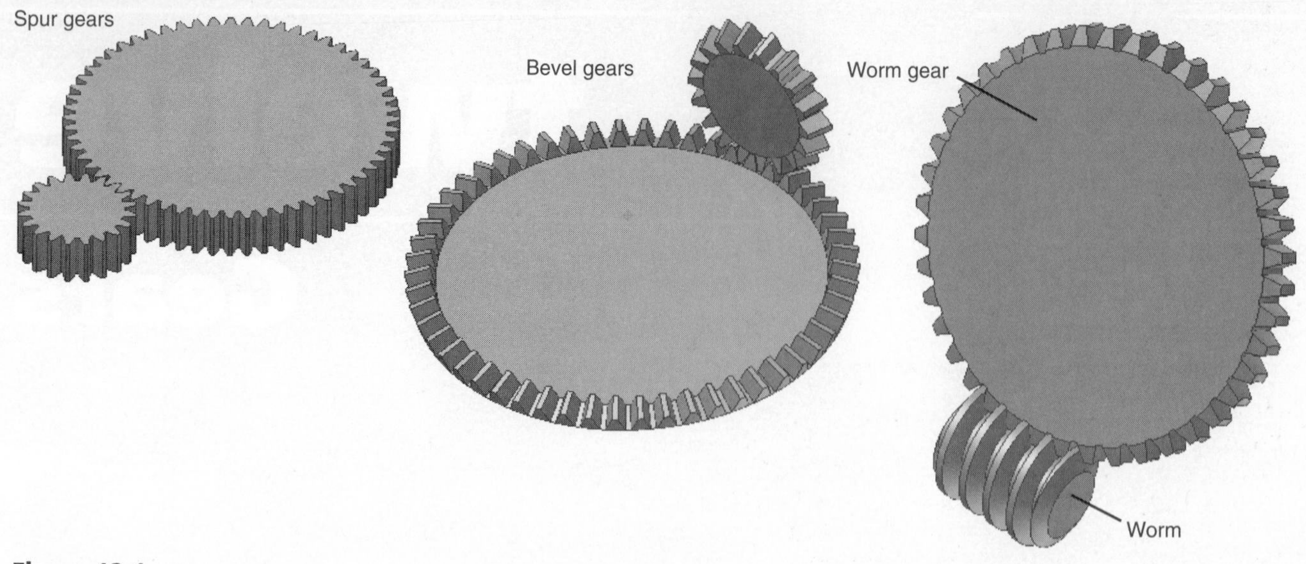

Figure 12-1

Module (*M*): The pitch diameter divided by the number of teeth. The metric equivalent of diametral pitch.

Number of Teeth (*N*): The number of teeth of a gear.

Circular Pitch (*CP*): The circular distance from a fixed point on one tooth to the same position on the next tooth as measured along the pitch circle. The circumference of the pitch circle divided by the number of teeth.

Preferred Pitches: The standard sizes available from gear manufacturers. Whenever possible, use preferred gear sizes.

Center Distance (*CD*): The distance between the center points of two meshing gears.

Backlash: The difference between a tooth width and the engaging space on a meshing gear.

Addendum (*a*): The height of a tooth above the pitch diameter.

Dedendum (*d*): The depth of a tooth below the pitch diameter.

Whole Depth: The total depth of a tooth. The addendum plus the dedendum.

Working Depth: The depth of engagement of one gear into another. Equal to the sum of the two gears' addendeums.

Circular Thickness: The distance across a tooth as measured along the pitch circle.

Face Width (*F*): The distance from front to back along a tooth as measured perpendicular to the pitch circle.

Outside Diameter: The largest diameter of the gear. Equal to the pitch diameter plus the addendum.

Root Diameter: The diameter of the base of the teeth. The pitch diameter minus the dedendum.

Clearance: The distance between the addendum of the meshing gear and the dedendum of the mating gear.

Pressure Angle: The angle between the line of action and a line tangent to the pitch circle. Most gears have pressure angles of either 14.5° or 20°.

See Figure 12-2.

Figure 12-2

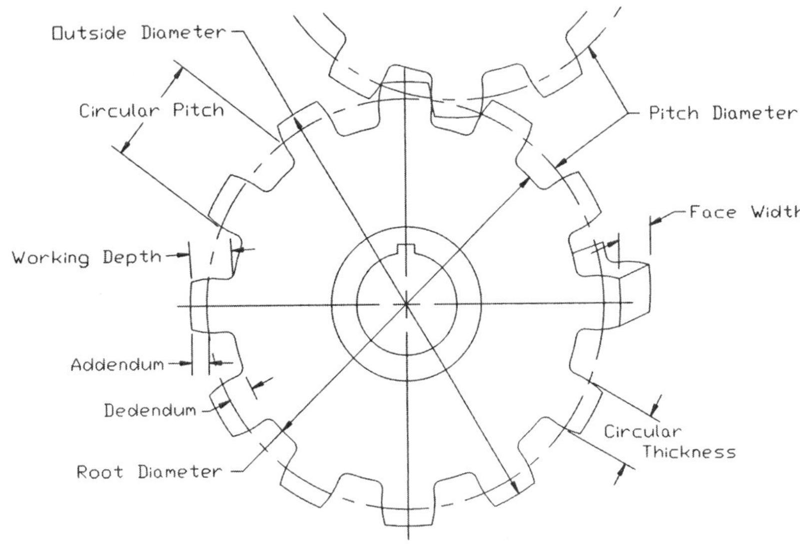

Gear Formulas

Figure 12-3 shows a chart of formulas commonly associated with gears. The formulas are for spur gears.

Figure 12-3

	Diametral pitch (P)	$P = \frac{N}{D}$
	Pitch diameter (D)	$D = \frac{N}{P}$
	Number of teeth (N)	$N = DP$
	Addendum (a)	$a = \frac{1}{P}$
Metric		
	Module (M)	$M = \frac{D}{N}$

Drawing Gears Using the Gear Tool

Draw two gears with 60 and 120 teeth, respectively, a diametral pitch of 24, a pressure angle of 14.5°, and a face width of 0.75 in.

1. Start a new drawing using the **Standard (in).iam** format.
2. Click the **Design** tab.

   See Figure 12-4.
3. Click the **Spur Gear** tool.

Figure 12-4

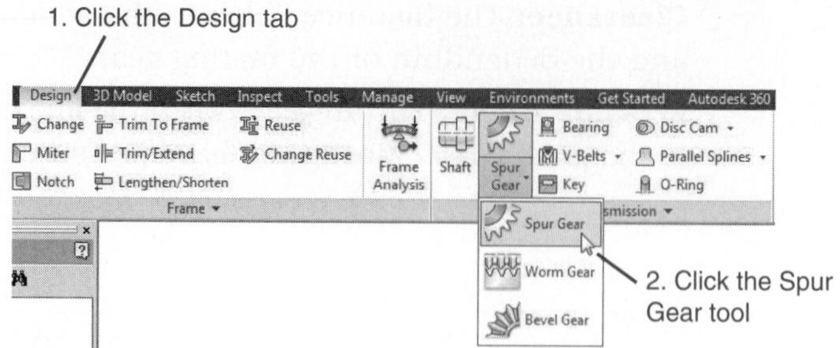

> **NOTE**
>
> In metric units diametral pitch is called *module*.

The **Spur Gears Component Generator** dialog box will appear. See Figure 12-5. There are several options available under the **Design Guide** option. The **Center Distance** option will be used for this example.

Figure 12-5

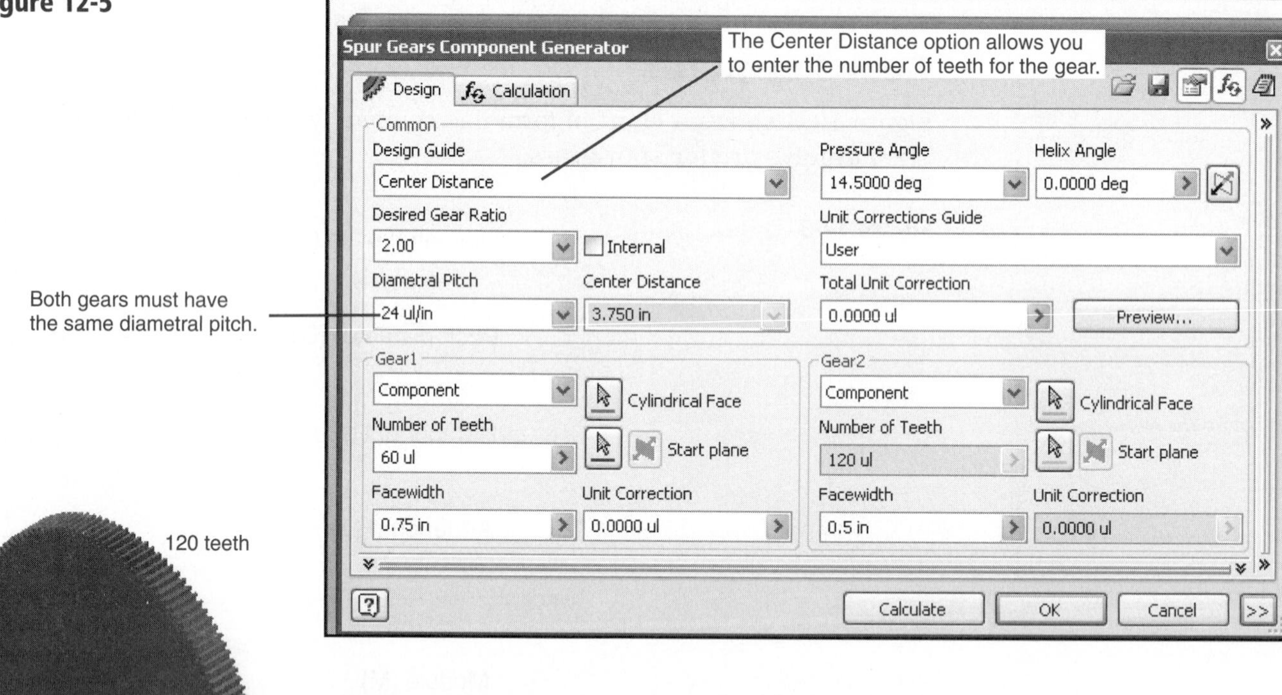

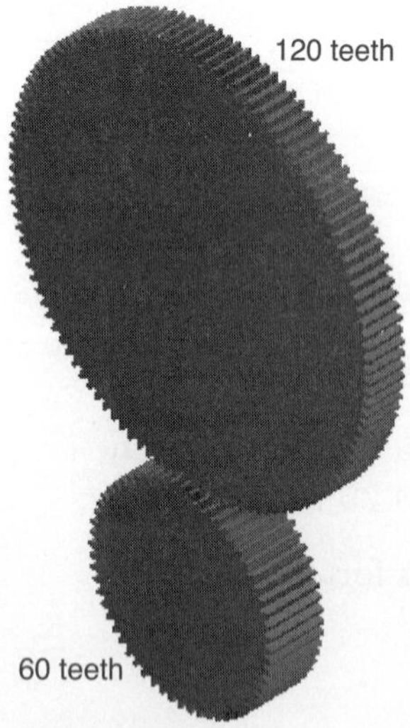

120 teeth

60 teeth

Figure 12-6

4 Enter the following values.

Gear1: 60 teeth, facewidth = 0.75

Gear2: 120 teeth, facewidth = 0.500

Diametral pitch = 24 ul/in

A gear ratio of 2.00 will be calculated automatically.

5 Click **Calculate,** then click **OK.**

The two gears will appear on the screen. See Figure 12-6.

6 Click **Spur Gears:1** in the browser box, right-click **Spur Gear1:1,** and click the **Isolate** option.

See Figure 12-7. Figure 12-8 shows the finished gear.

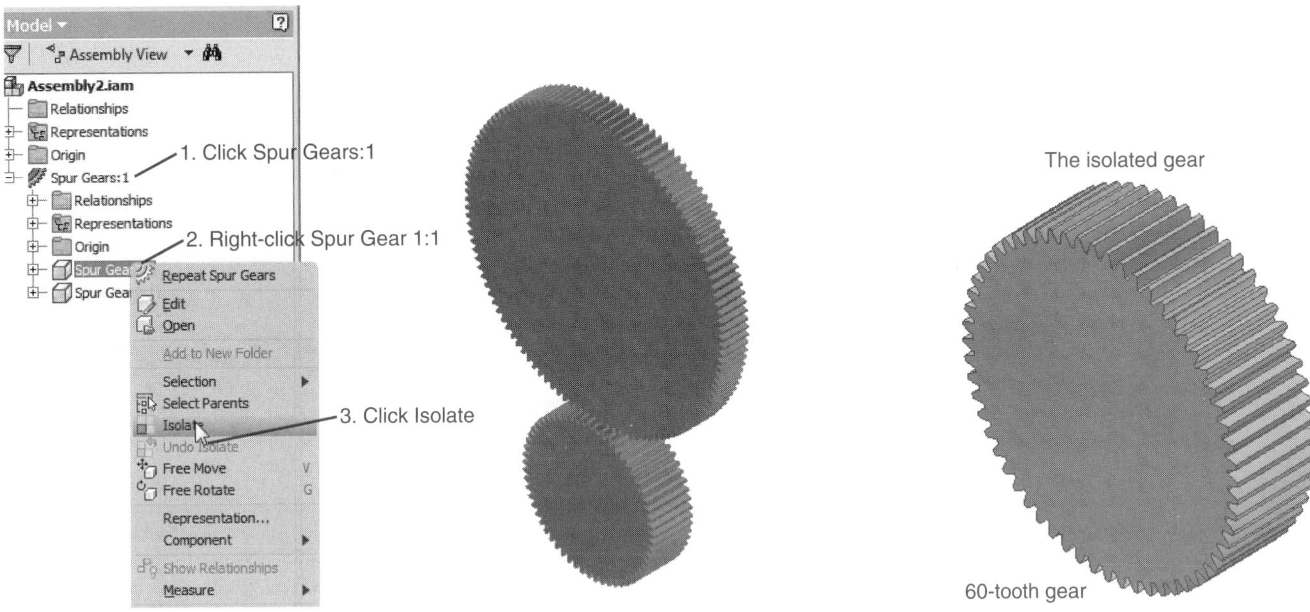

Figure 12-7

Figure 12-8

Gear Hubs

This section will show how to add a gear hub to a gear. The gear created and isolated in the last section and shown in Figure 12-8 will be used. A hub will be added, a hole will be created through the hub and gear, and a threaded hole for a setscrew will also be added.

EXERCISE 12-1 Adding a Hub to a Gear

1 Move the cursor into the browser area, right-click **Spur Gear1:1,** and select the **Edit** option.

See Figure 12-9.

Figure 12-9

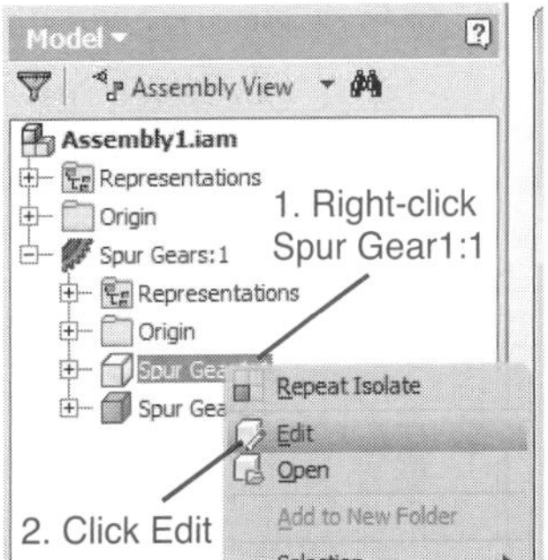

2 Move the cursor onto the front face of the gear, right-click the mouse, and select the **New Sketch** option.

See Figure 12-10.

3 Sketch a **Ø1.250** circle on the gear.

See Figure 12-11.

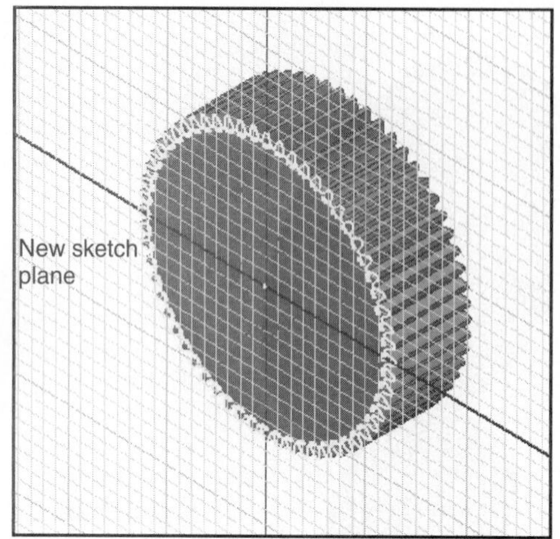

Figure 12-10

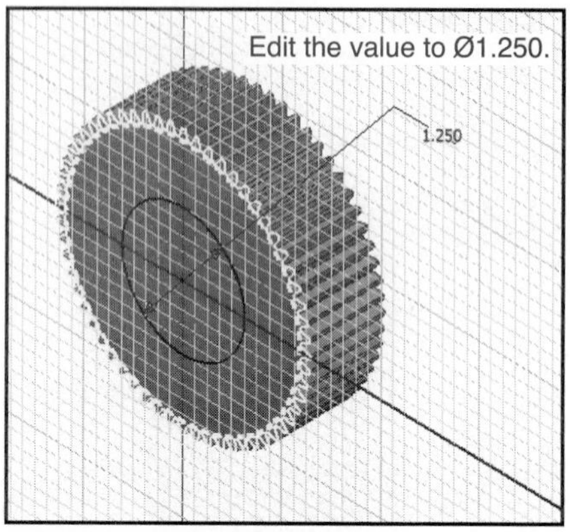

Edit the value to Ø1.250.

1.250

Figure 12-11

4 Right-click the mouse and select the **Finish 2D Sketch** option.

See Figures 12-12 and 12-13.

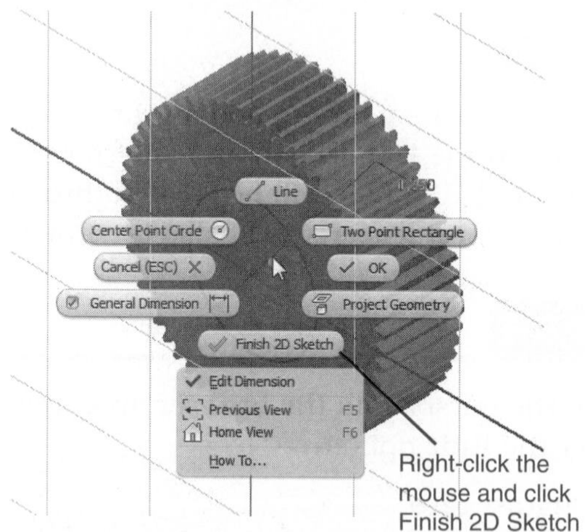

Figure 12-12

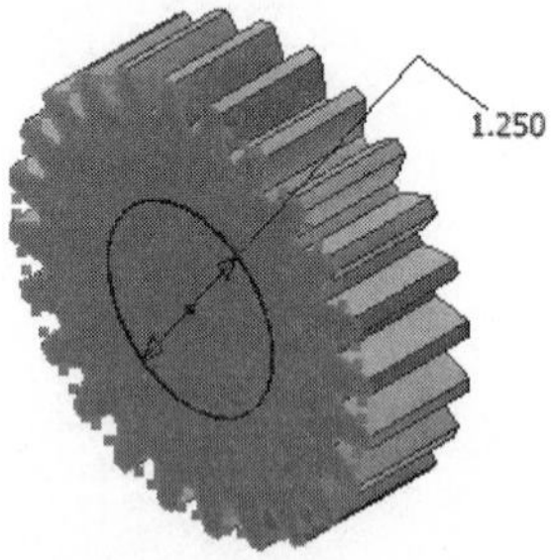

1.250

Figure 12-13

5 Click the **Extrude** tool and extrude the Ø1.250 circle **0.75.**

See Figure 12-14.

6 Create a new sketch plane on the top surface of the extrusion and locate a **Point, Center Point.**

See Figure 12-15.

7 Right-click the mouse and select the **Done** option. Right-click the mouse again and select the **Finish 2D Sketch** option.

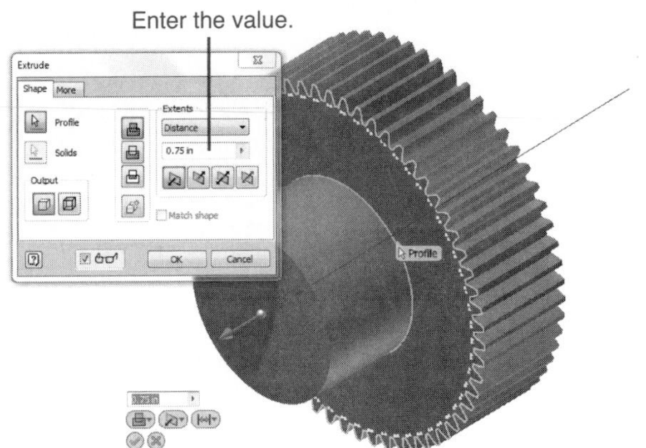

Figure 12-14

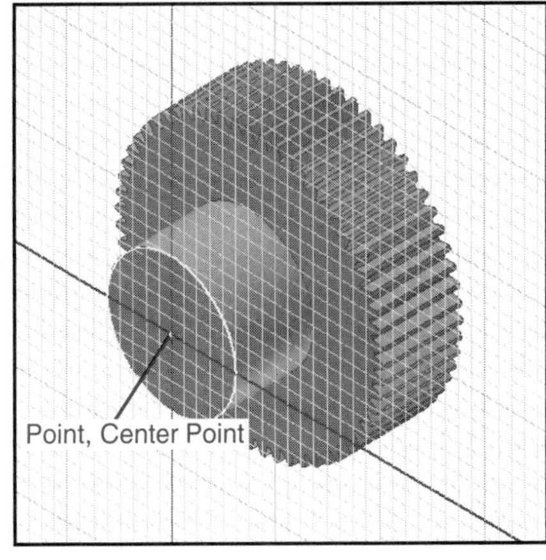

Figure 12-15

8 Click the **Hole** tool and add a **Ø.500** hole through the hub and gear **(Through All).**

See Figure 12-16.

9 Create a work plane tangent to the hub by clicking the **Work Plane** tool located on the flyout from the **Plane** tool on the **Work Features** panel under the **3D Model** tab, clicking the **XZ Plane** under the **Origin** heading of **Spur Gear1:1** in the browser box, and touching the edge of the hub's outer surface with the cursor.

See Figure 12-17.

Figure 12-16

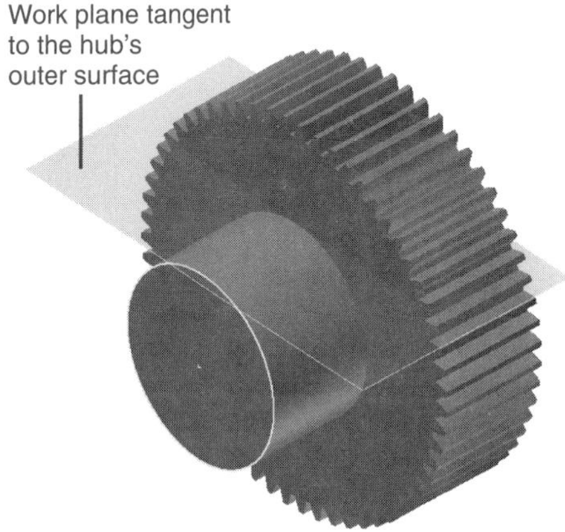

Figure 12-17

> **TIP**
>
> The work plane is created by clicking the **Work Plane** tool then the **XZ plane** under the **Origin** heading in the browser box. Move the cursor to the outside surface of the hub. A work plane will appear tangent to the hub.

10 Create a new sketch on the work plane and locate a **Point, Center Point .375** from the top edge of the hub.

See Figure 12-18.

11 Right-click the mouse and select the **Finish 2D Sketch** option.

12 Click the **Hole** tool and create a **10-24 UNC** threaded hole in the hub. Hide the work plane.

Figure 12-18

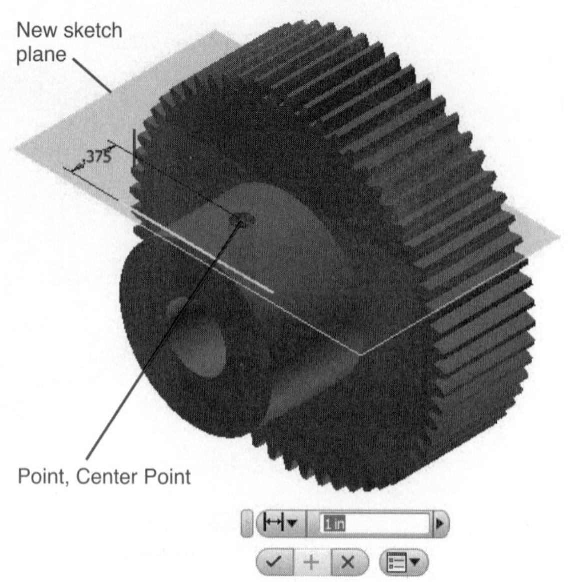

TIP

Do not use the **Through All** distance, as this will create two holes.

See Figure 12-19. The #10 is a hole size. It is a Ø.19 hole.

13 Right-click the mouse and select the **Finish Edit** option.

Hide the XZ work plane.

Figure 12-19

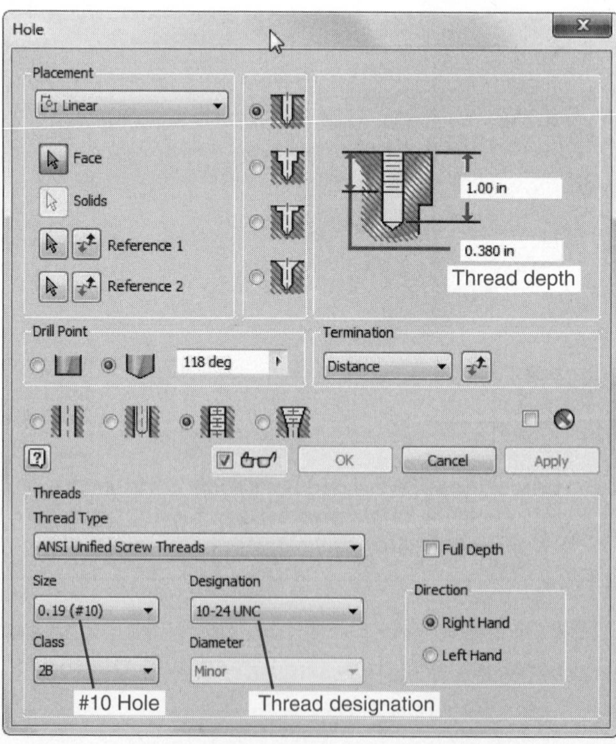

14 Click the **Assemble** tab and click the **Design** tab.

See Figure 12-20. Figure 12-21 shows the finished gear.

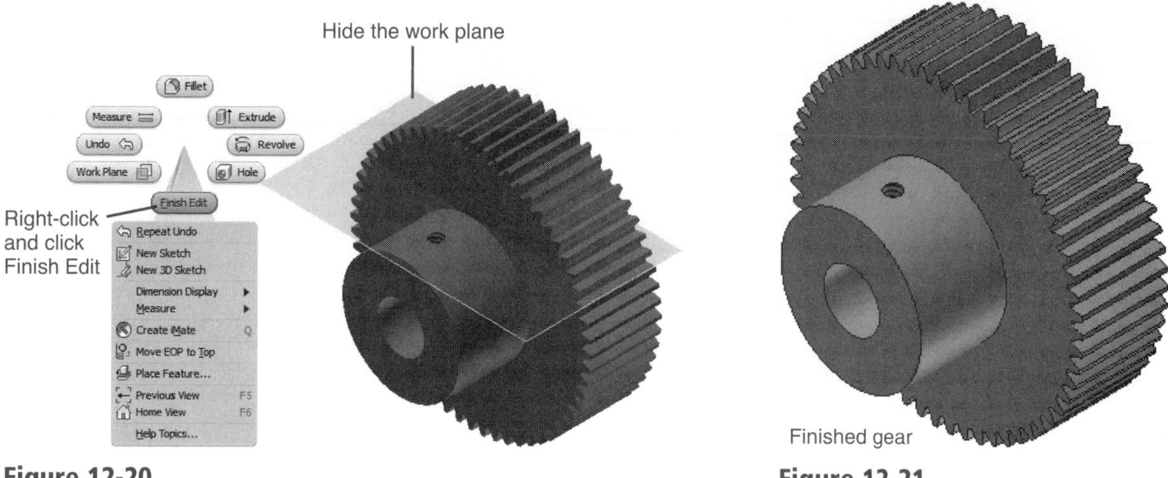

Figure 12-20

Figure 12-21

15 Click on the **Place from Content Center** tool.

16 Access the **Set Screws** and select **Hexagon Socket Set Screw – Half Dog Point – Inch.** Click **OK.**

See Figure 12-22.

Figure 12-22

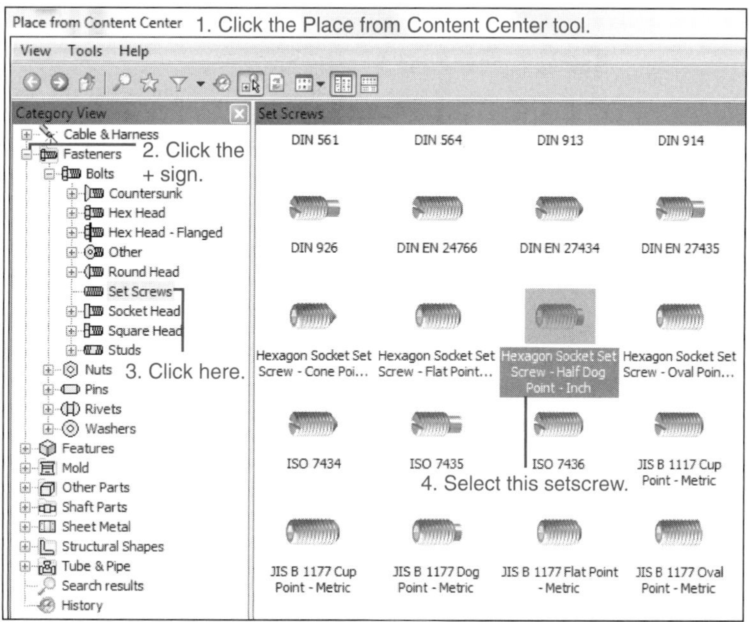

17 Select a **#10 UNC** thread that is **0.38** long. Click **OK.**

See Figure 12-23.

18 Use the **Constrain** tool and position the setscrew.

See Figure 12-24.

NOTE

In this example a work axis was created to help align the setscrew and the hole.

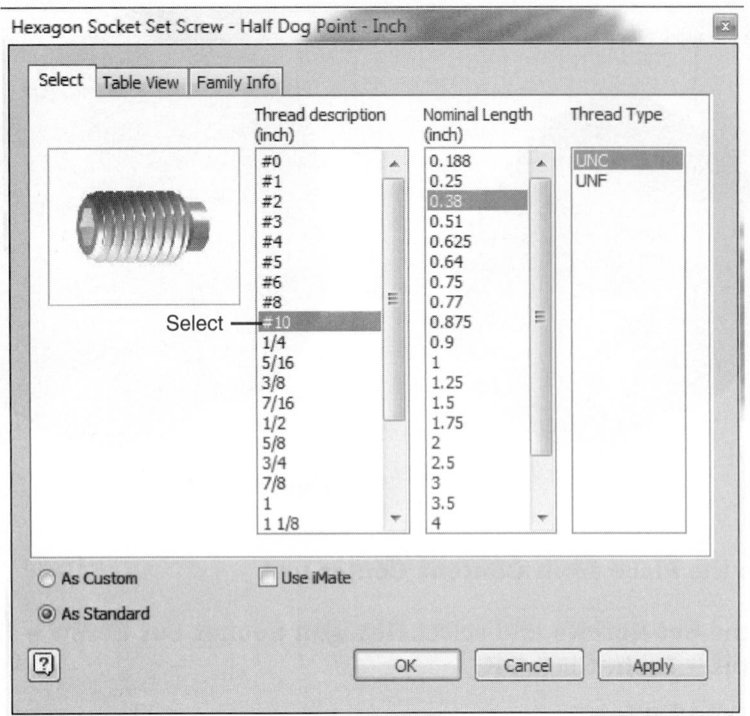

Figure 12-23

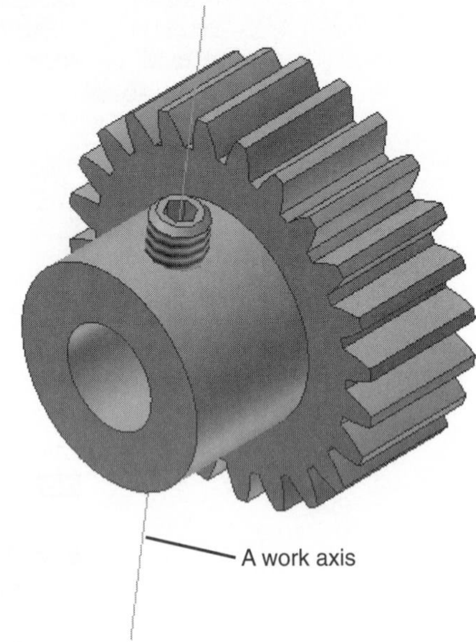

Figure 12-24

Gear Ratios

gear ratio: The ratio of number of teeth on the larger of two meshing gears to the number on the smaller gear.

The speed ratio between two gears is determined by the number of teeth on each gear. For example, if two spur gears have 60 teeth and 20 teeth, respectively, the ***gear ratio*** is 60/20 = 3/1. See Figure 12-25. Thus, if the larger gear with 60 teeth is rotating at 30 revolutions per minute (RPM), the smaller gear with 20 teeth will rotate at 90 RPM. It is a general rule of thumb not to use spur gear ratios greater than 5:1.

Bevel gear ratios are also determined by the number of teeth on each gear. Again, the general rule of thumb is not to use gear ratios greater than 5:1. See Figure 12-26.

The gear ratio for a worm and worm gear combination is determined by the number of teeth on the worm gear. The cylinder-shaped gear is

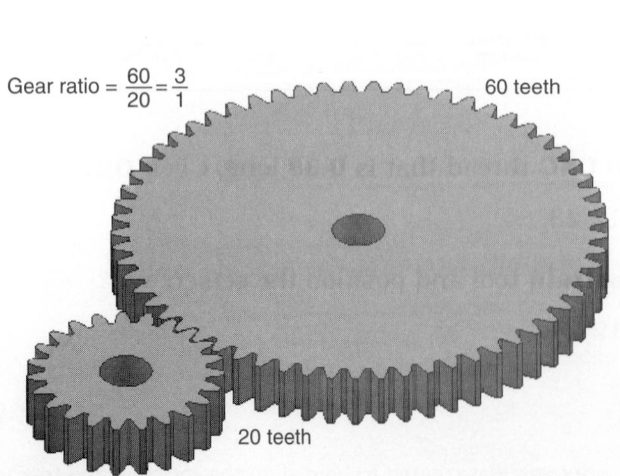

Figure 12-25

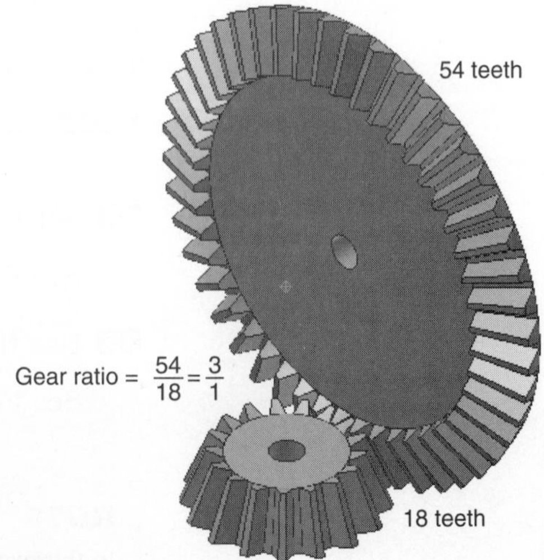

Figure 12-26

called a *worm*, and the round gear is called a *worm gear*. The worm is assumed to be 1 tooth. If a worm gear is meshed with a worm, and the worm gear has 42 teeth, the gear ratio will be 42:1. See Figure 12-27.

Figure 12-27

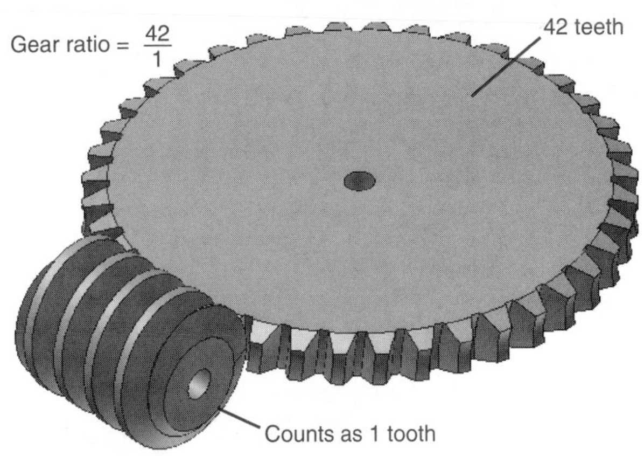

Gear Trains

gear train: The combination of more than two meshing gears.

When more than two gears are used in a design the combination is called a ***gear train***. Figure 12-28 shows a gear train that contains four gears: two 20-tooth gears, one 40-tooth, and one 60-tooth. The speed ratio between the input RPM and output RPM is determined by multiplying the individual gear ratios together. Observe that the 40-tooth gear and one of the 20-tooth gears are mounted on the same shaft. There is no speed ratio between these two gears, as they have the same angular velocity. The speed ratio is

$$\left(\frac{40}{20}\right)\left(\frac{60}{20}\right) = \frac{6}{1}$$

For an input speed of 1750 RPM, the output speed would be

$$1750\left(\frac{1}{6}\right) = 292 \text{ RPM}$$

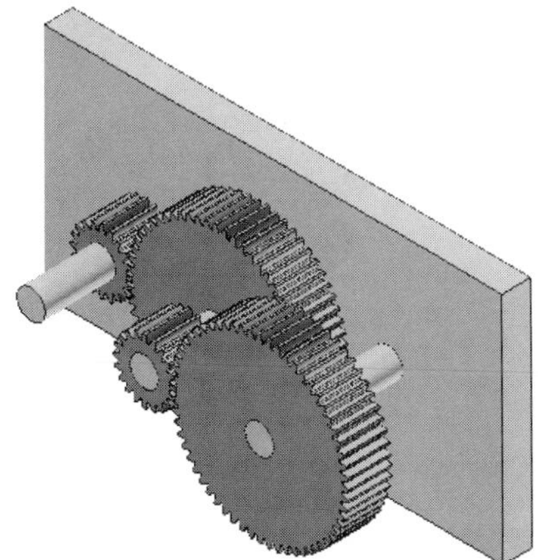

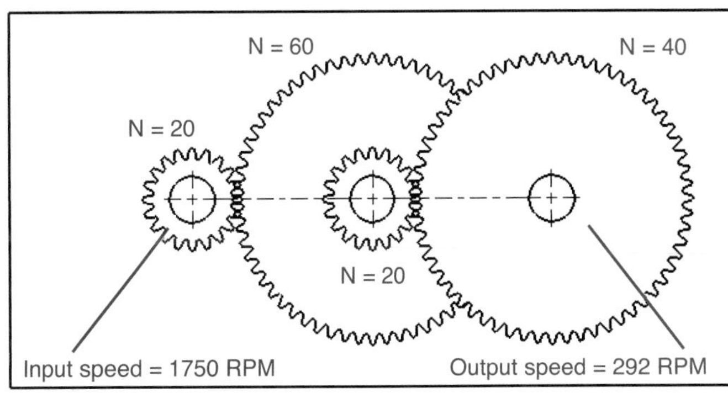

Figure 12-28

Figure 12-29 shows another gear train that includes six gears: three 20-tooth gears and three 60-tooth gears. The speed ratio between input and output speeds is

$$\left(\frac{60}{20}\right)\left(\frac{60}{20}\right)\left(\frac{60}{20}\right) = \frac{27}{1}$$

For an input speed of 1750 RPM, the output speed would be

$$\frac{1750}{27} = 64.8 \text{ RPM}$$

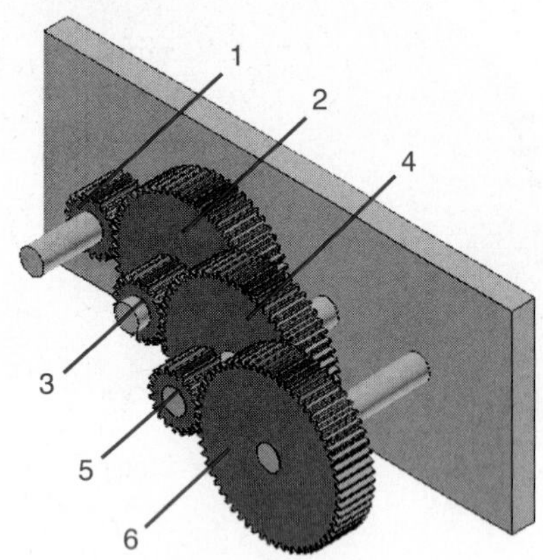

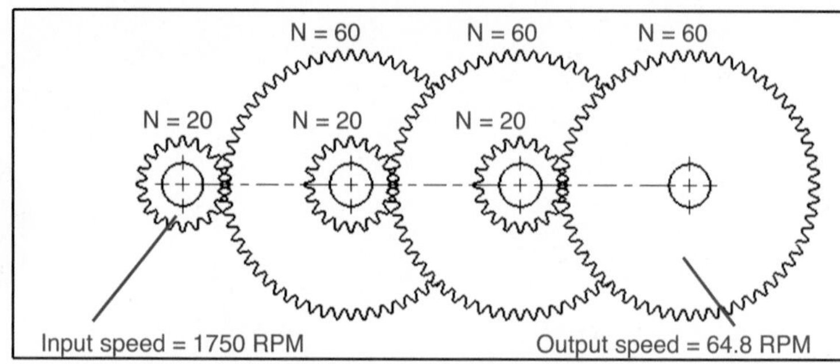

Figure 12-29

Gear Direction

Meshing gears always rotate in opposite directions. If gear 1 in Figure 12-29 were to rotate clockwise (CW), then gear 2 would rotate counterclockwise (CCW). Gear 3 would also rotate CCW, driving gear 4 in a CW direction. Gear 5 would rotate in the CW direction and drive gear 6 in the CCW direction. A gear called an ***idler*** may be added to a gear train for the sole purpose of changing the direction of the final rotation. Idler gears are usually identical with one of the gears with which they are meshing so as not to affect the final speed ratio. See Figure 12-30.

idler gear: A gear added to a gear train for the sole purpose of changing the direction of the final rotation.

Figure 12-30

An idler gear. An idler does not change the speed ratio between input and output but does affect the output direction.

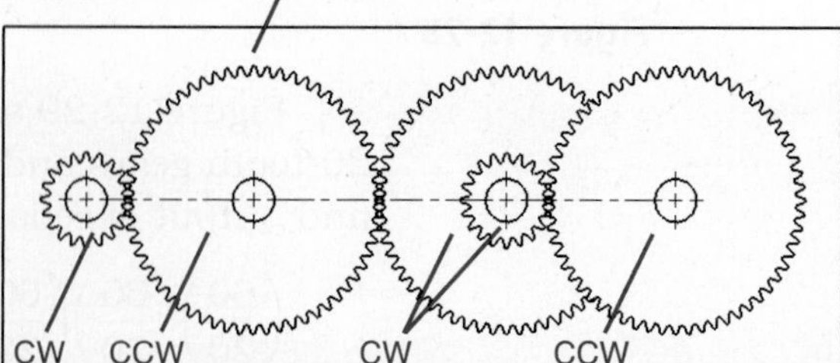

Gears with Keyways

Figure 12-31 shows a hubless spur gear. It is to be joined to a shaft using a square key. A keyway must be added to the gear. The gear has a bore of Ø.5000 in. and a face width of .500 in. Its pressure angle is 14.5, it has 48 teeth, and a diametral pitch of 24. It was drawn using the **Gear** tools on the **Power Transmission** panel under the **Design** tab, on a **Standard (in).iam** format drawing.

Figure 12-31

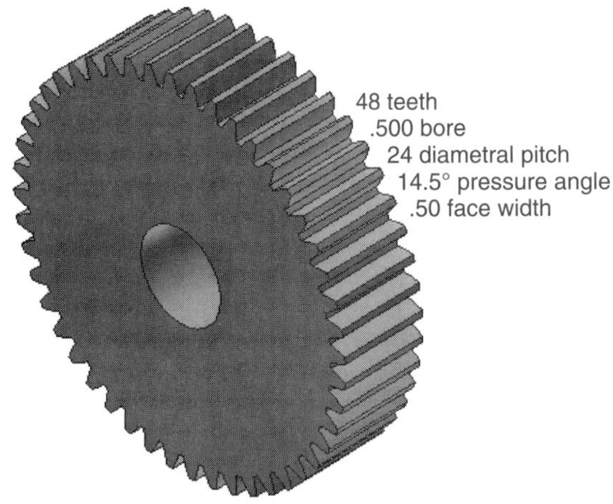

EXERCISE 12-2 **Adding a Keyway to a Gear**

First, determine the key that will be inserted into the gear. The information about the gear contained in the **Content Center** will specify the size of the keyway.

1 Click the **Place from Content Center** tool.

See Figure 12-32.

Figure 12-32

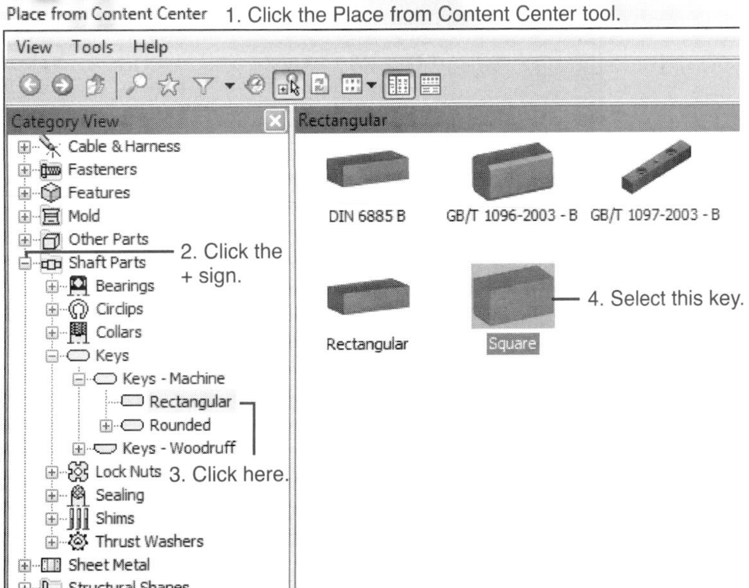

2 Select a **Square** key.

See Figure 12-33.

The bore of the gear is Ø.500, so the **0.4375 – 0.5625** shaft diameter range was selected, yielding a 1/8 × 1/8 width and height for the key. A nominal key length of **0.625** was selected.

3 Click the **Table View** tab and access **All Columns.**

See Figure 12-34. The table specifies a **Parallel Key Width** of **0.125** and a **Parallel Key Groove** of **0.0625.**

Figure 12-33

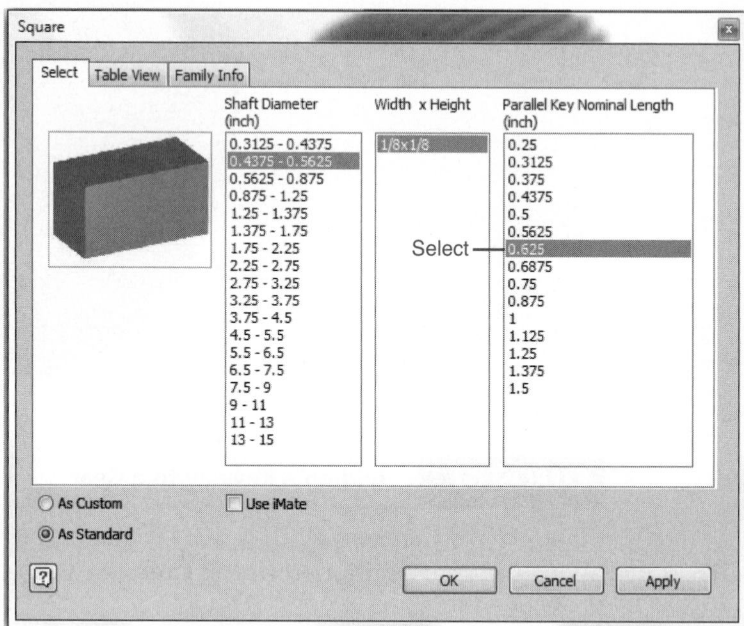

Figure 12-34

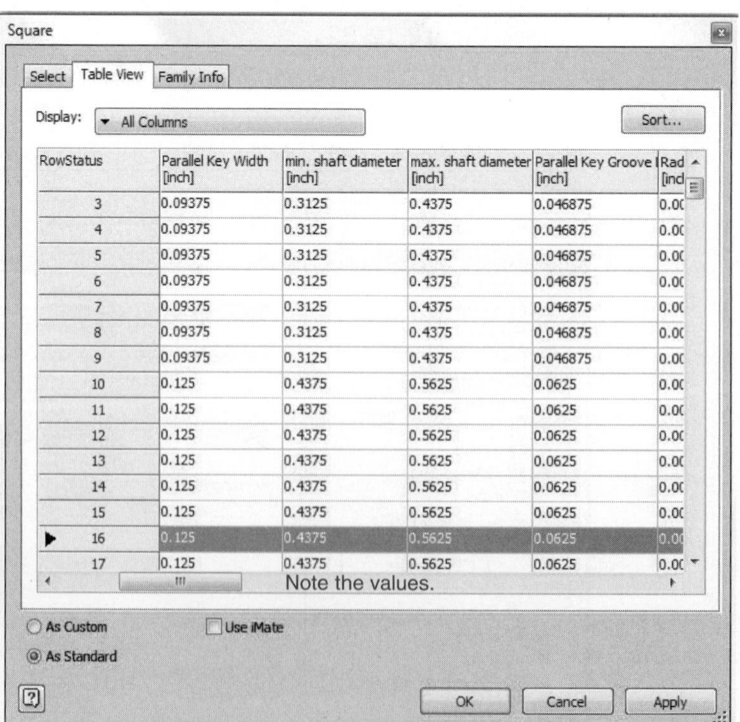

RowStatus	Parallel Key Width [inch]	min. shaft diameter [inch]	max. shaft diameter [inch]	Parallel Key Groove [inch]	Rad [incl
3	0.09375	0.3125	0.4375	0.046875	0.0(
4	0.09375	0.3125	0.4375	0.046875	0.0(
5	0.09375	0.3125	0.4375	0.046875	0.0(
6	0.09375	0.3125	0.4375	0.046875	0.0(
7	0.09375	0.3125	0.4375	0.046875	0.0(
8	0.09375	0.3125	0.4375	0.046875	0.0(
9	0.09375	0.3125	0.4375	0.046875	0.0(
10	0.125	0.4375	0.5625	0.0625	0.0(
11	0.125	0.4375	0.5625	0.0625	0.0(
12	0.125	0.4375	0.5625	0.0625	0.0(
13	0.125	0.4375	0.5625	0.0625	0.0(
14	0.125	0.4375	0.5625	0.0625	0.0(
15	0.125	0.4375	0.5625	0.0625	0.0(
16	0.125	0.4375	0.5625	0.0625	0.0(
17	0.125	0.4375	0.5625	0.0625	0.0(

4 Right-click the **Gear** callout in the browser box, select the **Edit** option, create a new sketch plane on the front surface of the gear, and draw a two-point rectangle as shown.

See Figure 12-35. The **.3130** value was derived from the bore radius of **.250** and the **Parallel Key Groove** requirement of **0.0625.** The width of the keyway is .125 (1/8).

5 Right-click the mouse and select the **Finish 2D Sketch** option.

See Figure 12-36.

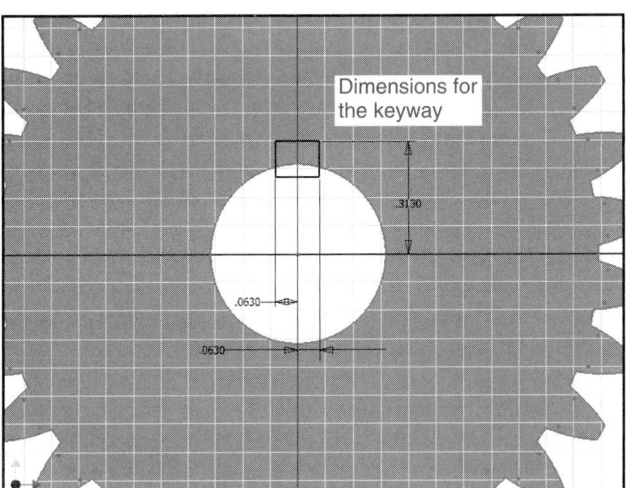

Figure 12-35

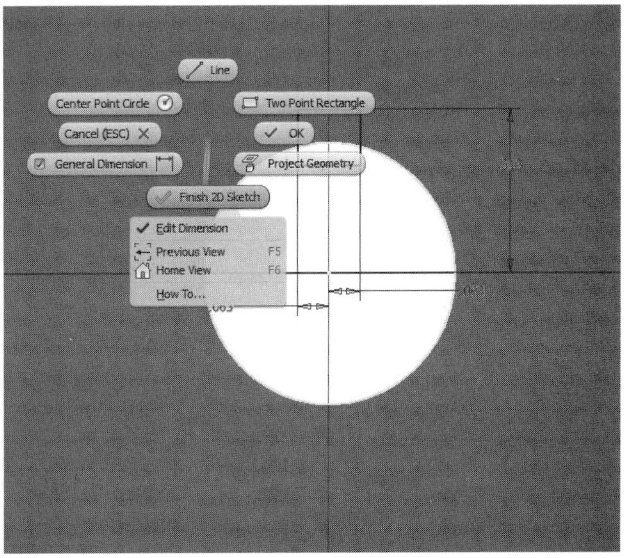

Figure 12-36

6 Use the **Extrude** tool and cut out the rectangle, producing the keyway.

See Figure 12-37.

7 Right-click the mouse and select the **Finish Edit** option.

8 Assemble the square key into the gear's keyway.

See Figure 12-38.

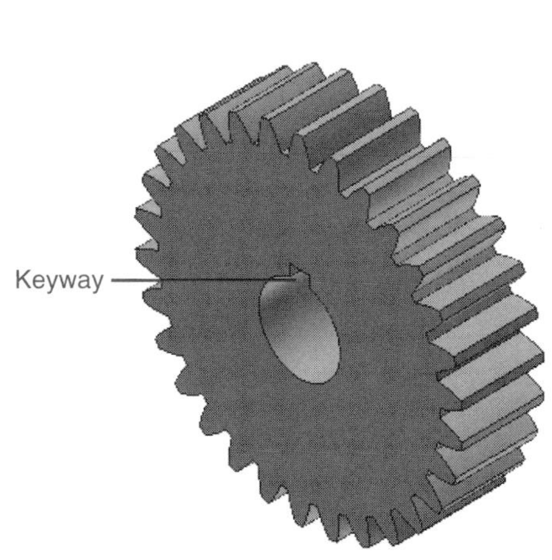

Figure 12-37

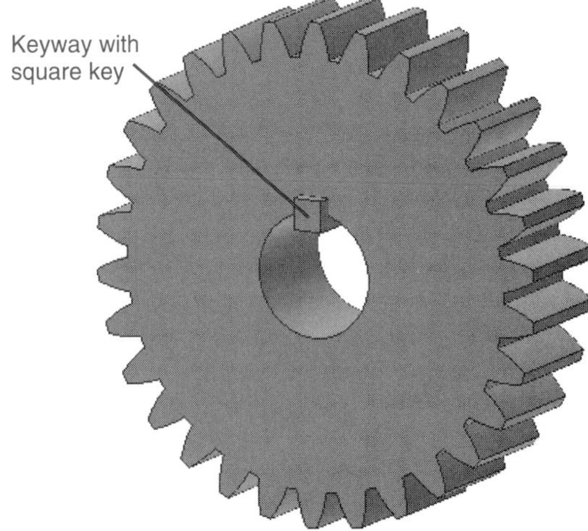

Figure 12-38

NOTE

Chapter 10 shows how to draw a keyway in a shaft.

Gear Assemblies

This section shows how to draw gear assemblies. Two meshing gears will be mounted onto a support plate using two shafts. The support plate and gear shafts were drawn using the dimensions shown in Figure 12-39.

Figure 12-39

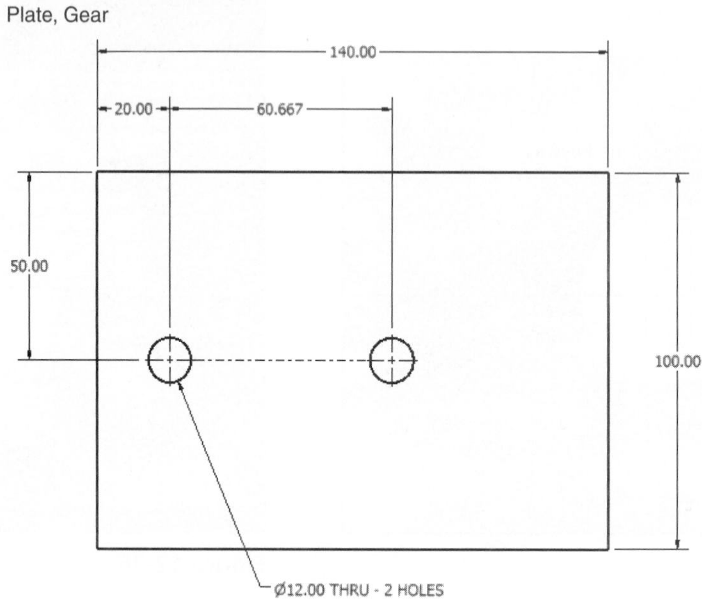

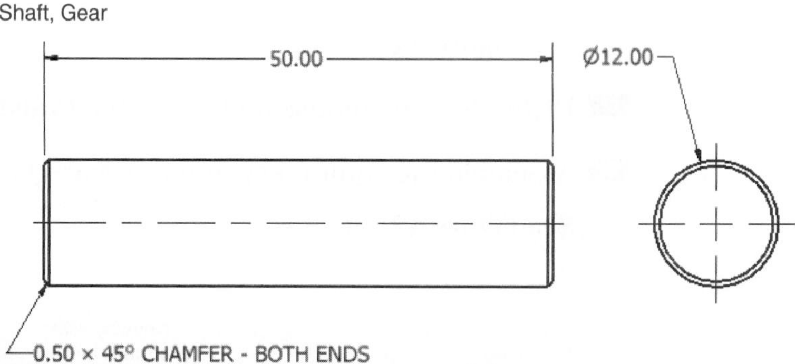

EXERCISE 12-3 **Drawing a Gear Assembly**

1. Create a new drawing using the **Standard (mm).iam** format.
2. Access the **Spur Gear** tool on the **Power Transmission** panel under the **Design** tab.
3. Set the values for the dimensions of the gear in the **Spur Gears Component Generator** dialog box as shown in Figure 12-40.
4. Finish the gear drawing.

   Figure 12-41 shows the finished gears.

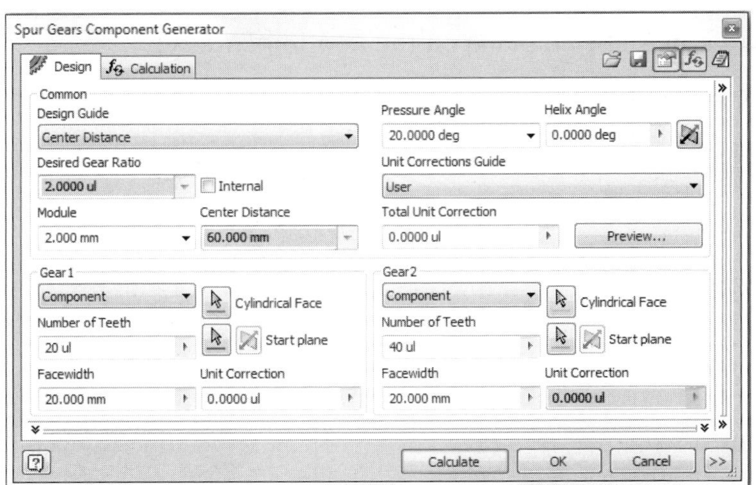

Figure 12-40

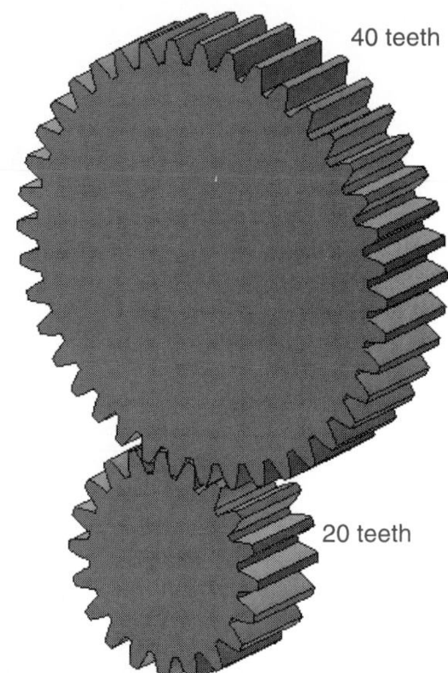

Figure 12-41

5 Click the + **sign** next to the **Spur Gears:1** heading in the browser box, right-click **Spur Gear1:1,** and select the **Isolate** option.

6 Right-click the **20-tooth** gear and select the **Edit** option.

7 Create a new sketch plane and draw a **Ø20.00** circle. Right-click the mouse and select the **Finish 2D Sketch** option.

See Figure 12-42.

Figure 12-42

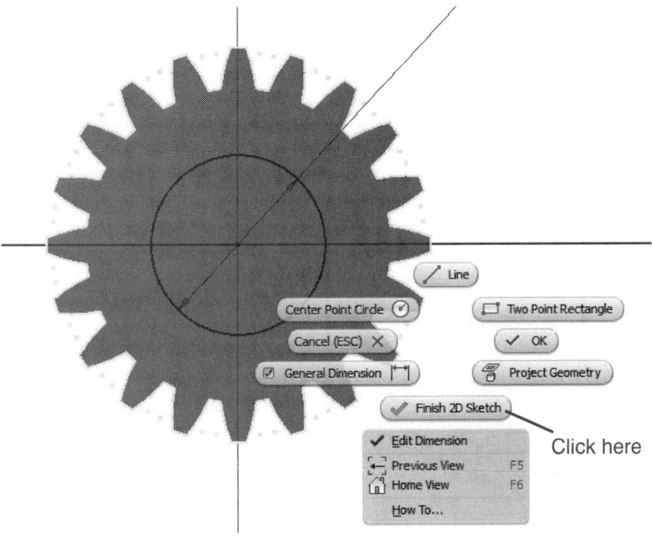

NOTE

See Exercise 12-1 for a more detailed explanation of how to draw gear hubs.

TIP

Do not select the **Finish Edit** option. Remember that you are working only on the 20-tooth gear, so you want to stay in that sketch mode.

8 Extrude the Ø**20** circle through a distance of **12 mm.**

See Figure 12-43.

9 Create a new sketch plane on the new top surface.

See Figure 12-44.

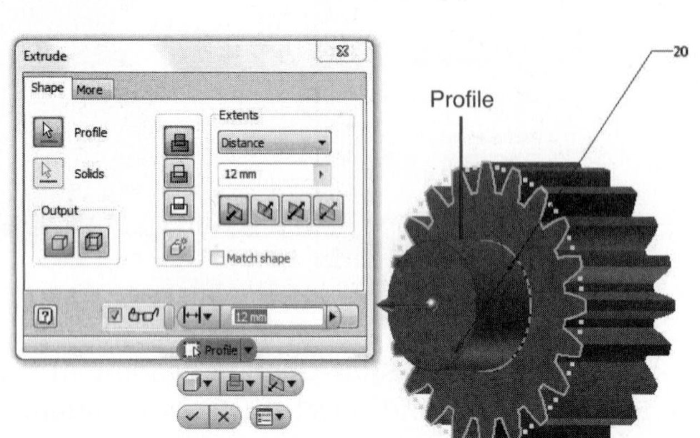

Figure 12-43

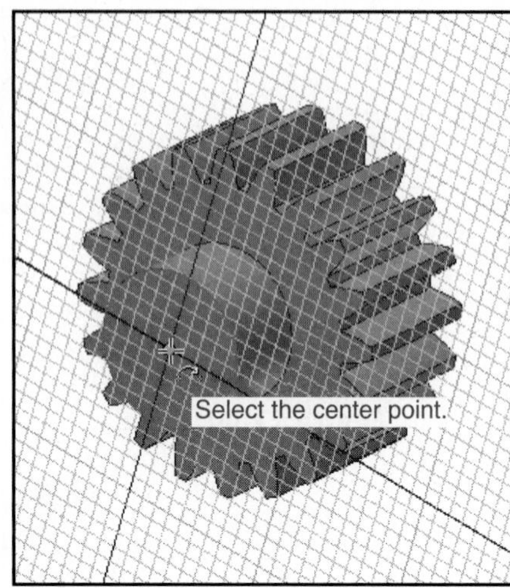

Figure 12-44

10 Draw a Ø**12.00** hole through the gear.

See Figure 12-45.

11 Create a new work plane tangent to the gear's hub. Locate a center point for a hole.

See Figure 12-46.

Figure 12-45

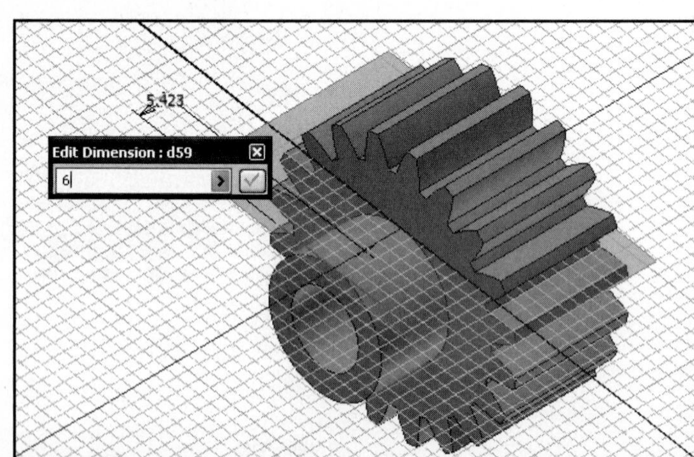

Figure 12-46

12 Draw a threaded hole. Use an **M4** × **0.7** metric thread.

See Figure 12-47.

13 Right-click the mouse and select the **Finish Edit** option.

Figure 12-47

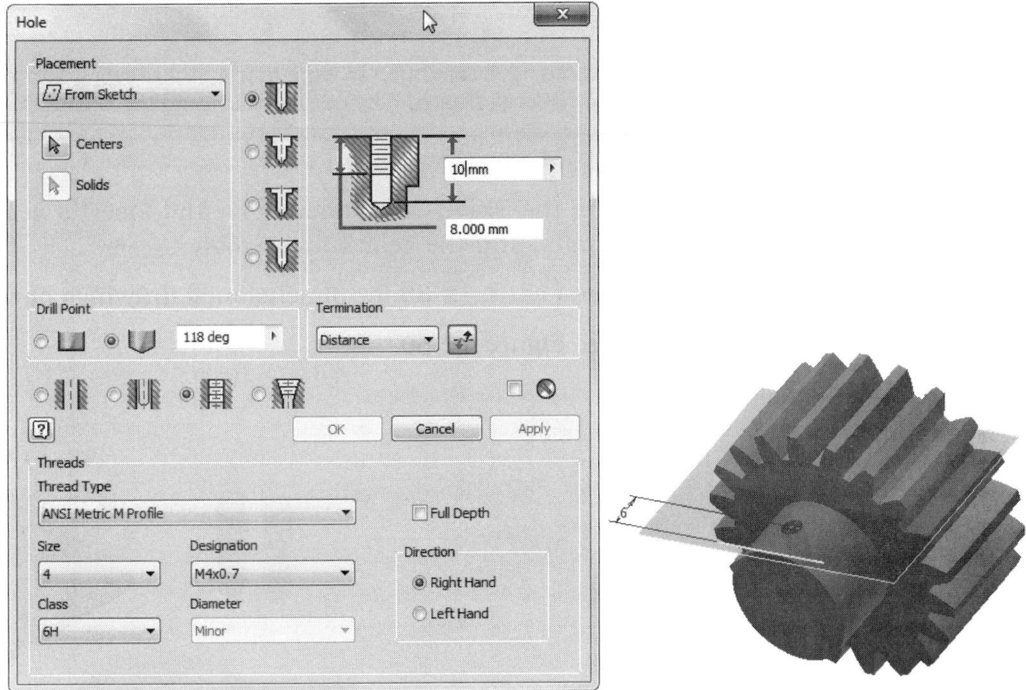

14 Right-click the mouse and click the **Undo Isolate** option.

The 40-tooth gear will reappear.

See Figure 12-48.

15 Isolate the 40-tooth gear and create a hub, bore, and **M4 × 0.7** threaded hole using the procedure described for the 20-tooth gear.

See Figure 12-49.

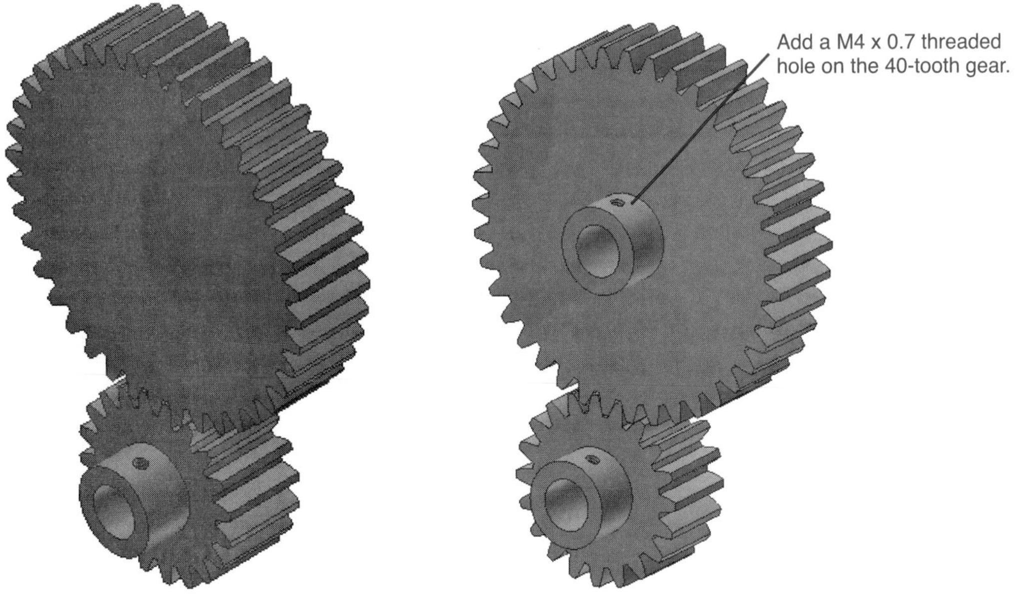

Figure 12-48

Figure 12-49

16 Save the two gears as **Gear Subassembly.**

17 Start a new drawing using the **Standard (mm).iam** format.

> **TIP**
>
> You could have continued to work in the gear drawing and either inserted the existing components or created new components using the **Create Component** tool.

18 Use the **Place Component** tool and insert a gear support plate, two shafts, and the gear subassembly.

See Figure 12-39 for dimensioned drawings of the plate and shafts.

See Figure 12-50.

Figure 12-50

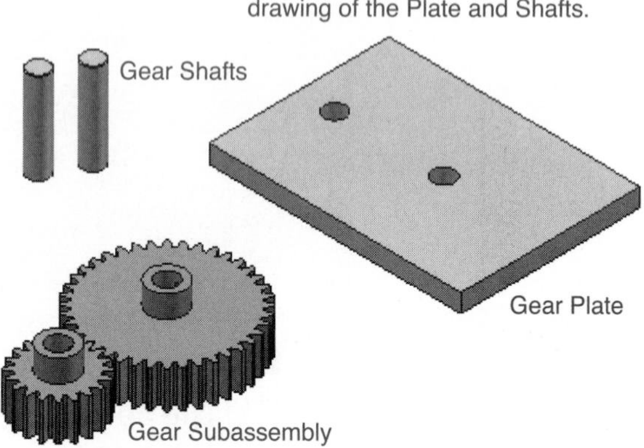
See Figure 12-39 for a dimensioned drawing of the Plate and Shafts.
Gear Shafts
Gear Plate
Gear Subassembly

19 Use the **Constrain** tool and assemble the gears, shafts, and support plate.

See Figure 12-51.

Figure 12-51

20 Click the **Place from Content Center** tool located under the **Assemble** tab.

Right-click the mouse and access the **Content Center,** click the **Set Screws** option, and select a **DIN EN 24766 Set Screw** with a point that has an **M4** thread and is **4** long.

See Figures 12-52 and 12-53.

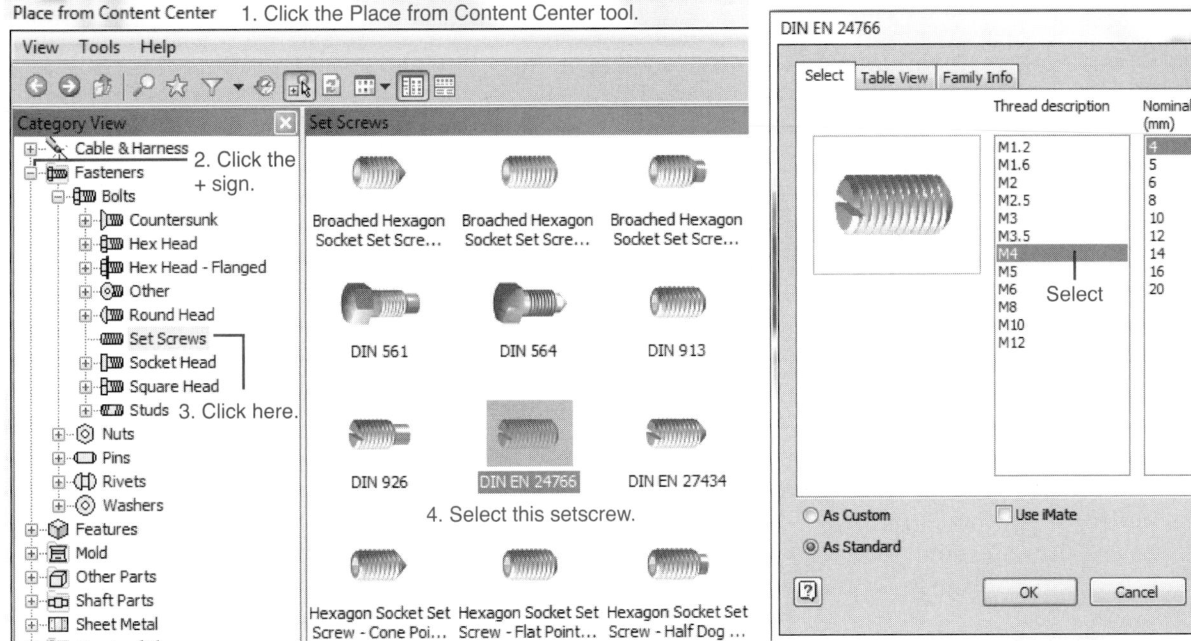

Figure 12-52

Figure 12-53

21 Click **OK.**

Add two setscrews to the assembly drawing. See Figure 12-54.

22 Insert the setscrews into the gear hubs.

Figure 12-55 shows the finished assembly.

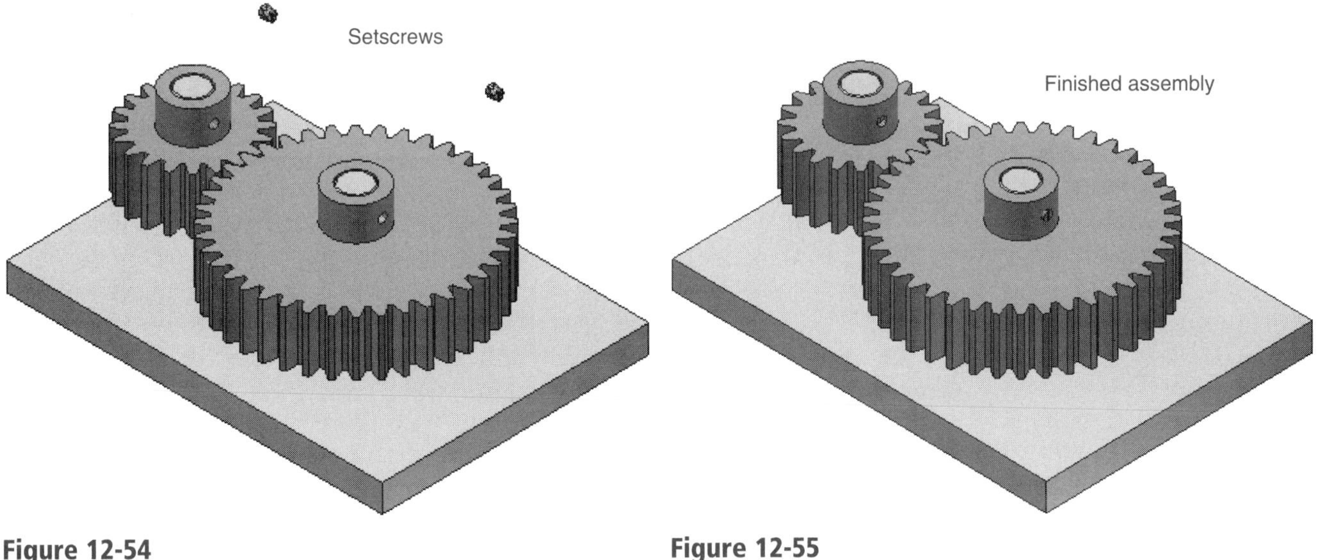

Figure 12-54

Figure 12-55

> **TIP**
>
> Use work planes and work axes if necessary to insert the setscrews into the hubs. Remember to assemble in the assembly drawing and not on individual components.

Bevel Gears

bevel gears: Conical-shaped gears that have intersecting axes.

Bevel gears are conical-shaped gears that have intersecting axes. Spur gears have parallel axes. Figure 12-56 shows a set of meshing bevel gears.

Figure 12-56

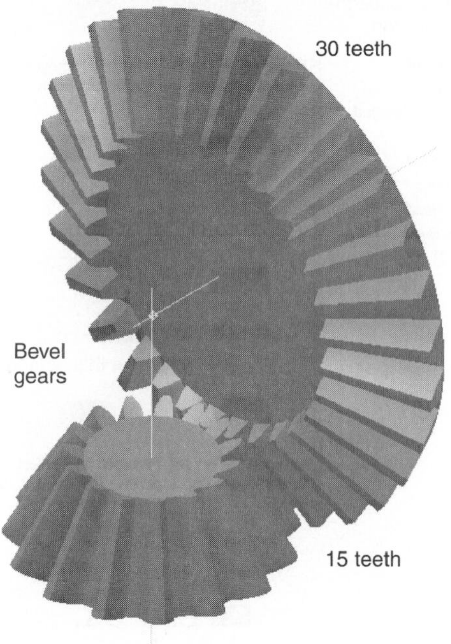

EXERCISE 12-4 **Drawing Bevel Gears**

Draw a set of bevel gears with 15 and 30 teeth, respectively. Define the module as 3.000, the shaft angle as 90°, the pressure angle as 20, and the face width as 20.00.

1. Start a new drawing using the **Standard (mm).iam** format.
2. Click the **Design** tab, and click the **Gear** tool located on the **Power Transmission** panel.

   See Figure 12-57.
3. Click the **Bevel Gear** tool.

   The **Bevel Gear** option is a flyout from the **Spur Gear** tool. The **Bevel Gear Component Generator** dialog box will appear. See Figure 12-57.
4. Enter **15** teeth for the small gear, **30** teeth for the large gear, and a **Module** value of **3.00.**
5. Click **Calculate.**
6. Click **OK.**

   The **File Naming** box will appear.
7. Click **OK.**

   The gears will appear on the screen.

Figure 12-57

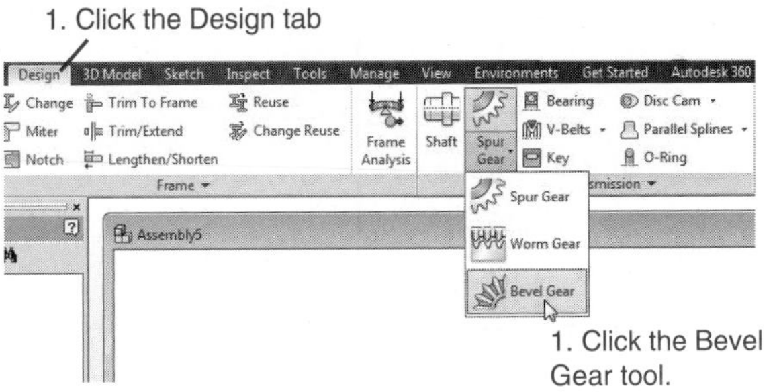

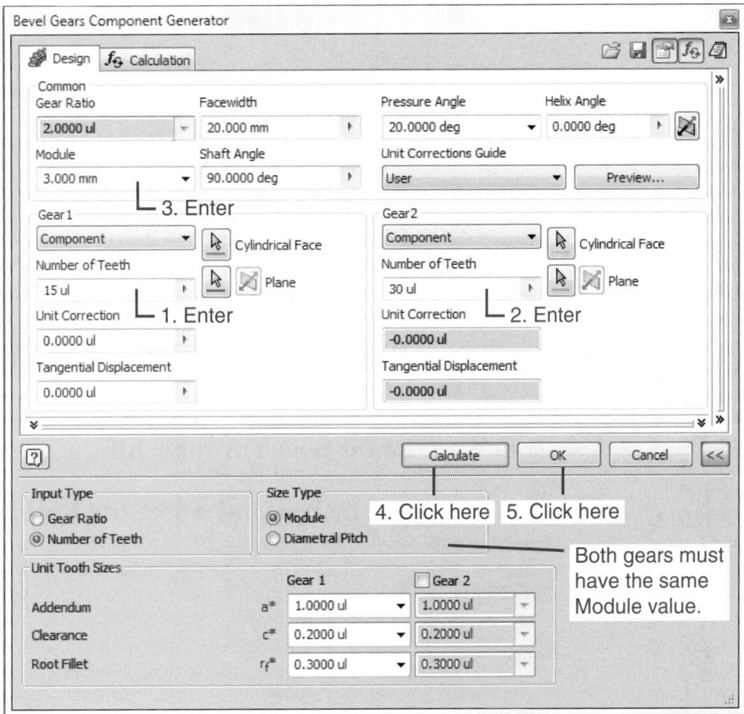

EXERCISE 12-5 Adding Hubs to Bevel Gears

1. Rotate the gears so that the bottom surface of the small gear is in view.

2. Click the + **sign** to the left of the **Bevel Gears1:1** heading in the browser box, right-click on the **Bevel Gear1:1** heading, and select the **Edit** option.

   See Figure 12-58.

3. Create a new sketch plane on the bottom surface of the smaller gear.

   See Figure 12-59.

4. Draw a **Ø24** circle on the new sketch plane.

5. Right-click the mouse and select the **Finish 2D Sketch** option.

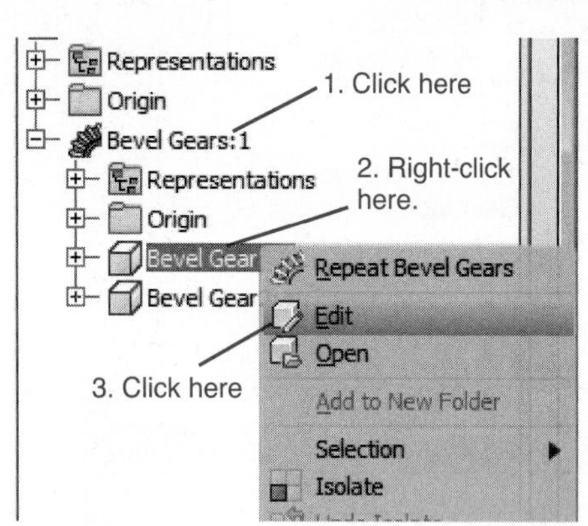

Figure 12-58

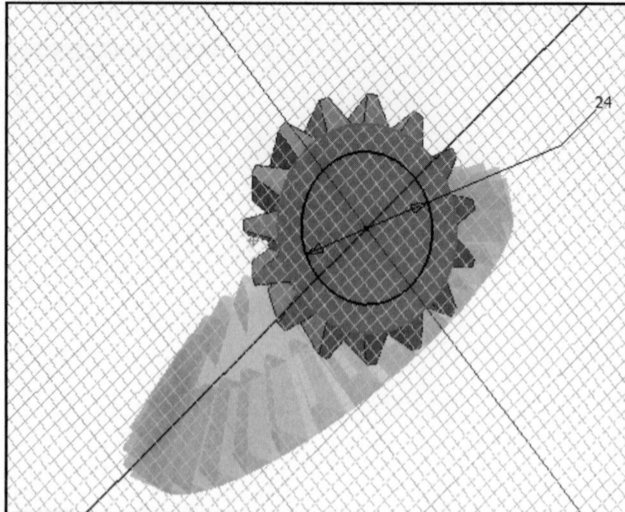

Figure 12-59

See Figure 12-60. Do not click the **Finish Edit** option. You are still working on the smaller gear, so continue working on the sketch.

6 Extrude the circle **12 mm.**

7 Right-click the mouse and create a new sketch plane on the top surface of the hub.

8 Create a **Ø12.00** hole **Through All.**

9 Right-click the mouse and select the **Finish Edit** option. Click the **Isometric** tool.

See Figure 12-61.

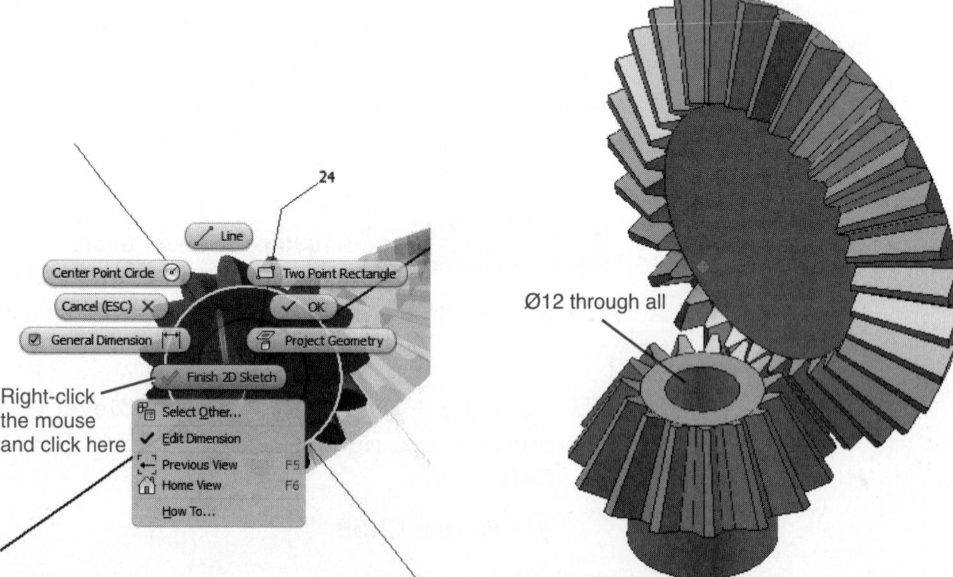

Figure 12-60

Figure 12-61

10 Repeat the procedure for the larger gear. Make the hub **15 mm** high with a **Ø15** hole.

See Figure 12-62. Figure 12-63 shows the finished bevel gears.

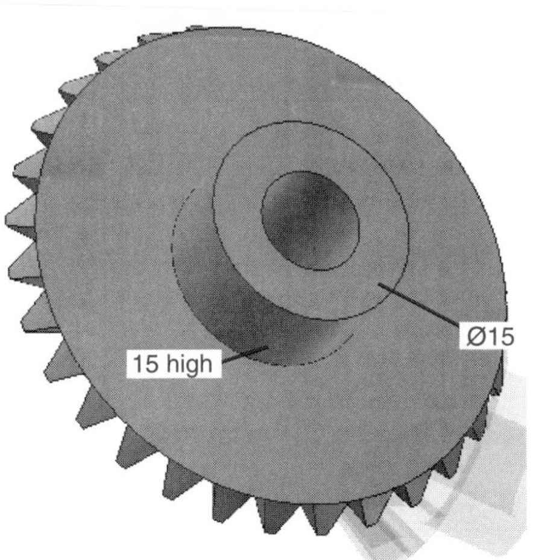

Figure 12-62

Figure 12-63

EXERCISE 12-6 Adding Keyways to Bevel Gears

Before we can draw keyways, we must determine the size of the key. In the previous section the gears were given bore diameters of Ø12 and Ø15, so keys must be selected to match those diameters.

1. Click the **Assemble** tab and click the **Place from Content Center** tool.
2. Select the **Shaft Parts** option, **Keys, Keys – Machine,** and **Rectangular.**
3. Select the **DIN 6885 B** square key.

   See Figure 12-64.

Figure 12-64

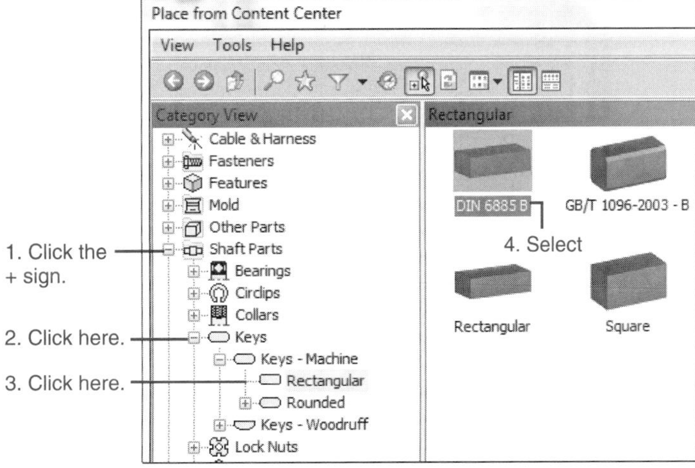

4. Select the **10 - 12** shaft diameter input.

   This will generate a key size of **4 × 4.**
5. Select a **20** nominal length.

   See Figure 12-65.

Figure 12-65

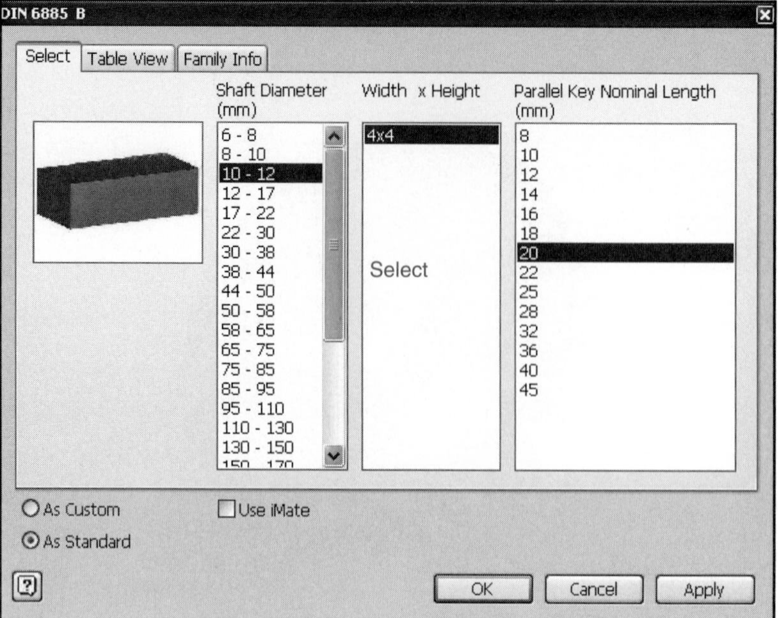

6 Click the **Table View** tab.

See Figure 12-66.

7 Scroll the table to determine the recommended **Hub Depth.**

In this example the value is **1.8.**

8 Close the **Content Center** and return to the bevel gear drawing.

9 Select the smaller gear, right-click the mouse, and select the **Edit** option. Create a new sketch plane on the top surface of the gear.

See Figure 12-67. Create a 2D view of the sketch plane if necessary.

Figure 12-66

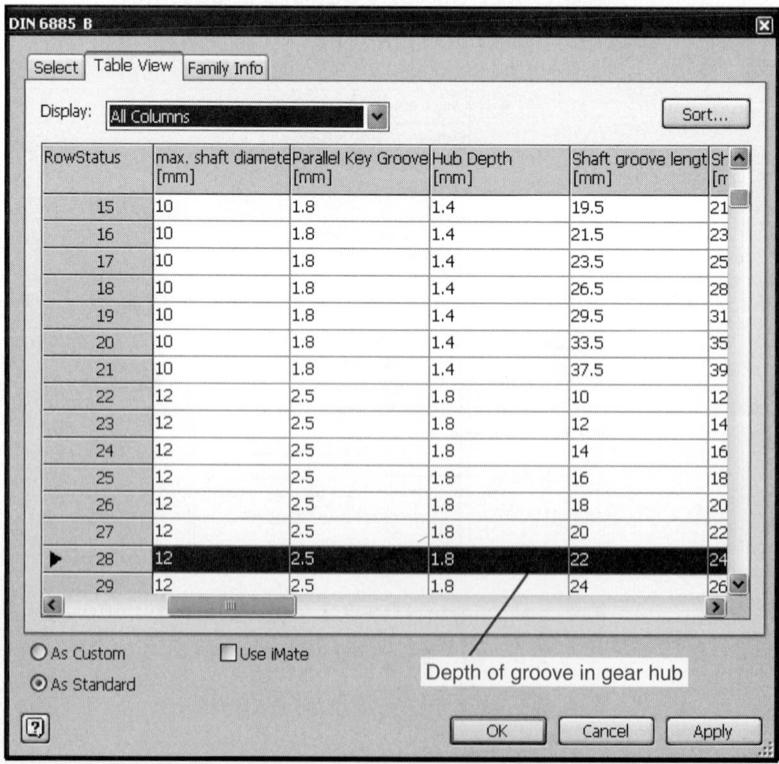

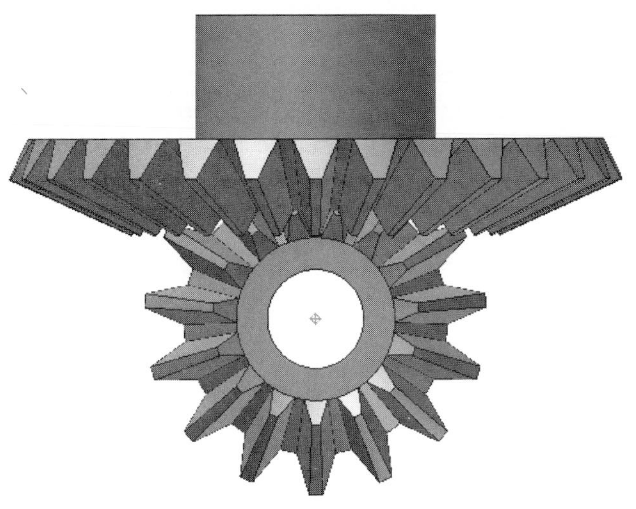

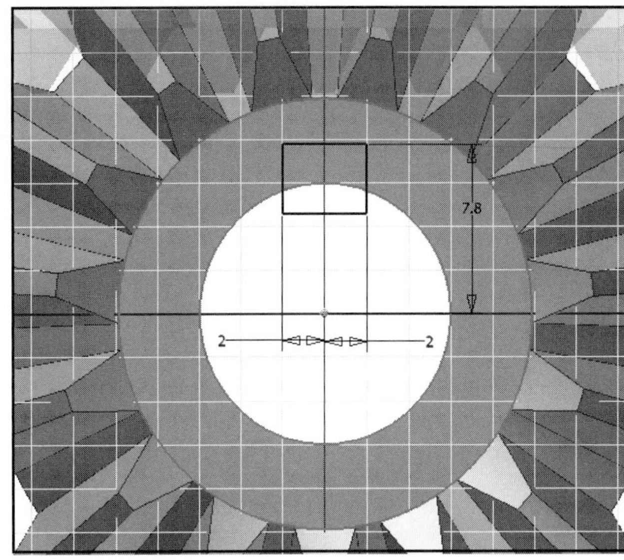

Figure 12-67

10 Draw a rectangle as shown.

The width value of **4** matches the key width of 4 (no tolerances were factored in). The **7.8** value was derived from adding the bore's radius (6.0) to the **Hub Depth** value given in the table (1.8). The keyway depth in the shaft is **2.5,** for a total key size of **4.3,** or slightly larger than the key height of 4. These values will vary when tolerances are considered.

11 Right-click the mouse, click the **Finish 2D Sketch** option, click the **Extrude** tool, and cut the rectangle through the gear, producing a keyway.

12 Repeat the procedure for the larger gear.

13 Right-click the mouse and select the **Finish Edit** option.

Figure 12-68 shows the finished keyways in the bevel gears.

Figure 12-68

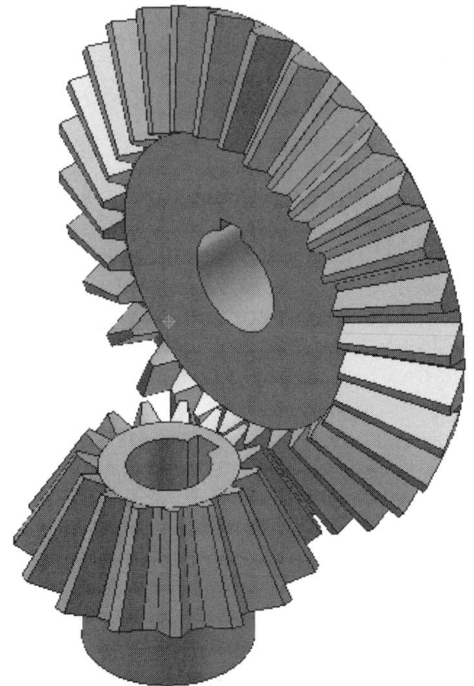

Supports for Bevel Gears

Figure 12-69 shows a dimensioned drawing of the bevel gears created in the last section. The drawing was created using the **ANSI (mm).idw** format. The two distances, 50.66 and 36.80, are needed to position the holes in the support structure.

Figure 12-69

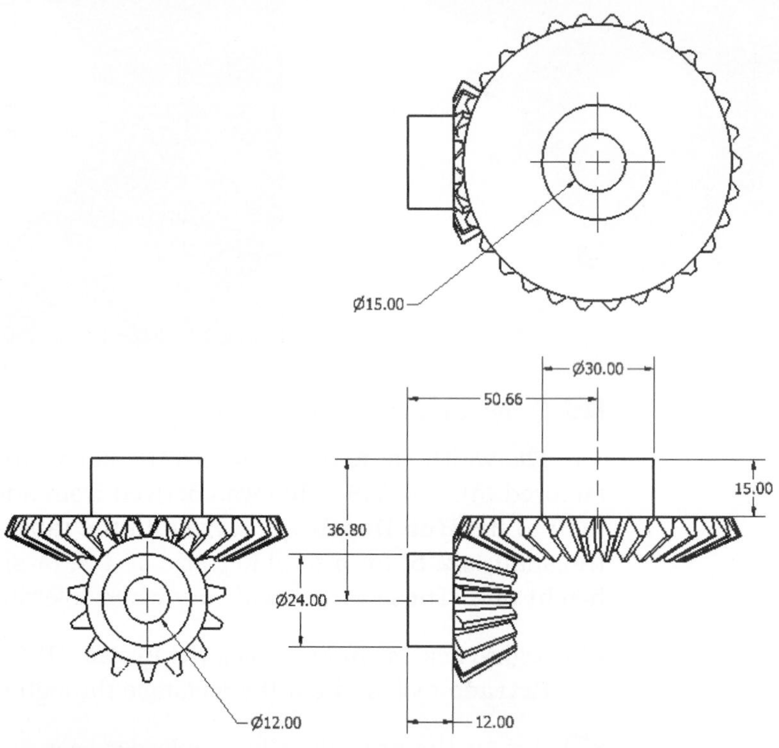

EXERCISE 12-7 **Designing a Support Structure**

1 Draw an L-shaped bracket.

In this example the horizontal portion of the bracket is **120 × 120 × 10,** and the vertical portion is **100 × 120 × 10.**

2 Draw a **Ø40** circle located as shown.

See Figure 12-70. The **55.66** value is derived from the 50.66 center distance and an additional 5 for a boss.

Figure 12-70

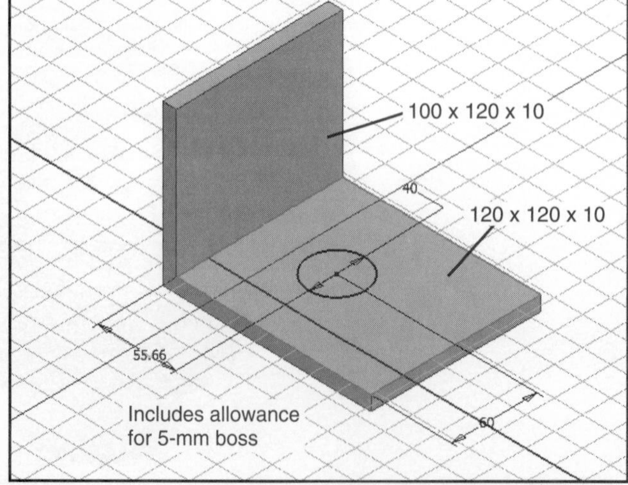

boss: A raised area that is used to add additional support for gears and shafts.

A ***boss*** is a raised area that is used to add additional support for gears and shafts. It also saves machining time, as only the surface of the boss is machined, not the entire surface.

3 Draw a boss **5** high with a **Ø15** hole.

See Figure 12-71.

4 Draw a boss **5** high with a **Ø12** hole on the vertical portion of the bracket.

See Figures 12-72 and 12-73.

5 Add any fillets required.

6 Use the **Constrain** tool and add the bevel gears to the support bracket.

See Figure 12-74.

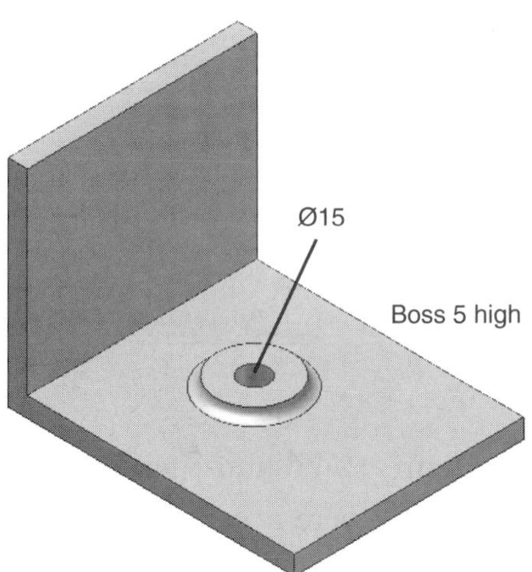

Figure 12-71

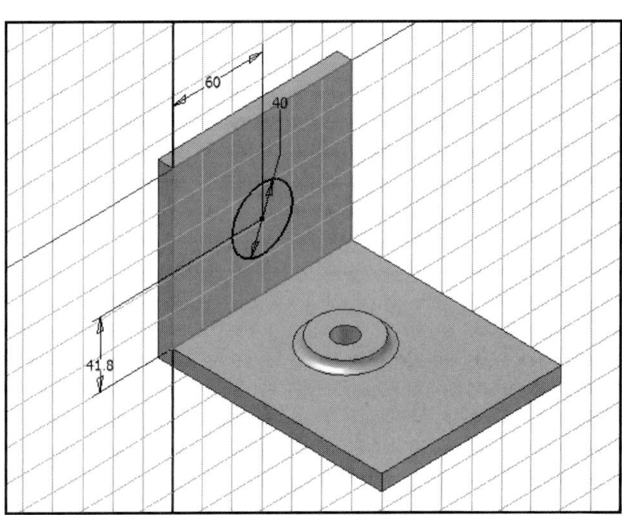

Figure 12-72

Figure 12-73

Figure 12-74

Worm Gears

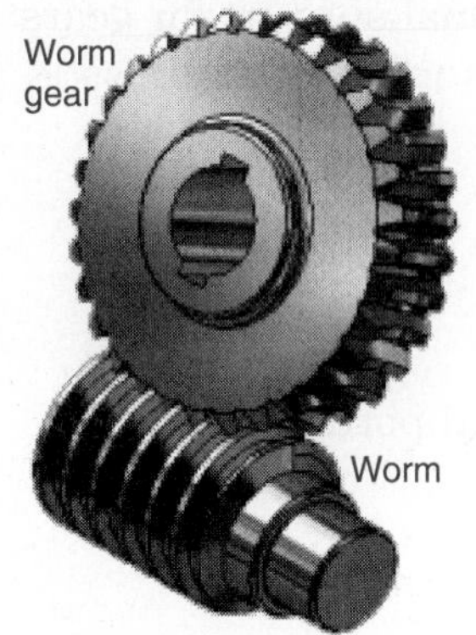

Figure 12-75

Figure 12-75 shows a worm and a worm gear. Worm gear ratios are based on the fact that the worm has a value of 1. If the worm gear has 40 teeth, the gear ratio is 40 to 1.

EXERCISE 12-8 Drawing a Worm and a Worm Gear

1. Create a new drawing using the **Standard (mm).iam** format.
2. Click the **Design** tab, then click the **Worm Gear** tool on the **Power Transmission** panel.

   The **Worm Gear** tool is a flyout from the **Spur Gear** tool.

   See Figure 12-76.

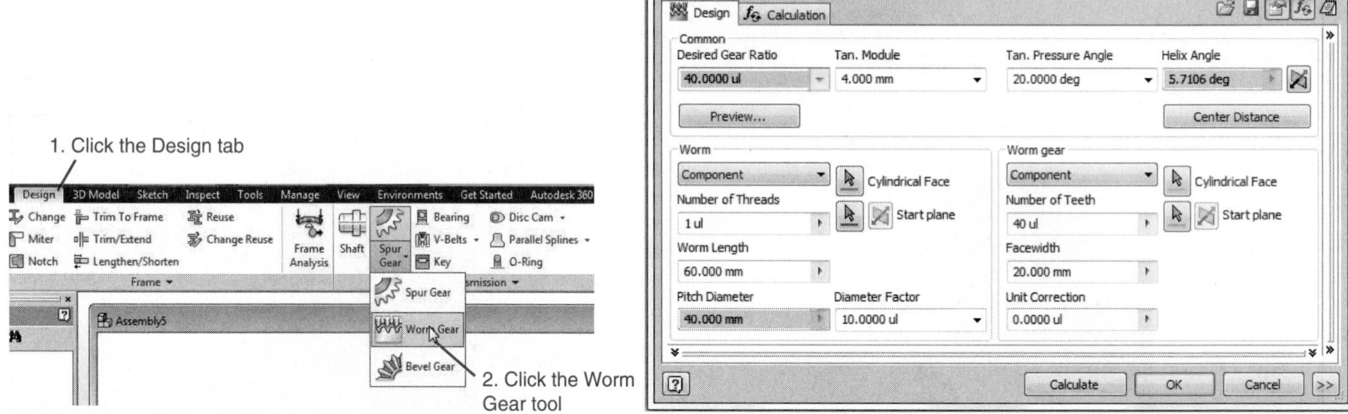

Figure 12-76

3. Enter the values shown in Figure 12-76 and click **OK.**

   In this example the default values were accepted.

   The **File Naming** dialog box will appear.
4. Click **OK.**

   Figure 12-77 shows the finished worm gears.
5. Click the **+ sign** to the left of the **Worm Gears1:1** heading in the browser box, right-click the **Worm:1** heading, and select the **Edit** option.

   See Figure 12-78.

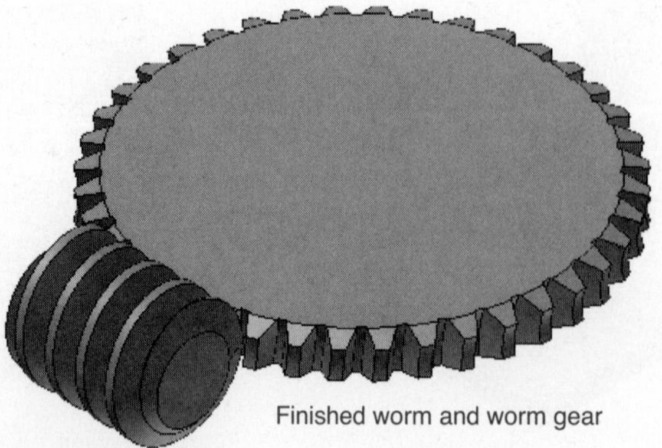

Figure 12-77

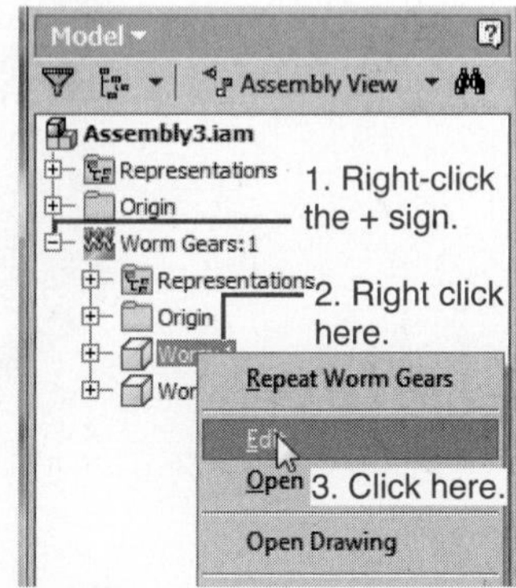

Figure 12-78

6 Right-click the end plane of the worm, create a new sketch plane on the end of the worm, and draw a **Ø20** circle.

See Figure 12-79.

7 Right-click the mouse and select the **Finish 2D Sketch** option.

8 Extrude the Ø20 circle a distance of **24** to create a hub. Create a new sketch plane on the top surface of the hub and create a **Ø12.0** hole.

9 Create a work plane tangent to the extruded hub.

See Figure 12-80.

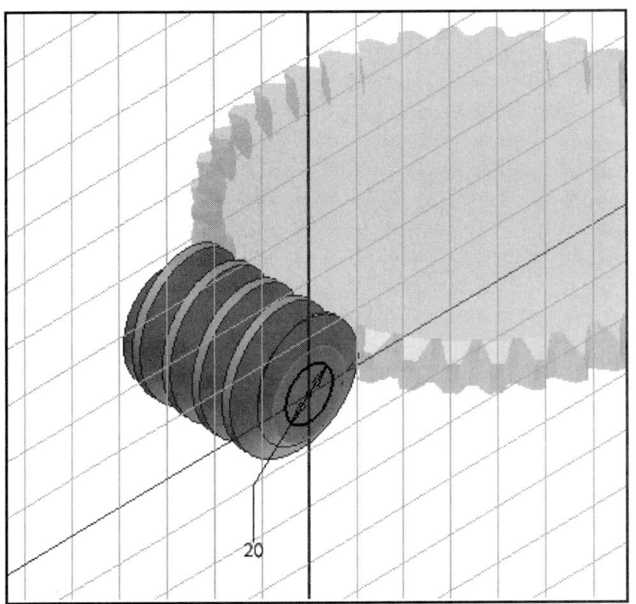

Figure 12-79

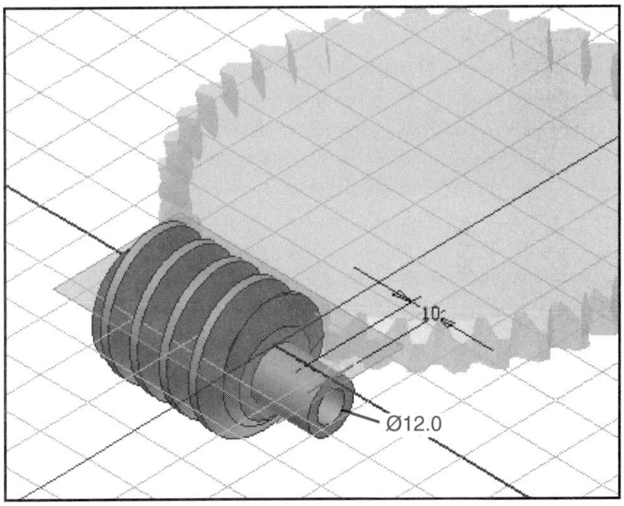

Figure 12-80

10 Add a threaded hole to the hub **10** from the top edge of the hub.

In this example an **M4** thread was added.

11 Right-click the worm and select the **Finish Edit** option.

12 Add a **Ø40** hub with a height of **20** and a **Ø20.0** hole to the worm gear using the same procedure as was used for the worm. Create an **M4** hole **10** from the top edge of the hub.

See Figure 12-81.

Figure 12-81

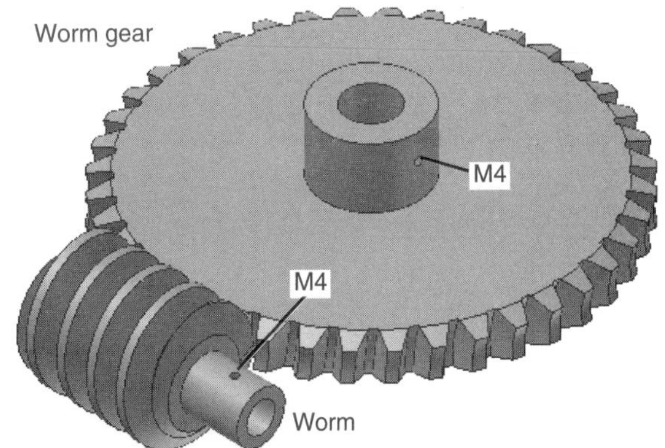

Supports for Worm Gears

Figure 12-82 shows orthographic views of a worm and worm gear drawn using the **ANSI (mm).idw** format. The **Dimension** tool was used to determine that the distance between their centers is 100.00 mm. A three-sided corner bracket was drawn. Each side of the bracket is 200 × 200 × 10, and the holes were located to match the 100.00 worm center requirement. Remember, the worm and the gear can be moved along the shafts to ensure correct alignment.

The corner bracket shown was created for clarity. It would be better to support the shafts at both ends with a boxlike structure.

Figure 12-82

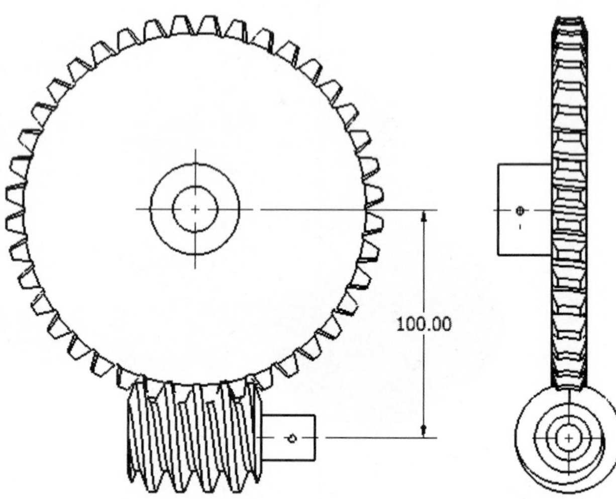
100.00

EXERCISE 12-9 **Drawing Worm Gear Supports**

1 Draw a corner bracket with two holes **100.00** apart.

Figure 12-83 shows two plates with two holes 100.00 apart vertically. Each side of the bracket is **200 × 200 × 10.**

Figure 12-83

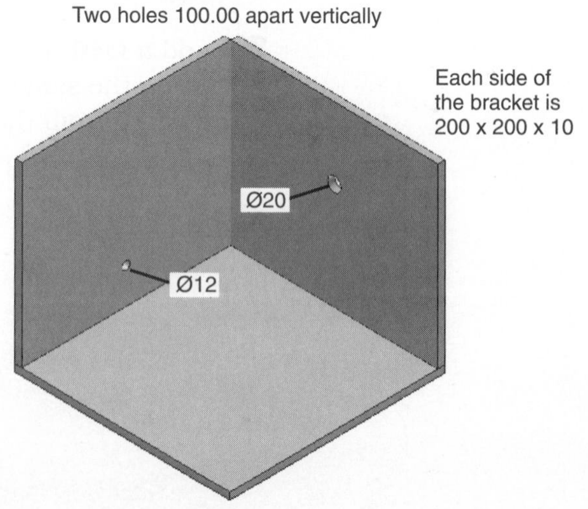
Two holes 100.00 apart vertically
Each side of the bracket is 200 x 200 x 10
Ø20
Ø12

2 Add the appropriate shafts.

See Figure 12-84.

3 Use the **Constrain** tool and insert the worm gear assembly.

Figure 12-85 shows the finished worm gear assembly support.

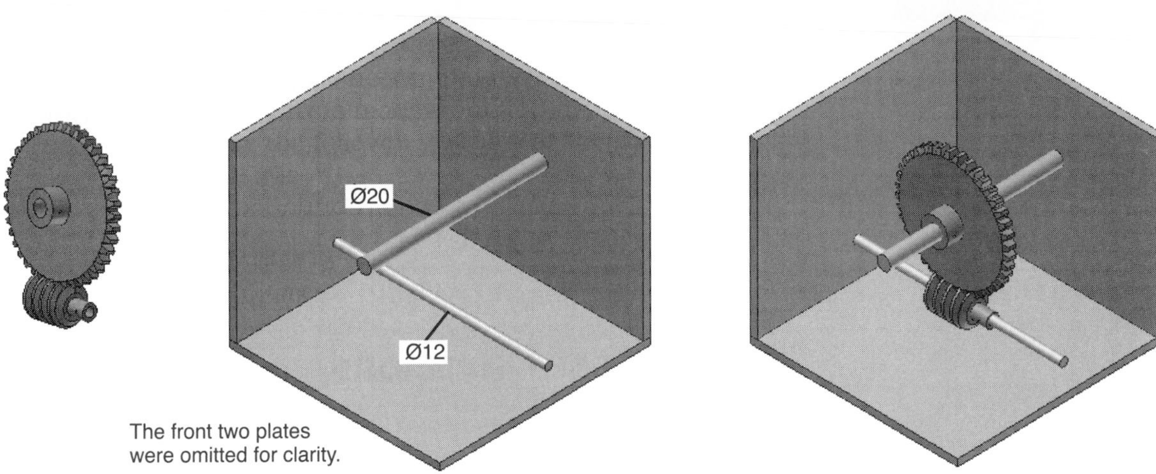
Ø20
Ø12
The front two plates
were omitted for clarity.

Figure 12-84

Figure 12-85

Chapter Summary

This chapter explained the differences among three types of gears: spur, bevel, and worm. Gear terminology, formulas, and ratios were explained.

The addition of hubs, splines, and keyways to gears was illustrated, and gears were assembled into gear trains.

Chapter Test Questions

Multiple Choice

Circle the correct answer.

1. What are the two components needed to create a worm gear assembly called?

a. Worm gear and spur gear

b. Worm and worm gear

c. Rack and worm gear

2. If a spur gear with 20 teeth is meshed with another spur gear with 50 teeth, what is the gear ratio?

a. 2:1
b. 1.75:1
c. 2.5:1
d. 3:1

3. If a worm is meshed with a worm gear that has 46 teeth, what is the gear ratio?

a. 46:1
b. 23:1
c. 192:1
d. 11:1

4. If four gears are meshed together and the first gear is turning clockwise, the fourth gear is turning

a. Clockwise

b. Counterclockwise

c. Dwelling

5. The hole in the center of a gear is called the

a. Center hole
b. Dedendum
c. Bore
d. Root diameter

6. The height of a tooth above the pitch diameter is called the

a. Face width
b. Dedendum
c. Addendum
d. Module

7. The diameter used to define the spacing of gears is called the

a. Pitch diameter
b. Diametral pitch
c. Preferred pitch
d. Center distance

8. A gear train is
   a. A group of at least six gears
   b. Any group of two or more meshing gears
   c. A group of gears with different pitches

9. The portion of a gear that protrudes from the gear's center and includes a bore is called the gear's
   a. Face
   b. Hub
   c. Backlash
   d. Root fillet

10. A listing of setscrews and keys used with gears is found in the
   a. **Design** tab
   b. **Assemble** panel
   c. **Content Center**

Matching

Match the terms in column A with the definition in column B.

Column A

a. Backlash ___________
b. Circular thickness ___________
c. Module ___________
d. Pressure angle ___________
e. Face width ___________

Column B

1. The angle between the line of action and a line tangent to the pitch circle
2. The pitch diameter divided by the number of teeth
3. The distance across a tooth as measured along the pitch circle
4. The distance from front to back along a tooth as measured perpendicular to the pitch circle
5. The difference between a tooth width and the engaging space on a meshing gear

True or False

Circle the correct answer.

1. **True or False:** A library of gears in included in the **Content Center.**
2. **True or False:** Gears must have the same pitch to mesh properly.
3. **True or False:** The term *pitch*, as used with gears manufactured using English units, is called *module* for gears manufactured using metric units.
4. **True or False:** Diametral pitch is the outside diameter of a gear.
5. **True or False:** Gear ratios are determined by the number of teeth on each gear.
6. **True or False:** The axes of beveled gears are located at 90° to each other.
7. **True or False:** Bevel gears can mesh with spur gears.
8. **True or False:** Worms have a ratio value of 1 when meshing with worm gears.
9. **True or False:** A boss is a turretlike shape usually added to castings.
10. **True or False:** The face width of a gear is its thickness.

Chapter Projects

Project 12-1: Inches

Draw two spur gears based on the following parameters. Locate the two threaded holes 90° apart as shown. On both gears the threaded hole is located halfway up the hub height.

	Gear 1	Gear 2
Number of teeth	16	48
Face width	.50	.50
Diametral pitch	24	24
Pressure angle	20	20
Hub Ø	.50	.50
Hub height	.375	.750
Bore	.250	.500
Threaded hole	0.138(#6)UNC	0.164(#8)UNC

Figure P12-1

Project 12-2: Inches

A. Draw two spur gears based on the following parameters. On both gears the threaded hole is located halfway up the hub height.

B. Design Exercise

Create a support plate and shafts for the two gears drawn in part A of this project. See the section on gear assemblies. Create two shafts, Ø.500 and Ø1.000 with .05 chamfers at each end. The shafts should be long enough to allow for at least .25 clearance between the gears and the support plate. The shafts should extend from the bottom surface of the plate to the top surface of each gear's hub.

Create a plate .375 thick. Size the plate so that it extends at least .25 beyond the edges on either gear. Mount the shafts in SKF Series RLS ball bearings, and create holes in the plate that will accommodate the outside diameters of the bearings.

	Gear 1	Gear 2
Number of teeth	20	50
Face width	1.00	1.00
Diametral pitch	12	12
Pressure angle	20	20
Hub Ø	1.00	1.00
Hub height	.375	.750
Bore	.500	1.000
Threaded hole	0.216(#12)UNC	0.216(#12)UNC

Figure P12-2

Project 12-3: Millimeters

Draw two spur gears based on the following parameters. On both gears the threaded hole is located halfway up the hub height.

	Gear 1	Gear 2
Number of teeth	30	90
Face width	20	20
Diametral pitch	1.5	1.5
Pressure angle	20	20
Hub Ø	30	50
Hub height	20	20
Bore	12	16
Threaded hole	M3	M5

Figure P12-3

Project 12-4: Millimeters

A. Draw two spur gears based on the following parameters. On both gears the threaded hole is located halfway up the hub height.

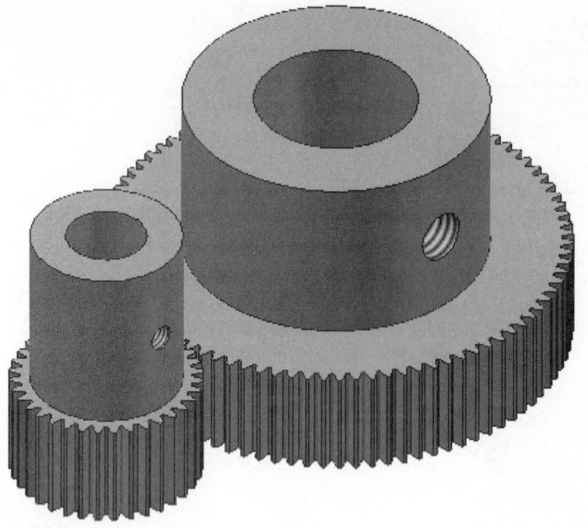

	Gear 1	Gear 2
Number of teeth	36	90
Face width	10	20
Diametral pitch	0.5	1.5
Pressure angle	20	20
Hub Ø	15	30
Hub height	16	16
Bore	8	16
Threaded hole	M3	M5

Figure P12-4

B. Design Exercise

Create a support plate and shafts for the two gears drawn in part A of this project. See the section on gear assemblies. Create two shafts, Ø8.0 and Ø16.0 with 0.50 chamfers at each end. The shafts should be long enough to allow for at least 5.0 clearance between the gears and the support plate. The shafts should extend from the bottom surface of the plate to the top surface of each gear's hub.

Create a plate 10 thick. Size the plate so that it extends at least 5.0 beyond the edges on either gear. Mount the shafts in DIN 1854-4 M plain bearings and create holes in the plate that will accommodate the outside diameters of the bearings.

Project 12-5: Inches

A. Draw two bevel gears based on the following parameters. On both gears the threaded hole is located halfway up the hub height.

	Gear 1	Gear 2
Shaft angle	90	90
Pressure angle	20	20
Helix angle	20	20
Number of teeth	24	60
Face width	.50	.50
Diametral pitch	16	16
Hub Ø	1.00	1.00
Hub height	.75	.75
Bore	.50	.50
Threaded hole	#8-UNC	#8-UNC

B. Design Exercise

Create an L-bracket support plate and shafts for the two gears drawn in part A of this project. See the section on supports for bevel gears. Create two shafts, Ø.50 with .05 chamfers at each end. The shafts should be long enough to allow for at least .25 clearance between the gears and the support plate. The shafts should extend from the bottom surface of the plate to the top surface of each gear's hub.

Create an L-bracket .375 thick. Size the bracket so that it extends at least .25 beyond the edges on either gear. Make the outside diameters of the bosses at least .025 greater than the outside diameter of the bearings. Make the bosses .25 high.

Mount the shafts in SKF Series RLS ball bearings, and create holes in the plate that will accommodate the outside diameters of the bearings.

Project 12-6: Millimeters

A. Draw two bevel gears based on the following parameters. On both gears the threaded hole is located halfway up the hub height.

	Gear 1	Gear 2
Shaft angle	90	90
Pressure angle	20	20
Helix angle	20	20
Number of teeth	20	40
Face width	12	12
Tangential module	2	2
Hub Ø	20	40
Hub height	16	20
Bore	12	20
Threaded hole	M5	M6

B. Design Exercise

Create an L-bracket support plate and shafts for the two gears drawn in part A of this project. See the section on supports for bevel gears. Create two shafts, Ø12.0 and Ø20.0 with 0.50 chamfers at each end. The shafts should be long enough to allow for at least 5.0 clearance between the gears and the support plate. The shafts should extend from the bottom surface of the plate to the top surface of each gear's hub.

Create an L-bracket 10 thick. Size the plate so that it extends at least 5.0 beyond the edges on either gear. Make the outside diameters of the bosses at least 5 greater than the outside diameter of the bearings. Make the bosses 5.00 high.

Mount the shafts in DIN 1854-4 M plain bearings, and create holes in the plate that will accommodate the outside diameters of the bearings.

Project 12-7: Inches

A. Draw a worm and a worm gear based on the following parameters. On both gears the threaded hole is located halfway up the hub height.

Worm gear: Number of teeth = 48		
Worm: Number of threads = 1		
	Worm	**Worm Gear**
Face width		.75
Diametral pitch	12	12
Pressure angle	20	20
Worm length	2.50	
Hub Ø	.50	.50
Hub height	.500	.750
Bore	.375	.500
Threaded hole	0.138(#6)UNC	0.164(#8)UNC

B. Design Exercise

Create a corner-bracket support plate and shafts for the two gears drawn in part A of this project. See the section on worm gear supports. Create two shafts, Ø.375 and Ø.500 with .05 chamfers at each end. The shafts should be long enough to allow for at least .25 clearance between the gears and the support plate. The shafts should extend from the back surface of the plate to the top surface of each gear's hub.

Create a corner bracket .375 thick. Size the bracket so that it extends at least .25 beyond the edges on either gear. Mount the shafts in SKF Series RLS ball bearings and create holes in the sides that will accommodate the outside diameters of the bearings.

Project 12-8: Inches

A. Draw a worm and a worm gear based on the following parameters. On both gears the threaded hole is located halfway up the hub height.

Worm gear: Number of teeth = 60		
Worm: Number of threads = 1		
	Worm	**Worm Gear**
Face width		24
Module	4	4
Pressure angle	14.5	14.5
Worm length	65	
Hub Ø	16	24
Hub height	12	16
Bore	8.0	10.0
Threaded hole	M4	M4

B. Design Exercise

Create a corner-bracket support plate and shafts for the two gears drawn in part A of this project. See the section on worm gear supports. Create two shafts, Ø8.0 and Ø10.0 with .50 chamfers at each end. The shafts should be long enough to allow for at least 5.0 clearance between the gears and the support plate. The shafts should extend from the back surface of the plate to the top surface of each gear's hub.

Create a corner bracket 8.0 thick. Size the bracket so that it extends at least 6.0 beyond the edges on either gear. Mount the shafts in DIN 1854-4 M plain bearings, and create holes in the sides that will accommodate the outside diameters of the bearings.

Project 12-9: Inches

A. Prepare an assembly drawing of the 2-Gear Assembly shown in Figure P12-9. The gears have the following parameters:

Center distance = 3.00		
	Gear 1	**Gear 2**
Number of teeth	48	96
Face width	.50	.50
Diametral pitch	24	24
Pressure angle	20	20
Hub Ø	.750	1.000
Hub height	.500	.500
Bore	.5.00	.625

B. Prepare a presentation drawing.

C. Animate the presentation drawing.

D. Prepare an exploded isometric drawing with assembly numbers and a parts list.

E. Prepare a detailed dimensioned drawing of each part.

2-Gear Assembly

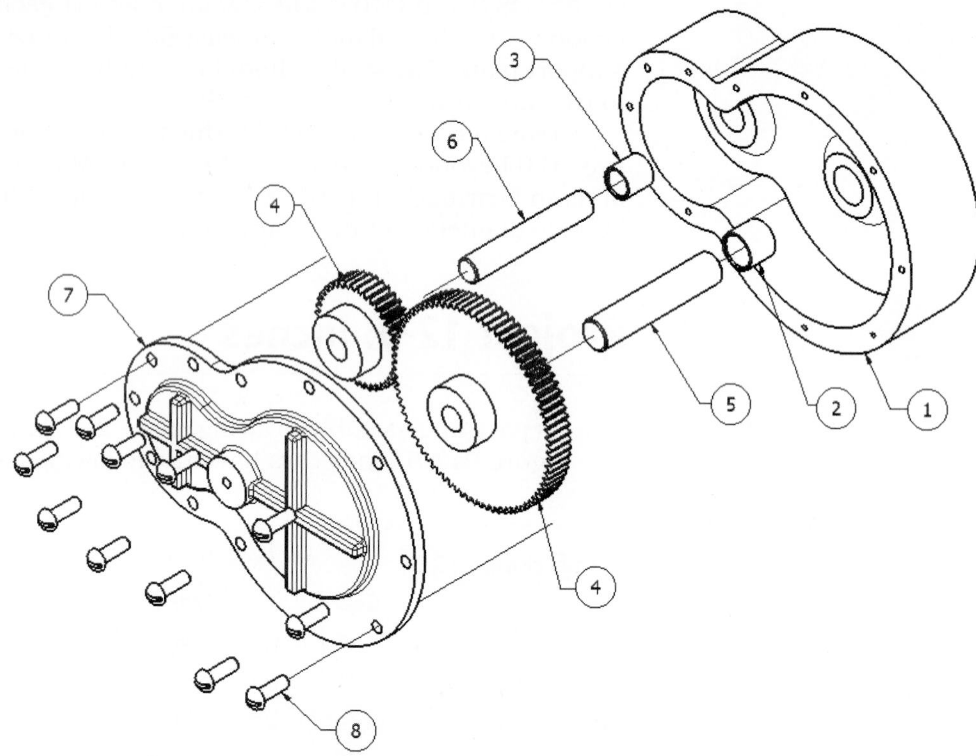

Figure P12-9A

Parts List				
ITEM	PART NUMBER	DESCRIPTION	MATERIAL	QTY
1	ENG-453-A	GEAR, HOUSING	CAST IRON	1
2	BU-1123	BUSHING Ø0.75	Delrin, Black	1
3	BU-1126	BUSHING Ø0.625	Delrin, Black	1
4	ASSEMBLY-6	GEAR ASSEMBLY	STEEL	1
5	AM-314	SHAFT, GEAR Ø.625	STEEL	1
6	AM-315	SHAFT, GEAR Ø.0.500	STEEL	1
7	ENG -566-B	COVER, GEAR	CAST IRON	1
8	ANSI B18.6.2 - 1/4-20 UNC - 0.75	Slotted Round Head Cap Screw	Steel, Mild	12

Figure P12-9B

Gear, Housing
P/N ENG-453-A
Cast Iron

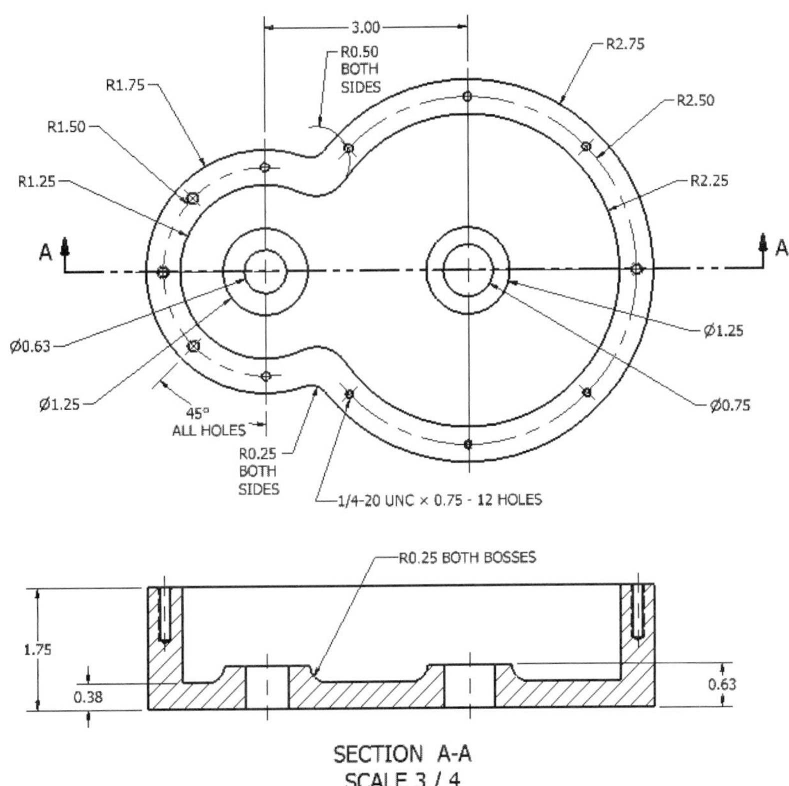

Figure P12-9C

Shaft, Gear Ø.625
P/N AM-314
Steel

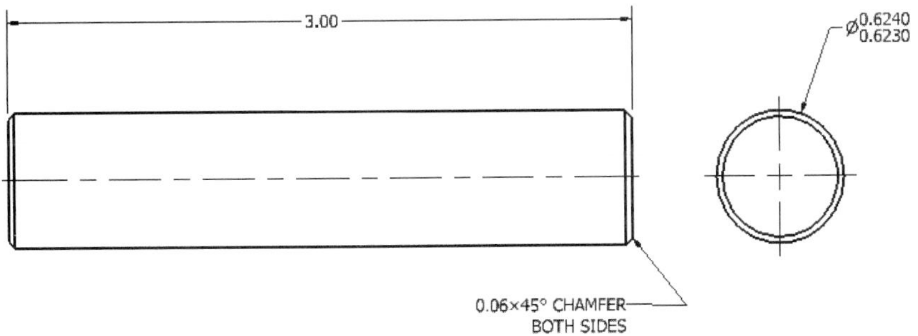

Figure P12-9D

Figure P12-9E

Shaft, Gear Ø.500
P/N AM-315
Steel

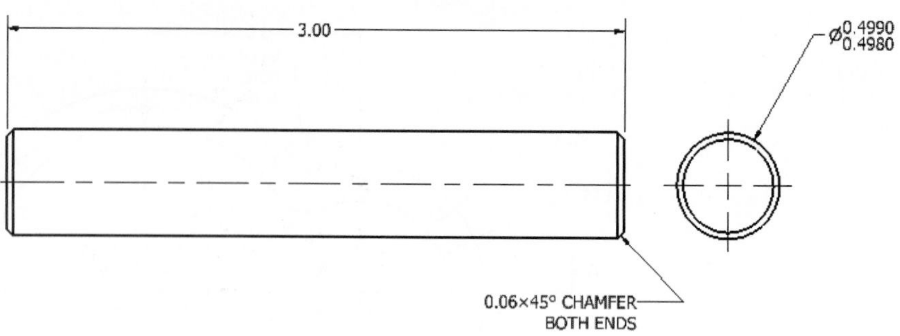

Figure P12-9F

Cover, Gear
P/N ENG-566-B
Cast Iron

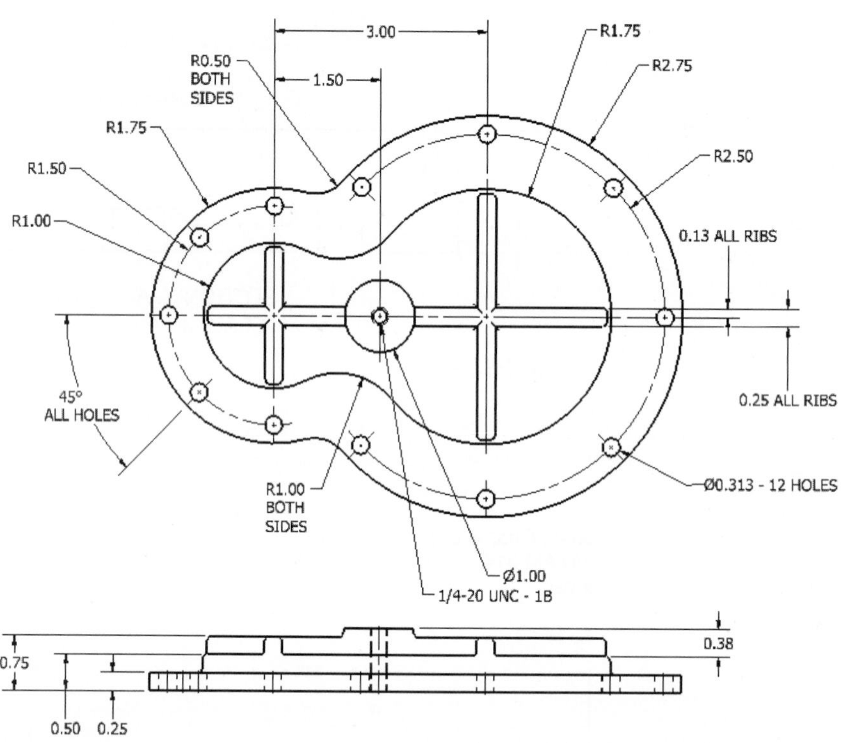

Figure P12-9G

Bushing Ø.625
P/N BU-1123
Delrin, Black

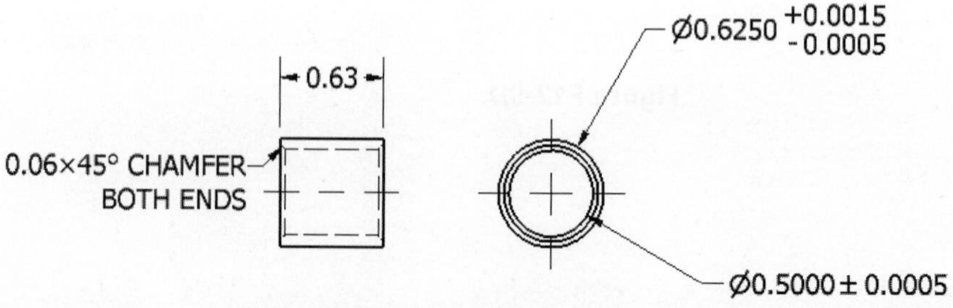

Bushing Ø0.750
P/N BU-1126
Delrin, Black

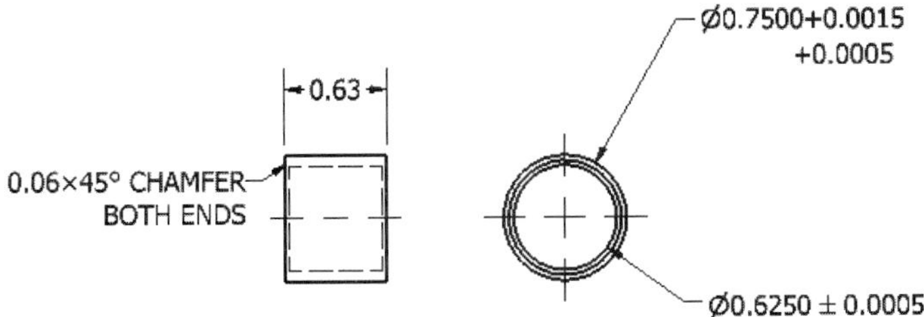

Figure P12-9H

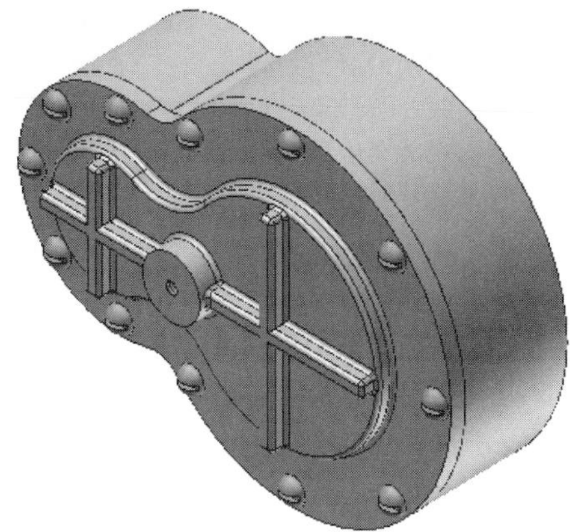

Figure P12-9I

Project 12-10: Millimeters

A. Prepare an assembly drawing of the 4-Gear Assembly shown in Figure P12-10. The gears have the following parameters:

Center distance = 3.00		
	Gears 1, 3	**Gears 2, 4**
Number of teeth	30	96
Face width	20.0	20.0
Module	2	2
Pressure angle	20	20
Hub Ø	50.0	60.0
Hub height	20.0	20.0
Bore	25.0	30.0

4-Gear Assembly

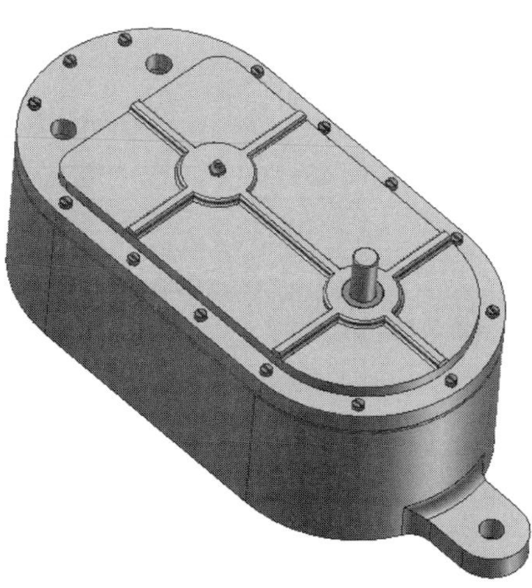

Figure P12-10A

Presentation Drawing

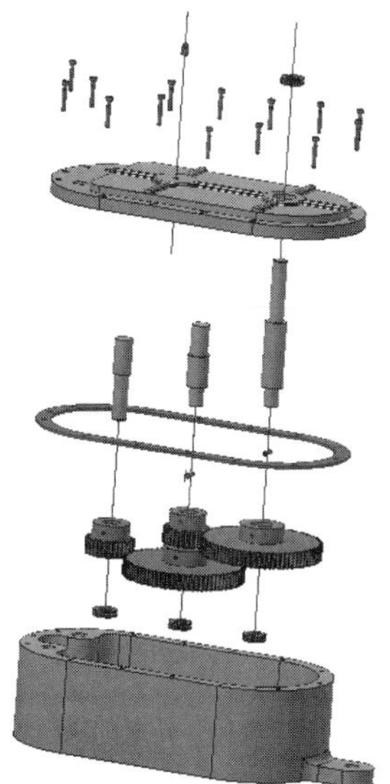

Figure P12-10B

B. Prepare a presentation drawing.

C. Animate the presentation drawing.

D. Prepare an exploded isometric drawing with assembly numbers and a parts list.

E. Prepare a detailed dimensioned drawing of each part.

Parts List

Parts List				
ITEM	PART NUMBER	DESCRIPTION	MATERIAL	QTY
1	ENG-311-1	4-GEAR HOUSING	CAST IRON	1
2	BS 5989: Part 1 - 0 10 - 20x32x8	Thrust Thrust Ball Bearing	Steel, Mild	4
3	SH-4002	SHAFT, NEUTRAL	STEEL	1
4	SH-4003	SHAFT, OUTPUT	STEEL	1
5	SH-4004A	SHAFT,INPUT	STEEL	1
6	4-GEAR-ASSEMBLY		STEEL	2
7	CSN 02 1181 - M6 x 16	Slotted Headless Set Screw - Flat Point	Steel, Mild	2
8	ENG-312-1	GASKET	Brass, Soft Yellow	1
9	COVER			1
10	CNS 4355 - M 6 x 35	Slotted Cheese Head Screw	Steel, Mild	14
11	CSN 02 7421 - M10 x 1coned short	Lubricating Nipple, coned Type A	Steel, Mild	1

Figure P12-10C

Housing, Gear
P/N ENG-311-1
Cast Iron

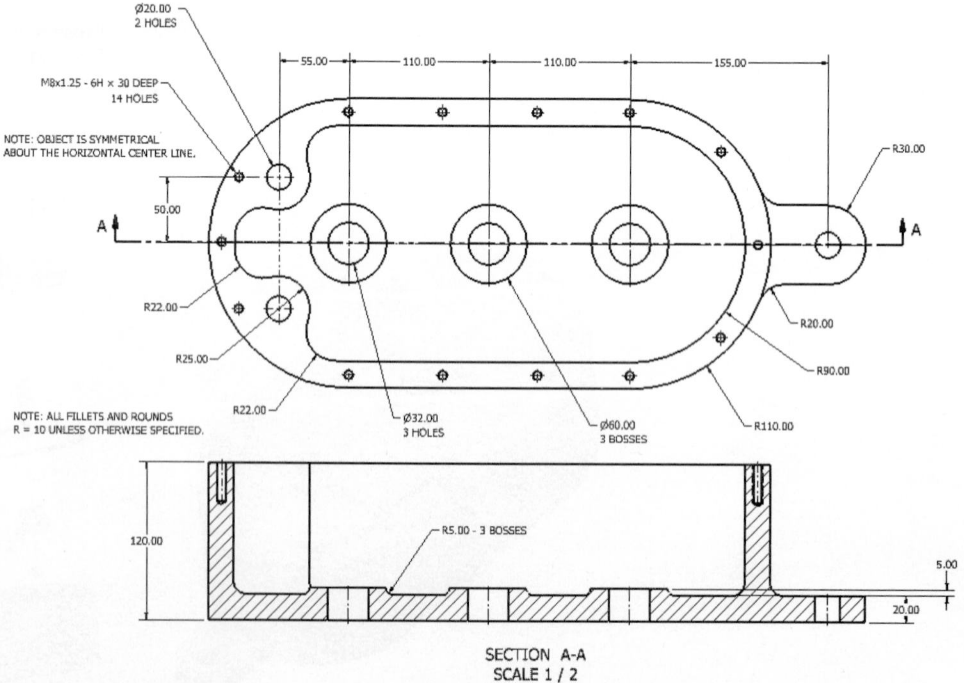

Figure P12-10D

Cover
P/N AM-311-2

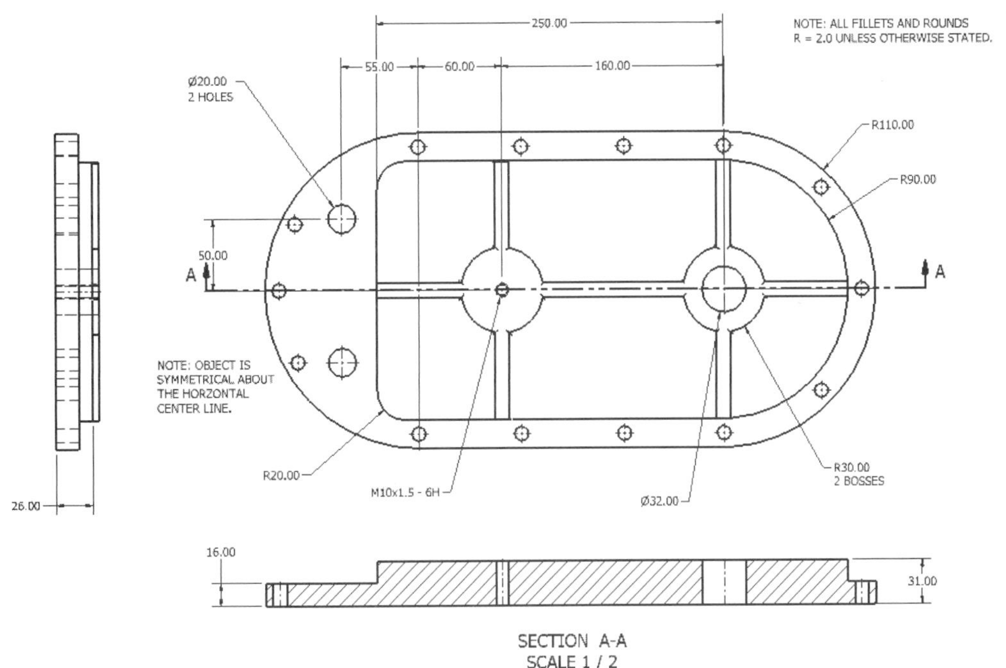

Figure P12-10E

Gasket
P/N ENG-312-1
Brass, soft yellow

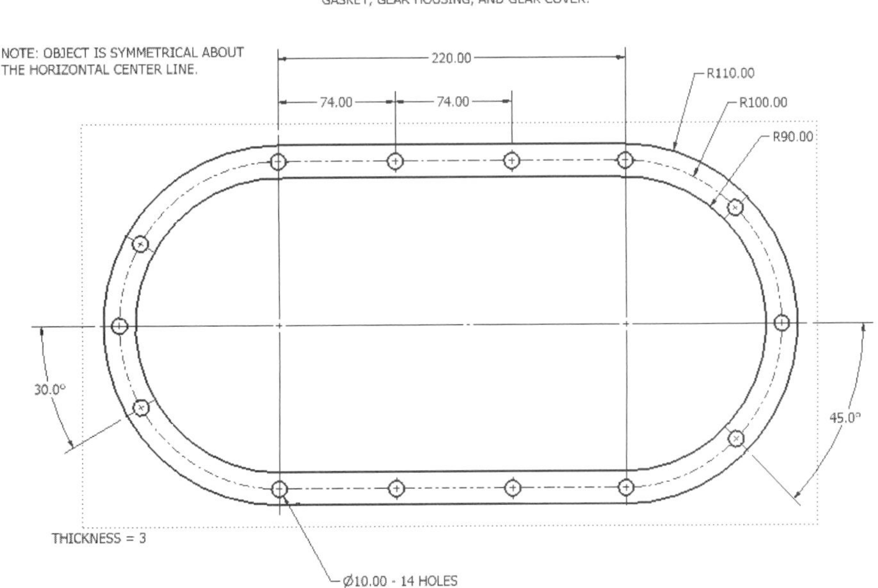

Figure P12-10F

Figure P12-10G Gear Subassembly

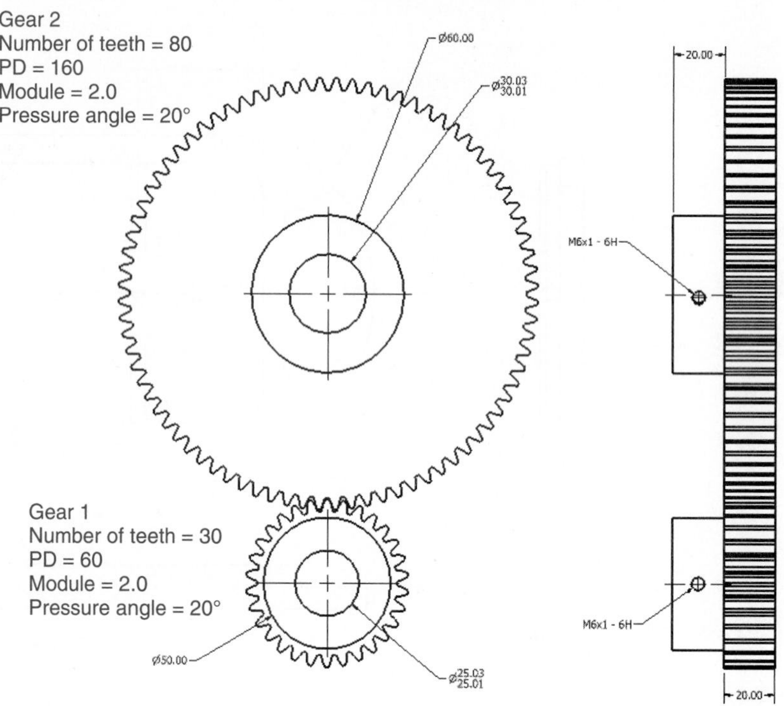

Figure P12-10H Shaft, Output
P/N SH-4003
Steel

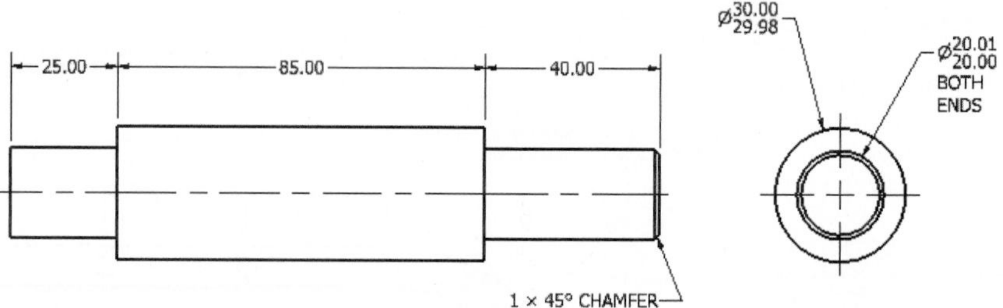

Figure P12-10I Shaft, Input
P/N SH-4004-A
Steel

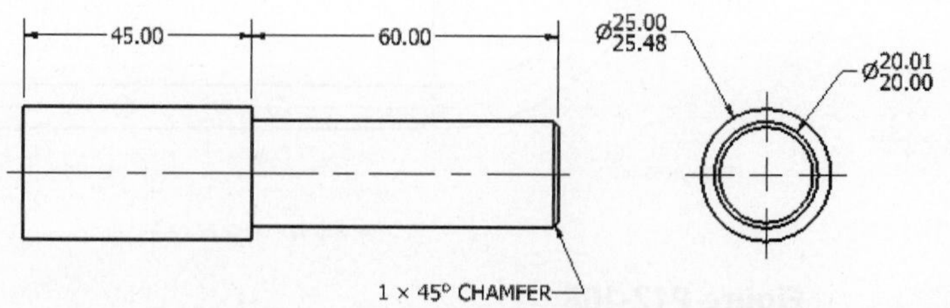

Figure P12-10J

Shaft, Neutral
P/N SH-4002
Steel

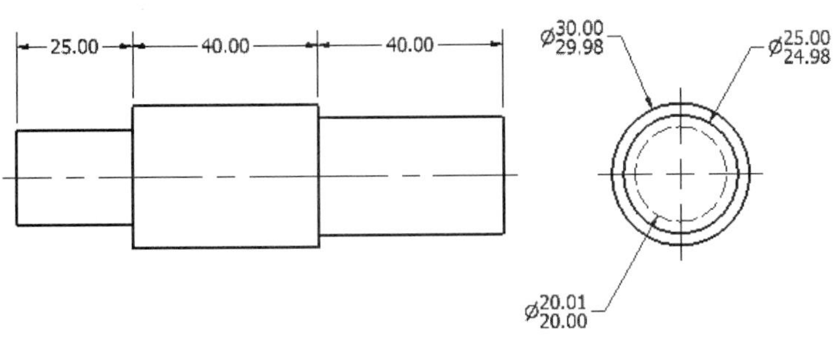

Figure P12-10K

Exploded Isometric Drawing

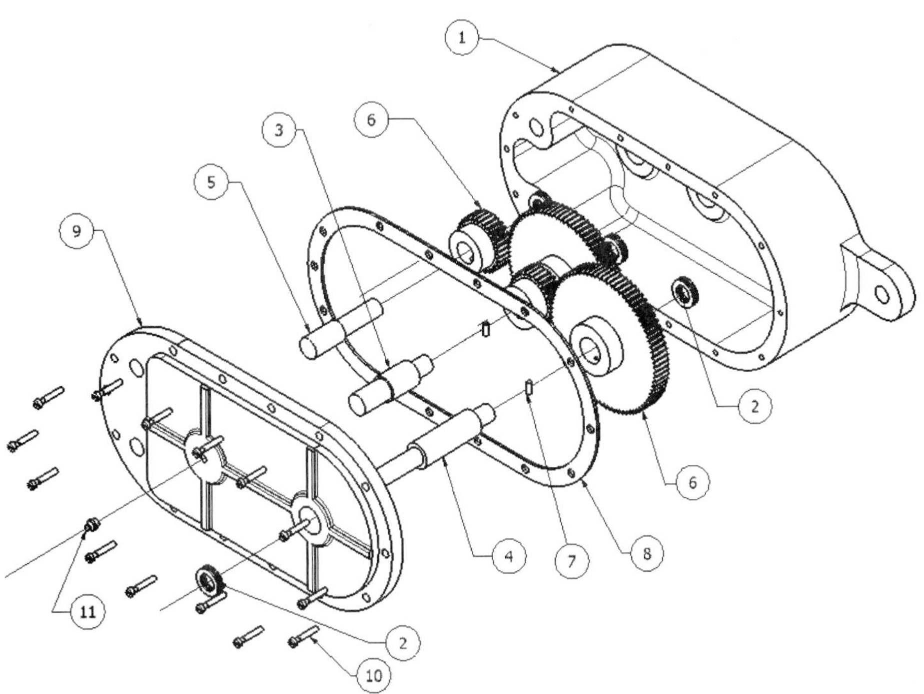

chapter thirteen

Sheet Metal Drawings

CHAPTER OBJECTIVES

- Learn how to create sheet metal drawings
- Learn about sheet metal gauges
- Understand sheet metal terminology

Introduction

This chapter explains how to create sheet metal drawings. Gauges for sheet metal are presented along with bend radii, flanges, tabs, reliefs, and flat patterns.

Sheet Metal Drawings

Figure 13-1 shows a 3D solid model of a sheet metal part and a dimensioned orthographic drawing of that part. The orthographic drawing was created from the 3D model. The following sections explain how to create the 3D sheet metal drawing.

EXERCISE 13-1 Creating a 3D Sheet Metal Drawing

1 Create a new drawing using the **Sheet Metal (mm).ipt** format.

See Figure 13-2. The **Sketch** panels will appear. See Figure 13-3. Sheet metal drawings are initiated as 2D sketches, then developed using a combination of **Sketch** and **Sheet Metal** panel tools.

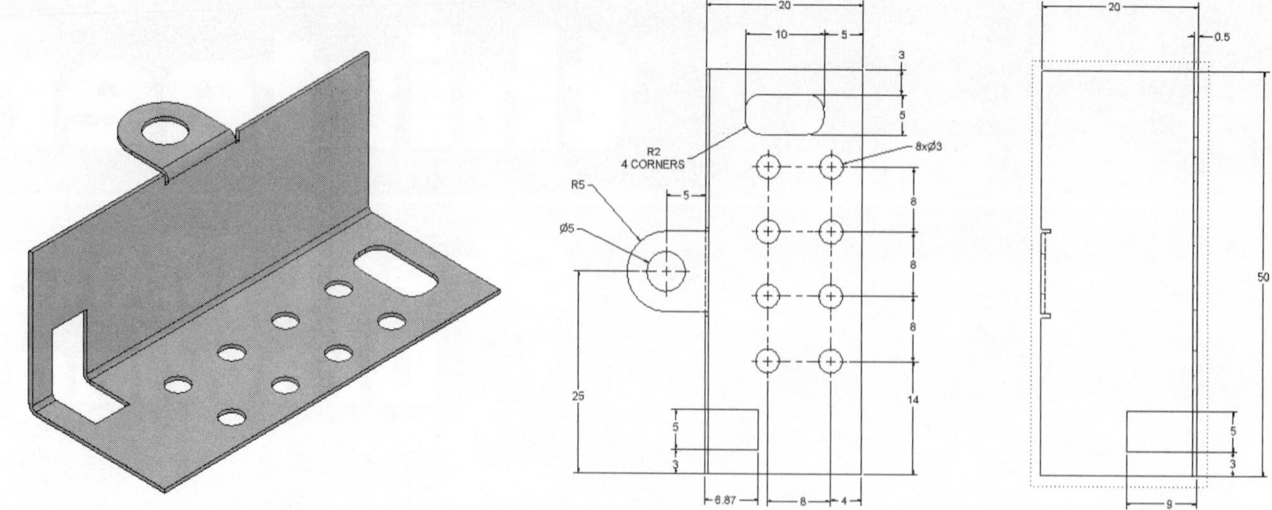

Figure 13-1

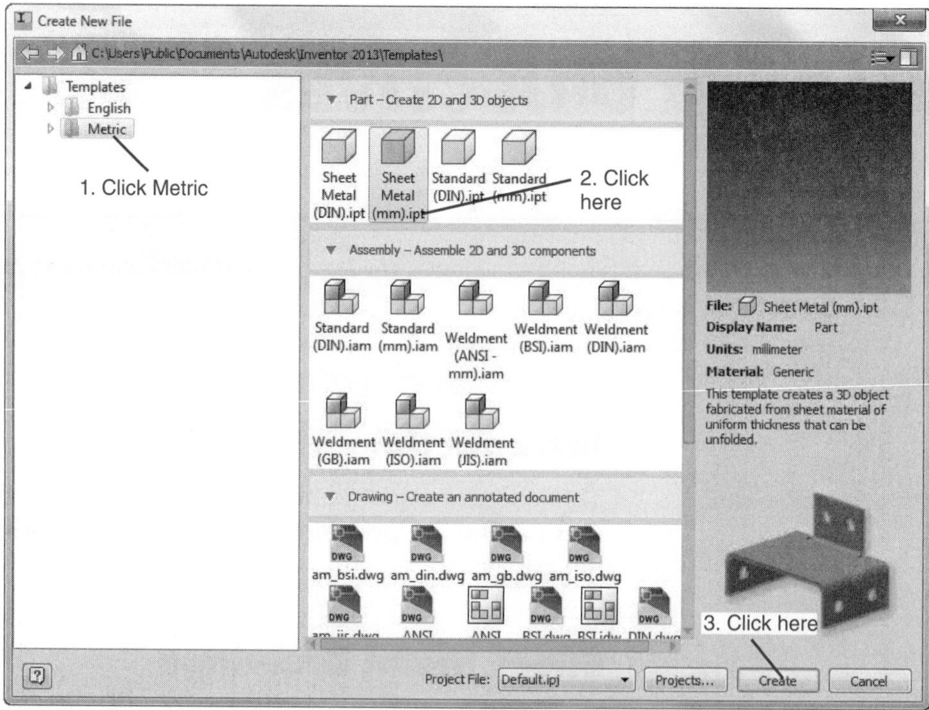

Figure 13-2

2 Click the **Start 2D Sketch** tool, select the XY plane, and use the **Two Point Rectangle** tool and draw a **20** × **50** rectangle.

3 Move the cursor into the area of the ViewCube and click the icon that looks like a house (the **Home** tool).

See Figure 13-3.

Figure 13-3

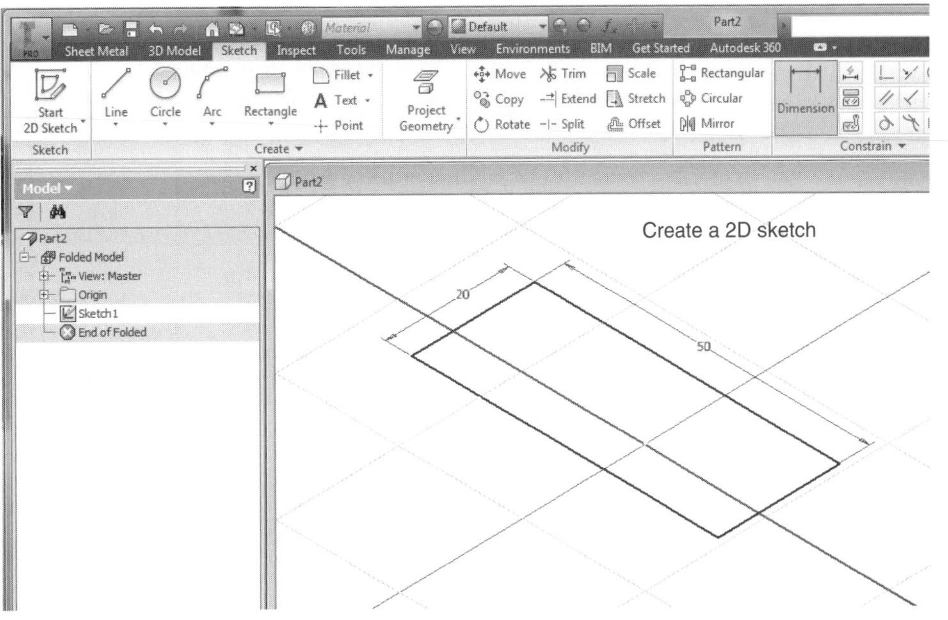

EXERCISE 13-2 Adding Thickness

1 Right-click the mouse again and select the **Finish 2D Sketch** option.

The **Sheet Metal** panels will appear. See Figure 13-4. Not all tools will be active at this time, but they will become active as the drawing progresses.

Figure 13-4

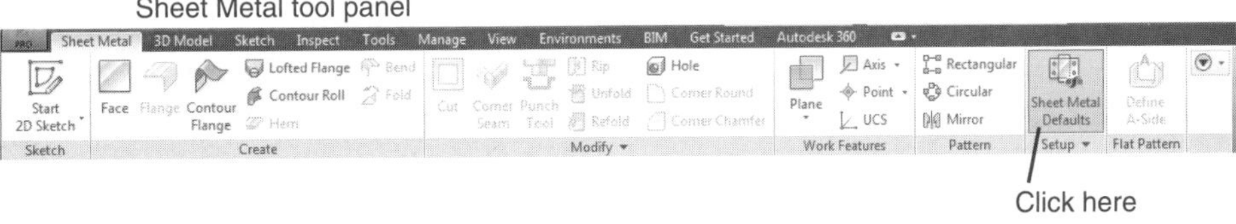

2 Select the **Sheet Metal Defaults** option located on the **Setup** panel under the **Sheet Metal** tab.

The **Sheet Metal Defaults** dialog box will appear. See Figure 13-5. The **Sheet Metal Defaults** dialog box is used to define the thickness, material, and bend characteristics of the part.

Figure 13-5

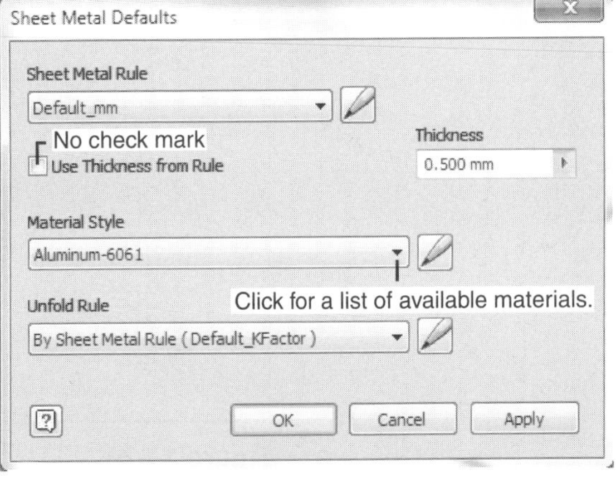

Inches

Gauge	Thickness (in.)	Gauge	Thickness (in.)
000 000	0.5800	18	0.0403
00 000	0.5165	19	0.0359
0 000	0.4600	20	0.0320
000	0.4096	21	0.0285
00	0.3648	22	0.0253
0	0.3249	23	0.0226
1	0.2893	24	0.0201
2	0.2576	25	0.0179
3	0.2294	26	0.0159
4	0.2043	27	0.0142
5	0.1819	28	0.0126
6	0.1620	29	0.0113
7	0.1443	30	0.0100
8	0.1285	31	0.0089
9	0.1144	32	0.0080
10	0.1019	33	0.0071
11	0.0907	34	0.0063
12	0.0808	35	0.0056
13	0.0720	36	0.0050
14	0.0641	37	0.0045
15	0.0571	38	0.0040
16	0.0508	39	0.0035
17	0.0453	40	0.0031

Wire and Sheet Metal Gauges

Figure 13-6

Inventor has many default values already in place. Figure 13-5 shows that the default thickness is 0.500 mm. Sheet metal is manufactured in standard thicknesses. Figure 13-6 is a partial listing of available standard sheet metal thicknesses in inches, and Figure 13-7 is a partial listing of sheet metal thicknesses in millimeters.

Figure 13-5 lists 0.500 mm as a standard thickness, so this default value will be used for this example. Click the check mark in the **Use Thickness from Rule** box. There should be no check mark in the box. Set the **Material Style** for **Aluminum-6061.**

Millimeters

0.050	0.50	4.0
0.060	0.60	5.0
0.080	0.80	6.0
0.10	1.0	8.0
0.12	1.2	10.0
0.16	1.6	
0.20	2.0	
0.25	2.5	
0.30	3.0	
0.40	3.5	

Figure 13-7

3 Accept the **0.500 mm Thickness** value, then click **OK.**

4 Select the **Face** tool from the **Create** panel located under the **Sheet Metal** tab.

5 Select the sketch, and click **OK.**

The 20 × 50 panel will now have a thickness of 0.50. See Figure 13-8.

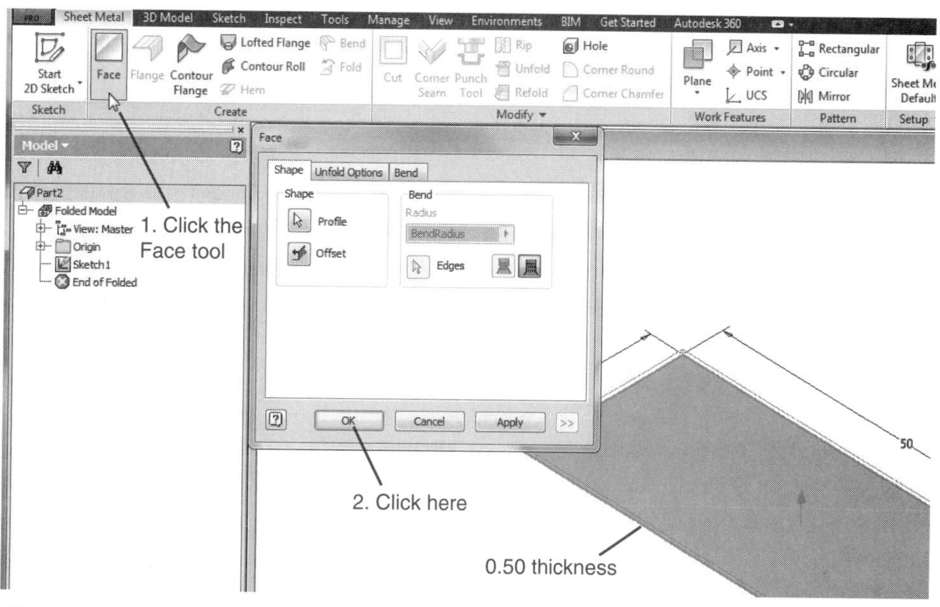

Figure 13-8

Bend Radii

As sheet metal is bent, the inside surface is subjected to compression, and the outside surface to tension. These forces cause the material to stretch slightly.

To edit the bend radius for a sheet metal part access the **Style and Standard Editor** dialog box by first clicking the **Edit Sheet Metal Rule** box located on the **Sheet Metal Defaults** dialog box. See Figure 13-9.

The **Relief Depth** and **Minimum Remnant** values shown on the **Style and Standard Editor** dialog box are calculated based on the thickness value specified in the **Sheet Metal Defaults** dialog box. This defines the **Relief Depth** as **0.50 mm.**

The default values for **Bend Radius** and **Relief Shape** will be accepted for this example.

1 Click the **Done** box.

NOTE

Reliefs are added to bends in sheet metal parts to prevent tearing as the bend is created.

The **Sheet Metal Defaults** dialog box will appear.

Figure 13-9

Sheet Metal Defaults
Sheet Metal Rule
Default_mm
Click here
Thickness
Use Thickness from Rule
0.500 mm
Material
Aluminum 6061
Unfold Rule
Default_KFactor
OK
Cancel
Apply

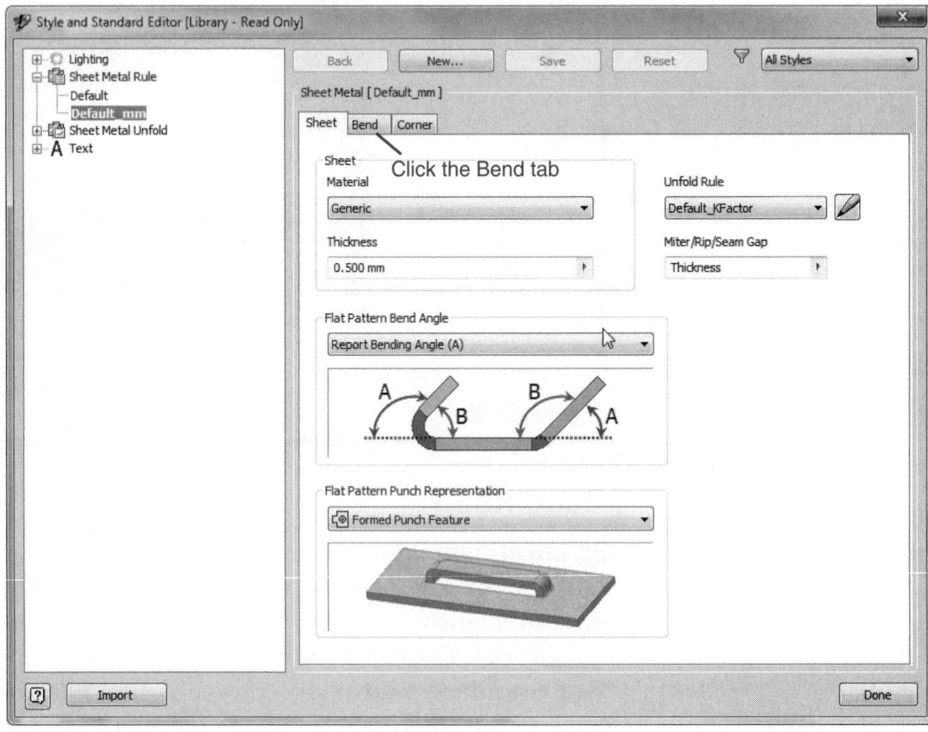
Style and Standard Editor [Library - Read Only]
Lighting
Sheet Metal Rule
Default
Default_mm
Sheet Metal Unfold
Text
Back
New...
Save
Reset
All Styles
Sheet Metal [Default_mm]
Sheet
Bend
Corner
Click the Bend tab
Sheet
Material
Generic
Thickness
0.500 mm
Unfold Rule
Default_KFactor
Miter/Rip/Seam Gap
Thickness
Flat Pattern Bend Angle
Report Bending Angle (A)
A
B
B
A
Flat Pattern Punch Representation
Formed Punch Feature
Import
Done

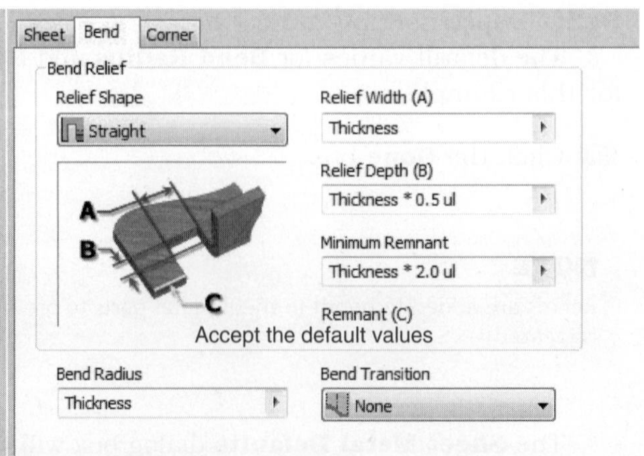
Sheet
Bend
Corner
Bend Relief
Relief Shape
Straight
Relief Width (A)
Thickness
Relief Depth (B)
Thickness * 0.5 ul
Minimum Remnant
Thickness * 2.0 ul
A
B
C
Remnant (C)
Accept the default values
Bend Radius
Thickness
Bend Transition
None

2 Click the **Cancel** box. (No changes were made to the bend parameters.)

Flanges

flange: A rim formed on the edge of sheet metal for strength.

A ***flange*** is a rim formed on the edge of sheet metal for strength.

1 Right-click the mouse and select the **Flange** tool.

The **Flange** dialog box will appear. See Figure 13-10.

Figure 13-10

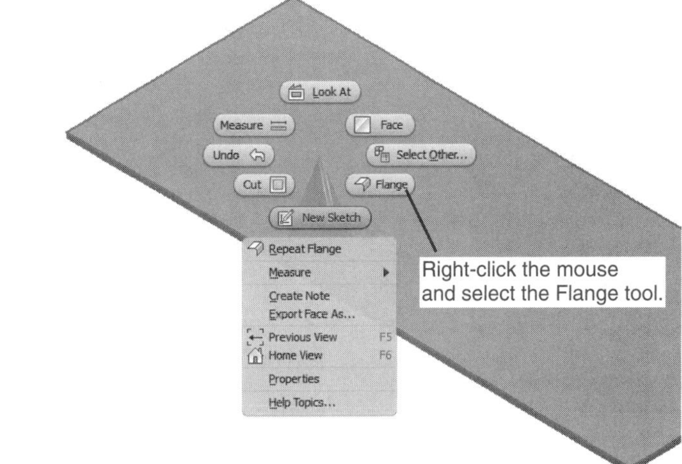

2 Set the length of the flange for **20 mm,** accept the **90.0° Flange Angle,** and select the lower rear edge of the sketch.

The lower edge was chosen because the flange total height is to be 20 mm. If the upper edge was chosen, the total height would be 20.5, the flange height plus the material thickness. Figure 13-11 shows the flange orientation resulting from edge selection.

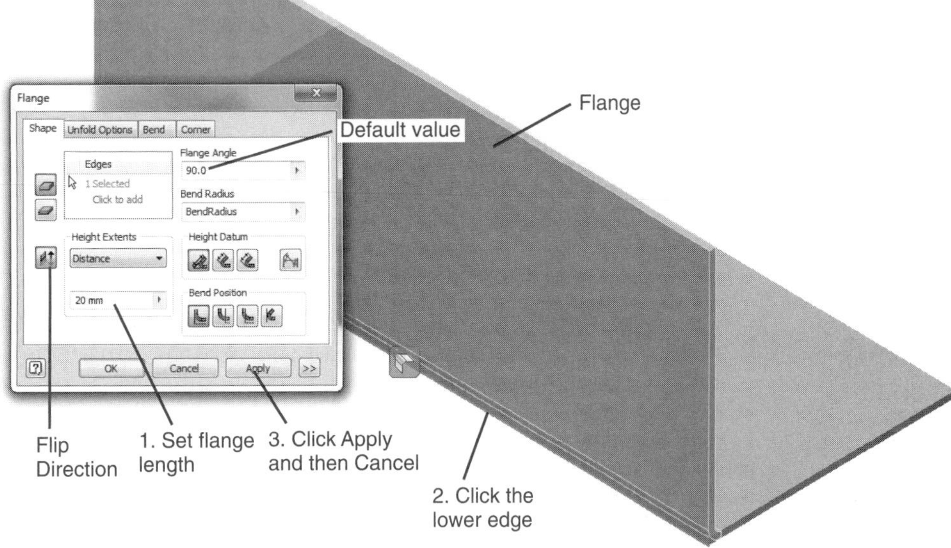

Figure 13-11

3. Use the **Flip Direction** button to change the flange orientation if necessary.

4. Click **Apply** and then **Cancel.**

   Figure 13-12 shows the resulting flange.

Figure 13-12

Tabs

tab: A feature similar to a flange but that does not run the entire length of the edge.

Tabs are similar to flanges, but tabs do not run the entire length of the edge, as flanges do. Tabs are created using a new sketch plane, and the **Two Point Rectangle** tool located under the **Sketch** tab. See Figure 13-13.

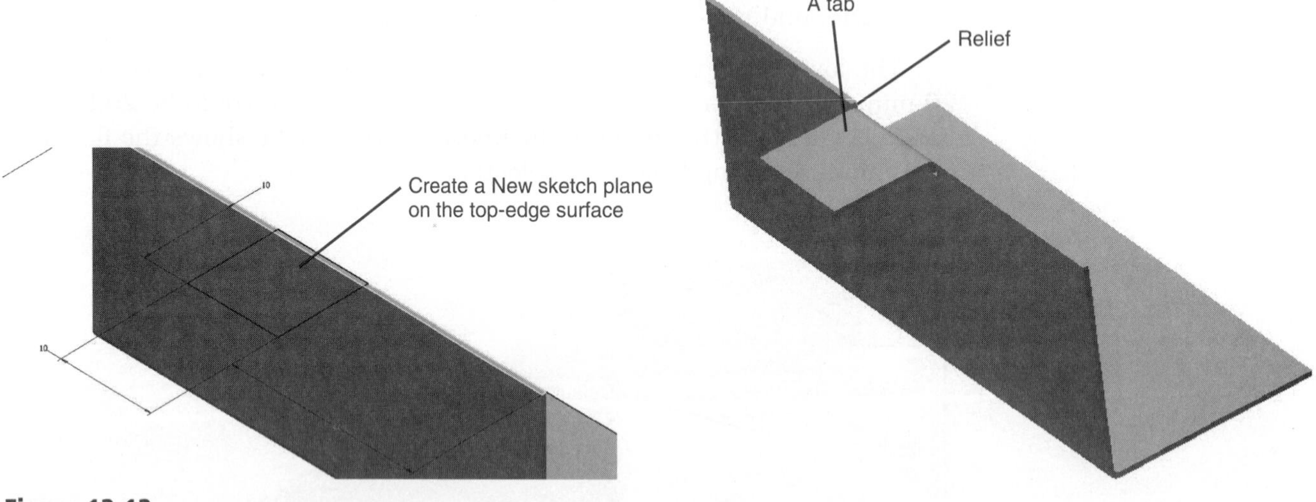

Figure 13-13

1. Zoom the part so that the top-edge surface is identifiable; select the top-edge surface of the vertical flange, right-click the mouse, and select the **New Sketch** option.

2. Use the **Two Point Rectangle** tool located on the **Draw** panel under the **Sketch** tab to extend from the edge of the vertical flange as shown.

3. Use the **Dimension** tool to size and locate the tab in accordance with the dimensions given in Figure 13-13.

4 Right-click the mouse and select the **Finish 2D Sketch** option.

5 Select the **Face** tool, and define the tab as the **Profile.**

6 Select **OK.**

Figure 13-13 shows the resulting tab.

Reliefs

relief: An area cut out of material to allow it to be bent.

Reliefs are cut out in material to allow it to be bent. If the material were not relieved, it would tear uncontrollably as the bend was formed.

Inventor's default relief value is equal to the thickness of the sheet metal material. Figure 13-13 shows the relief that was automatically created as the tab was formed.

Holes

Holes are added to sheet metal parts in the same manner as they are added to 3D models. See Figure 13-14.

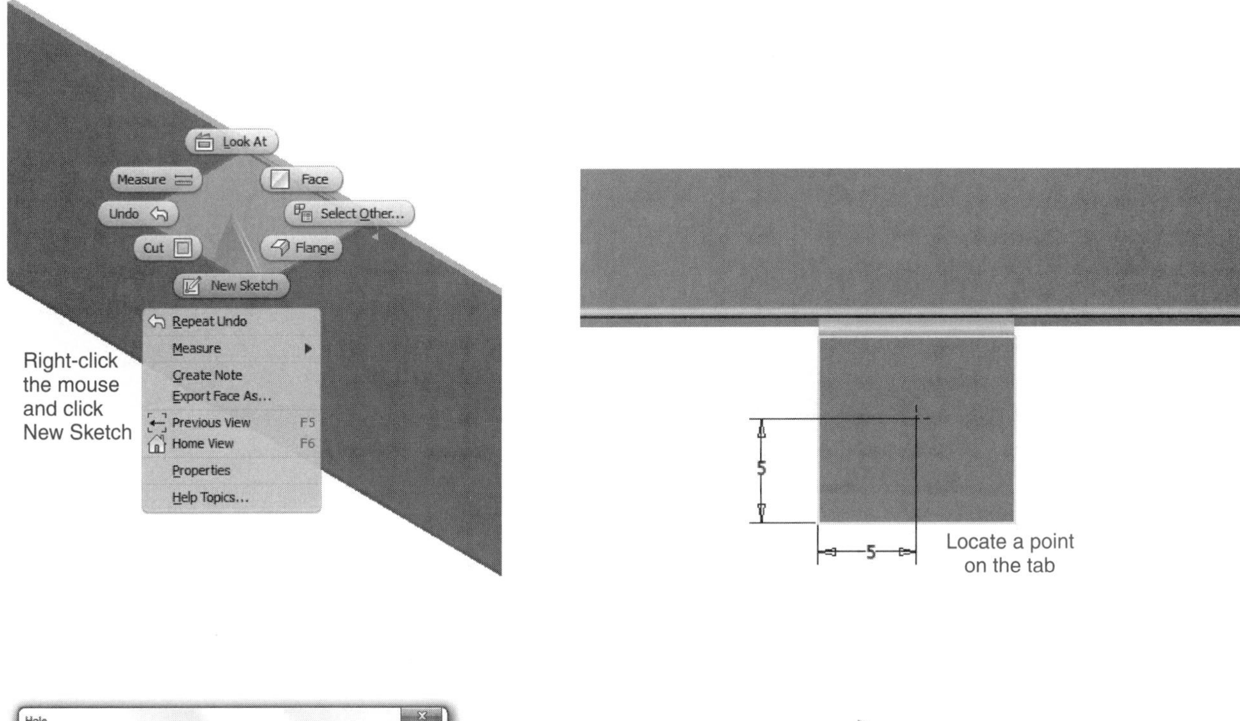

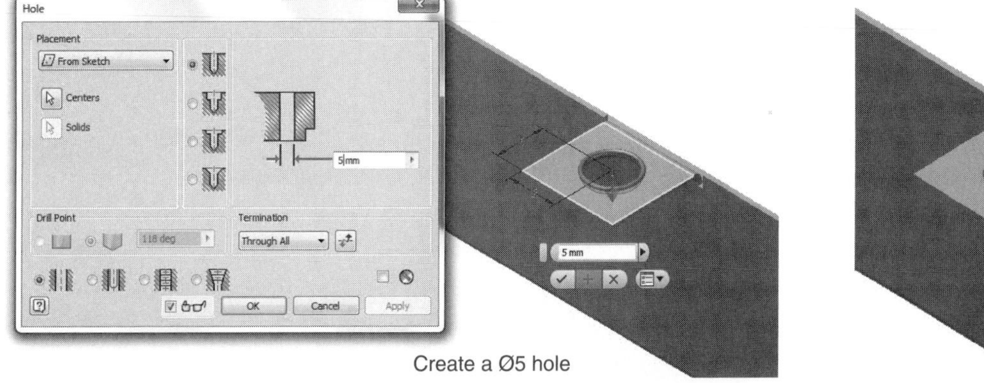

Figure 13-14

1 Create a new sketch plane on the top surface of the tab.

2 Use the **Point, Center Point** tool to define a center point.

3 Use the **Dimension** tool to dimension the center point location.

4 Right-click the mouse, and select the **Finish 2D Sketch** option.

5 Use the **Hole** tool on the **Modify** panel under the **3D Model** tab panel to create the hole.

In this example a **Ø5.00** hole was created located **5** from each edge of the tab.

Corners

Both internal and external corners are created using the **Corner Round** tool found on the **Sheet Metal** panel bar.

1 Click the **Corner Round** tool on the **Modify** panel under the **Sheet Metal** tab.

2 The **Corner Round** dialog box will appear. See Figure 13-15.

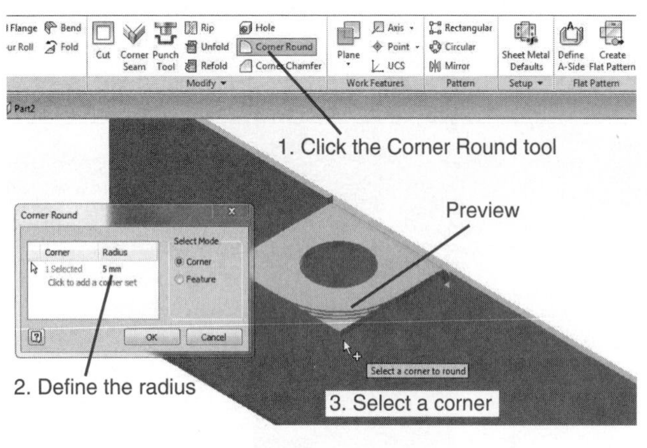

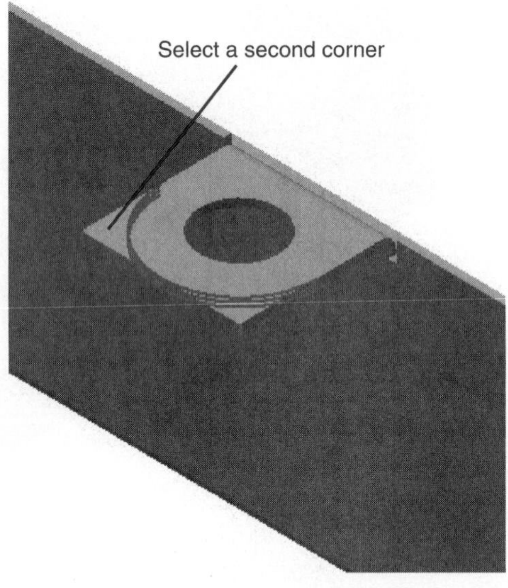

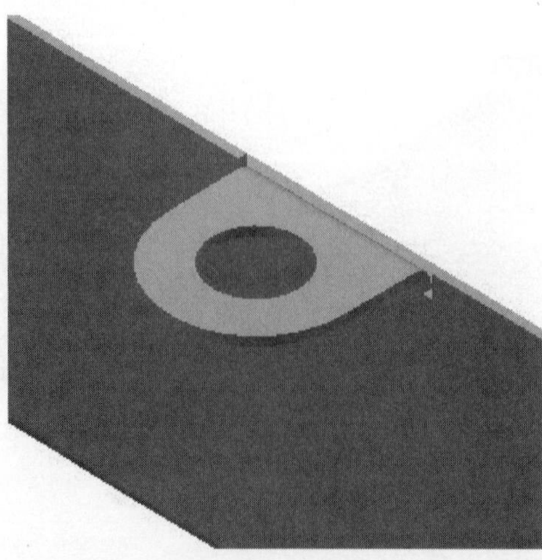

Figure 13-15

3. Set the **Radius** value for **5 mm.**

4. Select the two outside corners of the tab.

5. Click **OK.**

   Figure 13-15 shows the resulting rounded tab.

Cuts

Cuts may be any shape, other than a hole, that passes through the sheet metal. In this example a rectangular shape is used. See Figure 13-16.

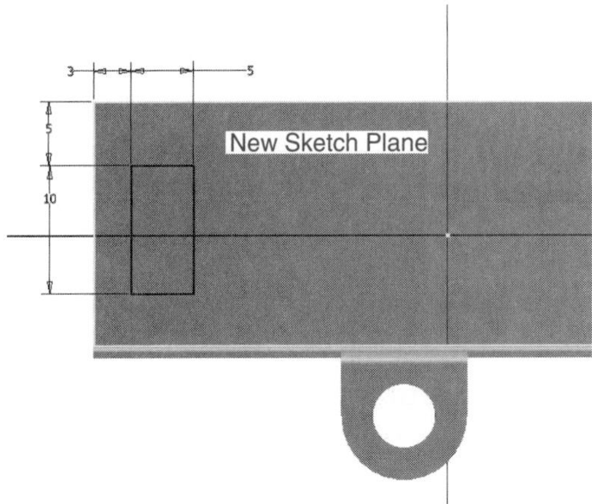

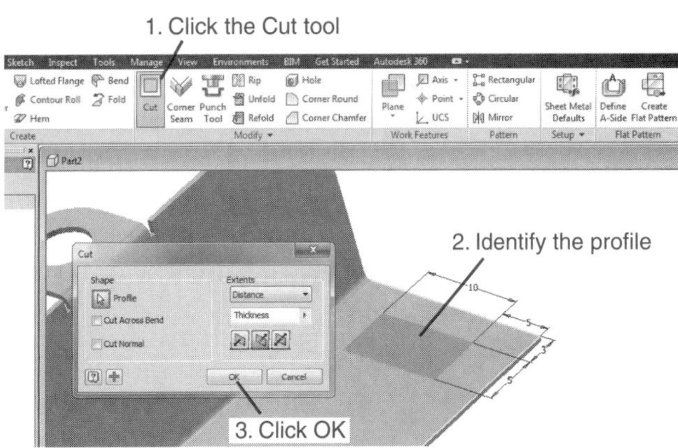

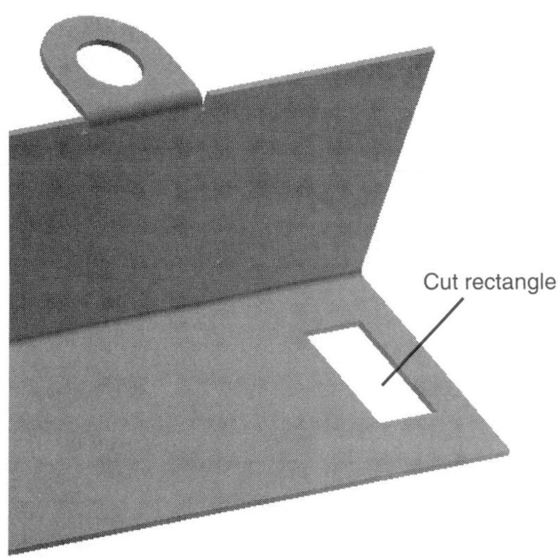

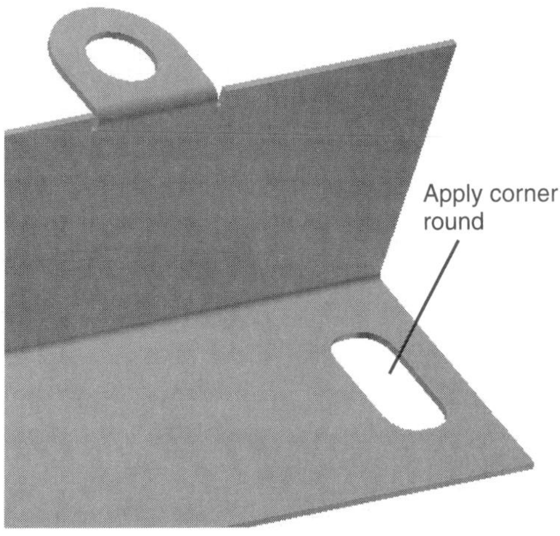

Figure 13-16

1. Reorient the part and create a new sketch plane and sketch a rectangle as shown. Use the **Dimension** tool to size and locate the rectangle.

2. Right-click the mouse and select the **Finish 2D Sketch** option.

   The **Sheet Metal** panel bar will appear.

3. Select the **Cut** tool from the **Modify** panel.

   The **Cut** dialog box will appear.

4. Select the rectangle as the **Profile.**

5. Ensure that the direction of the cut is correct, and click the **OK** box.

   The rectangular area will be removed. The depth of the cut will automatically be set for the thickness value.

6. Select the **Corner Round** tool and set the **Radius** value for **2 mm.**

7. Select the four inside corners of the rectangular cut.

8. Click the **OK** button on the **Corner Round** dialog box.

Cuts Through Normal Surfaces

Normal surfaces are surfaces that are perpendicular to each other. Cuts in normal sufaces are made by making intersecting cuts in both surfaces. See Figure 13-17.

1. Create a new sketch plane on the vertical flange as shown, and sketch a rectangle.

   Ensure that the rectangle extends beyond the rounded edge of the surface.

2. Use the **Dimension** tool to locate and size the rectangle.

3. Right-click the mouse and select the **Finish 2D Sketch** option.

4. Use the **Cut** tool to remove the rectangle.

5. Create another new sketch plane on the horizontal flange, and use the **Dimension** tool to size and locate the rectangle.

6. Right-click the mouse and select the **Finish 2D Sketch** option.

7. Click the **Cut** tool.

   The **Cut** dialog box will appear.

8. Click the **Cut Across Bend** box, then click **OK.**

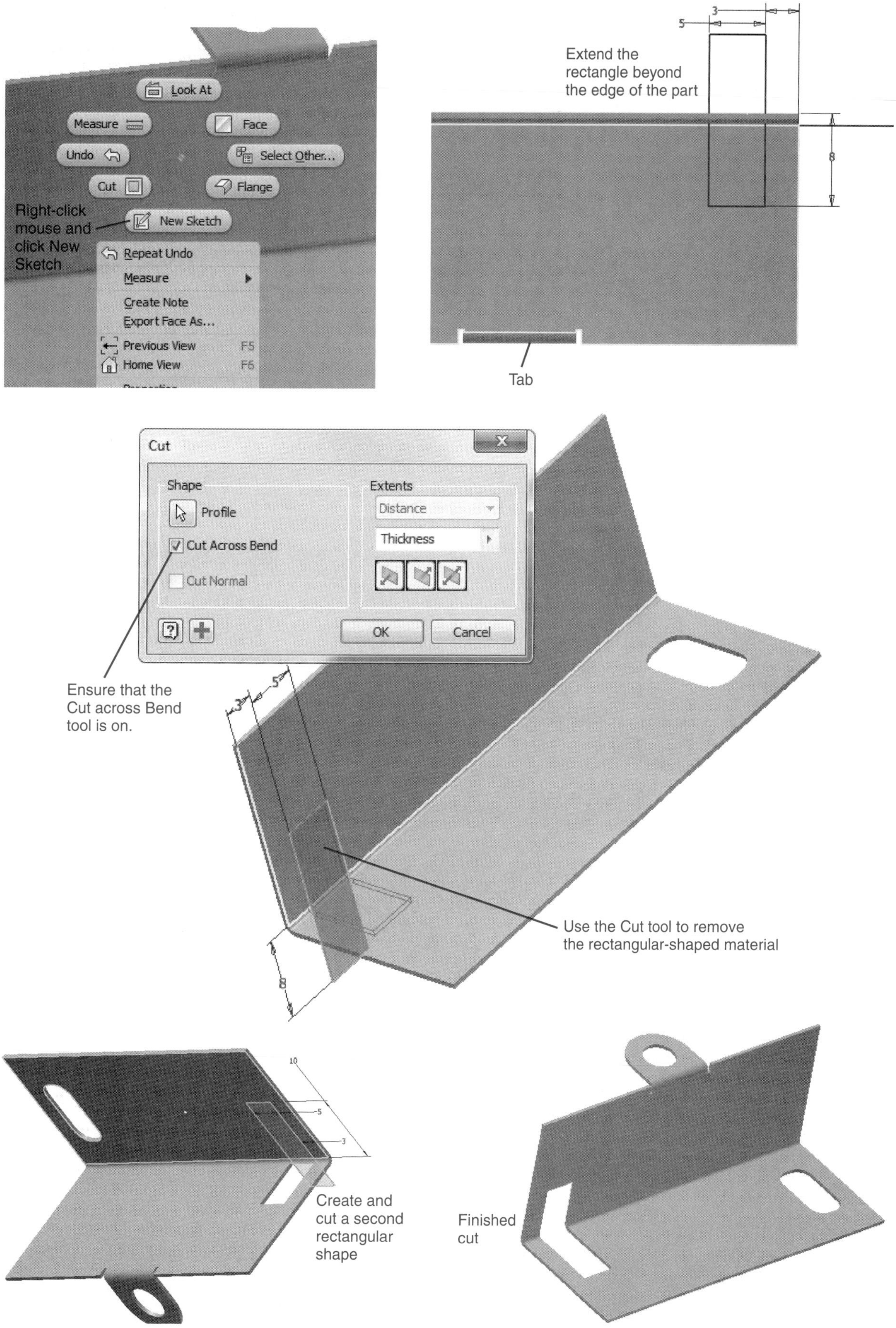

Look At
Measure
Face
Undo
Select Other...
Cut
Flange
Right-click mouse and click New Sketch
New Sketch
Repeat Undo
Measure
Create Note
Export Face As...
Previous View F5
Home View F6

Extend the rectangle beyond the edge of the part
5
3
8
Tab

Cut
Shape
Profile
Cut Across Bend
Cut Normal
Extents
Distance
Thickness
OK
Cancel

Ensure that the Cut across Bend tool is on.
3
5
8
Use the Cut tool to remove the rectangular-shaped material

10
5
3
Create and cut a second rectangular shape

Finished cut

Figure 13-17

Hole Patterns

A hole pattern is created from an existing hole. See Figure 13-18.

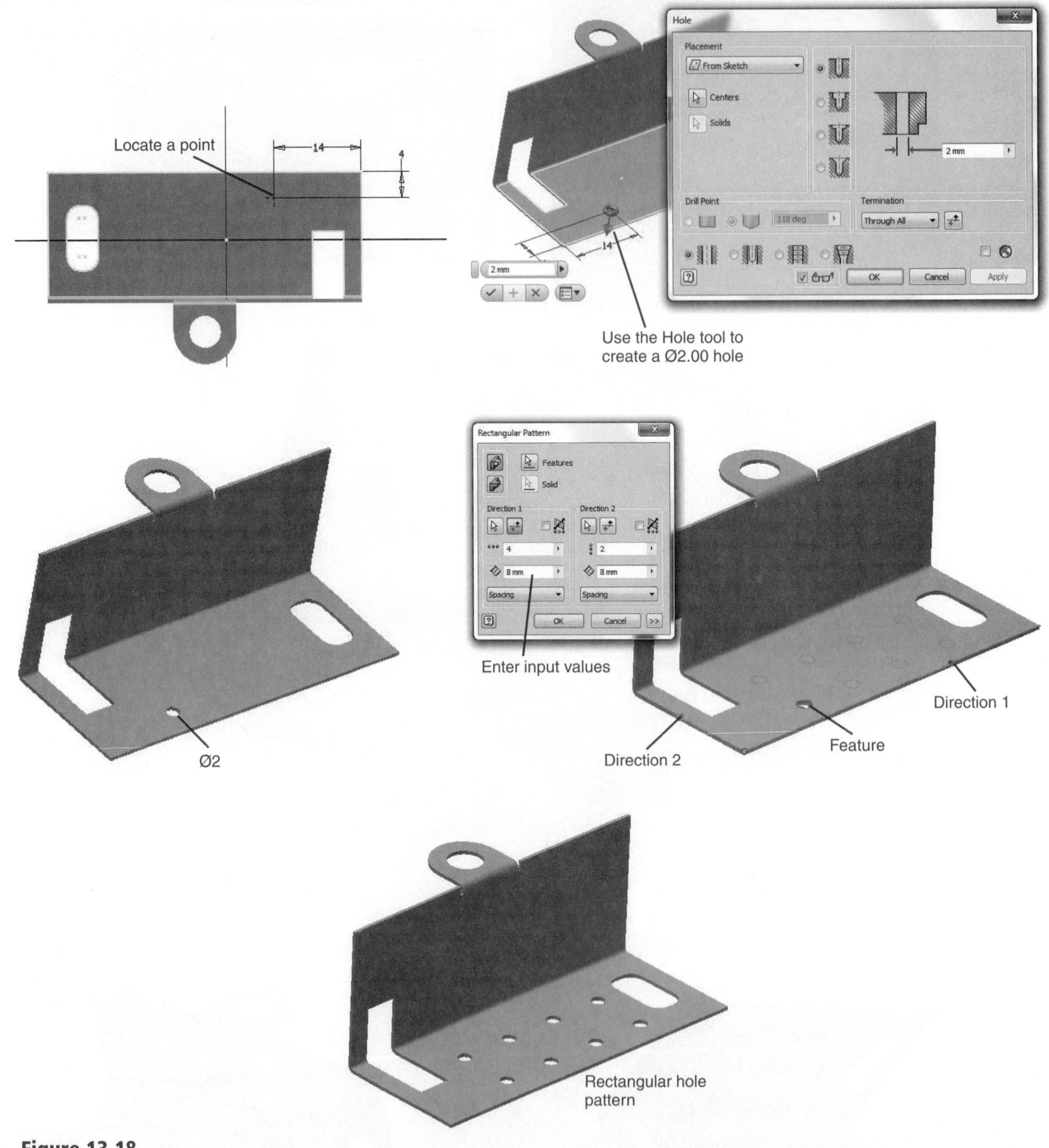

Figure 13-18

1. Create a new sketch plane on the horizontal flange.

2. Use the **Point, Center Point** tool and create a hole on the flange.

3. Use the **Dimension** tool to locate the center point.

4 Right-click the mouse and click the **Finish 2D Sketch** option.

5 Click the **Hole** tool on the **Sheet Metal** panel.

The **Hole** dialog box will appear.

6 Set the **Termination** for **Through All** and the hole's diameter for **2.**

7 Click **OK.**

The dimensions for the hole come from the dimensions given in Figure 13-1.

8 Click the **Rectangular** tool located on the **Pattern** panel under the **Sheet Metal** tab.

The **Rectangular Pattern** dialog box will appear.

9 Define the **Ø2 hole** as the **Feature.**

10 Click the arrow under the **Direction 1** heading, then the top front edge of the part. Use the **Flip Direction** button to change direction if necessary.

11 Set the number of holes under **Direction 1** for **4** and the spacing for **8 mm.**

12 Click the arrow under the **Direction 2** heading, and click the left front edge of the part to define the direction.

13 Set the number of holes for **2** and the distance for **8 mm.**

14 Click **OK.**

Flat Patterns

Flat patterns of 3D sheet metal parts can be created using the **Flat Pattern** tool. See Figure 13-19.

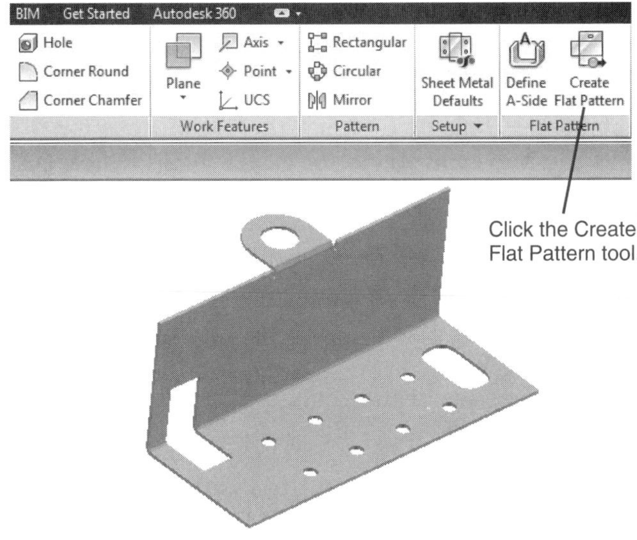

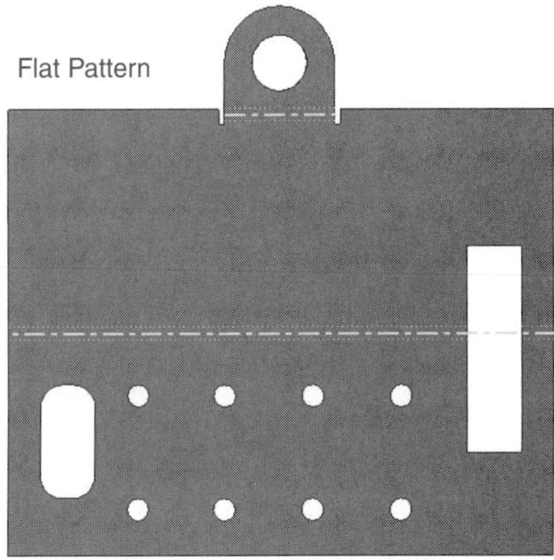

Figure 13-19

1 Click the **Flat Pattern** tool.

A flat pattern will automatically be created.

Punch Tool

The **Punch Tool** is used to create various shapes in sheet metal parts. Because sheet metal parts are thin, many shapes are created by punching through the material. Sheet metal is placed in a press and a tool with the desired shape is inserted. The press then presses down quickly, piercing the sheet metal with the punch tool to create the desired shape.

To Use the Punch Tool

1. Draw a **4-in.** × **6-in.** rectangle using the **Sheet Metal (in).ipt** format. It is presented in an isometric orientation.

   See Figure 13-20.

Figure 13-20

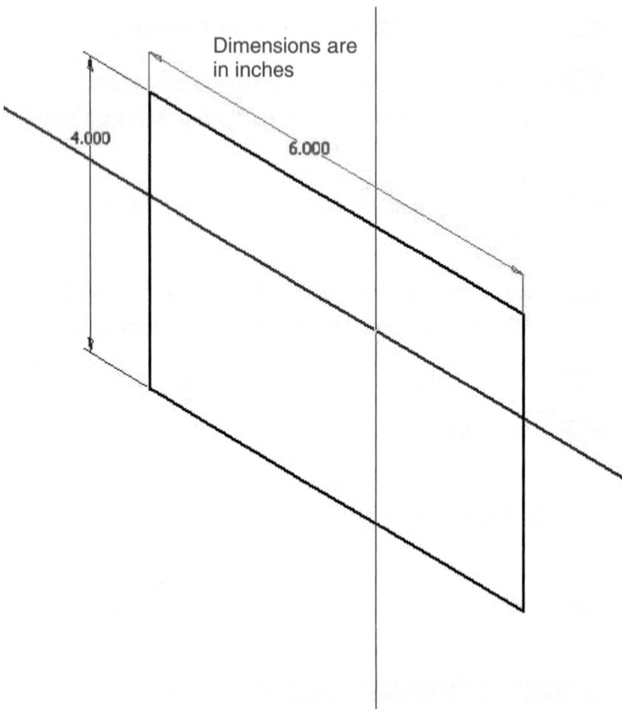

2. Right-click the mouse and select the **Finish 2D Sketch** option.
3. Click the **Sheet Metal Default** tool; set the **Thickness** for **0.1019** (#10 gauge) and the **Material Style** for **Steel, Mild.** Click **OK.**
4. Click the **Face** tool.

   There is only one shape on the screen, so the **Face** tool will automatically select the rectangle as the profile. See Figure 13-21.
5. Click **OK.**
6. Right-click the front surface of the rectangular part and create a new sketch plane.
7. Use the **Point, Center Point** tool and the **Dimension** tool and locate a center mark **1.25** from the left edge and **2.00** from the top edge.
8. Right-click the mouse and select the **Finish 2D Sketch** option.

   See Figure 13-22.
9. Click the **Punch Tool** on the **Sheet Metal** panel bar.

   The **PunchTool Directory** dialog box will appear. See Figure 13-23.

Figure 13-21

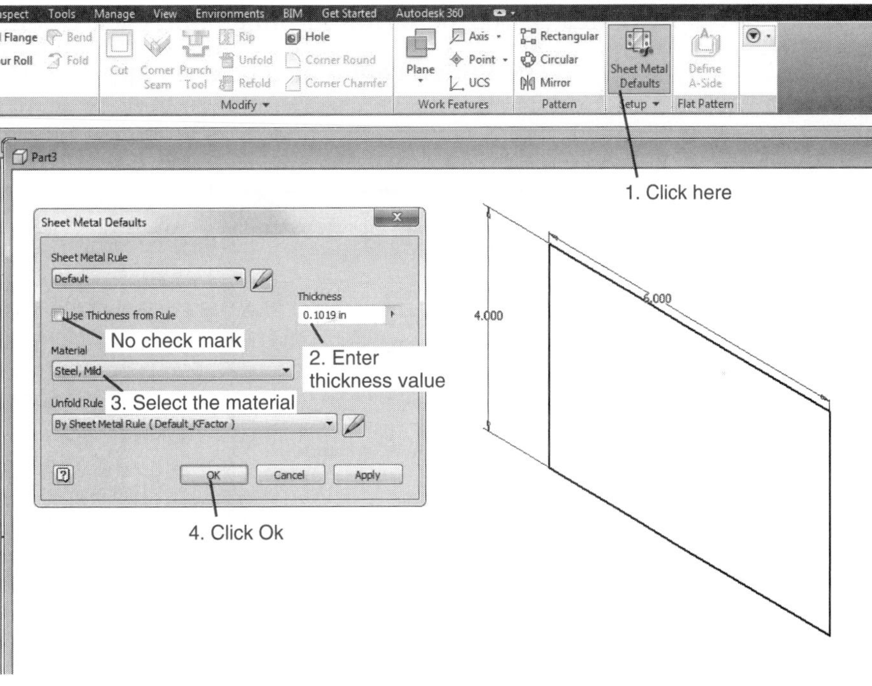

Figure 13-22

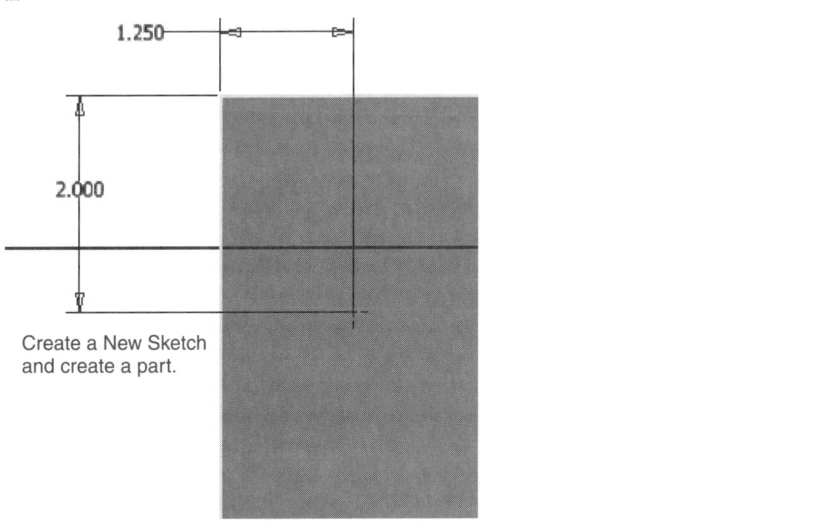

Figure 13-23

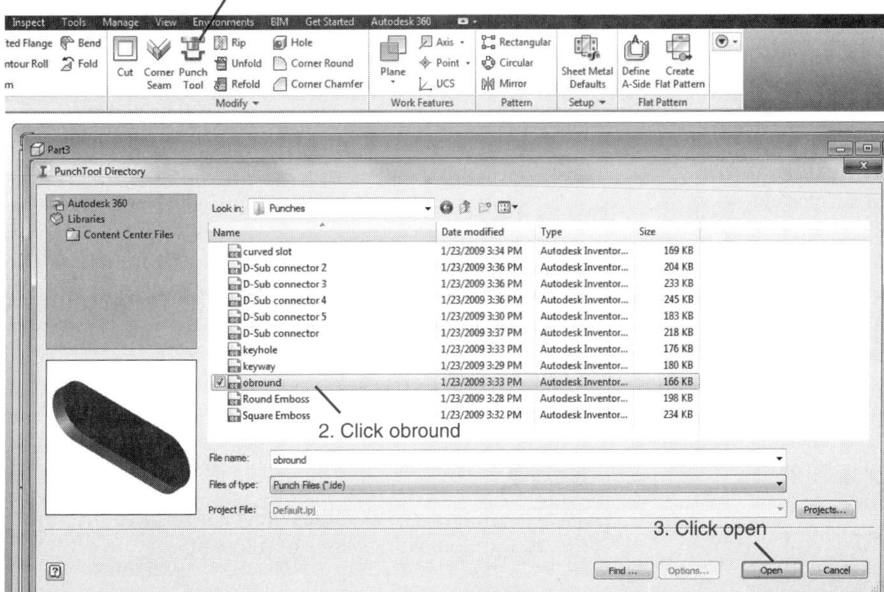

10 Select the **obround** punch and click **Open.**

The **PunchTool** dialog box will appear. See Figure 13-24.

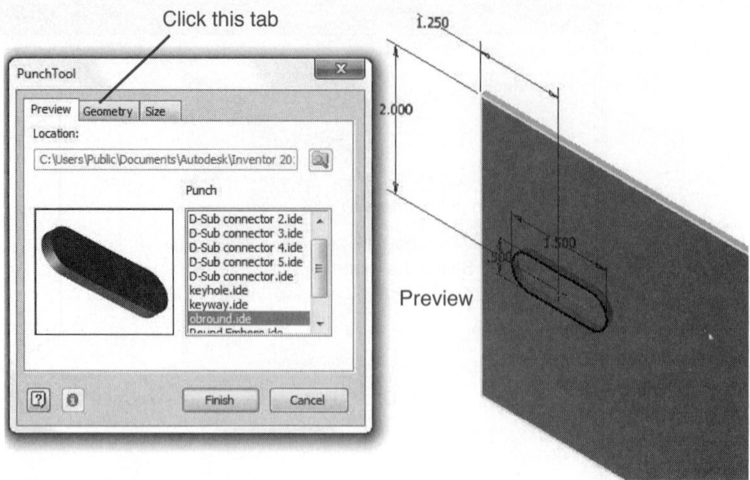

Figure 13-24

11 Click the **Geometry** tab and set the **Angle** value for **90°.**

See Figure 13-25.

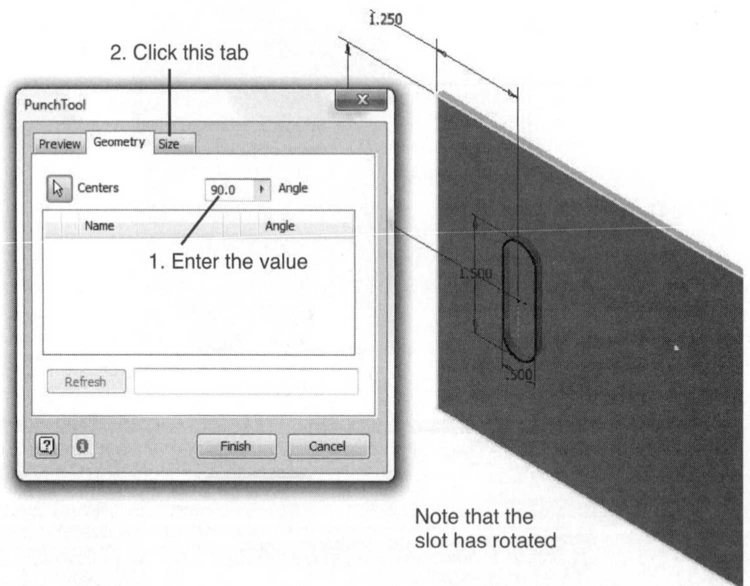

Figure 13-25

12 Click the **Size** tab and set the **length** and **width** values for **1.25** and **0.75,** respectively.

See Figure 13-26.

NOTE

The length and width values must conform to the standard sizes listed under the arrowheads. Random values are not allowed.

Figure 13-26

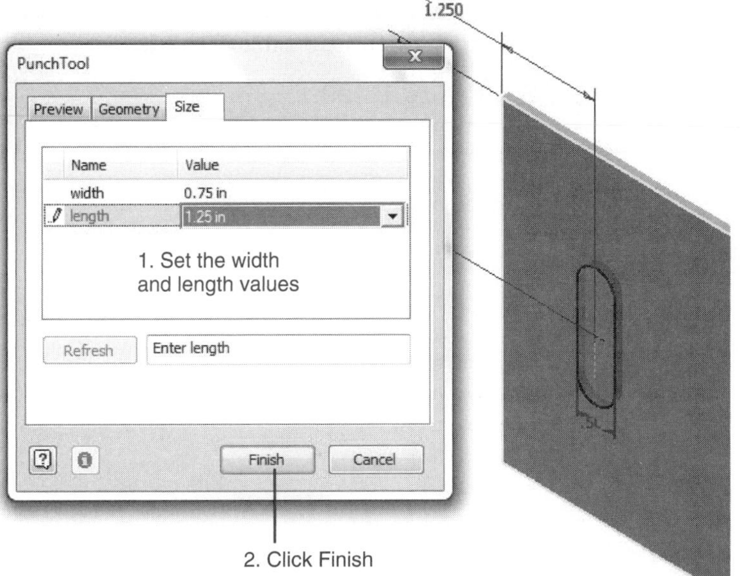

13 Click **Finish.**

See Figure 13-27.

14 Create another **New Sketch** plane on the front surface of the rectangular part and locate a center mark **3.00** from the left edge and **2.00** from the top edge.

15 Right-click the mouse and select the **Finish 2D Sketch** option.

See Figure 13-28.

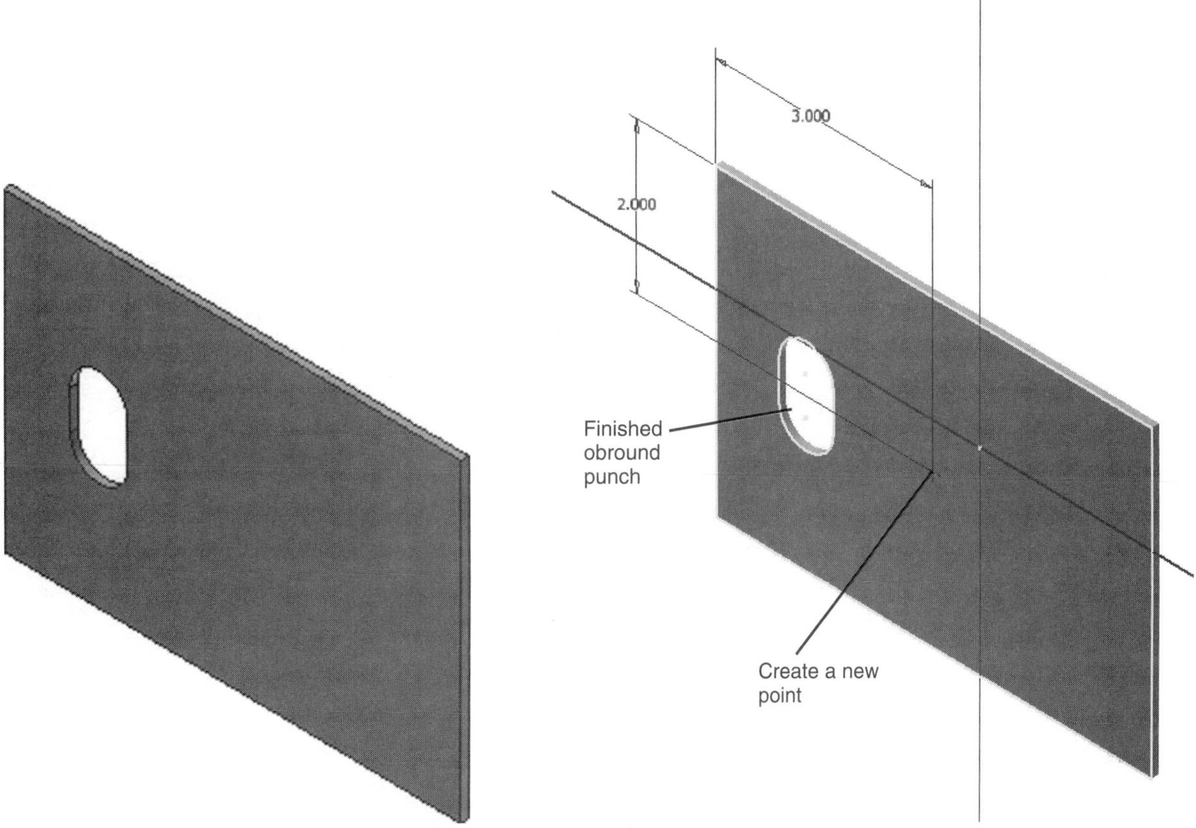

Figure 13-27

Figure 13-28

16 Click the **Punch Tool** and select the **keyhole.ide** option. Accept the default values and click **Finish.**

See Figure 13-29.

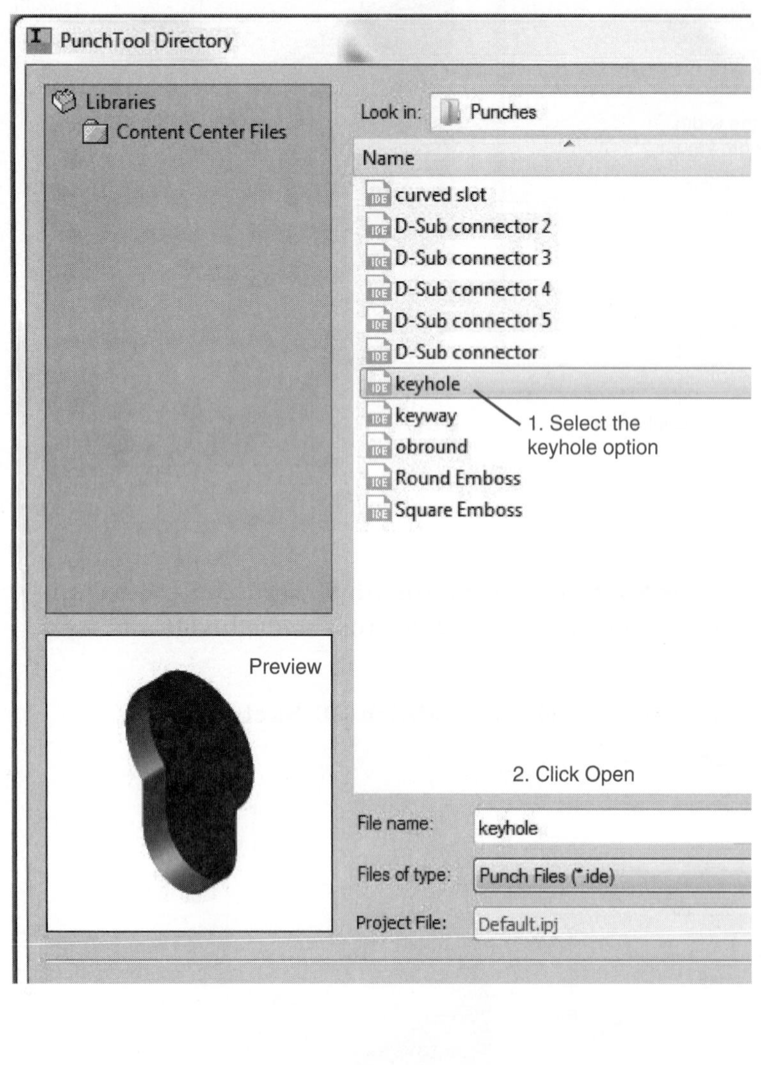

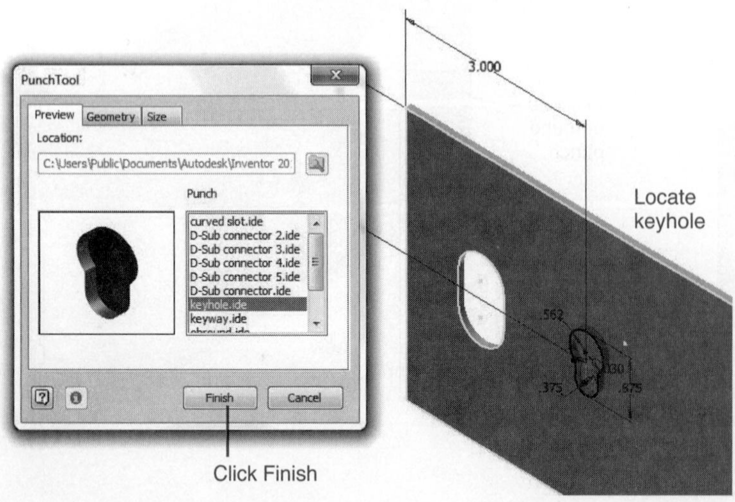

Figure 13-29

17 Create and locate a center mark **1.25** from the right edge and **2.00** from the top edge.

See Figure 13-30.

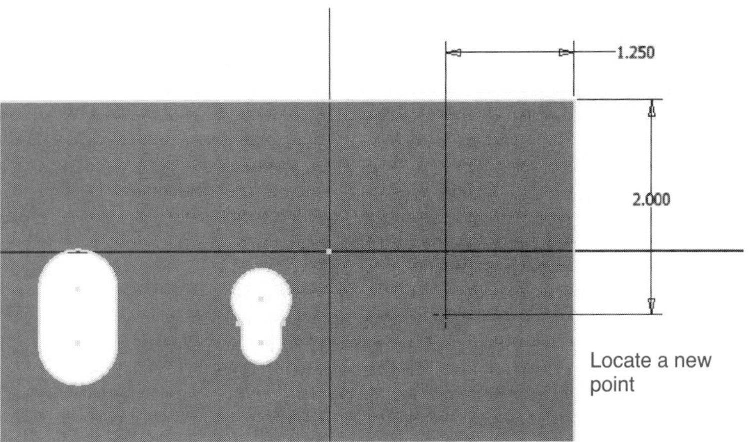

Figure 13-30

18 Access the **PunchTool Directory,** select the **Square Emboss** option, and click **Open.**

The **PunchTool** dialog box will appear. See Figure 13-31.

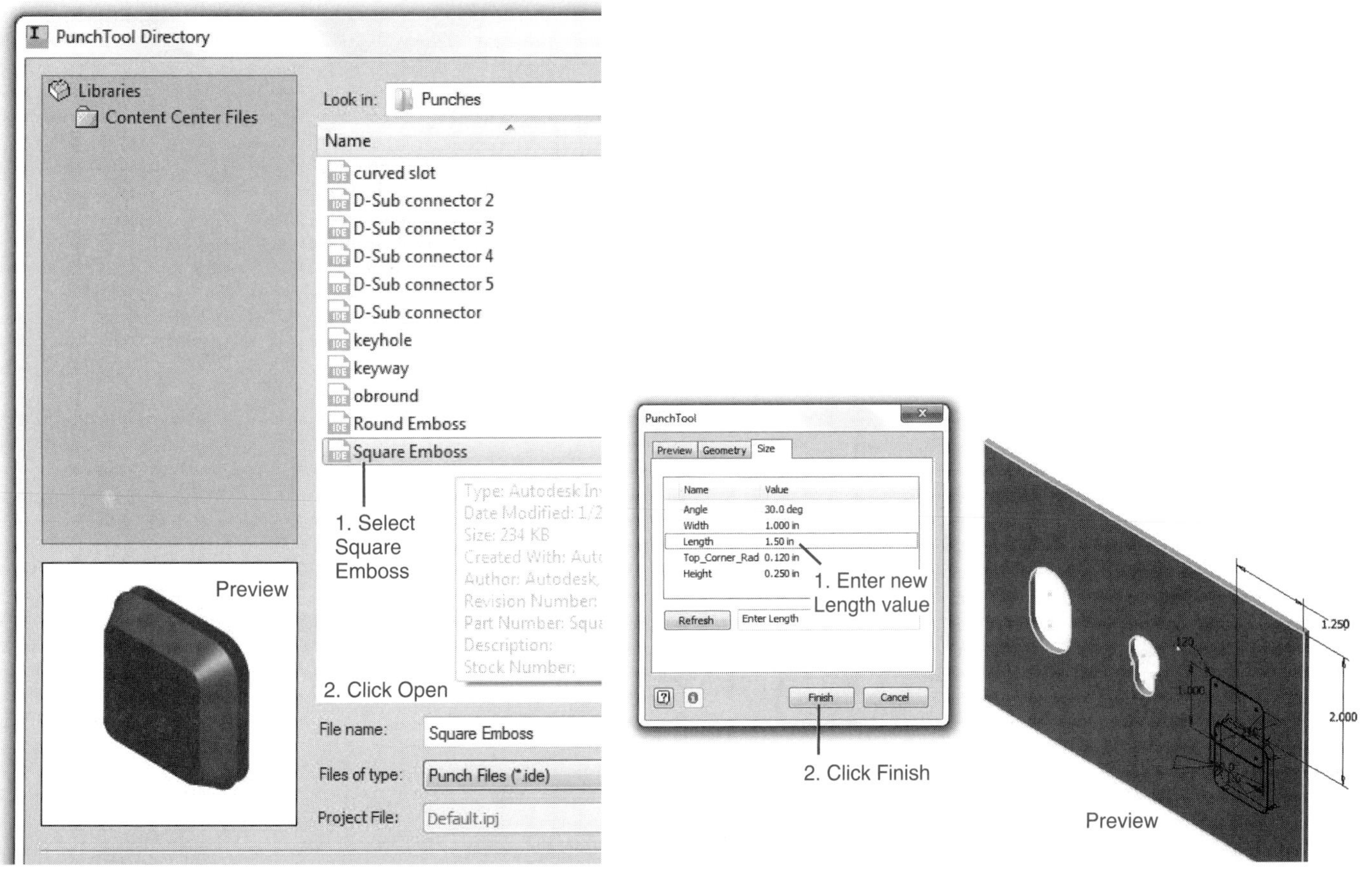

Figure 13-31

19 Change the **Length** value to **1.50.**

20 Click **Finish.**

Figure 13-32 shows the finished rectangular part.

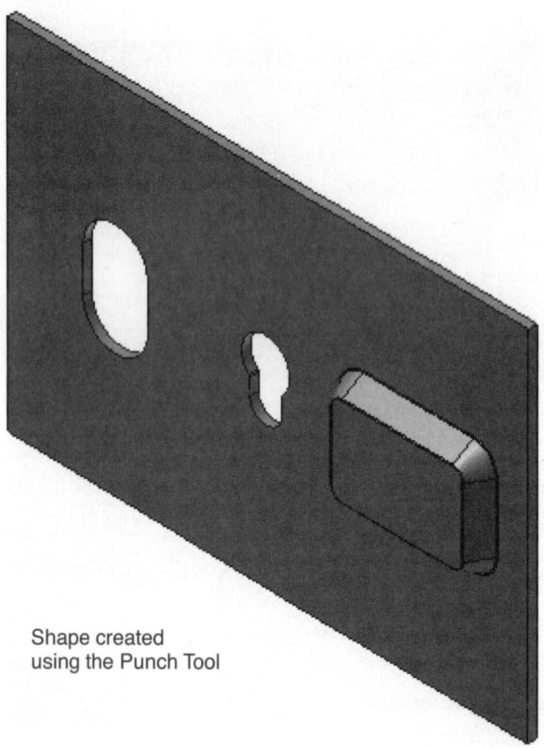

Figure 13-32

13 chapterthirteen

Chapter Summary

This chapter defined and illustrated how to create sheet metal drawings from 3D models and orthographic drawings. Features of sheet metal parts such as bend radii, flanges, tabs, and reliefs were presented, and flat patterns were created.

The use of the **Punch Tool** was also illustrated.

Chapter Test Questions

Multiple Choice

Circle the correct answer.

1. How thick is a piece of #12 gauge sheet metal?
   a. 0.4600 in.
   b. 0.0808 in.
   c. 0.0571 in.
   d. 0.0050 in.
2. How thick is a piece of #30 gauge sheet metal?
   a. 0.1443 in.
   b. 1/8 in.
   c. 0.0100 in.
   d. 0.0031 in.
3. The thickness of a piece of sheet metal is defined using which tool?
   a. **Sheet Metal Defaults**
   b. **Face**
   c. **Fold**
   d. **Hem**
4. A small piece of bent material that does not run the entire length of an edge is called a
   a. Flange
   b. Tab
   c. Relief
   d. Contour
5. Which of the following materials is not available in the **Material** option of the **Sheet Metal Defaults** dialog box?
   a. Steel, Mild
   b. Aluminum-6061
   c. Brass, Soft
   d. Plexiglas

True or False

Circle the correct answer.

1. **True or False:** In the English unit system the higher the sheet metal gauge number, the thinner the material.
2. **True or False:** A cut made next to a tab to allow for smooth bending is called a relief.
3. **True or False:** As sheet metal is bent, the inside surface is subjected to compression, and the outside surface to tension.
4. **True or False:** Normal surfaces are surfaces located 60° apart.
5. **True or False:** Punch tools can be used to create slots and keyholes.

Chapter Projects

Project 13-1

Redraw the sheet metal parts in Figures P13-1A through P13-1F using the given dimensions. Use the default values for all bend radii and reliefs.

Figure P13-1A
MILLIMETERS

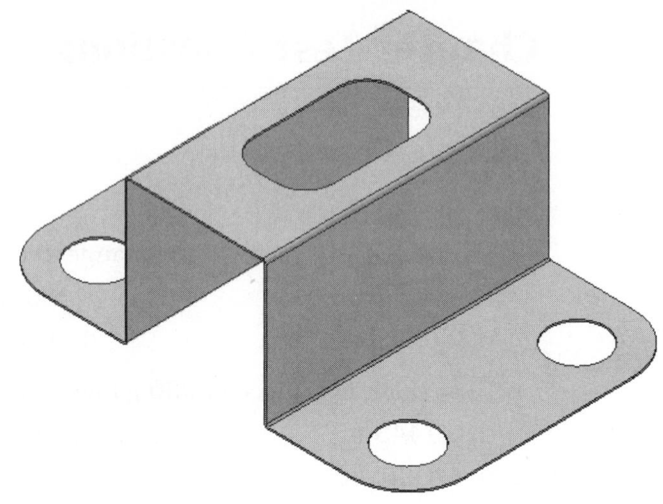

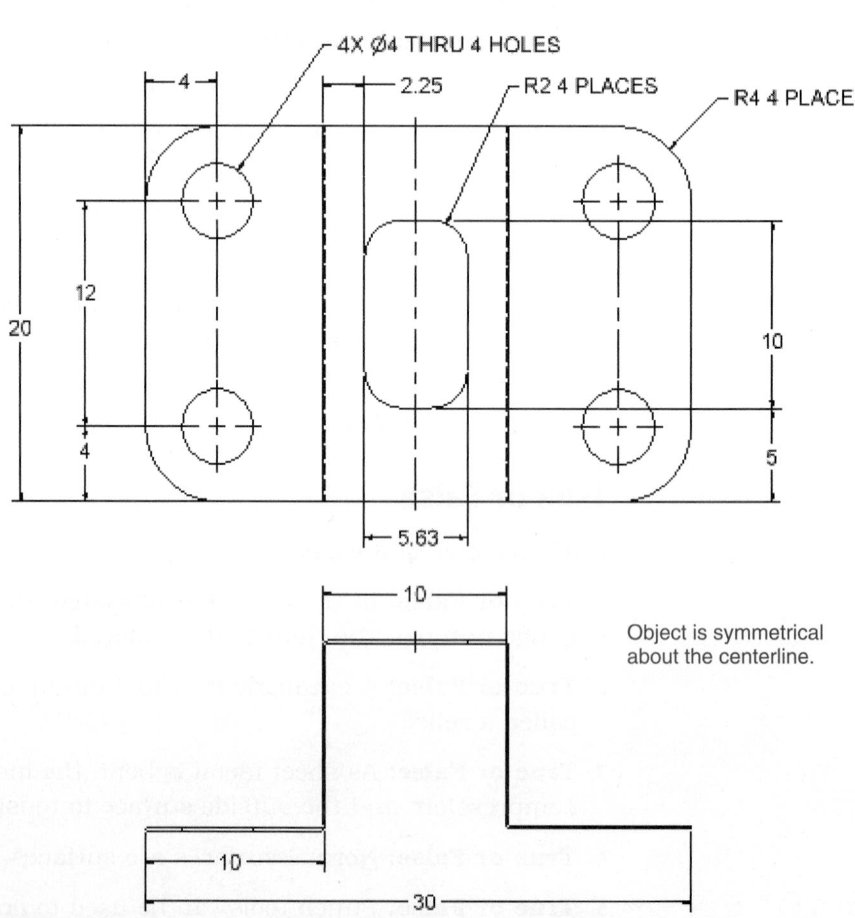

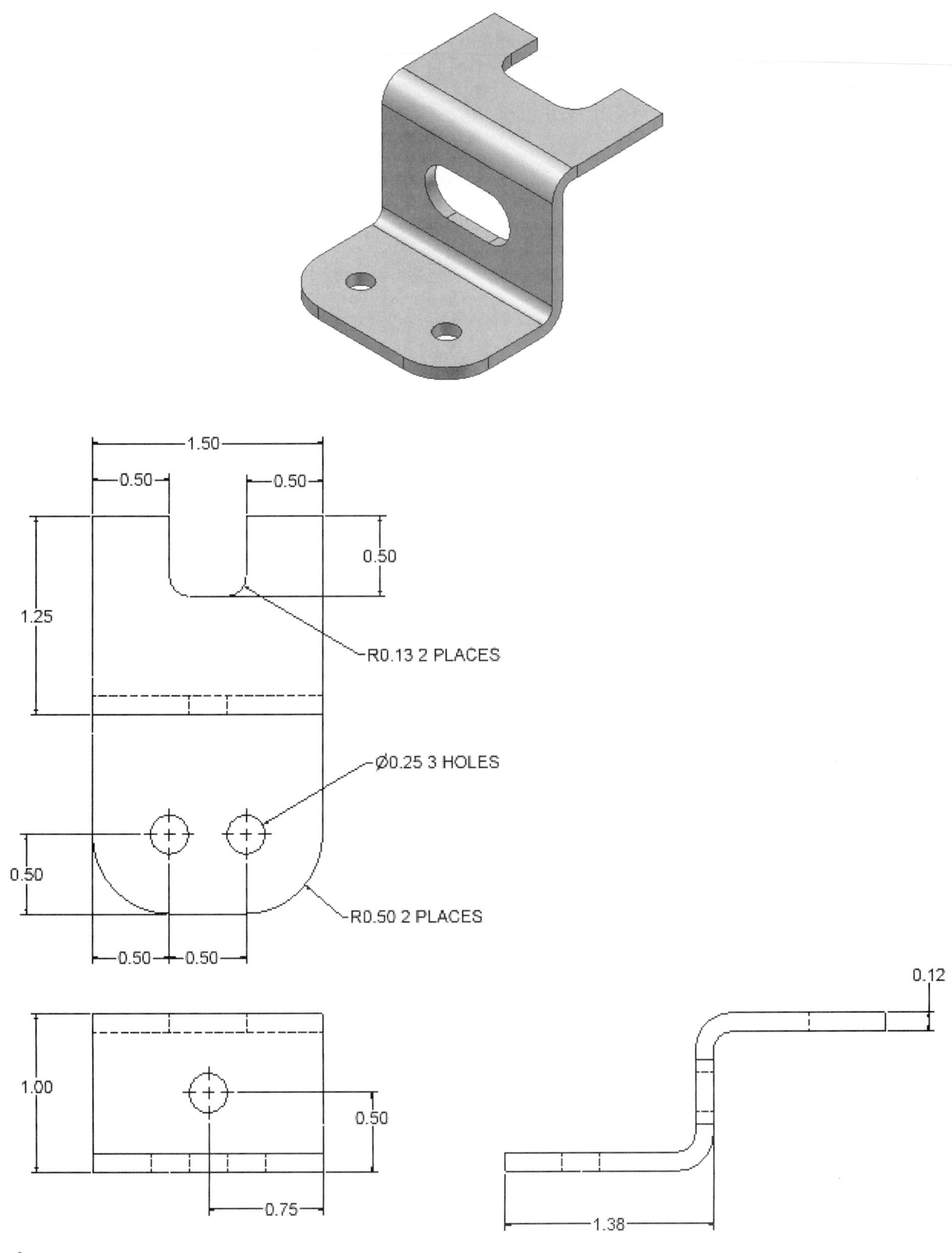

Figure P13-1B
INCHES

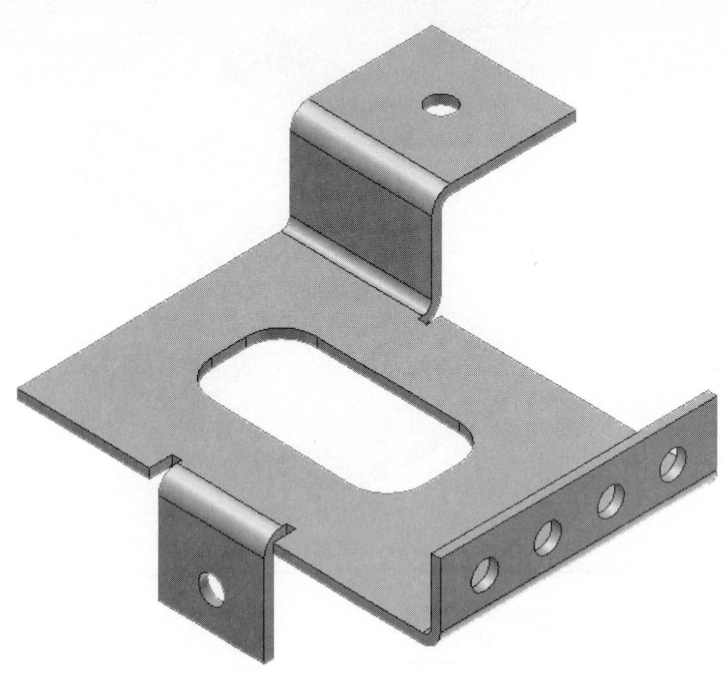

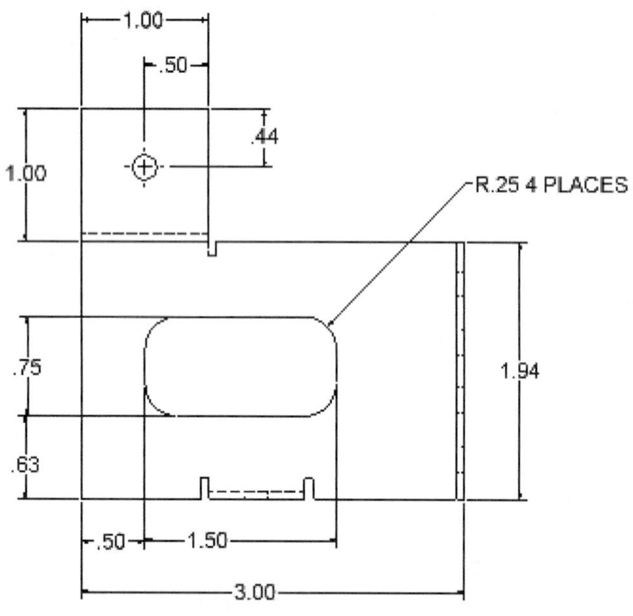

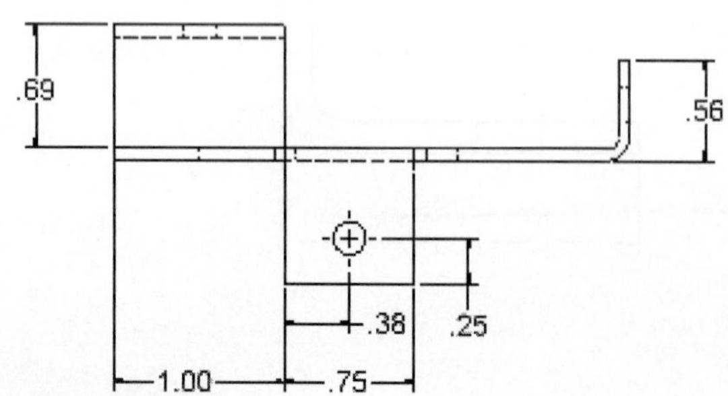

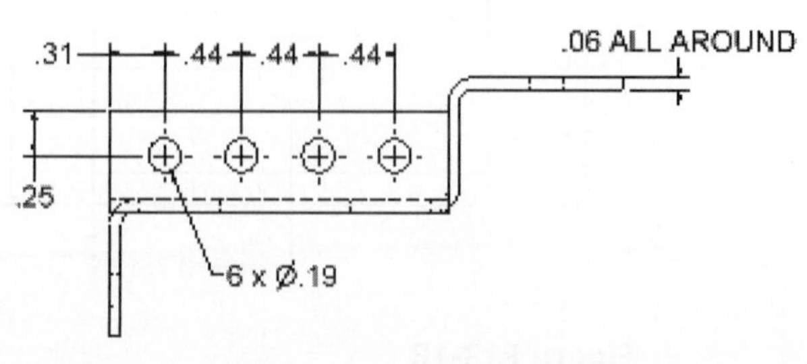

Figure P13-1C
INCHES

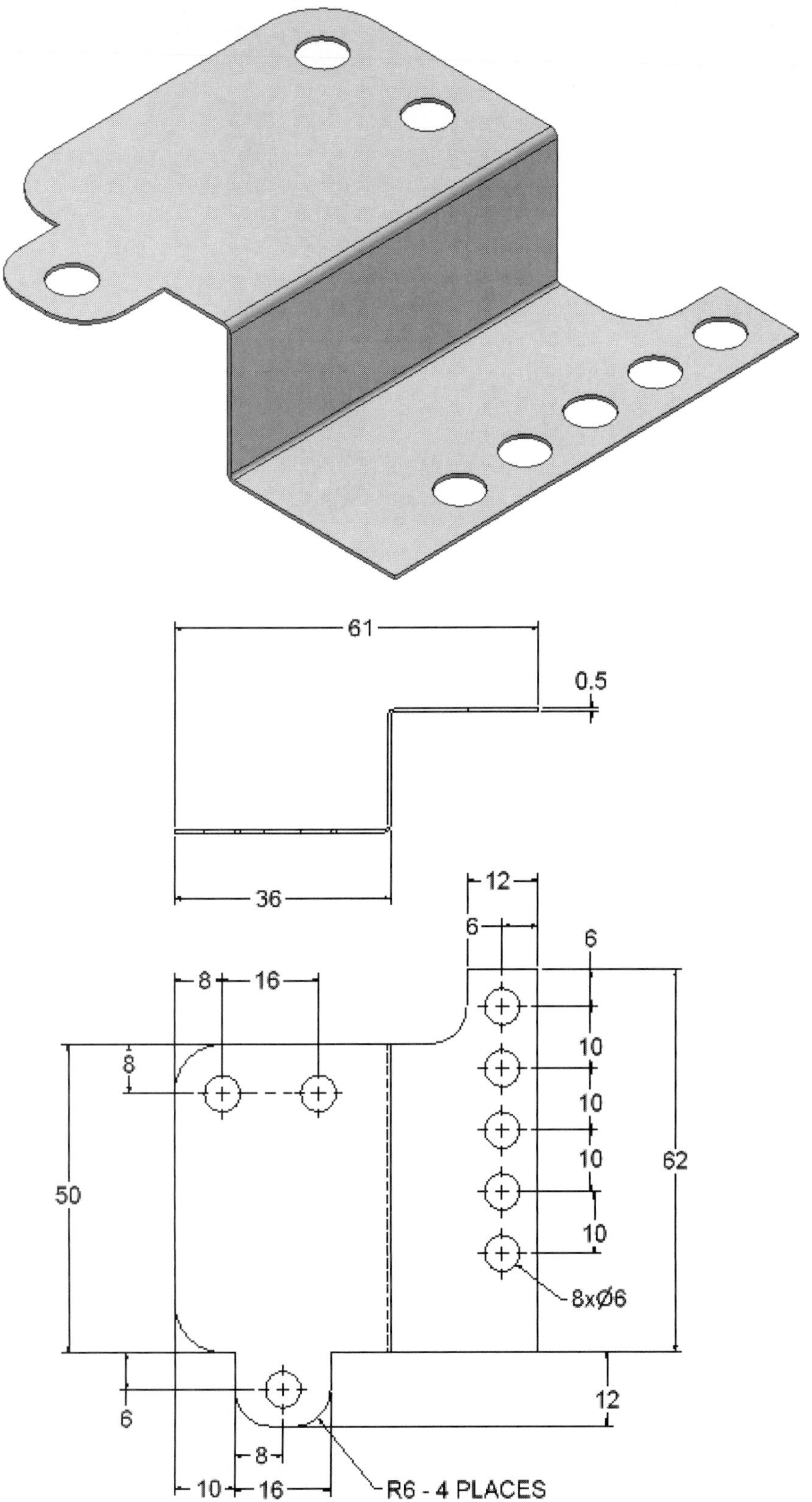

Figure P13-1D
MILLIMETERS

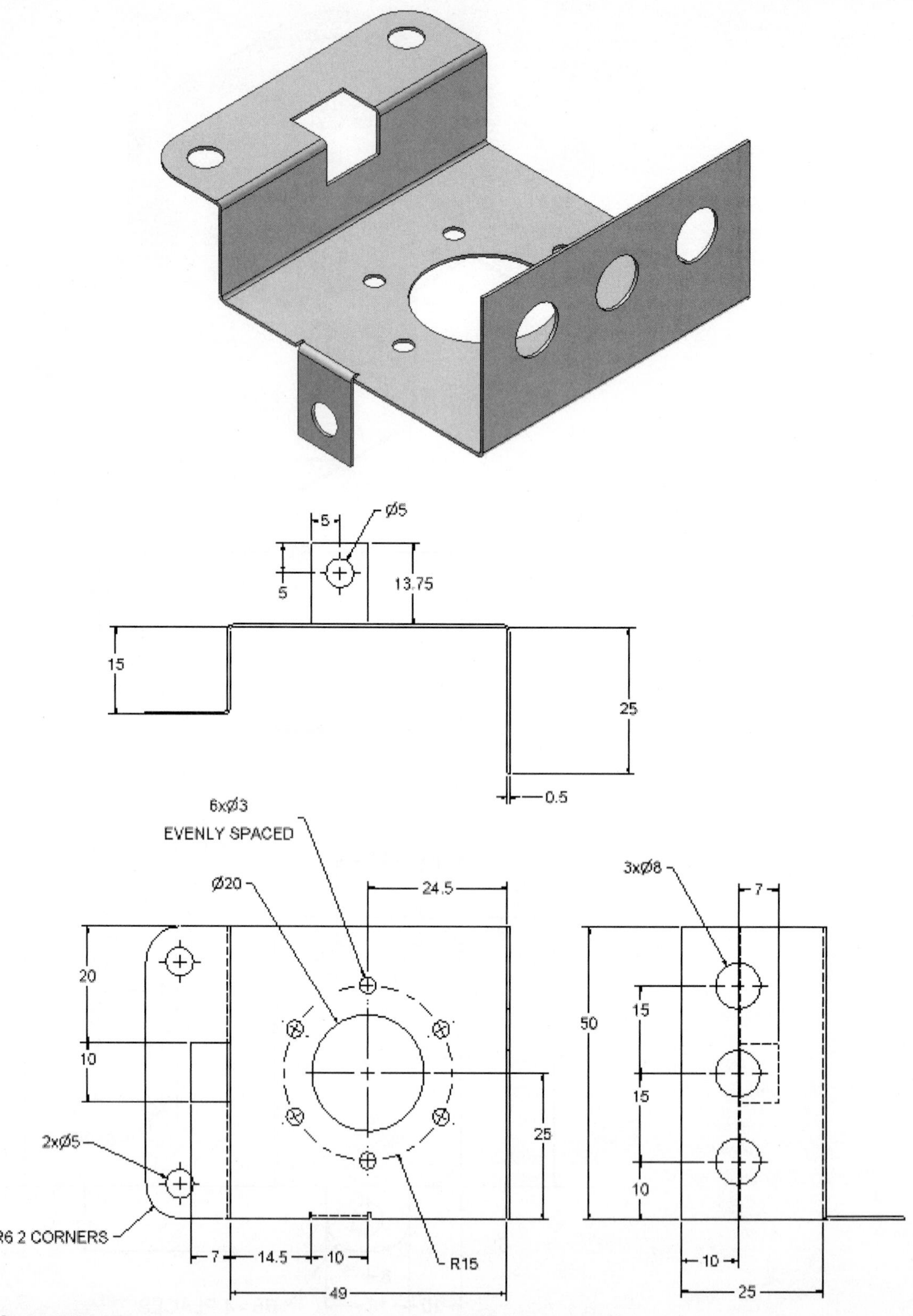

Figure P13-1E
MILLIMETERS

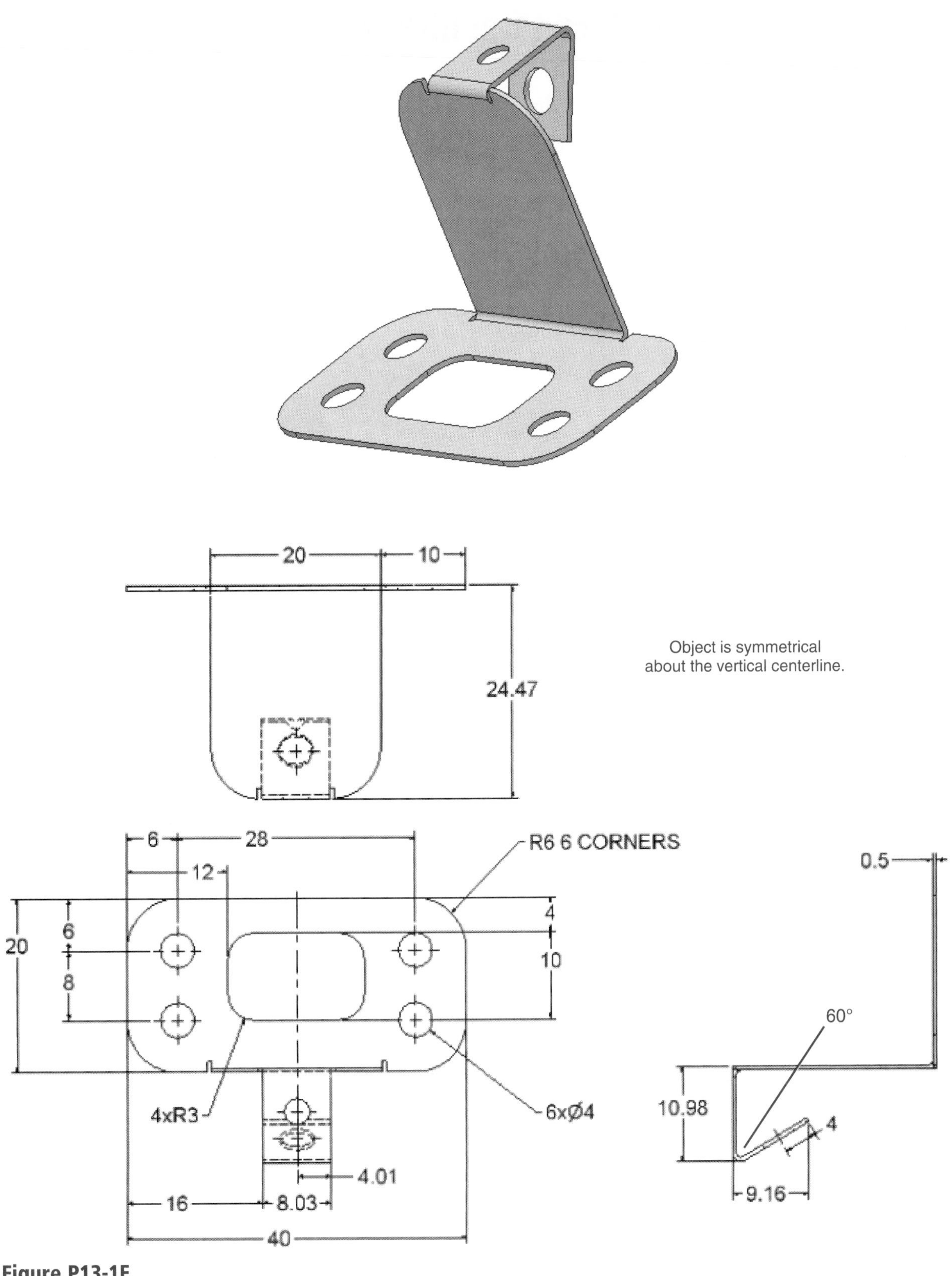

Figure P13-1F
MILLIMETERS

Project 13-2: Inches

Design and draw a box similar to that shown that has a capacity of

A. 100 cubic centimeters and is a cube.
B. 4 fluid ounces.
C. 100 cubic centimeters and is rectangular with the length of one side 2 times the length of the other.
D. 125 cubic inches and is a cube.
E. 125 cubic inches and is rectangular with the length of one side 1.5 times the length of the other.
F. 8 fluid ounces.

Figure P13-2

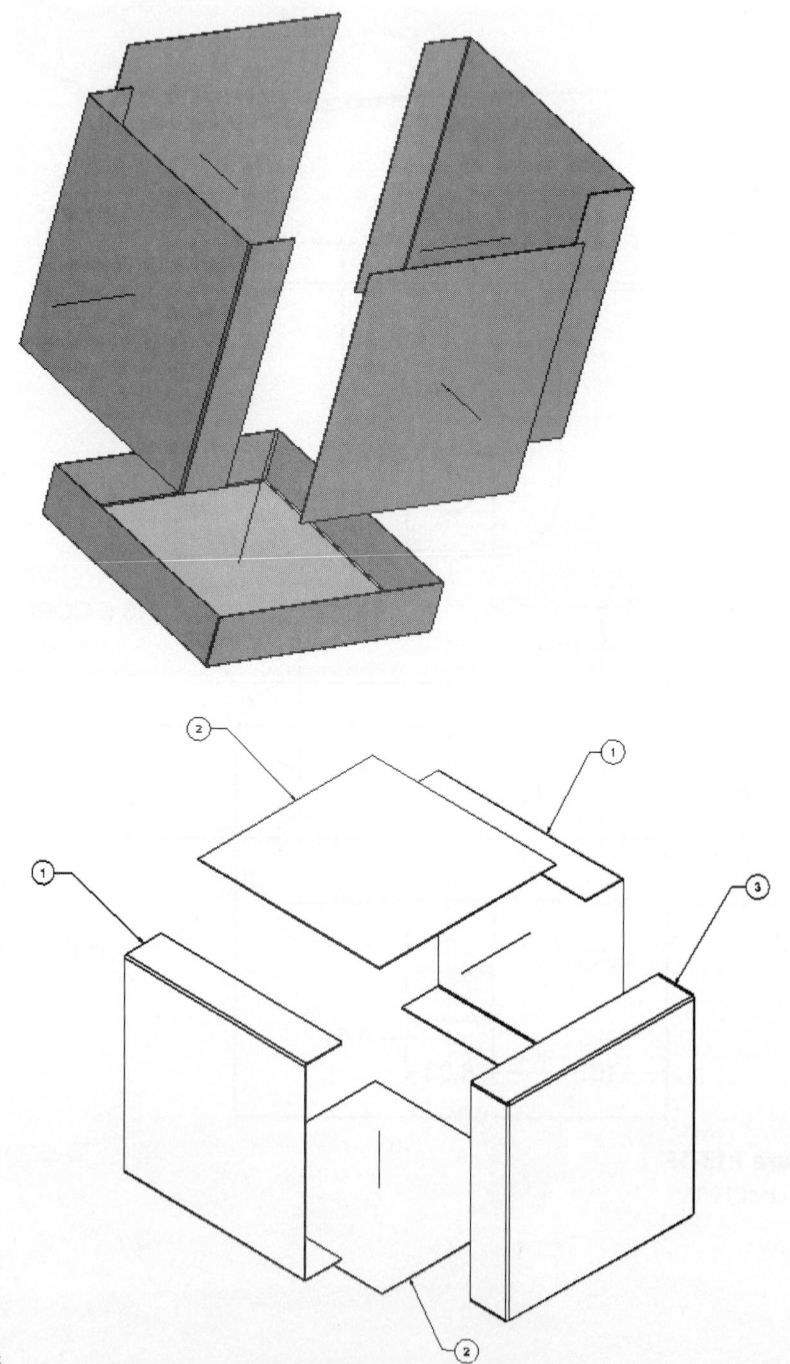

Project 13-3: Inches

Draw the sheet metal part shown in Figure P13-3.
The part is made from #16 gauge mild steel.
Use the following values for the **Punch Tool** shapes.

The Round Emboss:

Height = 0.125
Diameter = 1.250
Angle = 30°

The Slot (obround):

Length = 1.50
Width = 0.50

The Curved Slot:

Outer radius = 0.625
Inner radius = 0.250
Angle = 180°

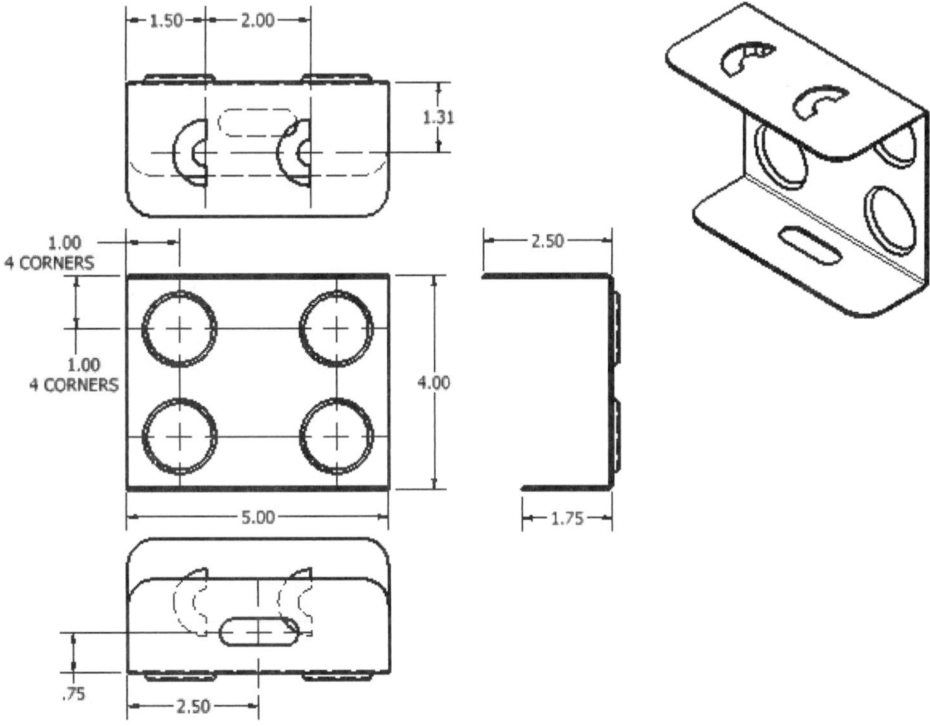

Figure P13-3

chapter fourteen

Weldment Drawings

CHAPTER OBJECTIVES

- Understand how to design and draw weldments
- Learn about fillet and groove welds
- Learn how to create weld symbols

Introduction

weldment: An assembly made from several smaller parts that have been welded together.

Weldments are assemblies made from several smaller parts that have been welded together. Weldments are often less expensive to manufacture because they save extensive machining time or replace expensive castings.

Fillet Welds

fillet weld: A weld usually created at 45° to join pieces that are perpendicular to each other; may be continuous or intermittent.

Figure 14-1 shows a simple weldment. It was created from two 0.375-in. thick plates and joined by a ***fillet weld.*** The base plate is 2.00 × 4.00 in., and the vertical plate is 1.25 × 4.00 in. Both parts are made from low-carbon steel.

EXERCISE 14-1 **Creating the Components**

1. Click the **New** tool, then the **English** tab, **Weldment (ANSI).iam,** and **Create.**
2. Click the **Assemble** tab and click the **Create** tool located on the **Component** panel.

   The **Create In-Place Component** dialog box will appear. See Figure 14-2.

Figure 14-1

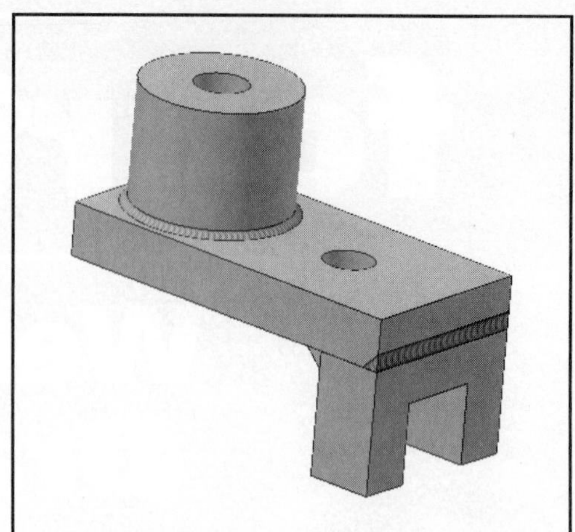

Figure 14-2

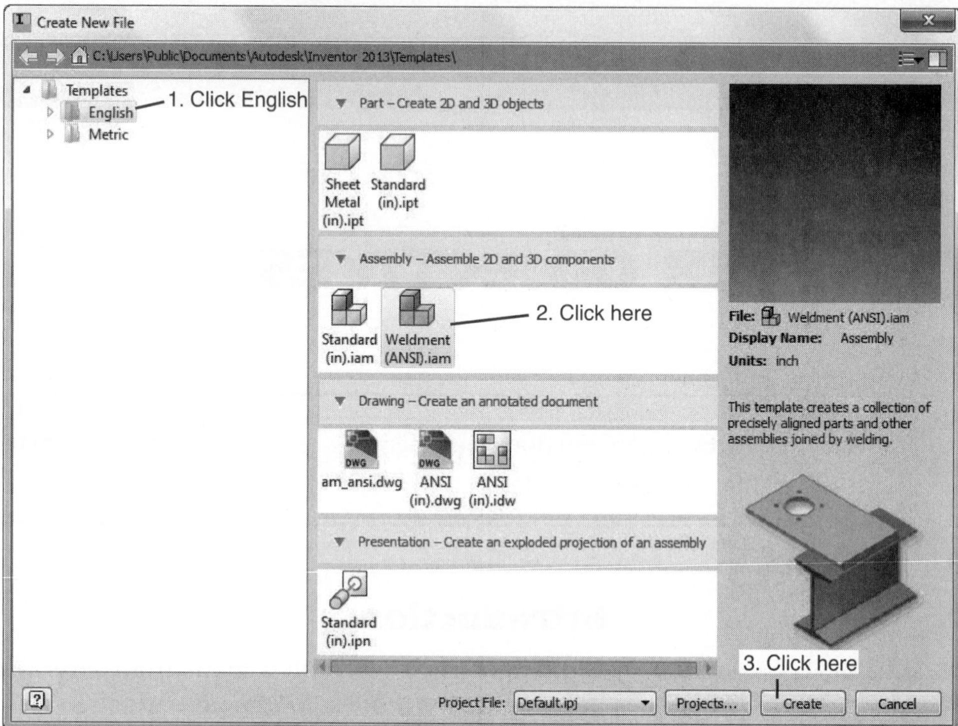

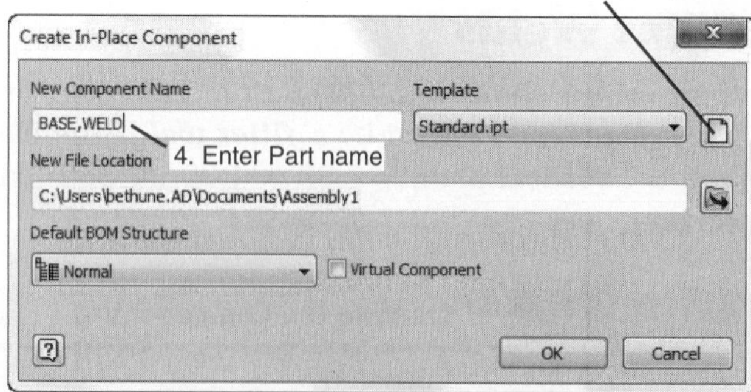

3 Define a new component named **BASE,WELD.**

4 Click the **Browse Templates** box.

The **Open Template** dialog box will appear. See Figure 14-3.

Figure 14-3

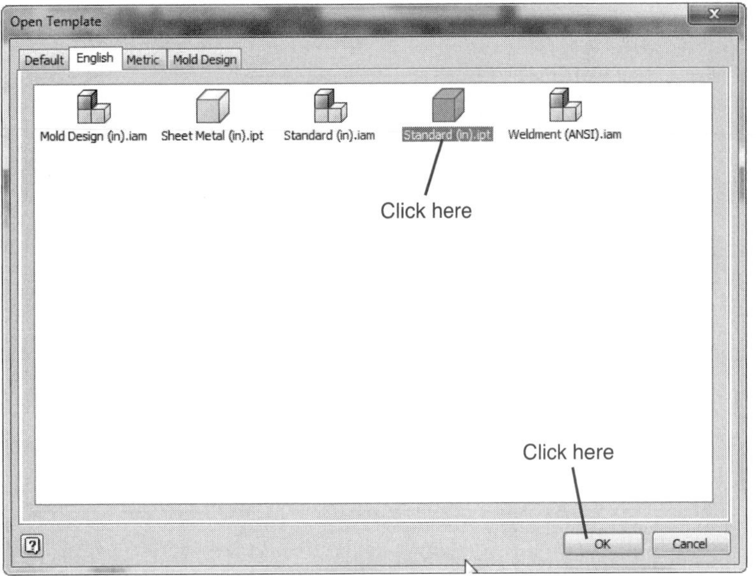

5 Select the **Standard (in).ipt** format. Click **OK.**

The **Create In-Place Component** dialog box will reappear. Note that **English\Standard (in).ipt** is the new template. See Figure 14-4.

Figure 14-4

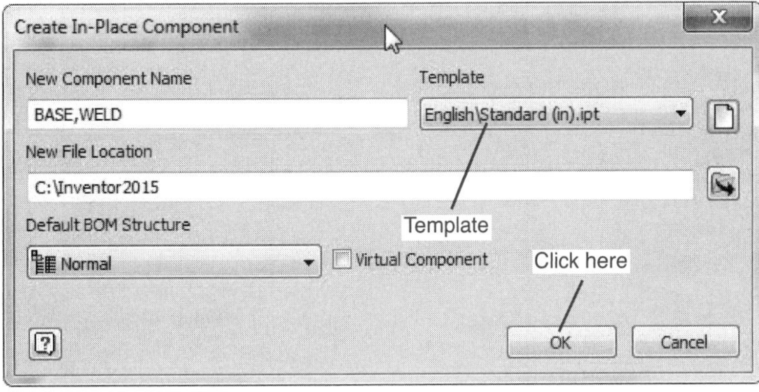

6 Click **OK.**

A small icon will appear next to the cursor.

7 Click the drawing screen.

The **Sketch** tab tools will appear.

8 Click the **Start 2D Sketch** tool, select the XY plane, and sketch a **2.00 × 4.00** rectangle.

9 Right-click the mouse and click **Finish 2D Sketch.**

The **3D Model** tab tools will appear.

10 Click the **Extrude** tool and define a thickness of **0.375** for the rectangle.

See Figure 14-5.

Figure 14-5

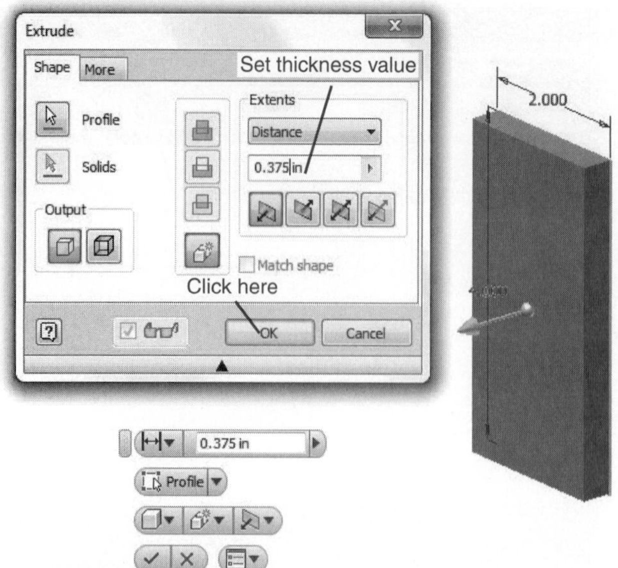

EXERCISE 14-2 Creating a Second Plate

1 Right-click the mouse and select the **Finish Edit** option.

See Figure 14-6.

Figure 14-6

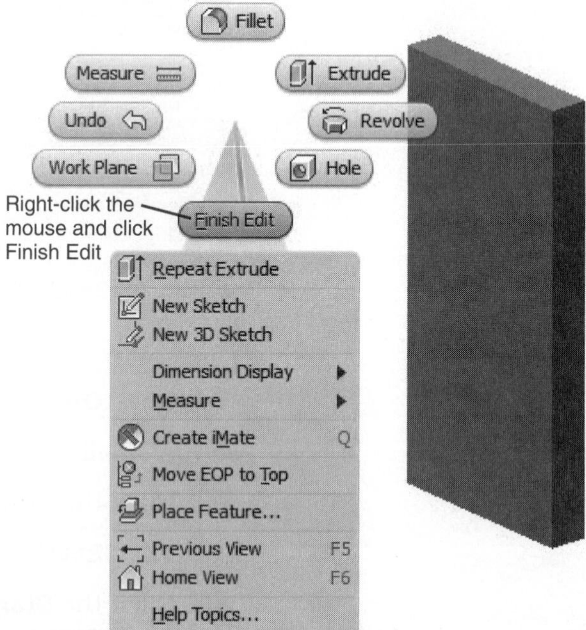

2 Select the **Copy** tool. The **Copy** tool is located on the **Pattern** panel under the **Assemble** tab.

See Figure 14-7. The **Copy Components: Status** dialog box will appear.

Figure 14-7

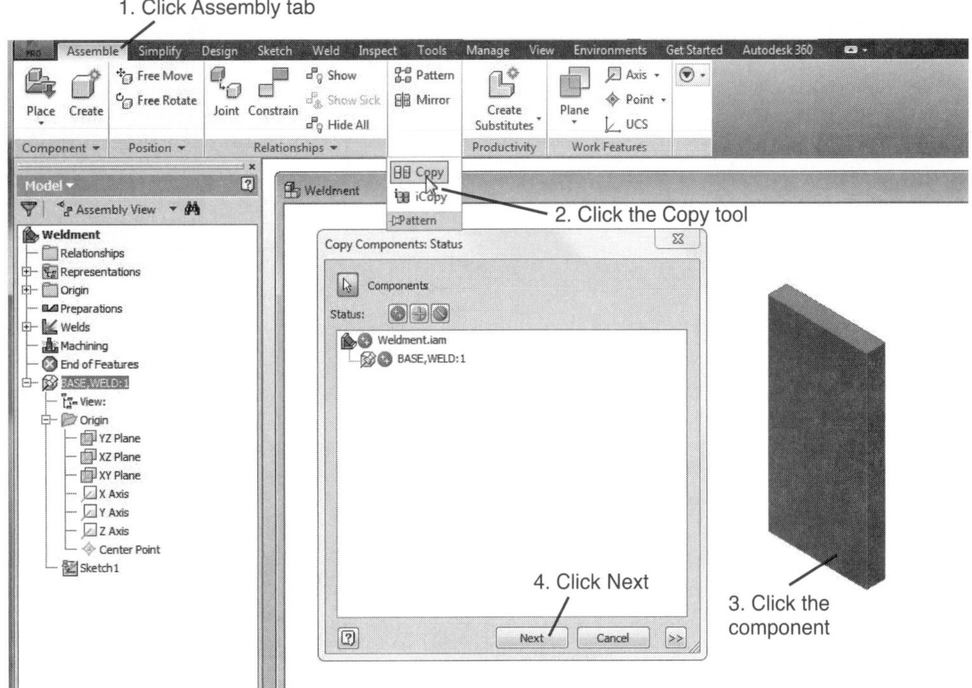

3 Click the **BASE,WELD:1** component.

The component file name will appear in the dialog box.

4 Click **Next.**

The **Copy Components: File Names** dialog box will appear. See Figure 14-8.

Figure 14-8

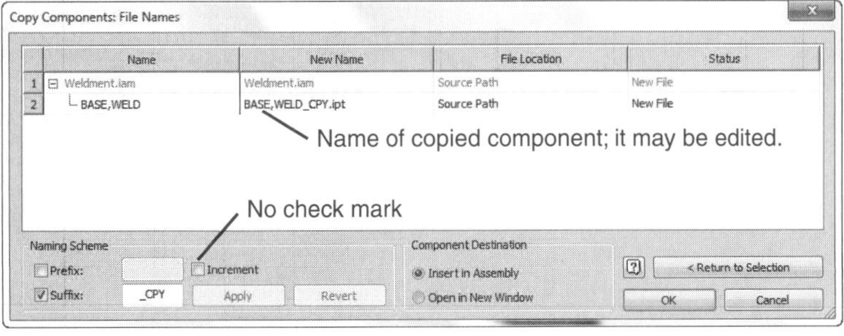

The copied component will be assigned a new file name. In this example the new name is **BASE,WELD_CPY.ipt.** Another name may be entered.

5 Click the **Increment** box and remove the check mark. Click the **OK** box.

The component copy will appear on the screen. See Figure 14-9.

Figure 14-9

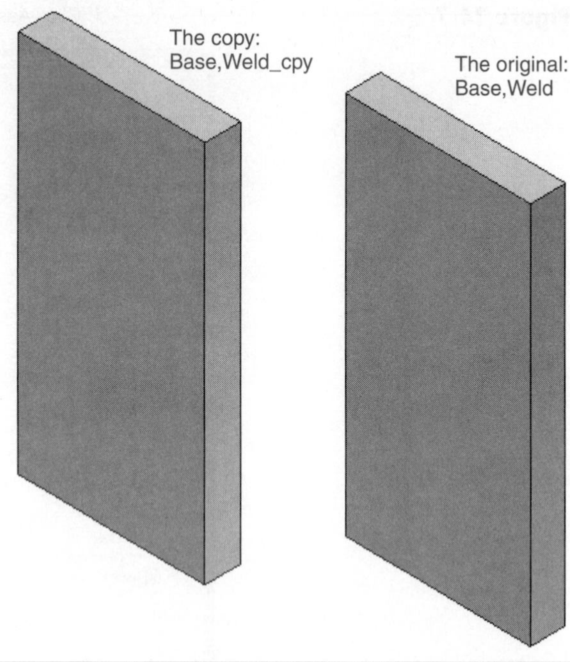

EXERCISE 14-3 Creating a T-Bracket

1 Right-click **BASE,WELD_CPY** in the browser box and click the **Grounded** option.

See Figure 14-10. This will remove the grounded constraint. The push-pin icon will disappear from the browser box.

Figure 14-10

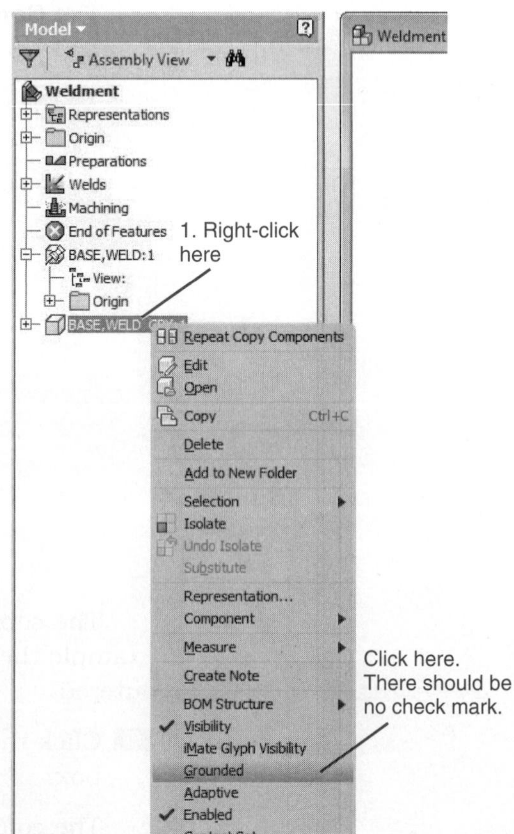

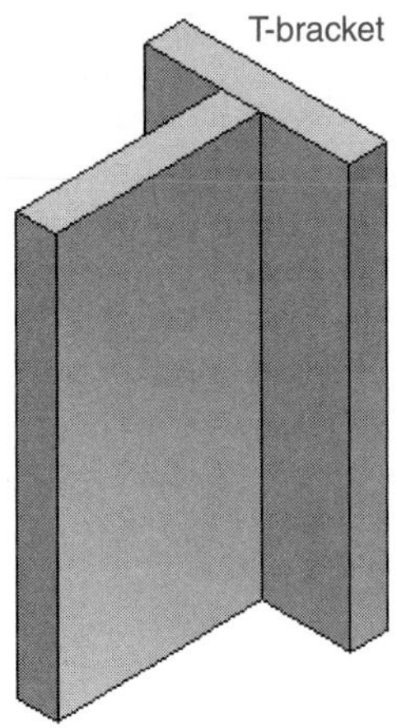

Figure 14-11

2 Use the **Free Rotate** and **Constrain** tools and assemble the two plates to form a T-bracket.

See Figure 14-11.

3 Right-click the mouse and select the **Finish Edit** option.

EXERCISE 14-4 Creating the Welds

1 Click the **Welds** tool on the **Process** panel under the **Weld** tab.

See Figure 14-12. The **Weld** panel tools will appear. (The **Weld** panel can also be accessed by double-clicking the **Welds** heading in the browser box.)

2 Click the **Fillet** tool on the **Weld** panel under the **Weld** tab.

The **Fillet Weld** dialog box will appear. See Figure 14-13.

Figure 14-12

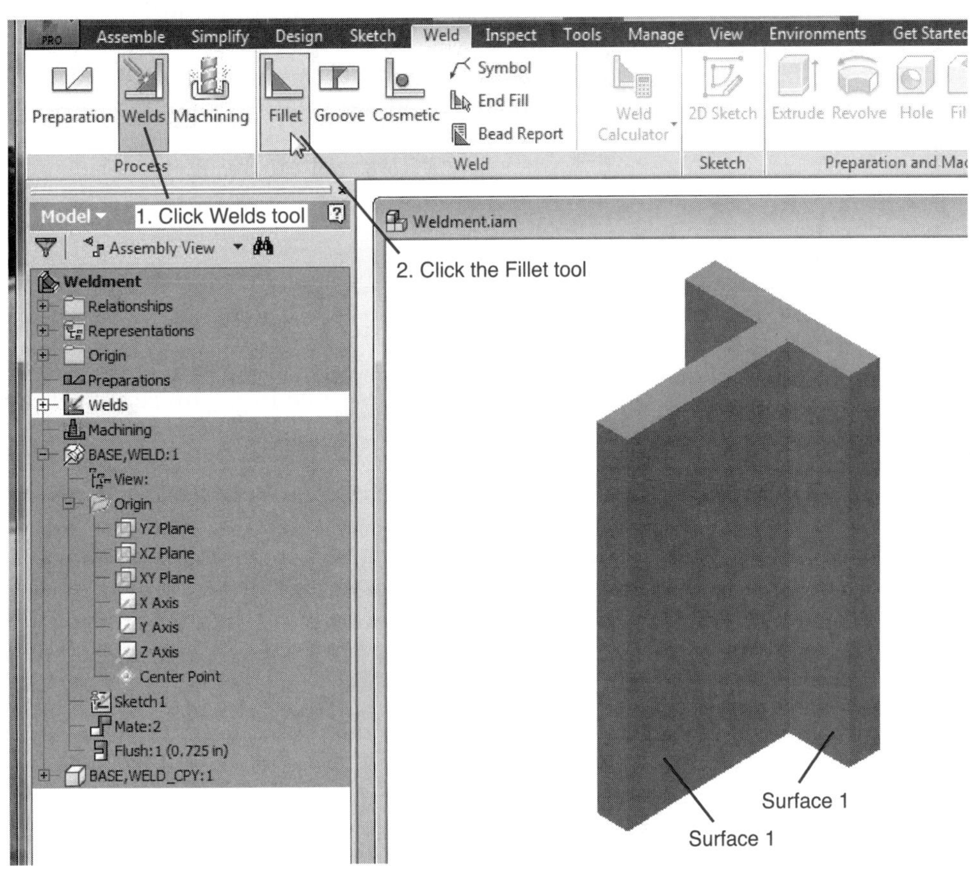

Figure 14-13

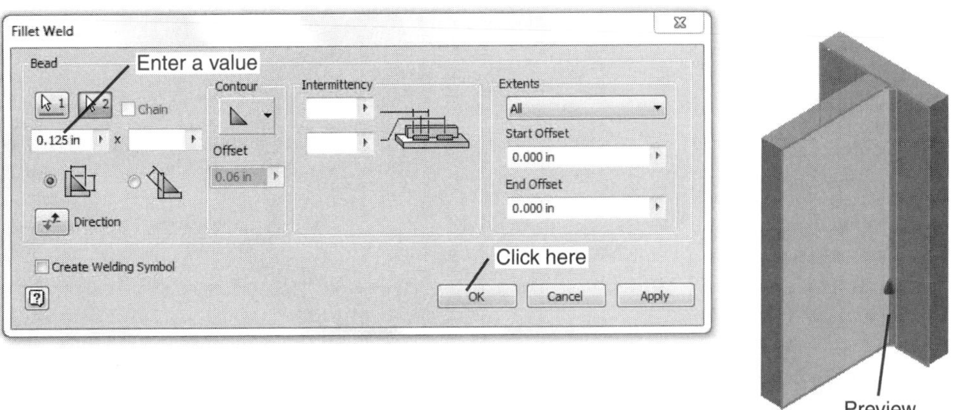

Figure 14-14

3. Set the weld size to **0.125,** as shown.

4. Click the **1** box, then click the front vertical surface of the part **BASE,WELD.**

5. Click the **2** box, then click the right vertical surface of the part **BASE,WELD_CPY.**

   A preview of the weld will appear as small right triangles. See Figure 14-13.

6. Click **OK.**

   Figure 14-14 shows the finished weld.

Intermittent Fillet Welds

Fillet welds may be located intermittently along a weld line.

1. Rotate the T-bracket created in the previous section so that the side opposite the fillet weld is expanded.

2. Click the **Fillet** tool on the **Weld** panel.

   The **Fillet Weld** dialog box will appear. See Figure 14-15.

3. Enter the appropriate **Intermittency** values.

   Note that the values **1.00** and **1.50** have been entered in the **Intermittency** box. The 1.00 value is the length of each weld, and the 1.50 value is the distance from the center of one weld to the center of the next.

4. Define surfaces **1** and **2** as before.

5. Click **Apply.**

   Figure 14-16 shows the finished intermittent welds.

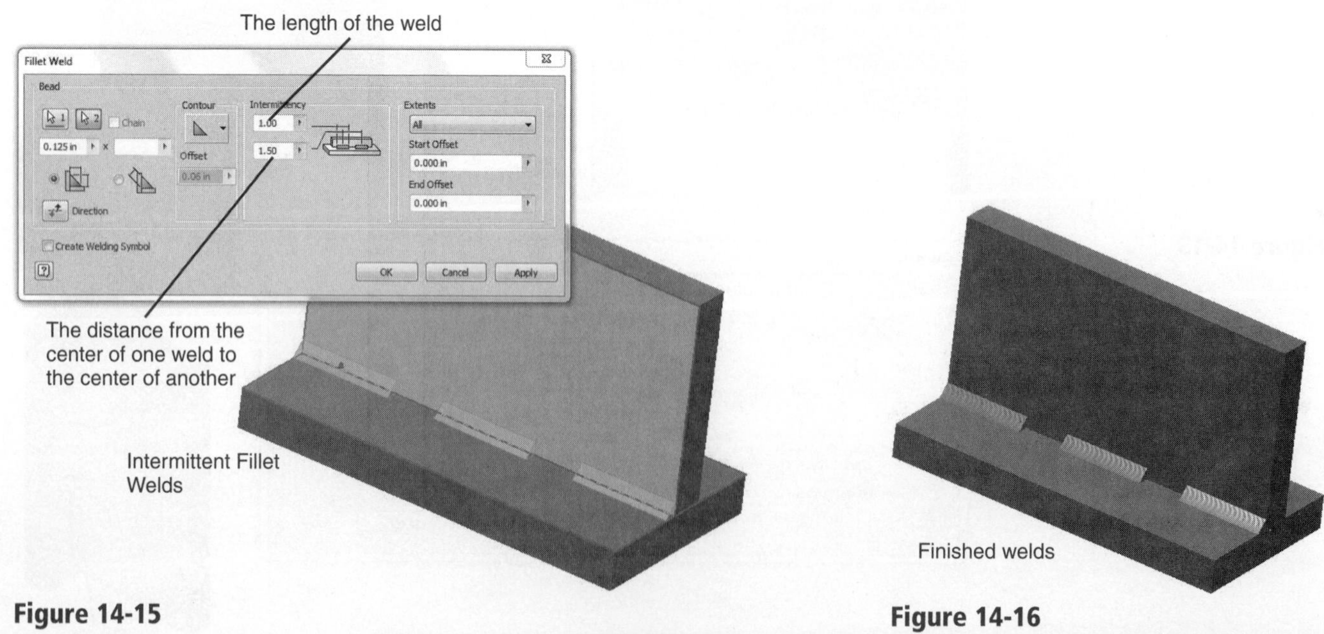

Figure 14-15

Figure 14-16

Weld Symbols

Welds are defined on drawings using symbols. The symbol for a fillet weld is shown in Figure 14-17. Note that the location of the flag-like portion of the symbol defines the location of the weld. It is not always possible to point directly at a weld location, so the **Other side** symbol is very useful.

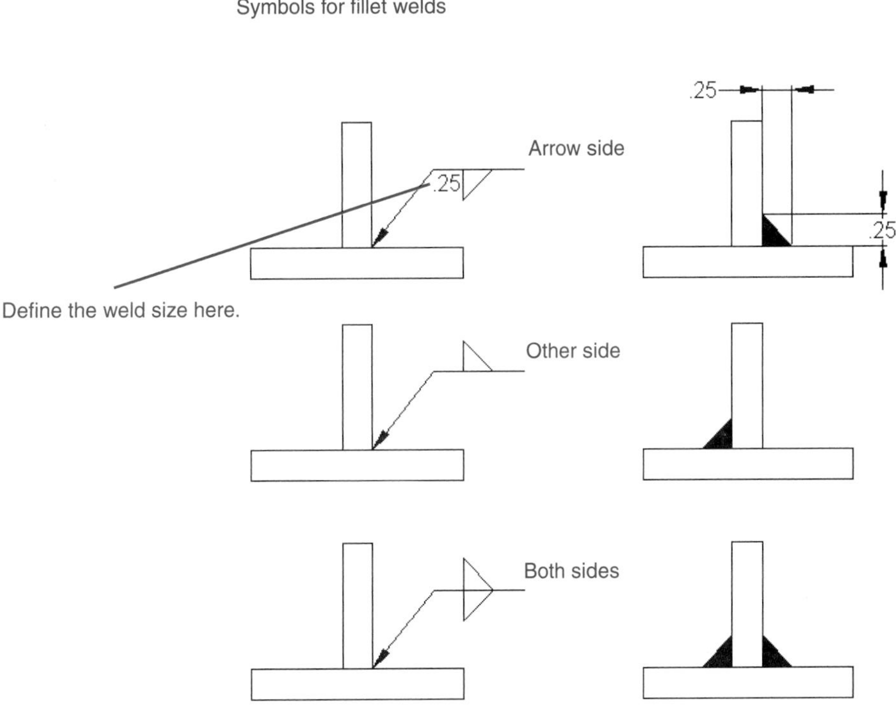

Figure 14-17

The size of the weld is defined as shown. Most fillet welds are created at 45°, although other angles are possible. A fillet weld defined by .25 indicates that the 45° weld is defined by two sides, both .25 long. Metric values are used to define a weld size in the same manner.

EXERCISE 14-5 **Adding a Weld Symbol to a Drawing**

This example will use the T-bracket shown in Figure 14-11.

1. Start a new drawing using English units and the **Weldment (ANSI) .iam** format.
2. Click the **Assemble** tab, select the **Place Component** tool on the **Component** panel, and locate the T-bracket on the drawing screen.
3. Rotate the T-bracket so that the portion of the bracket that does not have a weld is visible.

   See Figure 14-18.

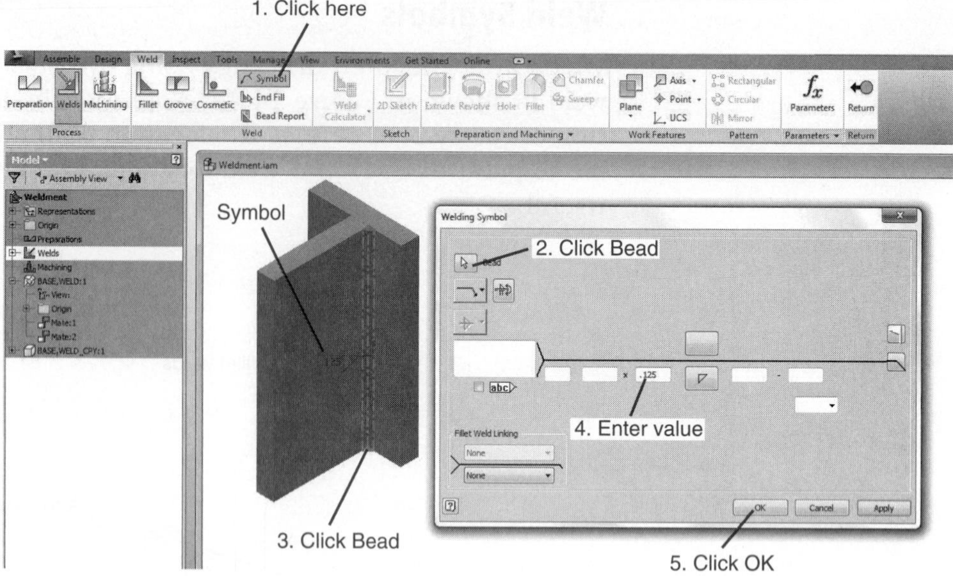

Figure 14-18

4 Click the **Weld** tool, click the **Fillet** tool, and add a **.125** fillet weld; click **OK.**

5 Click the **Symbol** tool located on the **Weld** panel.

The **Welding Symbol** dialog box will appear. See Figure 14-19.

6 Enter the **.125** value, click the **Bead** box, and click the fillet weld on the T-bracket.

The weld symbol will appear.

7 Click **OK.**

Figure 14-19 shows a weld drawing with a weld symbol. The weldment has been reoriented to better show the weld symbol.

NOTE

The fillet weld and symbol can be created together by first clicking the **Create Welding Symbol** box on the **Fillet Weld** dialog box. When the box is clicked the **Fillet Weld** dialog box will expand to include welding symbols.

Figure 14-20 shows the fillet weld symbol for an intermittent fillet weld. It was created using the same procedure as for the continuous weld.

Figure 14-19

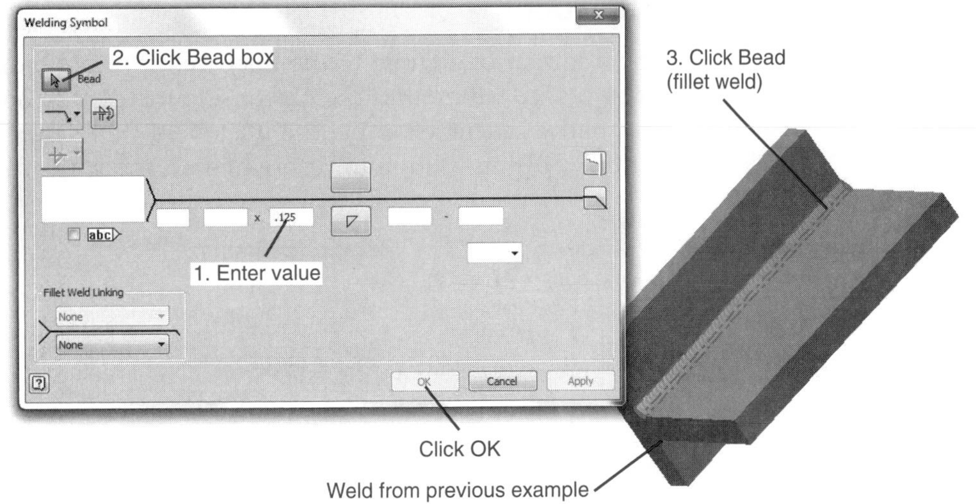

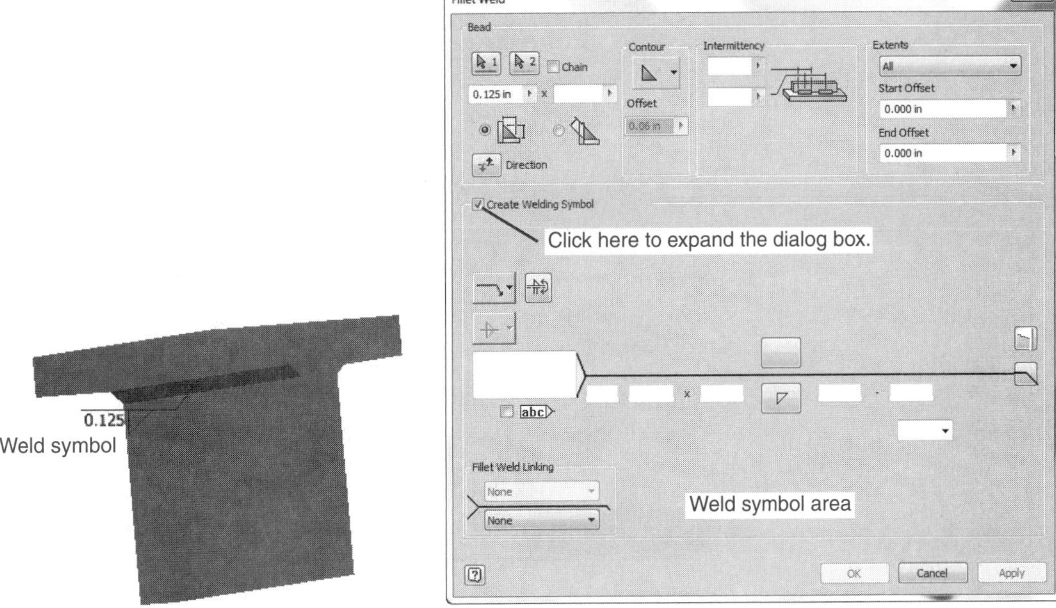

Figure 14-20

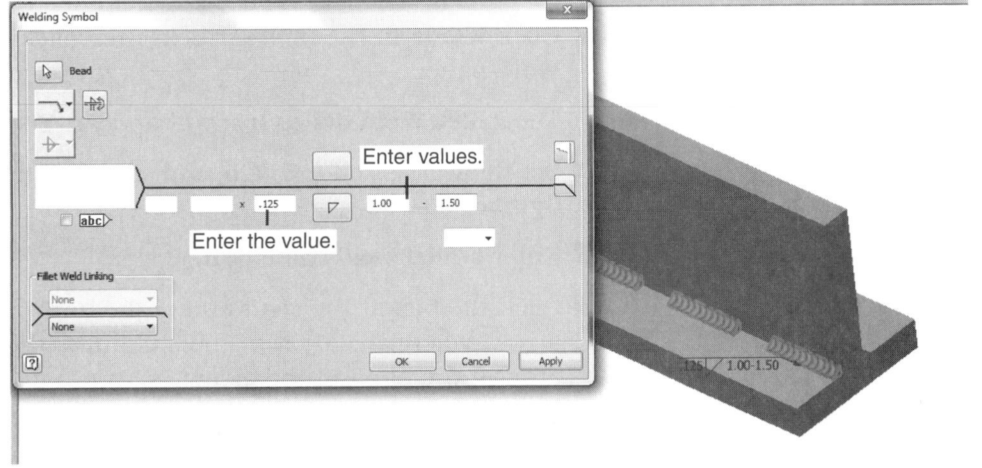

All Around

The addition of a circle to the fillet weld symbol indicates that the weld is to be placed *all around* the object. Figure 14-21 shows a cylinder welded to a plate. In this example the fillet weld was defined using millimeters. A 5-mm × 5-mm weld is to be created all the way around the cylinder.

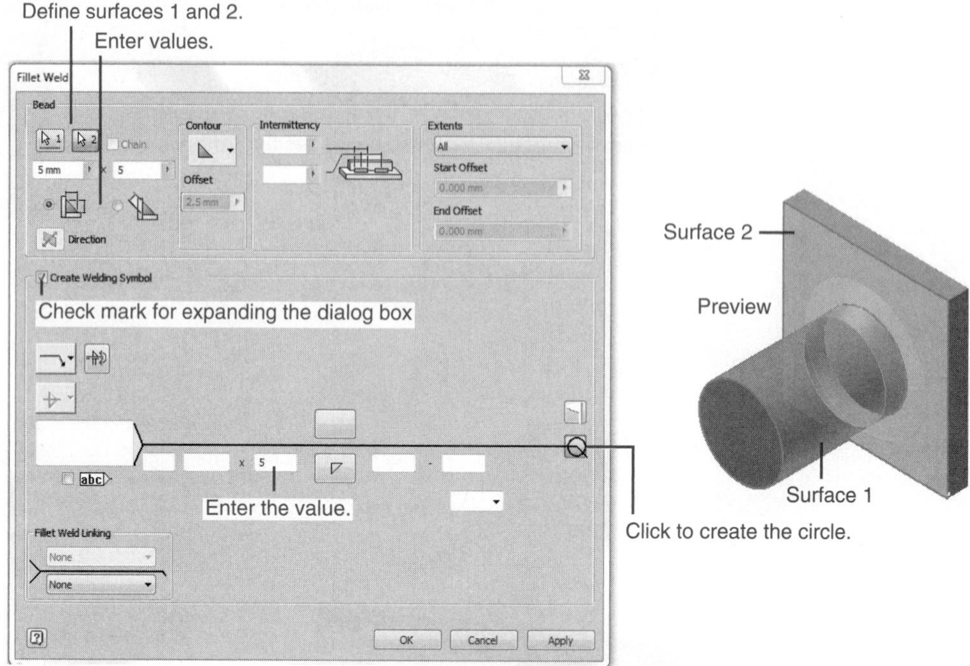

Figure 14-21

EXERCISE 14-6 **Creating an All-Around Fillet Weld**

1. Start a new drawing, click the **Metric** tab, and create a new drawing using the **Weldment (ISO).iam** format.
2. Create a **5 × 40 × 40-mm** plate and a **Ø20 × 30** cylinder.
3. Assemble the parts so that the cylinder is centered on the plate.
4. Click **Welds** in the browser box, then click the **Fillet Weld** tool on the **Weld** panel.

   The **Fillet Weld** dialog box will appear. See Figure 14-21.
5. Set the weld size for **5 × 5 mm,** then click the **Create Welding Symbol** box.
6. Enter a fillet weld value of **5.**
7. Click the box at the right end of the horizontal segment of the symbol arrow, where the arrow symbol changes direction, to create a circle around the bend in the arrow.
8. Click **OK.**

   Figure 14-22 shows the finished weld and symbol.

Figure 14-22

Finished all-around
weld and symbol
5

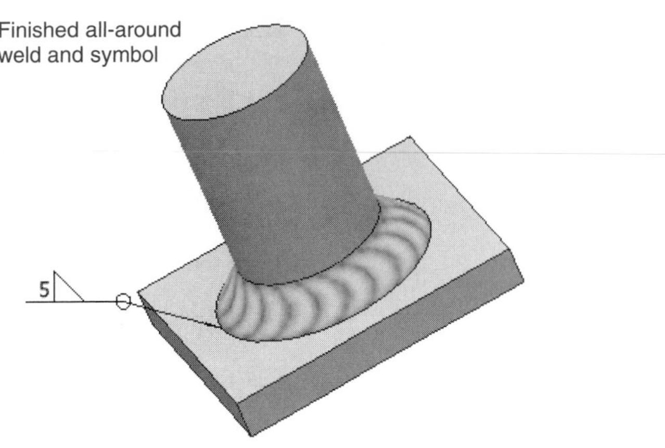

Weldments—Groove Welds

groove weld: A weld used when two parts abut, placed in the groove formed when a chamfer is cut into each part.

Groove welds are used when two parts abut. A chamfer is cut into each part and the weld is placed in the resulting groove. Figure 14-23 shows an L-bracket created as a weldment.

It was created as follows.

1. Draw a **.375 × 2.00 × 4.00-in.** plate and cut a **.19 × .19** chamfer as shown.

2. Create a weldment drawing using the **Weldment (ANSI).iam** format.

.19 x .19 Chamfer
.375 x 2.00 x 4.00 Plate
Create two plates and
assemble them as shown.
Use the Weldment (ANSI).
.iam format
4. Click Face Set 2
2. Click Face Set 1
Groove Weld
Bead
Face Set 1
Face Set 2
Fill Direction
Full Face Weld
Full Face Weld
Radial Fill
Chain Faces
Ignore Internal Loops
Create Welding Symbol
OK
Cancel
Apply
5. Click Face 2
3. Click Face 1
Preview
Groove weld
1. Click Full Face
Weld boxes
6. Click OK

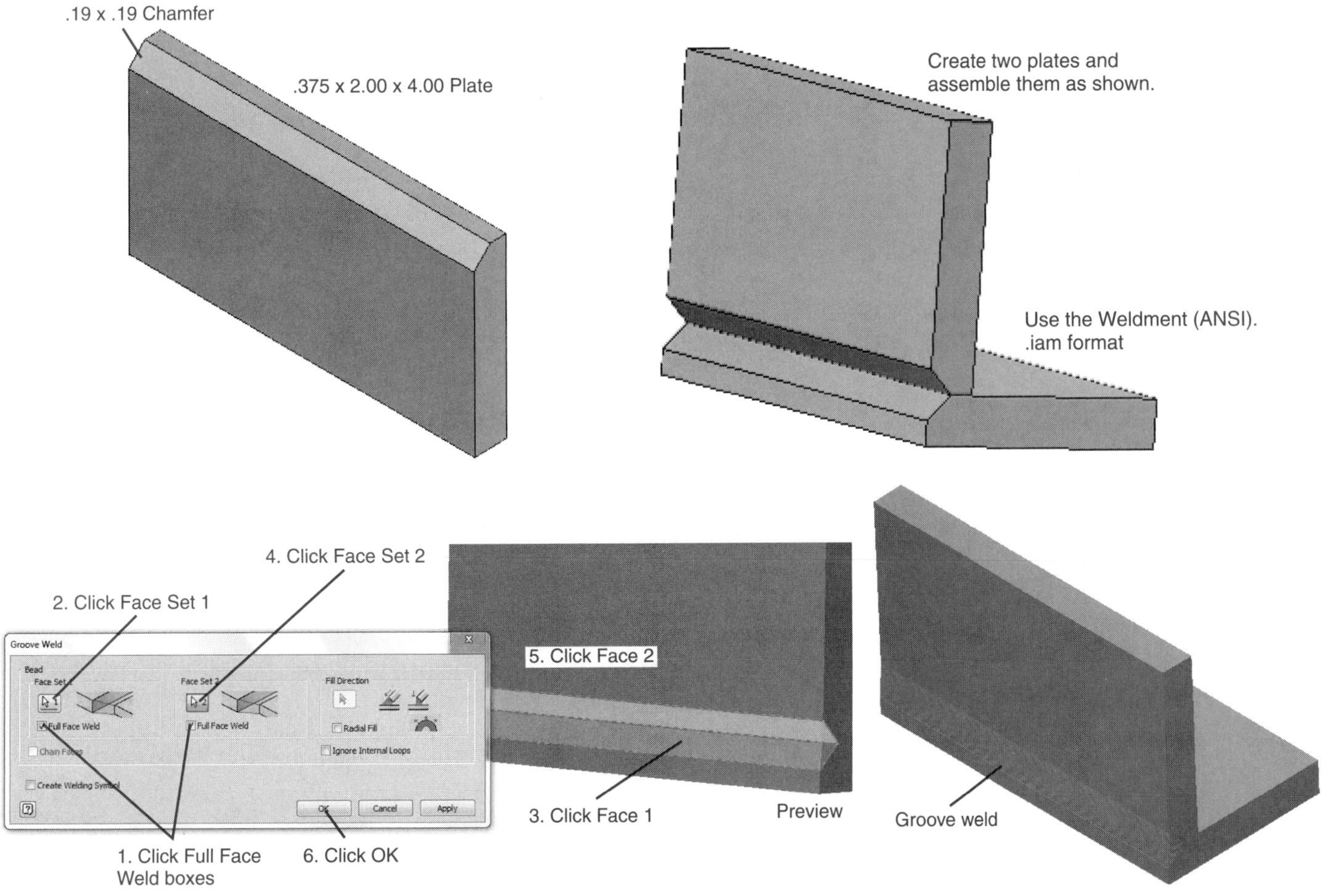

Figure 14-23

3 Click the **Assemble** tab, click the **Place Component** tool located on the **Component** panel, and add the two plates to the drawing. Use the **Contsrain** tool and assemble the plates as shown.

4 Click the **Weld** tab, click the **Welds** tool, and click the **Groove Weld** tool on the **Weld** panel.

The **Groove Weld** dialog box will appear.

5 Click the **Full Face Weld** boxes for both **Face Set 1** and **Face Set 2.**

6 Define **Face 1** and **Face 2** as shown.

7 Click **OK.**

Sample Problem SP14-1

Figure 14-24 shows an object that is to be manufactured as a weldment created from three parts. The three parts are the barrel, the center plate, and the front plate.

Figure 14-24

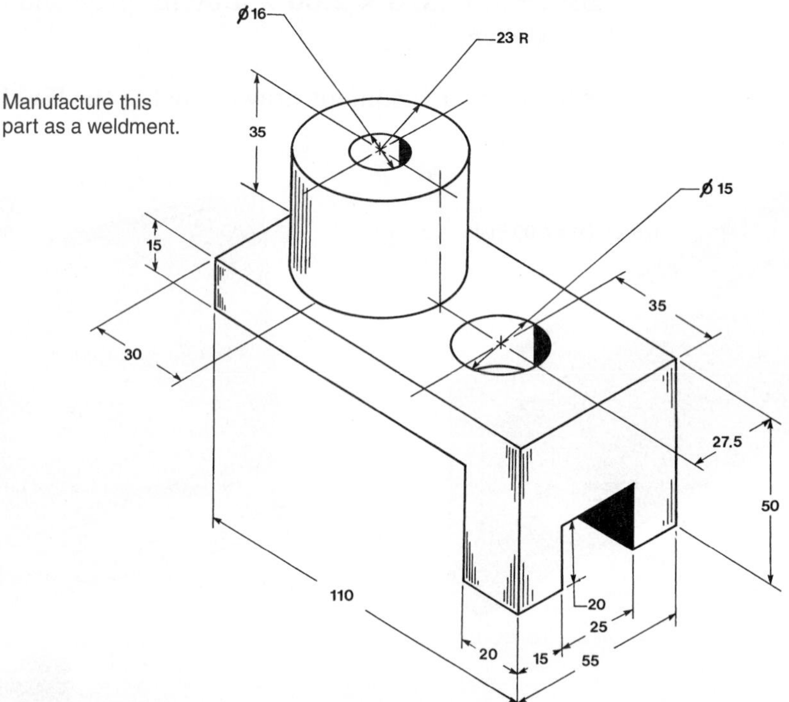

EXERCISE 14-7 Creating the Weldment

Use the **Weldment (ISO).iam** format. The dimensions for each part were derived from those given in Figure 14-24.

1 Start a new drawing, click the **Metric** tab, and use the **Weldment (ISO).iam** format.

2 Use the **Create** tool (use the **Standard (mm).ipt** format) and create a **15 × 55 × 110** plate with a **5 × 5** chamfer and a **Ø15** as shown.

3 Right-click and select the **Finish Edit** option.

4 Use the **Create** tool and create a **Ø46** × **35** barrel with a **Ø16** through-all hole. Use the **Constrain** tool to position the barrel.

5 Reorient the model and use the **Create** tool to create a **20** × **55** × **35** flange with a **25** × **20** cutout as shown.

6 Add a groove weld as shown.

7 Add a **5** × **5** fillet weld as shown.

8 Add a **2** × **2** fillet weld around the barrel as shown.

Figure 14-25 shows the weldment.

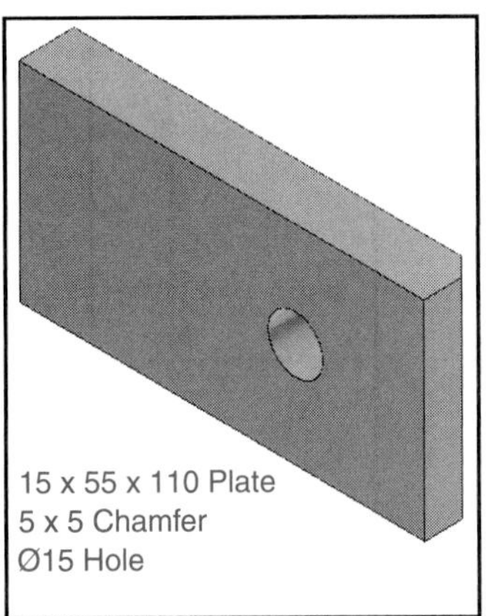

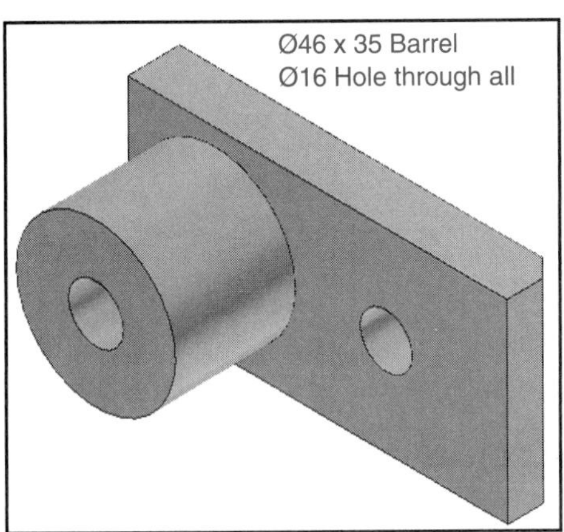

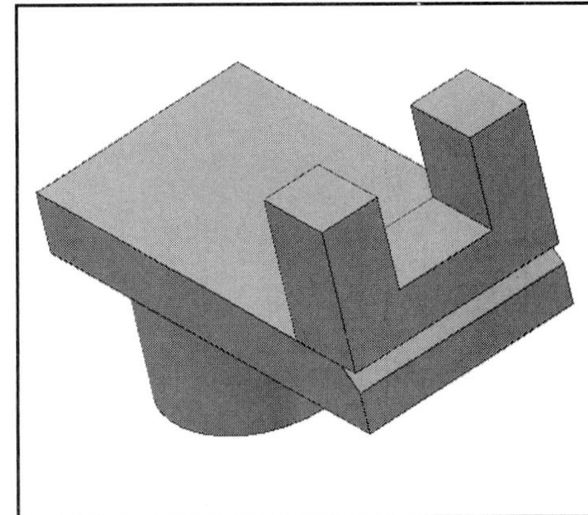

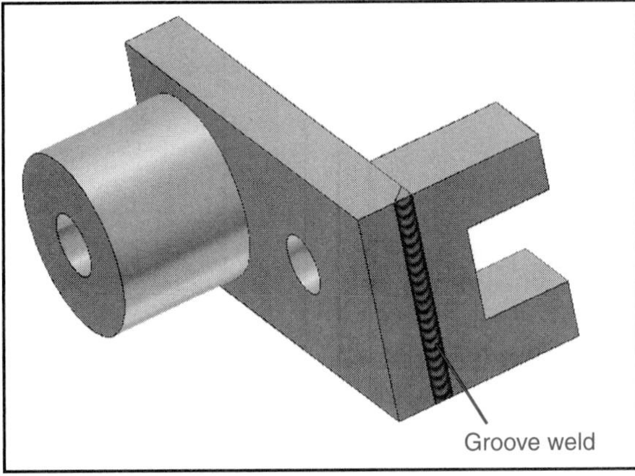

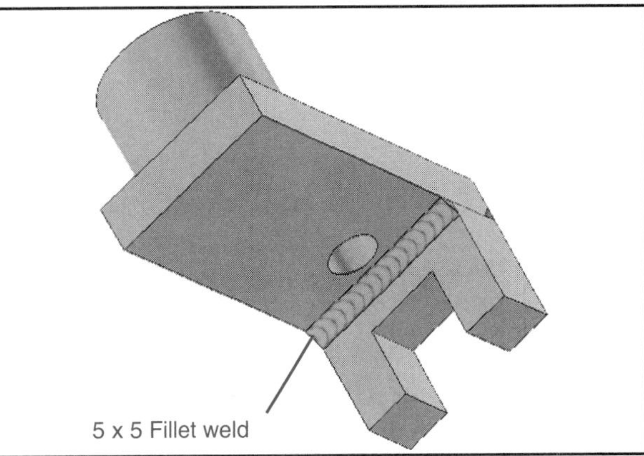

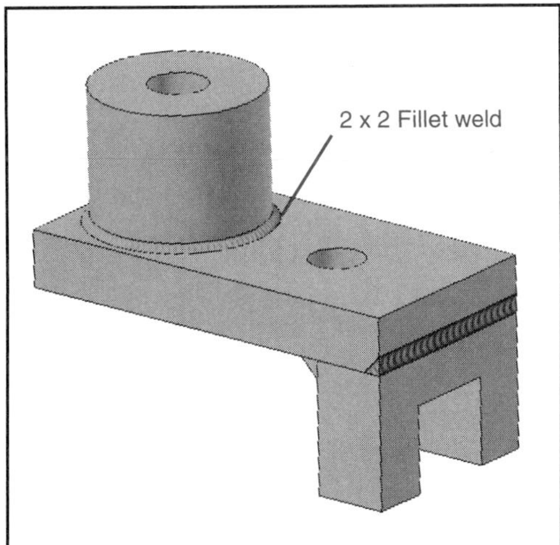

Figure 14-25

Chapter Summary

This chapter illustrated how to create and draw weldments, which are assemblies of several parts welded together. Fillet welds, both continuous and intermittent as well as all around, were introduced; welding symbols were added to drawings; and groove welds were illustrated.

Chapter Test Questions

Multiple Choice

Circle the correct answer.

1. What shape is the symbol for a fillet weld?
   a. Circle
   b. Flag-like
   c. Square
   d. Hexagon
2. A circle added to the fillet weld symbol means
   a. Weld all around
   b. Use a round weld bead
   c. Use a cosmetic weld
3. Groove welds are generally associated with
   a. Fillets
   b. Holes
   c. Chamfers
   d. Cutouts
4. The weld tools are accessed by clicking the **Welds** tool located in the
   a. **Weldment Assembly Panel**
   b. **Standard** toolbar
   c. **Tools** pull-down menu
   d. Browser
5. A weld symbol that has a symbol both above and below the horizontal segment of the symbol indicates
   a. Weld the other side
   b. Weld both sides
   c. Weld all around
   d. Weld the closest side

True or False

Circle the correct answer.

1. **True or False:** A weldment is an assembly made from several smaller parts that have been welded together.
2. **True or False:** Inventor can draw both continuous and intermittent welds.
3. **True or False:** The symbol for a fillet weld is a circle.
4. **True or False:** A groove weld is used when two parts abut.
5. **True or False:** Fillet welds can be defined in two ways: by two edge distances or the distance from front to back.

Chapter Project

Project 14-1

For Figures P14-1 through P14-10, redesign the given parts as weldments. Use either 5-mm or .20-in. fillet welds.

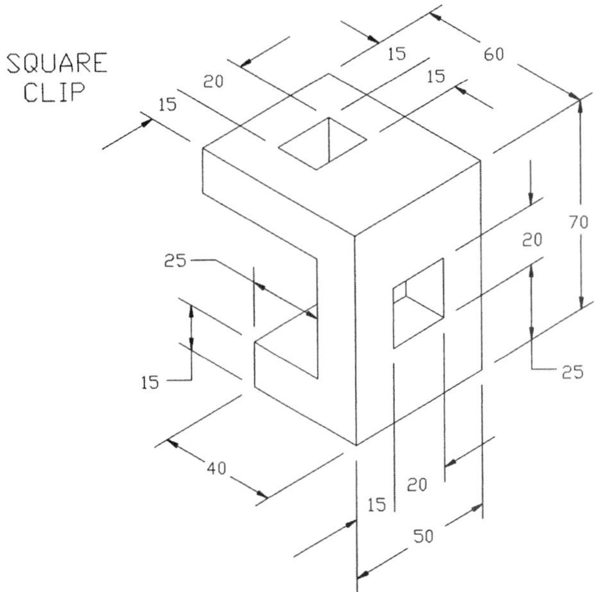

Figure P14-1
MILLIMETERS

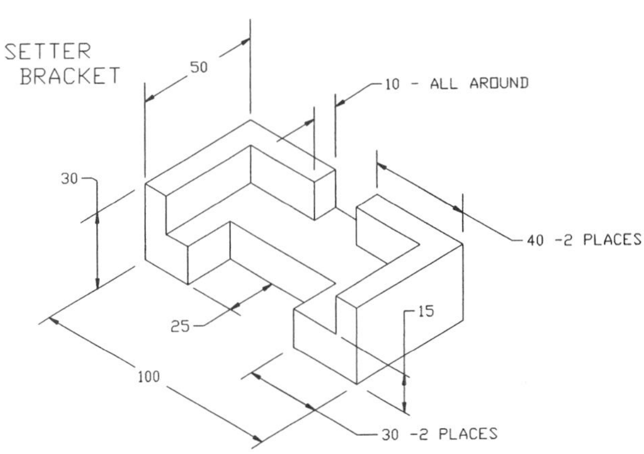

Figure P14-2
MILLIMETERS

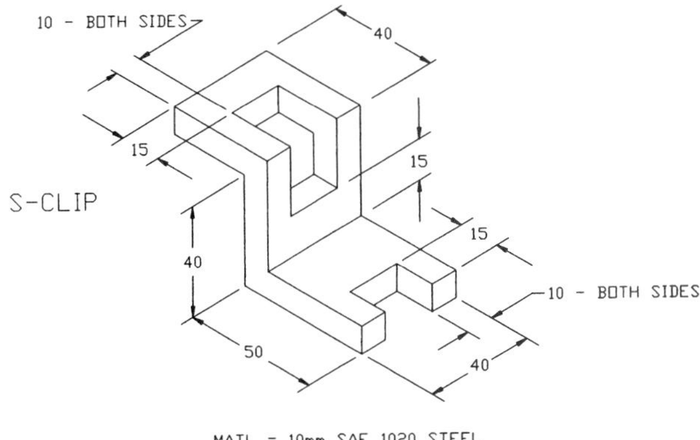

Figure P14-3
MILLIMETERS

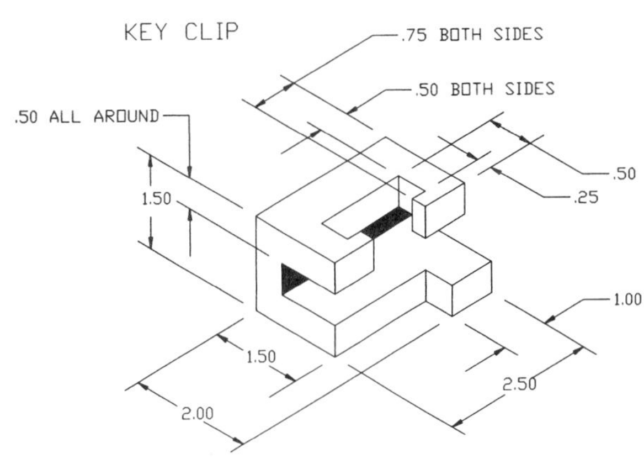

Figure P14-4
INCHES

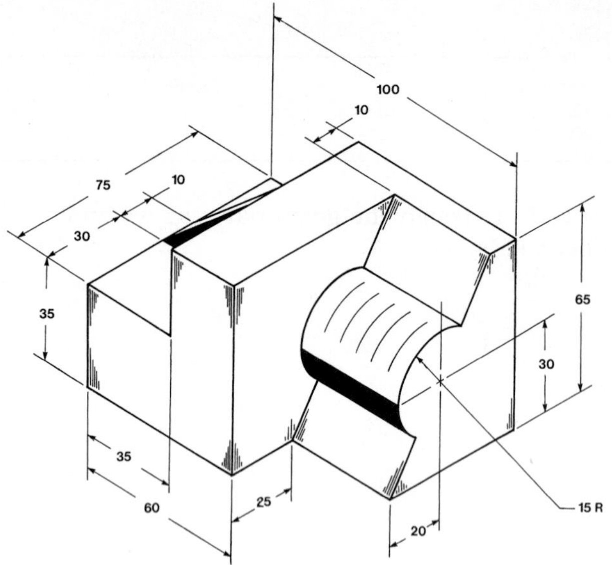

Figure P14-5
MILLIMETERS

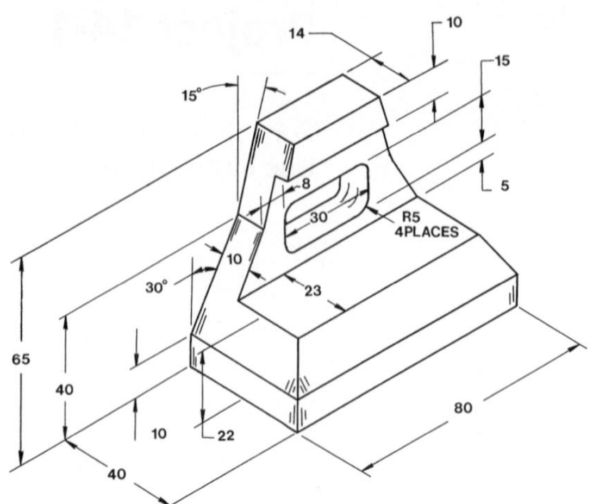

Figure P14-6
MILLIMETERS

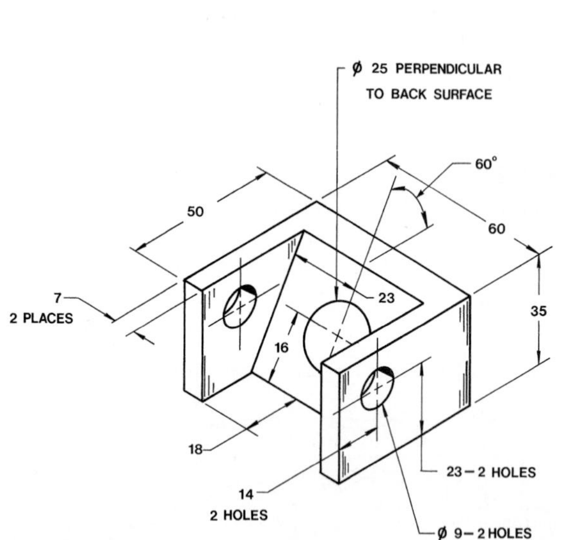

Figure P14-7
MILLIMETERS

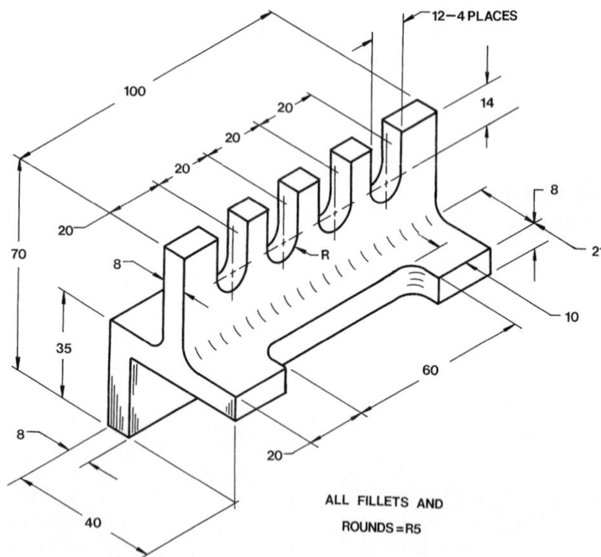

Figure P14-8
MILLIMETERS

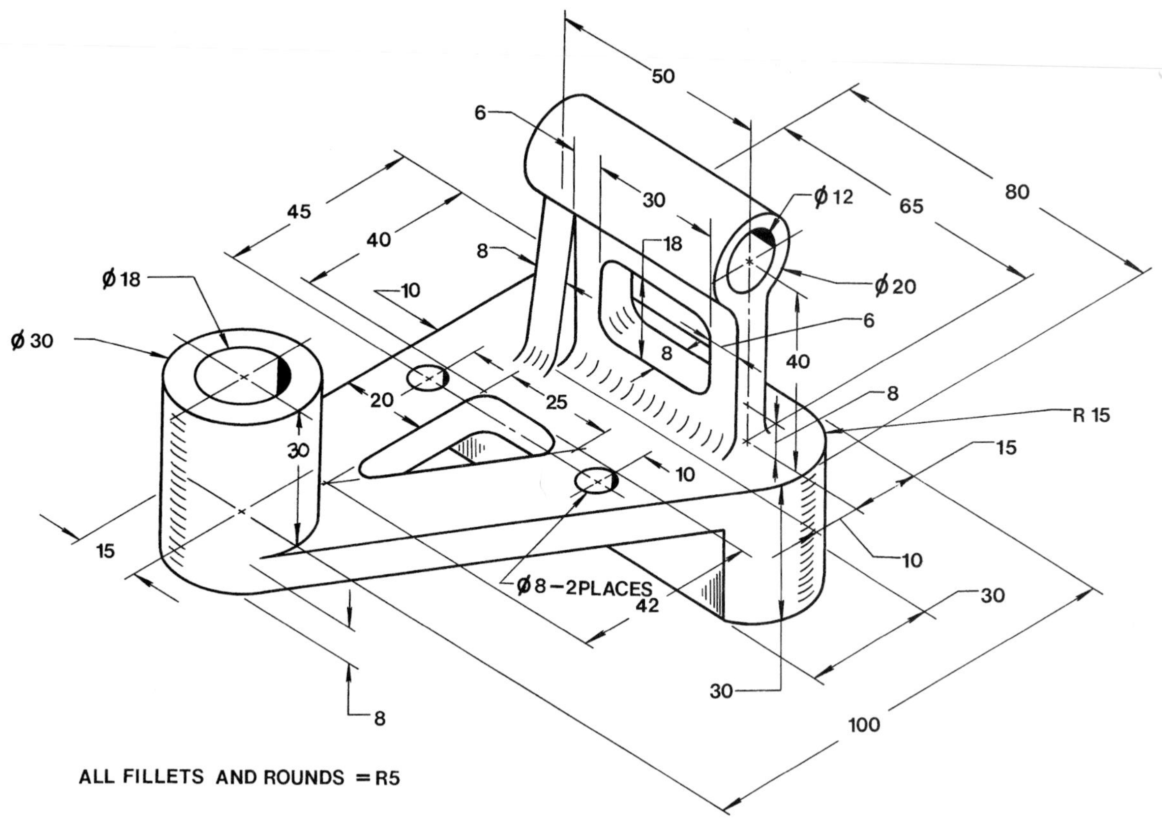

Figure P14-9
MILLIMETERS

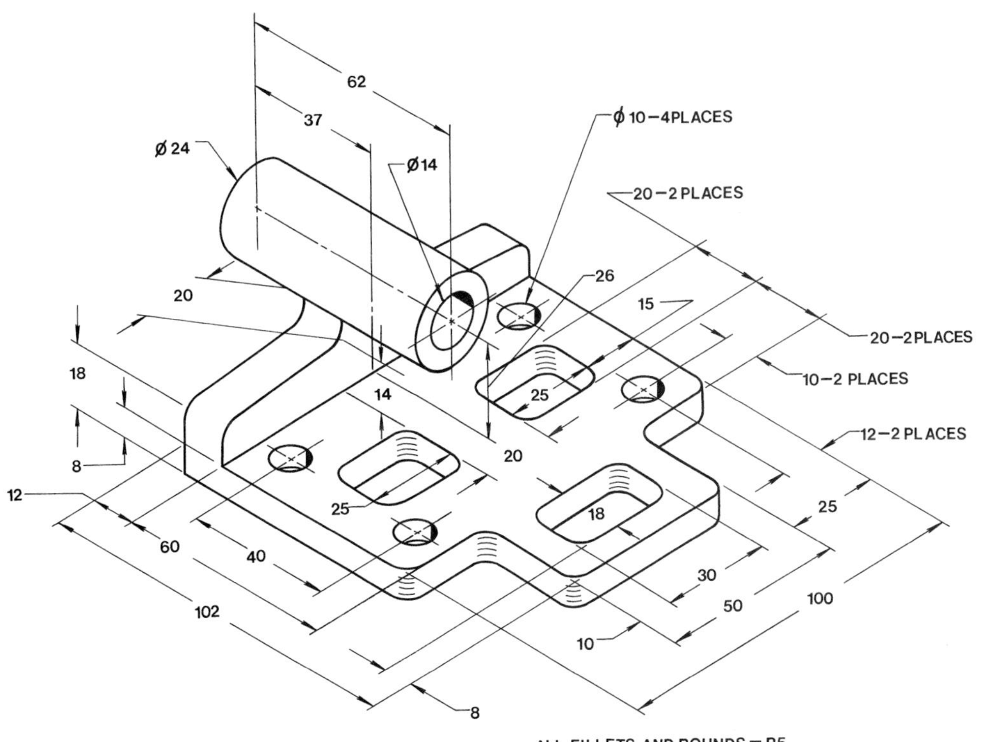

Figure P14-10
MILLIMETERS

chapter fifteen

Cams

CHAPTER OBJECTIVES

- Learn how to use the **Disc Cam** tool located on the **Power Transmission** panel under the **Design** tab
- Learn how to create and use displacement diagrams
- Learn how to insert cams into assembly drawings

Introduction

cam: An eccentric object that converts rotary motion into linear motion.

This chapter explains how to draw and design cams. ***Cams*** are eccentric objects that convert rotary motion into linear motion. Cams are fitted onto rotating shafts and lift and lower followers as they rotate.

Cams can be designed and drawn using the **Disc Cam** tool. Displacement diagrams are defined and cams are generated from the displacement diagrams. The shape of the cam's profile causes the follower to rise and fall as the cam rotates. Changes in the cam's displacement cause the follower to accelerate. Excessive acceleration can generate excessive forces.

Figure 15-1 shows a cam drawn using Inventor. The cam bore includes a keyway.

Figure 15-1

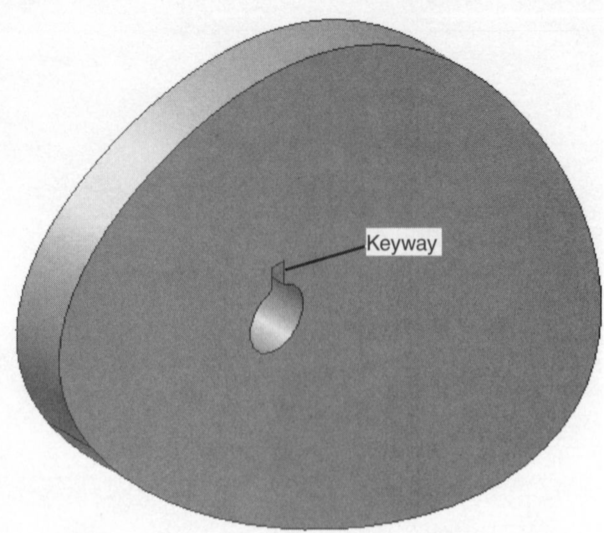

Displacement Diagrams

displacement diagram: A linear diagram used to define the motion of a cam.

Displacement diagrams are used to define the motion of a cam using a linear diagram. The distances are then transferred to a base circle to create the required cam shape. Inventor will automatically create a cam from given displacement information.

Figure 15-2 shows a displacement diagram and a cam shape generated from the information on the diagram. The displacement diagram shown is drawn using only straight lines, which is called *uniform motion.* This diagram results in points of discontinuity that can result in erratic follower motion. Several different shapes can be used to smooth out these areas and create smoother follower motion.

Figure 15-2

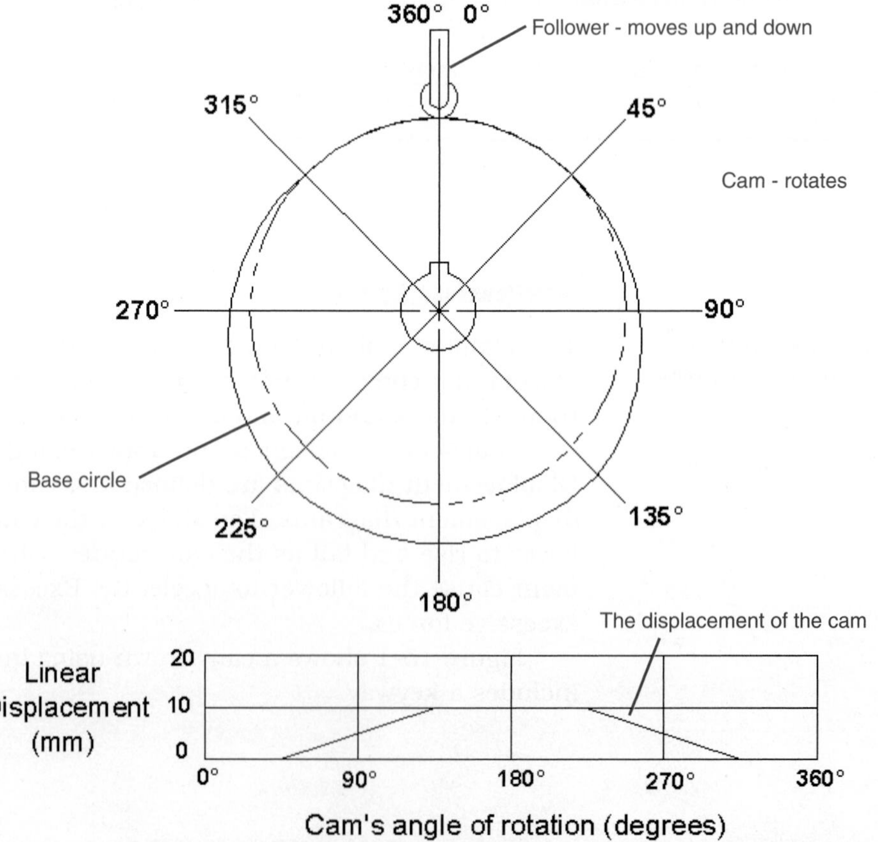

Drawing a Cam Using Inventor

A cam is drawn using the **Disc Cam** tool by first defining the physical characteristics of the cam then defining the cam's displacement diagram. A drawing of the cam will then automatically be generated.

EXERCISE 15-1 Drawing a Cam

1 Start a new drawing, click the **Metric** tab, and use the **Standard (mm).iam** format.

2 Click the **Design** tab and click the **Disc Cam** tool located on the **Power Transmission** panel.

The **Disc Cam Component Generator** dialog box will appear. See Figure 15-3. The first cam position segment, the one between 0° and 90°, will have a broken line around it. The broken line indicates that the segment is currently active.

Figure 15-3

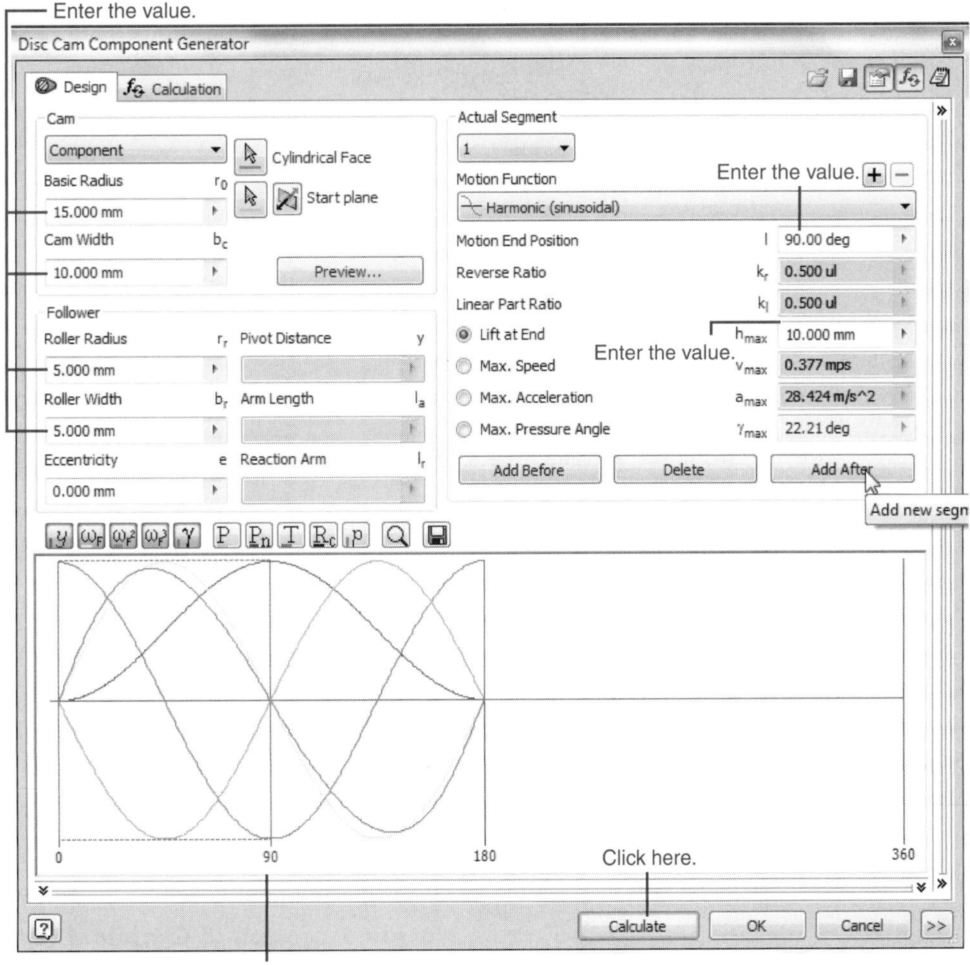

3 Enter a **Basic Radius** of **15.00, a Cam Width** of **10.00,** a **Roller Radius** of **5.00,** and a **Roller Width** of **5.00.**

These values are unique to this example.

4. Enter a **Lift at End** value of **10.00.**

   This value is noted as $\mathbf{h}_{max}$.

5. Click the **Add After** box.

   A third segment will appear to the right of the leftmost segment.

6. Click the rightmost of the three segments and drag the rightmost vertical line of the segment to the 360° line. This movement can also be accomplished by clicking the **Motion End Position** box, entering a value of **360,** and clicking the **Calculate** box.

7. Set the end position of the leftmost segment for **90** and the end position of the middle segment for **270** by clicking the individual segment, entering the appropriate value in the **Motion End Position** box, and clicking the **Calculate** box.

   See Figure 15-4.

Figure 15-4

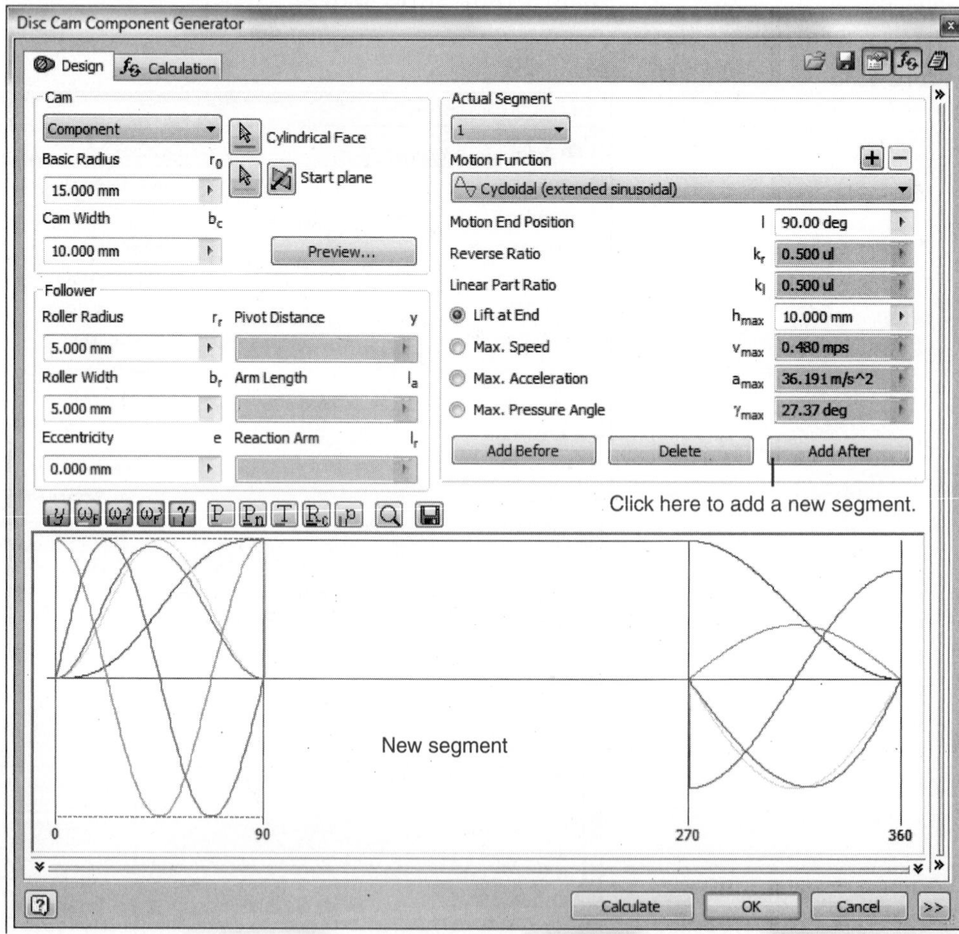

8. Enter a **Motion Function** of **Cycloidal (extended sinusoidal)** for the first and third segments.

9. Click the **Calculate** button to implement the new **Motion Functions**.

The middle segment of the cam, segment 2, is a dwell motion. Dwell portions of a cam have a constant radius so the follower will not move up or down during the dwell. Dwell segments are created, in this example, by having the **Lift at End** value of the first segment set at 10 and the **Lift at**

End value for the second segment also set for 10. The **Lift at End** value for the third segment is 0.000.

Figure 15-5 shows the **Calculation** portion of the **Disc Cam Component Generator** dialog box. Note that the $\mathbf{h}_{max}$ value is **10.000 mm** and the $\mathbf{h}_{min}$ value is **0.000 mm.**

Figure 15-5

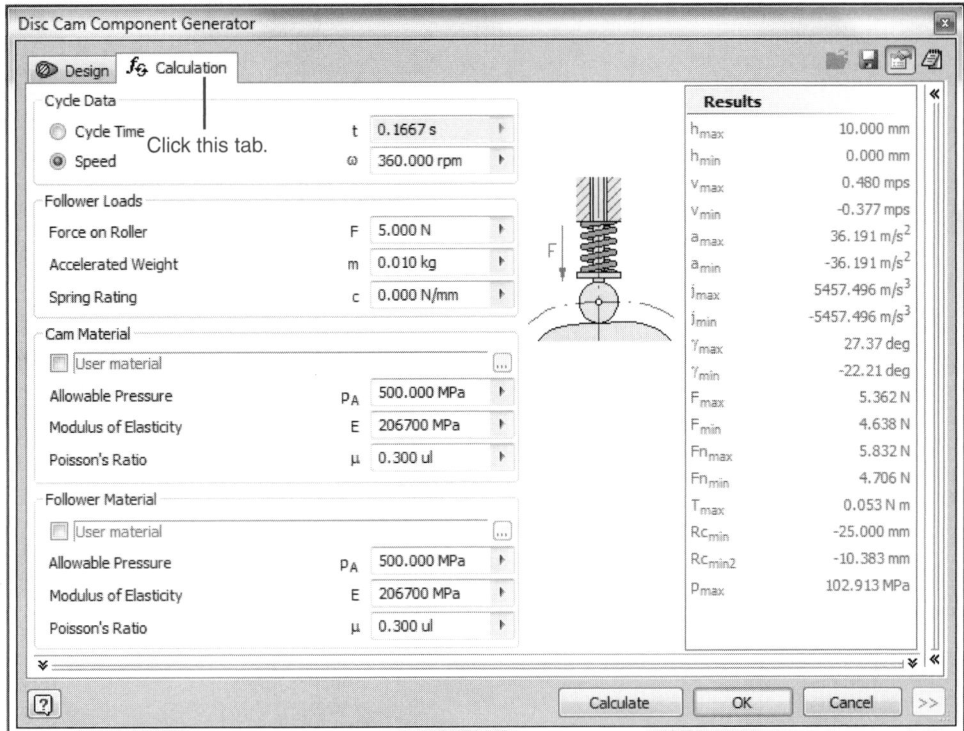

10 Click **OK.**

The **File Naming** dialog box will appear.

11 Click **OK.**

Figure 15-6 shows the resulting cam.

Figure 15-6

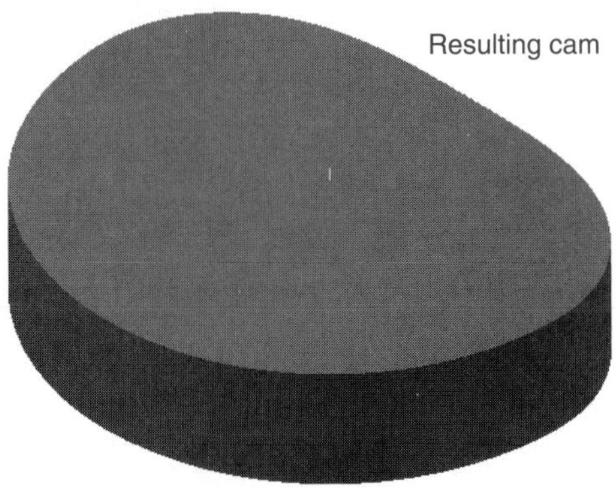

EXERCISE 15-2 Adding a Hole to a Cam

1 Locate the cursor on the cam, right-click the mouse, and click the **Open** option.

See Figure 15-7.

Figure 15-7

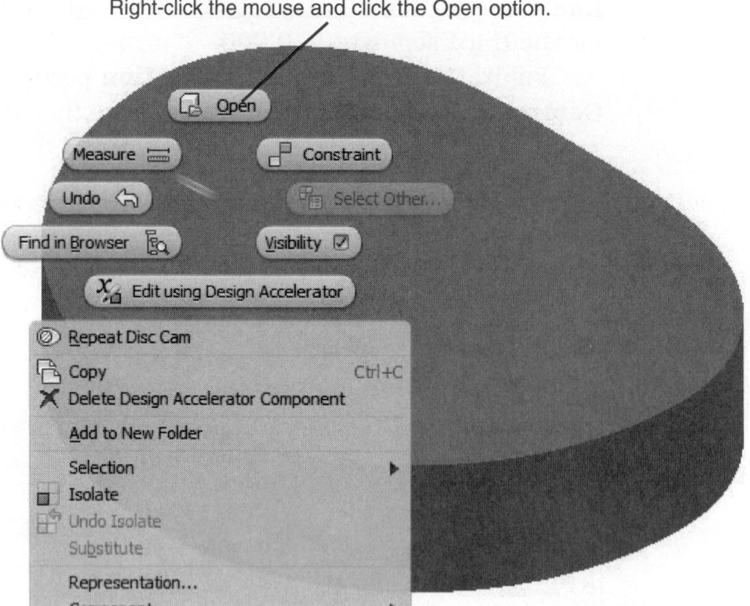

2 Right-click the mouse and click the **Edit** option.

See Figure 15-8.

Figure 15-8

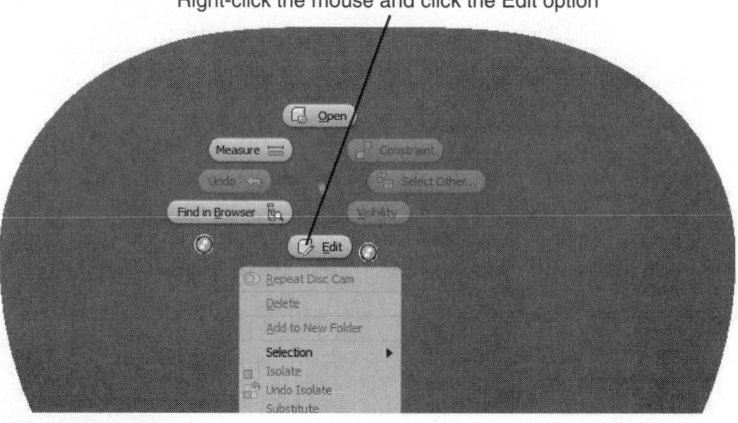

3 Click the house-shaped icon, the **Home** tool, above the ViewCube and create an **Isometric View**, then right-click again and click the **New Sketch** option.

See Figure 15-9.

4 Create a **Point, Center Point.**

See Figure 15-10.

5 Right-click the mouse and click the **Finish 2D Sketch** option.

6 Click the **Hole** tool and enter the values shown.

7 Click **OK.**

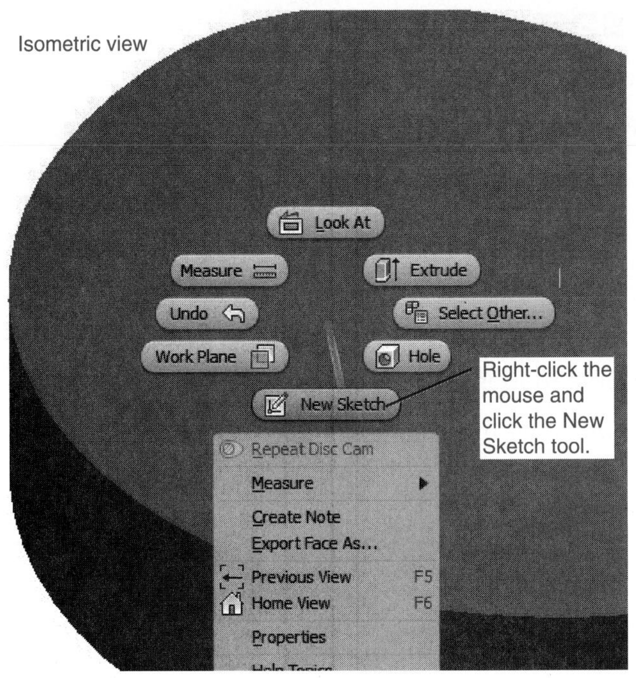

Figure 15-9

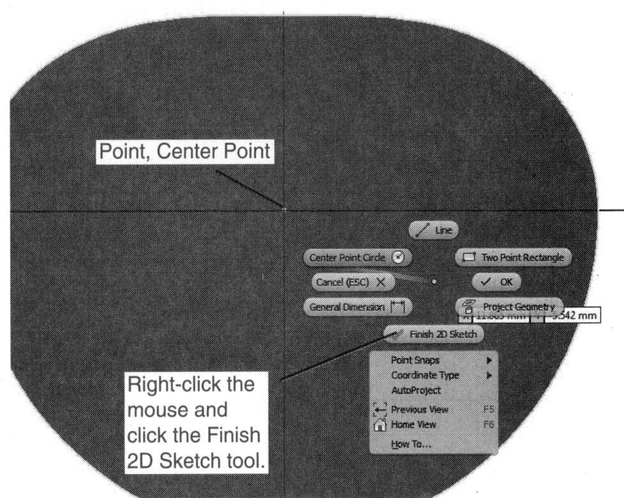

Figure 15-10

8 Right-click the mouse and click the **Finish Edit** option.

See Figure 15-11.

The cam will appear. See Figure 15-12.

Figure 15-11

Figure 15-12

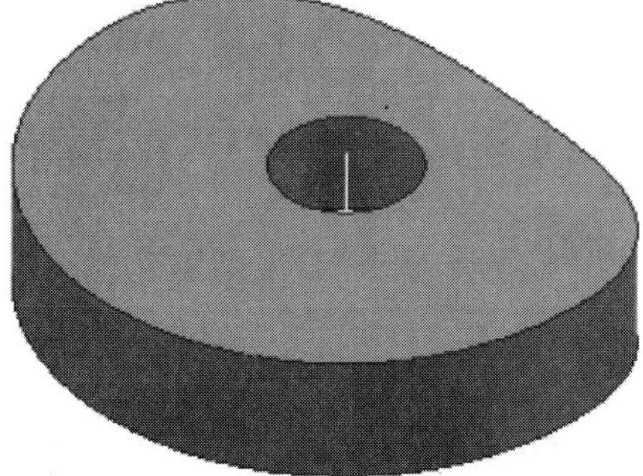

Sample Problem SP15-1

Design and draw a cam that meets the following specifications:

All linear dimensions are in millimeters.

Base circle = Ø80

Dwell = 45°

Rise 5 mm using harmonic motion over 90°

Dwell 45°

Rise 5 mm using harmonic motion over 90°

Drop 10 mm using parabolic with linear part motion in 90°

Bore = Ø12

Cam width = 10

Roller radius = 8

Roller width = 10

1 Create a drawing using the **Standard(mm).iam** format.

2 Click the **Design** tab.

3 Select the **Disc Cam** tool.

The **Disc Cam Component Generator** dialog box will appear. See Figure 15-13.

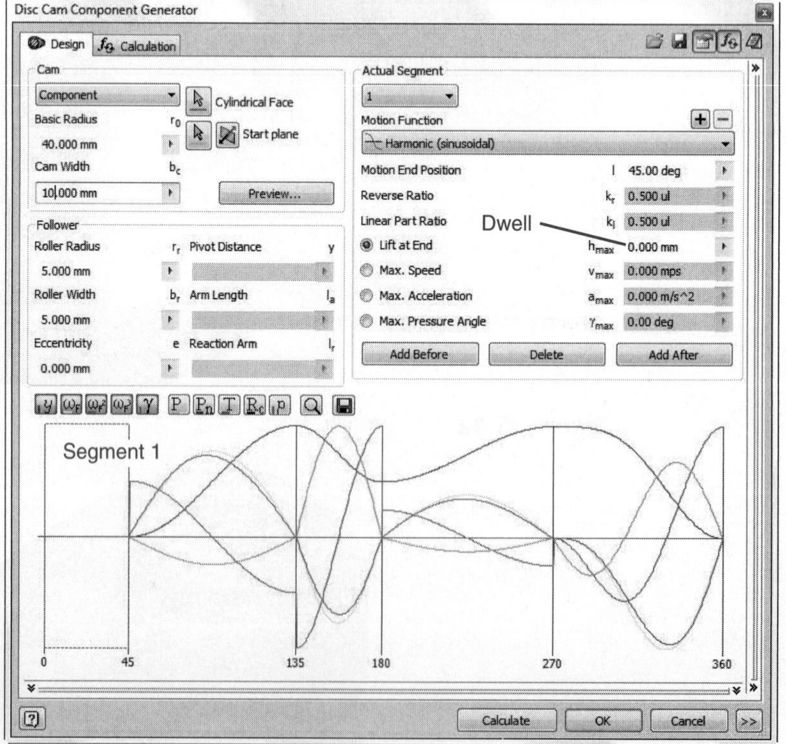

Figure 15-13

FIGURE 15-13
(Continued)

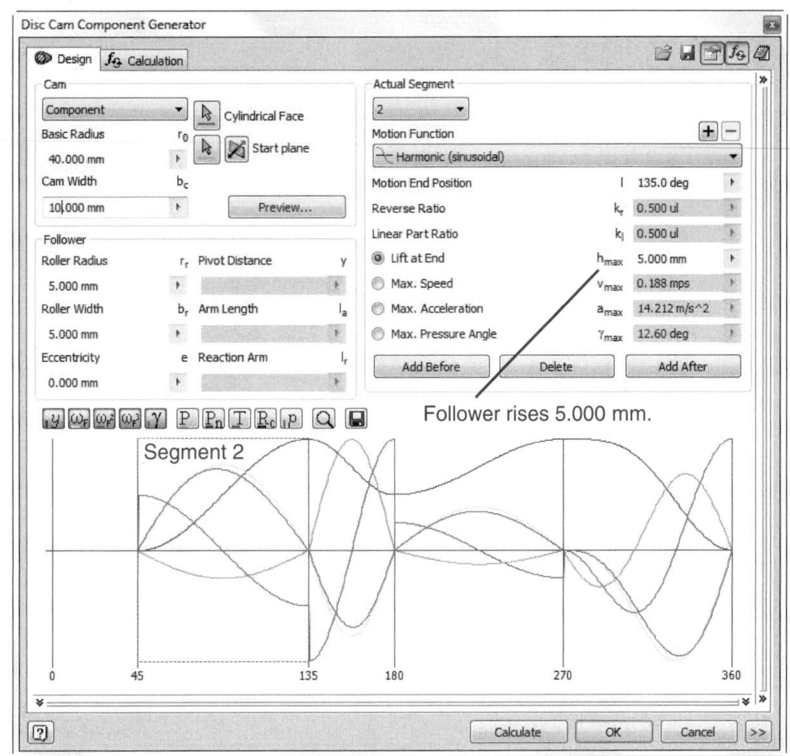

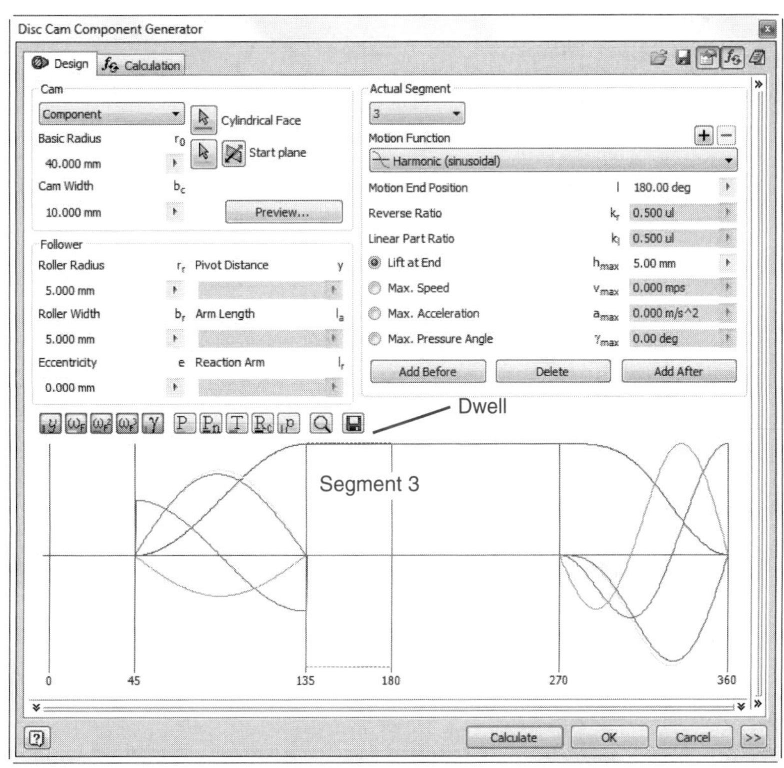

FIGURE 15-13
(Continued)

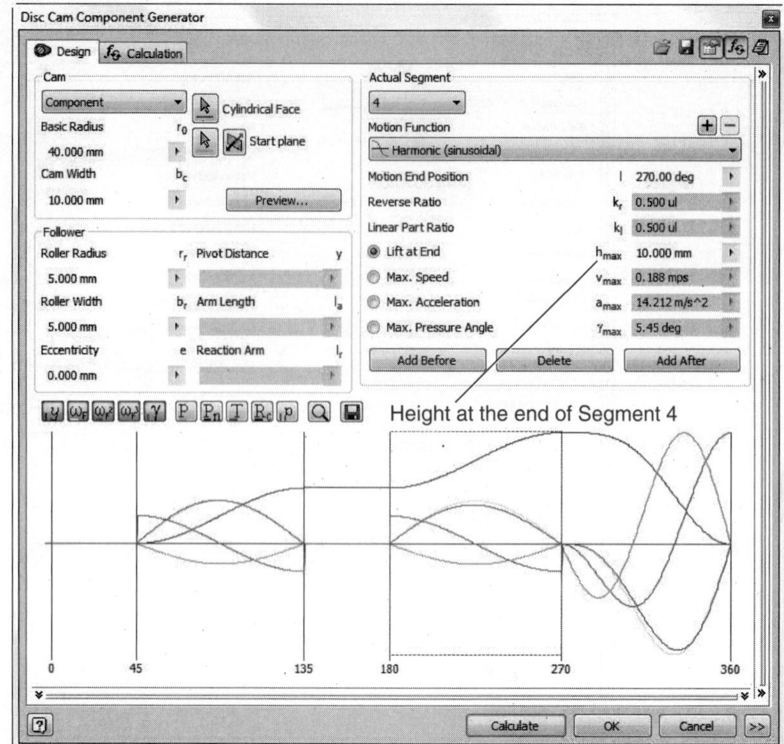

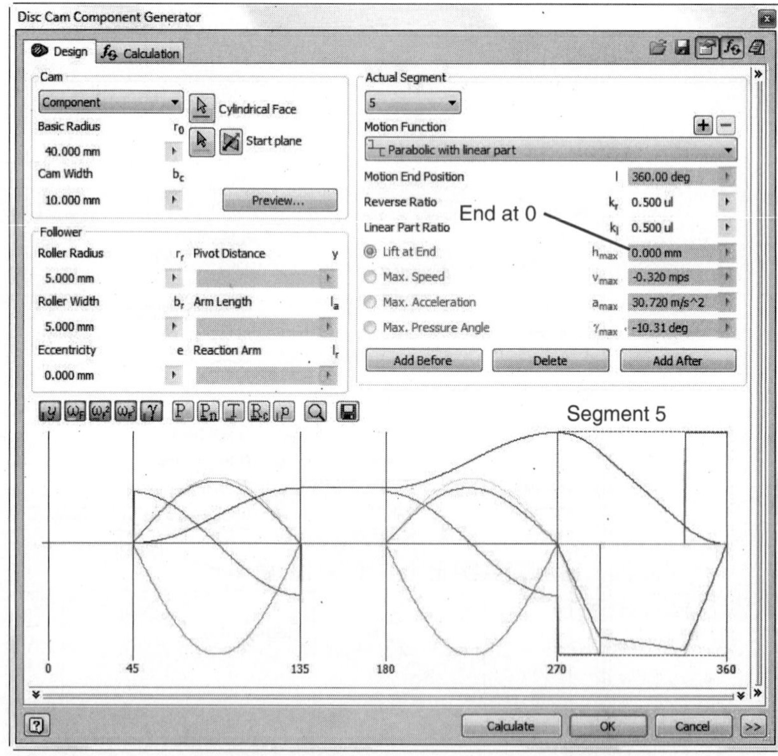

4 Enter the cam values.

5 Click the **Add After** box.

The cam will require five segments to define its motion. Segments can be added by using the **Add Before** or **Add After** option on the **Disc Cam Component Generator.**

6 Complete the displacement diagram using the given specifications.

Isolate each of the five segments and enter the appropriate values. For example, the first segment is a dwell for 45°. Set the **Motion End Position** for **24.00 deg** and **Lift at the End** for **00.00.**

7 Click **Calculate** after each segment input has been completed, then click **OK.**

Figure 15-14 shows the finished cam.

Figure 15-14

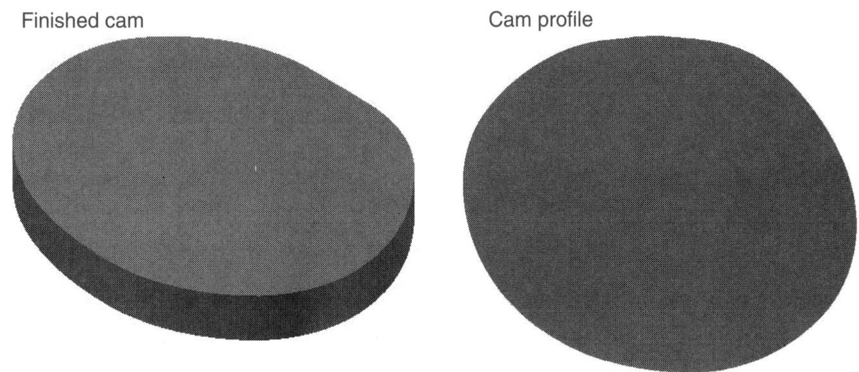

Cams and Followers

Figure 15-15 shows a cam that was previously created in **SP15-1**.

Figure 15-15

Start a New Assembly

1 Create a new **Standard (mm).iam** drawing.

2 Use the **Place Component** tool and add the cam created in **SP15-1**.

3 Click the **Create** tool located on the **Component** panel.

See Figure 15-16.

Figure 15-16

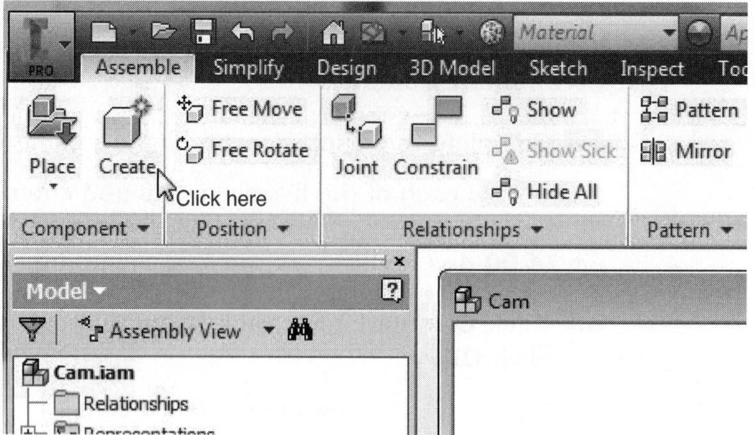

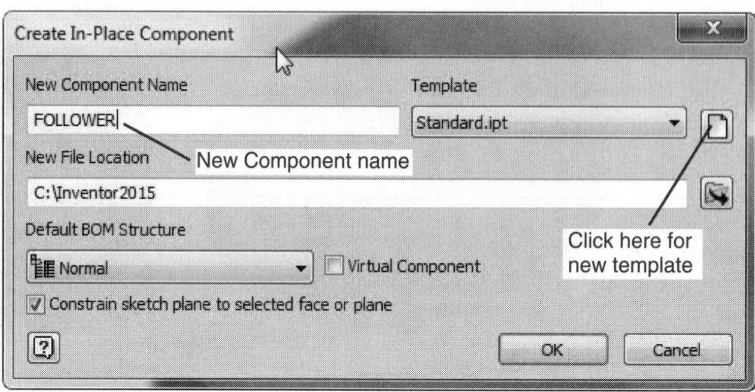

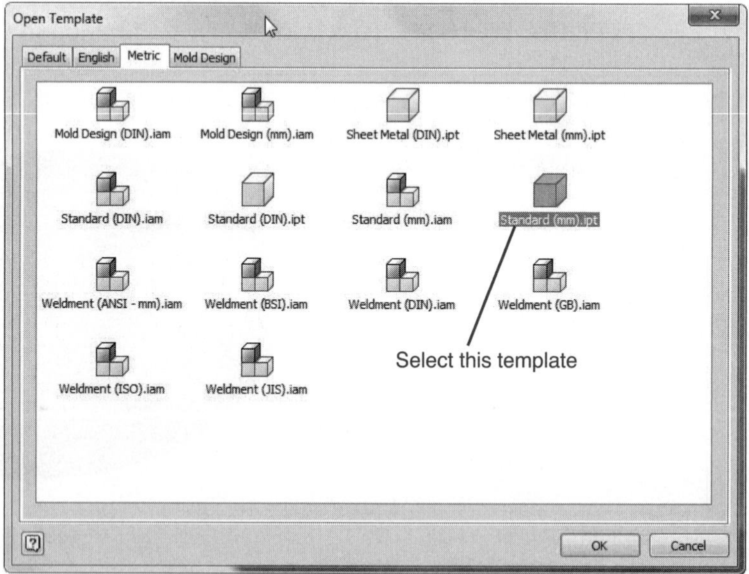

EXERCISE 15-3 Creating a Follower

1 Name the new component **FOLLOWER.**

The **Create In-Place Component** dialog box will appear. See Figure 15-16.

2. Click the **Browse Templates** box.

   The **Open Template** dialog box will appear.

3. Click the **Metric** tab.

   The **Open Template** dialog box will change.

4. Select the **Standard (mm).ipt** format; click **OK.**

   The **Create In-Place Component** dialog box will appear.

5. Click **OK.**

6. Move the cursor into the drawing area and click the left mouse button.

   You are now in **Sketch** mode. See Figure 15-17. The sketch plane is aligned with the sketch plane used to create the cam.

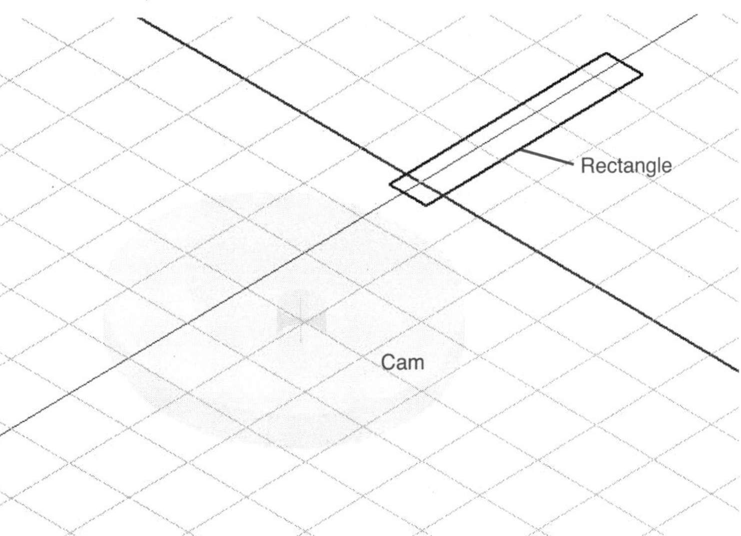

Figure 15-17

7. Use the **Rectangle** tool under the **Sketch** tab and sketch the approximate shape of the FOLLOWER.

8. Use the **Dimension** tool and size a rectangular follower to **10 × 60** as shown.

9. Click the right mouse button and select **Finish 2D Sketch,** right-click the mouse again, and select the **Finish Edit** option.

10. Right-click the rectangle and select the **Edit** option. Use the **Extrude** tool and extrude the rectangle **10 mm.**

11. Right-click the mouse and select **Finish Edit.**

    See Figure 15-18.

12. Use the **Constrain** tools **Tangent** and **Flush** to constrain the follower to the cam.

    See Figure 15-19.

The **Move and Rotate Component** tool may have to be used to position the follower so it can be constrained.

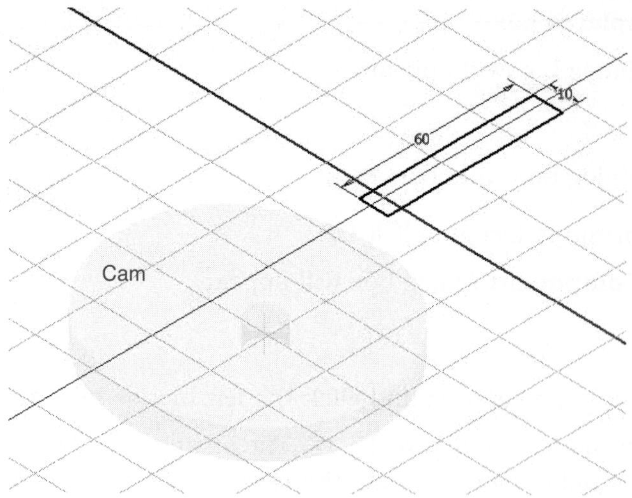

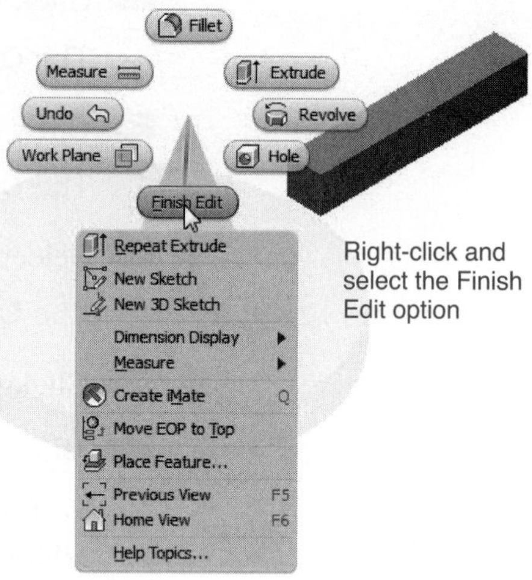

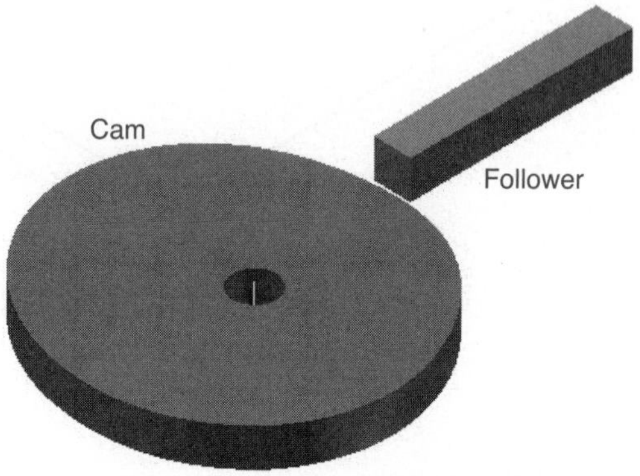

Figure 15-18

Figure 15-19

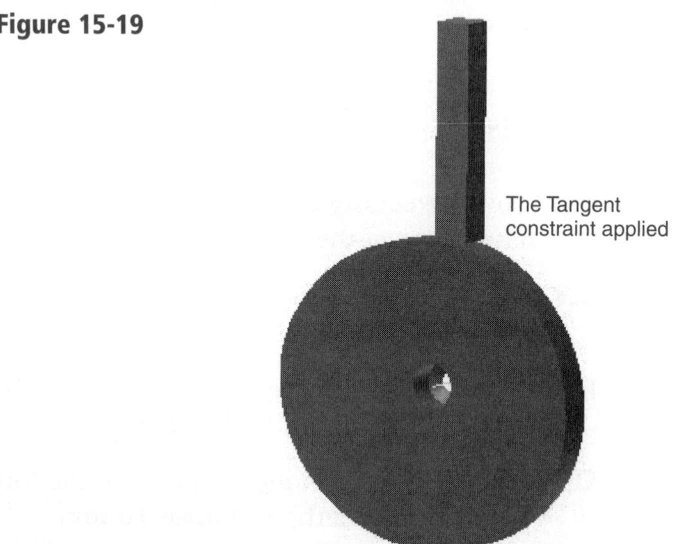

EXERCISE 15-4 **Creating a Follower Guide**

1. Right-click the mouse, click the **Create Component** tool, and create a new component named **GUIDE.** Use the **Standard (mm).ipt** format.

   See Figure 15-20.

2. Click the top surface of the follower to create a new sketch plane.

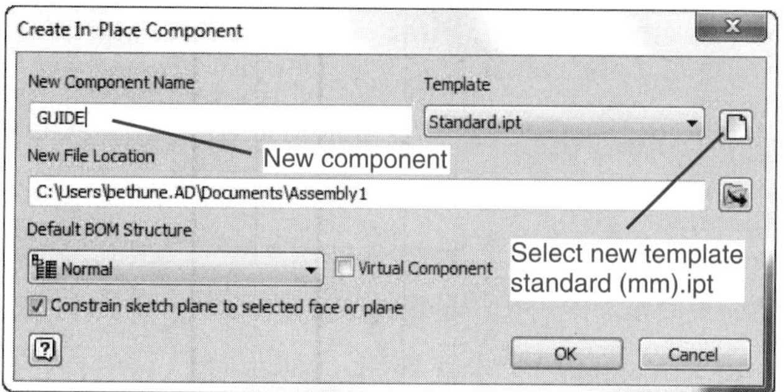
Create In-Place Component
New Component Name
GUIDE
Template
Standard.ipt
New File Location
C:\Users\bethune.AD\Documents\Assembly1
Default BOM Structure
Normal
Virtual Component
Constrain sketch plane to selected face or plane
OK
Cancel
New component
Select new template
standard (mm).ipt

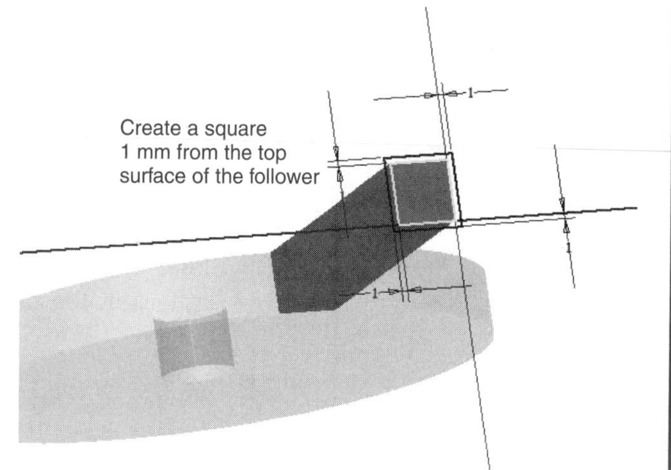
Create a square
1 mm from the top
surface of the follower

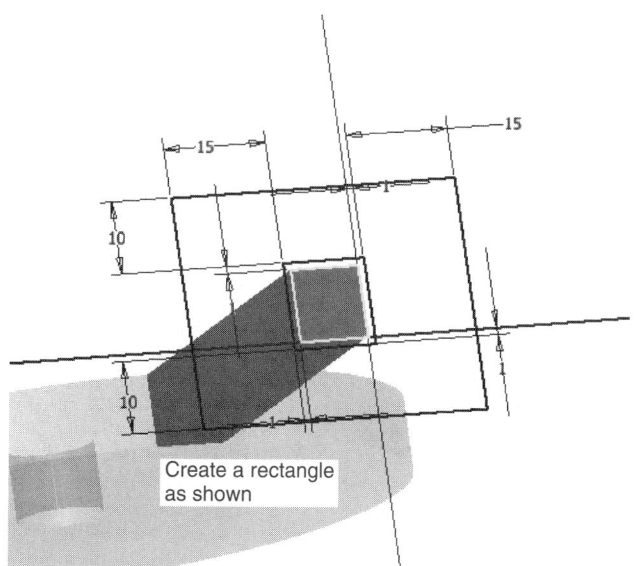
15
15
10
10
Create a rectangle
as shown

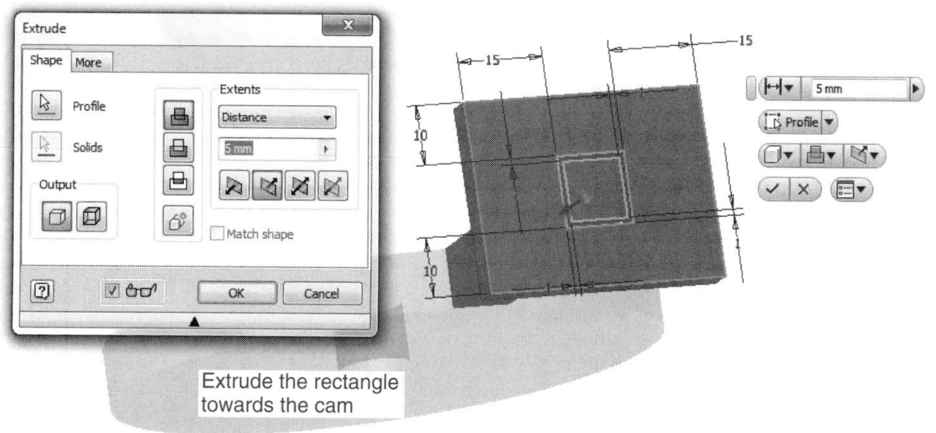
Extrude
Shape
More
Profile
Solids
Output
Extents
Distance
5 mm
Match shape
OK
Cancel
Profile
Extrude the rectangle
towards the cam

Figure 15-20

3 Click the **Project Geometry** tool located on the **Draw** panel under the **Sketch** tab, and click the four lines of the top surface of the follower. Right-click the mouse and select the **Done** option.

4 Sketch a rectangle around the projected top surface of the follower, then use the **Dimension** tool to define a **1-mm** clearance between the guide and the follower.

5 Sketch a second rectangle around the first rectangle as shown, then right-click the mouse and select the **Done** option, then the **Finish Sketch** option.

6 Use the **Extrude** tool to add a **5-mm** thickness to the guide.

7 Right-click the mouse and select the **Finish Edit** option.

See Figure 15-21.

There are many different types of cam followers. Figure 15-22 shows four different types. As a cam turns, the follower is pushed up and down. A spring is often used to force the follower to stay in contact with the cam surface.

Figure 15-21

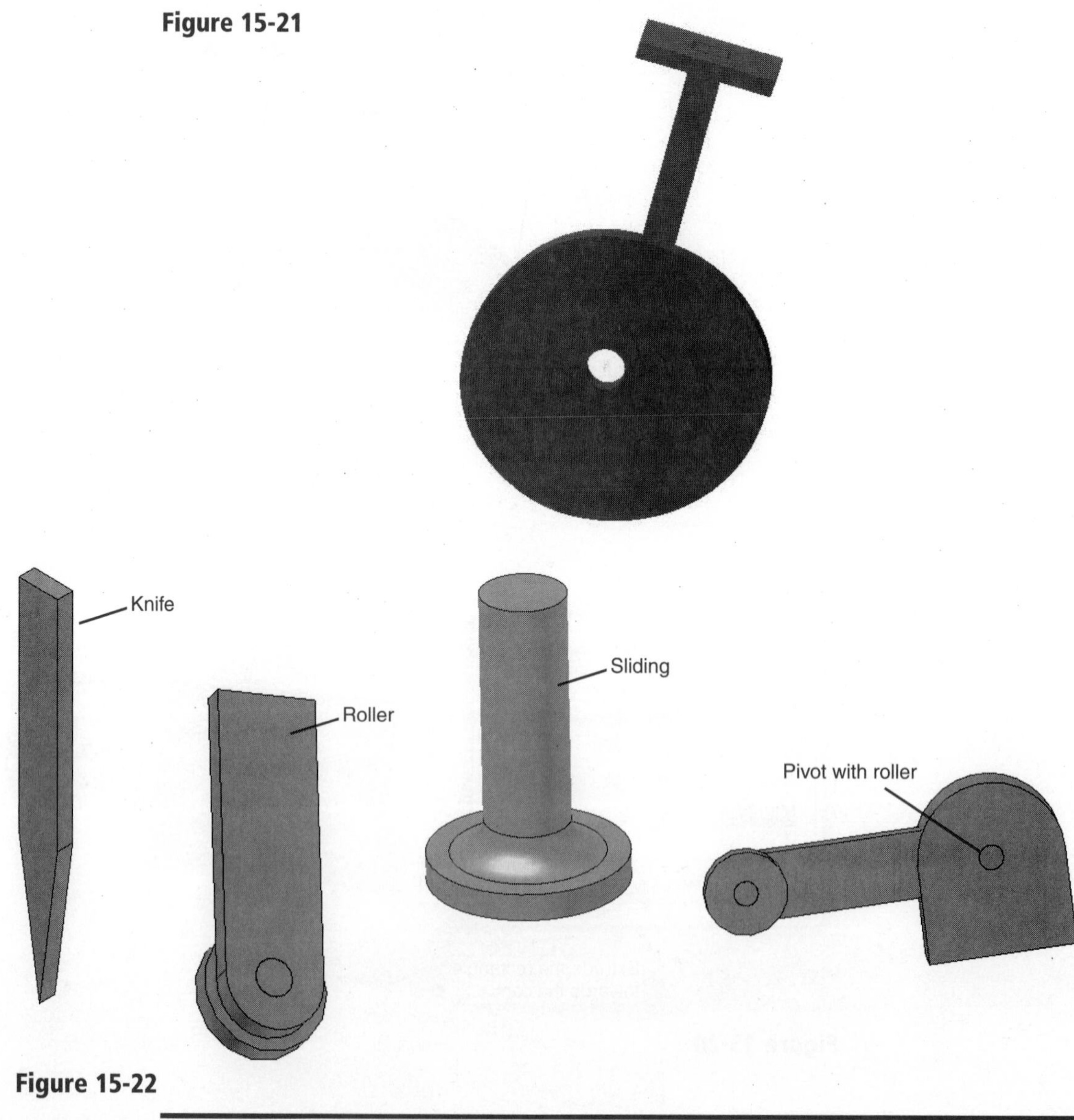

Figure 15-22

Chapter Summary

This chapter explained and illustrated how to draw and design cams, which are eccentric objects that convert rotary motion into linear motion. Displacement diagrams were defined and used to generate cams with the **Disc Cam** tool located under the **Design** tab. A follower was created, and the cam and follower assembly was animated.

Chapter Test Questions

Multiple Choice

Circle the correct answer.

1. Which of the following is not a type of cam follower?
   a. Rectangular c. Sliding
   b. Knife d. Roller
2. The basic shape about which a cam is created is called the
   a. Roller radius c. Base circle
   b. Cylindrical face d. Eccentricity
3. The acceleration and forces in a cam follower are affected by
   a. The cam's material
   b. The shape of the cam's profile surface
   c. Poisson's ratio
4. Why are keys used with cams?
   a. To control their pressure angles
   b. To transfer rotary motion for the driving shaft
   c. To change the effects of the basic radius
5. Which of the following is not a standard cam material available in Inventor?
   a. Steel SAE 1030 c. Aluminum Alloy Alclad 7075-T6
   b. Malleable cast iron 33-8 d. #2 Pine

True or False

Circle the correct answer.

1. **True or False:** Cams are eccentric objects that convert rotary motion into linear motion.
2. **True or False:** When a cam has a constant radius it creates a dwell motion.
3. **True or False:** The base circle of a cam is its outside diameter.
4. **True or False:** The part that contacts the profile surface of a cam is called a follower.
5. **True or False:** Cams can be animated.

Chapter Projects

Project 15-1: Millimeters

Draw the Cam Support Assembly shown in Figure P15-1. For this example the nominal dimensions for the bearings are as follows. More detailed dimensions can be found in the **Content Center.**

DIN625—SKF 6203 (ID × OD × THK) 17 × 40 × 10
DIN625—SKF 634 4 × 13 × 4
GB 2273.2-87—7/70 8 × 18 × 5

The nominal dimensions for the rectangular key are 5 × 5 × 16.
The values for the compression spring are as follows:

Wire diameter = 1.5
Inside diameter = 9.0
Loose spring length = 24
Preload spring length = 23
Fully loaded = 20
Working spring length = 21
Right coil direction
Active coils = 10.125

Figure P15-1A

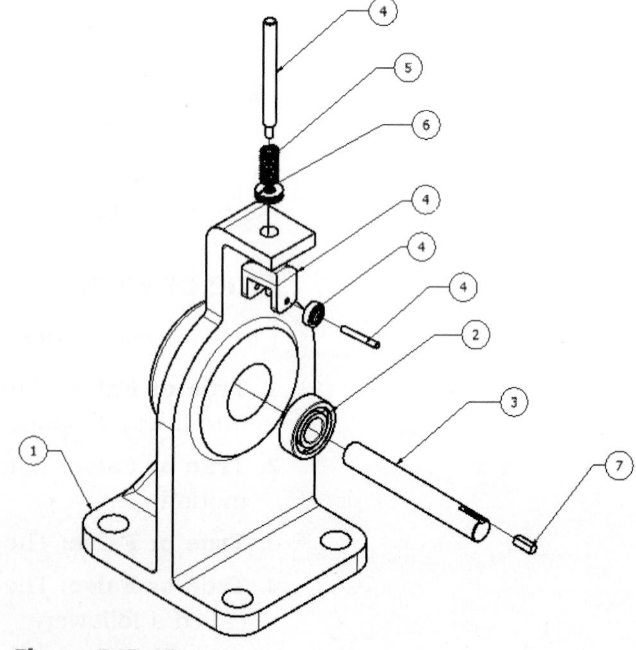

Figure P15-1B

Figure P15-1C

Parts List			
ITEM	QTY	PART NUMBER	DESCRIPTION
1	1	ENG-2008-A	BASE, CAST
2	1	DIN625 - SKF 6203	Single row ball bearings
3	1	SHF-4004-16	SHAFT: Ø16×120,WITH 2.3×5×16 KEYWAY
4	1		SUB-ASSEMBLY, FOLLOWER
5	1	SPR-C22	SPRING,COMPRESSION
6	1	GB 273.2-87 - 7/70 - 8 x 18 x 5	Rolling bearings - Thrust bearings - Plan of boundary dimensions
7	1	IS 2048 - 1983 - Specification for Parallel Keys and Keyways B 5 x 5 x 16	Specification for Parallel Keys and Keyways

Cast Base
P/N ENG-2008-A
MATL = Steel

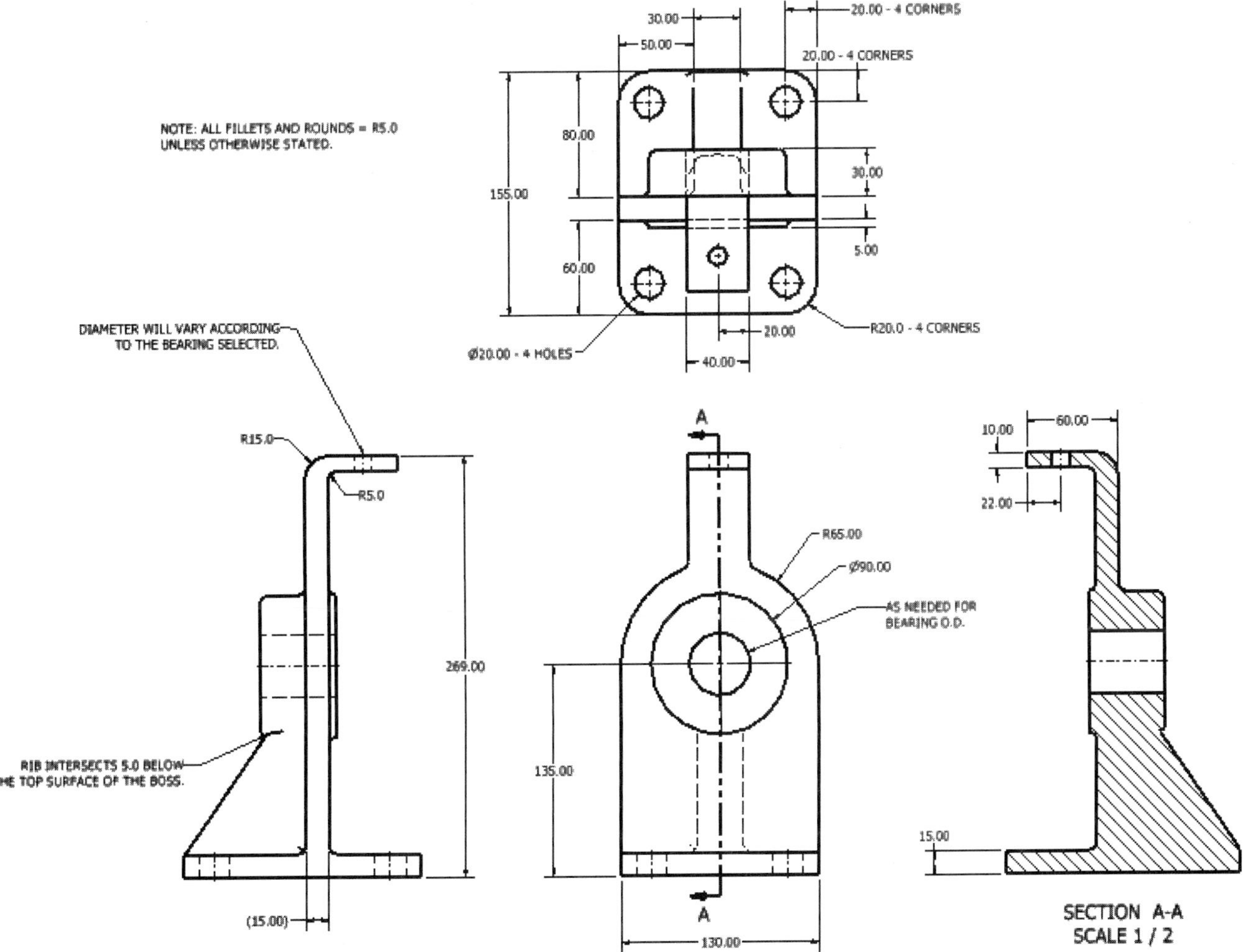

Figure P15-1D

Figure P15-1E Follower Subassembly

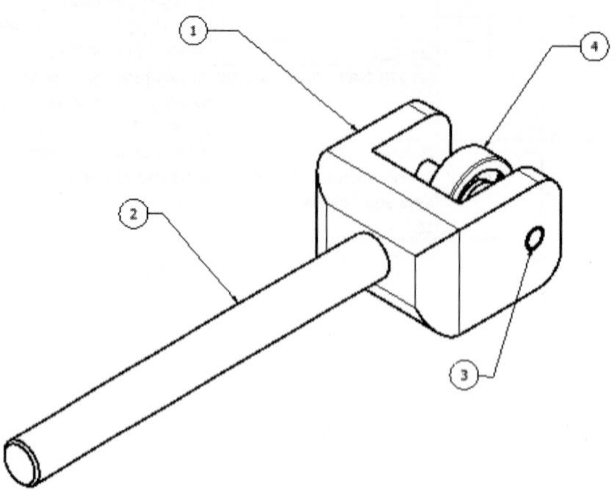

Parts List				
ITEM	PART NUMBER	DESCRIPTION	MATERIAL	QTY
1	AM-232	HOLDER	STEEL	1
2	AM-256	POST, FOLLOWER	STEEL	1
3	BS 1804-2 - 4 x 30	Parallel steel dowel pins - metric series	Steel, Mild	1
4	DIN625- SKF 634	Single row ball bearings	Steel, Mild	1

Figure P15-1F Holder
P/N AM-232
MATL = Steel

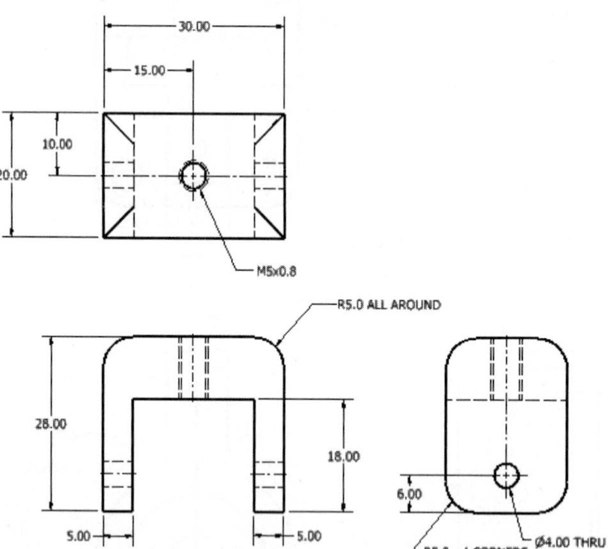

Figure P15-1G Follower Post
P/N AM-256
MATL = Steel

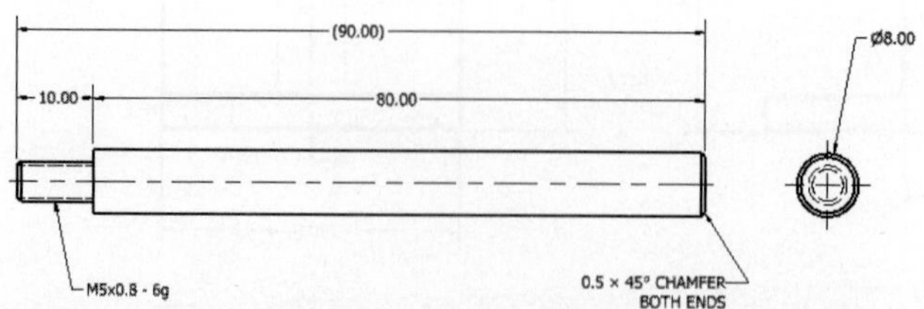

Shaft
P/N SHF-4004-16
MATL=Steel

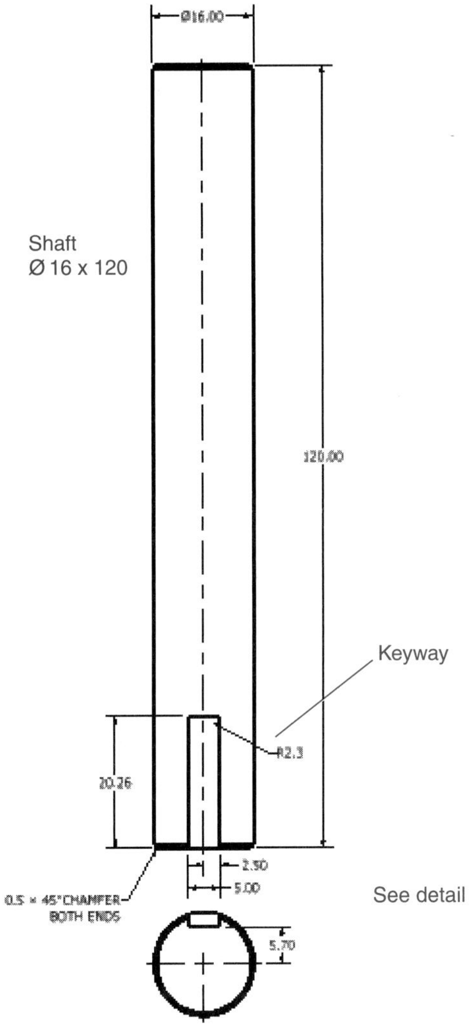

Figure P15-1H

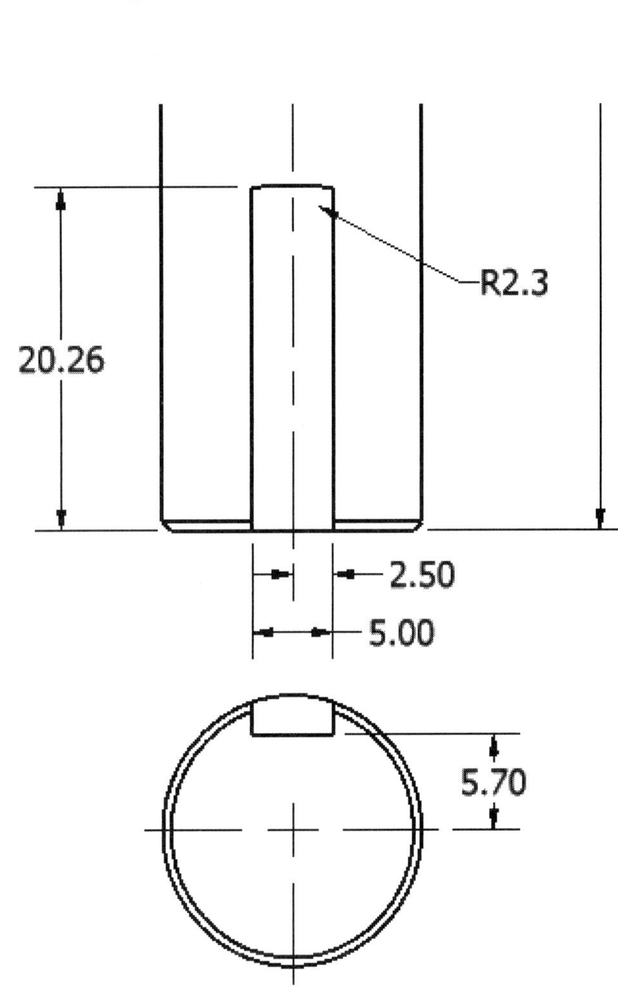

Figure P15-1I

Project 15-2: Millimeters

A. Draw the following cam:

Base circle = R73.0
Face width = 16.0
Rise 10.0 using harmonic motion in 90°
Dwell for 180°
Fall 10.0 using harmonic motion in 90°
Bore = Ø16.0
Keyway = 2.3 × 5 × 16 with a radius value of 2.3
Follower diameter = 16
Follower width = 4

B. Mount the cam into the Cam Support Assembly defined in Project 15-1. See Figure P15-2.

Figure P15-2

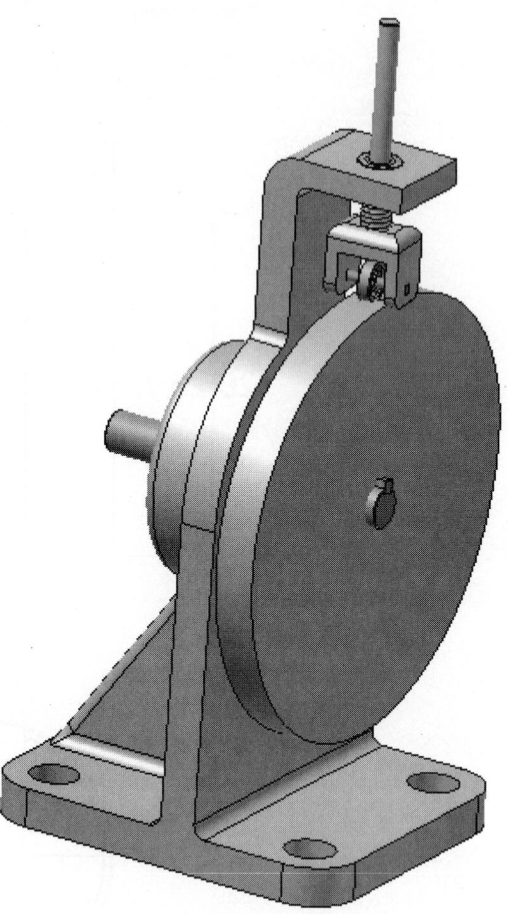

Project 15-3: Millimeters

A. Draw the following cam:

Base circle = Ø73.0
Face width = 16.0
Rise 8.0 using double harmonic motion—Part 1 motion in 90°
Dwell for 45°
Rise 6.0 using double harmonic motion—Part 1 motion in 90°
Dwell for 45°
Fall 14.0 using double harmonic motion—Part 1 motion in 90°
Bore = Ø16.0
Keyway = 2.3 × 5 × 16 with a radius value of 2.3
Follower diameter = 16
Follower width = 4

B. Mount the cam into the Cam Support Assembly defined in Project 15-1.

Project 15-4: Millimeters

A. Draw the following cam and add a Ø24.0 × 16.0 hub. Place a Ø16 hole through the hub and cam. Add an M4 hole though the hub located 10.0 from the top surface of the hub, and insert an M4 × 4.0 CSN 02 1181 Set Screw.

   Base circle = Ø73.0
   Face width = 16.0
   Rise 6.0 using cycloidal motion in 45°
   Dwell for 45°
   Rise 6.0 using cycloidal motion in 90°
   Dwell for 45°
   Fall 12.0 using cycloidal motion in 90°
   Bore = Ø16.0
   Keyway = 2.3 × 5 × 16 with a radius value of 2.3
   Follower diameter = 16
   Follower width = 4

B. Design
   Mount the cam into the Cam Support Assembly defined in Project 15-1. Modify the shaft presented in Project 15-1 by removing the keyway. Assign the modified shaft a new part number, SHT-466E.

Project 15-5: Millimeters

A. Draw the following cam:

   Base circle = Ø73.0
   Face width = 16.0
   Rise 12.0 using harmonic motion in 90°
   Dwell for 180°
   Fall 12.0 using harmonic motion in 90°
   Follower diameter = 16
   Follower width = 4

B. Design
   Make the following modifications, then mount the cam into the Cam Support Assembly defined in Project 15-1.
   Make the cam's bore 20.0.
   Modify the shaft to have a Ø20.0.
   Select a new key based on the Ø20.0 shaft, and add the appropriate keyway to the shaft and cam.
   Select a new bearing to accept the Ø20.0 shaft.
   Modify the hole in the Cast Base to accept the outside diameter of the selected bearing.
   Consider using a counterbored hole to mount the bearing.

appendix

Wire and Sheet Metal Gauges

Gauge	Thickness	Gauge	Thickness
000 000	0.5800	18	0.0403
00 000	0.5165	19	0.0359
0 000	0.4600	20	0.0320
000	0.4096	21	0.0285
00	0.3648	22	0.0253
0	0.3249	23	0.0226
1	0.2893	24	0.0201
2	0.2576	25	0.0179
3	0.2294	26	0.0159
4	0.2043	27	0.0142
5	0.1819	28	0.0126
6	0.1620	29	0.0113
7	0.1443	30	0.0100
8	0.1285	31	0.0089
9	0.1144	32	0.0080
10	0.1019	33	0.0071
11	0.0907	34	0.0063
12	0.0808	35	0.0056
13	0.0720	36	0.0050
14	0.0641	37	0.0045
15	0.0571	38	0.0040
16	0.0508	39	0.0035
17	0.0453	40	0.0031

Figure A-1

American Standard Clearance Locational Fits

Nominal Size Range Inches Over – To	Limits of Clearance	Class LC1 Standard Limits Hole H6	Class LC1 Standard Limits Shaft h5	Limits of Clearance	Class LC2 Standard Limits Hole H7	Class LC2 Standard Limits Shaft h6	Limits of Clearance	Class LC3 Standard Limits Hole H8	Class LC3 Standard Limits Shaft h7	Limits of Clearance	Class LC4 Standard Limits Hole H10	Class LC4 Standard Limits Shaft h9
0 - 0.12	0 0.45	+0.25 0	0 -0.2	0 0.65	+0.4 0	0 -0.25	0 1	+0.6 0	0 -0.4	0 2.6	+1.6 0	0 -1.0
0.12 - 0.24	0 0.5	+0.3 0	0 -0.2	0 0.8	+0.5 0	0 -0.3	0 1.2	+0.7 0	0 -0.5	0 3.0	+1.8 0	0 -1.2
0.24 - 0.40	0 0.65	+0.4 0	0 -0.25	0 1.0	+0.6 0	0 -0.4	0 1.5	+0.9 0	0 -0.6	0 3.6	+2.2 0	0 -1.4
0.40 - 0.71	0 0.7	+0.4 0	0 -0.3	0 1.1	+0.7 0	0 -0.4	0 1.7	+1.0 0	0 -0.7	0 4.4	+2.8 0	0 -1.6
0.71 - 1.19	0 0.9	+0.5 0	0 -0.4	0 1.3	+0.8 0	0 -0.5	0 2	+1.2 0	0 -0.8	0 5.5	+3.5 0	0 -2.0
1.19 - 1.97	0 1.0	+0.6 0	0 -0.4	0 1.6	+1.0 0	0 -0.6	0 2.6	+1.6 0	0 -1.0	0 6.5	+4.0 0	0 -2.5

Figure A-2A

Nominal Size Range Inches Over – To	Limits of Clearance	Class LC5 Standard Limits Hole H7	Class LC5 Standard Limits Shaft g6	Limits of Clearance	Class LC6 Standard Limits Hole H9	Class LC6 Standard Limits Shaft f8	Limits of Clearance	Class LC7 Standard Limits Hole H10	Class LC7 Standard Limits Shaft e9	Limits of Clearance	Class LC8 Standard Limits Hole H10	Class LC8 Standard Limits Shaft d9
0 - 0.12	0.1 0.75	+0..4 0	-0.1 -0.35	0.3 1.9	+1.0 0	-0.3 -0.9	0.6 3.2	+1.6 0	-0.6 -1.6	1.0 3.6	+1.6 0	-1.0 -2.0
0.12 - 0.24	0.15 0.95	+0.5 0	-0.15 -0.45	0.4 2.3	+1.2 0	-0.4 -1.1	0.8 3.8	+1.8 0	-0.8 -2.0	1.2 4.2	+1.8 0	-1.2 -2.4
0.24 - 0.40	0.2 1.2	+0.6 0	-0.2 -0.6	0.5 2.8	+1.4 0	-0.5 -1.4	1.0 4.6	+2.2 0	-1.0 -2.4	1.6 5.2	+2.2 0	-1.6 -3.0
0.40 - 0.71	0.25 1.35	+0.7 0	-0.25 -0.65	0.6 3.2	+1.6 0	-0.6 -1.6	1.2 5.6	+2.8 0	-1.2 -2.8	2.0 6.4	+2.8 0	-2.0 -3.6
0.71 - 1.19	0.3 1.6	+0.8 0	-0.3 -0.8	0.8 4.0	+2.0 0	-0.8 -2.0	1.6 7.1	+3.5 0	-1.6 -3.6	2.5 8.0	+3.5 0	-2.5 -4.5
1.19 - 1.97	0.4 2.0	+1.0 0	-0.4 -1.0	1.0 5.1	+2.5 0	-1.0 -2.6	2.0 8.5	+4.0 0	-2.0 -4.5	3.0 9.5	+4.0 0	-3.0 -5.5

Figure A-2B

American Standard Running and Sliding Fits

(Hole Basis)

Nominal Size Range Inches		Class RC1			Class RC2			Class RC3			Class RC4		
		Limits of Clearance	Standard Limits		Limits of Clearance	Standard Limits		Limits of Clearance	Standard Limits		Limits of Clearance	Standard Limits	
Over	To		Hole H5	Shaft g4		Hole H6	Shaft g5		Hole H7	Shaft f6		Hole H8	Shaft f7
0	0.12	0.1 0.45	+0.2 0	-0.1 -0.25	0.1 0.55	+0.25 0	-0.1 -0.3	0.3 0.95	+0.4 0	-0.3 -0.55	0.3 1.3	+0.6 0	-0.3 -0.7
0.12	0.24	0.15 0.5	+0.2 0	-0.15 -0.3	0.15 0.65	+0.3 0	-0.15 -0.35	0.4 1.12	+0.5 0	-0.4 -0.7	0.4 1.5	+0.7 0	-0.4 -0.0
0.24	0.40	0.2 0.6	+0.25 0	-0.2 -0.35	0.2 0.85	+0.4 0	-0.2 -0.45	0.5 1.5	+0.6 0	-0.5 -0.9	0.5 2.0	+0.9 0	-0.5 -1.1
0.40	0.71	0.25 0.75	+0.3 0	-0.25 -0.45	0.25 0.95	+0.4 0	-0.25 -0.55	0.6 1.7	+0.7 0	-0.6 -1.0	0.6 2.3	+1.0 0	-0.6 -1.3
0.71	1.19	0.3 0.95	+0.4 0	-0.3 -0.55	0.3 1.2	+0.5 0	-0.3 -0.7	0.8 2.1	+0.8 0	-0.8 -1.3	0.8 2.8	+1.2 0	-0.8 -1.6
1.19	1.97	0.4 1.1	+0.4 0	-0.4 -0.7	0.4 1.4	+0.6 0	-0.4 -0.8	1.0 2.6	+1.0 0	-1.0 -1.6	1.0 3.6	+1.6 0	-1.0 -2.0

Figure A-3A

Nominal Size Range Inches		Class RC5			Class RC6			Class RC7			Class RC8		
		Limits of Clearance	Standard Limits		Limits of Clearance	Standard Limits		Limits of Clearance	Standard Limits		Limits of Clearance	Standard Limits	
Over	To		Hole H8	Shaft e7		Hole H9	Shaft e8		Hole H9	Shaft d8		Hole H10	Shaft c9
0	0.12	0.6 1.6	+0.6 0	-0.6 -1.0	0.6 2.2	+1.0 0	-0.6 -1.2	1.0 2.6	+1.0 0	-1.0 -1.6	2.5 5.1	+1.6 0	-2.5 -3.5
0.12	0.24	0.8 2.0	+0.7 0	-0.8 -1.3	0.8 2.7	+1.2 0	-0.8 -1.5	1.2 3.1	+1.2 0	-1.2 -1.9	2.8 5.8	+1.8 0	-2.8 -4.0
0.24	0.40	1.0 2.5	+0.9 0	-1.0 -1.6	1.0 3.3	+1.4 0	-1.0 -1.9	1.6 3.9	+1.4 0	-1.6 -2.5	3.0 6.6	+2.2 0	-3.0 -4.4
0.40	0.71	1.2 2.9	+1.0 0	-1.2 -1.9	1.2 3.8	+1.6 0	-1.2 -2.2	2.0 4.6	+1.6 0	-2.0 -3.0	3.5 7.9	+2.8 0	-3.5 -5.1
0.71	1.19	1.6 3.6	+1.2 0	-1.6 -2.4	1.6 4.8	+2.0 0	-1.6 -2.8	2.5 5.7	+2.0 0	-2.5 -3.7	4.5 10.0	+3.5 0	-4.5 -6.5
1.19	1.97	2.0 4.6	+1.6 0	-2.0 -3.0	2.0 6.1	+2.5 0	-2.0 -3.6	3.0 7.1	+2.5 0	-3.0 -4.6	5.0 11.5	+4.0 0	-5.0 -7.5

Figure A-3B

American Standard Transition Locational Fits

Nominal Size Range Inches	Class LT1			Class LT2			Class LT3		
	Fit	Standard Limits		Fit	Standard Limits		Fit	Standard Limits	
Over To		Hole H7	Shaft js6		Hole H8	Shaft js7		Hole H7	Shaft k6
0 - 0.12	-0.10 +0.50	+0.4 0	+0.10 -0.10	-0.2 +0.8	+0.6 0	+0.2 -0.2			
0.12 - 0.24	-0.15 -0.65	+0.5 0	+0.15 -0.15	-0.25 +0.95	+0.7 0	+0.25 -0.25			
0.24 - 0.40	-0.2 +0.5	+0.6 0	+0.2 -0.2	-0.3 +1.2	+0.9 0	+0.3 -0.3	-0.5 +0.5	+0.6 0	+0.5 +0.1
0.40 - 0.71	-0.2 +0.9	+0.7 0	+0.2 -0.2	-0.35 +1.35	+1.0 0	+0.35 -0.35	-0.5 +0.6	+0.7 0	+0.5 +0.1
0.71 - 1.19	-0.25 +1.05	+0.8 0	+0.25 -0.25	-0.4 +1.6	+1.2 0	+0.4 -0.4	-0.6 +0.7	+0.8 0	+0.6 +0.1
1.19 - 1.97	-0.3 +1.3	+1.0 0	+0.3 -0.3	-0.5 +2.1	+1.6 0	+0.5 -0.5	+0.7 +0.1	+1.0 0	+0.7 +0.1

Figure A-4A

Nominal Size Range Inches	Class LT4			Class LT5			Class LT6		
	Fit	Standard Limits		Fit	Standard Limits		Fit	Standard Limits	
Over To		Hole H8	Shaft k7		Hole H7	Shaft n6		Hole H7	Shaft n7
0 - 0.12				-0.5 +0.15	+0.4 0	+0.5 +0.25	-0.65 +0.15	+0.4 0	+0.65 +0.25
0.12 - 0.24				-0.6 +0.2	+0.5 0	+0.6 +0.3	-0.8 +0.2	+0.5 0	+0.8 +0.3
0.24 - 0.40	-0.7 +0.8	+0.9 0	+0.7 +0.1	-0.8 +0.2	+0.6 0	+0.8 +0.4	-1.0 +0.2	+0.6 0	+1.0 +0.4
0.40 - 0.71	-0.8 +0.9	+1.0 0	+0.8 +0.1	-0.9 +0.2	+0.7 0	+0.9 +0.5	-1.2 +0.2	+0.7 0	+1.2 +0.5
0.71 - 1.19	-0.9 +1.1	+1.2 0	+0.9 +0.1	-1.1 +0.2	+0.8 0	+1.1 +0.6	-1.4 +0.2	+0.8 0	+1.4 +0.6
1.19 - 1.97	-1.1 +1.5	+1.6 0	+1.1 +0.1	-1.3 +0.3	+1.0 0	+1.3 +0.7	-1.7 +0.3	+1.0 0	+1.7 +0.7

Figure A-4B

American Standard Interference Locational Fits

Nominal Size Range Inches Over – To	Limits of Interference	Class LN1 Standard Limits Hole H6	Class LN1 Standard Limits Shaft n5	Limits of Interference	Class LN2 Standard Limits Hole H7	Class LN2 Standard Limits Shaft p6	Limits of Interference	Class LN3 Standard Limits Hole H7	Class LN3 Standard Limits Shaft r6
0 - 0.12	0 0.45	+0.25 0	+0.45 +0.25	0 0.65	+0.4 0	+0.63 +0.4	0.1 0.75	+0.4 0	+0.75 +0.5
0.12 - 0.24	0 0.5	+0.3 0	+0.5 +0.3	0 0.8	+0.5 0	+0.8 +0.5	0.1 0.9	+0.5 0	+0.9 +0.6
0.24 - 0.40	0 0.65	+0.4 0	+0.65 +0.4	0 1.0	+0.6 0	+1.0 +0.6	0.2 1.2	+0.6 0	+1.2 +0.8
0.40 - 0.71	0 0.8	+0.4 0	+0.8 +0.4	0 1.1	+0.7 0	+1.1 +0.7	0.3 1.4	+0.7 0	+1.4 +1.0
0.71 - 1.19	0 1.0	+0.5 0	+1.0 +0.5	0 1.3	+0.8 0	+1.3 +0.8	0.4 1.7	+0.8 0	+1.7 +1.2
1.19 - 1.97	0 1.1	+0.6 0	+1.1 +0.6	0 1.6	+1.0 0	+1.6 +1.0	0.4 2.0	+1.0 0	+2.0 +1.4

Figure A-5

American Standard Force and Shrink Fits

Nominal Size Range Inches Over – To	Limits of Interference	Class FN 1 Standard Limits Hole	Class FN 1 Standard Limits Shaft	Limits of Interference	Class FN 2 Standard Limits Hole	Class FN 2 Standard Limits Shaft	Limits of Interference	Class FN 3 Standard Limits Hole	Class FN 3 Standard Limits Shaft	Limits of Interference	Class FN 4 Standard Limits Hole	Class FN 4 Standard Limits Shaft
0 - 0.12	0.05 0.5	+0.25 0	+0.5 +0.3	0.2 0.85	+0.4 0	+0.85 +0.6				0.3 0.95	+0.4 0	+0.95 +0.7
0.12 - 0.24	0.1 0.6	+0.3 0	+0.6 +0.4	0.2 1.0	+0.5 0	+1.0 +0.7				0.4 1.2	+0.5 0	+1.2 +0.9
0.24 - 0.40	0.1 0.75	+0.4 0	+0.75 +0.5	0.4 1.4	+0.6 0	+1.4 +1.0				0.6 1.6	+0.6 0	+1.6 +1.2
0.40 - 0.56	0.1 0.8	+0.4 0	+0.8 +0.5	0.5 1.6	+0.7 0	+1.6 +1.2				0.7 1.8	+0.7 0	+1.8 +1.4
0.56 - 0.71	0.2 0.9	+0.4 0	+0.9 +0.6	0.5 1.6	+0.7 0	+1.6 +1.2				0.7 1.8	+0.7 0	+1.8 +1.4
0.71 - 0.95	0.2 1.1	+0.5 0	+1.1 +0.7	0.6 1.9	+0.8 0	+1.9 +1.4				0.8 2.1	+0.8 0	+2.1 +1.6
0.95 - 1.19	0.3 1.2	+0.5 0	+1.2 +0.8	0.6 1.9	+0.8 0	+1.9 +1.4	0.8 2.1	+0.8 0	+2.1 +1.6	1.0 2.3	+0.8 0	+2.1 +1.8
1.19 - 1.58	0.3 1.3	+0.6 0	+1.3 +0.9	0.8 2.4	+1.0 0	+2.4 +1.8	1.0 2.6	+1.0 0	+2.6 +2.0	1.5 3.1	+1.0 0	+3.1 +2.5
1.58 - 1.97	0.4 1.4	+0.6 0	+1.4 +1.0	0.8 2.4	+1.0 0	+2.4 +1.8	1.2 2.8	+1.0 0	+2.8 +2.2	1.8 3.4	+1.0 0	+3.4 +2.8

Figure A-6

Preferred Clearance Fits — Cylindrical Fits
(Hole Basis; ANSI B4.2)

Basic Size		Loose Running			Free Running			Close Running			Sliding			Locational Clear.		
		Hole H11	Shaft c11	Fit	Hole H9	Shaft d9	Fit	Hole H8	Shaft f7	Fit	Hole H7	Shaft g6	Fit	Hole H7	Shaft h6	Fit
4	Max	4.075	3.930	0.220	4.030	3.970	0.090	4.018	3.990	0.040	4.012	3.996	0.024	4.012	4.000	0.020
	Min	4.000	3.855	0.070	4.000	3.940	0.030	4.000	3.978	0.010	4.000	3.988	0.004	4.000	3.992	0.000
5	Max	5.075	4.930	0.220	5.030	4.970	0.090	5.018	4.990	0.040	5.012	4.996	0.024	5.012	5.000	0.020
	Min	5.000	4.855	0.070	5.000	4.940	0.030	5.000	4.978	0.010	5.000	4.988	0.004	5.000	4.992	0.000
6	Max	6.075	5.930	0.220	6.030	5.970	0.090	6.018	5.990	0.040	6.012	5.996	0.024	6.012	6.000	0.020
	Min	6.000	5.885	0.070	6.000	5.940	0.030	6.000	5.978	0.010	6.000	5.988	0.004	6.000	5.992	0.000
8	Max	8.090	7.920	0.260	8.036	7.960	0.112	8.022	7.987	0.050	8.015	7.995	0.029	8.015	8.000	0.024
	Min	8.000	7.830	0.080	8.000	7.924	0.040	8.000	7.972	0.013	8.000	7.986	0.005	8.000	7.991	0.000
10	Max	10.090	9.920	0.026	10.036	9.960	0.112	10.022	9.987	0.050	10.015	9.995	0.029	10.015	10.000	0.024
	Min	10.000	9.830	0.080	10.000	9.924	0.040	10.000	9.972	0.013	10.000	9.986	0.005	10.000	9.991	0.000
12	Max	12.112	11.905	0.315	12.043	11.950	0.136	12.027	11.984	0.061	12.018	11.994	0.035	12.018	12.000	0.029
	Min	12.000	11.795	0.095	12.000	11.907	0.050	12.000	11.966	0.016	12.000	11.983	0.006	12.000	11.989	0.000
16	Max	16.110	15.905	0.315	16.043	15.950	0.136	16.027	15.984	0.061	16.018	15.994	0.035	16.018	16.000	0.029
	Min	16.000	15.795	0.095	16.000	15.907	0.050	16.000	15.966	0.016	16.000	19.983	0.006	16.000	15.989	0.000
20	Max	20.130	19.890	0.370	20.052	19.935	0.169	20.033	19.980	0.074	20.021	19.993	0.041	20.021	20.000	0.034
	Min	20.000	19.760	0.110	20.000	19.883	0.065	20.000	19.959	0.020	20.000	19.980	0.007	20.000	19.987	0.000
25	Max	25.130	24.890	0.370	25.052	24.935	0.169	25.033	24.980	0.074	25.021	24.993	0.041	25.021	25.000	0.034
	Min	25.000	24.760	0.110	25.000	24.883	0.065	25.000	24.959	0.020	25.000	24.980	0.007	25.000	24.987	0.000
30	Max	30.130	29.890	0.370	30.052	29.935	0.169	30.033	29.980	0.074	30.021	29.993	0.041	30.021	30.000	0.034
	Min	30.000	29.760	0.110	30.000	29.883	0.065	30.000	29.959	0.020	30.000	29.980	0.007	30.000	29.987	0.000

Figure A-7

Preferred Transition and Interference Fits — Cylindrical Fits
(Hole Basis; ANSI B4.2)

Basic Size		Locational Trans.			Locational Trans.			Locational Inter.			Medium Drive			Force		
		Hole H7	Shaft k6	Fit	Hole H7	Shaft n6	Fit	Hole H7	Shaft p6	Fit	Hole H7	Shaft s6	Fit	Hole H7	Shaft u6	Fit
4	Max	4.012	4.009	0.011	4.012	4.016	0.004	4.012	4.020	0.000	4.012	4.027	-0.007	4.012	4.031	-0.011
	Min	4.000	4.001	-0.009	4.000	4.008	-0.016	4.000	4.012	-0.020	4.000	4.019	-0.027	4.000	4.023	-0.031
5	Max	5.012	5.009	0.011	5.012	5.016	0.004	5.012	5.020	0.000	5.012	5.027	-0.007	5.012	5.031	-0.011
	Min	5.000	5.001	-0.009	5.000	5.008	-0.016	5.000	5.012	-0.020	5.000	5.019	-0.027	5.000	5.023	-0.031
6	Max	6.012	6.009	0.011	6.012	6.016	0.004	6.012	6.020	0.000	6.012	6.027	-0.007	6.012	6.031	-0.011
	Min	6.000	6.001	-0.009	6.000	6.008	-0.016	6.000	6.012	-0.020	6.000	6.019	-0.027	6.000	6.023	-0.031
8	Max	8.015	8.010	0.014	8.015	8.019	0.005	8.015	8.024	0.000	8.015	8.032	-0.008	8.015	8.037	-0.013
	Min	8.000	8.001	-0.010	8.000	8.010	-0.019	8.000	8.015	-0.024	8.000	8.023	-0.032	8.000	8.028	-0.037
10	Max	10.015	10.010	0.014	10.015	10.019	0.005	10.015	10.024	0.000	10.015	10.032	-0.008	10.015	10.037	-0.013
	Min	10.000	10.001	-0.010	10.000	10.010	-0.019	10.000	10.015	-0.024	10.000	10.023	-0.032	10.000	10.028	-0.037
12	Max	12.018	12.012	0.017	12.018	12.023	0.006	12.018	12.029	0.000	12.018	12.039	-0.010	12.018	12.044	-0.015
	Min	12.000	12.001	-0.012	12.000	12.012	-0.023	12.000	12.018	-0.029	12.000	12.028	-0.039	12.000	12.033	-0.044
16	Max	16.018	16.012	0.017	16.018	16.023	0.006	16.018	16.029	0.000	16.018	16.039	-0.010	16.018	16.044	-0.015
	Min	16.000	16.001	-0.012	16.000	16.012	-0.023	16.000	16.018	-0.029	16.000	16.028	-0.039	16.000	16.033	-0.044
20	Max	20.021	20.015	0.019	20.021	20.028	0.006	20.021	20.035	-0.001	20.021	20.048	-0.014	20.021	20.054	-0.020
	Min	20.000	20.002	-0.015	20.000	20.015	-0.028	20.000	20.022	-0.035	20.000	20.035	-0.048	20.000	20.041	-0.054
25	Max	25.021	25.015	0.019	25.021	25.028	0.006	25.021	25.035	-0.001	25.021	25.048	-0.014	25.021	25.061	-0.027
	Min	25.000	25.002	-0.015	25.000	25.015	-0.028	25.000	25.022	-0.035	25.000	25.035	-0.048	25.000	25.048	-0.061
30	Max	30.021	30.015	0.019	30.021	30.028	0.006	30.021	30.035	-0.001	30.021	30.048	-0.014	30.021	30.061	-0.027
	Min	30.000	30.002	-0.015	30.000	30.015	-0.028	30.000	30.022	-0.035	30.000	30.035	-0.048	30.000	30.048	-0.061

Figure A-8

Preferred Clearance Fits — Cylindrical Fits
(Shaft Basis; ANSI B4.2)

Basic Size		Loose Running			Free Running			Close Running			Sliding			Locational Clear.		
		Hole C11	Shaft h11	Fit	Hole D9	Shaft h9	Fit	Hole F8	Shaft h7	Fit	Hole G7	Shaft h6	Fit	Hole H7	Shaft h6	Fit
4	Max	4.145	4.000	0.220	4.060	4.000	0.090	4.028	4.000	0.040	4.016	4.000	0.024	4.012	4.000	0.020
	Min	4.070	3.925	0.070	4.030	3.970	0.030	4.010	3.988	0.010	4.004	3.992	0.004	4.000	3.992	0.000
5	Max	5.145	5.000	0.220	5.060	5.000	0.090	5.028	5.000	0.040	5.016	5.000	0.024	5.012	5.000	0.020
	Min	5.070	4.925	0.070	5.030	4.970	0.030	5.010	4.988	0.010	5.004	4.992	0.004	5.000	4.992	0.000
6	Max	6.145	6.000	0.220	6.060	6.000	0.090	6.028	6.000	0.040	6.016	6.000	0.024	6.012	6.000	0.020
	Min	6.070	5.925	0.070	6.030	5.970	0.030	6.010	5.988	0.010	6.004	5.992	0.004	6.000	5.992	0.000
8	Max	8.170	8.000	0.260	8.076	8.000	0.112	8.035	8.000	0.050	8.020	8.000	0.029	8.015	8.000	0.024
	Min	8.080	7.910	0.080	8.040	7.964	0.040	8.013	7.985	0.013	8.005	7.991	0.005	8.000	7.991	0.000
10	Max	10.170	10.000	0.260	10.076	10.000	0.112	10.035	10.000	0.050	10.020	10.000	0.029	10.015	10.000	0.024
	Min	10.080	9.910	0.080	10.040	9.964	0.040	10.013	9.985	0.013	10.005	9.991	0.005	10.000	9.991	0.000
12	Max	12.205	12.000	0.315	12.093	12.000	0.136	12.043	12.000	0.061	12.024	12.000	0.035	12.018	12.000	0.029
	Min	12.095	11.890	0.095	12.050	11.957	0.050	12.016	11.982	0.016	12.006	11.989	0.006	12.000	11.989	0.000
16	Max	16.205	16.000	0.315	16.093	16.000	0.136	16.043	16.000	0.061	16.024	16.000	0.035	16.018	16.000	0.029
	Min	16.095	15.890	0.095	16.050	15.957	0.050	16.016	15.982	0.016	06.006	15.989	0.006	16.000	15.989	0.000
20	Max	20.240	20.000	0.370	20.117	20.000	0.169	20.053	20.000	0.074	20.028	20.000	0.041	20.021	20.000	0.034
	Min	20.110	19.870	0.110	20.065	19.948	0.065	20.020	19.979	0.020	20.007	19.987	0.007	20.000	19.987	0.000
25	Max	25.240	25.000	0.370	25.117	25.000	0.169	25.053	25.000	0.074	25.028	25.000	0.041	25.021	25.000	0.034
	Min	25.110	24.870	0.110	25.065	24.948	0.065	25.020	24.979	0.020	25.007	24.987	0.007	25.000	24.987	0.000
30	Max	30.240	30.000	0.370	30.117	30.000	0.169	30.053	30.000	0.074	30.028	30.000	0.041	30.021	30.000	0.034
	Min	30.110	29.870	0.110	30.065	29.948	0.065	29.979	29.979	0.020	30.007	29.987	0.007	30.000	29.987	0.000

Figure A-9

Preferred Transition and Interference Fits — Cylindrical Fits

(Shaft Basis; ANSI B4.2)

Basic Size		Locational Trans.			Locational Trans.			Locational Inter.			Medium Drive			Force		
		Hole K7	Shaft h6	Fit	Hole N7	Shaft h6	Fit	Hole F7	Shaft h6	Fit	Hole S7	Shaft h6	Fit	Hole U7	Shaft h6	Fit
4	Max	4.003	4.000	0.011	3.996	4.000	0.004	3.992	4.000	0.000	3.985	4.000	-0.007	3.981	4.000	-0.011
	Min	3.991	3.992	-0.009	3.984	3.992	-0.016	3.980	3.992	-0.020	3.973	3.992	-0.027	3.969	3.992	-0.031
5	Max	5.003	5.000	0.011	4.996	5.000	0.004	4.992	5.000	0.000	4.985	5.000	-0.007	4.981	5.000	-0.011
	Min	4.991	4.992	-0.009	4.984	4.992	-0.016	4.980	4.992	-0.020	4.973	4.992	-0.027	4.969	4.992	-0.031
6	Max	6.003	6.000	0.011	5.996	6.000	0.004	5.992	6.000	0.000	5.985	6.000	-0.007	5.981	6.000	-0.011
	Min	5.991	5.991	-0.009	5.984	5.992	-0.016	5.980	5.992	-0.020	5.973	5.992	-0.027	5.969	5.992	-0.031
8	Max	8.005	8.000	0.014	7.996	8.000	0.005	7.991	8.000	0.000	7.983	8.000	-0.008	7.978	8.000	-0.013
	Min	7.990	7.991	-0.010	7.981	7.991	-0.019	7.976	7.991	-0.024	7.968	7.991	-0.032	7.963	7.991	-0.037
10	Max	10.005	10.000	0.014	9.996	10.000	0.005	9.991	10.000	0.000	9.983	10.000	-0.008	9.978	10.000	-0.013
	Min	9.990	9.991	-0.010	9.981	9.991	-0.019	9.976	9.991	-0.024	9.968	9.991	-0.032	9.963	9.991	-0.037
12	Max	12.006	12.000	0.017	11.995	12.000	0.006	11.989	12.000	0.000	11.979	12.000	-0.010	11.974	12.000	-0.015
	Min	11.988	11.989	-0.012	11.977	11.989	-0.023	11.971	11.989	-0.029	11.961	11.989	-0.039	11.956	11.989	-0.044
16	Max	16.006	16.000	0.017	15.995	16.000	0.006	15.989	16.000	0.000	15.979	16.000	-0.010	15.974	16.000	-0.015
	Min	15.988	15.989	-0.012	15.977	15.989	-0.023	15.971	15.989	-0.029	15.961	15.989	-0.039	15.956	15.989	-0.044
20	Max	20.006	20.000	0.019	19.993	20.000	0.006	19.986	20.000	-0.001	19.973	20.000	-0.014	19.967	20.000	-0.020
	Min	19.985	19.987	-0.015	19.972	19.987	-0.028	19.965	19.987	-0.035	19.952	19.987	-0.048	19.946	19.987	-0.054
25	Max	25.006	25.000	0.019	24.993	25.000	0.006	24.986	25.000	-0.001	24.973	25.000	-0.014	24.960	25.000	-0.027
	Min	24.985	24.987	-0.015	24.972	24.987	-0.028	24.965	24.987	-0.035	24.952	24.987	-0.048	24.939	24.987	-0.061
30	Max	30.006	30.000	0.019	29.993	30.000	0.006	29.986	30.000	-0.001	29.973	30.000	-0.014	29.960	30.000	-0.027
	Min	29.985	29.987	-0.015	29.972	29.987	-0.028	29.987	29.987	-0.035	29.952	29.987	-0.048	29.939	29.987	-0.061

Figure A-10

Index

T

U

V

W